Biochemistry

A Short Course

Third Edition

John L. Tymoczko

Jeremy M. Berg

Lubert Stryer

W. H. FREEMAN
& COMPANY

A Macmillan Education Imprint

Publisher:	Kate Ahr Parker
Director of Marketing:	Sandy Lindelof
Senior Acquisitions Editor:	Lauren Schultz
Developmental Editor:	Heidi Bamatter
Media and Supplements Editors:	Amanda Dunning, Heidi Bamatter
Editorial Assistant:	Nandini Ahuja
Marketing Assistant:	Bailey James
Photo Editor:	Christine Buese
Photo Researcher:	Jacquelyn Wong
Cover and Interior Designer:	Vicki Tomaselli
Senior Project Editor:	Elizabeth Geller
Manuscript Editor:	Teresa Wilson
Illustrations:	Jeremy Berg, Gregory Gatto, Adam Steinberg, and Network Graphics
Illustration Coordinator:	Janice Donnola
Production Manager:	Paul Rohloff
Composition:	codeMantra
Printing and Binding:	Quad/Graphics
Cover Photo:	Fabio Ferrari/LaPresse/Icon Sportswire

Library of Congress Control Number: 2015934516

ISBN-10: 1-4641-2613-5
ISBN-13: 978-1-4641-2613-0

Printed in the United States of America
First printing

W. H. Freeman and Company
41 Madison Avenue
New York, NY 10010
www.whfreeman.com

**To our teachers
and students**

About the Authors

John L. Tymoczko is Towsley Professor of Biology at Carleton College, where he has taught since 1976. He currently teaches Biochemistry, Biochemistry Laboratory, Oncogenes and the Molecular Biology of Cancer, and Exercise Biochemistry and co-teaches an introductory course, Energy Flow in Biological Systems. Professor Tymoczko received his B.A. from the University of Chicago in 1970 and his Ph.D. in Biochemistry from the University of Chicago with Shutsung Liao at the Ben May Institute for Cancer Research. He then had a postdoctoral position with Hewson Swift of the Department of Biology at the University of Chicago. The focus of his research has been on steroid receptors, ribonucleoprotein particles, and proteolytic processing enzymes.

Jeremy M. Berg received his B.S. and M.S degrees in Chemistry from Stanford (where he did research with Keith Hodgson and Lubert Stryer) and his Ph.D. in Chemistry from Harvard with Richard Holm. He then completed a postdoctoral fellowship with Carl Pabo in Biophysics at Johns Hopkins University School of Medicine. He was an Assistant Professor in the Department of Chemistry at Johns Hopkins from 1986 to 1990. He then moved to Johns Hopkins University School of Medicine as Professor and Director of the Department of Biophysics and Biophysical Chemistry, where he remained until 2003. From 2003 to 2011, he served as Director of the National Institute of General Medical Sciences at the National Institutes of Health. In 2011, he moved to the University of Pittsburgh, where he is now Professor of Computational and Systems Biology and Pittsburgh Foundation Professor and Director of the Institute for Personalized Medicine. He served as President of the American Society for Biochemistry and Molecular Biology from 2011 to 2013. He is a Fellow of the American Association for the Advancement of Science and a member of the Institute of Medicine of the National Academy of Sciences. He is a recipient of the American Chemical Society Award in Pure Chemistry (1994), the Eli Lilly Award for Fundamental Research in Biological Chemistry (1995), the Harrison Howe Award (1997), and the Howard Schachman Public Service Award (2011), was named Maryland Outstanding Young Scientist of the Year (1995), and received public service awards from the Biophysical Society, the American Society for Biochemistry and Molecular Biology, the American Chemical Society, and the American Society for Cell Biology. He also received numerous teaching awards, including the W. Barry Wood Teaching Award (selected by medical students), the Graduate Student Teaching Award, and the Professor's Teaching Award for the Preclinical Sciences. He is coauthor, with Stephen J. Lippard, of the textbook *Principles of Bioinorganic Chemistry*.

Lubert Stryer is Winzer Professor of Cell Biology, Emeritus, in the School of Medicine and Professor of Neurobiology, Emeritus, at Stanford University, where he has been on the faculty since 1976. He received his M.D. from Harvard Medical School. Professor Stryer has received many awards for his research on the interplay of light and life, including the Eli Lilly Award for Fundamental Research in Biological Chemistry, the Distinguished Inventors Award of the Intellectual Property Owners' Association, and election to the National Academy of Sciences and the American Philosophical Society. He was awarded the National Medal of Science in 2006. The publication of his first edition of *Biochemistry* in 1975 transformed the teaching of biochemistry.

Preface

As humans, we are adept learning machines. Long before a baby learns that she can change a sheet of paper by crumpling it, she is absorbing vast amounts of information. This learning continues throughout life in myriad ways: learning to ride a bike and to take social cues from friends; learning to drive a car and balance a checkbook; learning to solve a quadratic equation and to interpret a work of art.

Of course, much of learning is necessary for survival, and even the simplest organisms learn to avoid danger and recognize food. However, humans are especially gifted in that we also acquire skills and knowledge to make our lives richer and more meaningful. Many students would agree that reading novels and watching movies enhance the quality of our lives because we can expand our horizons by vicariously being in situations we would never experience, reacting sympathetically or unsympathetically to characters who remind us of ourselves or are very different from anyone we have ever known. Strangely, at least to us as science professors, science courses are rarely thought of as being enriching or insightful into the human condition. Larry Gould, a former president of Carleton College, was also a geologist and an Arctic explorer. As a scientist, teacher, and administrator, he was very interested in science education especially as it related to other disciplines. In his inaugural address when he became president he said, "Science is a part of the same whole as philosophy and the other fields of learning. They are not mutually exclusive disciplines but they are independent and overlapping." Our goal was to write a book that encourages students to appreciate biochemistry in this broader sense, as a way to enrich their understanding of the world.

New to this Edition

This third edition takes into account recent discoveries and advances that have changed how we think about the fundamental concepts in biochemistry and human health. To meet the needs of instructors and students alike, particular attention has been paid to the topics outlined below.

Expanded Physiological Focus

A hallmark feature of *Biochemistry: A Short Course* is its physiological perspective on biochemical processes and its integration of clinical examples to apply and reinforce concepts. In the third edition, we build on this aspect of the book with:

- A **NEW** section: "Mutations in Genes Encoding Hemoglobin Subunits Can Result in Disease" (Chapter 9)

- 17 new Clinical Insights, demonstrating the relevance of biochemistry to human health and disease.

Features highlighting the physiological aspect of biochemistry have been expanded, and include the following:

 CLINICAL INSIGHT

Premature Aging Can Result from the Improper Attachment of a Hydrophobic Group to a Protein

Farnesyl is a hydrophobic group that is often attached to proteins, usually so that the protein is able to associate with a membrane (Figure 11.10). Inappropriate farnesylation has been shown to result in Hutchinson–Gilford progeria syndrome (HGPS), a rare disease of premature aging. Early postnatal development is normal, but the children fail to thrive, develop bone abnormalities, and have a small beaked nose, a receding jaw, and a complete loss of hair (Figure 11.11). Affected children usually die at an average age of 13 years of severe atherosclerosis, a cause of death more commonly seen in the elderly.

The cause of HGPS appears to be a mutation in the gene for the nuclear protein *lamin*, a protein that forms a scaffold for the nucleus and may take part in the regulation of gene expression. The folded polypeptide that will eventually become lamin is modified and processed many times before the mature protein is produced. One key processing event is the removal of a farnesyl group that had been added to the nascent protein earlier in processing. In HGPS patients, the farnesyl group is not removed, owing to a mutation in the lamin. The incorrectly processed lamin results in a deformed nucleus (Figure 11.11) and aberrant nuclear function that results in HGPS. Much research remains to determine precisely how the failure to remove the farnesyl group leads to such dramatic consequences.

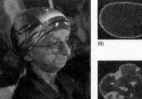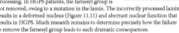

Figure 11.11 Hutchinson–Gilford progeria syndrome (HGPS). (A) A 15-year-old boy suffering from HGPS. (B) A normal nucleus. (C) A nucleus from a HGPS patient. [(A) AP Photo/Gerald Herbert; (B and C) Scaffidi, P., Gordon, L. and Misteli, T. (2005). The cell nucleus and aging: Tantalizing clues and hopeful promises. *PLoS Biol* 3 (11): e395. Courtesy of Paola Scaffidi.]

CLINICAL INSIGHTS

In the Clinical Insights, students see how the concepts most recently considered affect an aspect of a disease or its cure. By exploring biochemical concepts in the context of a disease, students learn how these concepts are relevant to human life and what happens when biochemistry goes awry.

BIOLOGICAL INSIGHTS

Biochemistry affects every aspect of our world, sometimes in strange and amazing ways. Like Clinical Insights, Biological Insights bolster students' understanding of biochemical concepts as they learn how simple changes in biochemical processes can have dramatic effects.

For a complete list of clinical and biological insights see pages xi–xii.

NUTRITIONAL EXAMPLES

Examples of the underlying relationship between nutrition and biochemistry abound.

Increased Coverage of the Fundamentals

The third edition features a greater emphasis on the fundamentals of biochemistry, specifically where metabolism is concerned (Chapters 14 and 15). In an effort to explain metabolism more fully, we've expanded on the following areas within Chapters 14 and 15:

- Digestive enzymes
- Protein digestion
- Celiac disease
- Energy
- Phosphates in biochemical processes

BIOLOGICAL INSIGHT

The Dead Zone: Too Much Respiration

Some marine organisms perform so much cellular respiration, and therefore consume so much molecular oxygen, that the oxygen concentration in the water is decreased to a level that is too low to sustain other organisms. One such hypoxic (low levels of oxygen) zone is in the northern Gulf of Mexico, off the coast of Louisiana where the Mississippi River flows into the Gulf (Figure 20.16). The Mississippi is extremely nutrient rich due to agricultural runoff; so plant microorganisms, called phytoplankton, proliferate so robustly that they exceed the amount that can be consumed by other members of the food chain. When the phytoplankton die, they sink to the bottom and are consumed by aerobic bacteria. The aerobic bacteria thrive to such a degree that other bottom-dwelling organisms, such as shrimp and crabs, cannot obtain enough O_2 to survive. The term "dead zone" refers to the inability of this area to support fisheries.

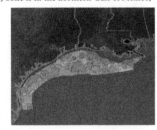

Figure 20.16 The Gulf of Mexico dead zone. The size of the dead zone in the Gulf of Mexico off Louisiana varies annually but may extend from the Louisiana and Alabama coasts to the westernmost coast of Texas. Reds and oranges represent high concentrations of phytoplankton and river sediment. [NASA/Goddard Space Flight Center/Scientific Visualization Studio.]

||||||||||||||||||||||||||||||||||||||||||||||||||||||||||||||||||||||||||||||||||||||||||||||||||||||||||||||||

Teaching and Learning Tools

In addition to providing an engaging contextual framework for the biochemistry throughout the book, we have created several opportunities for students to check their understanding, reinforce connections across the book, and practice what they have learned. These opportunities present themselves both in features throughout the text and in the many resources offered in LaunchPad.

ACTIVE LEARNING RESOURCES

In this new edition, we've responded to instructor requests to provide resources that aid in creating an active classroom environment. All of the new media resources for *Biochemistry: A Short Course* will be available in our new Macmillan Education LaunchPad system. For more information on LaunchPad see page ix. To help students adapt to an interactive course, we've added the following resources:

NEW **Case Studies** are a series of online biochemistry case studies that are assignable and assessable. Authored by Justin Hines, Assistant Professor of Chemistry at Lafayette College, each case study gives students practice in working with data, developing critical thinking skills, connecting topics, and applying knowledge to real scenarios. We also provide instructional guidance with each case study (with suggestions on how to use the case in the classroom) and aligned assessment questions for quizzes and exams.

NEW **Clicker Questions** are aligned with key concepts and misconceptions in each chapter so instructors can assess student understanding in real time during lectures.

END-OF-CHAPTER PROBLEMS

Each chapter includes a robust set of practice problems. We have revised and added to the total number of questions in the third edition.

- **Data Interpretation Problems** train students to analyze data and reach scientific conclusions.

- **Chapter Integration Problems** draw connections between concepts across chapters.

- **Challenge Problems** require calculations, understanding of chemical structures, and other concepts that are challenging for most students.

Brief solutions to all the end-of-chapter problems are provided in the "Answers to Problems" section in the back of the textbook. We are also pleased to offer expanded solutions in the accompanying *Student Companion*, by Frank Deis, Nancy Counts Gerber, Richard Gumport, and Roger Koeppe. (For more details on this supplement see page x.)

MARGIN FEATURES

We use the margin features in the textbook in several ways to help engage students, emphasize the relevance of biochemistry to their lives, and make it more accessible. We have given these features a new look to make them clearer and more easily identifiable.

- **Learning Objectives** are used in many different ways in the classroom. To help reinforce key concepts while the student is reading the chapter we have indicated them with a ✓ and number and integrated them on a chapter level as well as in the section introductions. They are also tied to the end-of-chapter problems to assist students in developing problem-solving skills and instructors in assessing students' understanding of some of the key concepts in each chapter.

6.4 Enzymes Facilitate the Formation of the Transition State

✓ 2 Explain the relation between the transition state and the active site of an enzyme, and list the characteristics of active sites.

The free-energy difference between reactants and products accounts for the equilibrium of a reaction, but enzymes accelerate how quickly this equilibrium is attained. How can we explain the rate enhancement in terms of thermodynamics? To do so, we have to consider not the end points of the reaction but the chemical pathway between the end points.

A chemical reaction of substrate S to form product P goes through a *transition state* X‡ that has a higher free energy than does either S or P. The double dagger denotes the transition state. The transition state is a fleeting molecular structure that is no longer the substrate but is not yet the product. The transition state is the least-stable and most-seldom-occurring species along the reaction pathway because it is the one with the highest free energy.

? QUICK QUIZ Explain why a person who has a trypsinogen deficiency will suffer from more digestion difficulties than will a person lacking most other zymogens.

- **Quick Quizzes** emulate that moment in a lecture when a professor asks, "Do you get it?" These questions allow students to check their understanding of the material as they read it so they can immediately gauge whether they need to review a discussion or can advance to the next topic. Answers are given at the end of each chapter.

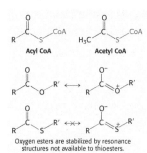

Oxygen esters are stabilized by resonance structures not available to thioesters.

- **Margin Structures** provide a quick reminder of a molecule or group that students may have seen earlier in the book or in another course. This allows students to understand the topic at hand without needing to look up a basic structure or organic chemistry principle elsewhere.

- **Did You Know?** features are short asides to the biochemical topic being discussed. They put a personal face on science, or, in the vein of Biological Insights, provide glimpses of how we use biochemistry in everyday life.

DID YOU KNOW?

Interestingly, digitalis was used effectively long before the discovery of the Na⁺–K⁺ ATPase. In 1785, William Withering, a British physician, heard tales of an elderly woman, known as "the old woman of Shropshire," who cured people of "dropsy" (which today would be recognized as congestive heart failure) with an extract of foxglove. Withering conducted the first scientific study of the effects of foxglove on congestive heart failure and documented its effectiveness.

NUTRITION FACTS

Niacin Also called vitamin B₃, niacin is a component of coenzymes NAD⁺ and NADP⁺ (pp. 268–270), which are used in electron-transfer reactions. There are many sources of niacin, including chicken breast. Niacin deficiency results in the potentially fatal disease pellagra, a condition characterized by dermatitis, dementia, and diarrhea. [Brand X Pictures]

- **Nutrition Facts** highlight essential vitamins in the margin next to where they are discussed as part of an enzyme mechanism or metabolic pathway. In these boxes, students will discover how we obtain vitamins from our diets and what happens if we do not have enough of them. These important molecules and their structures are listed in table form in the appendix of the book as well, to help students easily find where each vitamin is discussed in the book.

Media and Supplements

All of the new media resources for *Biochemistry: A Short Course* are available in our new LaunchPad system.

www.macmillanhighered.com/launchpad/tymoczko3e

LaunchPad is a dynamic, fully integrated learning environment that brings together all of our teaching and learning resources in one place. It includes easy-to-use, powerful assessment tracking and grading tools, a personalized calendar, an announcement center, and communication tools to help you manage your course. This learning system also contains the fully interactive **e-Book** and other newly updated resources for students and instructors, including the following:

For Students

- **Case Studies** are a series of online biochemistry case studies that are assignable and assessable. Authored by Justin Hines, Assistant Professor of Chemistry at Lafayette College, each case study gives students practice in working with data, developing critical thinking skills, connecting topics, and applying knowledge to real scenarios.

- **e-Book** allows students to read the online version of the textbook, which combines the contents of the printed book, electronic study tools, and a full complement of student media specifically created to support the text.

- **Hundreds of Self-Graded Practice Problems** allow students to test their understanding of concepts explained in the text, with immediate feedback.

- **Metabolic Map** helps students understand the principles and applications of the core metabolic pathways. Students can work through guided tutorials with embedded assessment questions, or explore the Metabolic Map on their own using the dragging and zooming functionality of the map.

- **Problem-Solving Videos**, created by Scott Ensign of Utah State University, provide 24/7 online problem-solving help to students. Through a two-part approach, each 10-minute video covers a key textbook problem representing a topic that students traditionally struggle to master. Dr. Ensign first describes a proven problem-solving strategy and then applies the strategy to the problem at hand in clear, concise steps. Students can easily pause, rewind, and review any steps they wish until they firmly grasp not just the solution but also the reasoning behind it. Working through the problems in this way is designed to make students better and more confident at applying key strategies as they solve other textbook and exam problems.

- **Living Figures** allow students to view textbook illustrations of protein structures online in interactive 3-D using Jmol. Students can zoom and rotate 54 "live" structures to get a better understanding of their three-dimensional nature and can experiment with different display styles (space-filling, ball-and-stick, ribbon, backbone) by means of a user-friendly interface.

- **Self-Assessment Tool** allows students to test their understanding by taking an online multiple-choice quiz provided for each chapter, as well as a general chemistry review.

- **Animated Techniques** illustrate laboratory techniques described in the text.

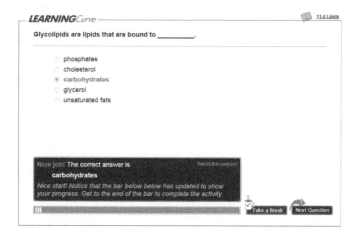

- **Learning Curve** is a self-assessment tool that helps students evaluate their progress. Students can test their understanding by taking an online multiple-choice quiz provided for each chapter, as well as a general chemistry review.

For Instructors

All the features listed above for students plus:

- **e-Book** Instructors teaching from the e-Book can assign either the entire textbook or a custom version that includes only the chapters that correspond to their syllabi. They can choose to add notes to any page of the e-Book and share these notes with their students. These notes may include text, animations, or photographs.

- **Clicker Questions** are aligned with key concepts and misconceptions in each chapter so instructors can assess student understanding in real time during lectures.

- **Newly Updated Lecture PowerPoint Files** have been developed to minimize preparation time for new users of the book. These files offer suggested lectures including key illustrations and summaries that instructors can adapt to their teaching styles.

- **Updated Textbook Images and Tables** are offered as high-resolution JPEG files. The JPEGs are also offered in separate PowerPoint files.

- **Test Bank**, by Harvey Nikkel of Grand Valley State University, Susan Knock of Texas A&M University at Galveston, and Joseph Provost of Minnesota State University Moorhead, offers more than 1500 questions in editable Word format.

Student Companion

(1-319-03295-8)

For each chapter of the textbook, the *Student Companion* includes:

- Chapter Learning Objectives and Summary

- Self-Assessment Problems, including multiple-choice, short-answer, matching questions, and challenge problems, and their answers

- Expanded Solutions to the end-of-chapter problems in the textbook

CLINICAL INSIGHTS   *This icon signals the beginning of a Clinical Insight in the text.*

BIOLOGICAL INSIGHTS   *This icon signals the beginning of a Biological Insight in the text.*

Acknowledgments

Our thanks go to the instructors and professors who have reviewed the chapters of this book. Their sharp eyes and keen insights strongly influenced us as we wrote and shaped the various drafts of each chapter to create this completed work.

Tabitha Amora,
  Ball State University
Bynthia Anose,
  Bethel University
Kimberly Bagley,
  SUNY Buffalo State
David Baker,
  Delta College
Michael Barbush,
  Baker University
Ellen Batchelder,
  Unity College
Moriah Beck,
  Wichita State University
Nina Bernstein,
  MacEwan University
Veronic Bezaire,
  Carleton University
Mary Bruno,
  University of Connecticut
John Cannon,
  Trinity International University
James Cheetham,
  Carleton University
Silvana Constantinescu,
  Marymount California University
Peter DiMaria,
  Delaware State University
Caryn Evilia,
  Idaho State University
Brenda Fredette,
  Medaille College
Scott Gabriel,
  Viterbo University
Ratna Gupta,
  Our Lady of the Lake College
Sarah Hosch,
  Oakland University
Kelly Johanson,
  Xavier University of Louisiana
Marjorie Jones,
  Illinois State University
Susan Knock,
  Texas A&M University at Galveston

Kris Koudelka,
  Point Loma Nazarene University
Ramaswamy Krishnamoorthi,
  Kansas State University
Isabel Larraza,
  North Park University
Linda Luck,
  SUNY Plattsburgh
Kumaran Mani,
  University of Wyoming
Jairam Menon,
  University of Michigan Medical School
David Mitchell,
  College of Saint Benedict & Saint John's University
Mautusi Mitra,
  University of West Georgia
Ashvin Mohindra,
  Fleming College
William Newton,
  Virginia Tech
Brian Nichols,
  University of Illinois at Chicago
Carleitta Paige-Anderson,
  Virginia Union University
Janice Pellino,
  Carthage College
Ivana Peralta,
  Vincennes University
Elizabeth Roberts-Kirchhoff,
  University of Detroit Mercy
John Rose,
  University of Georgia
Martina Rosenberg,
  University of New Mexico
Tricia Scott,
  Dalton State College
Richard Sheardy,
  Texas Woman's University
Kevin Siebenlist,
  Marquette University
Matt Thomas,
  State College of Florida
Jennifer Tsui,
  Marygrove College

Timothy Vail,
Northern Arizona University

Todd Weaver,
University of Wisconsin–La Crosse

Korin Wheeler,
Santa Clara University

Harvey Wiener,
Manchester Community College

Marc Wold,
University of Iowa

Adele Wolfson,
Wellesley College

The German scientist, writer, and statesman Johann Wolfgang von Goethe once remarked, "Thinking is easy, acting is difficult, and to put one's thoughts into action is the most difficult thing in the world." While we may disagree with Goethe's assertion that thinking is easy, we emphatically agree with the rest of the quotation. Thinking about biochemistry and then putting those thoughts into a book that is clear, welcoming, stimulating, and challenging is, if not the most difficult thing in the world, still very demanding. This task would be utterly impossible without our wonderful colleagues at W. H. Freeman. They are intelligent, dedicated, caring people who have taught us much about how to present science to students and, in the process, brought out the best in us. Although we have had the pleasure of working with our collaborators at W. H. Freeman on a number of projects, our appreciation of and gratefulness for their efforts and guidance are as sincere now as they were when we were inexperienced authors. Our experiences with this edition have been as delightful and rewarding as our past projects. We have many people to thank for this experience, some of whom we have worked with previously and some new to the effort. First, we would like to acknowledge the encouragement, patience, excellent advice, and good humor of our Publisher, Kate Ahr Parker. Kate can suggest difficult challenges with such grace and equanimity that we readily accept the challenge. New to our book team is our Senior Acquisitions Editor, Lauren Schultz. Her unfailing enthusiasm was a source of support and energy for the author team. New to our book team for this edition is Heidi Bamatter, our Developmental Editor. Heidi is another in a line of outstanding development editors that we have had the pleasure to work with at Freeman. Her insight, patience, and guidance made this effort successful and enjoyable. Elizabeth Geller, Senior Project Editor, managed the flow of the project with admirable efficiency. Teresa Wilson, our Manuscript Editor, enhanced the literary consistency and clarity of the text. Vicki Tomaselli, Design Manager, produced a design and layout that made the book welcoming and accessible. Christine Buese and Jacquelyn Wong, Photo Editor and Photo Researcher, respectively, found the photographs that helped to achieve one of our main goals—linking biochemistry to the everyday world of the student while making the text a visual treat. Janice Donnola, Illustration Coordinator, deftly directed the rendering of new illustrations. Paul Rohloff, Production Manager, made sure the difficulties of scheduling, composition, and manufacturing were readily overcome. We are more appreciative of the sales staff at W. H. Freeman for their enthusiastic support than we can put into words. Without the efforts of the sales force to persuade professors to examine our book, all of our own excitement and enthusiasm for this text would be meaningless. We also thank Susan Winslow. Her vision for science textbooks and her skill at gathering exceptional personnel make working with W. H. Freeman a true pleasure.

Thanks also to our many colleagues at our own institutions as well as throughout the country who patiently answered our questions and encouraged us on our quest. Finally, we owe a debt of gratitude to our families. Without their support, comfort, and understanding, this project could never have been undertaken, let alone successfully completed.

Brief Contents

Contents

SECTION 15
RNA Synthesis, Processing, and Regulation   657

Chapter 36 RNA Synthesis and Regulation in Bacteria   659

Chapter 37 Gene Expression in Eukaryotes   675

Chapter 38 RNA Processing in Eukaryotes   691

SECTION 16
Protein Synthesis and Recombinant DNA Techniques   705

Chapter 39 The Genetic Code   707

Biochemistry Helps Us to Understand Our World

The ultimate goal of all scientific endeavors is to develop a deeper, richer understanding of ourselves and the world in which we live. Biochemistry has had and will continue to have an extensive role in helping us to develop this understanding. *Biochemistry*, the study of living organisms at the molecular level, has shown us many of the details of the most fundamental processes of life. For instance, biochemistry has shown us how information flows from genes to molecules that have functional capabilities. In recent years, biochemistry has also unraveled some of the mysteries of the molecular generators that provide the energy that powers living organisms. The realization that we can understand such essential life processes has significant philosophical implications. What does it mean, biochemically, to be human? What are the biochemical differences between a human being, a chimpanzee, a mouse, and a fruit fly? Are we more similar than we are different?

The understanding achieved through biochemistry is greatly influencing medicine and other fields. Although we may not be accustomed to thinking of illness in relation to molecules, illness is ultimately some sort of malfunction at the molecular level. The molecular lesions causing sickle-cell anemia, cystic fibrosis, hemophilia, and many other genetic diseases have been elucidated at the biochemical level. Biochemistry is also contributing richly to clinical diagnostics. For example, elevated levels of heart enzymes in the blood reveal whether a patient has recently had a myocardial infarction (heart attack). Agriculture, too, is employing biochemistry to develop more effective, environmentally safer herbicides and pesticides and to create genetically engineered plants that are, for example, more resistant to insects.

1

In this section, we will learn some of the key concepts that structure the study of biochemistry. We begin with an introduction to the molecules of biochemistry, followed by an overview of the fundamental unit of biochemistry and life itself—the cell. Finally, we examine the weak reversible bonds that enable the formation of biological structures and permit the interplay between molecules that makes life possible.

✓ **By the end of this section, you should be able to:**

✓ 1 Describe the key classes of biomolecules and differentiate between them.

✓ 2 List the steps of the central dogma.

✓ 3 Identify the key features that differentiate eukaryotic cells from prokaryotic cells.

✓ 4 Describe the chemical properties of water and explain how water affects biochemical interactions.

✓ 5 Describe the types of noncovalent, reversible interactions and explain why reversible interactions are important in biochemistry.

✓ 6 Define pH and explain why changes in pH may affect biochemical systems.

Biochemistry and the Unity of Life

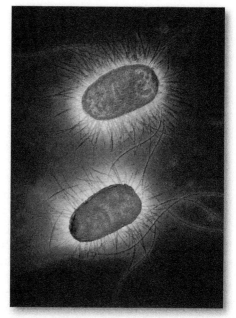

Despite their vast differences in mass—the African elephant has a mass 3×10^{18} times as great as that of the bacterium *E. coli*—and complexity, the biochemical workings of these two organisms are remarkably similar. [*E. coli*: Eye of Science/Science Source. Elephant: John Michael Evan Potter/Shutterstock.]

A key goal of biochemistry, one that has been met with striking success, is to understand what it means to be alive at the molecular level. Another goal is to extend this understanding to the organismic level—that is, to understand the effects of molecular manipulations on the life that an organism leads. For instance, understanding how the hormone insulin works at the molecular level illuminates how the organism controls the levels of common fuels—glucose and fats—in its blood. Often, such understanding facilitates an understanding of disease states, such as diabetes, which results when insulin signaling goes awry. In turn, this knowledge can be a source of insight into how the disease can be treated.

Biochemistry has been an active area of research for more than a century. Much knowledge has been gained about how a variety of organisms manipulate energy and information. However, one of the most exciting outcomes of biochemical research has been the realization that all organisms have much in common biochemically. *Organisms are remarkably uniform at the molecular level.* This observation is frequently referred to as the *unity of biochemistry,* but, in reality, it illustrates the unity of life. French biochemist Jacques Monod encapsulated this idea in 1954 with the phrase "Anything found to be true of [the bacterium] *E. coli* must also be true of elephants." This uniformity reveals that all organisms on Earth have arisen from a common ancestor. A core of essential biochemical processes, common to all organisms, appeared early in the evolution of life. The diversity of life in the modern world has been generated by evolutionary processes acting on these core processes through millions or even billions of years.

We begin our study of biochemistry by looking at commonalities. We will examine the molecules and molecular constituents that are used by all life forms and will then consider the rules that govern how biochemical information is accessed and how it is passed from one generation to the next. Finally, we will take an overview of the fundamental unit of life—the cell. This is just the beginning. All of the molecules and structures that we see in this chapter we will meet again and again as we explore the chemical basis of life.

1.1   Living Systems Require a Limited Variety of Atoms and Molecules

Ninety naturally occurring elements have been identified, yet only three—oxygen, hydrogen, and carbon—make up 98% of the atoms in an organism. Moreover, the abundance of these three elements in life is vastly different from their abundance in Earth's crust (Table 1.1). What can account for the disparity between what is available and what organisms are made of?

One reason that oxygen and hydrogen are so common is the ubiquity of water, or "the matrix of life," as biochemist Albert Szent-Györgi called it. This tiny molecule—consisting of only three atoms—makes life on Earth possible. Indeed, current belief is that all life requires water, which is why so much effort has been made in recent decades to determine whether Mars had water in the past and whether it still does. The importance of water for life is so crucial that its presence is tantamount to saying that life could be present. We will consider the properties of water and how these properties facilitate biochemistry in Chapter 2.

After oxygen and hydrogen, the next most-common element in living organisms is carbon. Most large molecules in living systems are made up predominantly of carbon. Fuel molecules are made entirely of carbon, hydrogen,

Table 1.1 Chemical compositions as percentage of total number of atoms

| Element | Composition in | | |
	Human beings (%)	Seawater (%)	Earth's crust (%)
Hydrogen	63	66	0.22
Oxygen	25.5	33	47
Carbon	9.5	0.0014	0.19
Nitrogen	1.4	<0.1	<0.1
Calcium	0.31	0.006	3.5
Phosphorus	0.22	<0.1	<0.1
Chloride	0.03	0.33	<0.1
Potassium	0.06	0.006	2.5
Sulfur	0.05	0.017	<0.1
Sodium	0.03	0.28	2.5
Magnesium	0.01	0.003	2.2
Silicon	<0.1	<0.1	28
Aluminum	<0.1	<0.1	7.9
Iron	<0.1	<0.1	4.5
Titanium	<0.1	<0.1	0.46
All others	<0.1	<0.1	<0.1

Note: Because of rounding, total percentages do not equal 100%.
Source: Data from E. Frieden, The chemical elements of life, *Sci. Am.* 227(1), 1972, p. 54.

and oxygen. Biological fuels, like the fuels that power machinery, react with oxygen to produce carbon dioxide and water. In regard to biological fuels, this reaction, called combustion, provides the energy to power the cell. As a means of seeing why carbon is uniquely suited for life, let us compare it with silicon, its nearest elemental relative. Silicon is much more plentiful than carbon in Earth's crust (Table 1.1), and, like carbon, can form four covalent bonds—a property crucial to the construction of large molecules. However, carbon-to-carbon bonds are stronger than silicon-to-silicon bonds. This difference in bond strength has two important consequences. First, large molecules can be built with the use of carbon–carbon bonds as the backbone because of the stability of these bonds. Second, more energy is released when carbon–carbon bonds undergo combustion than when silicon reacts with oxygen. Thus, carbon-based molecules are stronger construction materials and are better fuels than silicon-based molecules. Carbon even has an advantage over silicon after it has undergone combustion. Carbon dioxide is readily soluble in water and can exist as a gas; thus, it remains in biochemical circulation, given off by one tissue or organism to be used by another tissue or organism. In contrast, silicon is essentially insoluble after reaction with oxygen. After it has combined with oxygen, it is permanently out of circulation. Quartz is a common form of silicon dioxide.

Other elements have essential roles in living systems—notably, nitrogen, phosphorus, and sulfur. Moreover, some of the trace elements, although present in tiny amounts compared with oxygen, hydrogen, and carbon, are absolutely vital to a number of life processes. We will see specific uses of these elements as we proceed with our study of biochemistry.

1.2 There Are Four Major Classes of Biomolecules

Living systems contain a dizzying array of biomolecules. However, these biomolecules can be divided into just four classes: proteins, nucleic acids, lipids, and carbohydrates.

✓ 1 Describe the key classes of biomolecules and differentiate between them.

Proteins Are Highly Versatile Biomolecules

Much of our study of biochemistry will revolve around proteins. *Proteins* are constructed from 20 building blocks, called amino acids, linked by peptide bonds to form long unbranched polymers (**Figure 1.1**). These polymers fold into precise three-dimensional structures that facilitate a vast array of biochemical functions. Proteins serve as signal molecules (e.g., the hormone insulin signals that fuel is in the blood) and as receptors for signal molecules. Receptors convey to the cell that a signal has been received and initiates the cellular response. Thus, for example, insulin binds to its particular receptor, called the insulin receptor, and initiates the biological response to the presence of fuel in the blood. Proteins also play structural roles, allow mobility, and provide defenses against environmental

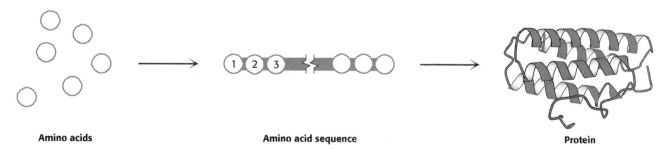

Amino acids   **Amino acid sequence**   **Protein**

Figure 1.1 Protein folding. The three-dimensional structure of a protein is dictated by the sequence of amino acids that constitute the protein.

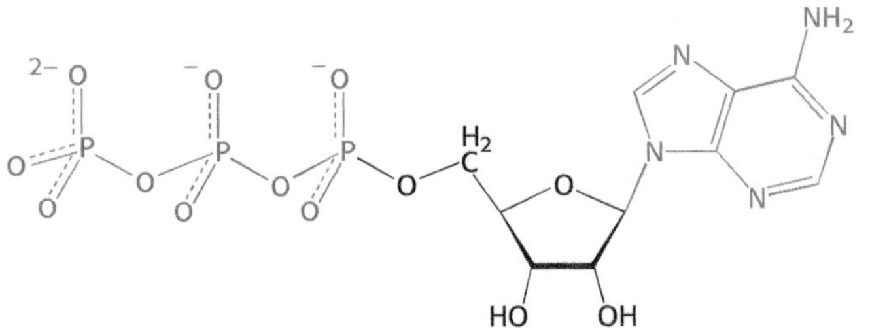

Adenosine triphosphate (ATP)

Figure 1.2 The structure of a nucleotide. A nucleotide (in this case, adenosine triphosphate) consists of a base (shown in blue), a five-carbon sugar (black), and at least one phosphoryl group (red).

dangers. Perhaps the most prominent role of proteins is that of *catalysts*—agents that enhance the rate of a chemical reaction without being permanently affected themselves. Protein catalysts are called *enzymes.* Every process that takes place in living systems depends on enzymes.

Nucleic Acids Are the Information Molecules of the Cell

As information keepers of the cell, the primary function of *nucleic acids* is to store and transfer information. They contain the instructions for all cellular functions and interactions. Like proteins, nucleic acids are linear molecules. However, nucleic acids are constructed from only four building blocks called *nucleotides.* A nucleotide is made up of a five-carbon sugar, either a deoxyribose or a ribose, attached to a heterocyclic ring structure called a base and at least one phosphoryl group (**Figure 1.2**).

There are two types of nucleic acid: *deoxyribonucleic acid* (DNA) and *ribonucleic acid* (RNA). Genetic information is stored in DNA—the "parts list" that determines the nature of an organism. DNA is constructed from four deoxyribonucleotides, differing from one another only in the ring structure of the bases—adenine (A), cytosine (C), guanine (G), and thymine (T). The information content of DNA is the sequence of nucleotides linked together by phosphodiester linkages. DNA in all higher organisms exists as a double-stranded helix (**Figure 1.3**). In the double helix, the bases interact with one another—A with T and C with G.

Figure 1.3 The double helix. Two individual chains of DNA interact to form a double helix. The sugar–phosphate backbone of one of the two chains is shown in red; the other is shown in blue. The bases are shown in green, purple, orange, and yellow.

RNA is a single-stranded form of nucleic acid. Some regions of DNA are copied as a special class of RNA molecules called messenger RNA (mRNA). mRNA is a template for the synthesis of proteins. Unlike DNA, mRNA is frequently broken down after use. RNA is similar to DNA in composition with two exceptions: the base thymine (T) is replaced by the base uracil (U), and the sugar component of the ribonucleotides contains an additional hydroxyl (—OH) group.

Lipids Are a Storage Form of Fuel and Serve as a Barrier

Among the key biomolecules, *lipids* are much smaller than proteins or nucleic acids. Whereas proteins and nucleic acids can have molecular weights of thousands to millions, a typical lipid has a molecular weight of 1300 g mol^{-1}. Moreover, lipids are not polymers made of repeating units, as are proteins and nucleic acids. A key characteristic of many biochemically important lipids is their dual chemical nature: part of the molecule is *hydrophilic,* meaning that it can dissolve in water, whereas the other part, made up of one or more hydrocarbon chains, is *hydrophobic* and cannot dissolve in water (**Figure 1.4**). This dual nature allows lipids to form barriers that delineate the cell from its environment and to establish intracellular compartments. In other words, lipids allow the development of "inside" and "outside" at a biochemical level. The hydrocarbon chains cannot interact with water and, instead, interact with those of other lipids to form a barrier, or membrane, whereas the water-soluble components interact with the aqueous environment on either side of the membrane. Lipids are also an important storage form of energy. As we will see, the hydrophobic component of lipids can undergo combustion to provide large amounts of cellular energy. Lipids are crucial signal molecules as well.

(A)

Hydrophobic tail

Hydrophilic head

(B)

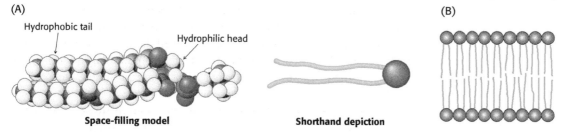

Space-filling model   **Shorthand depiction**

Figure 1.4 The dual properties of lipids. (A) One part of a lipid molecule is hydrophilic; the other part is hydrophobic. (B) In water, lipids can form a bilayer, constituting a barrier that separates two aqueous compartments.

Carbohydrates Are Fuels and Informational Molecules

Most of us already know that *carbohydrates* are an important fuel source for most living creatures. The most-common carbohydrate fuel is the simple sugar glucose. Glucose is stored in animals as *glycogen,* which consists of many glucose molecules linked end-to-end and has occasional branches (**Figure 1.5**). In plants, the storage form of glucose is starch, which is similar to glycogen in molecular composition.

There are thousands of different carbohydrates. They can be linked together in chains, and these chains can be highly branched, much more so than in glycogen or starch. Such chains of carbohydrates play important roles in helping cells to recognize one another. Many of the components of the cell exterior are decorated with various carbohydrates that can be identified by other cells and serve as sites of cell-to-cell interactions.

CH_2OH

H   O OH

H

HO   OH H

H

H   OH

Glucose

? QUICK QUIZ 1 Name the four classes of biomolecules, and state an important function of each class.

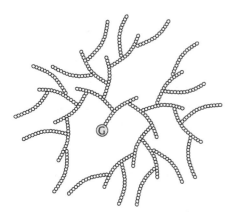

Figure 1.5 The structure of glycogen. Glycogen is a branched polymer composed of glucose molecules. The protein identified by the letter G at the center of the glycogen molecule is required for glycogen synthesis (Chapter 25).

1.3 The Central Dogma Describes the Basic Principles of Biological Information Transfer

✓ 2 List the steps of the central dogma.

Information processing in all cells is quite complex. It increases in complexity as cells become parts of tissues and as tissues become components of organisms. The scheme that underlies information processing at the level of gene expression was first proposed by Francis Crick in 1958.

Replication

DNA $\xrightarrow{\text{Transcription}}$ RNA $\xrightarrow{\text{Translation}}$ Protein

Crick called this scheme the *central dogma:* information flows from DNA to RNA and then to protein. Moreover, DNA can be replicated. The basic tenets of this dogma are true, but, as we will see later, this scheme is not as simple as depicted.

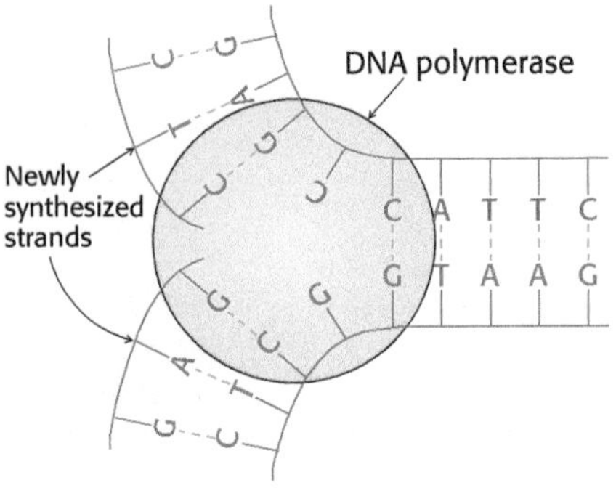

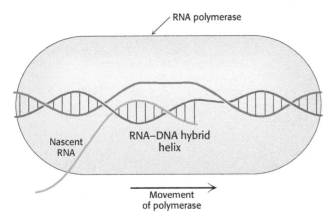

Figure 1.6 DNA replication. When the two strands of a DNA molecule are separated, each strand can serve as a template for the synthesis of a new partner strand. DNA polymerase catalyzes replication.

DID YOU KNOW?

As defined in the Oxford English Dictionary, to transcribe means to make a copy of (something) in writing; to copy out from an original; to write (a copy).

DNA constitutes the heritable information—the *genome*. This information is packaged into discrete units called *genes*. It is this collection of genes that determines the physical nature of the organism. When a cell duplicates, DNA is copied and identical genomes are then present in the newly formed daughter cells. The process of copying the genome is called *replication*. A group of enzymes, collectively called *DNA polymerase*, catalyze the replication process (**Figure 1.6**).

Genes are useless in and of themselves. The information must be rendered accessible. This accessibility is achieved in the process of *transcription* through which one form of nucleic acid, DNA, is transcribed into another form, RNA. The enzyme *RNA polymerase* catalyzes this process (**Figure 1.7**). Which genes are transcribed, as well as when and where they are transcribed, is crucial to the fate of the cell. For instance, although each cell in a human body has the DNA information that encodes the instructions to make all tissues, this information is parceled out. The genes expressed in the liver are different from those expressed in the muscles and brain. *Indeed, it is this selective expression that defines the function of a cell or tissue.*

A key aspect of the selective expression of genetic information is the transcription of genes into mRNA. The information encoded in mRNA is realized in the process of *translation* because information is literally translated from one chemical form (nucleic acid) into another (protein). Proteins have been described as the workhorses of the cell, and *translation renders the genetic information into a functional form.* Translation takes place on large macromolecular complexes called *ribosomes*, consisting of RNA and protein (**Figure 1.8**).

Now that you have been introduced to the key biomolecules and have briefly examined the central dogma of information transfer, let us look at the platform—the cell—that contains and coordinates the biochemistry required for life.

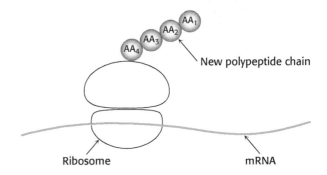

Figure 1.7 The transcription of RNA. Transcription, catalyzed by RNA polymerase, makes an RNA copy of one of the strands of DNA.

Figure 1.8 Translation takes place on ribosomes. A ribosome decodes the information in mRNA and translates it into the amino acid sequence of a protein.

✓ 3 Identify the key features that differentiate eukaryotic cells from prokaryotic cells.

1.4 Membranes Define the Cell and Carry Out Cellular Functions

The cell is the basic unit of life. Cells grow, replicate, and interact with their environment. Living organisms can be as simple as a single cell or as complex as a human body, which is composed of approximately 100 trillion cells. Every cell is delineated by a membrane that separates the inside of the cell from its environment. A *membrane* is a *lipid bilayer:* two layers of lipids organized with their hydrophobic chains interacting with one another and the hydrophilic head groups

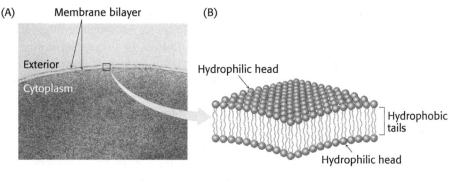

(A)

Membrane bilayer

Exterior

Cytoplasm

(B)

Hydrophilic head

Hydrophobic tails

Hydrophilic head

Figure 1.9 The bilayer structure of a membrane. (A) Membranes are composed of two layers or sheets. (B) The hydrophobic parts of the layers interact with each other, and the hydrophilic parts interact with the environment. [J. D. Robertson. "Discovery in Cell Biology: Membrane Structure." *Journal of Cell Biology* 91(1981): 189s–204s. Courtesy of J.D. Robertson.]

interacting with the environment (**Figure 1.9**).

There are two basic types of cells: eukaryotic cells and prokaryotic cells (**Figure 1.10**). The main difference between the two is the existence of membrane-enclosed compartments in *eukaryotes* and the absence of such compartments in *prokaryotes*. Prokaryotic cells, exemplified by the human gut

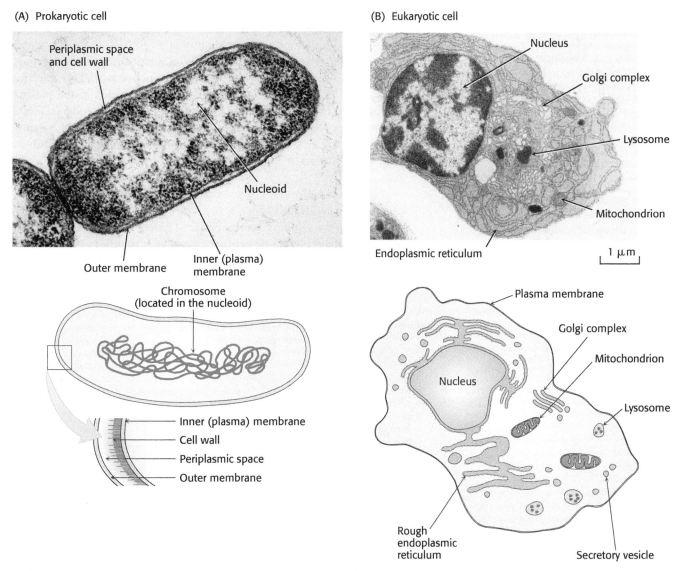

(A) Prokaryotic cell

Periplasmic space and cell wall

Nucleoid

Outer membrane

Inner (plasma) membrane

Chromosome (located in the nucleoid)

Inner (plasma) membrane
Cell wall
Periplasmic space
Outer membrane

(B) Eukaryotic cell

Nucleus

Golgi complex

Lysosome

Mitochondrion

Endoplasmic reticulum

1 μm

Plasma membrane

Golgi complex

Mitochondrion

Nucleus

Lysosome

Rough endoplasmic reticulum

Secretory vesicle

Figure 1.10 Prokaryotic and eukaryotic cells. Eukaryotic cells display more internal structure than do prokaryotic cells. Components within the interior of a eukaryotic cell, most notably the nucleus, are defined by membranes. [Micrographs: (A) ©Biology Pics/Science Source; (B) from P. C. Cross and K. L. Mercer, *Cell Tissue Ultrastructure: A Functional Perspective* (W. H. Freeman and Company, 1993), p. 199. Diagrams: (A and B) Information from H. Lodish et al., *Molecular Cell Biology*, 6th ed. (W. H. Freeman and Company, 2008), p. 3.]

bacterium *Escherichia coli*, have a relatively simple structure. They are surrounded by two membranes separated by the periplasmic space. Although human beings are composed of 100 trillion cells, we carry more than 10 times that number of bacteria in us and on us. For the most part, our attitude toward our prokaryotic colleagues is to "live and let live." For example, the prokaryotes living in our intestines assist us in the process of digestion. Prokaryotes are responsible for making our lives richer in other ways. Various prokaryotes provide us with buttermilk, yogurt, and cheese. Nevertheless, prokaryotes also cause a wide array of diseases.

Regardless of cell type—eukaryotic or prokaryotic, plant or animal—two biochemical features minimally constitute a cell: there must be (1) a barrier that separates the cell from its environment and (2) an inside that is chemically different from the environment and that accommodates the biochemistry of living. The barrier is called the *plasma membrane,* and the fundamental intracellular material is called the *cytoplasm.*

The plasma membrane The plasma membrane separates the inside of the cell from the outside, one cell from another cell. This membrane is impermeable to most substances, even to substances such as fuels, building blocks, and signal molecules that must enter the cell. Consequently, the barrier function of the membrane must be mitigated to permit the entry and exit of molecules and information. In other words, the membrane must be rendered semipermeable but in a very selective way. This *selective permeability* is the work of proteins that are embedded in the plasma membrane or associated with it (**Figure 1.11**). These proteins facilitate the entrance of fuels, such as glucose, and building blocks, such as amino acids, and they transduce information—for example, that insulin is in the blood stream.

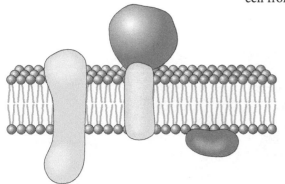

Figure 1.11 Membrane proteins. Proteins, embedded (yellow) in membranes and attached (blue) to them, permit the exchange of material and information with the environment.

The plant cell wall The plasma membrane of a plant is itself surrounded by a cell wall (**Figure 1.12**). The cell wall is constructed largely from cellulose, a long, linear polymer of glucose molecules. Cellulose molecules interact with one another as well as with other cell-wall components to form a sturdy protective wall for the cell.

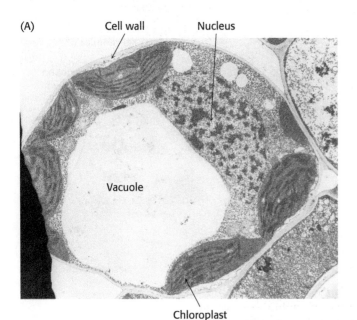

(A)

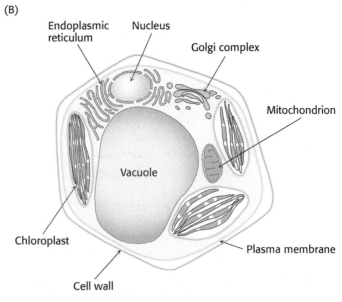

(B)

Figure 1.12 A plant cell. (A) Photomicrograph of a leaf cell. (B) Diagram of a typical plant cell. [(A) Biophoto Associates/Science Source.]

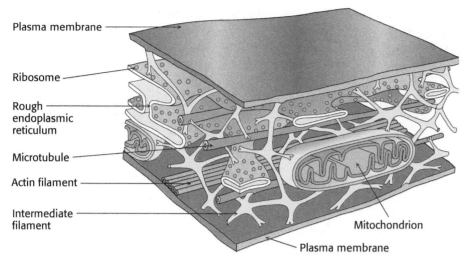

Plasma membrane

Ribosome

Rough endoplasmic reticulum

Microtubule

Actin filament

Intermediate filament

Mitochondrion

Plasma membrane

Figure 1.13 The cytoskeleton. Actin filaments, intermediate filaments, and microtubules are components of the cytoskeleton, which provides cell shape and contributes to cell movement. These components course throughout the cytoplasm, associating with all other cellular organelles. [Information from W. K. Purves et al., *Life*, 7th ed. (W. H. Freeman and Company, 2004), p. 80.]

The cytoplasm The inner substance of the cell, the material that is surrounded by the plasma membrane, is called the cytoplasm. The cytoplasm is the site of a host of biochemical processes, including the initial stage of glucose metabolism, fatty acid synthesis, and protein synthesis. Formerly, the cytoplasm was believed to be a "soup" of important biomolecules, but it is becoming increasingly clear that the biochemistry of the cytoplasm is highly organized by a network of structural filaments called the *cytoskeleton*. In many eukaryotes, the cytoskeleton is a network of three kinds of protein fibers—actin filaments, intermediate filaments, and microtubules—that support the structure of the cell, help to localize certain biochemical activities, and even serve as "molecular highways" by which molecules can be shuttled around the cell (**Figure 1.13**).

Biochemical Functions Are Sequestered in Cellular Compartments

A key difference between eukaryotic cells and prokaryotic cells is the presence of a complex array of intracellular, membrane-bounded compartments called *organelles* in eukaryotes (Figure 1.10B). We will now tour the cell to investigate prominent organelles, which we will see many times in our study of biochemistry.

The nucleus The largest organelle is the *nucleus,* which is a double-membrane-bounded organelle (**Figure 1.14**). *The nucleus is the information center of the cell*, the location of an organism's genome. The nuclear membrane is punctuated with pores that allow transport into and out of the nucleus. For instance, the molecular machines that synthesize DNA and RNA are formed in the cytoplasm, but function in the nucleus. The nucleus is where the genomic information is selectively expressed at the proper time and in the proper amount.

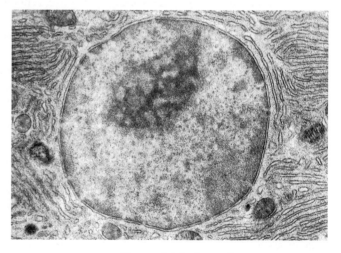

Figure 1.14 The nucleus. [Don W. Fawcett/Science Source.]

The mitochondrion The *mitochondrion* (plural, *mitochondria*) has two membranes—an outer mitochondrial membrane that is in touch with the cytoplasm and an inner mitochondrial membrane that defines the matrix of the mitochondrion—the mitochondrial equivalent of the cytoplasm. The inner mitochondrial membrane is highly invaginated (**Figure 1.15**). The space between the two membranes is the intermembrane space. In mitochondria, fuel molecules undergo combustion into carbon dioxide and water with the generation of cellular energy, adenosine triphosphate (ATP). Approximately 90% of the energy used by a typical cell is produced in this organelle. Poisons such as cyanide and

(A)

(B)

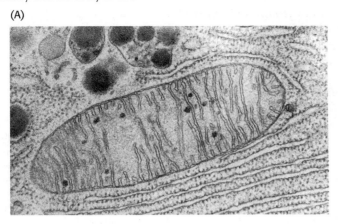

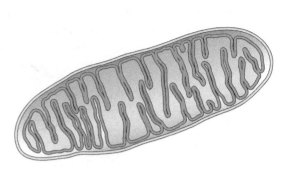

Figure 1.15 The mitochondrion. The mitochondrial matrix is shown in light blue in part B. [(A) Keith R. Porter/Science Source.]

carbon monoxide are so deadly precisely because they shut down the functioning of mitochondria. When we study the biochemistry of this organelle in detail, we will see that the structure of the mitochondrion plays an intimate role in its biochemical functioning.

The chloroplast Another double-membrane-bounded organelle vital to all life, but found only in plant cells, is the *chloroplast* (Figure 1.12). Chloroplasts power the plant cell, the plant, and the rest of the living world. A chloroplast is the site of a remarkable biochemical feat: the conversion of sunlight into chemical energy, a process called *photosynthesis*. Every meal that we consume, be it a salad or a large cut of juicy steak, owes its existence to photosynthesis. If photosynthesis were to halt, life on Earth would cease in about 25 years. The mass extinction of the Cretaceous period (65.1 million years ago), in which the dinosaurs met their demise, is believed to have been caused by a large meteor strike that propelled so much debris into the atmosphere that sunlight could not penetrate and photosynthesis ceased.

Some Organelles Process and Sort Proteins and Exchange Material with the Environment

Let us briefly examine other eukaryotic organelles (Figure 1.10B) in the context of how they cooperate with one another to perform vital biochemical tasks.

The endoplasmic reticulum The *endoplasmic reticulum* is a series of membranous sacs. Many biochemical reactions take place on the cytoplasmic surface of these sacs as well as in their interiors, or lumens. The endoplasmic reticulum comes in two types: the *smooth endoplasmic reticulum* (smooth ER, or SER) and the *rough endoplasmic reticulum* (rough ER, or RER), as illustrated in **Figure 1.16** (also Figure 1.10). The smooth endoplasmic reticulum plays a variety of roles, but an especially notable role is the processing of exogenous chemicals (chemicals originating outside the cell) such as drugs. The more drugs, including alcohol, ingested by an organism, the greater the quantity of smooth endoplasmic reticulum in the liver.

The rough endoplasmic reticulum appears rough because ribosomes are attached to the cytoplasmic side (Figure 1.14). Ribosomes that are free in the cytoplasm take part in the synthesis of proteins for use inside the cell. Ribosomes attached to the rough endoplasmic reticulum synthesize proteins that will either be inserted into cellular membranes or be secreted from the cell.

Proteins synthesized on the rough endoplasmic reticulum are transported into the lumen of the endoplasmic reticulum during the process of translation.

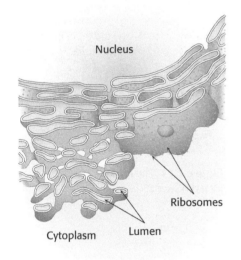

Nucleus

Ribosomes

Cytoplasm   Lumen

Figure 1.16 The endoplasmic reticulum (ER). Smooth ER lacks ribosomes. Rough ER has ribosomes attached to it. [Information from D. Sadava et al., *Life*, 8th ed. (W. H. Freeman and Company, 2008), p. 77.]

Inside the lumen of the rough endoplasmic reticulum, a protein folds into its final three-dimensional structure, with the assistance of other proteins called *chaperones,* and is often modified, for instance, by the attachment of carbohydrates. The folded, modified protein then becomes sequestered into regions of the rough endoplasmic reticulum that lack ribosomes. These regions bud off the rough endoplasmic reticulum as *transport vesicles.*

The Golgi complex The transport vesicles from the rough endoplasmic reticulum are carried to the *Golgi complex*—a series of stacked membranes—and fuse with it (**Figure 1.17**). Further processing of the proteins that were contained in the transport vesicles takes place in the Golgi complex. In particular, a different set of carbohydrates is added. Proteins with different fates are sorted in the Golgi complex.

Secretory granules A *secretory granule,* or zymogen granule, is formed when a vesicle filled with the proteins destined for secretion buds off the Golgi complex. The granule is directed toward the cell membrane. When the proper signal is received, the secretory granule fuses with the plasma membrane and dumps its cargo into the extracellular environment, a process called *exocytosis* (**Figure 1.18**).

The endosome Material is taken into the cell when the plasma membrane invaginates and buds off to form an *endosome* (not shown in Figure 1.10B). This process is called *endocytosis,* which is the opposite of exocytosis. Endocytosis is used to bring important biochemicals such as iron ions, vitamin B_{12}, and cholesterol into the cell. Endocytosis takes place through small regions of the membrane, such as when a protein is taken into the cell (**Figure 1.19**). Alternatively, large amounts of material can also be taken into the cell. When large amounts of material are taken into the cell, the process is called *phagocytosis*. **Figure 1.20** shows an immune-system cell, called a macrophage, phagocytizing bacteria. Macrophages phagocytize bacteria as a means of protecting an organism from infection. What is the fate of the vesicles formed by endocytosis or phagocytosis?

Lysosomes The *lysosome* is an organelle that contains a wide array of digestive enzymes. Lysosomes form in a manner analogous to the formation of secretory granules, but lysosomes fuse with endosomes instead of the cell membrane. After fusion has taken place, the lysosomal enzymes digest the material, releasing small molecules that can be used as building blocks or fuel by the cell.

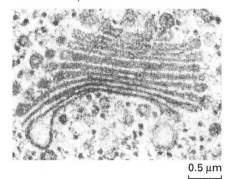

0.5 µm

Figure 1.17 The Golgi complex. [Courtesy of L. Andrew Staehelin, University of Colorado.]

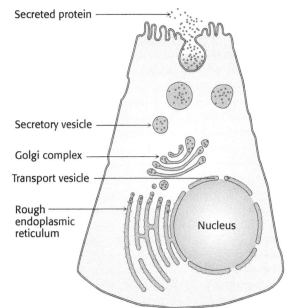

Figure 1.18 Exocytosis. The secretory pathway. [Information from H. Lodish et al., *Molecular Cell Biology*, 5th ed. (W. H. Freeman and Company, 2004), p. 169.]

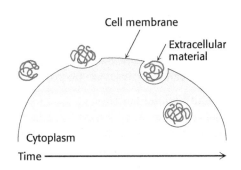

Figure 1.19 Endocytosis.

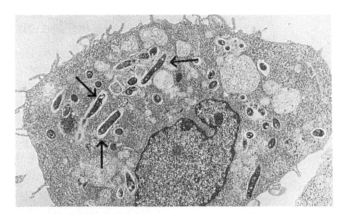

Figure 1.20 Phagocytosis. Bacteria (indicated by arrows) are phagocytosized by a macrophage. [Courtesy of Dr. Stanley Falkow.]

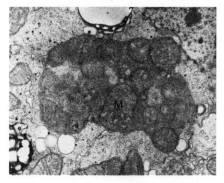

Figure 1.21 A lysosome. A micrograph of a lysosome in the process of digesting several mitochondria, one of which is labeled (M). [©Biophoto Associates/Science Source.]

? QUICK QUIZ 2 Name three organelles or structures found in plant cells but not in animal cells.

Lysosomes do not just degrade extracellular material, however. Another role is the digestion of damaged intracellular organelles (**Figure 1.21**).

Plant vacuoles Another organelle unique to plant cells, in addition to the chloroplast, is a large vacuole. In some plant cells, this single-membrane-bounded organelle may occupy as much as 80% of a cell's volume (Figure 1.12). These vacuoles store water, ions, and various nutrients. For instance, the vacuoles of citrus fruits are rich in citric acid, which is responsible for the tart taste of these fruits. Proteins transport the molecules across the vacuolar membrane.

⚕ CLINICAL INSIGHT

Defects in Organelle Function May Lead to Disease

Many pathological conditions arise due to malfunctions in various organelles. For instance, familial hypercholesterolemia, a disease in which children as young as 6 years old die of heart attacks, is caused by the inefficient endocytosis of cholesterol from the blood. The resulting high levels of cholesterol in the blood result in heart attacks. Tay–Sachs disease, characterized by muscle weakness, dementia, and death at an early age, usually before the age of 3, results from improper lysosome function. We will revisit these disorders and examine many others as we progress in our study of biochemistry.

Cellular organization attests to the high information content of the cell. But this brief overview has only touched the surface of the information processing that must take place to construct something as sophisticated as a cell. In the rest of this textbook, we will examine the biochemical energy and information pathways that construct and maintain living systems.

SUMMARY

1.1 Living Systems Require a Limited Variety of Atoms and Molecules
Oxygen, hydrogen, and carbon make up 98% of the atoms in living organisms. Hydrogen and oxygen are prevalent because of the abundance of water, and carbon is the most common atom in all biomolecules.

1.2 There Are Four Major Classes of Biomolecules
Proteins, nucleic acids, lipids, and carbohydrates constitute the four major classes of biomolecules. Proteins are the most versatile with an especially prominent role as enzymes. Nucleic acids are primarily information molecules: DNA is the genetic information in most organisms, whereas RNA plays a variety of roles, including serving as a link between DNA and proteins. Lipids serve as fuels and as membranes. Carbohydrates are key fuel molecules that also play a role in cell-to-cell interactions.

1.3 The Central Dogma Describes the Basic Principles of Biological Information Transfer
The central dogma of biology states that DNA is replicated to form new DNA molecules. DNA can also be transcribed to form RNA. Some information in the form of RNA, called messenger RNA, can be translated into proteins.

1.4 Membranes Define the Cell and Carry Out Cellular Functions
Membranes, formed of lipid bilayers, are crucial for establishing boundaries between cells and their environment and for establishing boundaries within internal regions of many cells. There are two structurally distinct types of cells: eukaryotic cells and prokaryotic cells. Eukaryotic cells are characterized by a complex array of intracellular membrane-bounded compartments called organelles. The nucleus is the largest organelle and

houses the genetic information of the cell. Other organelles play roles in energy transformation, in protein processing and secretion, and in digestion. In contrast, prokaryotic cells are smaller and less complex, and they lack membrane-bounded compartments.

KEY TERMS

unity of biochemistry (p. 3)
proteins (p. 5)
nucleic acids (p. 6)
nucleotides (p. 6)
deoxyribonucleic acid (DNA) (p. 6)
ribonucleic acid (RNA) (p. 6)
lipids (p. 6)
carbohydrates (p. 7)
glycogen (p. 7)
central dogma (p. 7)
replication (p. 8)

transcription (p. 8)
translation (p. 8)
membrane (p. 8)
lipid bilayer (p. 8)
eukaryotes (p. 9)
prokaryotes (p. 9)
plasma membrane (p. 10)
cytoplasm (p. 10)
cytoskeleton (p. 11)
nucleus (p. 11)
mitochondria (p. 11)

chloroplasts (p. 12)
endoplasmic reticulum (ER) (p. 12)
Golgi complex (p. 13)
secretory (zymogen) granules (p. 13)
exocytosis (p. 13)
endosomes (p. 13)
endocytosis (p. 13)
phagocytosis (p. 13)
lysosomes (p. 13)

? Answers to QUICK QUIZZES

1. Proteins: catalysts. Nucleic acids: information transfer. Lipids: fuel and structure. Carbohydrates: fuel and cell-to-cell communication.

2. Chloroplasts, vacuoles, and cell wall.

PROBLEMS

1. *E. coli and elephants.* What is meant by the phrase "unity of biochemistry"? What are the implications of the unity of biochemistry?

2. *Similar, but not the same.* Describe the structural differences between DNA and RNA. ✓ 1

3. *Polymers.* Differentiate between proteins and glycogen in regard to their polymeric structure. ✓ 1

4. *An authoritative belief.* Define the central dogma of biology. ✓ 2

5. *Information processing.* Define replication, transcription, and translation in regard to the central dogma. ✓ 2

6. *Hurry it up.* What is an enzyme? ✓ 1

7. *Complex and less so.* Differentiate between eukaryotic cells and prokaryotic cells. ✓ 3

8. *A diminutive organ?* What is an organelle? ✓ 3

9. *Double bounded.* Which organelles are surrounded by two membranes?

10. *Perforated.* How does the nuclear membrane differ from other membranes?

11. *An exit strategy.* Trace the pathway of the formation of a secretory protein from its gene to its exocytosis from the cell.

12. *Function and structure.* Match the function with the appropriate organelle in the column at right.

(a) Endoplasmic reticulum
_____

(b) Smooth endoplasmic reticulum _____

(c) Rough endoplasmic reticulum _____

(d) Golgi complex _____

(e) Transport vesicles
_____

(f) Secretory granules
_____

(g) Endosome _____

(h) Lysosome _____

(i) Mitochondrion
_____

(j) Chloroplast _____

(k) Nucleus _____

(l) Plasma membrane
_____

1. Location of most of the cell's DNA

2. Site of fuel oxidation

3. Separates the inside of the cell from the outside

4. Carries important biochemicals into the cell

5. Membrane with ribosomes attached

6. Site of photosynthesis

7. Contains digestive enzymes

8. Destined for fusion with the plasma membrane

9. Common form of cytoplasmic membrane

10. Site of carbohydrate addition to proteins

11. Facilitate communication between the rough endoplasmic reticulum and the Golgi complex

12. Processes exogenous chemicals

Selected Readings for this chapter can be found online at www.whfreeman.com/tymoczko3e.

CHAPTER 2

Water, Weak Bonds, and the Generation of Order Out of Chaos

Our senses—vision, taste, smell, hearing, and touch—allow us to experience the world. We delight in the softness of a kitten's fur and the loudness of its purr through touch and hearing. Remarkably, these sensuous pleasures depend on weak, reversible chemical bonds. [Marc Hill/Alamy.]

Cells, as shown in Chapter 1, present a remarkable display of functional order. Millions of individual molecules are the cell's building blocks, consisting of the four key biomolecules of life—proteins, nucleic acids, lipids, and carbohydrates. These molecules are, for the most part, stable because they are constructed with strong covalent bonds—bonds in which the electrons are shared by the participating atoms. However, the remarkable structure and function of the cell itself are stabilized by weak interactions that have only a fraction of the strength of covalent bonds.

Two questions immediately come to mind: How is such stabilization possible? And why is it advantageous? The answer to the first question is that there is stability in numbers. Many weak bonds can result in large stable structures. The answer to the second question is that weak bonds allow transient interactions. A substrate can bind to an enzyme, and the product can leave the enzyme. A hormone can bind to its receptor and then dissociate from the receptor after the signal has been received. Weak bonds allow for dynamic interactions and permit energy and information to move about the cell and organism. *Transient chemical interactions form the basis of biochemistry and life itself.*

Water is the solvent of life and greatly affects weak bonds, making some weaker and powering the formation of others. For instance, hydrophobic molecules, such as fats, cannot interact with water at all. Yet, this chemical antipathy is put to use. The formation of membranes and the intricate three-dimensional structure of biomolecules, most notably proteins, are powered by an energetic solution to the chemical opposition between water and hydrophobic molecules.

Our experience of life happens at a distance of 4 angstroms (4 Å, or 0.4 nm), the typical length of noncovalent bonds. The pressure of a held hand, the feeling of a kiss, the reading of the words on this page—all of these sensations are the result of large, covalently bonded molecules interacting noncovalently with a vast array of other large molecules or with sodium ions, photons, or an assortment of other signal molecules, all at a distance of approximately 4 Å.

In this chapter, we will focus on transient interactions between molecules—weak, reversible but essential interactions. We will see how molecules must meet before they can interact and will then examine the chemical foundations for the various weak interactions. Finally, we will examine especially prominent weak, reversible interactions—the ionization of water and weak acids.

> **DID YOU KNOW?**
>
> One angstrom (Å) = 0.1 nanometer (nm) = 1×10^{-10} meter (m). It is named after Swedish physicist Anders Jonas Ångström (1814–1874) who expressed wavelengths as multiples of 1×10^{-10} meter. That length was subsequently named an angstrom.

2.1 Thermal Motions Power Biological Interactions

In 1827, English botanist Robert Brown observed, under a microscope, pollen granules suspended in water. He noted that the granules darted randomly about and thought that he was observing the life force inherent in the pollen granules. He dismissed that idea when he observed the same behavior with dye particles in water or dust particles in air. The movement that he observed, subsequently referred to as *Brownian motion,* is a vital energy source for life. The movement of the particles that Brown observed is due to the random fluctuation of the energy content of the environment—thermal noise. The water and gas molecules of the environment are bouncing randomly about at a rate determined only by the temperature. When these molecules collide with pollen granules or dust motes, the particles move randomly themselves.

Brownian motion is responsible for initiating many biochemical interactions. In the context of the cell, water is the most common medium for the thermal noise of Brownian motion. Water is the lubricant that facilitates the flow of energy and information transformations through Brownian motion. Enzymes find their substrates; fuels can be progressively modified to yield energy, and signal molecules can diffuse from their sites of origin to their sites of effect, all through Brownian motion.

To be sure, the environment inside the cell is not as simple as just implied. Cells are not simply water-filled sacks with biomolecules bouncing about. As described in Chapter 1, a great deal of organization facilitates the Brownian-motion-driven exchange of metabolites and signal molecules. Examples of this organization will come up again many times in the course of our study of biochemistry.

Water is the medium for which Brownian motion provides the motive force for biochemical interactions. What are the properties of water that make it the perfect environment for life?

✓ 4 Describe the chemical properties of water and explain how water affects biochemical interactions.

2.2 Biochemical Interactions Take Place in an Aqueous Solution

Water is the solvent of life. Human beings are 65% water, tomatoes are 90% water, and a typical cell is about 70% water. Indeed, most organisms are mostly water, be they bacteria, cacti, whales, or elephants. Many of the organic molecules required for the biochemistry of living systems dissolve in water. In essence, water renders molecules mobile and permits Brownian-motion-powered interactions between molecules. What is the chemical basis of water's ability to dissolve so many biomolecules?

Water is a simple molecule, composed of two hydrogen atoms linked by covalent bonds to a single atom of oxygen. The important properties of water are due to the fact that oxygen is an electronegative atom. That is, although the bonds joining the hydrogen atoms to the oxygen atom are covalent, the electrons of the bond spend more time near the oxygen atom. Because the charge distribution is not uniform, the water molecule is said to be *polar*. The oxygen atom is slightly negatively charged (designated δ^-), and the hydrogen atoms are correspondingly slightly positively charged (δ^+).

This polarity has important chemical ramifications. The partially positively charged hydrogen atoms of one molecule of water can interact with the partially negatively charged oxygen atoms of another molecule of water. This interaction is called a *hydrogen bond* (**Figure 2.1**). As we will see, hydrogen bonds are not unique to water molecules and in fact are common weak bonds in biomolecules. Liquid water has a partly ordered structure in which hydrogen-bonded clusters of molecules are continually forming and breaking apart, with each molecule of water hydrogen-bonding to an average of 3.4 neighboring molecules. Hence, water is cohesive. The polarity of water and its ability to form hydrogen bonds renders it a solvent for any charged or polar molecule.

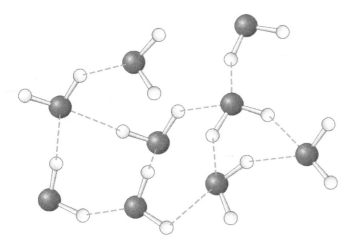

Figure 2.1 Hydrogen bonding in water. Hydrogen bonds (shown as dashed green lines) are formed between water molecules to produce a highly ordered and open structure.

The giant trees of the redwood forests—trees that can be hundreds of feet tall—are a remarkable demonstration of the cohesive power of water imparted by hydrogen bonds. Water rises to the treetops by transpiration, the evaporation of water from the topmost leaves. Transpiration pulls the water up from the roots. This column of water is maintained by hydrogen bonds between water molecules. Indeed, the strength of the hydrogen bond may play a limiting role in the ultimate height attained by a redwood tree. When the bonds break, an embolism (air bubble) forms, preventing further water flow to the top of the tree (**Figure 2.2**).

Although water's ability to dissolve many biochemicals is vitally important, the fact that water cannot dissolve certain compounds is equally important. A certain class of molecules termed *nonpolar*, or *hydrophobic*, cannot dissolve in water. These molecules, in the presence of water, behave exactly as the oil in an oil-and-vinegar salad dressing does: they sequester themselves away from the

Figure 2.2 Redwood forest. Hydrogen bonding allows water to travel from the roots to the top leaves of the giant redwoods. [Jorg Hackemann/Shutterstock.]

water, a process termed the hydrophobic effect (p. 23). However, living systems take great advantage of this chemical animosity to power the creation of many elaborate structures required for their continued existence. Indeed, cell and organelle membranes form because of the hydrophobic effect.

✓ 5 Describe the types of noncovalent, reversible interactions and explain why reversible interactions are important in biochemistry.

2.3 Weak Interactions Are Important Biochemical Properties

Readily reversible, noncovalent molecular interactions are essential interactions in the flow of energy and information. Such weak, noncovalent forces play roles in the faithful replication of DNA, the folding of proteins into elaborate three-dimensional forms, the specific recognition of reactants by enzymes, and the detection of molecular signals. *The three fundamental noncovalent bonds are (1) ionic bonds, or electrostatic interactions; (2) hydrogen bonds; and (3) van der Waals interactions.* They differ in geometry, strength, and specificity. Furthermore, these bonds are greatly affected in different ways by the presence of water. Let us consider the characteristics of each type.

Electrostatic Interactions Are Between Electrical Charges

Electrostatic interactions, also called *ionic bonds* or *salt bridges*, are the interactions between distinct electrical charges on atoms. They usually take place between atoms bearing a completely negative charge and a completely positive charge. The energy of an electrostatic interaction between two ions is given by *Coulomb's law*:

$$E = \frac{kq_1q_2}{Dr}$$

where E is the force, q_1 and q_2 are the charges on the two atoms (in units of the electronic charge), r is the distance between the two atoms (in angstroms), D is the dielectric constant (which accounts for the effects of the intervening medium), and k is the proportionality constant. Thus, the electrostatic interaction between two atoms bearing single opposite charges varies inversely with the square of the distance separating them as well as with the nature of the intervening medium. Electrostatic interactions are strongest in a vacuum, where $D = 1$. The distance for maximal bond strength is about 3 Å. Because of its polar characteristics, water (which has a dielectric constant of 80) weakens electrostatic interactions. Conversely, electrostatic interactions are maximized in an uncharged environment. For instance, the electrostatic interaction between two ions bearing single opposite charges separated by 3 Å in water has an energy of -5.8 kJ mol^{-1} (-1.4 kcal mol^{-1}), whereas that between the same two ions separated by 3 Å in a nonpolar solvent such as hexane (which has a dielectric constant of 2) has an energy of -231 kJ mol^{-1} (-55 kcal mol^{-1}). (Note: One kilojoule, abbreviated kJ, is equivalent to 0.239 kilocalorie, abbreviated kcal.)

Why does water weaken electrostatic interactions? Consider what happens when a grain of salt, NaCl, is added to water. Even in its crystalline form, salt is more appropriately represented in the ionic form, Na^+Cl^-. The salt dissolves—the ionic bond between Na^+ and Cl^- is destroyed—because the individual ions now bind to the water molecules rather than to each other (**Figure 2.3**). Water

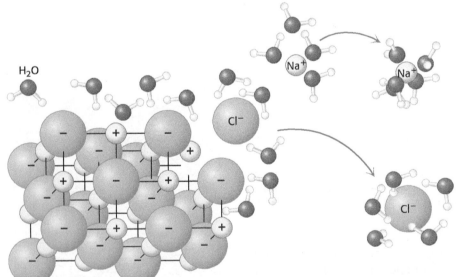

H_2O

Na^+ Na^+

Cl^-

Cl^-

Na^+Cl^-

Figure 2.3 Sodium chloride dissolves in water. As the sodium ions disperse, their positive charges are neutralized by the partially negative charges of the oxygen atoms of water. The chloride ions are surrounded by the partially positive charges on the hydrogen atoms. [Information from D. L. Nelson and M. M. Cox, *Lehninger Principles of Biochemistry*, 5th ed. (W. H. Freeman and Company, 2008), p. 47.]

can dissolve virtually any molecule that has sufficient partial or complete charges on the molecule to interact with water. This power to dissolve is crucial. Brownian motion powers collisions among the dissolved molecules, and many of these collisions result in fleeting but productive interactions.

Hydrogen Bonds Form Between an Electronegative Atom and Hydrogen

Hydrogen bonds are not unique to water molecules; the unequal distribution of charges that permit hydrogen-bond formation can arise whenever hydrogen is covalently bound to an electronegative atom. In biochemistry, the two most common electronegative atoms included in hydrogen bonds are oxygen and nitrogen. Typical hydrogen bonds that include these atoms are shown in **Figure 2.4**. Hydrogen bonds are much weaker than covalent bonds. They have energies ranging from 8 to 20 kJ mol^{-1} (from 2 to 5 kcal mol^{-1}) compared with approximately 418 kJ mol^{-1} (100 kcal mol^{-1}) for a carbon–hydrogen covalent bond. Hydrogen bonds are also somewhat longer than covalent bonds; their bond distances (measured from the hydrogen atom) range from 1.5 to 2.6 Å; hence, distances ranging from 2.4 to 3.5 Å separate the two nonhydrogen atoms in a hydrogen bond. Hydrogen bonds between two molecules will be disrupted by water, inasmuch as water itself forms hydrogen bonds with the molecules (**Figure 2.5**). Conversely, hydrogen bonding between two molecules is stronger in the absence of water.

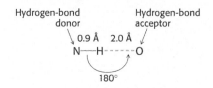

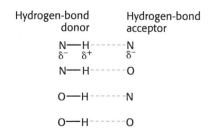

Figure 2.4 Hydrogen bonds that include nitrogen and oxygen atoms. The positions of the partial charges (δ^+ and δ^-) are shown.

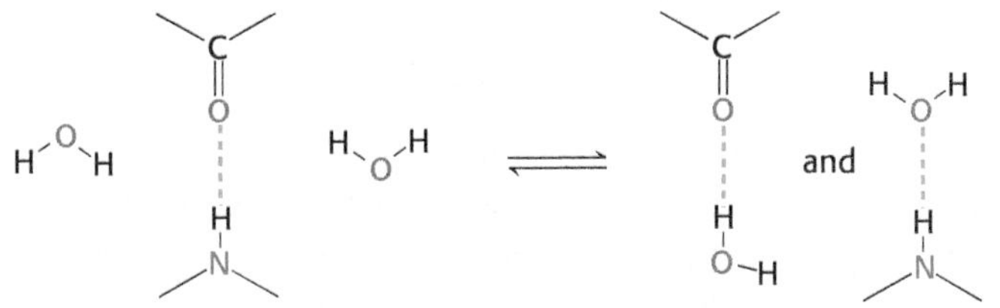

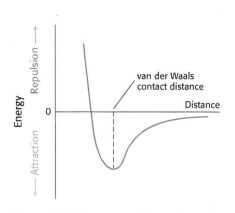

Figure 2.5 Disruption of hydrogen bonds. Competition from water molecules disrupts hydrogen bonds in other molecules.

van der Waals Interactions Depend on Transient Asymmetry in Electrical Charge

Many important biomolecules are neither polar nor charged. Nonetheless, such molecules can interact with each other electrostatically by a *van der Waals interaction*. The basis of a van der Waals interaction is that the distribution of electronic charge around an atom changes with time, and, at any instant, the charge distribution is not perfectly symmetric: there will be regions of partial positive charge and partial negative charge. This transient asymmetry in the electronic charge around an atom acts through electrostatic interactions to induce a complementary asymmetry in the electron distribution around its neighboring atoms. The resulting attraction between two atoms increases as they come closer to each other, until they are separated by the *van der Waals contact distance*, which corresponds to 3 to 4 Å, depending on the participating atoms (**Figure 2.6**). At a shorter distance, very strong repulsive forces become dominant because the outer electron clouds overlap. Energies associated with van der Waals interactions are quite small; typical interactions contribute from 2 to 4 kJ mol^{-1} (from 0.5 to 1.0 kcal mol^{-1}) per atom pair. However, when the surfaces of two large molecules with complementary shapes come together, a large number of atoms are in van der Waals contact, and the net effect, summed over many atom pairs, can be substantial. Although the motto of all weak electrostatic interactions might be "stability in numbers," the motto especially applies to van der Waals interactions.

A remarkable example of the power of van der Waals interactions is provided by geckos (**Figure 2.7**). These creatures can walk up walls and across ceilings,

Figure 2.6 The energy of a van der Waals interaction as two atoms approach each other. The energy is most favorable at the van der Waals contact distance. The energy rises rapidly owing to electron–electron repulsion as the atoms move closer together than this distance.

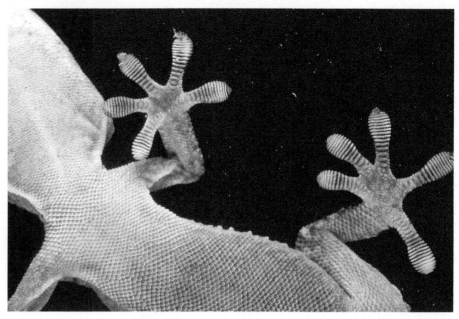

Figure 2.7 The power of van der Waals interactions. Geckos can cross a ceiling, held only by weak bonds called van der Waals forces. [Stephen Dalton/Science Source.]

defying gravity because of the van der Waals interactions between their feet and the surface of the wall or ceiling.

Weak Bonds Permit Repeated Interactions

An important feature of weak bonds is that they can be easily broken. DNA provides an excellent example of why the breakage of weak bonds is essential. Hydrogen bonds between base pairs stabilize the double helix and keep the coding information—the base sequence—inside the helix away from potential harmful reactions (**Figure 2.8**; also Figure 1.3). However, as stated in Chapter 1, if the information is to be at all useful, it must be accessible. Consequently, the double helix can be opened up—the strands separated—so that the DNA can be replicated or so that the genes on the DNA can be expressed. The weak interactions are strong enough to stabilize and protect the DNA but weak enough to allow access to the information of the base sequences under appropriate circumstances.

? QUICK QUIZ 1 All weak interactions can be said to be fundamentally electrostatic interactions. Explain.

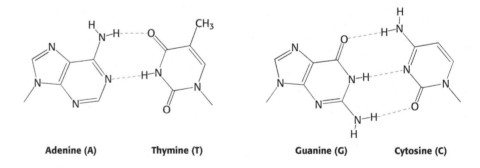

Figure 2.8 Stabilization of the double helix. Hydrogen bonds between adenine and thymine and between guanine and cytosine base pairs stabilize the double helix.

Adenine (A)   **Thymine (T)**   **Guanine (G)**   **Cytosine (C)**

2.4 Hydrophobic Molecules Cluster Together

The existence of life on Earth depends critically on the capacity of water to dissolve polar or charged molecules. However, not all biomolecules are polar or ionic. Such molecules are called nonpolar or hydrophobic molecules because these molecules simply cannot interact with water. Oil-and-vinegar salad dressing exemplifies the properties of hydrophobic molecules in the presence of water. Unless shaken vigorously, the oil and vinegar, the latter of which is predominantly

(A)

(B)

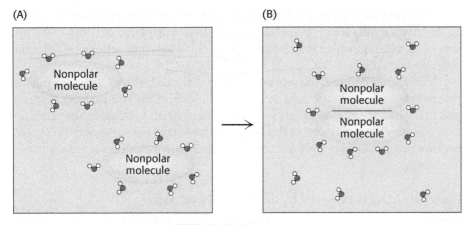

Figure 2.9 The hydrophobic effect. The aggregation of nonpolar groups in water leads to an increase in entropy owing to the release of water molecules into bulk water.

water, form two distinct layers. Even after shaking, the layers quickly re-form. What is the basis of this organization?

To understand how this organization takes place, we need to refer to the Second Law of Thermodynamics:

> *The total entropy of a system and its surroundings always increases in a spontaneous process.*

Entropy is a measure of randomness. The system may be a chemical reaction, a cell, or a bottle of salad dressing. Consider the introduction of a single nonpolar molecule, such as benzene, into some water (**Figure 2.9**).

A cavity in the water is created because the benzene has no chemical means of interacting with the water molecule. The cavity temporarily disrupts some hydrogen bonds between water molecules. The displaced water molecules then reorient themselves to form the maximum number of new hydrogen bonds. However, there are many fewer ways of forming hydrogen bonds around the benzene molecule than there are in pure water. *The water molecules around the benzene molecule are much more ordered than elsewhere in the solution.* The introduction of the nonpolar molecule into water has resulted in *a decrease in the entropy of water*. Now consider the arrangement of two benzene molecules in water. They do not reside in separate small cavities (Figure 2.9A); instead, they coalesce into a single larger one (Figure 2.9B). They become organized. The energetic basis for the formation of this order is that the association of the benzene molecules releases some of the ordered water molecules around the separated benzenes, increasing the entropy of the system. *Nonpolar solute molecules are driven together in water not primarily because they have a high affinity for each other but because, when they do associate, they release water molecules.* This entropy-driven association is called the *hydrophobic effect,* and the resulting interactions are called *hydrophobic interactions.* Hydrophobic interactions form spontaneously—no input of energy is required—because, when they form, the entropy of water increases.

Benzene

Membrane Formation Is Powered by the Hydrophobic Effect

The biological significance of the hydrophobic effect is more apparent when we consider molecules more complex than benzene, such as a phospholipid (Chapter 11). Recall that the structure of the phospholipid reveals two distinct chemical properties (Figure 1.4). The top of the molecule, called the head group, is hydrophilic, consisting of polar and charged species. However, the remainder of the molecule, consisting of two large hydrophobic chains, cannot interact with water. Such a molecule, with two distinct chemical personalities, is called an *amphipathic* or *amphiphilic molecule.* When exposed to water, the molecules orient themselves such that the hydrophilic head groups interact with the aqueous medium, whereas the hydrophobic tails are sequestered away from the water and interact only

Unfolded ensemble

↓

Folded ensemble

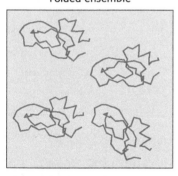

Figure 2.10 The folding of proteins. Protein folding entails the transition from a disordered mixture of unfolded molecules to a comparatively uniform solution of folded protein molecules.

? QUICK QUIZ 2 Explain how the following statement applies to biochemistry: Order can be generated by an increase in randomness.

with one another. Under the right conditions, they can form *membranes*. The lipids form a contiguous, closed bilayer, with two hydrophilic outsides and a hydrophobic interior, which is stabilized by van der Waals interactions between the hydrophobic tails.

Membranes define the inside and outside of the cell, as well as separating the components of eukaryotic cells into distinct biochemical compartments (pp. 8 and 11). Paradoxically, order has been introduced by an increase in the randomness of water. We will return to the topic of membranes many times in this book, because membranes are vital to many aspects of energy and information transformation.

Protein Folding Is Powered by the Hydrophobic Effect

Proteins, which we will consider in Chapters 3 and 4, are the true workhorses of biochemistry, playing prominent roles in all aspects of energy and information manipulation. Proteins play these roles because they are capable of forming complex three-dimensional structures that allow specific interactions with other biomolecules. These interactions define a protein's function. How does the hydrophobic effect favor protein folding? Consider a system consisting of identical unfolded protein molecules in aqueous solution (Figure 2.10). Each unfolded protein molecule can adopt a unique conformation—no two molecules will be in the same conformation—and so the system is quite disordered and the entropy of the collection of molecules is high. Yet, protein folding proceeds spontaneously under appropriate conditions, with all of the molecules assuming the same conformation, a clear decrease in entropy. To avoid violation of the Second Law of Thermodynamics, entropy must be increasing elsewhere in the system or in the surroundings.

How can we reconcile the apparent contradiction that proteins spontaneously assume an ordered structure, and yet entropy increases? We can again call on the hydrophobic effect to introduce order. Some of the amino acids that make up proteins have nonpolar groups (p. 39). These nonpolar amino acids have a strong tendency to associate with one another in the interior of the folded protein. The increased entropy of water resulting from the interaction of these hydrophobic amino acids helps to compensate for the entropy losses inherent in the folding process. Thus, the same thermodynamic principles that permit the formation of membranes facilitate the formation of the intricate three-dimensional structures. Although the hydrophobic effect powers the folding of proteins, many weak bonds, including hydrogen bonds and van der Waals interactions, are formed in the protein-folding process to stabilize the three-dimensional structure. These interactions replace interactions with water that take place in the unfolded protein.

Functional Groups Have Specific Chemical Properties

As we progress in our study of biochemistry, we will see a dizzying number of biomolecules. However, there are generalizations that make dealing with the large number of molecules easier. All biomolecules interact with one another and their environment by using the three types of reversible interactions and the hydrophobic effect, as heretofore discussed. As stated in Chapter 1, there are only four major classes of biomolecules. We can make the chemical basis of biochemistry even more manageable by noting that a limited number of groups of atoms with distinct chemical properties, called *functional groups*, are found in all biomolecules, including the four classes considered in Chapter 1 (Table 2.1). Each of the eight common functional groups listed in Table 2.1 confers similar chemical properties on the molecules of which it is a component. These groups are called functional groups because the chemical properties conferred are necessary for the biochemical function of the molecules. Note that all of these groups have hydrogen-bonding potential or the ability to form ionic bonds or both, except for the hydrophobic

Table 2.1   Some key functional groups in biochemistry

Functional group	Class of compounds	Structural formula	Example
Hydrophobic	Hydrocarbon chains (aliphatic)	$R—CH_3$	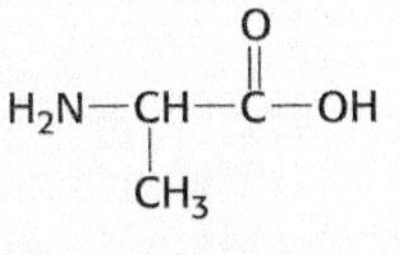 Alanine
	Aromatic (hydrocarbons in a ring structure with multiple double bonds)	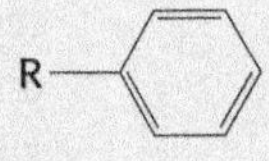	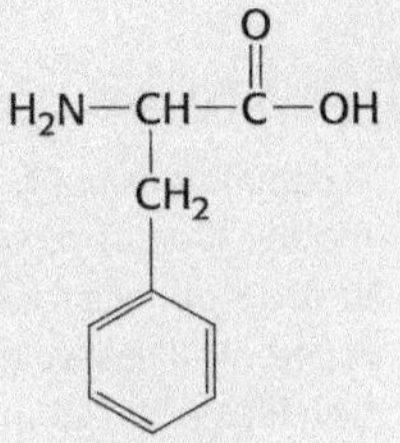 Phenylalanine
Hydroxyl	Alcohol	$R—OH$	Ethanol
Aldehyde	Aldehydes	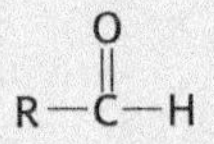	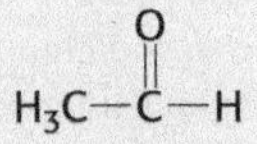 Acetaldehyde
Keto	Ketones	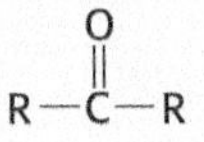	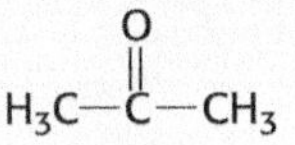 Acetone
Carboxyl	Carboxylic acid	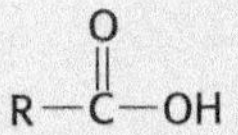	 Acetic acid
Amino	Amines	$R—NH_2$	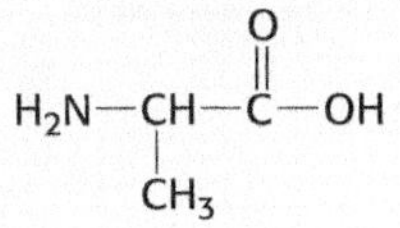 Alanine
Phosphate	Organic phosphates	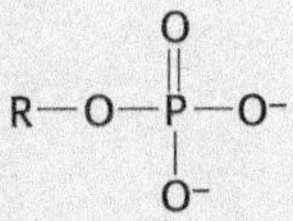	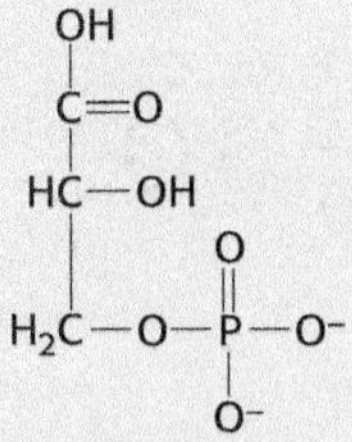 3-Phosphoglyceric acid
Sulfhydryl	Thiols	$R—SH$	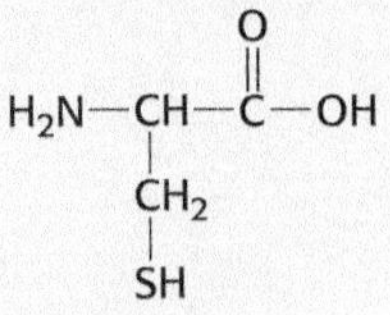 Cysteine

Note: There are many aliphatic (hydrocarbon chains) and aromatic groups. The methyl group and benzyl groups are shown as examples. Notice also that many of the examples have more than one functional group. The letter R stands for the remainder of the molecule. Finally, note that a *carbon atom double-bonded to an oxygen atom*, C═O, called a *carbonyl group*, is present in aldehydes, ketones, and carboxylic acids, including amino acids. Carbonyl groups are common in biochemicals.

groups, which can interact with other hydrophobic groups through van der Waals interactions. Essentially, every biomolecule that we encounter in our study of biochemistry will have one or more of these functional groups.

✓ 6 Define pH and explain why changes in pH may affect biochemical systems.

DID YOU KNOW?

A common example of a pathological modification of environmental pH is gastroesophageal reflux disease, or GERD. A chronic digestive disease, GERD develops when stomach acid refluxes into the esophagus. The backwash of acid, frequently experienced as heartburn, irritates the lining of the esophagus by exposing the tissue to very acidic conditions (pH 1 to 2). GERD can cause chronic inflammation in the esophagus that can lead to complications, including esophageal ulcers and esophageal cancer. Risk factors for GERD include smoking and obesity.

DID YOU KNOW?

In chemistry, equilibrium is the condition in which the concentrations of reactants and products have no net change over time.

2.5 pH Is an Important Parameter of Biochemical Systems

Especially important examples of reversible reactions are those that include the release or binding of a hydrogen ion, H^+, also called a proton. The *pH* of a solution is a measure of the hydrogen ion concentration, with values ranging from 0 to 14. The smaller numbers denote an acidic environment, and the larger numbers denote a basic environment. Indeed, pH is an important parameter of living systems. For example, the pH of human blood is about 7.4, and a deviation of $+/- 0.5$ units can result in coma or death. Why is maintaining the proper pH so vital? Alterations in pH can drastically affect the internal electrostatic environment of an organism, which can alter the weak bonds that maintain the structure of biomolecules. Altered structure usually means loss of function. For instance, ionic bonds may be weakened or disappear with a change in pH, and hydrogen bonds may or may not form, depending on the pH. Given how crucial maintaining proper pH is to the correct functioning of biochemical systems, it is important to have means of describing pH. We will now investigate the quantitative nature of pH, the effect of acids and bases on pH, and the means by which cells maintain an approximately constant pH.

Water Ionizes to a Small Extent

Very small amounts of pure water dissociate and form hydronium (H_3O^+) and hydroxyl (OH^-) ions, with the concentration ion of each being 10^{-7} M. For simplicity, we refer to the hydronium ion simply as a hydrogen ion (H^+) and write the equilibrium as

$$H_2O \rightleftharpoons H^+ + OH^-$$

The equilibrium constant K_{eq} of this dissociation is given by

$$K_{eq} = [H^+][OH^-]/[H_2O] \tag{1}$$

in which the brackets denote molar concentrations (M) of the molecules. Substituting the concentration of H^+ and OH^- (10^{-7} M, each) and the concentration of water (55.5 M), we see that the equilibrium constant of water is

$$K_{eq} = 10^{-7} M \times 10^{-7} M/55.5 M = 1.8 \times 10^{-16} M$$

Because the concentration of water is essentially unchanged by the small amount of ionization, we can ignore any change in the concentration of water and define a new constant:

$$K_w = K_{eq} \times [H_2O]$$

which then simplifies to

$$K_w = [H^+][OH^-] \tag{2}$$

K_w is the *ion product of water*. At 25°C, K_w is $1.0 \times 10^{-14} M^2$. The pH of any solution is quantitatively defined as

$$pH = \log_{10}(1/[H^+]) = -\log_{10}[H^+] \tag{3}$$

Consequently, the pH of pure water, which contains equal amounts of H^+ and OH^-, is equal to 7. Note that the concentrations of H^+ and OH^- are reciprocally related; thus, $pH + pOH = 14$, where pOH is determined by substituting the hydroxide ion concentration for the proton concentration in equation 3. If the concentration of H^+ is high, then the concentration of OH^- must be low, and vice versa. For example, if $[H^+] = 10^{-2}$ M, then $[OH^-] = 10^{-12}$ M.

Let us consider a problem that is a bit more complex. If the $[H^+]$ equals 2.5×10^{-4} M in a solution, what would the $[OH^-]$ be? We solve this problem by first remembering the equation for the ion product of water:

$$K_w = [H^+][OH^-] = 1.0 \times 10^{-14} \text{ M}^2$$

We can determine the $[OH^-]$ by rearranging the equation to solve for $[OH^-]$ and inserting the proton concentration:

$$[OH^-] = \frac{K_w}{[H^+]} = \frac{1.0 \times 10^{-14} \text{ M}^2}{2.5 \times 10^{-4} \text{ M}} = 4 \times 10^{-9} \text{ M}$$

Thus, if the concentration of protons or hydroxyl ions is known, the concentration of the unknown ion can be determined.

An Acid Is a Proton Donor, Whereas a Base Is a Proton Acceptor

Organic acids are prominent biomolecules. These acids will ionize to produce a proton and a base:

$$\text{Acid} \rightleftharpoons H^+ + \text{base}$$

The species formed by the ionization of an acid is its *conjugate base* and is distinguished from the ionized acid by having the suffix "ate." Conversely, the protonation of a base yields its *conjugate acid*. Let's consider acetic acid, a carboxylic acid. Carboxylic acids are key functional groups found in a host of biochemicals (Table 2.1). Acetic acid and acetate ion are a conjugate acid–base pair:

$$\text{CH}_3\text{COOH} \rightleftharpoons H^+ + \text{CH}_3\text{COOH}^-$$
Acetic acid   Acetate

Acids Have Differing Tendencies to Ionize

How can we measure the strength of an acid? For instance, how can we determine whether an acid will dissociate in a given biochemical environment, such as the blood? Let us examine weak acids, inasmuch as weak acids are the type found in biochemical systems. The ionization equilibrium of a weak acid (HA) is given by

$$\text{HA} \rightleftharpoons H^+ + A^-$$

The equilibrium constant K_a for this ionization is

$$K_a = \frac{[H^+][A^-]}{[HA]} \tag{4}$$

The larger the value of K_a is, the stronger the acid (**Figure 2.11**).

What is the relation between pH and the ratio of acid to base? In other words, how dissociated will an acid be at a particular pH? A useful expression establishing the relation between pH and the acid/base ratio can be obtained by rearrangement of equation 4:

$$\frac{1}{[H^+]} = \frac{1}{K_a} \frac{[A^-]}{[HA]} \tag{5}$$

Taking the logarithm of both sides of equation 5 gives

$$\log\left(\frac{1}{[H^+]}\right) = \log\left(\frac{1}{K_a}\right) + \log\left(\frac{[A^-]}{[HA]}\right) \tag{6}$$

We define $\log(1/K_a)$ as the pK_a of the acid.

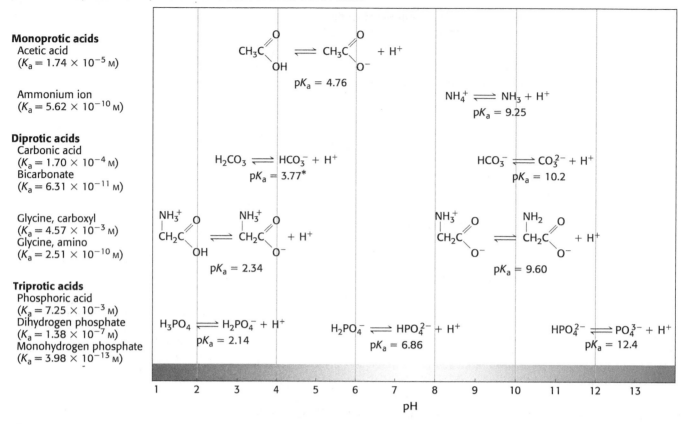

Monoprotic acids
Acetic acid
($K_a = 1.74 \times 10^{-5}$ M)

Ammonium ion
($K_a = 5.62 \times 10^{-10}$ M)

Diprotic acids
Carbonic acid
($K_a = 1.70 \times 10^{-4}$ M)
Bicarbonate
($K_a = 6.31 \times 10^{-11}$ M)

Glycine, carboxyl
($K_a = 4.57 \times 10^{-3}$ M)
Glycine, amino
($K_a = 2.51 \times 10^{-10}$ M)

Triprotic acids
Phosphoric acid
($K_a = 7.25 \times 10^{-3}$ M)
Dihydrogen phosphate
($K_a = 1.38 \times 10^{-7}$ M)
Monohydrogen phosphate
($K_a = 3.98 \times 10^{-13}$ M)

Figure 2.11 A variety of conjugate acid–base pairs. Many conjugate acid–base pairs are important in biochemistry. A sampling of such pairs is shown. Notice that some acids can release more than one proton. The dissociation reaction and the pK_a for each pair are shown where they lie along a pH gradient. The equilibrium constants for the acids are given at the left of the illustration. *Note the pK_a of carbonic acid. Because of the large reservoir of carbon dioxide in the blood and its rapid equilibration with carbonic acid, the pK_a of carbonic acid in blood is taken to be 6.1. [Data from D. L. Nelson and M. M. Cox, *Lehninger Principles of Biochemistry*, 5th ed. (W. H. Freeman and Company, 2008), Fig. 2.15.]

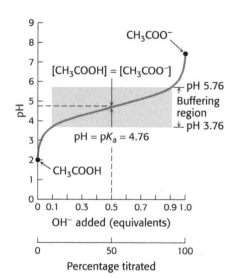

Figure 2.12 The titration curve for acetic acid. Notice that, near the pK_a of acetic acid, the pH does not change much with the addition of more base. [Data from D. L. Nelson and M. M. Cox, *Lehninger Principles of Biochemistry*, 5th ed. (W. H. Freeman and Company, 2008), p. 58.]

Substituting pH for log $1/[H^+]$ and pK_a for log $1/K_a$ in equation 6 yields

$$pH = pK_a + \log\left(\frac{[A^-]}{[HA]}\right) \qquad (7)$$

which is commonly known as the *Henderson–Hasselbalch equation*. Note that, when the concentration of ionized acid molecules equals that of unionized acid molecules, or $[A^-] = [HA]$, the $\log([A^-]/[HA]) = 0$, and so pK_a *is simply the pH at which the acid is half dissociated.*

Above pK_a, $[A^-]$ predominates, whereas below pK_a, $[HA]$ predominates. As a reference to the strength of an acid, pK_a is more useful than the ionization constant (K_a) because pK_a does not require the use of the sometimes cumbersome scientific notation. Note from Figure 2.11 that the most important biochemical acids will be predominately dissociated at physiological pH (~7.4) Thus, these molecules are usually referred to as the conjugate base (e.g., pyruvate) and not as the acid (e.g., pyruvic acid).

Buffers Resist Changes in pH

An acid–base conjugate pair (such as acetic acid and acetate ion) has an important property: it resists changes in the pH of a solution. In other words, it acts as a *buffer*. Consider the addition of OH$^-$ to a solution of acetic acid (HA):

$$HA + OH^- \rightleftharpoons A^- + H_2O$$

A plot of how the pH of this solution changes with the amount of OH$^-$ added is called a *titration curve* (**Figure 2.12**). Note that there is an inflection point in the

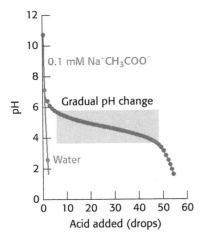

Figure 2.13 **Buffer action.** The addition of a strong acid–say, 1 M HCl–to pure water results in an immediate drop in pH to near 2, as the blue line shows. In contrast, the addition of acid to 0.1 mM sodium acetate ($Na^+CH_3COO^-$) results in a much more gradual change in pH until the pH drops below 3.5, as shown by the red line.

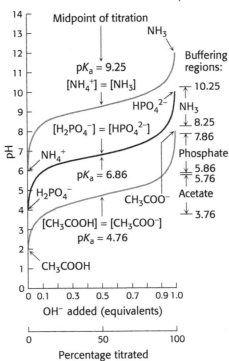

Figure 2.14 **The titration curves of three important weak acids.** Notice that the regions of buffering capacity differ. [Data from D. L. Nelson and M. M. Cox, *Lehninger Principles of Biochemistry*, 5th ed. (W. H. Freeman and Company, 2008), p. 59.]

curve at pH 4.8, which is the pK_a of acetic acid. In the vicinity of this pH, the addition of a relatively large amount of OH^- to the buffer produces little change in the pH of the solution. In other words, the buffer maintains the value of pH near the pK_a of the acid component of the buffer, despite the addition of either protons or hydroxide ions. **Figure 2.13** compares the changes in pH when a strong acid is added to pure water with the changes in pH when that same strong acid is added to a buffered solution. The pH of the buffered solution does not change nearly as rapidly as that of pure water. In general, a weak acid is most effective in buffering against pH changes in the vicinity of its pK_a value. **Figure 2.14** shows the buffering range of three different weak acids.

Buffers Are Crucial in Biological Systems

Knowledge of the workings of buffers is important for two reasons. First, much of biochemistry is experimentally investigated in vitro (in glass, or, in effect, in a test tube). Because the biomolecules that are being investigated are sensitive to pH, biochemists must use buffers to maintain the proper pH during the experiments. Choice of the proper buffer is often crucial to designing a successful experiment.

Second, we need to understand buffers so as to understand how an organism controls the pH of its internal environment in vivo—that is, in the living organism. As mentioned earlier, many biochemical processes generate acids. How is the pH of the organism maintained in response to the production of acid? What are the physiologically crucial buffers?

We will examine the buffering of blood pH as an example of a physiological system. In the aerobic biochemical consumption of fuels, carbon dioxide (CO_2) is produced (Chapter 19). The carbon dioxide reacts with water to produce a weak acid, carbonic acid:

$$CO_2 + H_2O \rightleftharpoons H_2CO_3 \qquad (8)$$

The carbonic acid readily dissociates into a proton and bicarbonate ion:

$$H_2CO_3 \rightleftharpoons H^+ + HCO_3^- \qquad (9)$$

The conjugate acid–base pair of H_2CO_3/HCO_3^- acts as a buffer. Protons released by an added acid will combine with bicarbonate ion, thus having little effect on the pH. The effectiveness of the H_2CO_3/HCO_3^- buffer system is enhanced by the fact that the quantity of the buffer in the blood can be rapidly adjusted. For instance, if there is an influx of acid into the blood, reaction 9 will proceed to the left, driving reaction 8 to the left. The newly generated CO_2 can then be expired from the lungs. In essence, the ratio of H_2CO_3/HCO_3^- is maintained, which keeps the pH constant. This mechanism of blood-pH control is called *compensatory respiratory alkalosis*.

Making Buffers Is a Common Laboratory Practice

Let's consider an exercise that is common in biological laboratories: making a buffer. Imagine that we have a protein and want to see how the activity changes as a function of pH. To do so, we need buffers covering a range of pH values, and we examine the protein's activity at each value. As an example, let's examine how to make 1 liter of 0.3 M acetate buffer, pH 4.46. We have on hand a 2 M solution of acetic acid (a weak acid, $pK_a = 4.76$) and a 2.5 M solution of potassium hydroxide (KOH; a strong base). These solutions are referred to as stock solutions. Remember that an acetate buffer will consist of a mixture of acetate ion and acetic acid; so we must first determine the ratio of acetate to acetic acid that will yield pH 4.46 and then calculate the amount of each constituent needed to yield a total molarity (concentration of acid plus concentration of base) equal to 0.3 M.

Essentially, all buffer problems can be solved with a little introductory chemistry and the Henderson–Hasselbalch equation:

$$pH = pK_a + \log\left(\frac{[A^-]}{[HA]}\right)$$

Insert the given parameters:

$$4.46 = 4.76 + \log\left(\frac{[\text{acetate}]}{[\text{acetic acid}]}\right)$$

or

$$-0.3 = \log\left(\frac{[\text{acetate}]}{[\text{acetic acid}]}\right)$$

Finding the antilog of 0.3 yields

$$\frac{[\text{acetate}]}{[\text{acetic acid}]} = 0.5$$

This equation tells us that, to obtain pH = 4.46 with an acetate buffer, two-thirds of the acetate will be contributed by the acid and one-third by the base. Because we need a total molarity of 0.3 M, we will need 0.2 M of acetic acid and 0.1 M of acetate. Now for the simple chemistry.

First of all, we do not have any acetate solution; we only have acetic acid and potassium hydroxide. Therefore, all of the acetate in the buffer will come from acetic acid, and we will then convert one-third of the acetic acid into acetate. So, how much of our 2 M acetic acid stock solution do we need to add to obtain 0.3 M acetic acid? Remember that a 0.3 M solution contains 0.3 mol per liter by definition. Now we must determine what volume of the 2 M acetic acid stock solution contains 0.3 mol. If a 2 M solution contains 2 mol per liter, then 0.15 liter, or 150 ml, of the stock solution will contain 0.3 mol of acetic acid. We pour this amount into a bottle labeled 0.3 M acetate buffer, pH 4.46. What about the acetate? The only source of acetate ion is acetic acid, and so we must generate the acetate by the addition of KOH to the 150 ml of acetic acid that we have set aside. Potassium hydroxide is a strong base: when mixed with acetic acid, it will convert an equivalent amount of the acetic acid into acetate, or, precisely, potassium acetate. We need to generate 0.1 M acetate ion from the 0.3 mol of acetic acid that we already have in the bottle. By using our earlier reasoning, we know that 0.04 liter, or 40 ml, of KOH will convert 40 ml of acetic acid into acetate. The complete 0.3 M acetate buffer, pH 4.46, solution is generated by adding together 150 ml of 2 M acetic acid, 40 ml of 2.5 M KOH, and 810 ml of water to make a final volume of 1 liter. A note of caution: it is safer to add the acid and the base to water. So, after the calculations are completed, add the water first, and then add the acid and base. Don't forget to mix well. Add your initials to the label so that we know whom to blame if the experiment doesn't work.

SUMMARY

2.1 Thermal Motions Power Biological Interactions

Brownian motion, the random movement of fluids and gases, is powered by the background thermal noise. Brownian motion inside the cell supplies the energy for many of the interactions required for a functioning biochemical system.

2.2 Biochemical Interactions Take Place in an Aqueous Solution

Most biochemical interactions take place in aqueous solutions. Water is a polar molecule, with the oxygen atom bearing a partial negative charge and the hydrogen atoms a partial positive charge. The charges on water molecules interact with opposite charges on other water molecules to form hydrogen bonds.

2.3 Weak Interactions Are Important Biochemical Properties

There are three common types of weak interactions found in biochemical systems. Electrostatic interactions take place between ions having opposite charges. The strength of electrostatic interactions depends on the nature of the medium. The basis of the hydrogen bond is the unequal distribution of charge that results whenever a hydrogen atom is covalently bonded to an electronegative atom, such as oxygen or nitrogen. Hydrogen bonds in biomolecules are weakened in the presence of water because water readily forms hydrogen bonds. Fleeting electrostatic interactions, termed van der Waals interactions, take place when the transient asymmetry of charges on one nonpolar molecule induce complementary asymmetry in nearby nonpolar molecules.

2.4 Hydrophobic Molecules Cluster Together

The Second Law of Thermodynamics states that the entropy of the universe is always increasing. This law is the basis of the hydrophobic effect: nonpolar molecules in aqueous solutions are driven together because of the resulting increase in entropy of water molecules. The hydrophobic effect is one of the most important energy considerations in biological systems, accounting for much of the structure of life, including membrane formation and the specific folding of proteins. Functional groups are groups of atoms found in many different biomolecules that confer specific chemical properties.

2.5 pH Is an Important Parameter of Biochemical Systems

The pH of a solution is a measure of hydrogen ion concentration and is an important parameter in biochemical systems, both in vivo and in vitro. Buffers are acid–base conjugate pairs that resist changes in pH. Buffers are crucial in biological systems because changes in pH can have drastic effects on the structure of biomolecules and can even result in death.

KEY TERMS

Brownian motion (p. 18)
hydrogen bond (p. 19)
electrostatic interactions (ionic bonds) (p. 20)

van der Waals interactions (p. 21)
entropy (p. 23)
hydrophobic effect (p. 23)

pH (p. 26)
buffers (p. 28)

? Answers to QUICK QUIZZES

1. Ionic bonds, hydrogen bonds, and van der Waals interactions all depend on the unequal distribution of electrons, resulting in an unequal distribution of charge.

2. The statement essentially describes the hydrophobic effect. Specific complicated biochemical structures can form, powered by the increase in entropy that results when hydrophobic groups are removed from aqueous solution.

PROBLEMS

1. *A random walk.* Define Brownian motion.

2. *Inequality.* Water is said to be polar but uncharged. How is it possible? ✓ 4

3. *United we stand.* Why are weak bonds important in biochemistry? ✓ 4

4. *Bond types.* What are the common types of weak bonds important in biochemistry? How does water affect these bonds? ✓ 4

5. *Temperature and bonds.* In liquid water, each molecule is hydrogen bonded to approximately 3.4 molecules of water. What effect would freezing water have on the number of hydrogen bonds? Heating water?

6. *Context matters.* What would be the effect of an organic solvent on electrostatic interactions? ✓ 5

7. *Some atoms don't share well.* What is an electronegative atom, and why are such atoms important in biochemistry? ✓ 5

8. *Oil and vinegar.* Define the hydrophobic effect. ✓ 5

9. *Laws are important.* How does the Second Law of Thermodynamics allow for the formation of biochemical order?

10. *Fourteen once.* If an aqueous solution has a hydrogen ion concentration of 10^{-5} M, what is the concentration of hydroxyl ion? ✓ 6

11. *Fourteen twice.* If an aqueous solution has a hydroxyl ion concentration of 10^{-2} M, what is the concentration of hydrogen ion? ✓ 6

12 *Acid–base chemistry.* Using the Henderson–Hasselbalch equation, show that, for a weak acid, the pK_a is the pH at which the concentration of the acid equals the concentration of the conjugate base. ✓ 6

13. *Acid strength.* What is the relation between the pK_a of an acid and the strength of the acid? ✓ 6

14. *Warts beware.* The pK_a of acetic acid is 4.76 and the pK_a of trichloroacetic acid, which is used to remove warts, is

0.7. Calculate the dissociation constant of each acid. Which is the stronger acid? ✓ 6

15. *Acids from the health food store?* Many important biochemicals are organic acids, such as pyruvic acid (pK_a = 2.50) and lactic acid (pK_a = 3.86). The conjugate bases are pyruvate and lactate, respectively. Determine, for each acid, which form—the acid or the conjugate base—predominates at pH 7.4. ✓ 6

Pyruvic acid   **Lactic acid**

Challenge Problems

16. *Find the pK_a.* For an acid HA, the concentrations of HA and A$^-$ are 0.075 and 0.025, respectively, at pH 6.0. What is the pK_a value for HA? ✓ 6

17. *pH indicator.* A dye that is an acid and that appears as different colors in its protonated and deprotonated forms can be used as a pH indicator. Suppose that you have a 0.001 M solution of a dye with a pK_a of 7.2. From the color, the concentration of the protonated form is found to be 0.0002 M. Assume that the remainder of the dye is in the deprotonated form. What is the pH of the solution? ✓ 6

18. *What's the ratio?* An acid with a pK_a of 8.0 is present in a solution with a pH of 6.0. What is the ratio of the protonated to the deprotonated form of the acid? ✓ 6

19. *Buffer capacity.* Two solutions of sodium acetate are prepared, one having a concentration of 0.1 M and the other having a concentration of 0.01 M. Calculate the pH values when the following concentrations of HCl have been added to each of these solutions: 0.0025 M, 0.005 M, 0.01 M, and 0.05 M. ✓ 6

20. *Another ratio.* Calculate the concentration of acetic acid and acetate ion in a 0.2 M acetate buffer at pH 5. The pK_a of acetic acid is 4.76. ✓ 6

21. *Buffers STAT!* You are working in a high-powered clinical biochemistry lab. The chief scientist rushes in and announces, "I need 500 ml of 0.2 M acetate, pH 5.0. STAT! Who is the best and brightest in this room?" All eyes turn toward you. You have solid anhydrous sodium acetate (MW = 82 g mol^{-1}) and a solution of 1 M acetic acid. Describe how you would make the buffer. ✓ 6

22. *Blood pH.* Following a bout of intense exercise, the pH of the exerciser's blood was found to be 7.1. If the HCO_3^- concentration is 8 mM, and the pK_a for HCO_3^- is 6.1, what is the concentration of CO_2 in the blood? ✓ 6

23. *Metabolic acidosis.* As we will see later in the course, pathological conditions arise where the blood pH falls because of excess acid production, a condition called metabolic acidosis. Excess protons in the blood decrease the amount of HCO_3^- and thus reduce the buffering capacity of blood. A rapid drop in pH could lead to death. Normal values for blood are: pH = 7.4, $[HCO_3^-]$ = 24.0 mM, $[CO_2]$ = 1.20 mM. ✓ 6

(a) If a patient has a blood pH = 7.03 and $[CO_2]$ = 1.2 mM, what is the $[HCO_3^-]$ in the patient's blood? The pKa of HCO_3^- = 6.1.

(b) Suggest a possible treatment for metabolic acidosis.

(c) Why might the suggestion for part (b) be of benefit to middle-distance runners?

CHAPTER 3
Amino Acids

CHAPTER 4
Protein Three-Dimensional
Structure

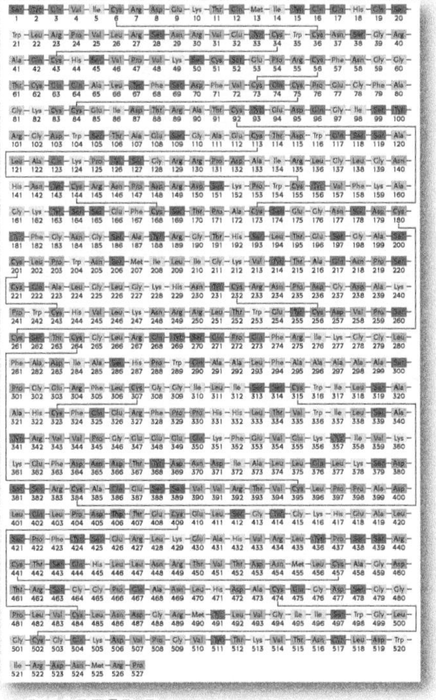

CHAPTER 5
Techniques in Protein
Biochemistry

SECTION **2**

Protein Composition and Structure

P roteins are the most versatile macromolecules in living systems and serve crucial functions in essentially all biological processes. They function as catalysts, provide mechanical support, generate movement, control growth and differentiation, and much more. Indeed, much of this book will focus on understanding what proteins do and how they perform these functions.

Several key properties enable proteins to participate in a wide range of functions:

1. Proteins are linear polymers built of monomers called amino acids, which are linked end to end. Remarkably, the sequence of amino acids determines the three-dimensional shape of the protein. Protein function directly depends on this three-dimensional structure.

2. Proteins contain a wide range of functional groups, which account for the broad spectrum of protein function.

3. Proteins can interact with one another and with other macromolecules to form complex assemblies. The proteins within these assemblies can act synergistically to generate capabilities that individual proteins may lack.

4. Some proteins are quite rigid, whereas others display a considerable flexibility. Rigid proteins can function as structural elements in cells and tissues. Proteins with some flexibility can act as hinges, springs, or levers that are crucial to protein function or to the assembly of protein complexes.

In this section, we will explore the structure of proteins from the ground up. We begin with an investigation of the chemical properties of the amino acids.

35

We will then see how the amino acids are linked together to form a polypeptide chain. The shape of the amino acids informs the shape and flexibility of the polypeptide chain. A functional protein is a complex three-dimensional structure, and we will examine the various levels of protein structure. We will see that the final three-dimensional structure of proteins forms spontaneously. Finally, we will examine how proteins are purified and characterized. Researchers take advantage of often slight differences in the physical characteristics of similar proteins to separate them from one another. We will also explore powerful immunological techniques that can be used to purify and further investigate proteins as well as other biomolecules.

✓ **By the end of this section, you should be able to:**

✓ 1 Identify the main classes of amino acids.

✓ 2 Compare and contrast the different levels of protein structure and how they relate to one another.

✓ 3 Describe the biochemical information that determines the final three-dimensional structure, and explain what powers the formation of this structure.

✓ 4 Explain how proteins can be purified.

✓ 5 Explain how immunological techniques can be used to purify and identify proteins.

CHAPTER 3

Amino Acids

In Lewis Carroll's wonderful tale *Through the Looking Glass,* Alice enters a world of delightful curiosities. Had Alice been a biochemist, she might have wondered if the stereochemistry of the amino acids in *Looking Glass* proteins were mirror images of the amino acids in her normal world. [The Granger Collection.]

The role of amino acids as the building blocks of proteins is readily apparent in health food stores. Indeed, their shelves are stocked with powdered amino acids for use as dietary supplements, enabling body builders to enhance muscle growth. However, amino acids are key biochemicals in their own right. Some amino acids function as signal molecules, such as neurotransmitters, and all amino acids are precursors to other biomolecules, such as hormones, nucleic acids, lipids, and, as just mentioned, proteins. In this chapter, we examine the fundamental chemical properties of amino acids.

$$^+H_3N—\overset{\overset{\displaystyle CH_3}{|}}{\underset{\underset{\displaystyle H}{|}}{C}}—COO^-$$

**Fischer projection
of alanine**

**Stereochemical rendering
of alanine**

Two Different Ways of Depicting Biomolecules Will Be Used

Proteins, as well as biochemicals in general, derive their amazing array of functions from the ability to form three-dimensional structures. How do we view these biomolecules on two-dimensional pages? In this book, molecules are depicted in two different ways, depending on the aspect of the molecule being emphasized.

When visualizing the *constituent atoms* of a molecule (such as carbons and hydrogens) is more important than seeing the shape of the molecule, we use a depiction called a Fischer projection. In a *Fischer projection,* every atom is identified and the bonds to the central carbon atom are represented by horizontal and vertical lines. By convention, the horizontal bonds are assumed to project out of the page toward the viewer, whereas the vertical bonds are assumed to project behind the page away from the viewer.

When emphasis is on a molecule's *function,* visualization of the shape of the molecule is more important. In such instances, stereochemical renderings are used because they convey an immediate sense of the molecule's structure and, therefore, a hint about its function. Stereochemical renderings also simplify the diagram, thereby making the function of a molecule clearer. Carbon and hydrogen atoms are not explicitly shown unless they are important to the activity of the molecule. In this way, the functional groups are easier to identify.

To illustrate the correct *stereochemistry* of tetrahedral carbon atoms, wedges are used to depict the direction of a bond into or out of the plane of the page. A solid wedge denotes a bond coming out of the plane of the page toward the viewer. A dashed wedge represents a bond going away from the viewer and behind the plane of the page. The remaining two bonds are depicted as straight lines.

✓ 1 Identify the main classes of amino acids.

3.1 Proteins Are Built from a Repertoire of 20 Amino Acids

Amino acids are the building blocks of proteins. An *α-amino acid* consists of a central carbon atom, called the *α carbon*, which is bonded to an amino group, a carboxylic acid group, a hydrogen atom, and a *side chain,* called the R group. Each kind of amino acid has a different R group.

Most Amino Acids Exist in Two Mirror-Image Forms

With four different groups connected to the tetrahedral α-carbon atom, α-amino acids are *chiral:* they may exist in one or the other of two mirror-image forms, called the L isomer and the D isomer (**Figure 3.1**).

Only L *amino acids are constituents of proteins.* What is the basis for the preference for L *amino acids*? The answer is not known, but evidence shows that pure L or D amino acids are slightly more soluble than a stable DL crystal. Consequently, if simply by chance, there were a small excess of the L amino acid, this small solubility difference could have been amplified over time so that the L isomer became dominant in solution.

All Amino Acids Have at Least Two Charged Groups

Free amino acids in solution at neutral pH exist predominantly as *dipolar ions* (also called *zwitterions*). In the dipolar form, the amino group is protonated (NH_3^+) and the carboxyl group is deprotonated (COO^-). The ionization state of an amino acid varies with pH (**Figure 3.2**). In acid solution (e.g., pH 1), the amino group is protonated (NH_3^+) and the carboxyl group is not dissociated (— COOH). As the pH is raised, the carboxylic acid is the first group to give up a proton, because its pK_a is near 2. The dipolar form persists until the pH approaches 9, when the protonated amino group loses a proton. Under physiological conditions, amino acids exist in the dipolar form.

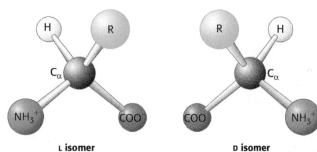

Figure 3.1 The L and D isomers of amino acids. The letter R refers to the side chain. The L and D isomers are mirror images of each other.

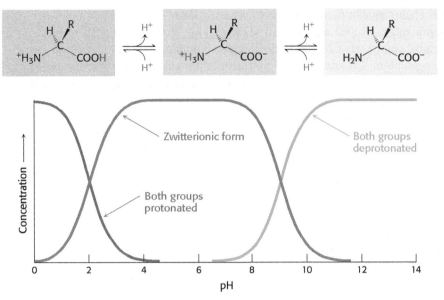

Figure 3.2 Ionization state as a function of pH. The ionization state of amino acids is altered by a change in pH. The zwitterionic form predominates near physiological pH, approximately 7.4.

3.2 Amino Acids Contain a Wide Array of Functional Groups

Twenty kinds of side chains varying in *size, shape, charge, hydrogen-bonding capacity, hydrophobic character,* and *chemical reactivity* are commonly found in proteins. Many of these properties are conferred by functional groups (Table 2.1). The amino acid functional groups include alcohols, thiols, thioethers, carboxylic acids, carboxamides, and a variety of basic groups. Most of these groups are chemically reactive.

All proteins in all species—bacterial, archaeal, and eukaryotic—are constructed from the same set of 20 amino acids with only a few exceptions. This fundamental alphabet for the construction of proteins is several billion years old. The remarkable range of functions mediated by proteins results from the diversity and versatility of these 20 building blocks.

Although there are many ways to classify amino acids, we will assort these molecules into four groups, on the basis of the general chemical characteristics of their R groups:

1. Hydrophobic amino acids with nonpolar R groups

2. Polar amino acids with neutral R groups but the charge is not evenly distributed

3. Positively charged amino acids with R groups that have a positive charge at physiological pH (pH ≈ 7.4)

4. Negatively charged amino acids with R groups that have a negative charge at physiological pH

Hydrophobic Amino Acids Have Mainly Hydrocarbon Side Chains

The amino acids having side chains consisting only of hydrogen and carbon are hydrophobic. The simplest amino acid is *glycine,* which has a single hydrogen atom as its side chain. With two hydrogen atoms bonded to the α-atom, glycine is unique in being achiral. *Alanine,* the next simplest amino acid, has a methyl group ($-CH_3$) as its side chain (**Figure 3.3**). The three-letter abbreviations and one-letter symbols under the names of the amino acids depicted in Figure 3.3 and in subsequent illustrations are an integral part of the vocabulary of biochemists.

Larger aliphatic side chains are found in the branched-chain amino acids *valine, leucine,* and *isoleucine. Methionine* contains a largely aliphatic side chain that includes a *thioether* ($-S-$) group. The different sizes and shapes of these hydrocarbon side chains enable them to pack together to form compact structures with little empty space. *Proline* also has an aliphatic side chain, but it differs from other members of the set of 20 in that its side chain is bonded to both the α-carbon and the nitrogen atom. Proline markedly influences protein architecture because its ring structure makes it more conformationally restricted than the other amino acids.

Two amino acids with simple *aromatic side chains* are also classified as hydrophobic (Figure 3.3). *Phenylalanine,* as its name indicates, contains a phenyl

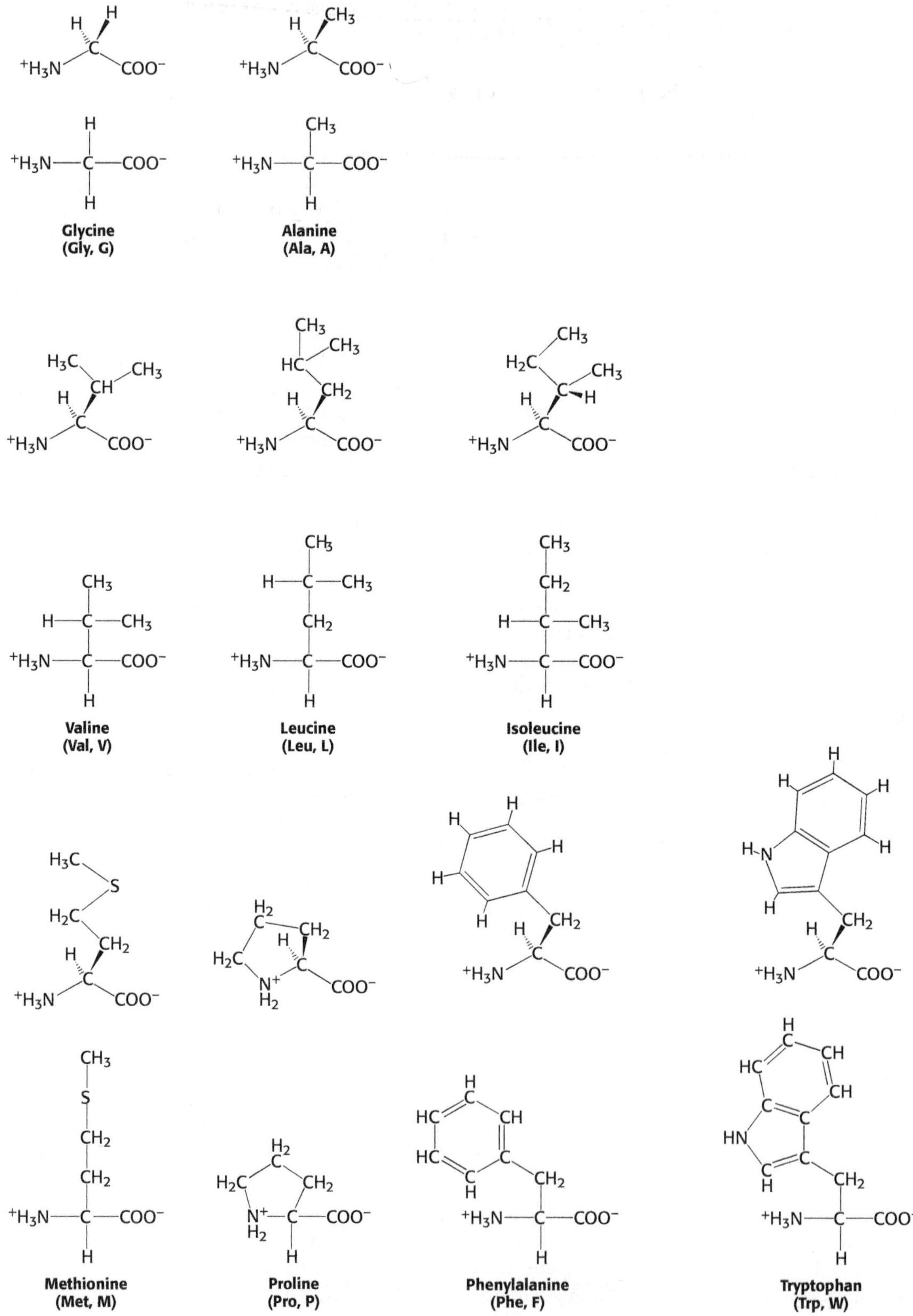

Figure 3.3
Hydrophobic amino acids.

ring attached in place of one of the methyl hydrogen atoms of alanine. *Tryptophan* has an indole ring joined to a methylene ($-CH_2-$) group; the indole group comprises two fused rings and an NH group.

The hydrophobic amino acids, particularly the larger aliphatic and aromatic ones, tend to cluster together inside the protein away from the aqueous environment of the cell. This tendency of hydrophobic groups to come together is called *the hydrophobic effect* (pp. 22–23) and is the driving force for the formation of the unique three-dimensional architecture of water-soluble proteins. The different sizes and shapes of these hydrocarbon side chains enable them to pack together to form compact structures with little empty space.

Polar Amino Acids Have Side Chains That Contain an Electronegative Atom

The next group of amino acids that we will consider are those that are neutral overall, yet they are polar because the R group contains an electronegative atom that hoards electrons. Three amino acids, *serine, threonine,* and *tyrosine,* contain hydroxyl ($-OH$) groups (**Figure 3.4**). The electrons in the O—H bond are attracted to the oxygen atom, making it partly negative, which in turn makes the

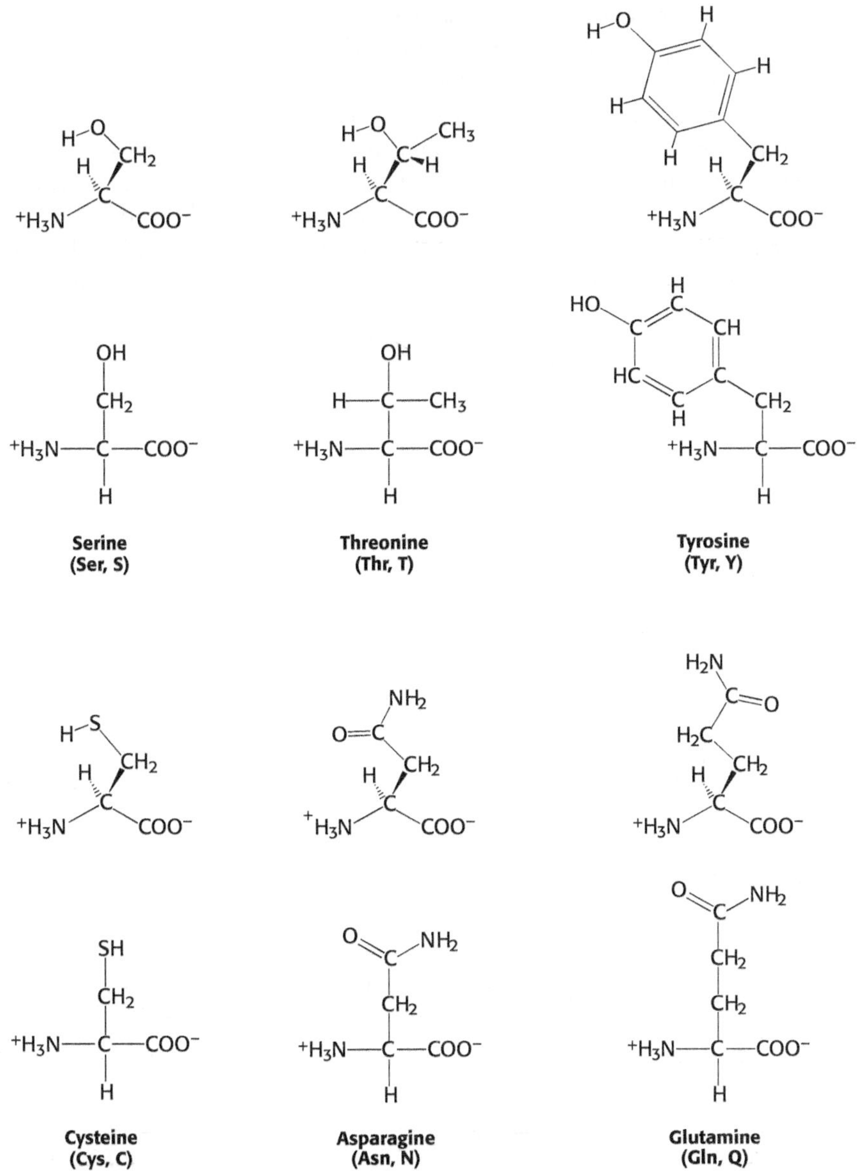

Serine (Ser, S)   **Threonine (Thr, T)**   **Tyrosine (Tyr, Y)**

Cysteine (Cys, C)   **Asparagine (Asn, N)**   **Glutamine (Gln, Q)**

Figure 3.4 Polar amino acids.

hydrogen partly positive. Serine can be thought of as a version of alanine with a hydroxyl group attached to the methyl group, whereas threonine resembles valine with a hydroxyl group in place of one of the valine methyl groups. Tyrosine is similar to phenylalanine but contains a hydrophilic hydroxyl group attached to the large aromatic ring. The hydroxyl groups on serine, threonine, and tyrosine make them more *hydrophilic* (water loving) and *reactive* than their respective nonpolar counterparts, alanine, valine, and phenylalanine. *Cysteine* is structurally similar to serine but contains a *sulfhydryl,* or *thiol* (—SH), group in place of the hydroxyl group. The sulfhydryl group is much more reactive than a hydroxyl group and can completely lose a proton at slightly basic pH to form the reactive thiolate group. Pairs of sulfhydryl groups in close proximity may form disulfide bonds—covalent bonds that are particularly important in stabilizing some proteins, as will be discussed in Chapter 4. In addition, the set of polar amino acids includes *asparagine* and *glutamine,* which contain a terminal *carboxamide.*

DID YOU KNOW?

Lysine is an essential amino acid (p. 44), which means that human beings cannot synthesize lysine and must obtain it in the diet. In experimental animals, kept on a cereal-based diet, inadequate dietary lysine increased stress-induced anxiety. Recent studies suggest that this response to lysine deprivation may be true for human beings, too.

Positively Charged Amino Acids Are Hydrophilic

We now turn to amino acids having positively charged side chains that render these amino acids highly hydrophilic (**Figure 3.5**). *Lysine* and *arginine* have long side chains that terminate with groups that are *positively charged* at neutral pH. Lysine is topped by an amino group and arginine by a guanidinium group. Note that the R groups of lysine and arginine have dual properties—the carbon chains constitute a hydrocarbon backbone, similar to the amino acid leucine, but the chain is terminated with a positive charge. Such combinations of characteristics contribute to the wide array of chemical properties of amino acids.

Histidine contains an imidazole group, an aromatic ring that also can be positively charged. With a pK_a value near 6, the imidazole group of histidine is unique in that it can be uncharged or positively charged near neutral pH, depending on its local environment (**Figure 3.6**). Indeed, histidine is often found in the active sites of enzymes, where the imidazole ring can bind and release protons in the course of enzymatic reactions.

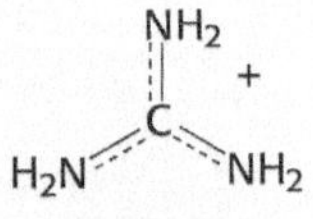

Guanidinium

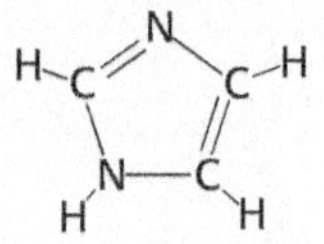

Imidazole

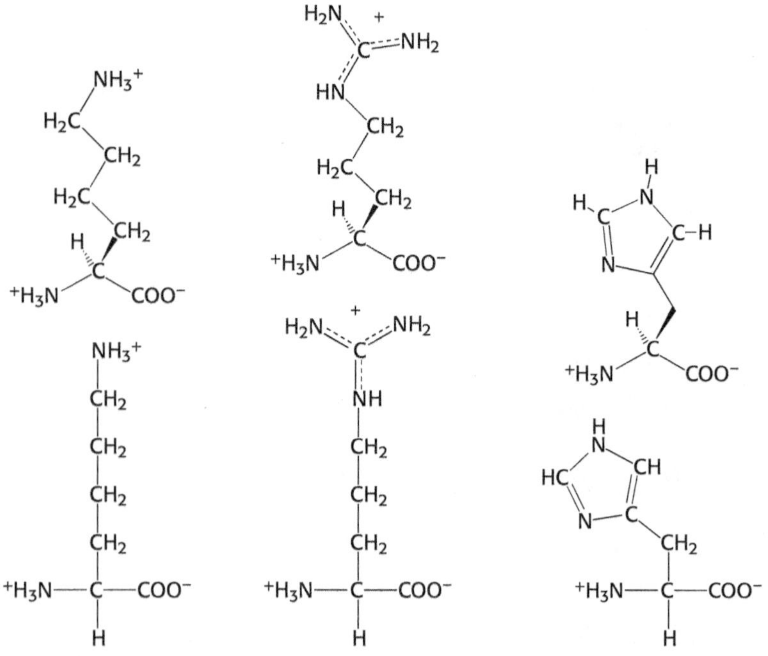

Lysine (Lys, K) **Arginine (Arg, R)** **Histidine (His, H)**

Figure 3.5 Positively charged amino acids.

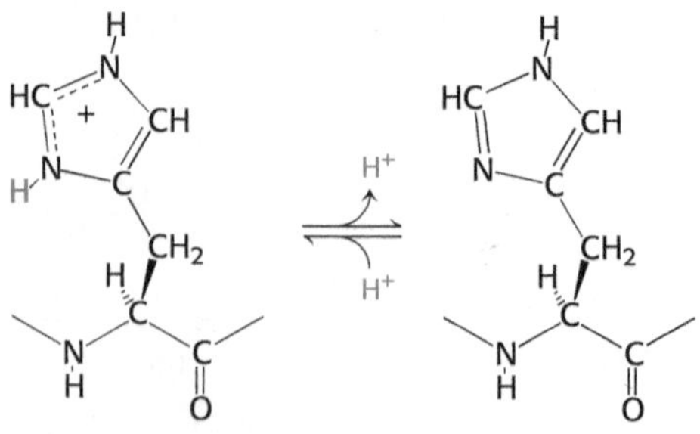

Figure 3.6 Histidine ionization. Histidine can bind or release protons near physiological pH.

Negatively Charged Amino Acids Have Acidic Side Chains

The two amino acids in this group, *aspartic acid* and *glutamic acid*, have *acidic side chains* that are usually negatively charged under intracellular conditions (**Figure 3.7**). These amino acids are often called *aspartate* and *glutamate* to emphasize the presence of the negative charge on their side chains. In some proteins, these side chains accept protons, which neutralize the negative charge. This ability is often functionally important. Aspartate and glutamate are related to asparagine and glutamine in which a carboxylic acid group in the former pair replaces a carboxamide in the latter pair.

The Ionizable Side Chains Enhance Reactivity and Bonding

Seven of the 20 amino acids—tyrosine, cysteine, arginine, lysine, histidine, and aspartic and glutamic acids—have readily ionizable side chains. These seven amino acids are able to form ionic bonds as well as to donate or accept protons to facilitate reactions. The ability to donate or accept protons is called acid–base catalysis and is an important chemical reaction for many enzymes. We will see the importance of histidine as an acid–base catalyst when we examine the protein-digesting enzyme chymotrypsin in Chapter 8. **Table 3.1** gives the equilibria and typical pK_a values for the ionization of the side chains of these seven amino acids.

QUICK QUIZ Name three amino acids that are positively charged at neutral pH. Name three amino acids that contain hydroxyl groups.

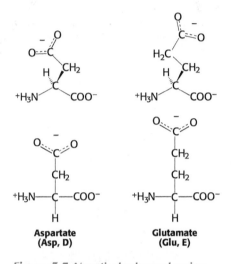

Aspartate (Asp, D)   **Glutamate (Glu, E)**

Figure 3.7 Negatively charged amino acids.

Table 3.1 Typical pK_a values of ionizable groups in proteins

Group	Acid	Base	Typical pK_a
Terminal α-carboxyl group			3.1
Aspartic acid Glutamic acid			4.1
Histidine			6.0
Terminal α-amino group			8.0
Cysteine			8.3
Tyrosine			10.9
Lysine			10.8
Arginine			12.5

Note: Values of pK_a depend on temperature, ionic strength, and the microenvironment of the ionizable group.

Table 3.2 Basic set of 20 amino acids

Nonessential	Essential
Alanine	Histidine
Arginine	Isoleucine
Asparagine	Leucine
Aspartate	Lysine
Cysteine	Methionine
Glutamate	Phenylalanine
Glutamine	Threonine
Glycine	Tryptophan
Proline	Valine
Serine	
Tyrosine	

3.3 Essential Amino Acids Must Be Obtained from the Diet

Most microorganisms can synthesize the entire basic set of 20 amino acids, whereas human beings can make only 11 of them. Amino acids that cannot be generated in the body must be supplied by the diet and are termed *essential amino acids*. The others are called nonessential amino acids (Table 3.2). These designations refer to an organism under a particular set of conditions. For example, a human adult can synthesize enough arginine to meet his or her needs, but a growing child requires more arginine than the body can provide to meet the protein-synthesis needs of rapid growth.

 CLINICAL INSIGHT

Pathological Conditions Result If Protein Intake Is Inadequate

The importance of amino acids in the diet (usually ingested as proteins) is illustrated by *kwashiorkor*, a particular form of malnutrition. This condition was first defined in the 1930s in Ghana. *Kwashiorkor* means "the disease of the displaced child" in the Ga language, a Ghanaian dialect; that is, the condition arises when a child is weaned because of the birth of a sibling. It is a form of malnutrition that results when protein intake is not sufficient. Initial symptoms of the disease are generalized lethargy, irritability, and stunted growth. If treated early enough, the effects of the disease are reversible. However, if not corrected early enough, many physiological systems fail to develop properly, including the brain. For instance, children will suffer from various infectious diseases because their immune systems, composed of many different proteins, cannot be constructed to function adequately. Likewise, the lack of protein prevents the complete development of the central nervous system, with resulting neurological problems.

The most common characteristic of a child suffering from kwashiorkor is a large protruding belly (Figure 3.8). The large belly is a sign not of excess calories but of edema, another result of the lack of protein. Edema is swelling that results from too much water in tissue. Insufficient protein in a child's blood distorts the normal distribution of water between plasma and surrounding capillaries. Although the swollen belly is most obvious, the limbs of a child suffering from kwashiorkor also are often swollen. Such suffering children are a devastating display of the centrality of protein to life.

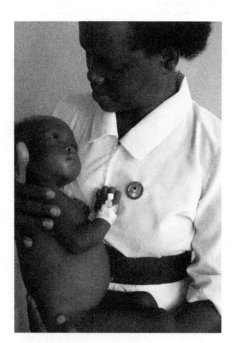

Figure 3.8 A child suffering from kwashiorkor. Note the swollen belly. This swelling (edema) is due to fluid collecting in the tissues because there is not enough protein in the blood. [Mauro Fermariello/Science Source.]

SUMMARY

3.1 Proteins Are Built from a Repertoire of 20 Amino Acids

Proteins are linear polymers of amino acids. Each amino acid consists of a central tetrahedral carbon atom that is bonded to an amino group, a carboxylic acid group, a distinctive side chain, and a hydrogen atom. These tetrahedral centers, with the exception of that of glycine, are chiral; only the L isomer exists in natural proteins. All natural proteins are constructed from the same set of 20 amino acids.

3.2 Amino Acids Contain a Wide Array of Functional Groups

The side chains of these 20 building blocks vary tremendously in size, shape, and the presence of functional groups. They can be grouped as follows on the basis of the chemical properties of the side chains: (1) hydrophobic side chains—glycine, alanine, valine, leucine, isoleucine, methionine, proline, and the aromatic amino acids phenylalanine and tryptophan; (2) polar amino acids—serine, threonine, tyrosine, asparagine, and glutamine; (3) positively charged amino acids—lysine, arginine, and histidine; and (4) negatively charged amino acids—aspartic acid and glutamic acid.

3.3 Essential Amino Acids Must Be Obtained from the Diet

Most microorganisms are capable of making all 20 of the amino acids from simpler molecules. Although human beings can make 11 amino acids, 9 must be acquired from the diet. These 9 amino acids are called essential amino acids because they are required for healthy growth and development.

KEY TERMS

side chain (R group) (p. 38)
L amino acid (p. 38)

dipolar ion (zwitterion) (p. 38)

essential amino acids (p. 44)

? Answer to QUICK QUIZ

Lysine, arginine, and histidine are positively charged at neutral pH. Three amino acids containing hydroxyl groups are serine, threonine, and tyrosine.

PROBLEMS

1. *Where's Elvis?* Translate the following amino acid sequence into one-letter code: Glu-Leu-Val-Ile-Ser-Ile-Ser-Leu-Ile-Val-Ile-Asn-Gly-Ile-Asn-Leu-Ala-Ser-Val-Glu-Gly-Ala-Ser.

2. *A nervous polecat.* Pyrrolysine (Pyl, O) and selenocysteine (Sec, U) are two uncommon amino acids. Knowing that these amino acids exist, translate the following amino acid sequence into one-letter code: Thr-Trp-Ile-Thr-Cys-His-Tyr-Leu-Ile-Thr-Thr-Ile-Glu-Phe-Glu-Arg-Arg-Glu-Thr-Ala-Arg-Glu-Asn-Thr-Tyr-Pyl-Sec-Met-Ala-Leu-Phe-Pyl-Tyr.

3. *Identify.* Examine the following four amino acids. What are their names, three-letter abbreviations, and one-letter symbols?

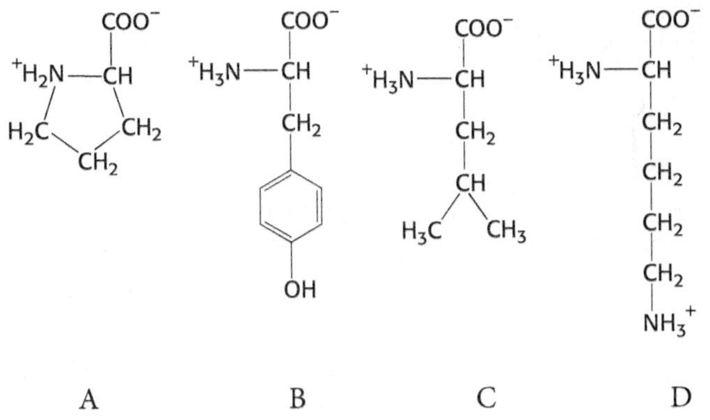

A   B   C   D

4.   *Properties.* In reference to the amino acids shown in problem 2, which are associated with the characteristics (a)–(e)? ✓ 1

(a) Hydrophobic side chain_____
(b) Basic side chain_____
(c) Three ionizable groups_____
(d) pK_a of approximately 10 in proteins_____
(e) Modified form of phenylalanine_____

5.   *Match 'em.* Match each amino acid in the left-hand column with the appropriate side-chain type in the right-hand column. ✓ 1

(a) Leu _____   1. hydroxyl-containing
(b) Glu _____   2. acidic
(c) Lys _____   3. basic
(d) Ser _____   4. sulfur-containing
(e) Cys _____   5. nonpolar aromatic
(f) Trp _____   6. nonpolar aliphatic

6.   *Solubility.* In each of the following pairs of amino acids, identify which amino acid would be most soluble in water: (a) Ala, Leu; (b) Tyr, Phe; (c) Ser, Ala; (d) Trp, His.

7.   *Charge and pH.* What is the net charge on the amino acid glycine at pH 7? At pH 12?

8.   *Isoelectric point.* The isoelectric point (pI) is the pH at which a molecule has no net charge. The amino acid glycine has two ionizable groups: (1) a carboxylic acid group with a pK_a of 2.72 and (2) an amino group with a pK_a of 9.60. Calculate the pI of glycine.

9.   *Positive R.* Which amino acids have positively charged R groups at pH 7?

10.   *Crucial versus noncrucial.* Differentiate between a nonessential and an essential amino acid.

11.   *Aromatic, not romantic.* What three amino acids have aromatic components in their side chains? ✓ 1

12.   *Getting a charge out of it.* Which amino acid side chains are capable of ionization? ✓ 1

13.   *Different states.* An amino acid with two ionizable groups can exist in four possible ionization states. Draw the four possible ionization states for alanine, and indicate which state predominates at pH 1, pH 7, and pH 11. ✓ 1

14.   *Carbolic acid-like.* Which amino acid contains a weakly acidic phenol-like R group?

15.   *Bonding is good.* Which of the following amino acids have R groups that have hydrogen-bonding potential? Ala, Gly, Ser, Phe, Glu, Tyr, Ile, and Thr.

Challenge Problems

16.   *Minor species.* For an amino acid such as alanine, the major species in solution at pH 7 is the zwitterionic form. Assume a pK_a value of 8 for the amino group and a pK_a value of 3 for the carboxylic acid. Estimate the ratio of the concentration of the neutral amino acid species (with the carboxylic acid protonated and the amino group neutral) to that of the zwitterionic species at pH 7.

17.   *Half-ionized.* Figure 3.2 shows the titration curve for a typical amino acid lacking a charged R group. Examine the graph, and determine the pK_a for the carboxyl group and the amino group of this amino acid.

18.   *+/0.* The R group of lysine has an amino group that can be positively charged or lose a proton to become neutral. The pK_a of the amino group is 10.8. Determine the fraction of amino group that is protonated at pH = 9.8 and at pH = 11.8.

19.   *0/−.* Glutamic acid has an R group that contains a carboxyl group whose pK_a = 4.1. What fraction of the R group carboxylic acid is protonated at pH = 2.1 and pH = 7.1?

Selected Readings for this chapter can be found online at www.whfreeman.com/tymoczko3e.

Protein Three-Dimensional Structure

A spider's web is a device built by the spider to trap prey. Spider silk, a protein, is the main component of the web. Many proteins have β sheets, a fundamental unit of protein structure, but silk is composed almost entirely of β sheets. [Javier Larrea/age fotostock.]

Proteins are the embodiment of the transition from the one-dimensional world of DNA sequences to the three-dimensional world of molecules capable of diverse activities. DNA encodes the sequence of amino acids that constitute a protein. The amino acid sequence is called the *primary structure*, and proteins typically consist of 50 to 300 amino acids. Functioning proteins, however, are not simply long polymers of amino acids. These polymers fold to form discrete three-dimensional structures with specific biochemical functions. Three-dimensional structure resulting from a regular pattern of hydrogen bonds between the NH and the CO components of the amino acids in the polypeptide chain is called *secondary structure*. The three-dimensional structure becomes more complex when the R groups of amino acids that are far apart in the primary structure bond with one another. This level of structure is called *tertiary structure* and is the highest level of structure that an individual polypeptide can attain. However, many proteins require more than one chain to function. Such proteins display *quaternary structure*, which can be as simple as a functional protein consisting of two identical polypeptide chains or as complex as one consisting of dozens of different polypeptide chains. Remarkably, the final three-dimensional structure of a protein is determined simply by the amino acid sequence of the protein.

In this chapter, we will examine the properties of the various levels of protein structure. Then, we will investigate how primary structure determines the final three-dimensional structure.

4.1   Primary Structure: Amino Acids Are Linked by Peptide Bonds to Form Polypeptide Chains

Proteins are complicated three-dimensional molecules, but their three-dimensional structure depends simply on their *primary structure*—the *linear polymers* formed by linking the α-carboxyl group of one amino acid to the α-amino group of another amino acid. The linkage joining amino acids in a protein is called a *peptide bond* (also called an *amide bond*). The formation of a dipeptide from two amino acids is accompanied by the loss of a water molecule (**Figure 4.1**). The equilibrium of this reaction lies on the side of hydrolysis rather than synthesis under most conditions. Hence, the biosynthesis of peptide bonds requires an input of free energy. Nonetheless, peptide bonds are quite stable kinetically because the rate of hydrolysis is extremely slow; the lifetime of a peptide bond in aqueous solution in the absence of a catalyst approaches 1000 years.

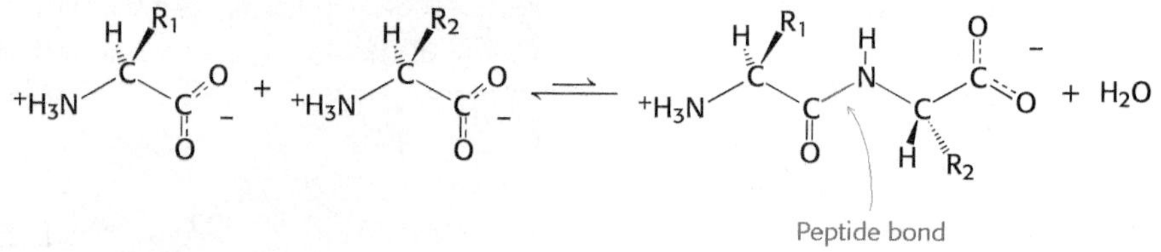

Figure 4.1 Peptide-bond formation. The linking of two amino acids is accompanied by the loss of a molecule of water.

A series of amino acids joined by peptide bonds form a *polypeptide chain*, and each amino acid unit in a polypeptide is called a *residue*. *A polypeptide chain has directionality*, sometimes called *polarity*, because its ends are different: an α-amino group is at one end, and an α-carboxyl group is at the other. By convention, the amino end is taken to be the beginning of a polypeptide chain, and so the sequence of amino acids in a polypeptide chain is written starting with the amino-terminal residue. Thus, in the pentapeptide Tyr-Gly-Gly-Phe-Leu (YGGFL), tyrosine is the amino-terminal (N-terminal) residue and leucine is the carboxyl-terminal (C-terminal) residue (**Figure 4.2**). The reverse sequence, Leu-Phe-Gly-Gly-Tyr (LFGGY), is a different pentapeptide, with different chemical properties. Note that the two peptides in question have the same *amino acid composition* but differ in primary structure.

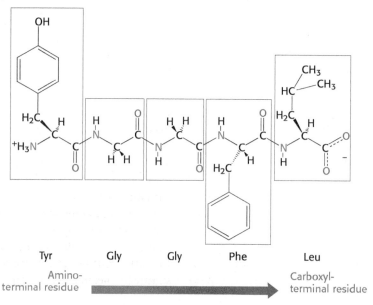

Figure 4.2 Amino acid sequences have direction. This illustration of the pentapeptide Tyr-Gly-Gly-Phe-Leu (YGGFL) shows the sequence from the amino terminus to the carboxyl terminus. This pentapeptide, Leu–enkephalin, is an opioid peptide that modulates the perception of pain.

Tyr   Gly   Gly   Phe   Leu
Amino-terminal residue   Carboxyl-terminal residue

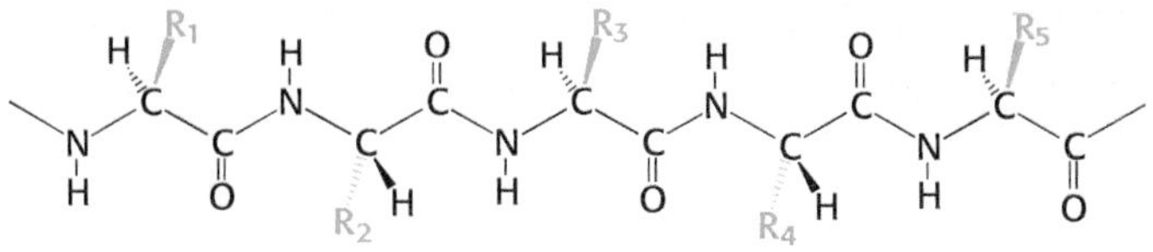

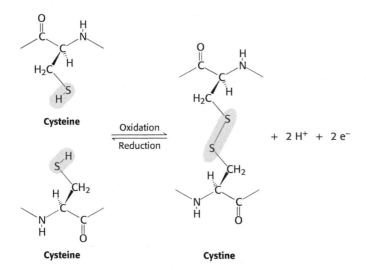

Figure 4.3 Components of a polypeptide chain. A polypeptide chain consists of a constant backbone (shown in black) and variable side chains (shown in green).

Figure 4.3 Components of a polypeptide chain. A polypeptide chain consists of a constant backbone (shown in black) and variable side chains (shown in green).

Carbonyl group

A polypeptide chain consists of a regularly repeating part, called the *main chain* or *backbone*, and a variable part, comprising the distinctive *side chains* (**Figure 4.3**). The polypeptide backbone is rich in hydrogen-bonding potential. Each residue contains a carbonyl group (C=O), which is a good hydrogen-bond acceptor, and, with the exception of proline, an amino group (N—H), which is a good hydrogen-bond donor. These groups interact with each other and with the functional groups of side chains to stabilize particular structures.

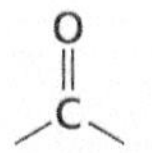

Most natural polypeptide chains contain between 50 and 2000 amino acid residues and are commonly referred to as proteins. The largest protein known is the muscle protein *titin*, which serves as a scaffold for the assembly of the contractile proteins of muscle. Titin consists of almost 27,000 amino acids. Peptides made of small numbers of amino acids are called *oligopeptides* or simply *peptides*. The mean molecular weight of an amino acid residue is about 110 g mol^{-1}, and so the molecular weights of most proteins are between 5500 and 22,000 g mol^{-1}. We can also refer to the mass of a protein in units of daltons; a *dalton* is a unit of mass very nearly equal to that of a hydrogen atom. A protein with a molecular weight of 50,000 g mol^{-1} has a mass of 50,000 daltons, or 50 kDa (kilodaltons).

In some proteins, the linear polypeptide chain is covalently cross-linked. The most common cross-links are *disulfide bonds*, formed by the oxidation of a pair of cysteine residues (**Figure 4.4**). The resulting unit of two linked cysteines is called *cystine*. Disulfide bonds can form between cysteine residues in the same polypeptide chain, or they can link two separate chains together. Rarely, nondisulfide cross-links derived from other side chains are present in proteins.

Cysteine

Oxidation ⇌ Reduction

$+ 2 H^+ + 2 e^-$

Cysteine

Cysteine

Cystine

Figure 4.4 Cross-links. The formation of a disulfide bond between two cysteine residues is an oxidation reaction.

Proteins Have Unique Amino Acid Sequences Specified by Genes

In 1953, Frederick Sanger determined the amino acid sequence of insulin, a protein hormone (**Figure 4.5**). *This work is a landmark in biochemistry because it showed for the first time that a protein has a precisely defined amino acid sequence* consisting only of L amino acids linked by peptide bonds. Sanger's accomplishment stimulated other scientists to carry out sequence studies of a wide variety of proteins. The complete primary acid sequences of millions of proteins are now known.

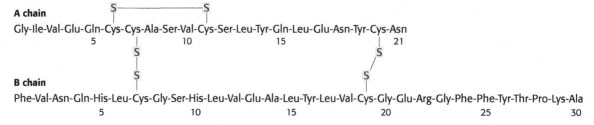

A chain

Gly-Ile-Val-Glu-Gln-Cys-Cys-Ala-Ser-Val-Cys-Ser-Leu-Tyr-Gln-Leu-Glu-Asn-Tyr-Cys-Asn

B chain

Phe-Val-Asn-Gln-His-Leu-Cys-Gly-Ser-His-Leu-Val-Glu-Ala-Leu-Tyr-Leu-Val-Cys-Gly-Glu-Arg-Gly-Phe-Phe-Tyr-Thr-Pro-Lys-Ala

Figure 4.5 Amino acid sequence of bovine insulin.

? QUICK QUIZ (a) What is the amino terminus of the tripeptide Gly–Ala–Asp? (b) What is the approximate molecular weight of a protein composed of 300 amino acids? (c) Approximately how many amino acids are required to form a protein with a molecular weight of 110,000?

Knowing amino acid sequences is important for several reasons. First, amino acid sequences determine the three-dimensional structures of proteins. Second, knowledge of the sequence of a protein is usually essential to elucidating its function (e.g., the catalytic mechanism of an enzyme). Third, alterations in amino acid sequence can produce abnormal function and disease. Severe and sometimes fatal diseases, such as sickle-cell anemia (Chapter 9) and cystic fibrosis, can result from a change in a single amino acid within a protein. Fourth, the sequence of a protein reveals much about its evolutionary history. Proteins resemble one another in amino acid sequence only if they have a common ancestor. Consequently, molecular events in evolution can be traced from amino acid sequences; molecular paleontology is a flourishing area of research.

Polypeptide Chains Are Flexible Yet Conformationally Restricted

Primary structure determines the three-dimensional structure of a protein, and the three-dimensional structure determines the protein's function. What are the rules governing the relation between an amino acid sequence and the three-dimensional structure of a protein? This question is very difficult to answer, but we know that certain characteristics of the peptide bond itself are important. First, *the peptide bond is essentially planar* (**Figure 4.6**). Thus, for a pair of amino acids linked by a peptide bond, six atoms lie in the same plane: the α-carbon atom and CO group of the first amino acid and the NH group and α-carbon atom of the second amino acid. Second, the peptide bond has considerable *double-bond character* owing to resonance structures: the electrons resonate between a pure single bond and a pure double bond.

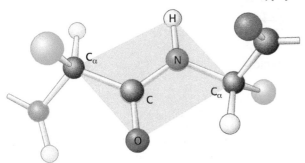

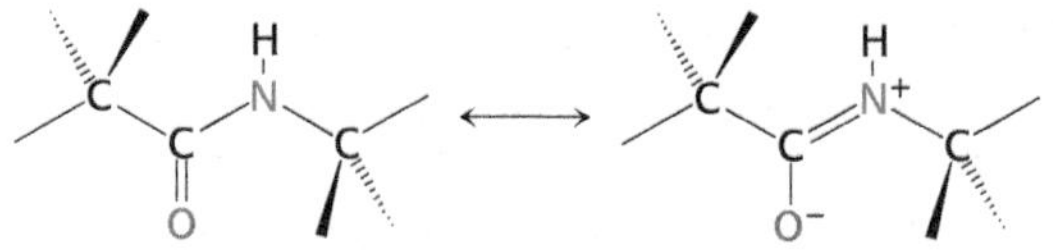

Figure 4.6 Peptide bonds are planar. In a pair of linked amino acids, six atoms (C_α, C, O, N, H, and C_α) lie in a plane. Side chains are shown as green balls.

Peptide-bond resonance structures

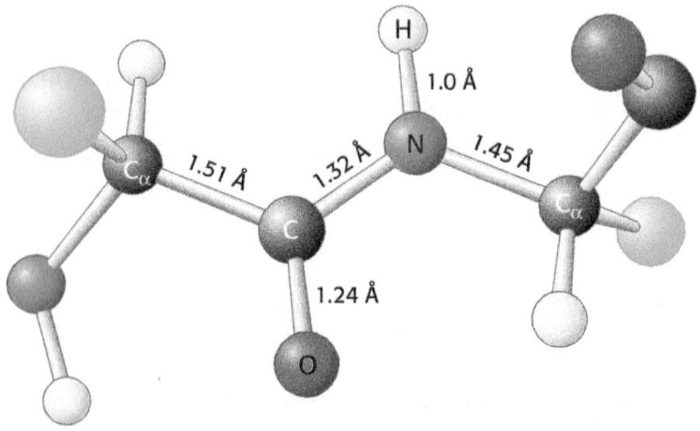

This partial double-bond character prevents rotation about this bond and thus constrains the conformation of the peptide backbone. The double-bond character is also expressed in the length of the bond between the CO and the NH groups. The C—N distance in a peptide bond is typically 1.32 Å (**Figure 4.7**), which is between the values expected for a C—N single bond (1.45 Å) and a C=N double bond (1.27 Å). Finally, *the peptide bond is uncharged*, allowing polymers of amino acids linked by peptide bonds to form tightly packed globular structures that would otherwise be inhibited by charge repulsion.

Figure 4.7 Typical bond lengths within a peptide unit. The peptide unit is shown in the trans configuration.

Two configurations are possible for a planar peptide bond. In the *trans* configuration, the two α-carbon atoms are on opposite sides of the peptide bond. In the *cis* configuration, these groups are on the same side of the peptide bond.

Almost all peptide bonds in proteins are trans. This preference for trans over cis can be explained by the fact that there are steric clashes between R groups in the cis configuration but not in the trans configuration (**Figure 4.8**).

In contrast with the peptide bond, the bonds between the amino group and the α-carbon atom and between the α-carbon atom and the carbonyl group are pure single bonds. The two adjacent rigid peptide units may rotate about these bonds, taking on various orientations. *This freedom of rotation about two bonds of each amino acid allows proteins to fold in many different ways.* The rotations about these bonds can be specified by *torsion angles* (**Figure 4.9**). The angle of rotation about the bond between the nitrogen atom and the α-carbon atom is called *phi (φ)*. The angle of rotation about the bond between the α-carbon atom and the carbonyl carbon atom is called *psi (ψ)*. A clockwise rotation about either bond as viewed toward the α-carbon atom corresponds to a positive value. The φ and ψ angles determine the path of the polypeptide chain.

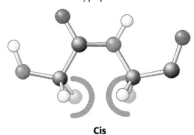

Trans　　　　　　**Cis**

Figure 4.8 Trans and cis peptide bonds. The trans form is strongly favored because of steric clashes in the cis form.

DID YOU KNOW?

Torsion angle, which is a measure of rotation about a bond, is usually taken to lie between −180 and +180 degrees. Torsion angles are sometimes called dihedral angles.

(A)　　　　　　　　　　(B)　　　　　　　　　　(C)

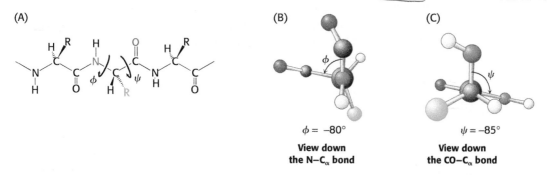

φ = −80°　　　　　ψ = −85°

View down the N–C$_\alpha$ bond　　　　**View down the CO–C$_\alpha$ bond**

Figure 4.9 Rotation about bonds in a polypeptide. The structure of each amino acid in a polypeptide can be adjusted by rotation about two single bonds. (A) Phi (φ) is the angle of rotation about the bond between the nitrogen and the α-carbon atoms, whereas psi (ψ) is the angle of rotation about the bond between the carbonyl carbon and the α-carbon atoms. (B) A view down the bond between the nitrogen and the α-carbon atoms. The angle φ is measured as the rotation of the carbonyl carbon attached to the α-carbon atom: positive if to the right, negative if to the left. (C) The angle ψ is measured by the rotation of the amino group as viewed down the bond from the carbonyl carbon to the α-carbon atom: positive if to the right, negative if to the left. Note that the view shown is the reverse of how the rotation is measured and consequently the angle has a negative value.

Are all combinations of φ and ψ possible? The Indian biophysicist Gopalasamudram Ramachandran recognized that many combinations are not found in nature because of steric clashes between atoms. He generated a two-dimensional plot, now called a *Ramachandran plot*, of the φ and ψ values of possible conformations (**Figure 4.10**). Three-quarters of the possible (φ, ψ) combinations are

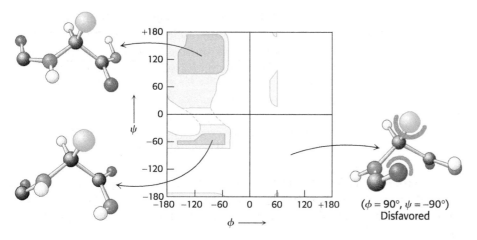

(φ = 90°, ψ = −90°)
Disfavored

Figure 4.10 A Ramachandran diagram showing the values of φ and ψ. Not all φ and ψ values are possible without collisions between atoms. The most favorable regions are shown in dark green on the graph; borderline regions are shown in light green. The structure on the right is disfavored because of steric clashes.

excluded simply by local steric clashes. *Steric exclusion, the fact that two atoms cannot be in the same place at the same time, restricts the number of possible peptide conformations and is thus a powerful organizing principle.*

4.2 Secondary Structure: Polypeptide Chains Can Fold into Regular Structures

Can a polypeptide chain fold into a regularly repeating structure? In 1951, Linus Pauling and Robert Corey proposed that certain polypeptide chains have the ability to fold into two periodic structures called the *α helix* (alpha helix) and the *β pleated sheet* (beta pleated sheet). Subsequently, other structures such as *turns* and *loops* were identified. Alpha helices, β pleated sheets, and turns are formed by a regular pattern of hydrogen bonds between the peptide NH and CO groups of amino acids that are often near one another in the linear sequence, or primary structure. Such regular folded segments are called *secondary structure*.

The Alpha Helix Is a Coiled Structure Stabilized by Intrachain Hydrogen Bonds

The first of Pauling and Corey's proposed secondary structures was the *α helix*, a rodlike structure with a tightly coiled backbone. The side chains of the amino acids composing the structure extend outward in a helical array (**Figure 4.11**).

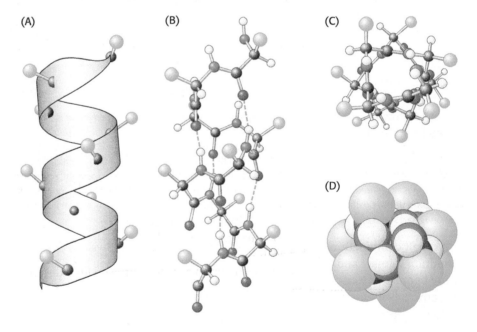

(A) (B) (C) (D)

Figure 4.11 The structure of the α helix. (A) A ribbon depiction shows the α-carbon atoms and side chains (green). (B) A side view of a ball-and-stick version depicts the hydrogen bonds (dashed lines) between NH and CO groups. (C) An end view shows the coiled backbone as the inside of the helix and the side chains (green) projecting outward. (D) A space-filling view of part C shows the tightly packed interior core of the helix.

The α helix is stabilized by hydrogen bonds between the NH and CO groups of the main chain. The CO group of each amino acid forms a hydrogen bond with the NH group of the amino acid that is situated four residues ahead in the sequence (**Figure 4.12**). Thus, except for amino acids near the ends of an α helix,

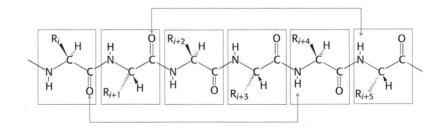

Figure 4.12 The hydrogen-bonding scheme for an α helix. In the α helix, the CO group of residue i forms a hydrogen bond with the NH group of residue $i + 4$.

all *the main-chain CO and NH groups are hydrogen bonded.* Each residue is related to the next one by a *rise*, also called *translation*, of 1.5 Å along the helix axis and a rotation of 100 degrees, which gives 3.6 amino acid residues per turn of helix. Thus, amino acids spaced three and four apart in the sequence are spatially quite close to one another in an α helix. In contrast, amino acids spaced two apart in the sequence are situated on opposite sides of the helix and so are unlikely to make contact. The *pitch* of the α helix is the length of one complete turn along the helix axis and is equal to the product of the translation (1.5 Å) and the number of residues per turn (3.6), or 5.4 Å. The *screw sense* of a helix can be right-handed (clockwise) or left-handed (counterclockwise). Right-handed helices are energetically more favorable because there are fewer steric clashes between the side chains and the backbone. *Essentially all α helices found in proteins are right-handed.* In schematic representations of proteins, α helices are depicted as twisted ribbons or rods (Figure 4.13).

Not all amino acids can be readily accommodated in an α helix. Branching at the β-carbon atom, as in valine, threonine, and isoleucine, tends to destabilize α helices because of steric clashes. Serine, aspartate, and asparagine also tend to disrupt α helices because their side chains contain hydrogen-bond donors or acceptors in close proximity to the main chain, where they compete for main-chain NH and CO groups. Proline also is an α helix breaker because it lacks an NH group and because its ring structure prevents it from assuming the ϕ value to fit into an α helix.

The α-helical content of proteins ranges widely, from none to almost 100%. For example, about 75% of the residues in ferritin, an iron-storage protein, are in α helices (Figure 4.14). Indeed, about 25% of all soluble proteins are composed of α helices connected by loops and turns of the polypeptide chain. Single α helices are usually less than 45 Å long. Many proteins that span biological membranes also contain α helices.

Beta Sheets Are Stabilized by Hydrogen Bonding Between Polypeptide Strands

Pauling and Corey named their other proposed periodic structural motif the β *pleated sheet* (β because it was the second structure that they elucidated). The β pleated sheet (more simply, the β sheet) differs markedly from the rodlike α helix in appearance and bond structure.

Instead of a single polypeptide strand, the β sheet is composed of two or more polypeptide chains called β *strands*. A β strand is almost fully extended rather than being tightly coiled as in the α helix. The distance between adjacent amino acids along a β strand is approximately 3.5 Å, in contrast with a distance of 1.5 Å along an α helix. The side chains of adjacent amino acids point in opposite directions (Figure 4.15).

A β sheet is formed by linking two or more β strands lying next to one another through hydrogen bonds. Adjacent chains in a β sheet can run in opposite directions (antiparallel β sheet) or in the same direction (parallel β sheet), as

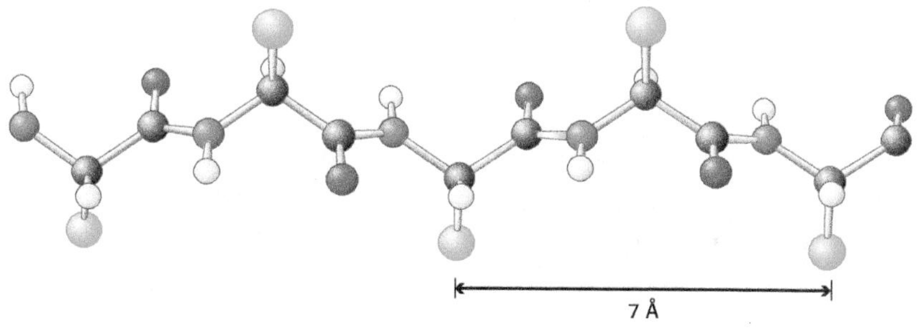

(A)   (B)

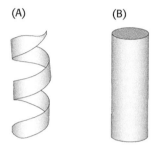

Figure 4.13 Schematic views of α helices. (A) A ribbon depiction. (B) A cylindrical depiction.

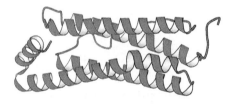

Figure 4.14 A largely α-helical protein. Ferritin, an iron-storage protein, is built from a bundle of a helices. [Drawn from 1AEW.pdb.]

Figure 4.15 The structure of a β strand. The side chains (green) are alternately above and below the plane of the strand. The bar shows the distance between two residues.

7 Å

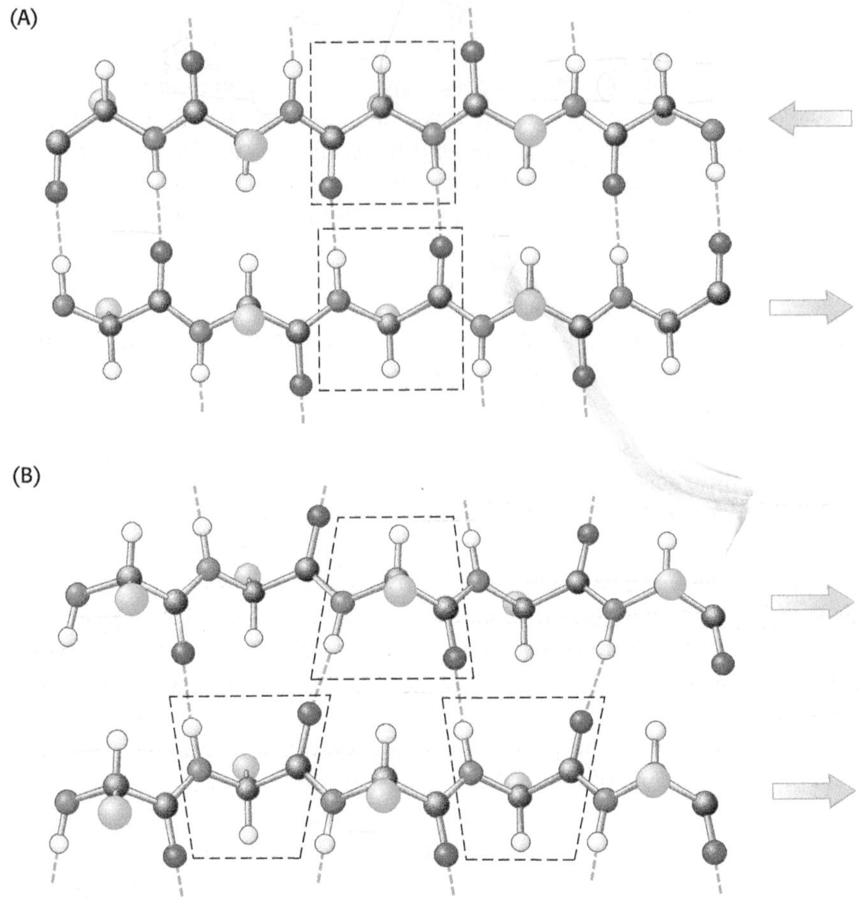

(A)

(B)

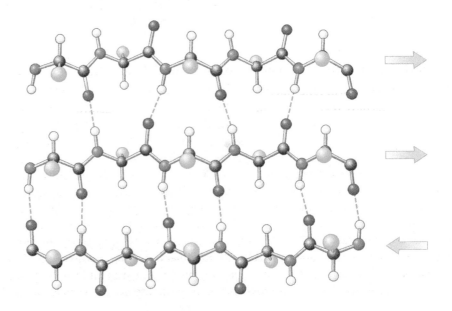

Figure 4.16 Antiparallel and parallel β sheets. (A) Adjacent β strands run in opposite directions. Hydrogen bonds (green dashes) between NH and CO groups connect each amino acid to a single amino acid on an adjacent strand, stabilizing the structure. (B) Adjacent β strands run in the same direction. Hydrogen bonds connect each amino acid on one strand with two different amino acids on the adjacent strand.

shown in **Figure 4.16**. Many strands, typically 4 or 5 but as many as 10 or more, can come together in a β sheet. Such β sheets can be purely antiparallel, purely parallel, or mixed (**Figure 4.17**). Unlike α helices, β sheets can consist of sections of a polypeptide that are not near one another. That is, in two β strands that lie next to each other, the last amino acid of one strand and the first amino acid of the adjacent strand are not necessarily neighbors in the amino acid sequence.

In schematic representations, β strands are usually depicted by broad arrows pointing in the direction of the carboxyl-terminal end to indicate the type of β sheet formed—parallel or antiparallel. Beta sheets can be almost flat, but most

Figure 4.17 The structure of a mixed β sheet.

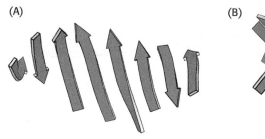

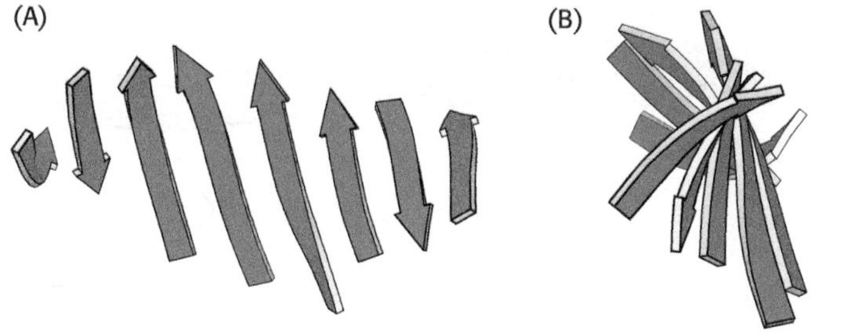

(A)

(B)

Figure 4.18 A twisted β sheet. (A) A schematic model. (B) The schematic view rotated by 90 degrees to illustrate the twist more clearly.

adopt a somewhat twisted shape (**Figure 4.18**). The β sheet is an important structural element in many proteins. For example, fatty-acid-binding proteins, which are important for lipid metabolism, are built almost entirely from β sheets (**Figure 4.19**).

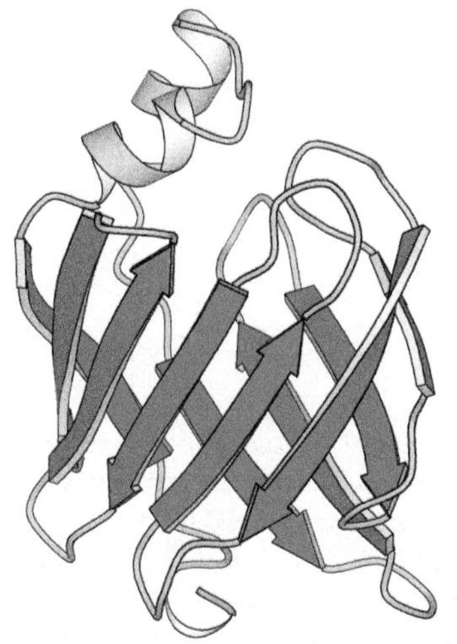

Polypeptide Chains Can Change Direction by Making Reverse Turns and Loops

Most proteins have compact globular shapes, requiring reversals in the direction of their polypeptide chains. Many of these reversals are accomplished by common structural elements called *reverse turns* and *loops* (**Figure 4.20**). Turns and loops invariably lie on the surfaces of proteins and thus often participate in interactions between other proteins and the environment. Loops exposed to an aqueous environment are usually composed of amino acids with hydrophilic R groups.

Figure 4.19 A protein rich in β sheets. The structure of a fatty-acid–binding protein. [Drawn from 1FTP.pdb.]

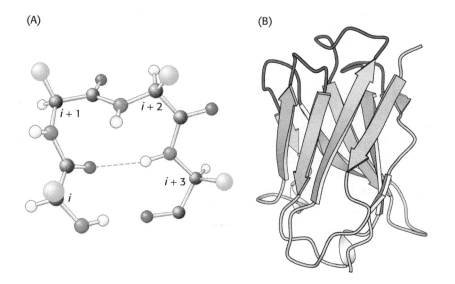

(A)

$i + 1$

$i + 2$

$i + 3$

i

(B)

Figure 4.20 The structure of a reverse turn. (A) The CO group of residue i of the polypeptide chain is hydrogen bonded to the NH group of residue $i + 3$ to stabilize the turn. (B) A part of an antibody molecule has surface loops (shown in red). [Drawn from 7FAB.pdb.]

Fibrous Proteins Provide Structural Support for Cells and Tissues

Fibrous proteins form long fibers that serve a structural role. Although some of these proteins have regions of complex three-dimensional structure, for the most part, the three-dimensional structure of fibrous proteins is relatively simple, consisting of extensive stretches of secondary structure. Special types of helices are present in two common fibrous proteins, α-keratin and collagen. α-Keratin, which is the primary component of wool and hair, consists of two right-handed α helices intertwined to form a type of left-handed superhelix called a *coiled coil*. α-Keratin is a member of a superfamily of proteins referred to as *coiled-coil proteins* (**Figure 4.21**). In these proteins, two or more helices can entwine to form a very stable structure that can have a length of 1000 Å (100 nm) or more. Human beings have approximately 60 members of this family, including intermediate filaments (p. 11) and the muscle proteins myosin and tropomyosin. The two helices in α-keratin are cross-linked by weak interactions such as van der

(A)

(B)

Figure 4.21 An α-helical coiled coil. (A) Space-filling model. (B) Ribbon diagram. The two helices wind around each other to form a superhelix. Such structures are found in many proteins, including keratin in hair, quills, claws, and horns. [Drawn from 1C1G.pdb.]

13
-Gly-Pro-Met-Gly-Pro-Ser-Gly-Pro-Arg-
22
-Gly-Leu-Hyp-Gly-Pro-Hyp-Gly-Ala-Hyp-
31
-Gly-Pro-Gln-Gly-Phe-Gln-Gly-Pro-Hyp-
40
-Gly-Glu-Hyp-Gly-Glu-Hyp-Gly-Ala-Ser-
49
-Gly-Pro-Met-Gly-Pro-Arg-Gly-Pro-Hyp-
58
-Gly-Pro-Hyp-Gly-Lys-Asn-Gly-Asp-Asp-

Figure 4.22 The amino acid sequence of a part of a collagen chain. Every third residue is glycine. Proline and hydroxyproline also are abundant.

Waals forces and ionic interactions. In addition, the two helices may be linked by disulfide bonds formed by neighboring cysteine residues.

A different type of helix is present in collagen, the most abundant mammalian protein. Collagen is the main fibrous component of skin, bone, tendon, cartilage, and teeth. It contains three helical polypeptide chains, each nearly 1000 residues long. Glycine appears at every third residue in the amino acid sequence, and the sequence glycine-proline-proline recurs frequently (**Figure 4.22**).

Hydrogen bonds within each peptide chain are absent in this type of helix. Instead, *the helices are stabilized by steric repulsion of the pyrrolidine rings of the proline residues* (**Figure 4.23**). The pyrrolidine rings keep out of each other's way

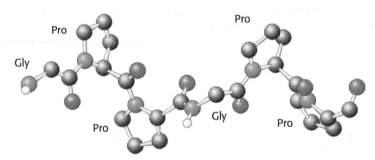

Figure 4.23 The conformation of a single strand of a collagen triple helix.

when the polypeptide chain assumes its helical form, which has about three residues per turn. Three strands wind around each other to form a *superhelical cable* that is stabilized by hydrogen bonds between strands. The hydrogen bonds form between the peptide NH groups of glycine residues and the CO groups of residues on the other chains. The inside of the triple-stranded helical cable is very crowded and explains why glycine has to be present at every third position on each strand: *the only residue that can fit in an interior position is glycine* (**Figure 4.24**A). The amino acid residue on either side of glycine is located on the outside of the cable, where there is room for the bulky rings of proline residues (Figure 4.24B).

(A)

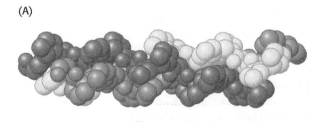

(B)

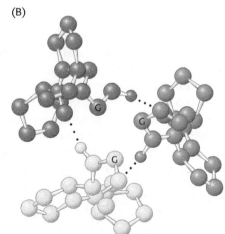

Figure 4.24 The structure of the protein collagen. (A) Space-filling model of collagen. Each strand is shown in a different color. (B) Cross section of a model of collagen. Each strand is hydrogen bonded to the other two strands. The atom of a glycine residue is identified by the letter G. Every third residue must be glycine because there is no space in the center of the helix. Notice that the pyrrolidine rings are on the outside.

✚ CLINICAL INSIGHT

Defects in Collagen Structure Result in Pathological Conditions

The importance of the positioning of glycine inside the triple helix is illustrated in the disorder *osteogenesis imperfecta*, also known as brittle bone disease. In this condition, which can vary from mild to very severe, other amino acids replace the internal glycine residue. This replacement leads to a delayed and improper folding of collagen, and the accumulation of defective collagen results in cell death. The most serious symptom is severe bone fragility. Defective collagen in the eyes causes the whites of the eyes to have a blue tint (blue sclera).

As we have seen, proline residues are important in creating the superhelical cable structure of collagen. Hydroxyproline is a modified version of proline, with a hydroxyl group replacing a hydrogen atom in the pyrrolidine ring. It is a common element of collagen, appearing in the glycine-proline-proline sequence as the second proline. Hydroxyproline is essential for stabilizing collagen, and its formation illustrates our dependence on vitamin C.

Vitamin C is required for the formation of stable collagen fibers because it assists in the formation of hydroxyproline from proline. Less stable collagen can result in scurvy. The symptoms of scurvy include skin lesions and blood-vessel fragility. Most notable are bleeding gums, the loss of teeth, and periodontal infections. Gums are especially sensitive to a lack of vitamin C because the collagen in gums turns over rapidly. Vitamin C is required for the continued activity of *prolyl hydroxylase,* which synthesizes hydroxyproline.

NUTRITION FACTS

Vitamin C
Human beings are among the few mammals unable to synthesize vitamin C. Citrus products are the most common source of this vitamin. Vitamin C functions as a general antioxidant to reduce the presence of reactive oxygen species throughout the body. In addition, it serves as a specific antioxidant by maintaining metals, required by certain enzymes such as the enzyme that synthesizes hydroxyproline, in the reduced state. [Photograph from Don Farrell/Digital Vision/Getty Images.]

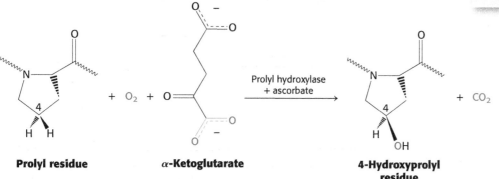

| **Prolyl residue** | **α-Ketoglutarate** | | **4-Hydroxyprolyl residue** | **Succinate** |

This reaction requires an Fe^{2+} ion to activate O_2. This iron ion, embedded in prolyl hydroxylase, is susceptible to oxidation, which inactivates the enzyme. How is the enzyme made active again? *Ascorbate* (vitamin C) comes to the rescue by reducing the Fe^{3+} of the inactivated enzyme. Thus, ascorbate serves here as a specific *antioxidant*.

4.3 Tertiary Structure: Water-Soluble Proteins Fold into Compact Structures

As already discussed, primary structure is the sequence of amino acids, and secondary structure is the simple repeating structures formed by hydrogen bonds between hydrogen and oxygen atoms of the peptide backbone. Another level of structure, *tertiary structure,* refers to the spatial arrangement of amino acid residues that are far apart in the sequence and to the pattern of disulfide bonds. This level of structure is the result of interactions between the R groups of the peptide chain. To explore the principles of tertiary structure, we will examine *myoglobin,* the first protein to be seen in atomic detail.

Myoglobin Illustrates the Principles of Tertiary Structure

Myoglobin is an example of a globular protein (**Figure 4.25**). In contrast with fibrous proteins such as keratin, globular proteins have a compact three-dimensional

(A)

Heme group
Iron atom

(B)

Heme group

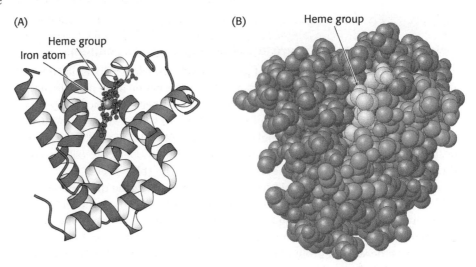

Figure 4.25 The three-dimensional structure of myoglobin. (A) A ribbon diagram shows that the protein consists largely of α helices. (B) A space-filling model in the same orientation shows how tightly packed the folded protein is. Notice that the heme group is nestled into a crevice in the compact protein with only an edge exposed. One helix is blue to allow comparison of the two structural depictions. [Drawn from 1A6N.pdb.]

structure and are water soluble. Globular proteins, with their more intricate three-dimensional structure, perform most of the chemical transactions in the cell.

Myoglobin, a single polypeptide chain of 153 amino acids, is an oxygen-binding protein found predominantly in heart and skeletal muscle; it appears to facilitate the diffusion of oxygen from the blood to the mitochondria, the primary site of oxygen utilization in the cell. The capacity of myoglobin to bind oxygen depends on the presence of heme, a prosthetic (helper) group containing an iron atom. *Myoglobin is an extremely compact molecule.* Its overall dimensions are 45 × 35 × 25 Å, an order of magnitude less than if it were fully stretched out. About 70% of the main chain is folded into eight α helices, and much of the rest of the chain forms turns and loops between helices.

Myoglobin, like most other proteins, is asymmetric because of the complex folding of its main chain. A unifying principle emerges from the distribution of side chains. The striking fact is that *the interior consists almost entirely of nonpolar residues* (**Figure 4.26**). The only polar residues on the interior are two histidine residues, which play critical roles in binding the heme iron and oxygen. The outside of myoglobin, on the other hand, consists of both nonpolar and polar residues, which can interact with water and thus render the molecule water soluble. The space-filling model shows that there is very little empty space inside.

(A)

(B)

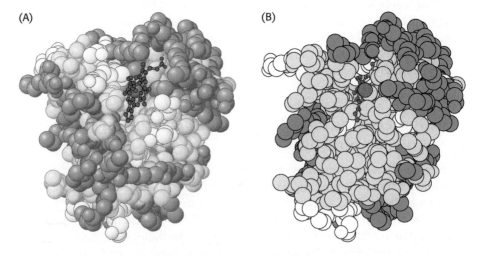

Figure 4.26 The distribution of amino acids in myoglobin. (A) A space-filling model of myoglobin, with hydrophobic amino acids shown in yellow, charged amino acids shown in blue, and others shown in white. Notice that the surface of the molecule has many charged amino acids, as well as some hydrophobic amino acids. (B) In this cross-sectional view, notice that mostly hydrophobic amino acids are found on the inside of the structure, whereas the charged amino acids are found on the protein surface. [Drawn from 1MBD.pdb.]

This contrasting distribution of polar and nonpolar residues reveals a key facet of protein architecture. In an aqueous environment such as the interior of a cell, protein folding is driven by the hydrophobic effect—the strong tendency of hydrophobic residues to avoid contact with water. *The polypeptide chain therefore folds so that its hydrophobic side chains are buried and its polar, charged chains are on the surface.* Similarly, an unpaired peptide NH or CO group of the main chain markedly prefers water to a nonpolar milieu. The only way to bury a segment of main chain in a hydrophobic environment is to pair all the NH and CO groups by hydrogen bonding. This pairing is neatly accomplished in an α helix or β sheet. Van der Waals interactions between tightly packed hydrocarbon side chains also contribute to the stability of proteins. We can now understand why the set of 20 amino acids contains several that differ subtly in size and shape. They provide a palette of shapes that can fit together tightly to fill the interior of a protein neatly and thereby maximize van der Waals interactions, which require intimate contact.

Some proteins that span biological membranes are "the exceptions that prove the rule" because they have the reverse distribution of hydrophobic and hydrophilic amino acids. For example, consider porins, proteins found in the outer membranes of many bacteria. Membranes are built largely of the hydrophobic hydrocarbon chains of lipids (Chapter 12). Thus, porins are covered on the outside largely by hydrophobic residues that interact with the hydrophobic environment. In contrast, the center of the protein contains many charged and polar amino acids that surround a water-filled channel running through the middle of the protein. Thus, because porins function in hydrophobic environments, they are "inside out" relative to proteins that function in aqueous solution.

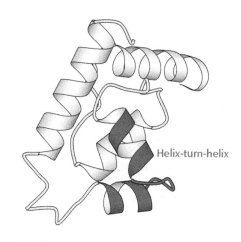

The Tertiary Structure of Many Proteins Can Be Divided into Structural and Functional Units

Certain combinations of secondary structure are present in many proteins and frequently exhibit similar functions. These combinations are called *motifs* or *supersecondary structures*. For example, an α helix separated from another α helix by a turn, called a *helix-turn-helix* unit, is found in many proteins that bind DNA (**Figure 4.27**).

Figure 4.27 The helix-turn-helix motif, a supersecondary structural element. Helix-turn-helix motifs are found in many DNA-binding proteins. [Drawn from 1LMB.pdb.]

Some polypeptide chains fold into two or more compact regions that may be connected by a flexible segment of polypeptide chain, rather like pearls on a string. These compact globular units, called *domains*, range in size from about 30 to 400 amino acid residues. For example, the extracellular part of CD4, a cell-surface protein on certain cells of the immune system, comprises four similar domains of approximately 100 amino acids each (**Figure 4.28**). Different proteins may have domains in common even if their overall tertiary structures are different.

Figure 4.28 Protein domains. The cell-surface protein CD4 consists of four similar domains. [Drawn from 1WIO.pdb.]

4.4 Quaternary Structure: Multiple Polypeptide Chains Can Assemble into a Single Protein

Many proteins consist of more than one polypeptide chain in their functional states. Each polypeptide chain in such a protein is called a *subunit*. *Quaternary structure* refers to the arrangement of subunits and the nature of their interactions. The interactions among subunits of proteins displaying quaternary structure are usually the weak interactions discussed in Chapter 2: hydrogen bonds, ionic bonds, and van der Waals interactions.

The simplest sort of quaternary structure is a *dimer* consisting of two identical subunits. This organization is present in Cro, a DNA-binding protein found in a bacterial virus called lambda (λ) that regulates gene expression (**Figure 4.29**). Quaternary structure can be as simple as two identical subunits or as complex as dozens of different polypeptide chains. More than one type of subunit can be present, often in variable numbers. For example, human hemoglobin, the oxygen-carrying protein in blood, consists of two subunits of one type (designated α) and two subunits of another type (designated β), as

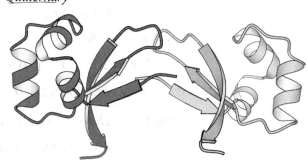

Figure 4.29 Quaternary structure. The Cro protein of bacteriophage λ is a dimer of identical subunits. [Drawn from 5CRO.pdb.]

illustrated in **Figure 4.30**. Thus, the hemoglobin molecule exists as an $\alpha_2\beta_2$ *tetramer*. Note that the hemoglobin subunits are called α and β for historical reasons, and the α and β designations have no relation to the α helix or the β strand.

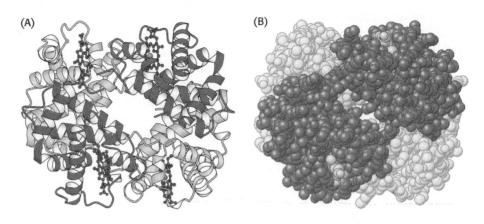

Figure 4.30 The $\alpha_2\beta_2$ tetramer of human hemoglobin. The structure of the two identical α subunits (red) and the two identical β subunits (yellow). (A) The ribbon diagram shows that they are composed mainly of α helices. (B) The space-filling model illustrates the close packing of the atoms and shows that the heme groups (gray) occupy crevices in the protein. [Drawn from 1A3N.pdb.]

✓ 3 Describe the biochemical information that determines the final three-dimensional structure of proteins, and explain what powers the formation of this structure.

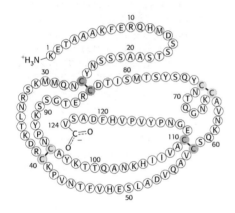

Figure 4.31 Amino acid sequence of bovine ribonuclease. The four disulfide bonds are shown in color. [Information from C. H. W. Hirs, S. Moore, and W. H. Stein, *J. Biol. Chem.* 235(1960):633–647.]

4.5 The Amino Acid Sequence of a Protein Determines Its Three-Dimensional Structure

How is the elaborate three-dimensional structure of proteins attained? The classic work of Christian Anfinsen in the 1950s on the enzyme ribonuclease revealed the relation between the amino acid sequence of a protein and its conformation. Ribonuclease is a single polypeptide chain consisting of 124 amino acid residues cross-linked by four disulfide bonds (**Figure 4.31**). Anfinsen's plan was to destroy the three-dimensional structure of the enzyme and to then determine the conditions required to restore the tertiary structure. Chaotropic agents, such as *urea*, which disrupt all of the noncovalent bonds in a protein, were added to a solution of the ribonuclease. The disulfide bonds where then cleaved reversibly with a sulfhydryl reagent such as *β-mercaptoethanol* (**Figure 4.32**). In the presence of a large excess of β-mercaptoethanol, the disulfides (cystines) are fully converted into sulfhydryls (cysteines).

When ribonuclease was treated with β-mercaptoethanol in 8 M urea, the product was a randomly coiled polypeptide chain *devoid of enzymatic activity*. When a protein is converted into a randomly coiled peptide without its normal activity, it is said to be *denatured* (**Figure 4.33**).

Anfinsen then made the critical observation that the denatured ribonuclease, freed of urea and β-mercaptoethanol by dialysis, slowly regained enzymatic activity. He immediately perceived the significance of this finding: the enzyme spontaneously refolded into a catalytically active form with all of the correct disulfide bonds re-forming. All the measured physical and chemical properties of

Figure 4.32 The role of β-mercaptoethanol in reducing disulfide bonds. Notice that, as the disulfides are reduced, the β-mercaptoethanol is oxidized and forms dimers.

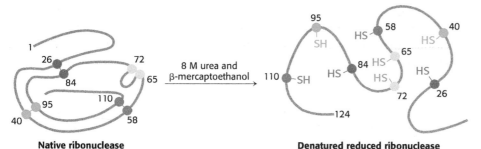

Native ribonuclease 8 M urea and β-mercaptoethanol → **Denatured reduced ribonuclease**

Figure 4.33 The reduction and denaturation of ribonuclease.

the refolded enzyme were virtually identical with those of the native enzyme. These experiments showed that *the information needed to specify the catalytically active three-dimensional structure of ribonuclease is contained in its amino acid sequence.* In other words, *the native structure is the thermodynamically most stable structure.* Subsequent studies have established the generality of this central principle of biochemistry: *sequence specifies conformation.* The dependence of conformation on sequence is especially significant because conformation determines function.

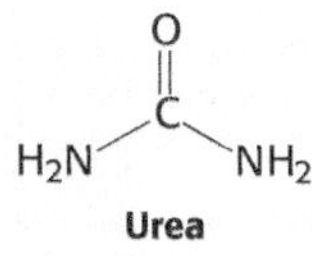

Similar refolding experiments have been performed on many other proteins. In many cases, the native structure can be generated under suitable conditions. For other proteins, however, refolding does not proceed efficiently. In these cases, the unfolded protein molecules usually become tangled up with one another to form aggregates. Inside cells, proteins called *chaperones* block such illicit interactions.

Proteins Fold by the Progressive Stabilization of Intermediates Rather Than by Random Search

How does a protein make the transition from an unfolded structure to a unique conformation in the native form? One possibility is that all possible conformations are tried out to find the energetically most favorable one. How long would such a random search take? Cyrus Levinthal calculated that, if each residue of a 100-residue protein can assume three different conformations, the total number of structures would be 3^{100}, which is equal to 5×10^{47}. If the conversion of one structure into another were to take 10^{-13} seconds (s), the total search time would be $5 \times 10^{47} \times 10^{-13}$ s, which is equal to 5×10^{34} s, or 1.6×10^{27} years. Clearly, it would take much too long for even a small protein to fold properly by randomly trying out all possible conformations. Moreover, Anfinsen's experiments showed that proteins do fold on a much more limited time scale. The enormous difference between calculated and actual folding times is called *Levinthal's paradox.* Levinthal's paradox and Anfinsen's results suggest that proteins do not fold by trying every possible conformation; rather, they must follow at least a partly defined folding pathway consisting of intermediates between the fully denatured protein and its native structure.

The way out of this paradox is to recognize the power of *cumulative selection.* Richard Dawkins, in his book *The Blind Watchmaker,* asked how long it would take a monkey poking randomly at a typewriter to reproduce Hamlet's remark to Polonius, "Methinks it is like a weasel" (**Figure 4.34**). An astronomically large number of keystrokes, of the order of 10^{40}, would be required. However, suppose that we preserved each correct character and allowed the monkey to retype only the wrong ones. In this case, only a few thousand keystrokes, on average, would be needed. The crucial difference between these cases is that the first employs a completely random search, whereas, in the second, partly correct intermediates are retained.

The essence of protein folding is the tendency to retain partly correct intermediates because they are slightly more stable than unfolded regions. However, the protein-folding problem is much more difficult than the one presented to our simian Shakespeare. First, the criterion of correctness is not a residue-by-residue

```
 200  ?T(\G{+s  x[A.N5~,#ATxSGpn`e□@
 400  oDr'Jh7s  DFR:W4l'u+^v6zpJseOi
 600  e2ih'8zs  n527x8l8d_ih=Hldseb.
 800  S#dh>}/s  ]tZqC%lP%DK<|!^aseZ.
1000  VOth>nLs  ut/isjl_kwojjwMasef.
1200  juth+nvs  it is[lukh?SCw=ase5.
1400  Iithdn4s  it isOl/ks/IxwLase~.
1600  M?thinrs  it is lXk?T"_woasel.
1800  MSthinWs  it is lwkN7□Kw(asel.
2000  Mhthin`s  it is likv,aww_asel.
2200  MMthinns  it is lik+5avwlasel.
2400  MethinXs  it is likydaqw)asel.
2600  Methin4s  it is lik2dasweasel.
2800  MethinHs  it is like□aTweasel.
2883  Methinks  it is like a weasel.
```

```
 200  )z~hg)W4{{cu!kO{d6jS!NlEyUx}p
 400  "W hi\kR.<&CfA%4-Y1G!iT$6({|6
 600  .L=hinkm4(uMGP^lAWoE6klwW=yiS
 800  AthinkaPa_vYH liR\Hb,Uo4\-"(
1000  OFthinksP)@fZO li8v] /+Eln26B
1200  6ithinksMVt -V likm+gl#K~}BFk
1400  vxthinksaEt □w like.SlGeutks.
1600  :Othinks<it MC likesN2[eaVe4.
1800  uxthinkqit Or likeQh)weaoeW.
2000  Y/thinks it id like7alwea)e&.
2200  Methinks it iW like a[weaWel.
2400  Methinks it is like a;weasel.
2431  Methinks it is like a weasel.
```

Figure 4.34 Typing-monkey analogy. A monkey randomly poking a typewriter could write a line from Shakespeare's *Hamlet,* provided that correct keystrokes were retained. In the two computer simulations shown, the cumulative number of keystrokes is given at the left of each line.

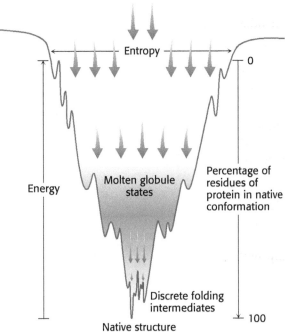

Beginning of helix formation and collapse

Energy

Entropy

0

Molten globule
states

Percentage of
residues of
protein in native
conformation

Discrete folding
intermediates

100

Native structure

Figure 4.35 Folding funnel. The folding funnel depicts the thermodynamics of protein folding. The top of the funnel represents all possible denatured conformations—that is, maximal conformational entropy. Depressions on the sides of the funnel represent semistable intermediates that may facilitate or hinder the formation of the native structure, depending on their depth. Secondary structures, such as helices, form and collapse onto one another to initiate folding. [Data from D. L. Nelson and M. M. Cox, *Lehninger Principles of Biochemistry,* 5th ed. (W. H. Freeman and Company, 2008), p. 143.]

scrutiny of conformation by an omniscient observer but rather the total free energy of the folding intermediate. Second, even correctly folded proteins are only marginally stable. The free-energy difference between the folded and the unfolded states of a typical 100-residue protein is 42 kJ mol^{-1} (10 kcal mol^{-1}); thus, each residue contributes on average only 0.42 kJ mol^{-1} (0.1 kcal mol^{-1}) of energy to maintain the folded state. This amount is less than the amount of thermal energy, which is 2.5 kJ mol^{-1} (0.6 kcal mol^{-1}) at room temperature. This meager stabilization energy means that correct intermediates, especially those formed early in folding, can be lost. Nonetheless, the interactions that lead to folding can stabilize intermediates as structure builds up. The analogy is that the monkey would be somewhat free to undo its correct keystrokes.

The folding of proteins is sometimes visualized as a *folding funnel,* or energy landscape (**Figure 4.35**). The breadth of the funnel represents all possible conformations of the unfolded protein. The depth of the funnel represents the energy difference between the unfolded and the native protein. Each point on the surface represents a possible three-dimensional structure and its energy value. The funnel suggests that there are alternative pathways to the native, or most stable, structure.

One model pathway postulates that local interactions take place first—in other words, secondary structure forms—and these secondary structures facilitate the long-range interactions leading to tertiary-structure formation. Another model pathway proposes that the hydrophobic effect brings together hydrophobic amino acids that are far apart in the amino acid sequence. The drawing together of hydrophobic amino acids in the interior leads to the formation of a globular structure. Because the hydrophobic interactions are presumed to be dynamic, allowing the protein to form progressively more stable interactions, the structure is called a *molten globule.* Another, more general model, called the nucleation–condensation model, is essentially a combination of the two preceding models. In the nucleation–condensation model, both local and long-range interactions take place to lead to the formation of the native state.

Some Proteins Are Inherently Unstructured and Can Exist in Multiple Conformations

The discussion of protein folding thus far is based on the paradigm that a given protein amino acid sequence will fold into a particular three-dimensional structure. This paradigm holds well for many proteins, such as enzymes and transport proteins. However, it has been known for some time that some proteins can adopt two different structures, one of which results in protein aggregation and pathological conditions (p. 63). Such alternate structures originating from a unique amino acid sequence were thought to be rare: the exception to the paradigm. However, recent work has called into question the universality of the idea that each amino acid sequence gives rise to one structure for certain proteins, even under normal cellular conditions.

Our first example is a class of proteins referred to as *intrinsically unstructured proteins* (IUPs). As the name suggests, these proteins, completely or in part, do not have a discrete three-dimensional structure under physiological conditions. Indeed, an estimated 50% of eukaryotic proteins have at least one unstructured region greater than 30 amino acids in length. These proteins assume a defined structure on interaction with other proteins. This molecular versatility means that one protein can assume different structures and interact with different partners, yielding different biochemical functions. IUPs appear to be especially important in signaling and regulatory pathways.

Another class of proteins that do not adhere to the paradigm are *metamorphic proteins*. These proteins appear to exist in an ensemble of structures of approximately equal energy that are in equilibrium. Small molecules or other proteins may bind to a particular member of the ensemble, resulting in a complex having a biochemical function that differs from that of another complex formed by the same metamorphic protein bound to a different partner. An especially clear example of a metamorphic protein is the cytokine lymphotactin. Cytokines are signal molecules in the immune system that bind to receptor proteins on the surface of immune-system cells, instigating an immunological response. Lymphotactin exists in two very different structures that are in equilibrium (**Figure 4.36**). One structure is a characteristic of chemokines, consisting of a three-stranded β sheet and a carboxyl-terminal helix. This structure binds to its receptor and activates it. The alternative structure is an identical dimer of all β sheets. When in this structure, lymphotactin binds to glycosaminglycan, a complex carbohydrate (Chapter 10). The biochemical activities of each structure are mutually exclusive: the cytokine structure cannot bind the glycosaminoglycan, and the β-sheet structure cannot activate the receptor. Yet, remarkably, both activities are required for full biochemical activity of the cytokine.

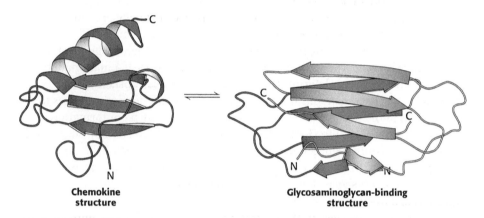

Chemokine structure   **Glycosaminoglycan-binding structure**

Figure 4.36 Lymphotactin exists in two conformations, which are in equilibrium. [R. L. Tuinstra, F. C. Peterson, S. Kutlesa, E. S. Elgin, M. A. Kron, and B. F. Volkman, *Proc. Natl. Sci. U. S. A.* 105:5057–5062, 2008, Fig. 3A.]

Note that IUPs and metamorphic proteins effectively expand the protein-encoding capacity of the genome. In some cases, a gene can encode a single protein that has more than one structure and function. These examples also illustrate the dynamic nature of the study of biochemistry and its inherent excitement: even well-established ideas are often subject to modifications.

 CLINICAL INSIGHT

Protein Misfolding and Aggregation Are Associated with Some Neurological Diseases

Understanding protein folding and misfolding is of more than academic interest. A host of diseases, including Alzheimer disease, Parkinson disease, Huntington disease, and transmissible spongiform encephalopathies (prion disease), are associated with improperly folded proteins. All of these diseases result in the deposition of protein aggregates, called *amyloid fibrils* or *plaques* (**Figure 4.37**). These diseases are consequently referred to as *amyloidoses*. A common feature of amyloidoses is that normally soluble proteins are converted into insoluble fibrils rich in β sheets. The correctly folded protein is only marginally more stable than the incorrect form. But the incorrect forms aggregates, pulling more correct forms into the incorrect form. We will focus on the transmissible spongiform encephalopathies.

One of the great surprises in modern medicine was the discovery by Stanley Prusiner that certain infectious neurological diseases were found to be

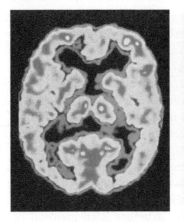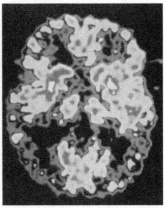

Figure 4.37 Alzheimer disease. Colored positron emission tomography (PET) scans of the brain of a normal person (left) and that of a patient who has Alzheimer disease (right). Color coding: high brain activity (red and yellow); low activity (blue and black). The Alzheimer patient's scan shows severe deterioration of brain activity. [Dr. Robert Friedland/Science Source.]

transmitted by agents that were similar in size to viruses but consisted only of protein. These diseases include *bovine spongiform encephalopathy* (commonly referred to as *mad cow disease*) and the analogous diseases in other organisms, including *Creutzfeldt–Jakob disease* (CJD) in human beings and *scrapie* in sheep. The agents causing these diseases are termed *prions*. Prions are composed largely or completely of a cellular protein called PrP, which is normally present in the brain. The prions are aggregated forms of the PrP protein termed PrP^{SC}.

The structure of the normal protein PrP contains extensive regions of α helix and relatively little β-strand structure. The structure of a mammalian PrP^{SC} has not yet been determined, because of challenges posed by its insoluble and heterogeneous nature. However, various evidence indicates that some parts of the protein that had been in α-helical or turn conformations have been converted into β-strand conformations. This conversion suggests that the PrP is only slightly more stable than the β-strand-rich PrP^{SC}; however, after the PrP^{SC} has formed, the β strands of one protein link with those of another to form β sheets, joining the two proteins and leading to the formation of aggregates, or amyloid fibrils.

With the realization that the infectious agent in prion diseases is an aggregated form of a protein that is already present in the brain, a model for disease transmission emerges (Figure 4.38). Protein aggregates built of abnormal forms of PrP act as nuclei to which other PrP molecules attach. Prion diseases can thus be transferred from one individual organism to another through the transfer of an aggregated nucleus, as likely happened in the mad cow disease outbreak in the United Kingdom in the 1990s: cattle given animal feed containing material from diseased cows developed the disease in turn. Amyloid fibers are also seen in the brains of patients with certain noninfectious neurodegenerative diseases such as Alzheimer and Parkinson diseases. How such aggregates lead to the death of the cells that harbor them is an active area of research.

Figure 4.38 The protein-only model for prion-disease transmission. A nucleus consisting of proteins in an abnormal conformation grows by the addition of proteins from the normal pool.

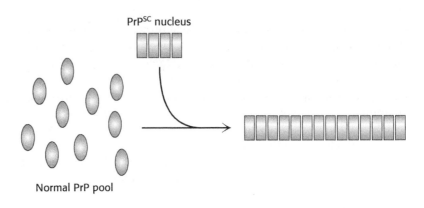

SUMMARY

4.1 Primary Structure: Amino Acids Are Linked by Peptide Bonds to Form Polypeptide Chains

The amino acids in a polypeptide are linked by amide bonds formed between the carboxyl group of one amino acid and the amino group of the next. This linkage, called a peptide bond, has several important properties. First, it is resistant to hydrolysis, and so proteins are remarkably stable kinetically. Second, each peptide bond has both a hydrogen-bond donor (the NH group) and a hydrogen-bond acceptor (the CO group). Because they are linear polymers, proteins can be described as sequences of amino acids. Such sequences are written from the amino to the carboxyl terminus.

4.2 Secondary Structure: Polypeptide Chains Can Fold into Regular Structures

Two major elements of secondary structure are the α helix and the β strand. In the helix, the polypeptide chain twists into a tightly packed rod. Within the helix, the CO group of each amino acid is hydrogen bonded to the NH group of the amino acid, four residues farther along the polypeptide chain. In the β strand, the polypeptide chain is nearly fully extended. Two or more β strands connected by NH-to-CO hydrogen bonds come together to form β sheets. The strands in β sheets can be antiparallel, parallel, or mixed.

4.3 Tertiary Structure: Water-Soluble Proteins Fold into Compact Structures

The compact, asymmetric structure that individual polypeptides attain is called tertiary structure. The tertiary structures of water-soluble proteins have features in common: (1) an interior formed of amino acids with hydrophobic side chains and (2) a surface formed largely of hydrophilic amino acids that interact with the aqueous environment. The driving force for the formation of the tertiary structure of water-soluble proteins is the hydrophobic interactions between the interior residues. Some proteins that exist in a hydrophobic environment, in membranes, display the inverse distribution of hydrophobic and hydrophilic amino acids. In these proteins, the hydrophobic amino acids are on the surface to interact with the environment, whereas the hydrophilic groups are shielded from the environment in the interior of the protein.

4.4 Quaternary Structure: Multiple Polypeptide Chains Can Assemble into a Single Protein

Proteins consisting of more than one polypeptide chain display quaternary structure; each individual polypeptide chain is called a subunit. Quaternary structure can be as simple as two identical subunits or as complex as dozens of different subunits. In most cases, the subunits are held together by noncovalent bonds.

4.5 The Amino Acid Sequence of a Protein Determines Its Three-Dimensional Structure

The amino acid sequence completely determines the three-dimensional structure and, hence, all other properties of a protein. Some proteins can be unfolded completely yet refold efficiently when placed under conditions in which the folded form is stable. The amino acid sequence of a protein is determined by the sequences of bases in a DNA molecule. This one-dimensional sequence information is extended into the three-dimensional world by the ability of proteins to fold spontaneously.

KEY TERMS ||||||||||||||||||||||||||||||||||||||||||||||||||||||||||||||||||||||||||||||||||||||||||||||||||||||||||||||||||||||||||||

primary structure (p. 48)
peptide (amide) bond (p. 48)
disulfide bond (p. 49)
phi (ϕ) angle (p. 51)
psi (ψ) angle (p. 51)
Ramachandran plot (p. 51)
secondary structure (p. 52)
α helix (p. 52)

β pleated sheet (p. 53)
β strand (p. 53)
tertiary structure (p. 57)
motif (supersecondary structure)
 (p. 59)
domain (p. 59)
subunit (p. 59)
quaternary structure (p. 59)

folding funnel (p. 62)
molten globule (p. 62)
intrinsically unstructured protein
 (IUP) (p. 62)
metamorphic protein (p. 63)
prion (p. 64)

? Answer to QUICK QUIZ

(a) Glycine is the amino terminus. (b) The average molecular weight of amino acids is 110. Therefore, a protein consisting of 300 amino acids has a molecular weight of approximately 33,000. (c) A protein with a molecular weight of 110,000 consists of approximately 1000 amino acids.

PROBLEMS |||||||||||||||||||||||||||||||||||||||||||||||||||||||||||||||||||||||||||||||||||||||||||||||||||||||||||||||||||||||||

1. *Matters of stability.* Proteins are quite stable. The lifetime of a peptide bond in aqueous solution is nearly 1000 years. However, the free energy of hydrolysis of proteins is negative and quite large. How can you account for the stability of the peptide bond in light of the fact that hydrolysis releases considerable energy?

2. *Name those components.* Examine the segment of a protein shown here. ✓ 2

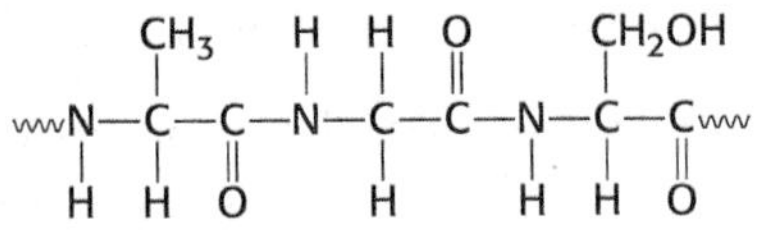

(a) What three amino acids are present?
(b) Of the three, which is nearest the N-terminal amino acid?
(c) Identify the peptide bonds.
(d) Identify the α-carbon atoms.

3. *Who's charged?* Draw the structure of the dipeptide Gly-His. What is the charge on the peptide at pH 4.0? At pH 7.5?

4. *First abbreviate, then charge.* Examine the following peptide, and answer parts *a* through *c.*

 Thr-Glu-Pro-Ile-Val-Ala-Pro-Met-Glu-Tyr-Gly-Lys

(a) Write the sequence using one-letter abbreviations.
(b) Estimate the net charge at pH 7.
(c) Estimate the net charge at pH 12.

5. *Neighborhood peer pressure?* Table 3.1 gives the typical pK_a values for ionizable groups in proteins. However, more than 500 pK_a values have been determined for individual groups in folded proteins. Account for this discrepancy.

6. *Prohibitions.* Why is rotation about the peptide bond prohibited, and what are the consequences of the lack of rotation? ✓ 2

7. *Don't they make a lovely pair?* Match the terms with the descriptions. ✓ 2

(a) Primary structure
 _____

(b) Peptide (amide) bond
 _____

(c) Disulfide bond
 _____

(d) phi (ϕ) angle _____
(e) psi (ψ) angle
 _____

(f) Ramachandran plot
 _____

(g) α helix _____
(h) β pleated sheet
 _____

(i) β strand _____

(j) Secondary structure
 _____

1. Forms between two cysteine amino acids
2. A rodlike structure with a tightly coiled backone
3. Angle of rotation about the bond between the N atom and the α-carbon atom
4. Fully extended polypeptide chain
5. Formed by hydrogen bonds between parallel or antiparallel chains
6. Regular repeating three-dimensional structures
7. The bond responsible for primary structure
8. Sequence of amino acids in a protein
9. Angle of rotation between the α-carbon atom and the carbonyl carbon atom
10. A plot of phi and psi angles

8. *Helix length.* Calculate the axial length of an α helix that is 120 amino acids long. How long would the polypeptide be if it were fully extended?

9. *Alphabet soup.* How many different polypeptides of 50 amino acids in length can be made from the 20 common amino acids? ✓ 2

10. *Sweet tooth, but calorie conscious.* Aspartame (Nutra-Sweet), an artificial sweetener, is a dipeptide composed of Asp-Phe in which the carboxyl terminus is modified by the attachment of a methyl group. Draw the structure of aspartame at pH 7.

11. *Vertebrate proteins?* What is meant by the term *polypeptide backbone?*

12. *Not a sidecar.* Define the term *side chain* in the context of amino acid or protein structure.

13. *One from many.* Differentiate between *amino acid composition* and *amino acid sequence.* ✓ 2

14. *Knowledge is good.* List some of the benefits of knowing the primary structure of a protein. ✓ 2

15. *Compare and contrast.* List some of the differences between an α helix and a β strand. ✓ 2

16. *Degrees of complication.* What are the levels of protein structure? Describe the type of bonds characteristic of each level. ✓ 2

17. *Two by two.* Match the terms with the descriptions. ✓ 2

(a) Tertiary structure
_____

(b) Supersecondary structure _____

(c) Domain _____

(d) Subunit _____

(e) Quaternary structure
_____

(f) Folding funnel
_____

(g) Molten globule
_____

(h) Metamorphic protein
_____

(i) Intrinsically unstructured protein
_____

(j) Prion _____

1. Basic component of quaternary structure
2. An energy landscape
3. Structure characterized by dynamic hydrophobic interactions
4. Proteins that in whole or in part lack discrete three-dimensional structure under physiological conditions
5. Refers to the spatial arrangement of amino acid residues that are far apart in the sequence
6. Proteins that exist in an ensemble of structures of approximately equal energy that are in equilibrium
7. Compact regions that may be connected by a flexible segment of polypeptide chain
8. Refers to the arrangement of subunits and the nature of their interactions
9. Cause of spongiform encephalopathies
10. Combinations of secondary structure are present in many proteins

18. *Who goes first?* Would you expect Pro–X peptide bonds to tend to have cis conformations like those of X–Pro bonds? Why or why not? ✓ 2

19. *Contrasting isomers.* Poly-L-leucine in an organic solvent such as dioxane is α helical, whereas poly-L-isoleucine is not. Why do these amino acids with the same number and kinds of atoms have different helix-forming tendencies? ✓ 2

20. *Active again.* A mutation that changes an alanine residue in the interior of a protein into valine is found to lead to a loss of activity. However, activity is regained when a second mutation at a different position changes an isoleucine residue into glycine. How might this second mutation lead to a restoration of activity? ✓ 3

21. *Exposure matters.* Many of the loops on the proteins that have been described are composed of hydrophilic amino acids. Why? ✓ 2 ✓ 3

22. *Goodbye native state. Hello chaos.* How would each of the following treatments contribute to protein denaturation? ✓ 3

(a) Heat
(b) Addition of the hydrophobic detergents
(c) Large changes in pH

23. *Often irreplaceable.* Glycine is a highly conserved amino acid residue in the evolution of proteins. Why? ✓ 3

24. *Potential partners.* Identify the groups in a protein that can form hydrogen bonds or electrostatic bonds with an arginine side chain at pH 7.

25. *Permanent waves.* The shape of hair is determined in part by the pattern of disulfide bonds in keratin, its major protein. How can curls be induced?

26. *Location is everything 1.* Most proteins have hydrophilic exteriors and hydrophobic interiors. Would you expect this structure to apply to proteins embedded in the hydrophobic interior of a membrane? Explain. ✓ 3

27. *Location is everything 2.* Proteins that span biological membranes often contain α helices. Given that the insides of membranes are highly hydrophobic, predict what type of amino acids will be in such a helix. Why is an α helix particularly suitable for existence in the hydrophobic environment of the interior of a membrane? ✓ 3

28. *Greasy patches.* The α and β subunits of hemoglobin bear a remarkable structural similarity to myoglobin. However, in the subunits of hemoglobin, residues that are hydrophilic in myoglobin are hydrophobic. Why? ✓ 3

29. *Maybe size does matter.* Osteogenesis imperfecta displays a wide range of symptoms, from mild to severe. On the basis of your knowledge of amino acid and collagen structure, propose a biochemical basis for the variety of symptoms.

Challenge Problems

30. *Scrambled ribonuclease.* When performing his experiments on protein refolding, Christian Anfinsen obtained a quite different result when reduced ribonuclease was reoxidized while it was still in 8 M urea and the preparation was then dialyzed to remove the urea. Ribonuclease reoxidized in this way had only 1% of the enzymatic activity of the native protein. Why were the outcomes so different when reduced ribonuclease was reoxidized in the presence and absence of urea? ✓ 3

31. *A little help.* Anfinsen found that scrambled ribonuclease spontaneously converted into fully active, native ribonuclease when trace amounts of β-mercaptoethanol were added to an aqueous solution of the protein. Explain these results. ✓ 3

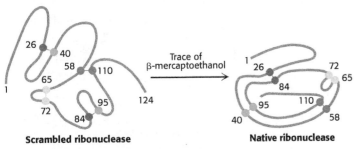

Scrambled ribonuclease   **Native ribonuclease**

32. *Shuffle test.* An enzyme called protein disulfide isomerase (PDI) catalyzes disulfide–sulfhydryl exchange reactions. PDI rapidly converts inactive scrambled ribonuclease into enzymatically active ribonuclease. In contrast, insulin is rapidly inactivated by PDI. What does this important observation imply about the relation between the amino acid sequence of insulin and its three-dimensional structure? ✓ 3

33. *Stretching a target.* A protease is an enzyme that catalyzes the hydrolysis of the peptide bonds of target proteins. How might a protease bind a target protein so that its main chain becomes fully extended in the vicinity of the vulnerable peptide bond?

34. $V = 4/3 \ \pi r^3$. Consider two proteins, one having a molecular weight of 10,000 and the other having a molecular weight of 100,000. Both are globular and of similar spherical shape. How will the ratio of hydrophilic to hydrophobic amino acids differ between the two proteins?

35. *Shape and dimension.* Tropomyosin, a 70-kDa muscle protein, is a two-stranded α-helical coiled coil. Estimate the length of the molecule.

36. *Concentration is crucial.* The concentration of RNase that Anfinsen used when performing his renaturation experiments was approximately 1 mg ml^{-1}. Inside a cell, the protein concentration is estimated to be more than 100 mg ml^{-1}. Predict the outcome of Anfinsen's experiments had he used a 100 mg ml^{-1} RNase.

Selected Readings for this chapter can be found online at www.whfreeman.com/tymoczko3e.

CHAPTER 5

Techniques in Protein Biochemistry

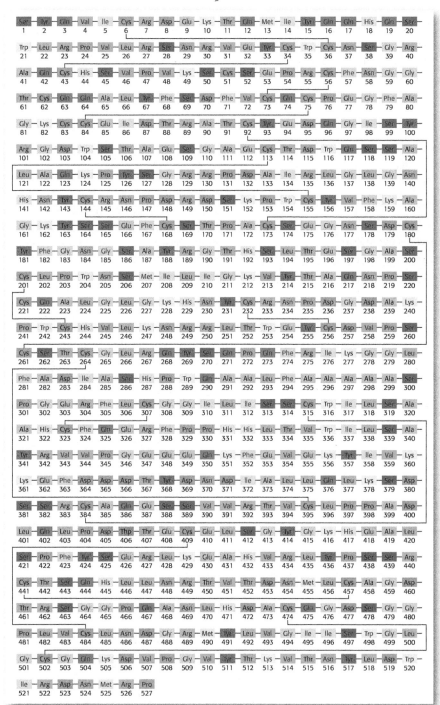

The amino acid sequence of tenecteplase, a fibrinolytic agent for the acute treatment of myocardial infarction. [Information from X. Rabasseda, *Drugs Today* 37(11):749, 2001.]

5.1 The Proteome Is the Functional Representation of the Genome

5.2 The Purification of Proteins Is the First Step in Understanding Their Function

5.3 Immunological Techniques Are Used to Purify and Characterize Proteins

5.4 Determination of Primary Structure Facilitates an Understanding of Protein Function

Earlier in this section, we learned some of the basic principles of protein structure and function, and proteins will continue to hold our attention as we examine them in their metabolic context. This focus on proteins raises an interesting question: How do we know what we know about proteins? The first step toward learning how proteins work in the cell is to study their behavior outside the cell, in vitro. To do so, the proteins must be separated from all of the other constituents of the cell so that their biochemical properties can be identified and characterized. In other words, the protein must be purified.

In this chapter, we will examine some of the key techniques of protein purification, including powerful immunological techniques. All of these techniques take advantage of biochemical properties unique to each protein. Then, we will learn how one crucial property of proteins—amino acid sequence, or primary structure—is elucidated.

5.1 The Proteome Is the Functional Representation of the Genome

Every year, the genomes of more organisms are being elucidated, revealing the exact DNA base sequences and the number of genes encoded. For example, researchers concluded that the roundworm *Caenorhabditis elegans* has a genome of 97 million bases and about 19,000 protein-encoding genes, whereas that of the fruit fly *Drosophila melanogaster* contains 180 million bases and about 14,000 protein-encoding genes. The completely sequenced human genome contains 3 billion bases and about 23,000 protein-encoding genes. But this genomic knowledge is analogous to a list of parts for a car: it does not explain which parts are present in different components or how the parts work together. A new word, the *proteome,* has been coined to signify a more complex level of information content—the level of *functional* information, which encompasses the types, functions, and interactions of proteins that yield a functional unit.

The term *proteome* is derived from *prote*ins expressed by the gen*ome*. The genome provides a list of gene products that *could* be present, but only a subset of these gene products will actually be expressed in a given biological context. The proteome tells us what proteins are functionally present. Unlike the genome, the proteome is not a fixed characteristic of the cell. Rather, because it represents the functional expression of information, it varies with cell type, developmental stage, and environmental conditions, such as the presence of hormones. Moreover, proteins can be enzymatically modified in a variety of ways. Furthermore, these proteins do not exist in isolation; they often interact with one another to form complexes with specific functional properties.

An understanding of the proteome is acquired by isolating, characterizing, and cataloging proteins. In some, but not all, cases, this process begins by separating a particular protein from all other biomolecules in the cell.

✓ 4 Explain how proteins can be purified.

5.2 The Purification of a Protein Is the First Step in Understanding Its Function

To understand a protein—its amino acid sequence, its three-dimensional structure, and its function in normal and pathological states—we need to purify the protein. In other words, we need to isolate the protein of interest from the thousands of other proteins in the cell. This protein sample may be only a fraction of 1% of the starting material, whether that starting material consists of cells in culture or a particular organ from a plant or animal. We will examine the purification of two proteins: an enzyme that is purified by standard biochemical techniques, and a hormone-binding protein, called a *receptor*, that proved refractory to standard biochemical techniques and thus required an immunological approach.

Proteins Can Be Purified on the Basis of Differences in Their Chemical Properties

Protein purification requires much ingenuity and patience, but, before we can even undertake the task, we need a test that identifies the protein that we are interested in. We will use this test after each stage of purification to see if the purification is working. Such a test is called an *assay,* and it is based on some unique identifying property of the protein. For enzymes, which are protein catalysts (Chapter 6), the assay is usually based on the reaction catalyzed by the enzyme in the cell. For instance, the enzyme lactate dehydrogenase, an important enzyme in glucose metabolism, carries out the following reaction:

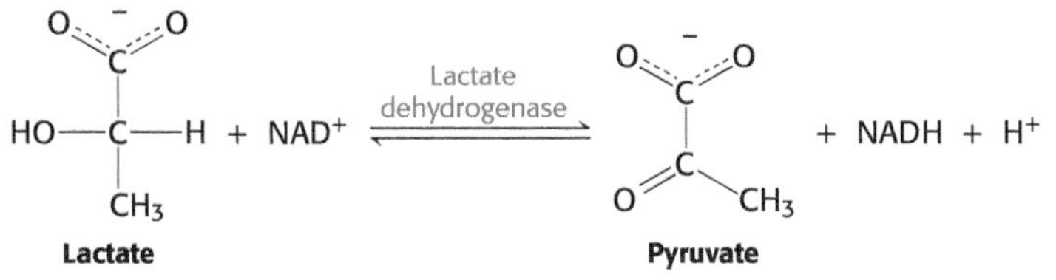

The product, reduced nicotinamide adenine dinucleotide (NADH), in contrast with the other reaction components, absorbs light at 340 nm. Consequently, we can follow the progress of the reaction by measuring the light absorbance at 340 nm in unit time—for instance, within 1 minute after the addition of the sample that contains the enzyme. Our assay for enzyme activity during the purification of lactate dehydrogenase is thus the increase in absorbance of light at 340 nm observed in 1 minute. Note that the assay tells us how much enzyme *activity* is present, not how much enzyme protein is present.

To be certain that our purification scheme is working, we need one additional piece of information—the amount of total protein present in the mixture being assayed. This measurement of the total amount of protein includes the enzyme of interest as well as all the other proteins present, but it is *not* a measure of enzyme activity. After we know both how much enzyme activity is present and how much protein is present, we can assess the progress of our purification by measuring the *specific activity,* the ratio of enzyme activity to the amount of protein in the enzyme assay at each step of our purification. The specific activity will rise as the protein of interest comprises a greater portion of the protein mixture used for the assay. In essence, the point of the purification is to remove all proteins except the protein in which we are interested. Quantitatively, it means that we want to maximize specific activity.

QUICK QUIZ 1 Why is an assay required for protein purification?

Proteins Must Be Removed from the Cell to Be Purified

Having found an assay, we must now break open the cells, releasing the cellular contents, so that we can gain access to our protein. The disruption of the cell membranes yields a *homogenate,* a mixture of all of the components of the cell but no intact cells. This mixture is centrifuged at low centrifugal force, yielding a pellet of heavy material at the bottom of the centrifuge tube and a lighter solution above, called the supernatant. The pellets are enriched in a particular organelle (**Figure 5.1**). The pellet and supernatant are referred to as *fractions* because we are *fractionating* the homogenate. The supernatant is centrifuged again at a greater force to yield yet another pellet and supernatant. This procedure, called *differential centrifugation,* yields several fractions of decreasing density, each still containing hundreds of different proteins, which are assayed for the activity being purified. Usually, one fraction will have more enzyme activity than any other fraction, and it then serves as the source of material to which more-discriminating purification techniques are applied. The fraction that is used as a source for further purification is often called the crude extract.

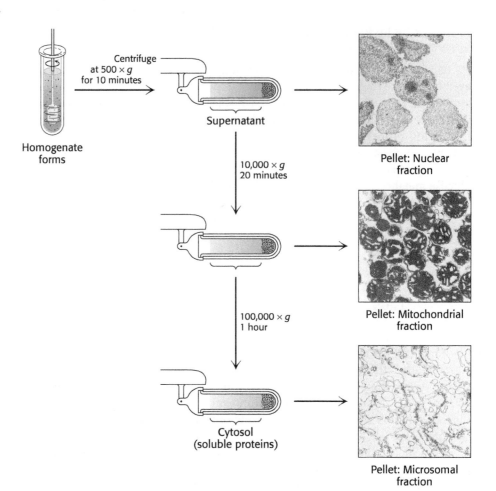

Figure 5.1 Differential centrifugation. Cells are disrupted in a homogenizer, and the resulting mixture, called the homogenate, is centrifuged in a step-by-step fashion of increasing centrifugal force. The denser material will form a pellet at lower centrifugal force than the less dense material. The isolated fractions can be used for further purification. [Photographs courtesy of S. Fleischer and B. Fleischer.]

Proteins Can Be Purified According to Solubility, Size, Charge, and Binding Affinity

Proteins are purified on the basis of differences in *solubility, size, charge,* and *specific binding affinity*. Usually, protein mixtures are subjected to a series of separations, each based on a different property.

Salting out Most proteins require some salt to dissolve, a process called *salting in*. However, most proteins precipitate out of solution at high salt concentrations, an effect called *salting out* (**Figure 5.2**). Salting out is due to competition between the salt ions and the protein for water to keep the protein in solution (water of solvation). The salt concentration at which a protein precipitates differs from one protein to another. Hence, salting out can be used to fractionate a mixture of proteins. Unfortunately, many proteins lose their activity in the presence of such high concentrations of salt. However, the salt can be removed by the process of *dialysis*. The protein–salt solution is placed in a small bag made of a semipermeable membrane, such as a cellulose membrane, with pores (**Figure 5.3**). Proteins are too large to fit through the pores of the membrane, whereas smaller molecules and ions such as salts can escape through the pores and emerge in the medium outside the bag (the dialysate).

Separation by size *Molecular exclusion chromatography,* also called gel-filtration chromatography, separates proteins on the basis of size. The sample is applied to the top of a column consisting of porous beads made of an insoluble polymer such as dextran, agarose, or polyacrylamide (**Figure 5.4**). Small molecules can enter these beads, but large ones cannot, and so those larger molecules

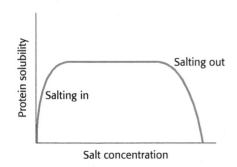

Figure 5.2 The dependency of protein solubility on salt concentration. The graph shows how altering the salt concentration affects the solubility of a hypothetical protein. Different proteins will display different curves.

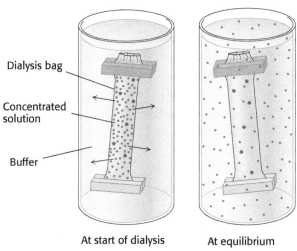

Dialysis bag

Concentrated solution

Buffer

At start of dialysis

At equilibrium

Figure 5.3 Dialysis. Protein molecules (red) are retained within the dialysis bag, whereas small molecules (blue) diffuse into the surrounding medium.

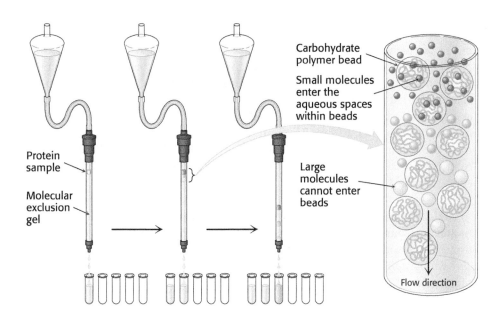

Protein sample

Molecular exclusion gel

Carbohydrate polymer bead

Small molecules enter the aqueous spaces within beads

Large molecules cannot enter beads

Flow direction

Figure 5.4 Gel-filtration chromatography. A mixture of proteins in a small volume is applied to a column filled with porous beads. Because large proteins cannot enter the internal volume of the beads, they emerge sooner than do small ones.

follow a shorter path to the bottom of the column and emerge first. Molecules that are of a size to occasionally enter a bead will flow from the column at an intermediate position, and small molecules, which take a longer, more circuitous path, will exit last.

Ion-exchange chromatography Proteins can be separated on the basis of their net charge by *ion-exchange chromatography*. If a protein has a net positive charge at pH 7, it will usually bind to a column of beads containing negatively charged carboxylate groups, whereas a negatively charged protein will not bind to the column (**Figure 5.5**). A positively charged protein bound to such a column can then be released by increasing the concentration of salt in the buffer poured over the column. The positively charged ions of the salt compete with positively charged groups on the protein for binding to the column. Likewise, a protein with a net negative charge will be bound to ion-exchange beads carrying positive charges and can be eluted from the column with the use of a buffer containing salt.

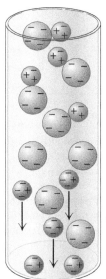

Positively charged protein binds to negatively charged bead

Negatively charged protein flows through

Figure 5.5 Ion-exchange chromatography. This technique separates proteins mainly according to their net charge.

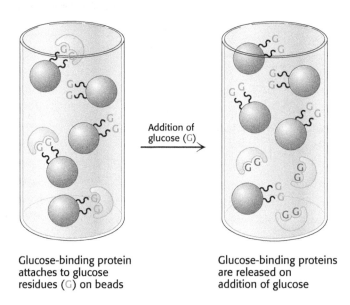

Glucose-binding protein attaches to glucose residues (G) on beads

Glucose-binding proteins are released on addition of glucose

Figure 5.6 Affinity chromatography. Affinity chromatography of concanavalin A (shown in yellow) on a solid support containing covalently attached glucose residues (G).

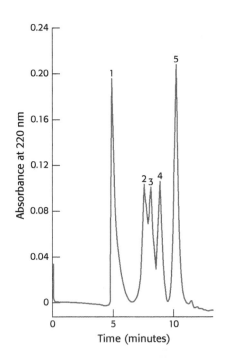

Figure 5.7 High-pressure liquid chromatography (HPLC). Gel filtration by HPLC clearly defines the individual proteins because of its greater resolving power. Proteins are detected by their absorbance of 220-nm light waves: (1) thyroglobulin (669 kDa), (2) catalase (232 kDa), (3) bovine serum albumin (67 kDa), (4) ovalbumin (43 kDa), and (5) ribonuclease (13.4 kDa). [Data from K. J. Wilson and T. D. Schlabach. In *Current Protocols in Molecular Biology*, vol. 2, suppl. 41, F. M. Ausubel, R. Brent, R. E. Kingston, D. D. Moore, J. G. Seidman, J. A. Smith, and K. Struhl, Eds. (Wiley, 1998), p. 10.14.1.]

Affinity chromatography *Affinity chromatography* is another powerful means of purifying proteins. This technique takes advantage of the fact that some proteins have a high affinity for specific chemical groups or specific molecules. For example, the plant protein concanavalin A, which binds to glucose reversibly, can be purified by passing a crude extract through a column of beads containing covalently attached glucose residues. Concanavalin A binds to such beads, whereas most other proteins do not. The bound concanavalin A can then be released from the column by adding a concentrated solution of glucose. The glucose in solution displaces the column-attached glucose residues from binding sites on concanavalin A (**Figure 5.6**).

High-pressure liquid chromatography The ability of column techniques to separate individual proteins, called the resolving power, can be improved substantially through the use of a technique called *high-pressure liquid chromatography* (HPLC), which is an enhanced version of the column techniques already discussed. The beads that make up the column material themselves are much more finely divided and, as a consequence, there are more interaction sites and thus greater resolving power. Because the column is made of finer material, pressure must be applied to the column to obtain adequate flow rates. The net result is high resolution as well as rapid separation (**Figure 5.7**).

Proteins Can Be Separated by Gel Electrophoresis and Displayed

How can we tell whether a purification scheme is effective? One way is to demonstrate that the specific activity rises with each purification step. Another is to visualize the number of proteins present at each step. The technique of *gel electrophoresis* makes the latter method possible.

A molecule with a net charge will move in an electric field, a phenomenon termed *electrophoresis*. The distance and speed that a protein moves in electrophoresis depends on the electric-field strength, the net charge on the protein, which is a function of the pH of the electrophoretic solution, and the shape of the protein. Electrophoretic separations are nearly always carried out in gels, such as polyacrylamide, because the gel serves as a molecular sieve that enhances separation. Molecules that are small compared with the pores in the gel readily move through the gel, whereas molecules much larger than the pores are almost immobile. Intermediate-size molecules move through the gel with various degrees

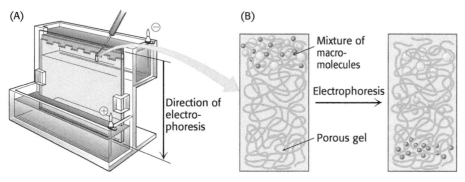

Figure 5.8 Polyacrylamide-gel electrophoresis. (A) Gel-electrophoresis apparatus. Typically, several samples undergo electrophoresis on one flat polyacrylamide gel. A microliter pipette is used to place solutions of proteins in the wells of the slab. A cover is then placed over the gel chamber and voltage is applied. The negatively charged SDS (sodium dodecyl sulfate)–protein complexes migrate in the direction of the anode, at the bottom of the gel. (B) The sieving action of a porous polyacrylamide gel separates proteins according to size, with the smallest moving most rapidly.

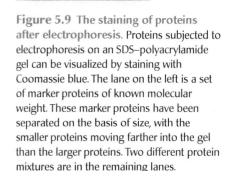

Figure 5.9 The staining of proteins after electrophoresis. Proteins subjected to electrophoresis on an SDS–polyacrylamide gel can be visualized by staining with Coomassie blue. The lane on the left is a set of marker proteins of known molecular weight. These marker proteins have been separated on the basis of size, with the smaller proteins moving farther into the gel than the larger proteins. Two different protein mixtures are in the remaining lanes. [Wellcome Photo Library, Wellcome Images.]

of ease. The electrophoresis of proteins is performed in a thin, vertical slab of polyacrylamide. The pH of the electrophoretic solution is adjusted so that all proteins are negatively charged. The direction of flow is from the cathode (negative charge) to the anode (positive charge) (**Figure 5.8**).

Proteins can be separated largely on the basis of mass by electrophoresis in a polyacrylamide gel in the presence of the detergent sodium dodecyl sulfate (SDS), a technique called SDS–PAGE (SDS–polyacrylamide-gel electrophoresis). The negatively charged SDS denatures proteins and binds to the denatured protein at a constant ratio of one SDS molecule for every two amino acids in the protein. The negative charges on the many SDS molecules bound to the protein "swamp" the normal charge on the protein and cause all proteins to have the same charge-to-mass ratio. Thus, proteins will differ only in their mass. Finally, a sulfhydryl agent such as mercaptoethanol is added to reduce disulfide bonds and completely linearize the proteins. The SDS–protein complexes are then subjected to electrophoresis. When the electrophoresis is complete, the proteins in the gel can be visualized by staining them with silver or a dye such as Coomassie blue, which reveals a series of bands (**Figure 5.9**). *Small proteins move rapidly through the gel, whereas large proteins stay at the top, near the point of application of the mixture.*

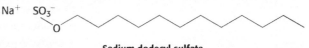

Sodium dodecyl sulfate (SDS)

Isoelectric focusing Proteins can also be separated electrophoretically on the basis of their relative contents of acidic and basic residues. The *isoelectric point* (pI) of a protein is the pH at which its net charge is zero. At this pH, the protein will not migrate in an electric field. If a mixture of proteins is subjected to electrophoresis in a pH gradient in a gel in the absence of SDS, each protein will move until it reaches a position in the gel at which the pH is equal to the pI of the protein. This method of separating proteins is called *isoelectric focusing*. Proteins differing by one net charge can be separated (**Figure 5.10**).

Two-dimensional electrophoresis Isoelectric focusing can be combined with SDS–PAGE to obtain very high resolution separations. A single sample is first subjected to isoelectric focusing. This

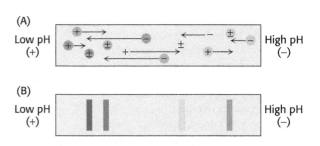

Figure 5.10 The principle of isoelectric focusing. A pH gradient is established in a gel before the sample has been loaded. (A) The sample is loaded, and voltage is applied. The proteins will migrate to their isoelectric pH, the location at which they have no net charge. (B) The proteins form bands that can be excised and used for further experimentation.

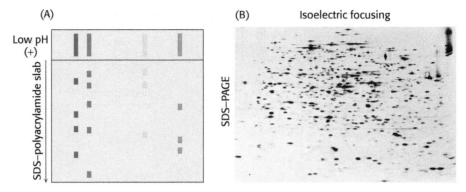

Figure 5.11 Two-dimensional gel electrophoresis. (A) A protein sample is initially fractionated in one direction by isoelectric focusing as described in Figure 5.10. The isoelectric-focusing gel is then attached to an SDS–polyacrylamide gel, and electrophoresis is performed in the second direction, perpendicular to the original separation. Proteins with the same pI value are now separated on the basis of mass. (B) Proteins from *E. coli* were separated by two-dimensional gel electrophoresis, resolving more than a thousand different proteins. The proteins were first separated according to their isoelectric pH in the horizontal direction and then by their apparent mass in the vertical direction. [(B) Courtesy of Dr. Patrick H. O'Farrell.]

single-lane gel is then placed horizontally on top of an SDS–polyacrylamide slab and subjected to electrophoresis again, in a direction perpendicular to the iso-electric focusing, to yield a two-dimensional pattern of spots. In such a gel, proteins have been separated in the horizontal direction on the basis of pI and in the vertical direction on the basis of mass. More than a thousand different proteins in the bacterium *Escherichia coli* can be resolved in a single experiment by *two-dimensional electrophoresis* (**Figure 5.11**).

Proteins isolated from cells under different physiological conditions can be subjected to two-dimensional electrophoresis. The intensities of individual spots on the gels can then be compared, which indicates that the concentrations of specific proteins have changed in response to the physiological state (**Figure 5.12**). How can we discover the identity of a protein that is showing such responses? Although many proteins are displayed on a two-dimensional gel, they are not identified. It is now possible to identify proteins by coupling two-dimensional gel electrophoresis with mass spectrometry, a highly sensitive technique for the determination of the precise mass of the proteins in a given sample (p. 88).

Figure 5.12 Alterations in protein levels detected by two-dimensional gel electrophoresis. Samples of (A) normal colon mucosa and (B) colorectal tumor tissue from the same person were analyzed by two-dimensional gel electrophoresis. In the gel section shown, changes in the intensity of several spots are evident, including a dramatic increase in levels of the protein indicated by the arrow, corresponding to the enzyme glyceraldehyde-3-phosphate dehydrogenase. [Courtesy of Qingsong Lin © 2006, The American Society for Biochemistry and Molecular Biology.]

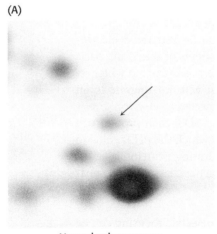

Normal colon mucosa

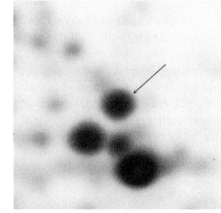

Colorectal tumor tissue

Table 5.1 Quantification of a purification protocol for a hypothetical protein

Step	Total protein (mg)	Total activity (units)	Specific activity (unites mg^{-1})	Yield (%)	Purification level
Homogenization	15,000	150,000	10	100	1
Salt fractionation	4,600	138,000	30	92	3
Ion-exchange chromatography	1,278	115,500	90	77	9
Gel-filtration chromatography	68.8	75,000	1,100	50	110
Affinity chromatography	1.75	52,500	30,000	35	3,000

A Purification Scheme Can Be Quantitatively Evaluated

Some combination of purification techniques will usually yield a pure protein. To determine the success of a protein-purification scheme, we monitor the procedure at each step by determining specific activity and by performing an SDS-PAGE analysis. Consider the results for the purification of a hypothetical protein, summarized in Table 5.1 and Figure 5.13. At each step, the following parameters are measured:

- *Total Protein.* The quantity of protein present in a fraction is obtained by determining the protein concentration of a portion of the fraction and multiplying by the fraction's total volume.

- *Total Activity.* The enzyme activity for the fraction is obtained by measuring the enzyme activity in the volume of fraction used in the assay and multiplying by the fraction's total volume.

- *Specific Activity.* This parameter, obtained by dividing total activity by total protein, enables us to measure the degree of purification by comparing specific

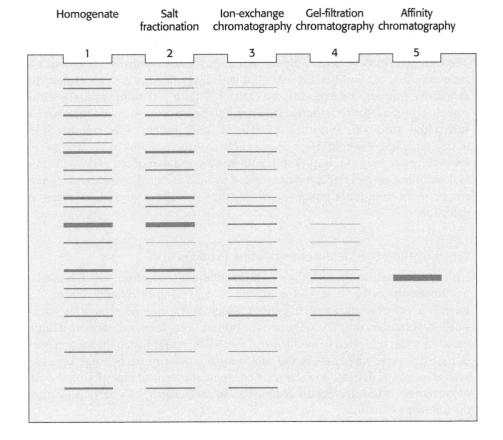

Homogenate   Salt fractionation   Ion-exchange chromatography   Gel-filtration chromatography   Affinity chromatography

1   2   3   4   5

Figure 5.13 Electrophoretic analysis of a protein purification. The purification scheme in Table 5.1 was analyzed by SDS-PAGE. Each lane contained 50 μg of sample. The effectiveness of the purification can be seen as the band for the protein of interest becomes more prominent relative to other bands.

activities after each purification step. Recall that the goal of a purification scheme is to maximize specific activity.

- *Yield.* This parameter is a measure of the total activity retained after each purification step as a percentage of the activity in the crude extract. The amount of activity in the initial extract is taken to be 100%.

- *Purification Level.* This parameter is a measure of the increase in purity and is obtained by dividing the specific activity, calculated after each purification step, by the specific activity of the initial extract.

As we see in Table 5.1, several purification steps can lead to several thousand-fold purification. Inevitably, in each purification step, some of the protein of interest is lost, and so our overall yield is 35%. A good purification scheme takes into account purification levels as well as yield.

The SDS-PAGE depicted in Figure 5.13 shows that, if we load the same amount of protein onto each lane after each step, the number of bands decreases in proportion to the level of purification and the amount of protein of interest increases as a proportion of the total protein present.

QUICK QUIZ 2 What physical differences among proteins allow for their purification?

✓ 5 Explain how immunological techniques can be used to purify and identify proteins.

5.3 Immunological Techniques Are Used to Purify and Characterize Proteins

For enzymes, the assay is a measure of enzyme activity—the disappearance of substrate or the appearance of product. Let us examine the purification of another type of protein, the estrogen receptor. In so doing, we will learn several more biochemical characterization techniques and we will see the power of immunological techniques.

The estrogen-receptor protein binds the female steroid hormone estradiol, an estrogen, and then regulates the expression of genes that play a role in the development of the female phenotype. But the estrogen receptor has no enzyme activity: How can we test for its presence? We can approach this question by asking another one: What is the most distinctive property of the estrogen receptor? The estrogen receptor is the only protein in estrogen-responsive tissues such as rat uterus that can bind to the estradiol with high affinity. We can exploit this distinctive property by exposing the cytosol (Figure 5.1) from rat uteri, which contains the receptor, to radiolabeled estradiol. Because the estrogen receptor has such a high affinity for estradiol, it will be the only protein in the cell that binds to this radioactive steroid. How do we know that the receptor has bound to this steroid? The answer to this question requires a second part of our assay—a means to detect the estradiol–receptor complex. A technique called *zonal, density gradient,* or, more commonly, *gradient centrifugation* provides a convenient means of detection.

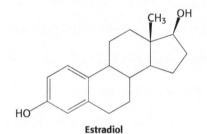

Estradiol

Centrifugation Is a Means of Separating Proteins

Earlier, we examined the technique of differential centrifugation, which is used to fractionate the cell into several components consisting of different organelles. Here, we examine ultracentrifugation, which is capable of separating much smaller molecular complexes. Proteins or protein complexes will move in a liquid medium when subjected to a centrifugal force. The rate at which these complexes or particles move when exposed to such a force is determined by three key characteristics: mass, density, and shape. A convenient means of quantifying the rate of movement is to calculate the *sedimentation coefficient(s)* of a particle by using the following equation:

$$s = m(1 - \bar{v}\rho)/f$$

Table 5.2 S values and molecular weights of sample proteins

Protein	S value (Svedberg units)	Molecular weight
Pancreatic trypsin inhibitor	1	6,520
Cytochrome c	1.83	12,310
Ribonuclease A	1.78	13,690
Myoglobin	1.97	17,800
Trypsin	2.5	23,200
Carbonic anhydrase	3.23	28,800
Concanavalin A	3.8	51,260
Malate dehydrogenase	5.76	74,900
Lactate dehydrogenase	7.54	146,200

where m is the mass of the particle, $\bar{v}$ is the partial specific volume (the reciprocal of the particle density), ρ is the density of the medium, and f is the frictional coefficient (a measure of the shape of the particle). The $(1 - \bar{v}\rho)$ term is the buoyant force exerted by the liquid medium.

Sedimentation coefficients are usually expressed in *Svedberg units* (S), equal to 10^{-13} s. The smaller the S value, the more slowly a molecule moves in a centrifugal field. The S values for a number of proteins are listed in Table 5.2.

Several important conclusions can be drawn from the preceding equation:

1. The sedimentation velocity of a particle depends in part on its mass. A more massive particle sediments more rapidly than does a less massive particle of the same shape and density.

2. Shape, too, influences the sedimentation velocity because it affects the viscous drag. The frictional coefficient f of a compact particle is smaller than that of an extended particle of the same mass. Hence, elongated particles sediment more slowly than do spherical ones of the same mass.

3. A dense particle moves more rapidly than does a less dense one because the opposing buoyant force $(1 - \bar{v}\rho)$ is smaller for the denser particle.

4. The sedimentation velocity also depends on the density of the solution (ρ). Particles sink when $\bar{v}\rho < 1$, float when $\bar{v}\rho > 1$, and do not move when $\bar{v}\rho = 1$.

Gradient Centrifugation Provides an Assay for the Estradiol–Receptor Complex

How do we analyze the rate of a protein's movement in a centrifugal field? The first step is to form a density gradient in a centrifuge tube. Differing proportions of a low-density solution (such as 5% sucrose) and a high-density solution (such as 20% sucrose) are mixed to create a linear gradient of sucrose concentration ranging from 20% at the bottom of the tube to 5% at the top (Figure 5.14). The role of the gradient is twofold. First, the gradient stabilizes the liquid and prevents

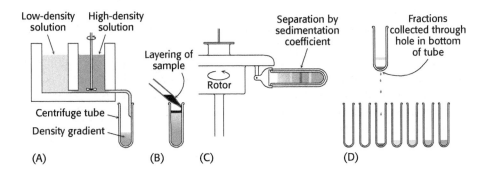

Figure 5.14 Zonal centrifugation. The steps are as follows: (A) form a density gradient, (B) layer the sample on top of the gradient, (C) place the tube in a swinging-bucket rotor and centrifuge it, and (D) collect the samples. [Information from D. Freifelder, *Physical Biochemistry*, 2d ed. (W. H. Freeman and Company, 1982), p. 397.]

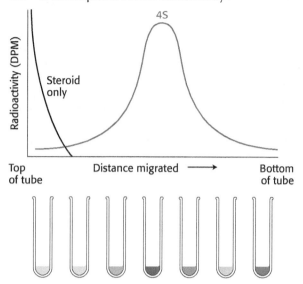

Figure 5.15 Gradient-centrifugation analysis of the estradiol–receptor complex. The receptor protein bound to radioactive estradiol migrates into the gradient on centrifugation. Although some unbound estradiol diffuses into the gradient, the steroid does not migrate to any significant extent. Abbreviation: DPM, disintegrations per minute.

movement of the liquid due to convection. Second, the resistance to movement resulting from the increasing density of the gradient helps to improve separation of the sample components.

Next, a small volume of a solution containing the mixture of proteins to be separated, which includes the radiolabeled estradiol–receptor complex, is placed on top of the density gradient. When the rotor is spun, proteins move through the gradient and separate according to their sedimentation coefficients. The time and speed of the centrifugation is determined empirically. The separated bands, or zones, of protein can be harvested by making a hole in the bottom of the tube and collecting the solution drop by drop. The drops can be measured for protein content and catalytic activity or another functional property. In regard to the estrogen receptor, the fractions are measured for radioactivity. **Figure 5.15** shows the radioactivity of each fraction after centrifugation. Radiolabeled estradiol alone is far too small to move under centrifugal force. Consequently, the radioactivity must be associated with the receptor. A comparison with standards established that the S value of the receptor was 4. However, hundreds of proteins have an S value near 4. Thus, although the radioactivity profile shows only the receptor, there are in fact many other proteins in the same region of the centrifuge tube. Although we can "see" the receptor only, it is still an impure collection of proteins.

Now that we have an assay for the receptor, we need to determine a strategy for purifying the receptor. Because this protein proved difficult to purify with the use of the methods described earlier, we will use immunological techniques. These techniques provide a powerful tool for examining all aspects of biochemical processes and are often the preferred means of protein purification. We will begin by first considering some general immunological properties and then move on to the technique of developing a monoclonal antibody to the estrogen receptor.

Antibodies to Specific Proteins Can Be Generated

Immunological techniques begin with the generation of antibodies to a particular protein. An *antibody* (also called an *immunoglobulin*, Ig; **Figure 5.16**) is itself a protein; it is synthesized by an animal in response to the presence of a foreign substance, called an *antigen*. Antibodies have specific and high affinity for the antigens that elicited their synthesis. The binding of antibody and antigen is a step in the immune response that protects the animal from infection. Proteins, polysaccharides, and nucleic acids can be effective antigens. When a protein is the antigen, an antibody recognizes a specific group or cluster of amino acids on the target molecule called an *antigenic determinant* or *epitope* (**Figure 5.17**). Animals have a very large repertoire of antibody-producing cells, each synthesizing an antibody of a single specificity. The binding of antigen to antibody stimulates the proliferation of the small number of cells that had already been forming an antibody capable of recognizing the antigen.

Figure 5.16 Antibody structure. (A) Immunoglobulin G (IgG) consists of four chains, two heavy chains (blue) and two light chains (red), linked by disulfide bonds. The heavy and light chains come together to form F$_{ab}$ domains, which have the antigen-binding sites at the ends. The two heavy chains form the F$_c$ domain. *Notice* that the F$_{ab}$ domains are linked to the F$_c$ domain by flexible linkers. (B) A more schematic representation of an IgG molecule. [Drawn from 1IGT.pdb.]

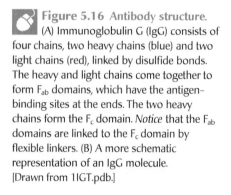
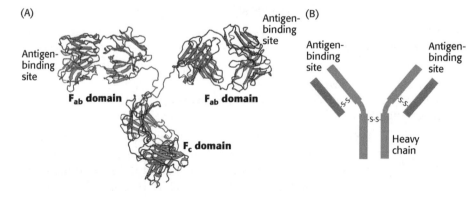

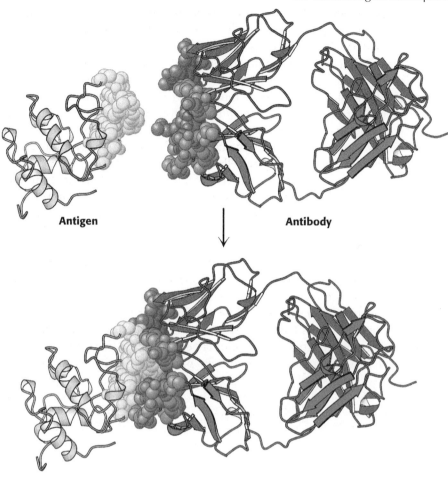

Figure 5.17 Antigen–antibody interactions. A protein antigen–in this case, lysozyme–binds to the end of an F_{ab} domain of an antibody. Notice that the end of the antibody and the antigen have complementary shapes, allowing a large amount of surface to be buried on binding. [Drawn from 3HFL.pdb.]

Polyclonal antibodies

Antigen

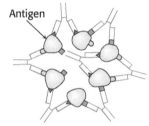

Monoclonal antibodies

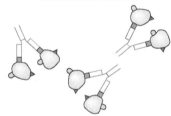

Immunological techniques depend on the ability to generate antibodies to a specific antigen. To obtain antibodies that recognize a particular protein, a biochemist injects the protein into a rabbit. The injected protein stimulates the reproduction of cells producing antibodies that recognize it. Blood is drawn from the immunized rabbit several weeks later and centrifuged to separate blood cells from the supernatant, which is called the serum (blood from which the cells have been removed). The serum, called an *antiserum*, contains antibodies to all antigens to which the rabbit has been exposed. Only some of them will be antibodies to the injected protein. Moreover, in many cases, many different types of antibody can bind to a single antigen. For instance, consider what occurs when 2,4-dinitrophenol (DNP) is used as an antigen to generate antibodies. Analyses of anti-DNP antibodies revealed a wide range of binding affinities: the dissociation constants ranged from about 0.1 nM to 1 μM. Correspondingly, a large number of bands were evident when anti-DNP antibody was subjected to isoelectric focusing (p. 75). These results indicate that animals produce many different antibodies, each recognizing a different surface feature of the same antigen. The antibodies are heterogeneous, or *polyclonal* (**Figure 5.18**). The heterogeneity of polyclonal antibodies can be advantageous for certain applications, such as the detection of a protein of low abundance, because each protein molecule can be bound by more than one antibody at multiple distinct antigenic sites.

Figure 5.18 Polyclonal and monoclonal antibodies. Most antigens have several epitopes. Polyclonal antibodies are heterogeneous mixtures of antibodies, each specific for one of the various epitopes on an antigen. Monoclonal antibodies are all identical, produced by clones of a single antibody-producing cell. They recognize one specific epitope. [Information from R. A. Goldsby, T. J. Kindt, and B. A. Osborne, *Kuby Immunology*, 4th ed. (W. H. Freeman and Company, 2000), p. 154.]

Monoclonal Antibodies with Virtually Any Desired Specificity Can Be Readily Prepared

The discovery of a means of producing *monoclonal antibodies* of virtually any desired specificity was a major breakthrough that intensified the power of immunological approaches. Just like working with impure proteins, working with an impure mixture of antibodies makes it difficult to interpret data. Rather than

2,4-Dinitrophenol (DNP)

separating the target antibody from the mixture, as we might separate a target protein from a mixture of molecules, we will find it more effective to isolate a group of identical cells producing a single kind of antibody for our experiment. The problem is that antibody-producing cells die in a short time in vitro, leaving us little product with which to work.

Immortal cell lines that produce monoclonal antibodies do exist. These cell lines are derived from a type of cancer, *multiple myeloma,* a malignant disorder of antibody-producing cells. In this cancer, a single transformed plasma cell divides uncontrollably, generating a very large number of cells of a single kind. Such a group of cells is a *clone* because the cells are descended from the same cell and have identical properties. The identical cells of the myeloma secrete large amounts of normal *antibodies of a single kind* generation after generation. However, although these antibodies were useful for elucidating antibody structure, nothing is known about their specificity, making them useless for the immunological methods described in the next pages.

César Milstein and Georges Köhler discovered that large amounts of a homogeneous antibody of nearly any desired specificity can be obtained by fusing a short-lived antibody-producing cell with an immortal myeloma cell. An antigen is injected into a mouse, and its spleen, an antibody-producing tissue, is removed several weeks later (**Figure 5.19**). A mixture of plasma cells from the spleen is fused in vitro with myeloma cells. Each of the resulting hybrid cells, called *hybridoma cells,* indefinitely produces the identical antibody

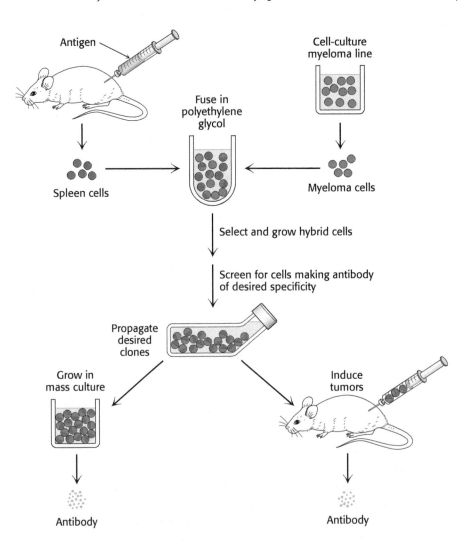

Figure 5.19 The preparation of monoclonal antibodies. Hybridoma cells are formed by the fusion of antibody-producing cells and myeloma cells. The hybrid cells are allowed to proliferate by growing them in selective medium. They are then screened to determine which ones produce antibody of the desired specificity. [Information from C. Milstein. Monoclonal antibodies. Copyright © 1980 by Scientific American, Inc. All rights reserved.]

specified by the parent cell from the spleen. Hybridoma cells will be produced for all of the proteins that were injected into the mouse. The cells can then be screened, by using some sort of specific assay for the antigen–antibody interaction, to determine which ones produce antibodies having the desired specificity. This process is repeated until a pure cell line, a clone producing a single antibody, is isolated.

Our goal now is to generate a monoclonal antibody to the estrogen receptor and then to use the antibody to isolate the receptor. To do so, we inject a small amount of cytosol from the rat uterus and generate hybridoma cells as just described. We now have a population of antibodies, some of which are specific for the estrogen receptor. How can we detect estrogen-receptor-specific antibodies? We will use the technique of gradient centrifugation once again. Our reasoning is as follows: if we mix the antibody preparation with the cytosol containing the radiolabeled estradiol–receptor complex, an antibody will bind to the estrogen receptor. This binding should alter the sedimentation profile obtained when we centrifuged the radiolabeled estradiol–receptor complex. Indeed, it is the case. When estradiol–receptor complex is incubated with antibodies, the complex of radioactive steroid and estrogen receptor undergoes sedimentation at 9S in a sucrose gradient, in association with the antibody (Figure 5.20). The population of antibody-producing cells is screened for those that are generating antibodies that bind the estradiol–receptor complex. After a certain number of screens, we will have the pure cell line—a monoclonal cell line—that is producing only one antibody, called a *monoclonal antibody*, to the estrogen receptor.

The Estrogen Receptor Can Be Purified by Immunoprecipitation

How can a pure monoclonal antibody to the estrogen receptor be used to isolate pure estrogen receptors from the morass of proteins in the cell cytosol? In essence, we can fish for the estrogen receptor by using these monoclonal antibodies as bait. The monoclonal antibody can be covalently linked to insoluble beads made of polyacrylamide or polysaccharides. The antibody-bound beads are added to a small amount of cytosol and then gently mixed for several hours (Figure 5.21). During this incubation, only the estrogen receptor will bind to the antibody. The mixture is then centrifuged and the beads, along with the attached antibody–antigen complex, will form a pellet at the bottom of the centrifuge tube. The supernatant, which contains all of the non-estrogen-receptor proteins, is discarded. The pellet is then resuspended in buffer and recentrifuged so as to wash away proteins that happened to have been trapped among the beads during the centrifugation. Finally, the addition of a protein denaturant to the mixture causes the estrogen receptor to detach from the antibody. The mixture is recentrifuged, but this time the estrogen receptor remains in the supernatant. The supernatant is collected and used as a source of receptor. This technique, called *immunoprecipitation*, is similar to affinity chromatography discussed earlier, except here the material bound to the matrix—the antibody—has high affinity for the proteins moving over the matrix.

Let us take a moment to admire what we have accomplished. We used an impure mixture of proteins containing the estrogen receptor to generate a heterogeneous mixture of antibody-producing cells, a small fraction of which are producing antibodies to the estrogen receptor. Then, we took advantage of two highly specific biochemical interactions—between the receptor and estradiol, and between the receptor and its antibody—to purify, first, a monoclonal antibody and, then, the receptor itself.

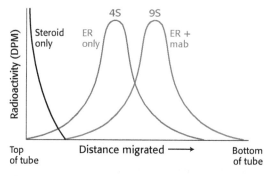

Figure 5.20 Alteration of the sedimentation profile when antibody binds to the receptor protein. The estradiol–receptor–antibody complex migrates farther into the centrifuge tube because it is larger than the estradiol–receptor complex alone. Abbreviations: DPM, disintegrations per minute; ER, estradiol–receptor complex; mab, monoclonal antibody.

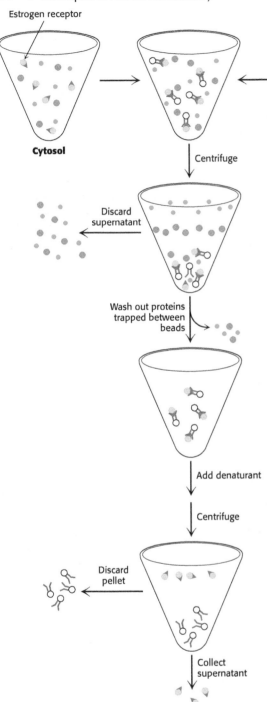

Figure 5.21 Purification by immunoprecipitation. Monoclonal antibody to the estrogen receptor is added to a preparation of cytosol containing the receptor. The antibody is attached to an insoluble bead. The mixture is then gently stirred for several hours to allow the interaction of the receptor and antibody. The mixture is centrifuged and the supernatant is discarded. The pellet is resuspended in buffer to wash out any trapped cytosolic components. The pellet is again collected and the supernatant discarded. The antibody–receptor complex is treated with a denaturant. On centrifugation, the supernatant contains pure estradiol receptor.

Proteins Can Be Detected and Quantified with the Use of an Enzyme-Linked Immunosorbent Assay

Antibodies not only aid in the purification of proteins, but also can be used as reagents to quantify the amount of a protein or other antigen. The technique is the *enzyme-linked immunosorbent assay* (ELISA). This method makes use of an enzyme that reacts with a colorless substrate to produce a colored product. For instance, peroxidase reacts with hydrogen peroxide and a reduced chromogenic substrate to yield an oxidized colored product. The enzyme is covalently linked to a specific antibody that recognizes a target antigen. If the antigen is present, the antibody–enzyme complex will bind to it; on the addition of the substrate, the enzyme will catalyze the reaction, generating the colored product. Thus, the presence of the colored product indicates the presence of the antigen. If no antigen is present, the enzyme-linked antibody cannot bind to anything and will be washed out of the reaction tube; no color will be generated when the substrate is added. ELISA is rapid and convenient, and can detect less than a nanogram (10^{-9} g) of a protein. ELISA can be performed with either polyclonal or monoclonal antibodies, but the use of monoclonal antibodies yields more reliable results.

We will consider two among the several types of ELISA. *The indirect ELISA is used to detect the presence of antibody* and is the basis of the test for HIV infection. The HIV test detects the presence of antibodies that recognize viral core proteins, the antigen. Purified viral core proteins are adsorbed to the bottom of a well. Antibodies from the person being tested are then added to the coated well. Only someone infected with HIV will have antibodies that bind to the antigen. Finally, enzyme-linked antibodies to human antibodies (e.g., enzyme-linked goat antibodies whose antigens are human antibodies) are allowed to react in the well, and unbound antibodies are removed by washing. Substrate is then applied. An enzyme reaction suggests that the enzyme-linked antibodies were bound to human antibodies, which in turn implies that the patient has antibodies to the viral antigen (**Figure 5.22**).

The sandwich ELISA is used to detect antigen rather than antibody. Antibody to a particular antigen is first adsorbed to the bottom of a well. This antibody is called the capture antibody. Next, blood or urine containing the antigen is added to the well and binds to the antibody. Finally, a second, different antibody to the antigen, the detection antibody, is added. This antibody is enzyme linked and is processed as described for indirect ELISA. In this case, the extent of color formation is directly proportional to the amount of antigen present. Consequently, it permits the measurement of small quantities of antigen (Figure 5.22). The sandwich ELISA is the basis of home pregnancy tests.

Western Blotting Permits the Detection of Proteins Separated by Gel Electrophoresis

ELISA is a powerful technique for screening large numbers of samples for the presence of particular antibodies or antigens. Another immunoassay technique called *western blotting* or *immunoblotting* allows the detection of very small quantities of a particular protein in a cell or in body fluid (**Figure 5.23**). Western blotting also allows the determination of the size of the target protein. A key step in western blotting is to subject the sample

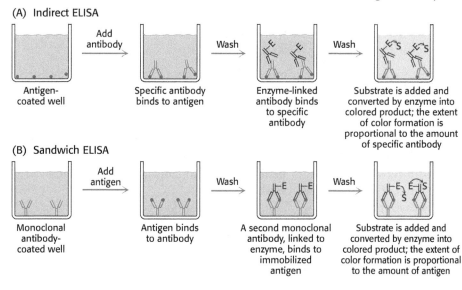

Figure 5.22 Indirect ELISA and sandwich ELISA. (A) In indirect ELISA, the production of color indicates the amount of an antibody to a specific antigen. (B) In sandwich ELISA, the production of color indicates the quantity of antigen. [Information from R. A. Goldsby, T. J. Kindt, and B. A. Osborne, *Kuby Immunology*, 4th ed. (W. H. Freeman and Company, 2000), p. 162.]

under investigation to electrophoresis on an SDS–polyacrylamide gel to separate the individual proteins (p. 74). How can we detect the presence of the protein of interest? In principle, we could soak the gel in a solution of antibody and detect the protein. However, antibodies are too large to penetrate the gel. To make the proteins accessible to the antibody, the resolved proteins on the gel are transferred to the surface of a polymer sheet (blotted) by an electrical current. An antibody that is specific for the protein of interest is added to the sheet and reacts with the antigen. The antibody–antigen complex on the sheet can then be detected by rinsing the sheet with a fluorescently labeled antibody specific for the first (e.g., goat antibody that recognizes mouse antibody). The fluorescently labeled second antibody reveals the presence of the first antibody. Alternatively, an enzyme on the second antibody generates a colored product, as in the ELISA method. Western blotting makes it possible to find a protein in a complex mixture, the proverbial needle in a haystack. It is the basis for the test for infection by hepatitis C, where it is used to detect a core protein of the virus. This technique is also very useful in monitoring protein purification and in the cloning of genes.

? QUICK QUIZ 3 What is the biochemical basis for the power of immunological techniques?

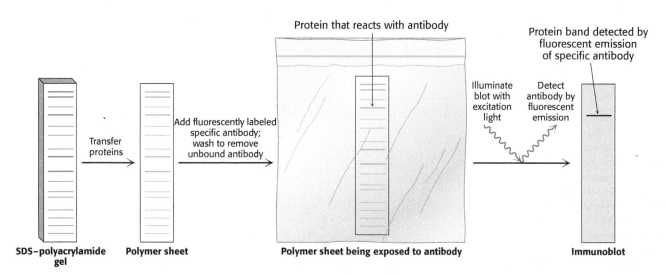

Figure 5.23 Western blotting. Proteins on an SDS–polyacrylamide gel are transferred to a polymer sheet and stained with fluorescent antibody. The fluorescent antibody is excited by light, and the band corresponding to the protein to which the antibody binds is visualized with an appropriate detector.

5.4 Determination of Primary Structure Facilitates an Understanding of Protein Function

Now that we have purified our protein, be it lactate dehydrogenase or the estrogen receptor, what is the next step in learning about the protein? An important means of characterizing a pure protein is to determine its primary structure, which can tell us much about the protein. Recall that the primary structure of a protein is the determinant of its three-dimensional structure, which ultimately determines the protein's function. Comparison of the sequence of normal proteins with those isolated from patients with pathological conditions allows an understanding of the molecular basis of diseases.

Let us examine first how we can sequence a simple peptide, such as

Ala-Gly-Asp-Phe-Arg-Gly

The first step is to determine the *amino acid composition* of the peptide. The peptide is hydrolyzed into its constituent amino acids by heating it in strong acid. The individual amino acids can then be separated by ion-exchange chromatography and visualized by treatment with *fluorescamine,* which reacts with the α-amino group to form a highly fluorescent product (**Figure 5.24**). The concentration of an amino acid in solution is proportional to the fluorescence of the solution. The solution is then run through a column. The amount of buffer required to remove the amino acid from the column is compared with the elution pattern of a standard mixture of amino acids, revealing the identity of the amino acid in the solution (**Figure 5.25**). The composition of our peptide is

(Ala, Arg, Asp, Gly₂, Phe)

The parentheses denote that this is the amino acid composition of the peptide, not its sequence.

The sequence of a protein can then be determined by a process called the Edman degradation. The *Edman degradation* sequentially removes one residue at a time from the amino end of a peptide (**Figure 5.26**). *Phenyl isothiocyanate* reacts with the terminal amino group of the peptide, which then cyclizes and breaks off the peptide, yielding an intact peptide shortened by one amino acid. The cyclic compound is a phenylthiohydantoin (PTH)–amino acid, which can be identified by chromatographic procedures. The Edman procedure can then be repeated sequentially to yield the amino acid sequence of the peptide.

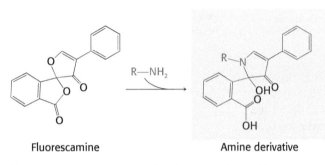

Fluorescamine   Amine derivative

Figure 5.24 Fluorescent derivatives of amino acids. Fluorescamine reacts with the α-amino group of an amino acid to form a fluorescent derivative.

Figure 5.25 Determination of amino acid composition. Different amino acids in a peptide hydrolysate can be separated by ion-exchange chromatography on a sulfonated polystyrene resin (such as Dowex-50). Buffers (in this case, sodium citrate) of increasing pH are used to elute the amino acids from the column. The amount of each amino acid present is determined from the absorbance. Aspartate, which has an acidic side chain, is the first to emerge, whereas arginine, which has a basic side chain, is the last. The original peptide is revealed to be composed of one aspartate, one alanine, one phenylalanine, one arginine, and two glycine residues.

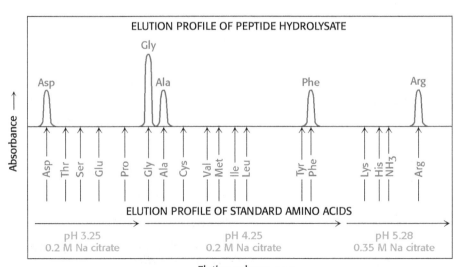

EDMAN DEGRADATION

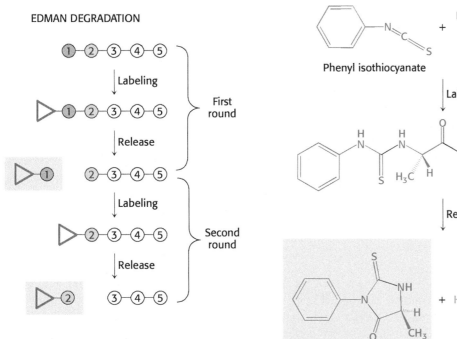

PTH-alanine   Peptide shortened by one residue

In principle, we should be able to sequence an entire protein by using the Edman method. In practice, the peptides cannot be much longer than about 50 residues, because the reactions of the Edman method are not 100% efficient and, eventually, the sequencing reactions are out of order. We can circumvent this obstacle by cleaving the original protein at specific amino acids into smaller peptides that can be sequenced independently. In essence, the strategy is to divide and conquer.

Specific cleavage can be achieved by chemical or enzymatic methods. Table 5.3 gives several ways of specifically cleaving polypeptide chains. The peptides obtained by specific chemical or enzymatic cleavage are separated, and the sequence of each purified peptide is then determined by the Edman method. At this point, the amino acid sequences of segments of the protein are known, but the order of these segments is not yet defined. How can we order the peptides to

Figure 5.26 The Edman degradation. The labeled amino-terminal residue (PTH-alanine in the first round) can be released without hydrolyzing the rest of the peptide. Hence, the amino-terminal residue of the shortened peptide (Gly-Asp-Phe-Arg-Gly) can be determined in the second round. Three more rounds of the Edman degradation reveal the complete sequence of the original peptide.

Table 5.3 Specific cleavage of polypeptides

Reagent	Cleavage site
Chemical cleavage	
Cyanogen bromide	Carboxyl side of methionine residues
O-Iodosobenzoate	Carboxyl side of tryptophan residues
Hydroxylamine	Asparagine-glycine bonds
2-Nitro-5-thiocyanobenzoate	Amino side of cysteine residues
Enzymatic cleavage	
Trypsin	Carboxyl side of lysine and arginine residues
Clostripain	Carboxyl side of arginine residues
Staphylococcal protease	Carboxyl side of aspartate and glutamate residues (glutamate only under certain conditions)
Thrombin	Carboxyl side of arginine
Chymotrypsin	Carboxyl side of tyrosine, tryptophan, phenylalanine, leucine, and methionine
Carboxypeptidase A	Amino side of carboxyl-terminal amino acid (not arginine, lysine, or proline)

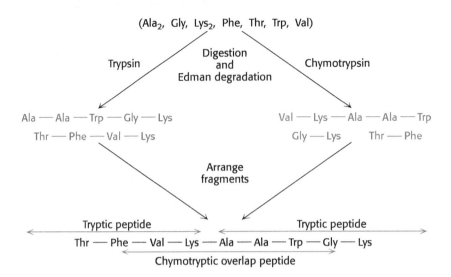

Figure 5.27 Overlap peptides. The peptide obtained by chymotryptic digestion overlaps two tryptic peptides, establishing their order.

obtain the primary structure of the original protein? The necessary additional information is obtained from *overlap peptides* (**Figure 5.27**). A second cleavage technique is used to split the polypeptide chain at different sites. Some of the peptides from the second cleavage will overlap two or more peptides from the first cleavage, and they can be used to establish the order of the peptides. The entire amino acid sequence of the polypeptide chain is then known.

Mass Spectrometry Can Be Used to Determine a Protein's Mass, Identity, and Sequence

Although Edman degradation has provided a wealth of sequence information, it has largely been supplanted by the powerful technique of mass spectrometry. Before we can examine how mass spectrometry can be used to sequence a protein, we will investigate how it can be used to determine a protein's mass and identity.

Mass spectrometry is a technique for analyzing ionized forms of molecules in the gas phase. It is most readily applied to gases or to volatile liquids that easily release gas-phase ions. Mass measurements are obtained by determining how readily an ion is accelerated in an applied electric field. Consider two ions with the same overall charge but with different masses. In a given electric field, the same force will act on each ion. However, the acceleration of the more massive ion due to this force will be less, according to Newton's third law, $F = ma$, where F is the force, m is the mass, and a is the acceleration. Thus, a measurement of the acceleration in a known applied force provides the mass.

Protein Mass   Two widely used methods, *matrix-assisted laser desorption–ionization* (MALDI) and *electrospray ionization* (ESI), have been developed to determine a protein's mass. We will focus on MALDI. In MALDI, the protein or peptide under study is coprecipitated with an organic compound that absorbs laser light of an appropriate wavelength (the "matrix"). The flash of a laser on the preparation expels molecules from the surface. These molecules capture electrons as they exit the matrix and hence leave as negatively charged ions.

After gas-phase ions have been generated, several approaches may be used to determine their mass. In *time of flight* (TOF) analysis, the ions are accelerated in an electric field toward a detector (**Figure 5.28**). The lighter ions are accelerated more, travel faster, and arrive at the detector first. Tiny amounts of biomolecules, as small as a few picomoles (pmol) to femtomoles (fmol), can be analyzed in this manner. A MALDI-TOF mass spectrum for a mixture of the proteins insulin and β-lactoglobulin is shown in **Figure 5.29**. The masses determined by MALDI-TOF are 5733.9 kDa and 18,364 kDa, respectively, com-

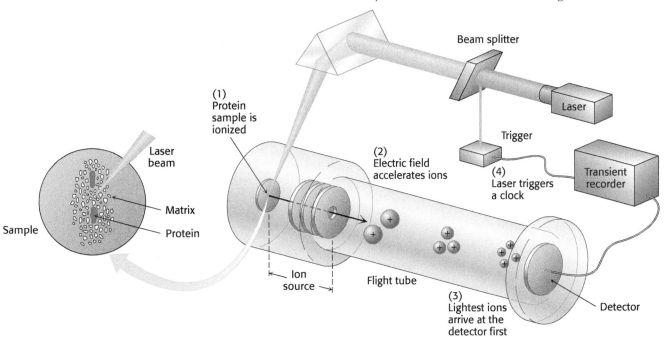

Figure 5.28 MALDI-TOF mass spectrometry. (1) The protein sample, embedded in an appropriate matrix, is ionized by the application of a laser beam. (2) An electric field accelerates the ions through the flight tube toward the detector. (3) The lightest ions arrive first. (4) The ionizing laser pulse also triggers a clock that measures the time of flight (TOF) for the ions. [Information from J. T. Watson, *Introduction to Mass Spectrometry*, 3d ed. (Lippincott-Raven, 1997), p. 279.]

pared with calculated values of 5733.5 kDa and 18,388 kDa. MALDI-TOF is indeed an accurate means of determining protein mass.

Protein Identity Although protein masses serve as convenient name tags for distinguishing proteins, the mass of a given protein is usually not enough to uniquely identify it among all possible proteins within a cell. However, the mass of the parent protein along with the masses of several protein fragments produced by a specific cleavage method can provide unique identification. Suppose we wish to identify proteins within a two-dimensional gel such as that described on page 75. After gel electrophoresis, the molecules in individual spots can be cleaved, often in the gel matrix itself, by using a protease such as trypsin. The

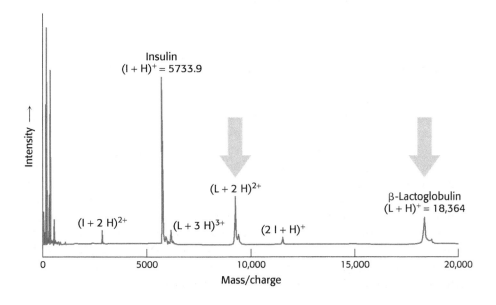

Figure 5.29 MALDI-TOF mass spectrum of insulin and β-lactoglobulin. A mixture of 5 pmol each of insulin (I) and β-lactoglobulin (L) was ionized by MALDI, which produces predominately singly charged molecular ions from peptides and proteins—the insulin ion $(I + H)^+$ and the lactoglobulin ion $(L + H)^+$. Molecules with multiple charges, such as those for β-lactoglobulin indicated by the blue arrows, as well as small quantities of a singly charged dimer of insulin $(2I + H)^+$, also are produced. [Data from J. T. Watson, *Introduction to Mass Spectrometry*, 3d ed. (Lippincott-Raven, 1997), p. 282.]

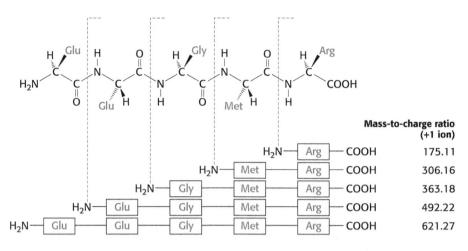

			Mass-to-charge ratio (+1 ion)
H₂N—Arg—COOH			175.11
H₂N—Met—Arg—COOH			306.16
H₂N—Gly—Met—Arg—COOH			363.18
H₂N—Glu—Gly—Met—Arg—COOH			492.22
H₂N—Glu—Glu—Gly—Met—Arg—COOH			621.27

Figure 5.30 Peptide sequencing by tandem mass spectrometry. Within the mass spectrometer, peptides can be fragmented by bombardment with inert gaseous ions to generate a family of product ions in which individual amino acids have been removed from one end. As drawn here, the carboxyl fragment of the cleaved peptide bond is ionized. [Data from H. Steen and M. Mann. *Nat. Rev. Mol. Cell Biol.* 5:699–711, 2004.]

mixture of fragments produced can then be analyzed by MALDI-TOF mass spectrometry. These peptide masses are matched against proteins in a database that have been "electronically cleaved" by a computer simulating the same fragmentation technique used for the experimental sample. In this way, the proteome within a given cell type or other sample can be analyzed in considerable detail.

Protein Sequence How can we employ mass spectrometry to sequence a protein? The use of mass spectrometry for protein sequencing takes advantage of the fact that ions of proteins that have been analyzed by a mass spectrometer, the *precursor ions,* can be broken into smaller peptide chains by bombardment with atoms of an inert gas such as helium or argon. These new fragments, or *product ions,* can be passed through a second mass analyzer for further mass characterization. The utilization of two mass analyzers arranged in this manner is referred to as *tandem mass spectrometry.* Importantly, the product-ion fragments are formed in chemically predictable ways that can provide clues to the amino acid sequence of the precursor ion. For peptide analytes, product ions can be formed such that individual amino acid residues are cleaved from the precursor ion (**Figure 5.30**). Hence, a family of ions is detected; each ion represents a fragment of the original peptide with one or more amino acids removed from one end.

Amino Acid Sequences Are Sources of Many Kinds of Insight

A protein's amino acid sequence is a valuable source of insight into the protein's function, structure, and history.

QUICK QUIZ 4 Differentiate between amino acid composition and amino acid sequence.

1. *The sequence of a protein of interest can be compared with all other known sequences to ascertain whether significant similarities exist. Does this protein belong to an established family?* A search for kinship between a newly sequenced protein and the millions of previously sequenced ones takes only a few seconds on a computer. If the newly isolated protein is a member of an established family of proteins, we can infer information about the protein's structure and function. For instance, chymotrypsin and trypsin are members of the serine protease family, a clan of proteolytic enzymes that have a common catalytic mechanism based on a reactive serine residue. If the sequence of the newly isolated protein shows sequence similarity with trypsin or chymotrypsin, the result suggests that it, too, may be a serine protease.

2. *Comparison of sequences of the same protein in different species yields a wealth of information about evolutionary pathways.* Genealogical relations between species can be established from sequence differences between their proteins. We can even estimate the time at which two evolutionary lines diverged, thanks to the clocklike nature of random mutations. For example, a comparison of serum albumins found in primates indicates that human beings and African apes diverged 5 million years ago, not 30 million years ago as was once thought. Sequence analyses have opened a new perspective on the fossil record and the pathway of human evolution.

3. *Amino acid sequences can be searched for the presence of internal repeats.* Such internal repeats can reveal the history of an individual protein itself. Many proteins apparently have arisen by the duplication of primordial genes. For example, calmodulin, a ubiquitous calcium sensor in eukaryotes (Chapter 13), contains four similar calcium-binding modules that arose by gene duplication (**Figure 5.31**).

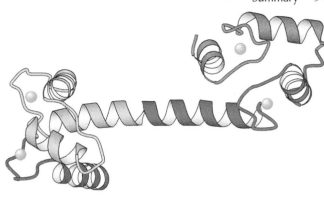

Figure 5.31 Repeating motifs in a protein chain. Calmodulin, a calcium sensor, contains four similar units (shown in red, yellow, blue, and orange) in a single polypeptide chain. Notice that each unit binds a calcium ion (shown in green). [Drawn from 1CLL.pdb.]

4. *Many proteins contain amino acid sequences that serve as signals designating their destinations or controlling their processing.* For example, a protein destined for export from a cell or for location in a membrane contains a *signal sequence,* a stretch of about 20 hydrophobic residues near the amino terminus that directs the protein to the appropriate membrane (Chapter 40). Another protein may contain a stretch of amino acids that functions as a *nuclear localization signal,* directing the protein to the nucleus.

5. *Sequence data allow a molecular understanding of diseases.* Many diseases are caused by mutations in DNA that result in alterations in the amino acid sequence of a particular protein. These alterations often compromise the protein's function. For instance, sickle-cell anemia is caused by a change in a single amino acid in the primary structure of the β chain of hemoglobin (Chapter 9). Approximately 70% of the cases of cystic fibrosis are caused by the deletion of one particular amino acid out of the 1480 amino acids in the protein that controls chloride transport across cell membranes. Indeed, a major goal of biochemistry is to elucidate the molecular basis of disease with the hope that this understanding will lead to effective treatment.

6. *Protein sequence is a guide to nucleic acid information.* Knowledge of a protein's primary structure allows access to genomic information. DNA sequences that correspond to a part of the amino acid sequence can be synthesized on the basis of the genetic code. These DNA sequences can be used as probes to isolate the gene encoding the protein or the DNA corresponding to the mRNA, called the cDNA or complementary DNA (Chapter 41).

SUMMARY

5.1 The Proteome Is the Functional Representation of the Genome

The rapid progress in gene sequencing has advanced another goal of bio-chemistry—the elucidation of the proteome. The proteome is the complete set of proteins expressed and includes information about how they are modified, how they function, and how they interact with other molecules. Unlike the genome, the proteome is not static and varies with cell type, developmental stage, and environmental conditions.

5.2 The Purification of Proteins Is the First Step in Understanding Their Function

Proteins can be separated from one another and from other molecules on the basis of such characteristics as solubility, size, charge, and binding affinity. SDS–PAGE separates the polypeptide chains of proteins under denaturing conditions largely according to mass. Proteins can also be separated electrophoretically on the basis of net charge by isoelectric focusing in a pH gradient.

5.3 Immunological Techniques Are Used to Purify and Characterize Proteins

Proteins can be detected and quantified by highly specific antibodies; monoclonal antibodies are especially useful because they are homogeneous. Monoclonal antibodies to the estrogen receptor can be generated and then used as a tool to purify the receptor with the use of the technique of immunoprecipitation. Enzyme-linked immunosorbent assays and western blots of SDS–polyacrylamide gels are used extensively.

5.4 Determination of Primary Structure Facilitates an Understanding of Protein Function

The amino acid composition of a protein can be ascertained by hydrolyzing the protein into its constituent amino acids. The amino acids can be separated by ion-exchange chromatography and quantitated by their reaction with fluorescamine. Amino acid sequences can be determined by Edman degradation, which removes one amino acid at a time from the amino end of a peptide. Longer polypeptide chains are broken into shorter ones for analysis by specifically cleaving them with a reagent that breaks the peptide at specific sites. Mass spectrometry allows determination of protein mass, identity, and primary structure. Amino acid sequences are rich in information concerning the kinship of proteins, their evolutionary relationships, and diseases produced by mutations. Knowledge of a sequence provides valuable clues to conformation and function.

KEY TERMS

proteome (p. 70)
assay (p. 71)
homogenate (p. 71)
differential centrifugation (p. 71)
salting out (p. 72)
dialysis (p. 72)
molecular exclusion chromatography (p. 72)
ion-exchange chromatography (p. 73)
affinity chromatography (p. 74)

high-pressure liquid chromatography (HPLC) (p. 74)
gel electrophoresis (p. 74)
isoelectric focusing (p. 75)
two-dimensional electrophoresis (p. 75)
gradient centrifugation (p. 78)
sedimentation coefficient (Svedberg units, S value) (p. 78)
antibody (immunoglobin) (p. 80)
antigen (p. 80)

antigenic determinant (epitope) (p. 80)
monoclonal antibody (p. 81)
immunoprecipitation (p. 83)
enzyme-linked immunosorbent assay (ELISA) (p. 84)
western blotting (p. 84)
Edman degradation (p. 86)
mass spectrometry (p. 88)
MALDI-TOF (p. 88)

Answers to QUICK QUIZZES

1. An assay, which should be based on some unique biochemical property of the protein that is being purified, allows the detection of the protein of interest.

2. Differences in size, solubility, charge, and the specific binding of certain molecules.

3. The texquisite specificity of the antibody–antigen interaction. This specificity allows for the detection of the antigen in the presence of large amounts of extraneous material.

4. Amino acid composition is simply the amino acids that are present in the protein. Many proteins can have the same amino acid composition. Amino acid sequence is the sequence of amino acids, or the primary structure, of the protein. Each protein has a unique amino acid sequence.

PROBLEMS

1. *Knowing that it's there.* Why is an assay necessary to purify a protein? ✓ 4

2. *Pair bonding.* Match the terms with the descriptions. ✓ 4

(a) Assay _____
(b) Molecular exclusion chromatography _____
(c) Ion-exchange chromatography _____
(d) Affinity chromatography _____
(e) High-pressure liquid chromatography (HPLC) _____
(f) Isoelectric focusing _____
(g) Sedimentation coefficient _____
(h) Antigenic determinant (epitope) _____
(i) Monoclonal antibodies _____
(j) Western blotting _____

1. Separating proteins on the basis of size differences
2. Allows high resolution and rapid separation
3. Produced by hybridoma cells
4. An immunoassay technique preceded by gel electrophoresis
5. A measure of the rate of movement due to centrifugal force
6. Separating proteins on the basis of net charge
7. Specific site recognized by an antibody
8. Based on the fact that proteins have a pH at which the net charge is zero
9. Based on attraction to a specific chemical group or molecule
10. A means of identifying a protein based on a unique property of the protein

3. *Salting out.* Why do proteins precipitate at high salt concentrations? ✓ 4

4. *Salting in.* Although many proteins precipitate at high salt concentrations, some proteins require salt to dissolve in water. Explain why some proteins require salt to dissolve.

5. *Competition for water.* What types of R groups would compete with salt ions for water of solvation?

6. *Column choice.* (a) The octapeptide AVGWRVKS was digested with the enzyme trypsin. Would ion-exchange or molecular exclusion chromatography be most appropriate for separating the products? Explain. ✓ 4

(b) Suppose that the peptide had, instead, been digested with chymotrypsin. What would be the optimal separation technique? Explain.

7. *Frequently used in shampoos.* The detergent sodium dodecyl sulfate (SDS) denatures proteins. Suggest how SDS destroys protein structure.

8. *Making more enzyme?* In the course of purifying an enzyme, a researcher performs a purification step that results in an *increase* in the total activity to a value greater than that present in the original crude extract. Explain how the amount of total activity might increase. ✓ 4

9. *Protein-purification problem.* Complete the following table. ✓ 4

Purification procedure	Total protein (mg)	Total activity (units)	Specific activity (units mg^{-1})	Purification level	Yield (%)
Crude extract	20,000	4,000,000		1	100
(NH$_4$)$_2$SO$_4$ precipitation	5000	3,000,000			
DEAE–cellulose chromatography	1500	1,000,000			
Gel-filtration chromatography	500	750,000			
Affinity chromatography	45	675,000			

Note: DEAE (diethylaminoethyl) bears a positive charge, and so DEAE-cellulose chromatography is anion-exchange chromatography.

10. *Charge to mass.* (a) Proteins treated with a sulfhydryl reagent such as β-mercaptoethanol and dissolved in sodium dodecyl sulfate have the same charge-to-mass ratio. Explain. ✓ 4

(b) Under what conditions might the statement in part *a* be incorrect?

(c) Some proteins migrate anomalously in SDS–PAGE gels. For instance, the molecular weight determined from an SDS–PAGE gel is sometimes very different from the molecular weight determined from the amino acid sequence. Suggest an explanation for this discrepancy.

11. *Push back.* During ultracentrifugation, the force opposing centrifugation is the buoyant force, equal to $(1 - \bar{v}\rho)$. Using the equation for buoyant force, show why a dense particle moves more rapidly than does a less dense one in ultracentrifugation.

12. *A special attraction.* What unique property of the estrogen receptor allows for its identification and purification? ✓ 5

13. *Many or one.* Differentiate between polyclonal and monoclonal antibodies. ✓ 5

14. *A falling out.* Explain how immunoprecipitation can be used to purify proteins. ✓ 5

15. *Don't you mean who?* What is an ELISA, and how is it used? ✓ 5

16. *John Wayne's favorite.* Describe western blotting. ✓ 5

17. *A question of efficiency.* The Edman method of protein sequencing can be used to determine a sequence of proteins no longer than approximately 50 amino acids. Why is this length limitation the case? ✓ 5

18. *Divide and conquer.* The determination of the mass of a protein by mass spectrometry often does not allow its unique identification among possible proteins within a complete proteome, but determination of the masses of all fragments produced by digestion with trypsin almost always allows unique identification. Explain.

Chapter Integration Problem

19. *Quaternary structure.* A protein was purified to homogeneity. Determination of the mass by gel-filtration chromatography yields 60 kDa. Chromatography in the presence of urea yields a 30-kDa species. When the chromatography is repeated in the presence of urea and β-mercaptoethanol, a single molecular species of 15 kDa results. Describe the structure of the molecule.

Data Interpretation Problems

20. *Protein solubility.* The isoelectric point (pI) of a protein is the pH at which a protein has no net charge. The pI is also the pH at which a protein is least soluble. The graph below shows the solubility of a protein as a function of pH in solutions of different ionic strengths. ✓ 4

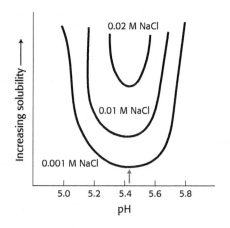

(a) Why is the protein least soluble at its pI (indicated by the red arrow in the figure)?

(b) Why does increasing the salt concentration increase the protein's solubility?

21. *Protein sequencing 1.* Determine the sequence of hexapeptide on the basis of the following data. Note: When the sequence is not known, a comma separates the amino acids. (Table 5.3)

Amino acid composition: (2R,A,S,V,Y)

Amino-terminal analysis of the hexapeptide: A

Trypsin digestion: (R,A,V) and (R,S,Y)

Carboxypeptidase A digestion: no digestion

Chymotrypsin digestion: (A,R,V,Y) and (R,S)

22. *Protein sequencing 2.* Determine the sequence of a peptide consisting of 14 amino acids on the basis of the following data.

Amino acid composition: (4S,2L,F,G,I,K,M,T,W,Y)

Amino-terminal analysis: S

Carboxypeptidase A digestion: L

Trypsin digestion: (3S,2L,F,I,M,T,W) (G,K,S,Y)

Chymotrypsin digestion: (F,I,S) (G,K,L) (L,S) (M,T) (S,W) (S,Y)

Amino-terminal analysis of (F,I,S) peptide: S

Cyanogen bromide treatment: (2S,F,G,I,K,L,M*,T,Y) (2S,L,W)

M*, methionine detected as homoserine

Challenge Problem

23. *Dialysis.* Suppose that you precipitate a protein with 1 M $(NH_4)_2SO_4$, and you wish to reduce the concentration of the $(NH_4)_2SO_4$. You take 1 ml of your sample and dialyze it in 1000 ml of buffer. At the end of dialysis, what is the concentration of $(NH_4)_2SO_4$ in your sample? How could you further lower the $(NH_4)_2SO_4$ concentration?

Selected Readings for this chapter can be found online at www.whfreeman.com/tymoczko3e.

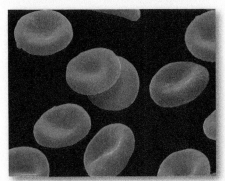

Basic Concepts and Kinetics of Enzymes

In Section 2, we considered the chemical workhorses of life—proteins. In this section, we examine an important class of proteins called enzymes. Protein enzymes are the most prominent catalysts of biological systems. *Catalysts* are chemicals that enhance the rate of reactions without being permanently altered themselves. The role of *enzymes*, then, is to make biochemically required reactions take place at a rate compatible with life.

Proteins as a class of macromolecules are well suited to be catalysts because of their capacity to form complex three-dimensional structures that can recognize one or a few molecules with high specificity. Collectively, the range of molecules on which enzymes can act is virtually unlimited. Because of their specificity, enzymes bring substrates (reactants) together at a particular site on the enzyme, called the *active site*, where they are oriented to facilitate the making and breaking of chemical bonds. The most striking characteristics of enzymes are their *catalytic power* and *specificity*.

Some enzymes are information sensors as well as catalysts. In addition to active sites, *allosteric enzymes* have distinct regulatory sites that bind to environmental signals. This binding modifies the activity of the active site.

We begin this section with a look at the basic properties of enzymes, with special emphasis on the energetics of enzyme-catalyzed reactions. We then move on to a kinetic analysis of enzymes. Next, we see how enzyme activity is modified by environmental conditions and examine the mechanism of action of chymotrypsin, a protein-digesting enzyme. We end the section with an examination of hemoglobin. This oxygen-transporting protein is a source of insight into the properties of allosteric proteins.

95

✓ **By the end of this section, you should be able to:**

✓ 1 Describe the relations between the enzyme catalysis of a reaction, the thermodynamics of the reaction, and the formation of the transition state.

✓ 2 Explain the relation between the transition state and the active site of an enzyme, and list the characteristics of active sites.

✓ 3 Explain what reaction velocity is.

✓ 4 Explain how reaction velocity is determined and how reaction velocities are used to characterize enzyme activity.

✓ 5 Identify the key properties of allosteric proteins, and describe the structural basis for these properties.

✓ 6 List environmental factors that affect enzyme activity, and describe how these factors exert their effects on enzymes.

✓ 7 Explain how allosteric properties contribute to hemoglobin function.

✓ 8 Identify the key regulators of hemoglobin function.

Basic Concepts of Enzyme Action

The activity of an enzyme is responsible for the glow of the luminescent jellyfish. The enzyme aequorin catalyzes the oxidation of a compound by oxygen in the presence of calcium to release CO_2 and light. [chain45154/Getty Images.]

The energy and information processing that takes place inside a cell consists of thousands of individual chemical reactions. For these reactions to take place in a physiologically useful fashion, they must occur at a rate that meets the cell's needs, and they must display specificity; that is, a particular reactant should always yield a particular product. Side reactions leading to the formation of useless or hazardous by-products must be minimized. In this chapter, we consider the key properties of enzymes, with a special look at the energetics of enzyme-catalyzed reactions.

6.1 Enzymes Are Powerful and Highly Specific Catalysts

Enzymes accelerate the rate of reactions by factors of as much as a million or more (Table 6.1). Indeed, most reactions in biological systems do not take place at perceptible rates in the absence of enzymes. Even a reaction as simple as adding water to carbon dioxide is catalyzed by an enzyme—namely, carbonic anhydrase. This reaction facilitates the transport of carbon dioxide from the tissues where it is produced to the lungs where it is exhaled. Carbonic anhydrase is one of the fastest known enzymes. Each enzyme molecule can hydrate 10^6 molecules of CO_2 *per second*. This catalyzed reaction is 10^7 times as fast as the uncatalyzed one. The transfer of CO_2 from the tissues to the blood and then to the lungs would be less complete in the absence of this enzyme.

Table 6.1   Rate enhancement by selected enzymes

Enzyme	Nonenzymatic half-life	Uncatalyzed rate (k_{un} s^{-1})	Catalyzed rate (k_{cat} s^{-1})	Rate enhancement (k_{cat} s^{-1}/k_{un} s^{-1})
OMP decarboxylase	78,000,000 years	2.8×10^{-16}	39	1.4×10^{17}
Staphylococcal nuclease	130,000 years	1.7×10^{-13}	95	5.6×10^{14}
AMP nucleosidase	69,000 years	1.0×10^{-11}	60	6.0×10^{12}
Carboxypeptidase A	7.3 years	3.0×10^{-9}	578	1.9×10^{11}
Ketosteroid isomerase	7 weeks	1.7×10^{-7}	66,000	3.9×10^{11}
Triose phosphate isomerase	1.9 days	4.3×10^{-6}	4,300	1.0×10^{9}
Chorismate mutase	7.4 hours	2.6×10^{-5}	50	1.9×10^{6}
Carbonic anhydrase	5 seconds	1.3×10^{-1}	1×10^{6}	7.7×10^{6}

Abbreviations: OMP, orotidine monophosphate; AMP, adenosine monophosphate.
Source: Data from A.Radzicka and R.Wolfenden, *Science* 267:90–93, 1995.

Proteolytic Enzymes Illustrate the Range of Enzyme Specificity

Enzymes are highly specific both in the reactions that they catalyze and in their choice of reactants, which are called *substrates*. An enzyme usually catalyzes a single chemical reaction or a set of closely related reactions. Let us consider *proteolytic enzymes* as an example. The biochemical function of these enzymes is to catalyze *proteolysis,* the hydrolysis of a peptide bond:

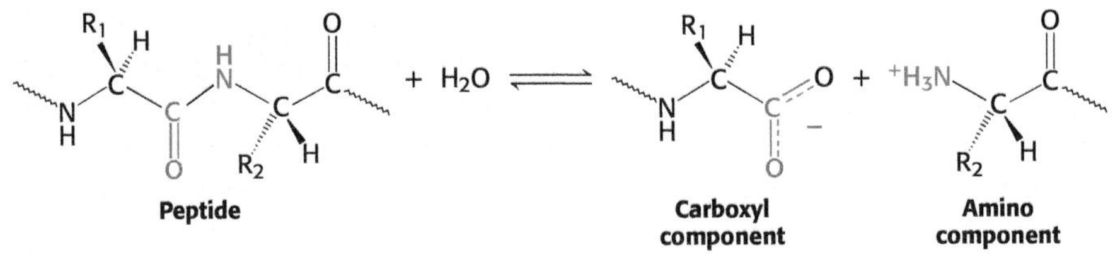

Peptide   **Carboxyl component**   **Amino component**

Proteolytic enzymes differ markedly in their degree of substrate specificity. Papain, which is found in papaya plants, is quite undiscriminating: it will cleave any peptide bond with little regard to the identity of the adjacent side chains. This lack of specificity accounts for its use in meat-tenderizing sauces. Trypsin, a digestive enzyme, is quite specific and catalyzes the splitting of peptide bonds only on the carboxyl side of lysine and arginine residues (**Figure 6.1**A). Thrombin, an enzyme that participates in blood clotting, is even more specific than trypsin. It catalyzes the hydrolysis of Arg–Gly bonds in particular peptide sequences only (Figure 6.1B). *The specificity of an enzyme is due to the precise interaction of the substrate with the enzyme. This precision is a result of the intricate three-dimensional structure of the enzyme protein.*

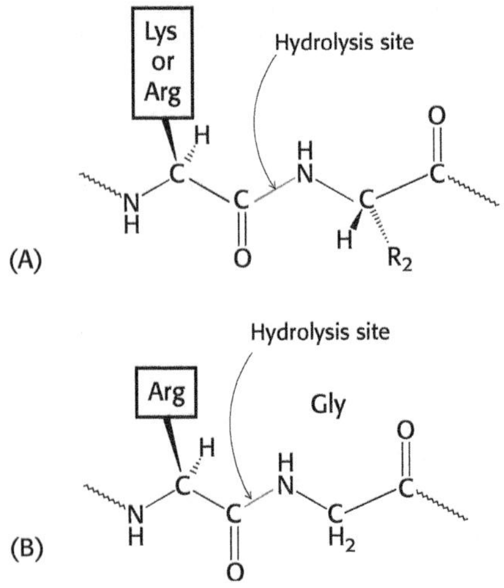

There Are Six Major Classes of Enzymes

More than a thousand enzymes have been identified—a daunting number for a student of biochemistry. Despite this large number, however, there are only six major classes of enzymes, which makes recognizing the function of enzymes much simpler. We will see many members of these classes as we progress in our study of biochemistry.

(A)

(B)

Figure 6.1 Enzyme specificity. (A) Trypsin cleaves on the carboxyl side of arginine and lysine residues, whereas (B) thrombin cleaves Arg–Gly bonds in particular sequences only.

1. *Oxidoreductases.* These enzymes transfer electrons between molecules. In other words, these enzymes catalyze oxidation–reduction reactions. We will first meet a member of this class, lactate dehydrogenase, when we consider glycolysis, the first pathway in glucose degradation.

2. *Transferases*. These enzymes transfer functional groups between molecules. Aminotransferases are prominent in amino acid synthesis and degradation, where they shuffle amine groups between donor and acceptor molecules.

3. *Hydrolyases*. A hydrolyase cleaves molecules by the addition of water. Trypsin, the proteolytic enzyme already discussed, is a hydrolyase.

4. *Lyases*. A lyase adds atoms or functional groups to a double bond or removes them to form double bonds. The lyase fumarase is crucial to aerobic fuel metabolism.

5. *Isomerases*. These enzymes move functional groups within a molecule. We will meet triose phosphate isomerase in glycolysis.

6. *Ligases*. Ligases join two molecules in a reaction powered by ATP hydrolysis. DNA ligase, an important enzyme in DNA replication, is representative of this class.

The classification of all enzymes into these six classes also allowed the development of a standard nomenclature for enzymes. Many enzymes have common names that provide little information about the reactions that they catalyze. Trypsin exemplifies this lack of information. Most other enzymes are named for their substrates and for the reactions that they catalyze, with the suffix "ase" added. Thus, a peptide hydrolase is an enzyme that hydrolyzes peptide bonds, whereas ATP synthase is an enzyme that synthesizes ATP. Common names will be used in this book, but let's examine the more accurate nomenclature.

The six groups (classes) of enzymes were subdivided and further subdivided so that a four-digit number preceded by the letters *EC* for Enzyme Commission, the entity that developed the classification system, could precisely identify all enzymes.

Consider trypsin as an example. Trypsin cleaves bonds by the addition of water; consequently, it is a member of group 3: *Hydrolyases*. Trypsin cleaves only peptide bonds, and hydrolyases that cleave peptide bonds are classified as 3.4. Trypsin employs a serine residue to facilitate hydrolysis and cleaves the protein chain internally (in contrast with the removal of amino acids from the end of the polypeptide chain). Such enzymes are placed in sub-sub-group 21 and identified as 3.4.21. Finally, trypsin cleaves peptide bonds in which the amino acid donating the carboxyl group to the peptide bond is either lysine or arginine. Thus, the number uniquely identifying trypsin is EC 3.4.21.4. Although the common names are used routinely, the classification number is used when the precise identity of the enzyme might be ambiguous.

DID YOU KNOW?

Hydrolysis reactions, the breaking of a chemical bond by the addition of a water molecule, are prominent in biochemistry.

6.2 Many Enzymes Require Cofactors for Activity

Although the chemical repertoire of amino acid functional groups is quite varied (pp. 39–43), they often cannot meet the chemical needs required for catalysis to take place. Thus, the catalytic activity of many enzymes depends on the presence of small molecules termed *cofactors*. The precise role varies with the cofactor and the enzyme. An enzyme without its cofactor is referred to as an *apoenzyme*; the complete, catalytically active enzyme is called a *holoenzyme*. Cofactors can be subdivided into two groups: (1) small organic molecules, derived from vitamins, called *coenzymes* and (2) metals (Table 6.2). Tightly bound coenzymes are called *prosthetic (helper) groups*. Loosely associated coenzymes are more like cosubstrates because, like substrates and products, they bind to the enzyme and are released from it. Coenzymes are distinct from normal substrates not only because they are often derived from vitamins but also because they are used by a variety of enzymes. Different enzymes that use the same coenzyme usually carry out similar chemical transformations.

Table 6.2 Enzyme cofactors

Cofactor	Enzyme*
Coenzyme[†]	
Thiamine pyrophosphate (TPP)	Pyruvate dehydrogenase
Flavin adenine dinucleotide (FAD)	Monoamine oxidase
Nicotinamide adenine dinucleotide (NAD^+)	Lactate dehydrogenase
Pyridoxal phosphate (PLP)	Glycogen phosphorylase
Coenzyme A (CoA)	Acetyl CoA carboxylase
Biotin	Pyruvate carboxylase
6'-Deoxyadenosyl cobalamin	Methylmalonyl mutase
Tetrahydrofolate	Thymidylate synthase
Metal	
Zn^{2+}	Carbonic anhydrase
Mg^{2+}	*Eco*RV
Ni^{2+}	Urease
Mo	Nitrogenase
Se	Glutathione peroxidase
$Mn^{2+ \leftrightarrow 3+}$	Superoxide dismutase
K^+	Acetoacetyl CoA thiolase

*The enzymes listed are examples of enzymes that employ the indicated cofactor.
[†]Often derived from vitamins, coenzymes can be either tightly or loosely bound to the enzyme.

✓ 1 Describe the relations between the enzyme catalysis of a reaction, the thermodynamics of the reaction, and the formation of the transition state.

6.3 Gibbs Free Energy Is a Useful Thermodynamic Function for Understanding Enzymes

Enzymes speed up the rate of chemical reactions, but the properties of the reaction—whether it can take place at all—depends on free-energy differences. *Gibbs free energy*, or more simply *free energy* (G), is a thermodynamic property that is a measure of useful energy, or energy that is capable of doing work. To understand how enzymes operate, we need to consider only two thermodynamic properties of the reaction: (1) the free-energy difference (ΔG) between the products and the reactants and (2) the free energy required to initiate the conversion of reactants into products. The former determines whether the reaction will take place spontaneously, whereas the latter determines the rate of the reaction. Enzymes affect only the latter. Let us review some of the principles of thermodynamics as they apply to enzymes.

The Free-Energy Change Provides Information About the Spontaneity but Not the Rate of a Reaction

The free-energy change of a reaction (ΔG) tells us whether the reaction can take place spontaneously:

1. *A reaction can take place spontaneously only if ΔG is negative.* "Spontaneously" in the context of thermodynamics means that the reaction will take place without the input of energy and, in fact, the reaction releases energy. Such reactions are said to be *exergonic*.

2. A reaction cannot take place spontaneously if ΔG is positive. An input of free energy is required to drive such a reaction. These reactions are termed *endergonic*.

3. In a system at equilibrium, there is no *net* change in the concentrations of the products and reactants, and ΔG is zero.

4. The ΔG of a reaction depends only on the free energy of the products (the final state) minus the free energy of the reactants (the initial state). *The ΔG of a reaction is independent of the path (or molecular mechanism) of the transformation.* The mechanism of a reaction has no effect on ΔG. For example, the ΔG for the transformation of glucose into CO_2 and H_2O is the same whether it takes place by combustion or by a series of enzyme-catalyzed steps in a cell.

5. *The ΔG provides no information about the rate of a reaction.* A negative ΔG indicates that a reaction *can* take place spontaneously, but it does not signify whether it will proceed at a perceptible rate. As will be discussed shortly (p. 103), the rate of a reaction depends on the *free energy of activation* ($\Delta G^{\ddagger}$), which is largely unrelated to the ΔG of the reaction.

The Standard Free-Energy Change of a Reaction Is Related to the Equilibrium Constant

As for any reaction, we need to be able to determine ΔG for an enzyme-catalyzed reaction to know whether the reaction is spontaneous or requires an input of energy. To determine the free-energy change of the reaction, we need to take into account the nature of both the reactants and the products as well as their concentrations.

Consider the reaction

$$A + B \rightleftharpoons C + D$$

The ΔG of this reaction is given by

$$\Delta G = \Delta G^{\circ} + RT \ln \frac{[C][D]}{[A][B]} \qquad (1)$$

in which ΔG° is the *standard free-energy change*, R is the gas constant, T is the absolute temperature, and [A], [B], [C], and [D] are the molar concentrations of the reactants. ΔG° is the free-energy change for this reaction under standard conditions—that is, when each of the reactants A, B, C, and D is present at a concentration of 1.0 M (for a gas, the standard state is usually chosen to be 1 atmosphere) before the initiation of the reaction, and the temperature is 298 K (298 kelvins, or 25°C). Thus, the ΔG of a reaction depends on the *nature* of the reactants (expressed in the ΔG° term of equation 1) and on their *concentrations* (expressed in the logarithmic term of equation 1).

A convention has been adopted to simplify free-energy calculations for biochemical reactions. The standard state is defined as having a pH of 7. Consequently, when H^+ is a reactant, its concentration has the value 1 (corresponding to a pH of 7) in the numbered equations that follow. The concentration of water also is taken to be 1 in these equations. The *standard free-energy change at pH 7*, denoted by the symbol $\Delta G^{\circ\prime}$, will be used throughout this book. The *kilojoule* (kJ) and the *kilocalorie* (kcal) will be used as the units of energy. As stated in Chapter 2, 1 kJ is equivalent to 0.239 kcal.

A simple way to determine the $\Delta G^{\circ\prime}$ is to measure the concentrations of reactants and products when the reaction has reached equilibrium. At equilibrium, there is no net change in the concentrations of reactants and products; in essence, the reaction has stopped and $\Delta G = 0$. At equilibrium, equation 1 then becomes

$$0 = \Delta G^{\circ\prime} + RT \ln \frac{[C][D]}{[A][B]} \qquad (2)$$

and so

$$\Delta G^{\circ\prime} = -RT \ln \frac{[C][D]}{[A][B]} \qquad (3)$$

DID YOU KNOW?

A *kilojoule* (kJ) is equal to 1000 J.
A *joule* (J) is the amount of energy needed to apply a 1-newton force over a distance of 1 meter.
A *kilocalorie* (kcal) is equal to 1000 cal.
A *calorie* (cal) is equivalent to the amount of heat required to raise the temperature of 1 gram of water from 14.5°C to 15.5°C.
1 kJ = 0.239 kcal

The equilibrium constant under standard conditions, K'_{eq}, is defined as

$$K'_{eq} = \frac{[C][D]}{[A][B]} \tag{4}$$

Substituting equation 4 into equation 3 gives

$$\Delta G^{\circ\prime} = -RT \ln K'_{eq} \tag{5}$$

which can be rearranged to give

$$K'_{eq} = e^{-\Delta G^{\circ\prime}/RT} \tag{6}$$

Substituting $R = 8.315 \times 10^{-3}\ kJ\ mol^{-1}\ K^{-1}$ and $T = 298\ K$ (corresponding to 25°C) gives

$$K'_{eq} = e^{-\Delta G^{\circ\prime}/2.47} \tag{7}$$

where $\Delta G^{\circ\prime}$ is here expressed in kilojoules per mole because of the choice of the units for R in equation 7. Thus, the standard free energy and the equilibrium constant of a reaction are related by a simple expression. For example, an equilibrium constant of 10 gives a standard free-energy change of $-5.69\ kJ\ mol^{-1}$ ($-1.36\ kcal\ mol^{-1}$) at 25°C (Table 6.3). Note that, for each 10-fold change in the equilibrium constant, the $\Delta G^{\circ\prime}$ changes by $5.69\ kJ\ mol^{-1}$ ($1.36\ kcal\ mol^{-1}$).

QUICK QUIZ Which of the following two reactions will take place spontaneously? What are the $\Delta G^{\circ\prime}$ values for the reverse reactions?

A → B $\Delta G^{\circ\prime} = -10\ kJ\ mol^{-1}$
C → D $\Delta G^{\circ\prime} = +10\ kJ\ mol^{-1}$

Table 6.3 Relation between $\Delta G^{\circ\prime}$ and K'_{eq} (at 25°C)

K'_{eq}	$\Delta G^{\circ\prime}$	
	kJ mol^{-1}	kcal mol^{-1}
10^{-5}	28.53	6.82
10^{-4}	22.84	5.46
10^{-3}	17.11	4.09
10^{-2}	11.42	2.73
10^{-1}	5.69	1.36
1	0	0
10	−5.69	−1.36
10^{2}	−11.42	−2.73
10^{3}	−17.11	−4.09
10^{4}	−22.84	−5.46
10^{5}	−28.53	−6.82

It is important to stress that whether the ΔG for a reaction is larger, smaller, or the same as $\Delta G^{\circ\prime}$ depends on the concentrations of the reactants and products. The criterion of spontaneity for a reaction is ΔG, not $\Delta G^{\circ\prime}$. This point is important because reactions that are not spontaneous, on the basis of $\Delta G^{\circ\prime}$, can be made spontaneous by adjusting the concentrations of reactants and products. This principle is the basis of the coupling of reactions to form metabolic pathways (Chapter 15).

Enzymes Alter the Reaction Rate but Not the Reaction Equilibrium

Because enzymes are such superb catalysts, it is tempting to ascribe to them powers that they do not have. *An enzyme cannot alter the laws of thermodynamics and consequently cannot alter the equilibrium of a chemical reaction.* Consider an enzyme-catalyzed reaction, the conversion of substrate, S, into product, P. **Figure 6.2** graphs the rate of product formation with time in the presence and absence of enzyme. Note that the amount of product formed is the same whether or not the enzyme is present, but, in the present example, the amount of product formed in seconds when the enzyme is present might take hours or centuries to form if the enzyme were absent (Table 6.1).

Why does the rate of product formation level off with time? The reaction has reached equilibrium. Substrate S is still being converted into product P, but P is being converted into S at a rate such that the amount of P remains constant. *Enzymes accelerate the attainment of equilibria but do not shift their positions. The equilibrium position is a function only of the free-energy difference between reactants and products.*

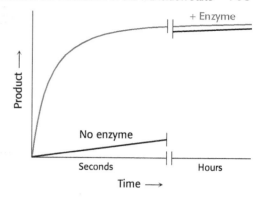

Figure 6.2 Enzymes accelerate the reaction rate. The same equilibrium point is reached but much more quickly in the presence of an enzyme.

6.4 Enzymes Facilitate the Formation of the Transition State

✓ 2 Explain the relation between the transition state and the active site of an enzyme, and list the characteristics of active sites.

The free-energy difference between reactants and products accounts for the equilibrium of a reaction, but enzymes accelerate how quickly this equilibrium is attained. How can we explain the rate enhancement in terms of thermodynamics? To do so, we have to consider not the end points of the reaction but the chemical pathway between the end points.

A chemical reaction of substrate S to form product P goes through a *transition state* $X^{\ddagger}$ that has a higher free energy than does either S or P. The double dagger denotes the transition state. The transition state is a fleeting molecular structure that is no longer the substrate but is not yet the product. The transition state is the least-stable and most-seldom-occurring species along the reaction pathway because it is the one with the highest free energy.

$$S \rightleftharpoons X^{\ddagger} \rightarrow P$$

The difference in free energy between the transition state and the substrate is called the *free energy of activation* or simply the *activation energy*, symbolized by $\Delta G^{\ddagger}$ (**Figure 6.3**):

$$\Delta G^{\ddagger} = G_X^{\ddagger} - G_S$$

Note that the energy of activation, or $\Delta G^{\ddagger}$, does not enter into the final ΔG calculation for the reaction, because the energy that had to be added to reach the transition state is released when the transition state becomes the product. The activation energy immediately suggests how enzymes accelerate the reaction rate without altering ΔG of the reaction: enzymes function to lower the activation energy. In other words, *enzymes facilitate the formation of the transition state.*

The combination of substrate and enzyme creates a reaction pathway whose transition-state energy is lower than what it would be without the enzyme (Figure 6.3). Because the activation energy is lower, more molecules have the energy required to reach the transition state and more product will be formed faster. Decreasing the activation barrier is analogous to lowering the height of a high-jump bar; more athletes will be able to clear the bar. *The essence of catalysis is stabilization of the transition state.*

The Formation of an Enzyme–Substrate Complex Is the First Step in Enzymatic Catalysis

Much of the catalytic power of enzymes comes from their binding to and then altering the structure of the substrate to promote the formation of the transition state. Thus, the first step in catalysis is the formation of an *enzyme–substrate* (ES)

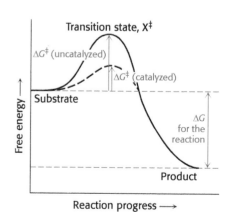

Figure 6.3 Enzymes decrease the activation energy. Enzymes accelerate reactions by decreasing $\Delta G^{\ddagger}$, the free energy of activation.

complex. Substrates bind to a specific region of the enzyme called the *active site*. Most enzymes are highly selective in the substrates that they bind. Indeed, the catalytic specificity of enzymes depends in part on the specificity of binding.

The Active Sites of Enzymes Have Some Common Features

The active site of an enzyme is the region that binds the substrates (and the cofactor, if any). It also contains the amino acid residues that directly participate in the making and breaking of bonds. These residues are called the *catalytic groups*. In essence, *the interaction of the enzyme and substrate at the active site promotes the formation of the transition state*. The active site is the region of the enzyme that most directly lowers the $\Delta G^{\ddagger}$ of the reaction, thus providing the rate-enhancement characteristic of enzyme action. Recall from Chapter 4 that proteins are not rigid structures, but are flexible and exist in an array of conformations. Thus, the interaction of the enzyme and substrate at the active site and the formation of the transition state is a dynamic process. Although enzymes differ widely in structure, specificity, and mode of catalysis, a number of generalizations concerning their active sites can be stated:

1. *The active site is a three-dimensional cleft or crevice* formed by groups that come from different parts of the amino acid sequence: indeed, amino acids near to one another in the primary structure are often sterically constrained from adopting the structural relations necessary to form the active site. In lysozyme, the important groups in the active site are contributed by residues numbered 35, 52, 62, 63, 101, and 108 in the sequence of 129 amino acids (Figure 6.4). Lysozyme, found in a variety of organisms and tissues including human tears, degrades the cell walls of some bacteria.

2. *The active site takes up a small part of the total volume of an enzyme.* Although most of the amino acid residues in an enzyme are not in contact with the substrate, the cooperative motions of the enzyme as a whole help to correctly position the catalytic residues at the active site. Experimental attempts to reduce the size of a catalytically active enzyme show that the minimum size requires about 100 amino acid residues. In fact, nearly all enzymes are made up of more than 100 amino acid residues, which gives them a mass greater than 10 kDa and a diameter of more than 25 Å, suggesting that all amino acids in the protein, not just those at the active site, are ultimately required to form a functional enzyme.

3. *Active sites are unique microenvironments.* The close association between the active site and the substrate means that water is usually excluded from the active site unless it is a reactant. The nonpolar microenvironment of the cleft enhances the binding of substrates as well as catalysis. Nevertheless, the cleft may also contain polar residues, some of which may acquire special properties essential for substrate binding or catalysis. The internal positions of these polar residues are biologically crucial exceptions to the general rule that polar residues are located on the surface of proteins, exposed to water.

4. *Substrates are bound to enzymes by multiple weak attractions.* The noncovalent interactions between the enzyme and the substrate in ES complexes are much weaker than covalent bonds. These weak reversible interactions are mediated by electrostatic interactions, hydrogen bonds, and van der Waals forces, powered by the hydrophobic effect. Van der Waals forces become significant in binding only when numerous substrate atoms simultaneously come close to many enzyme atoms. Hence, to bind as strongly as possible, the enzyme and substrate should have complementary shapes.

(A)

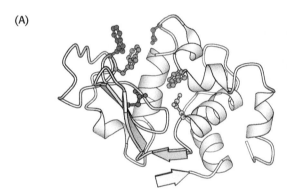

(B) N

C

1   35   52 62,63   101 108   129

Figure 6.4 Active sites may include distant residues. (A) Ribbon diagram of the enzyme lysozyme with several components of the active site shown in color. (B) A schematic representation of the primary structure of lysozyme shows that the active site is composed of residues that come from different parts of the polypeptide chain. [Drawn from 6LYZ.pdb.]

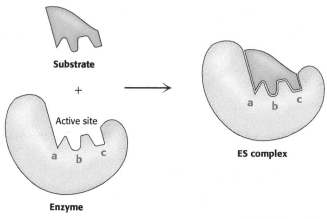

Figure 6.5 Lock-and-key model of enzyme–substrate binding. In this model, the active site of the unbound enzyme is complementary in shape to the substrate.

5. *The specificity of binding depends on the precisely defined arrangement of atoms in an active site.* Because the enzyme and the substrate interact by means of short-range forces that require close contact, a substrate must have a matching shape to fit into the site. Emil Fischer's analogy of the lock and key (**Figure 6.5**), expressed in 1890, has proved to be highly stimulating and fruitful. However, we now know that enzymes are flexible and that the shapes of the active sites can be markedly modified by the binding of substrate, a process of dynamic recognition called *induced fit* (**Figure 6.6**). Moreover, the substrate may bind to only certain conformations of the enzyme, in what is called *conformation selection.* Thus, the mechanism of catalysis is dynamic, involving structural changes with multiple intermediates of both reactants and the enzyme.

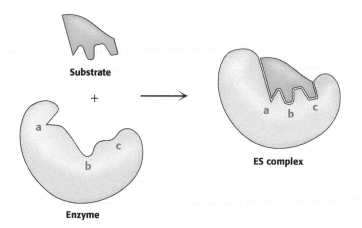

Figure 6.6 Induced-fit model of enzyme–substrate binding. In this model, the enzyme changes shape on substrate binding. The active site forms a shape complementary to the substrate only after the substrate has been bound.

The Binding Energy Between Enzyme and Substrate Is Important for Catalysis

Enzymes lower the activation energy, but where does the energy to lower the activation energy come from? Free energy is released by the formation of a large number of weak interactions between a complementary enzyme and substrate. The free energy released on binding is called the *binding energy.* Only the correct substrate can participate in most or all of the interactions with the enzyme and thus maximize binding energy, accounting for the exquisite substrate specificity exhibited by many enzymes. Furthermore, *the full complement of such interactions is formed only when the substrate is in the transition state.* Thus, the maximal binding energy is released when the enzyme facilitates the formation of the transition state. The energy released by the interactions between the enzyme and the substrate can be thought of as lowering the activation energy. The interaction of the enzyme with the substrate and reaction intermediates is fleeting, with molecular

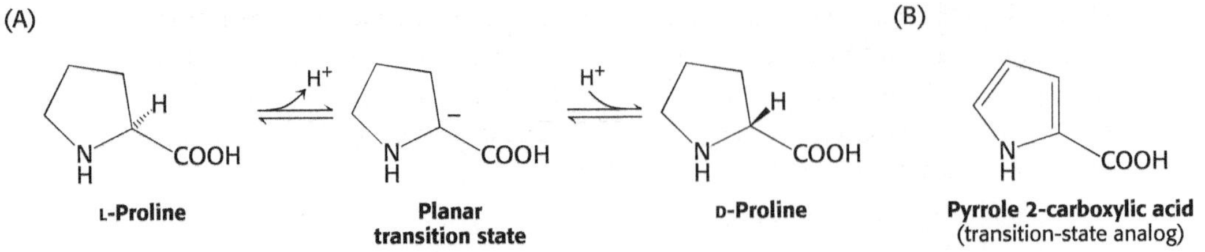

(A)   (B)

L-Proline   **Planar**   **D-Proline**   **Pyrrole 2-carboxylic acid**
                    transition state   (transition-state analog)

Figure 6.7 Inhibition by transition-state analogs. (A) The isomerization of L-proline to D-proline by proline racemase, a bacterial enzyme, proceeds through a planar transition state in which the α-carbon atom is trigonal rather than tetrahedral. (B) Pyrrole 2-carboxylate, a transition-state analog because of its trigonal geometry, is a potent inhibitor of proline racemase.

movements resulting in the optimal alignment of functional groups at the active site so that maximum binding energy occurs only between the enzyme and the transition state, the least-stable reaction intermediate. However, the transition state is too unstable to exist for long. It collapses to either substrate or product, but which of the two accumulates is determined only by the energy difference between the substrate and the product—that is, by the ΔG of the reaction.

Transition-State Analogs Are Potent Inhibitors of Enzymes

The importance of the formation of the transition state to enzyme catalysis is demonstrated by the study of compounds that resemble the transition state of a reaction but are not capable of being acted on by the enzyme. These mimics are called *transition-state analogs*. The inhibition of the enzyme proline racemase is an instructive example. The racemization of proline proceeds through a transition state in which the tetrahedral α-carbon atom has become trigonal (**Figure 6.7**). This picture is supported by the finding that the inhibitor pyrrole 2-carboxylate binds to the racemase 160 times as tightly as does proline. *The α-carbon atom of this inhibitor, like that of the transition state, is trigonal.* An analog that also carries a negative charge on α-carbon would be expected to bind even more tightly. In general, highly potent and specific inhibitors of enzymes can be produced by synthesizing compounds that more closely resemble the transition state than the substrate itself. The inhibitory power of transition-state analogs underscores the essence of catalysis: *selective binding of the transition state*.

If our understanding of the importance of the transition state to catalysis is correct, then antibodies that recognize transition states should function as catalysts. Antibodies have been generated that recognize the transition states of certain reactions, and these antibodies, called *catalytic antibodies* or *abzymes*, do indeed function as enzymes.

DID YOU KNOW?

Racemization is the conversion of one enantiomer into another; in regard to proline, the interconversion of the L and D isomers.

SUMMARY

6.1 Enzymes Are Powerful and Highly Specific Catalysts

The catalysts in biological systems are enzymes, and nearly all enzymes are proteins. Enzymes are highly specific and have great catalytic power. They can enhance reaction rates by factors of 10^6 or more.

6.2 Many Enzymes Require Cofactors for Activity

Cofactors required by enzymes for activity can be small, vitamin-derived organic molecules called coenzymes, or they can be metals.

6.3 Gibbs Free Energy Is a Useful Thermodynamic Function for Understanding Enzymes

Free energy (G) is the most valuable thermodynamic function for determining whether a reaction can take place and for understanding the energetics of catalysis. A reaction can take place spontaneously only if the change in free energy (ΔG) is negative. The free-energy change of a reaction that takes place when reactants and products are at unit activity is called the standard free-energy change ($\Delta G°$). Biochemists usually use $\Delta G°'$, the standard free-energy change at pH 7. Enzymes do not alter reaction equilibria; rather, they increase reaction rates.

6.4   Enzymes Facilitate the Formation of the Transition State

Enzymes serve as catalysts by decreasing the free energy of activation of chemical reactions. Enzymes accelerate reactions by providing a reaction pathway in which the transition state (the highest-energy species) has a lower free energy and, hence, is more rapidly formed than in the uncatalyzed reaction.

The first step in catalysis is the formation of an enzyme–substrate complex. Substrates are bound to enzymes at active-site clefts from which water is largely excluded when the substrate has been bound. The recognition of substrates by enzymes is often accompanied by conformational changes at active sites, and such changes facilitate the formation of the transition state.

KEY TERMS

enzyme (p. 97)
substrate (p. 98)
cofactor (p. 99)
apoenzyme (p. 99)

holoenzyme (p. 99)
coenzyme (p. 99)
free energy (p. 100)
transition state (p. 103)

free energy of activation (p. 103)
enzyme–substrate (ES) complex (p. 103)
active site (p. 104)
induced fit (p. 105)

❓ Answer to QUICK QUIZ

The reaction with the negative $\Delta G°'$ will be spontaneous (exergonic), whereas the reaction with the positive $\Delta G°'$ will not be spontaneous (endergonic). For the reverse reactions, the numerical values will be the same, but the signs of the reactions will be opposite.

PROBLEMS

1. *Raisons d'être.* What are the two properties of enzymes that make them especially useful catalysts? ✓ 1

2. *Shared properties.* What are the general characteristics of enzyme active sites? ✓ 2

3. *Partners.* What does an apoenzyme require to become a holoenzyme?

4. *Different partners.* What are the two main types of cofactors?

5. *One a day.* Why are vitamins necessary for good health?

6. *A function of state.* What is the fundamental mechanism by which enzymes enhance the rate of chemical reactions? ✓ 1

7. *Nooks and crannies.* What is the structural basis for enzyme specificity? ✓ 2

8. *Mutual attraction.* What is meant by the term *binding energy*? ✓ 2

9. *Catalytically binding.* What is the role of binding energy in enzyme catalysis? ✓ 2

10. *Sticky situation.* What would be the result of an enzyme having a greater binding energy for the substrate than for the transition state? ✓ 2

11. *Made for each other.* Match the term with the proper description.

(a) Enzyme _____
(b) Substrate _____
(c) Cofactor _____
(d) Apoenzyme _____
(e) Holoenzyme _____
(f) Coenzymes _____
(g) $\Delta G°'$ _____
(h) Transition state _____
(i) Active site _____
(j) Induced fit _____

1. The least-stable reaction intermediate
2. Site on the enzyme where catalysis takes place
3. Enzyme minus its cofactor
4. Protein catalyst
5. Function of K'_{eq}
6. Change in enzyme structure
7. Reactant in an enzyme-catalyzed reaction
8. A coenzyme or metal
9. Enzyme plus cofactor
10. Small vitamin-derived organic cofactors

12. *Give with one hand, take with the other.* Why does the activation energy of a reaction not appear in the final ΔG of the reaction? ✓ 1

13. *Making progress.* The illustrations below show the reaction-progress curves for two different reactions. Indicate the activation energy as well as the ΔG for each reaction. Which reaction is endergonic? Exergonic? ✓ 1

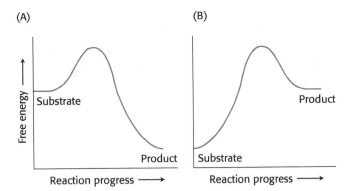

(A)   (B)

14. *Mountain climbing.* Proteins are thermodynamically unstable. The ΔG of the hydrolysis of proteins is quite negative, yet proteins can be quite stable. Explain this apparent paradox. What does it tell you about protein synthesis?

15. *Protection.* Suggest why the enzyme lysozyme, which degrades cell walls of some bacteria, is present in tears.

16. *Stability matters.* Transition-state analogs, which can be used as enzyme inhibitors and to generate catalytic antibodies, are often difficult to synthesize. Suggest a reason. ✓ 2

17. *Match'em.* Match the K'_{eq} values with the appropriate $\Delta G^{\circ\prime}$ values. ✓ 1

K'_{eq}	$\Delta G^{\circ\prime}$ (kJ mol^{-1})
(a) 1	28.53
(b) 10^{-5}	-11.42
(c) 10^{4}	5.69
(d) 10^{2}	0
(e) 10^{-1}	-22.84

18. *Free energy!* Consider the following reaction:

$$\text{Glucose 1-phosphate} \rightleftharpoons \text{glucose 6-phosphate}$$

After the reactants and products were mixed and allowed to reach equilibrium at 25°C, the concentration of each compound was measured:

$$[\text{Glucose 1-phosphate}]_{eq} = 0.01 \text{ M}$$
$$[\text{Glucose 6-phosphate}]_{eq} = 0.19 \text{ M}$$

Calculate K_{eq} and $\Delta G^{\circ\prime}$.

19. *More free energy!* The isomerization of dihydroxyacetone phosphate (DHAP) to glyceroldehyde 3-phosphate (GAP) has an equilibrium constant of 0.0475 under standard conditions (298 K, pH 7). Calculate $\Delta G^{\circ\prime}$ for the isomerization. Next, calculate ΔG for this reaction when the initial concentration of DHAP is 2×10^{-4} M and the initial concentration of GAP is 3×10^{-6} M. What do these values tell you about the importance of ΔG compared with that of $\Delta G^{\circ\prime}$ in understanding the thermodynamics of intracellular reactions? ✓ 1

20. *A tenacious mutant.* Suppose that a mutant enzyme binds a substrate 100 times as tightly as does the native enzyme. What is the effect of this mutation on catalytic rate if the binding of the transition state is unaffected? ✓ 2

21. *A question of stability.* Pyridoxal phosphate (PLP) is a coenzyme for the enzyme ornithine aminotransferase. The enzyme was purified from cells grown in PLP-deficient medium as well as from cells grown in medium that contained pyridoxal phosphate. The stability of the enzyme was then measured by incubating the enzyme at 37°C and assaying for the amount of enzyme activity remaining. The following results were obtained:

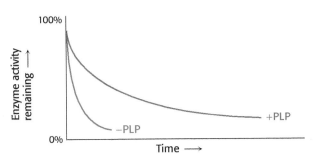

(a) Why does the amount of active enzyme decrease with the time of incubation?
(b) Why does the amount of enzyme from the PLP-deficient cells decline more rapidly?

Challenge Problems

22. *Free energy, yet again.* Assume that you have a solution of 0.1 M glucose 6-phosphate. To this solution, you add the enzyme phosphoglucomutase, which catalyzes the reaction. ✓ 1

$$\text{Glucose 6-phosphate} \xrightleftharpoons{\text{Phosphoglucomutase}} \text{glucose 1-phosphate}$$

The $\Delta G^{\circ\prime}$ for the reaction is +7.5 kJ mol^{-1} (+1.8 kcal mol^{-1}).

(a) Does the reaction proceed as written, and, if so, what are the final concentrations of glucose 6-phosphate and glucose 1-phosphate?
(b) Under what cellular conditions could you produce glucose 1-phosphate at a high rate?

23. *Potential donors and acceptors.* The hormone progesterone contains two ketone groups. At pH 7, which side chains of a protein might form hydrogen bonds with progesterone?

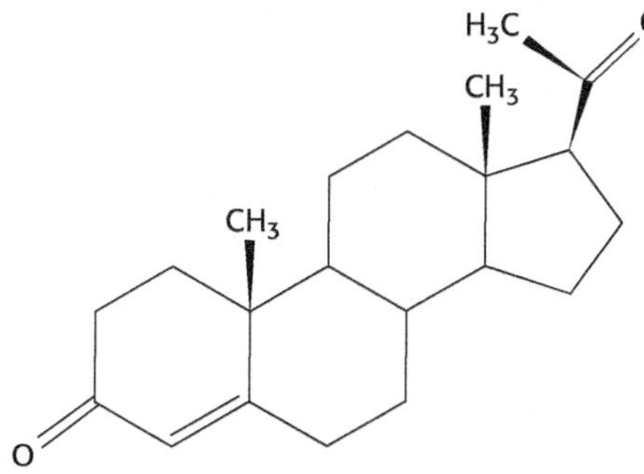

Progesterone

24. *The more things change, the more they stay the same.* Suppose that, in the absence of enzyme, the forward rate constant (k_F) for the conversion of S into P is $10^{-4}\ s^{-1}$ and the reverse rate constant (k_R) for the conversion of P into S is $10^{-6}\ s^{-1}$:

$$S \underset{10^{-6}\,s^{-1}}{\overset{10^{-4}\,s^{-1}}{\rightleftharpoons}} P$$

(a) What is the equilibrium for the reaction? What is the $\Delta G^{\circ\prime}$?

(b) Suppose an enzyme enhances the rate of the reaction 100-fold. What are the rate constants for the enzyme-catalyzed reaction? The equilibrium constant? The $\Delta G^{\circ\prime}$?

Selected Readings for this chapter can be found online at www.whfreeman.com/tymoczko3e.

CHAPTER 7

Kinetics and Regulation

7.1 Kinetics Is the Study of Reaction Rates

7.2 The Michaelis–Menten Model Describes the Kinetics of Many Enzymes

7.3 Allosteric Enzymes Are Catalysts and Information Sensors

7.4 Enzymes Can Be Studied One Molecule at a Time

Much of life is motion, whether at the macroscopic level of our daily life or at the molecular level of a cell. Studying motion is what motivated Eadweard James Muybridge to use stop-motion photography to analyze the gallop of a horse in 1878. In biochemistry, kinetics (derived from the Greek *kinesis*, meaning "movement") is used to capture the dynamics of enzyme activity. [Image Select/Art Resource, NY]

The primary function of enzymes is to accelerate the rates, or velocities, of reactions so that they are compatible with the needs of the organism. Thus, to understand how enzymes function, we need a kinetic description of their activity. This description will help us to quantify such kinetic parameters as how fast an enzyme can operate, how fast it will operate at substrate concentrations found in a cell, and what substrate is most readily operated on by the enzyme.

Some enzymes, however, need to do more than enhance the velocity of reactions. Metabolism in the cell is a complex array of dozens of metabolic pathways composed of thousands of different reactions, each catalyzed by a different enzyme. If all of these reactions were to take place in an unregulated fashion, metabolic chaos would result. An important class of enzymes called *allosteric enzymes* prevents this chaos and allows for the efficient integration of metabolism. These remarkable enzymes are not only catalysts, but also information sensors. They sense signals in the environment that allow them to adjust the rates of their reactions to meet the metabolic needs of the cell and facilitate the efficient coordination of the various metabolic pathways.

In this chapter, we will examine a kinetic model that describes the activity of a class of enzymes called *Michaelis–Menten enzymes* that do not display sophisticated regulatory properties. Then, we will consider another class of enzymes, the allosteric enzymes, that are highly regulated. We will see many examples of these information-sensing enzymes throughout our study of biochemistry.

✓ 3 Explain what reaction velocity is.

7.1 Kinetics Is the Study of Reaction Rates

The study of the rates of chemical reactions is called *kinetics,* and the study of the rates of enzyme-catalyzed reactions is called *enzyme kinetics.* We begin by briefly examining some of the basic principles of reaction kinetics.

What do we mean when we say the "velocity" or "rate" of a chemical reaction? Consider a simple reaction:

$$A \rightarrow P$$

The velocity of the reaction, V (for velocity), is the quantity of reactant A that disappears in a specified unit of time t. It is equal to the velocity of the appearance of product P, or the quantity of P that appears in a specified unit of time:

$$V = -d[A]/dt = d[P]/dt \qquad (1)$$

where d is the decrease in substrate concentration or the increase in product concentration. If A is yellow and P is colorless, we can follow the decrease in the concentration of A by measuring the decrease in the intensity of yellow color with time. Consider only the change in the concentration of A for now. The velocity of the reaction is directly related to the concentration of A by a proportionality constant k, called the *rate constant*:

$$V = k[A] \qquad (2)$$

Reactions in which the velocity is directly proportional to the reactant concentration are called *first-order reactions*. First-order rate constants have the unit of s^{-1} (per second).

Many important biochemical reactions are biomolecular; that is, they include two reactants. Such reactions are called *second-order reactions*. For example,

$$2 A \rightarrow P$$

or

$$A + B \rightarrow P$$

The corresponding rate equations often take the form

$$V = k[A]^2 \qquad (3)$$

and

$$V = k[A][B] \qquad (4)$$

The rate constants, called second-order rate constants, have the units $M^{-1} s^{-1}$ (per mole per second).

Sometimes, second-order reactions can appear to be first-order reactions. For instance, in reaction 4, if the concentration of B greatly exceeds that of A and if A is present at low concentrations, the reaction rate will be first order with respect to A and will not appear to depend on the concentration of B. These reactions are called *pseudo-first-order reactions*, and we will see them a number of times in our study of biochemistry.

Interestingly enough, under some conditions, a reaction can be zero order. In these cases, the rate is independent of reactant concentrations. As we will see shortly, enzyme-catalyzed reactions can approximate zero-order reactions under some circumstances.

7.2 The Michaelis–Menten Model Describes the Kinetics of Many Enzymes

The initial velocity of catalysis, which is defined as the number of moles of product formed per second shortly after the reaction has begun, varies with the substrate concentration [S] when enzyme concentration is constant, in the manner shown in **Figure 7.1**. Examining how velocity changes in response to changes in substrate concentration is a reasonable way to study enzyme activity because variation in substrate concentration depends on environmental circumstances (for instance, after a meal), but enzyme concentration is relatively constant, especially on the time scale of reaction rates.

Before we can fully interpret this graph, we need to examine some of the parameters of the reaction. Consider an enzyme E that catalyzes the conversion of S into P by the following reaction:

$$E + S \underset{k_{-1}}{\overset{k_1}{\rightleftharpoons}} ES \underset{k_{-2}}{\overset{k_2}{\rightleftharpoons}} E + P \tag{5}$$

where k_1 is the rate constant for the formation of the enzyme–substrate (ES) complex, k_2 is the rate constant for the formation of product P, and k_{-1} and k_{-2} are the constants for the respective reverse reactions. **Figure 7.2** shows that the amount of product formed is determined as a function of time for a series of substrate concentrations. As expected, in each case, the amount of product formed increases with time, although eventually a time is reached when there is *no net change* in the concentration of S or P. The enzyme is still actively converting substrate into product and vice versa, but the reaction equilibrium has been attained.

We can simplify the entire discussion of enzyme kinetics if we ignore the reverse reaction of product forming substrate. We can define the rate of catalysis V_0, or the initial rate of catalysis, as the number of moles of product formed per second when the reaction is just beginning—that is, when $t \approx 0$ and thus $[P] \approx 0$. Thus, for the graph in Figure 7.2, V_0 is determined for each substrate concentration by measuring the rate of product formation at early times before P accumulates. With the use of this approximation, reaction 5 can be simplified to the following reaction scheme:

$$E + S \underset{k_{-1}}{\overset{k_1}{\rightleftharpoons}} ES \xrightarrow{k_2} E + P \tag{6}$$

Enzyme E combines with substrate S to form an ES complex, with a rate constant k_1. The ES complex has two possible fates. It can dissociate to E and S, with a rate constant k_{-1}, or it can proceed to form product P, with a rate constant k_2.

In 1913, on the basis of this reaction scheme and with some simple assumptions, Leonor Michaelis and Maud Menten proposed a simple model to account for these kinetic characteristics. The critical feature in their treatment is that *a specific ES complex is a necessary intermediate in catalysis.* The notion that an enzyme needed to bind substrate before catalysis could take place was not evident to the early biochemists. Some believed that enzymes might release emanations that converted substrate into product.

The *Michaelis–Menten equation* describes the variation of enzyme activity as a function of substrate concentration. The derivation of the equation from the terms just described can be found in the Appendix at the end of this chapter. This equation is

$$V_0 = V_{max} \frac{[S]}{[S] + K_M} \tag{7}$$

✓ 4 Explain how reaction velocity is determined and how reaction velocities are used to characterize enzyme activity.

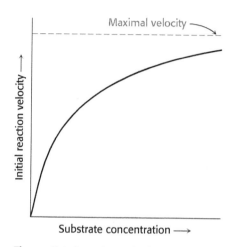

Figure 7.1 Reaction velocity versus substrate concentration in an enzyme-catalyzed reaction. An enzyme-catalyzed reaction approaches a maximal velocity.

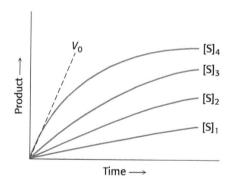

Figure 7.2 Determining initial velocity. The amount of product formed at different substrate concentrations is plotted as a function of time. The initial velocity (V_0) for each substrate concentration is determined from the slope of the curve at the beginning of a reaction, when the reverse reaction is insignificant. The initial velocity is illustrated for substrate concentration $[S]_4$.

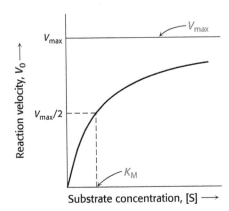

Figure 7.3 Michaelis–Menten kinetics.
A plot of the reaction velocity V_0 as a function of the substrate concentration [S] for an enzyme that obeys Michaelis–Menten kinetics shows that the maximal velocity V_{max} is approached asymptotically. The Michaelis constant K_M is the substrate concentration yielding a velocity of $V_{max}/2$.

? QUICK QUIZ 1 What value of [S], as a fraction of K_M, is required to obtain 80% V_{max}?

where

$$K_M = \frac{k_{-1} + k_2}{k_1} \qquad (8)$$

K_M, a compilation of rate constants called the *Michaelis constant*, is unique to each enzyme and is independent of enzyme concentration. K_M describes the properties of the enzyme–substrate interaction and thus will vary for enzymes that can use different substrates. The maximal velocity possible, V_{max}, can be attained only when all of the enzyme (total enzyme, or E_T) is bound to substrate (S):

$$V_{max} = k_2[E]_T \qquad (9)$$

V_{max} is directly dependent on enzyme concentration. Equation 7 describes the typical Michaelis–Menten curve as illustrated in **Figure 7.3**. At very low substrate concentrations, when [S] is much less than the value of K_M, the velocity is directly proportional to the substrate concentration; that is, $V_0 = (V_{max}/K_M)[S]$. At high substrate concentrations, when [S] is much greater than K_M, $V_0 \approx V_{max}$; that is, the velocity is maximal, independent of substrate concentration. When an enzyme is operating at V_{max}, all of the available enzyme is bound to substrate; the addition of more substrate will not affect the velocity of the reaction. The enzyme is displaying zero-order kinetics. Under these conditions, the enzyme is said to be *saturated*.

Consider the circumstances when $V_0 = V_{max}/2$. Under these conditions, the practical meaning of K_M is evident from equation 7. Using $V_0 = V_{max}/2$, and solving for [S], we see that [S] $= K_M$ at $V_0 = V_{max}/2$. Thus, *K_M is equal to the substrate concentration at which the reaction velocity is half its maximal value.* K_M is an important characteristic of an enzyme-catalyzed reaction and is significant for its biological function. Determination of V_{max} and K_M for an enzyme-catalyzed reaction is often one of the first characterizations of an enzyme undertaken.

⚕ CLINICAL INSIGHT

Variations in K_M Can Have Physiological Consequences

The physiological consequence of K_M is illustrated by the sensitivity of some persons to ethanol (**Figure 7.4**). Such persons exhibit facial flushing and rapid heart rate (tachycardia) after ingesting even small amounts of alcohol. In the liver, alcohol dehydrogenase converts ethanol into acetaldehyde:

Figure 7.4 Ethanol in alcoholic beverages is converted into acetaldehyde. [Angela Hampton Picture Library/Alamy.]

$$\underset{\text{Ethanol}}{CH_3CH_2OH} + NAD^+ \underset{\text{dehydrogenase}}{\overset{\text{Alcohol}}{\rightleftharpoons}} \underset{\text{Acetaldehyde}}{CH_3CHO} + NADH + H^+$$

Normally, the acetaldehyde, which is the cause of the symptoms when present at high concentrations, is processed to acetate by aldehyde dehydrogenase:

$$CH_3CHO + NAD^+ + H_2O \underset{\text{dehydrogenase}}{\overset{\text{Aldehyde}}{\rightleftharpoons}} CH_3COO^- + NADH + 2H^+$$

Most people have two forms of the aldehyde dehydrogenase, a low K_M mitochondrial form and a high K_M cytoplasmic form. In susceptible persons, the mitochondrial enzyme is less active owing to the substitution of a single amino acid, and so acetaldehyde is processed only by the cytoplasmic enzyme. Because the cytoplasmic enzyme has a high K_M, it achieves a high rate of catalysis only at very high concentrations of acetaldehyde. Consequently, less acetaldehyde is converted into acetate; excess acetaldehyde escapes into the blood and accounts for the physiological effects.

K_M and V_{max} Values Can Be Determined by Several Means

K_M is equal to the substrate concentration that yields $V_{max}/2$; however, V_{max}, like perfection, is approached but never attained. How, then, can we experimentally determine K_M and V_{max}, and how do these parameters enhance our understanding of enzyme-catalyzed reactions? The Michaelis constant K_M and the maximal velocity V_{max} can be readily derived from rates of catalysis measured at a variety of substrate concentrations if an enzyme operates according to the simple scheme given in equation 7. The derivation of K_M and V_{max} is most commonly achieved with the use of curve-fitting programs on a computer. However, an older method is a source of further insight into the meaning of K_M and V_{max}.

Before the availability of computers, the determination of K_M and V_{max} values required algebraic manipulation of the basic Michaelis–Menten equation. The Michaelis–Menten equation can be transformed into one that gives a straight-line plot. Taking the reciprocal of both sides of equation 7 gives the *Lineweaver–Burk equation*:

$$\frac{1}{V_0} = \frac{K_M}{V_{max}} \cdot \frac{1}{S} + \frac{1}{V_{max}} \tag{10}$$

A plot of $1/V_0$ versus $1/[S]$, called a *double-reciprocal plot*, yields a straight line with a y-intercept of $1/V_{max}$ and a slope of K_M/V_{max} (Figure 7.5). The intercept on the x axis is $-1/K_M$. This method is now rarely used because the data points at high and low concentrations are weighted differently and thus sensitive to errors.

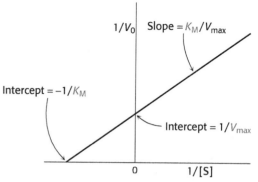

Figure 7.5 A double-reciprocal, or Lineweaver–Burk, plot. A double-reciprocal plot of enzyme kinetics is generated by plotting $1/V_0$ as a function $1/[S]$. The slope is K_M/V_{max}, the intercept on the vertical axis is $1/V_{max}$, and the intercept on the horizontal axis is $-1/K_M$.

K_M and V_{max} Values Are Important Enzyme Characteristics

The K_M values of enzymes range widely (Table 7.1). For most enzymes, K_M lies between 10^{-1} and 10^{-7} M. The K_M value for an enzyme depends on the particular substrate and on environmental conditions such as pH, temperature, and ionic strength. The Michaelis constant K_M, as already noted, is equal to the concentration of substrate at which half the active sites are filled. Thus, K_M provides a measure of the substrate concentration required for significant catalysis to take place. For many enzymes, experimental evidence suggests that the K_M value provides an approximation of substrate concentration in vivo, which in turn suggests that most enzymes evolved to have significant activity at the substrate concentration commonly available. We can speculate about why an enzyme might evolve to have a K_M value that corresponds to the substrate concentration normally available to the enzyme. If the normal concentration of substrate is approximately equal to K_M, the enzyme will display significant activity and yet the activity will be sensitive to changes in environmental conditions—that is, changes in substrate concentration. At values below K_M, enzymes are very sensitive to changes in substrate concentration but display little activity. At substrate values well above K_M, enzymes have great catalytic activity but are insensitive to changes in substrate concentration. Thus, with the normal substrate concentration being approximately K_M, the enzymes have significant activity ($1/2$ V_{max}) but are still sensitive to changes in substrate concentration.

Table 7.1 K_M values of some enzymes

Enzyme	Substrate	K_M (μM)
Chymotrypsin	Acetyl-L-tryptophanamide	5000
Lysozyme	Hexa-N-acetylglucosamine	6
β-Galactosidase	Lactose	4000
Carbonic anhydrase	CO_2	8000
Penicillinase	Benzylpenicillin	50

Table 7.2 Turnover numbers of some enzymes

Enzyme	Turnover number (per second)
Carbonic anhydrase	600,000
3-Ketosteroid isomerase	280,000
Acetylcholinesterase	25,000
Penicillinase	2,000
Lactate dehydrogenase	1,000
Chymotrypsin	100
DNA polymerase I	15
Tryptophan synthetase	2
Lysozyme	0.5

The maximal velocity V_{max} reveals the *turnover number* of an enzyme, which is *the number of substrate molecules that an enzyme can convert into product per unit time when the enzyme is fully saturated with substrate.* The turnover number is equal to the rate constant k_2, which is also called k_{cat}. If the total concentration of active sites, $[E]_T$, is known, then

$$V_{max} = k_2[E]_T \tag{11}$$

and thus

$$k_2 = V_{max}/[E]_T \tag{12}$$

For example, a 10^{-6} M solution of carbonic anhydrase catalyzes the formation of 0.6 M H_2CO_3 per second when the enzyme is fully saturated with substrate. Hence, k_2 is $6 \times 10^5 \text{ s}^{-1}$. This value is one of the largest known turnover numbers. Each catalyzed reaction takes place in a time equal to, on average, $1/k_2$, which is 1.7 μs for carbonic anhydrase. The turnover numbers of most enzymes with their physiological substrates fall in the range from 1 to 10^4 per second (Table 7.2).

k_{cat}/K_M Is a Measure of Catalytic Efficiency

When the substrate concentration is much greater than K_M, the velocity of catalysis approaches V_{max}. However, in the cell, most enzymes are not normally saturated with substrate. Under physiological conditions, the amount of substrate present is often between 10% and 50% K_M. Thus, the $[S]/K_M$ ratio is typically between 0.01 and 1.0. When $[S] \ll K_M$, the enzymatic rate is much less than k_{cat} (k_2) because most of the active sites are unoccupied. We can derive an equation that characterizes the kinetics of an enzyme under these cellular conditions:

$$V_0 = \frac{k_{cat}}{K_M}[E][S] \tag{13}$$

When $[S] \ll K_M$, almost all active sites are empty. In other words, the concentration of free enzyme, $[E]$, is nearly equal to the total concentration of enzyme, $[E]_T$; so

$$V_0 = \frac{k_{cat}}{K_M}[S][E]_T \tag{14}$$

Thus, when $[S] \ll K_M$, the enzymatic velocity depends on the values of k_{cat}/K_M, $[S]$, and $[E]_T$. Under these conditions, k_{cat}/K_M is the rate constant for the interaction of S and E. The rate constant k_{cat}/K_M, called the *specificity constant*, is a measure of catalytic efficiency because it takes into account both the rate of catalysis with a particular substrate (k_{cat}) and the nature of the enzyme–substrate interaction (K_M). For instance, by using k_{cat}/K_M values, we can compare an enzyme's preference for different substrates. Table 7.3 shows the k_{cat}/K_M values for several different substrates of chymotrypsin, a digestive enzyme secreted by the pancreas. Chymotrypsin clearly has a preference for cleaving next to bulky, hydrophobic side chains (p. 144).

Table 7.3 Substrate preferences of chymotrypsin

Amino acid in ester	Amino acid side chain	k_{cat}/K_M (s^{-1} M^{-1})
Glycine	—H	1.3×10^{-1}
Valine	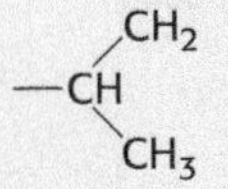 —CH with CH$_2$/CH$_3$	2.0
Norvaline	—$CH_2CH_2CH_3$	3.6×10^2
Norleucine	—$CH_2CH_2CH_2CH_3$	3.0×10^3
Phenylalanine	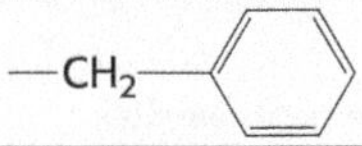 —CH_2—⟨benzene ring⟩	1.0×10^5

Source: Data from A. Fersht, *Structure and Mechanism in Protein Science: A Guide to Enzyme Catalysis and Protein Folding* (W.H. Freeman and Company, 1999), Table 6.3.

How efficient can an enzyme be? We can approach this question by determining whether there are any physical limits on the value of k_{cat}/K_M. Note that the k_{cat}/K_M ratio depends on k_1, k_{-1}, and k_{cat}, as can be shown by substituting for K_M:

$$k_{cat}/K_M = \frac{k_{cat}k_1}{k_{-1} + k_{cat}} = \left(\frac{k_{cat}}{k_{-1} + k_{cat}}\right)k_1 < k_1 \tag{15}$$

Suppose that the rate of formation of product (k_{cat}) is much faster than the rate of dissociation of the ES complex (k_{-1}). The value of k_{cat}/K_M then approaches k_1. Thus, the ultimate limit on the value of k_{cat}/K_M is set by k_1, the rate of formation of the ES complex. *This rate cannot be faster than the diffusion-controlled encounter of an enzyme and its substrate.* Diffusion limits the value of k_1, and so it cannot be higher than 10^8 to 10^9 s^{-1} M^{-1}. Hence, the upper limit on k_{cat}/K_M is between 10^8 and 10^9 s^{-1} M^{-1}.

The k_{cat}/K_M ratios of the enzymes superoxide dismutase, acetylcholinesterase, and triose phosphate isomerase are between 10^8 and 10^9 s^{-1} M^{-1}. Enzymes that have k_{cat}/K_M ratios at the upper limits have attained *kinetic perfection. Their catalytic velocity is restricted only by the rate at which they encounter substrate in the solution* (Table 7.4).

Most Biochemical Reactions Include Multiple Substrates

The simplest way to explain Michaelis–Menten kinetics is to use a one-substrate reaction as an example. However, most reactions in biological systems start with two substrates and yield two products. They can be represented by the following bisubstrate reaction:

$$A + B \rightleftharpoons P + Q$$

Table 7.4 Enzymes for which k_{cat}/K_M is close to the diffusion-controlled rate of encounter

Enzyme	k_{cat}/K_M (s^{-1} M^{-1})
Acetylcholinesterase	1.6×10^8
Carbonic anhydrase	8.3×10^7
Catalase	4×10^7
Crotonase	2.8×10^8
Fumarase	1.6×10^8
Triose phosphate isomerase	2.4×10^8
β-Lactamase	1×10^8
Superoxide dismutase	7×10^9

Source: Data from A. Fersht, *Structure and Mechanism in Protein Science: A Guide to Enzyme Catalysis and Protein Folding* (W. H. Freeman and Company, 1999), Table 4.5.

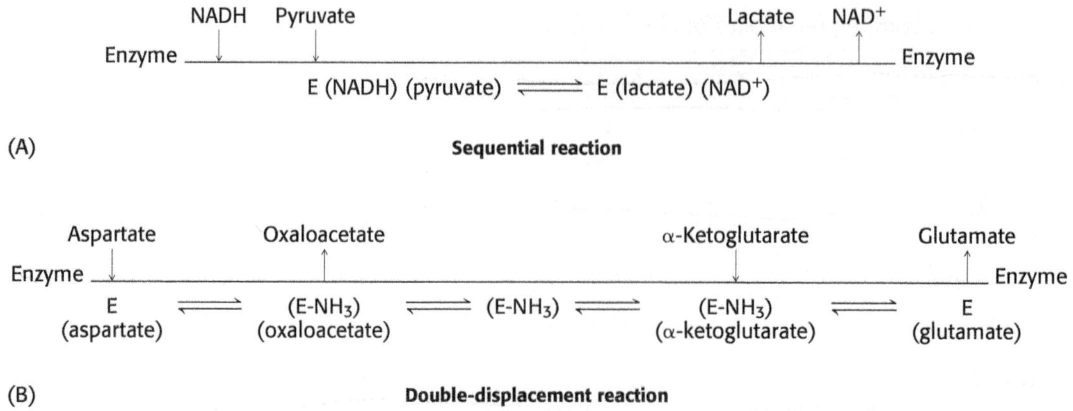

NADH   Pyruvate   Lactate   NAD$^+$

Enzyme ─┴──────┴───────────────────────────┴─────┴── Enzyme

E (NADH) (pyruvate) $\rightleftharpoons$ E (lactate) (NAD$^+$)

(A)   **Sequential reaction**

Aspartate   Oxaloacetate α-Ketoglutarate   Glutamate

Enzyme ─┴──────────┴───────────────────────┴──────────┴── Enzyme

E $\rightleftharpoons$ (E-NH$_3$) $\rightleftharpoons$ (E-NH$_3$) $\rightleftharpoons$ (E-NH$_3$) $\rightleftharpoons$ E
(aspartate)   (oxaloacetate)   (α-ketoglutarate)   (glutamate)

(B)   **Double-displacement reaction**

Figure 7.6 Cleland representations of bisubstrate reactions. (A) Sequential reaction. The first substrate (NADH) binds to the enzyme, followed by the second substrate (pyruvate) to form a ternary complex of two substrates and the enzyme. Catalysis then takes place, forming a ternary complex of two products and the enzyme. The products subsequently leave sequentially. (B) Double displacement. The first substrate (aspartate) binds, and the first catalytic step takes place, resulting in a substituted enzyme (E-NH$_3$). The first product (oxaloacetate) then leaves. The second substrate (α-ketoglutarate) binds to the substituted enzyme. The second catalytic step takes place, and the NH$_3$ is transferred to the substrate to form the final product glutamate, which departs the enzyme.

Many such reactions transfer a functional group, such as a phosphoryl or an ammonium group, from one substrate to the other. Those that are oxidation–reduction reactions transfer electrons between substrates. Although equations can be developed to describe the kinetics of multiple-substrate reactions, we will forego a kinetic description of such reactions and simply examine some general principles of bisubstrate reactions.

Multiple-substrate reactions can be divided into two classes: *sequential reactions* and *double-displacement reactions* (**Figure 7.6**). The depictions shown in Figure 7.6 are called Cleland representations.

Sequential reactions In sequential reactions (Figure 7.6A), all substrates must bind to the enzyme before any product is released. Consequently, in a bisubstrate reaction, *a ternary complex* consisting of the enzyme and both substrates forms. Sequential mechanisms are of two types: ordered, in which the substrates bind the enzyme in a defined sequence, and random.

Double-displacement (ping-pong) reactions In double-displacement, or ping-pong, reactions (Figure 7.6B), one or more products are released before all substrates bind the enzyme. The defining feature of double-displacement reactions is the existence of a *substituted enzyme intermediate,* in which the enzyme is temporarily modified. Reactions that shuttle amino groups between amino acids and α-ketoacids are classic examples of double-displacement mechanisms. As shown in Figure 7.6, the substrates and products appear to bounce on and off the enzyme just as a Ping-Pong ball bounces on and off a table.

✓ 5 Identify the key properties of allosteric proteins, and describe the structural basis for these properties.

7.3 Allosteric Enzymes Are Catalysts and Information Sensors

Enzymes that conform to simple Michaelis–Menten kinetics are called Michaelis–Menten enzymes. These enzymes are not regulated in the cell, and their activity is governed simply by mass action: if substrate is present, they catalyze. Most enzymes in the cell are Michaelis–Menten enzymes.

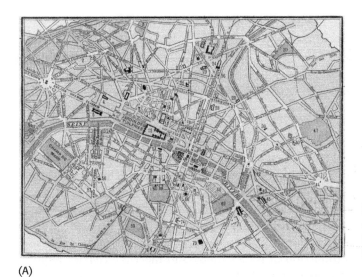

(A)

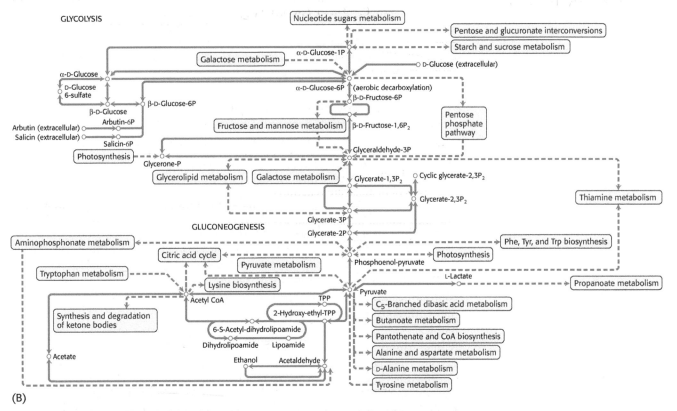

Figure 7.7 Complex interactions require regulation. (A) A part of a street map of Paris. (B) A schematic representation of several interconnecting metabolic pathways in plants. [(A) RoseOfSharon/Alamy Images.]

(B)

As important as catalysis is to the functioning of a cell, catalysis alone is not sufficient. The vast array of reaction pathways also need to be regulated so that the biochemical pathways function in a coherent fashion. **Figure 7.7**A is a street map of part of Paris. Every day, hundreds of thousands of vehicles travel these streets. Although the streets of Paris are notoriously difficult to navigate by car, imagine how much more difficult navigation would be without stop signs, stoplights, and traffic police. The metabolic map shown in Figure 7.7B shows that metabolic traffic also is immensely complex, and, like street traffic, metabolic traffic requires regulation. An effective means of regulating metabolic pathways is to regulate enzyme activity. Enzymes that regulate the flux of biochemicals through metabolic pathways are called *allosteric enzymes*, for reasons to be discussed shortly. Key features of allosteric enzymes include the regulation of catalytic activity by environmental signals, including the final product of the metabolic pathway regulated by the enzyme; kinetics that are more complex than those of Michaelis–Menten enzymes; and quaternary structure with multiple active sites in each enzyme.

Allosteric Enzymes Are Regulated by Products of the Pathways Under Their Control

Let us begin our consideration of allosteric enzymes by imagining what control processes would be useful for the efficient functioning of metabolic pathways.

We begin by examining a hypothetical metabolic pathway:

$$A \xrightarrow{e_1} B \xrightarrow{e_2} C \xrightarrow{e_3} D \xrightarrow{e_4} E \xrightarrow{e_5} F$$

There are five reactions in this metabolic pathway, catalyzed by five distinct enzymes, denoted e_1 through e_5. As is typical in real metabolic situations, the end product F is needed in limited amounts and cannot be stored. The initial reactant A is valuable and should be conserved unless F is needed. Finally, B, C, D, and E have no biological roles except as chemical intermediates in the synthesis of F. This last condition means that the first reaction, A → B, is the *committed step* in this pathway; after this reaction has taken place, B is committed to conversion into F. How can the production of F be regulated to meet the cellular requirements without making more than is needed?

In the simplest situation, when sufficient F is present, F can bind reversibly to e_1, the enzyme catalyzing the committed step, and inhibit the reaction, an effect called *feedback inhibition*.

$$A \xrightarrow[e_1]{(-)} B \xrightarrow{e_2} C \xrightarrow{e_3} D \xrightarrow{e_4} E \xrightarrow{e_5} F$$

Feedback inhibition is a common means of biochemical regulation. Feedback inhibitors usually bear no structural resemblance to the substrate or the product of the enzyme that they inhibit. Moreover, feedback inhibitors do not bind at the active site but rather at a distinct regulatory site on the allosteric enzyme. Allosteric (from the Greek *allos,* meaning "other," and *stereos,* meaning "structure") enzymes are so-named because they are regulated by molecules that bind to sites other than the active site. *Allosteric enzymes always catalyze the committed step of metabolic pathways.*

Another level of complexity is required when metabolic pathways must communicate with one another. Consider the pathway shown in **Figure 7.8** for the synthesis of K from two separate pathways producing F and I. How can these pathways be coordinated to produce the appropriate amount of K? Compound K might inhibit enzymes e_1 and e_{10} by feedback inhibition, inasmuch as these two enzymes control the committed step for each of the two required pathways. But, in this case, a second wasteful situation could arise: if more F were produced than I, some of the F molecules would be wasted because there would not be enough I to bind to F. Thus, how might the concentrations of F and I be balanced? Compound F could stimulate e_{10} but inhibit e_1 to balance the amount of I with F. Likewise, compound I could inhibit e_{10} but stimulate e_1. Thus, allosteric enzymes can recognize inhibitor molecules as well as stimulatory molecules.

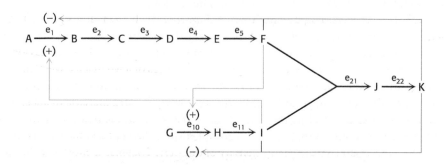

Figure 7.8 Two pathways cooperate to form a single product.

Allosterically Regulated Enzymes Do Not Conform to Michaelis–Menten Kinetics

Allosteric enzymes are distinguished by their response to changes in substrate concentration in addition to their susceptibility to regulation by other molecules. A typical velocity-versus-substrate curve for an allosteric enzyme is shown in **Figure 7.9**. The curve differs from what is expected of an enzyme that conforms to Michaelis–Menten kinetics because of the sharp increase in V_0 in the middle of the curve. The observed curve is referred to as sigmoidal because it resembles the letter S.

Allosteric Enzymes Depend on Alterations in Quaternary Structure

Thus far, we know two properties that are unique to allosteric enzymes: (1) regulation of catalytic activity and (2) sigmoidal kinetics. Is there a relation between these two unique characteristics of allosteric enzymes? Indeed there is, and the relation is best elucidated by first examining a model for how sigmoidal kinetics might be displayed.

We will consider one possible model, called the *concerted model* or the *MWC model* after Jacques Monod, Jefferies Wyman, and Jean-Pierre Changeux, the biochemists who developed it. The model is based on several premises. First, allosteric enzymes have multiple active sites on different polypeptide chains. Second, the enzyme can exist in two distinct conformations or states. One state, designated R for relaxed, is the active conformation, which catalyzes reactions. The other state, designated T for tense, is significantly less active. In the absence of substrate or signal molecules, R and T are in equilibrium, with T being the more stable state and thus the more common state. The T/R ratio, which is called *the allosteric constant* (L_0), is in the hundreds for a typical allosteric enzyme. However, the difference in stability is small enough that thermal jostling of the cellular environment provides enough energy to occasionally power the spontaneous conversion of a T form into an R form. Third, the concerted model also requires that all of the subunits or active sites of the enzyme must be in the same state; that is, all must be T or all must be R. No hybrids are allowed. This requirement is called the *symmetry rule*. Finally, substrate (S) binds more readily to the R form of the enzyme than to the T form.

How can we use this information to explain the sigmoidal nature of allosteric kinetics? Consider a population of allosteric enzymes with each enzyme containing four active sites on four subunits. Because it is more stable, most of the enzymes will be in the T state. The binding of S to the T form is difficult; thus, there will be little activity at low substrate concentrations (**Figure 7.10**). However, if the substrate concentration is increased, eventually enough S will be present so that, when a relaxed form of the enzyme spontaneously appears, S will bind to it. Because of the symmetry rule, if one S binds to the R form, all of the four potential active sites become trapped in the R form. Consequently, the next S to bind the enzyme will not have to unproductively collide with the many T forms, because R forms of the enzymes are accumulating owing to the binding of the initial S to the enzyme. *The binding of substrate disrupts the* T $\rightleftharpoons$ R *equilibrium in favor of* R. This behavior is called *cooperativity* and accounts for the sharp increase in V_0 of the velocity-versus-substrate concentration curve.

What is the physiological significance of cooperativity, seen as a sigmoidal kinetics? Allosteric enzymes transition from a less active state to a more active state within a narrow range of substrate concentration. The benefit of this behavior is illustrated in **Figure 7.11**, which compares the kinetics of a Michaelis–Menten enzyme (blue curve) to that of an allosteric enzyme (red curve). In this example, the Michaelis–Menten enzyme requires an approximately 27-fold increase in substrate concentration to increase V_0 from $0.1\ V_{max}$ to $0.8\ V_{max}$. In contrast, the allosteric enzyme requires only about a 4-fold increase in substrate concentration to attain the same increase in velocity. *The activity of allosteric enzymes is more sensitive to*

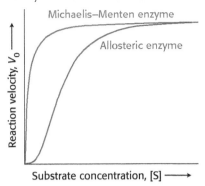

Figure 7.9 Kinetics for an allosteric enzyme. Allosteric enzymes display a sigmoidal dependence of reaction velocity on substrate concentration in contrast to the hyperbolic curve seen with Michaelis–Menten enzymes.

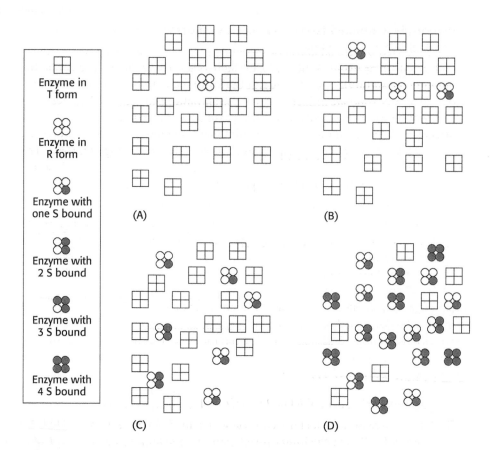

Figure 7.10 The concerted model for allosteric enzymes. (A) [T] >>> [R], meaning that L_0 is large. Consequently, it will be difficult for S to bind to an R form of the enzyme. (B) As the concentration of S increases, it will bind to one of the active sites on R, trapping all of the other active sites in the R state (by the symmetry rule.) (C) As more active sites are trapped in the R state, it becomes easier for S to bind to the R state. (D) The binding of S to R becomes easier yet as more of the enzyme is in the R form. In a velocity-versus-[S] curve, V_0 will be seen to rise rapidly toward V_{max}.

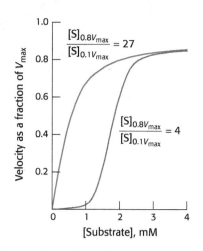

Figure 7.11 Allosteric enzymes display threshold effects. As the T-to-R transition occurs, the velocity increases over a narrower range of substrate concentration for an allosteric enzyme (red curve) than for a Michaelis–Menten enzyme (blue curve).

QUICK QUIZ 2 What would be the effect of a mutation in an allosteric enzyme that resulted in a T/R ratio of 0?

changes in substrate concentration near K_M than are Michaelis–Menten enzymes with the same V_{max}. This sensitivity is called a *threshold effect*: below a certain substrate concentration, there is little enzyme activity. However, after the threshold has been reached, enzyme activity increases rapidly. In other words, much like an "on or off" switch, cooperativity ensures that most of the enzyme is either on (R state) or off (T state). The vast majority of allosteric enzymes display sigmoidal kinetics.

Regulator Molecules Modulate the $T \rightleftharpoons R$ Equilibrium

How can we explain the signal-sensing capabilities of allosteric enzymes with the concerted model? *The regulatory molecules alter the equilibrium between T and R forms.* That is, they change the relative proportion of enzymes in the T and R states. A positive effector binds to the R form at a regulatory site, distinct from the active site, and stabilizes this form, thus increasing the concentration of R and making an R and S interaction more likely. A negative effector binds to T and stabilizes it; a negative effector increases the concentration of T and thereby decreases the likelihood of an R binding to an S. Figure 7.12 shows the effects of a positive and a negative regulator on the kinetics of an allosteric enzyme. The positive effector lowers the threshold concentration of substrate needed for activity, whereas the negative effector raises the threshold concentration. The effects of regulatory molecules on allosteric enzymes are referred to as *heterotropic effects*. Heterotropic effectors shift the sigmoidal curve to the left (activators) or right (inhibitors). In contrast, the effects of substrates on allosteric enzymes are referred to as *homotropic effects*. Homotropic effects account for the sigmoidal nature of the kinetics curve.

It is all but impossible to overestimate the importance of allosteric enzymes in biological systems. They respond immediately and specifically to chemical signals produced elsewhere in the cell. They are receivers and transducers of chemical information, allowing cross talk between metabolic pathways, not only within the cell, but also between cells and organs. The complexity of biological systems is due to the sensing of the environment by allosteric enzymes.

The Sequential Model Also Can Account for Allosteric Effects

In the concerted model, an allosteric enzyme can exist in only two states, T and R; no intermediate or hybrid states are allowed. An alternative model posits that the subunits of the allosteric enzyme undergo sequential changes in structure. The binding of substrate to one site influences the substrate binding to neighboring sites without necessarily inducing a transition encompassing the entire enzyme (Figure 7.13). The sequential model more readily accommodates negative cooperativity, in which the binding of one substrate decreases the affinity of other sites for the substrate. The results of studies on a number of allosteric proteins suggest that many behave according to some combination of the sequential and concerted models.

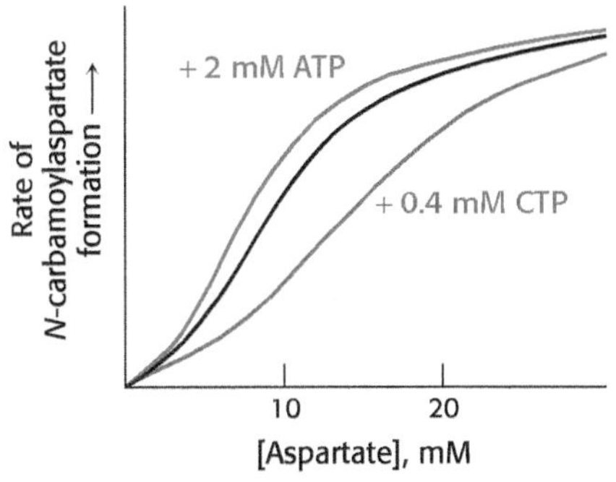

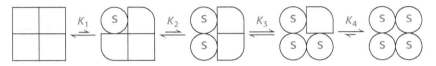

Figure 7.13 The sequential model. The binding of a substrate (S) changes the conformation of the subunit to which it binds. This conformational change induces changes in neighboring subunits of the allosteric enzyme that increase their affinity for the substrate. K_1, K_2, etc., are rate constants for the binding of substrate to the different states of the enzyme.

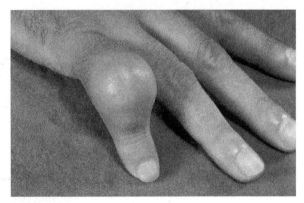

Figure 7.12 The effect of regulators on the allosteric enzyme aspartate transcarbamoylase. ATP is an allosteric activator of aspartate transcarbamoylase because it stabilizes the R state, making it easier for substrate to bind. As a result, the curve is shifted to the left, as shown in blue. Cytidine triphosphate (CTP) stabilizes the T state of aspartate transcarbamoylase, making it more difficult for substrate binding to convert the enzyme into the R state. As a result, the curve is shifted to the right, as shown in red.

🜊 CLINICAL INSIGHT

Loss of Allosteric Control May Result in Pathological Conditions

The importance of the allosteric control of enzymes can be seen in the pathway leading to the joint disease *gout*. In this disease, an excess of urate crystallizes in the fluid and lining of the joints, resulting in painful inflammation when cells of the immune system engulf the sodium urate crystals (Figure 7.14). Urate is a final product of the degradation of purines, the base components of the adenine and guanine nucleotides of DNA and RNA.

Although gout can be caused by a number of metabolic deficiencies (Chapter 31), one interesting cause is the loss of allosteric regulation by an important enzyme in purine synthesis. Phosphoribosylpyrophosphate synthetase (PRS) catalyzes the synthesis of phosphoribosylpyrophosphate (PRPP), which is a vital precursor for all nucleotide synthesis:

$$\text{Ribose 5-phosphate} + \text{ATP} \xrightarrow{\substack{\text{Phosphoribosylpyrophosphate} \\ \text{synthetase}}}$$
$$\text{5-phosphoribosyl-1-pyrophosphate} + \text{AMP}$$

PRS is normally feedback inhibited by purine nucleotides. However, in certain people, the regulatory site has undergone a mutation that renders PRS insensitive to feedback inhibition. The catalytic activity of PRS is unaffected and, in some cases, is increased, leading to a glut of purine nucleotides, which are converted into urate. The excess urate accumulates and causes gout.

Figure 7.14 A gout-inflamed joint. [Medical-on-Line/Alamy.]

7.4 Enzymes Can Be Studied One Molecule at a Time

Most experiments that are performed to determine an enzyme characteristic require an enzyme preparation in a buffered solution. Even a few microliters of such a solution will contain millions of enzyme molecules. Much that we have learned about enzymes thus far has come from such experiments, called *ensemble studies*. A basic assumption of ensemble studies is that all of the enzyme molecules

(A)

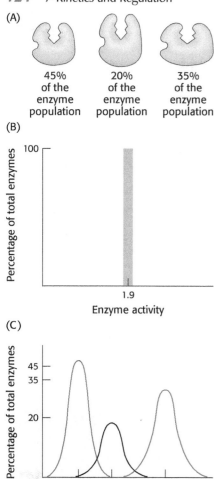

| 45% of the enzyme population | 20% of the enzyme population | 35% of the enzyme population |

(B)

(C)

Figure 7.15 Single-molecule studies can reveal molecular heterogeneity.

are the same or very similar. When we determine an enzyme property such as the value of K_M in ensemble studies, that value is of necessity an average value of all of the enzyme molecules present. However, we know that molecular heterogeneity—the ability of a molecule, with the passage of time, to assume several different structures that differ slightly in stability—is an inherent property of all large biomolecules (p. 62). How can we tell whether this molecular heterogeneity affects enzyme activity?

Consider this hypothetical situation. A Martian visits Earth to learn about higher education. The spacecraft hovers high above a university, and the Martian meticulously records how the student population moves about campus. Much information can be gathered from such studies: where students are likely to be at certain times on certain days; which buildings are used when and by how many. Now, suppose our visitor developed a high-magnification camera that could follow one student throughout the day. Such data would provide a much different perspective on college life: What does this student eat? To whom does she speak with? How much time does she spend studying? This new *in singulo* method, examining one individual at a time, yields a host of new information but also illustrates a potential pitfall of studying individuals, be they students or enzymes: How can we be certain that the student or molecule is representative and not an outlier? This pitfall can be avoided by studying enough individuals to satisfy statistical analysis for validity.

Let us leave the Martian to his observations and consider a more biochemical situation. **Figure 7.15**A shows an enzyme that displays molecular heterogeneity, with three active forms that catalyze the same reaction but at different rates. These forms have slightly different stabilities, but thermal noise is sufficient to power conversion among the forms. Each form is present as a fraction of the total enzyme population as indicated. If we were to perform an experiment to determine enzyme activity under a particular set of conditions with the use of ensemble methods, we would get a single value, which would represent the average of the heterogeneous assembly (Figure 7.15B). However, were we to perform a sufficient number of single-molecule experiments, we would discover that the enzyme has three different molecular forms with very different activities (Figure 7.15C). Moreover, these different forms would most likely correspond to important biochemical differences.

The development of powerful techniques has enabled biochemists to look into the workings of individual molecules. We are now able to observe events at a molecular level that reveal rare or transient structures and fleeting events in a reaction sequence, as well as to measure mechanical forces affecting or generated by an enzyme. Single-molecule studies open a new vista on the function of enzymes in particular and on all large biomolecules in general.

SUMMARY

7.1 Kinetics Is the Study of Reaction Rates

The velocity of a reaction is determined by the concentration of reactant and a proportionality constant called the rate constant. Reactions that are directly proportional to the reactant concentration are called first-order reactions. First-order rate constants have the units of s^{-1}. In many reactions in biochemistry, there are two reactants. Such reactions are called bimolecular reactions. The rate constants for biomolecular reactions, called second-order rate constants, have the units $M^{-1}s^{-1}$.

7.2 The Michaelis–Menten Model Describes the Kinetics of Many Enzymes

The kinetic properties of many enzymes are described by the Michaelis–Menten model. In this model, an enzyme (E) combines with a substrate (S) to form an enzyme–substrate (ES) complex, which can proceed to form a product (P) or to dissociate into E and S. The velocity V_0 of formation of product is given by the Michaelis–Menten equation:

$$V_0 = V_{max} \frac{[S]}{[S] + K_M}$$

in which V_{max} is the reaction velocity when the enzyme is fully saturated with substrate and K_M, the Michaelis constant, is the substrate concentration at which the reaction velocity is half maximal. The maximal velocity V_{max} is equal to the product of k_2, or k_{cat}, and the total concentration of enzyme. The kinetic constant k_{cat}, called the turnover number, is the number of substrate molecules converted into product per unit time at a single catalytic site when the enzyme is fully saturated with substrate. Turnover numbers for most enzymes are between 1 and 10^4 per second. The ratio of k_{cat}/K_M provides information about enzyme efficiency.

7.3 Allosteric Enzymes Are Catalysts and Information Sensors

Allosteric enzymes constitute an important class of enzymes whose catalytic activity can be regulated. These enzymes, which do not conform to Michaelis–Menten kinetics, have multiple active sites. These active sites display cooperativity, as evidenced by a sigmoidal dependence of reaction velocity on substrate concentration. Regulators of allosteric enzymes can stimulate enzyme activity or inhibit enzyme activity.

7.4 Enzymes Can Be Studied One Molecule at a Time

Many enzymes are now being studied in singulo, at the level of a single molecule. Such studies are important because they yield information that is difficult to obtain in studies of populations of molecules. Single-molecule methods reveal a distribution of enzyme characteristics rather than an average value as is acquired with the use of ensemble methods.

APPENDIX: Derivation of the Michaelis–Menten Equation

As already discussed, the key feature of the Michaelis–Menten treatment of enzyme kinetics is that a specific enzyme–substrate (ES) complex is a necessary intermediate in catalysis. The model proposed is

$$E + S \underset{k_{-1}}{\overset{k_1}{\rightleftharpoons}} ES \underset{k_{-2}}{\overset{k_2}{\rightleftharpoons}} E + P$$

An enzyme E combines with substrate S to form an ES complex, with a rate constant k_1. The ES complex has two possible fates. It can dissociate to E and S, with a rate constant k_{-1}, or it can proceed to form product P, with a rate constant k_2. The ES complex can also be re-formed from E and P by the reverse reaction with a rate constant k_{-2}. However, as already discussed, we simplify these reactions by considering the velocity of reaction at times close to zero, when there is negligible product formation and thus no reverse reaction ($k_{-2}[S][P] \approx 0$):

$$E + S \underset{k_{-1}}{\overset{k_1}{\rightleftharpoons}} ES \overset{k_2}{\rightarrow} E + P$$

We want an expression that relates the velocity of catalysis to the concentrations of substrate and enzyme and the rates of the individual steps. Our starting point is that the catalytic velocity is equal to the product of the concentration of the ES complex and k_2:

$$V_0 = k_2[ES] \tag{16}$$

Now we need to express [ES] in terms of known quantities. The rates of formation and breakdown of ES are given by

$$\text{Rate of formation of ES} = k_1[E][S] \tag{17}$$

$$\text{Rate of breakdown of ES} = (k_{-1} + k_2)[ES] \tag{18}$$

To simplify matters further, we use the *steady-state assumption*. In a steady state, the concentrations of intermediates—in this case, [ES]—stay the same even if the concentrations of starting materials and products are changing. This steady state is achieved when the rates of formation and breakdown of the ES complex are equal. Setting the right-hand sides of equations 17 and 18 equal gives

$$k_1[ES] = (k_{-1} + k_2)[ES] \tag{19}$$

By rearranging equation 19, we obtain

$$[E][S]/[ES] = (k_{-1} + k_2)/k_1 \tag{20}$$

Equation 20 can be simplified by defining a new constant, K_M, called the *Michaelis constant*:

$$K_M = \frac{k_{-1} + k_2}{k_1} \qquad (21)$$

Note that K_M has the units of concentration and is independent of enzyme and substrate concentrations.

Inserting equation 21 into equation 20 and solving for [ES] yields

$$[ES] = \frac{[E][S]}{K_M} \qquad (22)$$

Now, let us examine the numerator of equation 22. Because the substrate is usually present at much higher concentration than the enzyme, the concentration of uncombined substrate [S] is very nearly equal to the total substrate concentration. The concentration of uncombined enzyme [E] is equal to the total enzyme concentration $[E]_T$ minus the concentration of the ES complex:

$$[E] = [E]_T - [ES] \qquad (23)$$

Substituting this expression for [E] in equation 22 gives

$$[ES] = \frac{([E]_T - [ES])[S]}{K_M} \qquad (24)$$

Solving equation 24 for [ES] gives

$$[ES] = \frac{[E]_T[S]/K_M}{1 + [S]/K_M} \qquad (25)$$

or

$$[ES] = [E]_T \frac{[S]}{[S] + K_M} \qquad (26)$$

By substituting this expression for [ES] into equation 16, we obtain

$$V_0 = k_2[E]_T \frac{[S]}{[S] + K_M} \qquad (27)$$

The maximal velocity V_{max} is attained when the catalytic sites on the enzyme are saturated with substrate—that is, when $[ES] = [E]_T$. Thus,

$$V_{max} = k_2[E]_T \qquad (28)$$

Substituting equation 28 into equation 27 yields the *Michaelis–Menten equation*:

$$V_0 = V_{max} \frac{[S]}{[S] + K_M} \qquad (29)$$

KEY TERMS

kinetics (p. 112)
Michaelis–Menten equation (p. 113)
Michaelis constant (K_M) (p. 114)
Lineweaver–Burk equation (p. 115)

turnover number (p. 116)
k_{cat}/K_M (p. 116)
sequential reaction (p. 118)

double-displacement (ping-pong)
  reaction (p. 118)
allosteric enzyme (p. 118)

? Answers to QUICK QUIZZES

1. Use $0.8\ V_{max}$ as V_0 and solve for [S]. $S = 4\ K_M$.

2. All of the enzyme would be in the R form all of the time. There would be no cooperativity. The kinetics would look like that of a Michaelis–Menten enzyme.

PROBLEMS

1. *Different orders.* Differentiate between a first-order rate constant and a second-order rate constant. ✓ 3

2. *A fraudulent reaction?* What is a pseudo-first-order reaction? ✓ 3

3. *Changing in concert.* On the graph at the right, show how substrate and product concentrations of the simple enzyme-catalyzed reaction

$$S \longrightarrow P$$

change with time. ✓ 3

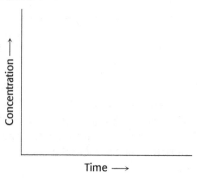

(a) For a reaction in which the equilibrium is in favor of product formation.

(b) For a reaction in which the equilibrium constant is 1.

4. *Active yet responsive.* What is the biochemical advantage of having a K_M approximately equal to the substrate concentration normally available to an enzyme? ✓ 4

5. *Defining attributes.* What is the defining characteristic for an enzyme catalyzing a sequential reaction? A double-displacement reaction? ✓ 4

6. *Affinity or not affinity? That is the question.* The affinity between a protein and a molecule that binds to the protein is frequently measured as a dissociation constant K_D.

Protein–small-molecule complex $\rightleftharpoons$
protein + small molecule

$$K_D = \frac{[\text{protein}][\text{small molecule}]}{[\text{protein–small-molecule complex}]}$$

Does K_M measure the affinity of the enzyme complex? Under what circumstances might K_M approximately equal K_D? ✓ 4

7. *Destroying the Trojan horse.* Penicillin is hydrolyzed and thereby rendered inactive by penicillinase (also known as β-lactamase), an enzyme present in some resistant bacteria. The mass of this enzyme in *Staphylococcus aureus* is 29.6 kDa. The amount of penicillin hydrolyzed in 1 minute in a 10-ml solution containing 10^{-9} g of purified penicillinase was measured as a function of the concentration of penicillin. Assume that the concentration of penicillin does not change appreciably during the assay. ✓ 4

[Penicillin] μM	Amount hydrolyzed (nanomoles)
1	0.11
3	0.25
5	0.34
10	0.45
30	0.58
50	0.61

(a) Plot V_0 versus [S] and $1/V_0$ versus $1/$[S] for these data. Does penicillinase appear to obey Michaelis–Menten kinetics? If so, what is the value of K_M?

(b) What is the value of V_{max}?

(c) What is the turnover number of penicillinase under these experimental conditions? Assume one active site per enzyme molecule.

8. *Hydrolytic driving force.* The hydrolysis of pyrophosphate to orthophosphate is important in driving forward biosynthetic reactions such as the synthesis of DNA. This hydrolytic reaction is catalyzed in *Escherichia coli* by a pyrophosphatase that has a mass of 120 kDa and consists of six identical subunits. For this enzyme, a unit of activity is defined as the amount of enzyme that hydrolyzes 10 μmol of pyrophosphate in 15 minutes at 37°C under standard assay conditions. The purified enzyme has a V_{max} of 2800 units per milligram of enzyme. ✓ 4

(a) How many moles of substrate are hydrolyzed per second per milligram of enzyme when the substrate concentration is much greater than K_M?

(b) How many moles of active sites are there in 1 mg of enzyme? Assume that each subunit has one active site.

(c) What is the turnover number of the enzyme? Compare this value with others mentioned in this chapter.

9. *A fresh view.* The plot of $1/V_0$ versus $1/$[S] is sometimes called a double-reciprocal or Lineweaver–Burk plot. Another way of expressing the kinetic data is to plot V_0 versus $V_0/$[S], which is known as an Eadie–Hofstee plot. ✓ 4

(a) Rearrange the Michaelis–Menten equation to give V_0 as a function of $V_0/$[S].

(b) What is the significance of the slope, the vertical intercept, and the horizontal intercept in a plot of V_0 versus $V_0/$[S]?

10. *More Michaelis–Menten.* For an enzyme that follows simple Michaelis–Menten kinetics, what is the value of V_{max} if V_0 is equal to 1 mmol minute^{-1} when [S] $= 1/10\, K_M$? ✓ 4

11. *Fractional occupation.* Starting with the Michaelis–Menten equation, derive an equation that shows the fraction of active sites filled (f_{ES}) at any V_0. ✓ 4

12. *They go together like spaghetti and meatballs.* Match the term with the proper description.

(a) Enzyme _____

(b) Kinetics _____

(c) Michaelis–Menten equation _____

(d) Michaelis constant (K_M) _____

(e) Lineweaver–Burk equation _____

(f) Turnover number _____

(g) k_{cat}/K_M _____

(h) Sequential reaction _____

(i) Double-displacement (ping-pong) reaction _____

(j) Allosteric enzyme _____

1. Substrate concentration that yields $1/2\,V_{max}$

2. k_2 or k_{cat}

3. Responds to environmental signals

4. The study of reaction rates

5. Describes the kinetics of simple one-substrate reactions

6. A ternary complex is formed

7. Protein catalyst

8. Double-reciprocal plot

9. A measure of enzyme efficiency

10. Includes a substituted enzyme intermediate

13. *A new view of cooperativity.* Draw a double-reciprocal plot for a typical Michaelis–Menten enzyme and an allosteric enzyme that have the same V_{max}. ✓ 5

14. *Angry biochemists.* Many biochemists go bananas, and justifiably, when they see a Michaelis–Menten plot like the one shown below. To see why they go bananas, determine the V_0 as a fraction of V_{max} when the substrate concentration is equal to 10 K_M and 20 K_M. Please control your outrage. ✓ 4

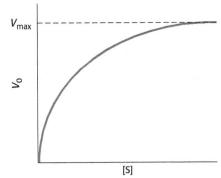

15. *Knowing when to say when.* What is feedback inhibition? Why is it a useful property? ✓ 5

16. *Turned upside down.* An allosteric enzyme that follows the concerted mechanism has a T/R ratio of 300 in the absence of substrate. Suppose that a mutation reversed the ratio. How would this mutation affect the relation between the velocity of the reaction and the substrate concentration? ✓ 5

17. *RT equilibrium.* Differentiate between homotropic and heterotropic effectors. ✓ 5

18. *Restoration project.* Aspartate transcarbamoylase (ATCase) is an allosteric enzyme that regulates the synthesis of uridine triphosphate (UTP) and cytidine triphosphate (CTP). It can be separated into regulatory subunits and catalytic subunits. If isolated regulatory subunits and catalytic subunits of ATCase are mixed, the native enzyme is reconstituted. What is the biological significance of the observation? ✓ 5

19. *Negative cooperativity.* You have isolated a dimeric enzyme that contains two identical active sites. The binding of substrate to one active site decreases the substrate affinity of the other active site. Can the concerted model account for this negative cooperativity? ✓ 5

Data Interpretation Problems

20. *A natural attraction, but more complicated.* You have isolated two versions of the same enzyme, a wild type and a mutant differing from the wild type at a single amino acid. Working carefully but expeditiously, you then establish the following kinetic characteristics of the enzymes. ✓ 4

	Maximum velocity	K_M
Wild type	100 μmol/min	10 mM
Mutant	1 μmol/min	0.1 mM

(a) With the assumption that a two-step reaction in which k_{-1} is much larger than k_2, which enzyme has the higher affinity for substrate?

(b) What is the initial velocity of the reaction catalyzed by the wild-type enzyme when the substrate concentration is 10 mM?

(c) Which enzyme alters the equilibrium more in the direction of product?

21. K_M *matters.* The amino acid asparagine is required by cancer cells to proliferate. Treating patients with the enzyme asparaginase is sometimes used as a chemotherapy treatment. Asparaginase hydrolyzes asparagine to aspartate and ammonia. The adjoining illustration shows the Michaelis–Menten curves for two asparaginases from different sources, as well as the concentration of asparagine in the environment (indicated by the arrow). Which enzyme would make a better chemotherapeutic agent? ✓ 4

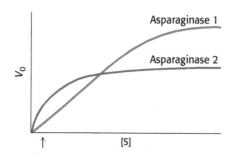

22. *Varying the enzyme.* For a one-substrate, enzyme-catalyzed reaction, double-reciprocal plots were determined for three different enzyme concentrations. Which of the following three families of curve would you expect to be obtained? Explain.

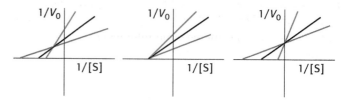

23. *Rate-limiting step.* In the conversion of A into D in the following biochemical pathway, enzymes E_A, E_B, and E_C have the K_M values indicated under each enzyme. If all of the substrates and products are present at a concentration of 10^{-4} M and the enzymes have approximately the same V_{max}, which step will be rate limiting and why? ✓ 4

$$A \underset{E_A}{\rightleftarrows} B \underset{E_B}{\rightleftarrows} C \underset{E_C}{\rightleftarrows} D$$

$$K_M = 10^{-2}\,M \quad 10^{-4}\,M \quad 10^{-4}\,M$$

24. *Enzyme specificity.* Catalysis of the cleavage of peptide bonds in small peptides by a proteolytic enzyme is described in the following table. ✓ 4

Substrate*	K_M (mM)	k_{cat} (s^{-1})
EMTA↓G	4.0	24
EMTA↓A	1.5	30
EMTA↓F	0.5	18

*See Chapter 2 for amino acid abbreviations.

The arrow indicates the peptide bond cleaved in each case.

(a) If a mixture of these peptides were presented to the enzyme with the concentration of each peptide being the same, which peptide would be digested most effectively? Least effectively? Briefly explain your reasoning, if any.

(b) The experiment is performed again on another peptide with the following results:

EMTI↓ F	9	18

On the basis of these data, suggest the features of the amino acid sequence that dictate the specificity of enzyme.

Challenge Problems

25. *Competing substrates.* Suppose that two substrates, A and B, compete for an enzyme. Derive an expression relating the ratio of the rates of utilization of A and B, V_A/V_B, to the concentrations of these substrates and their values of k_{cat} and K_M. (Hint: Express V_A as a function of k_{cat}/K_M for substrate A, and do the same for V_B.) Is specificity determined by K_M alone? ✓ 4

26. *Colored luminosity.* Tryptophan synthetase, a bacterial enzyme that contains a pyridoxal phosphate (PLP) prosthetic

group, catalyzes the synthesis of L-tryptophan from L-serine and an indole derivative. The addition of L-serine to the enzyme produces a marked increase in the fluorescence of the PLP group, as the adjoining graph shows. The subsequent addition of indole, the second substrate, reduces this fluorescence to a level even lower than that produced by the enzyme alone. How do these changes in fluorescence support the notion that the enzyme interacts directly with its substrates? ✓ 5

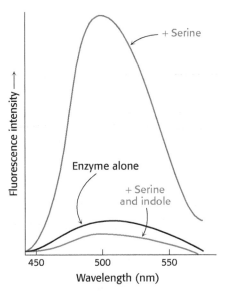

27. Too much of a good thing. A simple Michaelis–Menten enzyme, in the absence of any inhibitor, displayed the following kinetic behavior. The expected value of V_{max} is shown on the y axis. ✓ 4 ✓ 5

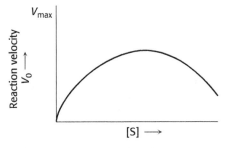

(a) Draw a double-reciprocal plot that corresponds to the velocity-versus-substrate curve.

(b) Provide an explanation for the kinetic results.

28. Paradoxical at first glance. Phosphonacetyl-L-aspartate (PALA) is a potent inhibitor of the allosteric enzyme aspartate transcarbamoylase (ATCase), which has six active sites, because PALA mimics the two physiological substrates. ATCase controls the synthesis of pyrimidine nucleotides. However, low concentrations of this unreactive bisubstrate analog *increase* the reaction velocity. On the addition of PALA, the reaction velocity increases until an average of three molecules of PALA are bound per molecule of enzyme. This maximal velocity is 17-fold greater than it is in the absence of PALA. The reaction velocity then decreases to nearly zero on the addition of three more molecules of PALA per molecule of enzyme. Why do low concentrations of PALA activate ATCase? ✓ 5

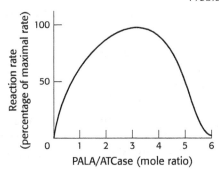

29. Distinguishing between models. The following graph shows the fraction of an allosteric enzyme in the R state (f_R) and the fraction of active sites bound to substrate (Y) as a function of substrate concentration. Which model, the concerted or sequential, best explains these results? ✓ 5

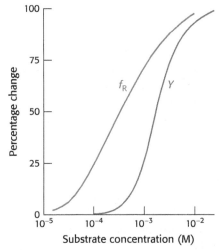

[Data from M. W. Kirschner and H. K. Schachman, *Biochemistry* 12:2997–3004, 1996.]

30. Reporting live from ATCase 1. The allosteric enzyme aspartate transcarbamoylase (ATCase) has six active sites, arranged as two catalytic trimers. ATCase was modified with tetranitromethane to form a colored nitrotyrosine group (λ_{max} = 430 nm) in each of its catalytic chains. The absorption by this reporter group depends on its immediate environment. An essential lysine residue at each catalytic site also was modified to block the binding of substrate. Catalytic trimers from this doubly modified enzyme were then combined with native trimers to form a hybrid enzyme. The absorption by the nitrotyrosine group was measured on addition of the substrate analog succinate. What is the significance of the alteration in the absorbance at 430 nm? ✓ 5

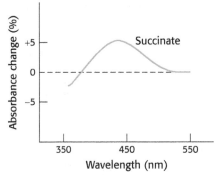

[Data from H. K. Schachman, *J. Biol. Chem.* 263:18583–18586, 1988.]

31. *Reporting live from ATCase 2.* A different ATCase hybrid was constructed to test the effects of allosteric activators and inhibitors. Normal regulatory subunits were combined with nitrotyrosine-containing catalytic subunits. The addition of ATP, an allosteric activator of ATCase, in the absence of substrate increased the absorbance at 430 nm, the same change elicited by the addition of succinate (see the graph in problem 30). Conversely, CTP, an allosteric inhibitor, in the absence of substrate decreased the absorbance at 430 nm. What is the significance of the changes in absorption of the reporter groups? ✓ 5

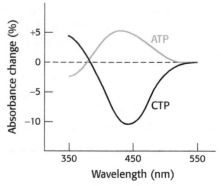

[Data from H. K. Schachman, *J. Biol. Chem.* 263:18583–18586, 1988.]

Selected Readings for this chapter can be found online at www.whfreeman.com/tymoczko3e.

CHAPTER 8

Mechanisms and Inhibitors

Chess and enzymes have in common the use of strategy, consciously thought out in the game of chess and selected by evolution for the action of an enzyme. In the chess match depicted here, each player struggles to develop strategies to checkmate the opponent. Likewise, enzymes have catalytic strategies, developed over evolutionary time, for binding their substrates and chemically acting on them. [Bridgeman-Giraudon/Art Resource, NY.]

Thus far, in our study of enzymes, we have learned that an enzyme binds a substrate at the active site to facilitate the formation of the transition state; we have developed a kinetic model for simple enzymes; and we learned that allosteric enzymes are not only catalysts, but also information sensors. We now turn our attention to the catalytic strategies of enzymes and then to an examination of how enzyme activity can be altered by environmental factors distinct from allosteric signals. Finally, we will consider the catalytic mechanism of the digestive enzyme chymotrypsin, one of the first enzymes understood in mechanistic detail.

8.1 A Few Basic Catalytic Strategies Are Used by Many Enzymes

As we will see in our study of biochemistry, enzymes catalyze a vast array of chemical reactions. Despite this diversity, all enzymes operate by facilitating the formation of the transition state. How the transition state is formed varies from

✓ 6 List environmental factors that affect enzyme activity, and describe how these factors exert their effects on enzymes.

enzyme to enzyme, but most enzymes commonly employ one or more of the following strategies to catalyze specific reactions:

1. *Covalent Catalysis.* In covalent catalysis, the active site contains a reactive group, usually a powerful nucleophile that becomes temporarily covalently modified in the course of catalysis. The proteolytic enzyme chymotrypsin provides an excellent example of this mechanism (p. 140).

2. *General Acid–Base Catalysis.* In general acid–base catalysis, a molecule other than water plays the role of a proton donor or acceptor. Chymotrypsin uses a histidine residue as a base catalyst to enhance the nucleophilic power of serine (p. 142).

3. *Metal Ion Catalysis.* Metal ions can function catalytically in several ways. For instance, a metal ion may serve as an electrophilic catalyst, stabilizing a negative charge on a reaction intermediate. Alternatively, a metal ion may generate a nucleophile by increasing the acidity of a nearby molecule, such as water. Finally, a metal ion may bind to the substrate, increasing the number of interactions with the enzyme and thus the binding energy. Metal ions are required cofactors for many of the enzymes that we will encounter in our study of biochemistry.

4. *Catalysis by Approximation and Orientation.* Many reactions include two distinct substrates. In such cases, the reaction rate may be considerably enhanced by bringing the two substrates into proximity and in the proper orientation on a single binding surface of an enzyme.

Recall from Chapter 6 our consideration of another key aspect of enzyme catalysis—*binding energy.* The full complement of binding interactions between an enzyme and a substrate is formed only when the substrate is in the transition state. The fact that binding energy is maximal when the enzyme binds to the transition state favors the formation of the transition state and thereby promotes catalysis.

8.2 Enzyme Activity Can Be Modulated by Temperature, pH, and Inhibitory Molecules

Regardless of which mechanism or mechanisms are employed by an enzyme to catalyze a reaction, the rate of catalysis is affected by the same environmental parameters that affect all chemical reactions, such as temperature and pH. Moreover, some chemicals that interact specifically with the elaborate three-dimensional structure of the enzyme also can affect enzyme activity.

Temperature Enhances the Rate of Enzyme-Catalyzed Reactions

As the temperature rises, the rate of most reactions increases. The rise in temperature increases the Brownian motion of the molecules, which makes interactions between an enzyme and its substrate more likely. For most enzymes, there is a temperature at which the increase in catalytic activity ceases and there is a precipitous loss of activity (**Figure 8.1**). What is the basis of this

Figure 8.1 The effect of heat on enzyme activity. Enzyme activity increases with temperature until the enzyme is denatured. The enzyme tyrosinase, which is part of the pathway that synthesizes the pigment that results in dark fur, has a low tolerance for heat in Siamese cats. It is inactive at normal body temperatures but functional at slightly lower temperatures. The extremities of a Siamese cat are cool enough for tyrosinase to gain function and produce pigment. [Photograph: Jane Burton/Dorling Kindersley/Getty Images.]

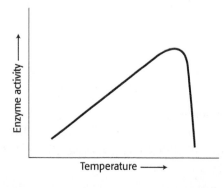

loss of activity? Recall from Chapter 4 that proteins have a complex three-dimensional structure that is held together by weak bonds. When the temperature increases beyond a certain point, the bonds maintaining the three-dimensional structure are not strong enough to withstand the polypeptide chain's thermal jostling and the protein loses the structure required for activity. The protein is said to be *denatured* (p. 60).

In organisms such as ourselves that maintain a constant body temperature (endotherms), the effect of outside temperature on enzyme activity is minimized. However, in organisms that assume the temperature of the ambient environment (ectotherms), temperature is an important regulator of biochemical and, indeed, biological activity. Lizards, for instance, are most active in warmer temperatures and relatively inactive in cooler temperatures, a behavioral manifestation of biochemical activity (**Figure 8.2**). Although endotherms are not as sensitive to ambient temperature as ectotherms, slight tissue temperature alterations are sometimes important. For instance, when athletes "warm-up," they are increasing the temperature of their muscles through exertion. This increase in temperature facilitates the biochemistry that will power the exercise.

Some organisms, such as thermophilic archaea, can live at temperatures of 80°C or higher, temperatures that would denature most proteins. The proteins in these organisms have evolved to be very resistant to thermal denaturation (**Figure 8.3**).

Most Enzymes Have an Optimal pH

Enzyme activity also often varies with pH, the H^+ concentration of the environment (p. 26). The activity of most enzymes displays a bell-shaped curve when examined as a function of pH. However, the optimal pH—the pH at which enzymes display maximal activity—varies with the enzyme and is correlated with the environment of the enzyme. For instance, the protein-digesting enzyme pepsin functions in the highly acidic environment of the stomach, where the pH is between 1 and 2. Most proteins are denatured at this pH, but pepsin functions very effectively. The enzymes of the pancreatic secretion of the upper small intestine, such as chymotrypsin, have pH optima near pH 8, in keeping with the pH of the intestine in this region (**Figure 8.4**).

How can we account for the pH effect on enzyme activity in regard to our understanding of enzymes in particular and proteins in general? Imagine an enzyme that requires the ionization of both a glutamic acid residue and a lysine residue at the active site for the enzyme to be functional. Thus, the enzyme would depend on the presence of a —COO^- group as well as an —NH_3^+ group (**Figure 8.5**). If the pH is lowered (the H^+ concentration increases), the —COO^-

Figure 8.2 A lizard basking in the sun. Ectotherms, such as the Namibian rock agama (*Agama planiceps*), adjust their body temperatures and, hence, the rate of biochemical reactions behaviorally. [Morales/age fotostock.]

Figure 8.3 Thermophilic archaea and their environment. Archaea can thrive in habitats as harsh as a volcanic vent. Here, the archaea form an orange mat surrounded by yellow sulfurous deposits. [Images & Volcanos/Science Source.]

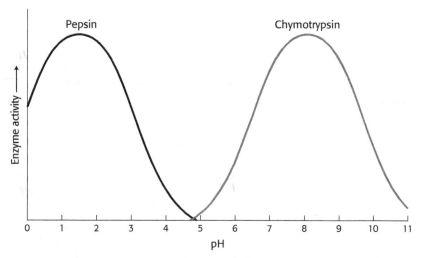

Figure 8.4 The pH dependence of the activity of the enzymes pepsin and chymotrypsin. Chymotrypsin and pepsin have different optimal pH values. The optimal pH for pepsin is noteworthy. Most proteins would be denatured at this acidic pH.

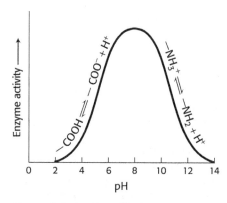

Figure 8.5 A pH profile for a hypothetical enzyme. The pH dependence of enzymes is due to the presence of ionizable R groups.

(A)

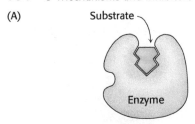

(B)

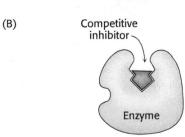

(C)

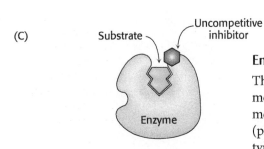

(D)

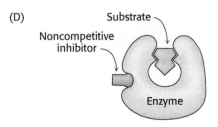

Figure 8.6 Reversible inhibitors. (A)
Substrate binds to an enzyme's active site to
form an enzyme–substrate complex. (B) A
competitive inhibitor binds at the active site
and thus prevents the substrate from binding.
(C) An uncompetitive inhibitor binds only to
the enzyme–substrate complex. (D) A
noncompetitive inhibitor does not prevent the
substrate from binding.

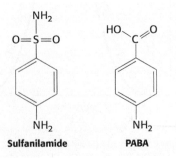

groups will gradually be converted into —COOH groups with a concomitant loss of enzyme activity. On the other side of the optimum, as the pH is raised (less H^+, more OH^-), the $-NH_3^+$ group loses an H^+ to OH^-, becoming a neutral $-NH_2$ group, and the enzyme activity is diminished. Often, the activity-versus-pH curves are due to several ionizable groups.

Is alteration of the pH of the cellular milieu ever used as a regulator device? The answer is yes, and a crucial enzyme in glucose metabolism in skeletal muscle provides an example. Phosphofructokinase (Chapter 16) controls the rate of metabolism of glucose under aerobic conditions (in the presence of oxygen) and under anaerobic conditions (in the absence of oxygen). A problem arises with the rapid processing of glucose in the absence of oxygen: the end product is lactic acid, which readily ionizes to lactate and a hydrogen ion. To prevent the muscle from becoming "pickled" by the high concentration of acid, the activity of phosphofructokinase decreases if the pH falls too drastically, which, in turn, reduces the metabolism of glucose to lactic acid. Phosphofructokinase is made up of multiple subunits, and the decrease in pH causes the subunits to dissociate, rendering the enzyme inactive and thus reducing lactic acid production.

Enzymes Can Be Inhibited by Specific Molecules

The activity of many enzymes can be inhibited by the binding of specific small molecules and ions. This means of enzyme inhibition serves as a major control mechanism in biological systems, typified by the regulation of allosteric enzymes (p. 118). In addition, many drugs and toxic agents act by inhibiting enzymes. This type of enzyme inhibition is not usually the result of evolutionary forces, as it is for allosteric enzymes, but rather due to design of inhibitors by scientists or simple chance discovery of inhibitory molecules. Examining inhibition can be a source of insight into the mechanism of enzyme action: specific inhibitors can often be used to identify residues critical for catalysis. Transition-state analogs are especially potent inhibitors.

Enzyme inhibition can be either reversible or irreversible. We begin the investigation of enzyme inhibition by first examining reversible inhibition. In contrast with irreversible inhibition, *reversible inhibition* is characterized by rapid dissociation of the enzyme–inhibitor complex. There are three common types of reversible inhibition: competitive inhibition, uncompetitive inhibition, and noncompetitive inhibition. These three types of inhibition differ in the nature of the interaction between the enzyme and the inhibitor and in the inhibitor's effect on enzyme kinetics (**Figure 8.6**). We will consider each of these types in turn.

In *competitive inhibition*, the inhibitor resembles the substrate and binds to the active site of the enzyme (Figure 8.6B). The substrate is thereby prevented from binding to the same active site. An enzyme can bind substrate (forming an ES complex) or inhibitor (EI), but not both (ESI). *A competitive inhibitor diminishes the rate of catalysis by reducing the proportion of enzyme molecules bound to a substrate.* At any given inhibitor concentration, competitive inhibition can be relieved by increasing the substrate concentration. Under these conditions, the substrate "outcompetes" the inhibitor for the active site.

Some competitive inhibitors are useful drugs. One of the earliest examples was the use of sulfanilamide as an antibiotic. Sulfanilamide is an example of a sulfa drug, a sulfur-containing antibiotic. Structurally, sulfanilamide mimics *p*-aminobenzoic acid (PABA), a metabolite required by bacteria for the synthesis of the coenzyme folic acid. Sulfanilamide binds to the enzyme that normally metabolizes PABA and competitively inhibits it, preventing folic acid synthesis. Human beings, unlike bacteria, absorb folic acid from the diet and are thus unaffected by the sulfa drug. Other competitive inhibitors commonly used as drugs include ibuprofen, which inhibits a cyclooxygenase that helps to generate the inflammatory response, and statins, which inhibit the key enzyme in cholesterol synthesis.

Uncompetitive inhibition is substrate-dependent inhibition in that the inhibitor binds only to the enzyme–substrate complex. The uncompetitive inhibitor's binding site is created only when the enzyme binds the substrate (Figure 8.6C). Uncompetitive inhibition cannot be overcome by the addition of more substrate. The herbicide glyphosate, also known as Roundup, is an uncompetitive inhibitor of an enzyme in the biosynthetic pathway for aromatic amino acids in plants. The plant dies because it lacks these amino acids.

In *noncompetitive inhibition*, the inhibitor and substrate can bind simultaneously to an enzyme molecule at different binding sites (Figure 8.6D). A noncompetitive inhibitor acts by decreasing the overall number of active enzyme molecules rather than by diminishing the proportion of enzyme molecules that are bound to substrate. Noncompetitive inhibition, in contrast with competitive inhibition, cannot be overcome by increasing the substrate concentration. Doxycycline, an antibiotic, functions at low concentrations as a noncompetitive inhibitor of a bacterial proteolytic enzyme (collagenase). Inhibition of this enzyme prevents the growth and reproduction of bacteria that cause gum (periodontal) disease.

Reversible Inhibitors Are Kinetically Distinguishable

How can we determine whether a reversible inhibitor acts by competitive, uncompetitive, or noncompetitive inhibition? Let us consider only enzymes that exhibit Michaelis–Menten kinetics—that is, enzymes that are not allosterically inhibited. Measurements of the rates of catalysis at different concentrations of substrate and inhibitor serve to distinguish the three types of reversible inhibition. In competitive inhibition, the inhibitor competes with the substrate for the active site. *The hallmark of competitive inhibition is that it can be overcome by a sufficiently high concentration of substrate* (Figure 8.7). The effect of a competitive inhibitor is to increase the apparent value of K_M, meaning that more substrate is needed to obtain the same reaction rate. This new apparent value of K_M is called K_M^{app}. In the presence of a competitive inhibitor, an enzyme will have the same V_{max} as in the absence of an inhibitor. At a sufficiently high concentration, virtually all the active sites are filled by substrate, and the enzyme is fully operative. The more inhibitor present, the more substrate is required to displace it and reach V_{max}.

In *uncompetitive inhibition*, the inhibitor binds only to the ES complex. This enzyme–substrate–inhibitor complex, ESI, does not proceed to form any product. Because some unproductive ESI complex will always be present, V_{max} will be lower in the presence of an inhibitor than in its absence (Figure 8.8). The uncompetitive inhibitor also lowers the apparent value of K_M, because the inhibitor binds to ES to form ESI, depleting ES. To maintain the equilibrium between E and ES, more S binds to E, increasing the apparent value of k_1 and thereby reducing the apparent value of K_M (see equation 8 on p. 114). Thus, a lower concentration of S is required to form half of the maximal concentration of ES, resulting in a reduction of the apparent value of K_M, now called K_M^{app}. Likewise, the value of V_{max} is decreased to a new value called V_M^{app}.

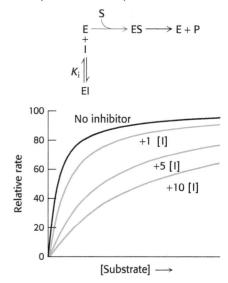

Figure 8.7 Kinetics of a competitive inhibitor. As the concentration of a competitive inhibitor increases, higher concentrations of substrate are required to attain a particular reaction velocity. The reaction pathway suggests how sufficiently high concentrations of substrate can completely relieve competitive inhibition.

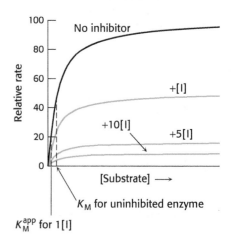

Figure 8.8 Kinetics of an uncompetitive inhibitor. The reaction pathway shows that the inhibitor binds only to the enzyme–substrate complex. Consequently, V_{max} cannot be attained, even at high substrate concentrations. Notice that the apparent value for $K_M (K_M^{app})$ is lowered, becoming smaller as more inhibitor is added.

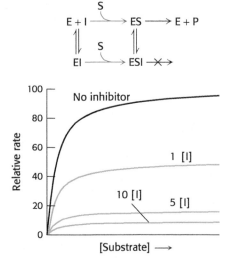

Figure 8.9 Kinetics of a noncompetitive inhibitor. The reaction pathway shows that the inhibitor binds both to free enzyme and to the enzyme–substrate complex. Consequently, as with uncompetitive competition, V_{max} cannot be attained. K_M remains unchanged, and so the reaction rate increases more slowly at low substrate concentrations than is the case for uncompetitive competition.

In *noncompetitive inhibition* (**Figure 8.9**), a substrate can bind to the enzyme–inhibitor complex as well as to the enzyme alone. In either case, the enzyme–inhibitor–substrate complex does not proceed to form product. The value of V_{max} is decreased to the new value V_{max}^{app}, whereas the value of K_M is unchanged. Why is V_{max} lowered although K_M remains unchanged? In essence, the inhibitor simply lowers the concentration of functional enzyme. The resulting solution behaves as a more dilute solution of enzyme. *Noncompetitive inhibition cannot be overcome by increasing the substrate concentration.*

Double-reciprocal plots are especially useful for distinguishing competitive, uncompetitive, and noncompetitive inhibitors. In competitive inhibition, the intercept on the y axis, $1/V_{max}$, is the same in the presence and in the absence of inhibitor, although the slope (K_M/V_{max}) is increased (**Figure 8.10**). The intercept is unchanged because a competitive inhibitor does not alter V_{max}. The increase in the slope of the $1/V_0$ versus $1/[S]$ plot indicates the strength of binding of competitive inhibitor.

In uncompetitive inhibition (**Figure 8.11**), the inhibitor combines only with the enzyme–substrate complex. The slope of the line is the same as that for the uninhibited enzyme, but the intercept on the y axis, $1/V_{max}$, is increased. Consequently, for uncompetitive inhibition, the lines in double-reciprocal plots are parallel.

In noncompetitive inhibition (**Figure 8.12**), the inhibitor can combine with either the enzyme or the enzyme–substrate complex. In pure noncompetitive inhibition, the values of the dissociation constants of the inhibitor and enzyme and of the inhibitor and enzyme–substrate complex are equal. The value of V_{max} is decreased to the new value V_{max}^{app}, and so the intercept on the vertical axis is increased. The slope when the inhibitor is present, which is equal to K_M/V_{max}^{app}, is larger by the same factor. In contrast with V_{max}, K_M is not affected by pure noncompetitive inhibition.

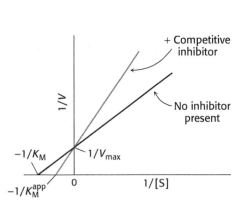

Figure 8.10 Competitive inhibition illustrated on a double-reciprocal plot. A double-reciprocal plot of enzyme kinetics in the presence and absence of a competitive inhibitor illustrates that the inhibitor has no effect on V_{max} but increases K_M.

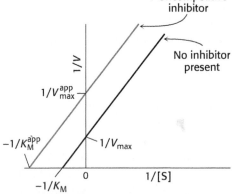

Figure 8.11 Uncompetitive inhibition illustrated on a double-reciprocal plot. An uncompetitive inhibitor does not affect the slope of the double-reciprocal plot. V_{max} and K_M are reduced by equivalent amounts.

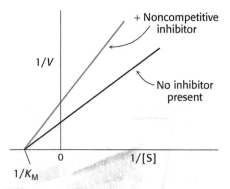

Figure 8.12 Noncompetitive inhibition illustrated on a double-reciprocal plot. A double-reciprocal plot of enzyme kinetics in the presence and absence of a noncompetitive inhibitor shows that K_M is unaltered and V_{max} is decreased.

Irreversible Inhibitors Can Be Used to Map the Active Site

Whereas a reversible inhibitor will both bind to an enzyme and dissociate from it rapidly, an *irreversible inhibitor* dissociates very slowly from its target enzyme because it has become tightly bound to the enzyme, either covalently or noncovalently. Some irreversible inhibitors are important drugs. Penicillin acts by covalently modifying the enzyme transpeptidase, thereby preventing the synthesis of bacterial cell walls and thus killing the bacteria (p. 138). Aspirin acts by covalently modifying the enzyme cyclooxygenase (the same enzyme inhibited by ibuprofen), reducing the synthesis of inflammatory signals.

Irreversible inhibitors that covalently bind to an enzyme are tools for elucidating the mechanism of enzymes. The first step in determining the chemical mechanism of an enzyme is to determine which functional groups are required for enzyme activity. Irreversible inhibitors modify functional groups, which can then be identified. If treatment with an irreversible inhibitor results in a loss of enzyme activity, then this loss suggests that the modified group is required for enzyme activity. Irreversible inhibitors can be assorted into four categories: group-specific reagents, affinity labels (substrate analogs), suicide inhibitors, and transition-state analogs.

Group-specific reagents modify specific R groups of amino acids. An example of a group-specific reagent is diisopropylphosphofluoridate (DIPF). DIPF inhibits the proteolytic enzyme chymotrypsin by modifying only 1 of the 28 serine residues in the protein, implying that this serine residue is especially reactive. As we will see on page 141, this serine residue is indeed at the active site. DIPF also revealed a reactive serine residue in acetylcholinesterase, an enzyme important in the transmission of nerve impulses (Figure 8.13). Thus, DIPF and similar compounds that bind and inactivate acetylcholinesterase are potent nerve gases.

? QUICK QUIZ 1 In the following graph, identify the curve that corresponds to each of the following conditions: no inhibition, competitive inhibition, noncompetitive inhibition, uncompetitive inhibition.

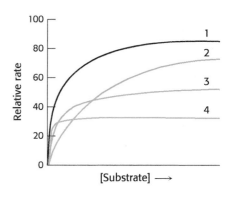

Figure 8.13 Enzyme inhibition by diisopropylphosphofluoridate (DIPF), a group-specific reagent. DIPF can inhibit an enzyme by covalently modifying a crucial serine residue.

Affinity labels, also called *substrate analogs*, are molecules that covalently modify active-site residues and are structurally similar to an enzyme's substrate. They are thus more specific for an enzyme's active site than are group-specific reagents. Tosyl-L-phenylalanine chloromethyl ketone (TPCK) is an affinity label for chymotrypsin (Figure 8.14). TPCK binds at the active site and then reacts irreversibly with a histidine residue at that site, inhibiting the enzyme.

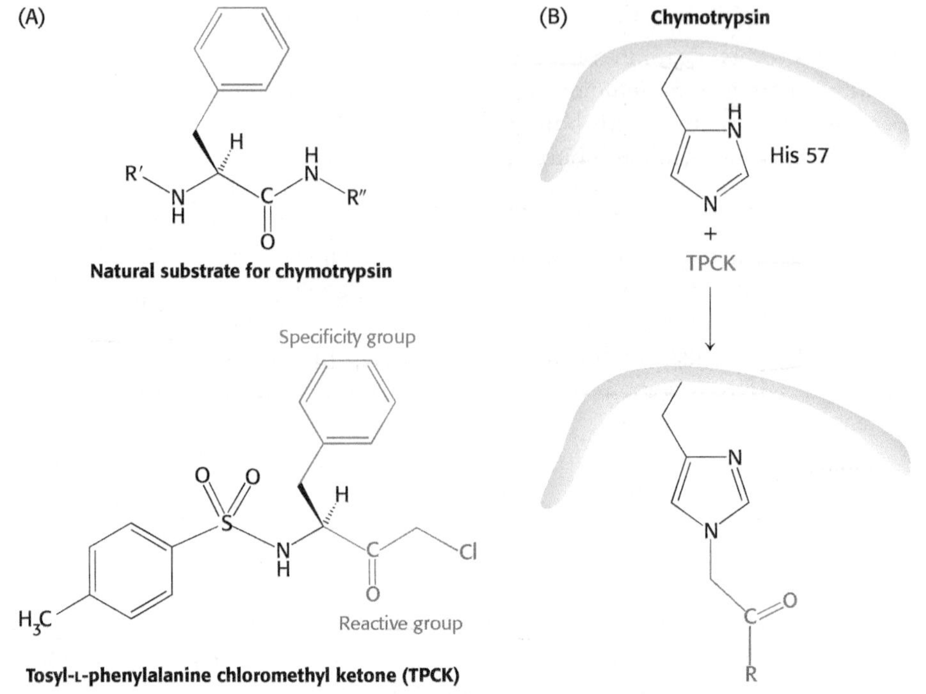

(A)

Natural substrate for chymotrypsin

Specificity group

Reactive group

Tosyl-L-phenylalanine chloromethyl ketone (TPCK)

(B) Chymotrypsin

His 57

TPCK

Figure 8.14 Affinity labeling. (A) Tosyl-L-phenylalanine chloromethyl ketone (TPCK) is a reactive analog of the normal substrate for the enzyme chymotrypsin. (B) TPCK binds at the chymotrypsin active site and modifies an essential histidine residue.

Suicide inhibitors, or *mechanism-based inhibitors*, are chemically modified substrates. These molecules provide researchers with the most specific means of modifying an enzyme's active site. The inhibitor binds to the enzyme as a substrate and is initially processed by the normal catalytic mechanism. The mechanism of catalysis then generates a chemically reactive intermediate that inactivates the enzyme through covalent modification. The fact that the enzyme participates in its own irreversible inhibition strongly suggests that the covalently modified group on the enzyme is catalytically vital. The antibiotic penicillin is a suicide inhibitor of the enzyme that synthesizes bacterial cell walls.

Transition-state analogs are potent inhibitors of enzymes (p. 106). As discussed earlier, the formation of the transition state is crucial to enzyme catalysis (p. 103). An important piece of evidence supporting the role of transition-state formation in catalysis is the inhibitory power of transition-state analogs.

🩺 CLINICAL INSIGHT

Penicillin Irreversibly Inactivates a Key Enzyme in Bacterial Cell-Wall Synthesis

Penicillin, the first antibiotic discovered, consists of a thiazolidine ring fused to a β-lactam ring to which a variable R group is attached by a peptide bond (**Figure 8.15**). This structure can undergo a variety of rearrangements, and, in particular, the β-lactam ring is very unstable. Indeed, this instability is closely tied to the antibiotic action of penicillin, as will be evident shortly.

How does penicillin inhibit bacterial growth? Let us consider the bacterium *Staphylococcus aureus*, the most common cause of staph infections. Penicillin works by interfering with the synthesis of the *S. aureus* cell walls. The *S. aureus* cell wall is made up of a macromolecule, called a *peptidoglycan* (**Figure 8.16**), which consists of linear polysaccharide chains that are cross-linked by short peptides (pentaglycines and tetrapeptides). The enormous bag-shaped peptidoglycan confers mechanical support and prevents bacteria from bursting in response to their high internal osmotic pressure. *Glycopeptide transpeptidase* catalyzes the formation of the cross-links that make the peptidoglycan so

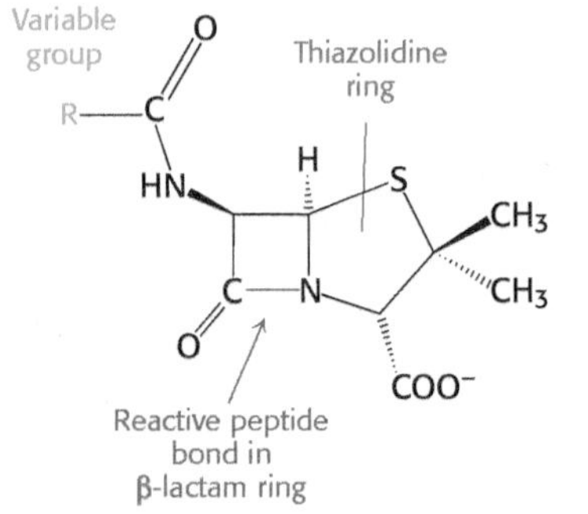

Variable group

Thiazolidine ring

Reactive peptide bond in β-lactam ring

Figure 8.15 The structure of penicillin. The reactive site of penicillin is the peptide bond of its β-lactam ring.

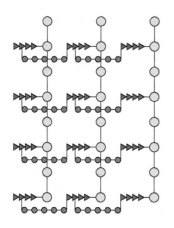

Figure 8.16 A schematic representation of the peptidoglycan in *Staphylococcus aureus*. The sugars are shown in yellow, the tetrapeptides in red, and the pentaglycine bridges in blue. The cell wall is a single, enormous, bag-shaped macromolecule because of extensive cross-linking.

stable (**Figure 8.17**). Bacterial cell walls are distinctive in containing D amino acids, which form cross-links by a mechanism different from that used to synthesize proteins.

Terminal glycine residue of pentaglycine bridge   **Terminal D-Ala-D-Ala unit**   **Gly-D-Ala cross-link**   **D-Ala**

Figure 8.17 The formation of cross-links in *S. aureus* peptidoglycan. The terminal amino group of the pentaglycine bridge in the cell wall attacks the peptide bond between two D-alanine residues to form a cross-link.

Penicillin inhibits the cross-linking transpeptidase by the Trojan horse stratagem: it mimics a normal substrate to enter the active site. To create cross-links between the tetrapeptides and pentaglycines, the transpeptidase normally forms an *acyl intermediate* with the penultimate D-alanine residue of the tetrapeptide. This covalent acyl-enzyme intermediate then reacts with the amino group of the terminal glycine in another peptide to form the cross-link (**Figure 8.18**). Penicillin is welcomed into the active site of the transpeptidase because it mimics the D-Ala-D-Ala moiety of the normal substrate. On binding to the transpeptidase, the serine residue at the active site attacks the carbonyl carbon atom of the lactam

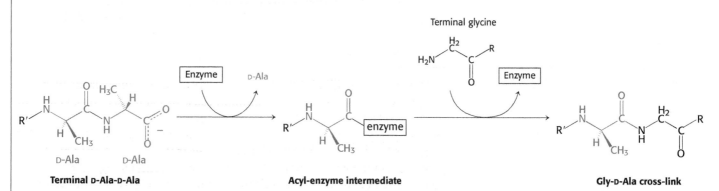

Terminal D-Ala-D-Ala   **Acyl-enzyme intermediate**   **Gly-D-Ala cross-link**

Figure 8.18 A transpeptidation reaction. An acyl-enzyme intermediate is formed in the transpeptidation reaction leading to cross-link formation.

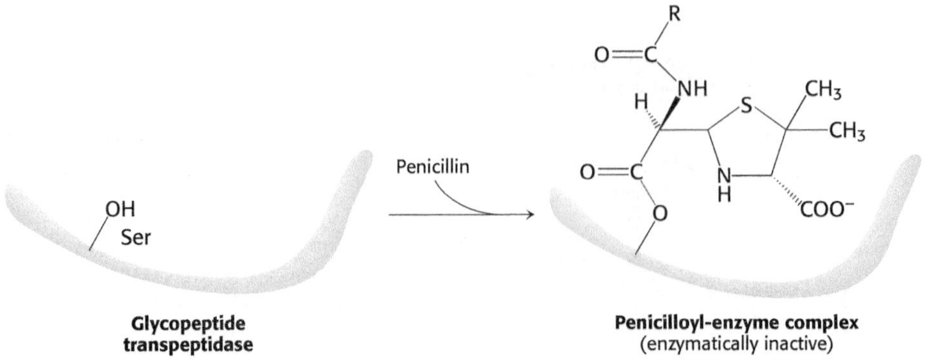

Figure 8.19 The formation of a penicilloyl-enzyme complex. Penicillin reacts with the transpeptidase to form an inactive complex, which is indefinitely stable.

ring to form the penicilloyl-serine derivative (**Figure 8.19**). This penicilloyl-enzyme does not react further. Hence, the transpeptidase is irreversibly inhibited and cell-wall synthesis cannot take place. Because the peptidase participates in its own inactivation, *penicillin acts as a suicide inhibitor.*

8.3 Chymotrypsin Illustrates Basic Principles of Catalysis and Inhibition

A detailed examination of the mechanism of action of the protein-degrading enzyme chymotrypsin will illustrate some of the basic principles of catalysis. It will also be useful as a case study for showing how enzyme mechanisms can be investigated, including the use of kinetics and enzyme inhibitors.

Protein turnover is an important process in living systems. After they are no longer needed in the cell, proteins must be degraded so that their constituent amino acids can be recycled for the synthesis of new proteins. Additionally, proteins ingested in the diet must be broken down into small peptides and amino acids for absorption in the intestine. Protein breakdown is catalyzed by a large class of enzymes called *proteolytic enzymes* or *proteases*. These enzymes cleave proteins by a hydrolysis reaction—the addition of a molecule of water to a peptide bond (p. 98).

One such proteolytic enzyme is chymotrypsin, which is secreted by the pancreas in response to a meal. Chymotrypsin cleaves peptide bonds selectively on the carboxyl-terminal side of the large hydrophobic amino acids such as tryptophan, tyrosine, phenylalanine, and methionine (**Figure 8.20**).

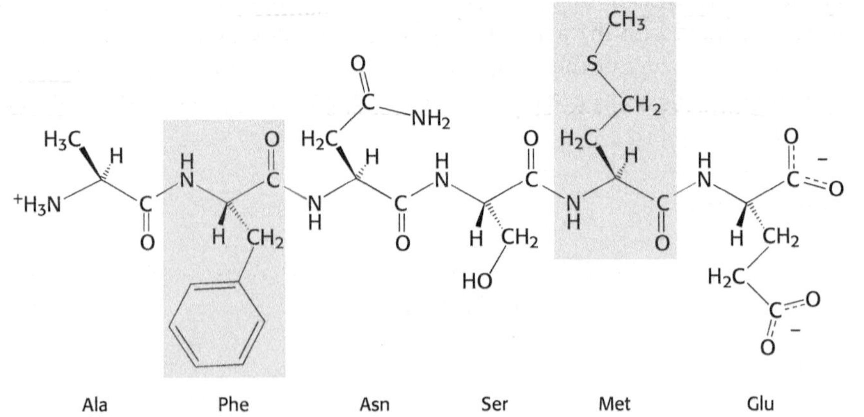

Figure 8.20 The specificity of chymotrypsin. Chymotrypsin cleaves proteins on the carboxyl side of aromatic or large hydrophobic amino acids (shaded orange). The red bonds indicate where chymotrypsin most likely acts.

Serine 195 Is Required for Chymotrypsin Activity

Chymotrypsin is a good example of the use of *covalent modification* as a catalytic strategy. The enzyme employs a powerful nucleophile to attack the unreactive carbonyl group of the substrate. This nucleophile becomes covalently attached to the substrate briefly in the course of catalysis.

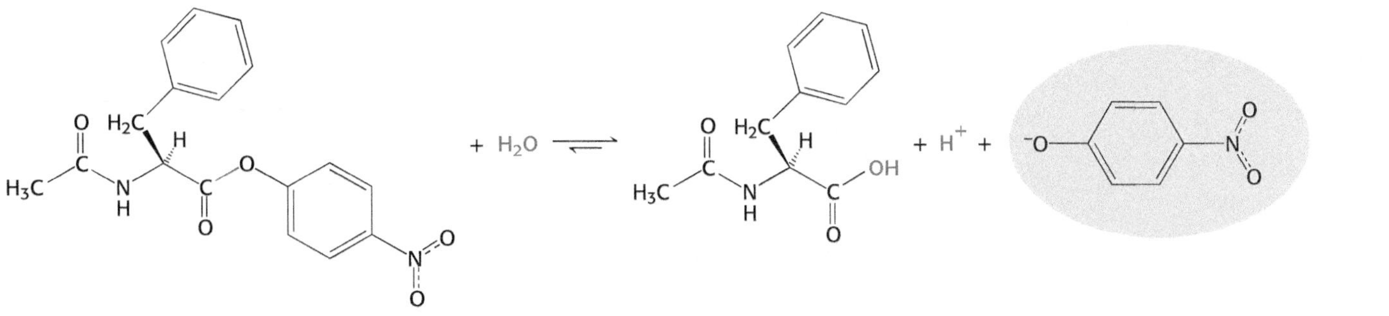

N-Acetyl-L-phenylalanine p-nitrophenyl ester

p-Nitrophenolate

Figure 8.21 A chromogenic substrate. N-Acetyl-L-phenylalanine p-nitrophenyl ester yields a yellow product, p-nitrophenolate, on cleavage by chymotrypsin. p-Nitrophenolate forms by deprotonation of p-nitrophenol at pH 7.

What is the nucleophile that chymotrypsin employs to attack the substrate carbonyl group? A clue came from the fact that chymotrypsin contains an extraordinarily reactive serine residue. Treatment with organofluorophosphates that modify serine residues, such as DIPF, was found to inactivate the enzyme irreversibly (Figure 8.13). Despite the fact that the enzyme possesses 28 serine residues, only one of them, serine 195, was modified, resulting in a total loss of enzyme activity. The use of the group-specific reagent DIPF alerted researchers to the importance of one particular serine residue in catalysis.

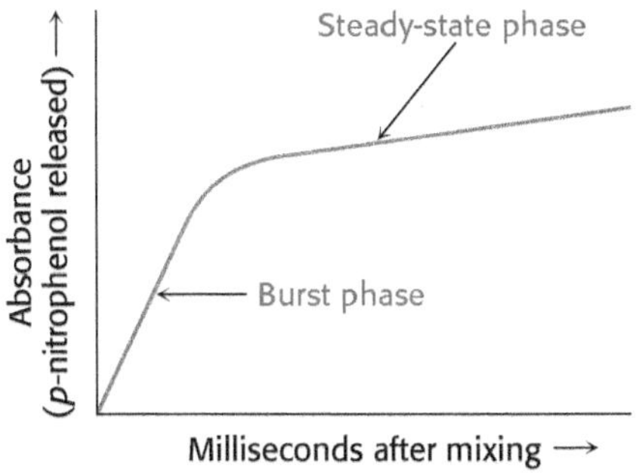

Chymotrypsin Action Proceeds in Two Steps Linked by a Covalently Bound Intermediate

A study of the enzyme's kinetics suggested a role for serine 195. The kinetics of enzyme action are often easily monitored by having the enzyme act on a substrate analog that forms a colored product. For chymotrypsin, such a *chromogenic substrate* is N-acetyl-L-phenylalanine p-nitrophenyl ester. One of the products formed by chymotrypsin's cleavage of this substrate is p-nitrophenolate, which has a yellow color (**Figure 8.21**). Measurements of the absorbance of light revealed the amount of p-nitrophenolate being produced and thus provided a facile means of investigating chymotrypsin activity.

Under steady-state conditions, the cleavage of the substrate obeys Michaelis–Menten kinetics. More insightful results were obtained by examining product formation within milliseconds of mixing the enzyme and substrate. The reaction between chymotrypsin and N-acetyl-L-phenylalanine p-nitrophenyl ester produced an initial rapid burst of colored product, followed by its slower formation as the reaction reached the steady state (**Figure 8.22**). These results suggest that hydrolysis proceeds in two steps. The burst is observed because the first step is more rapid than the second step.

The two steps are explained by the formation of a covalently bound enzyme–substrate intermediate (**Figure 8.23**). First, the acyl group of the substrate becomes covalently attached to serine 195 of the enzyme as p-nitrophenolate is released. The enzyme–acyl-group complex is called the *acyl-enzyme intermediate*. Second, the acyl-enzyme intermediate is hydrolyzed to release the carboxylic acid

Figure 8.22 Kinetics of chymotrypsin catalysis. Two stages are evident in the cleaving of N-acetyl-L-phenylalanine p-nitrophenyl ester by chymotrypsin: a rapid burst phase (pre-steady state) and a steady-state phase.

> **DID YOU KNOW?**
>
> In a steady-state system, the concentrations of the intermediates stay the same, even though the concentrations of substrate and products are changing. A sink filled with water that has the tap open just enough to match the loss of water down the drain is in a steady state. The level of the water in the sink never changes even though water is constantly flowing from the faucet through the sink and out through the drain.

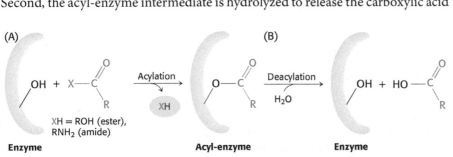

(A)   (B)

XH = ROH (ester), RNH₂ (amide)

Enzyme   **Acyl-enzyme**   **Enzyme**

Figure 8.23 Covalent catalysis. Hydrolysis by chymotrypsin takes place in two stages: (A) acylation to form the acyl-enzyme intermediate followed by (B) deacylation to regenerate the free enzyme.

component of the substrate and regenerate the free enzyme. Thus, one molecule of *p*-nitrophenolate is produced rapidly from each enzyme molecule as the acyl-enzyme intermediate is formed. However, it takes longer for the enzyme to be "reset" by the hydrolysis of the acyl-enzyme intermediate, and both steps are required for enzyme turnover.

The Catalytic Role of Histidine 57 Was Demonstrated by Affinity Labeling

The importance of a second residue in catalysis was shown by affinity labeling. The strategy was to have chymotrypsin react with a molecule that (1) specifically binds to the active site because it resembles a substrate and then (2) forms a stable covalent bond with a group on the enzyme that is in proximity. These criteria are met by TPCK (Figure 8.14). The phenylalanine side chain of TPCK enables it to bind specifically to chymotrypsin. The reactive group in TPCK is the chloromethyl ketone, which covalently modifies one of the ring nitrogens of histidine 57. TPCK is positioned to react with this residue because of its specific binding to the active site of the enzyme. The TPCK derivative of chymotrypsin is enzymatically inactive.

Serine Is Part of a Catalytic Triad That Includes Histidine and Aspartic Acid

Thus far, we have learned that serine 195 and histidine 57 are required for chymotrypsin activity, and the reaction proceeds through a substituted-enzyme intermediate. How can we integrate this information to elucidate the mechanism of chymotrypsin action? The side chain of serine 195 is hydrogen bonded to the imidazole ring of histidine 57. The —NH group of this imidazole ring is, in turn, hydrogen bonded to the carboxylate group of aspartate 102, another key component of the active site. This constellation of residues is referred to as the *catalytic triad.*

How does this arrangement of residues lead to the high reactivity of serine 195? The histidine residue serves to position the serine side chain and to polarize its hydroxyl group so that it is poised for deprotonation. In the presence of the substrate, histidine 57 accepts the proton from the serine-195 hydroxyl group. In doing so, histidine acts as a general base catalyst. The withdrawal of the proton from the hydroxyl group generates an alkoxide ion, which is a much more powerful nucleophile than a hydroxyl group is. The aspartate residue helps orient the histidine residue and make it a better proton acceptor through hydrogen bonding and electrostatic effects (Figure 8.24).

These observations suggest a mechanism for peptide hydrolysis (Figure 8.25). After substrate binding (step 1), the reaction begins with the oxygen atom of the side chain of serine 195 making a nucleophilic attack on the carbonyl carbon atom of the target peptide bond (step 2). There are now four atoms bonded to the carbonyl carbon atom, arranged as a tetrahedron, instead of three atoms in a

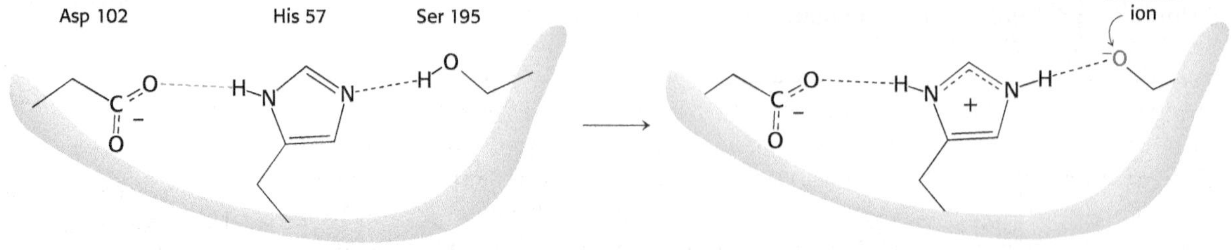

Figure 8.24 The catalytic triad. The catalytic triad, shown on the left, converts serine 195 into a potent nucleophile, as illustrated on the right.

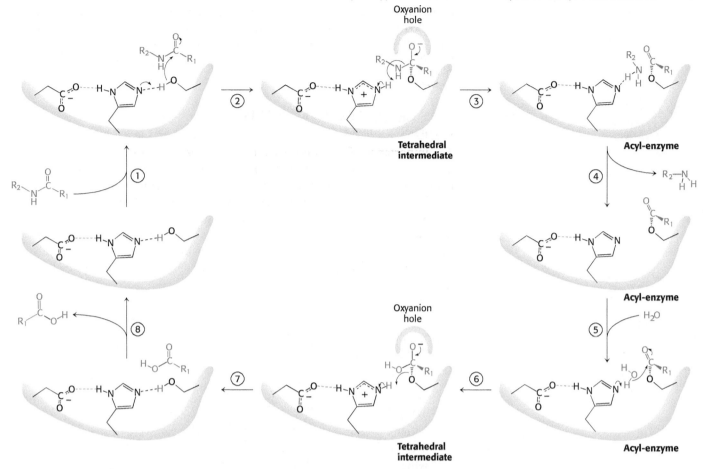

Figure 8.25 Peptide hydrolysis by chymotrypsin. The mechanism of peptide hydrolysis illustrates the principles of covalent and acid–base catalysis. The reaction proceeds in eight steps: (1) substrate binding, (2) serine's nucleophilic attack on the peptide carbonyl group, (3) collapse of the tetrahedral intermediate, (4) release of the amine component, (5) water binding, (6) water's nucleophilic attack on the acyl-enzyme intermediate, (7) collapse of the tetrahedral intermediate, and (8) release of the carboxylic acid component. The dashed green lines represent hydrogen bonds.

planar arrangement. The inherently unstable *tetrahedral-intermediate* form bears a negative charge on the oxygen atom derived from the carbonyl group. This charge is stabilized by a site termed the *oxyanion hole,* as shown in Figure 8.25. Interactions with NH groups in the oxyanion hole help to stabilize the tetrahedral intermediate (**Figure 8.26**). These interactions contribute to the binding energy that helps stabilize the transition state that precedes the formation of the tetrahedral intermediate (Figure 8.25). This tetrahedral intermediate collapses to generate the acyl-enzyme (step 3). This step is facilitated by the transfer of the proton from the positively charged histidine residue, now acting as a general acid catalyst, to the amino group of the substrate formed by cleavage of the peptide bond. The amine component is now free to depart from the enzyme (step 4), completing the first stage of the hydrolytic reaction—that is, acylation of the enzyme.

The next stage—deacylation of the enzyme—begins when a water molecule takes the place occupied earlier by the amine component of the substrate (step 5). The ester group of the acyl-enzyme is now hydrolyzed by a process that essentially repeats steps 2 through 4. Again acting as a general base catalyst, histidine 57 draws a proton away from the water molecule. The resulting OH⁻ ion attacks the carbonyl carbon atom of the acyl group, forming a tetrahedral intermediate (step 6). This structure breaks down to form the carboxylic acid product (step 7). Finally, the release of the carboxylic acid product readies the enzyme for another round of catalysis (step 8).

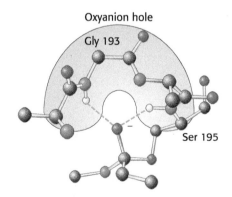

Figure 8.26 The oxyanion hole. The structure stabilizes the tetrahedral intermediate of the chymotrypsin reaction. Notice that hydrogen bonds (shown in green) link peptide NH groups and the negatively charged oxygen atom of the intermediate.

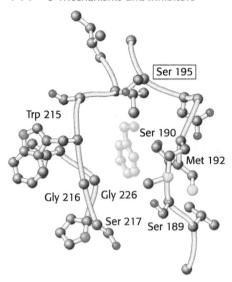

Figure 8.27 The specificity pocket of chymotrypsin. Notice that many hydrophobic groups line the deep specificity pocket. The structure of the pocket favors the binding of residues with long hydrophobic side chains such as phenylalanine (shown in green). Also notice that the active-site serine residue (serine 195) is positioned to cleave the peptide backbone between the residue bound in the pocket and the next residue in the sequence. The key amino acids that constitute the binding site are identified.

This mechanism accounts for all characteristics of chymotrypsin action except the observed preference for cleaving the peptide bonds just past residues with large, hydrophobic side chains. Examination of the three-dimensional structure of chymotrypsin with substrate analogs and enzyme inhibitors revealed the presence of a deep, relatively hydrophobic pocket, called the S1 pocket, into which the long, uncharged side chains of residues such as phenylalanine and tryptophan can fit. *The binding of an appropriate side chain into this pocket positions the adjacent peptide bond into the active site for cleavage* (Figure 8.27). The specificity of chymotrypsin depends almost entirely on which amino acid is directly on the amino-terminal side of the peptide bond to be cleaved.

❓ QUICK QUIZ 2 Using the Cleland representation (Figure 7.6), show the reaction progress of the hydrolysis of the tetrapeptide Ala-Phe-Gly-Ala by chymotrypsin.

SUMMARY

8.1 A Few Basic Catalytic Strategies Are Used by Many Enzymes

Although the detailed catalytic mechanisms of enzymes vary, many enzymes use one or more common strategies. They include (1) covalent catalysis, in which the enzyme becomes temporarily covalently modified; (2) general acid–base catalysis, in which some molecule other than water accepts or donates a proton; (3) metal ion catalysis, which functions in a variety of ways; and (4) catalysis by approximation and orientation, in which substrates are brought into proximity and oriented to facilitate the reaction.

8.2 Enzyme Activity Can Be Modulated by Temperature, pH, and Inhibitory Molecules

Specific small molecules or ions can inhibit even nonallosteric enzymes. Reversible inhibition is characterized by a rapid equilibrium between enzyme and inhibitor. A competitive inhibitor prevents the substrate from binding to the active site. It reduces the reaction velocity by diminishing the proportion of enzyme molecules that are bound to substrate. Competitive inhibition can be overcome by raising the substrate concentration. In uncompetitive inhibition, the inhibitor combines only with the enzyme–substrate complex. In noncompetitive inhibition, the inhibitor decreases the turnover number. Uncompetitive and noncompetitive inhibition cannot be overcome by raising the substrate concentration. In irreversible inhibition, the inhibitor is covalently linked to the enzyme or bound so tightly that its dissociation from the enzyme is very slow. Covalent inhibitors provide a means of mapping an enzyme's active site.

8.3 Chymotrypsin Illustrates Basic Principles of Catalysis and Inhibition

The cleavage of peptide bonds by chymotrypsin is initiated by the attack by a serine residue on the peptide carbonyl group. The attacking hydroxyl group is activated by interaction with the imidazole group of a histidine residue, which is, in turn, linked to an aspartate residue. This Ser-His-Asp catalytic triad generates a powerful nucleophile. The product of this initial reaction is a covalent intermediate formed by the enzyme and an acyl group derived from the bound substrate. The hydrolysis of this acyl-enzyme intermediate completes the cleavage process. The tetrahedral intermediates for these reactions have a negative charge on the peptide carbonyl oxygen atom. This negative charge is stabilized by interactions with peptide NH groups in a region on the enzyme termed the oxyanion hole.

KEY TERMS

covalent catalysis (p. 132)
general acid–base catalysis (p. 132)
metal ion catalysis (p. 132)
catalysis by approximation (p. 132)
binding energy (p. 132)
competitive inhibition (p. 134)

uncompetitive inhibition (p. 135)
noncompetitive inhibition (p. 135)
group-specific reagent (p. 137)
affinity label (substrate analog) (p. 137)
mechanism-based (suicide) inhibition
(p. 138)

transition-state analog (p. 138)
covalent modification (p. 140)
catalytic triad (p. 142)
oxyanion hole (p. 143)

? Answers to QUICK QUIZZES

1. Curve 1, no inhibition; curve 2, competitive inhibition; curve 3, noncompetitive inhibition; curve 4, uncompetitive inhibition.

2.

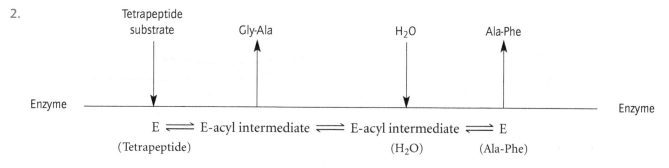

PROBLEMS

1. *Strategic solutions.* What are the four basic catalytic strategies used by many enzymes?

2. *Keeping busy.* Many isolated enzymes, if incubated at 37°C, will be denatured. However, if the enzymes are incubated at 37°C in the presence of substrate, the enzymes are catalytically active. Explain this apparent paradox.

3. *Controlled paralysis.* Succinylcholine is a fast-acting, short-duration muscle relaxant that is used when a tube is inserted into a patient's trachea or when a bronchoscope is used to examine the trachea and bronchi for signs of cancer. Within seconds of the administration of succinylcholine, the patient experiences muscle paralysis and is placed on a respirator while the examination proceeds. Succinylcholine is a competitive inhibitor of acetylcholinesterase, a nervous system enzyme, and this inhibition causes paralysis. However, succinylcholine is hydrolyzed by blood-serum cholinesterase, which shows broader substrate specificity than does the nervous system enzyme. Paralysis lasts until the succinylcholine is hydrolyzed by the serum cholinesterase, usually several minutes later. ✓ 6

(a) As a safety measure, serum cholinesterase is measured before the examination takes place. Explain why this measurement is good idea.

(b) What would happen to the patient if the serum cholinesterase activity were only 10 units of activity per liter rather than the normal activity of about 80 units?

(c) Some patients have a mutant form of the serum cholinesterase that displays a K_M of 10 mM, rather than the normal 1.4 mM. What will be the effect of this mutation on the patient?

4. *Specific action.* How can group-specific reagents be used to determine the mechanism of action of an enzyme? ✓ 6

5. *Like bread and butter.* Match the term with the description or compound. ✓ 6

(a) Competitive inhibition _____

(b) Uncompetitive inhibition _____

(c) Noncompetitive inhibition _____

1. Inhibitor and substrate can bind simultaneously

2. V_{max} remains the same but the K_M^{app} increases

3. Sulfanilamide

4. Binds to the enzyme–substrate complex only

5. Lowers V_{max} and K_M^{app}

6. Roundup

7. K_M remains unchanged but V_{max} is lower

8. Doxycycline

9. Inhibitor binds at the active site

6. *Mode of inhibition.* The kinetics of an enzyme are measured as a function of substrate concentration in the presence and in the absence of 2 μM inhibitor. ✓ 6

Velocity (μmol minute^{-1})

[S] (μM)	No inhibitor	Inhibitor
3	10.4	4.1
5	14.5	6.4
10	22.5	11.3
30	33.8	22.6
90	40.5	33.8

(a) What are the values of V_{max} and K_M in the absence of inhibitor? In its presence?

(b) What type of inhibition is it?

7. *A different mode.* The kinetics of the enzyme considered in problem 6 are measured in the presence of a different inhibitor. The concentration of this inhibitor is 100 μM. ✓ 6

Velocity (μmol minute^{-1})

[S] (μM)	No inhibitor	Inhibitor
3	10.4	2.1
5	14.5	2.9
10	22.5	4.5
30	33.8	6.8
90	40.5	8.1

(a) What are the values of V_{max} and K_M in the presence of this inhibitor? Compare them with those obtained in problem 6.

(b) What type of inhibition is it?

8. *Informative inhibition.* What are the four key types of irreversible inhibitors that can be used to study enzyme function? ✓ 6

9. *Self-preservation.* Some bacteria produce the enzyme β-lactamase, which cleaves and opens lactam rings. How would the presence of β-lactamase affect bacterial sensitivity to penicillin? ✓ 6

10. *One for all and all for one.* What is the catalytic triad, and what are the roles of the individual components in chymotrypsin activity?

11. *A burrow for oxyanions?* What is the purpose of the oxyanion hole in chymotrypsin?

12. *Burst.* What caused a "burst" of activity followed by a steady-state reaction when chymotrypsin was studied in the milliseconds subsequent to mixing the enzyme and substrate?

13. *Say no to cannibalism.* If chymotrypsin is such an effective protease, why doesn't it digest itself?

Chapter Integration Problems

14. *Mental experiment.* Picture in your mind the velocity-versus-substrate concentration curve for a typical Michaelis–Menten enzyme. Now, imagine that the experimental conditions are altered as described below. For each of the conditions described, fill in the table indicating precisely (when possible) the effect on V_{max} and K_M on the imagined Michaelis–Menten enzyme.

Experimental condition	V_{max}	K_M
(a) Twice as much enzyme is used.		
(b) Half as much enzyme is used.		
(c) A competitive inhibitor is present.		
(d) An uncompetitive inhibitor is present.		
(e) A pure noncompetitive inhibitor is present.		

15. *Type 1 and type 2.* In Section 7.2, we examined two types of bisubstrate reactions. Which of the two types best describes the action of chymotrypsin? Explain.

Challenge Problems

16. *Titration experiment.* The effect of pH on the activity of an enzyme was examined. At its active site, the enzyme has an ionizable group that must be negatively charged in order for substrate binding and catalysis to take place. The ionizable group has a pK_a of 6.0. The substrate is positively charged throughout the pH range of the experiment. ✓ 6

$$E^- + S^+ \rightleftharpoons E^-S^+ \longrightarrow E^- + P^+$$
$$+$$
$$H^+$$
$$\big\updownarrow$$
$$EH$$

(a) Draw the V_0-versus-pH curve when the substrate concentration is much greater than the K_M of the enzyme.

(b) Draw the V_0-versus-pH curve when the substrate concentration is much less than the K_M of the enzyme.

(c) At which pH will the velocity equal one-half of the maximal velocity attainable under the conditions described in (b)?

17. *Mutant.* Predict the effect of mutating the aspartic acid at the active site of chymotrypsin to asparagine.

18. *No burst.* Examination of the cleavage of the *amide* substrate A by chymotrypsin very early in the reaction reveals no burst. The reaction is monitored by noting the color produced by the release of the amino part of the substrate (highlighted in orange). Why is no burst observed?

A

19. *Variations on a theme.* Recall that TPCK, an affinity label, inactivates chymotrypsin by covalently binding to histidine 57. Trypsin is a protease very similar to chymotrypsin, except that it hydrolyzes peptide bonds on the carboxyl side of lysine or arginine. ✓ 6

(a) Name an affinity-labeling agent for trypsin.
(b) How would you test the agent's specificity?

Selected Readings for this chapter can be found online at www.whfreeman.com/tymoczko3e.

Hemoglobin, an Allosteric Protein

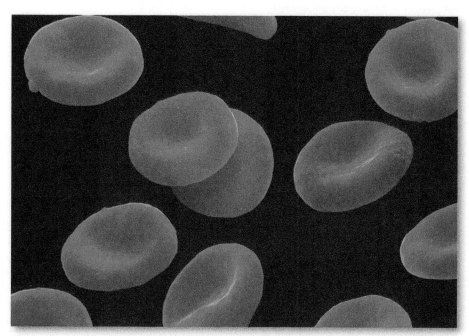

In the bloodstream, red blood cells carry oxygen from the lungs to tissues where oxygen is required. Hemoglobin, a four-subunit protein with an oxygen-binding pigment called heme, which gives blood its color, transports oxygen and releases it where it is needed. [Andrew Syred/Stone/Getty Images.]

A fascinating protein called *hemoglobin* is a component of red blood cells. This protein efficiently carries oxygen from the lungs to the tissues and contributes to the transport of carbon dioxide and hydrogen ions back to the lungs.

Hemoglobin is a wonderful example of the fact that allostery is not a property limited to enzymes. The basic principles of allostery are well illustrated by hemoglobin. Indeed, many of the principles of allosteric regulation were elucidated by the study of hemoglobin, the first allosteric protein known in atomic detail. The study of allosteric enzymes and hemoglobin is so intertwined that hemoglobin was designated an "honorary enzyme" by biochemist Jacques Monod.

In this chapter, we will examine the allosteric properties of hemoglobin, including how the oxygen-carrying capacity of this protein is regulated to meet the physiological requirements of an organism. We will also look at a close chemical relative of hemoglobin, the protein myoglobin. Located in muscle, *myoglobin* facilitates the diffusion of the oxygen to cellular sites that require oxygen and provides a reserve supply of oxygen in times of need.

✓ 7 Explain how allosteric properties contribute to hemoglobin function.

9.1 Hemoglobin Displays Cooperative Behavior

The binding of oxygen to hemoglobin isolated from red blood cells displays marked sigmoidal behavior, characteristic of cooperation between subunits, whereas myoglobin shows a hyperbolic curve (**Figure 9.1**). Indeed, the physiological significance of the *cooperative effect* in allosteric proteins is especially evident in the action of hemoglobin. Oxygen must be transported in the blood from the lungs, where the *partial pressure* of oxygen (pO_2) is high (approximately 100 torr), to the tissues, where the partial pressure of oxygen is much lower (typically 20 torr).

DID YOU KNOW?

Partial pressure is the fraction of the total pressure of a mixture of gases that is due to one component of the mixture—in this case, oxygen. For instance, 1 atmosphere of pressure = 760 torr, or the pressure of a column of mercury 760 mm high. Air is 21% oxygen in dry air at sea level, and so the partial pressure of oxygen is 159 torr. In the alveoli of the lungs, the partial pressure of oxygen is 100 torr owing to the addition of water vapor and to the gas exchange that takes place.

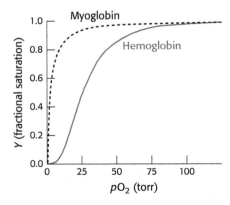

Figure 9.1 Oxygen binding by hemoglobin. This curve, obtained for hemoglobin in red blood cells, is shaped somewhat like the letter "S," indicating that distinct, but interacting, oxygen-binding sites are present in each hemoglobin molecule. For comparison, the binding curve for myoglobin is shown in black.

Let us consider how the cooperative behavior represented leads to efficient oxygen transport. In the lungs, hemoglobin becomes nearly saturated with oxygen, such that 98% of the oxygen-binding sites are occupied. When hemoglobin moves to the tissues, the oxygen saturation level drops to 32%. Thus, 66% (98 − 32 = 66) of the potential oxygen-binding sites release oxygen in the tissues. If myoglobin, with its high affinity for oxygen, were a transport protein for oxygen, it would release a mere 7% of its oxygen under these conditions (**Figure 9.2**). Why does hemoglobin release only part of the oxygen that it carries? As you might guess from the discussion of allosteric enzymes (Section 7.3), allosteric regulators released at the tissues can further enhance oxygen release. We will examine these regulators later in the chapter.

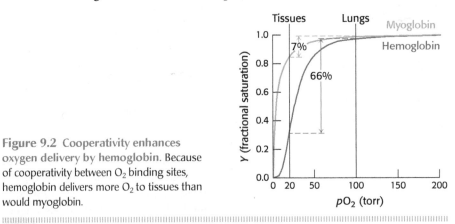

Figure 9.2 Cooperativity enhances oxygen delivery by hemoglobin. Because of cooperativity between O_2 binding sites, hemoglobin delivers more O_2 to tissues than would myoglobin.

9.2 Myoglobin and Hemoglobin Bind Oxygen in Heme Groups

Myoglobin, a single polypeptide chain, consists largely of α helices that are linked to one another by turns in the polypeptide chain to form a globular structure (**Figure 9.3**). The recurring structure is called a *globin fold* and is also seen in hemoglobin.

Myoglobin can exist in either of two forms: as *deoxymyoglobin*, an oxygen-free form, or as *oxymyoglobin*, a form having a bound oxygen molecule. The

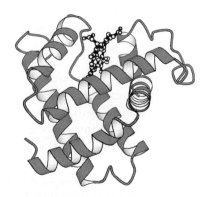

Myoglobin

Figure 9.3 The structure of myoglobin. Notice that myoglobin consists of a single polypeptide chain, formed of helices connected by turns, with one oxygen-binding site. [Drawn from 1MBD.pdb]

ability of myoglobin, as well as that of hemoglobin, to bind oxygen depends on the presence of a bound prosthetic group called *heme*.

Heme
(Fe-protoporphyrin IX)

Propionate group

Pyrrole ring

Methyl group

Vinyl group

The heme group gives muscle and blood their distinctive red color. It consists of an organic component and a central iron atom. The organic component, called *protoporphyrin,* is made up of four pyrrole rings linked by methine bridges to form a tetrapyrrole ring. Four methyl groups, two vinyl groups, and two propionate side chains are attached.

The iron atom lies in the center of the protoporphyrin, bonded to the four pyrrole nitrogen atoms. Under normal conditions, the iron is in the ferrous (Fe^{2+}) oxidation state. The iron ion can form two additional bonds, one on each side of the heme plane. These binding sites are called the fifth and sixth coordination sites. In hemoglobin and myoglobin, the fifth coordination site is occupied by the imidazole ring of a histidine residue of the protein. This histidine residue is referred to as the *proximal histidine.* In deoxyhemoglobin and deoxymyoglobin, the sixth coordination site remains unoccupied; this position is available for binding oxygen. The iron ion lies approximately 0.4 Å outside the porphyrin plane because an iron ion, in this form, is slightly too large to fit into the well-defined hole within the porphyrin ring (**Figure 9.4**, left). A second histidine, called the *distal histidine,* resides on the opposite side of the heme from the proximal histidine. The distal histidine prevents the oxidation of the heme iron to the ferric ion (Fe^{3+}), which cannot bind oxygen, and also reduces the ability of carbon monoxide to bind to the heme (problem 22).

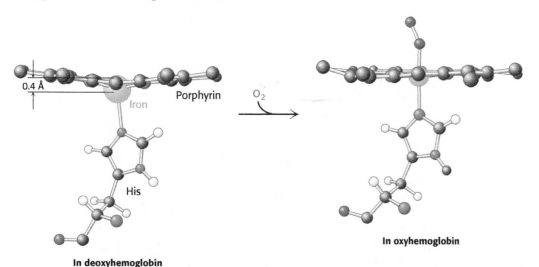

0.4 Å

Iron

Porphyrin

O_2

His

In deoxyhemoglobin

In oxyhemoglobin

Figure 9.4 Oxygen binding changes the position of the iron ion. The iron ion lies slightly outside the plane of the porphyrin in deoxyhemoglobin heme (left) but moves into the plane of the heme on oxygenation (right).

The binding of the oxygen molecule at the sixth coordination site of the iron ion substantially rearranges the electrons within the iron so that the ion becomes effectively smaller, allowing it to move into the plane of the porphyrin (Figure 9.4, right). The bound oxygen is stabilized by forming a hydrogen bond with the distal histidine.

 CLINICAL INSIGHT

Functional Magnetic Resonance Imaging Reveals Regions of the Brain Processing Sensory Information

The change in electronic structure that takes place when the iron ion moves into the plane of the porphyrin is paralleled by changes in the magnetic properties of hemoglobin; these changes are the basis for *functional magnetic resonance imaging* (fMRI), one of the most powerful methods for examining brain function. Nuclear magnetic resonance techniques detect signals that originate primarily from the protons in water molecules but are altered by the magnetic properties of hemoglobin. With the use of appropriate techniques, images can be generated that reveal differences in the relative amounts of deoxy- and oxyhemoglobin and thus the relative activity of various parts of the brain. When a specific part of the brain is active, blood vessels relax to allow more blood flow to the active region. Thus, a more active region of the brain will be richer in oxyhemoglobin.

These noninvasive methods reveal areas of the brain that process sensory information. For example, subjects have been imaged while breathing air that either does or does not contain odorants. When odorants are present, the fMRI technique detects an increase in the level of hemoglobin oxygenation (and, hence, brain activity) in several regions of the brain (Figure 9.5). Such regions include those in the primary olfactory cortex as well as other regions in which secondary processing of olfactory signals presumably takes place. Further analysis reveals the time course of activation of particular regions and other features. Functional MRI shows tremendous potential for mapping regions and pathways engaged in processing sensory information obtained from all the senses. Thus, *a seemingly incidental aspect of the biochemistry of hemoglobin has yielded the basis for observing the brain in action.*

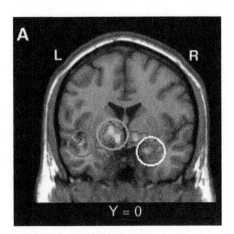

Figure 9.5 Brain response to odorants. A functional magnetic resonance image reveals brain response to odorants. The light spots indicate regions of the brain activated by odorants. [R. Osterbauer et al., *J. Neurophysiol. 93*(2005): 3434–3441.]

9.3 Hemoglobin Binds Oxygen Cooperatively

Like all allosteric proteins, hemoglobin displays quaternary structure. Human hemoglobin A, present in adults, consists of four subunits: two α *subunits* and two β *subunits*. The α and β subunits have similar three-dimensional structures and are similar to that of myoglobin. The three-dimensional structure of hemoglobin (Figure 9.6) is best described as a pair of identical $\alpha\beta$ *dimers* ($\alpha_1\beta_1$ and $\alpha_2\beta_2$) that

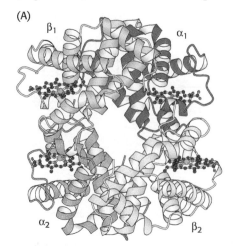
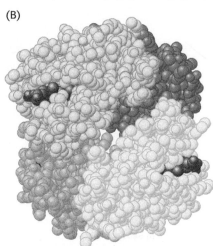

(A)   (B)

Figure 9.6 The quaternary structure of deoxyhemoglobin. Hemoglobin, which is composed of two α chains and two β chains, functions as a pair of $\alpha\beta$ dimers. (A) A ribbon diagram. (B) A space-filling model. [Drawn from 1A3N.pdb.]

associate to form the hemoglobin tetramer. In deoxyhemoglobin, these $\alpha\beta$ dimers are linked by an extensive interface, which includes, among other regions, the carboxyl terminus of each chain. Deoxyhemoglobin corresponds to the T state of hemoglobin, whereas oxyhemoglobin corresponds to the R state. Recall from the discussion of allosteric enzymes in Chapter 7 that the T state is less biochemically active than the R state. In regard to hemoglobin, the T state has a lower affinity for oxygen than does the R state.

How does oxygen binding lead to the structural transition from the T state to the R state? When the iron ion moves into the plane of the porphyrin, the histidine residue bound in the fifth coordination site moves with it. This histidine residue is part of an α helix, which also moves (**Figure 9.7**). The carboxyl terminal end of this α helix lies in the interface between the two $\alpha\beta$ dimers. Consequently, the structural transition at the iron ion is directly transmitted to the other subunits, resulting in substantial changes in quaternary structure that correspond to the T-to-R-state transition (**Figure 9.8**). The $\alpha_1\beta_1$ and $\alpha_2\beta_2$ dimers rotate approximately 15 degrees with respect to one another. The rearrangement of the dimer interface provides a pathway for communication between subunits: the presence of oxygen on one of the subunits is immediately communicated to the others so that the subunits change from T to R in concert.

Recall that we considered two models for cooperative binding (Section 7.3). Is the cooperative binding of oxygen by hemoglobin best described by the concerted or the sequential model? Neither model in its pure form fully accounts for the behavior of hemoglobin. Instead, a combined model is required. Hemoglobin behavior is concerted in that hemoglobin with three sites occupied by oxygen is almost always in the quaternary structure associated with the R state. The remaining open binding site has an affinity for oxygen more than 20-fold greater than that of fully deoxygenated hemoglobin binding its first oxygen atom. However, the behavior is not fully concerted, because hemoglobin with oxygen bound to only one of four sites remains primarily in the T-state quaternary structure. Yet this molecule binds oxygen three times as strongly as does fully deoxygenated hemoglobin, an observation consistent only with a sequential model. These results highlight the fact that the concerted and sequential models represent idealized cases, which real systems may approach but rarely attain.

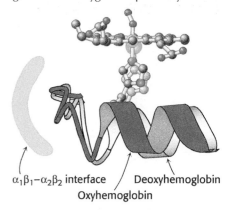

$\alpha_1\beta_1$–$\alpha_2\beta_2$ interface — Deoxyhemoglobin
Oxyhemoglobin

Figure 9.7 Conformational changes in hemoglobin. The movement of the iron ion on oxygenation brings the iron-associated histidine residue toward the porphyrin ring. The related movement of the histidine-containing α helix alters the interface between the $\alpha\beta$ dimers, instigating other structural changes. For comparison, the deoxyhemoglobin structure is shown in gray behind the oxyhemoglobin structure in color.

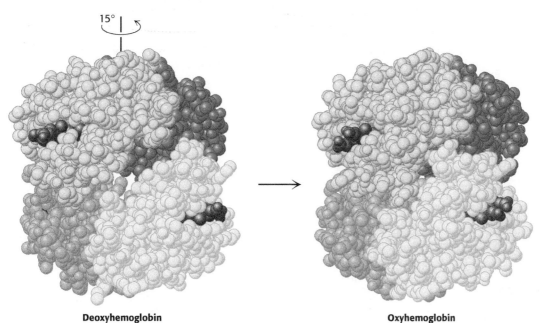

15°

Deoxyhemoglobin

Oxyhemoglobin

Figure 9.8 Quaternary structural changes on oxygen binding by hemoglobin. Notice that, on oxygenation, one $\alpha\beta$ dimer shifts with respect to the other by a rotation of 15 degrees. [Drawn from 1A3N.pdb and 1LFQ.pdb.]

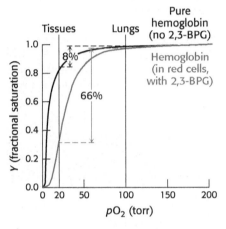

**2,3-Bisphosphoglycerate
(2,3-BPG)**

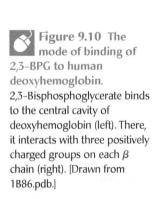

Figure 9.9 Oxygen binding by pure
hemoglobin compared with hemoglobin
in red blood cells. Pure hemoglobin binds
oxygen more tightly than does hemoglobin in
red blood cells. This difference is due to the
presence of 2,3-bisphosphoglycerate
(2,3-BPG) in red blood cells.

9.4   An Allosteric Regulator Determines the Oxygen Affinity of Hemoglobin

Oxygen binding by hemoglobin is analogous to substrate binding by an allosteric enzyme (Section 7.3). Is the oxygen-binding ability of hemoglobin also affected by regulatory molecules? A comparison of hemoglobin's behavior in the cell and out is a source of insight into the answer. Hemoglobin can be extracted from red blood cells and fully purified. Interestingly, the oxygen affinity of purified hemoglobin is much greater than that of hemoglobin within red blood cells. The affinity is so great that only 8% of the oxygen would be released to the tissues if hemoglobin in the cell behaved in the way that purified hemoglobin behaves, an amount insufficient to support aerobic tissues. What accounts for the difference between purified hemoglobin and hemoglobin in red blood cells? This dramatic difference is due to the presence, within these cells, of an allosteric regulator molecule—*2,3-bisphosphoglycerate* (2,3-BPG). This highly anionic compound is present in red blood cells at approximately the same concentration as hemoglobin (~2 mM). 2,3-BPG binds to hemoglobin, reducing its oxygen affinity so that 66% of the oxygen is released instead of a meager 8% (**Figure 9.9**).

How does 2,3-BPG affect oxygen affinity so significantly? Examination of the crystal structure of deoxyhemoglobin in the presence of 2,3-BPG reveals that a single molecule of 2,3-BPG binds in a pocket, present only in the T form, in the center of the hemoglobin tetramer. The interaction is facilitated by ionic bonds between the negative charges on 2,3-BPG and three positively charged groups of each β chain (**Figure 9.10**). Thus, 2,3-BPG binds preferentially to deoxyhemoglobin and stabilizes it, effectively reducing the oxygen affinity and leading to the release of oxygen. In order for the structural transition from T to R to take place, the bonds between hemoglobin and 2,3-BPG must be broken and 2,3-BPG must be expelled.

✚ CLINICAL INSIGHT

Hemoglobin's Oxygen Affinity Is Adjusted to Meet Environmental Needs

Let us consider another physiological example of the importance of 2,3-BPG binding to hemoglobin. Because a fetus obtains oxygen from the mother's hemoglobin rather than from the air, *fetal hemoglobin* must bind oxygen when the mother's hemoglobin is releasing oxygen. In order for the fetus to obtain enough oxygen to survive in a low-oxygen environment, its hemoglobin must have a high affinity for oxygen. How is the affinity of fetal hemoglobin for oxygen increased?

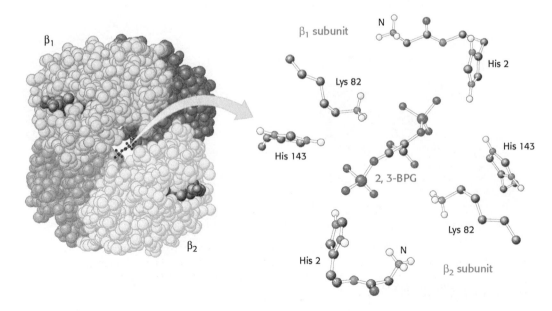

Figure 9.10 The mode of binding of 2,3-BPG to human deoxyhemoglobin. 2,3-Bisphosphoglycerate binds to the central cavity of deoxyhemoglobin (left). There, it interacts with three positively charged groups on each β chain (right). [Drawn from 1B86.pdb.]

The globin gene expressed by fetuses differs from that expressed by human adults; fetal hemoglobin tetramers include two α chains and two γ chains. The γ chain is 72% identical in amino acid sequence with the β chain. One noteworthy change is the substitution of a serine residue for histidine 143 in the γ chain of the 2,3-BPG-binding site. This change removes two positive charges from the 2,3-BPG-binding site (one from each chain) and reduces the affinity of 2,3-BPG for fetal hemoglobin. Reduced affinity for 2,3-BPG means that the oxygen-binding affinity of fetal hemoglobin is higher than that of the mother's (adult) hemoglobin (**Figure 9.11**). This difference in oxygen affinity allows oxygen to be effectively transferred from the mother's red blood cells to those of the fetus.

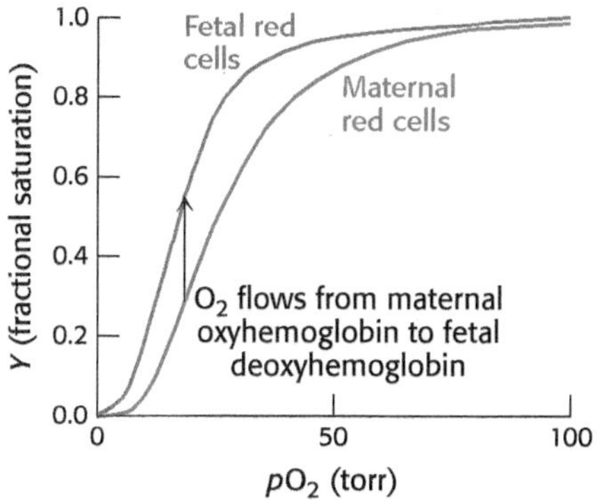

Figure 9.11 The oxygen affinity of fetal red blood cells. The oxygen affinity of fetal red blood cells is higher than that of maternal red blood cells because fetal hemoglobin does not bind 2,3-BPG as well as maternal hemoglobin does.

BIOLOGICAL INSIGHT

Hemoglobin Adaptations Allow Oxygen Transport in Extreme Environments

The bar-headed goose (**Figure 9.12**) provides another example of hemoglobin adaptations. This remarkable bird migrates over Mt. Everest at altitudes exceeding 9 km (5.6 miles), where the oxygen partial pressure is only 30% of that at sea level. In comparison, consider that human beings climbing Mt. Everest must take weeks to acclimate to the lower pO_2 and still usually require oxygen masks. Although the biochemical bases of the bird's amazing physiological feat are not clearly established, changes in hemoglobin may be important. Compared with its lower-flying cousins, the bar-headed goose has four amino acid changes in its hemoglobin, three in the α chain and one in the β chain. One of the changes in the α chain is proline for alanine, which disrupts a van der Waals contact and facilitates the formation of the R state, thus enabling the protein to more readily bind oxygen.

Figure 9.12 Bar-headed goose. [Tierbild Okapia/Science Source.]s

9.5 Hydrogen Ions and Carbon Dioxide Promote the Release of Oxygen

2,3-Bisphosphoglycerate is not the only allosteric regulator of hemoglobin activity. Indeed, actively metabolizing tissues—those most in need of oxygen, such as muscle—release signal molecules that further reduce the affinity of hemoglobin for oxygen. The signal molecules are hydrogen ion and carbon dioxide, heterotropic effectors that enhance oxygen release (**Figure 9.13**). The regulation of

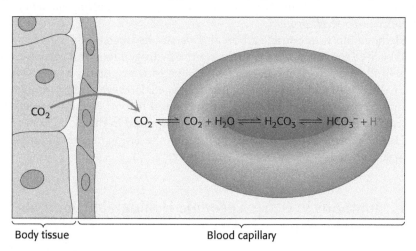

Figure 9.13 Carbon dioxide and pH. Carbon dioxide in the tissues diffuses into red blood cells. Inside a red blood cell, carbon dioxide reacts with water to form carbonic acid, in a reaction catalyzed by the enzyme carbonic anhydrase. Carbonic acid dissociates to form HCO_3^- and H^+, resulting in a drop in pH inside the red cell.

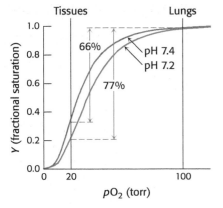

Figure 9.14 The effect of pH on the oxygen affinity of hemoglobin. Lowering the pH from 7.4 (red curve) to 7.2 (blue curve) results in the release of O_2 from oxyhemoglobin.

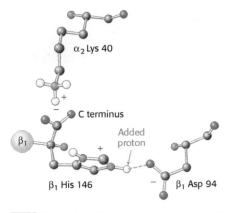

Figure 9.15 The chemical basis of the Bohr effect. In deoxyhemoglobin, three amino acid residues form two salt bridges that stabilize the T quaternary structure. The formation of one of the salt bridges depends on the presence of an added proton on histidine β146. The proximity of the negative charge on aspartate β94 in deoxyhemoglobin favors the protonation of this histidine. Notice that the salt bridge between histidine β146 and aspartate β94 is stabilized by a hydrogen bond (green dashed line). The green sphere represents the remainder of the β_1 chain.

QUICK QUIZ Name three factors that stabilize the deoxy form of hemoglobin.

oxygen binding by hydrogen ions and carbon dioxide is called the *Bohr effect* after Christian Bohr, who described this phenomenon in 1904.

The oxygen affinity of hemoglobin decreases as pH decreases from the value of 7.4 found in the lungs, at 100 torr of oxygen partial pressure, to pH 7.2 and an oxygen partial pressure of 20 torr found at active muscle (**Figure 9.14**). This difference in pH and partial pressure results in a release of oxygen amounting to 77% of total carrying capacity. Recall that only 66% of the oxygen would be released in the absence of any change in pH. What are the chemical and structural bases of the pH effect? Deoxyhemoglobin is stabilized by ionic bonds, or salt bridges, two of which are shown in **Figure 9.15**. Consider the ionic bond between aspartate β94 and the protonated side chain of histidine β146. At high pH, the side chain of histidine β146 is not protonated and the salt bridge does not form, thus favoring oxygen binding. As the pH drops, as in the vicinity of actively metabolizing tissues, the side chain of histidine β146 becomes protonated, the salt bridge with aspartate β94 forms, and the quaternary structure characteristic of deoxyhemoglobin is stabilized, leading to a greater release of oxygen at actively metabolizing tissues. The ionic bond between the carboxyl terminus of histidine β146 and lysine α140 also stabilizes the T state. This bond is disrupted by changes elsewhere in the β_1 chain that occur upon oxygen binding. Hemoglobin illustrates the fact that pH, as a cellular environmental factor, affects proteins other than enzymes.

In addition, hemoglobin responds to carbon dioxide with a decrease in oxygen affinity, thus facilitating the release of oxygen in tissues with a high carbon dioxide concentration, such as those in which fuel is actively undergoing combustion to form carbon dioxide and water. In the presence of carbon dioxide at a partial pressure of 40 torr, the amount of oxygen released approaches 90% of the maximum carrying capacity when the pH falls to 7.2 (**Figure 9.16**). Carbon dioxide stabilizes deoxyhemoglobin by reacting with the terminal amino groups to form *carbamate* groups, which are negatively charged.

Carbamate

The amino termini lie at the interface between the $\alpha\beta$ dimers. These negatively charged carbamate groups participate in salt-bridge interactions, characteristic of the T-state structure, which stabilize deoxyhemoglobin's structure and favor the release of oxygen. Thus, *the heterotropic regulation of hemoglobin by hydrogen ions and carbon dioxide further increases the oxygen-transporting efficiency of this magnificent allosteric protein.*

Hemoglobin with bound carbon dioxide and hydrogen ions is carried in the blood back to the lungs, where it releases the hydrogen ions and carbon dioxide and rebinds oxygen. Thus, hemoglobin helps to transport hydrogen ions and carbon dioxide in addition to transporting oxygen. However, transport by hemoglobin accounts for only about 14% of the total transport of these species; both hydrogen ions and carbon dioxide are also transported in the blood as bicarbonate (HCO_3^-) formed spontaneously or through the action of *carbonic anhydrase*, an enzyme abundant in red blood cells (**Figure 9.17**).

9.6 Mutations in Genes Encoding Hemoglobin Subunits Can Result in Disease

One of the first diseases understood at the molecular level was the blood disease *sickle-cell anemia*, which is caused by a single amino acid substitution in one hemoglobin chain. The name of the disorder comes from the abnormal sickle

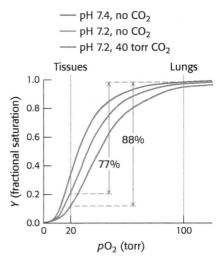

Figure 9.16 Carbon dioxide effects. The presence of carbon dioxide decreases the affinity of hemoglobin for oxygen even beyond the effect due to a decrease in pH, resulting in even more efficient oxygen transport from the tissues to the lungs.

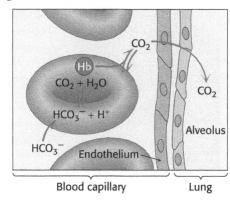
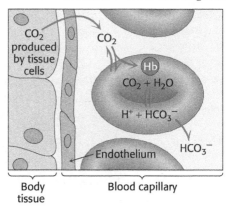

Figure 9.17 The transport of CO_2 from tissues to lungs. Most carbon dioxide is transported to the lungs in the form of HCO_3^- produced in red blood cells and then released into the blood plasma. A lesser amount is transported by hemoglobin in the form of an attached carbamate.

shape of red blood cells deprived of oxygen. Disorders due to mutations in hemoglobin (hemoglobinopathies) are not rare. In fact, approximately 7% of the world's population are carriers of some hemoglobin disorder caused by a variation in its amino acid sequence. In concluding this chapter, we will focus on the two most common hemoglobin disorders, sickle-cell anemia and thalassemia.

CLINICAL INSIGHT

Sickle-Cell Anemia Is a Disease Caused by a Mutation in Hemoglobin

In 1904, James Herrick, a Chicago physician, examined a 20-year-old black dental student who had been admitted to the hospital because of a cough and fever. The patient felt weak and dizzy and had a headache. For about a year, he had been having palpitations and shortness of breath. On physical examination, the patient appeared normal except that his heart was distinctly enlarged and he was markedly anemic.

The patient's blood smear contained unusual red cells, which Herrick described as sickle shaped (**Figure 9.18**). Other cases of this disease, called *sickle-cell anemia*, were found soon after the publication of Herrick's description. Indeed, sickle-cell anemia is not a rare disease, with an incidence among blacks of about 4 per 1000. In the past, it has usually been a fatal disease, often before age 30, as a result of infection, renal failure, cardiac failure, or thrombosis.

Sickle-cell anemia is genetically transmitted. Patients with sickle-cell anemia have two copies of the abnormal gene (are *homozygous*). Offspring who receive an abnormal gene from one parent and a normal gene from the other have *sickle-cell trait*. Such *heterozygous* people are usually not symptomatic. Only 1% of the red cells in a heterozygote's venous circulation are sickled, in contrast with about 50% in a homozygote.

Examination of the contents of sickled red blood cells reveals that hemoglobin molecules have bound together to form large fibrous aggregates that extend across the cell, deforming the red cells and giving them their sickle

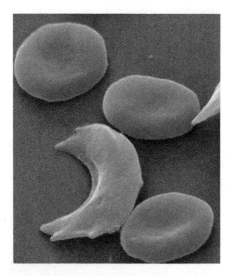

Figure 9.18 Sickled red blood cell. A micrograph showing a sickled red blood cell adjacent to normally shaped red blood cells. [Eye of Science/Science Source.]

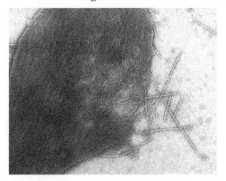

Figure 9.19 Sickle-cell hemoglobin fibers. An electron micrograph depicting a ruptured sickled red blood cell with fibers of sickle-cell hemoglobin emerging. [Courtesy of Dr. Robert Josephs and Dr. Thomas E. Wellems, University of Chicago.]

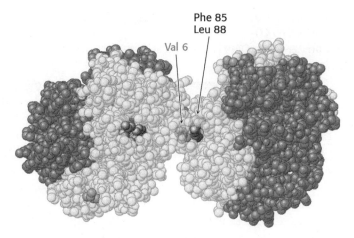

Figure 9.20 Deoxygenated hemoglobin S. The interaction between valine 6 (blue) on a β chain of one hemoglobin molecule and a hydrophobic patch formed by phenylalanine 85 and leucine 88 (gray) on a β chain of another deoxygenated hemoglobin molecule leads to hemoglobin aggregation. The exposed valine 6 residues of other β chains participate in other such interactions in HbS fibers. [Drawn from 2HBS.pdb.]

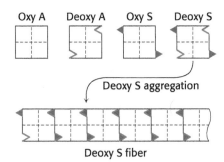

Figure 9.21 The formation of hemoglobin aggregates. The red triangle represents the sticky patch that is present on both oxy- and deoxyhemoglobin S but not on either form of hemoglobin A. The complementary site is represented by an indentation that can accommodate the triangle.

shape (Figure 9.19). Sickle-cell hemoglobin, referred to as *hemoglobin S* (HbS) to distinguish it from normal adult hemoglobin A (HbA), differs from HbA in a single amino acid substitution of valine for glutamate at position 6 of the β chains. This mutation places the nonpolar valine on the outside of hemoglobin S. *This alteration markedly reduces the solubility of the deoxygenated but not the oxygenated form of hemoglobin.* The exposed valine side chain of hemoglobin S interacts with a complementary hydrophobic patch on another hemoglobin molecule (Figure 9.20). The complementary site, formed by phenylalanine β85 and leucine β88, is exposed in deoxygenated but not in oxygenated hemoglobin. Thus, *sickling results when there is a high concentration of the deoxygenated form of hemoglobin S* (Figure 9.21). The oxygen affinity and allosteric properties of hemoglobin are virtually unaffected by the mutation, but large hemoglobin aggregates form that ultimately deform the cell.

A vicious cycle is set up when sickling takes place in a small blood vessel. The blockage of the vessel creates a local region of low oxygen concentration. Hence, more hemoglobin changes into the deoxy form and so more sickling takes place. Sickled red cells become trapped in the small blood vessels, which impairs circulation and leads to the damage of multiple organs. Sickled cells, which are more fragile than normal red blood cells, rupture (hemolyze) readily to produce severe anemia. Unfortunately, effective treatment of sickle-cell anemia has remained elusive. Note that sickle-cell anemia is another example of a pathological condition caused by inappropriate protein aggregation (p. 63).

Approximately 1 in 100 West Africans suffer from sickle-cell anemia. Given the often devastating consequences of the disease, why is the HbS mutation so prevalent in Africa and in some other regions? Recall that both copies of the HbA gene are mutated in people with sickle-cell anemia. However, if only one allele is mutated, the result is sickle-cell trait. People with sickle-cell trait are resistant to *malaria*, a disease carried by a parasite, *Plasmodium falciparum*, that lives within red blood cells at one stage in its life cycle. Because malaria is such a debilitating disease, people with the sickle-cell trait survive longer and have more children, increasing the prevalence of the HbS allele in regions where malaria is endemic.

A mother has her infant tested for sickle-call anemia at a medical center in Bamako, Mali. [FACELLY/SIPA/Newscom.]

⚕ CLINICAL INSIGHT

Thalassemia is caused by an imbalanced production of hemoglobin chains

Sickle-cell anemia is caused by the substitution of a single specific amino acid in one hemoglobin chain. *Thalassemia,* another prevalent inherited disorder of hemoglobin, is caused by the loss or substantial reduction of a single hemoglobin *chain*. The result is low levels of functional hemoglobin and a decreased production of red blood cells, which may lead to anemia, fatigue, pale skin, and spleen and liver malfunction. Thalassemia is a set of related diseases. In α-thalassemia, the α chain of hemoglobin is not produced in sufficient quantity. Consequently, hemoglobin tetramers form that contain only the β chain. These tetramers, referred to as *hemoglobin H* (HbH), bind oxygen with high affinity and no cooperativity. Thus, oxygen release in the tissues is poor. The most severe forms of α-thalassemia, in which the the α chain of hemoglobin is essentially absent, are usually fatal shortly before or just after birth. However, these forms are relatively rare.

In β-thalassemia, the β chain of hemoglobin is not produced in sufficient quantity. In the absence of β chains, the α chains form insoluble aggregates that precipitate inside immature red blood cells. The loss of red blood cells results in anemia. The severity of thalassemia depends on how much the gene is disrupted. The most severe form of β-thalassemia, called *thalassemia major* or *Cooley anemia,* results when genes from both parents are defective. Blood transfusions are the most common treatment for thalassemias.

SUMMARY

9.1   Hemoglobin Displays Cooperative Behavior

Hemoglobin is an allosteric protein that displays cooperative binding of molecular oxygen. This cooperativity is crucial to the functioning of hemoglobin because it allows rapid binding of O_2 in the lungs and easy release at the tissues where the O_2 is required to support metabolism.

9.2   Myoglobin and Hemoglobin Bind Oxygen in Heme Groups

Myoglobin is a largely α-helical protein that binds the prosthetic group heme. Heme consists of protoporphyrin, an organic component with four linked pyrrole rings, and a central iron ion in the ferrous (Fe^{2+}) state. The iron ion is coordinated to the side chain of a histidine residue in myoglobin, referred to as the proximal histidine. One of the oxygen atoms in O_2 binds to an open coordination site on the iron ion. Because of partial electron transfer from the iron ion to the oxygen atom, the iron ion moves into the plane of the porphyrin on oxygen binding.

9.3   Hemoglobin Binds Oxygen Cooperatively

Hemoglobin consists of four polypeptide chains: two α chains and two β chains. Each of these chains is similar in amino acid sequence to myoglobin and folds into a similar three-dimensional structure. The hemoglobin tetramer is best described as a pair of $\alpha\beta$ dimers. The oxygen-binding curve for hemoglobin has an "S"-like, or sigmoidal, shape, indicating that the oxygen binding is cooperative. Cooperative oxygen binding and release significantly increase the efficiency of oxygen transport.

The quaternary structure of hemoglobin changes on oxygen binding. The two $\alpha\beta$ dimers rotate by approximately 15 degrees with respect to each other in the transition from the T to the R state. Structural changes at the iron sites in response to oxygen binding are transmitted to the interface between $\alpha\beta$ dimers, influencing the T-to-R equilibrium.

9.4 An Allosteric Regulator Determines the Oxygen Affinity of Hemoglobin

Red blood cells contain 2,3-bisphosphoglycerate in concentrations approximately equal to that for hemoglobin. 2,3-BPG binds tightly to the T state but not to the R state, stabilizing the T state and lowering the oxygen affinity of hemoglobin. Fetal hemoglobin binds oxygen more tightly than does adult hemoglobin, owing to weaker 2,3-BPG binding. This difference allows oxygen transfer from maternal to fetal blood.

9.5 Hydrogen Ions and Carbon Dioxide Promote the Release of Oxygen

The oxygen-binding properties of hemoglobin are markedly affected by pH and by the presence of carbon dioxide, a phenomenon known as the Bohr effect. Increasing the concentration of hydrogen ions—that is, decreasing pH—decreases the oxygen affinity of hemoglobin, owing to the protonation of the amino termini and certain histidine residues. The protonated residues help stabilize the T state. Increasing concentrations of carbon dioxide decrease the oxygen affinity of hemoglobin by two mechanisms. First, carbon dioxide is converted into carbonic acid, which lowers the oxygen affinity of hemoglobin by decreasing the pH inside the red blood cell. Second, carbon dioxide adds to the amino termini of hemoglobin to form carbamates. These negatively charged groups stabilize deoxyhemoglobin through ionic interactions. Because hydrogen ions and carbon dioxide are produced in rapidly metabolizing tissues, the Bohr effect helps deliver oxygen to sites where it is most needed.

9.6 Mutations in Genes Encoding Hemoglobin Subunits Can Result in Disease

Sickle-cell disease is caused by a mutation in the β chain of hemoglobin that substitutes a valine residue for a glutamate residue. As a result, a hydrophobic patch forms on the surface of deoxy (T-state) hemoglobin that leads to the formation of fibrous polymers. These fibers distort red blood cells into sickle shapes. Sickle-cell disease was the first disease to be associated with a change in the amino acid sequence of a protein. Thalassemias are diseases caused by the reduced production of either the α or the β chain, yielding hemoglobin tetramers that contain only one type of hemoglobin chain. Such hemoglobin molecules are characterized by poor oxygen release and low solubility, leading to the destruction of red blood cells in the course of their development.

KEYTERMS

cooperative effect (p. 150)
heme (p. 151)
protoporphyrin (p. 151)
proximal histidine (p. 151)
distal histidine (p. 151)
α subunit (p. 152)

β subunit (p. 152)
$\alpha\beta$ dimer (p. 152)
2,3-bisphosphoglycerate (2,3-BPG) (p. 154)
fetal hemoglobin (p. 154)
Bohr effect (p. 156)

carbamate (p. 156)
carbonic anhydrase (p. 156)
sickle-cell anemia (p. 156)
thalassemia (p. 159)

❓ Answer to QUICK QUIZ

BPG binding; salt bridges between acidic and basic amino acids; salt bridges that include amino-terminal carbamate.

PROBLEMS

1. *Two by two.* Match each term with its description.

(a) Hemoglobin _____
(b) Myoglobin _____
(c) Heme _____
(d) Protoporphyrin
_____
(e) Proximal histidine
_____
(f) 2,3-Bisphosphoglycerate
_____
(g) Sickle-cell anemia
_____
(h) Bohr effect _____
(i) Carbonic anhydrase
_____
(j) Carbamate _____

1. Facilitates the formation of protons and bicarbonate
2. The regulation of oxygen binding by hydrogen ions and carbon dioxide
3. Binds in the center of the hemoglobin tetramer
4. Results from the change of a single amino acid in the β chain of hemoglobin
5. Oxygen-binding component of hemoglobin and myoglobin
6. Displays quaternary structure
7. Composed of four pyrrole rings
8. Amino termini structures that stabilize the T state
9. Binds the fifth coordination site in the heme
10. Displays tertiary structure only

2. *Hemoglobin content.* The average volume of a red blood cell is 87 mm^3. The mean concentration of hemoglobin in red cells is 0.34 g ml^{-1}.

(a) What is the weight of the hemoglobin contained in a red cell?
(b) How many hemoglobin molecules are there in a red cell?
(c) Could the hemoglobin concentration in red cells be much higher than the observed value? (Hint: Suppose that a red cell contained a crystalline array of hemoglobin molecules in a cubic lattice with 65-Å sides.)

3. *Iron content.* How much iron is there in the hemoglobin of a 70-kg adult? Assume that the blood volume is 70 ml kg^{-1} of body weight and that the hemoglobin content of blood is 0.16 g ml^{-1}.

4. *Oxygenating myoglobin.* The myoglobin content of some human muscles is about 8 g kg^{-1}. In sperm whale, the myoglobin content of muscle is about 80 g kg^{-1}.

(a) How much O$_2$ is bound to myoglobin in human muscle and in sperm-whale muscle? Assume that the myoglobin is saturated with O$_2$.

(b) The amount of oxygen dissolved in tissue water (in equilibrium with venous blood) at 37°C is about 3.5 × 10^{-5} M. What is the ratio of oxygen bound to myoglobin to that directly dissolved in the water of sperm-whale muscle?

5. *Cooperation is good.* What is the physiological significance of the cooperative binding of oxygen by hemoglobin? ✓ 7

6. *Suddenly breaking into pieces.* When crystals of deoxyhemoglobin are exposed to oxygen, the crystals shatter. Why? ✓ 7

7. *Hybrid vigor.* The oxygen-binding behavior of hemoglobin displays aspects of both the sequential model and the concerted model. Explain. ✓ 7

8. *Mom to baby.* What accounts for the fact that fetal hemoglobin has a higher oxygen affinity than maternal hemoglobin? ✓ 7

9. *Structural damage.* How does hemoglobin S cause tissue damage?

10. *Saving grace.* Hemoglobin A inhibits the formation of the long fibers of hemoglobin S and the subsequent sickling of the red cell on deoxygenation. Why does hemoglobin A have this effect?

11. *Screening the biosphere.* The first protein to have its structure determined was myoglobin from sperm whales. Propose an explanation for the observation that sperm-whale muscle is a rich source of this protein.

12. *Fits in the pocket.* Describe the role of 2,3-bisphosphoglycerate in the function of hemoglobin. ✓ 8

13. *High-altitude adaptation.* After a person spends a day or more at high altitude (with an oxygen partial pressure of 75 torr), the concentration of 2,3-bisphosphoglycerate in that person's red blood cells increases. What effect would an increased concentration of 2,3-BPG have on the oxygen-binding curve for hemoglobin? Explain why this adaptation would be beneficial for functioning well at high altitude. ✓ 8

14. *Blood doping.* Endurance athletes sometimes try an illegal method of blood doping called autologous transfusion. Some blood from the athlete is removed well before competition, and then transfused back into the athlete just before competition.

(a) Why might blood transfusion benefit the athlete?

(b) With time, stored red blood cells become depleted in 2,3-BPG. What might be the consequences of using such blood for a blood transfusion?

15. *A bad lecture?* What is the Bohr effect, and what is its chemical basis? ✓ 8

16. *I'll have the lobster.* Arthropods such as lobsters have oxygen carriers quite different from hemoglobin. The oxygen-binding sites do not contain heme but, instead,

are based on two copper(I) ions. The structural changes that accompany oxygen binding are shown below. How might these changes be used to facilitate cooperative oxygen binding?

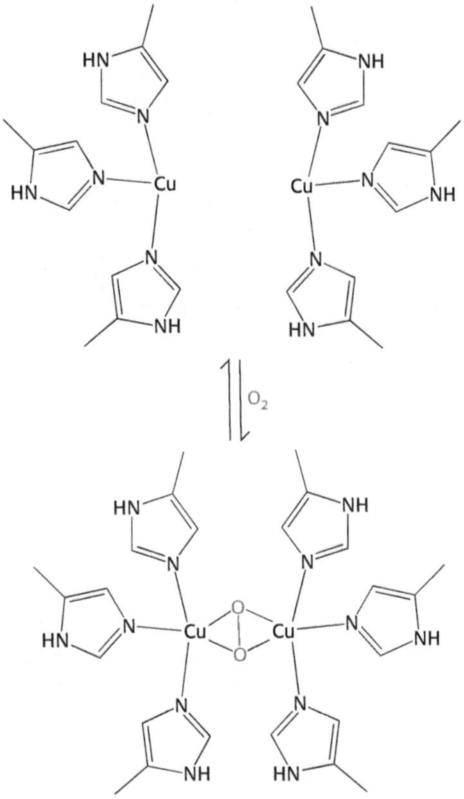

17. *Successful substitution.* Blood cells from some birds do not contain 2,3-bisphosphoglycerate; instead, they contain one of the compounds in parts *a* through *d*, which plays an analogous functional role. Which compound do you think is most likely to play this role? Explain briefly. ✓ 8

(a) 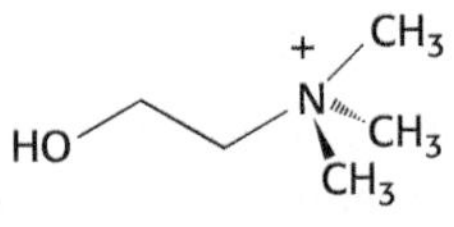 HO

Choline

(b) 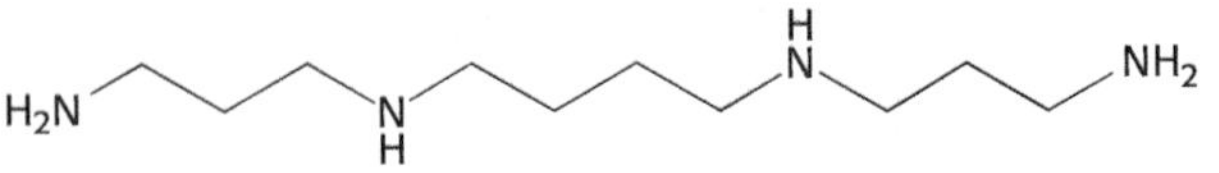 H₂N

Spermine

(c)

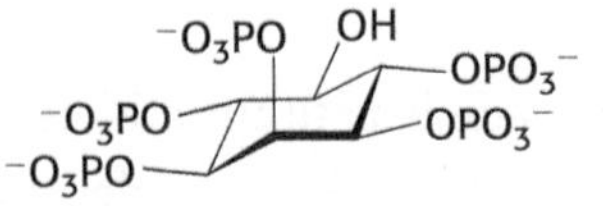

Inositol pentaphosphate

(d)

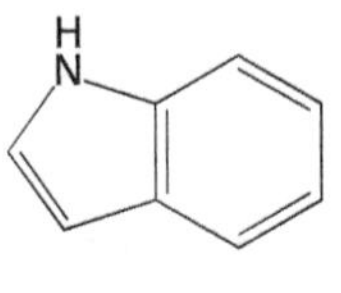

Indole

Data Interpretation Problem

18. *Leaning to the left or to the right.* The adjoining illustration shows several oxygen-dissociation curves. Assume that curve 3 corresponds to hemoglobin with physiological concentrations of CO_2 and 2,3-BPG at pH 7. Which curves represent each of the following perturbations? ✓ 8

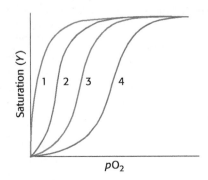

(a) Decrease in CO_2
(b) Increase in 2,3-BPG
(c) Increase in pH
(d) Loss of quaternary structure

Chapter Integration Problem

19. *Location is everything.* As shown in Figure 9.10, 2,3-bisphosphoglycerate lies in a central cavity, stabilizing the T state. What would be the effect of mutations that placed the BPG-binding site on the surface of hemoglobin? ✓ 8

Challenge Problems

20. *Release kinetics.* The dissociation constant K_D for the release of oxygen from oxymyoglobin is 10^{-6} M, where K_D is defined as

$$K_D = [Mb][O_2]/[MbO_2]$$

The rate constant for the combination of O_2 with myoglobin is 2×10^7 M^{-1} s^{-1}.

(a) What is the rate constant for the dissociation of O_2 from oxymyoglobin?
(b) What is the mean duration of the oxymyoglobin complex?

21. *Tuning proton affinity.* The pK_a of an acid depends partly on its environment. Predict the effect of each of the following environmental changes on the pK_a of a glutamic acid side chain. ✓ 8

(a) A lysine side chain is brought into proximity.
(b) The terminal carboxyl group of the protein is brought into close proximity.
(c) The glutamic acid side chain is shifted from the outside of the protein to a nonpolar site inside.

22. *Deadly gas.* Carbon monoxide is a colorless, odorless gas that binds to hemoglobin at an oxygen-binding site. Indeed, it binds 200 times as tightly as oxygen, accounting for its toxic nature. Even if only one of the four oxygen-binding sites on

hemoglobin is occupied by carbon monoxide, and the remaining three are bound to oxygen, oxygen is not released. Explain.

23. *Carrying a load.* Suppose that you are climbing a high mountain and the oxygen partial pressure in the air is reduced to 75 torr. Estimate the percentage of the oxygen-carrying capacity that will be utilized, assuming that the pH of both tissues and lungs is 7.4 and that the oxygen concentration in the tissues is 20 torr.

24. *A disconnect.* With the use of recombinant DNA techniques (Chapter 41), hemoglobin has been prepared in which the proximal histidine residues in both the α and the β subunits have been replaced by glycine. The imidazole

ring from the histidine residue can be replaced by adding free imidazole in solution.

Imidazole

Would you expect this modified hemoglobin to show cooperativity in oxygen binding? Why or why not?

25. *Parasitic effect.* When *Plasmodium falciparum,* a protozoan, lives inside red blood cells, the metabolism of the parasite tends to release acid. What effect is the presence of acid likely to have on the oxygen-carrying capacity of the red blood cells? On the likelihood that these cells will sickle? ✓ 8

Selected Readings for this chapter can be found online at www.whfreeman.com/tymoczko3e.

Carbohydrates and Lipids

CHAPTER 10
Carbohydrates

CHAPTER 11
Lipids

arbohydrates and lipids, along with proteins and nucleic acids, constitute the four major classes of biomolecules. We have already looked at proteins (Sections 2 and 3), with an emphasis on their roles as enzymes. Although carbohydrates and lipids play a diverse array of roles in living systems, their function as fuels is most obvious. This function is evident to anyone who examines the nutritional information on the labels of food packages. On such labels, lipids are designated as fats, for reasons that we will discover in Chapter 11.

Carbohydrates and lipids provide the energy to power all biochemical processes that take place inside a cell or organism. Their role as fuels is so paramount that the taste and texture of these molecules elicits gustatory pleasure, and most animals, including human beings, are behaviorally motivated to seek out foods rich in lipids and carbohydrates. For reasons to be considered in later chapters, lipids provide much more usable energy per gram than do carbohydrates. Yet most organisms maintain supplies of both types of fuel. Why have two sources of fuel if one of them provides so much more energy? The answer is that all lipids require oxygen to yield biologically useful energy. Although carbohydrates can release energy when they react with oxygen, they can also release energy when oxygen is scarce, as would be the case for the leg muscles of a runner sprinting to the finish line or for a bacterium growing in an oxygen-free environment. Thus, the use of both lipids and carbohydrates as fuel provides biochemical flexibility for meeting the various biological needs of an organism.

Carbohydrates and lipids also play important structural roles. For instance, carbohydrates provide the strength of plant cell walls, whereas lipids are ubiquitous in their role as membrane components. Indeed, carbohydrates and lipids can be joined together to form particular membrane components called *glycolipids*. Finally, these two classes of molecules play essential roles in signal-transduction pathways.

In Chapter 10, we will examine the biochemical properties of carbohydrates, highlighting some of their roles other than that of fuel. In Chapter 11, we will do the same for lipids, paying particular attention to the hydrophobic properties of these molecules.

✓ By the end of this section, you should be able to:

✓ 1 Differentiate between monosaccharides and polysaccharides in regard to structure and function.

✓ 2 Differentiate among the types of glycoproteins in regard to structure and function.

✓ 3 Describe the key chemical properties of fatty acids.

✓ 4 Identify the major lipids and describe their biochemical functions.

CHAPTER 10

Carbohydrates

Grains or cereal crops are an abundant source of carbohydrates worldwide. Katherine Lee Bates, a Wellesley College English professor, immortalized America's grain-producing capacity in her words to "America the Beautiful." Professor Bates was inspired to pen her words after experiencing the grandeur of America's West while visiting Colorado College. [Brent Winebrenner/Getty Images.]

Hardly a day goes by without reading or hearing something about "carbs" and diet. We will investigate carbohydrates as fuels in later chapters, but, before we do, let us examine what "carbs," or carbohydrates, are and see some non-fuel functions for these ubiquitous biomolecules. Carbohydrates are carbon-based molecules that are rich in hydroxyl (C—OH) groups. Indeed, the empirical formula for many carbohydrates is $(CH_2O)_n$—literally, a carbon hydrate. Simple carbohydrates are called monosaccharides. Complex carbohydrates—polymers of covalently linked monosaccharides—are called polysaccharides. A polysaccharide can be as simple as two identical monosaccharides linked together. Or it can be quite complex, consisting of dozens of different monosaccharides that are linked to form a polysaccharide composed of millions of monosaccharides. Monosaccharides are the monomers that make up polysaccharides, just as amino acids are the monomers that make up proteins. However, the nature of the covalent bonds linking the monosaccharides in a polysaccharide are much more varied than the canonical peptide bond of proteins. The variety of monosaccharides and the multiplicity of linkages forming polysaccharides mean that carbohydrates provide cells with a vast array of three-dimensional structures that can be used for a variety of purposes as simple as energy storage or as complex as cell–cell recognition signals.

✓ 1 Differentiate between monosaccharides and polysaccharides in regard to structure and function.

Carbonyl group

Aldehyde   **Ketone**

DID YOU KNOW?

Monosaccharides and other sugars are often represented by *Fischer projections*. Recall from Chapter 3 that, in a Fischer projection of a molecule, atoms joined to an asymmetric tetrahedral carbon atom by horizontal bonds are in front of the plane of the page and those joined by vertical bonds are behind the plane.

10.1 Monosaccharides Are the Simplest Carbohydrates

We begin our consideration of carbohydrates with monosaccharides, the simplest carbohydrates. These simple sugars serve not only as fuel molecules but also as fundamental constituents of living systems. For instance, DNA has a backbone consisting of alternating phosphoryl groups and deoxyribose, a cyclic five-carbon sugar.

Monosaccharides are aldehydes or ketones that have two or more hydroxyl groups. The smallest monosaccharides, composed of three carbons, are dihydroxyacetone and D- and L-glyceraldehyde.

Dihydroxyacetone
(a ketose)

D-Glyceraldehyde
(an aldose)

L-Glyceraldehyde
(an aldose)

Dihydroxyacetone is called a *ketose* because it contains a keto group (shown in red), whereas glyceraldehyde is called an *aldose* because it contains an aldehyde group (also shown in red). They are referred to as *trioses* (tri- for "three," referring to the three carbon atoms that they contain). Similarly, simple monosaccharides with four, five, six, or seven carbon atoms are called *tetroses, pentoses, hexoses,* or *heptoses,* respectively. Perhaps the monosaccharides of which we are most aware are the hexoses glucose and fructose. Glucose is an essential energy source for virtually all forms of life. Fructose is commonly used as a sweetener that is converted into glucose derivatives inside the cell.

Carbohydrates can exist in a dazzling variety of isomeric forms (**Figure 10.1**). Dihydroxyacetone and glyceraldehyde are *constitutional isomers* because they have identical molecular formulas but differ in how the atoms are ordered. *Stereoisomers* are isomers that differ in spatial arrangement. Glyceraldehyde has a single

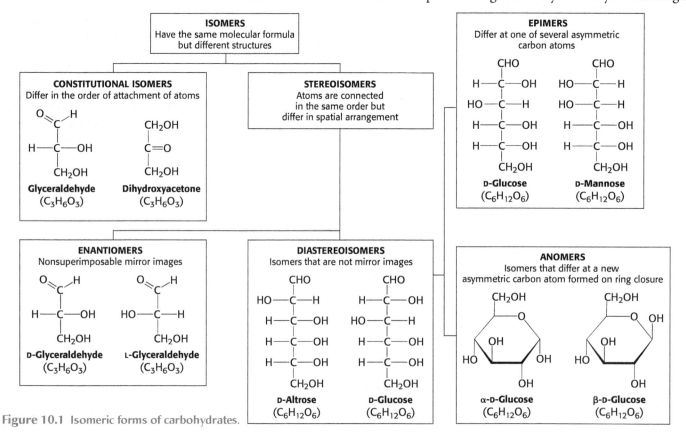

Figure 10.1 Isomeric forms of carbohydrates.

asymmetric carbon atom, and, thus, there are two stereoisomers of this sugar: D-glyceraldehyde and L-glyceraldehyde. These molecules are a type of stereoisomer called *enantiomers*, which are mirror images of each other.

Monosaccharides made up of more than three carbon atoms have multiple asymmetric carbon atoms, and so they exist not only as enantiomers but also as *diastereoisomers*, isomers that are not mirror images of each other. According to convention, the D and L isomers are determined by the configuration of the asymmetric carbon atom farthest from the aldehyde or keto group.

Figure 10.2 shows the common sugars that we will see most frequently in our study of biochemistry. D-Ribose, the carbohydrate component of RNA, is a five-carbon aldose, as is deoxyribose, the monosaccharide component of DNA. D-Glucose, D-mannose, and D-galactose are abundant six-carbon aldoses. Note that D-glucose and D-mannose differ in configuration only at C-2, the carbon atom in the second position. Sugars that are diastereoisomers differing in configuration at only a single asymmetric center are *epimers*. Thus, D-glucose and D-mannose are epimeric at C-2; D-glucose and D-galactose are epimeric at C-4. Note that ketoses have one less asymmetric center than aldoses with the same number of carbon atoms. D-Fructose is the most abundant ketohexose.

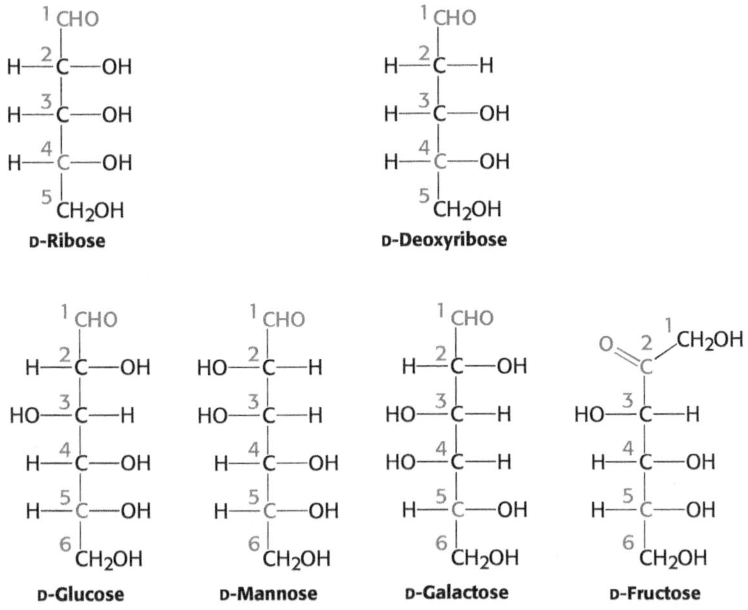

Figure 10.2 Common monosaccharides. Aldoses contain an aldehyde (shown in blue), whereas ketoses, such as fructose, contain an ketose (shown in blue). The asymmetric carbon atom farthest from the aldehyde or ketone (shown in red) designates the structures as being in the D configuration. The numbers are the standard designations for the positions of the carbon atoms (e.g., the number 2 identifies the carbon atom in the second position).

Many Common Sugars Exist in Cyclic Forms

The predominant forms of ribose, glucose, fructose, and many other sugars in solution are not open chains. Rather, the open-chain forms of these sugars cyclize into rings. The chemical basis for ring formation is that an aldehyde can react with an alcohol to form a *hemiacetal*.

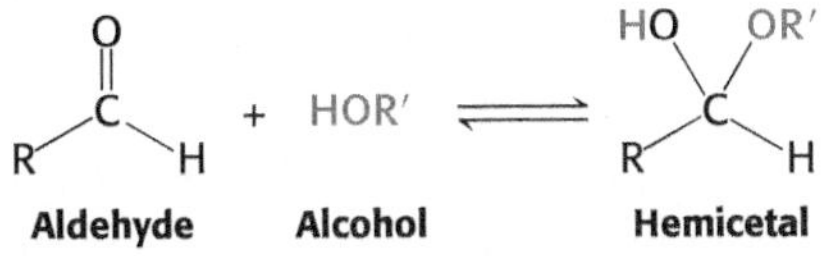

For an aldohexose such as glucose, the same molecule provides both the aldehyde and the alcohol: the C-1 aldehyde in the open-chain form of glucose reacts with the C-5 hydroxyl group to form an *intramolecular hemiacetal* (**Figure 10.3**). The resulting cyclic hemiacetal, a six-membered ring, is called *pyranose* because of its similarity to *pyran*.

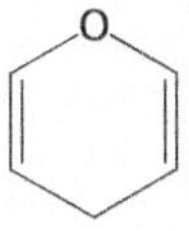

Pyran

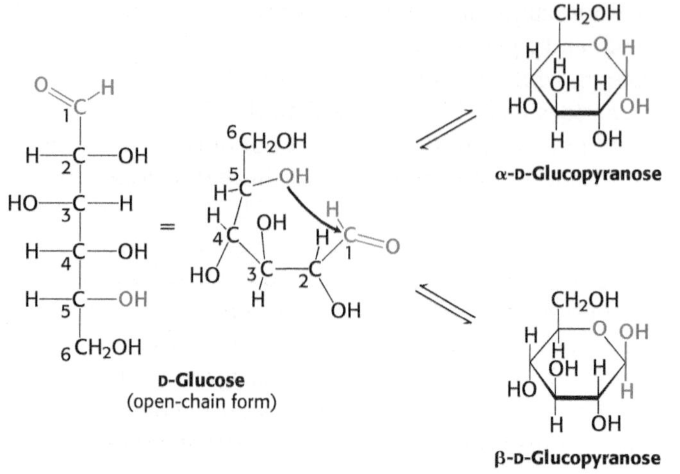

Figure 10.3 Pyranose formation. The open-chain form of glucose cyclizes when the C-5 hydroxyl group attacks carbon atom C-1 of the aldehyde group to form an intramolecular hemiacetal. Two anomeric forms, designated α and β, can result.

Similarly, a ketone can react with an alcohol to form a *hemiketal*.

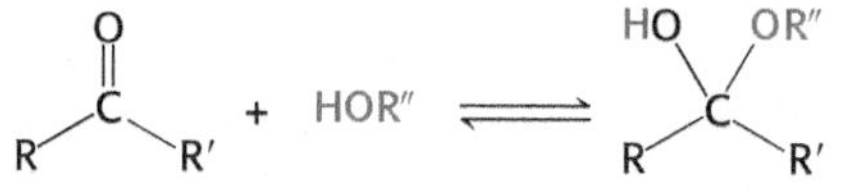

The C-2 keto group in the open-chain form of a ketohexose, such as fructose, can form an *intramolecular hemiketal* by reacting with either the C-6 hydroxyl group to form a six-membered cyclic hemiketal or the C-5 hydroxyl group to form a five-membered cyclic hemiketal (**Figure 10.4**). The five-membered ring is called a *furanose* because of its similarity to *furan*.

Furan

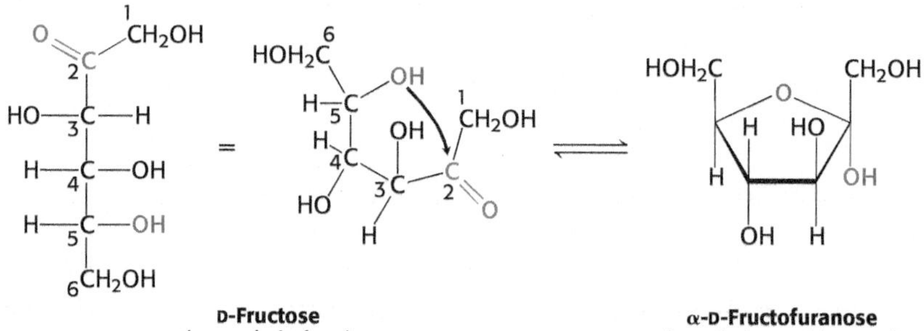

Figure 10.4 Furanose formation. The open-chain form of fructose cyclizes to a five-membered ring when the C-5 hydroxyl group attacks carbon C-2 of the ketone to form an intramolecular hemiketal. Two anomers are possible, but only the α anomer is shown.

The depictions of glucopyranose (glucose) and fructofuranose (fructose) shown in Figures 10.3 and 10.4 are *Haworth projections*. In such projections, the carbon atoms in the ring are not written out. The approximate plane of the ring is perpendicular to the plane of the paper, with the heavy line on the ring projecting toward the reader.

We have seen that carbohydrates may contain many asymmetric carbon atoms. An additional asymmetric center is created when a cyclic hemiacetal is formed, generating yet another diastereoisomeric form of sugars called *anomers*. In glucose, C-1 (the carbonyl carbon atom in the open-chain form) becomes an asymmetric center. Thus, two ring structures can be formed: α-D-glucopyranose and β-D-glucopyranose (Figure 10.3). For D sugars drawn as Haworth projections in the standard orientation as shown in Figure 10.3, *the designation α means that the hydroxyl group attached to C-1 is below the plane of the ring; β means that it is above the plane of the ring.* The C-1 carbon atom is called the *anomeric carbon atom*. An equilibrium mixture of glucose contains approximately one-third α anomer, two-thirds β anomer, and <1% of the open-chain form.

The furanose-ring form of fructose also has anomeric forms, in which α and β refer to the hydroxyl groups attached to C-2, the anomeric carbon atom (Figure 10.4). Fructose forms both pyranose and furanose rings. The pyranose α-D-**Fructofuranose**   β-D-**Fructofuranose**

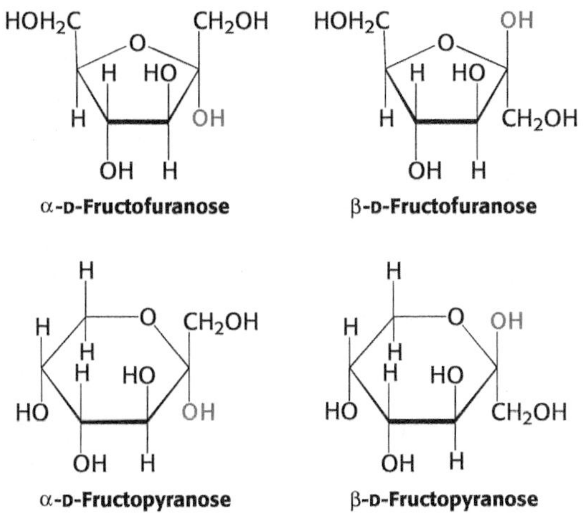

α-D-**Fructopyranose**   β-D-**Fructopyranose**

Figure 10.5 Ring structures of fructose. Fructose can form both five-membered furanose and six-membered pyranose rings. In each case, both α and β anomers are possible.

form predominates in fructose free in solution, and the furanose form predominates in many fructose derivatives (**Figure 10.5**).

β-D-Fructopyranose, found in honey, is one of the sweetest chemicals known. The β-D-fructofuranose form is not nearly as sweet. Heating converts β-fructopyranose into β-fructofuranose, reducing the sweetness of the solution. For this reason, corn syrup with a high concentration of fructose in the β-D-fructopyranose form is used as a sweetener in cold, but not hot, drinks.

Pyranose and Furanose Rings Can Assume Different Conformations

Hayworth projections provide simple means of depicting carbohydrates, but they are misleading in suggesting that the molecules are planar. For instance, stereochemical rendering of the six-membered pyranose ring shows that it is not planar, because of the tetrahedral geometry of its saturated carbon atoms. Instead, pyranose rings adopt two classes of conformations, termed "chair" and "boat" because of their resemblance to these objects (**Figure 10.6**). In the chair form, the substituents on the ring carbon atoms have two orientations: axial and equatorial. *Axial* bonds are nearly perpendicular to the average plane of the ring, whereas *equatorial* bonds are nearly parallel to this plane. Axial substituents sterically hinder each other if they emerge on the same side of the ring (e.g., 1,3-diaxial groups). In contrast, equatorial substituents are less crowded.

Although both the α and β chair form of D-glucopyranose exist in solution, the chair form of β-D-glucopyranose predominates because all axial positions are occupied by hydrogen atoms. The bulkier —OH and —CH$_2$OH groups emerge at the less hindered periphery. The boat form of glucose is disfavored because it is quite sterically hindered.

 CLINICAL INSIGHT

Glucose Is a Reducing Sugar

Because the α and β isomers of glucose are in an equilibrium that passes through the open-chain aldehyde form, glucose has some of the chemical properties of free aldehydes, such as the ability to react with oxidizing agents. For example, glucose can react with cupric ion (Cu^{2+}) reducing it to cuprous ion (Cu$^+$), while being oxidized to gluconic acid.

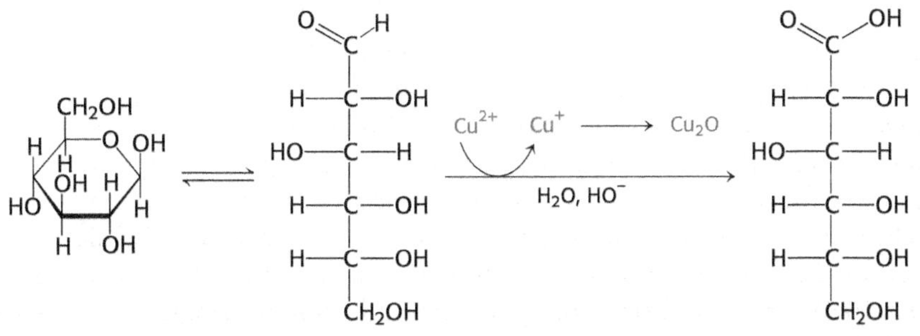

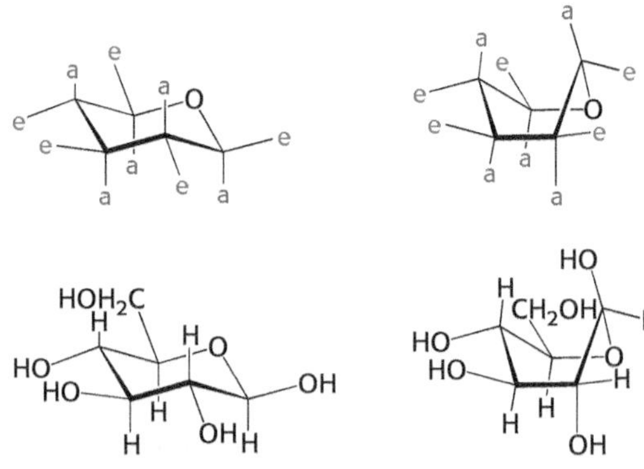

Figure 10.6 Chair and boat forms of β-D-glucose. The chair form is more stable owing to less steric hindrance because the axial positions are occupied by hydrogen atoms. Abbreviations: a, axial; e, equatorial.

Solutions of cupric ion (known as Fehling's solution) provide a simple test for the presence of sugars such as glucose. Sugars that react with solutions of cupric ion are called *reducing sugars*; those that do not are called *nonreducing sugars*. Reducing sugars can often nonspecifically react with amino groups on proteins not participating in a peptide bond. For instance, as a reducing sugar, glucose can react with hemoglobin to form glycosylated hemoglobin (hemoglobin A1c). Changes in the amount of glycosylated hemoglobin can be used to monitor the effectiveness of treatments for diabetes mellitus, a condition characterized by high levels of blood glucose (Chapter 25). Because the glycosylated hemoglobin remains in circulation, the amount of the modified hemoglobin corresponds to the long-term regulation—over several months—of glucose levels. In nondiabetic people, less than 6% of the hemoglobin is glycosylated, whereas in uncontrolled diabetics, almost 10% of the hemoglobin is glycosylated. Although the glycosylation of hemoglobin has no effect on oxygen binding and is thus benign, similar reducing reactions between sugars and other proteins are often detrimental to the body. These modifications, known as *advanced glycation end products* (AGEs), have been implicated in aging, arteriosclerosis, and diabetes, as well as other pathological conditions.

Monosaccharides Are Joined to Alcohols and Amines Through Glycosidic Bonds

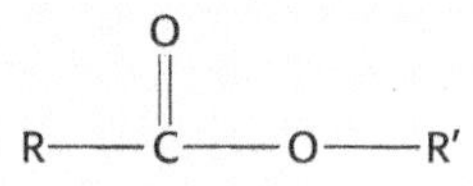

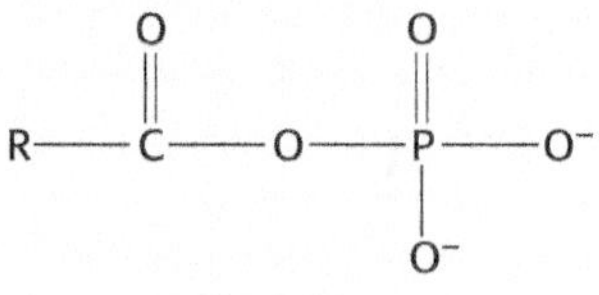

The biochemical properties of monosaccharides can be modified by reaction with other molecules. These modifications increase the biochemical versatility of carbohydrates, enabling them to serve as signal molecules or facilitating their metabolism. Three common reactants are alcohols, amines, and phosphates. A bond formed between the anomeric carbon atom of glucose and the oxygen atom of an alcohol is called a *glycosidic bond*—specifically, an *O-glycosidic bond*—and the resulting product is called a *glycoside*. O-Glycosidic bonds are prominent when carbohydrates are linked together to form long polymers and when they are attached to proteins. In addition, the anomeric carbon atom of a sugar can be linked to the nitrogen atom of an amine to form an *N-glycosidic bond*. Carbohydrates

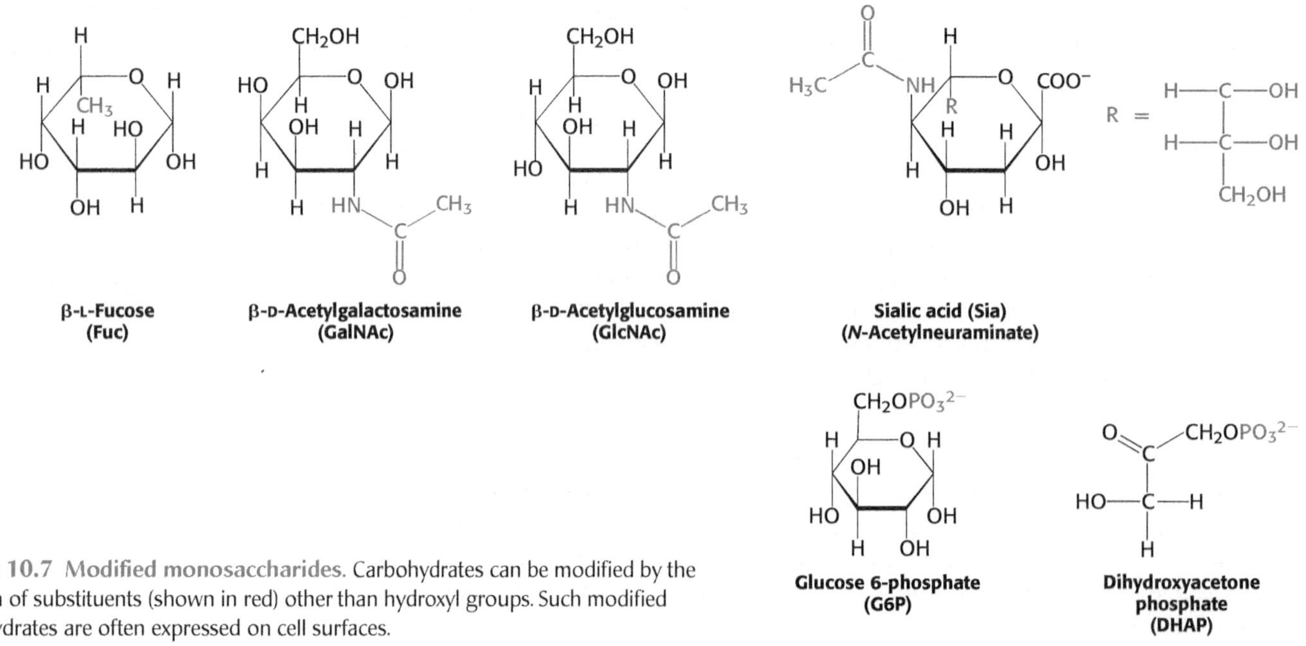

An ester

A phosphoester

Figure 10.7 Modified monosaccharides. Carbohydrates can be modified by the addition of substituents (shown in red) other than hydroxyl groups. Such modified carbohydrates are often expressed on cell surfaces.

β-L-Fucose (Fuc)

β-D-Acetylgalactosamine (GalNAc)

β-D-Acetylglucosamine (GlcNAc)

Sialic acid (Sia) (N-Acetylneuraminate)

Glucose 6-phosphate (G6P)

Dihydroxyacetone phosphate (DHAP)

can also form an ester linkage to phosphates, one of the most prominent modifications in carbohydrate metabolism. For instance, the phosphorylation of glucose is essential when glucose metabolizes to yield ATP (Chapter 16). Examples of modified carbohydrates are shown in **Figure 10.7**.

 BIOLOGICAL INSIGHT

Glucosinolates Protect Plants and Add Flavor to Our Diets

Plants, especially those of the order Brassicales, produce a special class of glycosides called *glucosinolates* as a defense against herbivory. When glucosinolate is hydrolyzed, isothiocyanate is released, generating a sharp taste that discourages further eating by an insect. The glucosinolate is stored separately from the activating enzyme, myrosinase; but on tissue damage, the enzyme and substrate combine for the hydrolysis reaction.

Many glucosinolates vary with the nature of the R component. The combination of the glucosinolate and myrosinase is sometimes called the "mustard-oil bomb." The mustard-oil bomb can also be activated by chewing, accounting for the sharp taste of mustard, relish, kale, broccoli, and other members of the order Brassicales. Interestingly, in human beings, certain glucosinolates stimulate the production of detoxifying enzymes that may play a role in cancer prevention, perhaps contributing to the cancer-protective effects of diets rich in Brassicales.

10.2 Monosaccharides Are Linked to Form Complex Carbohydrates

Because sugars contain many hydroxyl groups, glycosidic bonds can join one monosaccharide to another. *Oligosaccharides* are built by the linkage of two or more monosaccharides by *O*-glycosidic bonds (**Figure 10.8**). In the disaccharide maltose, for example, two D-glucose residues are joined by a glycosidic linkage between the α-anomeric form of C-1 on one sugar and the hydroxyl oxygen atom on C-4 of the adjacent sugar. Such a linkage is called an α-1,4-glycosidic bond. The fact that monosaccharides have multiple hydroxyl groups means that many different glycosidic linkages are possible. For example, consider three monosaccharides—glucose, mannose, and galactose. These molecules can be linked together in the laboratory to form more than 12,000 different structures. In this section, we will look at some of the most common oligosaccharides found in nature.

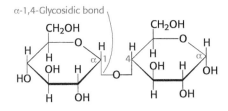

Figure 10.8 Maltose, a disaccharide. Two molecules of glucose are linked by an α-1,4-glycosidic bond to form the disaccharide maltose. The angles in the bonds to the central oxygen atom do not denote carbon atoms. The angles are added only for ease of illustration.

Specific Enzymes Are Responsible for Oligosaccharide Assembly

Oligosaccharides are synthesized through the action of specific enzymes, *glycosyltransferases*, which catalyze the formation of glycosidic bonds. Given the diversity of known glycosidic linkages, many different enzymes are required. Indeed, glycosyltransferases account for 1% to 2% of gene products in all organisms examined.

The general form of the reaction catalyzed by a glycosyltransferase is shown in **Figure 10.9**. The sugar to be added comes in the form of an activated (energy-rich) sugar nucleotide, such as UDP-glucose (UDP is the abbreviation for uridine diphosphate). The attachment of a nucleotide to enhance the energy content of a molecule is a common strategy in biosynthesis that we will see many times in our study of biochemistry (Chapters 16 and 25).

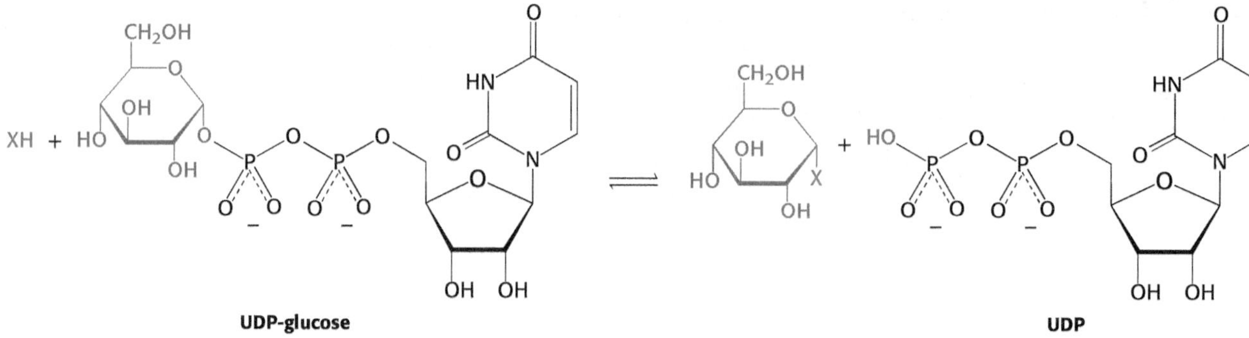

UDP-glucose

UDP

Figure 10.9 General form of a glycosyltransferase reaction. The sugar to be added comes from a sugar nucleotide—in this case, UDP-glucose. The acceptor, designated X in this illustration, can be a simple monosaccharide, a complex polysaccharide, or a serine, threonine, or asparagine residue of a protein.

Sucrose, Lactose, and Maltose Are the Common Disaccharides

A *disaccharide* consists of two sugars joined by an *O*-glycosidic bond. Three abundant disaccharides that we encounter frequently are sucrose, lactose, and maltose (**Figure 10.10**). *Sucrose* (common table sugar) is obtained commercially from sugar cane or sugar beets. The anomeric carbon atoms of a glucose unit and a fructose unit are joined in this disaccharide; the configuration of this glycosidic linkage is α for glucose and β for fructose. Sucrose can be cleaved into its component monosaccharides by the enzyme *sucrase*.

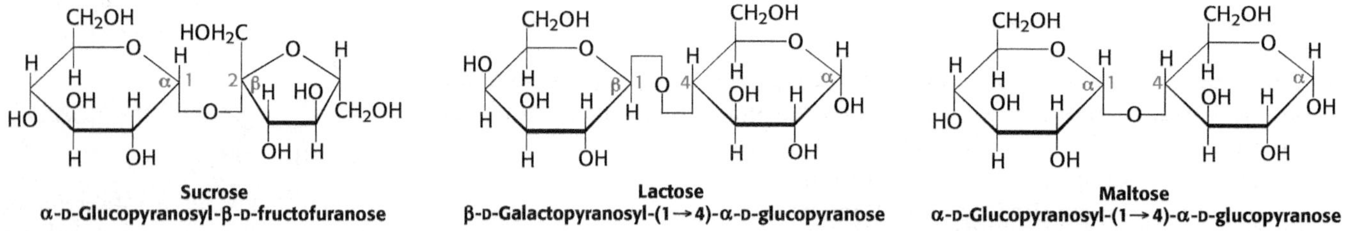

Sucrose
α-D-Glucopyranosyl-β-D-fructofuranose

Lactose
β-D-Galactopyranosyl-(1→4)-α-D-glucopyranose

Maltose
α-D-Glucopyranosyl-(1→4)-α-D-glucopyranose

Figure 10.10 Common disaccharides. Sucrose, lactose, and maltose are common dietary components. The angles in the bonds to the central oxygen atoms do not denote carbon atoms.

Lactose, the disaccharide of milk, consists of galactose joined to glucose by a β-1,4-glycosidic linkage. Lactose is hydrolyzed to these monosaccharides by *lactase* in human beings and by *β-galactosidase* in bacteria. In *maltose*, two glucose units are joined by an α-1,4-glycosidic linkage. Maltose comes from the hydrolysis of large polymeric oligosaccharides such as starch and glycogen and is in turn hydrolyzed to glucose by *maltase*. Sucrase, lactase, and maltase are located on the outer surfaces of epithelial cells lining the small intestine (**Figure 10.11**). The cleavage products of sucrose, lactose, and maltose can be further processed to provide energy in the form of ATP.

Glycogen and Starch Are Storage Forms of Glucose

Large polymeric oligosaccharides, formed by the linkage of multiple monosaccharides, are called *polysaccharides* and play vital roles in energy storage and in maintaining the structural integrity of an organism. The variety of different monosaccharides can be put together in any number of arrangements, creating a huge array of possible polysaccharides. If all of the monosaccharide units in a polysaccharide are the same, the polymer is called a *homopolymer*. The most common homopolymer in animal cells is *glycogen,* the storage form of glucose. Glycogen is present in most of our tissues but is most common in muscle and liver. As will be considered in detail in Chapters 24 and 25, glycogen is a large, branched polymer of glucose residues. Most of the glucose units in glycogen are linked by α-1,4-glycosidic bonds. The branches are formed by α-1,6-glycosidic bonds, present about once in 10 units (**Figure 10.12**).

Figure 10.11 Electron micrograph of microvilli. Lactase and other enzymes that hydrolyze carbohydrates are present on microvilli that project from the outer face of the plasma membrane of intestinal epithelial cells. [Courtesy of Louisa Howard, Dartmouth College.]

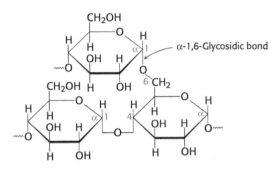

α-1,6-Glycosidic bond

Figure 10.12 Branch point in glycogen. Two chains of glucose molecules joined by α–1,4-glycosidic bonds are linked by an α–1,6-glycosidic bond to create a branch point. Such an α–1,6-glycosidic bond forms at approximately every 10 glucose units, making glycogen a highly branched molecule.

The nutritional reservoir in plants is the homopolymer *starch,* of which there are two forms. *Amylose,* the unbranched type of starch, consists of glucose residues in α-1,4 linkage. *Amylopectin,* the branched form, has about 1 α-1,6 linkage per 30 α-1,4 linkages, in similar fashion to glycogen except for its lower degree of branching. More than half the carbohydrate ingested by human beings is starch, found in wheat, potatoes, and rice, to name just a few sources (**Figure 10.13**). Amylopectin, amylose, and glycogen are rapidly hydrolyzed by α-*amylase,* an enzyme secreted by the salivary glands and the pancreas.

? QUICK QUIZ 1 Draw the structure of the disaccharide α-glycosyl–$(1 \rightarrow 6)$-galactose in the β anomeric form.

Cellulose, a Structural Component of Plants, Is Made of Chains of Glucose

Cellulose, the other major polysaccharide of glucose found in plants, serves a structural rather than a nutritional role as an important component of the plant cell wall. *Cellulose is among the most abundant organic compounds in the biosphere.* Some 10^{15} kg of cellulose is synthesized and degraded on Earth each year, an amount 1000 times as great as the combined weight of the human race. Cellulose is an unbranched polymer of glucose residues joined by β-1,4 linkages, in contrast with the α-1,4 linkage seen in starch and glycogen. This simple difference in stereochemistry yields two molecules with very different properties and biological functions. The β configuration allows cellulose to form very long, straight chains. Fibrils are formed by parallel chains that interact with one another through hydrogen bonds, generating a rigid, supportive structure. The straight chains formed by β linkages are optimal for the construction of fibers having a high tensile strength. The α-1,4 linkages in glycogen and starch produce

Figure 10.13 Pasta is a common source of starch. Pasta, thin pieces of unleavened dough molded into regular shapes, is made with flour, which has a high starch content. [Elena Schweitzer/Shutterstock.]

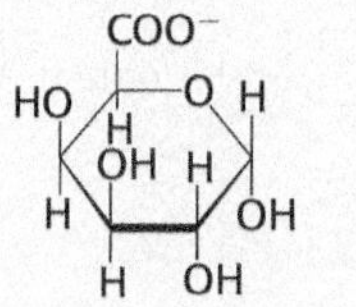

Galacturonate
(ionized galacturonic acid)

a very different molecular architecture: a hollow helix is formed instead of a straight chain (**Figure 10.14**). The hollow helix formed by α linkages is well suited to the formation of a more compact, accessible store of sugar. Although mammals lack cellulases and therefore cannot digest wood and vegetable fibers, cellulose and other plant fibers are still an important constituent of the mammalian diet as a component of dietary fiber. Soluble fiber such as *pectin* (polygalacturonate) slows the movement of food through the gastrointestinal tract, allowing improved digestion and the absorption of nutrients. Insoluble fibers, such as cellulose, increase the rate at which digestion products pass through the large intestine. This increase in rate can minimize exposure to toxins in the diet.

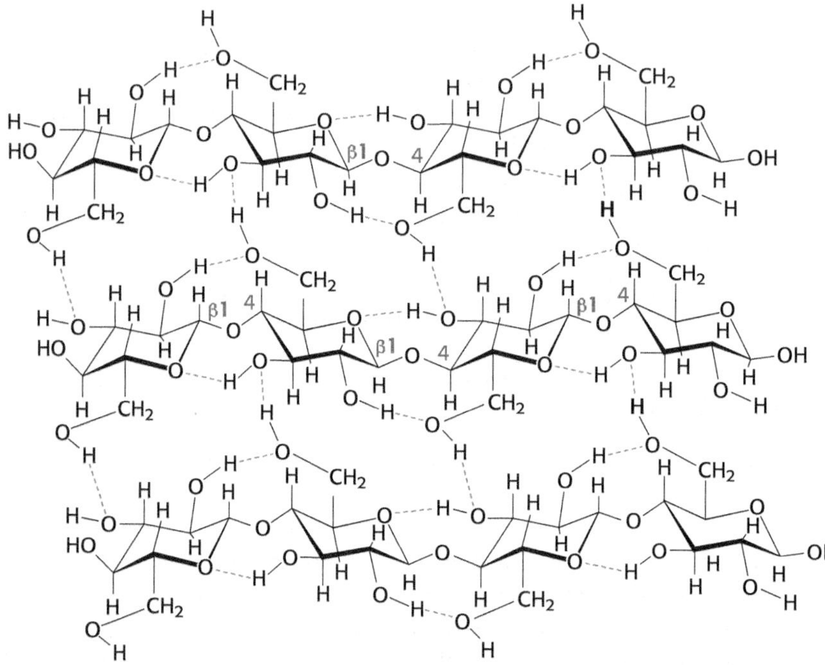

Cellulose
(β-1,4 linkages)

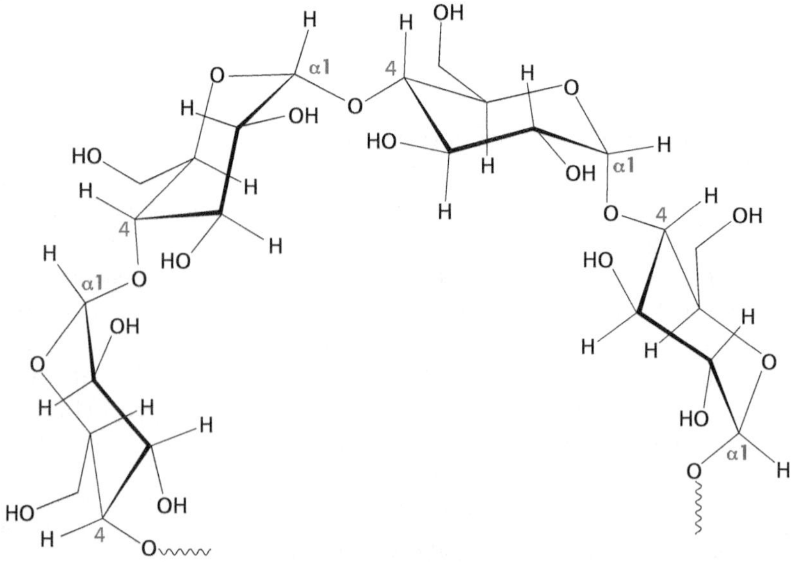

Starch and Glycogen
(α-1,4 linkages)

Figure 10.14 Glycosidic bonds determine polysaccharide structure. The β-1,4 linkages favor straight chains, which are optimal for structural purposes. The α-1,4 linkages favor bent structures, which are more suitable for storage.

10.3 Carbohydrates Are Attached to Proteins to Form Glycoproteins

✓ 2 Differentiate among the types of glycoproteins in regard to structure and function.

A carbohydrate group can be covalently attached to a protein to form a *glycoprotein*. Such modifications are not rare, as 50% of the proteome consists of glycoproteins. We will examine three classes of glycoproteins. In the first class, referred to simply as glycoproteins, the protein constituent is the largest component by weight. This versatile class plays a variety of biochemical roles. Many glycoproteins are components of cell membranes, where they take part in processes such as cell adhesion and the binding of sperm to eggs. Other glycoproteins are formed by linking carbohydrates to soluble proteins. Many of the proteins secreted from cells are glycosylated, or modified by the attachment of carbohydrates, including most proteins present in the serum component of blood.

The second class of glycoproteins comprises the *proteoglycans*. The protein component of proteoglycans is conjugated to a particular type of polysaccharide called a *glycosaminoglycan*. Carbohydrates make up a much larger percentage by weight of the proteoglycan compared with simple glycoproteins. Proteoglycans function as structural components and lubricants.

The third class of glycoproteins are the *mucins*, or *mucoproteins*, which, like proteoglycans, are predominately carbohydrate. Mucins, a key component of mucus, serve as lubricants. *N*-Acetylgalactosamine is usually the carbohydrate moiety bound to the protein in mucins. *N*-Acetylgalactosamine is an example of an *amino sugar*, so named because an amino group replaces a hydroxyl group.

Carbohydrates May Be Linked to Asparagine, Serine, or Threonine Residues of Proteins

In all classes of glycoproteins, sugars are attached either to the amide nitrogen atom in the side chain of asparagine (termed an *N-linkage*) or to the hydroxyl oxygen atom in the side chain of serine or threonine (termed an *O-linkage*), as shown in **Figure 10.15**, a process called *glycosylation*. All *N*-linked oligosaccharides have in common a pentasaccharide core consisting of three mannoses, a six-carbon sugar, and two *N*-acetylglucosamines, a glucosamine in which the nitrogen atom binds to an acetyl group. Additional sugars are attached to this core to form the great variety of oligosaccharide patterns found in glycoproteins (**Figure 10.16**).

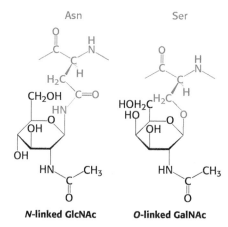

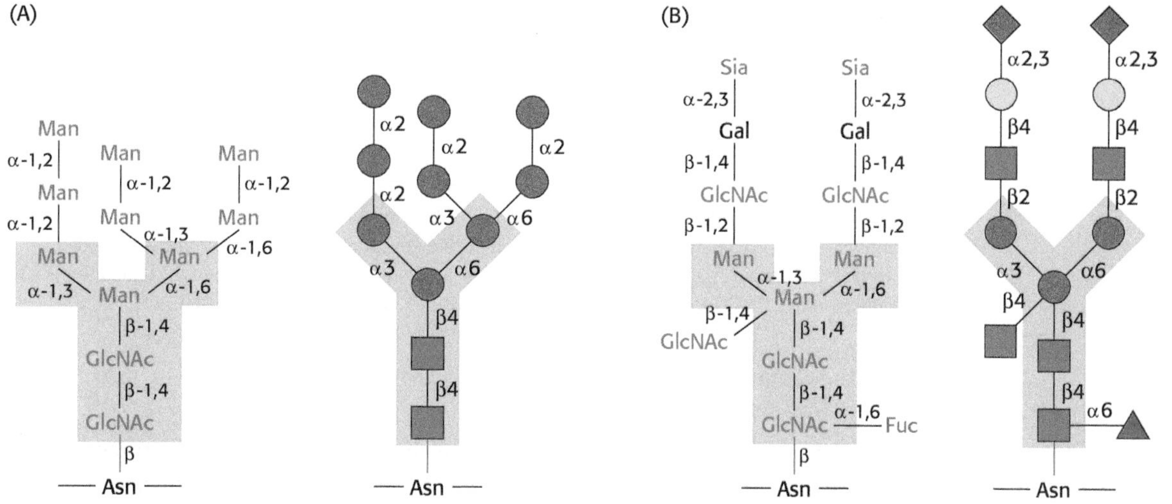

N-linked GlcNAc   **O-linked GalNAc**

Figure 10.15 Glycosidic bonds between proteins and carbohydrates. A glycosidic bond links a carbohydrate to the side chain of asparagine (*N*-linked) or to the side chain of serine or threonine (*O*-linked). The glycosidic bonds are shown in red. Abbreviations: GlcNAc, *N*-acetylglucosamine; GalNAc, *N*-acetylgalactosamine.

(A)

(B)

Figure 10.16 *N*-linked oligosaccharides. A pentasaccharide core (shaded gray) is common to all *N*-linked oligosaccharides and serves as the foundation for a wide variety of *N*-linked oligosaccharides, two of which are illustrated: (A) high-mannose type; (B) complex type. Detailed chemical formulas and schematic structures are shown for each type (for the key to the scheme, see Figure 10.17). Abbreviations for sugars: Fuc, fucose; Gal, galactose; GalNAc, *N*-acetylgalactosamine; GlcNAc, *N*-acetylglucosamine; Man, mannose; Sia, sialic acid.

CLINICAL INSIGHT

The Hormone Erythropoietin Is a Glycoprotein

Let us look at a glycoprotein present in the blood serum that has dramatically improved treatment for anemia, particularly that induced by cancer chemotherapy. The glycoprotein hormone *erythropoietin* (EPO) is secreted by the kidneys and stimulates the production of red blood cells by bone marrow, the tissue in interior of bones. EPO is composed of 165 amino acids and is *N*-glycosylated at three asparagine residues and *O*-glycosylated on a serine residue (**Figure 10.17**), making the mature EPO 40% carbohydrate by weight. Glycosylation enhances the stability of the protein in the blood; unglycosylated protein has only about 10% of the bioactivity of the glycosylated form because the protein is rapidly removed from the blood by the kidney. The availability of recombinant human EPO has greatly aided the treatment of anemias. Some endurance athletes have used recombinant human EPO to increase the red-blood-cell count and hence their oxygen-carrying capacity, a practice prohibited by most professional sports organizations. Drug-testing laboratories are able to distinguish some forms of prohibited recombinant EPO from natural EPO in athletes by detecting differences in the glycosylation patterns through the use of isoelectric focusing (p. 75).

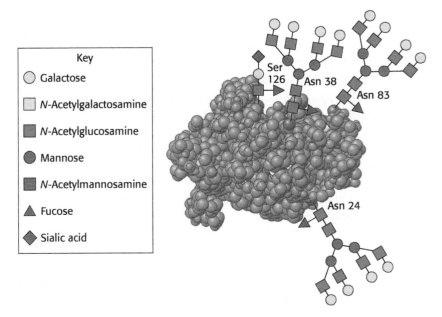

Key

○ Galactose

▢ *N*-Acetylgalactosamine

◼ *N*-Acetylglucosamine

● Mannose

◼ *N*-Acetylmannosamine

▲ Fucose

◆ Sialic acid

Figure 10.17 Oligosaccharides attached to erythropoietin. Erythropoietin has oligosaccharides linked to three asparagine residues and one serine residue. The structures shown are approximately to scale. The carbohydrate structures represented in the amino acid residues are depicted symbolically by employing a scheme (shown in the key, which also applies to Figures 10.16, 10.23, and 10.24) that is becoming widely used. [Drawn from 1BUY.pdf.]

Proteoglycans, Composed of Polysaccharides and Protein, Have Important Structural Roles

The second class of glycoproteins that we consider comprises the proteoglycans, proteins attached to a particular type of polysaccharide called glycosaminoglycans, as stated earlier. The glycosaminoglycan makes up as much as 95% of the biomolecule by weight, and so the proteoglycan resembles a polysaccharide more than a protein. Proteoglycans not only function as lubricants and structural components in connective tissue, but also mediate the adhesion of cells to the extracellular matrix, and bind factors that regulate cell proliferation.

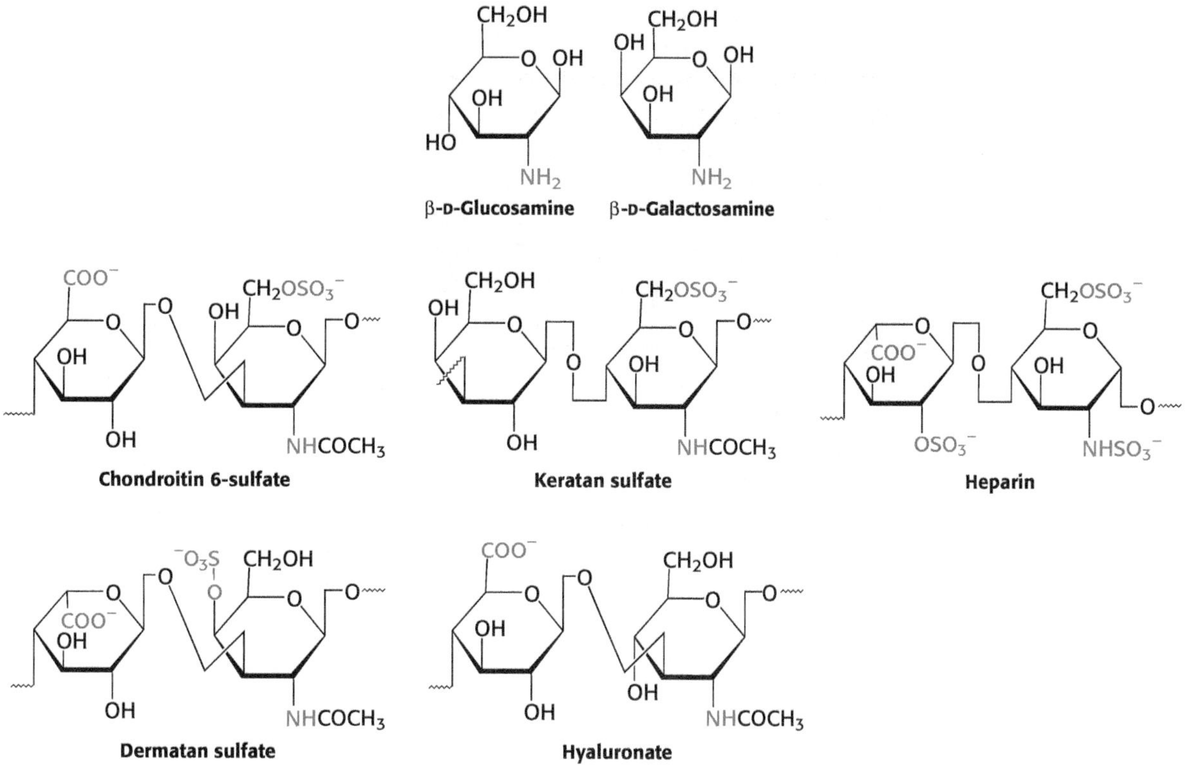

β-D-Glucosamine   β-D-Galactosamine

Chondroitin 6-sulfate   **Keratan sulfate**   **Heparin**

Dermatan sulfate   **Hyaluronate**

Figure 10.18 Repeating units in glycosaminoglycans. Structural formulas for five repeating units of important glycosaminoglycans illustrate the variety of modifications and linkages that are possible. Amino groups are shown in blue and negatively charged groups in red. Hydrogen atoms have been omitted for clarity. The right-hand structure is a glucosamine derivative in each case. The parent amino sugars, β-D-glucosamine and β-D-galactosamine, are shown for reference.

The properties of proteoglycans are determined primarily by the glycosaminoglycan component. Many glycosaminoglycans are made of repeating units of disaccharides containing a derivative of an amino sugar, either glucosamine or galactosamine (**Figure 10.18**). At least one of the two sugars in the repeating unit has a *negatively charged carboxylate or sulfate group*. The major glycosaminoglycans in animals are chondroitin sulfate, keratan sulfate, heparin, heparan sulfate, dermatan sulfate, and hyaluronate. *Mucopolysaccharidoses* are a collection of diseases, such as Hurler disease, that result from the inability to degrade glycosaminoglycans (**Figure 10.19**). Although precise clinical features vary with the disease, all mucopolysaccharidoses result in skeletal deformities and reduced life expectancies.

 CLINICAL INSIGHT

Proteoglycans Are Important Components of Cartilage

The proteoglycan *aggrecan* and the protein collagen are key components of cartilage. The triple helix of collagen (p. 56) provides structure and tensile strength, whereas aggrecan serves as a shock absorber (**Figure 10.20**). The protein component of aggrecan is a large molecule composed of 2397 amino acids (**Figure 10.21**). Many aggrecans are linked together by hyaluronate, a glycosaminoglycan. The aggrecan molecule itself is decorated with the glycosaminoglycans chondroitin sulfate and keratan sulfate. This combination of glycosaminoglycan and protein is especially suited to function as a shock absorber. Water is absorbed on the glycosaminoglycans, attracted by the many negative charges. Aggrecan can cushion compressive forces because the absorbed water enables the proteoglycan to spring back after having been deformed. When

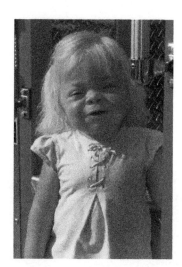

Figure 10.19 Hurler disease. Formerly called gargoylism, Hurler disease is a mucopolysaccharidosis having symptoms that include wide nostrils, a depressed nasal bridge, thick lips and earlobes, and irregular teeth. In Hurler disease, glycosaminoglycans cannot be degraded. The excess of these molecules is stored in the soft tissue of the facial regions, resulting in the characteristic facial features. [Courtesy of National MPS Society, www.mpssociety.org]

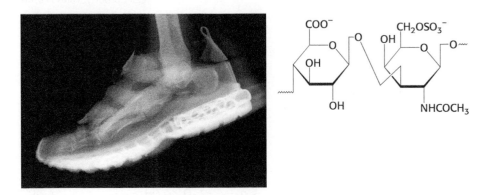

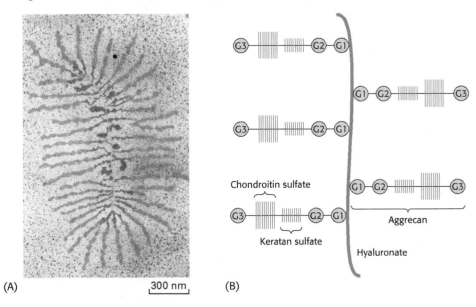

Figure 10.20 **Cartilage acts as a shock absorber.** The cartilage of a runner's foot cushions the impact of each step that she takes. The repeating unit of chondroitin sulfate, a glycosaminoglycan component of cartilage, is shown on the right. [Untitled x-ray/Nick Veasey/Getty Images.]

pressure is exerted, such as when the foot hits the ground in walking, water is squeezed from the glycosaminoglycan, cushioning the impact. When the pressure is released, the water rebinds. *Osteoarthritis*, the most common type of arthritis, can result from the proteolytic degradation of aggrecan and collagen in the cartilage in the joints due to inflammation. More typically, the cause is simple "wear and tear." Glucosamine and chondroitin (Figure 10.18) are widely promoted as a treatment for osteoarthritis, but the benefits to anyone but the producers and marketers of the supplement are very ambiguous.

Figure 10.21 **The structure of proteoglycan from cartilage.** (A) Electron micrograph of a proteoglycan from cartilage (with color added). Proteoglycan monomers emerge laterally at regular intervals from opposite sides of a central filament of hyaluronan. (B) Schematic representation in which G stands for globular domain. [(A) Reprinted from J. A. Buckwalter and L. Rosenberg, Structural changes during development in bovine fetal epiphyseal cartilage, *Collagen and Related Research* 3:489–504. Copyright Elsevier (1983). Courtesy of Dr. Lawrence Rosenberg.]

(A)   300 nm   (B)

Although glycosaminoglycans may not seem to be familiar molecules, they are common throughout the biosphere. *Chitin* is a glycosaminoglycan found in the exoskeleton of insects, crustaceans, and arachnids and is, next to cellulose, the second most abundant polysaccharide in nature (**Figure 10.22**). Cephalopods, such as squid, have razor sharp beaks, which are made of extensively cross-linked chitin, to disable and consume prey.

Figure 10.22 **Chitin, a glycosaminoglycan, is present in insect wings and the exoskeleton.** Glycosaminoglycans are components of the exoskeletons of insects, crustaceans, and arachnids. [FLPA/Alamy.]

CLINICAL INSIGHT

Mucins Are Glycoprotein Components of Mucus

The third class of glycoproteins that we examine comprises the mucins (mucoproteins). In mucins, the protein component is extensively glycosylated to serine or threonine residues by *N*-acetylgalactosamine (Figure 10.7). Mucins are capable of forming large polymeric structures and are common in mucous secretions. These glycoproteins are synthesized by specialized cells in the tracheobronchial, gastrointestinal, and urogenital tracts. Mucins are abundant in saliva where they function as lubricants.

(A)

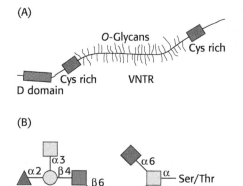

(B)

Figure 10.23 Mucin structure.
(A) A schematic representation of a
mucoprotein. The VNTR region is highly
glycosylated, forcing the molecule into an
extended conformation. The Cys–rich
domains and the D domain facilitate the
polymerization of many such molecules.
(B) An example of an oligosaccharide that is
bound to the VNTR region of the protein.
[Information from A. Varki, R. D. Cummings,
J. D. Esko, H. H. Freeze, P. Stanley, C. R.
Bertozzi, G. W. Hart, and M. E. Etzler, *Essentials
of Glycobiology*, 2d ed. (Cold Spring Harbor
Press, 2009), (Part A) p. 117, Fig. 9.1; (Part B)
p. 118, Fig. 9.2.]

A model of a mucin is shown in **Figure 10.23**A. The defining feature of the
mucins is a region of the protein backbone termed the *variable number of
tandem repeats* (VNTR) region, which is rich in serine and threonine residues
that are O-glycosylated. Indeed, the carbohydrate moiety can account for as
much as 80% of the molecule by weight. A number of core carbohydrate
structures are conjugated to the protein component of mucin. Figure 10.23B
shows one such structure.

Mucins adhere to epithelial cells and act as a protective barrier; they also
hydrate the underlying cells. In addition to protecting cells from
environmental insults, such as stomach acid, inhaled chemicals in the lungs,
and bacterial infections, mucins have roles in fertilization, the immune
response, and cell adhesion. Mucins are overexpressed in bronchitis and cystic
fibrosis, and the overexpression of mucins is also characteristic of
adenocarcinomas—cancers of the glandular cells of epithelial origin.

 BIOLOGICAL INSIGHT

Blood Groups Are Based on Protein Glycosylation Patterns

The human ABO blood groups illustrate the effects of
glycosyltransferases. Each blood group is designated by the
presence of one of the three different carbohydrates, termed A, B,
or O, on the surfaces of red blood cells (**Figure 10.24**). These
structures have in common an oligosaccharide foundation called
the O (or sometimes H) antigen. The A and B antigens differ from
the O antigen by the addition of one extra monosaccharide, either
N-acetylgalactosamine (for A) or galactose (for B) through an
α-1,3 linkage to a galactose moiety of the O antigen.

Specific glycosyltransferases add the extra monosaccharide to
the O antigen. There are three common genes encoding such
glycosyltransferases, and each person inherits the genes for each
type from each parent. The type A transferase specifically adds
N-acetylgalactosamine, whereas the type B transferase adds
galactose. These enzymes are identical in all but 4 of 354 positions.
The O phenotype is the result of a mutation in the O transferase
that results in the synthesis of inactive enzyme.

These structures have important implications for blood
transfusions and other transplantation procedures. If an antigen not
normally present in a person is introduced, the person's immune
system recognizes it as foreign. Adverse reactions can ensue, initiated
by the intravascular destruction of the incompatible red blood cells.

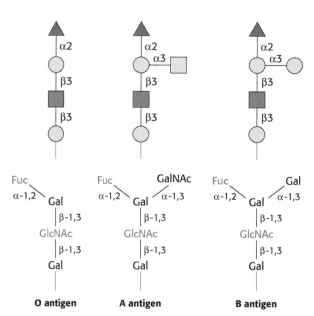

Figure 10.24 Structures of A, B, and O oligosaccharide
antigens. The carbohydrate structures represented in the upper
part of this illustration are depicted symbolically (refer to the key
in Figure 10.17).

QUICK QUIZ 2 Which amino acids are used for the attachment of carbohydrates to proteins?

Mannose residue

N-Acetylgalactosamine phosphotransferase — UDP-GlcNAc → UMP

GlcNAc

Mannose 6-phosphate residue

N-Acetylgalactosaminidase — H₂O → GlcNAc

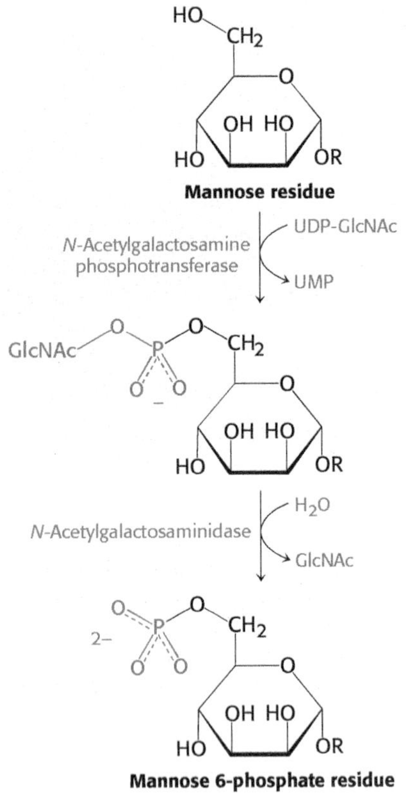

Figure 10.25 The formation of a mannose 6-phosphate marker. A glycoprotein destined for delivery to lysosomes acquires a phosphate marker in the Golgi complex in a two-step process. First, *N*-acetylgalactosamine phosphotransferase adds a phospho-*N*-acetylglucosamine unit to the 6-OH group of a mannose residue; then, *N*-acetylgalactosaminidase removes the added sugar to generate a mannose 6-phosphate residue in the core oligosaccharide.

Why are different blood types present in the human population? Suppose that a pathogenic organism such as a parasite expresses on its cell surface a carbohydrate antigen similar to one of the blood-group antigens. This antigen may not be readily detected as foreign in a person with the blood type that matches the parasite antigen, and the parasite will flourish. However, other people with different blood types will be protected. Hence, there will be selective pressure on human beings to vary blood type to prevent parasitic mimicry and a corresponding selective pressure on parasites to enhance mimicry. The constant "arms race" between pathogenic microorganisms and human beings drives the evolution of diversity of surface antigens within the human population.

🩺 CLINICAL INSIGHT

Lack of Glycosylation Can Result in Pathological Conditions

Although the role of carbohydrate attachment to proteins is not known in detail in most cases, data indicate that this glycosylation is important for the processing and stability of these proteins, as it is for EPO. Certain types of muscular dystrophy can be traced to improper glycosylation of dystroglycan, a membrane protein that links the extracellular matrix with the cytoskeleton (Section 1.4). Indeed, an entire family of severe inherited human diseases called *congenital disorders of glycosylation* has been identified. These pathological conditions reveal the importance of proper modification of proteins by carbohydrates and their derivatives.

An especially clear example of the role of glycosylation is provided by *I-cell disease* (also called *mucolipidosis II*), a lysosomal storage disease. *Lysosomes* are organelles that degrade and recycle damaged cellular components or material brought into the cell by endocytosis (p. 13). Normally, a carbohydrate marker directs certain digestive enzymes from the Golgi complex to lysosomes where they function. In patients with I-cell disease, lysosomes contain large *inclusions* of undigested glycosaminoglycans and glycolipids, the "I" in the name of the disease. These inclusions are present because the enzymes normally responsible for the degradation of glycosaminoglycans are missing from affected lysosomes. Remarkably, the enzymes are present at very high levels in the blood and urine. Thus, active enzymes are synthesized, but, in the absence of appropriate glycosylation, they are exported instead of being sequestered in lysosomes. In other words, *in I-cell disease, a whole series of enzymes is incorrectly addressed and delivered to the wrong location.* Normally, these enzymes contain a mannose 6-phosphate residue as a component of an *N*-oligosaccharide that serves as the marker directing the enzymes from the Golgi complex to lysosomes. In I-cell disease, however, the attached mannose lacks a phosphate. I-cell patients are deficient in the *N-acetylgalactosamine phosphotransferase* catalyzing the first step in the addition of the phosphoryl group; the consequence is the mistargeting of eight essential enzymes (**Figure 10.25**). I-cell disease causes the patient to suffer severe psychomotor retardation and skeletal deformities, similar to those in Hurler disease. Remarkably, mutations in the phosphotransferase have also been linked to stuttering. Why some mutations cause stuttering while others cause I-cell disease is a mystery.

10.4 Lectins Are Specific Carbohydrate-Binding Proteins

The diversity and complexity of the carbohydrate units and the variety of ways in which they can be joined in oligosaccharides and polysaccharides suggest that they are functionally important. Nature does not construct complex patterns when simple ones suffice. So why all this intricacy and diversity? It is now clear that these carbohydrate structures are the recognition sites for a special class of proteins. Such proteins, termed *glycan-binding proteins,* bind specific carbohydrate structures on neighboring cell surfaces. Originally discovered in plants, glycan-binding proteins are ubiquitous, and no living organisms have been found that lack these key proteins. We will focus on a particular class of glycan-binding proteins termed *lectins* (from Latin *legere,* "to select"). The interaction of lectins with their carbohydrate partners is another example of carbohydrates being information-rich molecules that guide many biological processes. The diverse carbohydrate structures displayed on cell surfaces are well suited to serving as sites of interaction between cells and their environments. Interestingly, the partners for lectin binding are often the carbohydrate moiety of glycoproteins.

Lectins Promote Interactions Between Cells

Cell–cell contact is a vital interaction in a host of biochemical functions, ranging from building a tissue from isolated cells to facilitating the transmission of information. The chief function of lectins is to facilitate cell–cell contact. A lectin usually contains two or more carbohydrate-binding sites. The lectins on the surface of one cell interact with arrays of carbohydrates displayed on the surface of another cell. Lectins and carbohydrates are linked by a number of weak noncovalent interactions that ensure specificity yet permit unlinking as needed.

We have already been introduced to a lectin obliquely. Recall that, in I-cell disease, lysosomal enzymes lack the appropriate mannose 6-phosphate, a molecule that directs the enzymes to the lysosome. Under normal circumstance, the *mannose 6-phosphate receptor,* a lectin, binds the enzymes in the Golgi apparatus and directs them to the lysosome.

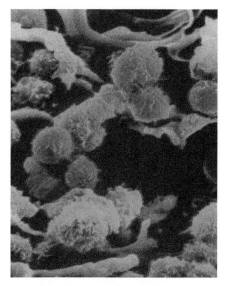

Figure 10.26 Selectins mediate cell–cell interactions. The scanning electron micrograph shows lymphocytes adhering to the endothelial lining of a lymph node. The L selectins on the lymphocyte surface bind specifically to carbohydrates on the lining of the lymph-node vessels. [CNRI/Science Source.]

 CLINICAL INSIGHT

Lectins Facilitate Embryonic Development

One class of lectins consists of proteins termed *selectins*, which bind immune-system cells to sites of injury in the inflammatory response (**Figure 10.26**). The L, E, and P forms of selectins bind specifically to carbohydrates on *l*ymph-node vessels, *e*ndothelium, or activated blood *p*latelets, respectively. L-Selectin, originally thought to participate only in the immune response, is produced by an embryo when it is ready to attach to the endometrium of the mother's uterus. For a short length of time, the endometrial cells present an oligosaccharide on the cell surface. When the embryo attaches through lectins, the attachment activates signal pathways in the endometrium to make implantation of the embryo possible.

 CLINICAL INSIGHT

Influenza Virus Binds to Sialic Acid Residues

Many pathogens gain entry into specific host cells by adhering to cell-surface carbohydrates. For example, influenza virus recognizes sialic acid residues linked to galactose residues that are present on cell-surface glycoproteins. The viral protein that binds to these sugars is called *hemagglutinin* (**Figure 10.27**A).

After binding to hemagglutinin, the virus is engulfed by the cell and begins to replicate. To exit the cell, a process essentially the reverse of viral entry occurs (Figure 10.27B). Viral assembly results in the budding of the viral particle from the cell. Upon complete assembly, the viral particle is still attached to sialic acid residues of the cell membrane by hemagglutinin on the surface of the new virions. Another viral protein, neuraminidase (sialidase), cleaves the

Figure 10.27 Viral receptors.
(A) Influenza virus targets cells by binding to sialic acid residues located at the termini of oligosaccharides present on cell-surface glycoproteins and glycolipids. These carbohydrates are bound by hemagglutinin, one of the major proteins expressed on the surface of the virus. (B) When viral replication is complete and the viral particle buds from the cell, the other major viral-surface protein, neuraminidase cleaves oligosaccharide chains to release the viral particle.

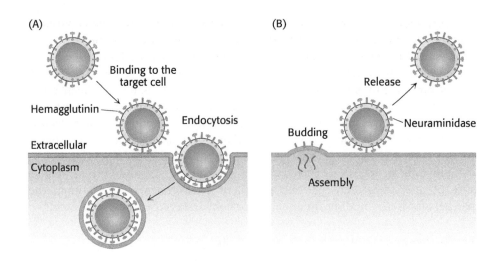

glycosidic bonds between the sialic acid residues and the rest of the cellular glycoprotein, freeing the virus to infect new cells, spreading the infection throughout the respiratory tract. Inhibitors of this enzyme such as oseltamivir (Tamiflu) and zanamivir (Relenza) are important anti-influenza agents.

Viral hemagglutinin's carbohydrate-binding specificity may play an important role in species specificity of infection and ease of transmission. For instance, avian influenza H5N1 (bird flu) is especially lethal and is readily spread from bird to bird. Although human beings can be infected by this virus, infection is rare and human-to-human transmission is rarer still. The biochemical basis of these characteristics is that the avian-virus hemagglutinin recognizes a different carbohydrate sequence from that recognized in human influenza. Although human beings have this sequence, it is located deep in the lungs. Infection is thus difficult, and, when it does occur, the avian virus is not readily transmitted by sneezing or coughing.

SUMMARY

10.1 Monosaccharides Are the Simplest Carbohydrates

An aldose is a carbohydrate with an aldehyde group (as in glyceraldehyde and glucose), whereas a ketose contains a keto group (as in dihydroxyacetone and fructose). A sugar belongs to the D series if the absolute configuration of its asymmetric carbon atom farthest from the aldehyde or keto group is the same as that of D-glyceraldehyde. Most naturally occurring sugars belong to the D series. The C-1 aldehyde in the open-chain form of glucose reacts with the C-5 hydroxyl group to form a six-membered pyranose ring. The C-2 keto group in the open-chain form of fructose reacts with the C-5 hydroxyl group to form a five-membered furanose ring. Pentoses such as ribose and deoxyribose also form furanose rings. An additional asymmetric center is formed at the anomeric carbon atom (C-1 in aldoses and C-2 in ketoses) in these cyclizations. Sugars are joined to alcohols and amines by glycosidic bonds at the anomeric carbon atom.

10.2 Monosaccharides Are Linked to Form Complex Carbohydrates

Sugars are linked to one another in disaccharides and polysaccharides by O-glycosidic bonds. Sucrose, lactose, and maltose are the common disaccharides. Sucrose (common table sugar) consists of α-glucose and β-fructose joined by a glycosidic linkage between their anomeric carbon atoms. Lactose (in milk) consists of galactose joined to glucose by a β-1,4 linkage. Maltose (in starch) consists of two glucoses joined by an α-1,4 linkage. Starch is a polymeric form of glucose in plants, and glycogen serves a similar role in animals. Most of the glucose units in starch and glycogen are in α-1,4 linkage. Glycogen has more branch points formed by α-1,6 linkages than does starch. Cellulose, the major structural polymer of plant cell walls, consists of glucose units joined by β-1,4 linkages. These β linkages give rise to long straight chains that form fibrils with high tensile strength. In contrast, the α linkages in starch and glycogen lead to open helices, in keeping with their roles as mobile energy stores.

10.3 Carbohydrates Are Attached to Proteins to Form Glycoproteins

Carbohydrates are commonly conjugated to proteins. Specific enzymes link the oligosaccharide units on proteins either to the side-chain oxygen atom of a serine or threonine residue or to the side-chain amide nitrogen atom of an asparagine residue. If the protein component is predominant, the conjugate of protein and carbohydrate is called a glycoprotein. Most secreted proteins are glycoproteins. The signal molecule erythropoietin is a glycoprotein. Glycoproteins are also prominent on the external surface of the plasma membrane. Proteins bearing covalently linked glycosaminoglycans are proteoglycans. Glycosaminoglycans are polymers of repeating disaccharides. One of the units in each repeat is a derivative of glucosamine or galactosamine. Mucoproteins, like proteoglycans, are predominantly carbohydrate by weight. The protein component is heavily O-glycosylated with N-acetylgalactosamine joining the oligosaccharide to the protein. Mucoproteins serve as lubricants.

10.4 Lectins Are Specific Carbohydrate-Binding Proteins

Carbohydrates on cell surfaces are recognized by proteins called lectins. In animals, the interplay of lectins and their sugar targets guides cell–cell contact. The viral protein hemagglutinin on the surface of the influenza virus recognizes sialic acid residues on the surfaces of cells invaded by the virus.

KEY TERMS

monosaccharide (p. 168)
ketose (p. 168)
aldose (p. 168)
enantiomer (p. 169)
diastereoisomer (p. 169)
epimer (p. 169)
anomer (p. 170)
reducing sugar (p. 171)

glycosidic bond (p. 172)
oligosaccharide (p. 173)
glycosyltransferase (p. 173)
disaccharide (p. 174)
polysaccharide (p. 175)
glycogen (p. 175)
starch (p. 175)
cellulose (p. 175)

glycoprotein (p. 177)
proteoglycan (p. 177)
glycosaminoglycan (p. 177)
mucin (mucoprotein) (p. 177)
lectin (p. 182)
selectin (p. 183)

? Answers to QUICK QUIZZES

1.

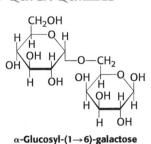

α-Glucosyl-(1→6)-galactose

2. Asparagine, serine, and threonine.

PROBLEMS

1. *Word origin.* Account for the origin of the term *carbohydrate.* ✓ 1

2. *Diversity.* How many different oligosaccharides can be made by linking one glucose, one mannose, and one galactose residue? Assume that each sugar is in its pyranose form. Compare this number with the number of tripeptides that can be made from three different amino acids. ✓ 1

3. *They go together like a horse and carriage.* Match each term with its description.

(a) Enantiomers _____
(b) Cellulose _____
(c) Lectins _____
(d) Glycosyltransferases
   _____
(e) Epimers _____
(f) Starch _____
(g) Carbohydrates
   _____
(h) Proteoglycan _____
(i) Mucoprotein _____
(j) Glycogen _____

1. Has the molecular formula of $(CH_2O)n$

2. Monosaccharides that differ at a single asymmetric carbon atom

3. The storage form of glucose in animals

4. The storage form of glucose in plants

5. Glycoprotein containing glycosaminoglycans

6. The most abundant organic molecule in the biosphere

7. N-Acetylgalactosamine is a key component of this glycoprotein

8. Carbohydrate-binding proteins

9. Enzymes that synthesize oligosaccharides

10. Stereoisomers that are mirror images of each other

4. *Couples.* Indicate whether each of the following pairs of sugars consists of anomers, epimers, or an aldose–ketose pair. ✓ 1

(a) D-glyceraldehyde and dihydroxyacetone
(b) D-glucose and D-mannose
(c) D-glucose and D-fructose
(d) α-D-glucose and β-D-glucose
(e) D-ribose and D-ribulose
(f) D-galactose and D-glucose

5. *Carbons and carbonyls.* To which classes of sugars do the monosaccharides shown here belong? ✓ 1

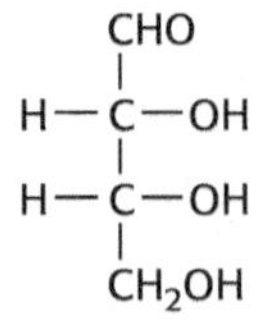

D-Erythrose 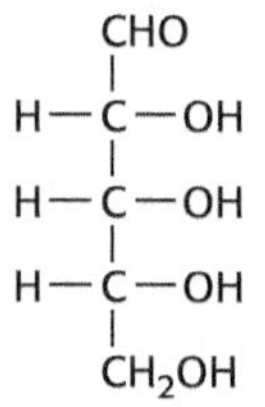   D-Ribose 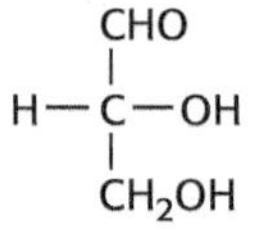   D-Glyceraldehyde

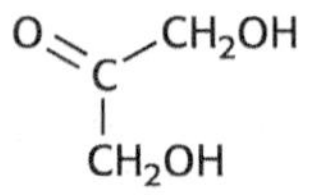

Dihydroxyacetone 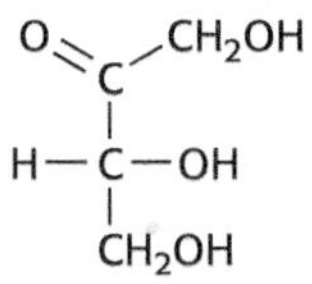   D-Erythrulose

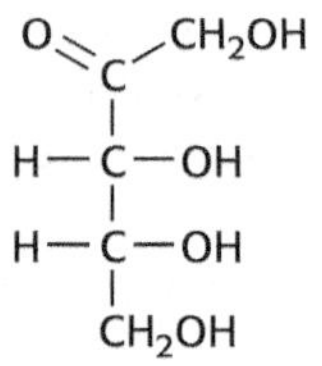

D-Ribulose 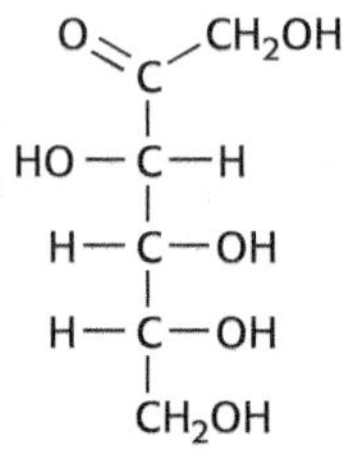   D-Fructose

6. *Chemical cousins.* Although an aldose with four asymmetric carbon atoms is capable of forming 16 diastereoisomers, only 8 of the isomers are commonly observed, including glucose. They are listed after the structure of D-glucose with their structural relation to glucose. Using the structure of glucose as a reference, draw the structures. ✓ 1

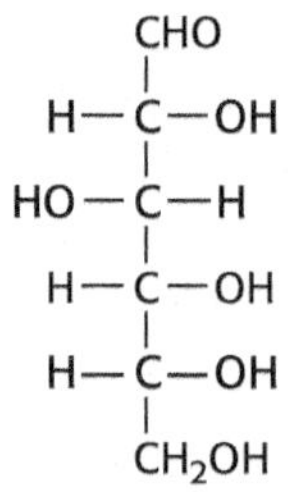

D-Glucose

D-Allose: Epimeric at C-3
D-Altrose: Isomeric at C-2 and C-3
D-Mannose: Epimeric at C-2
D-Gulose: Isomeric at C-3 and C-4
D-Idose: Isomeric at C-2, C-3, and C-4
D-Galactose: Epimeric at C-4
D-Talose: Isomeric at C-2 and C-4

7. *Telltale marker.* Glucose reacts slowly with hemoglobin and other proteins to form covalent compounds. Why is glucose reactive? What is the nature of the product formed? ✓ 1

8. *A lost property.* Glucose and fructose are reducing sugars. Sucrose, or table sugar, is a disaccharide consisting of both fructose and glucose. Is sucrose a reducing sugar? Explain. ✓ 1

9. *Oxygen source.* Does the oxygen atom attached to C-1 in methyl α-D-glucopyranoside come from glucose or methanol? ✓ 1

10. *Sugar lineup.* Identify the following four sugars. ✓ 1

(a)

(b)

(c)

(d)

11. *Cellular glue.* A trisaccharide unit of a cell-surface glycoprotein is postulated to play a critical role in mediating cell–cell adhesion in a particular tissue. Design a simple experiment to test this hypothesis. ✓ 2

12. *Component parts.* Raffinose, a minor constituent in sugar beets, is a trisaccharide. ✓ 2

Raffinose

(a) Is raffinose a reducing sugar? Explain.
(b) What are the monosaccharides that compose raffinose?
(c) β-Galactosidase is an enzyme that will remove galactose residues from an oligosaccharide. What are the products of β-galactosidase treatment of raffinose?

13. *Anomeric differences.* α-D-Mannose is a sweet-tasting sugar. β-D-Mannose, on the other hand, tastes bitter. A pure solution of α-D-mannose loses its sweet taste with time as it is converted into the β anomer. Draw the β anomer and explain how it is formed from the α anomer. ✓ 1

α-D-**Mannose**

14. *A taste of honey.* Fructose in its β-D-pyranose form accounts for the powerful sweetness of honey. The β-D-furanose form, although sweet, is not as sweet as the pyranose form. The furanose form is the more stable form. Draw the two forms and explain why it may not always be wise to cook with honey. ✓ 1

15. *Making ends meet.* (a) Compare the number of reducing ends to nonreducing ends in a molecule of glycogen. (b) As we will see in Chapter 24, glycogen is an important fuel-storage form that is rapidly mobilized. At which end—the reducing or nonreducing—would you expect most metabolism to take place? ✓ 1

16. *Meat and potatoes.* Compare the structures of glycogen and starch. ✓ 1

17. *Straight or with a twist?* Account for the different structures of glycogen and cellulose. ✓ 1

18. *Sweet proteins.* List the key classes of glycoprotein, their defining characteristics, and their biological functions. ✓ 2

19. *Life extender.* What is the function of the carbohydrate moiety that is attached to EPO? ✓ 2

20. *Carbohydrates—not just for breakfast anymore.* Differentiate between a glycoprotein and a lectin. ✓ 2

21. *Cushioning.* What is the role of the glycosaminoglycan in the cushioning provided by cartilage? ✓ 2

22. *Locks and keys.* What does the fact that all organisms contain lectins suggest about the role of carbohydrates? ✓ 2

23. *Undelivered mail. Not returned to sender.* I-cell disease results when proteins normally destined to the lysosomes lack the appropriate carbohydrate-addressing molecule (p. 182). Suggest another possible means by which I-cell disease might arise. ✓ 2

24. *Carbohydrates and proteomics.* Suppose that a protein contains six potential *N*-linked glycosylation sites. How many possible proteins can be generated, depending on which of these sites is actually glycosylated? Do not include the effects of diversity within the carbohydrate added. ✓ 2

25. *Many possibilities.* Why are polysaccharides considered information-rich molecules? ✓ 1

Chapter Integration Problem

26. *Stereospecificity.* Sucrose, a major product of photosynthesis in green leaves, is synthesized by a battery of enzymes. The substrates for sucrose synthesis, D-glucose and D-fructose, are a mixture of α and β anomers as well as acyclic compounds in solution. Nonetheless, sucrose consists of α-D-glucose linked by its C-1 atom to the C-2 atom of β-D-fructose. How can the specificity of sucrose be explained in light of the potential substrates?

Data Interpretation Problem

27. *Sore joints.* A contributing factor to the development of arthritis is the inappropriate proteolytic destruction of the aggrecan component of cartilage by the proteolytic enzyme aggrecanase. The immune-system signal molecule

interleukin 2 (IL-2) activates aggrecanase; in fact, IL-2 blockers are sometimes used to treat arthritis. Studies were undertaken to determine whether inhibitors of aggrecanase can counteract the effects of IL-2. Pieces of cartilage were incubated in media with various additions and the amount of aggrecan destruction was measured as a function of time. ✓ 1

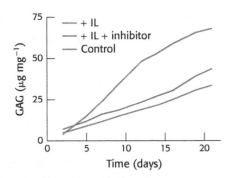

[Data from M. A. Pratta et al. *J. Biol. Chem.* 278:45539–45545, 2003, Fig. 7B.]

(a) Aggrecan degradation was measured by the release of glycosaminoglycan. What is the rational for this assay?
(b) Why might glycosaminoglycan release not indicate aggrecan degradation?
(c) What is the purpose of the control—cartilage incubated with no additions?
(d) What is the effect of adding IL-2 to the system?
(e) What is the response when an aggrecanase inhibitor is added in addition to IL-2?

(f) Why is there some aggrecan destruction in the control with the passage of time?

Challenge Problems

28. *Mutarotation.* The specific rotations of the α and β anomers of D-glucose are +112 degrees and +18.7 degrees, respectively. Specific rotation, $[\alpha]_D$, is defined as the observed rotation of light of wavelength 589 nm (the D line of a sodium lamp) passing through 10 cm of a 1 g ml^{-1} solution of a sample. When a crystalline sample of α-D-glucopyranose is dissolved in water, the specific rotation decreases from 112 degrees to an equilibrium value of 52.7 degrees. On the basis of this result, what are the proportions of the α and β anomers at equilibrium? Assume that the concentration of the open-chain form is negligible. ✓ 1

29. *Periodate cleavage.* Compounds containing hydroxyl groups on adjacent carbon atoms undergo carbon–carbon bond cleavage when treated with periodate ion (IO_4^-). How can this reaction be used to distinguish between pyranosides and furanosides if *cis*-glycans are cleaved more rapidly than *trans*-glycols? ✓ 1

30. *Mapping the molecule.* Each of the hydroxyl groups of glucose can be methylated with reagents such as dimethylsulfate under basic conditions. Explain how exhaustive methylation followed by the complete digestion of a known amount of glycogen would enable you to determine the number of branch points and reducing ends.

Selected Readings for this chapter can be found online at www.whfreeman.com/tymoczko3e.

CHAPTER 11

Lipids

The fruit of a small tree native to the eastern areas of the Mediterranean, olives are unusual in that they are an inedible fruit unless processed. Olive oil, which has been used in cooking since antiquity, is rich in monounsaturated fatty acids as well as antioxidants. These antioxidants may be one of the reasons for the positive health effects of the Mediterranean diet. [GoranStimac/Getty Images.]

*L*ipids are defined as water-insoluble molecules that are highly soluble in organic solvents. This important class of molecules has a variety of biochemical roles. For instance, lipids are widely used to store energy. They are also key components of membranes (Chapter 12) and play a variety of roles in signal-transduction pathways (Chapter 13). Unlike the three other classes of biomolecules (carbohydrates, amino acids, and nucleic acids), lipids do not form polymers. Their individual and collective properties—as noncovalent assemblies—make them extraordinarily important. We will examine five classes of lipids here:

1. *Free Fatty Acids (Nonesterified Fatty Acids).* This simplest type of lipid is most commonly used as a fuel. Fatty acids vary in hydrocarbon chain length, which has important ramifications when they are used as fuels and as components of membrane lipids.

2. *Triacylglycerols.* This class of lipid is the storage form of fatty acids.

3. *Phospholipids.* These membrane lipids consist of fatty acids attached to a scaffold that also bears a charged phosphoryl group, creating a macromolecule with a polar head and nonpolar tail.

4. *Glycolipids.* These lipids are bound to carbohydrates and are important membrane constituents.

5. *Steroids.* These lipids differ from the other classes in that they are polycyclic hydrocarbons. Steroids function as hormones that control a variety of physiological functions. The most common steroid is cholesterol, another vital membrane component.

✓ 3 Describe the key chemical properties of fatty acids.

11.1 Fatty Acids Are a Main Source of Fuel

Fats, or *fatty acids,* are chains of hydrogen-bearing carbon atoms, called hydrocarbons, which terminate with carboxylic acid groups. These hydrocarbon chains are of various lengths and may have one or more double bonds, depending on the fat. Two key roles for fatty acids are as fuels and as building blocks for membrane lipids. Fats are good fuels because they are more reduced than carbohydrates; that is, the carbon atoms are bonded to hydrogen atoms and other carbon atoms rather than to oxygen atoms, as is the case for carbohydrates. Because of this greater reduction, fats yield more energy than carbohydrates when undergoing combustion to carbon dioxide and water. We will return to this discussion in Chapter 15.

Fats have common names and systematic names. Although we will use the common names for fatty acids, familiarity with the systematic names is important because their use is sometimes required to prevent confusion. A fatty acid's systematic name is derived from the name of its parent hydrocarbon by the substitution of *oic* for the final *e*. For example, the C_{18} saturated fatty acid known familiarly as stearic acid is systematically called *octadecanoic acid* because the parent hydrocarbon is octadecane: *octadec-* for the 18 carbon atoms, and *–ane* because it is composed completely of single bonds. Fatty acids composed of single bonds only are called *saturated* fatty acids, because every carbon atom is attached to four other atoms. Fatty acids with one or more double or triple bonds are called *unsaturated* fatty acids. Oleic acid, a C_{18} fatty acid with one double bond, is called octadec*enoic* acid; whereas linoleic acid, with two double bonds, is octadeca*dienoic* acid; and linolenic acid, with three double bonds, is octadeca*trienoic* acid. The composition and structure of a fatty acid can also be designated by numbers. The notation 18:0 denotes a C_{18} fatty acid with no double bonds, whereas 18:2 signifies that there are two double bonds. The structure of the ionized forms of two common fatty acids—palmitic acid (16:0) and oleic acid (18:1)—are shown in **Figure 11.1**.

When considering fatty acids, we often need to distinguish the individual carbon atoms. Fatty acid carbon atoms are usually numbered starting at the carboxyl terminus but can also be numbered starting at the terminal carbon atom, as shown in the margin. Carbon atoms 2 and 3 are often referred to as α and β, respectively. The last carbon atom in the chain, which is almost always a methyl carbon atom, is called the *ω-carbon atom* (ω is the Greek symbol for

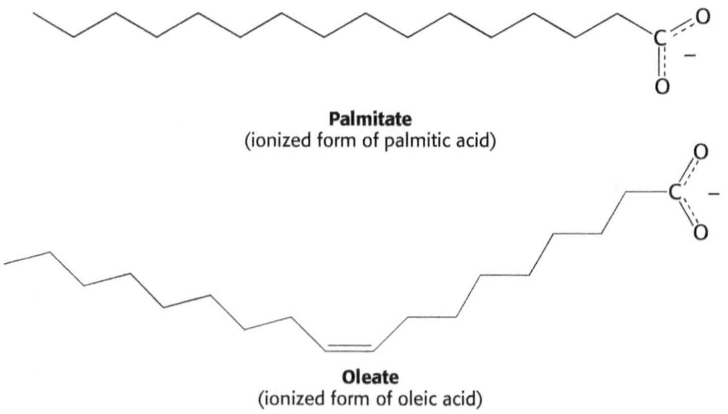

Palmitate
(ionized form of palmitic acid)

Oleate
(ionized form of oleic acid)

Figure 11.1 Structures of two fatty acids. Palmitate is a 16-carbon, saturated fatty acid, and oleate is an 18-carbon fatty acid with a single cis double bond.

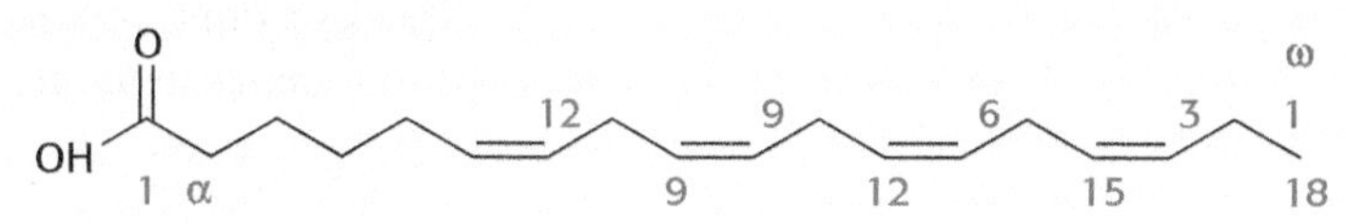

omega, the last letter of the Greek alphabet). The position of a double bond is represented by the symbol Δ followed by a superscript number. For example, *cis*-Δ^9 means that there is a cis double bond between carbon atoms 9 and 10; *trans*-Δ^2 means that there is a trans double bond between carbon atoms 2 and 3. *Cis* and *trans* designate the relative positions of substituents on either side of the double bond. Just like amino acids, fatty acids are ionized at physiological pH, and so it is preferable to refer to them according to their carboxylate form: for example, palmitate rather than palmitic acid.

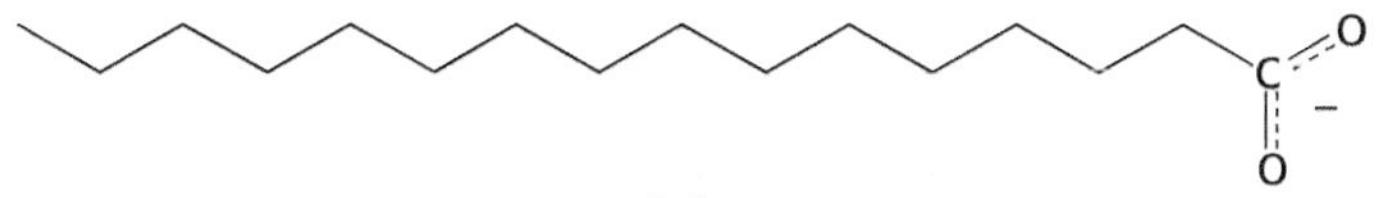

Palmitate
(ionized form of palmitic acid)

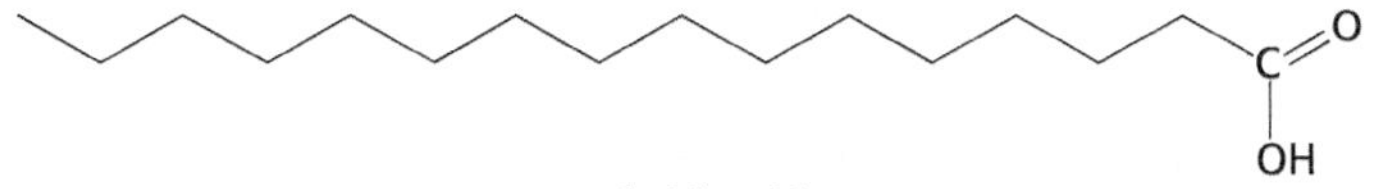

Palmitic acid

Fatty Acids Vary in Chain Length and Degree of Unsaturation

Fatty acids in biological systems usually contain an even number of carbon atoms, typically between 14 and 24 (Table 11.1), with the 16- and 18-carbon fatty acids being the most common. As heretofore described, the hydrocarbon chain may be saturated or it may contain one or more double bonds. The configuration of the double bonds in most unsaturated fatty acids is cis. The double bonds in polyunsaturated fatty acids are separated by at least one methylene group.

The properties of fatty acids and of lipids derived from them are markedly dependent on chain length and degree of saturation. Unsaturated fatty acids have lower melting points than those of saturated fatty acids of the same length. For example, the melting point of stearic acid is 69.6°C, whereas that of oleic acid (which contains one cis double bond) is 13.4°C. The melting points of polyunsaturated fatty acids of the C_{18} series are even lower. The presence of a cis double bond introduces a kink in the fatty acid and makes tight packing between

$—CH_2—$
Methylene group

Table 11.1 Some naturally occurring fatty acids in animals

Number of carbon atoms	Number of double bonds	Common name	Systematic name	Formula
12	0	Laurate	*n*-Dodecanoate	$CH_3(CH_2)_{10}COO^-$
14	0	Myristate	*n*-Tetradecanoate	$CH_3(CH_2)_{12}COO^-$
16	0	Palmitate	*n*-Hexadecanoate	$CH_3(CH_2)_{14}COO^-$
18	0	Stearate	*n*-Octadecanoate	$CH_3(CH_2)_{16}COO^-$
20	0	Arachidate	*n*-Eicosanoate	$CH_3(CH_2)_{18}COO^-$
22	0	Behenate	*n*-Docosanoate	$CH_3(CH_2)_{20}COO^-$
24	0	Lignocerate	*n*-Tetracosanoate	$CH_3(CH_2)_{22}COO^-$
16	1	Palmitoleate	*cis*-Δ^9-Hexadecenoate	$CH_3(CH_2)_5CH=CH(CH_2)_7COO^-$
18	1	Oleate	*cis*-Δ^9-Octadecenoate	$CH_3(CH_2)_7CH=CH(CH_2)_7COO^-$
18	2	Linoleate	*cis, cis*-Δ^9, Δ^{12}- Octadecadienoate	$CH_3(CH_2)_4(CH=CHCH_2)_2(CH)_6COO^-$
18	3	Linolenate	all-*cis*-Δ^9, Δ^{12}, Δ^{15}- Octadecatrienoate	$CH_3CH_2(CH=CHCH_2)_3(CH_2)_6COO^-$
20	4	Arachidonate	all-*cis* Δ^5, Δ^8, Δ^{11}, Δ^{14}- Eicosatetraenoate	$CH_3(CH_2)_4(CH=CHCH_2)_4(CH_2)_2COO^-$

the chains impossible. The lack of tight packing limits the van der Waals interactions between chains, lowering the melting temperature.

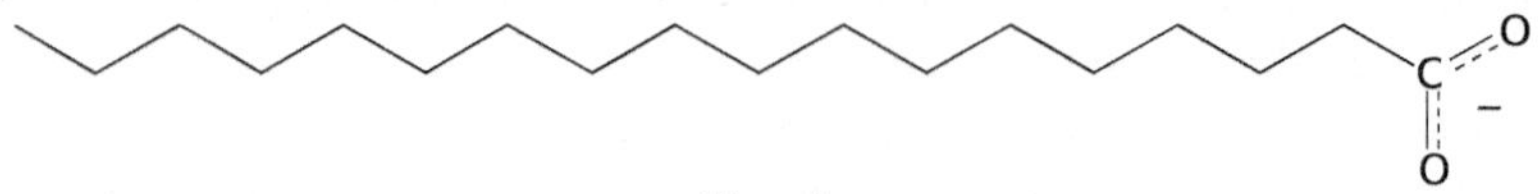

Stearate

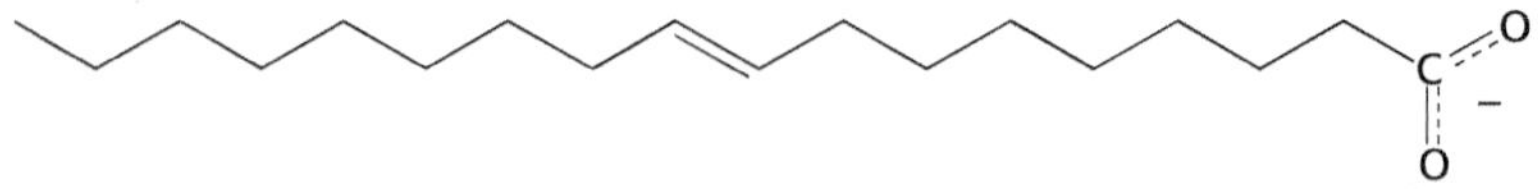

trans-**Oleate**

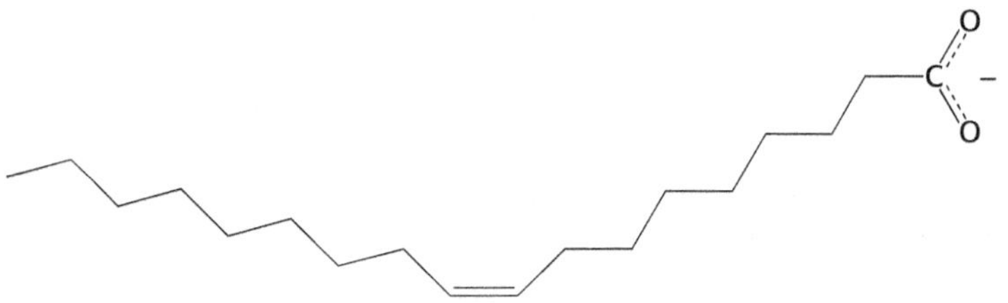

cis-**Oleate**

Chain length also affects the melting point, as illustrated by the fact that the melting temperature of palmitic acid (C_{16}) is 6.5 degrees lower than that of stearic acid (C_{18}). The fat that accumulates in the pan as bacon is fried is composed primarily of saturated fatty acids and solidifies soon after the burner is turned off. Olive oil, on the other hand, is composed of high concentrations of oleic acid and some polyunsaturated fatty acids and remains liquid at room temperature. The variability of melting points is not merely an arcane chemical insight. Melting temperatures of fatty acids are key elements in the control of the fluidity of cell membranes, and the proper degree of fluidity is essential for membrane function (Chapter 12). Thus, *short chain length and cis unsaturation enhance the fluidity of fatty acids and of their derivatives.*

The Degree and Type of Unsaturation Are Important to Health

Although fats are crucial biochemicals, too much saturated and trans-unsaturated fats in the diet are correlated with high blood levels of cholesterol and cardiovascular disease. The biochemical basis for this correlation remains to be determined, although trans-unsaturated fats appear to trigger inflammatory pathways in immune cells. In contrast, certain cis-polyunsaturated fatty acids are essential in our diets because we cannot synthesize them. Such fatty acids include the ω-3 fatty acids—polyunsaturated fatty acids common in cold-water fish such as salmon, which have been suggested to play a role in protection from cardiovascular disease. The important ω-3 fatty acids are α-linolenate, found in vegetable oils, and eicosapentaenoate (EPA, eicosapentaenoic acid) and docosahexaenoate (DHA, docosahexaenoic acid), both of which are found in fatty fish and shellfish.

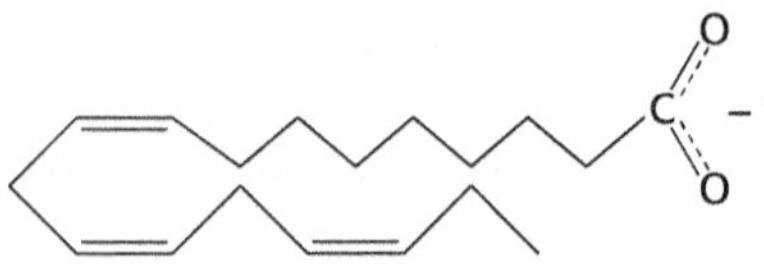

QUICK QUIZ 1 What factors determine the melting point of fatty acids?

α-Linolenate

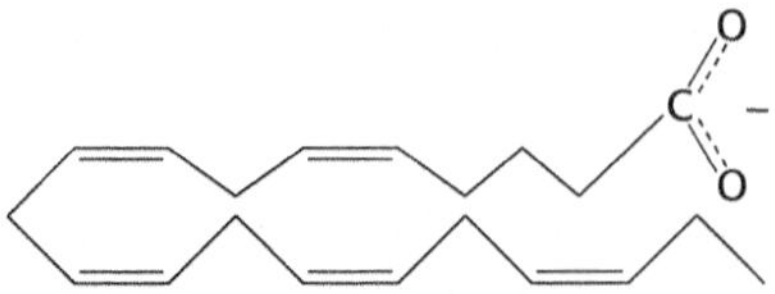

Eicosapentaenoate (EPA)

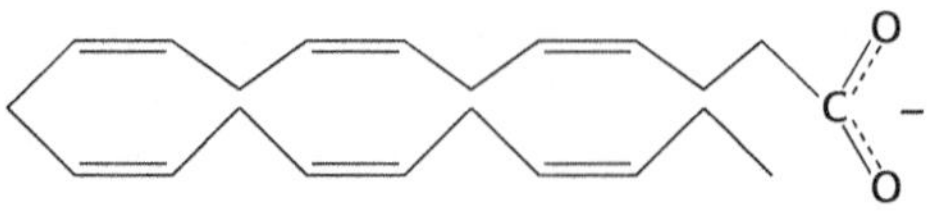

Docosahexaenoate (DHA)

These fatty acids are precursors to important hormones. However, the nature of their protective effect against heart disease is not known. Studies show that ω-3 fatty acids, especially those derived from marine organisms, can prevent sudden death from a heart attack, reduce triacylglycerols in the blood, lower blood pressure, and prevent thrombosis by slightly inhibiting blood clotting. Revealing how these molecules exert their effects is an active area of biochemical research.

11.2 Triacylglycerols Are the Storage Form of Fatty Acids

✓ 4 Identify the major lipids, and describe their biochemical functions.

Despite the fact that fatty acids are our principal energy source, the concentration of free fatty acids in cells or the blood is low because free fatty acids are strong acids. High concentrations of free fatty acids would disrupt the pH balance of the cells. Fatty acids required for energy generation are stored as *triacylglycerols*, which are formed by the attachment of three fatty acid chains to a glycerol molecule. The fatty acids are attached to the glycerol through ester linkages, in a process known as esterification. Common soaps are the sodium or potassium salts of fatty acids generated by treating triacylglycerols with strong bases (**Figure 11.2**).

$$H_2C-O-\overset{\overset{\displaystyle O}{\|}}{C}-CH_2-(CH_2)_n-CH_3$$
$$HC-O-\overset{\overset{\displaystyle O}{\|}}{C}-CH_2-(CH_2)_n-CH_3 \quad \Big\} \text{ Three fatty acid chains}$$
$$H_2C-O-\overset{\overset{\displaystyle O}{\|}}{C}-CH_2-(CH_2)_n-CH_3$$

Glycerol backbone

When energy is required during a fast (for instance, while sleeping), the fatty acids are cleaved from the triacylglycerol and carried to the cells. The ingestion of food replenishes the triacylglycerol stores.

As stated in Chapter 10, fatty acids are richer in energy than carbohydrates. Additionally, compared with carbohydrates, fatty acids store energy more efficiently in the form of triacylglycerols, which are hydrophobic and so are stored in a nearly anhydrous form. Polar carbohydrates, in contrast, bind to water molecules. Glycogen, for instance, binds to water. In fact, 1 g of dry glycogen binds about 2 g of water. Consequently, *a gram of nearly anhydrous fat stores more than six times as much energy as a gram of hydrated glycogen*. For this reason, triacylglycerols rather than glycogen were selected in evolution as the major energy reservoir. Consider a typical 70-kg man. Triacylglycerols constitute about 11 kg of his total body weight. If this amount of energy were stored in glycogen, his total body weight would be 55 kg greater. The glycogen stores provide enough energy to sustain biological function for about 18 to 24 hours, whereas the triacylglycerol stores allow survival for several weeks.

In mammals, the major site of accumulation of triacylglycerols is *adipose tissue*, which is distributed under the skin and elsewhere throughout the body. In adipose cells (fat cells), droplets of triacylglycerol coalesce to form a large globule in the

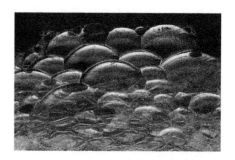

Figure 11.2 Soaps are the sodium and potassium salts of long chain fatty acids. They are derived from the treatment of triacylglycerols with a strong base, a process called saponification. Common sources of triacylglycerols are the animal lipids lard (from hogs) or tallow (from beef or sheep). [Tetra Images/Getty Images.]

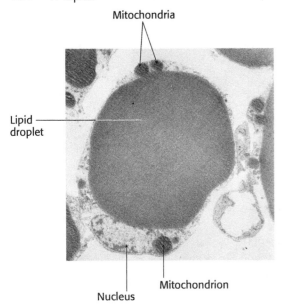

Figure 11.3 Electron micrograph of an adipose cell (adipocyte). A small band of cytoplasm surrounds the large deposit of triacylglycerols. [Biophoto Associates/ Science Source.]

cytoplasm, which may occupy most of the cell volume (**Figure 11.3**). Adipose cells are specialized for the synthesis and storage of triacylglycerols and for their mobilization into fuel molecules that are transported to other tissues by the blood. Adipose tissue also serves as a thermal insulator to help maintain body temperature. Polar bears provide an example of the insulating properties of adipose tissue. During fierce arctic storms, polar bears curl up into tight balls and are so well insulated that they are difficult to detect even with heat-sensing instruments.

The utility of triacylglycerols as an energy source is dramatically illustrated by the abilities of migratory birds, which can fly great distances without eating. Examples are the American golden plover and the ruby-throated hummingbird (**Figure 11.4**). The golden plover flies from Alaska to the southern tip of South America; a large segment of the flight (3800 km, or 2400 miles) is over open ocean, where the birds cannot feed. The ruby-throated hummingbird can fly nonstop across the Gulf of Mexico. Fatty acids derived from triacylglycerols provide the energy source for these prodigious feats.

Figure 11.4 Ruby-throated hummingbird.
[William Leaman/Alamy.]

11.3 There Are Three Common Types of Membrane Lipids

Thus far, we have considered only one type of complex lipid—triacylglycerols, the storage form of fatty acids. However, lipids can be used for more than just fuel and energy storage: they serve as hormones, messengers in signal-transduction pathways, and components of membranes. The first two of these functions of lipids will be considered in later chapters. We now consider the three major kinds of membrane lipids: *phospholipids, glycolipids,* and *cholesterol.*

Phospholipids Are the Major Class of Membrane Lipids

Phospholipids are abundant in all biological membranes. A phospholipid molecule is constructed from four components: one or more fatty acids, a platform to which the fatty acids are attached, a phosphate, and an alcohol attached to the phosphate.

The platform on which phospholipids are built may be *glycerol,* a three-carbon alcohol, or *sphingosine,* a more complex alcohol. Phospholipids derived from glycerol are called *phosphoglycerides* and are composed of a glycerol backbone to which are attached two fatty acid chains and a phosphorylated alcohol (**Figure 11.5**).

In phosphoglycerides, the hydroxyl groups at C-1 and C-2 of glycerol are esterified to the carboxyl groups of the two fatty acid chains. The C-3 hydroxyl group of the glycerol backbone is esterified to phosphoric acid. When no further additions are made, the resulting compound is *phosphatidate (diacylglycerol*

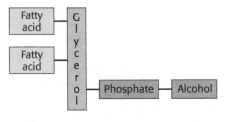

Figure 11.5 Schematic structure of a phospholipid. The platform is glycerol in this case.

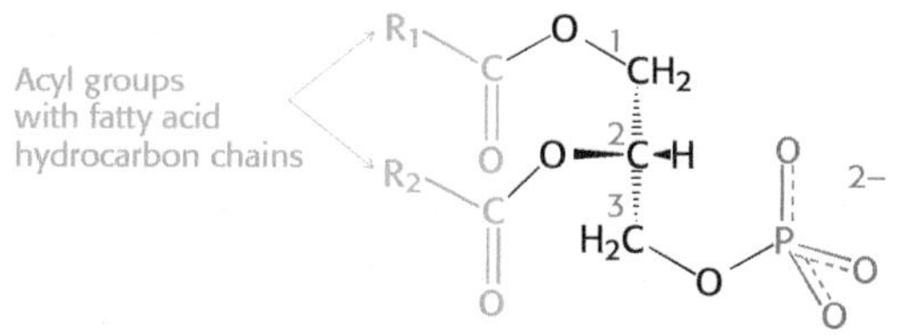

Phosphatidate
(Diacylglycerol 3-phosphate)

Figure 11.6 Structure of phosphatidate (diacylglycerol 3-phosphate). The absolute configuration of the central carbon atom (C-2) is shown.

3-phosphate), the simplest phosphoglyceride (**Figure 11.6**). Only small amounts of phosphatidate are present in membranes. However, the molecule is a key intermediate in the biosynthesis of the other phosphoglycerides as well as triacylglycerols (Chapter 29).

The major phosphoglycerides are derived from phosphatidate by the formation of an ester linkage between the phosphoryl group of phosphatidate and the hydroxyl group of one of several alcohols. The common alcohol moieties of phosphoglycerides are the amino acid serine, ethanolamine, choline, glycerol, and inositol. The structural formulas of phosphatidylcholine and the other principal phosphoglycerides are given in **Figure 11.7**.

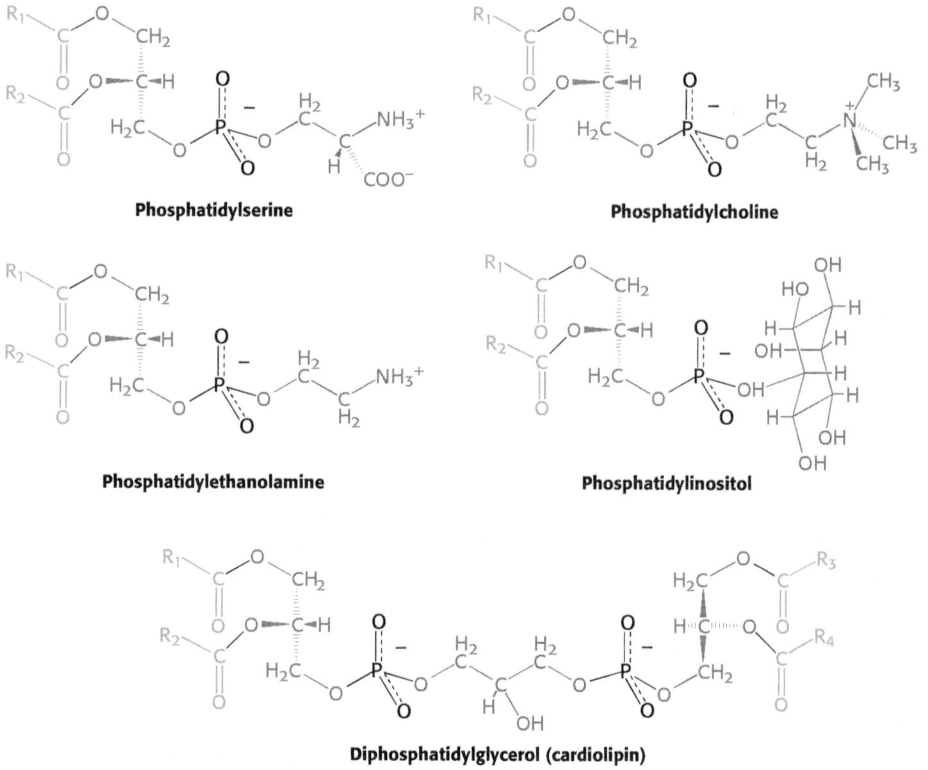

Phosphatidylserine

Phosphatidylcholine

Phosphatidylethanolamine

Phosphatidylinositol

Diphosphatidylglycerol (cardiolipin)

Figure 11.7 Some common phosphoglycerides found in membranes.

Phospholipids built on a sphingosine backbone are called *sphingolipids*. *Sphingosine* is an amino alcohol that contains a long, unsaturated hydrocarbon chain (**Figure 11.8**). *Sphingomyelin* is a common sphingolipid found in membranes. In sphingomyelin, the amino group of the sphingosine backbone is linked to a fatty acid by an amide bond. In addition, the primary hydroxyl group of sphingosine is attached to phosphorylcholine through an ester linkage. Sphingomyelin is found in the plasma membrane of many cells but is especially rich in the myelin sheath of nerve cells.

DID YOU KNOW?

Niemann–Pick disease can result from an accumulation of sphingomyelin owing to the lack of sphingomyelinase, an enzyme that degrades sphingomyelin. Symptoms of Niemann–Pick disease include mental retardation, seizures, eye paralysis, ataxia, and retarded growth.

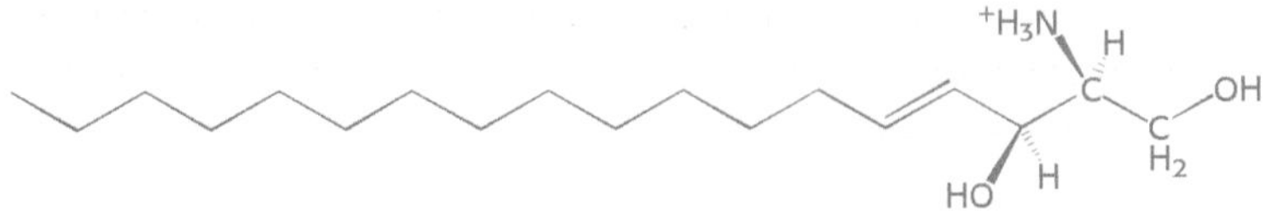

Figure 11.8 Structures of sphingosine and sphingomyelin. The sphingosine moiety of sphingomyelin is highlighted in blue.

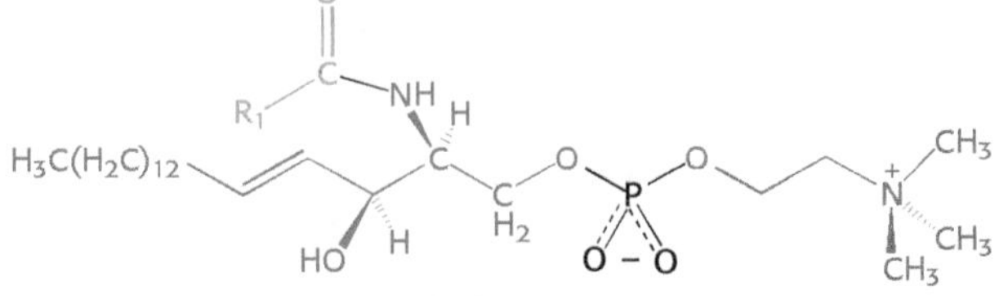

Membrane Lipids Can Include Carbohydrates

Glycolipids, as their name implies, are *sugar-containing lipids.* Glycolipids are ubiquitous in all cell membranes, where they play a role in cell–cell interactions. Like sphingomyelin, the glycolipids in animal cells are derived from sphingosine. The amino group of the sphingosine backbone is acylated by a fatty acid, as in sphingomyelin. Glycolipids differ from sphingomyelin in the identity of the unit that is linked to the primary hydroxyl group of the sphingosine backbone. In glycolipids, one or more sugars (rather than phosphorylcholine) are attached to this group. The simplest glycolipid, called a *cerebroside,* contains a single sugar residue, either glucose or galactose.

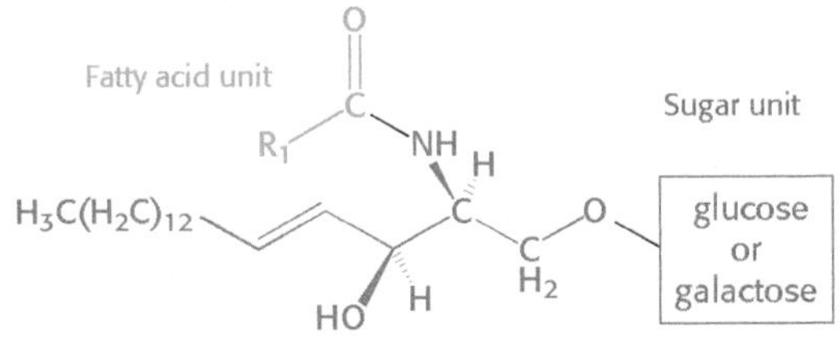

Cerebroside
(a glycolipid)

More complex glycolipids, such as *gangliosides,* contain a branched chain of as many as seven sugar residues. *Glycolipids are oriented in an asymmetric fashion in membranes with the sugar residues always on the extracellular side of the membrane.*

Steroids Are Lipids That Have a Variety of Roles

Steroids, the final class of lipids to be considered in this chapter, function as powerful hormones such as estradiol and testosterone, facilitate the digestion of lipids in the diet, and are key membrane constituents. Unlike the other classes of lipids, steroids exhibit a cyclical rather than linear structure. All steroids have a tetracyclic ring structure called the steroid nucleus. The steroid nucleus consists of three cyclohexane rings and a cyclopentane ring joined together.

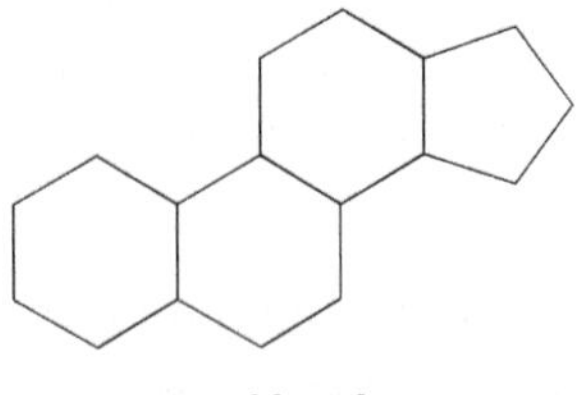

Steroid nucleus

All biochemically important steroids are modified versions of this basic structure.

Cholesterol is the most common steroid, and the precursor to biochemically active steroids. It is a member of a class of molecules called sterol—steroid with an alcohol functional group.

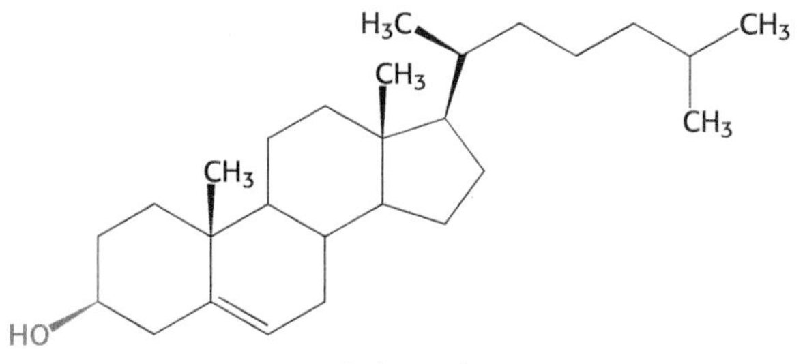

Cholesterol

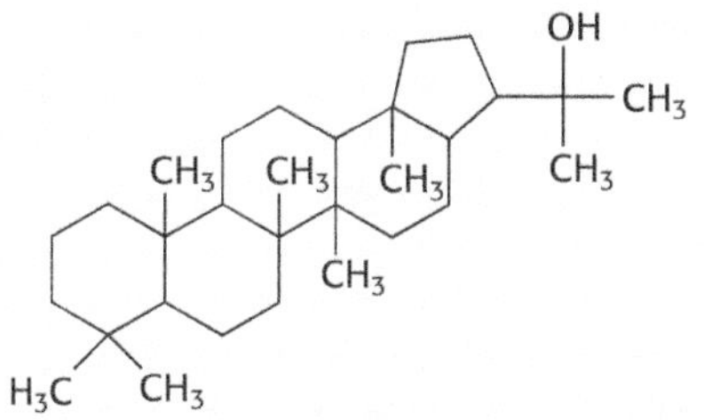

Diplopterol
(A hopanoid)

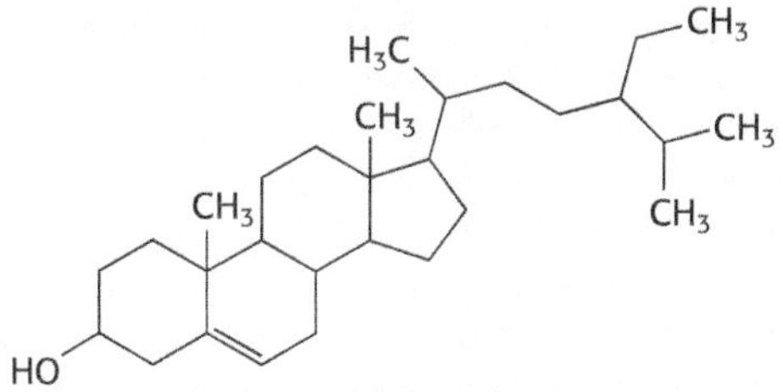

Sitosterol
(A plant sterol)

A hydrocarbon tail is linked to the steroid at one end, and a hydroxyl group is attached at the other end. As we will see in Chapter 12, cholesterol is important in maintaining proper membrane fluidity. Although cholesterol is absent from prokaryotes, all organisms require a similar molecule to moderate membrane fluidity. In prokaryotes, hopanoids play the role of cholesterol while various sterols serve the function in plants. Cholesterol is found in varying degrees in virtually all animal membranes. It constitutes almost 25% of the membrane lipids in certain nerve cells but is essentially absent from some intracellular membranes. Free cholesterol does not exist outside of membranes. Rather, it is esterified to a fatty acid for storage and transport.

 BIOLOGICAL INSIGHT

Membranes of Extremophiles Are Built from Ether Lipids with Branched Chains

The membranes of archaea differ in composition from those of eukaryotes or bacteria in two important ways. These differences clearly relate to the hostile living conditions of many archaea (Figure 8.3). First, the fatty acid chains are joined to a glycerol backbone by ether rather than ester linkages. The ether linkage is more resistant to hydrolysis than the ester linkage is. Second, the alkyl chains are branched rather than linear. They are built up from repeats of a fully saturated five-carbon fragment. These branched, saturated hydrocarbons pack more tightly than unbranched hydrocarbons, thereby protecting membrane integrity, and are more resistant to oxidation as well.

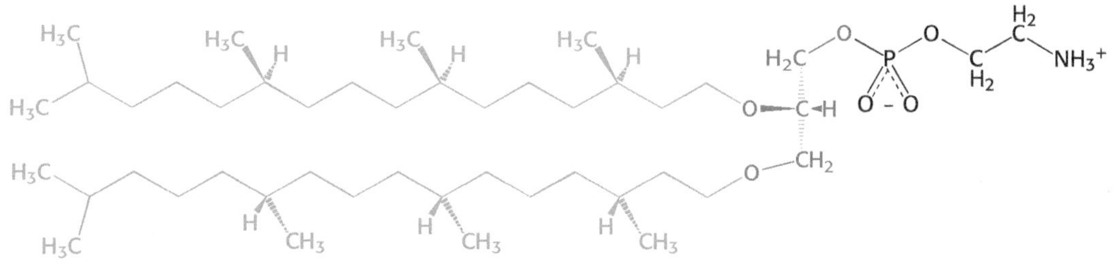

Membrane lipid from the archaeon *Methanococcus jannaschii*

The ability of these lipids to resist hydrolysis and oxidation may help these organisms to withstand the extreme conditions, such as high temperature, low pH, or high salt concentration, under which some of these archaea grow.

Membrane Lipids Contain a Hydrophilic and a Hydrophobic Moiety

The repertoire of membrane lipids is extensive. However, these lipids possess a critical common structural theme: *membrane lipids are amphipathic molecules* (amphiphilic molecules) containing both a *hydrophilic* and a *hydrophobic* moiety.

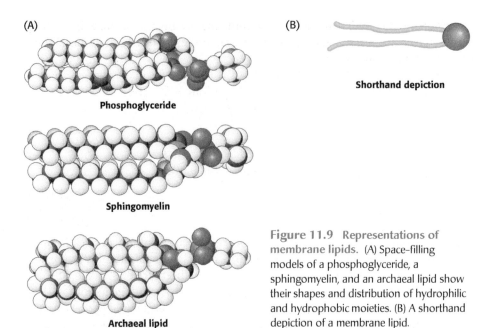

Phosphoglyceride

Sphingomyelin

Archaeal lipid

Shorthand depiction

Figure 11.9 Representations of membrane lipids. (A) Space-filling models of a phosphoglyceride, a sphingomyelin, and an archaeal lipid show their shapes and distribution of hydrophilic and hydrophobic moieties. (B) A shorthand depiction of a membrane lipid.

Let us look at a model of a phosphoglyceride, such as phosphatidylcholine. Its overall shape is roughly rectangular (**Figure 11.9**). The two hydrophobic fatty acid chains are approximately parallel to each other, whereas the hydrophilic phosphorylcholine moiety points in the opposite direction. Sphingomyelin has a similar conformation, as does the archaeal lipid depicted. Therefore, the following shorthand has been adopted to represent these membrane lipids: the hydrophilic unit, also called the *polar head group*, is represented by a circle, whereas the hydrocarbon tails are depicted by straight or wavy lines (Figure 11.9B).

? QUICK QUIZ 2 What are the three major types of membrane lipids?

Some Proteins Are Modified by the Covalent Attachment of Hydrophobic Groups

Just as carbohydrates are attached to proteins to modify the properties of proteins, lipids also can be attached to proteins to provide them with additional biochemical properties. Often, such attachments are necessary for a protein to associate with a hydrophobic environment such as a membrane. These hydrophobic attachments

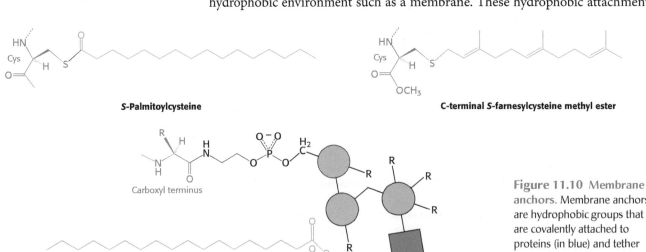

S-Palmitoylcysteine

C-terminal S-farnesylcysteine methyl ester

Carboxyl terminus

Glycosylphosphatidylinositol (GPI) anchor

Figure 11.10 Membrane anchors. Membrane anchors are hydrophobic groups that are covalently attached to proteins (in blue) and tether the proteins to the membrane. The green circles and blue square correspond to mannose and GlcNAc, respectively. R groups represent points of additional modification.

can insert into the hydrophobic interior of the membrane and localize the protein to the membrane surface. Such localization is required for protein function. Three such attachments are shown in **Figure 11.10**: (1) a palmitoyl group attached to a cysteine residue by a thioester bond, (2) a farnesyl group attached to a cysteine residue at the carboxyl terminus, and (3) a glycolipid structure termed a glycosylphosphatidylinositol (GPI) anchor attached to the carboxyl terminus.

 CLINICAL INSIGHT

Premature Aging Can Result from the Improper Attachment of a Hydrophobic Group to a Protein

Farnesyl is a hydrophobic group that is often attached to proteins, usually so that the protein is able to associate with a membrane (Figure 11.10). Inappropriate farnesylation has been shown to result in Hutchinson–Gilford progeria syndrome (HGPS), a rare disease of premature aging. Early postnatal development is normal, but the children fail to thrive, develop bone abnormalities, and have a small beaked nose, a receding jaw, and a complete loss of hair (**Figure 11.11**). Affected children usually die at an average age of 13 years of severe atherosclerosis, a cause of death more commonly seen in the elderly.

The cause of HGPS appears to be a mutation in the gene for the nuclear protein *lamin*, a protein that forms a scaffold for the nucleus and may take part in the regulation of gene expression. The folded polypeptide that will eventually become lamin is modified and processed many times before the mature protein is produced. One key processing event is the removal of a farnesyl group that had been added to the nascent protein earlier in processing. In HGPS patients, the farnesyl group is not removed, owing to a mutation in the lamin. The incorrectly processed lamin results in a deformed nucleus (Figure 11.11) and aberrant nuclear function that results in HGPS. Much research remains to determine precisely how the failure to remove the farnesyl group leads to such dramatic consequences.

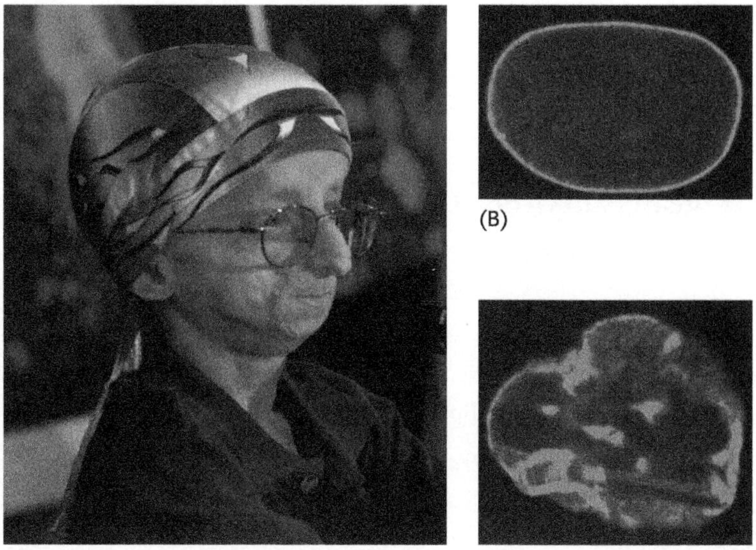

Figure 11.11 Hutchinson–Gilford progeria syndrome (HGPS). (A) A 15-year-old boy suffering from HGPS. (B) A normal nucleus. (C) A nucleus from a HGPS patient. [(A) AP Photo/Gerald Herbert; (B and C) Scaffidi, P., Gordon, L. and Misteli, T. (2005). The call nucleus and aging: Tantalizing clues and hopeful promises. *PLoS Biol* 3 (11): e395. Courtesy of Paola Scaffidi.]

SUMMARY

11.1 Fatty Acids Are a Main Source of Fuel

Lipids are defined as water-insoluble molecules that are soluble in organic solvents. Fatty acids are an important lipid in biochemistry. Fatty acids are hydrocarbon chains of various lengths and degrees of unsaturation that terminate with a carboxylic acid group. The fatty acid chains in membranes usually contain between 14 and 24 carbon atoms; they may be saturated or unsaturated. Short chain length and unsaturation enhance the fluidity of fatty acids and their derivatives by lowering the melting temperature.

11.2 Triacylglycerols Are the Storage Form of Fatty Acids

Fatty acids are stored as triacylglycerol molecules in adipose cells. Triacylglycerols are composed of three fatty acids esterified to a glycerol backbone. Triacylglycerols are stored in an anhydrous form.

11.3 There Are Three Common Types of Membrane Lipids

The major classes of membrane lipids are phospholipids, glycolipids, and cholesterol. Phosphoglycerides, a type of phospholipid, consist of a glycerol backbone, two fatty acid chains, and a phosphorylated alcohol. Phosphatidylcholine, phosphatidylserine, and phosphatidylethanolamine are major phosphoglycerides. Sphingomyelin, a different type of phospholipid, contains a sphingosine backbone instead of glycerol. Glycolipids are sugar-containing lipids derived from sphingosine. Cholesterol, which modulates membrane fluidity, is constructed from a steroid nucleus. A common feature of these membrane lipids is that they are amphipathic molecules, having one hydrophobic and one hydrophilic end.

KEY TERMS

fatty acid (p. 190)
triacylglycerol (p. 193)
phospholipid (p. 194)
sphingosine (p. 194)

phosphoglyceride (p. 194)
sphingomyelin (p. 195)
glycolipid (p. 196)
cerebroside (p. 196)

ganglioside (p. 196)
cholesterol (p. 197)
amphipathic molecule (p. 197)

? Answers to QUICK QUIZZES

1. Chain length and the degree of cis unsaturation.

2. Phospholipids, sphingolipids (of which glycolipids are a subclass), and cholesterol.

PROBLEMS

1. *Definition.* What are lipids? ✓ 3

2. *Like Lady and the Tramp.* Match the term with the proper description. ✓ 4

(a) Fatty acid
_____

(b) Triacylglycerol
_____

(c) Phospholipid
_____

(d) Sphingosine
_____

(e) Phosphoglyceride
_____

(f) Sphingomyelin
_____

(g) Glycolipid
_____

(h) Cerebroside
_____

(j) Cholesterol
_____

1. Membrane lipids with a glycerol backbone

2. Phospholipid especially common in nerve cells

3. The simplest glycolipid

4. Derived from sphingosine and found in all membranes

5. Chains of hydrogen-bearing carbon atoms

6. Complex glycolipids

7. A complex alcohol backbone for membrane lipids

8. Major class of membrane lipids

9. Steroid-based lipid

10. Storage form of fatty acids

3. *Structure and name.* Draw the structure of each of the following fatty acids, and give the structure its common name. ✓ 3

(a) *n*-Dodecanate
(b) *cis*-Δ^9-Hexadecenoate
(c) *cis*, *cis*-Δ^9, Δ^{12}-Octadecadienoate

4. *Fluidity matters.* Triacylglycerols are used for fuel storage in both plants and animals. The triacylglycerols from plants are often liquid at room temperature, whereas those from animals are solid. Suggest some chemical reasons for this difference. ✓ 3

5. *Contrast.* Distinguish between phosphoglycerides and triacylglycerols. ✓ 4

6. *Compare.* What structural features differentiate sphingolipids from phosphoglycerides? ✓ 4

7. *The head of the matter.* What are some molecules that form the polar head group of phospholipids? ✓ 4

8. *Depict a lipid.* Draw the structure of a triacylglycerol composed of equal amounts of palmitic acid, stearic acid, and oleic acid. ✓ 4

9. *Like finds like.* What structural characteristic of lipids accounts for their solubility in organic solvents? ✓ 4

10. *Like finds like, again.* Suppose that a small amount of phospholipid were exposed to an aqueous solution. What structure would the phospholipid molecules assume? What would be the driving force for the formation of this structure? ✓ 4

11. *Repetitive pattern.* How does the structure of steroids differ from the structure of other lipids? ✓ 4

12. *Linkages.* Platelet-activating factor (PAF) is a phospho-lipid that plays a role in allergic and inflammatory responses, as well as in toxic shock syndrome. The structure of PAF is shown here. How does it differ from the structures of the phospholipids discussed in this chapter? ✓ 4

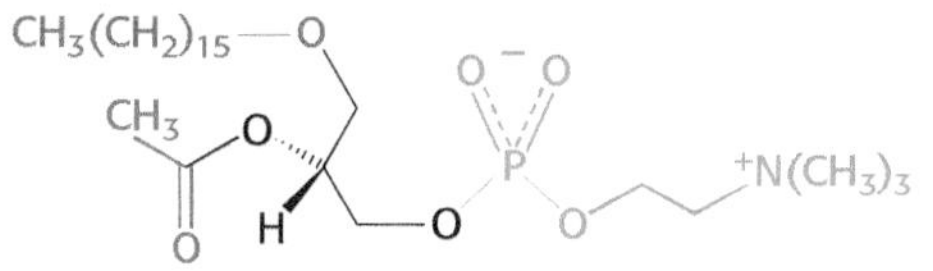

Platelet-activating factor (PAF)

13. *Coming clean.* Draw the structure of the sodium salt of stearic acid. How might it function to remove grease from your clothes or your hands? ✓ 3

14. *Hard-water problems.* Some metal salts of fatty acids are not as soluble as the sodium or potassium salts. For instance, magnesium or calcium salts of fatty acids are poorly soluble. How would taking a bath in water that is rich in magnesium or calcium affect the time you will need to spend cleaning the bathtub? ✓ 3

15. *Efficiency matters.* Why are lipids a more efficient storage form than glycogen? ✓ 4

16. *A sound diet.* Small mammalian hibernators can withstand body temperatures of 0° to 5°C without injury. However, the body fats of most mammals have melting temperatures of approximately 25°C. Predict how the composition of the body fat of hibernators might differ from that of their nonhibernating cousins. ✓ 3

Challenge Problems

17. *Molecules and melting points.* Match the fatty acid with the appropriate melting point. ✓ 3

(a) $CH_3(CH_2)_{18}CO_2H$ _____   1. 63°C

(b) $CH_3(CH_2)_{14}CO_2H$ _____   2. −5°C

(c) $CH_3(CH_2)_{10}CO_2H$ _____   3. 13°C

(d) $CH_3(CH_2)_7CH=CH(CH_2)_7CO_2H$ 4. 76°C
_____

(e) $CH_3(CH_2)_5CH=CH(CH_2)_7CO_2H$ 5. 45°C
_____

(f) $CH_3(CH_2)_4CH=CHCH_2CH=CH$ 6. −49°C
$(CH_2)_7CO_2H$ _____

(g) $CH_3(CH_2)_4(CH=CHCH_2)_4$ 7. 0°C
$(CH_2)_2CO_2H$ _____

18. *Melting point 1.* Explain why oleic acid (18 carbons, one cis bond) has a lower melting point than stearic acid, which has the same number of carbon atoms but is saturated. How would you expect the melting point of *trans*-oleic acid to compare with that of *cis*-oleic acid? ✓ 3

19. *Melting point 2.* Explain why the melting point of palmitic acid (C_{16}) is 6.5 degrees lower than that of stearic acid (C_{18}). ✓ 3

Selected Readings for this chapter can be found online at www.whfreeman.com/tymoczko3e.

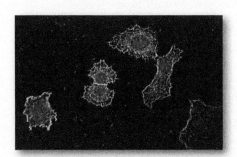

Cell Membranes, Channels, Pumps, and Receptors

The boundaries of cells are formed by membranes, the barriers that define the inside and the outside of a cell. These barriers are intrinsically impermeable, preventing molecules generated inside the cell from leaking out and unwanted molecules from diffusing in; yet they also contain transport systems that allow the cell to take up specific molecules and remove unwanted ones. Such transport systems provide membranes with the important property of *selective permeability*.

Selective permeability is conferred by two classes of membrane proteins: *pumps* and *channels*. Pumps use a source of free energy to drive the thermodynamically uphill transport of ions or molecules. Pump action is an example of *active transport*. Channels, in contrast, enable molecules, often ions, to flow rapidly through membranes in a thermodynamically downhill direction. Channel action illustrates *passive transport*, or *facilitated diffusion*.

The cell membrane is also impermeable to most information molecules, such as hormones, growth factors, and molecules in food or aromas that communicate taste or smell. Again, proteins serve to allow information to enter the cell from the environment. Membrane proteins called *receptors* sense information in the environment and transmit or transduce the information across the membrane to the cell interior.

First we will examine the basic structure of membranes and the proteins that are responsible for the property of selective permeability. Then, we will examine a special circumstance of transport across membranes—how environmental information is conveyed across the membrane to alter the biochemical workings of the cell.

✓ By the end of this section, you should be able to:

✓ 1 Identify the energetic force that powers the formation of membranes.

✓ 2 Explain why membranes are impermeable to most substances.

✓ 3 Describe the roles of proteins in making membranes selectively permeable.

✓ 4 Identify the basic components of all signal-transduction pathways.

✓ 5 Differentiate the various types of membrane receptors.

Membrane Structure and Function

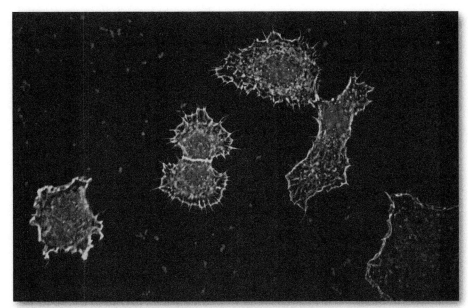

Cell membranes are not static structures. The membranes of these fibroblast cells show irregularities called ruffles and spikes. The ruffles and spikes are required for movement and phagocytosis. The nuclei are stained blue, and the actin component of the cytoskeleton is stained yellow. [Alex Gray/Wellcome Images.]

Membranes are the interface of two environments: (1) the cell interior and exterior in regard to the plasma membrane and (2) the cellular compartments and the cell interior in regard to internal membranes. Because these interfaces are unique, membranes of different cells and structures are as diverse in structure as they are in function. However, they do have in common a number of important attributes:

1. Membranes are *sheetlike structures,* only two molecules thick, that form *closed boundaries* between different compartments. The thickness of most membranes is between 60 Å (6 nm) and 100 Å (10 nm).

2. Membranes consist mainly of *lipids* and *proteins.* Membranes also contain *carbohydrates* that are linked to lipids and proteins (p. 196).

3. Membrane lipids are small molecules that have both *hydrophilic* and *hydrophobic* moieties. These lipids spontaneously form *closed bimolecular sheets* in aqueous media. These *lipid bilayers* are barriers to the flow of polar molecules.

4. Specific proteins mediate distinctive functions of membranes. Proteins serve as pumps, channels, receptors, energy transducers, and enzymes. Membrane proteins are embedded in lipid bilayers, which create suitable environments for their action.

5. Membranes are *noncovalent assemblies*. The constituent protein and lipid molecules are held together by many noncovalent interactions, which act cooperatively.

6. Membranes are *asymmetric*. The two faces of biological membranes always differ from each other.

7. Membranes are *fluid structures*. Lipid molecules diffuse rapidly in the plane of the membrane, as do proteins, unless they are anchored by specific interactions. In contrast, lipid molecules and proteins do not readily rotate across the membrane.

8. Most cell membranes are *electrically polarized* such that the inside is negative. Membrane potential plays a key role in transport, energy conversion, and excitability.

✓ 1 Identify the energetic force that powers the formation of membranes.

12.1 Phospholipids and Glycolipids Form Bimolecular Sheets

Recall from our earlier examination of membrane lipids that, although the repertoire of lipids is extensive, these lipids possess common structural elements: they are amphipathic molecules with a hydrophilic (polar) head group and a hydrophobic hydrocarbon tail (p. 198). *Membrane formation is a consequence of the amphipathic nature of the molecules.* Their polar head groups favor contact with water, whereas their hydrocarbon tails interact with one another in preference to water. How can molecules with these preferences arrange themselves in aqueous solutions? *The favored structure for most phospholipids and glycolipids in aqueous media is a lipid bilayer,* composed of two lipid sheets. The hydrophobic tails of each individual sheet interact with one another, forming a hydrophobic interior that acts as a permeability barrier. The hydrophilic head groups interact with the aqueous medium on each side of the bilayer. The two opposing sheets are called *leaflets* (**Figure 12.1**).

Lipid bilayers form spontaneously by a *self-assembly process*. In other words, the structure of a bimolecular sheet is inherent in the structure of the constituent lipid molecules. The growth of lipid bilayers from phospholipids is rapid and spontaneous in water. *The hydrophobic effect is the major driving force for the formation of lipid bilayers.* Recall that the hydrophobic effect also plays a dominant role in the folding of proteins (p. 62). Water molecules are released from the hydrocarbon tails of membrane lipids as these tails become sequestered in the nonpolar interior of the bilayer. Furthermore, *van der Waals attractive forces between the hydrocarbon tails favor close packing of the tails.* Finally, there are *electrostatic and hydrogen-bonding attractions between the polar head groups and water molecules.*

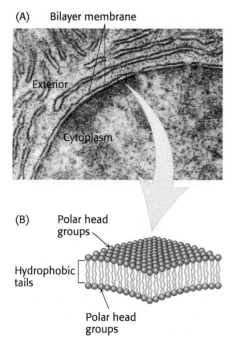

(A) Bilayer membrane

Exterior

Cytoplasm

(B) Polar head groups

Hydrophobic tails

Polar head groups

Figure 12.1 A section of phospholipid bilayer membrane. (A) Electron micrograph. (B) Space-filling model of an idealized view showing regular structures. [Don W. Fawcett/Science Source.]

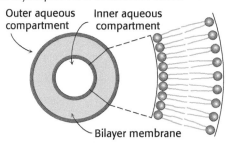

Figure 12.2 A liposome. A liposome, or lipid vesicle, is a small aqueous compartment surrounded by a lipid bilayer.

✚ CLINICAL INSIGHT

Lipid Vesicles Can Be Formed from Phospholipids

The propensity of phospholipids to form membranes has been used to create an important experimental and clinical tool. *Lipid vesicles,* or *liposomes,* are aqueous compartments enclosed by a lipid bilayer (Figure 12.2). These structures can be used to study membrane permeability or to deliver chemicals to cells. Liposomes are formed by suspending a membrane lipid in an aqueous medium and then *sonicating* (i.e., agitating by high-frequency sound waves) to give a dispersion of closed vesicles that are quite uniform in size. Vesicles formed by this method are nearly spherical in shape and have a diameter of about 500 Å (50 nm). Ions or molecules can be trapped in the aqueous compartments of lipid vesicles if the vesicles are formed in the presence of these substances (Figure 12.3).

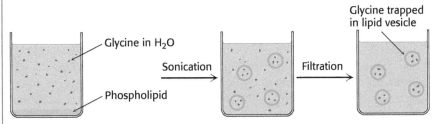

Figure 12.3 The preparation of glycine-containing liposomes. Liposomes containing glycine are formed by the sonication of phospholipids in the presence of glycine. Free glycine is removed by filtration.

Experiments are underway to develop clinical uses for liposomes. DNA-containing liposomes are currently being tested in more than 100 clinical trials for a variety of diseases. Liposomes may also contain drugs that can counter various pathological conditions. These liposomes fuse with the plasma membrane of many kinds of cells, introducing into the cells the molecules that they contain. The advantage of drug delivery by liposomes is that the drugs are more targeted than are systemic drugs, which means that less of the body is exposed to potentially toxic drugs. Liposomes are especially useful in targeting tumors and sites of inflammation, which have a high density of blood vessels. Liposomes concentrate in such areas of increased blood circulation, reducing the amount of drugs delivered to normal tissue. Currently, the largest use of liposomes is in the personal care industry where they are added to lotions and creams to deliver vitamins and other chemicals that are claimed to rejuvenate the skin.

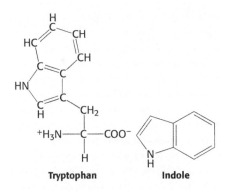

Tryptophan   **Indole**

Lipid Bilayers Are Highly Impermeable to Ions and Most Polar Molecules

The results of permeability studies of lipid bilayers have shown that *lipid bilayer membranes have a very low permeability for ions and most polar molecules.* The ability of molecules to move through a lipid environment, such as a membrane, is quite varied (Figure 12.4). For example, Na^+ and K^+ traverse these membranes

✓2 Explain why membranes are impermeable to most substances.

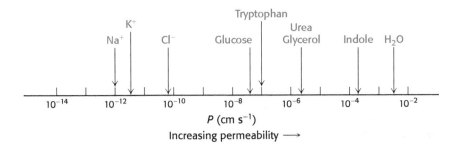

Figure 12.4 Permeability coefficients of ions and molecules in a lipid bilayer. The ability of molecules to cross a lipid bilayer spans a wide range of values. The permeability coefficient P, expressed in cm s^{-1}, provides a quantitative estimate of the rate of passage of a molecule across a membrane.

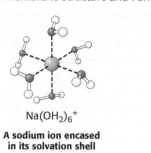

$Na(OH_2)_6^+$

A sodium ion encased in its solvation shell

10^9 times more slowly than H_2O. Tryptophan, a zwitterion at pH 7, crosses the membrane 10^3 times more slowly than indole, a structurally related molecule that lacks ionic groups. In fact, experiments show that *the permeability of small molecules is correlated with their relative solubilities in water and nonpolar solvents*. This relation suggests that a small molecule might traverse a lipid bilayer membrane in the following way: first, it sheds the water with which it is associated, called the solvation shell; then, it dissolves in the hydrocarbon core of the membrane; and, finally, it diffuses through this core to the other side of the membrane, where it is resolvated by water. An ion such as Na^+ cannot cross the membrane, because the replacement of its shell of polar water molecules by nonpolar interactions with the membrane interior is highly unfavorable energetically.

12.2 Membrane Fluidity Is Controlled by Fatty Acid Composition and Cholesterol Content

Many membrane processes, such as transport or signal transduction, depend on the fluidity of the membrane lipids, which is determined by the properties of fatty acid chains. Recall that the melting point of individual fatty acids depends on their length and the number of cis double bonds (p. 191). These same chemical properties affect the fluidity of the membranes of which the fatty acids are a component. The fatty acid chains in membrane bilayers may be arranged in an ordered, rigid state or in a relatively disordered, fluid state. The transition from the rigid to the fluid state takes place rather abruptly as the temperature is raised above T_m, the melting temperature (**Figure 12.5**). This *transition temperature depends on the length of the fatty acid chains and on their degree of unsaturation*. Long saturated fatty acids interact more strongly because of the increased number of van der Waals interactions than do short ones and thus favor the rigid state (**Figure 12.6**). On the other hand, *a cis double bond produces a bend in the hydrocarbon chain. This bend interferes with a highly ordered packing of fatty acid chains, and so T_m is lowered*.

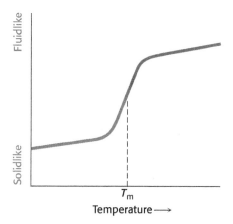

Figure 12.5 The phase-transition, or melting, temperature T_m for a phospholipid membrane. As the temperature is raised, the phospholipid membrane changes from a packed, ordered state to a more random one.

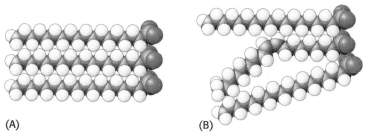

(A)   (B)

Figure 12.6 The packing of fatty acid chains in a membrane. The highly ordered packing of fatty acid chains is disrupted by the presence of cis double bonds. The space-filling models show the packing of (A) three molecules of stearate (C_{18}, saturated) and (B) a molecule of oleate (C_{18}, unsaturated) between two molecules of stearate.

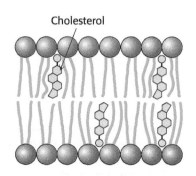

Cholesterol

Figure 12.7 Cholesterol disrupts the tight packing of the fatty acid chains. [Information from S. L. Wolfe, *Molecular and Cellular Biology* (Wadsworth, 1993).]

Bacteria regulate the fluidity of their membranes by varying the number of double bonds and the length of their fatty acid chains. In animals, *cholesterol* is the key modulator of membrane fluidity. Cholesterol contains a bulky steroid nucleus with a hydroxyl group at one end and a flexible hydrocarbon tail at the other end. The molecule inserts into bilayers with its long axis perpendicular to the plane of the membrane. Cholesterol's hydroxyl group forms a hydrogen bond with a carbonyl oxygen atom of a phospholipid head group, whereas its hydrocarbon tail is located in the nonpolar core of the bilayer. The different shape of cholesterol compared with that of phospholipids disrupts the regular interactions between fatty acid chains and helps maintain membrane fluidity (**Figure 12.7**).

In addition to its nonspecific effects on membrane fluidity, cholesterol can form specific complexes with saturated fatty acid components of lipids and specific proteins. These complexes concentrate within small (10–200 nm) and highly dynamic regions within membranes. The resulting structures are often referred to as *lipid rafts*. One result of these interactions is the *moderation of membrane fluidity*, making membranes less fluid but at the same time less subject to phase transitions. Although their small size and dynamic nature have made them very difficult to study, lipid rafts play a role in concentrating proteins required for signal-transduction pathways, regulate membrane curvature and budding, and facilitate the interaction between the extracellular matrix and the cytoskeleton.

? **QUICK QUIZ 1** Predict the effect on membrane-lipid composition if the temperature of a bacterial culture is raised from 37°C to 42°C.

12.3 Proteins Carry Out Most Membrane Processes

✓ 3 Describe the roles of proteins in making membranes selectively permeable.

We now turn to membrane proteins, which are responsible for most of the dynamic processes carried out by membranes. Membrane lipids form a permeability barrier and thereby establish compartments, whereas *specific proteins mediate nearly all other membrane functions*. In particular, proteins transport chemicals and information across a membrane. Membranes that differ in function differ in their protein content. Myelin, a membrane that serves as an electrical insulator around certain nerve fibers, has a low content of protein (18%). Membranes composed almost entirely of lipids are suitable for insulation because the hydrophobic components do not conduct currents well. In contrast, the plasma membranes, or exterior membranes, of most other cells must conduct the traffic of molecules into and out of the cells and so contain many pumps, channels, receptors, and enzymes. The protein content of these plasma membranes is typically 50%. Energy-transduction membranes, such as the internal membranes of mitochondria and chloroplasts, have the highest content of protein—typically, 75%. In general, *membranes performing different functions contain different kinds and amounts of proteins.*

Proteins Associate with the Lipid Bilayer in a Variety of Ways

Membrane proteins can be classified as being either *peripheral* or *integral* on the basis of their interaction with the hydrophobic interior of the membrane (**Figure 12.8**). *Integral membrane proteins* are embedded in the hydrocarbon chains of membrane lipids, and they can be released only when the membrane is physically disrupted. In fact, most integral membrane proteins span the lipid bilayer. In contrast, *peripheral membrane proteins* are bound to the head groups of lipids or the exposed portions of integral membrane proteins by electrostatic and hydrogen-bond interactions. These interactions may occur on either the cytoplasmic or the extracellular side of the membrane. Other proteins are anchored to the lipid bilayer by a covalently attached hydrophobic chain, such as a fatty acid (p. 198).

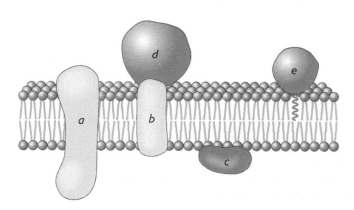

Figure 12.8 Integral and peripheral membrane proteins. Integral membrane proteins (*a* and *b*) interact extensively with the hydrocarbon region of the bilayer. Most known integral membrane proteins traverse the lipid bilayer. Some peripheral membrane proteins (*c*) interact with the polar head groups of the lipids. Other peripheral membrane proteins (*d*) bind to the surfaces of integral proteins. Some proteins (*e*) are tightly anchored to the membrane by a covalently attached lipid molecule.

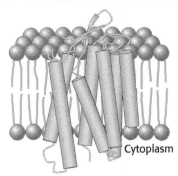

Figure 12.9 Structure of bacteriorhodopsin. Notice that bacteriorhodopsin consists largely of membrane-spanning α helices (represented by yellow cylinders). [Drawn from 1BRX.]

Proteins can span the membrane with α helices. For instance, consider *bacteriorhodopsin,* which uses light energy to transport protons from inside the bacterial cell to outside, generating a proton gradient used to form ATP (**Figure 12.9**). Bacteriorhodopsin is built almost entirely of α helices; seven closely packed α helices, arranged almost perpendicularly to the plane of the cell membrane, span its width. Just as the nonpolar hydrocarbon moieties of phospholipids associate with one another, the nonpolar α helices of bacteriorhodopsin associate with the hydrocarbon core of the lipid bilayer. *Membrane-spanning α helices are the most common structural motif in integral membrane proteins.*

The membrane-spanning parts of proteins can also be formed from β strands. Porin, a protein from the outer membranes of bacteria such as *E. coli,* represents a class of membrane proteins that are built from β strands and contain essentially no α helices (**Figure 12.10**). The arrangement of β strands is quite simple: each strand is hydrogen-bonded to its neighbor in an antiparallel arrangement, forming a single β sheet. The β sheet curls up to form a hollow cylinder that, as the protein's name suggests, forms a pore, or channel, in the membrane connecting the environment with the periplasmic space (p. 9). The outside surface of porin is appropriately nonpolar, given that it interacts with the hydrocarbon core of the membrane. In contrast, the inside of the channel is quite hydrophilic and is filled with water.

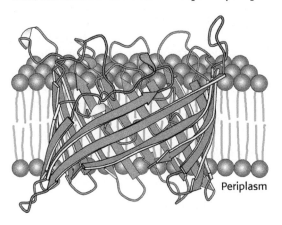

Figure 12.10 The structure of bacterial porin (from *Rhodopseudomonas blastica*). Notice that this membrane protein is built entirely of β strands. Only one monomer of the trimeric protein is shown. [Drawn from 1PRN.pdb.]

Another means by which a protein associates with a membrane is the embedding of just part of the protein into the membrane. The structure of the endoplasmic reticulum membrane-bound enzyme prostaglandin H_2 synthase-1 exemplifies this type of association. Prostaglandin H_2 synthase-1 lies along the outer surface of the membrane but is firmly bound by a set of α helices with hydrophobic surfaces that extend from the bottom of the protein into the membrane (**Figure 12.11**).

Figure 12.11 The attachment of prostaglandin H_2 synthase-1 to the membrane. Notice that prostaglandin H_2 synthase-1 is held in the membrane by a set of α helices (orange) coated with hydrophobic side chains. One monomer of the dimeric enzyme is shown. [Drawn from 1PTH.pdb.]

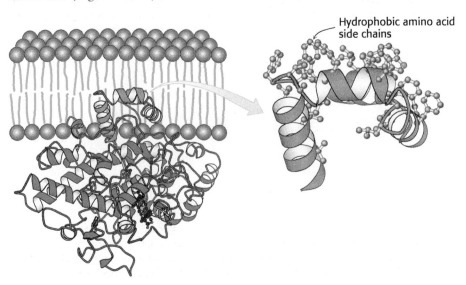

The Association of Prostaglandin H₂ Synthase-I with the Membrane Accounts for the Action of Aspirin

The localization of prostaglandin H_2 synthase-l in the membrane is crucial to its function. Prostaglandin H_2 synthase-1 catalyzes the conversion of arachidonic acid into prostaglandin H_2, which promotes inflammation and modulates gastric acid secretion, in a two-step process requiring a cyclooxygenase (COX) activity and a peroxidase activity. Arachidonic acid is a fatty acid generated by the hydrolysis of membrane lipids. Arachidonic acid moves from the nonpolar core of the lipid bilayer, where it is generated, to the active site of the enzyme, without entering an aqueous environment, by traveling through a hydrophobic channel in the protein (**Figure 12.12**). Indeed, nearly all of us have experienced the importance of this channel leading to the cyclooxygenase: drugs such as aspirin and ibuprofen block the channel and prevent prostaglandin synthesis, thereby reducing the inflammatory response. Aspirin donates an acetyl group to a serine residue (Ser 530) that lies along the path to the active site, thereby blocking the channel (**Figure 12.13**). Thus, aspirin and similar drugs are called COX inhibitors.

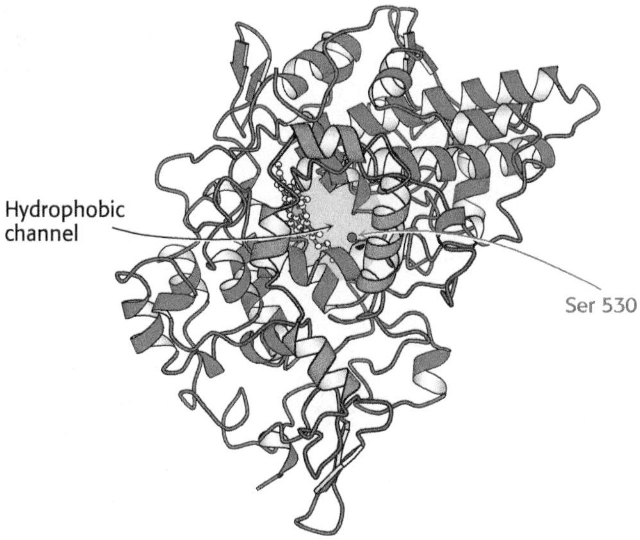

Hydrophobic channel

Ser 530

🖱 **Figure 12.12 The hydrophobic channel of prostaglandin H_2 synthase-1.** A view of prostaglandin H_2 synthase-1 from the membrane reveals the hydrophobic channel that leads to the active site. The membrane-anchoring helices are shown in orange. [Drawn from 1PTH.pdb.]

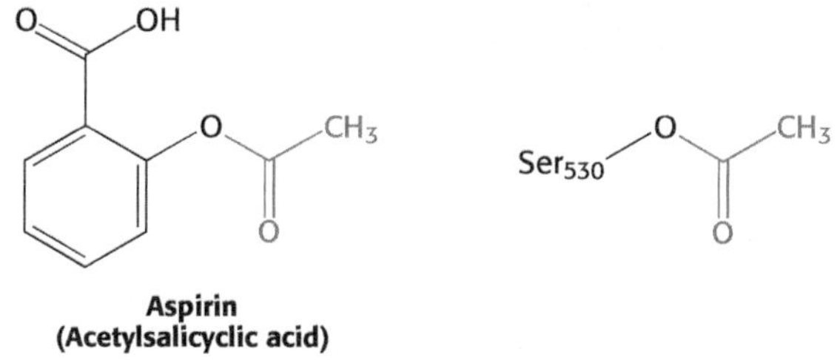

**Aspirin
(Acetylsalicyclic acid)**

Figure 12.13 Aspirin's effects on prostaglandin H_2 synthase-1. Aspirin acts by transferring an acetyl group to a serine residue in prostaglandin H_2 synthase-1.

12.4 Lipids and Many Membrane Proteins Diffuse Laterally in the Membrane

Membranes are not rigid, static structures. On the contrary, lipids and many membrane proteins are constantly in motion, a process called *lateral diffusion*. The rapid lateral movement of membrane constituents has been visualized with the use of fluorescence microscopy and the technique of *fluorescence recovery after photobleaching* (FRAP; **Figure 12.14**). First, a cell-surface component, such

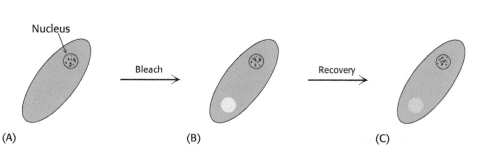

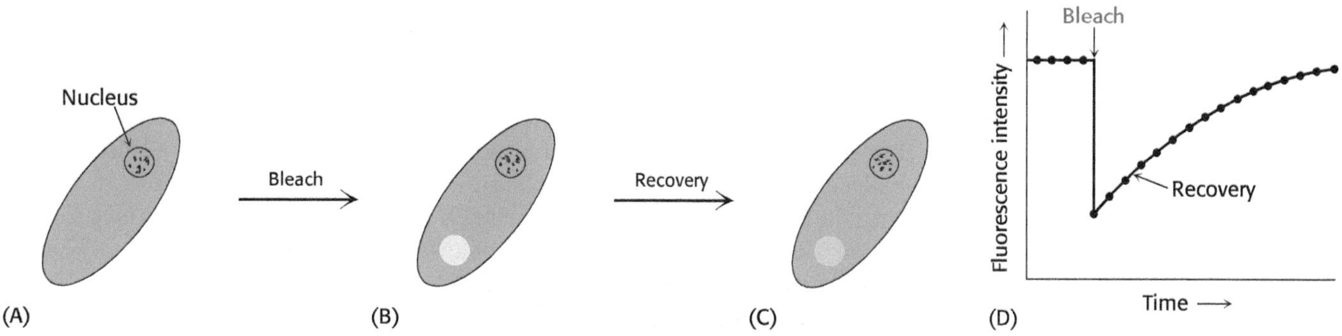

Nucleus

Bleach

Recovery

(A)   (B)   (C)   (D)

Bleach

Recovery

Fluorescence intensity →

Time →

Figure 12.14 The technique of fluorescence recovery after photobleaching (FRAP). (A) The cell surface fluoresces because of a labeled surface component. (B) The fluorescent molecules of a small part of the surface are bleached by an intense light pulse. (C) The fluorescence intensity recovers as bleached molecules diffuse out of the region and unbleached molecules diffuse into it. (D) The rate of recovery depends on the diffusion coefficient.

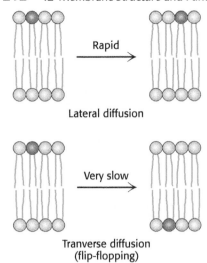

Figure 12.15 Lipid movement in membranes. Lateral diffusion of lipids is much more rapid than transverse diffusion (flip-flopping).

as a protein or the head group of a phospholipid, is fluorescently labeled. A small region of the cell surface (~3 μm^2) is viewed through a fluorescence microscope. The fluorescent molecules in this region are then destroyed (bleached) by an intense pulse of light from a laser. The fluorescence of this region is subsequently monitored as a function of time. If the labeled component is mobile, bleached molecules leave and unbleached molecules enter the affected region, which results in an increase in the fluorescence intensity. The rate of recovery of fluorescence depends on the lateral mobility of the fluorescence-labeled component. Such experiments show that a phospholipid molecule moves an average distance of 2 μm in 1 s. This rate means that *a lipid molecule can travel from one end of a bacterium to the other in a second*. In contrast, proteins vary markedly in their lateral mobility. *Some proteins are nearly as mobile as lipids, whereas others are virtually immobile.* Proteins are immobilized by attachment to the cytoskeleton or structural components outside the cell, called the extracellular matrix.

On the basis of the mobility of proteins in membranes, in 1972 S. Jonathan Singer and Garth Nicolson proposed a *fluid mosaic model* to describe the overall organization of biological membranes. The essence of their model is that *membranes are two-dimensional solutions of oriented lipids and globular proteins*. The lipid bilayer has a dual role: it is both a *solvent* for integral membrane proteins and a *permeability barrier*. Membrane proteins are free to diffuse laterally in the lipid matrix unless restricted by special interactions.

Although the lateral diffusion of membrane components can be rapid, the spontaneous rotation of lipids from one face of a membrane to the other is a very slow process. The transition of a molecule from one membrane surface to the other is called *transverse diffusion* or *flip-flopping*. The results of experiments show that *a phospholipid molecule flip-flops once in several hours*. Thus, a phospholipid molecule takes about 10^9 times as long to flip-flop across a membrane as it takes to diffuse a distance of 50 Å in the lateral direction (**Figure 12.15**). The free-energy barriers to flip-flopping are even larger for protein molecules than for lipids because proteins have more extensive polar regions. In fact, the flip-flopping of a protein molecule has not been observed. Hence, *membrane asymmetry can be preserved for long periods.*

12.5 A Major Role of Membrane Proteins Is to Function as Transporters

Transporter proteins are a specific class of pump or channel that facilitates the movement of molecules across a membrane. Each cell type expresses a specific set of transporters in its plasma membrane. The set of transporters expressed is crucial because these transporters largely determine the ionic composition inside a cell and the compounds that can be taken up from the cell's environment. In some senses, the array of transporters expressed by a cell determines its characteristics because a cell can execute only those biochemical reactions for which it has taken up the substrates.

Two factors determine whether a small molecule will cross a membrane: (1) the concentration gradient of the molecule across the membrane and (2) the molecule's solubility in the hydrophobic environment of the membrane. In accord with the Second Law of Thermodynamics, molecules spontaneously move from a region of higher concentration to one of lower concentration. For many molecules, the cell membrane is an obstacle to this movement, but, as discussed earlier (p. 207), some molecules can pass through the membrane because they dissolve in the lipid bilayer. Such molecules are called *lipophilic molecules*. The steroid hormones provide a physiological example. These cholesterol relatives can pass through a membrane in their path of movement in a process called *simple diffusion*.

Matters become more complicated when the molecule is highly polar. For example, sodium ions are present at 143 mM outside a typical cell and at 14 mM inside the cell, yet sodium does not freely enter the cell, because the charged ion cannot pass through the hydrophobic membrane interior. In some circumstances, such as in a nerve impulse, sodium ions must enter the cell. How are these ions able to cross the hydrophobic membrane interior? Sodium ions pass through specific channels in the hydrophobic barrier—channels that are formed by membrane proteins. This means of crossing the membrane is called *facilitated diffusion* because the diffusion across the membrane is facilitated by the channel. It is also called *passive transport* because the energy driving the ion movement originates in the ion gradient itself, without any contribution by the transport system. Channels, like enzymes, display substrate specificity in that they facilitate the transport of some ions but not other ions, not even those that are closely related.

How is the sodium gradient established in the first place? In this case, sodium must move, or be pumped, *against* a concentration gradient. Because moving the ion from a low concentration to a higher concentration results in a decrease in entropy, it requires an input of free energy. Protein *pumps* embedded in the membrane are capable of using an energy source to move the molecule up a concentration gradient in a process called *active transport*.

The Na⁺–K⁺ ATPase Is an Important Pump in Many Cells

Cells must control their intracellular salt concentrations to prevent unfavorable interactions with high concentrations of ions and to facilitate specific processes such as a nerve impulse. In particular, most animal cells contain a high concentration of K^+ and a low concentration of Na^+ relative to the extracellular fluid. These ionic gradients are generated by a specific transport system, an enzyme called the *Na^+–K^+ pump* or the *Na^+–K^+ ATPase*. The hydrolysis of ATP by the pump provides the energy needed for the active transport of three Na^+ ions out of the cell and two K^+ ions into the cell, generating the gradients (**Figure 12.16**). In other words, the Na^+–K^+ ATPase is an *ATP-driven pump*.

The active transport of Na^+ and K^+ is of great physiological significance. Indeed, more than a third of the ATP consumed by a resting animal is used to pump these ions. The Na^+–K^+ gradient in animal cells controls cell volume, renders neurons and muscle cells electrically excitable, and drives the active transport of sugars and amino acids. This third phenomenon is called secondary active transport (p. 214) because the sodium gradient generated by the Na^+–K^+ ATPase (the primary instance of active transport) can be used to power active transport of other molecules (the secondary instance of active transport) when the sodium flows down its gradient.

The pump is called the Na^+–K^+ ATPase because the hydrolysis of ATP takes place only when Na^+ and K^+ are present. The Na^+–K^+ ATPase is referred to as a *P-type ATPase* because it forms a key *p*hosphorylated intermediate. In the formation of this intermediate, a phosphoryl group obtained from ATP is linked to the side chain of a specific conserved aspartate residue in the pump to form phosphorylaspartate. Other examples of P-type ATPases include the *Ca^{2+} ATPase*, which transports calcium ions out of the cytoplasm and into the extracellular fluid, mitochondria, and sarcoplasmic reticulum of muscle cells, a key process in muscle contraction; the *gastric H^+–K^+ ATPase,* which is responsible for pumping protons into the stomach to lower the pH to near 1.0; and *P4-ATPase*, a flippase that moves membrane lipids from the outer leaflet to the inner leaflet. Indeed, P-type ATPases are found in all the kingdoms of life.

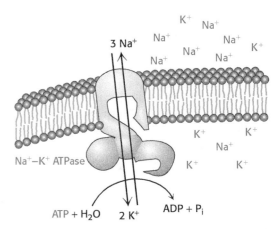

Figure 12.16 Energy transduction by membrane proteins. The Na^+–K^+ ATPase converts the free energy of phosphoryl transfer into the free energy of a Na^+ ion gradient.

Phosphorylaspartate

(A)

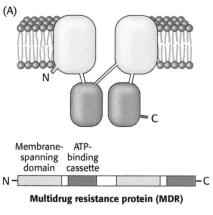

Multidrug resistance protein (MDR)

(B)

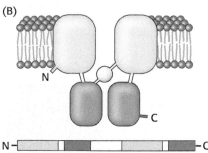

Cystic fibrosis transmembrane regulator (CFTR)

Figure 12.17 ABC transporters. The multidrug-resistance protein (MDR) and the cystic fibrosis transmembrane regulator (CFTR) are homologous proteins composed of two transmembrane domains and two ATP-binding domains, called ATP-binding cassettes (ABCs).

CLINICAL INSIGHT

Multidrug Resistance Highlights a Family of Membrane Pumps with ATP–Binding Domains

Studies of human disease revealed another large family of active-transport proteins, with structures and mechanisms quite different from those of the P-type ATPase family. Tumor cells often become resistant to drugs that were initially quite toxic to the cells. Remarkably, the development of resistance to one drug also makes the cells less sensitive to a range of other compounds. This phenomenon is known as *multidrug resistance*. In a significant discovery, the onset of multidrug resistance was found to correlate with the expression of a protein that acts as an ATP-dependent pump that extrudes a wide range of small molecules from cells that express it. The protein is called the *multidrug-resistance* (MDR) *protein* or *P-glycoprotein* ("glyco" because it includes a carbohydrate moiety). When cells are exposed to a drug, the MDR protein pumps the drug out of the cell before the drug can exert its effects. The MDR protein comprises four domains: two membrane-spanning domains and two ATP-binding domains (**Figure 12.17**A). The ATP-binding domains of these proteins are called *ATP-binding cassettes* (ABCs). Transporters that include these domains are called *ABC transporters*. ABC transporters are one of the largest protein superfamilies and are found in all forms of life.

Another example of an ABC transporter is the *cystic fibrosis transmembrane conductance regulator* (CFTR; Figure 12.17B). CFTR acts as an ATP-regulated chloride channel in the plasma membranes of epithelial cells. Mutations in the gene for CFTR cause a decrease in fluid and salt secretion by CFTR and result in cystic fibrosis. As a consequence of the malfunctioning CFTR, secretion from the pancreas is blocked and heavy, dehydrated mucus accumulates in the lungs, leading to chronic lung infections. Certain ABC transporters also function as flippases, but in contrast to the P-type flippase, these enzymes transport lipids from the inner leaflet to the outer leaflet of the cell membrane.

CLINICAL INSIGHT

Harlequin Ichthyosis Is a Dramatic Result of a Mutation in an ABC Transporter Protein

A number of human diseases in addition to cystic fibrosis result from defects in ABC transporter proteins. One especially startling disorder is harlequin ichthyosis, which results from a defective ABC transporter for lipids in keratinocytes, a common type of skin cell. Babies suffering from this very rare disorder are born encased in thick skin, which restricts their movement. As the skin dries out, hard diamond shaped plaques form, severely distorting facial features. The newborns usually die within a few days because of feeding difficulties, respiratory distress, or infections that are likely due to cracks in the skin.

Secondary Transporters Use One Concentration Gradient to Power the Formation of Another

Many active-transport processes are not directly driven by the hydrolysis of ATP. Instead, the thermodynamically uphill flow of one species of ion or molecule is coupled to the downhill flow of a different species. Membrane proteins that move ions or molecules uphill by this means are termed *secondary transporters* or *cotransporters.* These proteins can be classified as either *antiporters* or *symporters.* Antiporters couple the downhill flow of one species to the uphill flow of another in the *opposite direction* across the membrane; symporters use the flow of one species to drive the flow of a different species in the *same direction* across the membrane (**Figure 12.18**).

Glucose is moved into some animal cells by the sodium-glucose linked transporter (SGLT), a symporter powered by the simultaneous entry of Na^+. This free-energy input of Na^+ flowing down its concentration gradient is sufficient to generate a 66-fold concentration gradient of an uncharged molecule such as glucose (**Figure 12.19**). Recall that the sodium ion gradient was initially generated by the Na^+–K^+ ATPase, demonstrating that the action of the secondary active transporter depends on the primary active transporter.

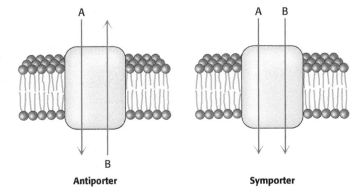

Figure 12.18 Antiporters and symporters. Secondary transporters can transport two substrates in opposite directions (antiporters) or two substrates in the same direction (symporters).

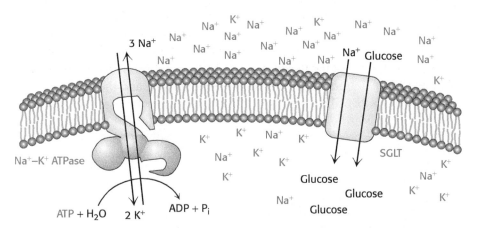

Figure 12.19 Secondary transport. The ion gradient set up by the Na^+–K^+ ATPase can be used to move materials into the cell, through the action of a secondary transporter such as the Na^+–glucose linked transporter, a symporter.

 CLINICAL INSIGHT

Digitalis Inhibits the Na^+–K^+ Pump by Blocking Its Dephosphorylation

The interplay between active transport and secondary active transport is especially well illustrated by the action of the cardiotonic steroids. Heart failure can result if the muscles in the heart are not able to contract with sufficient strength to effectively pump blood. Certain steroids derived from plants, such as digitalis and ouabain, are known as *cardiotonic steroids* because of their ability to strengthen heart contractions. Interestingly, cardiotonic steroids exert their effect by inhibiting the Na^+–K^+ pump.

Digitalis is a mixture of cardiotonic steroids derived from the dried leaf of the foxglove plant *Digitalis purpurea* (**Figure 12.20**). The compound increases the force of contraction of heart muscle and is consequently a choice drug in the treatment of congestive heart failure. Inhibition of the Na^+–K^+ pump by digitalis means that Na^+ is not pumped out of the cell, diminishing the Na^+ gradient. The reduced Na^+ gradient in turn affects the sodium–calcium exchanger. This exchanger, an example of secondary active transport, relies on Na^+ influx to simultaneously power the expulsion of Ca^+ from the cell. The diminished Na^+ gradient results in slower extrusion of Ca^{2+} by the sodium–calcium exchanger. The subsequent increase in the intracellular level of Ca^{2+} enhances the ability of cardiac muscle to contract.

Figure 12.20 Foxglove. Foxglove *(Digitalis purpurea)* is a highly poisonous plant due to the high concentration of potent cardiotonic steroids. Digitalis, one of the most widely used drugs, is obtained from foxglove. [Roger Hall/Shutterstock.]

DID YOU KNOW?

Interestingly, digitalis was used effectively long before the discovery of the Na^+–K^+ ATPase. In 1785, William Withering, a British physician, heard tales of an elderly woman, known as "the old woman of Shropshire," who cured people of "dropsy" (which today would be recognized as congestive heart failure) with an extract of foxglove. Withering conducted the first scientific study of the effects of foxglove on congestive heart failure and documented its effectiveness.

Specific Channels Can Rapidly Transport Ions Across Membranes

Pumps can transport ions at rates approaching several thousand ions per second. Other membrane proteins, the passive-transport systems called *ion channels,* are capable of ion-transport rates that are more than 1000 times as fast. These rates of transport through ion channels are close to rates expected for ions diffusing freely through aqueous solution. Yet ion channels are not simply tubes that span membranes through which ions can rapidly flow. Rather, they are highly sophisticated molecular machines that respond to chemical and physical changes in their environments and undergo precisely timed conformational changes to regulate when ions can flow into and out of the cell. Channels can be classified on the basis of the environmental signals that activate them. *Voltage-gated channels* are opened in response to changes in membrane potential, whereas *ligand-gated channels* are opened in response to the binding of small molecules (ligands), such as neurotransmitters.

Among the most important manifestations of ion-channel action is the nerve impulse, which is the fundamental means of communication in the nervous system. A *nerve impulse,* or *action potential,* is an electrical signal produced by the flow of ions across the plasma membrane of a neuron. In particular, Na^+ transiently flows into the cell and K^+ flows out. This ion traffic is through protein channels that are both specific and rapid. The importance of ion channels is illustrated by the effect of tetrodotoxin, which is produced by the pufferfish (**Figure 12.21**). Tetrodotoxin inhibits the Na^+ channel, and the lethal dose for human beings is just 10 ng.

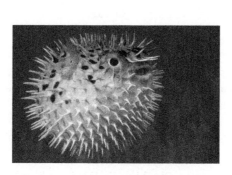

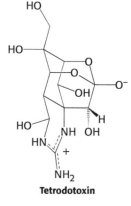

Tetrodotoxin

Figure 12.21 Pufferfish. The pufferfish is regarded as a culinary delicacy in Japan. Tetrodotoxin is produced by the pufferfish. Several people die every year in Japan from eating poorly prepared pufferfish. [Beth Swanson/Shutterstock.]

Visual image (left) and thermal image (right) of a mouse. [Left: Michiel de Wit/Shutterstock; right: Ted Kinsman/Science Source.]

❀ BIOLOGICAL INSIGHT

Venomous Pit Vipers Use Ion Channels to Generate a Thermal Image

A large family of cation channels comprises the TRP (*transient receptor potential*) channels. These channels serve a host of functions in vertebrates including detecting taste, pain, and temperature. Venomous pit vipers, such as the western diamondback rattlesnake (*Crotalus atrox*), possess TRP channels that are activated by infrared (750 nm–1 mm) radiation, thus enabling the snakes to create a thermal landscape that can be overlain by their visual landscape. This dual-image view enables the pit vipers to locate and track prey with great speed and accuracy, much to the consternation of a host of rodents.

The Structure of the Potassium Ion Channel Reveals the Basis of Ion Specificity

The K^+ channel is one of the most extensively studied ion channels and thus provides us with a clear example of how a channel function can be both specific

Cell exterior

3 Å

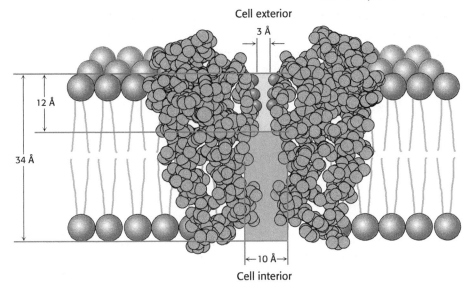

12 Å

34 Å

10 Å

Cell interior

Figure 12.22 A path through a channel. A potassium ion entering the K⁺ channel can pass a distance of 22 Å into the membrane while remaining solvated with water (blue). At this point, the pore diameter narrows to 3 Å (yellow), and potassium ions must shed their water and interact with carbonyl groups (red) of the pore amino acids.

and rapid. Beginning from the inside of the cell, the pore starts with a diameter of approximately 10 Å and then constricts to a smaller cavity with a diameter of 8 Å (**Figure 12.22**). Both the opening to the outside and the central cavity of the pore are filled with water, and a K^+ ion can fit in the pore without losing its shell of bound water molecules. Approximately two-thirds of the way through the membrane, the pore becomes more constricted (3-Å diameter). At that point, any K^+ ions must give up their water molecules and interact directly with groups from the protein. The channel structure effectively reduces the thickness of the membrane from 34 Å to 12 Å by allowing the solvated ions to penetrate into the membrane before the ions must directly interact with the channel.

The restricted part of the pore is built from residues contributed by the two transmembrane α helices. In particular, a stretch of five amino acids within this region function as the *selectivity filter* that determines the preference for K^+ over other ions (**Figure 12.23**). This region of the strand lies in an extended conformation and is oriented such that the peptide carbonyl groups are directed into the channel, in a good position to interact with the potassium ions. The potassium ion relinquishes its associated water molecules because it can bind with the oxygen atoms of the carbonyl groups of the selectivity filter.

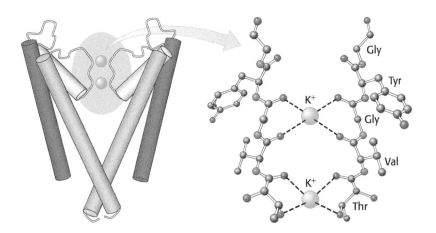

Gly

Tyr

K⁺

Gly

Val

K⁺

Thr

Figure 12.23 The selectivity filter of the potassium ion channel. Potassium ions interact with the carbonyl groups of the selectivity filter, located at the 3-Å-diameter pore of the K⁺ channel. Only two of the four channel subunits are shown.

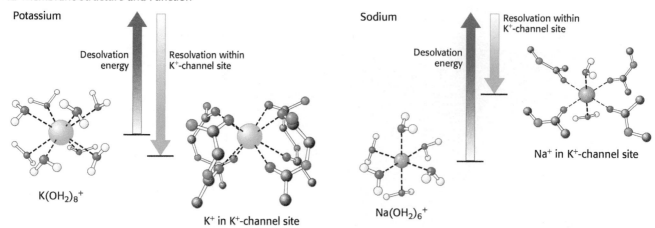

Figure 12.24 The energetic basis of ion selectivity. The energy cost of dehydrating a potassium ion is compensated by favorable interactions with the selectivity filter. Because a sodium ion is too small to interact favorably with the selectivity filter, the free energy of desolvation cannot be compensated, and the sodium ion does not pass through the channel.

Potassium ion channels are 100-fold more permeable to K^+ than to Na^+. How is this high degree of selectivity achieved? Ions having a radius larger than 1.5 Å cannot pass into the narrow diameter (3 Å) of the selectivity filter of the K^+ channel. However, a bare Na^+ is small enough to pass through the pore. Indeed, the ionic radius of Na^+ (0.95 Å) is substantially smaller than that of K^+ (1.33 Å). How, then, is Na^+ rejected?

Sodium ions are too small to react with the selectivity filter. For ions to relinquish their water molecules, other polar interactions—such as those that take place between the K^+ ion and the selectivity filter's carbonyl groups—must replace the interactions with water. The key point is that the free-energy costs of dehydrating these ions are considerable. *The channel pays the cost of dehydrating potassium ions by providing compensating interactions with the carbonyl oxygen atoms lining the selectivity filter* (**Figure 12.24**). Sodium ions are rejected because the energy required to dehydrate them would not be recovered. The K^+ channel does not closely interact with sodium ions, which must stay hydrated and, hence, cannot pass through the channel.

The Structure of the Potassium Ion Channel Explains Its Rapid Rate of Transport

The tight binding sites required for ion selectivity should slow the progress of ions through a channel, yet ion channels achieve rapid rates of ion transport. How is this apparent paradox resolved? A structural analysis of the K^+ channel at high resolution provides an appealing explanation. Four K^+-binding sites crucial for rapid ion flow are present in the constricted region of the K^+ channel. Consider the process of ion conductance starting from inside the cell (**Figure 12.25**). A hydrated potassium ion proceeds into the channel and through the relatively unrestricted part of the channel. The ion then gives up its coordinated water molecules and binds to a site within the selectivity-filter region. The ion can move between the four sites within the selectivity filter because they have similar energy levels and thus ion affinities. As each subsequent potassium ion moves into the selectivity filter, its positive charge will repel the potassium ion at the nearest site, causing it to shift to a site farther up the channel and in turn push upward any potassium ion already bound to a site farther up. Thus, each ion that binds anew favors the release of an ion from the other side of the channel. This multiple-binding-site mechanism solves the apparent paradox of high ion selectivity and rapid flow.

? QUICK QUIZ 2 What determines the direction of flow through an ion channel?

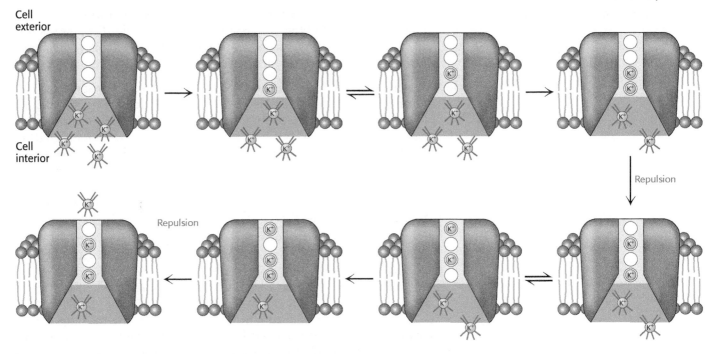

Figure 12.25 A model for K⁺-channel transport. The selectivity filter has four binding sites (white circles). Hydrated potassium ions can enter these sites, one at a time, losing their hydration shells (red lines). When two ions occupy adjacent sites, electrostatic repulsion forces them apart. Thus, as ions enter the channel from one side, other ions are pushed out the other side.

SUMMARY

12.1 Phospholipids and Glycolipids Form Bimolecular Sheets

Membrane lipids spontaneously form extensive bimolecular sheets in aqueous solutions. The driving force for membrane formation is the hydrophobic effect. The hydrophobic tails then interact with one another by van der Waals interactions, while the hydrophilic head groups interact with the aqueous medium. Lipid bilayers are cooperative structures, held together by many weak bonds. These lipid bilayers are highly impermeable to ions and most polar molecules, yet they are quite fluid, which enables them to act as a solvent for membrane proteins.

12.2 Membrane Fluidity Is Controlled by Fatty Acid Composition and Cholesterol Content

The degree of fluidity of a membrane partly depends on the chain length of its lipids and the extent to which their constituent fatty acids are unsaturated. In animals, cholesterol content also regulates membrane fluidity.

12.3 Proteins Carry Out Most Membrane Processes

Specific proteins mediate distinctive membrane functions such as transport, communication, and energy transduction. Many integral membrane proteins span the lipid bilayer, whereas others are only partly embedded in the membrane. Peripheral membrane proteins are bound to membrane surfaces or integral membrane proteins by electrostatic and hydrogen-bond interactions. Membrane-spanning proteins have regular structures, including β strands, although the α helix is the most common membrane-spanning structure.

12.4 Lipids and Many Membrane Proteins Diffuse Laterally in the Membrane

Membranes are dynamic structures in which proteins and lipids diffuse rapidly in the plane of the membrane (lateral diffusion), unless restricted by special interactions. In contrast, the rotation of lipids from one face of a membrane to the other (transverse diffusion, or flip-flopping) is usually very slow. Proteins do not rotate across bilayers; hence, membrane asymmetry can be preserved.

12.5 A Major Role of Membrane Proteins Is to Function as Transporters

Polar or charged molecules require proteins to form passages through the hydrophobic barrier. Passive transport or facilitated diffusion takes place when an ion or polar molecule moves down its concentration gradient. If a molecule moves against a concentration gradient, an external energy source is required; this movement is referred to as active transport and results in the generation of concentration gradients.

Active transport is often carried out at the expense of ATP hydrolysis. P-type ATPases pump ions against a concentration gradient and become transiently phosphorylated in the process of transport. Membrane proteins containing ATP-binding-cassette domains are another family of ATP-dependent pumps.

Many active-transport systems couple the uphill flow of one ion or molecule to the downhill flow of another. These membrane proteins, called secondary transporters or cotransporters, can be classified as antiporters or symporters. Antiporters couple the downhill flow of one type of ion in one direction to the uphill flow of another in the opposite direction. Symporters move both ions in the same direction.

Ion channels allow the rapid movement of ions across the hydrophobic barrier of the membrane. In regard to K^+ channels, hydrated potassium ions must transiently lose their coordinated water molecules as they move to the narrowest part of the channel, termed the selectivity filter.

KEY TERMS

lipid bilayer (p. 206)
liposome (p. 207)
cholesterol (p. 208)
lipid raft (p. 209)
integral membrane protein (p. 209)
peripheral membrane protein (p. 209)
simple diffusion (p. 212)
facilitated diffusion (passive
  transport) (p. 213)

active transport (p. 213)
Na^+–K^+ pump (Na^+–K^+ATPase)
  (p. 213)
P-type ATPase (p. 213)
multidrug-resistance (MDR) protein
  (P-glycoprotein) (p. 214)
ATP-binding cassette (ABC) domain
  (p. 214)

secondary transporter
  (cotransporter) (p. 214)
antiporter (p. 215)
symporter (p. 215)
ion channel (p. 216)
selectivity filter (p. 217)

? Answers to QUICK QUIZZES

1. The increase in temperature will increase the fluidity of the membrane. To prevent the membrane from becoming too fluid, the bacteria will incorporate longer-chain fatty acids into the membrane phospholipids. This alteration will increase van der Waals interactions among the chains and decrease fluidity.

2. Ion channels allow ion flow in either direction. In accordance with the Second Law of Thermodynamics, ions will flow down their concentration gradient.

PROBLEMS

1. *Shared traits.* Name some of the features common to all membranes. ✓ 1

2. *Simple diffusion.* What conditions are required for a small molecule to spontaneously pass through a membrane? ✓ 2

3. *Bread and jam.* Match each term with its description.

(a) Integral membrane protein _____

(b) Peripheral membrane protein _____

(c) Channel _____

(d) Passive transport _____

(e) Active transport _____

(f) Na$^+$–K$^+$ ATPase _____

(g) Secondary transporter _____

(h) Antiporter _____

(i) Symporter _____

(j) Ion channel _____

1. Facilitated diffusion
2. Uses the energy of one gradient to create another
3. Interacts tightly with the membrane interior
4. Molecules moving in opposite directions
5. Interacts with the border of a membrane
6. Allows rapid movement of molecules down a gradient across a membrane
7. Movement against a concentration gradient
8. Molecules moving in the same direction
9. Can be voltage-gated or ligand-gated
10. Inhibited by digitalis

4. *Solubility matters.* Arrange the following substances in order of increasing permeability through a lipid bilayer: (a) glucose; (b) glycerol; (c) Cl$^-$; (d) indole; (e) tryptophan. ✓ 2

5. *A helping hand.* Differentiate between simple diffusion and facilitated diffusion. ✓ 2

6. *Gradients.* Differentiate between passive transport and active transport. ✓ 3

7. *The golden mean.* Proper membrane fluidity is vital to membrane-protein function. Suggest how a loss of fluidity and how too much fluidity might affect membrane-protein function. ✓ 3

8. *Heart beats.* Outline the relation between the Na$^+$–K$^+$ ATPase and the strength of a heart contraction. Identify the relevant primary and secondary active-transport components. How do cardiotonic steroids affect the strength of a heartbeat? ✓ 3

9. *Hunting hippos.* Somali hunters use arrows that have been dipped in high concentrations of the cardiac glycoside ouabain to kill game. Indeed, there are reports that animals the size of a hippopotamus can be killed by ouabain-treated arrows. Suggest a biochemical basis for the lethal action of ouabain. ✓ 3

10. *Only so much energy.* Consider Figure 12.20, which illustrates the relation between the sodium–glucose symporter and the Na$^+$–K$^+$ ATPase. If the symporter were inhibited, what effect, if any, would such inhibition have on the ATPase? ✓ 3

11. *Commonalities.* What are two fundamental properties of all ion channels? ✓ 3

12. *Opening channels.* Differentiate between ligand-gated and voltage-gated channels. ✓ 3

13. *Behind the scenes.* Is the following statement true or false? Explain. ✓ 3

The sodium–glucose linked transporter does not depend on the hydrolysis of ATP.

14. *Powering movement.* List two forms of energy that can power active transport. ✓ 3

15. *Greasy patch.* A stretch of 20 amino acids is sufficient to form an α helix long enough to span the lipid bilayer of a membrane. How could this piece of information be used to search for membrane proteins in a data bank of primary sequences of proteins?

16. *Water-fearing.* Lipid bilayers are self-sealing. If a hole is introduced, the hole is filled in immediately. What is the energetic basis of this self-sealing? ✓ 1

17. *Embedded or not.* Differentiate between peripheral proteins and integral proteins.

18. *Water-loving.* All biological membranes are asymmetric. What is the energetic basis of this asymmetry? ✓ 1

19. *Pumping sodium.* Design an experiment to show that the action of the sodium-glucose linked transporter can be reversed in vitro to pump sodium ions across a membrane. ✓ 3

20. *Different directions.* The K$^+$ channel and the Na$^+$ channel have similar structures and are arranged in the same orientation in the cell membrane. Yet the Na$^+$ channel allows sodium ions to flow into the cell and the K$^+$ channel allows potassium ions to flow out of the cell. Explain. ✓ 3

Chapter Integration Problems

21. *Energy considerations.* Explain why an α helix is especially suitable for a transmembrane-protein segment.

22. *Speed and efficiency matter.* The neurotransmitter acetylcholine, which activates a ligand-gated ion channel, is rapidly destroyed by the enzyme acetylcholinesterase. This enzyme, which has a turnover number of 25,000 per second,

has attained catalytic perfection with a k_{cat}/K_M of 2×10^8 M^{-1} s^{-1}. Why is the efficiency of this enzyme physiologically crucial?

23. *Relief for sore joints.* Both aspirin and ibuprofen inhibit prostaglandin H_2 synthase-1 and relieve inflammation. Aspirin functions by blocking a channel in the enzyme, thereby preventing access to the substrate. Ibuprofen does not block this channel but still inhibits the synthase. How might ibuprofen function?

Data Interpretation and Challenge Problems

24. *Cholesterol effects.* The red curve on the following graph shows the fluidity of the fatty acids of a phospholipid bilayer as a function of temperature. The blue curve shows the fluidity in the presence of cholesterol. ✓ 2

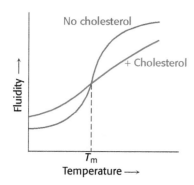

(a) What is the effect of cholesterol?

(b) Why might this effect be biologically important?

25. *Transport differences.* The rate of transport of two molecules, indole and glucose, across a cell membrane is shown in the following illustration. What are the differences between the transport mechanisms of the two molecules? Suppose that ouabain, a specific inhibitor of the Na^+–K^+ ATPase, inhibited the transport of glucose. What would this inhibition suggest about the mechanism of transport? ✓ 3

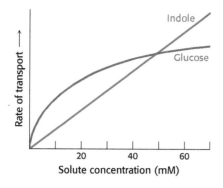

26. *Desert fish.* Certain fish living in desert streams alter their membrane-lipid composition in the transition from the heat of the day to the cool of the night. Predict the nature of the changes. ✓ 2

27. *A handy protective device.* The multidrug-resistance protein, which can lead to drug resistance in cancer and other pathological situations, is present in most normal cells. What purpose might this protein serve in normal cells? ✓ 3

28. *Looking-glass structures.* Phospholipids form lipid bilayers in water. What structure might form if phospholipids were placed in an organic solvent? ✓ 1

29. *Tarantula toxin.* Acid sensing is associated with pain, tasting, and other biological activities. Acid sensing is carried out by a ligand-gated channel that permits Na^+ influx in response to H^+. This family of acid-sensitive ion channels (ASICs) comprises a number of members. Psalmotoxin 1 (PcTX1), a venom from the tarantula, inhibits some members of this family. The following electrophysiological recordings of cells containing several members of the ASIC family were made in the presence of the toxin at a concentration of 10 nM. The channels were opened by changing the pH from 7.4 to the indicated values. PcTX1 was present for a short time (indicated by the black bar above the recordings), after which time it was rapidly washed from the system. ✓ 3

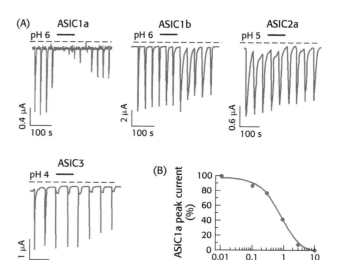

(A) Electrophysiological recordings of cells exposed to tarantula toxin. (B) Plot of peak current of a cell containing the ASIC1a protein versus the toxin concentration. [Data from P. Escoubas et al., *J. Biol. Chem.* 275:25116–25121, 2000.]

(a) Which of the ASIC family members—ASIC1a, ASIC1b, ASIC2a, or ASIC3—is most sensitive to the toxin?

(b) Is the effect of the toxin reversible? Explain.

(c) What concentration of PcTX1 yields 50% inhibition of the sensitive channel?

30. *Channel problems 1.* A number of pathological conditions result from mutations in the acetylcholine receptor channel, an ion channel that is activated by the binding of acetylcholine. One such mutation causes muscle weakness

and rapid fatigue. An investigation of the acetylcholine-generated currents through the acetylcholine receptor channel for both a control and a patient yielded the following results. What is the effect of the mutation on channel function? Suggest some possible biochemical explanations for the effect. ✓ 3

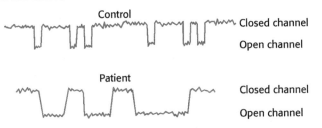

31. *Channel problems 2.* The acetylcholine receptor channel can undergo mutation leading to fast-channel syndrome (FCS), with clinical manifestations similar to those of slow-channel syndrome (SCS). What would the recordings of ion movement look like in this syndrome? Suggest a biochemical explanation. ✓ 3

Challenge Problems

32. *A dangerous snail.* Cone snails are carnivores that inject a powerful set of toxins into their prey, leading to rapid paralysis. Many of these toxins are found to bind to specific ion-channel proteins. Why are such molecules so toxic? How might such toxins be useful for biochemical studies? ✓ 3

33. *A perfect fit.* Provide an energetic explanation for how the potassium channel allows passage of the potassium ion but not the smaller sodium ion. ✓ 3

Selected Readings for this chapter can be found online at www.whfreeman.com/tymoczko3e.

Signal–Transduction Pathways

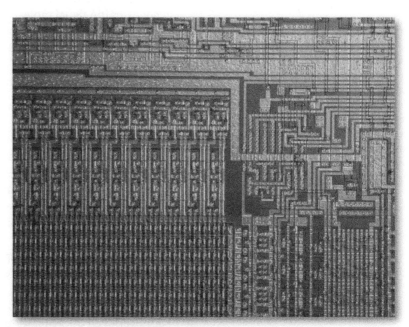

Signal transduction is an important capability in all life forms. It allows an organism to sense its environment and formulate the proper biochemical response. Just as the computer chip has "on–off" switches that allow the transmission of information, cells have molecular "on–off" switches that allow the transmission of information in the cell and between cells. [Astrid & Hanns-Frieder Michler/Science Source.]

T his chapter provides an overview of how cells receive, process, and respond to information from the environment, whether the information is in the form of light, smell, or blood-glucose concentration. *Signal-transduction* cascades mediate the sensing and processing of these stimuli. These molecular circuits detect, amplify, and integrate diverse external signals to generate responses such as changes in enzyme activity, gene expression, or ion-channel activity. This chapter introduces some of the basic principles of signal transduction and important classes of molecules that participate in common signal-transduction pathways.

✓ 4 Identify the basic components of all signal-transduction pathways.

13.1 Signal Transduction Depends on Molecular Circuits

Signal-transduction pathways follow a broadly similar course that can be viewed as a molecular circuit (**Figure 13.1**). All such circuits contain certain key steps:

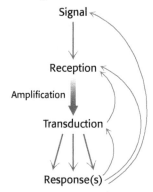

Figure 13.1 Principles of signal transduction. An environmental signal is first received by interaction with a cellular component, most often a cell-surface receptor. The information that the signal has arrived is then converted into other chemical forms, or *transduced*. The transduction process often comprises many steps. The signal is often amplified before evoking a response. Feedback pathways (blue arrows) regulate the entire signaling process.

1. *Release of the Primary Messenger.* A stimulus such as a wound or a digested meal triggers the release of the signal molecule, also called the *primary messenger.*

2. *Reception of the Primary Messenger.* Most signal molecules are too large and too polar to pass through the cell membrane or through transporters. Thus, the information presented by signal molecules must be transmitted across the cell membrane without the molecules themselves entering the cell. *Membrane receptors transfer information from the environment to a cell's interior.* Such receptors are integral membrane proteins that have both extracellular and intracellular domains. A binding site on the extracellular domain specifically recognizes the signal molecule (often referred to as the *ligand*). The formation of the receptor–ligand complex alters the tertiary or quaternary structure of the receptor, including the intracellular domain. However, structural changes in the few receptors that are bound to ligands are not sufficient to yield a response from the cell. The information conveyed by the receptor must be transduced into other forms of information that can alter the biochemistry of the cell.

3. *Relay of Information by the Second Messenger.* Structural changes in receptors lead to changes in the concentration of small molecules, called *second messengers,* that are used to relay information from the receptor–ligand complex. Particularly prominent second messengers include cyclic AMP (cAMP, or cyclic adenosine monophosphate) and cyclic GMP (cGMP, or cyclic guanosine monophosphate), calcium ion, inositol 1,4,5-trisphosphate (IP_3), and diacylglycerol (DAG; **Figure 13.2**).

   The use of second messengers has several consequences. One consequence is that second messengers are often free to diffuse to other compartments of the cell, such as the nucleus, where they can influence gene expression and other processes. Another consequence is that the signal may be amplified significantly in the generation of second messengers. Each activated receptor–ligand complex can lead to the generation of many second messengers within the cell. Thus, *a low concentration of signal molecules in the environment, even as little as a single molecule, can yield a large intracellular signal and response.*

4. *Activation of Effectors That Directly Alter the Physiological Response.* The ultimate effect of the signal pathway is to activate (or to inhibit) the pumps,

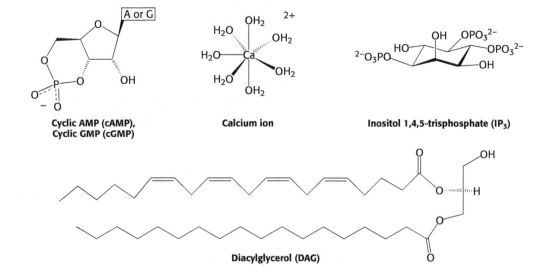

Figure 13.2 Common second messengers. Second messengers are intracellular molecules that change in concentration in response to environmental signals. That change in concentration conveys information inside the cell.

enzymes, and gene-transcription factors that directly control metabolic pathways, gene activation, and processes such as nerve transmission.

5. *Termination of the Signal.* After a signaling process has been initiated and the information has been transduced to affect other cellular processes, the signaling process must be terminated. Without such termination, cells lose their responsiveness to new signals. Moreover, signaling processes that fail to be terminated properly may lead to uncontrolled cell growth and cancer.

Essentially every biochemical process presented in the rest of this book either is a component of a signal-transduction pathway or can be affected by one.

13.2 Receptor Proteins Transmit Information into the Cell

Most receptor proteins that convey environmental information into the cell interior fall into three classes: seven-transmembrane-helix receptors, dimeric receptors that recruit protein kinases, and dimeric receptors that are protein kinases. We begin by considering the largest class of receptors, the seven-transmembrane-helix receptors.

Seven-Transmembrane-Helix Receptors Change Conformation in Response to Ligand Binding and Activate G Proteins

The *seven-transmembrane-helix* (7TM) *receptors* transmit information initiated by signals as diverse as photons, odorants, tastants, hormones, and neurotransmitters (Table 13.1). Several thousand such receptors are known, and the list continues to grow. Indeed, approximately 50% of the drugs that we use alter receptors of this class. Mutations in these receptors and their associated components cause a host of diseases, some of which are listed in Table 13.2. As the name indicates, these receptors contain seven helices that span the membrane bilayer (Figure 13.3). An example of a 7TM receptor that responds to chemical signals is the *β-adrenergic receptor*. This protein binds *epinephrine* (also called *adrenaline*), a hormone responsible for the "fight or flight" response. We will address the biochemical roles of this hormone in more detail later (Chapters 16 and 24).

✓ 5 Differentiate the various types of membrane receptors.

Table 13.1 Biological functions mediated by 7TM receptors

Hormone action
Hormone secretion
Neurotransmission
Chemotaxis
Exocytosis
Control of blood pressure
Embryogenesis
Cell growth and differentiation
Development
Smell
Taste
Vision
Viral infection

Source: Data from J. S. Gutkind, *J. Biol. Chem.* 273:1839–1842, 1998.

(A)

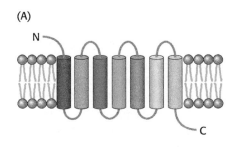

(B)

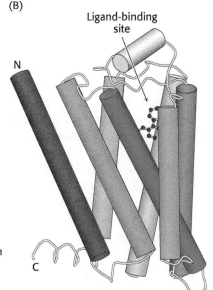

Ligand-binding site

Cytoplasmic loops

β₂-adrenergic receptor

Epinephrine

Figure 13.3 The structure of 7TM receptors. (A) Schematic representation of a 7TM receptor showing how it passes through the membrane seven times. (B) Three-dimensional structure of a subtype (β₂) of the β-adrenergic receptor, a 7TM receptor taking part in fuel mobilization. Notice the ligand-binding site near the extracellular surface. [Drawn from 2RH1.pdb.]

Table 13.2 Diseases resulting from defects in 7TM receptors

Color blindness
Familial hypogonadism
Short stature due to mutated growth hormone receptor
Extreme obesity
Congenital hypothyroidism
Incomplete bowel innervation (Hirschsprung disease)
Precocious puberty
Night blindness

The 7TM receptors undergo a change in conformation in response to ligand binding. *The binding of a ligand on the outside the cell induces a conformational change in the 7TM receptor that can be detected inside the cell.* As we shall see, 7TM receptors also have in common the next step in their signal-transduction cascades.

Ligand Binding to 7TM Receptors Leads to the Activation of G Proteins

Let us focus on the β-adrenergic receptor as a model of the 7TM receptor class. What is the next step in the pathway after the binding of epinephrine by the β-adrenergic receptor? The conformational change in the cytoplasmic domain of the receptor activates a GTP-binding protein. This signal-coupling protein is termed a *G protein* (*G* for guanyl nucleotide). The activated G protein stimulates the activity of adenylate cyclase, an enzyme that increases the concentration of the second messenger cAMP by forming it from ATP (**Figure 13.4**).

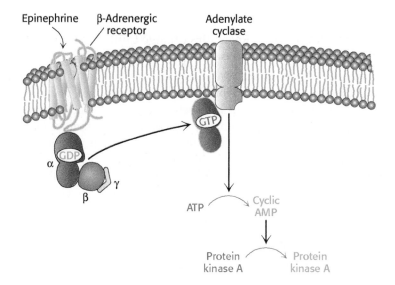

Figure 13.4 The activation of protein kinase A by a G-protein pathway. Hormone binding to a 7TM receptor initiates a signal-transduction pathway that acts through a G protein and cAMP to activate protein kinase A.

How do these G proteins operate? *In the unactivated state, the guanyl nucleotide bound to the G protein is GDP.* In this form, the G protein exists as a *heterotrimer* consisting of α, β, and γ subunits; the α subunit (referred to as G_α) binds the nucleotide (**Figure 13.5**). The α and γ subunits are usually anchored to the membrane by covalently attached fatty acids. *The exchange of the bound GDP for GTP is catalyzed by the ligand-bound receptor.* The ligand–receptor complex interacts with the heterotrimeric G protein and opens the nucleotide-binding site so that GDP (guanosine diphosphate) can depart and GTP (guanosine triphosphate) can bind. The α subunit simultaneously dissociates from the $\beta\gamma$ dimer ($G_{\beta\gamma}$; Figure 13.4). *The dissociation of the G-protein heterotrimer into G_α and $G_{\beta\gamma}$ units transmits the signal that the receptor has bound its ligand. The human genome encodes 15 genes for G_α, one of which ($G_{\alpha s}$) is expressed ubiquitously while the others are expressed in specific cell-types.*

A single ligand–receptor complex can stimulate nucleotide exchange in many G-protein heterotrimers. Thus, hundreds of G_α molecules are converted to their GTP-bound forms from their GDP-bound forms for each bound molecule of hormone, giving an *amplified response.* All 7TM receptors appear to be coupled to G proteins, and so the 7TM receptors are sometimes referred to as *G-protein-coupled receptors* or GPCRs.

(A)

(B)

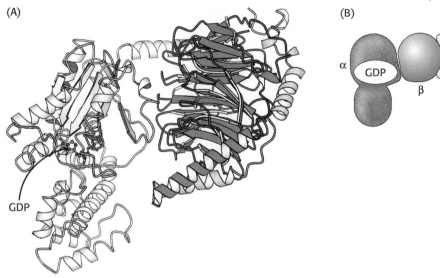

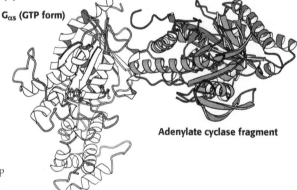

Gₐₛ (GTP form)

Adenylate cyclase fragment

Figure 13.5 A heterotrimeric G protein. (A) A ribbon diagram shows the relation between the three subunits. In this complex, the α subunit (gray and purple) is bound to GDP. Notice that GDP is bound in a pocket close to the surface at which the α subunit interacts with the βγ dimer (the β subunit is shown in blue and γ in yellow). (B) A schematic representation of the heterotrimeric G protein. [Drawn from 1GOT.pdb.]

Activated G Proteins Transmit Signals by Binding to Other Proteins

As just described, the formation of the ligand–receptor complex activates a G protein. How does the G protein propagate the message that the ligand is present? It does so by a variety of means, depending on the specific type of G protein. We will begin by examining one target of a G protein, the enzyme *adenylate cyclase* (**Figure 13.6**). The adenylate cyclase enzyme that is activated by the β-adrenergic signaling pathway is a membrane protein that contains 12 presumed membrane-spanning helices. The G_α protein binds to adenylate cyclase on the G_α surface that had bound the βγ dimer when the G_α protein was in its GDP form. $G_{\alpha s}$ ("s" stands for "stimulatory") stimulates adenylate cyclase activity, thus increasing cAMP production. *The net result is that the binding of epinephrine to the receptor on the cell surface increases the rate of cAMP production inside the cell.*

(A)

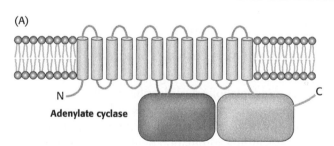

N

C

Adenylate cyclase

Figure 13.6 Adenylate cyclase activation. (A) Adenylate cyclase is a membrane protein with two large intracellular domains (red and orange) that contain the catalytic apparatus. (B) The structure of the complex between G_α in its GTP form bound to a catalytic fragment of adenylate cyclase. Notice that the surface of G_α that had been bound to the βγ dimer (Figure 13.5) now binds adenylate cyclase. [Drawn from 1AZS.pdb.]

Cyclic AMP Stimulates the Phosphorylation of Many Target Proteins by Activating Protein Kinase A

The increased concentration of cAMP can affect a wide range of cellular processes, depending on the cell type. For example, it enhances the degradation of storage fuels, increases the secretion of acid by the gastric mucosa in the cells of the stomach and intestines, leads to the dispersion of melanin pigment granules in skin cells, diminishes the aggregation of blood platelets, and induces the opening of chloride channels in the pancreas. How does cAMP affect so many cellular processes? Is there a common denominator for its diverse effects? Indeed there is. *Most effects of cAMP in eukaryotic cells are mediated by the activation of a single protein kinase.* This key enzyme is called *protein kinase A* (PKA). Kinases are enzymes that phosphorylate a substrate at the expense of a molecule of ATP. PKA consists of two regulatory (R) subunits and two catalytic (C) subunits. In the absence of cAMP, the R_2C_2 complex is catalytically inactive (**Figure 13.7**). The binding of cAMP to the regulatory subunits releases the catalytic subunits, which are enzymatically active on their own. Activated PKA then phosphorylates specific serine and threonine residues in many targets to alter their activity. The alteration in activity is due to structural and ionic changes that result from the introduction of the large negatively charged phosphate functional group. The cAMP cascade is turned off by *cAMP phosphodiesterase,* an enzyme that converts cAMP into AMP, which does not activate PKA. The C and R subunits subsequently rejoin to form the inactive enzyme.

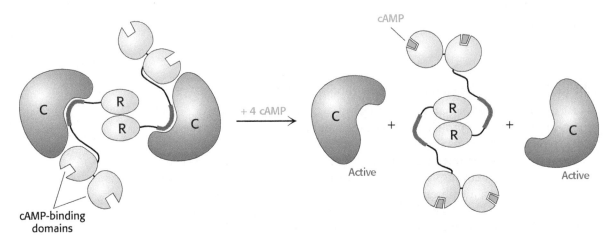

Figure 13.7 The regulation of protein kinase A. The binding of four molecules of cAMP activates protein kinase A by dissociating the inhibited holoenzyme (R_2C_2) into a regulatory subunit (R_2) and two catalytically active subunits (C).

CLINICAL INSIGHT

Mutations in Protein Kinase A Can Cause Cushing's Syndrome

Cushing's syndrome, a collection of diseases resulting from excess cortisol secretion by the adrenal cortex, is a metabolic disorder characterized by a variety of symptoms such as muscle weakness, thinning skin that is easily bruised, and osteoporosis. Cortisol, a steroid hormone (Section 29.5), has a number of physiological effects including stimulation of glucose synthesis, suppression the immune response, and inhibition of bone growth. The most common cause of Cushing's syndrome, called Cushing's disease, is a tumor of the pituitary gland that overstimulates cortisol secretion by the adrenal cortex. Recent work shows that a mutation that renders protein kinase A constitutively active also results in the syndrome. In these patients, the catalytic subunit of the enzyme is altered so that it no longer binds the regulatory subunit. Thus, the enzyme is active even in the absence of cAMP with the ultimate result being unregulated secretion of cortisol.

G Proteins Spontaneously Reset Themselves Through GTP Hydrolysis

The ability to shut down signal-transduction pathways is as critical as the ability to turn them on. How is the signaling pathway initiated by activated 7TM receptors switched off? G_α *subunits have intrinsic GTPase activity,* hydrolyzing bound GTP to GDP and P_i (inorganic orthophosphate) and thereby deactivating itself. This hydrolysis reaction is slow, however, requiring from seconds to minutes and thus allowing the GTP form of G_α to activate downstream components of the signal-transduction pathway before GTP hydrolysis deactivates the subunit. In essence, *the bound GTP acts as a built-in clock that spontaneously resets the G_α subunit after a short time period*. After GTP hydrolysis and the release of P_i, the GDP–bound form of G_α then reassociates with $G_{\beta\gamma}$ to reform the heterotrimeric protein (**Figure 13.8**).

The ligand-bound activated receptor must be reset as well to prevent the continuous activation of G proteins. A key step in the inactivation of the receptor rests on the fact that the receptor–ligand interaction is reversible (**Figure 13.9**). When the ligand dissociates, the receptor returns to its initial, unactivated state. The likelihood that the receptor remains in its unbound state depends on the concentration of ligand in the environment.

> **?** QUICK QUIZ 1 List the means by which the β-adrenergic pathway is terminated.

Adenylate cyclase

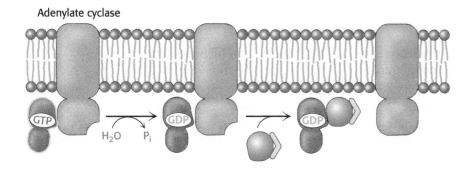

Figure 13.8 Resetting G_α. On hydrolysis of the bound GTP by the intrinsic GTPase activity of G_α, G_α reassociates with the $\beta\gamma$ dimer to form the heterotrimeric G protein, thereby terminating the activation of adenylate cyclase.

Dissociation

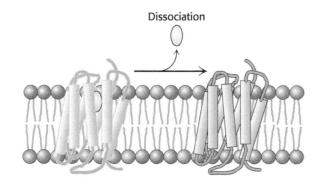

Figure 13.9 Signal termination. Signal transduction by the 7TM receptor is halted, in part, by dissociation of the signal molecule (yellow) from the receptor.

⚕ CLINICAL INSIGHT

Cholera and Whooping Cough Are Due to Altered G-Protein Activity

The alteration of G-protein-dependent signal pathways can result in pathological conditions. Let us first consider the mechanism of action of the cholera toxin, secreted by the intestinal bacterium *Vibrio cholerae*. Cholera is an acute diarrheal disease that can be life threatening. It causes a voluminous secretion of electrolytes and fluids from the intestines of infected persons (**Figure 13.10**). The cholera toxin, *choleragen*, is a protein composed of two functional units—a B subunit that binds to gangliosides (p. 196) on the surface of cells of the intestinal epithelium and a catalytic A subunit that enters the cell. The A subunit catalyzes the covalent modification of a $G_{\alpha s}$ protein. This modification stabilizes the active GTP-bound form of $G_{\alpha s}$, trapping the molecule in the active conformation. The active G protein, in turn, continuously activates protein kinase A. PKA opens a chloride channel (a CFTR channel, p. 214) and inhibits the $Na^+–H^+$ exchanger by phosphorylation. The net result of the phosphorylation of these channels is an excessive loss of NaCl and the loss of large amounts of water into the intestine. Patients suffering from cholera for 4 to 6 days may pass as much as twice their body weight in fluid. Treatment consists of rehydration with a glucose–electrolyte solution.

Figure 13.10 Death's dispensary. An 1866 cartoon illustrating that contaminated water is a frequent source of cholera infection. [The Granger Collection.]

Whereas cholera is a result of a G protein trapped in the active conformation, causing the signal-transduction pathway to be perpetually stimulated, pertussis, or whooping cough, is a result of the opposite situation. The toxin also modifies a G_α protein called $G_{\alpha i}$, which normally inhibits adenylate cyclase, closes Ca^{2+} channels, and opens K^+ channels. The effect of this modification is to lower the G protein's affinity for GTP, effectively trapping it in the "off" conformation. The symptoms of whooping cough, such as prolonged coughing that ends with a whoop as the patient gasps for air, have not yet been traced to the inhibition of any single target of the $G_{\alpha i}$ protein. Pertussis toxin is secreted by *Bordetella pertussis,* the bacterium responsible for whooping cough, one of the leading causes of infant mortality globally. Whooping cough is highly preventable through vaccination.

The Hydrolysis of Phosphatidylinositol Bisphosphate by Phospholipase C Generates Two Second Messengers

Cyclic AMP is not the only second messenger employed by 7TM receptors and the G proteins. We turn now to another ubiquitous second-messenger cascade used by many hormones to evoke a variety of responses. The *phosphoinositide cascade,* like the adenylate cyclase cascade, converts extracellular signals into intracellular ones. The intracellular messengers formed by activation of this pathway arise from the cleavage of *phosphatidylinositol 4,5-bisphosphate* (PIP_2), a membrane phospholipid. The binding of a hormone such as vasopressin, which regulates water retention, to its 7TM receptor, leads to the activation of *phospholipase C.* The G_α protein that activates phospholipase C is called $G_{\alpha q}$. The activated enzyme then hydrolyzes the phosphodiester linkage joining the phosphorylated inositol unit to the acylated glycerol moiety. The cleavage of PIP_2 produces two messengers: inositol 1,4,5-trisphosphate (IP_3), a soluble molecule that can diffuse from the membrane, and diacylglycerol (DAG), which stays in the membrane (**Figure 13.11**).

What are the biochemical effects of the second messenger IP_3? Unlike cAMP, IP_3 does not cause a cascade of phosphorylation to elicit a response from the cell.

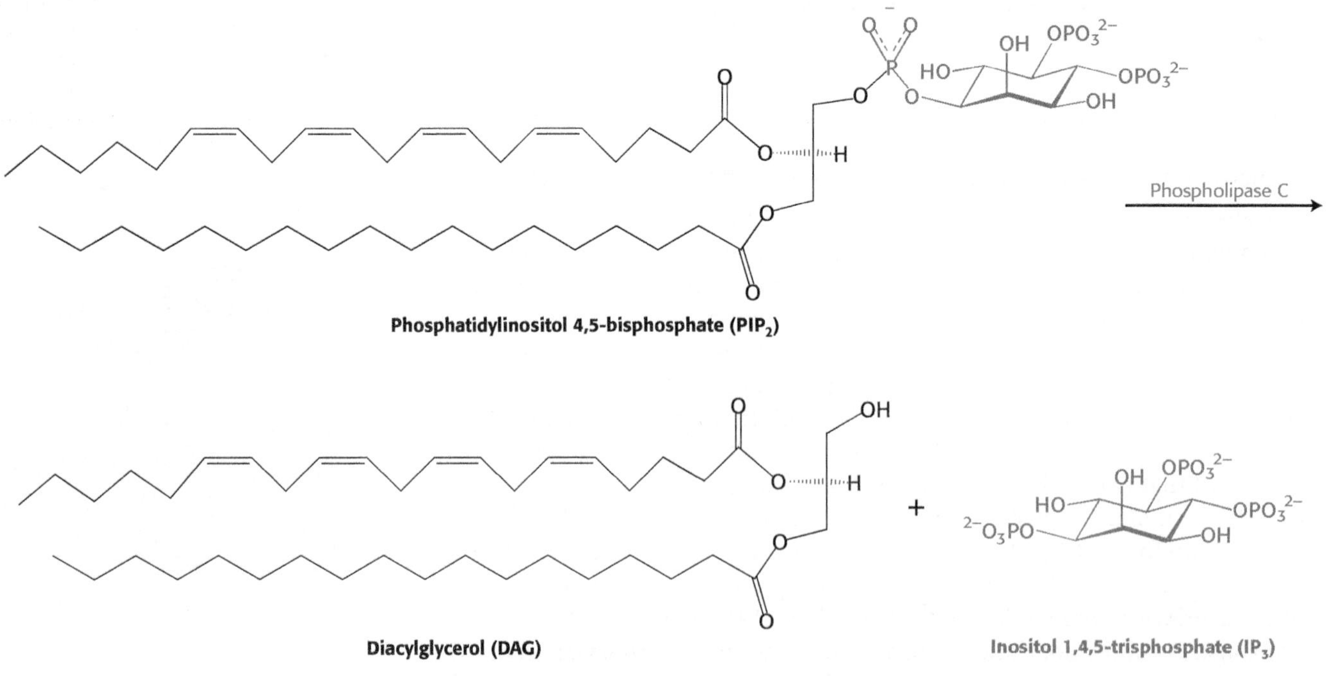

Phosphatidylinositol 4,5-bisphosphate (PIP₂)

Diacylglycerol (DAG)

Inositol 1,4,5-trisphosphate (IP₃)

Figure 13.11 The phospholipase C reaction. Phospholipase C cleaves the membrane lipid phosphatidylinositol 4,5-bisphosphate into two second messengers: diacylglycerol, which remains in the membrane, and inositol 1,4,5-trisphosphate, which diffuses away from the membrane.

IP$_3$ directly causes the rapid release of Ca^2+ from intracellular stores—the endoplasmic reticulum and, in muscle cells, the sarcoplasmic reticulum. IP$_3$ associates with a membrane protein called the *IP$_3$-gated channel* or *IP$_3$ receptor* to allow the flow of Ca^{2+} from the endoplasmic reticulum into the cell cytoplasm. The elevated level of Ca^{2+} in the cytoplasm then triggers a variety of biochemical processes such as smooth-muscle contraction, glycogen breakdown, and vesicle release.

The lifetime of IP$_3$ in the cell is very short—less than a few seconds. It is rapidly converted into derivatives that have no effect on the IP$_3$-gated channel. Lithium ion, widely used to treat bipolar affective disorder, may act by inhibiting the recycling of IP$_3$, although the details of lithium action remain to be determined.

Diacylglycerol, the other molecule formed by the receptor-triggered hydrolysis of PIP$_2$, also is a second messenger that, in conjunction with Ca^{2+}, activates *protein kinase C* (PKC), a protein kinase that phosphorylates serine and threonine residues in many target proteins (**Figure 13.12**).

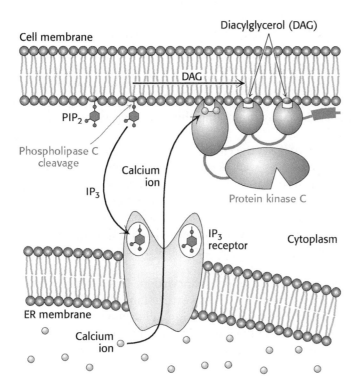

Figure 13.12 The phosphoinositide cascade. The cleavage of phosphatidylinositol 4,5-bisphosphate (PIP$_2$) into diacylglycerol (DAG) and inositol 1,4,5-trisphosphate (IP$_3$) results in the release of calcium ions (owing to the opening of the IP$_3$ receptor ion channels) and the activation of protein kinase C (owing to the binding of protein kinase C to free DAG in the membrane). Calcium ions bind to protein kinase C and help facilitate its activation.

13.3 Some Receptors Dimerize in Response to Ligand Binding and Recruit Tyrosine Kinases

The 7TM receptors initiate signal-transduction pathways through changes in tertiary structure that are induced by ligand binding. A fundamentally different mechanism is utilized by a number of other classes of receptors. For these receptors, ligand binding leads to changes in quaternary structures—specifically, the formation of receptor dimers.

Receptor Dimerization May Result in Tyrosine Kinase Recruitment

The growth-hormone receptor is a well-studied example of a receptor that dimerizes on ligand binding and subsequently recruits tyrosine kinases to propagate the signal. Growth hormone is a monomeric protein of 217 amino acids that is required for cell division and growth. Human growth hormone (HGH) is abused

(A)

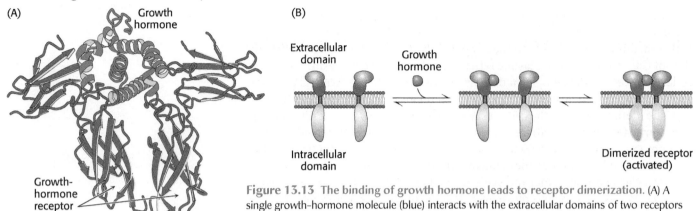

Growth hormone

Growth-hormone receptor (extracellular domains)

(B)

Extracellular domain

Growth hormone

Intracellular domain

Dimerized receptor (activated)

Figure 13.13 The binding of growth hormone leads to receptor dimerization. (A) A single growth-hormone molecule (blue) interacts with the extracellular domains of two receptors (red). (B) The binding of one hormone molecule to two receptors leads to the formation of a receptor dimer. Dimerization is a key step in this signal-transduction pathway.

DID YOU KNOW?

Overstimulation of the growth-hormone signal-transduction pathway can lead to pathological conditions. Acromegaly is a rare hormonal disorder resulting from the overproduction of growth hormone in middle age. A common characteristic of acromegaly is enlargement of the face, hands, and feet. Excessive production of growth hormone in children results in gigantism. André the Giant, who played the beloved giant Fezzik in Rob Reiner's classic film *The Princess Bride* suffered from gigantism.

by athletes who believe that HGH treatments will enhance their performance, although there is little evidence to support such beliefs. The growth-hormone receptor has an extracellular domain, a single membrane-spanning helix, and an intracellular domain. In the absence of bound hormone, the receptor is present as a monomer. Each hormone molecule binds to the extracellular domains of two receptor molecules, thus promoting the formation of a dimer of the receptor (**Figure 13.13**).

Dimerization of the extracellular domains of the receptor brings together the intracellular domains as well. Associated with each intracellular domain is a molecule of a tyrosine protein kinase, termed *Janus kinase 2* (JAK2), in an unactivated form (**Figure 13.14**). Tyrosine kinases phosphorylate proteins on the hydroxyl group of tyrosine residues. Dimerization of the growth-hormone receptors brings together the JAK2 proteins associated with each intracellular domain. Each of the kinases phosphorylates its partner, resulting in the activation of the kinases.

When activated by cross-phosphorylation, JAK2 can phosphorylate other substrates, such as a regulator of gene expression called STAT5 (STAT for *signal transducers and activators of transcription*). Phosphorylated STAT5 moves to the nucleus, where it binds to the DNA-binding sites to regulate gene expression. A signal received on the outside of the cell membrane is forwarded to the nucleus for action.

Figure 13.14 The cross-phosphorylation of two molecules of JAK2 induced by receptor dimerization. The binding of growth hormone (blue) leads to growth-hormone-receptor dimerization, which brings two molecules of Janus kinase 2 (JAK2; yellow) together in such a way that each phosphorylates key residues on the other. The activated JAK2 molecules remain bound to the receptor.

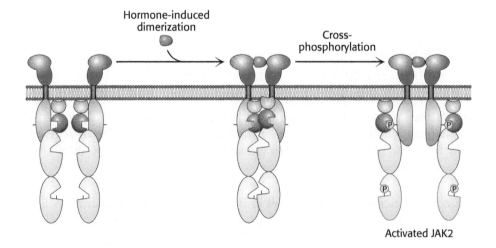

Hormone-induced dimerization

Cross-phosphorylation

Activated JAK2

CLINICAL INSIGHT

Some Receptors Contain Tyrosine Kinase Domains Within Their Covalent Structures

Some growth factors and hormones—such as epidermal growth factor (EGF), platelet-derived growth factor, and insulin—bind to the extracellular domains of transmembrane receptors that have tyrosine kinase domains within their intracellular domains. These *receptor tyrosine kinases* (RTKs) signal by mechanisms quite similar to those described for the pathway initiated by the growth-hormone receptor discussed in the preceding subsection. Humans have 58 known genes encoding receptor tyrosine kinases. Mutations in these receptors cause a range of pathologies, including arteriosclerosis, cancer, inflammation, and type 2 diabetes.

Consider, for example, *epidermal growth factor,* a 6-kDa polypeptide that stimulates the growth of epidermal and epithelial cells by binding to the *epidermal-growth-factor receptor,* a single polypeptide chain consisting of an extracellular growth hormone binding domain, a helix that spans the membrane, and an intracellular kinase domain. The receptor is monomeric and enzymatically inactive in the absence of the growth factor. *The binding of EGF to its extracellular domain causes the receptor to dimerize and undergo cross-phosphorylation and activation.*

How is the signal transferred beyond the receptor tyrosine kinase? A key *adaptor protein,* called *Grb-2,* links the phosphorylation of the EGF receptor to the stimulation of cell growth through a chain of protein phosphorylations (**Figure 13.15**). On phosphorylation of the receptor, Grb-2 binds to the phosphotyrosine residues of the receptor tyrosine kinase. Grb-2 then recruits a protein called *Sos.* Sos, in turn, binds to *Ras* and activates it. Ras is a very prominent signal-transduction component that we will consider shortly. Finally, Ras, in its activated form, binds to other components of the molecular circuitry, leading to the activation of the specific protein kinases that phosphorylate specific targets that promote cell growth. We see here another example of how a signal-transduction pathway is constructed. Specific protein–protein interactions link the original ligand-binding event to the final result—the stimulation of cell growth.

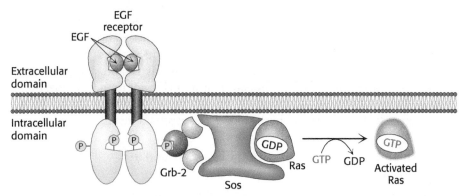

Figure 13.15 The EGF signaling pathway. The binding of epidermal growth factor (EGF) to its receptor leads to cross-phosphorylation of the receptor. The phosphorylated receptor binds Grb-2, which, in turn, binds Sos. Sos stimulates the exchange of GTP for GDP in Ras. Activated Ras binds to protein kinases and stimulates them (not shown).

Table 13.3 Ras superfamily of GTPases

Subfamily	Function
Ras	Regulates cell growth through serine or threonine protein kinases
Rho	Reorganizes cytoskeleton through serine or threonine protein kinases
Arf	Activates the ADP-ribosyltransferase of the cholera toxin A subunit; regulates vesicular trafficking pathways; activates phospholipase D
Rab	Plays a key role in secretory and endocytotic pathways
Ran	Functions in the transport of RNA and protein into and out of the nucleus

Ras Belongs to Another Class of Signaling G Proteins

The signal-transduction protein Ras is a member of a family of signal proteins—the *small G proteins*, or small GTPases. This large superfamily of proteins—grouped into subfamilies called Ras, Rho, Arf, Rab, and Ran—plays a major role in a host of cell functions, including growth, differentiation, cell motility, cytokinesis, and the transport of materials throughout the cell (Table 13.3). Like their relatives the heterotrimeric G proteins (p. 228), the small G proteins cycle between an active GTP-bound form and an inactive GDP-bound form. They differ from the heterotrimeric G proteins in being smaller (20–25 kDa compared with 30–35 kDa) and monomeric.

Recall that Sos is the immediate upstream link to Ras in the circuit conveying the EGF signal (Figure 13.15). How does Sos activate Ras? Sos binds to Ras, reaches into the nucleotide-binding pocket, and opens it up, allowing GDP to escape and GTP to enter in its place. Sos is referred to as a *guanine-nucleotide exchange factor* (GEF).

Like the G_α protein, Ras possesses an intrinsic GTPase activity, which serves to terminate the signal and return the system to the inactive state. This activity is slow but is augmented by helper proteins termed *GTPase-activating proteins* (GAPs). The guanine-nucleotide exchange factors and the GTPase-activating proteins allow the G-protein cycle to proceed with rates appropriate for a balanced level of downstream signaling.

13.4 Metabolism in Context: Insulin Signaling Regulates Metabolism

Insulin is among the principal hormones that regulate metabolism, and we will examine the effects of this hormone on metabolic pathways many times in the course of our study of biochemistry. This section presents an overview of its signal-transduction pathway. Insulin is the hormone released by the β cells of the pancreas after a meal has been eaten and is the biochemical signal designating the fed state. In all of its detail this multibranched pathway is quite complex, and so we will focus solely on the major branch. This branch leads to the mobilization of glucose transporters to the cell surface. These transporters allow the cell to take up the glucose that is plentiful in the bloodstream after a meal.

The Insulin Receptor Is a Dimer That Closes Around a Bound Insulin Molecule

Insulin is a peptide hormone that consists of two chains, linked by two disulfide bonds (Figure 13.16). The *insulin receptor* is a member of the receptor tyrosine kinase class of membrane proteins. Unlike the other members of the receptor tyrosine kinase class, however, the insulin receptor exists as a dimer even in the absence of insulin. Each subunit consists of one α chain and one β chain linked to one another by a single disulfide bond. Each α subunit lies completely outside the cell, whereas each β subunit starts in the extracellular domain and spans the

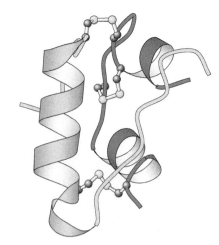

Figure 13.16 Insulin structure. Notice that insulin consists of two chains (shown in blue and yellow) linked by two interchain disulfide bonds. The α chain (blue) also has an intrachain disulfide bond. [Drawn from 1B2F.pdb.]

membrane with an α helix to the cytoplasmic side, where the kinase domain resides. The two α subunits are linked to one another with disulfide bonds (**Figure 13.17**).

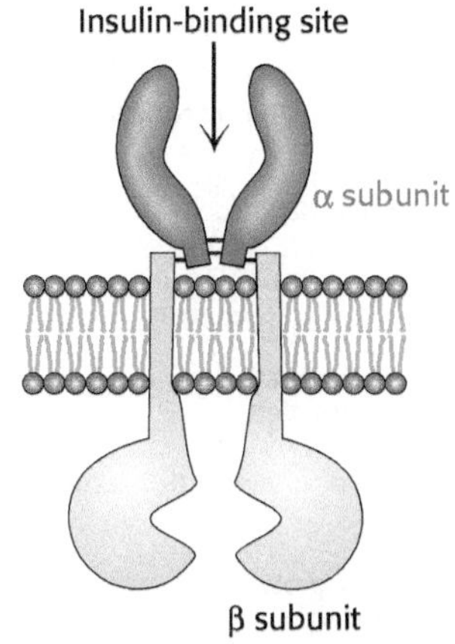

The Activated Insulin-Receptor Kinase Initiates a Kinase Cascade

Insulin binding causes a change in quaternary structure that results in cross-phosphorylation by the two kinase domains, activating the kinase. The activated kinase phosphorylates additional sites within the receptor. These phosphorylated sites act as docking sites for other substrates, including a class of molecules referred to as *insulin-receptor substrates* (IRS). The IRS proteins are subsequently phosphorylated by the tyrosine kinase activity of the insulin receptor.

In their phosphorylated form, the IRS molecules act as *adaptor proteins* (**Figure 13.18**). The phosphotyrosine residues in the IRS proteins are recognized by other proteins, including a lipid kinase that phosphorylates phosphatidylinositol 4,5-bisphosphate (PIP$_2$) to form phosphatidylinositol 3,4,5-trisphosphate (PIP$_3$), as depicted in **Figure 13.19**. This lipid product, in

Figure 13.17 The insulin receptor. The receptor consists of two units, each of which consists of an α subunit and a β subunit linked by a disulfide bond. The α subunit lies outside the cell, and two α subunits are linked by disulfide bonds to form a binding site for insulin. Each β subunit lies primarily inside the cell and includes a protein kinase domain.

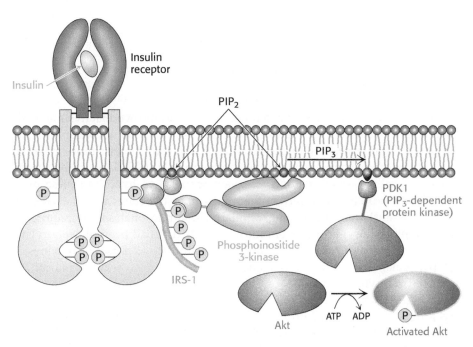

Figure 13.18 Insulin signaling. The binding of insulin results in the cross-phosphorylation and activation of the insulin receptor. Phosphorylated sites on the receptor act as binding sites for insulin-receptor substrates such as IRS-1. The lipid kinase phosphoinositide 3-kinase binds to phosphorylated sites on IRS-1 through its regulatory domain and then converts PIP$_2$ into PIP$_3$. Binding to PIP$_3$ activates PIP$_3$-dependent protein kinase (PDK1), which phosphorylates and activates kinases such as Akt. Activated Akt can then diffuse throughout the cell to continue the signal-transduction pathway.

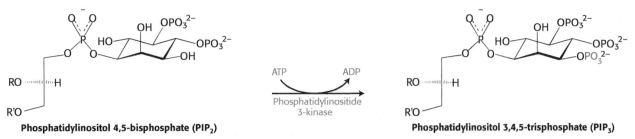

Figure 13.19 The action of a lipid kinase in insulin signaling. Phosphorylated IRS-1 and IRS-2 activate the enzyme phosphatidylinositide 3-kinase, an enzyme that converts PIP$_2$ into PIP$_3$.

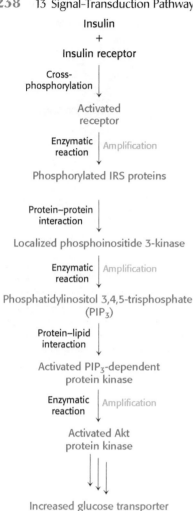

Insulin
+
Insulin receptor

Cross-
phosphorylation

Activated
receptor

Enzymatic | Amplification
reaction

Phosphorylated IRS proteins

Protein–protein
interaction

Localized phosphoinositide 3-kinase

Enzymatic | Amplification
reaction

Phosphatidylinositol 3,4,5-trisphosphate
(PIP_3)

Protein–lipid
interaction

Activated PIP_3-dependent
protein kinase

Enzymatic | Amplification
reaction

Activated Akt
protein kinase

Increased glucose transporter
on cell surface

Figure 13.20 Insulin signaling pathway. Key steps in the signal-transduction pathway initiated by the binding of insulin to the insulin receptor.

QUICK QUIZ 2 Why does it make good physiological sense for insulin to increase the number of glucose transporters in the cell membrane?

turn, activates PIP_3-dependent protein kinase (PDK1). This activated protein kinase phosphorylates and activates Akt (also called protein kinase B, PKB), another serine or threonine protein kinase (Figure 13.18). Akt is not membrane anchored and moves through the cell to phosphorylate enzymes that stimulate glycogen synthesis as well as components that control the trafficking of the glucose transporter GLUT4 to the cell surface. At the cell surface, GLUT4, one of a family of five glucose transporters, allows the entry of glucose down its concentration gradient into the cell, efficiently removing glucose from the blood. Thus, the signal is conveyed to the cell interior by the IRS protein through a series of membrane-anchored molecules to a protein kinase that finally leaves the membrane and elicits the cellular response to insulin.

The cascade initiated by the binding of insulin to the insulin receptor is summarized in **Figure 13.20**. The signal is amplified at several stages along this pathway. Because the activated insulin receptor itself is a protein kinase, each activated receptor can phosphorylate multiple IRS molecules. Activated enzymes further amplify the signal in at least two of the subsequent steps. Thus, a small increase in the concentration of circulating insulin can produce a robust intracellular response. Note that, as complicated as the pathway described here is, it is substantially less elaborate than the full network of pathways initiated by insulin.

Insulin Signaling Is Terminated by the Action of Phosphatases

We have seen that the activated G protein promotes its own inactivation by the release of a phosphoryl group from GTP. In contrast, proteins phosphorylated on serine, threonine, or tyrosine residues, such as insulin-receptor substrates, do not hydrolyze spontaneously; they are extremely stable kinetically. Specific enzymes, called *protein phosphatases,* are required to hydrolyze these phosphorylated proteins and convert them back into the states that they were in before the initiation of signaling. Similarly, lipid phosphatases are required to remove phosphoryl groups from inositol lipids that had been phosphorylated as part of a signaling cascade. In insulin signaling, three classes of enzymes are of particular note: protein tyrosine phosphatases that remove phosphoryl groups from tyrosine residues on the insulin receptor, lipid phosphatases that hydrolyze PIP_3 to PIP_2, and protein serine phosphatases that remove phosphoryl groups from activated protein kinases such as Akt. Many of these phosphatases are activated or recruited as part of the response to insulin. Thus, the binding of the initial signal sets the stage for the eventual termination of the response.

13.5 Calcium Ion Is a Ubiquitous Cytoplasmic Messenger

We have already seen that Ca^{2+} is a component of one signal-transduction circuit, the phosphoinositide cascade. Indeed, Ca^{2+} is itself an intracellular messenger in many eukaryotic signal-transducing pathways.

Calmodulin (CaM), a 17-kDa protein with four calcium-binding sites, serves as a calcium sensor in nearly all eukaryotic cells. *Calmodulin is activated by the binding of Ca^{2+} when the cytoplasmic calcium level is raised.* Calmodulin is a member of the *EF-hand protein family.* The *EF hand* is a Ca^{2+}-binding motif that consists of a helix, a loop, and a second helix. This motif, originally discovered in the protein parvalbumin, was named the EF hand because the helices designated E and F in parvalbumin that form the calcium-binding motif are positioned like the forefinger and thumb of the right hand (**Figure 13.21**).

The Ca²⁺–calmodulin complex stimulates a wide array of enzymes, pumps, and other target proteins. Two targets are especially noteworthy: one that propagates the Ca²⁺ signal and another that abrogates it. The binding of Ca²⁺–calmodulin to a *calmodulin-dependent protein kinase* (CaM kinase) activates the kinase and enables it to phosphorylate a wide variety of target proteins. CaM kinases regulate the metabolism of fuel, ionic permeability, neurotransmitter synthesis, and neurotransmitter release through the action of the Ca²⁺–calmodulin complex. The plasma-membrane *Ca²⁺–ATPase pump* is another important target of Ca²⁺–calmodulin. Stimulation of the pump by Ca²⁺–calmodulin drives the calcium level down to restore a low-calcium basal state to the cell, thus helping to terminate the signal.

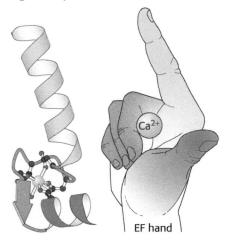

Figure 13.21 An EF hand. Formed by a helix–loop–helix unit, an EF hand is a binding site for Ca²⁺ in many calcium-sensing proteins. Here, the E helix is yellow, the F helix is blue, and calcium is represented by the green sphere. Notice that the calcium ion is bound in a loop connecting two nearly perpendicular helices. [Drawn from 1CLL.pdb.]

13.6 Defects in Signaling Pathways Can Lead to Diseases

In light of their complexity, it comes as no surprise that signal-transduction pathways occasionally fail, leading to pathological or disease states. Cancer, a set of diseases characterized by uncontrolled or inappropriate cell growth, is strongly associated with defects in signal-transduction proteins. Indeed, the study of cancer has contributed greatly to our understanding of signal-transduction proteins and pathways.

✚ CLINICAL INSIGHT

The Conversion of Proto-oncogenes into Oncogenes Disrupts the Regulation of Cell Growth

Genes that, when mutated, cause cancer often normally regulate cell growth. The unmutated, normally expressed versions of these genes are termed *proto-oncogenes,* and the proteins encoded by them are often signal-transduction proteins that regulate cell growth. If a proto-oncogene suffers a mutation that leads to unrestrained growth by the cell, the gene is then referred to as an *oncogene* (from the Greek *onco,* meaning "swelling," "mass," or "tumor").

The gene encoding Ras (p. 236), a component of the EGF-initiated pathway, is one of the genes most commonly mutated in human tumors. Mammalian cells contain three Ras proteins (H-, K-, and N-Ras), each of which cycles between inactive GDP and active GTP forms. The most common mutation in tumors is a loss of the intrinsic GTPase activity. Thus, the Ras protein is trapped in the "on" position and continues to stimulate cell growth, even in the absence of a continuing signal.

Mutated, or overexpressed, receptor tyrosine kinases also are frequently observed in tumors. For instance, the epidermal-growth-factor receptor (EGFR) is overexpressed in some colorectal and head and neck cancers. Because some small amount of the receptor can dimerize and activate the signaling pathway even without binding to EGF, the overexpression of the receptor increases the likelihood that a "grow and divide" signal will be inappropriately sent to the cell. This understanding of cancer-related signal-transduction pathways has led to a therapeutic approach that targets the EGFR. Antibodies, such as cetuximab (Erbitux), have been generated that inhibit the EGFR by competing with EGF for the binding site on the receptor while also blocking the change in conformation that causes dimerization. The result is that the EGFR-controlled pathway is not initiated.

Other genes can contribute to cancer development only when both copies of the gene normally present in a cell are deleted or otherwise damaged. Such genes are called *tumor-suppressor genes*. These genes encode proteins that either inhibit cell growth by turning off growth-promoting pathways or trigger the death of tumor cells. For example, genes for some of the phosphatases that participate in the termination of EGF signaling are tumor suppressors. Without any functional phosphatase present, EGF signaling persists after its initiation, stimulating inappropriate cell growth.

✠ CLINICAL INSIGHT

Protein Kinase Inhibitors May Be Effective Anticancer Drugs

The widespread occurrence of overactive protein kinases in cancer cells suggests that inhibitors of these enzymes might act as antitumor agents. For example, more than 90% of patients with chronic myelogenous leukemia (CML) have a specific chromosomal defect in affected cells (**Figure 13.22**). The translocation of genetic material between chromosomes 9 and 22 causes the *c-abl* ("c" for cellular) gene, which encodes a tyrosine kinase, to be inserted into the *bcr* gene on chromosome 22. The result is the production of a fusion protein called Bcr-Abl that consists primarily of sequences for the c-Abl kinase. However, the *bcr-abl* gene is expressed at higher levels than the gene encoding the normal c-Abl kinase, leading to an excess of signals for cell growth. In addition, the Bcr-Abl protein may have regulatory properties that are subtly different from those of the c-Abl kinase itself. Thus, leukemia cells express a unique target for drugs. A specific inhibitor of the Bcr-Abl kinase, imatinib mesylate (called Gleevec commercially), is now widely used to treat leukemia; more than 90% of patients responded well to the treatment. Thus, *our understanding of signal-transduction pathways is leading to conceptually new disease treatments.*

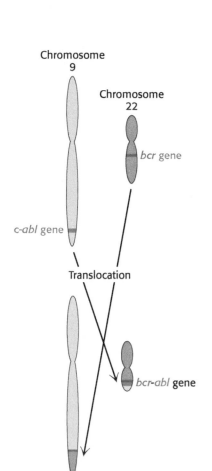

Figure 13.22 The formation of the *bcr-abl* **gene by translocation.** In chronic myelogenous leukemia, parts of chromosomes 9 and 22 are reciprocally exchanged, causing the *bcr* and *abl* genes to fuse. The protein kinase encoded by the *bcr-abl* gene is expressed at higher levels in cells having this translocation than is that encoded by the *c-abl* gene in normal cells.

SUMMARY

13.1 Signal Transduction Depends on Molecular Circuits

Most signal-transduction pathways are constructed with a similar set of components. A primary messenger, such as a hormone, binds to the extracellular part of a membrane-bound receptor. The messenger–receptor complex generates a second messenger inside the cell, which activates proteins that alter the biochemical environment inside the cell. Finally, means exist to terminate the signal-transduction pathway.

13.2 Receptor Proteins Transmit Information into the Cell

Seven-transmembrane-helix receptors operate in conjunction with heterotrimeric G proteins. The binding of a hormone to a 7TM receptor triggers the exchange of GTP for GDP bound to the α subunit of the G protein. G_α proteins can transmit information in a number of ways. $G_{\alpha s}$-GTP activates adenylate cyclase, an integral membrane protein that catalyzes the synthesis of cAMP. Cyclic AMP then activates protein kinase A by binding to its regulatory subunit, thus unleashing its catalytic chains. 7TM receptors activate $G_{\alpha q}$ proteins and the phosphoinositide pathway. The receptor-triggered activation of phospholipase C generates two intracellular messengers by hydrolysis of phosphatidylinositol 4,5-bisphosphate. Inositol trisphosphate opens calcium channels in the endoplasmic and sarcoplasmic reticulum membranes. Diacylglycerol activates protein kinase C.

13.3 Some Receptors Dimerize in Response to Ligand Binding and Recruit Tyrosine Kinases

Some ligands induce dimerization of the receptors to which they bind. Such a receptor contains an extracellular domain that binds the ligand, a transmembrane region, and a cytoplasmic domain that either binds a protein kinase or is a protein kinase. The growth-hormone receptor participates in an example of this type of signal-transduction pathway. Dimerization of the receptor activates Janus kinase 2, a protein kinase associated with the intracellular part of the receptor.

Receptor tyrosine kinases, such as the epidermal-growth-factor receptor and the insulin receptor, have intracellular domains that are kinases. When such receptor tyrosine kinases dimerize and are activated, cross-phosphorylation takes place. The phosphorylated tyrosines in activated receptor tyrosine kinases serve as docking sites for signaling proteins and permit further propagation of the signal. A prominent component of such pathways is the small GTPase Ras. The Ras protein, like the G_α subunit, cycles between an inactive form bound to GDP and an active form bound to GTP.

13.4 Metabolism in Context: Insulin Signaling Regulates Metabolism

The hormone insulin is secreted when blood concentration of glucose is high. Insulin binds to the insulin receptor, which is a receptor tyrosine kinase. The activated tyrosine kinase then phosphorylates insulin-receptor substrate. The signaling pathway continues, with the key result being an increase in glucose transporters in the cell membrane.

13.5 Calcium Ion Is a Ubiquitous Cytoplasmic Messenger

Calcium ion acts by binding to calmodulin and other calcium sensors. Calmodulin contains four calcium-binding modules called EF hands that recur in other proteins. Ca^{2+}–calmodulin activates target proteins by binding to them.

13.6 Defects in Signaling Pathways Can Lead to Diseases

If the genes encoding components of the signal-transduction pathways are altered by mutation, pathological conditions, most notably cancer, may result. In their mutated form, these genes are called oncogenes. The normal counterparts are called proto-oncogenes and function in pathways that control cell growth and replication. Mutated versions of Ras are frequently found in human cancers.

KEY TERMS

primary messenger (p. 226)
ligand (p. 226)
second messenger (p. 226)
seven-transmembrane-helix (7TM)
  receptor (p. 227)
G protein (p. 228)
G_α (p. 228)
$G_{\beta\gamma}$ (p. 228)

G-protein-coupled receptor (GCPR)
  (p. 228)
adenylate cyclase (p. 229)
protein kinase A (PKA) (p. 229)
phosphoinositide cascade (p. 232)
phospholipase C (p. 232)
receptor tyrosine kinase (RTK)
  (p. 235)

Ras (p. 235)
calmodulin (CaM) (p. 238)
EF hand (p. 238)
calmodulin-dependent protein (CaM)
  kinase (p. 239)
proto-oncogene (p. 239)
oncogene (p. 239)

? Answers to QUICK QUIZZES

1. Dissociation of epinephrine from the receptor. Conversion of cAMP into AMP by phosphodiesterase and the subsequent inhibition of PKA. Conversion of GTP into GDP by G_α and the subsequent reformation of the inactive heterotrimeric G protein.

2. Insulin signifies the fed state. Its presence leads to the removal of glucose from the blood for storage or metabolism. Increasing the number of glucose transporters available makes these biochemical processes more efficient.

PROBLEMS

1. *Like salt and pepper.* Match the terms with the descriptions. ✓ 4

(a) Primary messenger _____

(b) Second messenger _____

(c) G-protein coupled receptor _____

(d) Heterotrimeric G-protein _____

(e) $G_{\alpha s}$ _____

(f) Protein kinase A _____

(g) cAMP phosphodiesterase _____

(h) GTPase activity _____

(i) Cholera _____

(j) Pertussis _____

(k) Phospholipase C _____

(l) $G_{\alpha q}$ _____

(m) Inositol trisphosphate _____

(n) Protein kinase C _____

1. Activated by 7TM receptor
2. Stimulated by cAMP
3. Results in the reassociation of G_α and $G_{\beta\gamma}$
4. Generates two second messengers
5. Activated by diacylglycerol
6. Composed of seven transmembrane helices
7. Message received by the cell
8. Results from $G_{\alpha i}$ inhibition
9. Activates phospholipase C
10. Activates adenylate cyclase
11. Activates a Ca^{2+} channel
12. Due to persistent stimulation of $G_{\alpha s}$
13. Intracellular chemical that relays message from ligand receptor complex
14. Results in the inactivation of protein kinase A

2. *Class differences.* What are the three major classes of membrane receptors? ✓ 5

3. *Magnification.* Explain how a small number of hormones binding to the extracellular surface of a cell can have a large biochemical effect inside the cell. ✓ 4

4. *Common properties.* What are some of the structural features common to all membrane-bound receptors? ✓ 4 ✓ 5

5. *On–off.* Why is the GTPase activity of G proteins crucial to the proper functioning of a cell? ✓ 4

6. *Specificity.* Hormones affect the biochemistry of a distinct set of tissues. What accounts for the tissue specificity of hormone action? ✓ 4

7. *Like peanut butter and jelly.* Match the terms with the descriptions.

(a) Growth hormone binding _____

(b) Growth hormone receptor _____

(c) Receptor tyrosine kinase _____

(d) Grb-2 _____

(e) Sos _____

(f) Ras _____

(g) IRS _____

(h) Phosphoinositide kinase _____

(i) PIP$_3$-activated kinase _____

(j) Akt _____

1. Dimerization results in cross-phosphorylation
2. Activates Ras
3. Activates JAK
4. Adaptor protein in insulin signaling pathway
5. Activates Akt
6. Binds IRS and forms IP$_3$
7. Causes receptor dimerization
8. Promotes movement of glucose transporters to the cell membrane
9. Small G-protein
10. Binds receptor tyrosine kinase and Sos

8. *Chimeras.* In an elegant experiment on the nature of receptor tyrosine kinase signaling, a gene was synthesized that encoded a *chimeric receptor*—the extracellular part came from the insulin receptor, and the membrane-spanning and cytoplasmic parts came from the EGF receptor. The striking result was that the binding of insulin induced tyrosine kinase activity, as evidenced by rapid autophosphorylation. What does this result tell you about the signaling mechanisms of the EGF and insulin receptors? ✓ 5

9. *Antibodies mimicking hormones.* An antibody has two identical antigen-binding sites. Remarkably, antibodies to the extracellular parts of growth-factor receptors often lead to the same cellular effects as does exposure to growth factors. Explain this observation. ✓ 5

10. *Alike but different.* What is the difference between heterotrimeric G proteins and small G proteins? ✓ 4

11. *Facile exchange.* A mutated form of the α subunit of the heterotrimeric G protein has been identified; this form readily exchanges GDP for GTP even in the absence of an activated receptor. What would be the effect on a signaling pathway containing the mutated α subunit?

12. *Diffusion rates.* Normally, rates of diffusion vary inversely with molecular weights; so smaller molecules diffuse faster than do larger ones. In cells, however, calcium ion diffuses more slowly than does cAMP. Propose a possible explanation.

13. *Awash with glucose.* Glucose is mobilized for ATP generation in muscle in response to epinephrine, which activates $G_{\alpha s}$. Cyclic AMP phosphodiesterase is an enzyme that converts cAMP into AMP. How would inhibitors of cAMP phosphodiesterase affect glucose mobilization in muscle? ✓ 4

14. *Many defects.* Considerable effort has been directed toward determining the genes in which sequence variation contributes to the development of type 2 diabetes, a disease that results from a loss of sensitivity of cells to insulin. Approximately 800 genes have been implicated. Explain the significance of this observation.

15. *Growth-factor signaling.* Human growth hormone binds to a cell-surface membrane protein that is not a receptor tyrosine kinase. The intracellular domain of the receptor can bind other proteins inside the cell. Furthermore, studies indicate that the receptor is monomeric in the absence of hormone but dimerizes on hormone binding. Propose a possible mechanism for growth-hormone signaling. ✓ 5

16. *Different genes.* Differentiate among a proto-oncogene, an oncogene, and a tumor-repressor gene.

17. *Accelerator and brake.* Only one copy of a proto-oncogene need be mutated to enhance the development of cancer, whereas both copies of a tumor-suppressor gene must be mutated to contribute to cancer development. Explain this distinction.

18. *Redundancy.* Because of the high degree of genetic variability in tumors, typically no single anticancer therapy is universally effective for all patients, even within a given tumor type. Hence, it is often desirable to inhibit a particular pathway at more than one point in the signaling cascade. In addition to the EGFR-directed monoclonal antibody cetuximab (p. 239), propose alternative strategies for targeting the EGF signaling pathway for antitumor drug development.

Chapter Integration Problems

19. *A reappearance.* Ligand-gated channels can be thought of as receptors. Explain.

20. *Molecular exposure.* The binding of Ca^{2+} to calmodulin induces substantial conformational changes in its EF hands, exposing hydrophobic residues on the surface of the protein. How might this structural change help to propagate the calcium signal?

Challenge and Data Interpretation Problems

21. *Making connections.* Suppose that you were investigating a newly discovered growth-factor signal-transduction pathway. You found that, if you added a GTP analog in which the terminal phosphate was replaced by sulfate, the duration of the hormonal response was increased. What can you conclude?

22. *Vive la différence.* Why is the fact that a monomeric hormone binds to two identical receptor molecules, thus promoting the formation of a dimer of the receptor, considered remarkable? ✓ 5

23. *Active mutants.* Some protein kinases are inactive unless they are phosphorylated on key serine or threonine residues. In some cases, active enzymes can be generated by mutating these serine or threonine residues to glutamate. Explain.

24. *Signal-to-noise ratio.* At steady state, intracellular levels of Ca^{2+} must be kept low to prevent the precipitation of carboxylated and phosphorylated compounds, which form poorly soluble salts with Ca^{2+}. The cytoplasmic level of Ca^{2+} is approximately 100 nM, several orders of magnitude lower than the concentration in the extracellular medium. How might the cell maintain such low levels of intracellular Ca^{2+}? How does the cell take advantage of the difference in intracellular and extracellular Ca^{2+} concentrations?

25. *Receptor truncation.* You prepare a cell line that overexpresses a mutant form of EGFR in which the entire intracellular region of the receptor has been deleted. Predict the effect of overexpression of this construct on EGF signaling in this cell line. ✓ 5

26. *Total amplification.* Suppose that each β-adrenergic receptor bound to epinephrine converts 100 molecules of $G_{\alpha s}$ into their GTP-bound forms and that each molecule of activated adenylate cyclase produces 1000 molecules of cAMP per second. With the assumption of a full response, how many molecules of cAMP will be produced in 1 s after the formation of a single complex between epinephrine and the β-adrenergic receptor? ✓ 4

27. *Establishing specificity.* You wish to determine the hormone-binding specificity of a newly identified membrane receptor. Three different hormones, X, Y, and Z, were mixed with the receptor in separate experiments, and the percentage of binding capacity of the receptor was determined as a function of hormone concentration, as shown in graph A.

(A)

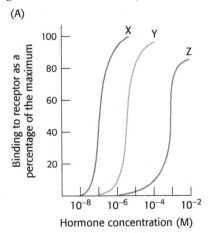

(a) What concentrations of each hormone yield 50% maximal binding?

(b) Which hormone shows the highest binding affinity for the receptor?

You next wish to determine whether the hormone–receptor complex stimulates the adenylate cyclase cascade. To do so, you measure adenylate cyclase activity as a function of hormone concentration, as shown in graph B.

(B)

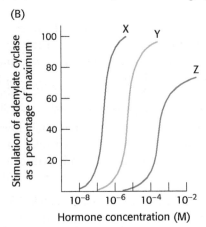

(c) What is the relation between the binding affinity of the hormone–receptor complex and the ability of the hormone to enhance adenylate cyclase activity? What can you conclude about the mechanism of action of the hormone–receptor complex?

(d) Suggest experiments that would determine whether a $G_{\alpha s}$ protein is a component of the signal-transduction pathway.

28. *Binding matters.* A scientist wishes to determine the number of receptors specific for a ligand X, which he has in both radioactive and nonradioactive form. In one experiment, he adds increasing amounts of the radioactive X and measures how much of it is bound to the cells. The result is shown as total activity in the adjoining graph. Next, he performs the same experiment, except that he includes a several hundredfold excess of nonradioactive X. This result is shown as nonspecific binding. The difference between the two curves is the specific binding.

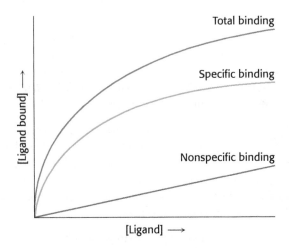

(a) Why is the total binding not an accurate representation of the number of receptors on the cell surface?

(b) What is the purpose of performing the experiment in the presence of excess nonradioactive ligand?

(c) What is the significance of the fact that specific binding attains a plateau?

29. *Counting receptors.* With the use of experiments such as those described in problems 27 and 28, the number of receptors in the cell membrane can be calculated. Suppose that the specific activity of the radioactive ligand is 10^{12} counts per minute (cpm) per millimole and that the maximal specific binding is 10^4 cpm per milligram of membrane protein. There are 10^{10} cells per milligram of membrane protein. Assume that one ligand binds per receptor. Calculate the number of receptor molecules present per cell.

Selected Readings for this chapter can be found online at www.whfreeman.com/tymoczko3e.

Basic Concepts and Design of Metabolism

CHAPTER 14

Digestion: Turning a Meal into Cellular Biochemicals

CHAPTER 15

Metabolism: Basic Concepts and Design

The concepts of conformation and dynamics developed in Sections 1 through 5—especially those dealing with the specificity and catalytic power of enzymes, the regulation of their catalytic activity, and the transport of molecules and ions across membranes—enable us to now ask questions that are fundamental to biochemistry:

1. *How does a cell extract energy and reducing power from its environment?*

2. *How does a cell synthesize the building blocks of its macromolecules and then the macromolecules themselves?*

These processes are carried out by a highly integrated network of chemical reactions that are collectively known as *metabolism* or *intermediary metabolism*. Metabolism can be subdivided into two categories: *catabolism* and *anabolism*. Catabolism is the set of reactions that extract biologically useful energy from environmental sources, such as meals. Anabolism is the set of reactions that use this energy and small molecules derived from the breakdown of food to synthesize new biomolecules, supramolecular complexes, and cells themselves.

More than a thousand chemical reactions take place, even in an organism as simple as the bacterium *Escherichia coli*. The array of reactions may seem overwhelming at first glance. However, closer scrutiny reveals that *metabolism has a coherent pattern containing many common motifs*. These motifs include the use of an energy currency and the repeated appearance of a limited number of activated intermediates. In fact, a group of about 100 molecules have central roles in all forms of life and are processed with the use of similar pathways. Moreover, these metabolic pathways are regulated in common ways.

Before a cell can begin the metabolism of a molecule for anabolic or catabolic purposes, the molecule must be made accessible to the cell. In higher organisms such as human beings, the conversion of meals into accessible biomolecules begins in the digestive tract with the biochemical process of digestion.

We begin this section with an investigation of how meals—dietary forms of the biochemicals required for survival—are converted into biochemicals that enter metabolic pathways. Next, we will examine some general principles and motifs of metabolism to provide a foundation for the more detailed studies to follow.

✓ By the end of this section, you should be able to:

✓ 1 Describe how dietary proteins, carbohydrates, and lipids are digested.

✓ 2 Explain how the release of pancreatic enzymes is coordinated with digestion in the stomach.

✓ 3 Identify the factors that make ATP an energy-rich molecule.

✓ 4 Explain how ATP can power reactions that would otherwise not take place.

✓ 5 Describe the relation between the oxidation state of a carbon molecule and its usefulness as a fuel.

CHAPTER 14

Digestion: Turning a Meal into Cellular Biochemicals

Growing requires vast amounts of energy and biochemical building blocks. These needs do not disappear as we age but are required to maintain our bodies against the wear and tear of living. The energy and building blocks come in the form of food, which must be converted into biochemicals in the process of digestion. [Stuart Pearce/Age Fotostock.]

Eating is a basic need of all organisms that relieves hunger and provides energy. For humans, however, eating is far more than consuming fuel. A well-cooked, well-presented meal is a gustatory, olfactory, and visual delight. Indeed, eating is one of the greatest pleasures that we as humans can experience. Sharing a meal, with friends in a playground or heads of state in a grand palace, is a time-honored means of building and strengthening bonds between individuals.

In this chapter, we will take a more prosaic view of eating. We will examine it as the first stage of biochemical fuel generation. Once the food is in our mouth, the process of digestion begins. *Digestion* is the biochemical set of reactions by which food is converted into molecules that will be further manipulated to yield biologically useful energy and biosynthetic building blocks.

✓ 1 Describe how dietary proteins, carbohydrates, and lipids are digested.

14.1 Digestion Prepares Large Biomolecules for Use in Metabolism

Let us begin our study of digestion by examining what happens after we take a bite of pizza, a delicious concoction of lipids, carbohydrates, and proteins

Figure 14.1 Pizza. Foods provide a pleasurable means of obtaining energy and building blocks for biological systems. [Mode/Ian O'Leary/Age Fotostock.]

Table 14.1 Gastric and pancreatic zymogens

Site of synthesis	Zymogen	Active enzyme
Stomach	Pepsinogen	Pepsin
Pancreas	Chymotrypsinogen	Chymotrypsin
Pancreas	Trypsinogen	Trypsin
Pancreas	Procarboxypeptidase	Carboxypeptidase
Pancreas	Proelastase	Elastase

✓ 2 Explain how the release of pancreatic enzymes is coordinated with digestion in the stomach.

(Figure 14.1). The components of our meal must be degraded into small molecules for absorption by the epithelial cells of the intestine and for transport in the blood. Proteins are digested to amino acids by *proteolytic enzymes (proteases)* secreted by the stomach and pancreas. Polysaccharides, such as starch, are cleaved into monosaccharides by α-amylase from the pancreas and to a lesser extent in saliva. Lipids are converted into fatty acids by lipases secreted by the pancreas. All of the digestive enzymes are hydrolases; that is, they cleave their substrates by the addition of a molecule of water.

Most Digestive Enzymes Are Secreted as Inactive Precursors

With the exception of α-amylase, all digestive enzymes are secreted as inactive forms, called *zymogens* or *proenzymes*, which are subsequently activated by proteolytic cleavage (Table 14.1). Before their secretion, zymogens exist in granules near the cell membrane. In response to signals discussed below, the granules fuse with the cell membrane, expelling their contents into the lumen of the intestine. The zymogens are activated when a part of the inactive precursor is proteolytically cleaved. The enzyme *enteropeptidase* (also called enterokinase), secreted by the epithelial cells of the small intestine, activates the pancreatic zymogen *trypsinogen* to form *trypsin*, which in turn activates the remaining pancreatic zymogens. Indeed, the stomach enzyme, *pepsin*, is itself secreted as a zymogen called *pepsinogen*. Pepsinogen has a small amount of enzyme activity and can activate itself to some degree in an acidic environment. The active pepsin activates the remaining pepsinogen.

14.2 Proteases Digest Proteins into Amino Acids and Peptides

Digestion begins in the mouth with the process of chewing, where teeth, tongue, and saliva are employed to homogenize a bite of pizza, converting it into an aqueous slurry that is more readily attacked by digestive enzymes than a piece of poorly chewed food would be.

Saliva, secreted by the salivary glands, is an aqueous solution of Na^+, K^+, Cl^-, and HCO_3^- that contains mucoproteins (p. 180). These components facilitate the homogenization of the food and lubricate the resulting slurry for swallowing. Saliva also contains α-amylase, which cleaves α-1,4 glycosidic bonds. Because of the short duration of time that food is in the mouth, polysaccharide breakdown in the mouth is minimal.

 CLINICAL INSIGHT

Protein Digestion Begins in the Stomach

Human beings ingest about 70 to 100 g of protein daily. In regard to our pizza, the meat and cheese provide most of the protein, which must be degraded so that the individual amino acids will be available for use in metabolic pathways. In addition to proteins in foods, 50 to 100 g of protein per day is sloughed off the cells of the intestine in the wear and tear of digestion; this protein is degraded, and the amino acids are salvaged.

Subsequent to homogenization, the food passes into the stomach, where two principal activities take place. First, the proteins are denatured by the acidic environment of the stomach, where the pH is maintained at values ranging from 1 to 2. This denaturation, caused by the breakage of ionic and hydrogen bonds due to the low pH, renders protein a better substrate for degradation. Second, the process of protein degradation begins in the stomach with the action of the proteolytic enzyme pepsin, as discussed above. The action of pepsin yields protein fragments that will be further degraded by the proteases of the intestine. Pepsin is a remarkable enzyme, displaying optimal activity in the pH range of 1 to 2, conditions so acidic that other proteins are denatured (p. 133).

How is the acid environment of the stomach generated? Specialized cells lining the stomach contain the membrane protein K^+/H^+ *ATPase* (gastric proton pump) that pumps protons into the stomach in exchange for K^+ at the expense of ATP hydrolysis. This proton pump is similar to the Na^+–K^+ ATPase discussed earlier (p. 213). In some individuals, the gastric proton pump can be too active, resulting in gastroesophageal reflux disease (GERD, p. 26), a condition where the stomach acid leaks back to the esophagus. In addition to being painful, GERD may eventually result in esophageal cancer if left untreated. A common treatment for GERD is to irreversibly inhibit the proton pump. One such inhibitor, omeprazole, is converted into sulfenic acid by the stomach acid, which rearranges to yield sulfonamide. Sulfonamide irreversibility modifies a cysteine residue on the pump (Figure 14.2). Omeprazole and other proton pump inhibitors are among the most commonly prescribed drugs.

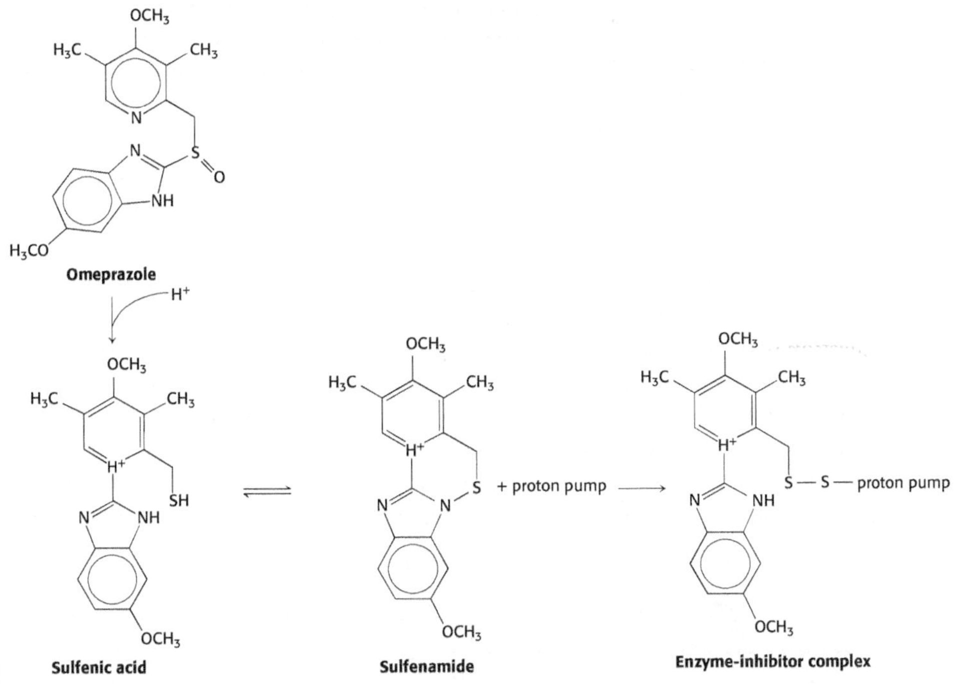

Figure 14.2 Omeprazole action.

Protein Digestion Continues in the Intestine

The partly digested proteins as well as carbohydrates and lipids move from the acidic environment of the stomach to the beginning of the small intestine. The low pH of the food stimulates the cells of the small intestine to release the hormone *secretin* (Figure 14.3). Secretin, in turn, promotes the release of sodium bicarbonate ($NaHCO_3$) from pancreatic cells, which neutralizes the pH of the food as it exits the stomach. The polypeptide products of pepsin digestion also stimulate the release of the hormone *cholecystokinin* (CCK) by intestinal cells. The pancreas responds to CCK (Figure 14.3) by releasing a host of digestive enzymes into the intestine, where the digestion of proteins continues and the digestion of lipids and carbohydrates begins.

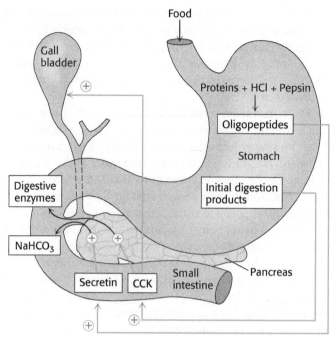

Figure 14.3 The hormonal control of digestion. Cholecystokinin (CCK) is secreted by specialized intestinal cells and causes the secretion of bile salts from the gall bladder and digestive enzymes from the pancreas. Secretin stimulates sodium bicarbonate ($NaHCO_3$) secretion from the pancreas, which neutralizes the stomach acid. [Information from D. Randall, W. Burggren, and K. French, *Eckert Animal Physiology*, 5th ed. (W. H. Freeman and Company, 2002), p. 658.]

The pancreatic proteases hydrolyze the proteins into small fragments called oligopeptides, but digestion is completed by enzymes called peptidases that are attached to the external surfaces of the intestinal cells. These enzymes cleave the oligopeptides into amino acids and di- and tripeptides that can be transported into an intestinal cell by transporters. At least seven different transporters exist, each specific to a different group of amino acids. A number of inherited disorders result from mutations in these transporters. For example, Hartnup disease, a rare disorder characterized by rashes, ataxia (lack of muscle control), delayed mental development, and diarrhea, results from a defect in the transporter for tryptophan and other nonpolar amino acids. The absorbed amino acids are subsequently released into the blood by a number of antiporters for use by other tissues (**Figure 14.4**).

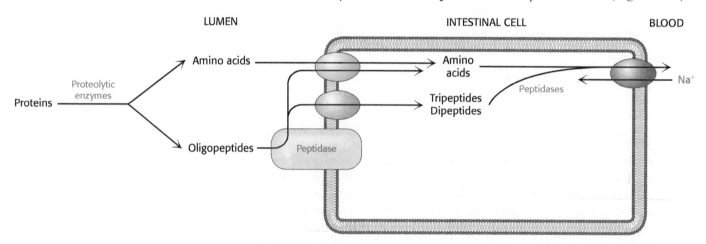

Figure 14.4 The digestion and absorption of proteins. Protein digestion is primarily a result of the activity of enzymes secreted by the pancreas. Peptidases associated with the intestinal epithelium further digest proteins. The amino acids and di- and tripeptides are absorbed into the intestinal cells by specific transporters. Free amino acids are then released into the blood through antiporters for use by other tissues.

Celiac Disease Results from the Inability to Properly Digest Certain Proteins

Celiac disease, also known as gluten enteropathy, is an intestinal inflammatory disorder that is triggered in susceptible individuals by protein in wheat, rye, and barley. Gluten is used as a general term referring to the proteins gliadin and glutenin in wheat and hordein and secalin in barley and rye, respectively. All glutens are rich in the amino acids proline and glutamine and are resistant to complete digestion by pepsin and the pancreatic proteases. Because of their composition, these proteins are sometimes called prolamins. Susceptible individuals are genetically disposed to mount an inflammatory response to gluten-derived peptides. The immune response damages the intestinal epithelial cells, which impairs nutrient absorption leading to a variety of problems, including weight loss, diarrhea, and anemia. The disease is quite common, with an estimated frequency of 1 in 250 Caucasians.

Glutens are storage proteins in plants, providing amino acids as well as carbon, nitrogen, and sulfur for growth and development. With regard to our pizza, the gluten in the dough gives it elasticity, helping it rise during heating and maintaining its final shape.

14.3 Dietary Carbohydrates Are Digested by Alpha-Amylase

How are the crust and the vegetables that topped our pizza converted into biochemicals? These ingredients are sources of carbohydrates, both complex, such as starch and any glycogen (p. 175) that might be present in the meat, and simple, such as sucrose (p. 174).

Like proteins, dietary carbohydrates are digested into molecules that can be readily absorbed by the intestine. The most common end products are the monosaccharides glucose, fructose, and galactose. Our primary source of carbohydrates is the complex carbohydrate starch, present in the crust. This branched homopolymer of glucose is digested primarily by the pancreatic enzyme *α-amylase*, which cleaves the α-1,4 bonds of starch but not the α-1,6 bonds (**Figure 14.5**). The products are the di- and trisaccharides maltose and maltotriose, and *limit dextrin*, which is not digestible by α-amylase because of the α-1,6 bonds. *Maltase* converts maltose into glucose, and *α-glucosidase* digests maltotriose and any other oligosaccharides that may have escaped digestion by the amylase. *α-Dextrinase* further digests limit dextrin into simple sugars. Maltase, α-glucosidase, and α-dextrinase are on the surfaces of the intestinal cells.

The digestion of disaccharides is simpler than the digestion of complex carbohydrates. Sucrose, a disaccharide consisting of glucose and fructose contributed by the vegetables, is digested by sucrase. Lactase degrades the milk-sugar lactose into glucose and galactose. Sucrase and lactase also reside on the surfaces of intestinal cells.

Glucose and galactose are transported into the intestinal epithelial cells by a secondary active-transport process carried out by the sodium–glucose linked transporter (SGLT) (p. 215) while fructose diffuses across the cell membrane through a transporter called GLUT5 (**Figure 14.6**). Another transporter, GLUT2, releases all three monosaccharides into the bloodstream, where they can travel to other tissues to be used as fuel. We will examine the GLUT proteins in more detail in Chapter 16.

DID YOU KNOW?

Baking pizza in an oven converts the dough into crust and solves a biochemical problem as well: starch is difficult to digest without first being hydrated, and heat allows the starch to absorb water.

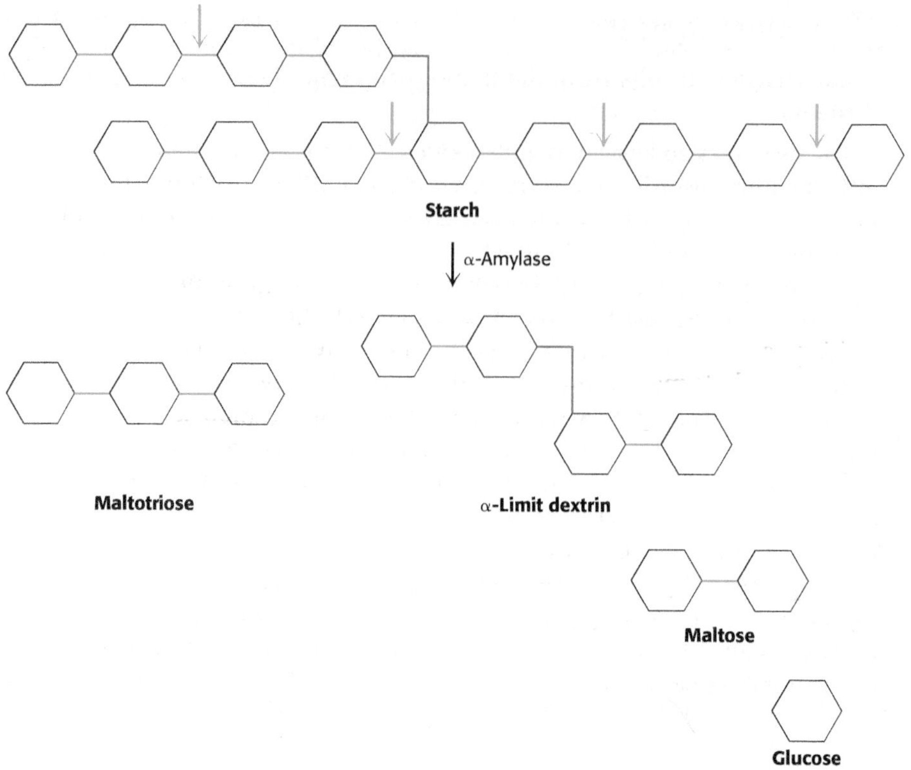

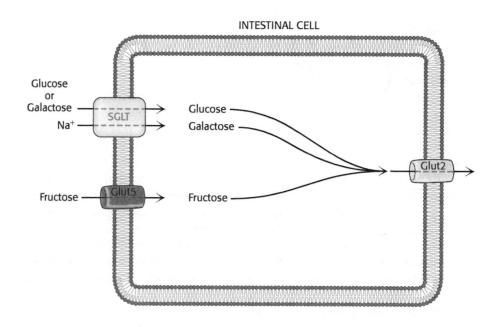

Starch

α-Amylase

Maltotriose

α-Limit dextrin

Maltose

Glucose

Figure 14.5 The digestion of starch by α-amylase. Amylase hydrolyzes starch into simple sugars. The α-1,4 bonds are shown in green. The α-1,6 bonds are red. The sites of α-amylase digestion are indicated by the small green arrows.

INTESTINAL CELL

Glucose or Galactose

Na^+

SGLT

Glucose

Galactose

Fructose

Glut5

Fructose

Glut2

Figure 14.6 Uptake of monosaccharides. The results of carbohydrate digestion, primarily glucose, galactose, and fructose, are transported into the intestinal cells by specific transport proteins. The carbohydrates also exit the cell with the assistance of transport proteins.

14.4 The Digestion of Lipids Is Complicated by Their Hydrophobicity

As we eat our pizza, we may notice that the box has greasy stains. These stains are caused by the lipids in our meal. The main sources of lipids in the pizza are the meat and the cheese, which, as already noted, are sources of protein as well. Most lipids are ingested in the form of triacylglycerols and must be degraded to fatty acids for absorption across the intestinal epithelium. Lipid digestion presents a problem because, unlike carbohydrates and proteins, lipids are not soluble in water. Recall that lipids are highly reduced molecules. This high degree of reduction accounts both for their high energy content and for their poor solubility in water (Chapter 11). How can the lipids be degraded to fatty acids if the lipids are not soluble in the same medium as the degradative enzymes are? Moreover, the lipid-digestion products—fatty acids—also are not water soluble; so, when digestion has taken place, how does fatty acid transport happen?

Lipids are prepared for digestion in the stomach. The grinding and mixing that takes place in the stomach converts lipids into an *emulsion*, a mixture of lipid droplets and water. Common emulsions include mayonnaise and shaken oil-and-vinegar salad dressing. After the lipids leave the stomach, emulsification is enhanced with the aid of *bile salts*, amphipathic molecules synthesized from cholesterol in the liver and secreted from the gall bladder in response to cholecystokinin (Figure 14.7). These molecules insert into the lipid droplets, making the triacylglycerols more readily digested. Triacylglycerols are degraded to free fatty acids and monoacylglycerol by enzymes secreted by the pancreas called *lipases* (Figure 14.8), which attach to the surface of a lipid droplet. Pancreatic lipases are also released into the intestine as zymogens that are subsequently activated. The final digestion products, free fatty acids and

Glycocholate

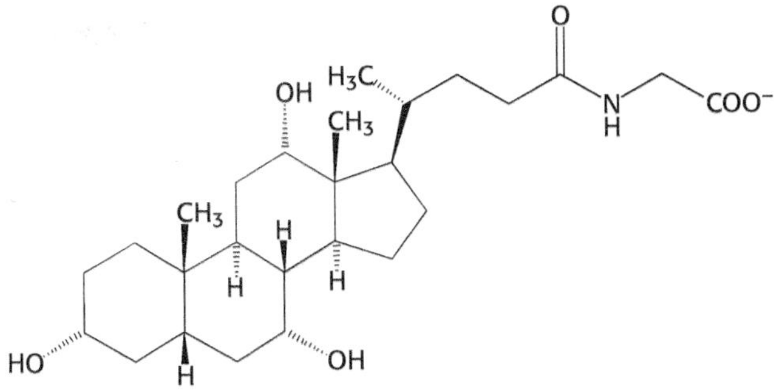

Figure 14.7 Glycocholate. Bile salts, such as glycocholate, facilitate lipid digestion in the intestine.

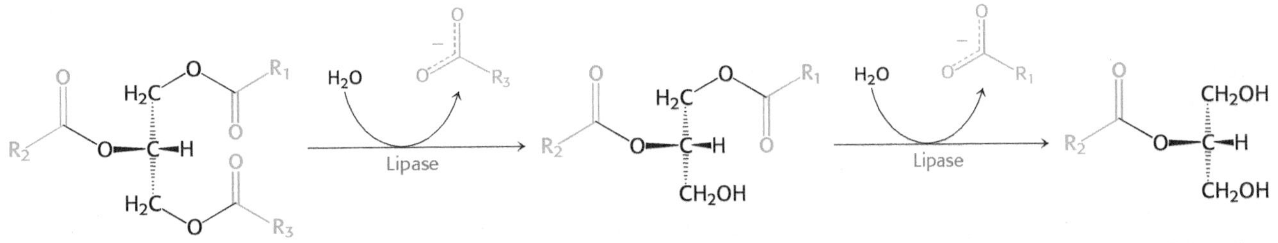

Triacylglycerol   **Diacylglycerol**   **Monoacylglycerol**

Figure 14.8 **The action of pancreatic lipases.** Lipases secreted by the pancreas convert triacylglycerols into fatty acids and monoacylglycerol for absorption into the intestine.

monoacylglycerol, are carried in *micelles* to the plasma membrane of the intestinal epithelial cells where they will subsequently be absorbed. Micelles are globular structures formed by small lipids in aqueous solutions (Figure 14.9). In a micelle, the polar head groups of the fatty acids and monoacylglycerol are in contact with the aqueous solution and the hydrocarbon chains are sequestered in the interior of the micelle. Micelle formation is also facilitated by bile salts. If the production of bile salts is inadequate due to liver disease, large amounts of fats (as much as 30 g per day) are excreted in the feces. This condition is referred to as *steatorrhea*, after stearic acid, a common fatty acid.

The fatty acids and monoacylglycerol are transported into the intestinal cells by membrane proteins such as the fatty-acid-binding protein (FABP) (Figure 14.10). Once inside the cell, fatty-acid-transport proteins (FATP) ferry them to the cytoplasmic face of the smooth endoplasmic reticulum (SER), where the triacylglycerols are resynthesized from fatty acids and monoacylglycerol. After transport into the lumen of the SER, the triacylglycerols associate with specific proteins and a small amount of phospholipid and cholesterol to form lipoprotein transport particles called *chylomicrons*, stable particles approximately 2000 Å (200 nm) in diameter. These particles are composed of 98% triacylglycerols with the proteins and phospholipid on the surface. The chylomicrons are released into the lymph system and then into the blood (Chapter 29).

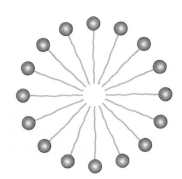

Figure 14.9 A diagram of a section of a micelle. Ionized fatty acids generated by the action of lipases readily form micelles.

LUMEN   INTESTINAL CELL

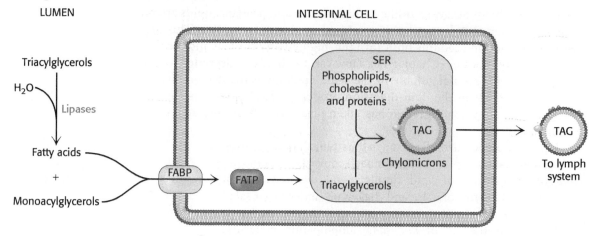

Figure 14.10 Chylomicron formation. Free fatty acids and monoacylglycerols are absorbed by intestinal epithelial cells. Triacylglycerols are resynthesized and packaged with other lipids and proteins to form chylomicrons, which are then released into the lymph system.

After a meal rich in lipids, the blood appears milky because of the high content of chylomicrons. These particles bind to membrane-bound lipoprotein lipases, primarily at adipose tissue and muscle, where the triacylglycerols are once again degraded into free fatty acids and monoacylglycerol for transport into the tissue. The triacylglycerols are then resynthesized and stored. In the muscle and other tissues, they can be oxidized to provide energy, as will be discussed in Chapter 27. Chylomicrons also function in the transport of fat-soluble vitamins and cholesterol.

QUICK QUIZ Explain why a person who has a trypsinogen deficiency will suffer from more digestion difficulties than will a person lacking most other zymogens.

BIOLOGICAL INSIGHT

Snake Venoms Digest from the Inside Out

Most animals ingest food and, in response to this ingestion, produce enzymes that digest the food. Many venomous snakes, on the other hand, do the opposite. They inject digestive enzymes into their prospective meals to begin the digestion process from the inside out, before they even consume the meals.

Snake venom, a highly modified form of saliva, consists of 50 to 60 different protein and peptide components that differ among species of snake and possibly even among individual snakes of the same species. Consider rattlesnakes (Figure 14.11). Rattlesnake venom contains a host of enzymes that digest the tissues of the victim. Phospholipases digest cell membranes at the site of the snakebite, causing a loss of cellular components. The phospholipases also disrupt the membranes of red blood cells, destroying them (a process called hemolysis). Collagenase digests the protein collagen, a major component of connective tissue (p. 56), whereas hyaluronidase digests hyaluronidate, a glycosaminoglycan (p. 178) component of connective tissue. The combined action of both collagenase and hyaluronidase is to destroy tissue at the site of the bite, enabling the venom to spread more readily throughout the victim.

Various proteolytic enzymes in the venom degrade basement membranes (a thin sheet of fibrous proteins, including collagen, that underlies the epithelial cells) and components of the extracellular matrix, leading to severe tissue damage. Some venoms contain proteolytic enzymes that stimulate the formation of blood clots as well as enzymes that digest blood clots. The net effect of these enzymes acting in concert may be to deplete all clotting factors

Figure 14.11 A rattlesnake poised to strike. Rattlesnakes inject digestive enzymes into their prospective meals. [Biosphoto/Daniel Heuclin.]

from the blood, and so clots do not form. Venoms also contain various peptides that have neurotoxic activities. The neurotoxins immobilize the prey, whereas the digestive enzymes reduce the size of the prey to make swallowing easier.

Despite the toxic nature of snake venoms, and venoms of all sorts, the study of the components of deadly concoctions have yielded a virtual pharmacopeia of clinically useful drugs. Drugs to combat hypertension (high blood pressure) have been developed following studies of venom components of the South African pit viper *Bothrops jararaca*. Drugs that reduce the likelihood of myocardial infarction (heart attack) resulted from the examination of the venom of the Southeastern pigmy rattlesnake *Sistrurus miliarius barbouri*. A component of the venom from the copperhead *Agkistrodon acutus* is showing promise as an anticancer drug.

SUMMARY

14.1 Digestion Prepares Large Biomolecules for Use in Metabolism

Digestion begins in the mouth, where food is homogenized into an aqueous slurry susceptible to enzyme digestion. The homogenized food then passes into the stomach, an acidic environment. The low pH of the stomach denatures proteins, thus preparing them for degradation.

14.2 Proteases Digest Proteins into Amino Acids and Peptides

Protein digestion begins in the stomach with the action of the proteolytic enzyme pepsin. The digestion products of pepsin stimulate the release of the hormone cholecystokinin from specialized cells in the upper intestine. Cholecystokinin stimulates the release of bile salts from the gall bladder and digestion enzymes from the pancreas in the form of zymogens or proenzymes. Enteropeptidase converts trypsinogen into trypsin, which, in turn, activates the other zymogens.

14.3 Dietary Carbohydrates Are Digested by Alpha-Amylase

Complex carbohydrates such as starch and glycogen are degraded by α-amylase, which cleaves the α-1,4 bonds of starch and glycogen. The products of α-amylase digestion are the disaccharide maltose and limit dextrin, a carbohydrate rich in α-1,6 bonds. Limit dextrin is digested by α-dextrinase. Sucrase and lactase digest the disaccharides sucrose and lactose, respectively.

14.4 The Digestion of Lipids Is Complicated by Their Hydrophobicity

Because dietary lipids, mostly triacylglycerols, are not water soluble, they must be converted into an emulsion, a mixture of lipid droplets and water, to be digested by lipases secreted by the pancreas. Bile salts, provided by the gall bladder, facilitate the formation of the emulsion. The products of lipase digestion—free fatty acids and monoacylglycerol—form micelles for absorption by the intestinal cells. In the cells of the intestine, triacylglycerols are re-formed and packaged into lipoprotein particles called chylomicrons for transport in the lymph system and the blood.

KEY TERMS

digestion (p. 247)
proteolytic enzymes (proteases) (p. 248)
zymogens (proenzymes) (p. 248)
enteropeptidase (p. 248)
trypsinogen (p. 248)

trypsin (p. 248)
pepsin (p. 248)
secretin (p. 249)
cholecystokinin (CCK) (p. 249)
α-amylase (p. 251)

emulsion (p. 253)
bile salt (p. 253)
lipase (p. 253)
micelle (p. 253)
chylomicron (p. 253)

? Answer to QUICK QUIZ

Trypsin, which is formed from trypsinogen, activates most of the other zymogens. Hence, a deficiency in trypsinogen would lead to a loss of activity of virtually all of the enzymes required for digestion. On the other hand, loss of a zymogen for lipase, for instance, would impair only lipid digestion, without affecting the digestion of other molecules.

PROBLEMS

1. *Fast and furious*. Match each term with its description. ✓ 1

(a) Digestion _____
(b) Cholecystokinin _____
(c) Zymogens _____
(d) Enteropeptidase _____
(e) α-Amylase _____
(f) Emulsion _____
(g) Bile salts _____
(h) Chylomicron _____
(i) K⁺/H⁺ ATPase _____
(j) Lipase _____

1. Proenzymes
2. Digests carbohydrates
3. Transports dietary lipids from intestinal cells to elsewhere in the body
4. Results in the secretion of digestion enzymes
5. Inserts into emulsions to facilitate lipase action
6. The conversion of food into simple biochemicals
7. Acidifies the stomach
8. Hydrolyzes lipids
9. Activates trypsin
10. Water–lipid droplets

2. *Necessary but not sufficient*. Why is digestion required for fuel metabolism even though no useful energy is harnessed in the process?

3. *Mother knows best*. When your mother told you to chew your food well, she had your best biochemical interests at heart. Explain. ✓ 1

4. *Accessibility matters 1*. Dietary proteins must be denatured for efficient digestion. Why? ✓ 1

5. *Deconstruction*. Outline how starch and glycogen are digested. ✓ 1

6. *Safeguard*. Trypsin inhibitor is a pancreatic polypeptide that binds trypsin with very high affinity, preventing it from digesting proteins. Why might a lack of trypsin inhibitor cause pancreatitus (inflammation of the pancreas)? ✓ 2

7. *Not too al dente*. The digestion of macaroni is more efficient after the pasta has been heated in water. Why? ✓ 1

8. *Phase problems*. What properties of lipids make their digestion more complicated than that of carbohydrates and proteins? How are lipids made accessible for digestion? ✓ 1

9. *Accessibility matters 2*. Why is emulsification required for efficient lipid digestion? ✓ 1

10. *Humors are necessary*. How would a lack of bile salts affect digestion?

11. *Doing their part*. What is the role of micelles in lipid digestion?

12. *Precautions*. Why are most digestive enzymes produced as zymogens? ✓ 2

13. *Dual roles*. What two biochemical roles does CCK play? ✓ 2

14. *Electrolyte disturbance*. Hyponatremia, or low blood sodium, results from excessive losses of fluid and sodium, or from dilution of sodium levels with excess fluid. Hyponatremia is treated by oral rehydration therapy, the administration of a solution of glucose and sodium. Suggest one role for the sodium.

Challenge Problem

15. *Zymogen activation*. When very low concentrations of pepsinogen are added to an acidic medium, how does the half-time of activation depend on zymogen concentration? ✓ 2

Selected Readings for this chapter can be found online at www.whfreeman.com/tymoczko3e.

Metabolism: Basic Concepts and Design

An infinite number of books can be written with only a limited number of letters, 26 in the case of English. Likewise, the complex biochemistry of a cell—intermediary metabolism—is constructed from a limited number of recurring motifs, reactions, and molecules. [MARKA/ Alamy.]

The generation of energy from the oxidation of food takes place in three stages (Figure 15.1). *In the first stage (top panel), large molecules in food are broken down into smaller units.* This process is *digestion.* As discussed in Chapter 14, digestion renders the macromolecules in our meals into biochemically more manageable fragments. Proteins are hydrolyzed to the 20 amino acids, polysaccharides are hydrolyzed to simple sugars such as glucose, and fats are hydrolyzed to fatty acids. This stage is strictly a preparation stage; no useful energy is captured at this point.

In the second stage (middle panel), these numerous small molecules are degraded to a few simple units that play a central role in metabolism. In fact, most of them—sugars, fatty acids, glycerol, and several amino acids—are converted into acetyl CoA, the activated two-carbon unit that is the fuel for the final stages of aerobic metabolism. Some adenosine triphosphate (ATP) is generated in the second stage, but the amount is small compared with that obtained in the third stage.

In the third stage (bottom panel), ATP is produced from the complete oxidation of acetyl CoA. The third stage consists of the citric acid cycle and oxidative

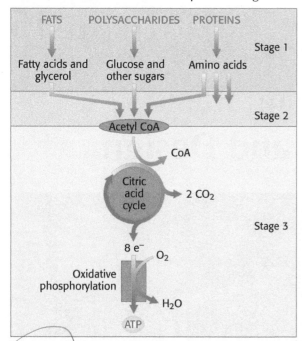

Figure 15.1 Stages of catabolism. The extraction of energy from fuels can be divided into three stages.

phosphorylation, which are the final common pathways in the oxidation of fuel molecules. Acetyl CoA brings the breakdown products of proteins, sugars, and fats into the citric acid cycle (also called the tricarboxylic acid (TCA) cycle or Krebs cycle), where they are completely oxidized to CO_2.

Stages 2 and 3 are the topic of Sections 7 through 13 and consist of many metabolic pathways and hundreds of reactions. In this chapter, we will see the basic principles that underlie not only the biochemistry of Sections 7 through 13, but also the remainder of the text. Just as a simple alphabet can be used to generate an unlimited number of books, the basic principles of biochemistry discussed in this chapter can be manipulated to allow a cell and an organism to respond to a wide range of physiological circumstances.

15.1 Energy Is Required to Meet Three Fundamental Needs

Living organisms require a continual input of free energy for three major purposes: (1) the performance of mechanical work in muscle contraction and cellular movements, (2) the active transport of molecules and ions, and (3) the synthesis of macromolecules and other biomolecules from simple precursors. The free energy used in these processes is derived from the environment. Photosynthetic organisms, or *phototrophs*, obtain this energy by trapping sunlight in a chemical form, whereas *chemotrophs*, which include humans, obtain energy through the oxidation of carbon fuels. In this chapter, we will examine some of the basic principles that underlie energy flow in all living systems. These principles are as follows:

1. Fuels are degraded and large molecules are constructed step by step in a series of linked reactions called *metabolic pathways*.

2. An energy currency common to all life forms, ATP, links energy-releasing pathways with energy-requiring pathways.

3. The oxidation of carbon fuels powers the formation of ATP.

4. Although there are many metabolic pathways, a limited number of types of reactions and particular intermediates are common to many pathways.

5. Metabolic pathways are highly regulated to allow the efficient use of fuels and to coordinate biosynthetic processes.

15.2 Metabolism Is Composed of Many Interconnecting Reactions

Metabolism is a linked series of chemical reactions that begins with a particular biomolecule and converts it into some other required biomolecule in a carefully defined fashion (**Figure 15.2**). These metabolic pathways process a biomolecule from a starting point (glucose, for instance) to an end point (carbon dioxide, water, and biochemically useful energy, in regard to glucose) without the generation of wasteful or harmful side products. There are many such defined pathways in the cell (**Figure 15.3**), together called *intermediary metabolism*, and we will examine many of them in some detail later. These pathways are interdependent—a biochemical ecosystem—and their activities are coordinated by exquisitely sensitive means of communication in which allosteric enzymes are predominant. We considered the principles of this communication in Chapters 7 and 13.

Metabolism Consists of Energy-Yielding Reactions and Energy-Requiring Reactions

We can divide metabolic pathways into two broad classes: (1) those that convert energy from fuels into biologically useful forms, such as ATP or ion gradients, and (2) those that require inputs of energy to proceed. Although this division is often imprecise, it is nonetheless a useful distinction in an examination of metabolism. Those reactions that transform fuels into cellular energy are called *catabolic reactions* or, more generally, *catabolism*.

$$\text{Fuel (carbohydrates, fats)} \xrightarrow{\text{Catabolism}} CO_2 + H_2O + \text{useful energy}$$

Those reactions that require energy—such as the synthesis of glucose, fats, or DNA—are called *anabolic reactions* or *anabolism*. The useful forms of energy that are produced in catabolism are employed in anabolism to generate complex structures from simple ones, or energy-rich states from energy-poor ones.

$$\text{Useful energy} + \text{simple precursors} \xrightarrow{\text{Anabolism}} \text{complex molecules}$$

Some pathways can be either anabolic or catabolic, depending on the energy conditions in the cell. They are referred to as *amphibolic pathways.*

An important general principle of metabolism is that, *although biosynthetic and degradative pathways often have reactions in common, the regulated, irreversible reactions of each pathway are almost always distinct from each other.*

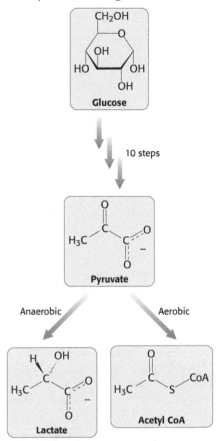

Figure 15.2 Glucose metabolism. Glucose is metabolized to pyruvate in 10 linked reactions. Under anaerobic conditions, pyruvate is metabolized to lactate and, under aerobic conditions, to acetyl CoA. The glucose-derived carbon atoms of acetyl CoA are subsequently oxidized to CO_2.

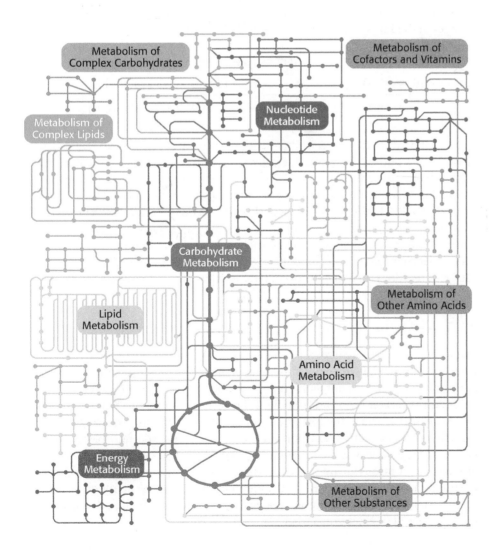

Figure 15.3 Metabolic pathways. Each node represents a particular biochemical, and the lines represent reactions linking the chemicals. [From the Kyoto Encyclopedia of Genes and Genomes (www.genome.ad.jp/kegg).]

This separation is necessary for energetic reasons, as will be evident in subsequent chapters. It also facilitates the control of metabolism.

A Thermodynamically Unfavorable Reaction Can Be Driven by a Favorable Reaction

How are specific pathways constructed from individual reactions? A pathway must satisfy minimally two criteria: (1) the individual reactions must be *specific,* and (2) the entire set of reactions that constitute the pathway must be *thermodynamically favored.* A reaction that is specific will yield only one particular product or set of products from its reactants. For example, glucose can undergo step-by-step conversion to yield carbon dioxide and water as well as useful energy. This conversion is extremely energy efficient because each step is facilitated by enzymes—highly specific catalysts (Section 3). The thermodynamics of metabolism is most readily approached in relation to free energy, which was discussed in Chapter 6. A reaction can take place spontaneously only if ΔG, the change in free energy, is negative. Recall that ΔG for the formation of products C and D from substrates A and B is given by

$$\Delta G = \Delta G^{\circ\prime} + RT \ln \frac{[C][D]}{[A][B]}$$

Thus, the ΔG of a reaction depends on the *nature* of the reactants and products (expressed by the $\Delta G^{\circ\prime}$ term, the standard free-energy change) and on their *concentrations* (expressed by the second term).

An important thermodynamic fact is that the overall free-energy change for a chemically coupled series of reactions is equal to the sum of the free-energy changes of the individual steps. Consider the following reactions:

$$A \rightleftharpoons B + C \qquad \Delta G^{\circ\prime} = +21 \text{ kJ mol}^{-1}(+5 \text{ kcal mol}^{-1})$$
$$B \rightleftharpoons D \qquad \Delta G^{\circ\prime} = -34 \text{ kJ mol}^{-1}(-8 \text{ kcal mol}^{-1})$$
$$\overline{A \rightleftharpoons C + D} \qquad \Delta G^{\circ\prime} = -13 \text{ kJ mol}^{-1}(-3 \text{ kcal mol}^{-1})$$

Under standard conditions, A cannot be spontaneously converted into B and C, because $\Delta G^{\circ\prime}$ is positive. However, the conversion of B into D under standard conditions is thermodynamically feasible. Because free-energy changes are additive, the conversion of A into C and D has a $\Delta G^{\circ\prime}$ of -13 kJ mol^{-1} (-3 kcal mol^{-1}), which means that it can take place spontaneously under standard conditions. Thus, *a thermodynamically unfavorable reaction can be driven by a thermodynamically favorable reaction to which it is coupled.* In this example, the reactions are coupled by the shared chemical intermediate B. Metabolic pathways are formed by the coupling of enzyme-catalyzed reactions such that the overall free energy of the pathway is negative.

✓ 3 Identify the factors that make ATP an energy-rich molecule.
✓ 4 Explain how ATP can power reactions that would otherwise not take place.

15.3 ATP Is the Universal Currency of Free Energy

Just as commerce is facilitated by the use of a common monetary currency, the commerce of the cell—metabolism—is facilitated by the use of a common energy currency, *adenosine triphosphate* (ATP). Part of the free energy derived from the oxidation of carbon fuels and from light is transformed into this readily available molecule, which acts as the free-energy donor in most energy-requiring processes such as motion, active transport, or biosynthesis. Indeed, most of catabolism consists of reactions that extract energy from fuels such as carbohydrates and fats and convert it into ATP. Interestingly, the other nucleoside triphosphates are as energy rich as ATP and are in fact sometimes used as free-energy donors. The reason why ATP, and not another nucleoside triphosphate, is the cellular energy currency is lost in evolutionary history.

ATP Hydrolysis Is Exergonic

ATP is a nucleotide consisting of adenine, a ribose, and a triphosphate unit (**Figure 15.4**). In considering the role of ATP as an energy carrier, we can focus on its triphosphate moiety. *ATP is an energy-rich molecule because its triphosphate unit contains two phosphoanhydride linkages.* Phosphoanhydride linkages are formed between two phosphoryl groups accompanied by the loss of a molecule of water. A large amount of free energy is liberated when ATP is hydrolyzed to adenosine diphosphate (ADP) and orthophosphate (P_i) or when ATP is hydrolyzed to adenosine monophosphate (AMP) and pyrophosphate (PP_i):

$$ATP + H_2O \rightleftharpoons ADP + P_i$$
$$\Delta G^{\circ\prime} = -30.5 \text{ kJ mol}^{-1}(-7.3 \text{ kcal mol}^{-1})$$

$$ATP + H_2O \rightleftharpoons AMP + PP_i$$
$$\Delta G^{\circ\prime} = -45.6 \text{ kJ mol}^{-1}(-10.9 \text{ kcal mol}^{-1})$$

The precise $\Delta G^{\circ\prime}$ for these reactions depends on the ionic strength of the medium and on the concentrations of Mg^{2+} and other metal ions in the medium (problems 27 and 32). Under typical cellular concentrations, the actual ΔG for these hydrolyses is approximately -50 kJ mol^{-1} ($-12 \text{ kcal mol}^{-1}$).

The free energy liberated in the hydrolysis of ATP is harnessed to drive reactions that require an input of free energy, such as muscle contraction. In turn, ATP is formed from ADP and P_i when fuel molecules are oxidized in chemotrophs or when light is trapped by phototrophs. *This ATP–ADP cycle is the fundamental mode of energy exchange in biological systems.*

ATP Hydrolysis Drives Metabolism by Shifting the Equilibrium of Coupled Reactions

An otherwise unfavorable reaction can be made possible by coupling to ATP hydrolysis. Consider an endergonic chemical reaction, a reaction that would not occur without an input of free energy, but yet is required for biosynthetic pathway.

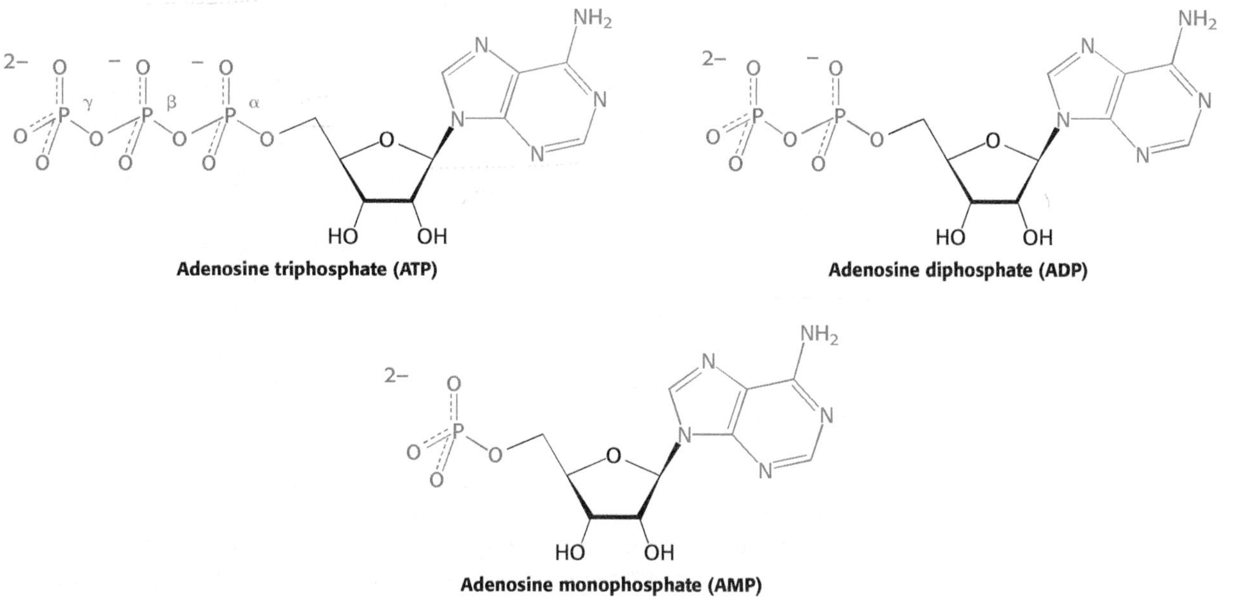

Figure 15.4 Structures of ATP, ADP, and AMP. These adenylates consist of adenine (blue), a ribose (black), and a tri-, di-, or monophosphate unit (red). The innermost phosphorus atom of ATP is designated P_α, the middle one P_β, and the outermost one P_γ.

Suppose that the standard free energy of the conversion of compound A into compound B is $+16.7$ kJ mol^{-1} ($+4.0$ kcal mol^{-1}):

$$A \rightleftharpoons B \qquad \Delta G^{\circ\prime} = +16.7 \text{ kJ mol}^{-1}(+4.0 \text{ kcal mol}^{-1})$$

The equilibrium constant K'_{eq} of this reaction at 25°C is related to $\Delta G^{\circ\prime}$ (in units of kilojoules per mole) by

$$K'_{eq} = [B]_{eq}/[A]_{eq} = e^{-\Delta G^{\circ\prime}/5.69} = 1.15 \times 10^{-3}$$

Thus, the net conversion of A into B cannot take place when the molar ratio of B to A is equal to or greater than 1.15×10^{-3}. However, A can be converted into B under these conditions if the reaction is coupled to the hydrolysis of ATP. Under standard conditions, the $\Delta G^{\circ\prime}$ of hydrolysis is approximately -30.5 kJ mol^{-1} (-7.3 kcal mol^{-1}). The new overall reaction is

$$A + ATP + H_2O \rightleftharpoons B + ADP + P_i$$
$$\Delta G^{\circ\prime} = -13.8 \text{ kJ mol}^{-1}(-3.3 \text{ kcal mol}^{-1})$$

Its free-energy change of -13.8 kJ mol^{-1} (-3.3 kcal mol^{-1}) is the sum of the value of $\Delta G^{\circ\prime}$ for the conversion of A into B [$+16.7$ kJ mol^{-1} ($+4.0$ kcal mol^{-1})] and the value of $\Delta G^{\circ\prime}$ for the hydrolysis of ATP [-30.5 kJ mol^{-1} (-7.3 kcal mol^{-1})]. At pH 7, the equilibrium constant of this coupled reaction is

$$K'_{eq} = \frac{[B]_{eq}}{[A]_{eq}} \times \frac{[ADP]_{eq}[P_i]_{eq}}{[ATP]_{eq}} = e^{13.8/2.47} = 2.67 \times 10^2$$

At equilibrium, the ratio of [B] to [A] is given by

$$\frac{[B]_{eq}}{[A]_{eq}} = K'_{eq} \frac{[ATP]_{eq}}{[ADP]_{eq}[P_i]_{eq}}$$

which means that the hydrolysis of ATP enables compound A to be converted into compound B until the [B]/[A] ratio reaches a value of 2.67×10^2. This equilibrium ratio is strikingly different from the value of 1.15×10^{-3} for the reaction A → B in the absence of ATP hydrolysis. In other words, coupling the hydrolysis of ATP with the conversion of A into B under standard conditions has changed the equilibrium ratio of B to A by a factor of about 10^5. If we were to use the ΔG° of hydrolysis of ATP under cellular conditions [-50.2 kJ mol^{-1} (-12 kcal mol^{-1})] in our calculations instead of $\Delta G^{\circ\prime}$, the change in the equilibrium ratio would be even more dramatic, of the order of 10^8 (problem 31).

We see here the thermodynamic essence of ATP's action as an *energy-coupling agent*. In the cell, the hydrolysis of an ATP molecule in a coupled reaction changes the equilibrium ratio of products to reactants by a very large factor. Thus, *a thermodynamically unfavorable reaction sequence can be converted into a favorable one by coupling it to the hydrolysis of ATP molecules in a new reaction.*

Note that A and B in the preceding coupled reaction may be interpreted very generally, not only as different metabolites. For example, A and B may represent activated and unactivated conformations of a protein that is activated by phosphorylation with ATP. Through such changes in protein conformation, muscle proteins convert the chemical energy of ATP into the mechanical energy of muscle contraction. Alternatively, A and B may refer to the concentrations of an ion or molecule on the outside and inside of a cell, as in the active transport of a nutrient.

The High Phosphoryl-Transfer Potential of ATP Results from Structural Differences Between ATP and Its Hydrolysis Products

What makes ATP an efficient energy currency? To answer this question, we need a means of comparing the tendency of various organic compounds bearing a phosphoryl group to transfer the phosphoryl group to an acceptor molecule. A common means of comparison is to determine the amount of energy released when the phosphorylated compound transfers the phosphoryl group to water under standard conditions. The energy released is called the *standard free energy of hydrolysis*. Let us compare the standard free energy of hydrolysis of ATP with that of a phosphate ester, such as glycerol 3-phosphate:

$$ATP + H_2O \rightleftharpoons ADP + P_i$$
$$\Delta G^{\circ\prime} = -30.5 \text{ kJ mol}^{-1} (-7.3 \text{ kcal mol}^{-1})$$

$$\text{Glycerol 3-phosphate} + H_2O \rightleftharpoons \text{glycerol} + P_i$$
$$\Delta G^{\circ\prime} = -9.2 \text{ kJ mol}^{-1} (-2.2 \text{ kcal mol}^{-1})$$

The magnitude of $\Delta G^{\circ\prime}$ for the hydrolysis of glycerol 3-phosphate is much smaller than that of ATP, which means that the tendency of ATP to transfer its terminal phosphoryl group to water is stronger than that of glycerol 3-phosphate. In other words, ATP has a larger *phosphoryl-transfer potential* (phosphoryl-group-transfer potential) than does glycerol 3-phosphate.

The high phosphoryl-transfer potential of ATP can be explained by features of the ATP structure. Because $\Delta G^{\circ\prime}$ depends on the *difference* in free energies of the products and reactants, we need to examine the structures of both ATP and its hydrolysis products, ADP and P_i, to understand the basis of high phosphoryl-transfer potential. Four factors differentiate the stability of the reactants and products: *electrostatic repulsion, resonance stabilization, an increase in entropy,* and *stabilization due to hydration*.

1. *Electrostatic Repulsion.* At pH 7, the triphosphate unit of ATP carries about four negative charges (Figure 15.4). These charges repel one another because they are in close proximity. The repulsion between them is reduced when ATP is hydrolyzed.

2. *Resonance Stabilization.* Orthophosphate (P_i), one of the products of ATP hydrolysis, has greater resonance stabilization than do any of the phosphates in ATP. Orthophosphate has a number of resonance forms of similar energy (**Figure 15.5**), whereas the γ phosphoryl group of ATP has a smaller number (**Figure 15.6**). Forms like that shown on the right in Figure 15.5 are unfavorable because a positively charged oxygen atom is adjacent to a positively charged phosphorus atom, an electrostatically unfavorable juxtaposition.

3. *Increase in Entropy.* The entropy of the products is greater, in that there are now two molecules instead of a single ATP molecule. We disregard the molecule of water used to hydrolyze the ATP because given the high concentration (55.5 M), there is effectively no change in the concentration of water during the reaction.

4. *Stabilization Due to Hydration.* Water binds to ADP and P_i, stabilizing these molecules and thereby rendering the reverse reaction, the synthesis of ATP, less favorable.

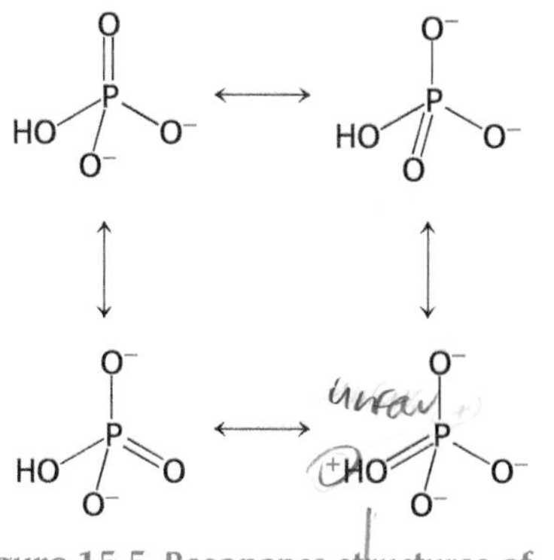

Figure 15.5 Resonance structures of orthophosphate.

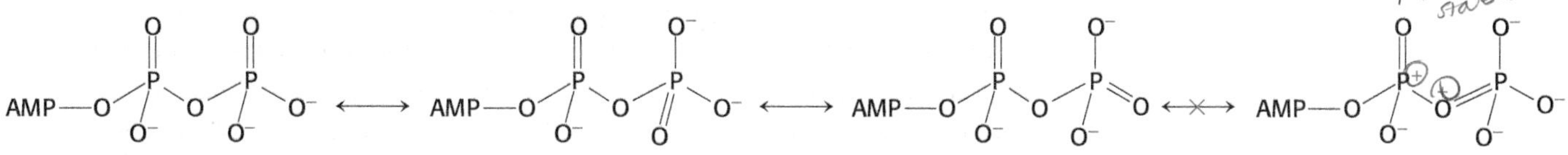

Figure 15.6 Improbable resonance structure. There are fewer resonance structures available to the γ-phosphate of ATP than to free orthophosphate.

ATP is often called a high-energy phosphate compound, and its phosphoan-hydride bonds are referred to as high-energy bonds. Indeed, a "squiggle" (~P) is often used to indicate such a bond. Nonetheless, there is nothing special about the bonds themselves. *They are high-energy bonds in the sense that much free energy is released when they are hydrolyzed*, for the reasons listed above.

Phosphoryl-Transfer Potential Is an Important Form of Cellular Energy Transformation

The standard free energies of hydrolysis provide a convenient means of comparing the phosphoryl-transfer potential of phosphorylated compounds. Such comparisons reveal that ATP is not the only compound with a high phosphoryl-transfer potential. In fact, some compounds in biological systems have a higher phosphoryl-transfer potential than that of ATP. These compounds include phosphoenolpyruvate (PEP), 1,3-bisphosphoglycerate (1,3-BPG), and creatine phosphate (Figure 15.7). Thus, PEP can transfer its phosphoryl group to ADP to form ATP. Indeed, this transfer is one of the ways in which ATP is generated in the breakdown of sugars (Chapter 16). Of significance is that ATP has a phosphoryl-transfer potential that is intermediate among the biologically important phosphorylated molecules (Table 15.1). *This intermediate position enables ATP to function efficiently as a carrier of phosphoryl groups.*

? QUICK QUIZ What properties of ATP make it an especially effective phosphoryl-transfer-potential compound?

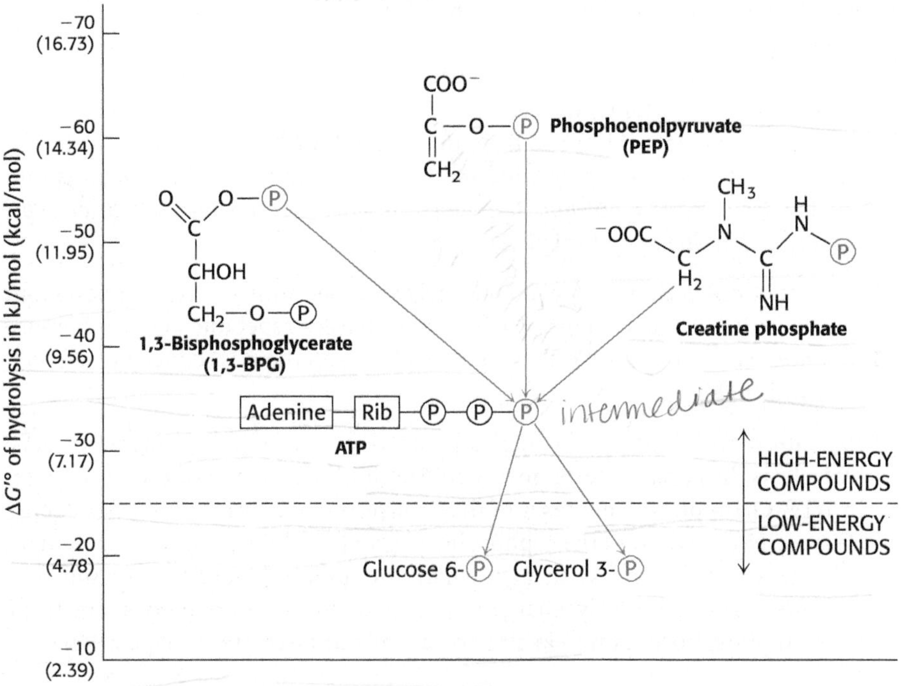

Figure 15.7 ATP has a central position in phosphoryl-transfer reactions. The role of ATP as the cellular energy currency is illustrated by its relation to other phosphorylated compounds. ATP has a phosphoryl-transfer potential that is intermediate among the biologically important phosphorylated molecules. High-phosphoryl-transfer-potential compounds derived from the metabolism of fuel molecules are used to power ATP synthesis. In turn, ATP donates a phosphoryl group to other biomolecules to facilitate their metabolism. [Data from D. L. Nelson and M. M. Cox, *Lehninger Principles of Biochemistry*, 5th ed. (W. H. Freeman and Company, 2009), Fig. 13–19.]

Table 15.1 Standard free energies of hydrolysis ($\Delta G^{\circ\prime}$) of some phosphorylated compounds

Compound	kJ mol^{-1}	kcal mol^{-1}
Phosphoenolpyruvate (PEP)	−61.9	−14.8
1,3-Bisphosphoglycerate (1,3-BPG)	−49.4	−11.8
Creatine phosphate	−43.1	−10.3
ATP (to ADP)	−30.5	−7.3
Glucose 1-phosphate	−20.9	−5.0
Pyrophosphate (PP$_i$)	−19.3	−4.6
Glucose 6-phosphate	−13.8	−3.3
Glycerol 3-phosphate	−9.2	−2.2

🩺 CLINICAL INSIGHT

Exercise Depends on Various Means of Generating ATP

At rest, muscle contains only enough ATP to sustain contractile activity for less than a second. Creatine phosphate, a high-phosphoryl-transfer-potential molecule in vertebrate muscle, serves as a reservoir of high-potential phosphoryl groups that can be readily transferred to ADP. This reaction is catalyzed by *creatine kinase*:

$$\text{Creatine phosphate} + \text{ADP} \xrightleftharpoons[]{\text{Creatine kinase}} \text{ATP} + \text{creatine}$$

In resting muscle, typical concentrations of these metabolites are [ATP] = 4 mM, [ADP] = 0.013 mM, [creatine phosphate] = 25 mM, and [creatine] = 13 mM. Because of its abundance and high phosphoryl-transfer potential relative to that of ATP (Table 15.1), creatine phosphate is a highly effective phosphoryl buffer. Indeed, creatine phosphate is the major source of phosphoryl groups for ATP regeneration for activities that require quick bursts of energy or short, intense sprints (**Figure 15.8**). The fact that creatine phosphate can replenish ATP pools is the basis of creatine's use as a dietary supplement by athletes in sports requiring short bursts of intense activity. After the creatine phosphate pool is depleted, ATP must be generated through metabolism (**Figure 15.9**).

Figure 15.8 Sprint to the finish. Creatine phosphate is an energy source for intense sprints. [David Stockman/AFP/Getty Images.]

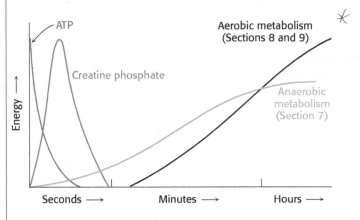

Figure 15.9 Sources of ATP during exercise. Exercise is initially powered by existing high–phosphoryl-transfer compounds (ATP and creatine phosphate). Subsequently, the ATP must be regenerated by metabolic pathways.

Phosphates Play a Prominent Role in Biochemical Processes

We have seen in Chapter 13 and in this chapter the prominence of phosphoryl group transfer from ATP to acceptor molecules. How is it that phosphate came to play such a prominent role in biology? Phosphate and its esters have several characteristics that render it useful for biochemical systems. First, phosphate esters have the important property of being *thermodynamically unstable* while being *kinetically stable.* Phosphate esters are thus molecules whose energy release can be manipulated by enzymes. Second, the stability of phosphate esters is due to the negative charges that make them resistant to hydrolysis in the absence of enzymes. This accounts for the presence of phosphate in the backbone of DNA. Third, because phosphate esters are so kinetically stable, they make ideal regulatory molecules, added to proteins by kinases and removed only by phosphatases. As we will see many times, phosphates are also frequently added to metabolites that might otherwise diffuse through the cell membrane.

No other ions have the chemical characteristics of phosphate. Citrate is not sufficiently charged to prevent hydrolysis. Arsenate forms esters that are unstable and susceptible to spontaneous hydrolysis. Indeed, arsenate is poisonous to cells because it can replace phosphate in reactions required for ATP synthesis, generating unstable compounds and preventing ATP synthesis. Silicate is more abundant than phosphate, but silicate salts are virtually insoluble, and in fact, are used for biomineralization. Only phosphate has the chemical properties to meet the needs of living systems.

✓ 5 Describe the relation between the oxidation state of a carbon molecule and its usefulness as a fuel.

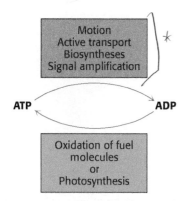

Figure 15.10 The ATP–ADP cycle. This cycle is the fundamental mode of energy exchange in biological systems.

15.4 The Oxidation of Carbon Fuels Is an Important Source of Cellular Energy

ATP serves as the principal *immediate donor of free energy* in biological systems rather than as a long-term storage form of free energy. In a typical cell, an ATP molecule is consumed within a minute of its formation. Although the total quantity of ATP in the body is limited to approximately 100 g, *the turnover of this small quantity of ATP is very high.* For example, a resting human being consumes about 40 kg (88 pounds) of ATP in 24 hours. During strenuous exertion, the rate of utilization of ATP may be as high as 0.5 kg (1.1 pounds) per minute. For a 2-hour run, 60 kg (132 pounds) of ATP is utilized. Clearly, having mechanisms for regenerating ATP is vital. Motion, active transport, signal amplification, and biosynthesis can take place only if ATP is continually regenerated from ADP (**Figure 15.10**). *The generation of ATP is one of the primary roles of catabolism.*

Carbon Oxidation Is Paired with a Reduction

The carbon in fuel molecules—such as glucose and fats—is oxidized to CO_2, and the energy released is used to regenerate ATP from ADP and P_i. As discussed earlier, the set of reactions that accomplish transformation are called catabolic reactions or, simply, catabolism. Understanding the transition from the energy inherent in reduced organic molecules—biological fuels—to ATP is a central theme in the study of biochemistry. Let's examine some of basic principles of carbon oxidation. We will return to this subject many times, most notably in Sections 8 through 10.

All oxidation reactions include the loss of electrons from the molecule being oxidized—carbon fuels in this case—and the gaining of those electrons by some other molecule, a process termed reduction. Such coupled reactions are called *oxidation–reduction* or *redox reactions.* In aerobic organisms, the ultimate electron acceptor in the oxidation of carbon is O_2 and the oxidation product is CO_2. Consequently, the more reduced a carbon is to begin with, the more free energy is released by its oxidation. **Figure 15.11** shows the $\Delta G^{\circ\prime}$ of oxidation for one-carbon compounds.

Although fuel molecules are more complex (**Figure 15.12**) than the single-carbon compounds depicted in Figure 15.11, when a fuel is oxidized, the oxidation takes place one carbon atom at a time. The carbon-oxidation energy is used in some cases to create a compound with high phosphoryl-transfer potential and in other cases to create an ion gradient (Section 9). In either case, the end point is the formation of ATP.

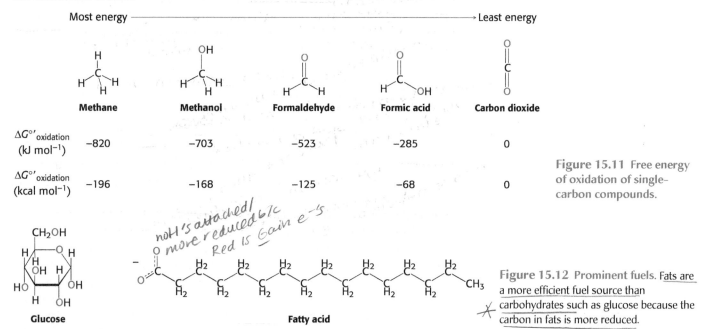

Most energy ——————————————————————→ Least energy

	Methane	Methanol	Formaldehyde	Formic acid	Carbon dioxide
$\Delta G^{\circ\prime}$ oxidation (kJ mol^{-1})	−820	−703	−523	−285	0
$\Delta G^{\circ\prime}$ oxidation (kcal mol^{-1})	−196	−168	−125	−68	0

Figure 15.11 Free energy of oxidation of single-carbon compounds.

Glucose   Fatty acid

Figure 15.12 Prominent fuels. Fats are a more efficient fuel source than carbohydrates such as glucose because the carbon in fats is more reduced.

Compounds with High Phosphoryl-Transfer Potential Can Couple Carbon Oxidation to ATP Synthesis

How is the energy released in the oxidation of a carbon compound converted into ATP? As an example, consider glyceraldehyde 3-phosphate, which is a metabolite of glucose formed in the oxidation of that sugar (Chapter 16). The C-1 carbon atom (shown in red) is at the aldehyde-oxidation level and is capable of further oxidation (Figure 15.11). Oxidation of the aldehyde to an acid will release energy.

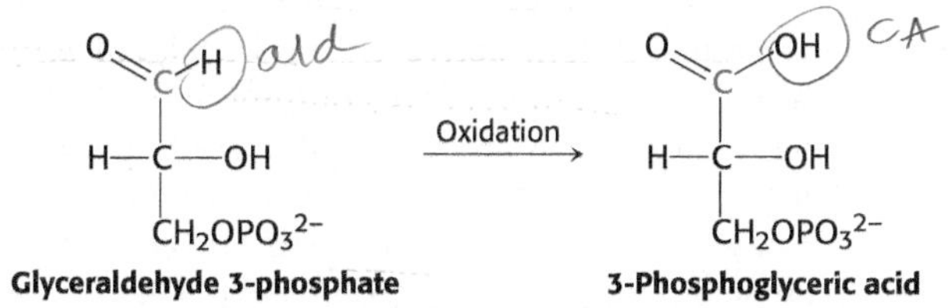

$$\text{Glyceraldehyde 3-phosphate} \xrightarrow{\text{Oxidation}} \text{3-Phosphoglyceric acid}$$

However, the oxidation does not take place directly. Instead, the carbon oxidation generates an acyl phosphate, 1,3-bisphosphoglycerate. The electrons released are captured by NAD$^+$, which we will consider shortly.

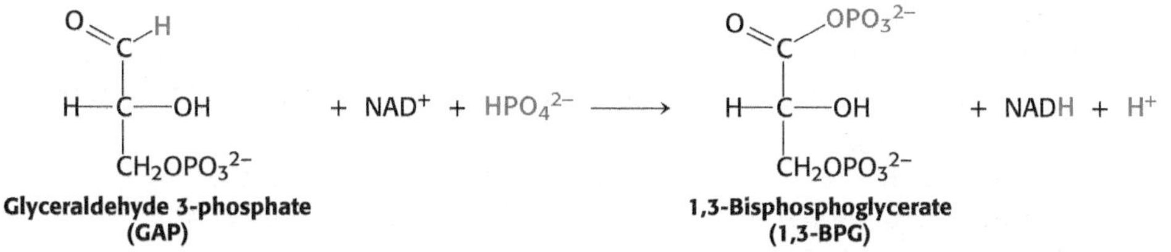

$$\text{Glyceraldehyde 3-phosphate (GAP)} + NAD^+ + HPO_4^{2-} \longrightarrow \text{1,3-Bisphosphoglycerate (1,3-BPG)} + NADH + H^+$$

For reasons similar to those discussed for ATP, 1,3-bisphosphoglycerate has a high phosphoryl-transfer potential. Thus, the cleavage of 1,3-BPG can be coupled to the synthesis of ATP.

$$
\text{1,3-Bisphosphoglycerate} \quad + \text{ADP} \longrightarrow \text{3-Phosphoglyceric acid} \quad + \text{ATP}
$$

The energy of oxidation is initially trapped as a high-phosphoryl-transfer-potential compound and then used to form ATP. The oxidation energy of a carbon atom is transformed into phosphoryl-transfer potential—first, as 1,3-bisphosphoglycerate and, ultimately, as ATP. We will consider these reactions in mechanistic detail in Chapter 16.

15.5 Metabolic Pathways Contain Many Recurring Motifs

At first glance, metabolism seems intimidating because of the sheer number of reactants and reactions. Nevertheless, there are unifying themes that make comprehending it more manageable. These themes include common metabolites and regulatory schemes that stem from a common evolutionary heritage.

Activated Carriers Exemplify the Modular Design and Economy of Metabolism

We have seen that phosphoryl transfer can be used to drive otherwise thermodynamically unfavorable reactions, alter the energy or conformation of a protein, or serve as a signal to alter the activity of a protein. The phosphoryl-group donor in all of these reactions is ATP. In other words, *ATP is an activated carrier of phosphoryl groups because phosphoryl transfer from ATP is an energetically favorable, or exergonic, process.* The use of *activated carriers* is a recurring motif in biochemistry, and we will consider several such carriers here. Many such activated carriers function as coenzymes—small organic molecules that serve as cofactors for enzymes (p. 99):

1. *Activated Carriers of Electrons for Fuel Oxidation.* In aerobic organisms, the ultimate electron acceptor in the oxidation of fuel molecules is O_2. However, electrons are not transferred directly to O_2. Instead, fuel molecules reduce or transfer electrons to special carriers, which are either *pyridine nucleotides* or *flavins.* The reduced forms of these carriers then transfer their high-potential electrons to O_2. In other words, these carriers have a higher affinity for electrons than do carbon fuels but a lower affinity for electrons than does O_2; consequently, the electrons flow from an unstable configuration (low affinity) to a stable one (high affinity). Indeed, the tendency of these electrons to flow toward a more stable arrangement accounts for their description as being activated. The energy that is released is converted into ATP by mechanisms to be discussed in Chapter 21.

Nicotinamide adenine dinucleotide (NAD^+), a pyridine nucleotide, is a major electron carrier in the oxidation of fuel molecules (**Figure 15.13**). The

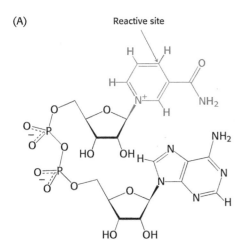

(A) Reactive site

(B)

$$
NAD^+ + 2\,H^+ + 2\,e^- \rightleftharpoons NADH + H^+
$$

Figure 15.13 A nicotinamide-derived electron carrier. (A) Nicotinamide adenine dinucleotide (NAD^+) is a prominent carrier of high-energy electrons derived from the vitamin niacin (nicotinamide) shown in red. (B) NAD^+ is reduced to NADH.

reactive part of NAD^+ is its nicotinamide ring, a pyridine derivative synthesized from the vitamin niacin. *In the oxidation of a substrate, the nicotinamide ring of NAD^+ accepts a hydrogen ion and two electrons, which are equivalent to a hydride ion ($H:^-$).* The reduced form of this carrier is *NADH*. In the oxidized form, the nitrogen atom carries a positive charge, as indicated by NAD^+. Nicotinamide adenine dinucleotide, or NAD^+, is the electron acceptor in many reactions of the following type:

pyridine

$$\underset{\substack{| \\ H}}{\overset{OH}{\underset{R}{\overset{|}{C}}}} {}_{R'} + NAD^+ \rightleftharpoons \overset{O}{\underset{R}{\overset{||}{C}}}{}_{R'} + NADH + H^+$$

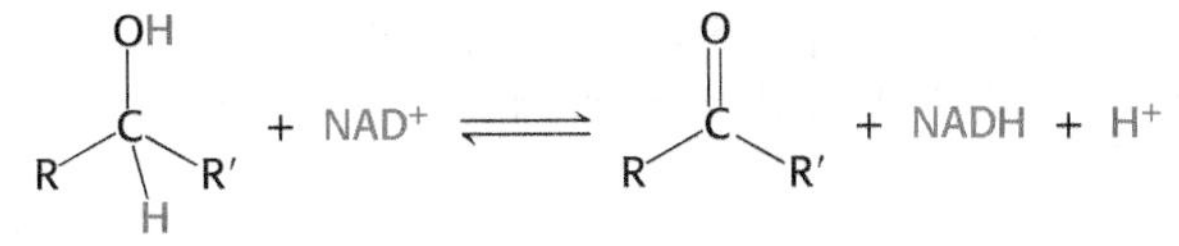

This redox reaction is often referred to as a dehydrogenation because protons accompany the electrons. One proton and two electrons of the substrate are directly transferred to NAD^+, whereas the other proton appears in the solvent as a proton.

The other major electron carrier in the oxidation of fuel molecules is the coenzyme *flavin adenine dinucleotide* (**Figure 15.14**). The abbreviations for the oxidized and reduced forms of this carrier are FAD and $FADH_2$, respectively. FAD is the electron acceptor in reactions of the following type:

flavin

$$\underset{\substack{| \ | \\ H \ H}}{\overset{H \ H}{\underset{R}{\overset{| \ |}{C-C}}}}{}_{R'} + FAD \rightleftharpoons \underset{\substack{| \ | \\ H \ H}}{\overset{R \quad R'}{C=C}} + FADH_2$$

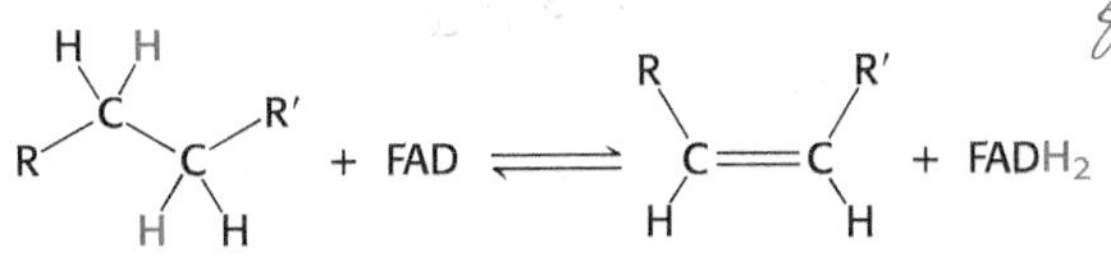

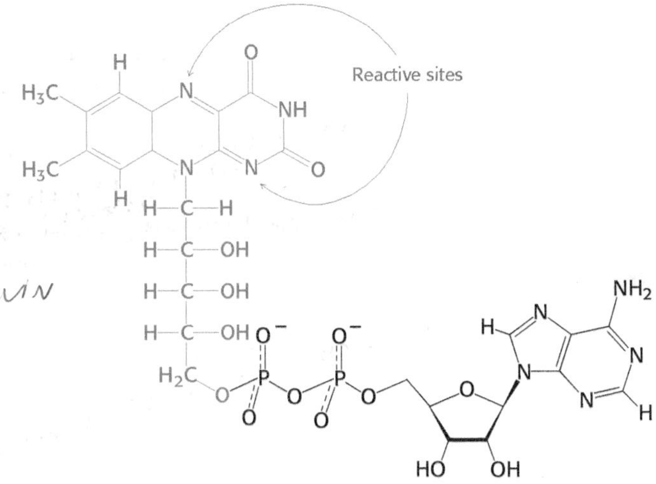

Figure 15.14 **The structure of the oxidized form of flavin adenine dinucleotide (FAD).** This electron carrier consists of the vitamin riboflavin (shown in blue) and an ADP unit (shown in black).

The reactive part of FAD is its isoalloxazine ring, a derivative of the vitamin riboflavin (**Figure 15.15**). FAD, like NAD^+, can accept two electrons. In doing so, FAD, unlike NAD^+, takes up two protons. These carriers of high-potential electrons as well as flavin mononucleotide (FMN), an electron carrier related to FAD, will be considered further in Chapters 20 and 21.

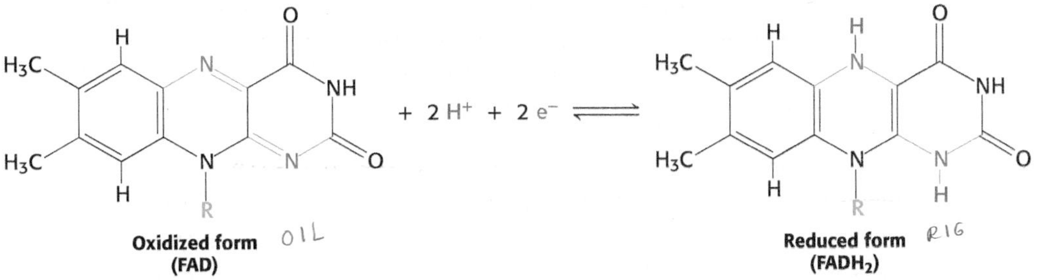

Figure 15.15 **Structures of the reactive parts of FAD and $FADH_2$.** The electrons and protons are carried by the isoalloxazine ring component of FAD and $FADH_2$.

2. *Activated Carriers of Electrons for the Synthesis of Biomolecules.* High-potential electrons are required for anabolic reactions. In most biosyntheses, the precursors are more oxidized than the products and, hence, reducing power is needed in addition to ATP. This process is called *reductive biosynthesis.* For example, in the biosynthesis of fatty acids, the keto group of a two-carbon unit

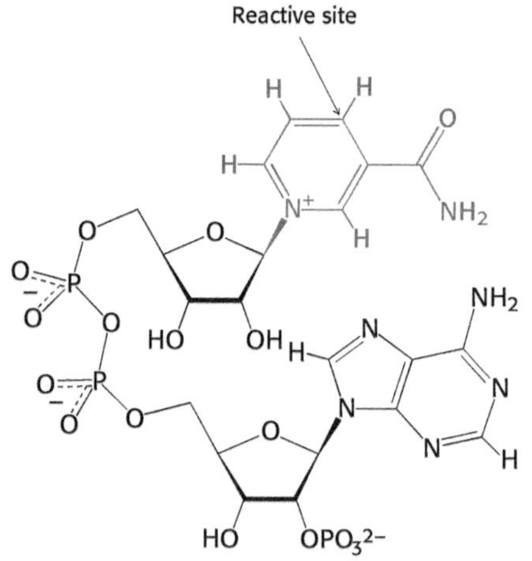

Figure 15.16 The structure of nicotinamide adenine dinucleotide phosphate (NADP$^+$). NADP$^+$ provides electrons for biosynthetic purposes. Notice that the reactive site is the same in NADP$^+$ and NAD$^+$.

is reduced to a methylene group in several steps. This sequence of reactions requires an input of four electrons:

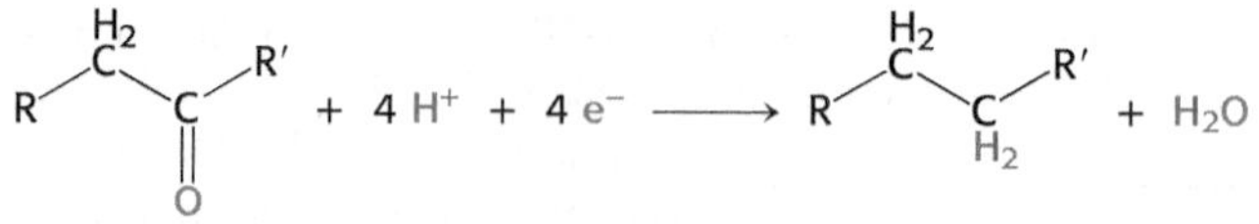

The electron donor in most reductive biosyntheses is NADPH, the reduced form of nicotinamide adenine dinucleotide phosphate (NADP$^+$). NADPH differs from NADH in that the 2′-hydroxyl group of its adenosine moiety is esterified with phosphate (**Figure 15.16**). NADPH carries electrons in the same way as NADH. However, *NADPH is used almost exclusively for reductive biosyntheses, whereas NADH is used primarily for the generation of ATP*. The extra phosphoryl group on NADPH is a tag that enables enzymes to distinguish between high-potential electrons to be used in anabolism and those to be used in catabolism.

3. *An Activated Carrier of Two-Carbon Fragments.* Coenzyme A (also called CoA-SH), another central molecule in metabolism, is a carrier of acyl groups (**Figure 15.17**). A key constituent of coenzyme A is the vitamin pantothenate. Acyl groups are important constituents both in catabolism, as in the oxidation of fatty acids, and in anabolism, as in the synthesis of membrane lipids. The terminal sulfhydryl group in CoA is the reactive site. Acyl groups are linked to the sulfhydryl group of CoA by thioester bonds. The resulting derivative is called an *acyl CoA*. An acyl group often linked to CoA is the acetyl unit; this derivative is called *acetyl CoA*. The $\Delta G°'$ for the hydrolysis of acetyl CoA has a large negative value:

$$\text{Acetyl CoA} + \text{H}_2\text{O} \rightleftharpoons \text{acetate} + \text{CoA} + \text{H}^+$$

$$\Delta G°' = -31.4 \text{ kJ mol}^{-1} (-7.5 \text{ kcal mol}^{-1})$$

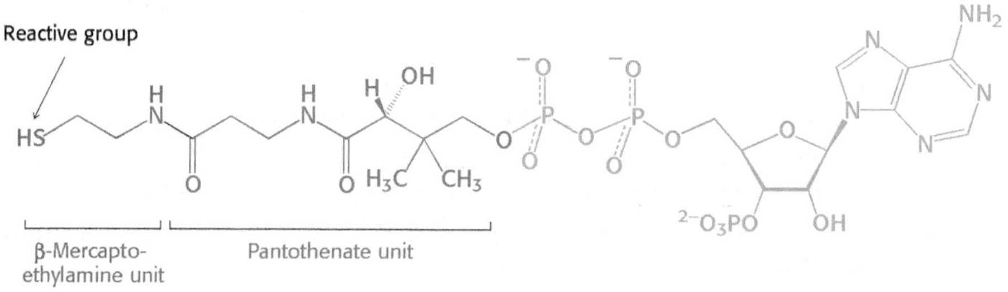

Figure 15.17 The structure of coenzyme A (CoA-SH).

β-Mercapto-ethylamine unit   Pantothenate unit

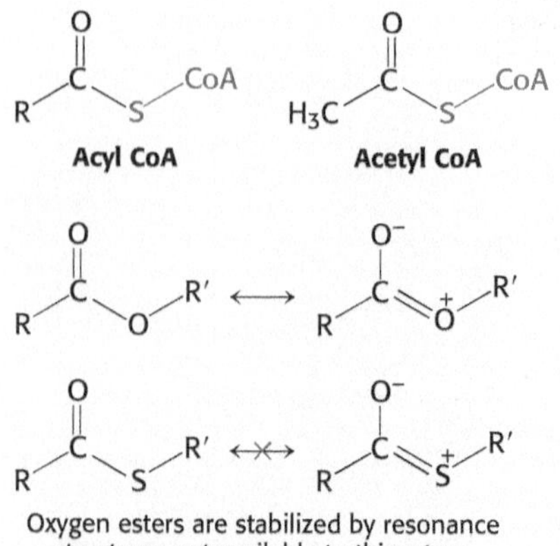

Oxygen esters are stabilized by resonance structures not available to thioesters.

The hydrolysis of a thioester is thermodynamically more favorable than that of an oxygen ester, such as those in fatty acids, because the electrons of the C=O bond form less stable resonance structures with the C—S bond than with the C—O bond. Consequently, *acetyl CoA has a high acetyl-group-transfer potential because transfer of the acetyl group is exergonic.* Acetyl CoA carries an activated acetyl group, just as ATP carries an activated phosphoryl group.

Additional features of activated carriers are responsible for two key aspects of metabolism. First, NADH, NADPH, and FADH$_2$ react slowly with O$_2$ in the absence of a catalyst. Likewise, ATP and acetyl CoA are hydrolyzed slowly (in times of many hours or even days) in the absence of a catalyst. These molecules are kinetically quite stable in the face of a large thermodynamic driving force for reaction with O$_2$ (in regard to the electron carriers) and H$_2$O (for ATP and acetyl CoA). *The kinetic stability of these molecules in the absence of specific catalysts is essential for their biological function because it enables enzymes to control the flow of free energy and reducing power.*

Second, *most interchanges of activated groups in metabolism are accomplished by a rather small set of carriers* (**Table 15.2**). The existence of a recurring set of activated carriers in all organisms is one of the unifying motifs of biochemistry.

Table 15.2 Some activated carriers in metabolism

Carrier molecule in activated form	Group carried	Vitamin precursor
ATP	Phosphoryl	
NADH and NADPH	Electrons	Nicotinate (niacin) (vitamin B_3)
$FADH_2$	Electrons	Riboflavin (vitamin B_2)
$FMNH_2$	Electrons	Riboflavin (vitamin B_2)
Coenzyme A	Acyl	Pantothenate (vitamin B_5)
Lipoamide	Acyl	
Thiamine pyrophosphate	Aldehyde	Thiamine (vitamin B_1)
Biotin	CO_2	Biotin (vitamin B_7)
Tetrahydrofolate	One-carbon units	Folate (vitamin B_9)
S-Adenosylmethionine	Methyl	
Uridine diphosphate glucose	Glucose	
Cytidine diphosphate diacylglycerol	Phosphatidate	
Nucleoside triphosphates	Nucleotides	

Note: Many of the activated carriers are coenzymes that are derived from water-soluble vitamins.

 CLINICAL INSIGHT

Lack of Activated Pantothenate Results in Neurological Problems

Pantothenate kinase associated degeneration, formerly called Hallervorden–Spatz syndrome, is a pathological condition characterized by neurodegeneration and iron accumulation in the brain. Specifically, a patient having this disorder displays abnormal postures and disrupted movement (dystonia), an inability to articulate words (dysarthria), involuntary writhing movements (choreathetosis), and muscle rigidity and resting tremors (parkinsonism). All patients with classic pantothenate kinase associated degeneration lack the enzyme pantothenate kinase.

An important regulatory enzyme in the biosynthetic pathway for coenzyme A, pantothenate kinase activates the vitamin pantothenate by phosphorylating it at the expense of ATP.

NUTRITION FACTS

Pantothenate Pantothenate is a component of coenzyme A and a cofactor required for fatty acid synthesis. It is plentiful in many foods, with egg yolk being a rich source. A deficiency of pantothenate alone has never been documented. [Photograph from FoodCollection/AgeFotostock.]

Pantothenate + ATP $\rightleftharpoons$ (Pantothenate kinase) 4′-Phosphopantothenate + ADP

In subsequent steps, pantothenate phosphate is converted into coenzyme A. Presumably, the symptoms of pantothenate kinase associated degeneration are predominately neurological because the nervous system is absolutely dependent on aerobic metabolism, and coenzyme A is an essential player in the aerobic metabolism of all fuels (as we will see in Sections 8 and 9).

Many Activated Carriers Are Derived from Vitamins

Almost all the activated carriers that act as coenzymes are derived from *vitamins*—organic molecules needed in small amounts in the diets of many higher animals. Table 15.3 lists the vitamins that act as coenzymes. The series of vitamins known as the vitamin B group is shown in Figure 15.18. Note that, in all cases, the

Table 15.3 The B vitamins

Vitamin	Coenzyme	Typical reaction type	Consequences of deficiency
Thiamine (B$_1$)	Thiamine pyrophosphate	Aldehyde transfer	Beriberi (weight loss, heart problems, neurological dysfunction)
Riboflavin (B$_2$)	Flavin adenine dinucleotide (FAD)	Oxidation–reduction	Cheliosis and angular stomatitis (lesions of the mouth), dermatitis
Pyridoxine (B$_6$)	Pyridoxal phosphate	Group transfer to or from amino acids	Depression, confusion, convulsions
Nicotinic acid (niacin, B$_3$)	Nicotinamide adenine dinucleotide (NAD$^+$)	Oxidation–reduction	Pellagra (dermatitis, depression, diarrhea)
Pantothenic acid (B$_5$)	Coenzyme A	Acyl-group transfer	Hypertension
Biotin (B$_7$)	Biotin–lysine adducts (biocytin)	ATP-dependent carboxylation and carboxyl-group transfer	Rash about the eyebrows, muscle pain, fatigue (rare)
Folic acid (B$_9$)	Tetrahydrofolate	Transfer of one-carbon components; thymine synthesis	Anemia, neural-tube defects in development
Cobalamin (B$_{12}$)	5′-Deoxyadenosyl cobalamin	Transfer of methyl groups; intramolecular rearrangements	Anemia, pernicious anemia, methylmalonic acidosis

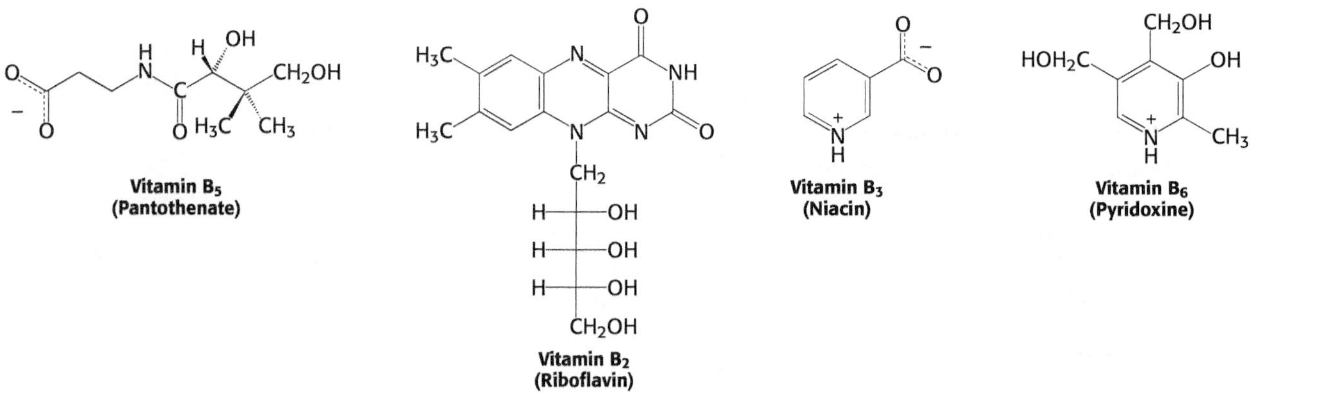

Figure 15.18 Structures of some of the B vitamins. These vitamins are often referred to as water-soluble vitamins because of the ease with which they dissolve in water.

vitamin must be modified before it can serve its function (Appendix D). We have already touched on the roles of niacin, riboflavin, and pantothenate. These three and the other B vitamins will appear many times in our study of biochemistry.

Vitamins serve the same important roles in nearly all forms of life, but higher animals lost the capacity to synthesize them in the course of evolution. For instance, whereas *E. coli* can thrive on glucose and organic salts, human beings require at least 12 vitamins in their diet. The biosynthetic pathways for vitamins can be complex; thus, it is biologically more efficient to ingest vitamins than to synthesize the enzymes required to construct them from simple molecules. This efficiency comes at the cost of dependence on other organisms for chemicals essential for life. Indeed, vitamin deficiency can generate diseases in all organisms requiring these molecules (Table 15.4; see also Table 15.3).

Not all vitamins function as coenzymes. Vitamins designated by the letters A, C, D, E, and K (Figure 15.19; Table 15.4) have a diverse array of functions. Vitamin A (retinol) is the precursor of retinal, the light-sensitive group in rhodopsin and other visual pigments, and retinoic acid, an important signaling molecule. A deficiency of this vitamin leads to night blindness. In addition, young animals require vitamin A for growth. Vitamin C, or ascorbate, acts as an antioxidant. A deficiency in vitamin C can lead to scurvy, a disease characterized

Table 15.4 Noncoenzyme vitamins

Vitamin	Function	Deficiency
A	Roles in vision, growth, reproduction	Night blindness, cornea damage, damage to respiratory and gastrointestinal tracts
C (ascorbate)	Antioxidant	Scurvy (swollen and bleeding gums, subdermal hemorrhaging)
D	Regulation of calcium and phosphate metabolism	Rickets (children): skeletal deformities, impaired growth Osteomalacia (adults): soft, bending bones
E	Antioxidant	Lesions in muscles and nerves (rare)
K	Blood coagulation	Subdermal hemorrhaging

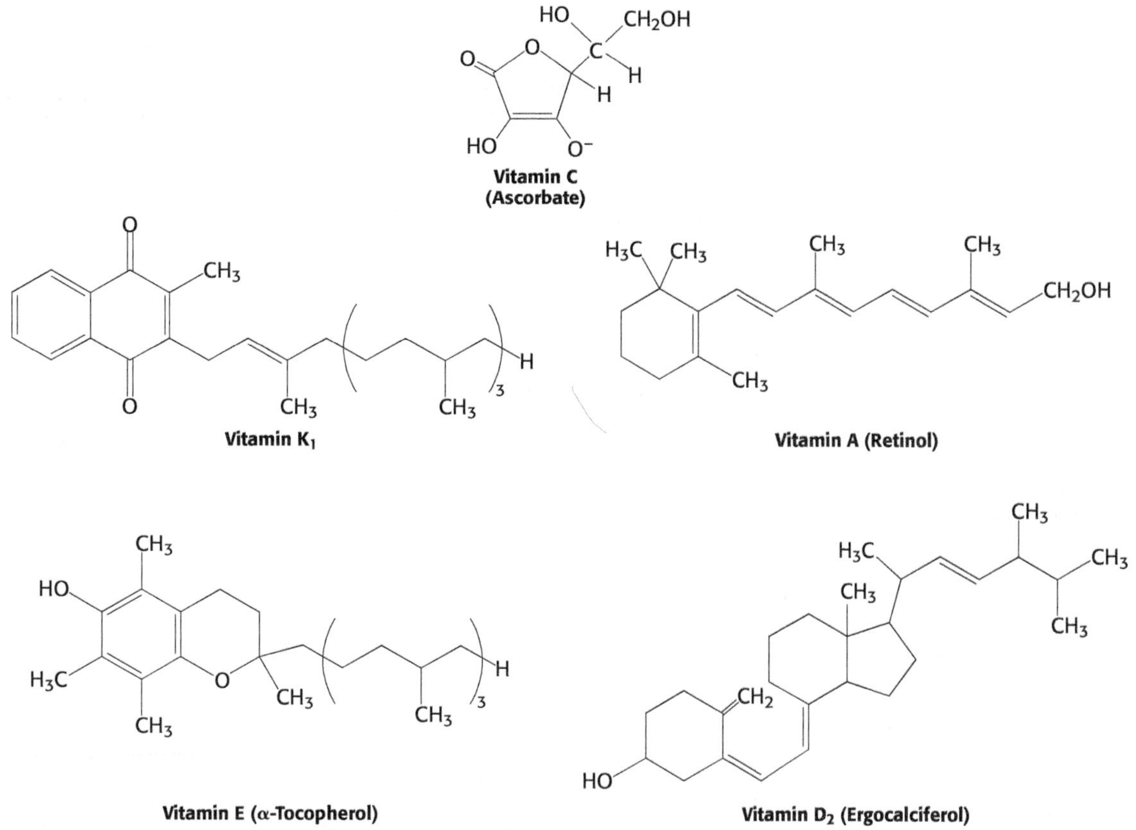

Figure 15.19 Structures of some vitamins that do not function as coenzymes. These vitamins, with the exception of vitamin C, are often called the fat-soluble vitamins because of their hydrophobic nature.

by skin lesions and blood-vessel fragility due to malformed collagen (p. 57). A metabolite of vitamin D is a hormone that regulates the metabolism of calcium and phosphorus. A deficiency in vitamin D impairs bone formation in growing animals. Vitamin E (α-tocopherol) deficiency causes a variety of neuromuscular pathologies. This vitamin reacts with reactive oxygen species, such as hydroxyl radicals, and inactivates them before they can oxidize unsaturated membrane lipids, damaging cell structures. Vitamin K is required for normal blood clotting.

15.6 Metabolic Processes Are Regulated in Three Principal Ways

It is evident that the complex network of metabolic reactions must be rigorously regulated. The levels of available nutrients must be monitored and the activity of metabolic pathways must be altered and integrated to create *homeostasis*, a stable biochemical environment. At the same time, metabolic control must be flexible, able to adjust metabolic activity to the changing external and internal environments of cells. **Figure 15.20** illustrates the nutrient pools and their connections that must be monitored and regulated. *Metabolism is regulated through control of (1) the amounts of enzymes, (2) their catalytic activities, and (3) the accessibility of substrates.*

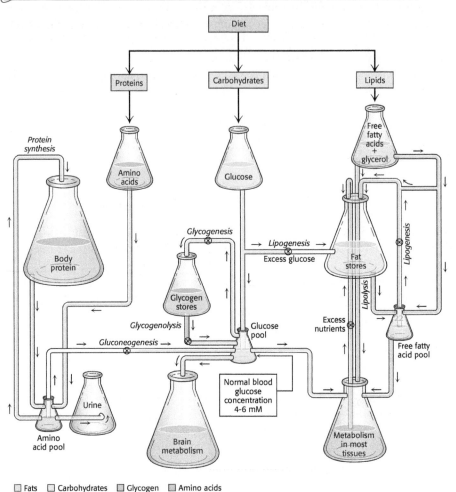

Figure 15.20 Homeostasis. Maintaining a constant cellular environment requires complex metabolic regulation that coordinates the use of nutrient pools.

☐ Fats   ☐ Carbohydrates   ☐ Glycogen   ▉ Amino acids

The Amounts of Enzymes Are Controlled

The amount of a particular enzyme depends on both its rate of synthesis and its rate of degradation. The level of many enzymes is adjusted primarily by a change in the *rate of transcription* of the genes encoding them (Section 15). In *E. coli*, for example, the presence of lactose induces within minutes a more than 50-fold increase in the rate of synthesis of β-galactosidase, an enzyme required for the breakdown of this disaccharide.

Catalytic Activity Is Regulated

The catalytic activity of enzymes is controlled in several ways. *Allosteric control* is especially important. For example, the first reaction in many biosynthetic pathways is allosterically inhibited by the ultimate product of the pathway, an example of *feedback inhibition*. This type of control can be almost instantaneous. Another recurring mechanism is the activation and deactivation of enzymes by *reversible covalent modification*. Reversible modification is often the end point of the signal-transduction cascades discussed in Chapter 13. For example, glycogen phosphorylase, the enzyme catalyzing the breakdown of glycogen, a storage form of glucose, is activated by the phosphorylation of a particular serine residue when glucose is scarce.

Hormones coordinate metabolic relations between different tissues, often by regulating the reversible modification of key enzymes. For instance, the hormone epinephrine triggers a signal-transduction cascade in muscle, resulting in the phosphorylation and activation of key enzymes and leading to the rapid degradation of glycogen to glucose, which is then used to supply ATP for muscle contraction. Glucagon has the same effect in liver, but the glucose is released into the blood for other tissues to use. As described in Chapter 13, many hormones act through intracellular messengers, such as cyclic AMP (cAMP) and calcium ion, which coordinate the activities of many target proteins.

Many reactions in metabolism are controlled by the *energy status* of the cell. One index of the energy status is the *energy charge*, which is the fraction of all of the adenine nucleotide molecules in the form of ATP plus half the fraction of adenine nucleotides in the form of ADP, given that ATP contains two phosphoanhydride linkages, whereas ADP contains one. Hence, the energy charge is defined as

$$\text{Energy charge} = \frac{[\text{ATP}] + 1/2[\text{ADP}]}{[\text{ATP}] + [\text{ADP}] + [\text{AMP}]}$$

The energy charge can have a value ranging from 0 (all AMP) to 1 (all ATP). A high-energy charge inhibits ATP-generating (catabolic) pathways because the cell has sufficient levels of ATP. ATP-utilizing (anabolic) pathways, on the other hand, are stimulated by a high-energy charge because ATP is available. In plots of the reaction rates of such pathways versus the energy charge, the curves are steep near an energy charge of 0.9, where they usually intersect (Figure 15.21). Evidently, the control of these pathways has evolved to maintain the energy charge within rather narrow limits. In other words, *the energy charge, like the pH of a cell, is buffered.* The energy charge of most cells ranges from 0.80 to 0.95.

An alternative index of the energy status is the *phosphorylation potential*, which is directly related to the free-energy storage available in the form of ATP. The phosphorylation potential is defined as

$$\text{Phosphorylation potential} = \frac{[\text{ATP}]}{[\text{ADP}] + [\text{P}_i]}$$

The phosphorylation potential, in contrast with the energy charge, depends on the concentration of inorganic orthophosphate (P_i).

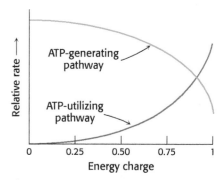

Figure 15.21 Energy charge regulates metabolism. High concentrations of ATP inhibit the relative rates of a typical ATP-generating (catabolic) pathway and stimulate the typical ATP-utilizing (anabolic) pathway.

The Accessibility of Substrates Is Regulated

Controlling the *availability of substrates* is another means of regulating metabolism in all organisms. For instance, glucose breakdown can take place in many cells only if insulin is present to promote the entry of glucose into the cell. In eukaryotes, metabolic regulation and flexibility are enhanced by compartmentalization. The transfer of substrates from one compartment of a cell to another can serve as a control point. For example, fatty acid oxidation takes place in mitochondria, whereas fatty acid synthesis takes place in the cytoplasm. *Compartmentalization segregates opposed reactions.*

SUMMARY

15.1 Energy Is Required to Meet Three Fundamental Needs

All organisms, from simple bacteria to humans, require energy for three major purposes: (1) the mechanical work of movement, (2) the active transport of molecules across membranes, and (3) the biosynthesis of biomolecules and, ultimately, the synthesis of new cells.

15.2 Metabolism Is Composed of Many Interconnecting Reactions

The process of energy transduction takes place through metabolism, a highly integrated network of chemical reactions. Metabolism can be subdivided into catabolism (reactions employed to extract energy from fuels) and anabolism (reactions that use this energy for biosynthesis). The most valuable thermodynamic concept for understanding bioenergetics is free energy. A reaction can take place spontaneously only if the change in free energy (ΔG) is negative. A thermodynamically unfavorable reaction can be driven by a thermodynamically favorable one, which is the hydrolysis of ATP in many cases.

15.3 ATP Is the Universal Currency of Free Energy

The energy derived from catabolism is transformed into adenosine triphosphate. ATP hydrolysis is exergonic, and the energy released can be used to power cellular processes, including motion, active transport, and biosynthesis. Under cellular conditions, the hydrolysis of ATP shifts the equilibrium of a coupled reaction by a factor of 10^8. ATP, the universal currency of energy in biological systems, is an energy-rich molecule because it contains two phosphoanhydride linkages.

15.4 The Oxidation of Carbon Fuels Is an Important Source of Cellular Energy

ATP formation is coupled to the oxidation of carbon fuels. Electrons are removed from carbon atoms and passed to O_2 in a series of oxidation–reduction reactions. Electrons flow down a stability gradient, a process that releases energy. The energy of carbon oxidation can be trapped as high-phosphoryl-transfer-potential compounds, which can then be used to power the synthesis of ATP.

15.5 Metabolic Pathways Contain Many Recurring Motifs

Metabolism is characterized by common motifs. A small number of recurring activated carriers, such as ATP, NADH, and acetyl CoA, transfer activated groups in many metabolic pathways. NADPH, which carries two electrons at a high potential, provides reducing power in the biosynthesis of cell components from more oxidized precursors. Many activated carriers are derived from vitamins, small organic molecules required in the diets of many animals.

15.6 Metabolic Processes Are Regulated in Three Principal Ways

Metabolism is regulated in a variety of ways, the principal ways among them being as follows. First, the amounts of some enzymes are controlled by regulation of the rate of synthesis and degradation. Second, the catalytic activities of many enzymes are regulated by allosteric interactions (as in feedback inhibition) and by covalent modification. Third, the movement of many substrates into cells and subcellular compartments also is controlled.

The energy charge, which depends on the relative amounts of ATP, ADP, and AMP, plays a role in metabolic regulation. A high-energy charge inhibits ATP-generating (catabolic) pathways, whereas it stimulates ATP-utilizing (anabolic) pathways.

KEY TERMS

phototroph (p. 258)
chemotroph (p. 258)
metabolism (p. 258)
intermediary metabolism (p. 258)
catabolism (p. 259)

anabolism (p. 259)
amphibolic pathway (p. 259)
adenosine triphosphate (ATP) (p. 260)
phosphoryl-transfer potential (p. 263)
activated carrier (p. 268)

vitamin (p. 271)
energy charge (p. 275)
phosphorylation potential (p. 275)

? Answer to QUICK QUIZ

Charge repulsion is reduced when a phosphoryl group is removed from ATP. The products of ATP hydrolysis have more resonance forms than does ATP. Entropy is increased when one molecule of ATP is converted into two product molecules, ADP and P_i. The products of ATP hydrolysis are more effectively stabilized by association with water than is ATP.

PROBLEMS

1. *Complex patterns.* What is meant by intermediary metabolism?

2. *Opposites.* Differentiate between anabolism and catabolism.

3. *Graffiti.* While walking to biochemistry class with a friend, you see the following graffiti spray painted on the wall of the science building: "When a system is in equilibrium, the Gibbs free energy is maximum." You are disgusted, not only at the vandalism, but at the ignorance of the vandal. Your friend asks you to explain. How do you respond?

4. *Why bother to eat?* What are the three primary uses for cellular energy?

5. *Like Antony and Cleopatra.* Match the terms in the two columns.

(a) Cellular energy currency _____
(b) Anabolic electron carrier _____
(c) Phototroph _____
(d) Catabolic electron carrier _____
(e) Oxidation–reduction reaction _____
(f) Activated carrier of two-carbon fragments _____
(g) Vitamin _____
(h) Anabolism _____
(i) Amphibolic reaction _____
(j) Catabolism _____

1. NAD^+
2. Coenzyme A
3. Precursor to coenzymes
4. Yields energy
5. Requires energy
6. ATP
7. Transfers electrons
8. $NADP^+$
9. Converts light energy into chemical energy
10. Used in anabolism and catabolism

6. *Energy to burn.* What factors account for the high phosphoryl-transfer potential of nucleoside triphosphates? ✓ 3

7. *Currency matters.* Why does it make good sense to have a single nucleotide, ATP, function as the cellular energy currency? ✓ 3

8. *Close, personal friends.* Why is ATP usually associated with magnesium or manganese ions? ✓ 3

9. *Environmental conditions.* The standard free energy of hydrolysis for ATP is -30.5 kJ mol^{-1} (-7.3 kcal mol^{-1}):

What conditions might be changed to alter the free energy of hydrolysis? ✓ 3

10. *Energy flow.* What is the direction of each of the following reactions when the reactants are initially present in equimolar amounts? Use the data given in Table 15.1. ✓ 4

(a) $ATP + H_2O \rightleftharpoons ADP + P_i$

(b) $ATP + glycerol \rightleftharpoons glycerol\ 3-phosphate + ADP$

(c) $ATP + pyruvate \rightleftharpoons phosphoenolpyruvate + ADP$

(d) $ATP + glucose \rightleftharpoons glucose\ 6-phosphate + ADP$

11. *A proper inference.* What information do the $\Delta G^{\circ\prime}$ data given in Table 15.1 provide about the relative rates of hydrolysis of pyrophosphate and acetyl phosphate?

12. *A potent donor.* Consider the following reaction. ✓ 4

$$ATP + pyruvate \rightleftharpoons phosphoenolpyruvate + ADP$$

(a) Calculate $\Delta G^{\circ\prime}$ and K'_{eq} at 25°C for this reaction by using the data given in Table 15.1.
(b) What is the equilibrium ratio of pyruvate to phosphoenolpyruvate if the ratio of ATP to ADP is 10?

13. *Isomeric equilibrium.* Calculate $\Delta G^{\circ\prime}$ for the isomerization of glucose 6-phosphate to glucose 1-phosphate. What is the equilibrium ratio of glucose 6-phosphate to glucose 1-phosphate at 25°C?

14. *Activated acetate.* The formation of acetyl CoA from acetate is an ATP-driven reaction:

$$Acetate + ATP + CoA \rightleftharpoons acetyl\ CoA + AMP + PP_i$$

(a) Calculate $\Delta G^{\circ\prime}$ for this reaction by using data given in this chapter.
(b) The PP_i formed in the preceding reaction is rapidly hydrolyzed in vivo because of the ubiquity of inorganic pyrophosphatase. The $\Delta G^{\circ\prime}$ for the hydrolysis of PP_i is -19.2 kJ mol^{-1} (-4.6 kcal mol^{-1}). Calculate the $\Delta G^{\circ\prime}$ for the overall reaction, including pyrophosphate hydrolysis. What effect does the hydrolysis of PP_i have on the formation of acetyl CoA?

15. *Brute force?* Metabolic pathways frequently contain reactions with positive standard free-energy values, yet the reactions still take place. How is it possible? ✓ 4

16. *Recurring motif.* What is the structural feature common to ATP, FAD, NAD^+, and CoA? ✓ 5

17. *Outsourcing.* Outsourcing, a common business practice, is contracting with another business to perform a particular

function. Higher organisms were the original outsourcers, frequently depending on lower organisms to perform key biochemical functions. Give an example from this chapter of biochemical outsourcing.

18. *High-energy electrons.* What are the activated electron carriers for catabolism? For anabolism?

19. *Less reverberation.* Thioesters, common in biochemistry, are more unstable (energy rich) than oxygen esters. Explain.

20. *Staying in control.* What are the three principal means of controlling metabolic reactions?

21. *Running downhill.* Glycolysis is a series of 10 linked reactions that convert one molecule of glucose into two molecules of pyruvate with the concomitant synthesis of two molecules of ATP (Chapter 16). The $\Delta G^{\circ\prime}$ for this set of reactions is -35.6 kJ mol^{-1} (-8.5 kcal mol^{-1}), whereas the ΔG° is -90 kJ mol^{-1} (-22 kcal mol^{-1}). Explain why the free-energy release is so much greater under intracellular conditions than under standard conditions.

22. *Oxidation matters.* Examine the pairs of molecules, and identify the more reduced molecule in each pair. ✓ 5

Chapter Integration Problems

23. *Breakdown products.* Digestion is the first stage in the extraction of energy from food, but no useful energy is acquired in this stage. Why is digestion considered a stage in energy extraction?

24. *Kinetic versus thermodynamic.* The reaction of NADH with oxygen to produce NAD$^+$ and H$_2$O is very exergonic, yet the reaction of NADH and oxygen takes place very slowly. Why does a thermodynamically favorable reaction not take place rapidly?

Chapter Integration and Challenge Problems

25. *Acid strength.* The pK_a of an acid is a measure of its proton-group-transfer potential.

(a) Derive a relation between $\Delta G^{\circ\prime}$ and pK_a.
(b) What is the $\Delta G^{\circ\prime}$ for the ionization of acetic acid, which has a pK_a of 4.8?

26. *Activated sulfate.* Fibrinogen, a precursor to the blood-clot protein fibrin, contains tyrosine-*O*-sulfate. Propose an activated form of sulfate that could react in vivo with the aromatic hydroxyl group of a tyrosine residue in a protein to form tyrosine-*O*-sulfate.

Data Interpretation and Challenge Problem

27. *Opposites attract.* The graph below shows how the ΔG for the hydrolysis of ATP varies as a function of the Mg^{2+} concentration (pMg $= -\log[\text{Mg}^{2+}]$). ✓ 3

(a) How does decreasing $[\text{Mg}^{2+}]$ affect the ΔG of hydrolysis for ATP?
(b) Explain this effect.

Challenge Problems

28. *Raison d'être.* The muscles of some invertebrates are rich in *arginine phosphate* (phosphoarginine). Propose a function for this amino acid derivative.

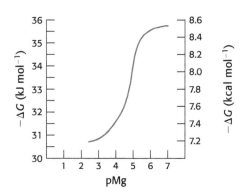

Arginine phosphate

29. *Ergogenic help or hindrance?* Creatine is a popular, but untested, dietary supplement.

(a) What is the biochemical rationale for the use of creatine?
(b) What type of exercise would most benefit from creatine supplementation?

30. *Standard conditions versus real life 1.* The enzyme aldolase catalyzes the following reaction in the glycolytic pathway:

$$\text{Fructose 1, 6-bisphosphate} \overset{\text{Aldolase}}{\rightleftharpoons}$$
dihydroxyacetone phosphate + glyceraldehyde 3-phosphate

The $\Delta G^{\circ\prime}$ for the reaction is $+23.8$ kJ mol^{-1} ($+5.7$ kcal mol^{-1}), whereas the ΔG° in the cell is -1.3 kJ mol^{-1} (-0.3 kcal mol^{-1}). Calculate the ratio of reactants to products under equilibrium and intracellular conditions. Using your results, explain how the reaction can be endergonic under standard conditions and exergonic under intracellular conditions.

31. *Standard conditions versus real life 2.* On page 262, a reaction, $A \rightleftharpoons B$, with a $\Delta G^{\circ\prime} = +16.7$ kJ mol^{-1} ($+4.0$ kcal mol^{-1}) is shown to have a K_{eq} of 1.15×10^{-3}. The K_{eq} is increased to 2.67×10^2 if the reaction is coupled to ATP hydrolysis under standard conditions. The ATP-generating system of cells maintains the [ATP]/[ADP][P$_i$] ratio at a high level, typically of the order of 500 M^{-1}. Calculate the ratio of B/A under cellular conditions. ✓ 4

32. *Not all alike.* The concentrations of ATP, ADP, and P$_i$ differ with cell type. Consequently, the release of free energy with the hydrolysis of ATP will vary with cell type. Use the following table to calculate the ΔG° for the hydrolysis of ATP in liver, muscle, and brain cells. In which cell type is the free energy of ATP hydrolysis most negative? ✓ 4

	ATP (mM)	ADP (mM)	P$_i$ (mM)
Liver	3.5	1.8	5.0
Muscle	8.0	0.9	8.0
Brain	2.6	0.7	2.7

Selected Readings for this chapter can be found online at www.whfreeman.com/tymoczko3e.

Glycolysis and Gluconeogenesis

CHAPTER 16
Glycolysis

CHAPTER 17
Gluconeogenesis

Wegin our study of metabolism by focusing on the processing of glucose, a fundamental fuel molecule for virtually all life forms. The first metabolic pathway that we encounter is glycolysis, an ancient pathway employed by a host of organisms. *Glycolysis is the sequence of reactions that converts one molecule of glucose into two molecules of pyruvate while generating ATP*. Glycolysis serves two major functions in the cell. First, this set of reactions generates ATP. Indeed, some tissues, such as the brain and red blood cells, rely solely on glucose as a fuel; consequently, glycolysis is especially important in these tissues. The second major function of glycolysis is to provide building blocks for biosynthesis. For instance, the molecules formed during glycolysis are used as precursors for amino acid and fatty acid synthesis.

Because glucose is such an important fuel, the end products of many biochemical pathways are salvaged to synthesize glucose in the process of *gluconeogenesis*. Gluconeogenesis is vital for ensuring that the brain and red blood cells have adequate supplies of glucose during a fast, such as a night's sleep. Although glycolysis and gluconeogenesis have some enzymes in common, the two pathways are not simply the reverse of each other. In particular, the highly exergonic, irreversible steps of glycolysis are bypassed in gluconeogenesis with reactions that render gluconeogenesis exergonic under cellular condition. The two pathways are reciprocally regulated so that glycolysis and gluconeogenesis do not take place in the same cell at the same time to a significant extent, thereby preventing the waste in energy that would result if glucose were being broken down at the same time as it is being synthesized.

CHAPTER 16

Glycolysis

Usain Bolt sprints to a world record in the 200-meter finals at the 2008 Beijing Olympics. Glucose metabolism can generate the ATP to power muscle contraction. During a sprint, when the ATP needs outpace oxygen delivery, as would be the case for Bolt, glucose is metabolized to lactate. When oxygen delivery is adequate, glucose is metabolized more efficiently to carbon dioxide and water. [Christophe Karaba/epa/Corbis.]

Earlier, we examined how complex carbohydrates are digested into biochemically useful molecules, such as glucose (Chapter 14). Glucose is the principal carbohydrate in living systems and an important fuel. In mammals, it is the only fuel that the brain uses under nonstarvation conditions and the only fuel that red blood cells can use at all. Indeed, almost all organisms use glucose, and most process it in a similar fashion. Why is glucose such a prominent fuel, rather than some other monosaccharide? We can speculate on the reasons. First, glucose is one of several monosaccharides formed from formaldehyde under prebiotic conditions, and so it may have been available as a fuel source for primitive biochemical systems. Second, glucose is the most stable hexose because the hydroxyl groups and the hydroxymethyl group are all in the equatorial position, minimizing steric clashes (p. 171). Third, glucose has a low tendency, relative to other monosaccharides, to nonenzymatically glycosylate proteins. In their open-chain forms, monosaccharides contain carbonyl groups that can covalently modify the amino groups of proteins. Such nonspecifically modified proteins often do not function effectively (p. 172). Glucose has a strong tendency to exist in the ring formation and, consequently, relatively little tendency to modify proteins.

Glucose

We start this section with glycolysis, paying special attention to the regulation of this pathway. We proceed to gluconeogenesis, again with a focus on regulation. The section ends with a discussion of the regulation of glycolysis and gluconeogenesis within a cell as well as between tissues.

In this chapter, we first examine how ATP is generated in glycolysis and how ATP can be generated in the absence of oxygen. We then see how sugars other than glucose are converted into glycolytic intermediates. The chapter ends with a discussion of the regulation of glycolysis.

✓ 1 Describe how ATP is generated in glycolysis.

16.1 Glycolysis Is an Energy-Conversion Pathway

We now begin our consideration of the glycolytic pathway. This pathway is common to virtually all cells, both prokaryotic and eukaryotic. In eukaryotic cells, glycolysis takes place in the cytoplasm. Glucose is converted into two molecules of pyruvate with the concomitant generation of two molecules of ATP.

Glycolysis can be thought of as comprising two stages (**Figure 16.1**). Stage 1 is the trapping and preparation phase. No ATP is generated in this stage. Stage 1 begins with the conversion of glucose into fructose 1,6-bisphosphate, which consists of three steps: a phosphorylation, an isomerization, and a second phosphorylation reaction. *The strategy of these initial steps in glycolysis is to trap the glucose in the cell and form a compound that can be readily cleaved into phosphorylated three-carbon units.* Stage 1 is completed with the cleavage of the fructose 1,6-bisphosphate into two phosphorylated three-carbon fragments. These resulting three-carbon units are readily interconvertible. In stage 2, ATP is harvested when the three-carbon fragments are oxidized to pyruvate.

Hexokinase Traps Glucose in the Cell and Begins Glycolysis

Glucose enters cells through specific transport proteins (p. 303) and has one principal fate inside the cell: *it is phosphorylated by ATP to form glucose 6-phosphate.* This step is notable for two reasons: (1) glucose 6-phosphate cannot pass through the membrane to the extracellular side, because it is not a substrate for the glucose transporters, and (2) the addition of the phosphoryl group facilitates the metabolism of glucose to phosphorylated three-carbon compounds with high phosphoryl-transfer potential. The transfer of the phosphoryl group from ATP to the hydroxyl group on carbon 6 of glucose is catalyzed by *hexokinase:*

Glucose   + ATP →(Hexokinase)→ **Glucose 6-phosphate (G-6P)** + ADP + H$^+$

Phosphoryl transfer is a fundamental reaction in biochemistry. *Kinases are enzymes that catalyze the transfer of a phosphoryl group from ATP to an acceptor.* Hexokinase, then, catalyzes the transfer of a phosphoryl group from ATP to a variety of six-carbon sugars *(hexoses)*, such as glucose and mannose. *Hexokinase, as well as all other kinases, requires Mg^{2+} (or another divalent metal ion such as Mn^{2+}) for activity.* The divalent metal ion forms a complex with ATP.

X-ray crystallographic studies of yeast hexokinase revealed that the binding of glucose induces a large conformational change in the enzyme. Hexokinase consists of two lobes, which move toward each other when glucose is bound (**Figure 16.2**). The cleft between the lobes closes, and the bound glucose becomes surrounded by protein, except for the carbon atom that will accept the phosphoryl

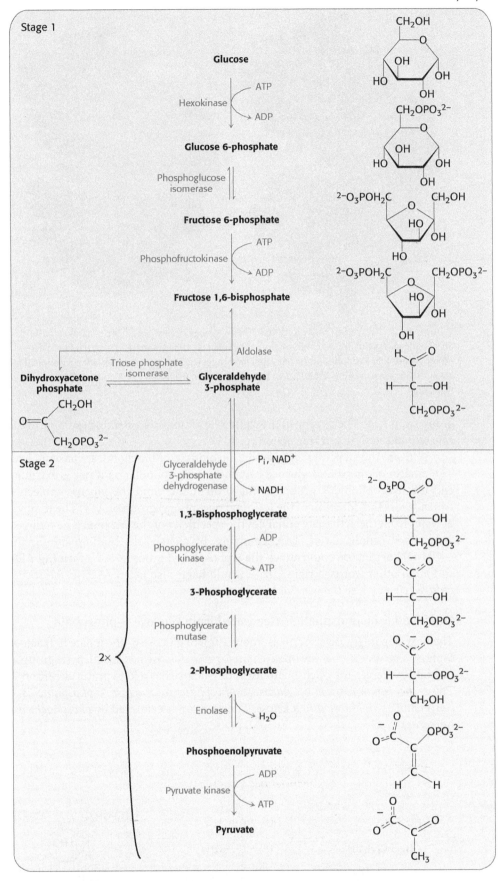

Figure 16.1 Stages of glycolysis. The glycolytic pathway can be divided into two stages: (1) glucose is trapped, destabilized, and cleaved into two interconvertible three-carbon molecules, generated by the cleavage of six-carbon fructose; and (2) the three-carbon units are oxidized to pyruvate, generating ATP.

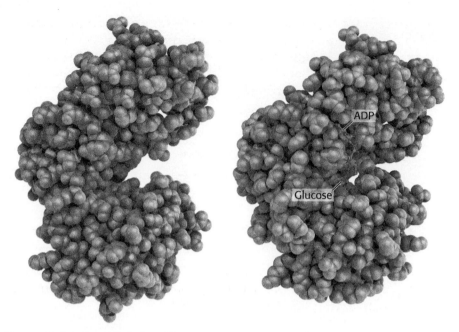

Figure 16.2 Induced fit in hexokinase. The two lobes of hexokinase are separated in the absence of glucose (left). The conformation of hexokinase changes markedly on binding glucose (right). Notice that two lobes of the enzyme come together, creating the necessary environment for catalysis. [Information from RSCB Protein Data Bank; drawn from PDB 2YHX and 1HKG by Adam Steinberg.]

group from ATP. The closing of the cleft in hexokinase is a striking example of the role of *induced fit* in enzyme action (p. 105).

Why are the structural changes in hexokinase of biochemical consequence? The environment around glucose becomes more nonpolar as water is extruded and the hydrophobic R groups of the protein surround the glucose molecule, which favors the donation of the terminal phosphoryl group of ATP. The removal of water from the active site enhances the specificity of the enzyme. If hexokinase were rigid, a molecule of H_2O occupying, by chance, the binding site for the —CH_2OH of glucose could attack the γ phosphoryl group of ATP, forming ADP and P_i. In other words, a rigid kinase would likely also be an ATPase. *Substrate-induced cleft closing is a general feature of kinases.*

Fructose 1,6-bisphosphate Is Generated from Glucose 6-phosphate

The next step in glycolysis is the isomerization of glucose 6-phosphate to fructose 6-phosphate. Recall that the open-chain form of glucose has an aldehyde group at carbon 1, whereas the open-chain form of fructose has a keto group at carbon 2. Thus, *the isomerization of glucose 6-phosphate to fructose 6-phosphate is a conversion of an aldose into a ketose.* The reaction is catalyzed by *phosphoglucose isomerase.*

**Glucose 6-phosphate
(G-6P)**

**Glucose 6-phosphate
(open-chain form)**

**Fructose 6-phosphate
(open-chain form)**

**Fructose 6-phosphate
(F-6P)**

This isomerization is crucial because only three-carbon molecules are metabolized in the later stages of glycolysis. Glucose 6-phosphate is not readily cleaved into two three-carbon fragments, while fructose 6-phosphate is. A second phosphorylation reaction follows the isomerization step, trapping the sugar as the fructose isomer. *Fructose 6-phosphate is phosphorylated by ATP to fructose 1,6-bisphosphate* (F-1,6-BP):

$$\text{Fructose 6-phosphate (F-6P)} + \text{ATP} \xrightleftharpoons[]{\text{Phosphofructokinase}} \text{Fructose 1,6-bisphosphate (F-1,6-BP)} + \text{ADP} + H^+$$

Fructose 6-phosphate
(F-6P)

Fructose 1,6-bisphosphate
(F-1,6-BP)

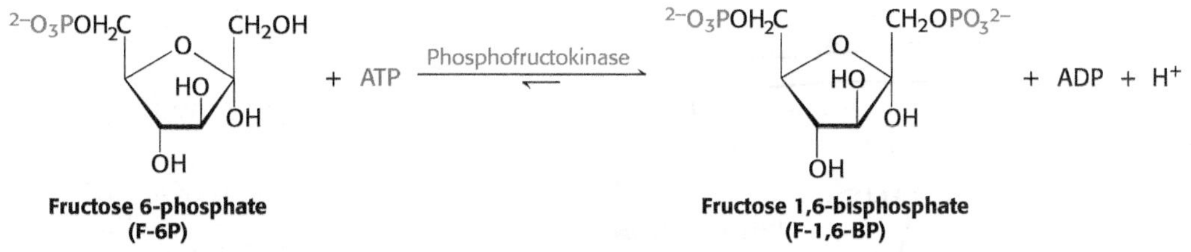

This reaction, which is irreversible under cellular conditions, is catalyzed by *phosphofructokinase* (PFK), an allosteric enzyme that is the key regulatory enzyme for glycolysis (p. 299).

 CLINICAL INSIGHT

The Six-Carbon Sugar Is Cleaved into Two Three-Carbon Fragments

The second stage of glycolysis begins with the cleavage of fructose 1,6-bisphosphate into two triose phosphates, *glyceraldehyde 3-phosphate* (GAP) and *dihydroxyacetone phosphate* (DHAP). The products of the remaining steps in glycolysis consist of three-carbon units rather than six-carbon units. This reaction, which is readily reversible, is catalyzed by *aldolase:*

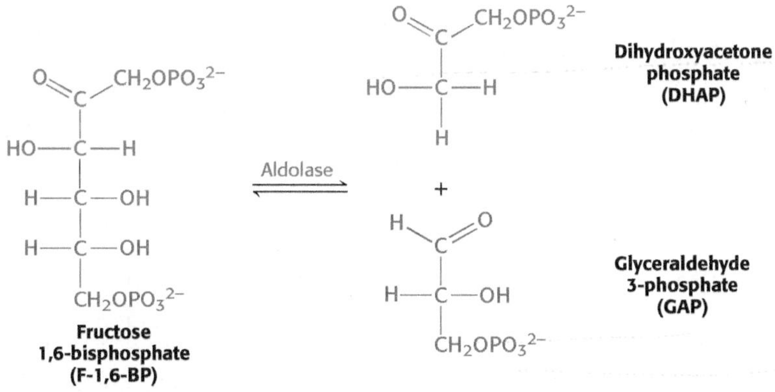

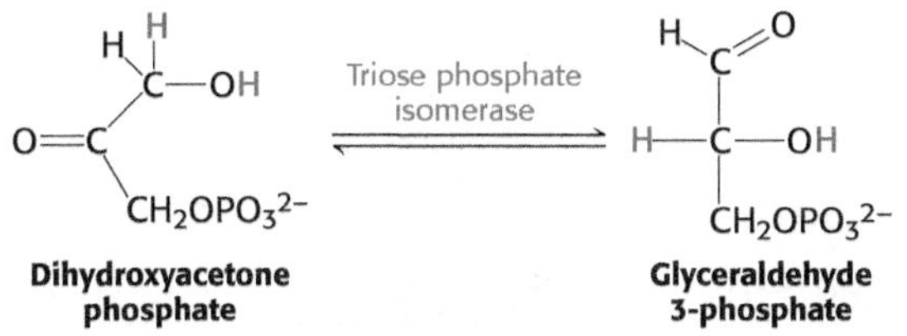

> **DID YOU KNOW?**
>
> The prefix *bis-* in bisphosphate means that two separate monophosphoryl groups are present, whereas the prefix *di-* in diphosphate (as in adenosine diphosphate) means that two phosphoryl groups are present and are connected by an anhydride linkage.

$$\text{Fructose 1,6-bisphosphate (F-1,6-BP)} \xrightleftharpoons[]{\text{Aldolase}} \text{Dihydroxyacetone phosphate (DHAP)} + \text{Glyceraldehyde 3-phosphate (GAP)}$$

Fructose
1,6-bisphosphate
(F-1,6-BP)

Dihydroxyacetone
phosphate
(DHAP)

Glyceraldehyde
3-phosphate
(GAP)

Glyceraldehyde 3-phosphate is on the direct pathway of glycolysis, whereas dihydroxyacetone phosphate is not. These compounds are isomers that can be readily interconverted so as not to waste the useful three-carbon fragment that would be lost if dihydroxyacetone remained in its unusable form. The isomerization of these three-carbon phosphorylated sugars is catalyzed by *triose phosphate isomerase* (TPI, sometimes abbreviated TIM):

$$\text{Dihydroxyacetone phosphate} \xrightleftharpoons[]{\text{Triose phosphate isomerase}} \text{Glyceraldehyde 3-phosphate}$$

Dihydroxyacetone
phosphate

Glyceraldehyde
3-phosphate

This reaction is rapid and reversible. At equilibrium, 96% of the triose phosphate is dihydroxyacetone phosphate. However, the reaction proceeds readily from dihydroxyacetone phosphate to glyceraldehyde 3-phosphate because the subsequent reactions of glycolysis remove this product. The importance of this reaction is demonstrated by the fact that TPI deficiency, a rare condition, is the only glycolytic enzymopathy that is lethal. This deficiency is characterized by severe hemolytic anemia and neurodegeneration.

The Oxidation of an Aldehyde Powers the Formation of a Compound Having High Phosphoryl-Transfer Potential

The preceding steps in glycolysis have transformed one molecule of glucose into two molecules of glyceraldehyde 3-phosphate, but no energy has yet been extracted. On the contrary, two molecules of ATP have been expended. We come now to the final stage of glycolysis, a series of steps that harvest some of the energy contained in glyceraldehyde 3-phosphate as ATP. The initial reaction in this sequence is the *conversion of glyceraldehyde 3-phosphate into 1,3-bisphosphoglycerate* (1,3-BPG), an oxidation–reduction reaction catalyzed by *glyceraldehyde 3-phosphate dehydrogenase*. Dehydrogenases are enzymes that catalyze oxidation–reduction reactions, often transferring a hydride ion from a donor molecule to NAD^+ or transferring a hydride ion from NADH to an acceptor molecule:

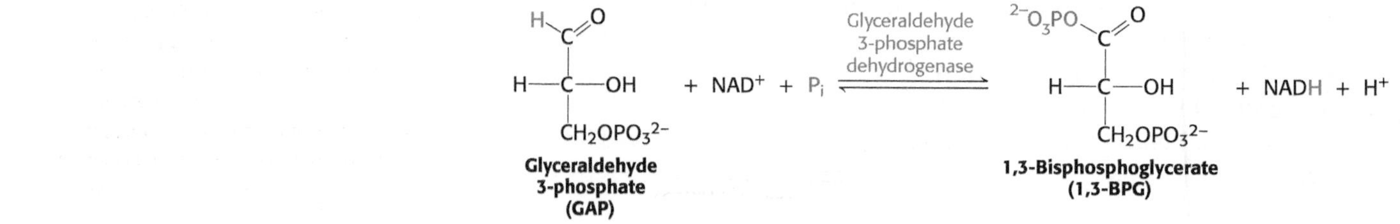

1,3-Bisphosphoglycerate is an acyl phosphate, which is a mixed anhydride of phosphoric acid and a carboxylic acid. Such compounds have a high phosphoryl-transfer potential (p. 264); one of its phosphoryl groups is transferred to ADP in the next step in glycolysis.

Let us consider this reaction in some detail because it illustrates the essence of energy transformation and metabolism itself: the energy of carbon oxidation is captured as high phosphoryl-transfer potential. The reaction catalyzed by glyceraldehyde 3-phosphate dehydrogenase can be viewed as the sum of two processes: the *oxidation* of the aldehyde (in this case, glyceraldehyde 3-phosphate) to a carboxylic acid by NAD^+ and the *joining* of the carboxylic acid (3-phosphoglycerate) and orthophosphate to form the acyl-phosphate product, 1,3-bisphosphoglycerate:

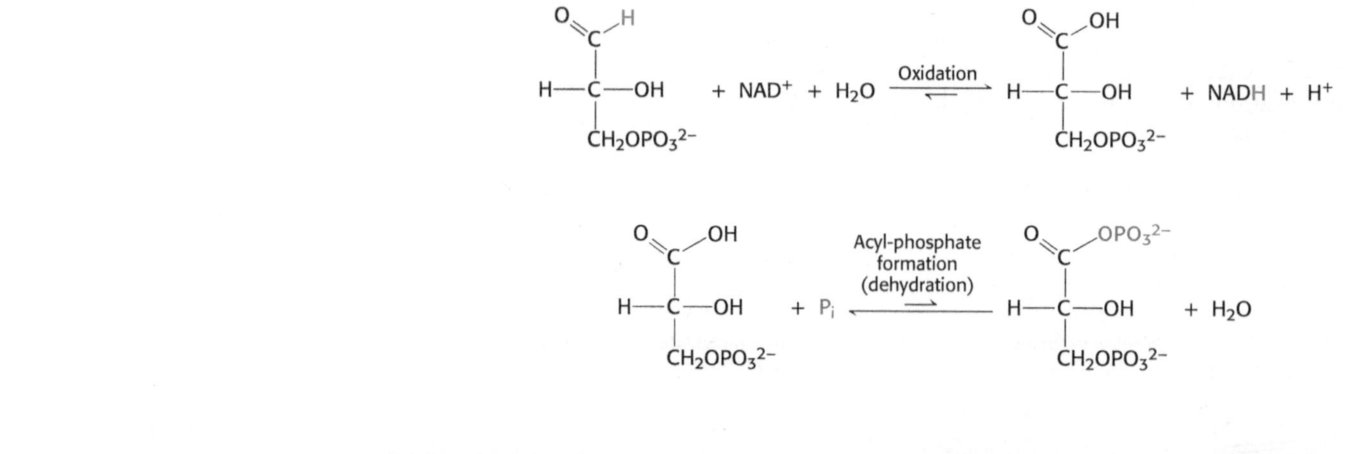

The first reaction is thermodynamically quite favorable, with a standard free-energy change, $\Delta G^{\circ\prime}$, of approximately -50 kJ mol^{-1} (-12 kcal mol^{-1}), whereas the second reaction is quite unfavorable, with a standard free-energy change of the same magnitude but the opposite sign. If these two reactions simply took place in succession, the second reaction would not take place at a biologically significant rate, because of its very large activation energy (Figure 16.3). These two processes must be coupled so that the favorable aldehyde oxidation can be used to drive the formation of the acyl phosphate. How are these reactions coupled? *The key is an intermediate that is linked to the enzyme by a thioester after the aldehyde has been oxidized.* This intermediate reacts with orthophosphate to form the high-energy compound 1,3-bisphosphoglycerate. The thioester is a free-energy intermediate between the aldehyde and the free carboxylic acid. The favorable oxidation and unfavorable phosphorylation reactions are coupled by the *thioester intermediate*, which preserves much of the free energy released in the oxidation reaction (Figure 16.3B).

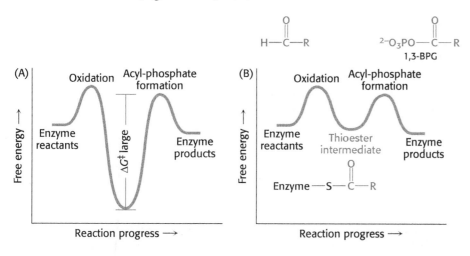

Figure 16.3 Free-energy profiles for glyceraldehyde oxidation followed by acyl-phosphate formation. (A) A hypothetical case with no coupling between the two processes. The second step must have a large activation barrier, making the reaction very slow. (B) The actual case with the two reactions coupled through a thioester intermediate. The thioester intermediate is more stable than the reactant, and, hence, its formation is spontaneous. However, the intermediate is less stable than the product, which forms spontaneously. Thus, the barrier separating oxidation from acyl-phosphate formation is eliminated.

ATP Is Formed by Phosphoryl Transfer from 1,3-Bisphosphoglycerate

1,3-Bisphosphoglycerate is an energy-rich molecule with a greater phosphoryl-transfer potential than that of ATP (p. 264). Thus, 1,3-BPG can be used to power the synthesis of ATP from ADP and orthophosphate. *Phosphoglycerate kinase* catalyzes the transfer of the phosphoryl group from the acyl phosphate of 1,3-bisphosphoglycerate to ADP. ATP and 3-phosphoglycerate are the products:

The formation of ATP in this manner is referred to as *substrate-level phosphorylation* because the phosphate donor, 1,3-BPG, is a kinase substrate with high phosphoryl-transfer potential. We will contrast this manner of ATP formation with the formation of ATP from ionic gradients in Chapters 20 and 21.

We now start to see a return on our initial investment of two molecules of ATP in stage 1. Going backward in the pathway one step to glyceraldehyde 3-phosphate, we find the outcomes of the reactions catalyzed by glyceraldehyde 3-phosphate dehydrogenase and phosphoglycerate kinase to be as follows:

1. Glyceraldehyde 3-phosphate, an aldehyde, is oxidized to 3-phosphoglycerate, a carboxylic acid.

2. NAD$^+$ is concomitantly reduced to NADH.
3. ATP is formed from P$_i$ and ADP at the expense of carbon-oxidation energy.

In essence, the energy released in the oxidation of glyceraldehyde 3-phosphate to 3-phosphoglycerate is temporarily trapped as 1,3-bisphosphoglycerate. This energy powers the transfer of a phosphoryl group from 1,3-bisphosphoglycerate to ADP to yield ATP. Keep in mind that, because of the actions of aldolase and triose phosphate isomerase on fructose 1,6-bisphosphate at the end of stage 1, two molecules of glyceraldehyde 3-phosphate were formed and, hence, two molecules of ATP were generated. These ATP molecules make up for the two molecules of ATP consumed in the first stage of glycolysis.

Additional ATP Is Generated with the Formation of Pyruvate

In the remaining steps of glycolysis, 3-phosphoglycerate is converted into pyruvate, and a second molecule of ATP is formed from ADP:

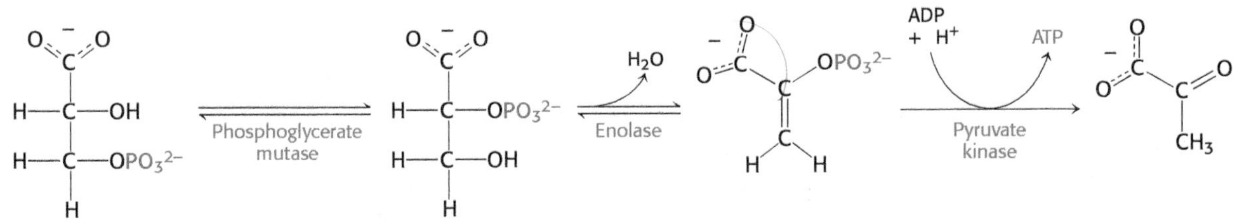

The first reaction is a rearrangement. *3-Phosphoglycerate is converted into 2-phosphoglycerate* by *phosphoglycerate mutase,* which shifts the position of the phosphoryl group. In general, a *mutase* is an enzyme that catalyzes the intramolecular shift of a chemical group, such as a phosphoryl group.

In the next reaction, the dehydration of 2-phosphoglycerate catalyzed by enolase introduces a double bond, creating an *enol phosphate,* an unstable class of molecule in relation to an alcohol such as 2-phosphoglycerate. *Enolase* catalyzes the formation of the enol phosphate *phosphoenolpyruvate* (PEP). This dehydration markedly elevates the transfer potential of the phosphoryl group.

Why does phosphoenolpyruvate have such a high phosphoryl-transfer potential? The phosphoryl group traps the molecule in its unstable enol form. When the phosphoryl group has been donated to ATP, the enol is able to undergo a conversion into the more stable ketone—namely, pyruvate. Hence, pyruvate is formed, and ATP is generated concomitantly. The irreversible transfer of a phosphoryl group from phosphoenolpyruvate to ADP is catalyzed by *pyruvate kinase:*

$$\text{Phosphoenolpyruvate} + \text{ADP} \xrightarrow{\text{Pyruvate kinase}} \text{pyruvate} + \text{ATP}$$

Because the two molecules of ATP used in forming fructose 1,6-bisphosphate were regenerated in the creation of two molecules of 3-phosphoglycerate, the two molecules of ATP generated from the two molecules of phosphoenolpyruvate are "profit."

What is the energy source for the formation of phosphoenolpyruvate? The answer to this question becomes clear when we compare the structures of 2-phosphoglycerate and pyruvate. The formation of pyruvate from 2-phosphoglycerate is, in essence, an internal oxidation–reduction reaction; carbon 3 takes electrons from carbon 2 in the conversion of 2-phosphoglycerate into pyruvate. Compared with 2-phosphoglycerate, C-3 is more reduced in pyruvate, whereas C-2 is more oxidized. Once again, carbon oxidation powers the synthesis of a compound with high phosphoryl-transfer potential—phosphoenolpyruvate here and 1,3-bisphosphoglycerate earlier—which allows the synthesis of ATP.

QUICK QUIZ 1 The gross yield of ATP from the metabolism of glucose to two molecules of pyruvate is four molecules of ATP. However, the net yield is only two molecules of ATP. Why are the gross and net values different?

Two ATP Molecules Are Formed in the Conversion of Glucose into Pyruvate

The net reaction in the transformation of glucose into pyruvate is

$$\text{Glucose} + 2P_i + 2\text{ ADP} + 2\text{ NAD}^+ \longrightarrow$$
$$2\text{ pyruvate} + 2\text{ ATP} + 2\text{ NADH} + 2\text{ H}^+ + 2\text{ H}_2\text{O}$$

Thus, *two molecules of ATP are generated in the conversion of glucose into two molecules of pyruvate.* The reactions of glycolysis are summarized in Table 16.1 below.

Note that the energy released in the anaerobic conversion of glucose into two molecules of pyruvate is about -90 kJ mol^{-1} (-22 kcal mol^{-1}). We shall see in Chapters 20 and 21 that much more energy can be released from glucose in the presence of oxygen.

16.2 NAD⁺ Is Regenerated from the Metabolism of Pyruvate

✓ 2 Explain why the regeneration of NAD⁺ is crucial to fermentations.

The conversion of glucose into two molecules of pyruvate results in the net synthesis of ATP. However, an energy-converting pathway that stops at pyruvate will not proceed for long, because redox balance has not been maintained. This imbalance is caused by the activity of glyceraldehyde 3-phosphate dehydrogenase, which leads to the reduction of NAD⁺ to NADH when glyceraldehyde 3-phosphate is oxidized. In the cell, there are limited amounts of NAD⁺, which is

Table 16.1 Reactions of glycolysis

Step	Reaction	Enzyme	Reaction type	$\Delta G^{\circ\prime}$ in kJ mol^{-1} (kcal mol^{-1})	ΔG in kJ mol^{-1} (kcal mol^{-1})
1	Glucose + ATP ⟶ glucose 6-phosphate + ADP + H⁺	Hexokinase	Phosphoryl transer	$-16.7\ (-4.0)$	$-33.5\ (-8.0)$
2	Glucose 6-phosphate ⇌ fructose 6-phosphate	Phosphoglucose isomerase	Isomerization	$+1.7\ (+0.4)$	$-2.5\ (-0.6)$
3	Fructose 6-phosphate + ATP ⟶ fructose 1,6-bisphosphate + ADP + H⁺	Phosphofructokinase	Phosphoryl transfer	$-14.2\ (-3.4)$	$-22.2\ (-5.3)$
4	Fructose 1,6-bisphosphate ⇌ dihydroxyacetone phosphate + glyceraldehyde 3-phosphate	Aldolase	Aldol cleavage	$+23.8\ (+5.7)$	$-1.3\ (-0.3)$
5	Dihydroxyacetone phosphate ⇌ glyceraldehyde 3-phosphate	Triose Phosphate isomerase	Isomerization	$+7.5\ (+1.8)$	$+2.5\ (+0.6)$
6	Glyceraldehyde 3-phosphate + P$_i$ + NAD⁺ ⇌ 1,3-bisphosphoglycerate + NADH + H⁺	Glyceraldehyde 3-phosphate dehydrogenase	Phosphorylation coupled to oxidation	$+6.3\ (+1.5)$	$-1.7\ (-0.4)$
7	1,3-Bisphosphoglycerate + ADP ⇌ 3-phosphoglycerate + ATP	Phosphoglycerate kinase	Phosphoryl transfer	$-18.8\ (-4.5)$	$+1.3\ (+0.3)$
8	3-Phosphoglycerate ⇌ 2-phoshoglycerate	Phosphoglycerate mutase	Phosphoryl shift	$+4.6\ (+1.1)$	$+0.8\ (+0.2)$
9	2-Phosphoglycerate ⇌ Phosphoenolpyruvate + H₂O	Enolase	Dehydration	$+1.7\ (+0.4)$	$-3.3\ (-0.8)$
10	Phosphoenolpyruvate + ADP + H⁺ ⟶ pyruvate + ATP	Pyruvate kinase	Phosphoryl transfer	$-31.4\ (-7.5)$	$-16.7\ (-4.0)$

Note: ΔG, the actual free-energy change, has been calculated from $\Delta G^{\circ\prime}$ and known concentrations of reactants under typical physiological conditions. Glycolysis can proceed only if the ΔG values of all reactions are negative. The small positive ΔG values of three of the above reactions indicate that the concentrations of metabolities in vivo in cells undergoing glycolysis are not precisely known.

NUTRITION FACTS

Niacin Also called vitamin B_3, niacin is a component of coenzymes NAD^+ and $NADP^+$ (pp. 268–270), which are used in electron-transfer reactions. There are many sources of niacin, including chicken breast. Niacin deficiency results in the potentially fatal disease pellagra, a condition characterized by dermatitis, dementia, and diarrhea. [Brand X Pictures]

Figure 16.4 Diverse fates of pyruvate. Ethanol and lactate can be formed by reactions that include NADH. Alternatively, a two-carbon unit from pyruvate can be coupled to coenzyme A (see Chapter 18) to form acetyl CoA.

derived from the vitamin niacin, a dietary requirement for human beings. Consequently, NAD^+ must be regenerated for glycolysis to proceed. Thus, the final process in the pathway is the regeneration of NAD^+ through the metabolism of pyruvate.

Fermentations Are a Means of Oxidizing NADH

The sequence of reactions from glucose to pyruvate is similar in most organisms and most types of cells. In contrast, the fate of pyruvate is variable. Three reactions of pyruvate are of primary importance: conversion into ethanol, lactate, or carbon dioxide and water (Figure 16.4). The first two reactions are fermentations that take place in the absence of oxygen. *Fermentations are ATP-generating processes in which organic compounds act as both donors and acceptors of electrons.* In the presence of oxygen, the most common situation in multicellular organisms and for many unicellular ones, pyruvate is metabolized to carbon dioxide and water through the citric acid cycle and the electron-transport chain (Sections 8 and 9). In these circumstances, oxygen accepts electrons and protons to form water. We now take a closer look at these three possible fates of pyruvate.

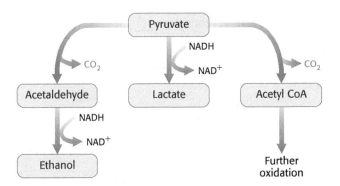

1. *Ethanol* is formed from pyruvate in yeast and several other microorganisms. The first step is the decarboxylation of pyruvate. This reaction is catalyzed by *pyruvate decarboxylase,* which requires the coenzyme thiamine pyrophosphate. This coenzyme is derived from the vitamin thiamine (B_1). The second step is the reduction of acetaldehyde to ethanol by NADH, in a reaction catalyzed by *alcohol dehydrogenase.* Acetaldehyde is thus the organic compound that accepts the electrons in this fermentation. This reaction regenerates NAD^+:

The conversion of glucose into ethanol is an example of *alcoholic fermentation.* The net result of this anaerobic process is

$$\text{Glucose} + 2P_i + 2\,\text{ADP} + 2\,H^+ \longrightarrow 2\,\text{ethanol} + 2\,CO_2 + 2\,\text{ATP} + 2\,H_2O$$

Note that NAD^+ and NADH do not appear in this equation, even though they are crucial for the overall process. NADH generated by the oxidation of glyceraldehyde 3-phosphate is consumed in the reduction of acetaldehyde to ethanol. Thus, *there is no net oxidation–reduction in the conversion of glucose into ethanol* (Figure 16.5). The ethanol formed in alcoholic fermentation is a key ingredient in brewing and winemaking, and the carbon dioxide formed accounts for some of the carbonation in beer and champagne.

Figure 16.5 Maintaining redox balance in alcoholic fermentation. The NADH produced by the glyceraldehyde 3-phosphate dehydrogenase reaction must be reoxidized to NAD^+ for the glycolytic pathway to continue. In alcoholic fermentation, alcohol dehydrogenase oxidizes NADH and generates ethanol.

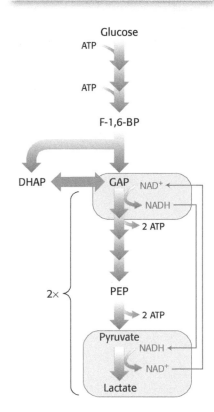

NUTRITION FACTS

Thiamine Also called vitamin B_1, thiamine is a component of the coenzyme thiamine pyrophosphate (TPP), which is used in decarboxylation reactions. Pork and legumes are good sources of thiamine. A deficiency of thiamine results in beriberi, the symptoms of which include muscle weakness, anorexia, and an enlarged heart. [Ruslan/age fotostock]

2. *Lactate* is formed from pyruvate in a variety of microorganisms in a process called *lactic acid fermentation*. Pyruvate accepts the electrons from NADH to form lactate in a reaction catalyzed by *lactate dehydrogenase*:

The overall reaction in the conversion of glucose into lactate is

$$\text{Glucose} + 2P_i + 2\,\text{ADP} \longrightarrow 2\,\text{lactate} + 2\,\text{ATP} + 2\,H_2O$$

As in alcoholic fermentation, there is no net oxidation–reduction. The NADH formed in the oxidation of glyceraldehyde 3-phosphate is consumed in the reduction of pyruvate (**Figure 16.6**). *The regeneration of NAD⁺ in the reduction of pyruvate to lactate or ethanol sustains the continued process of glycolysis under anaerobic conditions.*

Certain types of skeletal muscles in most animals can function anaerobically for short periods. For example, a specific type of muscle fiber, called *fast twitch* or *type IIb fibers*, perform short bursts of intense exercise. The ATP needs rise faster than the ability of the body to provide oxygen to the muscle. The muscle functions anaerobically until fatigue sets in, which is caused, in part, by lactate buildup. Indeed, the pH of resting type IIb muscle fibers, which is about 7.0, may fall to as low as 6.3 during the bout of exercise. The drop in pH inhibits phosphofructokinase (p. 299). A lactate/H^+ symporter (p. 214) allows the exit of lactate and H^+ from the muscle cell.

3. Only a fraction of the energy of glucose is released in its anaerobic conversion into ethanol or lactate. Much more energy can be extracted aerobically by means of the citric acid cycle and the electron-transport chain, which combust, or oxidize, glucose all the way to CO_2 and H_2O. The entry point to this oxidative pathway is *acetyl coenzyme A* (acetyl CoA), which is formed from pyruvate inside mitochondria:

$$\text{Pyruvate} + NAD^+ + \text{CoA} \longrightarrow \text{acetyl CoA} + CO_2 + NADH$$

This reaction will be considered in detail in Chapter 18. The NAD^+ required for this reaction and for the oxidation of glyceraldehyde 3-phosphate is regenerated in the electron-transport chain in mitochondria (Chapter 20).

Figure 16.6 Maintaining redox balance in lactic acid fermentation. In lactic acid fermentation, lactate dehydrogenase oxidizes NADH to produce lactic acid and regenerate NAD^+.

BIOLOGICAL INSIGHT

Fermentations Provide Usable Energy in the Absence of Oxygen

Fermentations yield only a fraction of the energy available from the complete combustion of glucose. Why is a relatively inefficient metabolic pathway so extensively used? The fundamental reason is that fermentations do not require oxygen. The ability to survive without oxygen affords a host of living accommodations such as soils, deep water, and skin pores. Some organisms, called *obligate anaerobes,* cannot survive in the presence of O_2, which is a highly reactive compound. The bacterium *Clostridium perfringens,* the cause of gangrene, is an example of an obligate anaerobe. Other pathogenic obligate anaerobes are listed in Table 16.2. Some organisms, such as yeast, are *facultative anaerobes* that metabolize glucose aerobically when oxygen is present and perform fermentation when oxygen is absent.

QUICK QUIZ 2 Lactic acid fermentation and alcoholic fermentation are oxidation–reduction reactions. Identify the ultimate electron donor and electron acceptor.

Table 16.2 Examples of pathogenic obligate anaerobes

Bacterium	Result of infection
Clostridium tetani	Tetanus (lockjaw)
Clostridium botulinum	Botulism (an especially severe type of food poisoning)
Clostridium perfringens	Gas gangrene (gas is produced as an end point of the fermentation, distorting and destroying the tissue)
Bartonella hensela	Cat scratch fever (flu-like symptoms)
Bacteroides fragilis	Abdominal, pelvic, pulmonary, and blood infections

Many food products, including sour cream, yogurt, various cheeses, beer, wine, and sauerkraut, result from fermentation. Yogurt is produced by the fermentation of lactose in milk to lactate by a mixed culture of *Lactobacillus acidophilus* and *Streptococcus thermophilus.* Sour cream begins with pasteurized light cream, which is fermented to lactate by *Streptococcus lactis.* The lactate is further fermented to ketones and aldehydes by *Leuconostoc citrovorum.* The second fermentation adds to the taste and aroma of sour cream. Yeast, *Saccharomyces cerevisiae,* ferments carbohydrates to ethanol and carbon dioxide, providing some of the ingredients for an array of alcohol beverages. Although we have considered only lactic acid and alcoholic fermentation, microorganisms are capable of generating a wide array of molecules as end points to fermentation.

DID YOU KNOW?

Fermentation is an ATP-generating process in which organic compounds act as both donors and acceptors of electrons. Fermentation can take place in the absence of O_2. Louis Pasteur, who discovered fermentation, described it as *la vie sans l'air* ("life without air").

16.3 Fructose and Galactose Are Converted into Glycolytic Intermediates

Although glucose is the monosaccharide most commonly used as an energy source, others also are important fuels. Let us consider how two common sugars—fructose and galactose—can be funneled into the glycolytic pathway (Figure 16.7). Fructose is a component of sucrose, or table sugar (p. 174), as well as high-fructose corn syrup, which is used as a sweetener in many foods and drinks. Galactose is a component of lactose, or milk sugar. There are no catabolic pathways dedicated to metabolizing fructose or galactose as there is for glucose, and so the strategy is to convert these sugars into an intermediate in glycolysis.

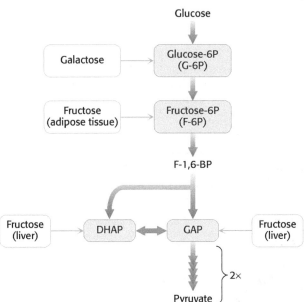

Glucose

Galactose → Glucose-6P (G-6P)

Fructose (adipose tissue) → Fructose-6P (F-6P)

F-1,6-BP

Fructose (liver) → DHAP ⇌ GAP ← Fructose (liver)

2×

Pyruvate

Figure 16.7 Entry points in glycolysis for fructose and galactose.

Fructose Is Converted into Glycolytic Intermediates by Fructokinase

Fructose can take one of two pathways to enter the glycolytic pathway. Much of the ingested fructose is metabolized by the liver, using the *fructose 1-phosphate pathway* (Figure 16.8). The first step is the phosphorylation of *fructose* to *fructose 1-phosphate* by *fructokinase*. Fructose 1-phosphate is then split into *glyceraldehyde* and *dihydroxyacetone phosphate,* an intermediate in glycolysis. This aldol cleavage is catalyzed by a specific *fructose 1-phosphate aldolase.* Dihydroxyacetone phosphate continues into stage 2 of glycolysis, whereas glyceraldehyde is then phosphorylated to *glyceraldehyde 3-phosphate,* a glycolytic intermediate, by *triose kinase.* In other tissues, such as adipose tissues, *fructose can be phosphorylated to the glycolytic intermediate fructose 6-phosphate by hexokinase.*

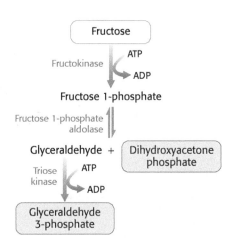

Fructose

↓ Fructokinase — ATP → ADP

Fructose 1-phosphate

↕ Fructose 1-phosphate aldolase

Glyceraldehyde + Dihydroxyacetone phosphate

↓ Triose kinase — ATP → ADP

Glyceraldehyde 3-phosphate

Figure 16.8 Fructose metabolism. Fructose enters the glycolytic pathway in the liver through the fructose 1-phosphate pathway.

 CLINICAL INSIGHT

Excessive Fructose Consumption Can Lead to Pathological Conditions

Fructose is a common sweetener, as mentioned above. Epidemiological as well as clinical studies have linked excessive fructose consumption to obesity, fatty liver, and insulin insensitivity. These conditions may eventually result in type 2 diabetes (non-insulin-dependent diabetes). Studies have shown that these disorders are not necessarily the result of simple excess energy consumption. What aspects of fructose metabolism are the contributing factors then? Note that, as shown in Figure 16.8, the actions of fructokinase and triose kinase bypass the most important regulatory step in glycolysis, the phosphofructokinase-catalyzed reaction. The fructose-derived glyceraldehyde 3-phosphate and dihydroxyacetone phosphate are processed by glycolysis to pyruvate and subsequently to acetyl CoA in an unregulated fashion. As we will see in Chapter 28, this excess acetyl CoA is converted to fatty acids, which can be transported to adipose tissue, resulting in obesity. The liver also begins to accumulate fatty acids, resulting in fatty liver. The activity of the fructokinase and triose kinase can deplete the liver of ATP and inorganic phosphate, compromising liver function.

Galactose Is Converted into Glucose 6-phosphate

Like fructose, *galactose* is an abundant sugar that must be converted into metabolites of glucose (Figure 16.7). Galactose is converted into *glucose 6-phosphate* in four steps. The first reaction in the *galactose–glucose interconversion pathway* is the phosphorylation of galactose to galactose 1-phosphate by *galactokinase*:

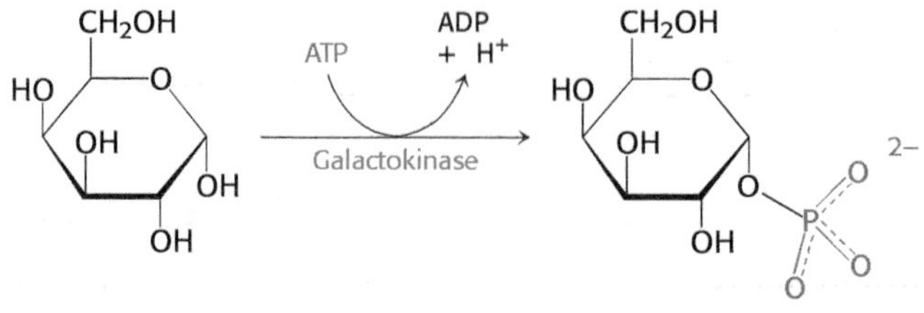

Galactose   **Galactose 1-phosphate**

Galactose 1-phosphate then acquires a uridyl group from uridine diphosphate glucose (UDP-glucose), which activates the sugar phosphate so that it can be converted into glucose (Figure 16.9). UDP-monosaccharides are another example of an activated intermediate (p. 174) and are used as an intermediate in the synthesis of glycosidic linkages, the bonds between monosaccharides. The products of this reaction, which is catalyzed by *galactose 1-phosphate uridyl transferase*, are UDP-galactose and glucose 1-phosphate. The galactose moiety of UDP-galactose is then epimerized to glucose by UDP-galactose 4-epimerase. The epimerase inverts the hydroxyl group at carbon 4 to create glucose.

The sum of the reactions catalyzed by galactokinase, the transferase, and the epimerase is

$$\text{Galactose} + \text{ATP} \longrightarrow \text{glucose 1-phosphate} + \text{ADP} + \text{H}^+$$

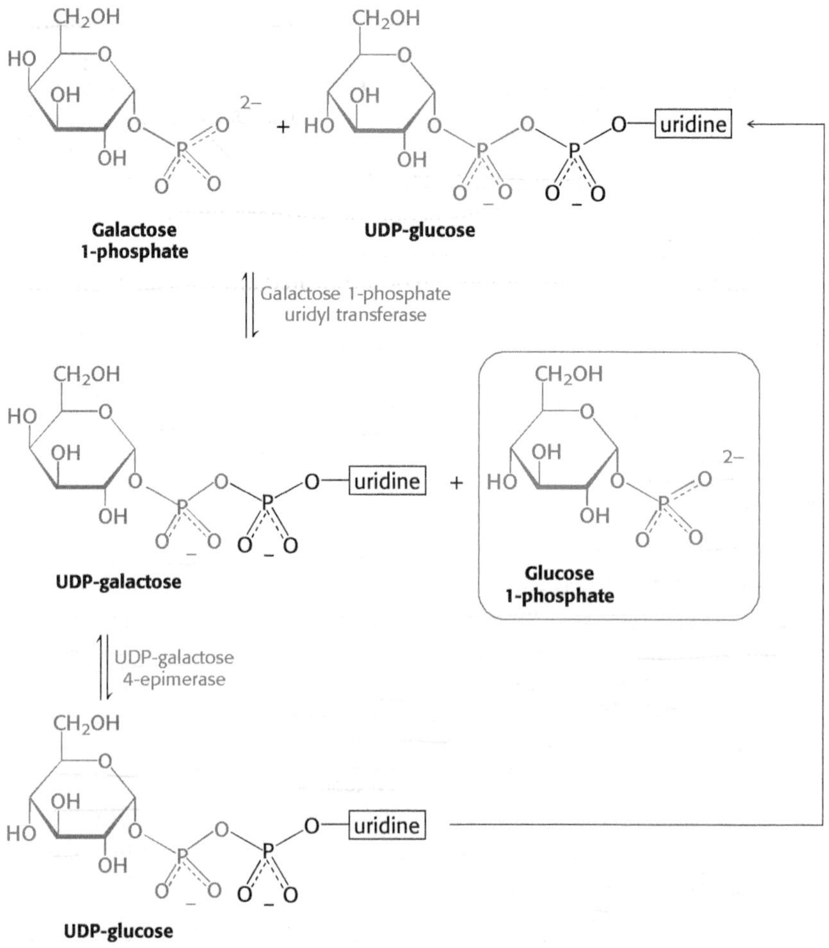

Figure 16.9 Galactose metabolism. Galactose 1-phosphate reacts with activated glucose (UDP-glucose) to form UDP-galactose, which is subsequently converted into UDP-glucose.

Note that UDP-glucose is not consumed in the conversion of galactose into glucose, because it is regenerated from UDP-galactose by the epimerase. This reaction is reversible, and the product of the reverse direction also is important. The conversion of UDP-glucose into UDP-galactose is essential for the synthesis of complex polysaccharides and glycoproteins (Chapter 10). Ordinarily, galactose fills this role, but, if the amount of galactose in the diet is inadequate to meet these needs, UDP-galactose is used.

Finally, glucose 1-phosphate, formed from galactose, is isomerized to glucose 6-phosphate by *phosphoglucomutase*:

$$\text{Glucose 1-phosphate} \underset{}{\overset{\text{Phosphoglucomutase}}{\rightleftharpoons}} \text{glucose 6-phosphate}$$

We shall return to this reaction when we consider the synthesis and degradation of glycogen, which proceeds through glucose 1-phosphate, in Chapters 24 and 25.

⚕ CLINICAL INSIGHT

Many Adults Are Intolerant of Milk Because They Are Deficient in Lactase

Many adults are unable to metabolize the milk sugar lactose and experience gastrointestinal disturbances if they drink milk. *Lactose intolerance*, or *hypolactasia*, is most commonly caused by a deficiency of the enzyme *lactase*, which cleaves lactose into glucose and galactose:

"Deficiency" is not quite the appropriate term, because a decrease in lactase is normal in the course of development in all mammals. As children are weaned and milk becomes less prominent in their diets, lactase activity normally declines to about 5 to 10% of the level at birth. This decrease is not as pronounced with some groups of people, most notably people of Northern European ancestry, who can continue to ingest milk without gastrointestinal difficulties (**Figure 16.10**). With the appearance of milk-producing domesticated animals, an adult with

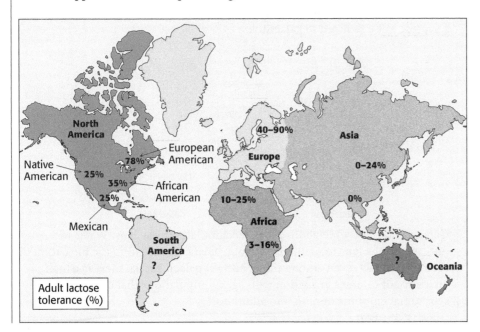

Figure 16.10 Lactose tolerance is most common in Europe. A mutation arose that prevented lactase activity from diminishing in adults. This mutation was beneficial because of the availability of milk from dairy farming. The question marks on the map indicate places where data are insufficient. [Data from J. W. Kalat, *Introduction to Psychology*, 8th ed. (Wadsworth, 2007), Fig. 10.11.]

active lactase would hypothetically have had a selective advantage in being able to consume calories from the readily available milk. Because dairy farming originated only about 10,000 years ago, the evolutionary selective pressure on lactase persistence must have been substantial, attesting to the biochemical value of being able to use milk as an energy source into adulthood.

What happens to the lactose in the intestine of a lactase-deficient person? The lactose is a good energy source for microorganisms in the colon, and they ferment it to lactic acid while also generating methane (CH_4) and hydrogen gas (H_2). The gas creates the uncomfortable feeling of gut distension and the annoying problem of flatulence. The lactate produced by the microorganisms draws water into the intestine, as does any undigested lactose, resulting in diarrhea. If severe enough, the gas and diarrhea hinder the absorption of other nutrients such as fats and proteins. The simplest treatment is to avoid the consumption of products containing much lactose. Alternatively, the enzyme lactase can be ingested with milk products.

 CLINICAL INSIGHT

Galactose Is Highly Toxic If the Transferase Is Missing

Less common than lactose intolerance are disorders that interfere with the metabolism of galactose. The disruption of galactose metabolism is referred to as *galactosemia*. The most common form, called *classic galactosemia*, is an inherited deficiency in galactose 1-phosphate uridyl transferase activity. Afflicted infants fail to thrive. They vomit or have diarrhea after consuming milk, and enlargement of the liver and jaundice are common, sometimes progressing to cirrhosis. Cataracts will form, and lethargy and retarded mental development also are common. The blood-galactose level is markedly elevated, and galactose is found in the urine. The absence of the transferase in red blood cells is a definitive diagnostic criterion.

The most common treatment is to remove galactose (and lactose) from the diet. An enigma of galactosemia is that, although elimination of galactose from the diet prevents liver disease and cataract development, most patients still suffer from central nervous system malfunction, most commonly a delayed acquisition of language skills. Female patients also display ovarian failure.

Cataract formation is better understood. A cataract is the clouding of the normally clear lens of the eye (**Figure 16.11**). If the transferase is not active in the lens of the eye, the presence of aldose reductase causes the accumulating galactose to be reduced to galactitol:

(A)   (B)

Figure 16.11 Cataracts are evident as the clouding of the lens. (A) A healthy eye. (B) An eye with a cataract. [(A) Tim Mainiero/ Shutterstock; (B) SPL/Science Source.]

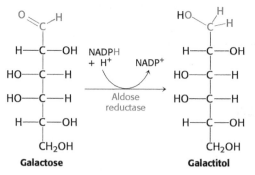

Galactitol is poorly metabolized and accumulates in the lens. Water will diffuse into the lens to maintain osmotic balance, instigating the formation of cataracts. In fact, even among those without galactosemia, there is a high incidence of cataract formation with age in populations that consume substantial amounts of milk into adulthood.

16.4 The Glycolytic Pathway Is Tightly Controlled

The glycolytic pathway has a dual role: it degrades glucose to generate ATP, and it provides building blocks for biosynthetic reactions, such as the formation of fatty acids and amino acids. The rate of conversion of glucose into pyruvate is regulated to meet these two major cellular needs. *In metabolic pathways, enzymes catalyzing irreversible reactions are potential sites of control.* In glycolysis, the reactions catalyzed by hexokinase, phosphofructokinase, and pyruvate kinase are irreversible, and each of them serves as a control site. These enzymes become more active or less so in response to the reversible binding of allosteric effectors or covalent modification. We will consider the control of glycolysis in two different tissues—skeletal muscle and liver.

Glycolysis in Muscle Is Regulated by Feedback Inhibition to Meet the Need for ATP

Glycolysis in skeletal muscle provides ATP primarily to power contraction. Consequently, the primary control of muscle glycolysis is the energy charge of the cell—the ratio of ATP to AMP. Glycolysis is stimulated as the energy charge falls—a signal that the cell needs more ATP. Let us examine how each of the key regulatory enzymes responds to changes in the amounts of ATP and AMP present in the cell.

   Phosphofructokinase *Phosphofructokinase is the most important control site in the mammalian glycolytic pathway.* High levels of ATP allosterically inhibit the enzyme (a 340-kDa tetramer). ATP binds to a specific regulatory site that is distinct from the catalytic site. The binding of ATP lowers the enzyme's affinity for fructose 6-phosphate. AMP reverses the inhibitory action of ATP. AMP competes with ATP for the binding site but, when bound, does not inhibit the enzyme. Consequently *the activity of the enzyme increases when the ATP/AMP ratio is lowered* (**Figure 16.12**). A decrease in pH also inhibits phosphofructokinase activity by augmenting the inhibitory effect of ATP. The pH might fall when fast-twitch muscle is functioning anaerobically, producing excessive quantities of lactic acid. The inhibition of glycolysis, and therefore of lactic acid fermentation, protects the muscle from damage that would result from the accumulation of too much acid.

   Why does AMP but not ADP stimulate the activity of phosphofructokinase? When ATP is being utilized rapidly, the enzyme *adenylate kinase* can form ATP from ADP by the following reaction:

$$ADP + ADP \rightleftharpoons ATP + AMP$$

Thus, some ATP is salvaged from ADP, and AMP becomes the signal for the low-energy state (problem 16.38).

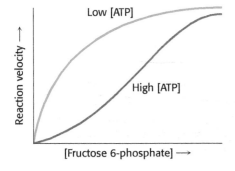

Figure 16.12 The allosteric regulation of phosphofructokinase. A high level of ATP inhibits the enzyme by decreasing its affinity for fructose 6-phosphate. AMP diminishes the inhibitory effect of ATP.

Hexokinase Phosphofructokinase is the primary regulatory enzyme in glycolysis, but it is not the only one. Hexokinase, the enzyme catalyzing the first step of glycolysis, is inhibited by its product, glucose 6-phosphate. High concentrations of glucose 6-phosphate signal that the cell no longer requires glucose for energy, and so no more glucose needs to be broken down. The glucose will then be left in the blood. A rise in glucose 6-phosphate concentration is a means by which phosphofructokinase communicates with hexokinase. When phosphofructokinase is inactive, the concentration of fructose 6-phosphate rises. In turn, the level of glucose 6-phosphate rises because it is in equilibrium with fructose 6-phosphate. Hence, *the inhibition of phosphofructokinase leads to the inhibition of hexokinase.*

Why is phosphofructokinase rather than hexokinase the pacemaker of glycolysis? The reason becomes evident on noting that glucose 6-phosphate is not solely a glycolytic intermediate. In muscle, for example, glucose 6-phosphate can also be converted into glycogen. The first irreversible reaction unique to the glycolytic pathway, the *committed step* (p. 120), is the phosphorylation of fructose 6-phosphate to fructose 1,6-bisphosphate. Thus, phosphofructokinase as the primary control site in glycolysis is highly appropriate. In general, *the enzyme catalyzing the committed step in a metabolic sequence is the most important control element in the pathway because it regulates flux down the pathway.*

Pyruvate kinase Pyruvate kinase, the enzyme catalyzing the third irreversible step in glycolysis, controls the efflux from this pathway. This final step yields ATP and pyruvate, a central metabolic intermediate that can be oxidized further or used as a building block. ATP allosterically inhibits pyruvate kinase to decrease the rate of glycolysis when the energy charge of the cell is high. When the pace of glycolysis increases, fructose 1,6-bisphosphate, the product of the preceding irreversible step in glycolysis, activates the kinase to enable it to keep pace with the oncoming high flux of intermediates. A summary of the regulation of glycolysis in resting and active muscle is shown in Figure 16.13.

The Regulation of Glycolysis in the Liver Corresponds to the Biochemical Versatility of the Liver

The liver has a greater diversity of biochemical functions than muscle. Significantly, the liver maintains blood-glucose concentration: it stores glucose as glycogen when glucose is plentiful, and it releases glucose when supplies are low. It also uses glucose to generate reducing power for biosynthesis (Chapter 26) as well as to synthesize a host of building blocks for other biomolecules. So, although the liver has many of the regulatory features of muscle glycolysis, the regulation of glycolysis in the liver is more complex.

Phosphofructokinase Liver phosphofructokinase can be regulated by ATP as in muscle, but such regulation is not as important since the liver does not experience the sudden ATP needs that a contracting muscle does. Likewise, low pH is not a metabolic signal for the liver enzyme, because lactate is not normally produced in the liver. Indeed, as we will see, lactate is converted into glucose in the liver.

Glycolysis in the liver furnishes carbon skeletons for biosyntheses, and so a signal indicating whether building blocks are abundant or scarce should also regulate phosphofructokinase. In the liver, *phosphofructokinase is inhibited by citrate,* an early intermediate in the citric acid cycle (Chapter 19). A high level of citrate in the cytoplasm means that biosynthetic precursors are abundant, and so there is no need to degrade additional glucose for this purpose. In this way, citrate enhances the inhibitory effect of ATP on phosphofructokinase.

The key means by which glycolysis in the liver responds to changes in blood glucose is through the signal molecule *fructose 2,6-bisphosphate* (F-2,6-BP), a

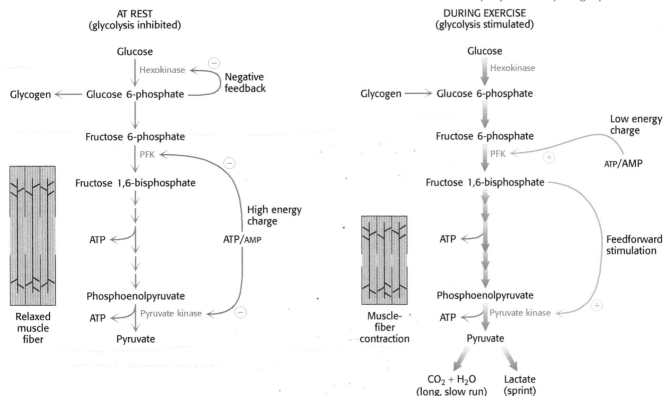

Figure 16.13 **The regulation of glycolysis in muscle.** At rest (left), glycolysis is not very active (thin arrows). The high concentration of ATP inhibits phosphofructokinase (PFK) and pyruvate kinase, while glucose 6-phosphate inhibits hexokinase. Glucose 6-phosphate is converted into glycogen (Chapter 25). During exercise (right), the decrease in the ATP/AMP ratio resulting from muscle contraction activates phosphofructokinase and hence glycolysis. The flux down the pathway is increased, as represented by the thick arrows.

potent activator of phosphofructokinase. After a meal rich in carbohydrates, the concentration of glucose in the blood rises. In the liver, the concentration of fructose 6-phosphate rises when blood-glucose concentration is high because of the action of hexokinase and phosphoglucose isomerase, and the abundance of fructose 6-phosphate accelerates the synthesis of F-2,6-BP (**Figure 16.14**). Hence _an abundance of fructose 6-phosphate leads to a higher concentration of F-2,6-BP._ Fructose 2,6-biphosphate stimulates glycolysis by increasing phosphofructokinase's affinity for fructose 6-phosphate and diminishing the inhibitory effect of ATP (**Figure 16.15**). Glycolysis is thus accelerated when glucose is abundant. Such a process is called _feedforward stimulation._ We will examine the synthesis and degradation of this regulatory molecule after we have considered gluconeogenesis (Chapter 17).

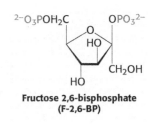

**Fructose 2,6-bisphosphate
(F-2,6-BP)**

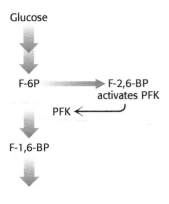

Figure 16.14 **The regulation of phosphofructokinase by fructose 2,6-bisphosphate.** High concentrations of fructose 6-phosphate (F-6P) activate the enzyme phosphofructokinase (PFK) through an intermediary, fructose 2,6-bisphosphate (F-2,6-BP).

Figure 16.15 The activation of phosphofructokinase by fructose 2,6-bisphosphate. (A) The sigmoidal dependence of velocity on substrate concentration becomes hyperbolic in the presence of 1 μM fructose 2,6-bisphosphate, indicating that more of the enzyme is active at lower substrate concentrations in the presence of fructose 2,6-bisphosphate. (B) ATP, acting as a substrate, initially stimulates the reaction. As the concentration of ATP increases, it acts as an allosteric inhibitor. The inhibitory effect of ATP is reduced by fructose 2,6-bisphosphate, which renders the enzyme less sensitive to ATP inhibition. [Data from E. Van Schaftingen, M. F. Jett, L. Hue, and H. G. Hers. *Proc. Natl. Acad. Sci. U. S. A.* 78:3483–3486, 1981.]

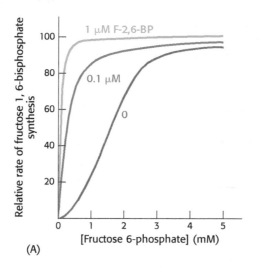
(A)

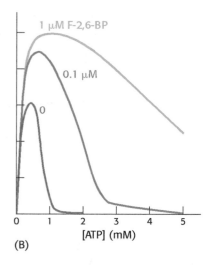
(B)

Hexokinase and glucokinase In the liver as well as in muscle, hexokinase is a regulatory enzyme. The hexokinase reaction is controlled in the liver as in muscle. However, the enzyme primarily responsible for phosphorylating glucose in the liver is not hexokinase, but *glucokinase* (hexokinase IV), an isozyme of hexokinase. *Isozymes,* or *isoenzymes,* are enzymes encoded by different genes with different amino acid sequences, yet they catalyze the same reaction. Isozymes usually differ in kinetic or regulatory properties. Glucokinase phosphorylates glucose only when glucose is abundant, as would be the case after a meal. The reason is that glucokinase's K_M for glucose is about 50-fold higher than that of hexokinase, which means that glucose 6-phosphate is formed only when glucose is abundant. Furthermore, glucokinase is not inhibited by its product, glucose 6-phosphate, as hexokinase is. The high K_M of glucokinase for glucose in the liver gives the brain and muscles first call for glucose when its supply is limited, and it ensures that glucose will not be wasted when it is abundant. The role of glucokinase is to provide glucose 6-phosphate for the synthesis of glycogen and for the formation of fatty acids. Drugs that activate liver glucokinase are being evaluated as a treatment for type 2 or insulin-insensitive diabetes.

Glucokinase is also present in the β cells of the pancreas, where the increased formation of glucose 6-phosphate by glucokinase when the blood-glucose concentration is high leads to the secretion of the hormone insulin. Insulin signals the need to remove glucose from the blood for storage as glycogen or conversion into fat.

Pyruvate kinase Several isozymic forms of pyruvate kinase (a tetramer of 57-kDa subunits) encoded by different genes are present in mammals: the L form predominates in liver, and the M form in muscle and brain. The L and M forms of pyruvate kinase have many properties in common. Indeed, the liver enzyme behaves much as the muscle enzyme does in regard to allosteric regulation, except that the liver enzyme is also inhibited by alanine (synthesized in one step from pyruvate), a signal that building blocks are available. Moreover, the isozymic forms differ in their susceptibility to phosphorylation. The catalytic properties of the L form—but not of the M form—are controlled by reversible phosphorylation (Figure 16.16). When the blood-glucose concentration is low, the glucagon-triggered cyclic AMP cascade (p. 321) leads to the phosphorylation of pyruvate kinase, which diminishes its activity. This hormone-triggered

Figure 16.16 The control of the catalytic activity of pyruvate kinase. Pyruvate kinase is regulated by allosteric effectors and covalent modification. Fructose 1,6-bisphosphate allosterically stimulates the enzyme, while ATP and alanine (in liver) are allosteric inhibitors. Glucagon, secreted in response to low blood glucose, promotes phosphorylation and inhibition of the enzyme. When blood-glucose levels are adequate, the enzyme is dephosphorylated and activated.

phosphorylation prevents the liver from consuming glucose when it is more urgently needed by brain and muscle. We see here a clear-cut example of how isoenzymes contribute to the metabolic diversity of different organs. We will return to the control of glycolysis after considering gluconeogenesis (Chapter 17).

A Family of Transporters Enables Glucose to Enter and Leave Animal Cells

Several glucose transporters mediate the thermodynamically downhill movement of glucose across the plasma membranes of animal cells. Each member of this protein family, named GLUT1 to GLUT5, consists of a single polypeptide chain of about 500 amino acids (Table 16.3).

Table 16.3 Family of glucose transporters

Name	Tissue location	K_M	Comments
GLUT1	All mammalian tissues	1 mM	Basal glucose uptake
GLUT2	Liver and pancreatic β cells	15–20 mM	In the pancreas, plays a role in the regulation of insulin In the liver, removes excess glucose from the blood
GLUT3	All mammalian tissues	1 mM	Basal glucose uptake
GLUT4	Muscle and fat cells	5 mM	Amount in muscle plasma membrane increases with endurance training
GLUT5	Small intestine	—	Primarily a fructose transporter

The members of this family have distinctive roles:

1. GLUT1 and GLUT3, present in nearly all mammalian cells, are responsible for transporting glucose into the cell under normal conditions. Like enzymes, transporters have K_M values, except that, for transporters, K_M is the concentration of the chemical transported that yields one-half maximal transport velocity. The K_M value for glucose for GLUT1 and GLUT3 is about 1 mM, significantly less than the normal serum-glucose concentration, which typically ranges from 4 mM to 8 mM. Hence, GLUT1 and GLUT3 continuously transport glucose into cells at an essentially constant rate.

2. GLUT2, present in liver and pancreatic β cells, is distinctive in having a very high K_M value for glucose (15–20 mM). Hence, glucose enters these tissues at a biologically significant rate only when there is much glucose in the blood. The pancreas can thereby sense the glucose level and adjust the rate of insulin secretion accordingly. Insulin signals the need to remove glucose from the blood for storage as glycogen or conversion into fat (Chapters 25 and 28). The high K_M value of GLUT2 also ensures that glucose rapidly enters liver cells only in times of plenty.

3. GLUT4, which has a K_M value of 5 mM, transports glucose into muscle and fat cells. The number of GLUT4 transporters in the plasma membrane increases rapidly in the presence of insulin, which signals the presence of glucose in the blood. Hence, insulin promotes the uptake of glucose by muscle and fat. Endurance exercise training also increases the amount of GLUT4 present in muscle membranes by a means independent of insulin.

4. GLUT5, present in the small intestine, functions primarily as a fructose transporter.

⚕ CLINICAL INSIGHT

Aerobic Glycolysis Is a Property of Rapidly Growing Cells

Tumors have been known for decades to display enhanced rates of glucose uptake and glycolysis. Indeed, rapidly growing tumor cells will metabolize glucose to lactate even in the presence of oxygen, a process called *aerobic glycolysis* or the "Warburg effect," after Otto Warburg, the biochemist who first noted this characteristic of cancer cells in the 1920s. In fact, tumors with a high glucose uptake are particularly aggressive, and the cancer is likely to have a poor prognosis. A nonmetabolizable glucose analog, $2\text{-}^{18}\text{F-2-D-deoxyglucose}$, detectable by a combination of positron emission tomography (PET) and computer-aided tomography (CAT), easily visualizes tumors (Figure 16.17).

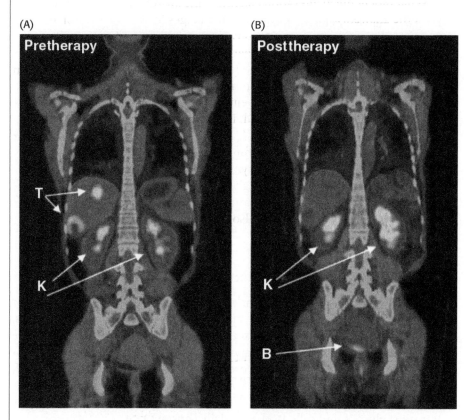

Figure 16.17 Tumors can be visualized with $2\text{-}^{18}\text{F-2-D-deoxyglucose}$ (FDG) and positron emission tomography. (A) A nonmetabolizable glucose analog (FDG) infused into a patient and detected by a combination of positron emission and computer-aided tomography reveals the presence of a malignant tumor (T). (B) After 4 weeks of treatment with a tyrosine kinase inhibitor, the tumor shows no uptake of FDG, indicating decreased metabolism. Excess FDG, which is excreted in the urine, also visualizes the kidney (K) and bladder (B). [Image courtesy of A. D. Van den Abbeele, Dana-Farber Cancer Institute, Boston.]

What selective advantage does aerobic glycolysis offer the tumor over the energetically more efficient oxidative phosphorylation? Research is being actively pursued to answer the question, but we can speculate on the benefits. First, aerobic glycolysis generates lactic acid that is then secreted. Acidification of the tumor environment has been shown to facilitate tumor invasion and inhibit the immune system from attacking the tumor. However, even leukemia cells perform aerobic glycolysis, and leukemia is not an invasive cancer. Second, and perhaps more importantly, the increased uptake of glucose and formation of glucose 6-phosphate provides substrates for another metabolic pathway, the pentose phosphate pathway (Chapter 20), that generates biosynthetic reducing power. Moreover, the pentose phosphate pathway, in cooperation with glycolysis, produces precursors for biomolecules necessary for growth, such as nucleotides. Finally, cancer cells grow more rapidly than the blood vessels that nourish them; thus, as solid tumors grow, the oxygen concentration in their environment falls. In other words, they begin to experience *hypoxia*, a deficiency of oxygen. The use of aerobic glycolysis reduces the dependence of cell growth on oxygen. Not all of the precursor needs are met by enhanced glucose metabolism. Cancer cells also require glutamine, which is channeled into the mitochondria to replenish citric acid cycle components used for biosynthesis.

What biochemical alterations facilitate the switch to aerobic glycolysis? Again, the answers are not complete, but changes in gene expression of isozymic forms of two glycolytic enzymes may be crucial. Tumor cells express an isozyme of hexokinase that binds to mitochondria. There, the enzyme has ready access to any ATP generated by oxidative phosphorylation and is not susceptible to feedback inhibition by its product, glucose 6-phosphate. More importantly, an embryonic isozyme of pyruvate kinase, pyruvate kinase M, is also expressed. Remarkably, this isozyme has a lower catalytic rate than normal pyruvate kinase and creates a bottleneck, allowing the use of glycolytic intermediates for biosynthetic processes required for cell proliferation. The need for biosynthetic precursors is greater than the need for ATP, suggesting that even glycolysis at a reduced rate produces sufficient ATP to allow cell proliferation. Although originally discovered in cancer cells, the Warburg effect is also observed in noncancerous rapidly dividing cells.

Table 16.4 Proteins in glucose metabolism encoded by genes regulated by hypoxia-inducible factor

GLUT1
GLUT3
Hexokinase
Phosphofructokinase
Aldolase
Glyceraldehyde 3-phosphate dehydrogenase
Phosphoglycerate kinase
Enolase
Pyruvate kinase

 CLINICAL INSIGHT

Cancer and Exercise Training Affect Glycolysis in a Similar Fashion

The hypoxia that some tumors experience with rapid growth activates a transcription factor, *hypoxia-inducible transcription factor* (HIF-1). HIF-1 increases the expression of most glycolytic enzymes and the glucose transporters GLUT1 and GLUT3 (Table 16.4). These adaptations by the cancer cells enable a tumor to survive until blood vessels can grow. HIF-1 also increases the expression of signal molecules, such as vascular endothelial growth factor (VEGF), that facilitate the growth of blood vessels that will provide nutrients to the cells (Figure 16.18). Without new blood vessels, a tumor would cease to grow and either die or remain harmlessly small. Efforts are underway to develop drugs that inhibit the growth of blood vessels in tumors.

Interestingly, anaerobic exercise training activates HIF-1, producing the same effects as those seen in the tumor—enhanced ability to generate ATP anaerobically and a stimulation of blood-vessel growth. These biochemical effects account for the improved athletic performance that results from training and demonstrate how behavior can affect biochemistry. Other signals from sustained muscle contraction trigger muscle mitochondrial biogenesis, allowing aerobic energy generation and forestalling the need to resort to aerobic glycolysis.

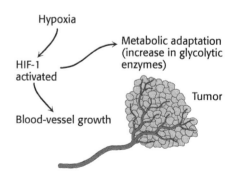

Figure 16.18 The alteration of gene expression in tumors because of hypoxia. The hypoxic conditions inside a tumor mass lead to the activation of the hypoxia-inducible transcription factor (HIF-1), which induces metabolic adaptation (an increase in glycolytic enzymes) and activates angiogenic factors that stimulate the growth of new blood vessels. [Information from C. V. Dang and G. L. Semenza, *Trends Biochem. Sci.* 24:68–72, 1999.]

16.5 Metabolism in Context: Glycolysis Helps Pancreatic Beta Cells Sense Glucose

Insulin is a polypeptide hormone secreted by the β cells of the pancreas in response to an increase in blood-glucose concentration, as would be the case after a meal. The function of insulin is to stimulate the uptake of glucose by tissues, notably muscle and adipose tissue. Although many of the details of how glucose stimulates insulin secretion remain to be elucidated, the general outline of the process is becoming clear.

Glucose enters the β cells of the pancreas through the glucose transporter GLUT2. As already discussed, GLUT2 will allow glucose entry only when blood glucose is plentiful. The β cell metabolizes glucose glycolytically to pyruvate, which is subsequently processed by cellular respiration to CO_2 and H_2O (Sections 8 and 9). This metabolism generates ATP, as we have already seen for glycolysis (Figure 16.19). The resulting increase in the ATP/ADP ratio closes an

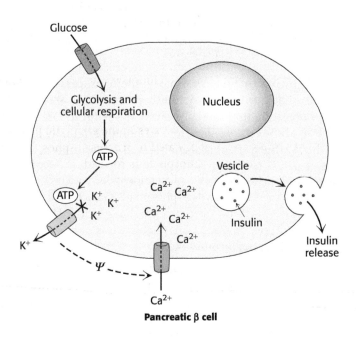

Figure 16.19 Insulin release is regulated by ATP. The metabolism of glucose by glycolysis increases the concentration of ATP, which causes an ATP-sensitive potassium channel to close. The closure of this channel alters the charge across the membrane (Ψ) and causes a calcium channel to open. The influx of calcium causes insulin-containing granules to fuse with the plasma membrane, releasing insulin into the blood.

ATP-sensitive K^+ channel that, when open, allows potassium to flow out of the β cell. The resulting alteration in the cellular ionic environment results in the opening of a Ca^{2+} channel. The influx of Ca^{2+} causes insulin-containing secretory vesicles to fuse with the cell membrane and release insulin into the blood. Thus, the increase in energy charge resulting from the metabolism of glucose has been translated by ion channels into a physiological response—the secretion of insulin and the subsequent removal of glucose from the blood.

SUMMARY

16.1 Glycolysis Is an Energy-Conversion Pathway

Glycolysis is the set of reactions that converts glucose into pyruvate. The 10 reactions of glycolysis take place in the cytoplasm. In the first stage, glucose is converted into fructose 1,6-bisphosphate by a phosphorylation, an isomerization, and a second phosphorylation reaction. Two molecules of ATP are consumed per molecule of glucose in these reactions, which are the prelude to the net synthesis of ATP. Fructose 1,6-bisphosphate is then cleaved by aldolase into dihydroxyacetone phosphate and glyceraldehyde 3-phosphate, which are readily interconvertible. In the second stage, ATP is generated. Glyceraldehyde 3-phosphate is oxidized and phosphorylated to form 1,3-bisphosphoglycerate, an acyl phosphate with a high phosphoryl-transfer potential. This molecule transfers a phosphoryl group to ADP to form ATP and 3-phosphoglycerate. A phosphoryl shift and a dehydration form phosphoenolpyruvate, a second intermediate with a high phosphoryl-transfer potential. Another molecule of ATP is generated as phosphoenol-pyruvate is converted into pyruvate. There is a net gain of two molecules of ATP in the formation of two molecules of pyruvate from one molecule of glucose.

16.2 NAD^+ Is Regenerated from the Metabolism of Pyruvate

The electron acceptor in the oxidation of glyceraldehyde 3-phosphate is NAD^+, which must be regenerated for glycolysis to continue. In aerobic organisms, the NADH formed in glycolysis transfers its electrons to O_2 through the electron-transport chain, which thereby regenerates NAD^+. Under anaerobic conditions and in some microorganisms, NAD^+ is regenerated by the reduction of pyruvate to lactate. In other microorganisms, NAD^+ is regenerated by the reduction of pyruvate to ethanol. These two processes are examples of fermentations.

16.3 Fructose and Galactose Are Converted into Glycolytic Intermediates

Although glucose is the most common monosaccharide fuel molecule, fructose and galactose also are frequently available. These monosaccharides must be converted into intermediates in the glycolytic pathway to be used as fuel. Fructose is phosphorylated to fructose 6-phosphate in most tissues. In the liver, fructose 1-phosphate is formed and cleaved to yield dihydroxyacetone phosphate and glyceraldehyde, which is subsequently phosphorylated to glyceraldehyde 3-phosphate. Galactose is converted into glucose 1-phosphate in a four-step process that includes UDP-glucose, the activated form of glucose. Glucose 1-phosphate is converted into glucose 6-phosphate by phosphoglucomutase.

16.4 The Glycolytic Pathway Is Tightly Controlled

The glycolytic pathway has a dual role: (1) it degrades glucose to generate ATP, and (2) it provides building blocks for the synthesis of cellular components. The rate of conversion of glucose into pyruvate is regulated to meet these two major cellular needs. Under physiological conditions, the reactions of glycolysis are readily reversible, except for those catalyzed by hexokinase, phosphofructokinase, and pyruvate kinase. Phosphofructokinase, the most important control element in glycolysis, is inhibited by high levels of ATP and citrate, and it is activated by AMP and fructose 2,6-bisphosphate. In the liver, this bisphosphate signals that glucose is abundant. Hence, phosphofructokinase is active when either energy or building blocks are needed. Hexokinase is inhibited by glucose 6-phosphate, which accumulates when phosphofructokinase is inactive. ATP and alanine allosterically inhibit pyruvate kinase, the other control site, and fructose 1,6-bisphosphate activates the enzyme. Consequently, pyruvate kinase is maximally active when the energy charge is low and glycolytic intermediates accumulate.

16.5 Metabolism in Context: Glycolysis Helps Pancreatic Beta Cells Sense Glucose

The increase in the ratio of ATP/ADP that results from the metabolism of glucose to pyruvate closes K^+ channels in the membranes of β cells of the pancreas. The resulting change in the cellular ionic environment causes an influx of Ca^{2+}, which, in turn, leads to insulin secretion. Insulin stimulates the uptake of glucose by muscle and adipose tissue.

KEY TERMS ||||||||||||||||||||||||||||||||||||||||||||||||||||||||||||||||||||||||||||||||||||||

glycolysis (p. 284)
hexokinase (p. 284)
kinase (p. 284)
phosphofructokinase (PFK) (p. 287)
thioester intermediate (p. 289)

substrate-level phosphorylation (p. 289)
mutase (p. 290)
enol phosphate (p. 290)
pyruvate kinase (p. 290)
alcoholic fermentation (p. 292)

lactic acid fermentation (p. 293)
obligate anaerobe (p. 294)
committed step (p. 300)
feedforward stimulation (p. 301)
isozyme (p. 302)

❓ Answers to QUICK QUIZZES

1. Two molecules of ATP are produced per molecule of glyceraldehyde 3-phosphate, and, because two molecules of GAP are produced per molecule of glucose, the total ATP yield is four. However, two molecules of ATP are required to convert glucose into fructose 1,6-bisphosphate. Thus, the net yield is only two molecules of ATP.

2. In both cases, the electron donor is glyceraldehyde 3-phosphate. In lactic acid fermentation, the electron acceptor is pyruvate, converting it into lactate. In alcoholic fermentation, acetaldehyde is the electron acceptor, forming ethanol.

PROBLEMS ||||||||||||||||||||||||||||||||||||||||||||||||||||||||||||||||||||||||||||||||||||||||||||||||||||||||||

1. *Just right.* Suggest some possible reasons why glucose is fuel used by all organisms.

2. *Like the owl and the pussycat.* Match each term with its description.

(a) Hexokinase _____
(b) Phosphoglucose isomerase _____
(c) Phosphofructokinase _____
(d) Aldolase _____
(e) Triose phosphate isomerase _____
(f) Glyceraldehyde 3-phosphate dehydrogenase _____
(g) Phosphoglycerate kinase _____
(h) Phosphoglycerate mutase _____
(i) Enolase _____
(j) Pyruvate kinase _____

1. Forms fructose 1,6-bisphosphate
2. Generates the first high-phosphoryl-transfer-potential compound that is not ATP
3. Converts glucose 6-phosphate into fructose 6-phosphate
4. Phosphorylates glucose
5. Generates the second molecule of ATP
6. Cleaves fructose 1,6-bisphosphate
7. Generates the second high-phosphoryl-transfer-potential compound that is not ATP
8. Catalyzes the interconversion of three-carbon isomers
9. Converts 3-phosphoglycerate into 2-phosphoglycerate
10. Generates the first molecule of ATP

3. *Who takes? Who gives?* Lactic acid fermentation and alcoholic fermentation are oxidation–reduction reactions. Identify the ultimate electron donor and electron acceptor.

4. *ATP yield.* Each of the following molecules is processed by glycolysis to lactate. How much ATP is generated from each molecule?

(a) Glucose 6-phosphate
(b) Dihydroxyacetone phosphate
(c) Glyceraldehyde 3-phosphate
(d) Fructose
(e) Sucrose

5. *Enzyme redundancy?* Why is it advantageous for the liver to have both hexokinase and glucokinase to phosphorylate glucose?

6. *Enzyme properties.* In the liver and the pancreas, hexokinase and glucokinase phosphorylate glucose. Glucokinase is active only when the blood concentration of glucose is high. How might glucokinase differ kinetically from hexokinase so as to function only at high glucose levels?

7. *Required isomerization.* Why is the isomerization of glucose 6-phosphate to fructose 6-phosphate an important step in glycolysis? How is the conversion of the fructose isomer back into the glucose isomer prevented?

8. *Magic?* The interconverison of DHAP and GAP greatly favors the formation of DHAP at equilibrium. Yet the conversion of DHAP by triose phosphate isomerase proceeds readily. Why? ✓ 1

9. *Between two extremes.* What is the role of a thioester in the formation of ATP in glycolysis?

10. *Corporate sponsors.* Some of the early research on glycolysis was supported by the brewing industry. Why would the brewing industry be interested in glycolysis? ✓ 2

11. *Recommended daily allowance.* The recommended daily allowance for the vitamin niacin is 15 mg per day. How would glycolysis be affected by niacin deficiency? ✓ 2

12. *Who's on first?* Although both hexokinase and phosphofructokinase catalyze irreversible steps in glycolysis and the hexokinase-catalyzed step is first, phosphofructokinase is nonetheless the pacemaker of glycolysis. What does this information tell you about the fate of the glucose 6-phosphate formed by hexokinase?

13. *The tortoise and the hare.* Why is the regulation of phosphofructokinase by energy charge not as important in the liver as it is in muscle? ✓ 1

14. *Running in reverse.* Why can't the reactions of the glycolytic pathway simply be run in reverse to synthesize glucose?

15. *Destiny.* What are the principle fates of pyruvate generated in glycolysis?

16. *Road blocks.* What reactions of glycolysis are irreversible under intracellular conditions?

17. *No pickling.* Why is it in the muscle's best interest to export lactic acid into the blood during intense exercise?

18. *Proper preparation.* Describe the pathways by which fructose is prepared for entry into glycolysis.

19. *Trouble ahead.* Suppose that an obligate anaerobe suffered a mutation that resulted in the loss of triose phosphate isomerase activity. How would this loss affect the ATP yield of fermentation? Could such an organism survive?

20. *High potential.* What is the equilibrium ratio of phosphoenolpyruvate to pyruvate under standard conditions when $[ATP]/[ADP] = 10$?

21. *Hexose–triose equilibrium.* What are the equilibrium concentrations of fructose 1,6-bisphosphate, dihydroxyacetone phosphate, and glyceraldehyde 3-phosphate when 1 mM fructose 1,6-bisphosphate is incubated with aldolase under standard conditions?

22. *Distinctive sugars.* The intravenous infusion of fructose into healthy volunteers leads to a two- to fivefold increase in the level of lactate in the blood, a far greater increase than that observed after the infusion of the same amount of glucose.

(a) Why is glycolysis more rapid after the infusion of fructose?
(b) Fructose has been used in place of glucose for intravenous feeding. Why is this use of fructose unwise?

23. *Arsenate poisoning.* Arsenate (AsO_4^{3-}) closely resembles P_i in structure and reactivity. In the reaction catalyzed by glyceraldehyde 3-phosphate dehydrogenase, arsenate can replace phosphate in attacking the energy-rich thioester intermediate. The product of this reaction, 1-arseno-3-phosphoglycerate, is unstable. It and other acyl arsenates are rapidly and spontaneously hydrolyzed. What is the effect of arsenate on energy generation in a cell?

24. *Reduce, reuse, recycle.* In the conversion of glucose into two molecules of lactate, the NADH generated earlier in the pathway is oxidized to NAD^+. Why is it not to the cell's advantage to simply make more NAD^+ so that the regeneration would not be necessary? After all, the cell would save much energy because it would no longer need to synthesize lactic acid dehydrogenase. ✓ 2

25. *Diverted resources.* Phosphofructokinase converts fructose 6-phosphate to fructose 1,6-bisphosphate, the committed step on the pathway that synthesizes ATP. However, some fructose 6-phosphate is converted into fructose 2,6-bisphosphate. Explain why depleting the substrate of PFK to form fructose 2,6-bisphosphate is not a wasteful use of substrate.

26. *Like Batman and Robin.* Match parts *a* through *j* with parts 1 through 10.

(a) Glucose 6-phosphate _____	1. Inhibits phosphofructokinase in the liver
(b) Low ATP/AMP ratio _____	2. Glucokinase
(c) Citrate _____	3. GLUT2
(d) Low pH _____	4. Inhibits hexokinase
(e) Fructose 1,6-bisphosphate _____	5. Inhibits phosphofructokinase in muscle
(f) Fructose 2,6-bisphosphate _____	6. Inhibits phosphofructokinase
(g) Insulin _____	7. Stimulates pyruvate kinase
(h) Has a high K_M for glucose _____	8. Stimulates phosphofructokinase in the liver
(i) Transporter specific to liver and pancreas _____	9. Causes the insertion of GLUT4 into cell membranes
(j) High ATP/AMP ratio _____	10. Stimulates phosphofructokinase

Chapter Integration Problems

27. *State function.* Fructose 2,6-bisphosphate is a potent stimulator of phosphofructokinase. Explain how fructose 2,6-bisphosphate might function in the concerted model for allosteric enzymes.

28. *Not just for energy.* People with galactosemia display central nervous system abnormalities even if galactose is eliminated from the diet. The precise reason for this is not known. Suggest a plausible explanation.

29. *Not MTV.* A ligand-gated ion channel and a voltage-gated ion channel play keys roles in the secretion of insulin by the pancreas. What are the channels, and what is their function in insulin secretion? ✓ 2

30. *Power to the rbc!* Hexokinase in red blood cells has a K_M of approximately 50 μM. Because life is hard enough as it is, let's assume that the hexokinase displays Michaelis–Menten kinetics. What concentration of blood glucose would yield v_o equal to 90% V_{max}? What does this result tell you if normal blood-glucose levels range between approximately 3.6 and 6.1 mM?

Data Interpretation and Challenge Problems

31. *Now that's unusual.* Phosphofructokinase has been isolated from the hyperthermophilic archaeon *Pyrococcus furiosus*. The kinase was subjected to standard biochemical analysis to determine basic catalytic parameters. The processes under study were of the form

Fructose 6-Phosphate $+ (x - P_i) \longrightarrow$
$$\text{fructose } 1, 6\text{-biphosphate} + (x)$$

The assay measured the increase in fructose 1,6-bisphosphate. Selected results are shown in the adjoining graph.

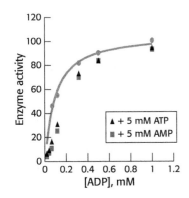

[Data from J. E. Tuininga et al., *J. Biol. Chem.* 274:21023–21028, 1999.]

(a) How does the *P. furiosus* phosphofructokinase differ from the phosphofructokinase considered in this chapter?
(b) What effects do AMP and ATP have on the reaction with ADP?

32. *Confused?* The adjoining graph shows muscle phosphofructokinase activity as a function of ATP concentration in the presence of a constant concentration of fructose 6-phosphate. Explain these results, and discuss how they relate to the role of phosphofructokinase in glycolysis. ✓ 1

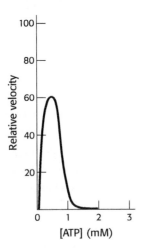

33. *Brewing.* The adjoining graph shows results of experiments on the alcoholic fermentation of glucose with the use of yeast extracts. The graph shows the volume of carbon dioxide released (*y* axis) as a function of time (*x* axis). ✓ 2

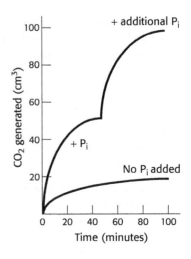

(a) Is measuring the rate of carbon dioxide release a reliable means to measure alcoholic fermentation? Explain.
(b) Why is glucose fermentation dependent on phosphate?
(c) Why is more carbon dioxide generated if the extract is supplemented with phosphate?
(d) During the fermentation, what would the ratio of carbon dioxide generated to phosphate consumed be expected to be?
(e) When fermentation slowed, a hexose bisphosphate accumulated. What is this compound, and why would it accumulate?

Challenge Problems

34. *Kitchen chemistry.* Sucrose is commonly used to preserve fruits. Why is glucose not suitable for preserving foods? ✓ 1

35. *Tracing carbon atoms.* Glucose labeled with ^{14}C at C-1 is incubated with the glycolytic enzymes and necessary cofactors.

(a) What is the location of ^{14}C in the pyruvate that is formed? (Assume that the interconversion of glyceraldehyde 3-phosphate and dihydroxyacetone phosphate is very rapid compared with the subsequent step.)
(b) If the specific activity of the glucose substrate is 10 mCi mmol^{-1}, what is the specific activity of the pyruvate that is formed?

36. *Lactic acid fermentation.* ✓ 2

(a) Write a balanced equation for the conversion of glucose into lactate.
(b) Calculate the standard free-energy change of the formation of lactate from glucose by using the data given in Table 16.1 and the fact that $\Delta G^{\circ\prime}$ is -25 kJ mol^{-1} (-6 kcal mol^{-1}) for the following reaction:

$$\text{Pyruvate} + \text{NADH} + \text{H}^+ \rightleftharpoons \text{lactate} + \text{NAD}^+$$

(c) What is the free-energy change (ΔG, not $\Delta G^{\circ\prime}$) of this reaction when the concentrations of reactants are glucose, 5 mM; lactate, 0.05 mM; ATP, 2 mM; ADP, 0.2 mM; and P$_i$, 1 mM?

37. *Après vous.* Why is it physiologically advantageous for the pancreas to use GLUT2, with a high K_M value, as the transporter that allows glucose entry into β cells?

38. *Bypass.* In the liver, fructose can be converted into glyceraldehyde 3-phosphate and dihydroxyacetone phosphate without passing through the phosphofructokinase-regulated reaction. Show the reactions that make this conversion possible. Why might ingesting high levels of fructose have deleterious physiological effects? ✓ 1

39. *An informative analog.* Xylose has the same structure as that of glucose except that xylose has a hydrogen atom at C-5 in place of a hydroxymethyl group. The rate of ATP hydrolysis by hexokinase is markedly enhanced by the addition of xylose. Why? ✓ 1

40. *Quick-change artist.* Adenylate kinase is responsible for interconverting the adenylate nucleotide pool:

$$ADP + ADP \rightleftharpoons ATP + AMP$$

The equilibrium constant for this reaction is close to 1, inasmuch as the number of phosphoanhydride bonds is the same on each side of the equation. Use the equation for the equilibrium constant for this reaction to show why changes in [AMP] are a more effective indicator of the adenylate pool than those in [ATP]. ✓ 1

Selected Readings for this chapter can be found online at www.whfreeman.com/tymoczko3e.

CHAPTER 17

Gluconeogenesis

Fasting is a part of many cultures and religions, including those of the Teton Sioux. Fasting is believed to cleanse the body and soul and to foster spiritual awakening. Gluconeogenesis is an important metabolic pathway during times of fasting because it supplies glucose to the brain and red blood cells, tissues that depend on this vital fuel.
[Library of Congress, Prints & Photographs Division, Edward S. Curtis Collection, LC-USZ62-99611.]

We now turn to *the synthesis of glucose from noncarbohydrate precursors, a process called gluconeogenesis*. Maintaining levels of glucose is important because the brain depends on glucose as its primary fuel and red blood cells use glucose as their only fuel, as stated earlier (p. 281). The daily glucose requirement of the brain in a typical adult human being is about 120 g, which accounts for most of the 160 g of glucose needed daily by the whole body. The amount of glucose present in body fluids is about 20 g, and that readily available from glycogen, the storage form of glucose, is approximately 190 g. Thus, the direct glucose reserves are sufficient to meet glucose needs for about a day. Gluconeogenesis is especially important during a longer period of fasting or starvation.

The major site of gluconeogenesis is the liver, with a small amount also taking place in the kidney. Little gluconeogenesis takes place in the brain, skeletal muscle, or heart muscle. Rather, *gluconeogenesis in the liver and kidney helps to maintain the glucose concentration in the blood, from which it can be extracted by the brain and muscle to meet their metabolic demands.*

In this chapter, we begin by examining the reactions that constitute the gluconeogenic pathway. We then investigate the reciprocal regulation of gluconeogenesis and glycolysis. The chapter ends with a look at how gluconeogenesis and glycolysis are coordinated between tissues.

✓ 3 Describe how gluconeogenesis is powered in the cell.

17.1 Glucose Can Be Synthesized from Noncarbohydrate Precursors

Although glucose is usually available in the environment of most organisms, this molecule is such a common fuel that a pathway exists in virtually all forms of life to synthesize it from simple precursors. *The gluconeogenic pathway converts pyruvate into glucose.* Noncarbohydrate precursors of glucose are first converted into pyruvate or enter the pathway at later intermediates (**Figure 17.1**). The major noncarbohydrate precursors are *lactate, amino acids,* and *glycerol.* Lactate is formed by active skeletal muscle through lactic acid fermentation when the rate of glycolysis exceeds the rate at which muscle can process pyruvate aerobically (p. 293). Lactate is readily converted into pyruvate in the liver by the action of lactate dehydrogenase. Amino acids are derived from proteins in the diet and, during starvation, from the breakdown of proteins in skeletal muscle (Chapter 32). The hydrolysis of triacylglycerols (Chapter 27) in fat cells yields glycerol and fatty acids. Glycerol is a precursor of glucose, but animals cannot convert fatty acids into glucose for reasons that will be given later (Chapter 27). Unlike lactate and amino acids, glycerol can be metabolized by glycolysis or converted into glucose by gluconeogenesis. Glycerol may enter either the gluconeogenic or the glycolytic pathway at dihydroxyacetone phosphate.

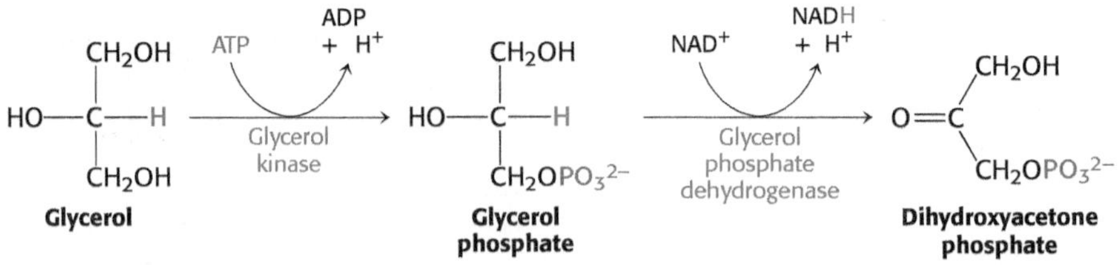

Gluconeogenesis Is Not a Complete Reversal of Glycolysis

In glycolysis, glucose is converted into pyruvate; in gluconeogenesis, pyruvate is converted into glucose. However, *gluconeogenesis is not a reversal of glycolysis.* Several reactions must differ because the equilibrium of glycolysis lies far on the side of pyruvate formation. The actual free energy change for the formation of pyruvate from glucose is about -90 kJ mol^{-1} ($-22 \text{ kcal mol}^{-1}$) under typical cellular

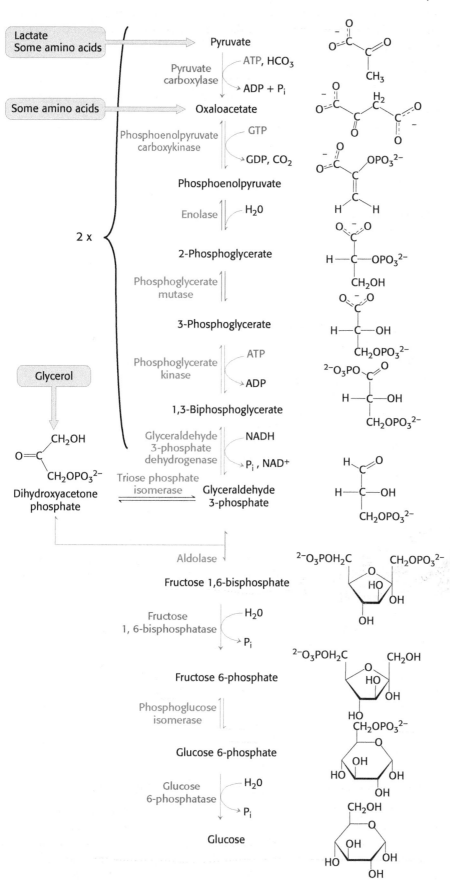

Figure 17.1 The pathway of gluconeogenesis. The distinctive reactions and enzymes of this pathway are shown in red. The other reactions are common to glycolysis. The enzymes for gluconeogenesis are located in the cytoplasm, except for pyruvate carboxylase (in the mitochondria) and glucose 6-phosphatase (membrane bound in the endoplasmic reticulum). The entry points for lactate, glycerol, and amino acids are shown.

conditions. Most of the decrease in free energy in glycolysis takes place in the three irreversible steps catalyzed by hexokinase, phosphofructokinase, and pyruvate kinase:

$$\text{Glucose} + \text{ATP} \xrightarrow{\text{Hexokinase}} \text{glucose 6-phosphate} + \text{ADP}$$

$$\Delta G = -33 \text{ kJ mol}^{-1} \ (-8.0 \text{ kcal mol}^{-1})$$

$$\text{Fructose 6-phosphate} + \text{ATP} \xrightarrow{\text{Phosphofructokinase}}$$

$$\text{fructose 1,6-bisphosphate} + \text{ADP}$$

$$\Delta G = -22 \text{ kJ mol}^{-1} \ (-5.3 \text{ kcal mol}^{-1})$$

$$\text{Phosphoenolpyruvate} + \text{ADP} \xrightarrow{\text{Pyruvate kinase}} \text{pyruvate} + \text{ATP}$$

$$\Delta G = -17 \text{ kJ mol}^{-1} \ (-4.0 \text{ kcal mol}^{-1})$$

In gluconeogenesis, these irreversible reactions of glycolysis must be bypassed.

The Conversion of Pyruvate into Phosphoenolpyruvate Begins with the Formation of Oxaloacetate

The first step in gluconeogenesis is the carboxylation of pyruvate to form oxaloacetate at the expense of a molecule of ATP, a reaction catalyzed by *pyruvate carboxylase*. This reaction occurs in the mitochondria.

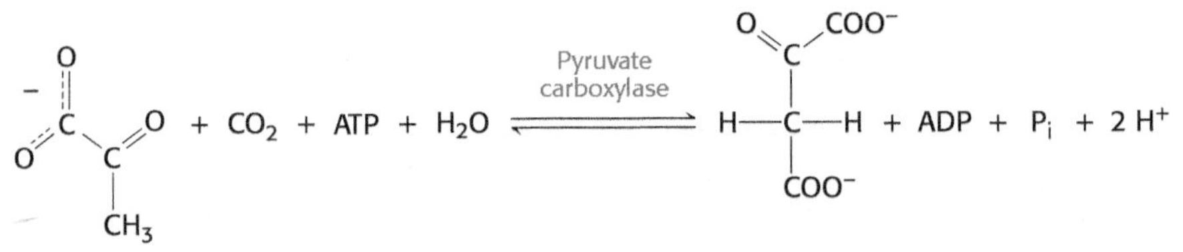

Pyruvate carboxylase requires *biotin, a covalently attached prosthetic group that serves as the carrier of activated* CO_2. The carboxylate group of biotin is linked to the ε-amino group of a specific lysine residue by an amide bond (Figure 17.2). Recall that, in aqueous solutions, CO_2 exists primarily as HCO_3^- with the aid of carbonic anhydrase (p. 155).

The carboxylation of pyruvate takes place in three stages:

$$HCO_3^- + \text{ATP} \rightleftharpoons HOCO_2\text{-}PO_3^{2-} + \text{ADP}$$

$$\text{Biotin–enzyme} + HOCO_2\text{-}PO_3^{2-} \longrightarrow CO_2\text{–biotin–enzyme} + P_i$$

$$CO_2\text{–biotin–enzyme} + \text{pyruvate} \rightleftharpoons \text{biotin–enzyme} + \text{oxaloacetate}$$

Pyruvate carboxylase functions as a tetramer composed of four identical subunits, and each subunit consists of four domains (Figure 17.3). The biotin carboxylase domain catalyzes the formation of carboxyphosphate and the subsequent attachment of CO_2 to the second domain, the biotin carboxyl carrier protein (BCCP), the site of the covalently attached biotin. Once bound to CO_2, BCCP leaves the biotin carboxylase active site and swings almost the entire length of the subunit ($\approx 75\text{Å}$) to the active site of the pyruvate carboxylase domain, which transfers the CO_2 to pyruvate to form oxaloacetate. BCCP in one subunit interacts with the active sites on an adjacent subunit. The fourth domain facilitates the formation of the tetramer.

The first partial reaction of pyruvate carboxylase, the formation of carboxybiotin, depends on the presence of acetyl CoA. *Biotin is not carboxylated unless acetyl CoA is bound to the enzyme*. Acetyl CoA has no effect on the second

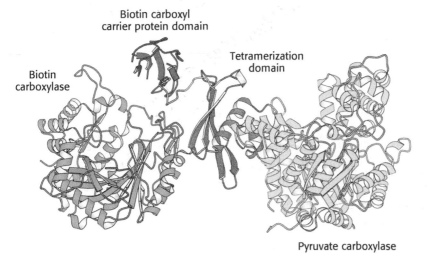

Biotin

Carboxybiotin covalently bound to
ε-amino group of a lysine

Figure 17.2 Structure of biotin and carboxybiotin.

Biotin carboxyl
carrier protein domain

Tetramerization
domain

Biotin
carboxylase

Biotin
carboxylase

Pyruvate carboxylase

Figure 17.3 A subunit of pyruvate carboxylase. Biotin, covalently attached to the biotin carboxyl carrier domain, transports CO_2 from the biotin carboxylase active site to the pyruvate carboxylase active site of an adjacent subunit. [Information from G. Lasso, L. P. C. Yu, D. Gil, S. Xiang, L. Tong, and M. Valle, *Structure* 18:1300–1310, 2010.]

NUTRITION FACTS

Biotin Also called vitamin B_7, biotin is used in CO_2 transfer and carboxylation reactions. Biotin deficiency is characterized by muscle pain, lethargy, anorexia, and depression. This vitamin is synthesized by microflora in the intestinal tract and can be obtained in the diet from liver, soybeans, nuts, and many other sources. [Photograph from Charles Bruglag/FeaturesPics.]

partial reaction. The allosteric activation of pyruvate carboxylase by acetyl CoA is an important physiological control mechanism that will be discussed in Section 17.4.

Oxaloacetate Is Shuttled into the Cytoplasm and Converted into Phosphoenolpyruvate

Oxaloacetate must be transported to the cytoplasm to complete the synthesis of phosphoenolpyruvate. Oxaloacetate is first reduced to malate by malate dehydrogenase. Malate is transported across the mitochondrial membrane and

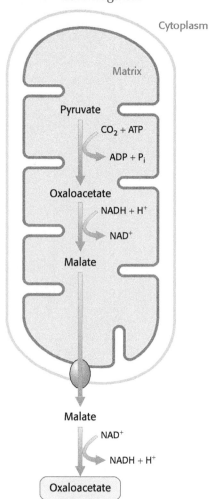

Figure 17.4 Compartmental cooperation. Oxaloacetate used in the cytoplasm for gluconeogenesis is formed in the mitochondrial matrix by the carboxylation of pyruvate. Oxaloacetate leaves the mitochondrion by a specific transport system in the form of malate, which is reoxidized to oxaloacetate in the cytoplasm.

reoxidized to oxaloacetate by a cytoplasmic NAD^+-linked malate dehydrogenase (Figure 17.4). The formation of oxaloacetate from malate also provides NADH for use in subsequent steps in gluconeogenesis. Finally, oxaloacetate is simultaneously *decarboxylated* and *phosphorylated* by *phosphoenolpyruvate carboxykinase* (PEPCK) to generate phosphoenolpyruvate. The phosphoryl donor is GTP. The CO_2 that was added to pyruvate by pyruvate carboxylase comes off in this step:

The sum of the reactions catalyzed by pyruvate carboxylase and phosphoenolpyruvate carboxykinase is

$$\text{Pyruvate} + \text{ATP} + \text{GTP} + \text{H}_2\text{O} \longrightarrow$$
$$\text{phosphoenolpyruvate} + \text{ADP} + \text{GDP} + \text{P}_i + 2\,\text{H}^+$$

This pair of reactions bypasses the irreversible reaction catalyzed by pyruvate kinase in glycolysis.

Why is a carboxylation and a decarboxylation required to form phosphoenolpyruvate from pyruvate? Recall that, in glycolysis, the presence of a phosphoryl group traps the unstable enol isomer of pyruvate as phosphoenolpyruvate (p. 290). However, the addition of a phosphoryl group to pyruvate is a highly unfavorable reaction: the $\Delta G^{\circ\prime}$ of the reverse of the glycolytic reaction catalyzed by pyruvate kinase is $+31$ kJ mol^{-1} ($+7.5$ kcal mol^{-1}). In gluconeogenesis, the use of the carboxylation and decarboxylation steps results in a much more favorable $\Delta G^{\circ\prime}$. The formation of phosphoenolpyruvate from pyruvate in the gluconeogenic pathway has a $\Delta G^{\circ\prime}$ of $+0.8$ kJ mol^{-1} ($+0.2$ kcal mol^{-1}). A molecule of ATP is used to power the addition of a molecule of CO_2 to pyruvate in the carboxylation step. That CO_2 is then removed to power the formation of phosphoenolpyruvate in the decarboxylation step. *Decarboxylations often drive reactions that are otherwise highly endergonic.* This metabolic motif is used in the citric acid cycle (p. 346), the pentose phosphate pathway (p. 474), and fatty acid synthesis (p. 511).

The Conversion of Fructose 1,6-bisphosphate into Fructose 6-phosphate and Orthophosphate Is an Irreversible Step

The newly formed phosphoenolpyruvate is then metabolized by the enzymes of glycolysis but in the reverse direction. These reactions are near equilibrium under intracellular conditions; so, when conditions favor gluconeogenesis, the reverse reactions will take place until the next irreversible step is reached. This step is the hydrolysis of fructose 1,6-bisphosphate to fructose 6-phosphate and P_i.

Fructose 1,6-bisphosphatase is the enzyme responsible for this step and, like its glycolytic counterpart, it is an allosteric enzyme that participates in regulation—in this case, of gluconeogenesis. We will return to its regulatory properties later in the chapter.

The Generation of Free Glucose Is an Important Control Point

The fructose 6-phosphate generated by fructose 1,6-bisphosphatase is readily converted into glucose 6-phosphate. In most tissues, gluconeogenesis ends here. Free glucose is not generated; rather glucose 6-phosphate is commonly converted into glycogen, the storage form of glucose. The final step in the generation of free glucose takes place primarily in the liver, a tissue whose metabolic duty is to maintain adequate levels of glucose in the blood for use by other tissues. Free glucose is not formed in the cytoplasm. Rather, glucose 6-phosphate is transported into the lumen of the endoplasmic reticulum, where it is hydrolyzed to glucose by *glucose 6-phosphatase*, which is bound to the ER membrane (**Figure 17.5**). Glucose and P_i are then shuttled back to the cytoplasm by a pair of transporters.

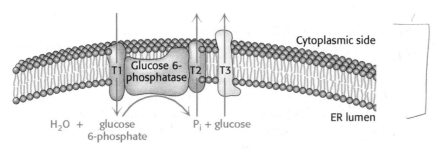

Figure 17.5 Generation of glucose from glucose 6-phosphate. Several endoplasmic reticulum (ER) proteins play a role in the generation of glucose from glucose 6-phosphate. T1 transports glucose 6-phosphate into the lumen of the ER, whereas T2 and T3 transport P_i and glucose, respectively, back into the cytoplasm. [Information from A. Buchell and I. D. Waddel. *Biochem. Biophys. Acta* 1092:129–137, 1991.]

Six High-Transfer-Potential Phosphoryl Groups Are Spent in Synthesizing Glucose from Pyruvate

As we have seen in the preceding sets of reactions, the formation of glucose from pyruvate is energetically unfavorable unless it is coupled to reactions that are favorable. Compare the stoichiometry of gluconeogenesis with that of the reverse of glycolysis. The stoichiometry of gluconeogenesis is

$$2\ \text{Pyruvate} + 4\ \text{ATP} + 2\ \text{GTP} + 2\ \text{NADH} + 2\ \text{H}^+ + 6\ \text{H}_2\text{O} \longrightarrow$$
$$\text{glucose} + 4\ \text{ADP} + 2\ \text{GDP} + 6\ \text{P}_i + 2\ \text{NAD}^+$$
$$\Delta G^{\circ\prime} = -38\ \text{kJ mol}^{-1}\ (-9\ \text{kcal mol}^{-1})$$

In contrast, the stoichiometry for the reversal of glycolysis is

$$\text{Pyruvate} + 2\ \text{ATP} + 2\ \text{NADH} + 2\ \text{H}^+ + 2\ \text{H}_2\text{O} \longrightarrow$$
$$\text{glucose} + 2\ \text{ADP} + 2\ \text{P}_i + 2\ \text{NAD}^+$$
$$\Delta G^{\circ\prime} = +90\ \text{kJ mol}^{-1}\ (+22\ \text{kcal mol}^{-1})$$

Note that *six* nucleoside triphosphate molecules are hydrolyzed to synthesize glucose from pyruvate in gluconeogenesis, whereas only *two* molecules of ATP are generated in glycolysis in the conversion of glucose into pyruvate. Thus, the extra cost of gluconeogenesis is four high-phosphoryl-transfer-potential molecules for

? QUICK QUIZ 1 What barrier prevents glycolysis from simply running in reverse to synthesize glucose? How is this barrier overcome in gluconeogenesis?

each molecule of glucose synthesized from pyruvate. The four additional high-phosphoryl-transfer-potential molecules are needed to turn an energetically unfavorable process (the reversal of glycolysis) into a favorable one (gluconeogenesis). Here, we have a clear example of the coupling of reactions: nucleoside triphosphate (NTP) hydrolysis is used to power an energetically unfavorable reaction.

✓ **4** Describe the coordinated regulation of glycolysis and gluconeogenesis.

17.2 Gluconeogenesis and Glycolysis Are Reciprocally Regulated

Gluconeogenesis and glycolysis are coordinated so that, within a cell, one pathway is relatively inactive while the other one is highly active. If both sets of reactions were highly active at the same time, the net result would be the hydrolysis of four nucleoside triphosphates (two ATP molecules plus two GTP molecules) per reaction cycle. Both glycolysis and gluconeogenesis are exergonic under cellular conditions, and so there is no thermodynamic barrier to such simultaneous activity. However, the activities of the distinctive enzymes of each pathway are controlled so that both pathways are not highly active at the same time. The rate of glycolysis is also determined by the concentration of glucose, and the rate of gluconeogenesis is controlled by the concentrations of lactate and other precursors of glucose. *The basic premise of the reciprocal regulation is that, when glucose is abundant, glycolysis will predominate. When glucose is scarce, gluconeogenesis will take over.*

Energy Charge Determines Whether Glycolysis or Gluconeogenesis Will Be More Active

The key regulation site in the gluconeogenesis pathway is the interconversion of fructose 6-phosphate and fructose 1,6-bisphosphate (**Figure 17.6**). Consider first a situation in which energy is needed. In this case, the concentration of AMP is high. Under this condition, AMP stimulates phosphofructokinase but inhibits fructose 1,6-bisphosphatase. Thus, glycolysis is turned on and gluconeogenesis is inhibited. Conversely, high levels of ATP and citrate indicate that the energy charge is high and that biosynthetic intermediates are abundant. ATP and citrate inhibit phosphofructokinase, whereas citrate activates fructose 1,6-bisphosphatase. Under these conditions, glycolysis is nearly switched off and gluconeogenesis is promoted. Why does citrate take part in this regulatory scheme? As we will see in Chapter 19, citrate reports on the status of the citric acid cycle, the primary pathway for oxidizing fuels in the presence of oxygen. High levels of citrate indicate an energy-rich situation and the presence of precursors for biosynthesis, most notably fatty acid synthesis (Chapter 28).

Glycolysis and gluconeogenesis are also reciprocally regulated at the interconversion of phosphoenolpyruvate and pyruvate in the liver. The glycolytic enzyme pyruvate kinase is inhibited by allosteric effectors ATP and alanine, which signal that the energy charge is high and that building blocks are abundant. Conversely, pyruvate carboxylase, which catalyzes the first step in gluconeogenesis from pyruvate, is inhibited by ADP. Likewise, ADP inhibits phosphoenolpyruvate carboxykinase. Pyruvate carboxylase is activated by acetyl CoA, which, like citrate, indicates that the citric acid cycle is producing energy and biosynthetic intermediates (Chapter 19). Hence, when the cell is rich in biosynthetic precursors and ATP, anabolic pathways are favored. In the liver, gluconeogenesis can produce glucose to replenish glycogen stores (Chapter 25) and fatty acid synthesis can take place.

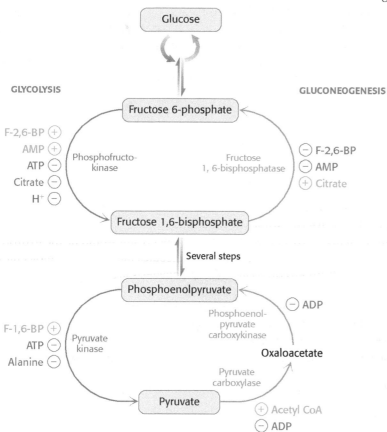

Figure 17.6 The reciprocal regulation of gluconeogenesis and glycolysis in the liver. The level of fructose 2,6-bisphosphate (F-2,6-BP) is high in the fed state and low in starvation. Another important control is the inhibition of pyruvate kinase by phosphorylation during starvation.

The Balance Between Glycolysis and Gluconeogenesis in the Liver Is Sensitive to Blood-Glucose Concentration

In the liver, rates of glycolysis and gluconeogenesis are adjusted to maintain blood-glucose concentration. Recall that fructose 2,6-bisphosphate is a potent activator of phosphofructokinase (PFK), the primary regulatory enzyme in glycolysis (p. 300). Fructose 2,6-bisphosphate is also an inhibitor of fructose 1,6-bisphosphatase. When blood glucose concentration is low, fructose 2,6-bisphosphate is dephosphorylated to form fructose 6-phosphate, which no longer binds to PFK. How is the amount of fructose 2,6-bisphosphate controlled to rise and fall with blood-glucose concentration? Two enzymes regulate the concentration of this molecule: one phosphorylates fructose 6-phosphate, and the other dephosphorylates fructose 2,6-bisphosphate. Fructose 2,6-bisphosphate is formed from fructose 6-phosphate in a reaction catalyzed by *phosphofructokinase 2 (PFK2)*, a different enzyme from phosphofructokinase. In the reverse direction, fructose 6-phosphate is formed through the hydrolysis of fructose 2,6-bisphosphate by a specific phosphatase, *fructose bisphosphatase 2* (FBPase2). The striking finding is that *both PFK2 and FBPase2 are present in a single 55-kDa polypeptide chain* (**Figure 17.7**). This *bifunctional enzyme* contains an N-terminal *regulatory domain*, followed by a *kinase domain* and a *phosphatase domain*.

What controls whether PFK2 or FBPase2 dominates the bifunctional enzyme's activities in the liver? The activities of PFK2 and FBPase2 are reciprocally controlled by *the phosphorylation of a single serine residue*. When glucose is scarce, as it is during a night's fast, a rise in the blood concentration of the hormone glucagon triggers a cyclic AMP signal cascade (p. 228), leading to the phosphorylation

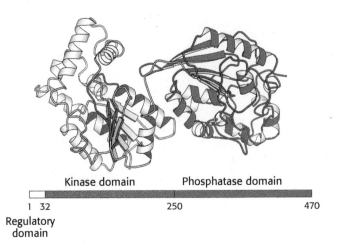

Kinase domain   Phosphatase domain

1 32   250   470

Regulatory
domain

Figure 17.7 The domain structure of the bifunctional regulatory enzyme phosphofructokinase 2/fructose 2,6-bisphosphatase. The kinase domain (purple) is fused to the phosphatase domain (red). The bar represents the amino acid sequence of the enzyme. [Drawn from 1BIF.pdb.]

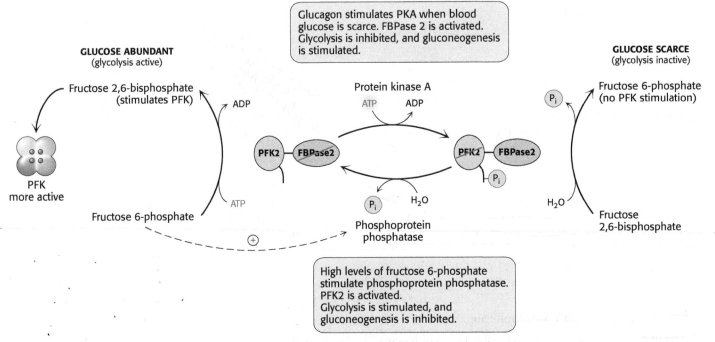

Glucagon stimulates PKA when blood glucose is scarce. FBPase 2 is activated. Glycolysis is inhibited, and gluconeogenesis is stimulated.

GLUCOSE ABUNDANT
(glycolysis active)

Fructose 2,6-bisphosphate
(stimulates PFK)

ADP

Protein kinase A

ATP   ADP

PFK
more active

ATP

PFK2   FBPase2

P_i H_2O

Phosphoprotein
phosphatase

Fructose 6-phosphate

$+$

GLUCOSE SCARCE
(glycolysis inactive)

Fructose 6-phosphate
(no PFK stimulation)

P_i

PFK2   FBPase2

P_i

H_2O

Fructose
2,6-bisphosphate

High levels of fructose 6-phosphate stimulate phosphoprotein phosphatase. PFK2 is activated. Glycolysis is stimulated, and gluconeogenesis is inhibited.

Figure 17.8 Control of the synthesis and degradation of fructose 2,6-bisphosphate. A low blood-glucose level as signaled by glucagon leads to the phosphorylation of the bifunctional enzyme and, hence, to a lower level of fructose 2,6-bisphosphate, slowing glycolysis. High levels of fructose 6-phosphate accelerate the formation of fructose 2,6-bisphosphate by facilitating the dephosphorylation of the bifunctional enzyme.

of this bifunctional enzyme by protein kinase A (Figure 17.8). This covalent modification activates FBPase2 and inhibits PFK2, lowering the concentration of F-2,6-BP. Gluconeogenesis predominates. Glucose formed by the liver under these conditions is essential for the viability of the brain. Glucagon stimulation of protein kinase A also inactivates pyruvate kinase in the liver (p. 302).

Conversely, when blood glucose is abundant, as it is after a meal, glucagon concentration in the blood falls and the insulin concentration rises. Now, gluconeogenesis is not needed and the phosphoryl group is removed from the bifunctional enzyme. This covalent modification activates PFK2 and inhibits FBPase2. The resulting increase in the concentration of F-2,6-BP accelerates glycolysis. The coordinated control of glycolysis and gluconeogenesis is facilitated by the location of the kinase and phosphatase domains on the same polypeptide chain as the regulatory domain.

QUICK QUIZ 2 What are the regulatory means that prevent high levels of activity in glycolysis and gluconeogenesis simultaneously? What would be the result if both pathways functioned rapidly at the same time?

CLINICAL INSIGHT

Insulin Fails to Inhibit Gluconeogenesis in Type 2 Diabetes

Insulin, the hormone that signals the presence of fuels in the blood, or the fed state, normally inhibits gluconeogenesis. When glucose is present in the blood after a meal, there is no need to synthesize glucose. Recall from Chapter 13 that insulin normally starts a signaling cascade that increases the expression of genes that encode the proteins of glycolysis. Insulin also turns off the gene that encodes phosphoenolpyruvate carboxykinase (PEPCK), thereby inhibiting gluconeogenesis. However, in type 2 diabetes (non-insulin-dependent diabetes), insulin, although present, fails to inhibit the expression of the gene that encodes PEPCK and other genes of gluconeogenesis. This metabolic circumstance is called *insulin resistance* and is the defining feature of type 2 diabetes. The higher-than-normal concentration of PEPCK results in an increased output of glucose by the liver even when glucose from the diet is present. Blood glucose rises to abnormally high levels (hyperglycemia). The high concentration of blood glucose alters blood osmolarity resulting in loss of water from tissues and the secretion of vasopressin (also called antidiuretic hormone, ADH), causing excessive thirst (polydipsia). Excess glucose, along with water, is excreted by the kidneys resulting in frequent urination (polyuria). Blurred vision, fatigue, and headache are caused by the lack of energy due to the inability to use glucose as a fuel. The cause of type 2 diabetes, the most common metabolic disease in the world, is unknown, although obesity may be a contributing factor. Untreated, type 2 diabetes can progress to insulin-dependent diabetes. The treatment of type 2 diabetes includes weight loss, a healthy diet, exercise, and drug treatment to enhance sensitivity to insulin (**Figure 17.9**).

Figure 17.9 Diet can help to prevent the development of type 2 diabetes. A healthy diet, one rich in fruits and vegetables, is an important step in preventing or treating type 2 diabetes. [Photodisc/Getty Images.]

CLINICAL INSIGHT

Substrate Cycles Amplify Metabolic Signals

A pair of reactions such as the phosphorylation of fructose 6-phosphate to fructose 1,6-bisphosphate in the glycolytic pathway and its hydrolysis back to fructose 6-phosphate in the gluconeogenic pathway is called a *substrate cycle*. As already mentioned, both reactions are not simultaneously fully active in most cells, because of reciprocal allosteric controls. However, the results of isotope-labeling studies have shown that some fructose 6-phosphate is phosphorylated to fructose 1,6-bisphosphate even in gluconeogenesis. There is also a limited degree of cycling in other pairs of opposed irreversible reactions. This cycling was regarded as an imperfection in metabolic control, and so substrate cycles have sometimes been called *futile cycles*. Indeed, there are pathological conditions, such as malignant hyperthermia, in which control is lost and opposing pathways proceed rapidly. In malignant hyperthermia, there is rapid, uncontrolled hydrolysis of ATP, which generates heat and can raise body temperature to 44°C (111°F). Muscles may become rigid and severely damaged as well.

Despite such extraordinary circumstances, substrate cycles now seem likely to be biologically important. One possibility is that *substrate cycles amplify metabolic signals*. Suppose that the rate of conversion of A into B is 100 and of B into A is 90, giving an initial net flux of 10. Assume that an allosteric effector increases the A → B rate by 20% to 120 and reciprocally decreases the

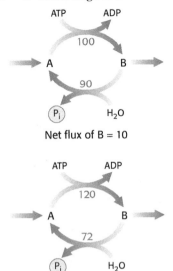

Net flux of B = 10

Net flux of B = 48

Figure 17.10 A substrate cycle. This ATP-driven cycle operates at two different rates. A small change in the rates of the two opposing reactions results in a large change in the *net* flux of product B.

B → A rate by 20% to 72. The new net flux is 48, and so a 20% change in the rates of the opposing reactions has led to a 380% increase in the net flux. In the example shown in **Figure 17.10**, this amplification is made possible by the rapid hydrolysis of ATP. The flux of each step of the glycolytic pathway has been suggested to increase as much as 1000-fold at the initiation of intense exercise, when a lot of ATP is needed. Because the allosteric activation of enzymes alone seems unlikely to explain this increased flux, the existence of substrate cycles may partly account for the rapid rise in the rate of glycolysis.

17.3 Metabolism in Context: Precursors Formed by Muscle Are Used by Other Organs

Lactate produced by active skeletal muscle and red blood cells is a source of energy for other organs. Red blood cells lack mitochondria and can never oxidize glucose completely. Recall that, in contracting type IIb skeletal muscle during vigorous exercise, the rate at which glycolysis produces pyruvate exceeds the rate at which the citric acid cycle oxidizes it. In these cells, lactate dehydrogenase reduces excess pyruvate to lactate to restore redox balance (p. 293). However, lactate is a dead end in metabolism. It must be converted back into pyruvate before it can be metabolized. Lactate and protons are transported out of these cells into the blood. *In contracting skeletal muscle, the formation and release of lactate lets the muscle generate ATP in the absence of oxygen and shifts the burden of metabolizing lactate from muscle to other organs.* The lactate in the bloodstream has two fates. In one fate, the plasma membranes of some cells, particularly cells in cardiac muscle and slow-twitch (type 1) skeletal muscle, contain carriers that make the cells highly permeable to lactate. The molecule diffuses from the blood into such permeable cells. Inside these well-oxygenated cells, lactate can be reverted back to pyruvate and metabolized through the citric acid cycle and oxidative phosphorylation to generate ATP. The use of lactate in place of glucose by these cells makes more circulating glucose available to the active muscle cells. In the other fate, excess lactate enters the liver and is converted first into pyruvate and then into glucose by the gluconeogenic pathway. Thus, *the liver restores the level of glucose necessary for active muscle cells, which derive ATP from the glycolytic conversion of glucose into lactate.* These reactions constitute the *Cori cycle* (**Figure 17.11**).

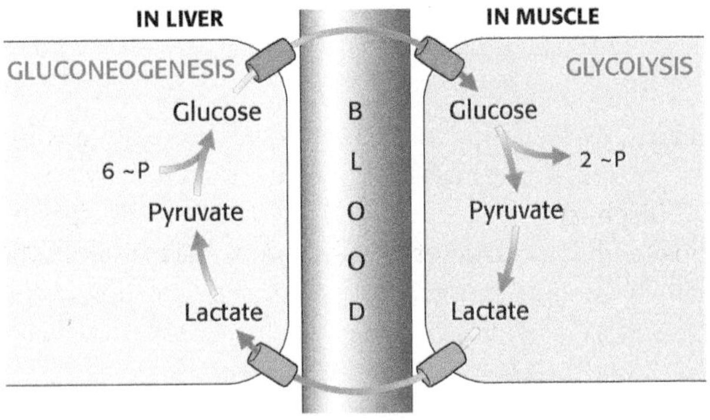

Figure 17.11 The Cori cycle. Lactate formed by active muscle is converted into glucose by the liver. This cycle shifts part of the metabolic burden of active muscle to the liver. The symbol ~P represents nucleoside triphosphates.

SUMMARY

17.1 Glucose Can Be Synthesized from Noncarbohydrate Precursors

Gluconeogenesis is the synthesis of glucose from noncarbohydrate sources, such as lactate, amino acids, and glycerol. Several of the reactions that convert pyruvate into glucose are common to glycolysis. Gluconeogenesis, however, requires four new reactions to bypass the three irreversible reactions in glycolysis. In two of the new reactions, pyruvate is carboxylated in mitochondria to oxaloacetate, which, in turn, is decarboxylated and phosphorylated in the cytoplasm to phosphoenolpyruvate. Two high phosphoryl-transfer potential molecules power these reactions, which are catalyzed by pyruvate carboxylase and phosphoenolpyruvate carboxykinase, respectively. The other distinctive reactions of gluconeogenesis are the hydrolyses of fructose 1,6-bisphosphate and glucose 6-phosphate, which are catalyzed by specific phosphatases. A major precursor for gluconeogenesis by the liver is lactate and produced by active skeletal muscle. The formation of lactate during intense muscular activity buys time and shifts part of the metabolic burden from muscle to the liver.

17.2 Gluconeogenesis and Glycolysis Are Reciprocally Regulated

Gluconeogenesis and glycolysis are reciprocally regulated so that one pathway is less active while the other is highly active. Phosphofructokinase and fructose 1,6-bisphosphatase are key control points. Fructose 2,6-bisphosphate, an intracellular signal molecule present at higher levels when glucose is abundant, activates glycolysis and inhibits gluconeogenesis by regulating these enzymes. Pyruvate kinase and pyruvate carboxylase are regulated by other effectors so that both are not maximally active at the same time. Allosteric regulation and reversible phosphorylation, which are rapid, are complemented by transcriptional control, which takes place in hours or days.

17.3 Metabolism in Context: Precursors Formed by Muscle Are Used by Other Organs

Lactate that is generated by glycolysis in contracting muscle is released into the bloodstream. This lactate is removed from the blood by the liver and is converted into glucose by gluconeogenesis. This metabolic cooperation between muscle and liver is called the Cori cycle.

KEY TERMS

gluconeogenesis (p. 314)
pyruvate carboxylase (p. 316)
biotin (p. 316)

fructose 1,6-bisphosphatase (p. 319)
glucose 6-phosphatase (p. 319)
bifunctional enzyme (p. 321)

substrate cycle (p. 323)
Cori cycle (p. 324)

? Answers to QUICK QUIZZES

1. The reverse of glycolysis is highly endergonic under cellular conditions. The expenditure of six NTP molecules in gluconeogenesis renders gluconeogenesis exergonic.

2. Reciprocal regulation at the key allosteric enzymes in the two pathways. For instance, PFK is stimulated by fructose 2,6-bisphosphate and AMP. The effect of these signals is opposite that of fructose 1,6-bisphosphatase. If both pathways were operating simultaneously, a futile cycle would result. ATP would be hydrolyzed, yielding only heat.

PROBLEMS

1. *It is not hard to meet expenses. They are everywhere.* What energetic barrier prevents glycolysis from simply running in reverse to synthesize glucose? What is the energetic cost of overcoming this barrier? ✓ 3

2. *Like Minneapolis and St. Paul.* Match each term with its description.

(a) Lactate _____

(b) Pyruvate carboxylase _____

1. Generates oxaloacetate

2. Readily converted into DHAP

(c) Acetyl CoA _____

(d) Phosphoenolpyruvate carboxykinase _____

(e) Glycerol _____

(f) Fructose1, 6-bisphosphatase _____

(g) Glucose 6-phosphatase _____

3. Generates a high-phosphoryl-transfer-potential compound

4. Found predominantly in liver

5. Gluconeogenic counterpart of PFK

6. Readily converted into pyruvate

7. Required for pyruvate carboxylase activity

3. *Road blocks.* What reactions of glycolysis are not reversible under intracellular conditions? How are these reactions bypassed in gluconeogenesis? ✓ 3

4. *Biotin snatcher.* Avidin, a 70-kDa protein in egg white, has very high affinity for biotin. In fact, it is a highly specific inhibitor of biotin enzymes. Which of the following conversions would be blocked by the addition of avidin to a cell homogenate?

(a) Glucose → pyruvate
(b) Pyruvate → glucose
(c) Oxaloacetate → glucose
(d) Malate → oxaloacetate
(e) Pyruvate → oxaloacetate
(f) Glyceraldehyde 3-phosphate →
$\qquad\qquad$ fructose 1,6-bisphosphate

5. *Working at cross-purposes?* Gluconeogenesis takes place during intense exercise, which seems counterintuitive. Why would an organism synthesize glucose and, at the same time, use glucose to generate energy? ✓ 3

6. *Different needs.* Liver is primarily a gluconeogenic tissue, whereas muscle is primarily glycolytic. Why does this division of labor make good physiological sense? ✓ 3

7. *Metabolic mutants.* What would be the effect on an organism's ability to use glucose as an energy source if a mutation inactivated glucose 6-phosphatase in the liver?

8. *Never let me go.* Why does the lack of glucose 6-phosphatase activity in the brain and muscle make good physiological sense?

9. *Which way to go?* Compare the roles of lactate dehydrogenase in gluconeogenesis and in lactic acid fermentation. ✓ 3

10. *Match 'em 1.* The following sequence is a part of the sequence of reactions in gluconeogenesis:

Pyruvate $\xrightarrow[A]{}$ oxaloacetate $\xrightarrow[B]{}$ malate $\xrightarrow[C]{}$

$\qquad$ oxaloacetate $\xrightarrow[D]{}$ phosphoenolpyruvate

Match the capital letters representing the reaction in the gluconeogenic pathway with parts *a, b, c,* and so on.

(a) Takes place in mitochondria
(b) Takes place in the cytoplasm
(c) Produces CO_2
(d) Consumes CO_2
(e) Requires NADH
(f) Produces NADH
(g) Requires ATP
(h) Requires GTP
(i) Requires thiamine
(j) Requires biotin
(k) Is regulated by acetyl CoA

11. *Salvaging resources.* In starvation, protein degradation takes place in muscle. Explain how this degradation might affect gluconeogenesis in the liver.

12. *Counting high-energy compounds 1.* How many NTP molecules are required for the synthesis of one molecule of glucose from two molecules of pyruvate? How many NADH molecules? ✓ 3

13. *Counting high-energy compounds 2.* How many NTP molecules are required to synthesize glucose from each of the following compounds? ✓ 3

(a) Glucose 6-phosphate
(b) Fructose 1,6-bisphosphate
(c) Two molecules of oxaloacetate
(d) Two molecules of dihydroxyacetone phosphate

14. *Two cycles.* What are the two potential substrate cycles in the glycolytic and gluconeogenic pathways? ✓ 4

15. *Useful cycles.* What is the regulatory role for the substrate cycles in glycolysis and gluconeogenesis? ✓ 4

16. *Not running at cross-purposes.* Describe the reciprocal regulation of gluconeogenesis and glycolysis. ✓ 4

Chapter Integration Problems

17. *Match 'em 2.* Indicate which of the conditions listed in the right-hand column increase the activity of the glycolytic and gluconeogenic pathways. ✓ 4

(a) Glycolysis_____
(b) Gluconeogenesis_____

1. Increase in ATP
2. Increase in AMP
3. Increase in fructose 2,6-bisphosphate
4. Increase in citrate
5. Increase in acetyl CoA
6. Increase in insulin
7. Increase in glucagon
8. Fasting
9. Fed

18. *Lending a hand.* How might enzymes that remove amino groups from alanine and aspartate contribute to gluconeogenesis?

19. *Even more metabolic mutants.* Predict the effect of each of the following mutations on the pace of glycolysis in liver cells. ✓ 4

(a) Loss of the allosteric site for ATP in phosphofructokinase
(b) Loss of the binding site for citrate in phosphofructokinase
(c) Loss of the phosphatase domain of the bifunctional enzyme that controls the level of fructose 2,6-bisphosphate
(d) Loss of the binding site for fructose 1,6-bisphosphate in pyruvate kinase

20. *A salvage operation.* Glycerol is released when lipids are used as a fuel. The released glycerol can be salvaged and can be used in glycolysis or gluconeogenesis in the liver. Show the reactions that are required for this conversion.

21. *Hungry yeast.* Yeast are facultative anaerobes—they can grow in the absence of oxygen (anaerobically) using alcoholic fermentation or in the presence of oxygen (aerobically) using cellular respiration. Interestingly, yeast cannot live anaerobically using glycerol as their only fuel source. Explain why yeast cannot survive metabolizing glycerol anaerobically.

Data Interpretation Problem

22. *Cool bees.* In principle, a futile cycle that includes phosphofructokinase and fructose 2,6-bisphosphatase could be used to generate heat. The heat could be used to warm tissues. For instance, certain bumblebees have been reported to use such a futile cycle to warm their flight muscles on cool mornings.

Scientists undertook a series of experiments to determine if a number of species of bumblebee use this futile cycle. Their approach was to measure the activity of PFK and F-1,6-BPase in flight muscle. ✓ 3

(a) What was the rationale for comparing the activities of these two enzymes?
(b) The following data show the activities of both enzymes for a variety of bumblebee species (genera *Bombus* and *Psithyrus*). Do these results support the notion that bumblebees use futile cycles to generate heat? Explain.

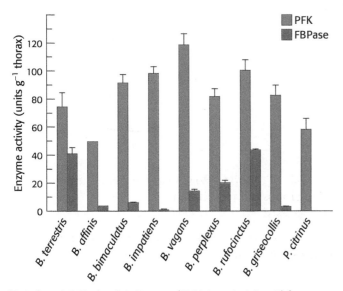

[Data from J. F. Staples, E. L. Koen, and T. M. Laverty, *J. Exp. Biol.* 207:749–754, 2004, p. 751.]

(c) In which species might futile cycling take place? Explain your reasoning.
(d) Do these results prove that futile cycling does not participate in heat generation?

Challenge Problems

23. *Waste not, want not.* Why is the conversion of lactic acid from the blood into glucose in the liver in an organism's best interest?

24. *More metabolic mutants.* What are the likely consequences of a genetic disorder rendering fructose 1,6-bisphosphatase in the liver less sensitive to regulation by fructose 2,6-bisphosphate? ✓ 4

25. *Tracing carbon atoms.* If cells synthesizing glucose from lactate are exposed to CO_2 labeled with ^{14}C, what will be the distribution of label in the newly synthesized glucose?

26. *Powering pathways.* Compare the stoichiometries of glycolysis and gluconeogenesis. Recall that the input of one ATP equivalent changes the equilibrium constant of a reaction by a factor of about 10^8 (p. 262). By what factor do the additional high-phosphoryl-transfer compounds alter the equilibrium constant of gluconeogenesis? ✓ 3

Selected Readings for this chapter can be found online at www.whfreeman.com/tymoczko3e.

The Citric Acid Cycle

CHAPTER 18
Preparation for the Cycle

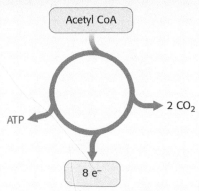

CHAPTER 19
Harvesting Electrons from the Cycle

You learned in Chapter 16 that glucose can be metabolized in glycolysis to pyruvate, yielding some ATP. However, the process of glycolysis is inefficient, capturing only a fraction of the energy inherent in a glucose molecule as ATP. More of the energy can be accessed if the pyruvate is completely oxidized into carbon dioxide and water. The combustion of fuels into carbon dioxide and water to generate ATP is called *cellular respiration* and is the source of more than 90% of the ATP required by human beings. Cellular respiration, unlike glycolysis, is an aerobic process, requiring molecular oxygen–O_2. In eukaryotes, cellular respiration takes place inside the double-membrane-bounded mitochondria, whereas glycolysis is cytoplasmic.

Cellular respiration can be divided into two parts. First, carbon fuels are completely oxidized with a concomitant generation of high-transfer-potential electrons in a series of reactions variously called the *citric acid cycle* (CAC), the *tricarboxylic acid* (TCA) *cycle*, or the *Krebs cycle*, after Sir Hans Krebs, the first to propose the existence of the cycle. In the second part of cellular respiration, referred to as *oxidative phosphorylation*, the high-transfer-potential electrons are transferred to oxygen to form water in a series of oxidation–reduction reactions. This transfer is highly exergonic, and the released energy is used to synthesize ATP. We will focus on the citric acid cycle in this section, leaving oxidative phosphorylation until Section 10.

The citric acid cycle is the central metabolic hub of the cell. It is the gateway to the aerobic metabolism of all fuel molecules. The cycle is also crucial for anabolism, serving as an important source of precursors for the building blocks of many other molecules such as amino acids, nucleotide bases, and porphyrin (the organic component of heme, p. 151). The citric acid cycle component oxaloacetate also is an important precursor to glucose (p. 316).

329

We begin this section by examining a crucial reaction in living systems: the conversion of glucose-derived pyruvate into acetyl CoA, an activated acetyl unit and the actual substrate for the citric acid cycle. This reaction links glycolysis and cellular respiration, thus allowing for the complete combustion of glucose, a fundamental fuel in all living systems. We will then study the citric acid cycle itself, the final common pathway for the oxidation of all fuel molecules, carbohydrates, fats, and amino acids.

✓ By the end of this section, you should be able to:

✓ 1 Explain why the reaction catalyzed by the pyruvate dehydrogenase complex is a crucial juncture in metabolism.

✓ 2 Identify the means by which the pyruvate dehydrogenase complex is regulated.

✓ 3 Identify the primary catabolic purpose of the citric acid cycle.

✓ 4 Explain the advantage of the oxidation of acetyl CoA in the citric acid cycle.

✓ 5 Describe how the citric acid cycle is regulated.

✓ 6 Describe the role of the citric acid cycle in anabolism.

CHAPTER 18

Preparation for the Cycle

One-way traffic facilitates traffic flow and directs flow toward specific locations. Pyruvate dehydrogenase is the one-way link between glycolysis and cellular respiration, directing pyruvate into the formation of acetyl CoA. [PhotoAlto/Alamy.]

A s you learned in Chapter 16, the pyruvate produced by glycolysis can have many fates. In the absence of oxygen (anaerobic conditions), the pyruvate is converted into lactic acid or ethanol, depending on the organism. In the

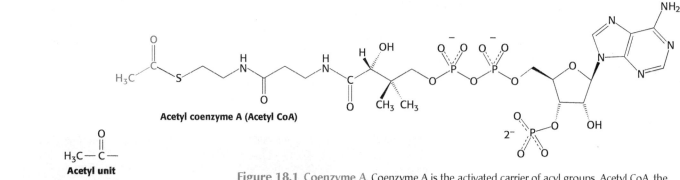

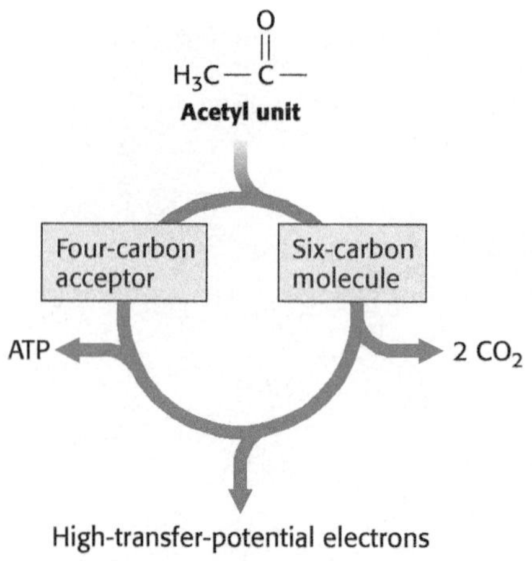

Acetyl coenzyme A (Acetyl CoA)

Figure 18.1 Coenzyme A. Coenzyme A is the activated carrier of acyl groups. Acetyl CoA, the fuel for the citric acid cycle, is formed by the pyruvate dehydrogenase complex.

Acetyl unit

Four-carbon acceptor

Six-carbon molecule

ATP

2 CO_2

High-transfer-potential electrons

Figure 18.2 An overview of the citric acid cycle. The citric acid cycle oxidizes two-carbon units, producing two molecules of CO_2, one molecule of ATP, and high-transfer-potential electrons.

presence of oxygen (aerobic conditions), it is converted into a molecule, called *acetyl coenzyme A* (acetyl CoA; **Figure 18.1**), that is able to enter the *citric acid cycle*. The path that pyruvate takes depends on the energy needs of the cell and the oxygen availability. In most tissues, pyruvate is processed aerobically because oxygen is readily available. For instance, in resting human muscle, most pyruvate is processed aerobically by first being converted into acetyl CoA. However, in very active muscle, for instance, the thigh muscles of a sprinter, much of the pyruvate is processed to lactate because the oxygen supply cannot meet the oxygen demand.

A schematic portrayal of the citric acid cycle is shown in **Figure 18.2**. The citric acid cycle accepts two-carbon acetyl units in the form of acetyl CoA. These two-carbon acetyl units are introduced into the cycle by binding to a four-carbon acceptor molecule. The two-carbon units are oxidized to CO_2, and the resulting high-transfer-potential electrons are captured. The acceptor molecule is regenerated, capable of processing another two-carbon unit. The cyclic nature of these reactions enhances their efficiency.

In this chapter, we examine the enzyme complex that catalyzes the formation of acetyl CoA from pyruvate, how this enzyme complex is regulated, and some pathologies that result if the function of the enzyme complex is impaired. However, the pyruvate dehydrogenase complex is not the only source of acetyl CoA. In particular, fatty acid degradation yields much acetyl CoA, as we will see in Chapter 27.

✓ 1 **Explain why the reaction catalyzed by the pyruvate dehydrogenase complex is a crucial juncture in metabolism.**

18.1 Pyruvate Dehydrogenase Forms Acetyl Coenzyme A from Pyruvate

Glycolysis takes place in the cytoplasm of the cell, but the citric acid cycle takes place in mitochondria (**Figure 18.3**). Pyruvate must therefore be transported into mitochondria to be aerobically metabolized. In the mitochondrial matrix, pyruvate is oxidatively decarboxylated by the *pyruvate dehydrogenase complex* to form acetyl CoA:

$$\text{Pyruvate} + \text{CoA} + \text{NAD}^+ \longrightarrow \text{acetyl CoA} + CO_2 + \text{NADH} + \text{H}^+$$

Figure 18.3 Mitochondrion. The double membrane of the mitochondrion is evident in this electron micrograph. The oxidative decarboxylation of pyruvate and the sequence of reactions in the citric acid cycle take place within the matrix. [(Left) Omikron/Science Source.]

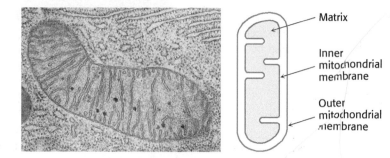

Matrix

Inner mitochondrial membrane

Outer mitochondrial membrane

Table 18.1   Pyruvate dehydrogenase complex of *E. coli*

Enzyme	Abbreviation	Number of chains	Prosthetic group	Reaction catalyzed
Pyruvate dehydrogenase component	E_1	24	TPP	Oxidative decarboxylation of pyruvate
Dihydrolipoyl transacetylase	E_2	24	Lipoamide	Transfer of acetyl group to CoA
Dihydrolipoyl dehydrogenase	E_3	12	FAD	Regeneration of the oxidized form of lipoamide

Abbreviations: TPP, thiamine pyrophosphate; FAD, flavin adenine dinucleotide.

Recall that glycolysis generates two molecules of pyruvate for each glucose molecule metabolized. *This irreversible conversion of pyruvate into acetyl CoA is the link between glycolysis and the citric acid cycle* (Figure 18.4). This reaction is a decisive reaction in metabolism: it commits the carbon atoms of carbohydrates to oxidation by the citric acid cycle or to the synthesis of lipids (Chapter 29). Note that the pyruvate dehydrogenase complex produces CO_2 and captures high-transfer-potential electrons in the form of NADH, thus foreshadowing the key features of the reactions of the citric acid cycle.

The pyruvate dehydrogenase complex is a large, highly integrated complex of three distinct enzymes (Table 18.1), each with its own active site. It is a member of a family of extremely large similar complexes with molecular masses ranging from 4 million to 10 million daltons (Figure 18.5). As we will see, their elaborate structures allow substrates to travel efficiently from one active site to another, connected by tethers to the core of the complex.

The Synthesis of Acetyl Coenzyme A from Pyruvate Requires Three Enzymes and Five Coenzymes

We will examine the mechanism of action of the pyruvate dehydrogenase complex in some detail because it catalyzes a key step in metabolism—the link between glycolysis and the citric acid cycle that allows the complete oxidation of glucose. The mechanism of the pyruvate dehydrogenase reaction is wonderfully complex, more so than is suggested by its simple stoichiometry. The reaction requires the participation of the three enzymes of the pyruvate dehydrogenase complex—pyruvate dehydrogenase, dihydrolipoyl transacetylase, and dihydrolipoyl dehydrogenase—as well as five coenzymes. The coenzymes *thiamine pyrophosphate* (TPP), *lipoic acid,* and *flavin adenine dinucleotide* (FAD) serve as catalytic coenzymes, and CoA and nicotinamide adenine dinucleotide (NAD^+) are stoichiometric coenzymes. Catalytic coenzymes, like enzymes, are not permanently altered by participation in the reaction. Stoichiometric coenzymes function as substrates.

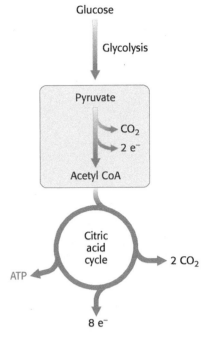

Figure 18.4 The link between glycolysis and the citric acid cycle. Pyruvate produced by glycolysis is converted into acetyl CoA, the fuel of the citric acid cycle. Fatty acid degradation is also an important source of acetyl CoA for the citric acid cycle (Chapter 27).

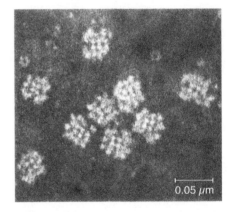

Figure 18.5 Electron micrograph of the pyruvate dehydrogenase complex from *E. coli*. [Courtesy of Dr. Lester Reed.]

Thiamine pyrophosphate (TPP)

Lipoic acid

The conversion of pyruvate into acetyl CoA consists of three steps: decarboxylation, oxidation, and the transfer of the resultant acetyl group to CoA:

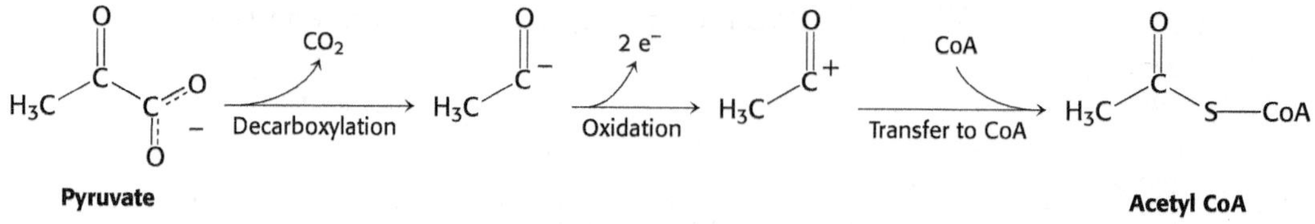

$$ \text{Pyruvate} \xrightarrow[\text{Decarboxylation}]{CO_2} H_3C\text{-C-} \xrightarrow[\text{Oxidation}]{2e^-} H_3C\text{-C+} \xrightarrow[\text{Transfer to CoA}]{CoA} \text{Acetyl CoA} $$

Pyruvate

Acetyl CoA

These steps must be coupled to preserve the free energy derived from the decarboxylation step to drive the formation of NADH and acetyl CoA.

1. *Decarboxylation.* Pyruvate combines with the ionized (carbanion) form of TPP and is then decarboxylated to yield hydroxyethyl-TPP:

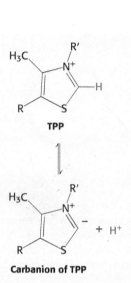

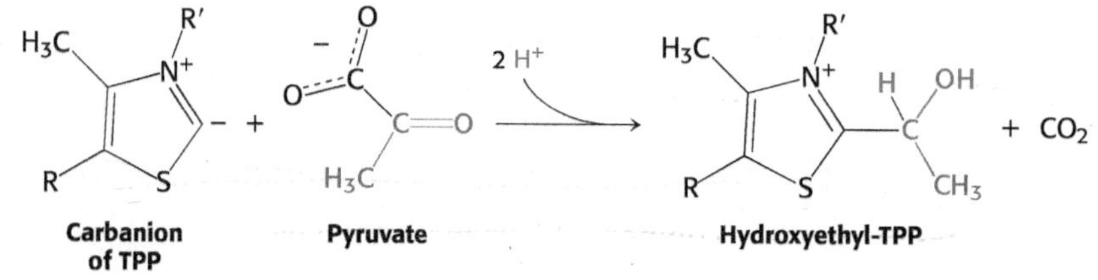

Carbanion of TPP

Carbanion of TPP   **Pyruvate**   **Hydroxyethyl-TPP**

This reaction, the rate-limiting step in the synthesis of acetyl CoA, is catalyzed by the *pyruvate dehydrogenase component* (E_1) of the multienzyme complex. TPP is the coenzyme of the pyruvate dehydrogenase component.

2. *Oxidation.* The hydroxyethyl group attached to TPP is *oxidized* to form an acetyl group while being simultaneously transferred to lipoamide, a derivative of lipoic acid. Note that this transfer results in the formation of an energy-rich thioester bond.

Hydroxyethyl-TPP   **Lipoamide**   **Carbanion of TPP**   **Acetyllipoamide**
(ionized form)

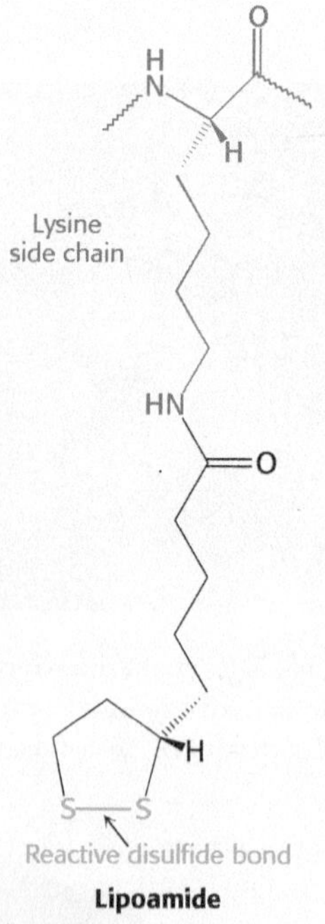

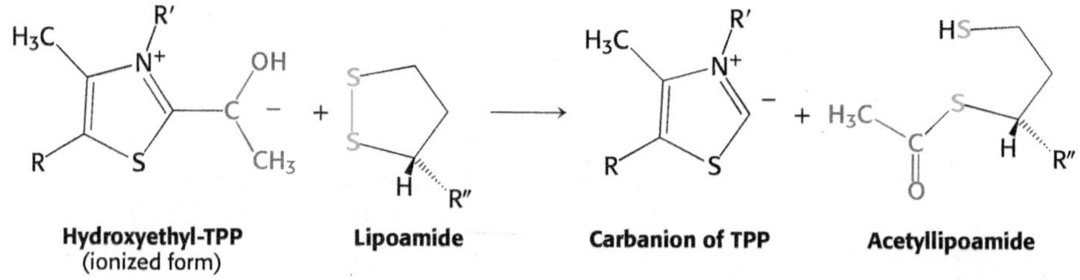

The disulfide group of lipoamide is reduced to its disulfhydryl form in this reaction. The reaction, also catalyzed by the pyruvate dehydrogenase component E_1, yields *acetyllipoamide*.

3. *Formation of acetyl CoA.* The acetyl group is transferred from acetyllipoamide to CoA to form acetyl CoA. *Dihydrolipoyl transacetylase* (E_2) catalyzes this reaction. The energy-rich thioester bond is preserved as the acetyl group is transferred to CoA. Acetyl CoA, the fuel for the citric acid cycle, has now been generated from pyruvate:

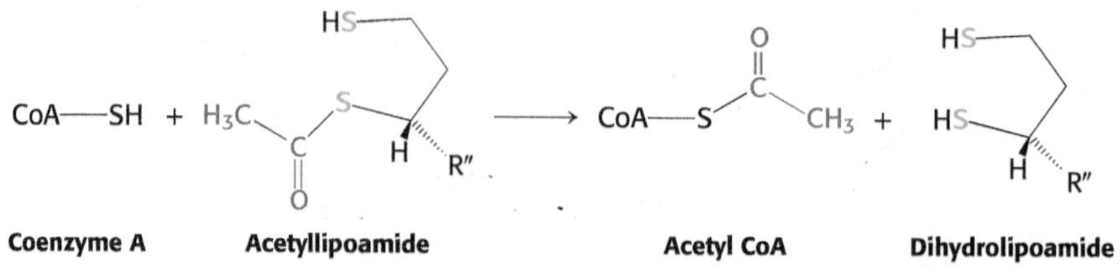

Coenzyme A   **Acetyllipoamide**   **Acetyl CoA**   **Dihydrolipoamide**

However, the pyruvate dehydrogenase complex cannot complete another catalytic cycle until the dihydrolipoamide is oxidized to lipoamide. In a fourth step, *the oxidized form of lipoamide is regenerated by dihydrolipoyl dehydrogenase* (E_3). Two electrons are transferred to an FAD prosthetic group of the enzyme and then to NAD^+:

This electron transfer from FAD to NAD^+ is unusual because the common role for FAD is to receive electrons from NADH. The electron-transfer potential of FAD is increased by its association with the enzyme, enabling it to transfer electrons to NAD^+. Proteins tightly associated with FAD are called *flavoproteins*.

Flexible Linkages Allow Lipoamide to Move Between Different Active Sites

The structures of all of the component enzymes of the pyruvate dehydrogenase complex are known, albeit from different complexes and species. Thus, it is now possible to construct an atomic model of the complex to understand its activity.

The core of the complex is formed by the transacetylase component E_2. Transacetylase consists of eight catalytic trimers assembled to form a hollow cube. Each of the three subunits forming a trimer has three major domains (**Figure 18.6**). At the amino terminus is a small domain that contains a flexible lipoamide cofactor covalently attached to a lysine side chain. The lipoamide domain is followed by a small domain that interacts with E_3 within the complex. A larger transacetylase domain completes an E_2 subunit. The eight E_2 trimers constitute the core of the complex and are surrounded by 24 copies of E_1 (an $\alpha_2\beta_2$ tetramer) and 12 copies of E_3 (an $\alpha\beta$ dimer). In mammals, this core contains another protein, E_3-binding

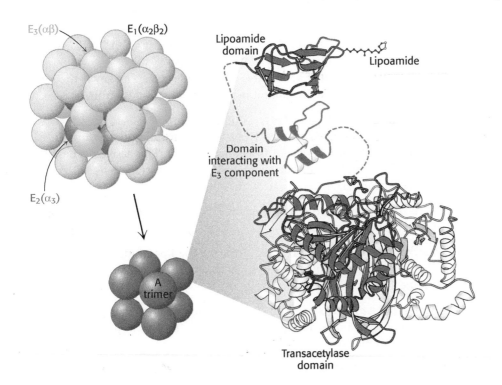

Figure 18.6 A schematic representation of the pyruvate dehydrogenase complex. The transacetylase core (E_2) is shown in red, the pyruvate dehydrogenase component (E_1) in yellow, and the dihydrolipoyl dehydrogenase (E_3) in green. The number and type of subunits of each enzyme are given parenthetically. Each red ball represents a trimer of three E_2 subunits. Notice that each subunit consists of three domains: a lipoamide-binding domain, a small domain for interaction with E_3, and a large transacetylase catalytic domain. The transacetylase domain has three subunits, with one subunit depicted in red and the other two in white in the ribbon representation.

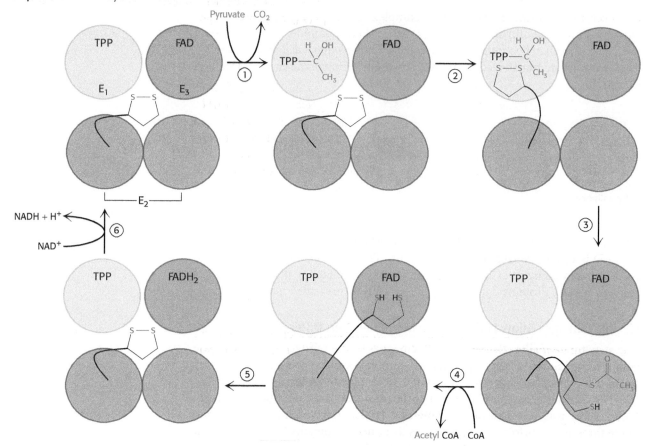

Figure 18.7 Reactions of the pyruvate dehydrogenase complex. At the top (left), the enzyme (represented by a yellow, a green, and two red spheres) is unmodified and ready for a catalytic cycle. (1) Pyruvate is decarboxylated to form hydroxyethyl–TPP by E_1. (2) The lipoamide arm of E_2 moves into the active site of E_1. (3) E_1 catalyzes the transfer of the two–carbon group to the lipoamide group to form the acetyl–lipoamide complex. (4) E_2 catalyzes the transfer of the acetyl moiety to CoA to form the product acetyl CoA. The dihydrolipoamide arm then swings to the active site of E_3. E_3 catalyzes (5) the oxidation of the dihydrolipoamide acid and (6) the transfer of the protons and electrons to NAD^+ to complete the reaction cycle.

protein (E_3-BP), which facilitates the interaction between E_2 and E_3. If E_3-BP is missing, the PDH complex has greatly reduced activity.

How do the three distinct active sites work in concert? The key is the long, flexible lipoamide arm of the E_2 subunit, which carries substrate from active site to active site (Figure 18.7).

1. Pyruvate is decarboxylated at the active site of E_1, forming the hydroxyethyl-TPP intermediate, and CO_2 leaves as the first product. This active site lies deep within the E_1 complex, connected to the enzyme surface by a 20-Å-long hydrophobic channel.

2. E_2 inserts the lipoamide arm of the lipoamide domain into the deep channel in E_1 leading to the active site.

3. E_1 catalyzes the transfer of the acetyl group to the lipoamide. The acetylated arm then leaves E_1 and enters the E_2 cube to visit the active site of E_2, located deep in the cube at the subunit interface.

4. The acetyl moiety is transferred to CoA, and the second product, acetyl CoA, leaves the cube. The reduced lipoamide arm then swings to the active site of the E_3 flavoprotein.

5. At the E_3 active site, the lipoamide is oxidized by coenzyme FAD. The reactivated lipoamide is ready to begin another reaction cycle.

6. The final product, NADH, is produced with the reoxidation of $FADH_2$ to FAD.

The structural integration of three kinds of enzymes and the long flexible lipoamide arm make the coordinated catalysis of a complex reaction possible. The proximity of one enzyme to another *increases the overall reaction rate and minimizes side reactions.* All the intermediates in the oxidative decarboxylation of pyruvate remain bound to the complex throughout the reaction sequence and are readily transferred as the flexible arm of E_2 calls on each active site in turn.

18.2 The Pyruvate Dehydrogenase Complex Is Regulated by Two Mechanisms

The pyruvate dehydrogenase complex is stringently regulated by multiple allosteric interactions and covalent modifications. As stated earlier, glucose can be formed from pyruvate through the gluconeogenic pathway (p. 314). However, *the formation of acetyl CoA from pyruvate is an irreversible step in animals and thus they are unable to convert acetyl CoA back into glucose.* The oxidative decarboxylation of pyruvate to acetyl CoA commits the carbon atoms of glucose to either of two principal fates: (1) oxidation to CO_2 by the citric acid cycle with the concomitant generation of energy or (2) incorporation into lipid, because acetyl CoA is a key precursor for lipid synthesis (Chapter 28 and **Figure 18.8**). High concentrations of reaction products inhibit the reaction: acetyl CoA inhibits the transacetylase component (E_2) by directly binding to it, whereas NADH inhibits the dihydrolipoyl dehydrogenase (E_3). High concentrations of NADH and acetyl CoA inform the enzyme that the energy needs of the cell have been met or that enough acetyl CoA and NADH have been produced from fatty acid degradation (Chapter 27). In either case, there is no need to metabolize pyruvate to acetyl CoA. This inhibition has the effect of sparing glucose, because most pyruvate is derived from glucose by glycolysis.

The key means of regulation of the complex in eukaryotes is covalent modification—in this case, phosphorylation (**Figure 18.9**). *Phosphorylation of the pyruvate dehydrogenase component (E_1) by pyruvate dehydrogenase (PDH) kinase switches off the activity of the complex. Deactivation is reversed by the action of PDH phosphatase.* In mammals, the kinase and the phosphatase are associated with the E_2-E_3-BP complex, again highlighting the structural and mechanistic importance of this core. Moreover, both the kinase and the phosphatase are themselves regulated.

✓ 2 Identify the means by which the pyruvate dehydrogenase complex is regulated.

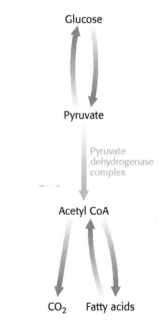

Figure 18.8 From glucose to acetyl CoA. The synthesis of acetyl CoA by the pyruvate dehydrogenase complex is a key irreversible step in the metabolism of glucose.

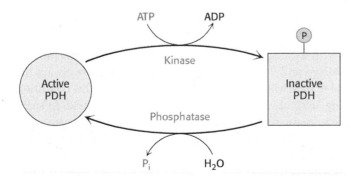

Figure 18.9 The regulation of the pyruvate dehydrogenase complex. PDH kinase phosphorylates and inactivates pyruvate dehydrogenase (PDH), and PDH phosphatase activates the dehydrogenase by removing the phosphoryl group. The kinase and the phosphatase are highly regulated enzymes.

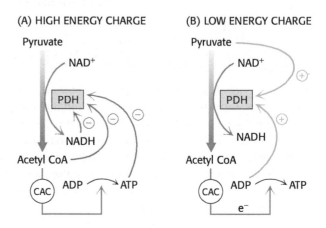

Figure 18.10 Response of the pyruvate dehydrogenase complex to the energy charge. The pyruvate dehydrogenase complex is regulated to respond to the energy charge of the cell. (A) The complex is inhibited by its immediate products, NADH and acetyl CoA, as well as by the ultimate product of cellular respiration, ATP. (B) The complex is activated by pyruvate and ADP, which inhibit the kinase that phosphorylates PDH.

To see how this regulation works under biological conditions, consider muscle that is becoming active after a period of rest (Figure 18.10). At rest, the muscle will not have significant energy demands. Consequently, the NADH/NAD$^+$, acetyl CoA/CoA, and ATP/ADP ratios will be high. These high ratios stimulate PDH kinase, promoting phosphorylation and, hence, deactivation of the pyruvate dehydrogenase complex. In other words, high concentrations of immediate (acetyl CoA and NADH) and ultimate (ATP) products of the pyruvate dehydrogenase complex inhibit its activity. Thus, *pyruvate dehydrogenase is switched off when the energy charge is high.*

As exercise begins, the concentrations of ADP and pyruvate will increase as muscle contraction consumes ATP and glucose is converted into pyruvate to meet the energy demands. Both ADP and pyruvate activate the dehydrogenase by inhibiting PDH kinase. Moreover, the phosphatase is stimulated by Ca^{2+}, a signal that also initiates muscle contraction. A rise in the cytoplasmic Ca^{2+} level to stimulate muscle contraction elevates the mitochondrial Ca^{2+} level. The rise in mitochondrial Ca^{2+} activates the phosphatase, enhancing pyruvate dehydrogenase activity.

In some tissues, the phosphatase is regulated by hormones. In liver, epinephrine binds to the α-adrenergic receptor to initiate the phosphatidylinositol pathway (p. 232), causing an increase in Ca^{2+} concentration that activates the phosphatase. In tissues capable of fatty acid synthesis (such as the liver and adipose tissue), insulin (the hormone that signifies the fed state) stimulates the phosphatase, increasing the conversion of pyruvate into acetyl CoA. In these tissues, the pyruvate dehydrogenase complex is activated to funnel glucose to pyruvate and then to acetyl CoA and ultimately to fatty acids.

 CLINICAL INSIGHT

Defective Regulation of Pyruvate Dehydrogenase Results in Lactic Acidosis

In people with a phosphatase deficiency, pyruvate dehydrogenase is always phosphorylated and thus inactive. Consequently, glucose always has to take the anaerobic path to lactate rather than acetyl CoA. This condition results in unremitting *lactic acidosis*—high blood levels of lactic acid. In such an acidic environment, many tissues malfunction, most notably the central nervous system. One treatment for the condition is to place the patient on a ketogenic (high fat, adequate protein, low carbohydrate) diet to minimize the need to metabolize glucose.

CLINICAL INSIGHT

Enhanced Pyruvate Dehydrogenase Kinase Activity Facilitates the Development of Cancer

Recall that cancer cells metabolize glucose to lactate even in the presence of oxygen, a phenomenon called aerobic glycolysis or the Warburg effect (p. 304). Under these conditions, the transcription factor hypoxia inducible factor-1 (HIF-1) increases the amount of the proteins required for glycolysis. In addition, HIF-1 stimulates the synthesis of pyruvate dehydrogenase kinase. The kinase inhibits the pyruvate dehydrogenase complex, preventing the conversion of pyruvate into acetyl CoA. The pyruvate remains in the cytoplasm, facilitating aerobic glycolysis. Moreover, even in the absence of increased synthesis of PDH kinase, mutations in PDH kinase have been identified that lead to enhanced activity, thereby contributing to increased aerobic glycolysis and the subsequent development of cancer as heretofore described. Enhanced lactate production resulting from aerobic glycolysis further enhances the activity of HIF-1.

? QUICK QUIZ List some of the advantages of organizing the enzymes that catalyze the formation of acetyl CoA from pyruvate into a single large complex.

 CLINICAL INSIGHT

The Disruption of Pyruvate Metabolism Is the Cause of Beriberi

The importance of the coordinated activity of the pyruvate dehydrogenase complex is illustrated by disorders that result from the absence of a key coenzyme. Recall that thiamine pyrophosphate is a coenzyme for the pyruvate dehydrogenase activity of the pyruvate dehydrogenase complex. *Beriberi*, a neurological and cardiovascular disorder, is caused by a dietary deficiency of thiamine (vitamin B_1). Thiamine deficiency results in insufficient pyruvate dehydrogenase activity because thiamine pyrophosphate cannot be formed. The disease has been and continues to be a serious health problem in the Far East because rice, an important food there, has a rather low content of thiamine. This deficiency is partly ameliorated if the whole rice grain is soaked in water before milling; some of the thiamine in the husk then leaches into the rice kernel (**Figure 18.11**). The problem is exacerbated if the rice is polished, a practice that prevents spoilage and extends the storage life of rice, because only the outer layer contains significant amounts of thiamine. A form of beriberi, called Wernicke's encephalopathy, is also occasionally seen in alcoholics who are severely malnourished and thus thiamine deficient. The disease is characterized by neurological and cardiac symptoms. Damage to the peripheral nervous system is expressed as pain in the limbs, weakness of the musculature, and distorted skin sensation. The heart may be enlarged and the cardiac output inadequate.

Thiamine pyrophosphate is not just crucial to the conversion of pyruvate to acetyl CoA. In fact, *this coenzyme is the prosthetic group of three important enzymes: pyruvate dehydrogenase, α-ketoglutarate dehydrogenase* (a citric acid cycle enzyme, Chapter 19), *and transketolase*. Transketolase functions in the pentose phosphate pathway, which will be considered in Chapter 26. *The common feature of enzymatic reactions utilizing TPP is the transfer of an activated aldehyde unit.* As expected in a body in which TPP is deficient, the levels of pyruvate and α-ketoglutarate in the blood of patients with beriberi are higher than normal. The increase in the level of pyruvate in the blood is especially pronounced after the ingestion of glucose. A related finding is that the activities of the pyruvate dehydrogenase complex and the α-ketoglutarate dehydrogenase complex in vivo are abnormally low. The low transketolase activity of red blood cells in beriberi is an easily measured and reliable diagnostic indicator of the disease.

Figure 18.11 Milled and polished rice. Brown rice is milled to remove only the outer husk. Further milling (polishing) removes the inner husk also, resulting in white rice. [Image Source/Age Fotostock.]

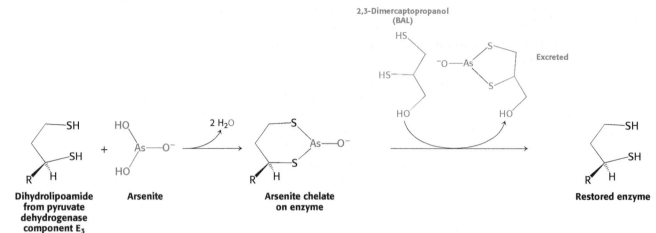

Figure 18.12 Arsenite poisoning. Arsenite inhibits the pyruvate dehydrogenase complex by inactivating the dihydrolipoamide component of the transacetylase. Some sulfhydryl reagents, such as 2,3-dimercaptopropanol, relieve the inhibition by forming a complex with the arsenite that can be excreted.

Figure 18.13 Mad Hatter. The Mad Hatter is one of the characters that Alice meets at a tea party in her journey through Wonderland. Real hatters worked with mercury, which inhibited an enzyme responsible for providing the brain with energy. The lack of energy would lead to peculiar behavior, often described as "mad." [The Granger Collection.]

Why does TPP deficiency lead primarily to neurological disorders? The nervous system relies essentially on glucose as its only fuel. The product of glycolysis—pyruvate—can enter the citric acid cycle only through the pyruvate dehydrogenase complex. With that enzyme inactive, the nervous system has no source of fuel. In contrast, most other tissues can use fats as a source of fuel for the citric acid cycle.

Symptoms similar to those of beriberi appear in organisms exposed to mercury or arsenite (AsO_3^{3-}). Both substances have a high affinity for sulfhydryls in close proximity to one another, such as those in the reduced dihydrolipoyl groups of the E_3 component of the pyruvate dehydrogenase complex (**Figure 18.12**). The binding of mercury or arsenite to the dihydrolipoyl groups inhibits the complex and leads to central nervous system pathologies. The proverbial phrase "mad as a hatter" refers to the strange behavior of poisoned hat makers who used mercury nitrate to soften and shape animal furs (**Figure 18.13**). This form of mercury is absorbed through the skin. Similar symptoms afflicted the early photographers, who used vaporized mercury to create daguerreotypes.

Treatment for these poisons is the administration of sulfhydryl reagents with adjacent sulfhydryl groups to compete with the dihydrolipoyl residues for binding with the metal ion. The reagent–metal complex is then excreted in the urine. Indeed, 2,3-dimercaptopropanol (Figure 18.12) was developed after World War I as an antidote to lewisite, an arsenic-based chemical weapon. This compound was initially called BAL, for *B*ritish *a*nti-*l*ewisite.

SUMMARY

18.1 Pyruvate Dehydrogenase Forms Acetyl Coenzyme A from Pyruvate

Most fuel molecules enter the citric acid cycle as acetyl CoA. The link between glycolysis and the citric acid cycle is the oxidative decarboxylation of pyruvate to form acetyl CoA. In eukaryotes, this reaction and those of the cycle take place inside mitochondria, in contrast with glycolysis, which takes place in the cytoplasm. The enzyme complex catalyzing this reaction, the pyruvate dehydrogenase complex, consists of three distinct enzyme activities. Pyruvate dehydrogenase catalyzes the decarboxylation of pyruvate and the formation of acetyllipoamide. Dihydrolipoyl transacetylase forms acetyl CoA, and dihydrolipoyl dehydrogenase regenerates the active transacetylase. The complex requires five cofactors: thiamine pyrophosphate, lipoic acid, coenzyme A, NAD^+, and FAD.

18.2 The Pyruvate Dehydrogenase Complex Is Regulated by Two Mechanisms

The irreversible formation of acetyl CoA from pyruvate is the regulatory point for the entry of glucose-derived pyruvate into the citric acid cycle. The pyruvate dehydrogenase complex is allosterically inhibited by acetyl CoA and NADH. The activity of the pyruvate dehydrogenase complex is stringently controlled by reversible phosphorylation by an associated kinase and phosphatase. High concentrations of ATP and NADH stimulate the kinase, which phosphorylates and inactivates the complex. ADP and pyruvate inhibit the kinase, whereas Ca^{2+} stimulates the phosphatase, which dephosphorylates and thereby activates the complex.

The importance of the pyruvate dehydrogenase complex to metabolism, especially to catabolism in the central nervous system, is illustrated by beriberi. Beriberi is a neurological condition that results from a deficiency of thiamine, the vitamin precursor of thiamine pyrophosphate. The lack of TPP impairs the activity of the pyruvate dehydrogenase component of the pyruvate dehydrogenase complex. Arsenite and mercury are toxic because of their effects on the complex. These chemicals bind to the lipoic acid coenzyme of dihydrolipoyl dehydrogenase, inhibiting the activity of this enzyme.

KEY TERMS

acetyl coenzyme A (acetyl CoA) (p. 332)
citric acid cycle (p. 332)
pyruvate dehydrogenase complex
   (p. 332)
thiamine pyrophosphate (TPP) (p. 333)
lipoic acid (p. 333)

flavin adenine dinucleotide (FAD)
   (p. 333)
pyruvate dehydrogenase (E_1) (p. 334)
acetyllipoamide (p. 334)
dihydrolipoyl transacetylase (E_2)
   (p. 334)

dihydrolipoyl dehydrogenase (E_3)
   (p. 335)
flavoprotein (p. 335)
beriberi (p. 339)

? Answer to QUICK QUIZ

The advantages are as follows:

1. The reaction is facilitated by having the active sites in proximity.

2. The reactants do not leave the enzyme until the final product is formed. Constraining the reactants minimizes loss due to diffusion and minimizes side reactions.

3. All of the enzymes are present in the correct amounts.

4. Regulation is more efficient because the regulatory enzymes—the kinase and phosphatase—are part of the complex.

PROBLEMS

1. *A one-way link.* What reaction serves to link glycolysis and the citric acid cycle, and what is the enzyme that catalyzes the reaction? ✓ 1

2. *Naming names.* What are the five enzymes (including regulatory enzymes) that constitute the pyruvate dehydrogenase complex? Which reactions do they catalyze? ✓ 1

3. *Waste and fraud?* Figure 18.7 shows the steps in the pyruvate dehydrogenase complex reaction cycle. A key product, acetyl CoA is released after the fourth step. What is the purpose of the remaining steps? ✓ 1

4. *The ol' two step plus one.* The conversion of pyruvate into acetyl CoA consists of three steps. What are these steps? ✓ 1

5. *Predetermined events.* The conversion of pyruvate into acetyl CoA commits the carbon atoms to either of two principal fates. What are the fates? ✓ 1

6. *Coenzymes.* What coenzymes are required by the pyruvate dehydrogenase complex, and what are their roles?

7. *More coenzymes.* Distinguish between catalytic coenzymes and stoichiometric coenzymes in the pyruvate dehydrogenase complex.

8. *Rolling uphill?* When lipoamide is reoxidized, what is the immediate electron acceptor and what is the final electron receptor of the reaction? Why is the observed electron transfer unusual?

9. *Like Watson and Holmes.* Match each term with its description.

(a) Acetyl CoA _____

(b) Citric acid cycle _____

(c) Pyruvate dehydrogenase complex _____

(d) Thiamine pyrophosphate _____

(e) Lipoic acid _____

(f) Pyruvate dehydrogenase _____

(g) Acetyllipoamide _____

(h) Dihydrolipoyl transacetylase _____

(i) Dihydrolipoyl dehydrogenase _____

(j) Beriberi _____

1. Catalyzes the link between glycolysis and the citric acid cycle

2. Coenzyme required by transacetylase

3. Final product of pyruvate dehydrogenase

4. Catalyzes the formation of acetyl CoA

5. Regenerates active transacetylase

6. Fuel for the citric acid cycle

7. Coenzyme required by pyruvate dehydrogenase

8. Catalyzes the oxidative decarboxylation of pyruvate

9. Due to a deficiency of thiamine

10. Central metabolic hub

10. *Alternative fates.* Compare the regulation of the pyruvate dehydrogenase complex in muscle and in liver. ✓ 2

11. *Mutations.* (a) Predict the effect of a mutation that enhances the activity of the kinase associated with the PDH complex. (b) Predict the effect of a mutation that reduces the activity of the phosphatase associated with the PDH complex. ✓ 2

12. *Flaking paint, green wallpaper.* Claire Boothe Luce, ambassador to Italy in the 1950s (and Connecticut congressperson, playwright, editor of *Vanity Fair,* and the wife of Henry Luce, founder of *Time* magazine and *Sports Illustrated*), became ill when she was staying at the ambassadorial residence in Rome. The arsenic-based paint on the dining-room ceiling was flaking; the wallpaper of her bedroom in the ambassadorial residence was colored a mellow green owing to the presence of cupric arsenite in the pigment. Suggest a possible cause of Ambassador Luce's illness. ✓ 2

13. *Energy rich.* What are the thioesters in the reaction catalyzed by PDH complex?

14. *Danbury shakes.* From 1850 until World War II, Danbury, Connecticut, was considered the "hat capital of the world." One popular product was the felted fur hat. In the process of

manufacturing the hats, fur was soaked in mercury nitrate. Many of the workers displayed neurological problems, including tremors, which came to be known as the "Danbury shakes." Suggest a biochemical explanation for the workers' problems.

Chapter Integration Problems

15. *Alternative fuels.* As we will see (Chapter 27), fatty acid breakdown generates a large amount of acetyl CoA. What will be the effect of fatty acid breakdown on pyruvate dehydrogenase complex activity? On glycolysis? ✓ 1

16. *Crucial intermediates.* Thioesters not only are important in the reaction catalyzed by the pyruvate dehydrogenase complex, but also are crucial for the generation of pyruvate itself. What is the thioester in glycolysis that helps to generate pyruvate?

Challenge Problems

17. *Lactic acidosis.* Patients in shock often suffer from lactic acidosis owing to a deficiency of O_2. ✓ 2

(a) Why does a lack of O_2 lead to lactic acid accumulation?

(b) One treatment for shock is to administer dichloroacetate (DCA), which inhibits the kinase associated with the pyruvate dehydrogenase complex. What is the biochemical rationale for this treatment?

18. *DCA again.* Patients with pyruvate dehydrogenase deficiency show high levels of lactic acid in the blood. However, in some cases, treatment with dichloroacetate (DCA) lowers lactic acid levels.

(a) How does dichloroacetate act to stimulate pyruvate dehydrogenase activity?

(b) What does this suggest about pyruvate dehydrogenase activity in patients who respond to DCA?

19. *Force feeding.* Inhibitors of pyruvate dehydrogenase kinase have been proposed as potential treatments for type 2 diabetes, which is characterized by high blood levels of glucose due to insulin resistance (p. 323). Suggest a biochemical rationale for this proposal. ✓ 2

20. *A potent inhibitor.* Thiamine thiazolone pyrophosphate binds to pyruvate dehydrogenase about 20,000 times more strongly than thiamine pyrophosphate does, and it competitively inhibits the enzyme. Why?

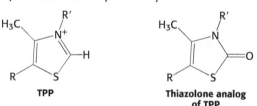

TPP   Thiazolone analog of TPP

Selected Readings for this chapter can be found online at www.whfreeman.com/tymoczko3e.

Harvesting Electrons from the Cycle

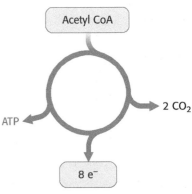

Roundabouts, or traffic circles, function as hubs to facilitate traffic flow. The citric acid cycle is the biochemical hub of the cell, oxidizing carbon fuels, usually in the form of acetyl CoA, and serving as a source of precursors for biosynthesis. [(Left) Lynn Saville/Getty Images.]

In Chapter 18, we learned how glucose is metabolized to acetyl CoA. This reaction, catalyzed by the *pyruvate dehydrogenase complex*, is the irreversible link between glycolysis and the citric acid cycle. Here, we will learn a common fate of acetyl CoA when oxygen is present and energy is required: the complete combustion of the acetyl group by the citric acid cycle.

What is the function of the citric acid cycle in transforming fuel molecules into ATP? Recall that fuel molecules are carbon compounds that are capable of being oxidized—of losing electrons (p. 266). These fuel molecules are first processed to acetyl CoA, the actual fuel for the citric acid cycle. The *citric acid cycle* includes a series of oxidation–reduction reactions that ultimately result in the oxidation of the acetyl group to two molecules of carbon dioxide. These oxidations generate high-transfer-potential or high-energy electrons that will be used to power the synthesis of ATP. *The function of the citric acid cycle is the harvesting of high-energy electrons from carbon fuels.*

We begin with an examination of the cycle itself, noting that it consists of two parts: one part oxidizes carbon atoms to CO_2 and the other regenerates oxaloacetate. We then see how this metabolic hub is regulated, and we end with an examination of the glyoxylate cycle, a cycle unique to plants and some microorganisms that uses reactions of the citric acid cycle.

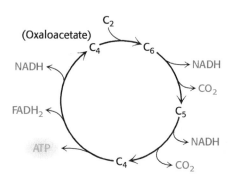

Figure 19.1 An overview of the citric acid cycle. The citric acid cycle oxidizes two-carbon units, producing two molecules of CO_2, one molecule of ATP, and high-energy electrons in the form of NADH and $FADH_2$.

19.1   The Citric Acid Cycle Consists of Two Stages

The overall pattern of the citric acid cycle is shown in **Figure 19.1**. The two-carbon acetyl unit condenses with a four-carbon component of the citric acid cycle oxaloacetate to yield the six-carbon tricarboxylic acid citrate. Citrate releases CO_2 twice, which yields high-energy electrons. A four-carbon compound remains. This four-carbon compound is further oxidized to regenerate oxaloacetate, which can initiate another round of the cycle. Two carbon atoms enter the cycle as an acetyl unit, and two carbon atoms leave the cycle in the form of two molecules of CO_2.

Note that the citric acid cycle itself neither generates much ATP nor includes oxygen as a reactant (Figure 19.1). Instead, the citric acid cycle captures high-energy electrons from citrate and its metabolites and uses these electrons to form NADH and $FADH_2$. These electron carriers yield nine molecules of ATP when they are oxidized by O_2 in *oxidative phosphorylation* (Section 9). Electrons released in the reoxidation of NADH and $FADH_2$ flow through a series of membrane proteins (referred to as the *electron-transport chain*) to generate a proton gradient across the membrane. This proton gradient is used to generate ATP from ADP and inorganic phosphate, as we will see in Section 9 (**Figure 19.2**).

The citric acid cycle occurs in essentially two stages. In the first stage, two carbon atoms are introduced into the cycle by coupling to oxaloacetate to form citrate, and two carbons are released as CO_2 as citrate is metabolized to a four-carbon molecule. In the second stage of the cycle, the resulting four-carbon molecule is metabolized to regenerate oxaloacetate, allowing continued functioning of the cycle.

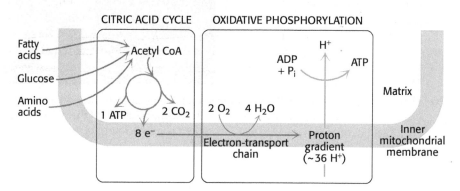

Figure 19.2 Cellular respiration. The citric acid cycle constitutes the first stage in cellular respiration, the removal of high-energy electrons from carbon fuels (blue pathway). These electrons reduce O_2 to generate a proton gradient (red pathway), which is used to synthesize ATP (green pathway). The reduction of O_2 and the synthesis of ATP constitute oxidative phosphorylation.

✓ 3 Identify the primary catabolic purpose of the citric acid cycle.

✓ 4 Explain the advantage of the oxidation of acetyl CoA in the citric acid cycle.

19.2   Stage One Oxidizes Two Carbon Atoms to Gather Energy-Rich Electrons

In the first part of the citric acid cycle, the four-carbon molecule oxaloacetate condenses with a two-carbon acetyl unit to yield citrate, a six-carbon tricarboxylic acid. As citrate moves through the first part of the cycle, two carbon atoms are lost as CO_2 in a process called *oxidative decarboxylation*. This decarboxylation yields a four-carbon molecule and high-transfer-potential electrons captured as two molecules of NADH.

Citrate Synthase Forms Citrate from Oxaloacetate and Acetyl Coenzyme A

The citric acid cycle begins with the condensation of a four-carbon unit, oxaloacetate, and a two-carbon unit, the acetyl group of acetyl CoA. Oxaloacetate reacts with acetyl CoA and H_2O to yield citrate and CoA.

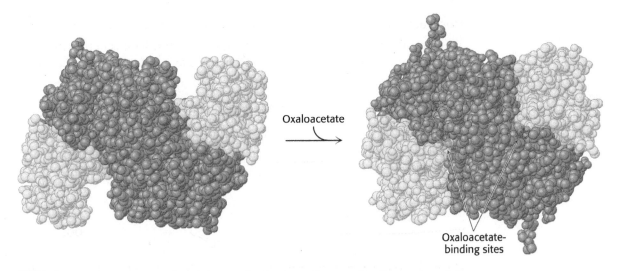

This reaction is catalyzed by *citrate synthase*. Oxaloacetate first condenses with acetyl CoA to form *citryl CoA*, a molecule that is energy rich because it contains the thioester that originated in acetyl CoA. The hydrolysis of citryl CoA thioester to citrate and CoA drives the overall reaction far in the direction of the synthesis of citrate. In essence, *the hydrolysis of the thioester powers the synthesis of a new molecule from two precursors.*

DID YOU KNOW?

A synthase is an enzyme that catalyzes a synthetic reaction in which two units are joined usually without the direct participation of ATP (or another nucleoside triphosphate).

The Mechanism of Citrate Synthase Prevents Undesirable Reactions

Because the condensation of acetyl CoA and oxaloacetate initiates the citric acid cycle, side reactions, such as the hydrolysis of acetyl CoA, must be minimized. Let us examine how citrate synthase prevents such wasteful reactions.

Mammalian citrate synthase is a dimer of identical 49-kDa subunits. Each active site is located in a cleft between the large and the small domains of a subunit, adjacent to the subunit interface. X-ray crystallographic studies revealed that the enzyme undergoes large conformational changes in the course of catalysis. Citrate synthase binds substrate in a sequential, ordered fashion: oxaloacetate binds first, followed by acetyl CoA. The reason for the ordered binding is that *oxaloacetate induces a major structural rearrangement leading to the creation of a binding site for acetyl CoA.* The binding of oxaloacetate converts the open form of the enzyme into a closed form (**Figure 19.3**). These structural changes create a binding site for acetyl CoA.

Figure 19.3 Conformational changes in citrate synthase on binding oxaloacetate. The small domain of each subunit of the homodimer is shown in yellow; the large domains are shown in blue. (Left) Open form of enzyme alone. (Right) Closed form of the liganded enzyme. [Drawn from 5CSC.pdb and 4CTS.pdb.]

Citrate synthase first catalyzes the condensation of citrate and acetyl CoA to form citryl CoA. The newly formed citryl CoA induces additional structural changes in the enzyme, causing the active site to become completely enclosed. The enzyme then cleaves the citryl CoA thioester by hydrolysis. CoA leaves the enzyme, followed by citrate, and the enzyme returns to the initial open conformation.

We can now understand how the wasteful hydrolysis of acetyl CoA is prevented. Citrate synthase is well suited to the hydrolysis of *citryl* CoA but not *acetyl* CoA. How is this discrimination accomplished? First, acetyl CoA does not bind to the enzyme until oxaloacetate is bound and ready for condensation. Second, the catalytic residues crucial for the hydrolysis of the thioester linkage are not appropriately positioned until citryl CoA is formed. *Induced fit prevents an undesirable side reaction.*

Citrate Is Isomerized into Isocitrate

Keep in mind that a major purpose of the citric acid cycle is the oxidation of carbon atoms leading to the capture of high-transfer-potential electrons. The newly formed citrate molecule is not optimally structured for the required oxidation reactions. In particular, the hydroxyl group is not properly located in the citrate molecule for the oxidative decarboxylations that follow. Thus, citrate is isomerized into isocitrate to enable the six-carbon unit to undergo oxidative decarboxylation. The isomerization of citrate is accomplished by a *dehydration* step followed by a *hydration* step. The result is an interchange of an H and an OH. The enzyme catalyzing both steps is called *aconitase* because *cis-aconitate* is an intermediate (**Figure 19.4**).

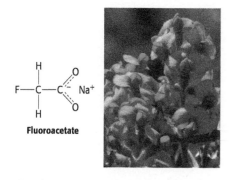

Fluoroacetate

Figure 19.4 Aconitase is inhibited by a metabolite of fluoroacetate. Fluoroacetate, a toxin, is activated to fluoroacetyl CoA, which reacts with citrate to form fluorocitrate, a suicide inhibitor (p. 138) of aconitase. After having been irreversibly inhibited, aconitase shuts down the citric acid cycle and cellular respiration, accomplishing its role as a pesticide. Members of the genus *Gastrolobium* in Australia contain fluoroacetate. [@Krystyna Szulecka/Alamy Images.]

Citrate   **cis-Aconitate**   **Isocitrate**

Isocitrate Is Oxidized and Decarboxylated to Alpha-Ketoglutarate

We come now to the first of four oxidation–reduction reactions in the citric acid cycle. The oxidative decarboxylation of isocitrate to α-ketoglutarate is catalyzed by *isocitrate dehydrogenase:*

$$\text{Isocitrate} + \text{NAD}^+ \longrightarrow \alpha\text{-ketoglutarate} + \text{CO}_2 + \text{NADH}$$

The intermediate in this reaction is oxalosuccinate, an unstable α-ketoacid. While bound to the enzyme, it loses CO_2 to form α-ketoglutarate. This oxidation generates the first high-transfer-potential electron carrier in the cycle, NADH:

Isocitrate   **Oxalosuccinate**   **α-Ketoglutarate**

Succinyl Coenzyme A Is Formed by the Oxidative Decarboxylation of Alpha-Ketoglutarate

The conversion of isocitrate into α-ketoglutarate is followed by a second oxidative decarboxylation reaction, removing CO_2 from the α-ketoglutarate (five carbon atoms) to form succinyl CoA (four carbon atoms):

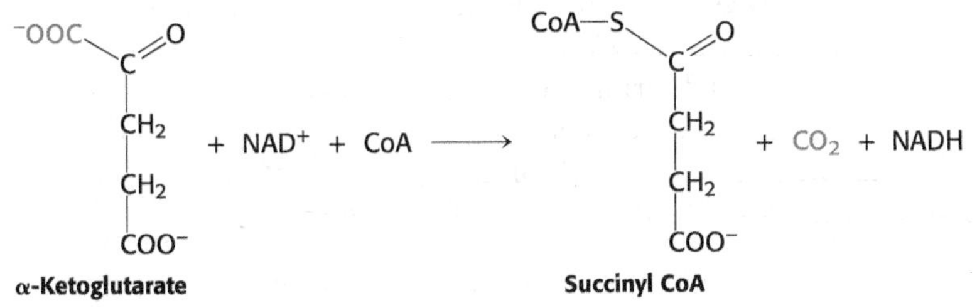

α-Ketoglutarate   Succinyl CoA

This reaction is catalyzed by the *α-ketoglutarate dehydrogenase complex,* an organized assembly of three kinds of enzymes that is structurally similar to the pyruvate dehydrogenase complex (p. 332). In fact, the E_3 component is identical in both enzymes. The oxidative decarboxylation of α-ketoglutarate closely resembles that of pyruvate, also an α-ketoacid:

$$\text{Pyruvate} + \text{CoA} + \text{NAD}^+ \xrightarrow{\text{Pyruvate dehdrogenase complex}} \text{acetyl CoA} + CO_2 + \text{NADH} + \text{H}^+$$

$$\alpha\text{-Ketoglutarate} + \text{CoA} + \text{NAD}^+ \xrightarrow{\alpha\text{-Ketoglutate dehdrogenase complex}} \text{succinyl CoA} + CO_2 + \text{NADH}$$

Both reactions include the decarboxylation of an α-ketoacid and the subsequent formation of a thioester linkage with CoA that has a high transfer potential. The reaction mechanisms are entirely analogous, employing the same coenzymes and reaction steps. At this point in the citric acid cycle, two carbon atoms have entered the cycle and two carbon atoms have been oxidized to CO_2. The electrons from the oxidations are captured in two molecules of NADH.

19.3 Stage Two Regenerates Oxaloacetate and Harvests Energy-Rich Electrons

The second part of the citric acid cycle consists of the regeneration of the starting material, oxaloacetate. This regeneration is accomplished by a series of reactions that begins with a four-carbon molecule and ends with a four-carbon molecule. However, the rearrangements within this set of reactions harvest energy in the form of high-energy electron carriers and a molecule of ATP.

A Compound with High Phosphoryl-Transfer Potential Is Generated from Succinyl Coenzyme A

The succinyl CoA produced by α-ketoglutarate dehydrogenase in the preceding step is an energy-rich thioester compound. The $\Delta G^{\circ\prime}$ for the hydrolysis of succinyl CoA is about -33.5 kJ mol^{-1} (-8.0 kcal mol^{-1}), which is comparable to that of ATP (-30.5 kJ mol^{-1}, or -7.3 kcal mol^{-1}). How is this energy utilized in the citric acid cycle? Recall that, in the citrate synthase reaction, the cleavage of the thioester powers the synthesis of the six-carbon citrate from the four-carbon oxaloacetate and the two-carbon fragment. In this case, *the cleavage of the thioester of succinyl CoA is coupled to the phosphorylation of a purine nucleoside diphosphate, usually ADP.* This reaction is catalyzed by *succinyl CoA synthetase* (succinate thiokinase):

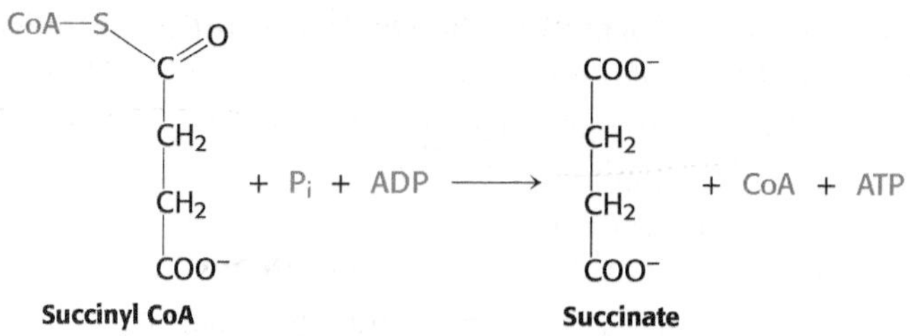

This reaction is the only step in the citric acid cycle that directly yields a compound with high phosphoryl-transfer potential. In mammals, there are two forms of the enzyme, one specific for ADP and the other for GDP. In tissues that perform large amounts of cellular respiration, such as skeletal and heart muscle, the ADP-requiring enzyme predominates. In tissues that perform many anabolic reactions, such as the liver, the GDP-requiring enzyme is common. The GDP-requiring enzyme is believed to work in reverse of the direction observed in the citric acid cycle; that is, GTP is used to power the synthesis of succinyl CoA, which is a precursor for heme synthesis. The *E. coli* enzyme uses either GDP or ADP as the phosphoryl-group acceptor.

Succinyl Coenzyme A Synthetase Transforms Types of Biochemical Energy

The mechanism of this reaction is a clear example of an energy transformation: energy inherent in the thioester is transformed into phosphoryl-group-transfer potential (**Figure 19.5**). In the first step the coenzyme A of succinyl CoA is displaced by orthophosphate. This displacement results in another energy-rich compound, succinyl phosphate, which is a mixed anhydride. Next, a histidine residue on the enzyme acts as a moving arm, detaching the phosphoryl group from succinyl phosphate, then swinging over to ADP bound to the enzyme, and transfers the phosphoryl group to form ATP. The participation of high-energy compounds in all the steps is evidenced by the fact that the reaction is readily reversible: $\Delta G^{\circ\prime} = -3.4 \text{ kJ mol}^{-1} (-0.8 \text{ kcal mol}^{-1})$. The generation of ATP in a reaction in which a high-phosphoryl-transfer-potential compound (succinyl phosphate)

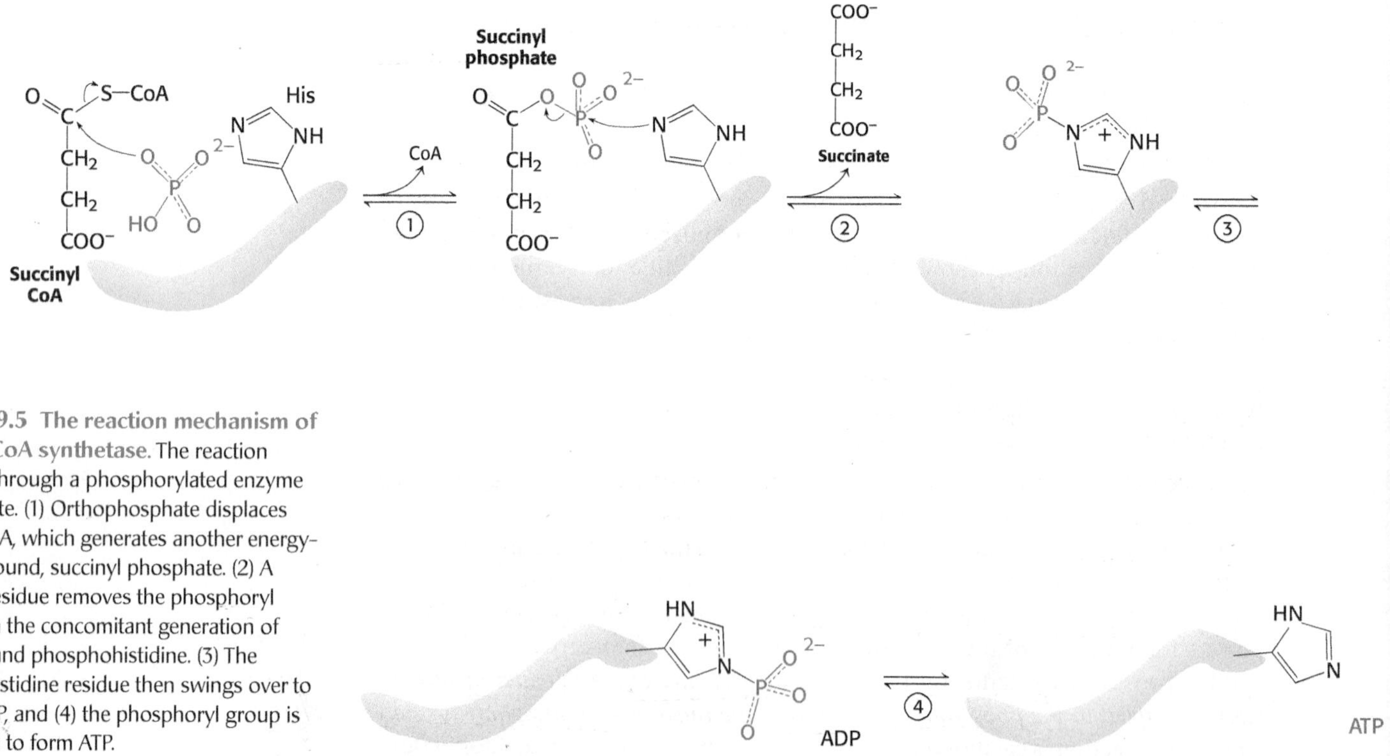

Figure 19.5 The reaction mechanism of succinyl CoA synthetase. The reaction proceeds through a phosphorylated enzyme intermediate. (1) Orthophosphate displaces coenzyme A, which generates another energy–rich compound, succinyl phosphate. (2) A histidine residue removes the phosphoryl group with the concomitant generation of succinate and phosphohistidine. (3) The phosphohistidine residue then swings over to bound ADP, and (4) the phosphoryl group is transferred to form ATP.

transfers the phosphate to ADP to generate ATP is called *substrate-level phos-phorylation*. Recall that glycolysis forms ATP with substrate-level phosphoryla-tion reactions (Chapter 16). In Section 9, we will examine how ion gradients can be used to power ATP formation.

Oxaloacetate Is Regenerated by the Oxidation of Succinate

Succinate is subsequently oxidized to regenerate oxaloacetate.

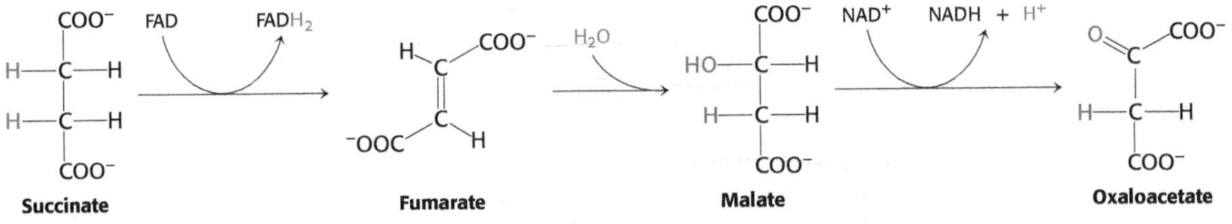

The reactions constitute a metabolic motif that we will see again in fatty acid degradation (Chapter 27) and synthesis (Chapter 28). A methylene group (CH_2) is converted into a carbonyl group ($C{=}O$) in three steps: an oxidation, a hydra-tion, and a second oxidation reaction. Oxaloacetate is thereby regenerated for another round of the cycle, and more energy is extracted in the form of $FADH_2$ and NADH.

The first step in this set of reactions, the oxidation of succinate to fumarate, is catalyzed by *succinate dehydrogenase*. The hydrogen acceptor is FAD rather than NAD^+, which is used in the other three oxidation reactions in the cycle. FAD is the hydrogen acceptor in this reaction because the free-energy change is insufficient to reduce NAD^+.

Succinate dehydrogenase differs from other enzymes in the citric acid cycle because it is embedded in the inner mitochondrial membrane in association with the electron-transport chain, which also is set in the inner mitochondrial membrane. The electron-transport chain is the link between the citric acid cycle and ATP formation (Section 9). $FADH_2$ produced by the oxidation of succinate does not dissociate from the enzyme, in contrast with NADH produced in other oxidation–reduction reactions. Rather, succinate dehydrogenase transfers two electrons directly from $FADH_2$ to coenzyme Q (CoQ). A component of the electron-transport chain, CoQ passes electrons to the ultimate acceptor, molecu-lar oxygen, as we shall see in Section 9.

After succinate has been oxidized to fumarate, the next step is the hydration of fumarate to form L-malate, a reaction catalyzed by *fumarase*. Finally, malate is oxidized to form oxaloacetate. This reaction is catalyzed by *malate dehydroge-nase*, and NAD^+ is again the hydrogen acceptor:

$$\text{Malate} + NAD^+ \rightleftharpoons \text{oxaloacetate} + NADH + H^+$$

The standard free energy for this reaction, unlike that for the other steps in the citric acid cycle, is significantly positive $\Delta G^{\circ\prime} = +29.7 \text{ kJ mol}^{-1}$, or $+7.1$ kcal mol^{-1}). The oxidation of malate is driven by the use of the products—oxaloacetate by citrate synthase and NADH by the electron-transport chain.

The Citric Acid Cycle Produces High-Transfer-Potential Electrons, an ATP, and Carbon Dioxide

The net reaction of the citric acid cycle is

$$\text{Acetyl CoA} + 3 \text{ NAD}^+ + \text{FAD} + \text{ADP} + P_i + 2 H_2O \longrightarrow$$
$$2 CO_2 + 3 \text{ NADH} + FADH_2 + \text{ATP} + 2 H^+ + \text{CoA}$$

DID YOU KNOW?

Apples are a rich source of malic acid, which used to be called "acid of apples." In fact, the word malic is derived from the Latin *malum*, meaning "apple."

Let us review the reactions that give this stoichiometry (**Figure 19.6** and **Table 19.1**):

1. Two carbon atoms enter the cycle in the condensation of an acetyl unit (from acetyl CoA) with oxaloacetate. Two carbon atoms leave the cycle in the form of CO_2 in the successive decarboxylations catalyzed by isocitrate dehydrogenase and α-ketoglutarate dehydrogenase.

2. Four pairs of hydrogen atoms leave the cycle in four oxidation reactions. Two NAD^+ molecules are reduced in the oxidative decarboxylations of isocitrate and α-ketoglutarate, one FAD molecule is reduced in the oxidation of succinate, and one NAD^+ molecule is reduced in the oxidation of malate. Recall also that one NAD^+ molecule is reduced in the oxidative decarboxylation of pyruvate to form acetyl CoA (p. 332).

3. One ATP is generated from the cleavage of the thioester linkage in succinyl CoA.

4. Two water molecules are consumed: one in the synthesis of citrate by the hydrolysis of citryl CoA and the other in the hydration of fumarate.

Isotope-labeling studies revealed that the two carbon atoms that enter each cycle are not the ones that leave. The two carbon atoms that enter the cycle as the acetyl group are retained in the initial two decarboxylation reactions (Figure 19.6)

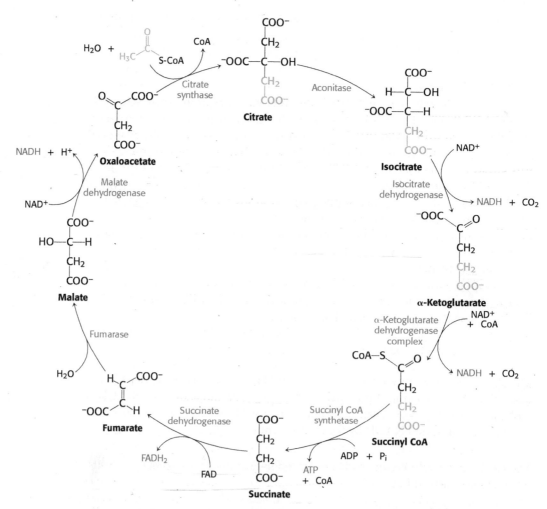

Figure 19.6 The citric acid cycle. Notice that, because succinate is a symmetric molecule, the identity of the carbon atoms from the acetyl unit is lost.

Table 19.1   Citric acid cycle

Step	Reaction	Enzyme	Prosthetic group	Type*	$\Delta G^{\circ\prime}$ kJ mol^{-1}	kcal mol^{-1}
1	Acetyl CoA + oxaloacetate + $H_2O \longrightarrow$ citrate + CoA + H^+	Citrate synthase		a	−31.4	−7.5
2a	Citrate $\rightleftharpoons$ cis-aconitate + H_2O	Aconitase	Fe-S	b	+8.4	+2.0
2b	cis-Aconitate + $H_2O \rightleftharpoons$ isocitrate	Aconitase	Fe-S	c	−2.1	−0.5
3	Isocitrate + NAD$^+$ $\rightleftharpoons$ α-ketoglutarate + CO_2 + NADH	Isocitrate dehydrogenase		d + e	−8.4	−2.0
4	α-ketoglutarate + NAD$^+$ + CoA $\rightleftharpoons$ succinyl CoA + CO_2 + NADH	α-ketoglutarate dehydrogenase complex	Lipoic acid, FAD, TPP	d + e	−30.1	−7.2
5	Succinyl CoA + P_i + ADP $\rightleftharpoons$ succinate + ATP + CoA	Succinyl CoA synthetase		f	−3.3	−0.8
6	Succinate + FAD (enzyme-bound) $\rightleftharpoons$ fumarate + FADH$_2$ (enzyme-bound)	Succinate dehydrogenase	FAD, Fe-S	e	0	0
7	Fumarate + H_2O $\rightleftharpoons$ L-malate	Fumarase		e	−3.8	−0.9
8	L-Malate + NAD$^+$ $\rightleftharpoons$ oxaloacetate + NADH + H^+	Malate dehydrogenase		e	+29.7	+7.1

*Reaction type: a, condensation; b, dehydration; c, hydration; d, decarboxylation; e, oxidation; f, substrate-level phosphorylation.

and then remain incorporated in the four-carbon acids of the cycle. Note that succinate is a symmetric molecule. Consequently, the two carbon atoms that enter the cycle can occupy any of the carbon positions in the subsequent metabolism of the four-carbon acids. The two carbon atoms that enter the cycle as the acetyl group will be released as CO_2 in *subsequent* trips through the cycle. To understand why citrate is not processed as a symmetric molecule, see problems 19 and 20 at the end of the chapter.

Evidence is accumulating that the enzymes of the citric acid cycle are physically associated with one another. The close arrangement of enzymes enhances the efficiency of the citric acid cycle because a reaction product can pass directly from one active site to the next through connecting channels, a process called *substrate channeling*.

The key catabolic function of the citric acid cycle is the production of high-energy electrons in the form of NADH and FADH$_2$. As will be considered in Section 9, the electron-transport chain oxidizes the NADH and FADH$_2$ formed in the citric acid cycle and ultimately results in the generation of 2.5 ATP per NADH, and 1.5 ATP per FADH$_2$. One complete turn through the citric acid cycle generates 3 NADH molecules, 1 FADH$_2$ molecule, and 1 ATP molecule, for a grand total of 10 molecules of ATP. In dramatic contrast, the anaerobic glycolysis of 1 glucose molecule generates only 2 ATP molecules (and 2 lactate molecules).

Recall that molecular oxygen does not participate directly in the citric acid cycle. However, the cycle operates only under aerobic conditions because NAD$^+$ and FAD can be regenerated in mitochondria only by the transfer of electrons to molecular oxygen. *Glycolysis has both an aerobic and an anaerobic mode, whereas the citric acid cycle is strictly aerobic.* As discussed in Chapter 16, glycolysis can proceed under anaerobic conditions because NAD$^+$ is regenerated in the conversion of pyruvate into lactate or ethanol.

DID YOU KNOW?

The manuscript proposing the citric acid cycle was submitted for publication to *Nature* but was rejected. Dr. Hans Krebs proudly displayed the rejection letter of June 1937 throughout his career as encouragement for young scientists. His work was subsequently published in *Enzymologia*.

? QUICK QUIZ 1 Malonate is a competitive inhibitor of succinate dehydrogenase. How will the concentrations of citric acid cycle intermediates change immediately after the addition of malonate? Why is malonate not a substrate for succinate dehydrogenase?

COO$^-$
|
CH$_2$
|
COO$^-$

Malonate

19.4 The Citric Acid Cycle Is Regulated

The citric acid cycle is the final common pathway for the aerobic oxidation of fuel molecules. Moreover, as we will see shortly (p. 353) and repeatedly elsewhere in our study of biochemistry, the cycle is an important source of building blocks for a host of biomolecules. As befits its role as the metabolic hub of the cell, entry into the cycle and the rate of the cycle itself are controlled at several stages. We have seen (p. 337) that the pyruvate dehydrogenase complex, the link between glycolysis and the citric acid cycle, is a regulatory site controlling the metabolism of pyruvate by the citric acid cycle.

The Citric Acid Cycle Is Controlled at Several Points

The regulation of pyruvate dehydrogenase is a crucial site of control, modulating the conversion of glucose-derived pyruvate into acetyl CoA. However, pyruvate is not the only source of acetyl CoA. Indeed, acetyl CoA derived from fat breakdown enters the cycle directly (Chapter 27). Consequently, the rate of the citric acid cycle itself also must be precisely controlled to meet an animal cell's needs for ATP (Figure 19.7). The primary control points are the allosteric enzymes *isocitrate dehydrogenase* and *α-ketoglutarate dehydrogenase.*

For many tissues, such as the liver, the ATP needs are approximately constant on the time scale of minutes to hours. Thus, for these tissues, the citric acid cycle is operating at a constant rate. But, just as driving at a constant speed usually requires touching the gas pedal and the break occasionally, the rate of the citric acid cycle must be "tuned" with increases and decreases. What signals regulate the rate of the cycle? *Isocitrate dehydrogenase is allosterically stimulated by ADP, which signifies the need for more energy.* In contrast, NADH, which signals the presence of high-transfer-potential electrons, inhibits isocitrate dehydrogenase. Likewise, ATP, the ultimate end product of fuel catabolism, is inhibitory.

A second control site in the citric acid cycle is α-ketoglutarate dehydrogenase, which catalyzes the rate-limiting step in the citric acid cycle. Some aspects of this enzyme's control are like those of the pyruvate dehydrogenase complex, as might be expected from the similarity of the two enzymes. α-Ketoglutarate dehydrogenase is inhibited by the products of the reaction that it catalyzes, succinyl CoA and NADH. In addition, α-ketoglutarate dehydrogenase is inhibited by high levels of ATP. Thus, *the rate of the cycle is decreased when the cell has high levels of ATP and NADH.* α-Ketoglutarate dehydrogenase deficiency is observed in a number of neurological disorders, including Alzheimer disease. Importantly, several other enzymes in the cycle require NAD^+ or FAD. These enzymes will function only when the energy charge is low, because only then will NAD^+ and FAD be available.

Although different regulatory mechanisms might be expected in skeletal muscle, where the energy needs can vary immensely and rapidly, this does not appear to be the case. While the rate of the citric acid cycle will increase 100-fold during intense exercise, the rate of the cycle in muscle is under the control of the same two enzymes as in liver.

The use of isocitrate dehydrogenase and α-ketoglutarate dehydrogenase as control points integrates the citric acid cycle with other pathways and highlights the central role of the citric acid cycle in metabolism. For instance, the inhibition of isocitrate dehydrogenase leads to a buildup of citrate, because the interconversion of isocitrate and citrate is readily reversible under intracellular conditions. Citrate can be transported to the cytoplasm where it signals phosphofructokinase to halt glycolysis (p. 300) and where it can serve as a source of acetyl CoA for fatty acid synthesis (Chapter 28). The α-ketoglutarate that accumulates when α-ketoglutarate dehydrogenase is inhibited can be used as a precursor for several amino acids and the purine bases (Chapter 30).

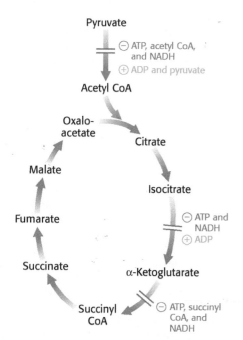

Pyruvate

⊖ ATP, acetyl CoA, and NADH
⊕ ADP and pyruvate

Acetyl CoA

Oxalo-acetate

Citrate

Malate

Isocitrate

Fumarate

⊖ ATP and NADH
⊕ ADP

Succinate

α-Ketoglutarate

Succinyl CoA

⊖ ATP, succinyl CoA, and NADH

Figure 19.7 Control of the citric acid cycle. The citric acid cycle is regulated primarily by the concentrations of ATP and NADH. The key control points are the enzymes isocitrate dehydrogenase and α-ketoglutarate dehydrogenase.

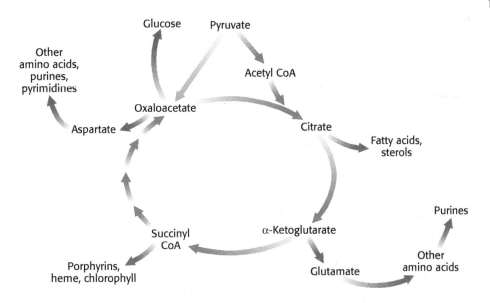

Figure 19.8 Biosynthetic roles of the citric acid cycle. Intermediates are drawn off for biosyntheses (shown by red arrows) when the energy needs of the cell are met. Intermediates are replenished by the formation of oxaloacetate from pyruvate (green arrow).

In many bacteria, the funneling of two-carbon fragments into the cycle also is controlled. *The synthesis of citrate from oxaloacetate and acetyl CoA carbon units is a key control point in these organisms*. ATP is an allosteric inhibitor of citrate synthase. The effect of ATP is to increase the value of K_M for acetyl CoA. Thus, as the level of ATP increases, less of this enzyme reacts with acetyl CoA and so less citrate is formed.

The Citric Acid Cycle Is a Source of Biosynthetic Precursors

Thus far, we have focused on the citric acid cycle as the *major degradative pathway for the generation of ATP*. As a major metabolic hub of the cell, the citric acid cycle also *provides intermediates for biosyntheses* (Figure 19.8). For example, most of the carbon atoms in porphyrins, the precursors to the heme groups in hemoglobin and myoglobin, come from *succinyl CoA*. Many of the amino acids are derived from *α-ketoglutarate* and *oxaloacetate*. These biosynthetic processes will be considered in subsequent chapters.

The Citric Acid Cycle Must Be Capable of Being Rapidly Replenished

The citric acid cycle is crucial for generating biological energy and is a source of building blocks for biosynthetic reactions. This dual use presents a problem. Suppose that much oxaloacetate is converted into glucose and, subsequently, the energy needs of the cell rise. The citric acid cycle will operate at a reduced capacity unless new oxaloacetate is formed, because acetyl CoA cannot enter the cycle unless it condenses with oxaloacetate. Even though oxaloacetate is recycled, a minimal level must be maintained to allow the cycle to function. *Citric acid cycle intermediates must be replenished if any are drawn off for biosyntheses.*

How is oxaloacetate replenished? Mammals lack the enzymes for the net conversion of acetyl CoA into oxaloacetate or any other citric acid cycle intermediate. Rather, oxaloacetate is formed by the carboxylation of pyruvate, in a reaction catalyzed by the biotin-dependent enzyme *pyruvate carboxylase* (Figure 19.9):

$$\text{Pyruvate} + CO_2 + \text{ATP} + H_2O \longrightarrow \text{oxaloacetate} + \text{ADP} + P_i + 2H^+$$

Recall that this enzyme plays a crucial role in gluconeogenesis (p. 316). It is active only in the presence of acetyl CoA, which signifies the need for more oxaloacetate. If the energy charge is high, oxaloacetate is converted into glucose. If the energy charge is low, oxaloacetate replenishes the citric acid cycle. The synthesis

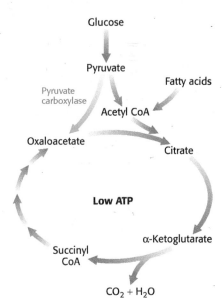

Figure 19.9 Pyruvate carboxylase replenishes the citric acid cycle. The rate of the citric acid cycle increases during exercise, requiring the replenishment of oxaloacetate and acetyl CoA. Oxaloacetate is replenished by its formation from pyruvate (green arrow). Acetyl CoA can be produced from the metabolism of both pyruvate and fatty acids.

of oxaloacetate by the carboxylation of pyruvate is an example of an *anaplerotic reaction* (of Greek origin, meaning to "fill up"), a reaction that leads to the net synthesis, or replenishment, of pathway components. Note that, because the citric acid cycle is a cycle, it can be replenished by the generation of any of the intermediates. For instance, the removal of the nitrogen groups from glutamate and aspartate yield α-ketoglutarate and oxaloacetate, respectively. Glutamine is an especially important source of citric acid cycle intermediates in rapidly growing cells, including cancer cells. Glutamine is converted into glutamate and then into α-ketoglutarate.

? QUICK QUIZ 2 Why is acetyl CoA an especially appropriate activator for pyruvate carboxylase?

🜍 CLINICAL INSIGHT

Defects in the Citric Acid Cycle Contribute to the Development of Cancer

Four enzymes crucial to cellular respiration are known to contribute to the development of cancer: succinate dehydrogenase, fumarase, pyruvate dehydrogenase kinase, and isocitrate dehydrogenase Mutations that alter the activity of the first three of these enzymes enhance aerobic glycolysis (p. 304). In aerobic glycolysis, cancer cells preferentially metabolize glucose to lactate even in the presence of oxygen. Defects in all of these enzymes have a common biochemical link: the transcription factor *hypoxia inducible factor 1* (HIF-1).

Normally, HIF-1 upregulates the enzymes and transporters that enhance glycolysis only when oxygen concentration falls, a condition called hypoxia. Under normal conditions, HIF-1 is hydroxylated by prolyl hydroxylase 2 and is subsequently destroyed by a large complex of proteolytic enzymes. The degradation of HIF-1 prevents the enhanced synthesis of glycolytic proteins. Prolyl hydroxylase 2 requires α-ketoglutarate, ascorbate (Vitamin C), and oxygen for activity. Thus, when oxygen concentration falls, prolyl hydroxylase 2 is inactive, HIF-1 is not hydroxylated and not degraded, and the synthesis of proteins required for glycolysis is stimulated. As a result, the rate of glycolysis is increased.

Defects in the enzymes of the citric acid cycle can significantly alter the regulation of prolyl hydroxylase 2. When either succinate dehydrogenase or fumarase is defective, succinate and fumarate accumulate in the mitochondria and spill over into the cytoplasm. Both succinate and fumarate are competitive inhibitors of prolyl hydroxylase 2. The inhibition of prolyl hydroxylase 2 results in the stabilization of HIF-1 because HIF-1 is no longer hydroxylated. Lactate, the end product of glycolysis, also appears to inhibit prolyl hydroxylase 2 by interfering with the action of ascorbate. In addition to increasing the amount of the proteins required for glycolysis, HIF-1 stimulates the production of pyruvate dehydrogenase kinase (PDH kinase), as discussed earlier (p. 337).

Mutations in isocitrate dehydrogenase result in the generation of an oncogenic metabolite, 2-hydroxyglutarate. The mutant enzyme catalyzes the conversion of isocitrate to α-ketoglutarate, but then reduces α-ketoglutarate to form 2-hydroxyglutarate. 2-Hydroxyglutarate alters the methylation patterns in DNA and reduces dependence on growth factors for proliferation. These changes alter gene expression and promote unrestrained cell growth.

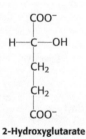

2-Hydroxyglutarate

These observations linking citric acid cycle enzymes to cancer suggest that cancer is also a metabolic disease, not simply a disease of mutant growth factors and cell-cycle-control proteins. The realization that there is a metabolic component to cancer opens the door to new thinking about the control of the disease. Indeed, preliminary experiments suggest that, if cancer cells undergoing aerobic glycolysis are forced by pharmacological manipulation to use oxidative phosphorylation, the cancer cells lose their malignant properties. It is also interesting to note that the citric acid cycle, which has been studied for decades, still has secrets to be revealed by future biochemists.

19.5 The Glyoxylate Cycle Enables Plants and Bacteria to Convert Fats into Carbohydrates

Acetyl CoA that enters the citric acid cycle has only one fate: oxidation to CO_2 and H_2O. Most organisms thus cannot convert acetyl CoA into glucose, because, although oxaloacetate, a key precursor to glucose, is formed in the citric acid cycle, the two decarboxylations that take place before the regeneration of oxaloacetate preclude the *net* conversion of acetyl CoA into glucose.

In plants and some microorganisms, a metabolic pathway does indeed exist that allows the conversion of acetyl CoA generated from fat stores into glucose. This reaction sequence, called the *glyoxylate cycle,* is similar to the citric acid cycle but bypasses the two decarboxylation steps of the cycle. Another important difference is that two molecules of acetyl CoA enter per turn of the glyoxylate cycle, compared with one molecule in the citric acid cycle.

The glyoxylate cycle (**Figure 19.10**), like the citric acid cycle, begins with the condensation of acetyl CoA and oxaloacetate to form citrate, which is then isomerized to isocitrate. Instead of being decarboxylated, as in the citric acid cycle, isocitrate is cleaved by *isocitrate lyase* into succinate and glyoxylate. The ensuing steps regenerate oxaloacetate from glyoxylate. First, acetyl CoA condenses with glyoxylate to form malate in a reaction catalyzed by *malate synthase,* and then malate is oxidized to oxaloacetate, as in the citric acid cycle. The sum of these reactions is

$$2\,\text{Acetyl CoA} + \text{NAD}^+ + 2\,H_2O \longrightarrow \text{succinate} + 2\,\text{CoA} + \text{NADH} + 2\,H^+$$

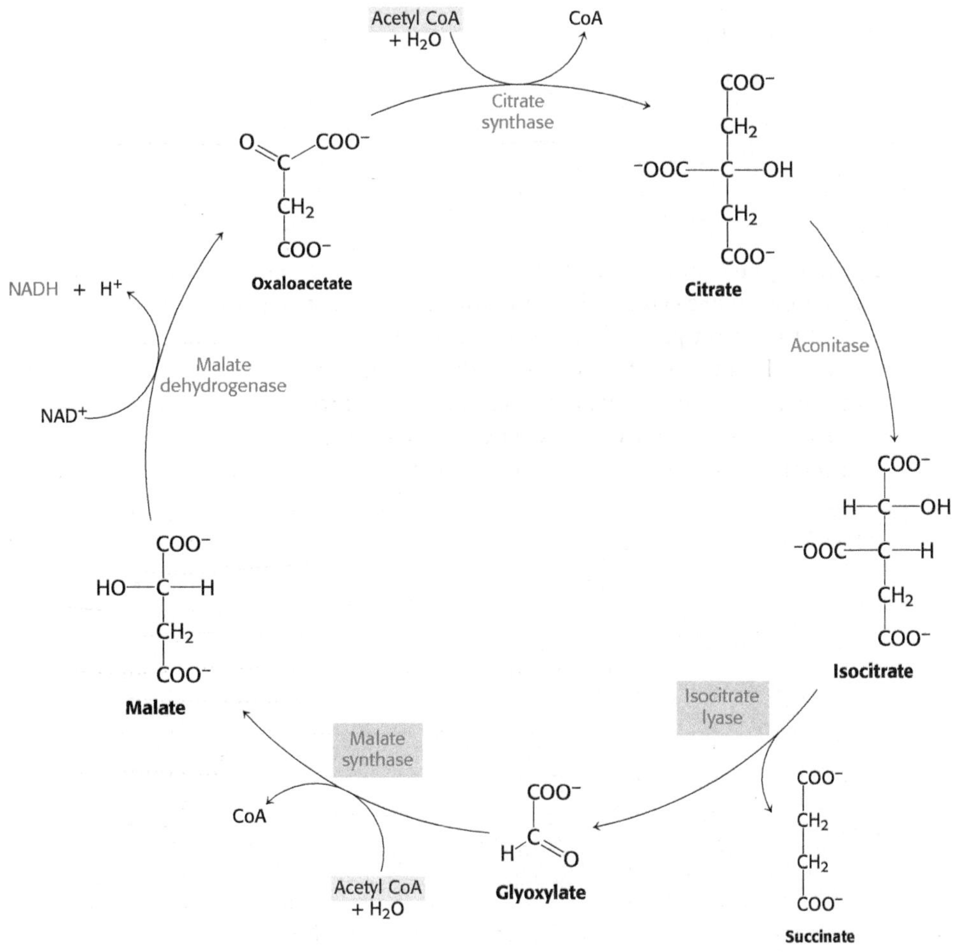

Figure 19.10 The glyoxylate pathway. The glyoxylate cycle allows plants and some microorganisms to grow on acetate because the cycle bypasses the decarboxylation steps of the citric acid cycle. The reactions of this cycle are the same as those of the citric acid cycle except for the ones catalyzed by isocitrate lyase and malate synthase, which are boxed in blue.

Figure 19.11 Sunflowers can convert acetyl CoA into glucose. Tour de France cyclists pass a field of sunflowers. The glyoxylate cycle is especially prominent in sunflowers. [Javier Soriano/AFP/Getty Images.]

In plants, these reactions take place in organelles called *glyoxysomes*. This cycle is especially prominent in oil-rich seeds, such as those from sunflowers, cucumbers, and castor beans (Figure 19.11). Succinate, released midcycle, can be converted into carbohydrates by a combination of the citric acid cycle and gluconeogenesis. The carbohydrates power seedling growth until the plant can begin photosynthesis. Thus, organisms with the glyoxylate cycle gain metabolic versatility because they can use acetyl CoA as a precursor of glucose and other biomolecules.

SUMMARY

19.1 The Citric Acid Cycle Consists of Two Stages

The first stage of the citric acid cycle consists of the condensation of acetyl CoA with oxaloacetate, followed by two oxidative decarboxylations. In the second stage of the cycle, oxaloacetate is regenerated, coupled with the formation of high-transfer-potential electrons and a molecule of ATP.

19.2 Stage One Oxidizes Two Carbon Atoms to Gather Energy-Rich Electrons

The cycle starts with the condensation of oxaloacetate (containing four carbon atoms, abbreviated as C_4) and acetyl CoA (C_2) to give citrate (C_6), which is isomerized to isocitrate (C_6). Oxidative decarboxylation of this intermediate gives α-ketoglutarate (C_5). The second molecule of carbon dioxide comes off in the next reaction, in which α-ketoglutarate is oxidatively decarboxylated to succinyl CoA (C_4).

19.3 Stage Two Regenerates Oxaloacetate and Harvests Energy-Rich Electrons

The thioester of succinyl CoA is cleaved by orthophosphate to yield succinate, and a high-phosphoryl-transfer-potential compound in the form of ATP is concomitantly generated. Succinate is oxidized to fumarate (C_4), which is then hydrated to form malate (C_4). Finally, malate is oxidized to regenerate oxaloacetate (C_4). Thus, two carbon atoms from acetyl CoA enter the cycle, and two carbon atoms leave the cycle as CO_2 in the successive decarboxylations catalyzed by isocitrate dehydrogenase and α-ketoglutarate dehydrogenase. In the four oxidation–reduction reactions in the cycle, three pairs of electrons are transferred to NAD^+ and one pair to FAD. These reduced electron carriers are subsequently oxidized by the electron-transport chain to generate approximately 9 molecules of ATP. In addition, 1 molecule of ATP is directly formed in the citric acid cycle. Hence, a total of 10 molecules of ATP are generated for each two-carbon fragment that is completely oxidized to H_2O and CO_2.

19.4 The Citric Acid Cycle Is Regulated

The citric acid cycle operates only under aerobic conditions because it requires a supply of NAD^+ and FAD. The electron acceptors are regenerated when NADH and $FADH_2$ transfer their electrons to O_2 through the electron-transport chain, with the concomitant production of ATP. Consequently, the rate of the citric acid cycle depends on the need for ATP. In eukaryotes, the regulation of two enzymes in the cycle also is important for control. A high energy charge diminishes the activities of isocitrate dehydrogenase and α-ketoglutarate dehydrogenase. These mechanisms complement each other in reducing the rate of formation of acetyl CoA when the energy charge of the cell is high and when biosynthetic intermediates are abundant.

When the cell has adequate energy available, the citric acid cycle can provide a source of building blocks for a host of important biomolecules, such as nucleotide bases, proteins, and heme groups. This use depletes the cycle of intermediates. When the cycle again needs to metabolize fuel, anaplerotic reactions replenish the cycle intermediates.

19.5 The Glyoxylate Cycle Enables Plants and Bacteria to Convert Fats into Carbohydrates

The glyoxylate cycle enhances the metabolic versatility of many plants and bacteria. This cycle, which uses some of the reactions of the citric acid cycle, enables these organisms to convert fats into glucose. Two molecules of acetyl CoA are converted into succinate, which can be used to synthesize glucose.

KEY TERMS

pyruvate dehydrogenase complex
  (p. 343)
citric acid cycle (p. 343)
citrate synthase (p. 344)

isocitrate dehydrogenase (p. 346)
α-ketoglutarate dehydrogenase
  (p. 347)
anaplerotic reaction (p. 354)

glyoxylate cycle (p. 355)
isocitrate lyase (p. 355)
malate synthase (p. 355)
glyoxysome (p. 356)

? Answers to QUICK QUIZZES

1. Succinate will increase in concentration, followed by α-ketoglutarate and the other intermediates "upstream" of the site of inhibition. Succinate has two methylene groups that are required for the dehydrogenation, whereas malonate has but one.

2. Pyruvate carboxylase should be active only when the acetyl CoA concentration is high. Acetyl CoA might accumulate if the energy needs of the cell are not being met because of a deficiency of oxaloacetate. Under these conditions, the pyruvate carboxylase catalyzes an anaplerotic reaction. Alternatively, acetyl CoA might accumulate because the energy needs of the cell have been met. In this circumstance, pyruvate will be converted back into glucose, and the first step in this conversion is the formation of oxaloacetate.

PROBLEMS

1. *The bottom line.* What is the net equation of the citric acid cycle? ✓ 3

2. *A hoax, perhaps?* The citric acid cycle is part of aerobic respiration, but no O_2 is required for the cycle. Explain this paradox. ✓ 3

3. *Thing 1 and Thing 2.* The citric acid cycle can be thought of as taking place in two stages. What are the two stages? ✓ 3

4. *Like Jack and Jill.* Match each enzyme with its description.

(a) Pyruvate dehydroge-
    nase complex _____

(b) Citrate synthase

    _____

(c) Aconitase _____

(d) Isocitrate
    dehydrogenase _____

(e) α-Ketoglutarate
    dehydrogenase _____

(f) Succinyl CoA synthe-
    tase _____

(g) Succinate
    dehydrogenase _____

(h) Fumarase _____

(i) Malate dehydrogenase

    _____

(j) Pyruvate carboxylase

    _____

1. Catalyzes the
   formation of isocitrate

2. Synthesizes succinyl
   CoA

3. Generates malate

4. Generates ATP

5. Converts pyruvate into
   acetyl CoA

6. Converts pyruvate into
   oxaloacetate

7. Condenses oxaloac-
   etate and acetyl CoA

8. Catalyzes the forma-
   tion of oxaloacetate

9. Synthesizes fumarate

10. Catalyzes the forma-
    tion of α-ketoglutarate

5. *One from two.* The synthesis of citrate from acetyl CoA and oxaloacetate is a biosynthetic reaction. What is the energy source that drives the formation of citrate? ✓ 3

6. *A penny saved . . .* How is the wasteful hydrolysis of acetyl CoA prevented by citrate synthase? ✓ 3

7. $C_2 + C_2 \longrightarrow C_4$.

(a) Which enzymes are required to get the *net synthesis* of oxaloacetate from acetyl CoA? ✓ 6
(b) Write a balanced equation for the net synthesis.
(c) Do mammalian cells contain the requisite enzymes?

8. *Driving force.* What is the value of $\Delta G^{\circ\prime}$ for the complete oxidation of the acetyl unit of acetyl CoA by the citric acid cycle? ✓ 4

9. *Acting catalytically.* The citric acid cycle itself, which is composed of enzymatically catalyzed steps, can be thought of as essentially a supramolecular enzyme. Explain. ✓ 4

10. *Seven o'clock roadblock.* Malonate is a competitive inhibitor of succinate dehydrogenase. How will the concentrations of citric acid cycle intermediates change immediately after the addition of malonate? Why is malonate not a substrate for succinate dehydrogenase? ✓ 4

$$\begin{array}{c} COO^- \\ | \\ CH_2 \\ | \\ COO^- \end{array}$$
Malonate

11. *One of a kind.* How is succinate dehydrogenase unique compared with the other enzymes in the citric acid cycle?

12. *Phosphate requirement.* What step in the citric acid cycle requires a molecule of inorganic phosphate? ✓ 3

13. *An added bonus.* What reaction in the citric acid cycle results in the direct formation of a molecule of ATP? ✓ 3

14. *Regulators.* Which enzymes are the key regulatory enzymes of the citric acid cycle itself? ✓ 5

15. *Versatility.* What is the chief benefit of being able to perform the glyoxylate cycle?

Chapter Integration Problem

16. *Kissin' cousins.* How does the decarboxylation of α-ketoglutarate resemble that of pyruvate decarboxylation?

17. *Fats into glucose?* Fats are usually metabolized into acetyl CoA and then further processed through the citric acid cycle. In Chapter 17, we learned that glucose could be synthesized from oxaloacetate, a citric acid cycle intermediate. Why, then, after a long bout of exercise depletes our carbohydrate stores, do we need to replenish those stores by eating carbohydrates? Why do we not simply replace them by converting fats into carbohydrates?

18. *No signal, no activity.* Why is acetyl CoA an especially appropriate activator for pyruvate carboxylase? ✓ 5

Mechanism Problems

19. *Symmetry problems.* In experiments carried out in 1941 to investigate the citric acid cycle, oxaloacetate labeled with ^{14}C in the carboxyl carbon atom (shown in red below) farthest from the keto group was introduced into an active preparation of mitochondria.

$$\begin{array}{c} O \diagdown \quad \diagup COO^- \\ C \\ | \\ CH_2 \\ | \\ COO^- \end{array}$$
Oxaloacetate

Analysis of the α-ketoglutarate formed showed that none of the radioactive label had been lost. Decarboxylation of α-ketoglutarate then yielded succinate devoid of radioactivity. All of the label was in the released CO_2. Why were the early investigators of the citric acid cycle surprised that *all* of the label emerged in the CO_2?

20. *Symmetric molecules reacting asymmetrically.* The interpretation of the experiments described in problem 19 was that citrate (or any other symmetric compound) cannot be an intermediate in the formation of α-ketoglutarate, because of the asymmetric fate of the label. This view seemed compelling until Alexander Ogston incisively pointed out in 1948 that "it is possible that *an asymmetric enzyme which attacks a symmetrical compound can distinguish between its identical groups.*" For simplicity, consider a molecule in which two hydrogen atoms, a group

X, and a different group Y are bonded to a tetrahedral carbon atom as a model for citrate. Explain how a symmetric molecule can react with an enzyme in an asymmetric way.

Challenge, Data Interpretation, and Chapter Integration Problems

21. *A little goes a long way.* As will become clearer in Section 9, the activity of the citric acid cycle can be monitored by measuring the amount of O_2 consumed. The greater the rate of O_2 consumption, the faster the rate of the cycle. Hans Krebs used this assay to investigate the cycle in 1937. He used as his experimental system minced pigeon-breast muscle, which is rich in mitochondria. In one set of experiments, Krebs measured the O_2 consumption in the presence of carbohydrate only and in the presence of carbohydrate and citrate. The results are shown in the following table. ✓ 4

Effect of citrate on oxygen consumption by minced pigeon-breast muscle

	Micromoles of oxygen consumed	
Time (min)	Carbohydrate only	Carbohydrate plus 3 μmol of citrate
10	26	28
60	43	62
90	46	77
150	49	85

(a) How much O_2 would be absorbed if the added citrate were completely oxidized to H_2O and CO_2?
(b) On the basis of your answer to part *a*, what do the results given in the table suggest?

22. *Arsenite poisoning.* The effect of arsenite on the experimental system of problem 21 was then examined. Experimental data (not presented here) showed that the amount of citrate present did not change in the course of the experiment in the absence of arsenite. However, if arsenite was added to the system, different results were obtained, as shown in the following table. ✓ 4

Disappearance of citric acid in pigeon-breast muscle in the presence of arsenite

Micromoles of citrate added	Micromoles of citrate found after 40 minutes	Micromoles of citrate used
22	00.6	21
44	20.0	24
90	56.0	34

(a) What is the effect of arsenite on the disappearance of citrate?
(b) How is the arsenite's action altered by the addition of more citrate?
(c) What do these data suggest about the site of action of arsenite?

23. *Isocitrate lyase and tuberculosis.* The bacterium *Mycobacterium tuberculosis*, the cause of tuberculosis, can invade the lungs and persist in a latent state for years. During this time, the bacteria reside in granulomas—nodular scars containing bacteria and host-cell debris in the center and surrounded by immune cells. The granulomas are lipid-rich, oxygen-poor environments. How these bacteria manage to persist is something of a mystery. Experimental evidence suggests that the glyoxylate cycle is required for the persistence. The following data show the amount of bacteria (presented as colony-forming units, or cfu) in mice lungs in the weeks after an infection.

In graph A, the black data points represent the results for wild-type bacteria and the red data points represent the results for bacteria from which the gene for isocitrate lyase was deleted.

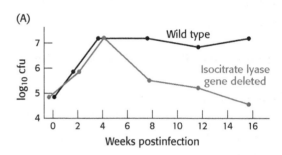

(A)

(a) What is the effect of the absence of isocitrate lyase?
The gene encoding isocitrate lyase was reinserted into bacteria from which it had previously been deleted (Chapter 41 describes the reinsertion technique.) In graph B, black data points represent bacteria into which the gene was reinserted and red data points represent bacteria in which the gene was still missing.

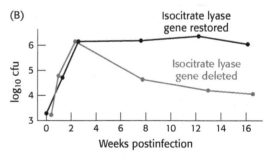

(B)

[Data for graphs A and B after J. D. McKinney, K. Höner zu Bentrup, E. J. Muñoz-Elias, et al., *Nature* 406:735–738, 2000.]

(b) Do these results support those obtained in part *a*?
(c) What is the purpose of the experiment in part *b*?
(d) Why do these bacteria perish in the absence of the glyoxylate cycle?

Challenge Problems

24. *Flow of carbon atoms.* What is the fate of the radioactive label when each of the following compounds is added to a cell extract containing the enzymes and cofactors of the glycolytic pathway, the pyruvate dehydrogenase complex, and the citric acid cycle? (The ^{14}C label is printed in red.) ✓ 3

(a)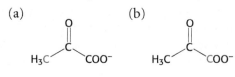

(b)

(c)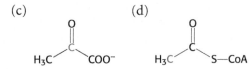

(d)

(e) Glucose 6-phosphate labeled at C-1.

25. *Coupling reactions.* The oxidation of malate by NAD^+ to form oxaloacetate is a highly endergonic reaction under standard conditions ($\Delta G^{\circ\prime} = 29$ kJ mol^{-1}, or 7 kcal mol^{-1}). The reaction proceeds readily under physiological conditions.

(a) Why?

(b) Assume an $[NAD^+]/[NADH]$ ratio of 8 and a pH of 7, and give the lowest $[malate]/[oxaloacetate]$ ratio at which oxaloacetate can be formed from malate.

26. *Power differentials.* As we will see in the next chapter, when NADH reacts with oxygen, 2.5 ATP are generated. When $FADH_2$ reduces oxygen, only 1.5 ATP are generated. Why then does succinate dehydrogenase produce $FADH_2$ and not NADH when succinate is reduced to fumarate?

27. *Back to Orgo.* Before any oxidation can take place in the citric acid cycle, citrate must be isomerized into isocitrate. Why?

28. *Synthesizing α-ketoglutarate.* With the use of the reactions and enzymes considered in this chapter and in Chapter 18, pyruvate can be converted into α-ketoglutarate without depleting any of the citric acid cycle components. Write a balanced reaction scheme for this conversion, showing cofactors and identifying the required enzymes. ✓ 6

Selected Readings for this chapter can be found online at www.whfreeman.com/tymoczko3e.

CHAPTER 20
The Electron-Transport Chain

CHAPTER 21
The Proton-Motive Force

Oxidative Phosphorylation

The amount of ATP that human beings need in order to go about their lives is staggering. A sedentary male of 70 kg (154 lb) requires about 8400 kJ (2000 kcal) for a day's worth of activity. This much energy requires 83 kg of ATP. However, human beings possess only about 250 g of ATP, less than 1% of the daily required amount. The disparity between the amount of ATP that we have and the amount that we require is solved by recycling spent ATP back to usable ATP. Each ATP molecule is recycled from ADP approximately 300 times per day. *This recycling takes place primarily through oxidative phosphorylation, in which ATP is formed as a result of the transfer of electrons from NADH or FADH$_2$ to O$_2$ by a series of electron carriers*. This process, which takes place in mitochondria, is the major source of ATP in aerobic organisms. For example, oxidative phosphorylation generates 26 of the 30 molecules of ATP that are formed when 1 molecule of glucose is completely oxidized to CO$_2$ and H$_2$O.

Oxidative phosphorylation is the culmination of the series of energy transformations as presented in Section 8, called *cellular respiration* or, simply, *respiration*, in their entirety. Carbon fuels are first oxidized in the citric acid cycle to yield high-transfer-potential electrons. In oxidative phosphorylation, high-transfer-potential electrons flow through a series of large protein complexes embedded in the inner mitochondrial membrane, called the *respiratory chain*, to reduce oxygen to water. The electron flow through these complexes is a series of highly exergonic oxidation–reduction reactions that power the pumping of protons from the inside of the mitochondria to the outside, establishing a proton gradient, called the *proton-motive force*. The final phase of oxidative phosphorylation is carried out by an ATP-synthesizing assembly that is driven by the flow of protons back into

361

the mitochondrial matrix. Oxidative phosphorylation vividly shows that *proton gradients are an interconvertible currency of free energy in biological systems.*

We begin this section with an examination of how the electron-transport chain harnesses the energy released when electrons flow from NADH and $FADH_2$ to oxygen to generate a proton gradient. We will then see how the energy inherent in the proton gradient is converted into ATP. Finally, we will see how the process of oxidative phosphorylation is regulated and how transporters facilitate the movement of biochemicals between the mitochondria and the cytoplasm.

✓ By the end of this section, you should be able to:

✓ 1 Describe the key components of the electron-transport chain and how they are arranged.

✓ 2 Explain the benefits of having the electron-transport chain located in a membrane.

✓ 3 Describe how the proton-motive force is converted into ATP.

✓ 4 Identify the ultimate determinant of the rate of cellular respiration.

CHAPTER 20

The Electron-Transport Chain

A bicycle chain converts the energy from the rider's legs into forward movement of the rider and the bike. Likewise, the electron-transport chain transfers the energy of the oxidation of carbon fuels to the energy of a proton gradient. [© Dattatreya/Alamy.]

We begin our study of *oxidative phosphorylation* by examining the oxidation–reduction reactions that allow the flow of electrons from NADH and $FADH_2$ to oxygen. The electron flow, which is very exergonic, takes place in four large protein complexes that are embedded in the inner

363

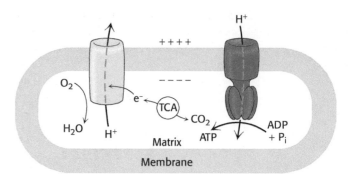

Figure 20.1 An overview of oxidative phosphorylation. Oxidation and ATP synthesis are coupled by transmembrane proton fluxes. The respiratory chain (yellow structure) transfers electrons from NADH and FADH₂ to oxygen and simultaneously generates a proton gradient. ATP synthase (red structure) converts the energy of the proton gradient into ATP.

DID YOU KNOW?

Respiration is an ATP-generating process in which an inorganic compound (such as molecular oxygen) serves as the ultimate electron acceptor. The electron donor can be either an organic compound or an inorganic one.

mitochondrial membrane, together called the *respiratory chain* or the *electron-transport chain*. Importantly, three of these complexes use the energy released by the electron flow to pump protons from the mitochondrial matrix into the space between the inner and outer mitochondrial membranes. The proton gradient is then used to power the synthesis of ATP by oxidative phosphorylation, a process that we will examine in Chapter 21 (**Figure 20.1**). Collectively, the generation of high-transfer-potential electrons by the citric acid cycle, their flow through the respiratory chain, and the accompanying synthesis of ATP is called *respiration* or *cellular respiration*.

20.1 Oxidative Phosphorylation in Eukaryotes Takes Place in Mitochondria

Like the citric acid cycle, the respiratory chain and ATP synthesis take place in mitochondria. Whereas the citric acid cycle takes place in the mitochondrial matrix, the flow of electrons through the respiratory chain and the process of ATP synthesis take place in the mitochondrial inner membrane, a location, as we will see, that is crucial to the process of oxidative phosphorylation.

Mitochondria Are Bounded by a Double Membrane

Electron microscopic studies reveal that mitochondria have two membrane systems: an *outer membrane* and an extensive, highly folded *inner membrane*. The inner membrane is folded into a series of internal ridges called *cristae*. Hence, there are two compartments in mitochondria: (1) the *intermembrane space* between the outer and the inner membranes and (2) the *matrix*, which is bounded by the inner membrane (**Figure 20.2**). The mitochondrial matrix is the site of the

Figure 20.2 Electron micrograph (A) and diagram (B) of a mitochondrion. [(A) Keith R. Porter/Science Source. (B) Information from *Biology of the Cell* by Stephen L. Wolfe. © 1972 by Wadsworth Publishing Company, Inc., Belmont, California 94002. Adapted by permission of the publisher.]

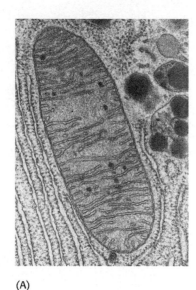

(A)

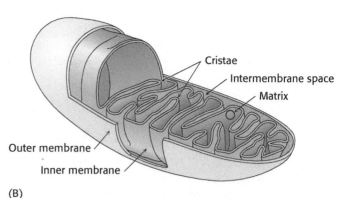

(B)

reactions of the citric acid cycle and fatty acid oxidation. In contrast, oxidative phosphorylation takes place in the inner mitochondrial membrane. The increase in surface area of the inner mitochondrial membrane provided by the cristae creates more sites for oxidative phosphorylation than would be the case with a simple, unfolded membrane. Human beings contain an estimated 14,000 m^2 of inner mitochondrial membrane, which is the approximate equivalent to the area of three football fields in the United States.

The outer membrane is quite permeable to most small molecules and ions because it contains many copies of *mitochondrial porin,* a 30- to 35-kDa pore-forming protein also known as VDAC, for *voltage-dependent anion channel.* VDAC regulates the flux of molecules crucial to the function of cellular respiration—usually anionic species such as phosphate, components of the citric acid cycle, and the adenine nucleotides—across the outer membrane. In contrast, the inner membrane is intrinsically impermeable to nearly all ions and polar molecules. A large family of transporters shuttle metabolites such as ATP, pyruvate, and citrate across the inner mitochondrial membrane. The two faces of this membrane will be referred to as the *matrix side* and the *cytoplasmic side* (the latter because it is freely accessible to most small molecules in the cytoplasm). In bacteria, the electron-driven proton pumps and ATP-synthesizing complex are located in the plasma membrane.

 BIOLOGICAL INSIGHT

Mitochondria Are the Result of an Endosymbiotic Event

Mitochondria are semiautonomous organelles that live in an endosymbiotic relationship with the host cell. These organelles contain their own DNA, which encodes a variety of different proteins and RNAs. However, mitochondria also contain many proteins encoded by nuclear DNA. Cells that contain mitochondria depend on these organelles for oxidative phosphorylation, and the mitochondria in turn depend on the cell for their very existence. How did this intimate symbiotic relationship come to exist?

An *endosymbiotic event* is thought to have taken place whereby a free-living organism capable of oxidative phosphorylation was engulfed by another cell. The double-membrane, circular DNA, and mitochondrial-specific transcription and translation machinery all point to this conclusion. Thanks to the rapid accumulation of sequence data for mitochondrial and bacterial genomes, speculation on the origin of the "original" mitochondrion with some authority is now possible. The most mitochondrial-like bacterial genome is that of *Rickettsia prowazekii,* the cause of louse-borne typhus. Sequence data suggest that all extant mitochondria are derived from an ancestor of *R. prowazekii* as the result of a single endosymbiotic event.

The evidence that modern mitochondria result from a single event comes from examination of the most bacteria-like mitochondrial genome, that of the protozoan *Reclinomonas americana.* Its genome contains 97 genes, of which 62 specify proteins. The genes encoding these proteins include all of the protein-coding genes found in all of the sequenced mitochondrial genomes (**Figure 20.3**). It seems unlikely that mitochondrial genomes resulting from several endosymbiotic events could have been independently reduced to the same set of genes found in *R. americana.*

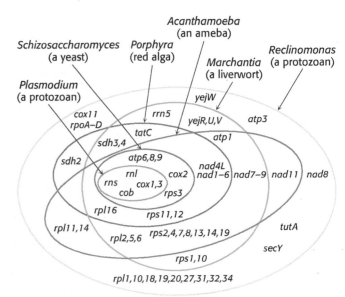

Figure 20.3 Overlapping gene complements of mitochondria. The genes within each oval are those present within the organism represented by the oval. The genome of *Reclinomonas* contains all the protein-coding genes found in all the sequenced mitochondrial genomes. [Information from M. W. Gray, G. Burger, and B. F. Lang, *Science* 283:1476–1481, 1999.]

Note that transient engulfment of prokaryotic cells by larger cells is not uncommon in the microbial world. In regard to mitochondria, such a transient relationship became permanent as the bacterial cell lost most of its DNA, making it incapable of independent living, and the host cell became dependent on the ATP generated by its tenant.

✓ 1 Describe the key components of the electron-transport chain and how they are arranged.

20.2 Oxidative Phosphorylation Depends on Electron Transfer

In Section 8, the primary catabolic function of the citric acid cycle was identified as the generation of NADH and $FADH_2$ by the oxidation of acetyl CoA. In oxidative phosphorylation, electrons from NADH and $FADH_2$ are used to reduce molecular oxygen to water. The highly exergonic reduction of molecular oxygen by NADH and $FADH_2$ is accomplished through a number of electron-transfer reactions, which take place in a set of membrane proteins known as the *electron-transport chain*. Electrons from NADH and $FADH_2$ flow through the components of the electron chain, ultimately resulting in the reduction of oxygen.

The Electron-Transfer Potential of an Electron Is Measured as Redox Potential

In oxidative phosphorylation, the *electron-transfer potential* of NADH or $FADH_2$ is converted into the *phosphoryl-transfer potential* of ATP. To better understand this conversion, we need quantitative expressions for these forms of free energy. The measure of phosphoryl-transfer potential is already familiar to us: it is given by $\Delta G^{\circ\prime}$ for the hydrolysis of the phosphoryl compound. The corresponding expression for the electron-transfer potential is E_0', the *reduction potential* (also called the *redox potential* or *oxidation–reduction potential*).

Consider a substance that can exist in an oxidized form, X, and a reduced form, X^-. Such a pair is called a *redox couple*. The reduction potential of this couple can be determined by measuring the electromotive force generated by a *sample half-cell* connected to a *standard reference half-cell* (**Figure 20.4**). The sample half-cell consists of an electrode immersed in a solution of 1 M oxidant (X) and 1 M reductant (X^-). The standard reference half-cell consists of an electrode immersed in a 1 M H^+ solution that is in equilibrium with H_2 gas at 1 atmosphere (1 atm) of pressure. The electrodes are connected to a voltmeter, and an agar (a gelatinous polysaccharide) bridge allows the flow of ions between the half-cells. Electrons then flow from one half-cell to the other through the wire connecting the two cells. If the reaction proceeds in the direction

$$X^- + H^+ \longrightarrow X + \tfrac{1}{2}H_2$$

the reactions in the half-cells (referred to as *half-reactions* or *couples*) must be

$$X^- \longrightarrow X + e^-$$
$$H^+ + e^- \longrightarrow \tfrac{1}{2}H_2$$

Thus, electrons flow from the sample half-cell to the standard reference half-cell, and the sample-cell electrode is taken to be negative with respect to the standard-cell electrode. *The reduction potential of the X:X^- couple is the observed voltage at the start of the experiment* (when X, X^-, and H^+ are 1 M with 1 atm of H_2). *The reduction potential of the H^+:H_2 couple is defined to be 0 volts.*

The meaning of the reduction potential is now evident. A negative reduction potential means that the oxidized form of a substance has lower affinity for electrons than does H_2, as in the preceding example. A positive reduction potential means that the oxidized form of a substance has higher affinity for electrons than does H_2. These comparisons refer to standard conditions—namely, 1 M oxidant, 1 M reductant, 1 M H^+, and 1 atm H_2. Thus, *a strong reducing agent (such as NADH)*

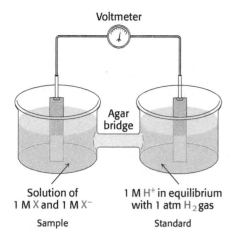

Voltmeter

Agar bridge

Solution of 1 M X and 1 M X^-

Sample

1 M H^+ in equilibrium with 1 atm H_2 gas

Standard

Figure 20.4 The measurement of redox potential. Apparatus for the measurement of the standard oxidation–reduction potential of a redox couple. Electrons flow through the wire connecting the cells, whereas ions flow through the agar bridge.

Table 20.1 Standard reduction potentials of some reactions

Oxidant	Reductant	n	E_0' (V)
Succinate + CO_2	α-Ketoglutarate	2	−0.67
Acetate	Acetaldehyde	2	−0.60
Ferredoxin (oxidized)	Ferredoxin (reduced)	1	−0.43
$2\,H^+$	H_2	2	−0.42
NAD^+	$NADH + H^+$	2	−0.32
$NADP^+$	$NADPH + H^+$	2	−0.32
Lipoate (oxidized)	Lipoate (reduced)	2	−0.29
Glutathione (oxidized)	Glutathione (reduced)	2	−0.23
FAD	$FADH_2$	2	−0.22
Acetaldehyde	Ethanol	2	−0.20
Pyruvate	Lactate	2	−0.19
$2\,H^+$	H_2	2	0.00[1]
Cytochrome *b* (+3)	Cytochrome *b* (+2)	1	+0.07
Dehydroascorbate	Ascorbate	2	+0.08
Ubiquinone (oxidized)	Ubiquinone (reduced)	2	+0.10
Cytochrome *c* (+3)	Cytochrome *c* (+2)	1	+0.22
Fe (+3)	Fe (+2)	1	+0.77
$\frac{1}{2}\,O_2 + 2\,H^+$	H_2O	2	+0.82

Note: E_0' is the standard oxidation-reduction potential (pH 7, 25°C, except where noted), and n is the number of electrons transferred. E_0' refers to the partial reaction written as Oxidant + $e^- \longrightarrow$ reductant
[1]Standard oxidation-reduction potential at pH = 0.

is poised to donate electrons and has a negative reduction potential, whereas a strong oxidizing agent (such as O_2) is ready to accept electrons and has a positive reduction potential. The reduction potentials of many biologically important redox couples are known (Table 20.1). Table 20.1 is like those presented in chemistry textbooks, except that a hydrogen ion concentration of 10^{-7} M (pH 7) instead of 1 M (pH 0) is the standard state adopted by biochemists. This difference is denoted by the prime in E_0'. Recall that the prime in $\Delta G^{\circ\prime}$ denotes a standard free-energy change at pH 7.

The standard free-energy change $\Delta G^{\circ\prime}$ is related to the change in reduction potential $\Delta E_0'$ by

$$\Delta G^{\circ\prime} = -nF\Delta E_0'$$

in which n is the number of electrons transferred, F is a proportionality constant called the *faraday* (96.48 kJ mol^{-1} V^{-1}, or 23.06 kcal mol^{-1} V^{-1}), $\Delta E_0'$ is in volts, and $\Delta G^{\circ\prime}$ is in kilojoules or kilocalories per mole.

DID YOU KNOW?

An oxidant (oxidizing agent) is the acceptor of electrons in an oxidation–reduction reaction.

A reductant (reducing agent) is the donator of electrons in an oxidation–reduction reaction.

Electron Flow Through the Electron-Transport Chain Creates a Proton Gradient

The driving force of oxidative phosphorylation is the electron-transfer potential of NADH or $FADH_2$ relative to that of O_2. Recall that NADH and $FADH_2$ are carriers of high-transfer-potential electrons generated in the citric acid cycle and elsewhere in metabolism. How much energy is released by the reduction of O_2 with NADH? Let us calculate $\Delta G^{\circ\prime}$ for this reaction. The pertinent half-reactions are

$$\tfrac{1}{2}\,O_2 + 2\,H^+ + 2\,e^- \longrightarrow H_2O \qquad E_0' = +0.82\text{ V} \qquad (A)$$

$$NAD^+ + H^+ + 2\,e^- \longrightarrow NADH \qquad E_0' = -0.32\text{ V} \qquad (B)$$

Subtracting reaction B from reaction A yields

$$\tfrac{1}{2}O_2 + NADH + H^+ \longrightarrow H_2O + NAD^+ \qquad \Delta E_0' = 1.14 \text{ V} \qquad \text{(C)}$$

Note that the sign of E_0' for the $NAD^+/NADH$ half-reaction is changed when NADH donates electrons. The standard free energy for this reaction is then given by the following equation:

$$\Delta G^{\circ\prime} = -nF\Delta E_0'$$

Substituting the appropriate values yields

$$\Delta G^{\circ\prime} = (-2 \times 96.48 \text{ kJ mol}^{-1} \text{ V}^{-1} \times 0.82 \text{ V})$$

$$+(-2 \times 96.48 \text{ kJ mol}^{-1} \text{ V}^{-1} \times -0.32 \text{ V})$$

$$= -158.2 \text{ kJ mol}^{-1} + (-61.9 \text{ kJ mol}^{-1})$$

$$= -220.1 \text{ kJ mol}^{-1} (-52.6 \text{ kcal mol}^{-1})$$

This release of free energy is substantial. Recall that $\Delta G^{\circ\prime}$ for the synthesis of ATP is $+30.5$ kJ mol^{-1} ($+7.3$ kcal mol^{-1}). The energy released by the reduction of each electron carrier generates a proton gradient that is then used for the synthesis of ATP and the transport of metabolites across the mitochondrial membrane. Indeed, each proton transported out of the matrix to the cytoplasmic side corresponds to 21.8 kJ mol^{-1} (5.2 kcal mol^{-1}) of free energy.

The Electron-Transport Chain Is a Series of Coupled Oxidation–Reduction Reactions

Electron flow from NADH to O_2 is accomplished by a series of intermediate electron carriers—a bucket brigade of electron carriers—that are coupled as members of sequential redox reactions. As we will see, NADH is oxidized by passing electrons to *flavin mononucleotide* (FMN), an electron carrier similar to flavin adenine dinucleotide (FAD) but lacking the nucleotide component (**Figure 20.5**). Reduced FMN is subsequently oxidized by the next electron carrier in the chain, and the process repeats itself as the electrons flow down the electron-transport chain until they finally reduce O_2. The members of the electron-transport chain are arranged so that the electrons always flow to components with more positive reduction potentials (a higher electron affinity).

Electrons are transferred from NADH to O_2 through a chain of three large protein complexes called NADH-Q oxidoreductase, Q-cytochrome c oxidoreductase,

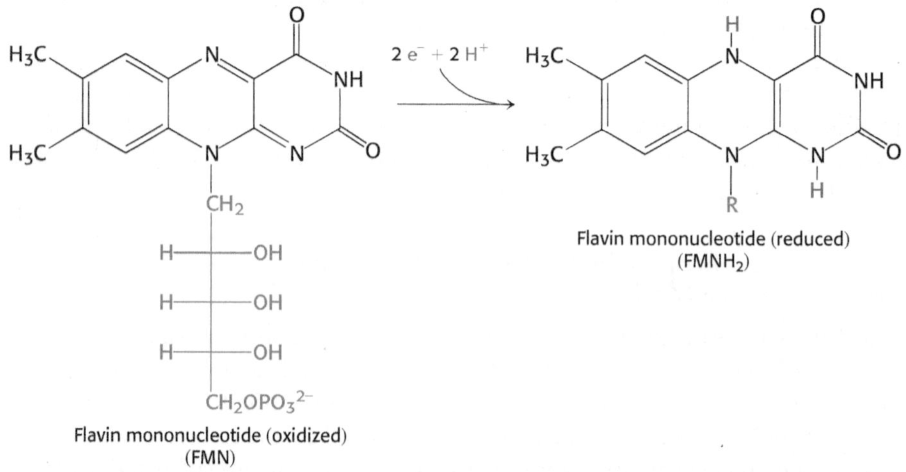

Figure 20.5 The oxidation states of flavin mononucleotide. The electron carrier component is the same in both flavin mononucleotide and flavin dinucleotide (FAD).

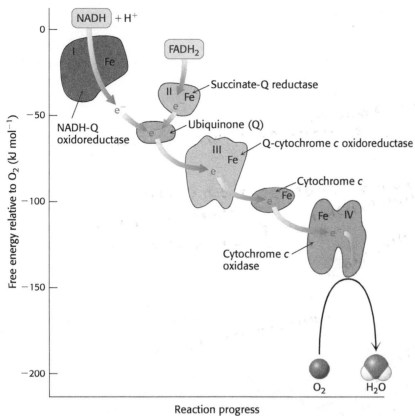

Figure 20.6 Components of the electron-transport chain. Electrons flow down an energy gradient from NADH to O_2. The flow is catalyzed by four protein complexes. Iron is a component of all of the complexes as well as cytochrome c. [Data from D. Sadava et al., *Life*, 8th ed. (Sinauer, 2008), p. 150.]

and cytochrome *c* oxidase (**Figure 20.6**). These complexes appear to be associated in a supramolecular complex termed the *respirasome*. As in the citric acid cycle, such supramolecular complexes facilitate the rapid transfer of substrate and prevent the release of reaction intermediates. A fourth large protein complex, called succinate-Q reductase, contains the succinate dehydrogenase that generates $FADH_2$ in the citric acid cycle (p. 349). Electrons from this $FADH_2$ enter the electron-transport chain at Q-cytochrome *c* oxidoreductase.

Before we look at how NADH and $FADH_2$ reduce O_2 in more detail, it's worth pointing out several features of the electron-transport chain that affect the energy yield of the high-transfer-potential electrons and facilitate the efficient flow of electrons (Figure 20.6). First of all, electrons from $FADH_2$ feed into the chain "downstream" of those from NADH because the electrons of $FADH_2$ have a lower reduction potential (Table 20.1). As a result, $FADH_2$-derived electrons pump fewer protons and thus yield fewer molecules of ATP. Second, note that iron is a prominent electron carrier, appearing in several places. Iron in the electron-transport chain appears in two fundamental forms: associated with sulfur as iron–sulfur clusters located in *iron–sulfur proteins* (also called *nonheme-iron proteins*; **Figure 20.7**), and as components of a heme-prosthetic group, which are

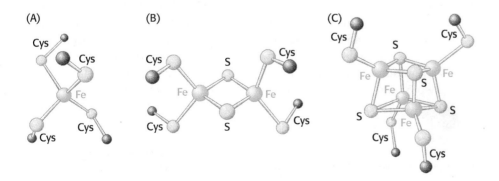

Figure 20.7 Iron–sulfur clusters. (A) A single iron ion bound by four cysteine residues. (B) 2Fe-2S cluster with iron ions bridged by sulfide ions. (C) 4Fe-4S cluster. Each of these clusters can undergo oxidation–reduction reactions.

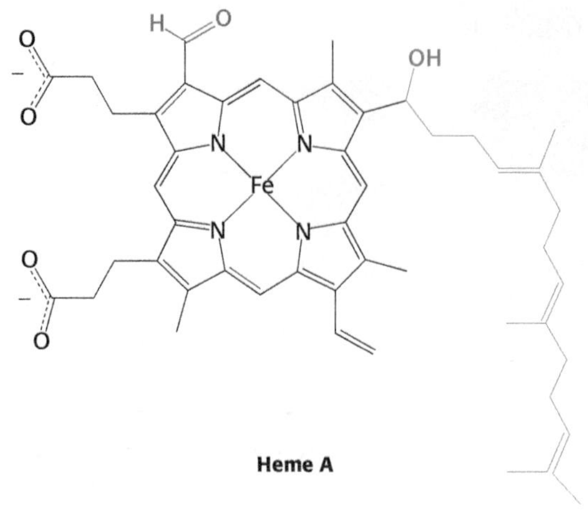

Heme A

Figure 20.8 A heme component of cytochrome *c* oxidase.

embedded in a special class of proteins called *cytochromes* (**Figure 20.8**). In both iron–sulfur proteins and cytochromes, iron shuttles between its reduced ferrous (Fe^{2+}) and its oxidized ferric state (Fe^{3+}):

$$Fe^{2+} \rightleftharpoons Fe^{3+} + e^-$$

The heme-prosthetic group of cytochromes is iron-protoporphyrin IX, the same heme present in hemoglobin and myoglobin (p. 151). The iron in hemoglobin and myoglobin, in contrast to the iron in cytochromes, remains in the Fe^{2+} oxidation state.

The fact that iron is an electron carrier in several places in the electron-transport chain raises a puzzling question. If the flow of electrons from NADH to O_2 is exergonic and iron has a reduction potential of $+0.77$ V (Table 20.1), how can iron participate in several places in the electron-transport chain if each step is exergonic? In other words, how can iron have several different reduction potentials? The answer to the puzzle is that the oxidation–reduction potential of iron ions can be altered by their environment. In regard to the electron-transport chain, the iron ion is not free; rather, it is embedded in different proteins, enabling iron to have various reduction potentials and to play a role at several different locations in the chain.

The metal copper also appears as a member of the final component of the electron-transport chain, where it is alternatively oxidized and reduced:

$$Cu^+ \rightleftharpoons Cu^{2+} + e^-$$

Another key feature of the electron-transport chain is the prominence of coenzyme Q (Q) as an electron carrier. *Coenzyme Q*, also known as *ubiquinone* because it is a *ubi*quitous *quinone* in biological systems, is a quinone derivative with a long isoprenoid tail, which renders the molecule hydrophobic and allows it to diffuse rapidly within the inner mitochondrial membrane, where it shuttles protons and electrons about.

Coenzyme Q consists of more than one five-carbon isoprene unit. The exact number depends on the species in which it is found. The most common form in mammals contains 10 isoprene units (coenzyme Q_{10}). For simplicity, the subscript will be omitted from this abbreviation because all varieties function in an identical manner. Quinones can exist in several oxidation states (**Figure 20.9**). In the fully oxidized state (Q), coenzyme Q has two keto groups. The addition of one electron yields a semiquinone radical anion ($Q^{\cdot-}$), whereas the addition of one electron and one proton results in the semiquinone form ($QH\cdot$). The addition of a second electron and proton generates ubiquinol (QH_2), the fully reduced form of coenzyme Q. Thus, *for quinones, electron-transfer reactions are coupled to proton binding and release,* a property that is key to transmembrane proton transport. A pool of Q and QH_2—the *Q pool*—is thought to exist in the inner mitochondrial membrane, although recent research suggests that it is confined to the respirasome.

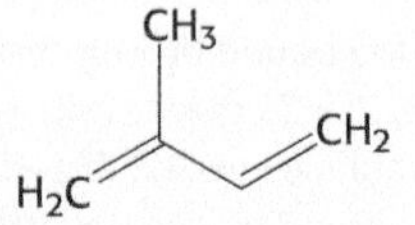

Isoprene

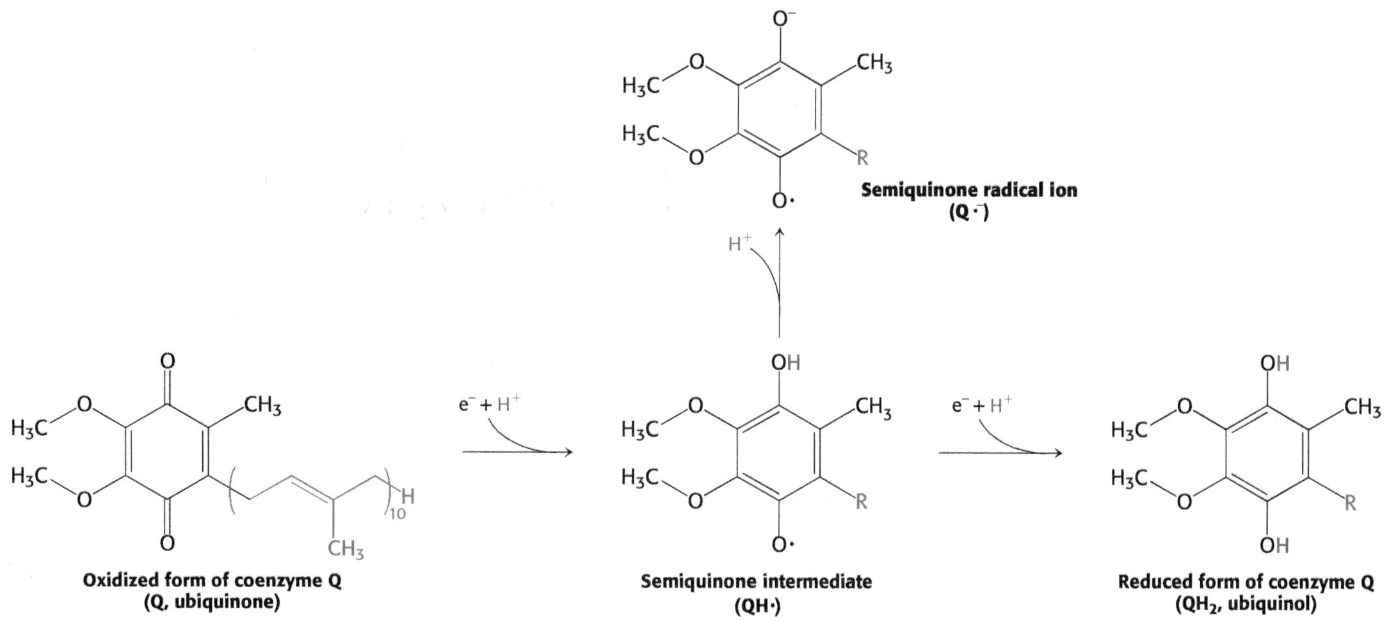

Figure 20.9 Oxidation states of quinones. The reduction of ubiquinone (Q) to ubiquinol (QH$_2$) proceeds through a semiquinone intermediate (QH•).

CLINICAL INSIGHT

Loss of Iron–Sulfur Cluster Results in Friedreich's Ataxia

The importance of Fe-S clusters is illustrated by the loss of function of the protein frataxin. Frataxin is a small (14.2 kDa) mitochondrial protein that is crucial for the synthesis of Fe-S clusters. Mutations in frataxin causes Friedreich's ataxia, a disease of varying severity that affects the central and peripheral nervous systems as well as the heart and skeletal systems. Severe cases lead to death in young adult life. The most common mutation is trinucleotide expansion (Section 35.1) in the gene for frataxin.

20.3 The Respiratory Chain Consists of Proton Pumps and a Physical Link to the Citric Acid Cycle

✓ 2 Explain the benefits of having the electron-transport chain located in a membrane.

How are the various electron carriers arranged to yield an exergonic reaction that ultimately results in the generation of a proton gradient? As discussed previously (Figure 20.6), electrons are transferred from NADH to O$_2$ through a chain of three protein complexes (Table 20.2 and Figure 20.10). *Electron flow within these transmembrane complexes leads to the transport of protons across the inner mitochondrial membrane.* A fourth protein complex, succinate-Q reductase, in contrast with the other complexes, does not pump protons.

The High-Potential Electrons of NADH Enter the Respiratory Chain at NADH-Q Oxidoreductase

The electrons of NADH enter the chain at *NADH-Q oxidoreductase* (also called *Complex I* and *NADH dehydrogenase*), an enormous enzyme (>900 kDa) consisting of approximately 46 polypeptide chains and two types of prosthetic groups: FMN and iron–sulfur clusters. This proton pump is L-shaped, with a hydrophobic horizontal arm lying in the membrane and a hydrophilic vertical arm that projects into the matrix. The electrons flow from NADH to FMN and then through a

Table 20.2 Components of the mitochondrial electron-transport chain

Enzyme complex	Mass (kDa)	Subunits	Prosthetic group	Oxidant or reductant		
				Matrix side	Membrane core	Cytoplasmic side
NADH-Q oxidoreductase	>900	46	FMN Fe-S	NADH	Q	
Succinate-Q reductase	140	4	FAD Fe-S	Succinate	Q	
Q-cytochrome c oxidoreductase	250	11	Heme b_H Heme b_L Heme c_1 Fe-S		Q	Cytochrome c
Cytochrome c oxidase	160	13	Heme a Heme a_3 Cu_A and Cu_B			Cytochrome c

Sources: J. W. DePierre and L. Ernster, *Annu. Rev. Biochem.* 46:215, 1977; Y. Hatefi, *Annu. Rev. Biochem.* 54:1015, 1985; and J. E. Walker, *Q. Rev. Biophys.* 25:253, 1992.

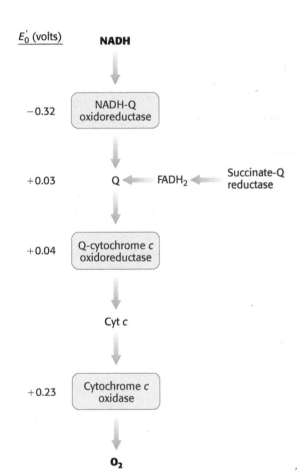

Figure 20.10 The components of the electron-transport chain are arranged in complexes. Notice that the electron affinity of the components increases as electrons move down the chain. The complexes shown in yellow boxes are proton pumps. Cyt c stands for cytochrome c.

series of seven iron–sulfur clusters to Q. Note that all of the redox reactions take place in the extramembranous part of NADH-Q oxidoreductase. Although the precise stoichiometry of the reaction catalyzed by this enzyme is not completely worked out, it appears to be

$$NADH + Q + 5H^+_{matrix} \longrightarrow NAD^+ + QH_2 + 4H^+_{cytoplasm}$$

Recent structural studies have suggested how Complex I acts as a proton pump. What are the structural elements required for proton pumping? The membrane-embedded part of the complex has four proton half-channels consisting, in part, of vertical helices. One set of half-channels is exposed to the matrix and the other to the intermembrane space (**Figure 20.11**). The vertical helices are linked on the matrix side by a long horizontal helix (HL) that connects the matrix half-channels, while the intermembrane space half-channels are joined by a series of β-hairpin-helix connecting elements (βH). An enclosed Q chamber, the site where Q accepts electrons from NADH, exists near the junction of the hydrophilic portion and the membrane-embedded portion. Finally, a hydrophilic funnel connects the Q chamber to a water-lined channel, into which the half-channels open, that extends the entire length of the membrane-embedded portion.

How do these structural elements cooperate to pump protons out of the matrix? When Q accepts two electrons from NADH, generating Q^{2-}, the negative charges on Q^{2-} interact electrostatically with negatively charged amino acid residues in the membrane-embedded arm, causing conformational changes in the long horizontal helix and the βH elements. These changes in turn alter the structures of the connected vertical helices that change the pK_a of amino acids, allowing protons from the matrix to first bind to the amino acids, then dissociate into the water-lined channel and finally enter the intermembrane space. Thus, *the flow of two electrons from NADH to coenzyme Q through NADH-Q oxidoreductase leads to the pumping of four hydrogen ions out of the matrix of the mitochondrion.* Q^{2-} subsequently takes up two protons from the matrix as it is reduced to QH_2. The removal of these protons from the matrix contributes to the formation of the proton-motive force. The QH_2 subsequently leaves the enzyme for the Q pool, allowing another reaction cycle to occur.

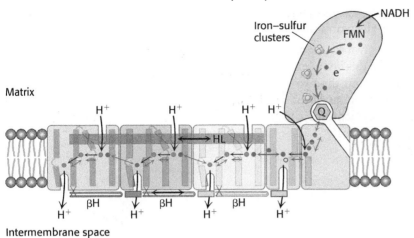

Figure 20.11 Coupled electron–proton transfer reactions through NADH-Q oxidoreductase. Electrons flow in Complex I from NADH through FMN and a series of iron–sulfur clusters to ubiquinone (Q), forming Q^{2-}. The charges on Q^{2-} are electrostatically transmitted to hydrophilic amino acid residues (shown as red and blue balls) that power the movement of HL and βH components. This movement changes the conformation of the transmembrane helices and results in the transport of four protons out of the mitochondrial matrix. [Information from R. Baradaran et al., *Nature* 494:443–448, 2013.]

Ubiquinol Is the Entry Point for Electrons from FADH₂ of Flavoproteins

Recall that $FADH_2$ is formed in the citric acid cycle in the oxidation of succinate to fumarate by succinate dehydrogenase (p. 349). This enzyme is part of the *succinate-Q reductase complex (Complex II)*, an integral membrane protein of the inner mitochondrial membrane. The electron carriers in this complex are FAD, iron–sulfur proteins, and Q. $FADH_2$ does not leave the complex. Rather, its electrons are transferred to Fe-S centers and then to Q for entry into the electron-transport chain. The succinate-Q reductase complex, in contrast with NADH-Q oxidoreductase, does not transport protons. Consequently, less ATP is formed from the oxidation of $FADH_2$ than from NADH.

Electrons Flow from Ubiquinol to Cytochrome *c* Through Q-Cytochrome *c* Oxidoreductase

The second of the three proton pumps in the respiratory chain is *Q-cytochrome c oxidoreductase* (also known as *Complex III* and *cytochrome c reductase*). The function of Q-cytochrome *c* oxidoreductase is to catalyze the transfer of electrons from QH_2 produced by NADH-Q oxidoreductase and the succinate-Q reductase complex to oxidized *cytochrome c (Cyt c)*, a water-soluble protein, and concomitantly pump protons out of the mitochondrial matrix. The flow of a pair of electrons through this complex leads to the effective net transport of $2\,H^+$ to the intermembrane space, half the yield obtained with NADH-Q oxidoreductase because of a smaller thermodynamic driving force.

$$QH_2 + 2\,\text{Cyt}\,c_{ox} + 2\,H^+_{matrix} \longrightarrow Q + 2\,\text{Cyt}\,c_{red} + 4\,H^+_{\text{intermembrane space}}$$

Q-cytochrome *c* oxidoreductase, a complex protein composed of 22 subunits, contains a total of three hemes, which themselves are contained within two cytochrome subunits: two hemes, termed heme b_L (L for low affinity) and heme b_H (H for high affinity) within cytochrome *b*, and one heme within cytochrome c_1. Because of these groups, this enzyme is also known as *cytochrome bc₁*. In addition to the hemes, the enzyme also contains an iron–sulfur protein with a 2Fe-2S center. This center, termed the *Rieske center*, is unusual because one of the iron ions is coordinated by two histidine residues rather than two cysteine residues.

The Q Cycle Funnels Electrons from a Two-Electron Carrier to a One-Electron Carrier and Pumps Protons

QH_2 passes two electrons to Q-cytochrome c oxidoreductase, but the acceptor of electrons in this complex, cytochrome c, can accept only one electron. How does the switch from the two-electron carrier ubiquinol to the one-electron carrier cytochrome c take place? The mechanism for the coupling of electron transfer from Q to cytochrome c to transmembrane proton transport is known as the Q cycle (**Figure 20.12**).

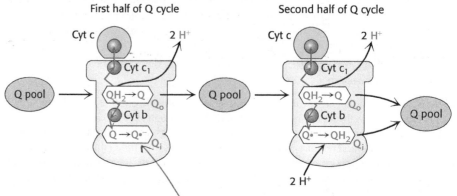

Figure 20.12 The Q cycle. In the first half of the cycle, two electrons of a bound QH_2 are transferred, one to cytochrome c and the other, after passing through cytochrome b, to a bound Q in a second binding site to form the semiquinone radical anion $Q\cdot^-$. The newly formed Q dissociates and enters the Q pool. In the second half of the cycle, a second QH_2 also gives up its electrons, one to a second molecule of cytochrome c and the other to reduce $Q\cdot^-$ to QH_2. This second electron transfer results in the uptake of two protons from the matrix. The path of electron transfer is shown in red.

The cycle begins when two QH_2 molecules bind to the complex consecutively, each giving up two electrons and two H^+. These protons are released to the intermembrane space. QH_2 binds to the first Q binding site (Q_o), and the two electrons travel through the complex to different destinations. One electron flows, first, to the Rieske 2Fe-2S cluster; then, to cytochrome c_1; and, finally, to a molecule of oxidized cytochrome c, converting it into its reduced form. The reduced cytochrome c molecule is free to move away from the enzyme to the final complex of the respiratory chain.

The second electron passes through two heme groups of cytochrome b to an oxidized ubiquinone in a second Q binding site (Q_i). The Q in the second binding site is reduced to a semiquinone radical anion ($Q\cdot^-$) by the electron from the first QH_2. The now fully oxidized Q leaves the first Q site, free to reenter the Q pool.

A second molecule of QH_2 binds to Q-cytochrome c oxidoreductase and reacts in the same way as the first. One of the electrons is transferred to cytochrome c. The second electron passes through the two heme groups of cytochrome b to the partly reduced ubiquinone bound in the second binding site. On the addition of the electron from the second QH_2 molecule, this quinone radical anion takes up two protons from the matrix side to form QH_2. *The removal of these two protons from the matrix contributes to the formation of the proton gradient.* This complex set of reactions can be summarized as follows: four protons are released into the intermembrane space, and two protons are removed from the mitochondrial matrix.

$$2\,QH_2 + Q + 2\,\text{Cyt } c_{ox} + 2\,H^+ \longrightarrow 2Q + QH_2 + 2\,\text{Cyt } c_{red} + 4\,H^+_{\text{intermembrane space}}$$

In one Q cycle, two QH_2 molecules are oxidized to form two Q molecules, and then one Q molecule is reduced to QH_2. The problem of how to efficiently funnel electrons from a two-electron carrier (QH_2) to a one-electron carrier (cytochrome c) is solved by the Q cycle. The cytochrome b component of the reductase is in essence a recycling device that enables both electrons of QH_2 to be used effectively.

Cytochrome *c* Oxidase Catalyzes the Reduction of Molecular Oxygen to Water

The last of the three proton-pumping assemblies of the respiratory chain is *cyto-chrome c oxidase (Complex IV)*. Cytochrome *c* oxidase catalyzes the transfer of electrons from the reduced form of cytochrome *c* to molecular oxygen, the final acceptor:

$$4 \text{ Cyt } c_{red} + 8 \text{ H}^+_{matrix} + \text{O}_2 \longrightarrow 4 \text{ Cyt } c_{ox} + 2 \text{ H}_2\text{O} + 4 \text{ H}^+_{intermembrane space}$$

The requirement of oxygen for this reaction is what makes "aerobic" organisms aerobic. Obtaining oxygen for this reaction is the primary reason that human beings must breathe. Four electrons are funneled to O_2 to completely reduce it to two molecules of H_2O, and, concurrently, protons are pumped from the matrix to the cytoplasmic side of the inner mitochondrial membrane. This reaction is quite thermodynamically favorable. From the reduction potentials in Table 20.1, the standard free-energy change for this reaction is calculated to be $\Delta G°' = -231.8$ kJ mol^{-1} (-55.4 kcal mol^{-1}). As much of this free energy as possible must be captured in the form of a proton gradient for subsequent use in ATP synthesis.

Cytochrome *c* oxidase, which consists of 13 subunits, contains two heme groups (*heme a* and *heme a₃*) and three copper ions, arranged as two copper centers (Cu_A and Cu_B), with Cu_A containing two copper ions. Four molecules of reduced cytochrome *c* generated by Q-cytochrome *c* oxidoreductase bind consecutively to cytochrome *c* oxidase and transfer an electron to reduce one molecule of O_2 to H_2O (**Figure 20.13**). Let us examine the steps required to generate water at the terminus of the electron-transport chain:

QUICK QUIZ 1 Why are the electrons carried by $FADH_2$ not as energy rich as those carried by NADH? What is the consequence of this difference?

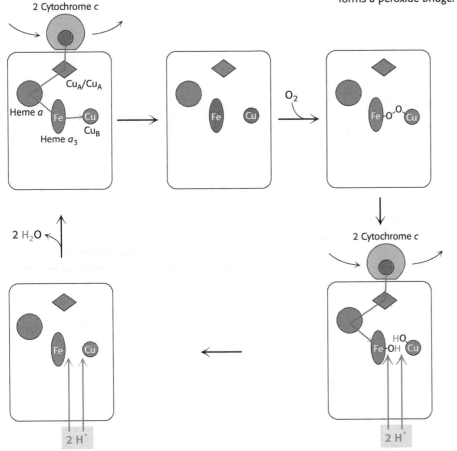

1. Two molecules of cytochrome *c* sequentially transfer electrons to reduce Cu_B and heme a_3.

2. Reduced Cu_B and Fe in heme a_3 bind O_2, which forms a peroxide bridge.

4. The addition of two more protons leads to the release of water.

3. The addition of two more electrons and two more protons cleaves the peroxide bridge.

Figure 20.13 The cytochrome *c* oxidase mechanism. The cycle begins and ends with all prosthetic groups in their oxidized forms (shown in blue). Reduced forms are in red. Four cytochrome *c* molecules donate four electrons, which, in allowing the binding and cleavage of an O_2 molecule, also makes possible the import of four H^+ from the matrix to form two molecules of H_2O, which are released from the enzyme to regenerate the initial state.

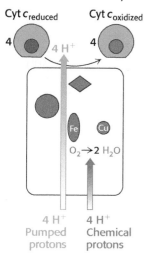

Figure 20.14 Proton transport by cytochrome *c* oxidase. Four protons are taken up from the matrix side to reduce one molecule of O_2 to two molecules of H_2O. These protons are called "chemical protons" because they participate in a clearly defined reaction with O_2. Four additional "pumped" protons are transported out of the matrix and released on the cytoplasmic side in the course of the reaction. The pumped protons double the efficiency of free-energy storage in the form of a proton gradient for this final step in the electron-transport chain.

1. Electrons from two molecules of reduced cytochrome *c* flow through the oxidation–reduction reactions, one stopping at Cu_B and the other at heme a_3. With both centers in the reduced state, they together can now bind an oxygen molecule.

2. As molecular oxygen binds, it removes an electron from each of the nearby ions in the active center to form a peroxide (O_2^{2-}) bridge between them.

3. Two more molecules of cytochrome *c* bind and release electrons that travel to the active center. The addition of an electron as well as H^+ to each oxygen atom reduces the two ion–oxygen groups to Cu_B^{2+}—OH and Fe^{3+}—OH.

4. Reaction with two more H^+ ions allows the release of two molecules of H_2O and resets the enzyme to its initial, fully oxidized form:

$$4 \text{ Cyt } c_{red} + 4 \text{ H}^+_{matrix} + O_2 \longrightarrow 4 \text{ Cyt } c_{ox} + 2 H_2O$$

The four protons in this reaction come exclusively from the matrix. Thus, the consumption of these four protons contributes directly to the proton gradient. Consuming these four protons requires 87.2 kJ mol^{-1} (19.8 kcal mol^{-1}), an amount substantially less than the free energy released from the reduction of oxygen to water. What is the fate of this missing energy? Remarkably, *cytochrome c oxidase uses this energy to pump four additional protons from the matrix to the cytoplasmic side of the membrane in the course of each reaction cycle for a total of eight protons removed from the matrix* (**Figure 20.14**). The details of how these protons are transported through the protein is still under study. Thus, the overall process catalyzed by cytochrome *c* oxidase is

$$4 \text{ Cyt } c_{red} + 8 \text{ H}^+_{matrix} + O_2 \longrightarrow 4 \text{ Cyt } c_{ox} + 2 H_2O + 4 \text{ H}^+_{intermembrane space}$$

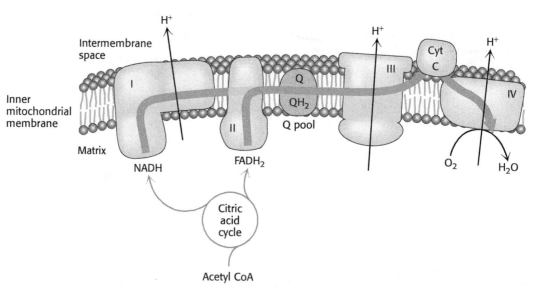

Figure 20.15 The electron-transport chain. High-energy electrons in the form of NADH and FADH₂ are generated by the citric acid cycle. These electrons flow through the respiratory chain, which powers proton pumping and results in the reduction of O_2.

Figure 20.15 summarizes the flow of electrons from NADH and FADH₂ through the respiratory chain. This series of exergonic reactions is coupled to the pumping of protons from the matrix. As we will see in Chapter 21, the energy inherent in the proton gradient will be used to synthesize ATP.

? QUICK QUIZ 2 Amytal is a barbiturate sedative that inhibits electron flow through Complex I. How would the addition of amytal to actively respiring mitochondria affect the relative oxidation–reduction states of the components of the electron-transport chain and the citric acid cycle?

✳ BIOLOGICAL INSIGHT

The Dead Zone: Too Much Respiration

Some marine organisms perform so much cellular respiration, and therefore consume so much molecular oxygen, that the oxygen concentration in the water is decreased to a level that is too low to sustain other organisms. One such hypoxic (low levels of oxygen) zone is in the northern Gulf of Mexico, off the coast of Louisiana where the Mississippi River flows into the Gulf (Figure 20.16). The Mississippi is extremely nutrient rich due to agricultural runoff; so plant microorganisms, called phytoplankton, proliferate so robustly that they exceed the amount that can be consumed by other members of the food chain. When the phytoplankton die, they sink to the bottom and are consumed by aerobic bacteria. The aerobic bacteria thrive to such a degree that other bottom-dwelling organisms, such as shrimp and crabs, cannot obtain enough O_2 to survive. The term "dead zone" refers to the inability of this area to support fisheries.

Figure 20.16 The Gulf of Mexico dead zone. The size of the dead zone in the Gulf of Mexico off Louisiana varies annually but may extend from the Louisiana and Alabama coasts to the westernmost coast of Texas. Reds and oranges represent high concentrations of phytoplankton and river sediment. [NASA/Goddard Space Flight Center/Scientific Visualization Studio.]

> **DID YOU KNOW?**
>
> In anaerobic respiration in some organisms, chemicals other than oxygen are used as the final electron acceptor in an electron-transport chain. Because none of these electron acceptors are as electropositive as O_2, not as much energy is released and, consequently, not as much ATP is generated.

Toxic Derivatives of Molecular Oxygen Such As Superoxide Radical Are Scavenged by Protective Enzymes

Molecular oxygen is an ideal terminal electron acceptor because its high affinity for electrons provides a large thermodynamic driving force. However, the *reduction of O_2 can result in dangerous side reactions.* The transfer of four electrons leads to safe products (two molecules of H_2O), but partial reduction generates hazardous compounds. In particular, *the transfer of a single electron to O_2 forms superoxide ion, whereas the transfer of two electrons yields peroxide:*

$$O_2 \xrightarrow{e^-} O_2^{\cdot -} \xrightarrow{e^-} O_2^{2-}$$

$$\underset{\text{ion}}{\underset{\text{Superoxide}}{\phantom{O_2^{\cdot-}}}} \qquad \underset{\text{Peroxide}}{\phantom{O_2^{2-}}}$$

From 2% to 4% of the oxygen molecules consumed by mitochondria are converted into superoxide ion, predominantly at Complexes I and III. Superoxide, peroxide, and species that can be generated from them such as the hydroxyl radical (OH·) are collectively referred to as *reactive oxygen species* or *ROS*. Reactive oxygen species react with essentially all macromolecules in the cell, including proteins, nucleotide bases, and membranes. Oxidative damage caused by ROS has been implicated in the aging process as well as in a growing list of diseases (Table 20.3).

What are the cellular defense strategies against oxidative damage by ROS? Chief among them is the enzyme *superoxide dismutase.* This enzyme scavenges

> **DID YOU KNOW?**
>
> In mammals, the mutation rate for mitochondrial DNA is 10- to 20-fold higher than that for nuclear DNA. This higher rate is believed to be due in large part to the inevitable generation of reactive oxygen species by oxidative phosphorylation in mitochondria.

Table 20.3   Pathological and other conditions that may be due to free-radical injury

Atherogenesis	Acute renal failure
Emphysema; bronchitis	Down syndrome
Parkinson disease	Retrolental fibroplasia (conversion of the retina into a fibrous mass in premature infants)
Duchenne muscular dystrophy	Cerebrovascular disorders
Cervical cancer	Ischemia; reperfusion injury
Alcoholic liver disease	
Diabetes	

Source: Data from D. B. Marks, A. D. Marks, and C. M. Smith, *Basic Medical Biochemistry: A Clinical Approach* (Williams & Wilkins, 1996), p. 331.

superoxide radicals by catalyzing the conversion of two of these radicals into hydrogen peroxide and molecular oxygen:

$$2O_2^{\cdot-} + 2H^+ \xrightleftharpoons{\text{Superoxide dismutase}} O_2 + H_2O_2$$

This reaction takes place in two steps. The oxidized form of the enzyme is reduced by superoxide to form oxygen (**Figure 20.17**). The reduced form of the enzyme then reacts with a second superoxide ion to form peroxide, which takes up two protons along the reaction path to yield hydrogen peroxide.

The hydrogen peroxide formed by superoxide dismutase and by other processes is scavenged by *catalase,* a ubiquitous heme protein that catalyzes the dismutation of hydrogen peroxide into water and molecular oxygen:

$$2H_2O_2 \xrightleftharpoons{\text{Catalase}} O_2 + 2H_2O$$

Superoxide dismutase and catalase are remarkably efficient, performing their reactions at or near the diffusion-limited rate (p. 117). Other cellular defenses against oxidative damage include the antioxidant vitamins—vitamins E and C as well as ubiquinol. Because it is lipophilic, vitamin E is especially useful in protecting membranes from lipid peroxidation. Ubiquinol is the only lipid-soluble antioxidant synthesized by human beings.

A long-term benefit of exercise may be to increase the amount of superoxide dismutase in the cell. The elevated aerobic metabolism during exercise causes more ROS to be generated. In response, the cell synthesizes more protective enzymes. The net effect is one of protection, because the increase in superoxide dismutase more effectively protects the cell during periods of rest.

Despite the fact that reactive oxygen species are known hazards, recent evidence suggests that the controlled generation of these molecules may be important components of signal-transduction pathways. For instance, growth factors have been shown to increase ROS as part of their signaling pathway, and ROS regulate channels and transcription factors. ROS have been implicated in the control of cell differentiation, the immune response, autophagy, circadian rhythms as well as other metabolic activities. The dual roles of ROS are an excellent example of the wondrous complexity of the biochemistry of living systems: even potentially harmful substances can be harnessed to play useful roles.

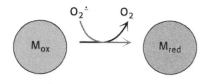
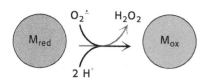

Figure 20.17 Superoxide dismutase mechanism. The oxidized form of superoxide dismutase (M_{ox}) reacts with one superoxide ion to form O_2 and generate the reduced form of the enzyme (M_{red}). The reduced form then reacts with a second superoxide ion and two protons to form hydrogen peroxide and regenerate the oxidized form of the enzyme.

SUMMARY

20.1 Oxidative Phosphorylation in Eukaryotes Takes Place in Mitochondria

Mitochondria generate most of the ATP required by aerobic cells through a joint endeavor of the reactions of the citric acid cycle, which take place in the mitochondrial matrix, and oxidative phosphorylation, which takes place in the inner mitochondrial membrane.

20.2 Oxidative Phosphorylation Depends on Electron Transfer

In oxidative phosphorylation, the synthesis of ATP is coupled to the flow of electrons from NADH or $FADH_2$ to O_2 by a proton gradient across the inner mitochondrial membrane. Electron flow through three asymmetrically oriented transmembrane complexes results in the pumping of protons out of the mitochondrial matrix and the generation of a membrane potential. ATP is synthesized when protons flow back to the matrix.

20.3 The Respiratory Chain Consists of Proton Pumps and a Physical Link to the Citric Acid Cycle

The electron carriers in the respiratory assembly of the inner mitochondrial membrane are quinones, flavins, iron–sulfur complexes, heme groups of cytochromes, and copper ions. Electrons from NADH are transferred to the FMN prosthetic group of NADH-Q oxidoreductase (Complex I), the first of four complexes. This oxidoreductase also contains Fe-S centers. The electrons emerge in QH_2, the reduced form of ubiquinone (Q). The citric acid cycle enzyme succinate dehydrogenase is a component of the succinate-Q reductase complex (Complex II), which donates electrons from $FADH_2$ to Q to form QH_2. This hydrophobic carrier transfers its electrons to Q-cytochrome c oxidoreductase (Complex III), a complex that contains cytochromes b and c_1 and an Fe-S center. This complex reduces cytochrome c, a water-soluble peripheral membrane protein. Cytochrome c, like Q, is a mobile carrier of electrons, which it then transfers to cytochrome c oxidase (Complex IV). This complex contains cytochromes a and a_3 and three copper ions. A heme iron ion and a copper ion in this oxidase transfer electrons to O_2, the ultimate acceptor, to form H_2O.

KEY TERMS

oxidative phosphorylation (p. 363)
electron-transport chain (p. 364)
cellular respiration (p. 364)
reduction (redox, oxidation–reduction,
  E'_0) potential (p. 366)
flavin mononucleotide (FMN) (p. 368)
iron–sulfur (nonheme-iron) protein
  (p. 369)

coenzyme Q (Q, ubiquinone) (p. 370)
Q pool (p. 370)
NADH-Q oxidoreductase (Complex I)
  (p. 371)
succinate-Q reductase (Complex II)
  (p. 373)
Q-cytochrome c oxidoreductase
  (Complex III) (p. 373)

cytochrome c (Cyt c) (p. 373)
Q cycle (p. 374)
cytochrome c oxidase (Complex IV)
  (p. 375)
superoxide dismutase (p. 377)
catalase (p. 378)

? Answers to QUICK QUIZZES

1. The reduction potential of $FADH_2$ is less than that of NADH (Table 20.1). As a result, when those electrons are passed along to oxygen, less energy is released. The consequence of the difference is that electron flow from $FADH_2$ to O_2 pumps fewer protons than does the electron flow from NADH.

2. Complex I would be reduced, whereas Complexes II, III, and IV would be oxidized. The citric acid cycle would become reduced because it has no way to oxidize NADH.

PROBLEMS

1. *Give and accept.* Distinguish between an oxidizing agent and a reducing agent. ✓ 1

2. *Like mac and cheese.* Match each term with its description. ✓ 1

(a) Respiration _____
(b) Redox potential _____
(c) Electron-transport chain _____
(d) Flavin mono-nucleotide (FMN) _____
(e) Iron–sulfur protein _____
(f) Coenzyme Q _____
(g) Cytochrome c _____
(h) Q cycle _____
(i) Superoxide dismutase _____
(j) Catalase _____

1. Converts reactive oxygen species into hydrogen peroxide
2. Electron flow from NADH and $FADH_2$ to O_2
3. Facilitates electron flow from FMN to coenzyme Q
4. An ATP-generating process in which an inorganic compound serves as the final electron acceptor
5. Measure of the tendency to accept or donate electrons
6. Converts hydrogen peroxide into oxygen and water
7. Funnels electrons from a two-electron carrier to a one-electron carrier
8. Lipid-soluble electron carrier
9. Donates electrons to Complex IV
10. Accepts electrons from NADH in Complex I

3. *Reference states.* The standard oxidation–reduction potential for the reduction of O_2 to H_2O is given as 0.82 V in Table 20.1. However, the value given in textbooks of chemistry is 1.23 V. Account for this difference. ✓ 1

4. *Thermodynamic constraint.* Compare the $\Delta G°'$ values for the oxidation of succinate by NAD^+ and by FAD. Use the data given in Table 20.1, and assume that E_0' for the FAD–$FADH_2$ redox couple is nearly 0 V. Why is FAD rather than NAD^+ the electron acceptor in the reaction catalyzed by succinate dehydrogenase? ✓ 1

5. *Hitchin' a ride.* What is the evidence that modern mitochondria arose from a single endosymbiotic event?

6. *Benefactor and beneficiary.* Identify the oxidant and the reductant in the following reaction:

$$\text{Pyruvate} + \text{NADH} + H^+ \longrightarrow \text{lactate} + NAD^+$$

7. *Location, location, location.* Iron is a component of many of the electron carriers of the electron-transport chain. How can it participate in a series of coupled redox reactions if the $\Delta E'_0$ value is +0.77 V as seen in Table 20.1? ✓ 1 ✓ 2

8. *Six of one, half dozen of the other.* How is the redox potential ($\Delta E_0'$) related to the free-energy change of a reaction ($\Delta G°'$)? ✓ 1

9. *Structural considerations.* Explain why coenzyme Q is an effective mobile electron carrier in the electron-transport chain. ✓ 1

10. *Line up.* Place the following components of the electron-transport chain in their proper order: (a) cytochrome c; (b) Q-cytochrome c oxidoreductase; (c) NADH-Q oxidoreductase; (d) cytochrome c oxidase; (e) ubiquinone. ✓ 2

11. *Like Dolce and Gabbana.* Match the terms on the left with those on the right.

(a) Complex I _____
(b) Complex II _____
(c) Complex III _____
(d) Complex IV _____
(e) Ubiquinone _____

1. Q-cytochrome c oxidoreductase
2. Coenzyme Q
3. Succinate-Q reductase
4. NADH-Q oxidoreductase
5. Cytochrome c oxidase

12. *ROS, not ROUS.* What are the reactive oxygen species, and why are they especially dangerous to cells?

13. *Inhibitors.* Rotenone inhibits electron flow through NADH-Q oxidoreductase. Antimycin A blocks electron flow between cytochromes b and c_1. Cyanide blocks electron flow through cytochrome oxidase to O_2. Predict the relative oxidation–reduction state of each of the following respiratory-chain components in mitochondria that are treated with each of the inhibitors: NAD^+; NADH-Q oxidoreductase; coenzyme Q; cytochrome c_1; cytochrome c; cytochrome a. ✓ 1

14. *Efficiency.* What is the advantage of having Complexes I, III, and IV associated with one another in the form of a respirasome? ✓ 2

Chapter Integration Problems

15. *Recycling device.* The cytochrome b component of Q-cytochrome c oxidoreductase enables both electrons of QH_2 to be effectively utilized in generating a proton-motive force. Cite another recycling device in metabolism that brings a potentially dead-end reaction product back into the mainstream.

16. *Maybe you shouldn't take your vitamins.* Exercise is known to increase insulin sensitivity and to ameliorate type 2 diabetes (p. 323). Recent research suggests that taking antioxidant vitamins might mitigate the beneficial effects of exercise with respect to ROS protection.

(a) What are antioxidant vitamins?
(b) How does exercise protect against ROS?

(c) Explain why vitamins might counteract the effects of exercise.

17. *Linked In.* What citric acid cycle enzyme is also a component of the electron-transport chain?

18. *Breathe or ferment?* Compare fermentation and respiration with respect to electron donors and electron acceptors.

Challenge Problems

19. *Weaker electrons.* Electrons from NADH pump more protons as a consequence of reaction with oxygen than do the electrons from $FADH_2$. Calculate the energy released by the reduction of O_2 with $FADH_2$. ✓ 2

20. *Crossover point.* The precise site of action of a respiratory-chain inhibitor can be revealed by the *crossover technique*. Britton Chance devised elegant spectroscopic methods for determining the proportions of the oxidized and reduced forms of each carrier. This determination is feasible because the forms have distinctive absorption spectra, as illustrated in the adjoining graph for cytochrome *c*.

You are given a new inhibitor and find that its addition to respiring mitochondria causes the carriers between NADH and QH_2 to become more reduced and those between cytochrome *c* and O_2 to become more oxidized. Where does your inhibitor act? ✓ 2

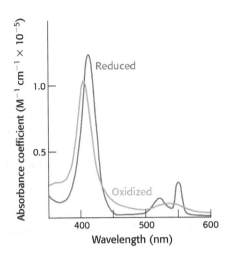

Selected Readings for this chapter can be found online at www.whfreeman.com/tymoczko3e.

The Proton-Motive Force

Itaipu Binacional, on the border between Brazil and Paraguay, is one of biggest hydroelectric dams in the world. The dam transforms the energy of falling water into electrical energy. Analogously, the mitochondrial enzyme ATP synthase transforms the energy of protons falling down an energy gradient into ATP. [Heeb Christian/AgeFotostock.]

In Chapter 20, we considered the flow of electrons from NADH to O_2, an exergonic process:

$$NADH + \tfrac{1}{2}O_2 + H^+ \longrightarrow H_2O + NAD^+$$

$$\Delta G^{\circ\prime} = -220.1 \text{ kJ mol}^{-1} \ (-52.6 \text{ kcal mol}^{-1})$$

During the electron flow, protons are pumped from the mitochondrial matrix to the outside of the inner mitochondrial membrane, creating a proton gradient. Energy is thus transformed as electron transfer potential is converted into a proton gradient, an energy-rich circumstance because the entropy of the protons is reduced. Next, we will see how the energy of the proton gradient powers the synthesis of ATP:

$$ADP + P_i + H^+ \rightleftharpoons ATP + H_2O$$

$$\Delta G^{\circ\prime} = +30.5 \text{ kJ mol}^{-1} \, (+7.3 \text{ kcal mol}^{-1})$$

✓ 3 Describe how the proton-motive force is converted into ATP.

DID YOU KNOW?

Some have argued that, along with the elucidation of the structure of DNA, the discovery that ATP synthesis is powered by a proton gradient is one of the two major advancements in biology in the twentieth century. However, Mitchell's initial postulation of the chemiosmotic theory was not warmly received by all. Efraim Racker, one of the early investigators of ATP synthase, recalls that some thought of Mitchell as a court jester, whose work was of no consequence. Peter Mitchell was awarded the Nobel Prize in chemistry in 1978 for his contributions to understanding oxidative phosphorylation.

21.1 A Proton Gradient Powers the Synthesis of ATP

A molecular assembly in the inner mitochondrial membrane carries out the synthesis of ATP. This enzyme complex was originally called the *mitochondrial ATPase* or F_1F_0 *ATPase* because it was discovered through its catalysis of the reverse reaction, the hydrolysis of ATP. *ATP synthase* or F_1F_0 *ATP synthase* are preferred names since these emphasize its actual role in the mitochondrion.

How is the oxidation of NADH coupled to the phosphorylation of ADP? This was one of the most perplexing questions biochemists ever faced, requiring decades of work to solve. Electron transfer was first suggested to lead to the formation of a covalent high-energy intermediate that serves as a compound having a high phosphoryl-transfer potential. Such a compound could, in a manner analogous to substrate-level phosphorylation in glycolysis (p. 289), transfer a phosphoryl group to ADP to form ATP. An alternative proposal was that electron transfer aids the formation of an activated protein conformation, which then drives ATP synthesis. The search for such intermediates for several decades proved fruitless.

In 1961, Peter Mitchell suggested a radically different mechanism, *the chemiosmotic hypothesis*. He proposed that *electron transport and ATP synthesis are coupled by a proton gradient across the inner mitochondrial membrane*. In his model, the transfer of electrons through the respiratory chain leads to the pumping of protons from the matrix to the cytoplasmic side of the inner mitochondrial membrane. The H^+ concentration becomes lower in the matrix, and an electric field with the matrix side negative is generated (**Figure 21.1**). Protons then flow back into the matrix to equalize the distribution. Mitchell's idea was that this flow of protons drives the synthesis of ATP by ATP synthase. The energy-rich, unequal distribution of protons is called the *proton-motive force*, which is composed of two components: a chemical gradient and a charge gradient. The chemical gradient for protons can be represented as a pH gradient. The charge gradient is created by the positive charge on the unequally distributed protons forming

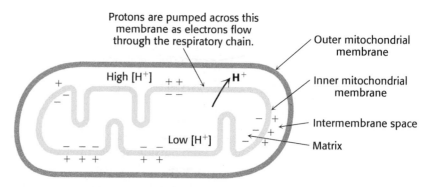

Figure 21.1 Chemiosmotic hypothesis. Electron transfer through the respiratory chain leads to the pumping of protons from the matrix to the cytoplasmic side of the inner mitochondrial membrane. The pH gradient and membrane potential constitute a proton–motive force that is used to drive ATP synthesis.

the chemical gradient. Mitchell proposed that both components power the synthesis of ATP:

Proton-motive force (Δp) = chemical gradient (ΔH) + charge gradient ($\Delta \psi$)

Mitchell's highly innovative hypothesis that oxidation and phosphorylation are coupled by a proton gradient is now supported by a wealth of evidence, including, importantly, that an intact proton-impermeable membrane is required for this coupling. Indeed, electron transport does generate a proton gradient across the inner mitochondrial membrane. The pH outside is 1.4 units lower than inside, and the voltage difference, or membrane potential, is 0.14 V, the outside being positive. This membrane potential is equivalent to a free energy of 20.8 kJ (5.2 kcal) per mole of protons.

An artificial system representing the cellular respiration system was created to elegantly demonstrate the basic principle of the chemiosmotic hypothesis. The role of the *electron-transport chain* was played by bacteriorhodopsin, a purple membrane protein from halobacteria that pumps protons when illuminated. Synthetic vesicles containing bacteriorhodopsin and mitochondrial ATP synthase purified from beef heart were created (**Figure 21.2**). When the vesicles were exposed to light, ATP was formed. This key experiment clearly showed that *the respiratory chain and ATP synthase are biochemically separate systems, linked only by a proton-motive force.*

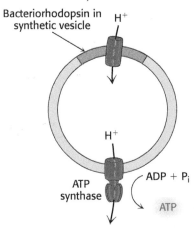

Figure 21.2 Testing the chemiosmotic hypothesis. ATP is synthesized when reconstituted membrane vesicles containing bacteriorhodopsin (a light-driven proton pump) and ATP synthase are illuminated. The orientation of ATP synthase in this reconstituted membrane is the reverse of that in the mitochondrion.

ATP Synthase Is Composed of a Proton-Conducting Unit and a Catalytic Unit

Two parts of the puzzle of how NADH oxidation is coupled to ATP synthesis are now evident: (1) electron transport generates a proton-motive force; (2) ATP synthesis by ATP synthase is powered by a proton-motive force. How is the proton-motive force converted into the high phosphoryl-transfer potential of ATP?

ATP synthase is a large, complex enzyme that looks like a ball on a stick, located in the inner mitochondrial membrane (**Figure 21.3**). Much of the "stick" part, called the F_0 subunit, is embedded in the inner mitochondrial membrane. The 85-Å-diameter ball, called the F_1 subunit, protrudes into the mitochondrial matrix. The F_1 subunit contains the catalytic activity of the synthase. In fact, isolated F_1 subunits display ATPase activity.

The F_1 subunit consists of five types of polypeptide chains (α_3, β_3, γ, δ, and ε). The three α and three β subunits, which make up the bulk of the F_1, are arranged alternately in a hexameric ring. The active sites reside on the β subunits. Beginning just above the α and β subunits is a central stalk consisting of the γ and ε proteins. The γ subunit includes a long helical coiled coil that extends into the center of the $\alpha_3\beta_3$ hexamer. *Each of the β subunits is distinct because each*

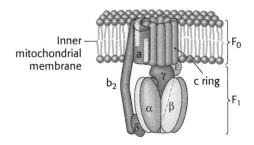

Figure 21.3 The structure of ATP synthase. A schematic structure of ATP synthase is shown. Notice that part of the enzyme complex (the F_0 subunit) is embedded in the inner mitochondrial membrane, whereas the remainder (the F_1 subunit) resides in the matrix. [Drawn from 1E79.pdb and 1C0V.pdb.]

interacts with a different face of the γ subunit. Distinguishing the three β subunits is crucial for understanding the mechanism of ATP synthesis.

The F_0 subunit is a hydrophobic segment that spans the inner mitochondrial membrane. *F_0 contains the proton channel of the complex.* This channel consists of a ring comprising from 8 to 15 **c** subunits, depending on the source of the enzyme, that are embedded in the membrane. A single **a** subunit binds to the outside of the ring. The F_0 and F_1 subunits are connected in two ways: by the central γε stalk and by an exterior column. The exterior column consists of one **a** subunit, two **b** subunits, and the δ subunit.

ATP synthases bind to one another to form dimers, which then associate to form large oligomers of dimers. This association stabilizes the individual enzymes to the rotational forces required for catalysis and facilitates the curvature of the inner mitochondrial membrane. The formation of the cristae allows the proton pumps of the electron transport chain to localize the proton gradient in the vicinity of the synthases, which are located at the tips of the cristae, thereby enhancing the efficiency of ATP synthesis (**Figure 21.4**).

Proton Flow Through ATP Synthase Leads to the Release of Tightly Bound ATP

ATP synthase catalyzes the formation of ATP from ADP and orthophosphate:

$$ADP^{3-} + HPO_4^{2-} + H^+ \rightleftharpoons ATP^{4-} + H_2O$$

There are three active sites on the enzyme, each performing one of three different functions at any instant. These functions are (1) trapping of ADP and P_i, (2) ATP synthesis, and (3) ATP release and ADP and P_i binding. The proton-motive force causes the three active sites to sequentially change functions as protons flow through the membrane-embedded component of the enzyme. Indeed, the enzyme consists of a moving part and a stationary part: (1) the moving unit, or *rotor*, consists of the **c** ring and the γε stalk; (2) the stationary unit, or *stator*, is composed of the remainder of the molecule.

? QUICK QUIZ 1 Why do isolated F_1 subunits of ATP synthase catalyze ATP hydrolysis?

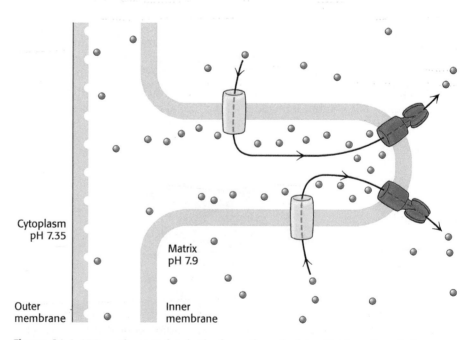

Cytoplasm
pH 7.35

Matrix
pH 7.9

Outer
membrane

Inner
membrane

Figure 21.4 ATP synthase assists in the formation of cristae. The formation of oligomers of dimers of ATP synthase facilitates the formation of cristae, creating an area where the protons have ready access to the ATP synthase.

How do the three active sites of ATP synthase respond to the flow of protons? A number of experimental observations suggested a *binding-change mechanism* for proton-driven ATP synthesis. This proposal states that a β subunit can perform each of three sequential steps—trapping, synthesis, and release and binding—in the process of ATP synthesis by changing conformation. As already noted, interactions with the γ subunit make the three β subunits structurally distinct (**Figure 21.5**). At any given moment, one β subunit will be in the L, or loose, conformation. This conformation binds ADP and Pᵢ. A second subunit will be in the T, or tight, conformation. This conformation binds ATP with great avidity, so much so that it will convert bound ADP and Pᵢ into ATP. Both the L and the T conformations are sufficiently constrained that they cannot release bound nucleotides. The final subunit will be in the O, or open, form. This form has a more open conformation and can bind or release adenine nucleotides.

The rotation of the γ subunit drives the interconversion of these three forms (**Figure 21.6**). ADP and Pᵢ bound in the subunit in the T-form combine to form ATP. Suppose that the γ subunit is rotated by 120 degrees in a counterclockwise direction (as viewed from the top). This rotation converts the T-form site into an O-form site with the nucleotide bound as ATP. Concomitantly, the L-form site is converted into a T-form site, enabling the transformation of an additional ADP and Pᵢ into ATP. The ATP in the O-form site can now depart from the enzyme to be replaced by ADP and Pᵢ. An additional 120-degree rotation converts this O-form site into an L-form site, trapping these substrates. Each subunit progresses from the T to the O to the L-form with no two subunits ever present in the same conformational form. This mechanism suggests that ATP can be synthesized and released by driving the rotation of the γ subunit in the appropriate direction. Note that the role of the proton gradient is not to directly participate in the formation of ATP but rather to drive the release of ATP from the enzyme.

Rotational Catalysis Is the World's Smallest Molecular Motor

Is it possible to observe the proposed rotation directly? Ingenious experiments have demonstrated rotation through the use of a simple experimental system consisting solely of cloned α₃β₃γ subunits (**Figure 21.7**). The β subunits were engineered to contain amino-terminal polyhistidine tags, which have a high affinity for nickel ions. This property of the tags allowed the α₃β₃ assembly to be immobilized on a glass surface that had been coated with nickel ions. The γ subunit was linked to a fluorescently labeled actin filament to provide a long

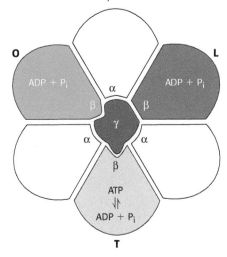

Figure 21.5 ATP synthase nucleotide-binding sites are not equivalent. The γ subunit passes through the center of the α₃β₃ hexamer and makes the nucleotide-binding sites in the β subunits distinct from one another.

Progressive alteration of the forms of the three active sites of ATP synthase
Subunit 1 L → T → O → L → T → O ⋯
Subunit 2 O → L → T → O → L → T ⋯
Subunit 3 T → O → L → T → O → L ⋯

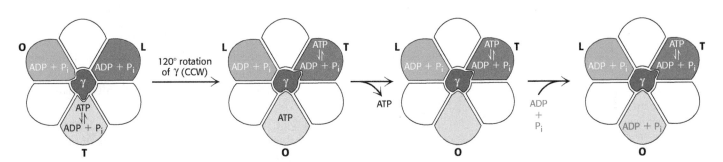

Figure 21.6 A binding–change mechanism for ATP synthase. The rotation of the γ subunit interconverts the three β subunits. The subunit in the T (tight) form converts ADP and Pᵢ into ATP but does not allow ATP to be released. When the γ subunit is rotated counterclockwise (CCW) 120 degrees, the T-form subunit is converted into the O form, allowing ATP release. New molecules of ADP and Pᵢ can then bind to the O-form subunit. An additional 120–degree rotation (not shown) traps these substrates in an L-form subunit.

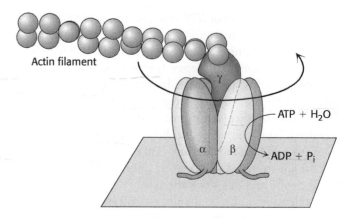

Figure 21.7 Direct observation of ATP-driven rotation in ATP synthase. The $\alpha_3\beta_3$ hexamer of ATP synthase is fixed to a surface, with the γ subunit projecting upward and linked to a fluorescently labeled actin filament. The addition and subsequent hydrolysis of ATP result in the counterclockwise rotation of the γ subunit, which can be directly seen under a fluorescence microscope.

segment that could be observed under a fluorescence microscope. Remarkably, the addition of ATP caused the actin filament to rotate unidirectionally in a counterclockwise direction. *The γ subunit was rotating, driven by the hydrolysis of ATP.* Thus, the catalytic activity of an individual molecule could be observed. The counterclockwise rotation is consistent with the predicted direction for hydrolysis because the molecule was viewed from below relative to the view shown in Figure 21.6.

More detailed analysis in the presence of lower concentrations of ATP revealed that the γ subunit rotates in 120-degree increments. Each increment corresponds to the hydrolysis of a single ATP molecule. In addition, from the results obtained by varying the length of the actin filament, thereby increasing the rotational resistance, and measuring the rate of rotation, the enzyme appears to operate near 100% efficiency; that is, essentially all of the energy released by ATP hydrolysis is converted into rotational motion.

Proton Flow Around the c Ring Powers ATP Synthesis

The direct observation of the γ subunit's rotary motion is strong evidence for the rotational mechanism for ATP synthesis. The last remaining question is: How does proton flow through F_0 drive the rotation of the γ subunit. An elegant mechanism has been postulated that provides a clear answer to this question. The mechanism depends on the structures of the **a** and **c** subunits of F_0 (**Figure 21.8**). The **a** subunit directly abuts the membrane-spanning ring formed by 8 to 15 **c** subunits. Evidence is consistent with a structure for the **a** subunit that includes two hydrophilic half-channels that do not span the membrane (Figure 21.8). Thus, protons can enter into either of these channels, but they cannot move completely across the membrane. The **a** subunit is positioned such that each half-channel directly interacts with one **c** subunit.

Each polypeptide chain of the **c** subunit forms a pair of α helices that span the membrane. Glutamic acid (or aspartic acid) is found in the middle of one of the helices. If the glutamate is charged (unprotonated), the **c** subunit will not move into the hydrophobic interior of the membrane. The key to proton movement across the membrane is that, in a proton-rich environment, such as the cytoplasmic side of the inner mitochondrial membrane, a proton will enter a channel and bind to the glutamate residue, while the glutamic acid in the other half-channel

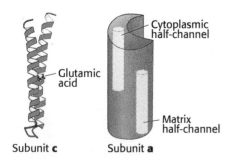

Figure 21.8 Components of the proton-conducting unit of ATP synthase. The **c** subunit consists of two helices that span the membrane. A glutamic acid residue in one of the helices lies on the center of the membrane. The structure of the **a** subunit appears to include two half-channels that allow protons to enter and pass part way but not completely through the membrane.

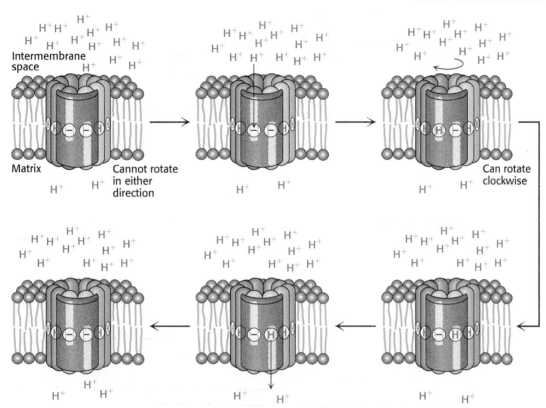

Figure 21.9 Proton motion across the membrane drives the rotation of the c ring. A proton enters from the intermembrane space into the cytoplasmic half-channel to neutralize the charge on a glutamate or aspartate residue in a **c** subunit. With this charge neutralized, the **c** ring can rotate clockwise by one **c** subunit, moving the amino acid residue out of the membrane into the matrix half-channel. This proton can move into the matrix, resetting the system to its initial state.

will release a proton to the proton-poor environment of the matrix (**Figure 21.9**). The **c** subunit with the bound proton then moves into the hydrophobic environment of the inner membrane as the ring rotates by one **c** subunit. This rotation brings the newly deprotonated **c** subunit from the matrix half-channel to the proton-rich cytoplasmic half-channel, where it can bind a proton. *The movement of protons through the half-channels from the high proton concentration of the inner membrane space to the low proton concentration of the matrix powers the rotation of the c ring.* The **a** unit remains stationary as the **c** ring rotates. Each proton that enters the cytoplasmic half-channel of the **a** unit moves through the membrane by riding around on the rotating **c** ring to exit through the matrix half-channel into the proton-poor environment of the matrix (**Figure 21.10**). The rate of **c** ring rotation is remarkable, nearly 100 revolutions per second.

How does the rotation of the **c** ring lead to the synthesis of ATP? The **c** ring is tightly linked to the γ and ε subunits. Thus, as the **c** ring turns, these subunits are turned inside the $\alpha_3\beta_3$ hexamer unit of F_1. The rotation of the γ subunit in turn promotes the synthesis of ATP through the binding-change mechanism (p. 387). The exterior column formed by the two **b** chains and the δ subunit prevents the $\alpha_3\beta_3$ hexamer from rotating in sympathy with the **c** ring. Each 360-degree rotation of the γ subunit leads to the synthesis and release of 3 molecules of ATP. The number of **c** subunits determines the efficiency with which the proton gradient is converted into ATP synthesis. For instance, if there are

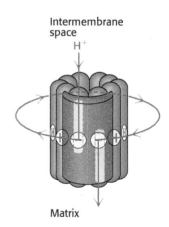

Figure 21.10 Proton path through the membrane. Each proton enters the cytoplasmic half-channel, follows a complete rotation of the **c** ring, and exits through the other half-channel into the matrix.

? QUICK QUIZ 2 ATP synthases isolated from different sources often have different numbers of **c** subunits. What effect would altering the number of **c** subunits have on the yield of ATP as a function of proton flow?

10 **c** subunits in the ring (as is the case for yeast mitochondrial ATP synthase), each molecule of ATP generated requires the transport of 10/3 = 3.33 protons. Recent evidence shows that the **c** rings of all vertebrates are composed of 8 subunits, making vertebrate ATP synthase the most efficient ATP synthase known, with the transport of only 2.7 protons required for ATP synthesis. For simplicity, we will assume that 3 protons must flow into the matrix for each molecule of ATP formed. As we will see, the electrons from NADH pump enough protons to generate 2.5 molecules of ATP, whereas those from $FADH_2$ yield 1.5 molecules of ATP.

Let us return for a moment to the example with which we began this section. If a resting human being requires 85 kg of ATP per day for bodily functions, then 3.3×10^{25} protons must flow through ATP synthase per day, or 3.3×10^{21} protons per second. Figure 21.11 summarizes the process of oxidative phosphorylation.

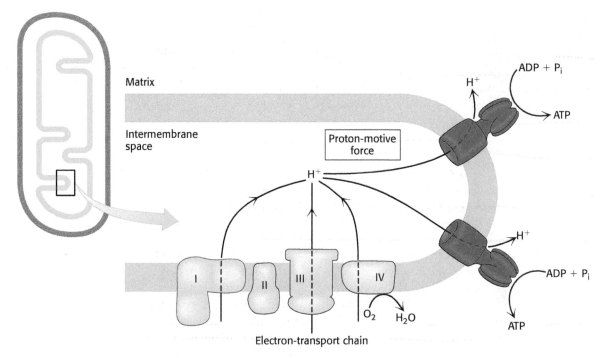

Figure 21.11 An overview of oxidative phosphorylation. The electron-transport chain generates a proton gradient, which is used to synthesize ATP.

21.2 Shuttles Allow Movement Across Mitochondrial Membranes

The inner mitochondrial membrane is impermeable to most molecules, yet much exchange has to take place between the cytoplasm and the mitochondria. This exchange is mediated by an array of membrane-spanning transporter proteins.

Electrons from Cytoplasmic NADH Enter Mitochondria by Shuttles

One function of the respiratory chain is to regenerate NAD^+ for use in glycolysis. NADH generated by the citric acid cycle and fatty acid oxidation is already in the mitochondrial matrix, but how is cytoplasmic NADH reoxidized to NAD^+ under aerobic conditions? NADH cannot simply pass into mitochondria for oxidation by the respiratory chain, because the inner mitochondrial membrane is impermeable to NADH and NAD^+. The solution is that *electrons from NADH, rather than NADH itself, are carried across the mitochondrial membrane*. One of several means of introducing electrons from NADH into the electron-transport chain is the *glycerol 3-phosphate shuttle* (Figure 21.12). The first step in this shuttle is the transfer of a pair of electrons from NADH to dihydroxyacetone phosphate, a glycolytic intermediate, to form glycerol 3-phosphate. This reaction is catalyzed by a glycerol 3-phosphate dehydrogenase in the cytoplasm. Glycerol 3-phosphate is reoxidized to dihydroxyacetone phosphate on the outer surface of the inner mitochondrial membrane by a membrane-bound isozyme of glycerol 3-phosphate dehydrogenase. An electron pair from glycerol 3-phosphate is transferred to an FAD prosthetic group in this enzyme to form $FADH_2$. This reaction also regenerates dihydroxyacetone phosphate.

The reduced flavin transfers its electrons to the electron carrier Q, which then enters the respiratory chain as QH_2. *When cytoplasmic NADH transported by the glycerol 3-phosphate shuttle is oxidized by the respiratory chain, 1.5 rather than 2.5 molecules of ATP are formed.* The yield is lower because the electrons from cytoplasmic NADH are carried to the electron-transport chain by FAD. The use of FAD enables electrons from cytoplasmic NADH to be transported into mitochondria against an NADH concentration gradient that is formed because the reduced cofactors build up in the mitochondria when oxygen demand is high. The price of this transport is one molecule of ATP per two electrons. This glycerol 3-phosphate shuttle is especially prominent in muscle and enables it to sustain a very high rate of oxidative phosphorylation. Indeed, some insects lack lactate dehydrogenase and are completely dependent on the glycerol 3-phosphate shuttle for the regeneration of cytoplasmic NAD^+.

In the heart and liver, electrons from cytoplasmic NADH are brought into mitochondria by the *malate–aspartate shuttle*, which is mediated by two membrane carriers and four enzymes (Figure 21.13). Electrons are transferred from NADH in the cytoplasm to oxaloacetate, forming malate, which traverses the inner mitochondrial membrane in exchange for α-ketoglutarate by an antiporter (p. 215). The malate is then reoxidized to oxaloacetate by NAD^+ in the matrix to

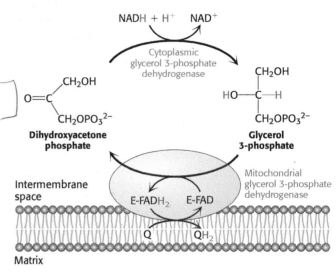

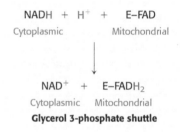

Figure 21.12 The glycerol 3-phosphate shuttle. Electrons from NADH can enter the mitochondrial electron-transport chain by reducing dihydroxyacetone phosphate to glycerol 3-phosphate. Electron transfer to an FAD prosthetic group in a membrane-bound glycerol 3-phosphate dehydrogenase reoxidizes glycerol 3-phosphate. Subsequent electron transfer to Q to form QH_2 allows these electrons to enter the electron-transport chain.

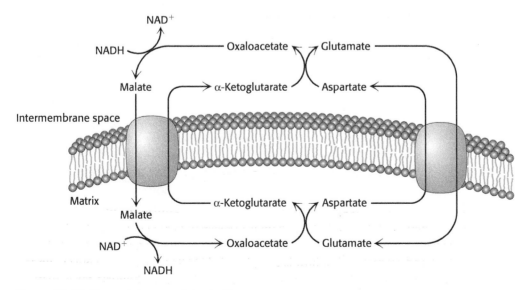

Figure 21.13 The malate–aspartate shuttle.

NADH   +   NAD$^+$
Cytoplasmic   Mitochondrial

NAD$^+$ +   NADH
Cytoplasmic   Mitochondrial

Malate–aspartate shuttle

form NADH in a reaction catalyzed by the citric acid cycle enzyme malate dehydrogenase. The resulting oxaloacetate does not readily cross the inner mitochondrial membrane, and so a transamination reaction (Chapter 30) is needed to form aspartate, which can be transported to the cytoplasmic side by another antiporter in exchange for glutamate. Glutamate donates an amino group to oxaloacetate, forming aspartate and α-ketoglutarate. In the cytoplasm, aspartate is then deaminated to form oxaloacetate and the cycle is restarted.

The Entry of ADP into Mitochondria Is Coupled to the Exit of ATP

The major function of oxidative phosphorylation is to generate ATP from ADP. However, these nucleotides do not diffuse freely across the inner mitochondrial membrane. How are these highly charged molecules moved across the inner membrane into the cytoplasm? A specific transport protein, *ATP-ADP translocase* (also called the *adenine nucleotide transporter, ANT*) enables these molecules to transverse this permeability barrier. Most importantly, the flows of ATP and ADP are coupled. *ADP enters the mitochondrial matrix only if ATP exits, and vice versa.* This process is carried out by the translocase, an antiporter:

$$ADP^{3-}_{cytoplasm} + ATP^{4-}_{matrix} \longrightarrow ADP^{3-}_{matrix} + ADP^{4-}_{cytoplasm}$$

ATP-ADP translocase is highly abundant, constituting about 15% of the protein in the inner mitochondrial membrane. The abundance is a manifestation of the fact that human beings exchange the equivalent of their weight in ATP each day. Despite the fact that the mitochondria are the sites of ATP synthesis, both the cytoplasm and the nucleus have more ATP than the mitochondria, a testament to the efficiency of transport.

The 30-kDa translocase contains a single nucleotide-binding site that alternately faces the matrix and the cytoplasmic sides of the membrane (**Figure 21.14**). The key to transport is that ATP has one more negative charge than that of ADP. Thus, in an actively respiring mitochondrion with a positive membrane potential, ATP transport out of the mitochondrial matrix and ADP transport into the matrix are favored. This ATP–ADP exchange is not without

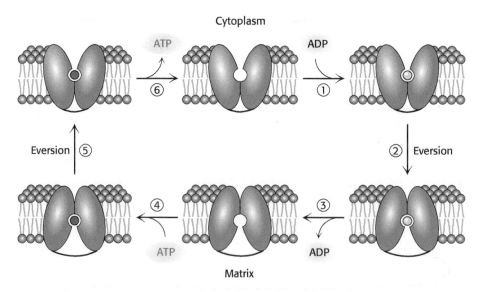

Figure 21.14 The mechanism of mitochondrial ATP-ADP translocase. The translocase catalyzes the coupled entry of ADP into the matrix and the exit of ATP from it. The binding of ADP (1) from the cytoplasm favors eversion of the transporter (2) to release ADP into the matrix (3). Subsequent binding of ATP from the matrix to the everted form (4) favors eversion back to the original conformation (5), releasing ATP into the cytoplasm (6).

significant energetic cost; about a quarter of the energy yield from electron transfer by the respiratory chain is consumed to regenerate the membrane potential that is tapped by this exchange process. The inhibition of ATP-ADP translocase leads to the subsequent inhibition of cellular respiration as well (p. 398).

Mitochondrial Transporters Allow Metabolite Exchange Between the Cytoplasm and Mitochondria

ATP-ADP translocase is only one of many mitochondrial transporters for ions and charged molecules (**Figure 21.15**). The *phosphate carrier,* which works in concert with ATP-ADP translocase, mediates the exchange of $H_2PO_4^-$ for OH^-. The combined action of these two transporters leads to the exchange of cytoplasmic ADP and P_i for matrix ATP at the cost of the influx of one H^+ (owing to the transport of one OH^- out of the matrix, which lowers the proton gradient by binding an H^+ to form water). These two transporters, which provide ATP synthase with its substrates, are associated with the synthase to form a large complex called the *ATP synthasome.*

Other carriers also are present in the inner mitochondrial membrane. The dicarboxylate carrier enables malate, succinate, and fumarate to be exported from the mitochondrial matrix in exchange for P_i. The tricarboxylate carrier exchanges citrate and H^+ for malate. Pyruvate in the cytoplasm enters the mitochondrial membrane in exchange for OH^- by means of the pyruvate carrier. In all, more than 40 such carriers are encoded in the human genome.

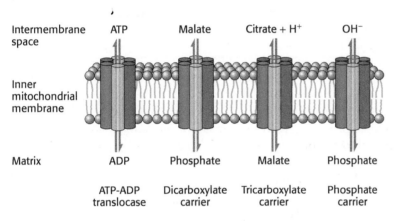

Figure 21.15 Mitochondrial transporters. Transporters (also called carriers) are transmembrane proteins that carry specific ions and charged metabolites across the inner mitochondrial membrane.

21.3 Cellular Respiration is Regulated by the Need for ATP

✓ 4 Identify the ultimate determinant of the rate of cellular respiration.

We have observed many times that most catabolic pathways are regulated in some fashion by the ATP concentration. Because ATP is the end product of *cellular respiration,* the ATP needs of the cell are the ultimate determinant of the rate of respiratory pathways and their components. Before we explore the nature of this regulation, let us calculate the ATP yield from the conversion of glucose to CO_2 and H_2O.

The Complete Oxidation of Glucose Yields About 30 Molecules of ATP

We can estimate how many molecules of ATP are formed when glucose is completely oxidized to CO_2. We say "estimate" because, in contrast with the ATP yield of glycolysis and the citric acid cycle (which yield 4 molecules of ATP per molecule of glucose and 1 molecule of ATP per molecule of pyruvate, respectively), the stoichiometries of proton pumping, ATP synthesis, and metabolite-transport processes need not be an integer or even have fixed values. As stated earlier, the best current estimates for the number of protons pumped out of the matrix by NADH-Q oxidoreductase, Q-cytochrome c oxidoreductase, and cytochrome c oxidase per electron pair are four, four, and two, respectively. The synthesis of a molecule of ATP is driven by the flow of about three protons through ATP synthase. An additional proton is consumed in transporting ATP from the matrix to the cytoplasm. Hence, about 2.5 molecules of cytoplasmic ATP are generated as a result of the flow of a pair of electrons from NADH to O_2. For electrons that enter at the level of Q-cytochrome c oxidoreductase, such as those from the oxidation of succinate or cytoplasmic NADH transferred by the glycerol-phosphate shuttle, the yield is about 1.5 molecules of ATP per electron pair. Hence, as tallied in Table 21.1, *about 30 molecules of ATP are formed when glucose is completely oxidized to CO₂*. Most of the ATP, 26 of 30 molecules formed, is generated by oxidative phosphorylation. Recall that the anaerobic metabolism of glucose yields only 2 molecules of ATP. The efficiency of cellular respiration is manifested in the observation that one of the effects of endurance exercise, a practice that calls for much ATP for an extended period of time, is to increase the number of mitochondria and blood vessels in muscle and therefore increase the extent of ATP generation by oxidative phosphorylation.

Table 21.1 ATP yield from the complete oxidation of glucose

Reaction sequence	ATP yield per glucose molecule
Glycolysis: Conversion of glucose into pyruvate (in the cytoplasm)	
Phosphorylation of glucose	−1
Phosphorylation of fructose 6-phosphate	−1
Dephosphorylation of 2 molecules of 1,3-BPG	+2
Dephosphorylation of 2 molecules of phosphoenolpyruvate	+2
2 molecules of NADH are formed in the oxidation of 2 molecules of glyceraldehyde 3-phosphate	
Conversion of pyruvate into acetyl CoA (inside mitochondria)	
2 molecules of NADH are formed	
Citric acid cycle (inside mitochondria)	
2 molecules of ATP are formed from 2 molecules of succinyl CoA	+2
6 molecules of NADH are formed in the oxidation of 2 molecules each of isocitrate, α-ketoglutarate, and malate	
2 molecules of FADH₂ are formed in the oxidation of 2 molecules of succinate	
Oxidative phosphorylation (inside mitochondria)	
2 molecules of NADH are formed in glycolysis; each yields 1.5 molecules of ATP (assuming transport of NADH by the glycerol 3-phosphate shuttle)	+3
2 molecules of NADH are formed in the oxidative decarboxylation of pyruvate; each yields 2.5 molecules of ATP	+5
2 molecules of FADH₂ are formed in the citric acid cycle; each yields 1.5 molecules of ATP	+3
6 molecules of NADH are formed in the citric acid cycle; each yields 2.5 molecules of ATP	+15
Net Yield Per Molecule of Glucose	+30

Source: The ATP yield of oxidative phosphorylation is based on values given in P. C. Hinkle, M. A. Kumar, A. Resetar, and D. L. Harris. *Biochemistry* 30:3576, 1991.
Note: The current value of 30 molecules of ATP per molecule of glucose supersedes the earlier one of 36 molecules of ATP. The stoichiometries of proton pumping, ATP synthesis, and metabolite transport should be regarded as estimates. About two more molecules of ATP are formed per molecule of glucose oxidized when the malate-aspartate shuttle rather than the glycerol 3-phosphate shuttle is used.

The Rate of Oxidative Phosphorylation Is Determined by the Need for ATP

How is the rate of the electron-transport chain controlled? Under most physiological conditions, electron transport is tightly coupled to phosphorylation. *Electrons do not usually flow through the electron-transport chain to O_2 unless ADP is simultaneously phosphorylated to ATP.* When ADP concentration rises, as would be the case in active muscle that is continuously consuming ATP, the rate of oxidative phosphorylation increases to meet the ATP needs of the cell. The regulation of the rate of oxidative phosphorylation by the ADP level is called *respiratory control* or *acceptor control.* Experiments on isolated mitochondria demonstrate the importance of ADP level (**Figure 21.16**). The rate of oxygen consumption by mitochondria increases markedly when ADP is added and then returns to its initial value when the added ADP has been converted into ATP.

As discussed in Chapter 19, the level of ADP likewise indirectly affects the rate of the citric acid cycle. At low concentrations of ADP, as in a resting muscle, NADH and $FADH_2$ produced by the citric acid cycle are not oxidized back to NAD^+ and FAD by the electron-transport chain. The citric acid cycle slows because there is less NAD^+ and FAD to feed the cycle. As the ADP level rises and oxidative phosphorylation speeds up, NADH and $FADH_2$ are oxidized, and the citric acid cycle becomes more active. *Electrons do not flow from fuel molecules to O_2 unless ATP needs to be synthesized.* We see here another example of the regulatory significance of the energy charge (**Figure 21.17**).

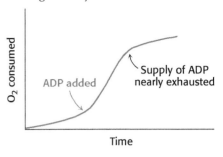

Figure 21.16 Respiratory control. Electrons are transferred to O_2 only if ADP is concomitantly phosphorylated to ATP.

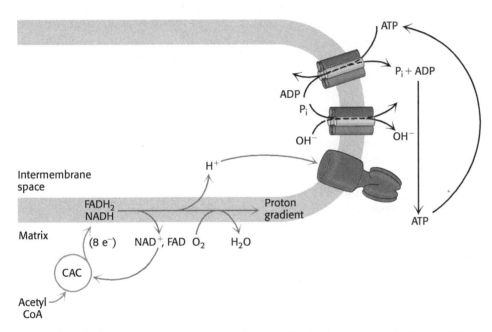

Figure 21.17 The energy charge regulates the use of fuels. The synthesis of ATP from ADP and P_i controls the flow of electrons from NADH and $FADH_2$ to oxygen. The availability of NAD^+ and FAD in turn control the rate of the citric acid cycle (CAC).

CLINICAL INSIGHT

ATP Synthase Can Be Regulated

Mitochondria contain an evolutionarily conserved protein, *inhibitory factor 1* (IF1), that specifically inhibits the potential hydrolytic activity of the F_0F_1 ATP synthase. What is the function of IF1? Consider a circumstance where tissues may be deprived of oxygen (ischemia). Without oxygen as the electron acceptor, the electron transport chain will be unable to generate the proton-motive force. The ATP in the mitochondria would be hydrolyzed by the synthase, working in reverse (problem 22). The role of IF1 is to prevent the wasteful hydrolysis of ATP by inhibiting the hydrolytic activity of the synthase.

IF1 is over-expressed in many types of cancer. This over-expression plays a role in the induction of the Warburg effect, the switch from oxidative phosphorylation to aerobic glycolysis as the principle means for ATP synthesis (p. 304).

✹ BIOLOGICAL INSIGHT

Regulated Uncoupling Leads to the Generation of Heat

Some organisms possess the ability to uncouple oxidative phosphorylation from ATP synthesis to generate heat. Such uncoupling is a means to maintain body temperature in hibernating animals, in some newborn animals (including human beings), and in mammals adapted to cold. In animals, *brown fat* or *brown adipose tissue* (BAT) is specialized tissue for this process of *nonshivering thermogenesis*. In contrast, *white adipose tissue* (WAT), which constitutes the bulk of adipose tissue, plays no role in thermogenesis but serves as an energy source.

Brown adipose tissue is very rich in mitochondria, often called *brown-fat mitochondria*. The inner mitochondrial membrane of these mitochondria contains a large amount of *uncoupling protein 1* (UCP-1), also called *thermogenin*. UCP-1 transports protons from the intermembrane space to the matrix with the assistance of fatty acids. In essence, *UCP-1 generates heat by short-circuiting the mitochondrial proton battery*. The energy of the proton gradient, normally captured as ATP, is released as heat as the protons flow through UCP-1 to the mitochondrial matrix (**Figure 21.18**). This dissipative proton pathway is activated when the core body temperature begins to fall. In response to a temperature drop, α adrenergic hormones stimulate the release of free fatty acids from triacylglycerols stored in cytoplasmic lipid droplets (p. 193). Long chain fatty acids bind to the cytoplasmic face of UCP-1, and the carboxyl group binds a proton. This causes a structural change in UCP-1 so that the protonated carboxyl now faces the proton-poor environment of the matrix, and the proton is released. Proton release resets UCP-1 to the initial state.

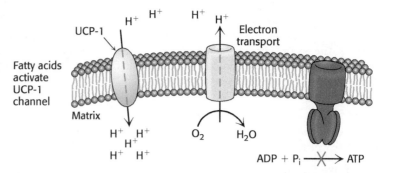

Figure 21.18 The action of an uncoupling protein. Uncoupling protein 1 (UCP-1) generates heat by permitting the influx of protons into the mitochondria without the synthesis of ATP.

We can witness the effects of a lack of nonshivering thermogenesis by examining pig behavior. Pigs are unusual mammals in that they have large litters and are the only ungulates (hoofed animals) that build nests for birth (Figure 21.19). These behavioral characteristics appear to be the result of a biochemical deficiency. Pigs lack UCP-1 and, hence, brown fat. Piglets must rely on other means of thermogenesis. Nesting, large litter size, and shivering are means by which pigs compensate for a lack of brown fat.

Until recently, adult humans were believed to lack brown fat tissue. However, new studies have established that adults, women especially, have brown adipose tissue on the neck and upper chest regions that is activated by cold (Figure 21.20). Obesity leads to a decrease in brown adipose tissue.

Figure 21.19 Nesting piglets. A wild boar, a member of the pig family, is shown with her nesting piglets. [Duncan Usher/Alamy Images.]

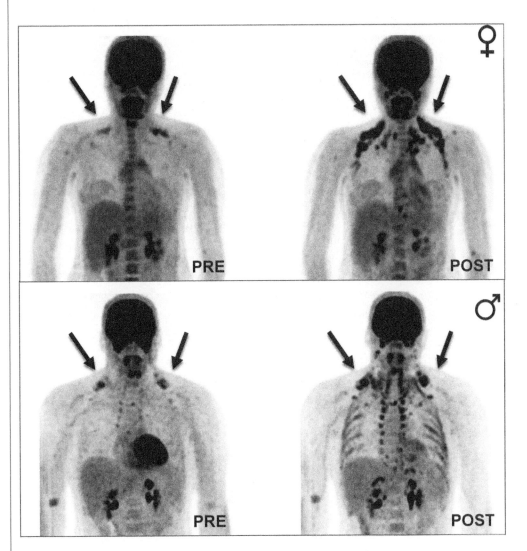

Figure 21.20 Brown adipose tissue is revealed on exposure to cold. The results of PET–CT (positron emission and computerized tomography) scanning show the uptake and distribution of ^{18}F-fluorodeoxyglucose (^{18}F-FDG), a nonmetabolizable glucose analog, in adipose tissue. The patterns of ^{18}F-FDG uptake in the same subject are dramatically different under thermoneutral conditions (left) and after exposure to cold (right). [Republished with permission of American Society for Clinical Investigation, from *J. Clin. Invest.* 123(8):3395-3403, 2013. doi:10.1172/JCI68993. Permission conveyed through Copyright Clearance Center, Inc.]

CLINICAL INSIGHT

Oxidative Phosphorylation Can Be Inhibited at Many Stages

Many potent and lethal poisons exert their effects by inhibiting oxidative phosphorylation at one of a number of different locations.

Inhibition of the electron-transport chain   The four complexes of the electron-transport chain can be inhibited by different compounds, blocking electron transfer downstream and thereby shutting down oxidative phosphorylation. NADH-Q oxidoreductase (Complex I) is inhibited by *rotenone*, which is used as a fish and insect poison, and *amytal*, a barbiturate sedative. Inhibitors of Complex I prevent the utilization of NADH as a substrate (Figure 21.21). Rotenone exposure, along with a genetic predisposition, has been implicated in the development of Parkinson disease, a condition characterized by resting tremor, slowness of movement, inability to initiate movement, rigidity, and postural instability. Inhibition of Complex I does not impair electron flow from $FADH_2$, because these electrons enter through QH_2, beyond the block. Q-cytochrome *c* oxidoreductase (Complex III) is inhibited by *antimycin A*, an antibiotic isolated from *Streptomyces* that is used as a fish poison. Furthermore, electron flow in cytochrome *c* oxidase (Complex IV) can be blocked by *cyanide* (CN^-), *azide* (N_3^-), and *carbon monoxide* (CO). Cyanide and azide react with the ferric form (Fe^{3+}) of heme a_3, whereas carbon monoxide inhibits the ferrous form (Fe^{2+}). Inhibition of the electron-transport chain also inhibits ATP synthesis because the proton-motive force can no longer be generated.

Inhibition of ATP synthase   *Oligomycin*, an antibiotic used as an antifungal agent, and dicyclohexylcarbodiimide (DCC), used in peptide synthesis in the laboratory, prevent the influx of protons through ATP synthase by binding to the carboxylate group of the **c** subunits required for proton binding. Modification of only one **c** subunit by DCC is sufficient to inhibit the rotation of the entire **c** ring and hence ATP synthesis. If actively respiring mitochondria

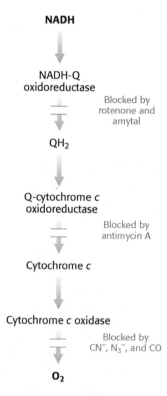

Figure 21.21 Sites of action of some inhibitors of electron transport.

are exposed to an inhibitor of ATP synthase, the electron-transport chain ceases to operate. This observation clearly illustrates that electron transport and ATP synthesis are normally tightly coupled.

Uncoupling electron transport from ATP synthesis The tight coupling of electron transport and phosphorylation in mitochondria can be uncoupled by *2,4-dinitrophenol* (DNP) and certain other acidic aromatic compounds. These substances carry protons across the inner mitochondrial membrane, down their concentration gradients. In the presence of these uncouplers, electron transport from NADH to O_2 proceeds in a normal fashion, but ATP is not formed by mitochondrial ATP synthase, because the proton-motive force across the inner mitochondrial membrane is continuously dissipated. This loss of respiratory control leads to increased oxygen consumption and the oxidation of NADH. Indeed, in the accidental ingestion of uncouplers, large amounts of metabolic fuels are consumed, but no energy is captured as ATP. Rather, energy is released as heat. DNP is the active ingredient in some herbicides and fungicides. Remarkably, some people consume DNP as a weight-loss drug, despite the fact that the U.S. Food and Drug Administration (FDA) banned its use in 1938. There are also reports that Soviet soldiers were given DNP to keep them warm during the long Russian winters. Chemical uncouplers are nonphysiological, unregulated counterparts of uncoupling proteins.

Drugs are being sought that would function as mild uncouplers, uncouplers not as potentially-lethal as DNP, for use in treatment of obesity and related pathologies. Xanthohumol, a prenylated chalcone found in hops and beer, shows promise in this regard. A chalcone is an aromatic ketone. Xanthohumol also scavenges free radicals and is used for treatment of certain types of cancers.

Inhibition of ATP export ATP-ADP translocase is specifically inhibited by very low concentrations of *atractyloside* (a plant glycoside) or *bongkrekic acid* (an antibiotic found in fermented coconut contaminated by the bacterium *Burkholderia gladioli*). Atractyloside binds to the translocase when its nucleotide site faces the cytoplasm, whereas bongkrekic acid binds when this site faces the mitochondrial matrix. Oxidative phosphorylation stops soon after either inhibitor is added, showing that ATP-ADP translocase is essential for maintaining adequate amounts of ADP to accept the energy associated with the proton-motive force.

Xanthohumol

CLINICAL INSIGHT

Mitochondrial Diseases Are Being Discovered in Increasing Numbers

The number of diseases that can be attributed to mitochondrial mutations is steadily growing in step with our growing understanding of the biochemistry and genetics of mitochondria. Mitochondrial diseases are estimated to affect from 10 to 15 per 100,000 people, roughly equivalent to the prevalence of the muscular dystrophies. The first mitochondrial disease to be understood was *Leber hereditary optic neuropathy* (LHON), a form of blindness that strikes in midlife as a result of mutations in Complex I. Some of these mutations impair NADH utilization, whereas others block electron transfer to Q. Mutations in Complex I are the most frequent cause of mitochondrial diseases. The accumulation of mutations in mitochondrial genes in a span of several decades may contribute to aging, degenerative disorders, and cancer.

A human egg harbors several hundred thousand molecules of mitochondrial DNA, whereas a sperm contributes only a few hundred and thus has little effect on the mitochondrial genotype. Because the maternally-inherited mitochondria are present in large numbers and not all of the mitochondria may be affected, the pathologies of mitochondrial mutants can be quite complex. Even within a single family carrying an identical mutation, chance fluctuations in the percentage of mitochondria with the mutation lead to large variations in the nature and severity of the symptoms of the pathological condition as well as the time of onset. As the percentage of defective mitochondria increases, energy-generating capacity diminishes until, at some threshold, the cell can no longer function properly. Defects in cellular respiration are doubly dangerous. Not only does energy transduction decrease, but also the likelihood that reactive oxygen species will be generated increases. Organs that are highly dependent on oxidative phosphorylation, such as the nervous system, the retina and the heart, are most vulnerable to mutations in mitochondrial DNA.

Power Transmission by Proton Gradients Is a Central Motif of Bioenergetics

The main concept presented in this chapter is that mitochondrial electron transfer and ATP synthesis are linked by a transmembrane proton gradient. ATP synthesis in bacteria and chloroplasts also is driven by proton gradients. In fact, proton gradients power a variety of energy-requiring processes such as the active transport of calcium ions by mitochondria, the entry of some amino acids and sugars into bacteria, the rotation of bacterial flagella, and the transfer of electrons from $NADP^+$ to NADPH. As we have already seen, proton gradients can also be used to generate heat. It is evident that *proton gradients are a central interconvertible currency of free energy in biological systems* (**Figure 21.22**). Peter Mitchell noted that the proton-motive force is a marvelously simple and effective store of free energy because it requires only a thin, closed lipid membrane between two aqueous phases.

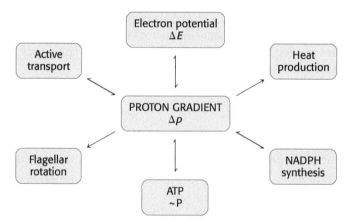

Figure 21.22 The proton gradient is an interconvertible form of free energy.

SUMMARY

21.1 A Proton Gradient Powers the Synthesis of ATP

The flow of electrons through Complexes I, III, and IV of the electron-transport chain leads to the transfer of protons from the matrix side to the cytoplasmic side of the inner mitochondrial membrane. A proton-motive force consisting of a pH gradient (matrix side basic) and a membrane potential (matrix side negative) is generated. The flow of protons back to the matrix side through ATP synthase drives ATP synthesis. The enzyme complex is a molecular motor made of two operational units: a rotating component and a stationary component. The rotation of the γ subunit induces structural changes in the β subunit that result in the synthesis and release of ATP from the enzyme. Proton influx provides the force for the rotation.

The flow of two electrons through NADH-Q oxidoreductase, Q-cytochrome c oxidoreductase, and cytochrome c oxidase generates a gradient sufficient to synthesize 1, 1, and 0.5 molecule of ATP, respectively. Hence, 2.5 molecules of ATP are formed per molecule of NADH oxidized in the mitochondrial matrix, whereas only 1.5 molecules of ATP are made per molecule of $FADH_2$ oxidized, because its electrons enter the chain at QH_2, after the first proton-pumping site.

21.2 Shuttles Allow Movement Across Mitochondrial Membranes

Mitochondria employ a host of transporters, or carriers, to move molecules across the inner mitochondrial membrane. The electrons of cytoplasmic NADH are transferred into the mitochondria by the glycerol 3-phosphate shuttle to form $FADH_2$ from FAD or by the malate–aspartate shuttle to form mitochondrial NADH. The entry of ADP into the mitochondrial matrix is coupled to the exit of ATP by ATP-ADP translocase, a transporter driven by membrane potential.

21.3 Cellular Respiration Is Regulated by the Need for ATP

About 30 molecules of ATP are generated when a molecule of glucose is completely oxidized to CO_2 and H_2O. Electron transport is normally tightly coupled to phosphorylation. NADH and $FADH_2$ are oxidized only if ADP is simultaneously phosphorylated to ATP, a form of regulation called acceptor or respiratory control. Proteins have been identified that uncouple electron transport and ATP synthesis for the generation of heat. Uncouplers such as 2,4-dinitrophenol also can disrupt this coupling; they dissipate the proton gradient by carrying protons across the inner mitochondrial membrane.

KEY TERMS

ATP synthase (F_1F_0 ATP synthase, Complex V) (p. 384)
proton-motive force (p. 384)
electron-transport chain (p. 385)

glycerol 3-phosphate shuttle (p. 391)
malate–aspartate shuttle (p. 391)
ATP-ADP translocase (adenine nucleotide translocase, ANT) (p. 392)

cellular respiration (p. 393)
respiratory (acceptor) control (p. 395)
uncoupling protein 1 (UCP-1) (p. 396)

? Answers to QUICK QUIZZES

1. Recall from the discussion of enzyme-catalyzed reactions that the direction of a reaction is determined by the ΔG difference between substrate and products. An enzyme speeds up the rate of both the forward and the backward reactions. The hydrolysis of ATP is exergonic, and so ATP synthase will enhance the hydrolytic reaction.

2. The number of **c** subunits is significant because it determines the number of protons that must be transported to generate a molecule of ATP. ATP synthase must rotate 360 degrees to synthesize three molecules of ATP; so, the more **c** subunits there are, the more protons are required to rotate the F_1 units 360 degrees.

1. *Reclaim resources.* Human beings have only about 250 g of ATP, but even a couch potato needs about 83 kg of ATP to open that bag of chips and use the remote. How is this discrepancy between requirements and resources reconciled?

2. *Like Barbie and Ken.* Match each term with its description.

(a)	ATP synthase _____	1.	Cytoplasmic NADH to mitochondrial FADH$_2$
(b)	Proton-motive force _____	2.	Results in heat instead of ATP
(c)	Electron-transport chain _____	3.	Catalytic subunit
(d)	Glycerol 3-phosphate shuttle _____	4.	Converts the proton-motive force into ATP
(e)	Malate–aspartate shuttle _____	5.	Proton channel
(f)	Respiratory (acceptor) control _____	6.	Composed of a chemical gradient and a charge gradient
(g)	Uncoupling protein _____	7.	A proton merry-go-round
(h)	F$_1$ subunit _____	8.	Generates the proton gradient
(i)	F$_0$ subunit _____	9.	ADP controls the rate of respiration
(j)	c ring _____	10.	Cytoplasmic NADH to mitochondrial NADH

3. *Energy harvest.* What is the yield of ATP when each of the following substrates is completely oxidized to CO$_2$ by a mammalian cell homogenate (p. 71)? Assume that glycolysis, the citric acid cycle, and oxidative phosphorylation are fully active.

(a) Pyruvate
(b) Lactate
(c) Fructose 1,6-bisphosphate
(d) Phosphoenolpyruvate
(e) Galactose
(f) Dihydroxyacetone phosphate

4. *Potent poisons.* What is the effect of each of the following inhibitors on electron transport and ATP formation by the respiratory chain?

(a) Azide
(b) Atractyloside
(c) Rotenone
(d) DNP
(e) Carbon monoxide
(f) Antimycin A

5. *A question of coupling.* What is the mechanistic basis for the observation that the inhibitors of ATP synthase also lead to an inhibition of the electron-transport chain? ✓ 3

6. *O$_2$ consumption.* Oxidative phosphorylation in mitochondria is often monitored by measuring oxygen consumption. When oxidative phosphorylation is proceeding rapidly, the mitochondria will rapidly consume oxygen. If there is little oxidative phosphorylation, only small amounts of oxygen will be used. You are given a suspension of isolated mitochondria and directed to add the following compounds in the order from *a* to *h*. With the addition of each compound, all of the previously added compounds remain present. Predict the effect of each addition on oxygen consumption by the isolated mitochondria.

(a) Glucose
(b) ADP + P$_i$
(c) Citrate
(d) Oligomycin
(e) Succinate
(f) 2,4-Dinitrophenol
(g) Rotenone
(h) Cyanide

7. *Runaway mitochondria 1.* The number of molecules of inorganic phosphate incorporated into organic form per atom of oxygen consumed, termed the *P:O ratio*, was frequently used as an index of oxidative phosphorylation. Suppose that the mitochondria of a patient oxidize NADH irrespective of whether ADP is present. The P:O ratio for oxidative phosphorylation by these mitochondria is less than normal. Predict the likely symptoms of this disorder.

8. *An essential residue.* The conduction of protons by the F$_0$ unit of ATP synthase is blocked by the modification of a single side chain by dicyclohexylcarbodiimide, which reacts readily with carboxyl groups. What are the most likely targets of action of this reagent? How might you use site-specific mutagenesis to determine whether this residue is essential for proton conduction? ✓ 3

9. *Runaway mitochondria 2.* Years ago, it was suggested that uncouplers would make wonderful diet drugs. Explain why this idea was proposed and why it was rejected. Why might the producers of antiperspirants be supportive of the idea? ✓ 3

10. *Coupled processes.* If actively respiring mitochondria are exposed to an inhibitor of ATP synthase, the electron-transport chain ceases to operate. Why?

11. *Gone with the flow.* What is the actual role of protons in the synthesis of ATP by F$_0$F$_1$ ATP synthase? ✓ 3 ✓ 4

12. *Opposites attract.* An arginine residue (Arg 210) in the a subunit of *E. coli* ATP synthase is near the aspartate residue (Asp 61) in the matrix-side proton channel. How might Arg 210 assist proton flow? ✓ 3

13. *Variable c subunits.* Recall that the number of c subunits in the c ring appears to range between 8 and 15. This

number is significant because it determines the number of protons that must be transported to generate a molecule of ATP. Each 360-degree rotation of the γ subunit leads to the synthesis and release of three molecules of ATP. Thus, if there are 10 **c** subunits in the ring (as was observed in a crystal structure of yeast mitochondrial ATP synthase), each ATP generated requires the transport of 10/3 = 3.33 protons. How many protons are required to form ATP if the ring has 8 **c** subunits? 12? 15? ✓ 3

14. *To each according to its needs.* It has been noted that the mitochondria of muscle cells often have more cristae than the mitochondria of liver cells. Provide an explanation for this observation. ✓ 4

15. *Everything is connected.* If actively respiring mitochondria are exposed to an inhibitor of ATP-ADP translocase, the electron-transport chain ceases to operate. Why? ✓ 4

16. *Exaggerating the difference.* Why must ATP-ADP translocase use Mg^{2+}-free forms of ATP and ADP? ✓ 4

17. *A Brownian ratchet wrench.* What causes the **c** subunits of ATP synthase to rotate? What determines the direction of rotation? ✓ 3

18. *Multiple uses.* Give an example of the use of the proton-motive force in ways other than for the synthesis of ATP.

19. *Connections.* How does the inhibition of ATP-ADP translocase affect the citric acid cycle? Glycolysis? ✓ 4

20. *Respiratory control.* The rate of oxygen consumption by mitochondria increases markedly when ADP is added and then decreases to its initial value when the added ADP has been converted into ATP (Figure 21.16). Why does the rate decrease? ✓ 4

21. *The same, but different.* Why is the electroneutral exchange of $H_2PO_4^-$ for OH^- indistinguishable from the electroneutral symport of $H_2PO_4^-$ and H^+?

22. *Counterintuitive.* Under some conditions, mitochondrial ATP synthase has been observed to run in reverse. How would that situation affect the proton-motive force?

23. *Not hearsay, but real evidence.* Describe the evidence supporting the chemiosmotic hypothesis. ✓ 3

24. *Imposing a gradient.* Mitoplasts are mitochondria that lack the outer membrane but are still capable of oxidative phosphorylation. Suppose that you were to soak mitoplasts in a pH 7 buffer for several hours. Then, you rapidly isolated the mitoplasts and mixed them in a pH 4 buffer containing ADP and P_i. Would ATP synthesis take place? Explain. ✓ 3

25. *With sympathy.* Predict the effect on ATP synthesis if the **b** and δ subunits of the ATPase were absent.

Chapter Integration Problems

26. *Obeying the laws of thermodynamics.* Why will isolated F_1 subunits display ATPase activity but not ATP synthase

activity? How can the enzyme then function as ATP synthase in mitochondria?

27. *Etiology? What does that mean?* What does the fact that rotenone increases the susceptibility to Parkinson disease indicate about the etiology of this disease?

28. *The right location.* Some cytoplasmic kinases, enzymes that phosphorylate substrates at the expense of ATP, bind to voltage-dependent anion channels (VDACs, p. 365). What might the advantage of this binding be?

29. *No exchange.* Mice that are completely lacking ATP-ADP translocase (ANT^-/ANT^-) can be made by the knockout technique. Remarkably, these mice are viable but have the following pathological conditions: (1) high serum levels of lactate, alanine, and succinate; (2) little electron transport; and (3) a 6- to 8-fold increase in the level of mitochondrial H_2O_2 compared with that in normal mice. Provide a possible biochemical explanation for each of these conditions. ✓ 4

30. *Alternative routes.* The most common metabolic sign of mitochondrial disorders is lactic acidosis. Why? ✓ 4

Data Interpretation and Challenge Problem

31. *Mitochondrial disease.* A mutation in a mitochondrial gene encoding a component of ATP synthase has been identified. People who have this mutation suffer from muscle weakness, ataxia (uncoordinated movement), and retinitis pigmentosa. A tissue biopsy was performed on each of three patients having this mutation, and submitochondrial particles were isolated that were capable of succinate-sustained ATP synthesis. First, the activity of the ATP synthase was measured on the addition of succinate and the following results were obtained.

ATP synthase activity (nmol of ATP formed min^{-1} mg^{-1})	
Controls	3.0
Patient 1	0.25
Patient 2	0.11
Patient 3	0.17

(a) What was the purpose of the addition of succinate?

(b) What is the effect of the mutation on succinate-coupled ATP synthesis?

Next, the ATPase activity of the enzyme was measured by incubating the submitochondrial particles with ATP in the absence of succinate.

ATP hydrolysis (nmol of ATP hydrolyzed min^{-1} mg^{-1})	
Controls	33
Patient 1	30
Patient 2	25
Patient 3	31

(c) Why was succinate omitted from the reaction?

(d) What is the effect of the mutation on ATP hydrolysis?

(e) What do these results, in conjunction with those obtained in the first experiment, tell you about the nature of the mutation?

Challenge Problems

32. *P:O ratios.* The P:O ratio can be used to monitor oxidative phosphorylation (problem 7).

(a) What is the relation of the P:O ratio to the ratio of the number of protons translocated per electron pair ($H^+/2e^-$) and the ratio of the number of protons needed to synthesize ATP and transport it to the cytoplasm (P/H^+)?

(b) What are the P:O ratios for electrons donated by matrix NADH and by succinate?

33. *Cyanide antidote.* The immediate administration of nitrite is a highly effective treatment for cyanide poisoning. What is the basis for the action of this antidote? (Hint: Nitrite oxidizes ferrohemoglobin to ferrihemoglobin.)

34. *Currency exchange.* For a proton-motive force of 0.2 V (matrix negative), what is the maximum [ATP]/[ADP][P_i] ratio compatible with ATP synthesis? Calculate this ratio three times, assuming that the number of protons translocated per ATP formed is two, three, and four and that the temperature is 25°C. ✓ 3

35. *Identifying the inhibition.* You are asked to determine whether a chemical is an electron-transport-chain inhibitor or an inhibitor of ATP synthase. Design an experiment to make this determination.

Selected Readings for this chapter can be found online at www.whfreeman.com/tymoczko3e.

The Light Reactions of Photosynthesis and the Calvin Cycle

CHAPTER 22
The Light Reactions

CHAPTER 23
The Calvin Cycle

In Chapter 2, we discussed Brownian motion, the movement of molecules due to background thermal energy. The speed of Brownian motion is proportional to the temperature, a measure of the background thermal energy. The temperature in the sun is approximately 15 million degrees Celsius. At this temperature, Brownian motion is so powerful that the collision of two hydrogen atoms results in the formation of a helium atom and the release of energy:

$$\text{Hydrogen} + \text{hydrogen} \longrightarrow \text{helium} + \text{energy}$$

An almost unfathomable amount of energy is released from the sun at a given instant. The majority of this energy will simply pass into outer space, forever useless, a dramatic example of the inevitability of entropy. However, Earth is immersed in this celestial glow as it makes its yearly journey about the sun. The sun's energy is in the form of broad-spectrum electromagnetic radiation, of which a particular segment is especially important to life on Earth—the region between 380 nm and 750 nm—that is, visible light.

On our planet are organisms capable of collecting the electromagnetic energy of the visible spectrum and converting it into chemical energy. Green plants are the most obvious of these organisms, though 60% of this conversion is carried out by algae and bacteria. This transformation is perhaps the most important of all of the energy transformations that we will see in our study of biochemistry; without it, life on our planet as we know it simply could not exist.

405

If plants stopped converting light energy into chemical energy, all higher forms of life would be extinct in about 25 years.

The process of converting light energy into chemical energy is called *photosynthesis,* which uses the energy of the sun to convert carbon dioxide and water into carbohydrates and oxygen. These carbohydrates not only provide the energy to run the biological world, but also provide the carbon molecules to make a wide array of biomolecules. Photosynthetic organisms are called *autotrophs* (literally, "self-feeders") because they can synthesize glucose from carbon dioxide and water, by using sunlight as an energy source, and then recover some of this energy from the synthesized glucose through the glycolytic pathway and aerobic metabolism. Organisms that obtain energy from chemical fuels only are called *heterotrophs,* which ultimately depend on autotrophs for their fuel.

Photosynthesis is composed of two parts: the light reactions and the dark reactions. The *light reactions* transform light energy into two forms of biochemical energy with which we are already familiar: reducing power and ATP. The *dark reactions* then use the products of the light reactions to drive the reduction of CO_2 and its conversion into glucose and other sugars. The dark reactions are also called the *Calvin cycle* or *light-independent reactions.*

Just as cellular respiration takes place in mitochondria, photosynthesis takes place in specialized organelles called *chloroplasts.* We will examine the chloroplast and then proceed to the light reactions of photosynthesis. Many of the biochemical principles that apply to oxidative phosphorylation also apply to photosynthesis. Finally, we will examine the Calvin cycle to learn how the products of the light reactions are used to synthesize glucose.

✓ By the end of this section, you should be able to:

✓ 1 Describe the light reactions.

✓ 2 Identify the key products of the light reactions.

✓ 3 Explain how redox balance is maintained during the light reactions.

✓ 4 Explain the function of the Calvin cycle.

✓ 5 Describe how the light reactions and the Calvin cycle are coordinated.

CHAPTER 22

The Light Reactions

A barley field. Throughout the day, the barley converts sunlight into chemical energy with the biochemical process of photosynthesis. Because barley is a major animal feed crop, the sun's energy will soon make its way up the food chain to human beings. [Brian Jannsen/Alamy.]

In Sections 8 and 9, we learned that ATP is generated through cellular respiration, the oxidation of the carbon atoms of carbohydrates, notably glucose, to CO_2 with the reduction of O_2 to water. In photosynthesis, this process must be reversed—reducing CO_2 and oxidizing H_2O to synthesize glucose:

$$\text{Energy} + 6\,H_2O + 6\,CO_2 \xrightarrow{\text{Photosynthesis}} C_6H_{12}O_6 + 6\,O_2$$

$$C_6H_{12}O_6 + 6\,O_2 \xrightarrow[\text{respiration}]{\text{Cellular}} 6\,CO_2 + 6\,H_2O + \text{energy}$$

Although the processes of respiration and photosynthesis are chemically opposite each other, the biochemical principles governing the two processes are nearly identical. The key to both processes is the generation of high-energy electrons. The citric acid cycle oxidizes carbon fuels to CO_2 to generate high-energy electrons. The flow of these high-energy electrons down an electron-transport chain generates a proton-motive force. This proton-motive force is then transduced by ATP synthase to form ATP. To synthesize glucose from CO_2, high-energy electrons are required for two purposes: (1) to provide reducing power to reduce

CO_2 and (2) to generate ATP to power the reduction. How can high-energy electrons be generated without using a chemical fuel? *Photosynthesis uses energy from light to boost electrons from a low-energy state to a high-energy state.* In the high-energy, unstable state, nearby molecules can abscond with the excited electrons. These electrons are then used directly to produce biosynthetic reducing power in the form of NADPH, and they are used indirectly, through an electron-transport chain that generates a proton-motive force across a membrane, to drive the synthesis of ATP. The reactions that are powered by sunlight are called the *light reactions* (**Figure 22.1**).

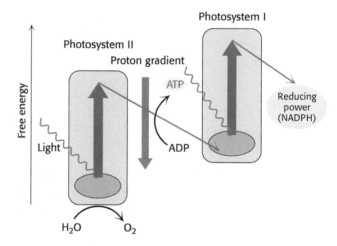

Figure 22.1 The light reactions of photosynthesis. Photosynthesis in plants consists of two photosystems. The two systems work in concert so that light is absorbed, and the energy is used to drive electrons from water to generate NADPH and to drive protons across a membrane. These protons return through ATP synthase to make ATP.

We begin with chloroplasts, the organelles in which photosynthesis takes place. We then examine the light-absorbing molecules that initiate the capture of light as high-energy electrons. Finally, we see how the high-energy electrons are used to form reducing power and ATP, the energy sources for the synthesis of glucose.

22.1 Photosynthesis Takes Place in Chloroplasts

In Section 9, we learned that oxidative phosphorylation is compartmentalized into mitochondria. Likewise, photosynthesis is sequestered into organelles called *chloroplasts,* typically 5 mm long. Like a mitochondrion, a chloroplast has an outer membrane and an inner membrane, with an intervening intermembrane space (**Figure 22.2**). The inner membrane surrounds a space called the *stroma,* which contains the enzymes that use reducing power and ATP to convert CO_2 into sugar. The stroma is akin to the matrix of the mitochondrion, which is the location of carbon chemistry during respiration. In the stroma are membranous discs called *thylakoids,* which are stacked to form a *granum.* Grana are linked by regions of thylakoid membrane called *stroma lamellae.* The thylakoid membranes separate the thylakoid lumen from the stroma. Thus, chloroplasts have three different membranes (*outer, inner,* and *thylakoid membranes*) and three separate spaces (*intermembrane, stroma,* and *thylakoid lumen*). The thylakoid membrane and the inner membrane, like the inner mitochondrial membrane, are impermeable to most molecules and ions. The outer membrane of a chloroplast, like that of a mitochondrion, is highly permeable to small molecules and ions. Like the mitochondrial cristae, the thylakoid membranes are the sites of coupled oxidation–reduction reactions that generate the proton-motive force. Plant leaf cells contain between 1 and 100 chloroplasts, depending on the species, cell type, and growth conditions.

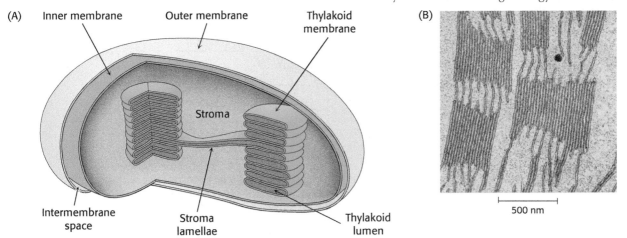

(A) Inner membrane   Outer membrane   Thylakoid membrane

Stroma

Intermembrane space   Stroma lamellae   Thylakoid lumen

(B)

500 nm

Figure 22.2 The structure of a chloroplast. (A) A diagram of a chloroplast. (B) An electron micrograph of a chloroplast from a spinach leaf shows the thylakoid membranes packed together to form grana. [(A) Information from S. L. Wolfe, *Biology of the Cell*, p. 130. © 1972 by Wadsworth Publishing Company, Inc. Adapted by permission of the publisher. (B) Courtesy of Dr. Kenneth Miller.]

BIOLOGICAL INSIGHT

Chloroplasts, Like Mitochondria, Arose from an Endosymbiotic Event

Chloroplasts contain their own DNA and the machinery for replicating and expressing it. However, chloroplasts are not autonomous: nuclear DNA encodes many proteins contained in chloroplasts. How did the intriguing relation between the cell and its chloroplasts develop? We now know that, in a manner analogous to the evolution of mitochondria (p. 365), chloroplasts are the result of endosymbiotic events in which a photosynthetic microorganism, most likely an ancestor of a cyanobacterium, was engulfed by a eukaryotic host.

The chloroplast genome is smaller than that of a cyanobacterium, but the two genomes have key features in common. Both are circular and have a single start site for DNA replication, and the genes of both are arranged in operons— sequences of functionally related genes under common control (Chapter 36). In the course of evolution, many of the genes of the chloroplast ancestor were transferred to the plant cell's nucleus or, in some cases, lost entirely, thus establishing a fully dependent relationship.

22.2 Photosynthesis Transforms Light Energy into Chemical Energy

The role of chloroplasts in plants is to capture light energy. The trapping of light energy—the conversion of light energy into chemical energy—is the key to photosynthesis and thus the key to life.

The first event in photosynthesis is the absorption of light by a photorecep-tor molecule. Photoreceptor molecules, capable of absorbing the energy of light of a specific wavelength, are also called pigments. What happens when a photo-receptor molecule absorbs a photon of light of the appropriate energy? The energy from the light excites an electron from its ground energy level to an excited energy level (**Figure 22.3**). For most compounds that absorb light, the excited electron simply returns to the ground state and the absorbed energy is converted into heat or light, as when a molecule fluoresces. In photoreceptor molecules, however, the excitation energy released when one electron returns to its ground state may be accepted by an electron in a neighboring molecule,

✓ 1 Describe the light reactions.

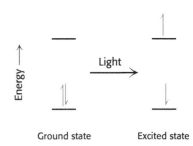

Ground state   Excited state

Figure 22.3 Light absorption. The absorption of light leads to the excitation of an electron from its ground state to a higher energy level. The horizontal lines represent different energy levels of the atoms in the pigment molecules.

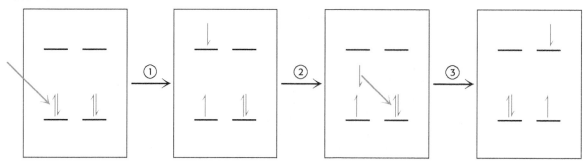

Figure 22.4 Resonance energy transfer. (1) An electron can accept energy from electromagnetic radiation of appropriate wavelength and jump to a higher energy state. (2) When the excited electron falls back to its lower energy state, the absorbed energy is released. (3) The released energy can be absorbed by an electron in a nearby molecule, and this electron jumps to a high-energy state. The horizontal lines represent different energy levels of the atoms in the pigment molecules.

which then jumps to a higher energy state. This process of moving energy from one electron to another is called *resonance energy transfer* (**Figure 22.4**). Resonance energy transfer allows energy to be passed from one molecule to another, which, as we will see later, is an important property for increasing the efficiency of photosynthesis (p. 411).

Alternatively, the excited electron itself may move to a nearby molecule that has a lower excited state, in a process called *electron transfer* (**Figure 22.5**). A positive charge forms on the initial molecule, owing to the loss of an electron, and a negative charge forms on the acceptor, owing to the gain of an electron. Hence, this process is referred to as *photoinduced charge separation*. The excited electron in its new molecule now has reducing power: it can reduce other molecules to store the energy originally obtained from light in chemical forms. The pair of electron carriers at which the charge separation takes place are called the *special pair* and are located at the *reaction center*.

DID YOU KNOW?

A photon is a discrete bundle, or quantum, of electromagnetic energy that travels at the speed of light (c). The energy of a photon (E) is directly proportional to the frequency (v) of its vibration or inversely proportional to its wavelength (λ). The proportionality constant (h) is Planck's constant.
$E = hv = hc/\lambda$

Figure 22.5 Photoinduced charge separation. If a suitable electron acceptor is nearby, an electron that has been moved to a high-energy level by light absorption can move from the excited molecule to the acceptor.

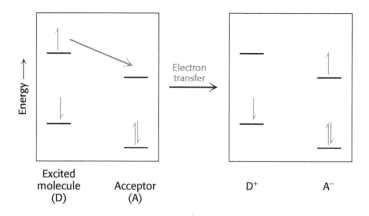

A pyrrole

Chlorophyll Is the Primary Light Acceptor in Most Photosynthetic Systems

The principal photoreceptor in the chloroplasts of most green plants is *chlorophyll*, a substituted tetrapyrrole (**Figure 22.6**). The four nitrogen atoms of the pyrroles are bound to a magnesium ion, just as a heme group is bound to iron (p. 150). A distinctive feature of chlorophyll is the presence of *phytol*, a highly hydrophobic 20-carbon alcohol, esterified to an acid side chain.

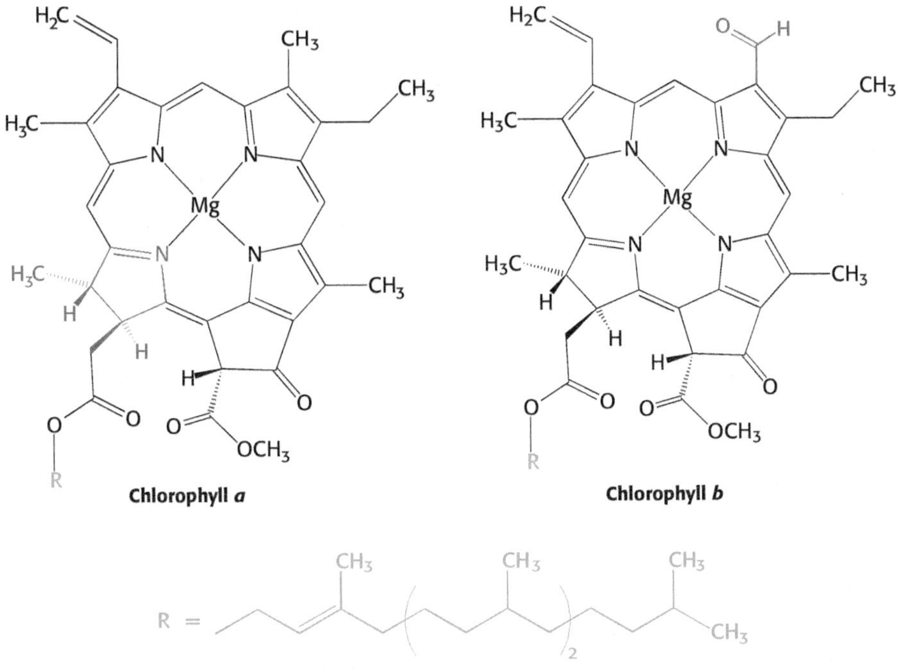

Chlorophyll a

Chlorophyll b

R =

Figure 22.6 Chlorophyll. Like heme, chlorophyll *a* is a cyclic tetrapyrrole (shown in red) with a magnesium ion at the center of the structure. Chlorophyll *b* has a formyl group in place of a methyl group in chlorophyll *a*. The hydrophobic phytol group (R) is shown in green.

Chlorophylls are very effective photoreceptors because they contain networks of alternating single and double bonds. Such compounds are called *polyenes* and display resonance structures. Because the electrons are not held tightly to a particular atom and thus can resonate within the larger ring structure, they can be readily excited by light of suitable energy.

There are two primary varieties of chlorophyll molecules, chlorophyll *a* (Chl$_a$) and chlorophyll *b* (Chl$_b$). Chlorophyll *a*, which is the principal chlorophyll in green plants, absorbs maximally at 420 nm and at 670 nm, leaving a large gap in the middle of the visible region. Chlorophyll *b* differs from chlorophyll *a* in having a formyl group in place of a methyl group. This small difference moves its two major absorption peaks toward the center of the visible region, where chlorophyll *a* does not absorb well (**Figure 22.7**).

The absorption of energy by chlorophyll in the reaction center, from electron transfer or directly from light, allows the chlorophyll molecule to donate an electron to a nearby molecule, its partner in the special pair, which results in photoinduced charge separation:

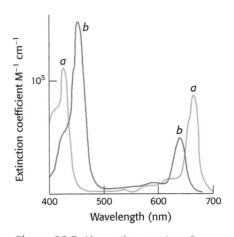

Figure 22.7 Absorption spectra of chlorophylls *a* and *b*. The extinction coefficient (*y* axis) is a measure of a molecule's ability to absorb light.

$$\text{Chl}_a + \text{acceptor} \xrightarrow[\text{absorption}]{\text{Light}} \text{Chl}_a{}^* + \text{acceptor} \xrightarrow[\text{transfer}]{\text{Electron}} \text{Chl}_a{}^+ + \text{acceptor}^-$$

Excited
chlorophyll

Separation of charge

The negatively charged acceptor now has a high electron-transfer potential. Thus, light energy is converted into chemical energy.

Light-Harvesting Complexes Enhance the Efficiency of Photosynthesis

A light-gathering system that relied only on the chlorophyll *a* molecules of the reaction center would be inefficient for two reasons. First, chlorophyll *a* molecules absorb only a part of the solar spectrum. Second, even in spectral regions in which chlorophyll *a* absorbs light, many photons would pass through without being absorbed, because of the relatively low density of chlorophyll *a* molecules in a reaction center. The absorption spectrum is expanded by the use of

DID YOU KNOW?

"Life has set itself the task to catch the light on the wing to earth and to store the most elusive of all powers in rigid form."

—Robert Julius Mayer, a German physician (1814–1878), one of the first to recognize the law of the conservation of energy

β-Carotene

Carotene absorbs blue and indigo light and thus looks orange. Breakdown products of the carotene of grass are responsible for the aroma of hay. β–Carotene is also a precursor for retinol, an important visual pigment in the eye.

antennae molecules: chlorophyll *a* molecules not in a reaction center, chlorophyll *b*, and other accessory pigments such as carotenoids. The *carotenoids* are polyenes that also absorb light between 400 and 500 nm. The accessory pigments are arranged in numerous *light-harvesting complexes* that completely surround the reaction center (**Figure 22.8**). *These pigments absorb light and deliver the energy to the reaction center by resonance energy transfer for conversion into chemical forms.* Accessory pigments, such as the carotenoids, are responsible for most of the yellow and red colors of fruits and flowers. They also provide the brilliance of fall colors, when the green chlorophyll molecules are degraded to reveal the carotenoids.

In addition to their role in transferring energy to reaction centers, the carotenoids and other accessory pigments serve a safeguarding function. Carotenoids suppress damaging photochemical reactions, particularly those including oxygen, that can be induced by bright sunlight. This protection may be especially important in the fall when the primary pigment chlorophyll is being degraded and thus not able to absorb light energy. Plants lacking carotenoids are quickly killed on exposure to light and oxygen.

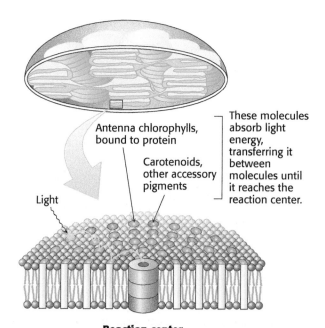

(A)

Figure 22.8 Energy transfer from accessory pigments to reaction centers. (A) Light energy absorbed by accessory chlorophyll molecules or other pigments can be transferred to reaction centers by resonance energy transfer (yellow arrows), where it drives photoinduced charge separation. (B) Accessory pigments increase the range of light absorption well beyond what is possible with chlorophyll *a* and *b* alone. [Data from D. L. Nelson and M. M. Cox, *Lehninger Principles of Biochemistry*, 5th ed. (W. H. Freeman and Company, 2008), (A) Fig. 19.52, (B) Fig. 19. 48.]

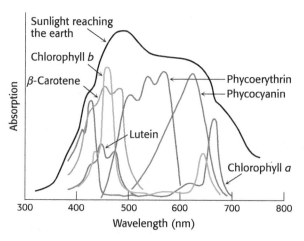

(B)

⬡ BIOLOGICAL INSIGHT

Chlorophyll in Potatoes Suggests the Presence of a Toxin

Chlorophyll synthesis is a warning sign when it comes to identifying poisonous potatoes. Light activates a noxious pathway in potatoes that leads to the synthesis of solanine, a toxic alkaloid. Plant alkaloids include such molecules as nicotine, caffeine, morphine, cocaine, and codeine.

Solanine

Solanine is toxic to animals because it inhibits acetylcholinesterase, an enzyme crucial for controlling the transmission of nerve impulses. Potato plants are thought to synthesize solanine to discourage insects from eating the potato. Light also causes potatoes to synthesize chlorophyll, which causes the tubers to turn green. Potatoes that are green have been exposed to light and are therefore probably also synthesizing solanine (**Figure 22.9**). For this reason, it is best not to eat green potatoes or potato chips with green edges.

Figure 22.9 Toxic potatoes. Potatoes that are exposed to light synthesize chlorophyll, resulting in greenish potatoes. Light also activates a pathway that results in the synthesis of solanine, a toxic alkaloid. Potato chips made from light-exposed potatoes have green edges. [Science Photo Library/Alamy.]

22.3 Two Photosystems Generate a Proton Gradient and NADPH

✓ **2** Identify the key products of the light reactions.

✓ **3** Explain how redox balance is maintained during the light reactions.

With an understanding of the principles of how photosynthetic organisms generate high-energy electrons, let us examine the biochemical systems that coordinate the electron capture and their use to generate reducing power and ATP, resources that will be used to power the synthesis of glucose from CO_2. Photosynthesis in green plants is mediated by two kinds of membrane-bound, light-sensitive complexes—*photosystem I* (PS I) and *photosystem II* (PS II), each with its own characteristic reaction center (**Figure 22.10**). Photosystem I responds to light with wavelengths shorter than 700 nm and is responsible for providing electrons to reduce $NADP^+$ to NADPH, a versatile reagent for driving biosynthetic processes requiring reducing power. Photosystem II responds to wavelengths shorter than 680 nm, sending electrons through a membrane-bound proton pump called *cytochrome b_6f* and then on to photosystem I to replace the electrons donated by PS I to $NADP^+$. The electrons in the reaction center of photosystem II are replaced when two molecules of H_2O are oxidized to generate a molecule of O_2. As we will soon see, *electrons flow from water through photosystem II, the cytochrome b_6f complex, and photosystem I and are finally accepted by $NADP^+$*. In the course of this flow, a

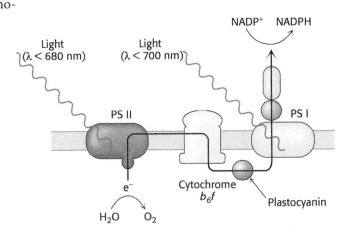

Figure 22.10 Two photosystems. The absorption of photons by two distinct photosystems (PS I and PS II) is required for complete electron flow from water to $NADP^+$.

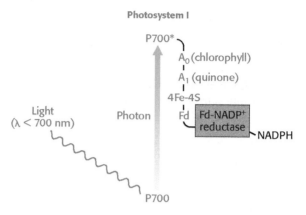

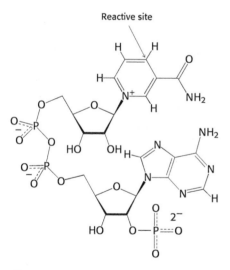

Figure 22.12 The structure of ferredoxin. In plants, ferredoxin contains a 2Fe-2S cluster. This protein accepts electrons from photosystem I and carries them to ferredoxin-NADP+ reductase. [Drawn from 1FXA.pdb.]

Figure 22.11 Photosystem I. Light absorption induces electron transfer from P700 down an electron-transport chain that includes a chlorophyll molecule, a quinone molecule, and three 4Fe-4S clusters to reach ferredoxin. The electron donating pigment is denoted as P700*.

proton gradient is established across the thylakoid membrane. This proton gradient is the driving force for ATP production.

Photosystem I Uses Light Energy to Generate Reduced Ferredoxin, a Powerful Reductant

The first stage of the light reactions is catalyzed by photosystem I (**Figure 22.11**). Photosystem I typically includes 14 polypeptide chains and multiple associated proteins and cofactors. The core of this system is a pair of similar subunits, PsaA (83 kDa) and PsaB (82 kDa), which bind 80 chlorophyll molecules as well as other redox factors. A special pair of chlorophyll a molecules lies at the center of the structure and absorbs wavelengths of light that are longer than those absorbed by chlorophyll a not associated with a photosystem. The special pair of the reaction center of photosystem I are called *P700*. When activated by light (designated P700*), the reaction center initiates photoinduced charge separation that generates the high-energy electrons. The electrons flow down an electron-transport chain through a chlorophyll called A_0 and a quinone called A_1 to a set of 4Fe-4S clusters. From there, the electrons are transferred to ferredoxin (Fd), a protein containing a 2Fe-2S cluster coordinated to four cysteine residues (**Figure 22.12**).

Although reduced ferredoxin is a strong reductant, it is not useful for driving many reactions, in part because ferredoxin carries only one available electron. *The currency of readily available biosynthetic reducing power in cells is NADPH,* which shuttles two electrons as a hydride ion. Indeed, NADPH functions exactly as NADH does (p. 268) and differs structurally only in that it contains a phosphoryl group on the 2′-hydroxyl group of one of the ribose units (**Figure 22.13**). This phosphoryl groups acts as a label identifying NADPH as a reductant for biosynthetic (anabolic) reactions, and so it is not oxidized by the respiratory chain as is NADH. This distinction holds true in all biochemical systems, including that of human beings.

How can reduced ferredoxin be used to drive the reduction of $NADP^+$ to NADPH? This reaction is catalyzed by *ferredoxin-NADP+ reductase*, a flavoprotein. The bound FAD component in this enzyme collects two electrons, one at a time, from two molecules of reduced ferredoxin as it proceeds from its oxidized form to its fully reduced form (**Figure 22.14**). The enzyme then transfers a hydride ion to $NADP^+$ to form NADPH. Note that this reaction takes place on the stromal side of the thylakoid membrane. The use of a stromal proton in the reduction of $NADP^+$ makes the stroma more basic than the thylakoid lumen and thus contributes to the formation of the proton-motive force across the thylakoid membrane that will be used to synthesize ATP. The first of the raw materials required to reduce CO_2—biosynthetic reducing power as NADPH—has been generated. However, photosystem I is now deficient in electrons. How can these electrons be replenished?

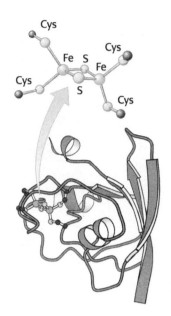

Figure 22.13 The structure of NADP+. Nicotinamide adenine dinucleotide phosphate ($NADP^+$) is a prominent carrier of high-energy electrons. $NADP^+$ binds a hydride ion (H^-) at the reactive site to form NADPH in a reaction entirely analogous to the reduction of NAD^+ (Figure 15.12).

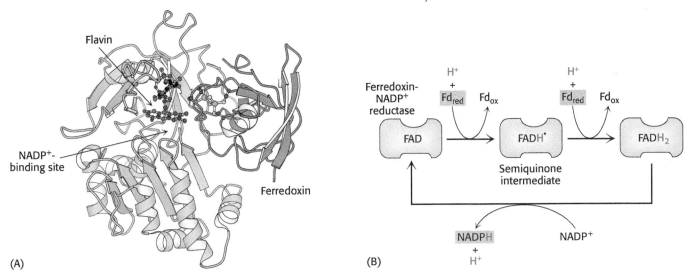

Figure 22.14 The structure and function of ferredoxin–NADP$^+$ reductase. (A) The structure of ferredoxin–NADP$^+$ reductase is shown in yellow. This enzyme accepts electrons, one at a time, from ferredoxin (shown in orange). (B) Ferredoxin–NADP$^+$ reductase first accepts two electrons and two protons from two molecules of reduced ferredoxin (Fd$_{red}$) to form FADH$_2$, which then transfers two electrons and a proton to NADP$^+$ to form NADPH. [(A) Drawn from 1EWY.pdb.]

Photosystem II Transfers Electrons to Photosystem I and Generates a Proton Gradient

Restocking the electrons of photosystem I requires another source of electrons. One of the roles of photosystem II is to provide electrons to photosystem I. Photosystem II catalyzes the light-driven transfer of electrons from water to photosystem I. In the process, protons are pumped into the thylakoid lumen to generate a proton-motive force. Photosystem II is an enormous assembly of more than 20 subunits. It is formed by D1 and D2, a pair of similar 32-kDa proteins that span the thylakoid membrane (**Figure 22.15**).

The photochemistry of photosystem II begins with the excitation of a special pair of chlorophyll molecules that are often called *P680*, generating P680* (**Figure 22.16**). P680* rapidly transfers electrons to a nearby pheophytin (designated Ph), a chlorophyll with two H$^+$ ions in place of the central Mg^{2+} ion. From there, the electrons are transferred first to a tightly bound plastoquinone at site Q$_A$ and then to a mobile plastoquinone at site Q$_B$. Plastoquinone is an electron carrier that closely resembles ubiquinone, a component of the mitochondrial electron-transport chain (p. 370). With the arrival of a second electron and the uptake of two protons, the mobile plastoquinone is reduced to plastoquinol, QH$_2$.

DID YOU KNOW?

Pheophytin is formed in the cooking of some green vegetables when the magnesium ion of chlorophyll is lost and replaced by protons. The absorption properties of pheophytin are different from those of the chlorophylls, often resulting in cooked vegetables that are a dull green compared with their uncooked counterparts.

Plastoquinone
(oxidized form, Q)

+2 electrons and 2 H$^+$

−2 electrons and 2 H$^+$

$(n = 6$ to $10)$

Plastoquinol
(reduced form, QH$_2$)

The site of quinone reduction in PS II is on the side of the stroma. Thus, the two protons that are taken up with the reduction of each molecule of plastoquinone come from the stroma, increasing the proton gradient. This gradient is further

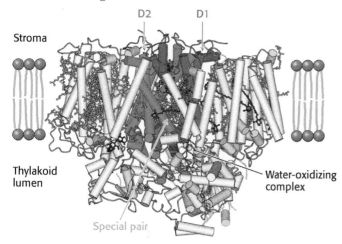

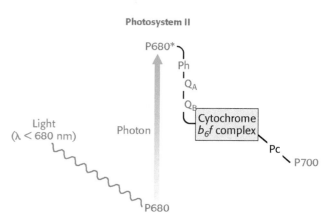

Figure 22.15 **The structure of photosystem II.** The D1 and D2 subunits are shown in red and blue, respectively, and the numerous bound chlorophyll molecules are shown in dark green. Notice that the special pair and the water-oxidizing complex (the site of oxygen evolution) lie toward the thylakoid-lumen side of the membrane. [Drawn from 1S5L.pdb.]

Figure 22.16 **Photosystem II provides electrons to photosystem I.** As electrons flow from photosystem II to photosystem I, they pass through the cytochrome b_6f complex, a proton pump. The electron donating pigment is denoted as P680*. Abbreviations: Ph, pheophytin; Q_A and Q_B, plastoquinones; Pc, plastocyanin.

augmented by the link between photosystem II and photosystem I, the *cytochrome b_6f complex.*

Cytochrome b_6f Links Photosystem II to Photosystem I

The QH_2 produced by photosystem II transfers its electrons to plastocyanin, which in turn donates the electrons to photosystem I, thereby replenishing the missing electrons in photosystem I. The chain starts when the electrons are transferred, one at a time from QH_2, to plastocyanin (Pc), a copper-containing protein in the thylakoid lumen (Figure 22.10). This reaction is catalyzed by cytochrome b_6f (plastoquinol-plastocyanin reductase), a proton pump:

$$QH_2 + 2 \text{ plastocyanin (Cu}^{2+}) \xrightarrow{\text{Cytochrome } b_6f} Q + 2 \text{ plastocyanin (Cu}^+) + 2 \text{ H}^+_{\text{thylakoid lumen}}$$

The cytochrome b_6f reaction is reminiscent of that catalyzed by ubiquinol cytochrome *c* oxidoreductase in oxidative phosphorylation (p. 374). Indeed, most components of the cytochrome b_6f complex are similar to those of Q-cytochrome *c* oxidoreductase.

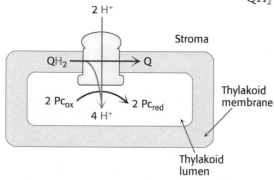

Figure 22.17 **Cytochrome b_6f's contribution to the proton gradient.** The cytochrome b_6f complex (yellow structure) oxidizes QH_2 to Q through the Q cycle. Four protons are released into the thylakoid lumen in each cycle.

The oxidation of plastoquinol by the cytochrome b_6f complex takes place by a mechanism nearly identical with the Q cycle of the electron-transport chain of cellular respiration. In photosynthesis, the net effect of the Q cycle is, first, to release protons from QH_2 into the lumen and, second, to pump protons from the stroma to the lumen, strengthening the proton-motive force (Figure 22.17). Thus, electrons from photosystem II are used to replace electrons in photosystem I and, in the process, augment the proton gradient. However, the electron-deficient P680 must now be replenished with electrons.

The Oxidation of Water Achieves Oxidation–Reduction Balance and Contributes Protons to the Proton Gradient

We experience the results of attaining photosynthetic redox balance with every breath that we take. The electron-deficient *P680$^+$ is a very strong oxidant and is*

(A)

(B)

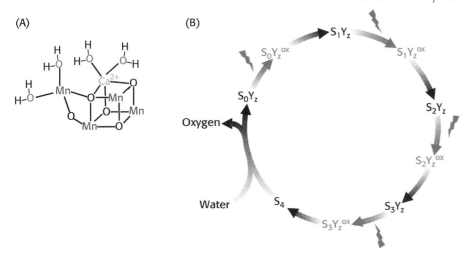

Figure 22.18 The core of the water-oxidizing complex. (A) The deduced core structure of the WOC including four manganese ions and one calcium ion is shown. The valence states of the individual manganese ions are not indicated because of uncertainty about the charge on the individual ions. The center is oxidized, one electron at a time, until two H_2O molecules are linked to form a molecule of O_2, which is then released from the complex. (B) The absorption of photons by the reaction center (red arrows) generates a tyrosine radical (Y), which then extracts electrons from the manganese ions. The structures are designated S_0 to S_4 to indicate the number of electrons that have been removed.

capable of extracting electrons from water molecules, resulting in the release of oxygen. This reaction, the *photolysis of water,* takes place at the *water-oxidizing complex* (WOC), also called the *manganese center.* The core of this complex includes a calcium ion, four manganese ions, and four water molecules (**Figure 22.18**A). Manganese is especially suitable for photolyzing water because of its ability to exist in multiple oxidation states and to form strong bonds with oxygen-containing species.

In its reduced form, the WOC oxidizes two molecules of water to form a single molecule of oxygen. Each time the absorbance of a photon powers the removal of an electron from P680, the positively charged special pair extracts an electron from a tyrosine residue of subunit D1 of the WOC, forming a tyrosine radical (Figure 22.18B). The tyrosine radical then removes an electron from a manganese ion. This process occurs four times, with the result that H_2O is oxidized to generate O_2 and H^+:

$$2\,H_2O \longrightarrow 4\,e^- + 4\,H^+ + O_2$$

Four photons must be absorbed to extract four electrons from a water molecule. The four electrons harvested from water are used to reduce two molecules of Q to QH_2. All oxygenic phototrophs, the most common type of photosynthetic organism, use the same inorganic core and protein components for capturing the energy of sunlight. A single solution to the biochemical problem of extracting electrons from water evolved billions of years ago, and has been conserved for use under a wide variety of phylogenetic and ecological circumstances. *The photolysis of water by photosynthetic organisms is essentially the sole source of oxygen in the biosphere.* This gas, so precious to our lives, is in essence a waste product of photosynthesis that results from the need to achieve redox balance (**Figure 22.19**).

The cooperation between photosystem I and photosystem II creates electron flow—an electrical current—from H_2O to $NADP^+$. The pathway of electron flow

Figure 22.19 *Elodea.* The production of oxygen is evident by the generation of bubbles in this aquatic plant. [Colin Milkens/ Oxford Scientific Films/Getty Images.]

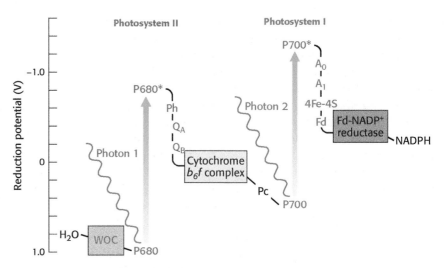

Figure 22.20 The pathway of electron flow from H_2O to $NADP^+$ in photosynthesis. This endergonic reaction is made possible by the absorption of light by photosystem II (P680) and photosystem I (P700). Abbreviations: Ph, pheophytin; Q_A and Q_B, plastoquinones; Pc, plastocyanin; A_0 and A_1, acceptors of electrons from P700*; Fd, ferredoxin; WOC, water-oxidizing complex.

QUICK QUIZ Photosystem I produces a powerful reductant, whereas photosystem II produces a powerful oxidant. Identify the reductant and oxidant and describe their roles.

is called the *Z scheme of photosynthesis* because the redox diagram from P680 to P700* looks like the letter Z (**Figure 22.20**).

Although we have focused on the green plants, which use H_2O as the electron donor in photosynthesis, **Table 22.1** shows that photosynthetic bacteria display a wide variety of electron donors.

Table 22.1 Major groups of photosynthetic bacteria

Bacteria	Photosynthetic electron donor	O_2 use
Green sulfur	H_2, H_2S, S	Anoxygenic
Green nonsulfur	Variety of amino acids and organic acids	Anoxygenic
Purple sulfur	H_2, H_2S, S	Anoxygenic
Purple nonsulfur	Usually organic molecules	Anoxygenic
Cyanobacteria	H_2O	Oxygenic

22.4 A Proton Gradient Drives ATP Synthesis

A crucial result of electron flow from water to $NADP^+$ is the formation of a proton gradient. Such a gradient can be maintained because the thylakoid membrane is impermeable to protons. As discussed in Chapter 21, energy inherent in the proton gradient, called the *proton-motive force* (Δp), is described as the sum of two components: a charge gradient and a chemical gradient. The dual nature of Δp is nicely illustrated in chloroplasts. In chloroplasts, nearly all of Δp arises from the proton gradient, whereas, in mitochondria, the contribution from the membrane potential is larger. The reason for this difference is that the thylakoid membrane, though impermeable to most ions, including H^+, is quite permeable to Cl^- and Mg^{2+} acquired from soil. The light-induced transfer of H^+ into the thylakoid lumen is accompanied by the transfer of either Cl^- in the same direction or Mg^{2+} (1 Mg^{2+} per 2 H^+) in the opposite direction. Consequently, electrical neutrality is maintained and no membrane potential is generated.

The ATP Synthase of Chloroplasts Closely Resembles That of Mitochondria

The proton-motive force generated by the light reactions is converted into ATP by the *ATP synthase* of chloroplasts, also called the *CF_1-CF_0 complex* (C stands for chloroplast and F for factor). CF_1-CF_0 ATP synthase closely resembles the F_1-F_0 complex of mitochondria (p. 385). CF_0 is analogous to F_0 in mitochondria and conducts protons across the thylakoid membrane, whereas CF_1, which is analogous to F_1, catalyzes the formation of ATP from ADP and P_i.

CF_0 is embedded in the thylakoid membrane. It consists of four different polypeptide chains known as I (17 kDa), II (16.5 kDa), III (8 kDa), and IV (27 kDa) with an estimated stoichiometry of 1:2:(10–14):1. Subunits I and II correspond to the **b** mitochondrial subunits, whereas subunits III and IV correspond to subunits **c** and **a,** respectively, of the mitochondrial F_0 subunit. CF_1, the site of ATP synthesis, has a subunit composition $\alpha_3\beta_3\gamma\delta\varepsilon$. The β subunits contain the catalytic sites, similar to the F_1 subunit of mitochondrial ATP synthase. Protons flow *out* of the thylakoid lumen through ATP synthase into the stroma. Because CF_1 is on the stromal surface of the thylakoid membrane, the newly synthesized ATP is released directly into the stroma (**Figure 22.21**). Recall that NADPH formed through the action of photosystem I and ferredoxin-NADP$^+$ reductase also is released into the stroma. Thus, *ATP and NADPH, the products of the light reactions of photosynthesis, are appropriately positioned for the subsequent dark reactions, in which CO_2 is converted into carbohydrate.*

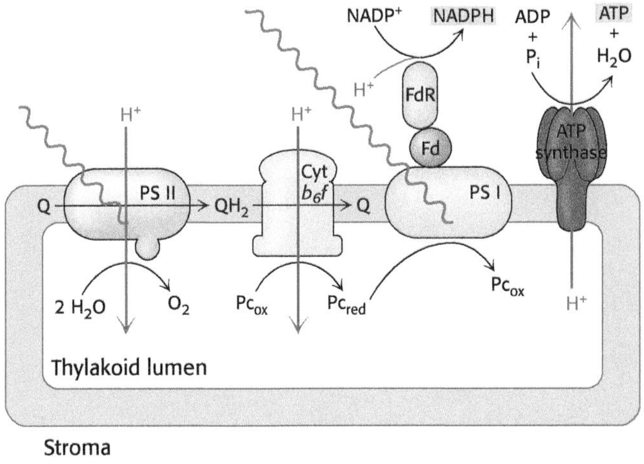

Figure 22.21 A summary of the light reactions of photosynthesis. The light-induced electron transfer in photosynthesis drives protons into the thylakoid lumen. The excess protons flow out of the lumen through ATP synthase to generate ATP in the stroma. Fd, ferredoxin; FdR, ferredoxin reductase.

The Activity of Chloroplast ATP Synthase Is Regulated

The activity of the ATP synthase is sensitive to the redox conditions in the chloroplast. For maximal activity, a specific disulfide bond in the γ subunit must be reduced to two cysteines. The reductant is reduced thioredoxin, which is formed from ferredoxin generated in photosystem I, by ferredoxin-thioredoxin reductase, an iron-sulfur containing enzyme:

$$2 \text{ Reduced ferredoxin } + \text{ thioredoxin disulfide } \underset{}{\overset{\text{Ferredoxin} - \text{thioredoxin reductase}}{\rightleftharpoons}}$$
$$2 \text{ oxidized ferredoxin } + \text{ reduced thioredoxin } + 2 \text{ H}^+$$

Conformational changes in the ε subunit also contribute to synthase regulation. The ε subunit appears to exist in two conformations. One conformation inhibits ATP hydrolysis by the synthase, while the other, which is generated by an increase in the proton-motive force, allows ATP synthesis and facilitates the reduction of the disulfide bond in the γ subunit. Thus, synthase activity is maximal when biosynthetic reducing power and a proton gradient are available. We will see in Chapter 23 that redox regulation is also important in photosynthetic carbon metabolism.

Cyclic Electron Flow Through Photosystem I Leads to the Production of ATP Instead of NADPH

On occasion, when the ratio of NADPH to NADP$^+$ is very high, NADP$^+$ may not be available to accept electrons from ferredoxin. In these circumstances, specific large protein complexes allow cyclic electron flow that powers ATP synthesis, a process called *cyclic photophosphorylation*. In this situation, the electrons in reduced ferredoxin can be transferred back to the cytochrome b_6f complex rather than to NADP$^+$. Each electron then flows back through the cytochrome b_6f complex to reduce plastocyanin, which can then be reoxidized by P700* to complete a cycle (**Figure 22.22**). The net outcome of this cyclic flow of electrons is the pumping of protons by the cytochrome b_6f complex. The resulting proton gradient then drives the synthesis of ATP. *In cyclic photophosphorylation, ATP is generated independently of the formation of NADPH.* Photosystem II does not participate in cyclic photophosphorylation, and so O_2 is not formed from H_2O.

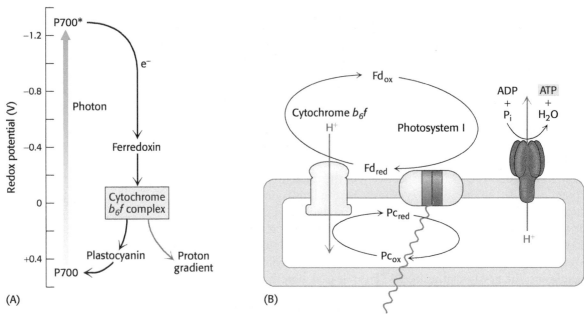

Figure 22.22 **Cyclic photophosphorylation.** (A) A scheme showing the energetic basis for cyclic photophosphorylation. (B) In this pathway, electrons from reduced ferredoxin are transferred to cytochrome b_6f rather than to ferredoxin-NADP$^+$ reductase. The flow of electrons through cytochrome b_6f pumps protons into the thylakoid lumen. These protons flow through ATP synthase to generate ATP. Neither NADPH nor O_2 is generated by this pathway.

The Absorption of Eight Photons Yields One O_2, Two NADPH, and Three ATP Molecules

We can now estimate the overall stoichiometry for the light reactions. The absorption of 4 photons by photosystem II generates one molecule of O_2 and releases 4 protons into the thylakoid lumen. The two molecules of plastoquinol are oxidized by the Q cycle of the cytochrome b_6f complex to release 8 protons into the lumen. The absorption of 4 additional photons by photosystem I generates four molecules of reduced plastocyanin, which subsequently reduce four molecules of ferredoxin. The four molecules of reduced ferredoxin generate two molecules of NADPH. Thus, the overall reaction is

$$2\,H_2O + 2\,NADP^+ + 10\,H^+_{stroma} \longrightarrow O_2 + 2\,NADPH + 12\,H^+_{lumen}$$

The 12 protons released in the lumen can then flow through ATP synthase. Let us assume that there are 12 subunit III components in CF_0. Twelve protons must pass through CF_0 to complete one full rotation of CF_1. Just as in mitochondrial ATP synthase, a single rotation generates three molecules of ATP. Given the ratio of three molecules of ATP to 12 protons, the overall reaction is

$$2\,H_2O + 2\,NADP^+ + 10\,H^+_{stroma} \longrightarrow O_2 + 2\,NADPH + 12\,H^+_{lumen}$$
$$\underline{3\,ADP^{3-} + 3\,P_i^{2-} + 3\,H^+ + 12\,H^+_{lumen} \longrightarrow 3\,ATP^{4-} + 3\,H_2O + 12\,H^+_{stroma}}$$
$$2\,NADP^+ + 3\,ADP^{3-} + 3\,P_i^{2-} + H^+ \longrightarrow O_2 + 2\,NADPH + 3\,ATP^{4-} + H_2O$$

Thus, 8 photons are required to yield three molecules of ATP (2.7 photons/ATP).

Cyclic photophosphorylation is a somewhat more productive way to synthesize ATP than noncyclic photophosphorylation (the Z scheme). The absorption of 4 photons by photosystem I leads to the release of 8 protons into the lumen by the cytochrome b_6f system. These protons flow through ATP synthase to yield two molecules of ATP. Thus, every 2 photons absorbed yield one molecule of ATP. No NADPH is produced.

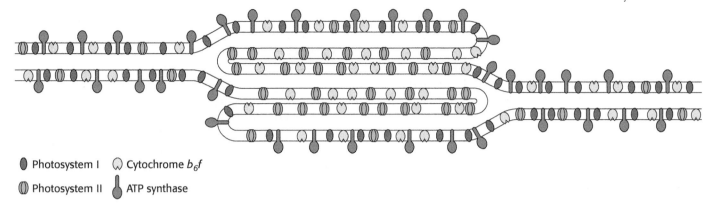

● Photosystem I   ◔ Cytochrome b_6f

◍ Photosystem II   ⌗ ATP synthase

Figure 22.23 Locations of photosynthesis components. Photosynthetic assemblies are differentially distributed in the stacked and unstacked regions of thylakoid membranes. [Information from a drawing kindly provided by Dr. Jan M. Anderson and Dr. Bertil Andersson.]

The Components of Photosynthesis Are Highly Organized

The complexity of photosynthesis, seen already in the elaborate interplay of multiple components, extends even to the placement of the components in the thylakoid membranes. *The thylakoid membranes of most plants are differentiated into stacked (appressed) and unstacked (nonappressed) regions.* Stacking increases the amount of thylakoid membrane in a given chloroplast volume. Both regions surround a common internal thylakoid lumen, but only unstacked regions make direct contact with the chloroplast stroma. Stacked and unstacked regions differ in the nature of their photosynthetic assemblies (**Figure 22.23**). Photosystem I and ATP synthase are located almost exclusively in unstacked regions, whereas photosystem II is present mostly in stacked regions. The cytochrome b_6f complex is found in both regions. Plastoquinone and plastocyanin are the mobile carriers of electrons between assemblies located in different regions of the thylakoid membrane. A common internal thylakoid lumen enables protons liberated by photosystem II in stacked membranes to be utilized by ATP synthase molecules that are located far away in unstacked membranes.

What is the functional significance of this spatial differentiation of the thylakoid membrane system? The positioning of photosystem I in the unstacked membranes gives it direct access to the stroma for the reduction of $NADP^+$. ATP synthase, too, is located in the unstacked region to provide space for its large CF_1 globule and so that it can bind stromal ADP and phosphate and release its ATP product into the stroma. The raw materials for CO_2 reduction and carbohydrate synthesis are thus accessible to the stromal enzymes. The tight quarters of the appressed region pose no problem for photosystem II, which interacts with two molecules found in the thylakoid lumen—a small polar electron donor (H_2O) and a highly lipid soluble electron carrier (plastoquinone).

(A)

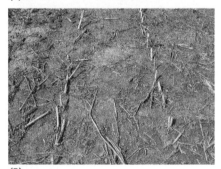

(B)

Figure 22.24 The effects of herbicide treatment. (A) A robust culture of dandelions. (B) The same plot of land after herbicide treatment. [Courtesy Dr. Chris Boerboom/University of Wisconsin, Madison.]

▦ BIOLOGICAL INSIGHT

Many Herbicides Inhibit the Light Reactions of Photosynthesis

Many commercial herbicides kill weeds by interfering with the action of one of the photosystems (**Figure 22.24**). Inhibitors of photosystem II block electron flow, whereas inhibitors of photosystem I divert electrons from ferredoxin, the terminal part of this photosystem. Photosystem II inhibitors include urea derivatives such as *diuron* and triazine derivatives such as *atrazine*, the most widely used weed killer in the United States. These chemicals bind to a site on the D1 subunit of photosystem II and block the formation of plastoquinol (QH_2) (p. 415).

Paraquat (1,1′-dimethyl-4,4′-bipyridinium), one of the most widely used herbicides in the world, is an inhibitor of photosystem I. Paraquat can accept electrons from photosystem I to become a radical. This radical reacts with O_2 to produce reactive oxygen species such as superoxide ion and hydroxyl radical (OH·). Such reactive oxygen species react with double bonds in membrane lipids, damaging the membrane. Paraquat displays acute oral toxicity in human beings, in whom it also leads to the generation of reactive oxygen species.

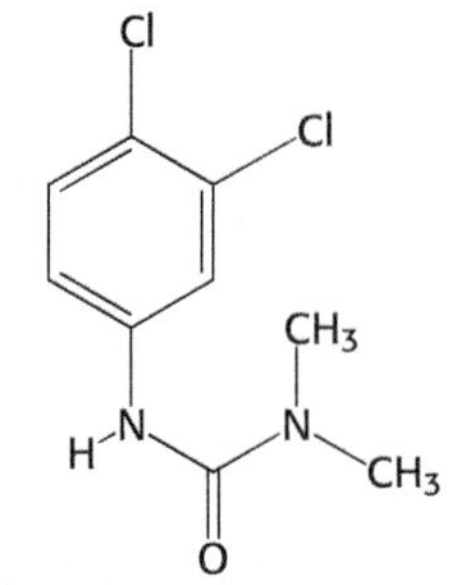

Diuron

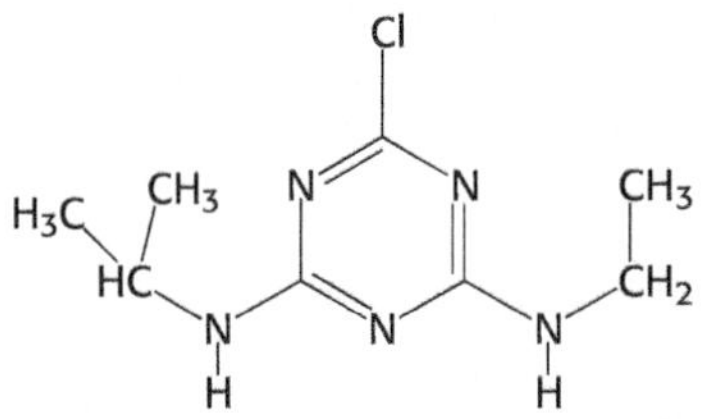

Atrazine

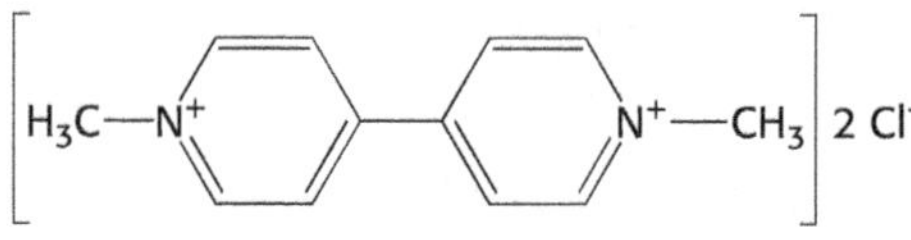

Paraquat

SUMMARY

22.1 Photosynthesis Takes Place in Chloroplasts

The proteins that participate in the light reactions of photosynthesis are located in the thylakoid membranes of chloroplasts. The light reactions result in (1) the formation of reducing power for the production of NADPH, (2) the generation of a transmembrane proton gradient for the formation of ATP, and (3) the production of O_2.

22.2 Photosynthesis Transforms Light Energy into Chemical Energy

Chlorophyll molecules—tetrapyrroles with a central magnesium ion—absorb light quite efficiently because they are polyenes. An electron excited to a high-energy state by the absorption of a photon can move to a nearby electron acceptor. In photosynthesis, an excited electron leaves a pair of associated chlorophyll molecules known as the special pair located at the reaction center. Light-harvesting complexes that surround the reaction centers contain additional molecules of chlorophyll *a*, as well as carotenoids and chlorophyll *b* molecules, which absorb light in the center of the visible spectrum. These accessory pigments increase the efficiency of light capture by absorbing light and transferring the energy to reaction centers through resonance energy transfer.

22.3 Two Photosystems Generate a Proton Gradient and NADPH

Photosynthesis in green plants is mediated by two linked photosystems. In photosystem I, the excitation of special-pair P700 releases electrons that flow to ferredoxin, a powerful reductant. Ferredoxin-NADP$^+$ reductase catalyzes the formation of NADPH. In photosystem II, the excitation of a special pair of chlorophyll molecules called P680 leads to electron transfer to plastocyanin, which replenishes the electrons removed from photosystem I. A proton gradient is generated as electrons pass through photosystem II, through the cytochrome b_6f complex and through ferredoxin-NADP$^+$ reductase. The electrons extracted from photosystem II are replenished by the extraction of electrons from a water molecule at a center containing four manganese ions. One molecule of O_2 is generated at this center for each four electrons transferred.

22.4 A Proton Gradient Drives ATP Synthesis

The proton gradient across the thylakoid membrane creates a proton-motive force used by ATP synthase to form ATP. The ATP synthase of chloroplasts (also called CF_1-CF_0) closely resembles the ATP-synthesizing assembly of mitochondria (F_1-F_0). If the $NADPH/NADP^+$ ratio is high, electrons transferred to ferredoxin by photosystem I can reenter the cytochrome b_6f complex. This process, called cyclic photophosphorylation, leads to the generation of a proton gradient by the cytochrome b_6f complex without the formation of NADPH or O_2.

KEY TERMS

light reactions (p. 408)
chloroplast (p. 408)
stroma (p. 408)
thylakoid (p. 408)
granum (p. 408)
photoinduced charge separation
  (p. 410)
special pair (p. 410)

reaction center (p. 410)
chlorophyll (p. 410)
light-harvesting complex (p. 411)
carotenoid (p. 412)
photosystem I (PS I) (p. 413)
photosystem II (PS II) (p. 413)
P700 (p. 414)
P680 (p. 415)

cytochrome b_6f complex (p. 416)
water-oxidizing complex (WOC)
  (p. 417)
Z scheme of photosynthesis (p. 418)
proton-motive force $\Delta(p)$ (p. 418)
ATP synthase (CF_1-CF_0 complex)
  (p. 418)
cyclic photophosphorylation (p. 419)

? Answer to QUICK QUIZ

Photosystem I generates ferredoxin, a powerful reductant that reduces $NADP^+$ to NADPH, a biosynthetic reducing power. Photosystem II activates the manganese complex, a powerful oxidant capable of oxidizing water, generating

O_2 and electrons that will ultimately be used to reduce $NADP^+$. In the process of oxidizing water, the manganese complex also generates protons to form the proton gradient used to generate ATP.

PROBLEMS

1. *A crucial prereq.* Human beings do not produce energy by photosynthesis, yet this process is critical to our survival. Explain. ✓ 2

2. *The accounting.* What is the overall reaction of photosynthesis? ✓ 1 ✓ 3

3. *Like a fife and drum.* Match each term with its description. ✓ 1

(a) Light reactions
  _____

(b) Chloroplasts _____

(c) Reaction center
  _____

(d) Chlorophyll _____

(e) Light-harvesting
  complex _____

(f) Photosystem I _____

(g) Photosystem II
  _____

(h) Cytochrome b_6f
  complex _____

(i) Water-oxidizing
  complex _____

(j) ATP synthase _____

1. Uses resonance energy transfer to reach the reaction center
2. Generates NADPH
3. Pumps protons
4. Site of photoinduced charge separation
5. Cellular location of photosynthesis
6. CF_1-CF_0 complex
7. Generates ATP, NADPH, and O_2
8. Site of oxygen generation
9. Transfers electrons from H_2O to P700
10. Primary photosynthetic pigment

4. *A single wavelength.* Photosynthesis can be measured by measuring the rate of oxygen production. When plants are exposed to light of wavelength 680 nm, more oxygen is evolved than if the plants are exposed to light of 700 nm. Explain. ✓ 2

5. *Combining wavelengths.* If plants described in problem 4 are illuminated by a combination of light of 680 nm and 700 nm, the oxygen production exceeds that of either wavelength alone. Explain. ✓ 2

6. *If a little is good.* What is the advantage of having an extensive set of thylakoid membranes in the chloroplasts? ✓ 2

7. *Separating charges.* What is the significance of photoinduced separation of charge in photosynthesis? ✓ 1

8. *Cooperation.* Explain how light-harvesting complexes enhance the efficiency of photosynthesis. ✓ 1

9. *One thing leads to another.* Identify the ultimate electron acceptor and the ultimate electron donor in photosynthesis. What powers the electron flow between the donor and the acceptor? ✓ 3

10. *Neutralization compensation.* In chloroplasts, a greater pH gradient across the thylakoid membrane is required to power the synthesis of ATP than is required across the mitochondrial inner membrane. Explain. ✓ 2

11. *Environmentally appropriate.* Chlorophyll is a hydrophobic molecule. Why is this property crucial for the function of chlorophyll? ✓ 1

12. *Networking.* Why is chlorophyll an effective light-absorbing pigment? ✓ 1

13. *Proton origins.* What are the various sources of protons that contribute to the generation of a proton gradient in chloroplasts? ✓ 3

14. *That's not right!* Explain the defect or defects in the hypothetical scheme for the light reactions of photosynthesis depicted here.

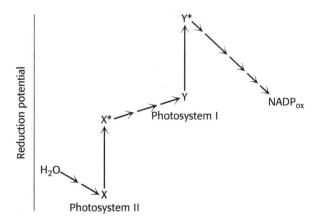

15. *Electron transfer.* Calculate the $\Delta E_0'$ and $\Delta G^{\circ\prime}$ for the reduction of $NADP^+$ by ferredoxin. Use the data given in Table 20.1. ✓ 3

16. *To boldly go.* (a) It can be argued that, if life were to exist elsewhere in the universe, it would require some process like photosynthesis. Why is this argument reasonable? (b) If the starship *Enterprise* were to land on a distant planet and find no measurable oxygen in the atmosphere, could the crew conclude that photosynthesis is not taking place?

17. *Weed killer 1.* Dichlorophenyldimethylurea (DCMU), a herbicide, interferes with photophosphorylation and O_2 evolution. However, it does not block O_2 evolution in the presence of an artificial electron acceptor such as ferricyanide. Propose a site for the inhibitory action of DCMU. ✓ 3

18. *Weed killer 2.* Predict the effect of the herbicide dichlorophenyldimethylurea (DCMU) on a plant's ability to perform cyclic photophosphorylation. ✓ 3

19. *Gedanken experiment.* Suppose you had a suspension of chloroplasts that lacked ADP and P_i. You exposed these chloroplasts to light for a period of time, after which you plunged them into darkness while simultaneously adding ADP and P_i. To what extent, if any, would you expect ATP synthesis to take place?

20. *Staunching the flow.* Venturicidin, an antibiotic isolated from a strain of *Streptomyces,* inhibits proton flow through the CF_0 subunit of chloroplast ATP synthase. What would be the effect of adding venturicidin to a suspension of chloroplasts that are robustly generating oxygen? ✓ 2

21. *Increasing the flow.* Consider again the situation described in problem 20. What could you add to the inhibited suspension of chloroplasts that would restore oxygen evolution? ✓ 2

Chapter Integration Problems

22. *Functional equivalents.* What structural feature of mitochondria corresponds to the thylakoid membranes?

23. *Compare and contrast.* Compare and contrast oxidative phosphorylation and photosynthesis.

24. *Backward.* In what way is the electron transfer in ferridoxin-$NADP^+$ reductase similar to that of the pyruvate dehydrogenase complex?

25. *Looking for a place to rest.* Albert Szent-Györgyi, Nobel Prize–winning biochemist, once said something to this effect: "Life is nothing more than an electron looking for a place to rest." Explain how this pithy statement applies to photosynthesis and cellular respiration. ✓ 3

Data Interpretation and Chapter Integration Problem

26. *The same, but different.* The $\alpha_3\beta_3\gamma$ complex of mitochondrial or chloroplast ATP synthase will function as an ATPase in vitro. The chloroplast enzyme (both synthase and ATPase activity) is sensitive to redox control, whereas the mitochondrial enzyme is not. To determine where the enzymes differ, a segment of the mitochondrial γ subunit was removed and replaced with the equivalent segment from the chloroplast γ subunit. The ATPase activity of the modified enzyme was then measured as a function of redox conditions. Dithiothreitol (DTT) is a small molecule used by biochemists to reduce disulfide bonds. Thioredoxin is a protein, common in all forms of life, that reduces disulfide bonds.

(a) What is the redox regulator of the chloroplast ATP synthase in vivo? The adjoining graph shows the ATPase activity of modified and control enzymes under various redox conditions.

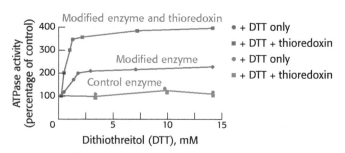

(b) What is the effect of increasing the reducing power of the reaction mixture for the control and the modified enzymes?

(c) What is the effect of the addition of thioredoxin? How do these results differ from those in the presence of DTT alone? Suggest a possible explanation for the difference.

(d) Did the researchers succeed in identifying the region of the γ subunit responsible for redox regulation?

(e) What is the biological rationale of regulation by high concentrations of reducing agents?

(f) What amino acids in the γ subunit are most likely affected by the reducing conditions?

(g) What experiments might confirm your answer to part *f*?

Challenge Problems

27. *Efficiency matters.* What fraction of the energy of 700-nm light absorbed by photosystem I is trapped as high-energy electrons? 700-nm photons have an energy content of 172 kJ mol^{-1} of photons (also called an einstein). ✓ 1

28. *Infrared harvest.* Consider the relation between the energy of a photon and its wavelength. ✓ 1

(a) Some bacteria are able to harvest 1000-nm light. What is the energy (in kilojoules or kilocalories) of a mole (also called an einstein) of 1000-nm photons?

(b) What is the maximum increase in redox potential that can be induced by a 1000-nm photon?

(c) What is the minimum number of 1000-nm photons needed to form ATP from ADP and P_i? Assume a ΔG of 50 kJ mol^{-1} (12 kcal mol^{-1}) for the phosphorylation reaction.

29. *Hill reaction.* In 1939, Robert Hill discovered that chloroplasts evolve O_2 when they are illuminated in the presence of an artificial electron acceptor such as ferricyanide $[Fe^{3+}(CN)_6]^{3-}$. Ferricyanide is reduced to ferrocyanide $[Fe^{2+}(CN)_6]^{4-}$ in this process. No NADPH or reduced plastocyanin is produced. Propose a mechanism for the Hill reaction. ✓ 3

30. *Energy accounts.* On page 430, the balance sheet for the cost of the synthesis of glucose is presented. Eighteen molecules of ATP are required. Yet, when glucose undergoes combustion in cellular respiration, 30 molecules of ATP are produced. Account for the difference. ✓ 2

Selected Readings for this chapter can be found online at www.whfreeman.com/tymoczko3e.

CHAPTER 23

The Calvin Cycle

Rain forests are primary sites of carbon fixation. Approximately 50% of terrestrial carbon fixation takes place in these forests. The logging of rain forests and the subsequent loss of carbon fixation account in part for the increase in carbon dioxide in the atmosphere. [Mattias Klum/National Geographic/Getty Images.]

The light reactions transform light energy into ATP and biosynthetic reducing power, NADPH. The second part of photosynthesis uses these raw materials to reduce carbon atoms from their fully oxidized state as carbon dioxide to the more reduced state as a hexose. These reactions are called the *dark reactions* or the light-independent reactions because light is not directly needed. The dark reactions are also called either the *Calvin–Benson cycle,* after Melvin Calvin and Andrew Benson, the biochemists who elucidated the pathway, or simply the *Calvin cycle*. The source of the carbon atoms in the Calvin cycle is the simple molecule carbon dioxide. The Calvin cycle brings into living systems the carbon atoms that will become biochemical constituents of all organisms.

✓ 4 Explain the function of the Calvin cycle.

23.1 The Calvin Cycle Synthesizes Hexoses from Carbon Dioxide and Water

The Calvin cycle takes place in the stroma of chloroplasts, the photosynthetic organelles. In this vital process, *carbon dioxide gas is trapped in an organic molecule, 3-phosphoglycerate, which has many biochemical fates.*

The Calvin cycle has three stages (**Figure 23.1**):

1. The fixation of CO_2 by ribulose 1,5-bisphosphate to form two molecules of 3-phosphoglycerate. Recall that 3-phosphoglycerate is an intermediate in glycolysis and gluconeogenesis.

2. The reduction of 3-phosphoglycerates to form hexose sugars.

3. The regeneration of ribulose 1,5-bisphosphate so that more CO_2 can be fixed (incorporated into an organic molecule).

Keep in mind that, owing to the stoichiometry of the conversion of six molecules of CO_2 into a hexose sugar, to synthesize a glucose molecule or any other hexose sugar, the three stages must take place six times. The regeneration process is a complicated one—hence the dotted arrow and simplification of the process depicted in Figure 23.1; the details will be revealed as we move through the process. For simplicity's sake, we will consider the reactions one molecule at a time or three molecules at a time.

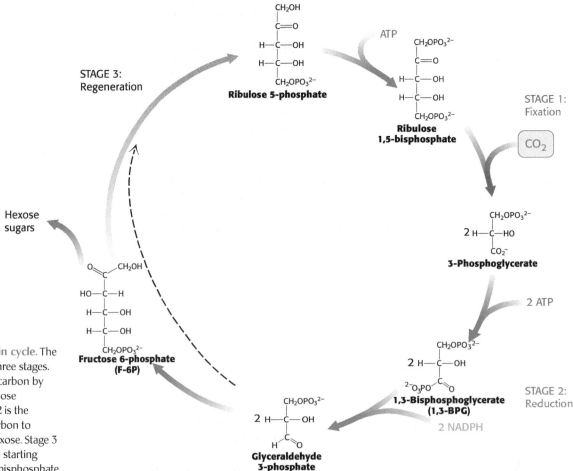

Figure 23.1 The Calvin cycle. The Calvin cycle consists of three stages. Stage 1 is the fixation of carbon by the carboxylation of ribulose 1,5-bisphosphate. Stage 2 is the reduction of the fixed carbon to begin the synthesis of hexose. Stage 3 is the regeneration of the starting compound, ribulose 1,5-bisphosphate.

Carbon Dioxide Reacts with Ribulose 1,5-bisphosphate to Form Two Molecules of 3-Phosphoglycerate

The first stage in the Calvin cycle is the *fixation of CO$_2$*. This highly exergonic reaction ($\Delta G^{\circ\prime} = -51.9$ kJ mol^{-1}, or -12.4 kcal mol^{-1}) is catalyzed by *ribulose 1,5-bisphosphate carboxylase/oxygenase* (usually called *rubisco*), an enzyme located on the stromal surface of the thylakoid membranes of chloroplasts. *This reaction is the rate-limiting step in hexose synthesis.*

Carbon fixation begins with the conversion of ribulose 1,5-bisphosphate into a highly reactive enediolate intermediate. The CO$_2$ molecule condenses with the enediolate intermediate to form an unstable six-carbon compound, which is rapidly hydrolyzed to two molecules of 3-phosphoglycerate:

Thus, carbon dioxide, a waste product of cellular respiration, is reintroduced to the biochemical world. Plants that use only the Calvin cycle to fix carbon dioxide are called C$_3$ plants, because the three-carbon molecule 3-phosphoglycerate is formed immediately after the rubisco reaction.

Rubisco in chloroplasts consists of eight large (L, 55-kDa) subunits and eight small (S, 13-kDa) ones (**Figure 23.2**). Each L chain contains a catalytic site and a regulatory site. The S chains enhance the catalytic activity of the L chains. This enzyme is very abundant in chloroplasts, constituting more than 16% of their total protein. In fact, rubisco is the most abundant enzyme in plants and probably the most abundant protein in the biosphere. Large amounts are required because rubisco is an inefficient enzyme; its maximal catalytic rate is only 3 s^{-1}. Rubisco also requires a bound CO$_2$ for catalytic activity in addition to the substrate CO$_2$. The bound CO$_2$ forms a carbamate with lysine 201 of the large subunit (margin on p. 430). The carbamate binds to a Mg^{2+} ion that is required for catalytic activity. We will see the regulatory significance of this requirement later.

In the absence of CO$_2$, rubisco binds to its substrate, ribulose 1,5-bisphosphate, so tightly that its enzyme activity is inhibited. The enzyme *rubisco activase* uses ATP to induce structural changes in rubisco that allow release of the bound substrate and formation of the required carbamate. The ATP requirement of rubisco activase coordinates rubisco activity with the light reactions.

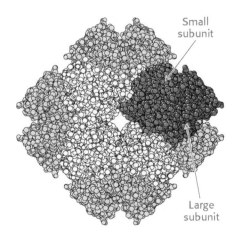

Figure 23.2 The structure of rubisco. The enzyme ribulose 1,5-bisphosphate carboxylase/oxygenase (rubisco) comprises eight large subunits (one shown in red and the others in yellow) and eight small subunits (one shown in blue and the others in white). The active sites lie in the large subunits. [Drawn from 1RXO.pdb.]

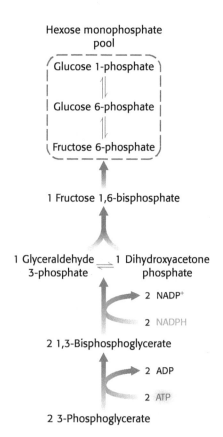

Lysine side chain

Carbamate

Hexose monophosphate pool

Glucose 1-phosphate

Glucose 6-phosphate

Fructose 6-phosphate

1 Fructose 1,6-bisphosphate

1 Glyceraldehyde 3-phosphate ⇌ 1 Dihydroxyacetone phosphate

2 NADP⁺

2 NADPH

2 1,3-Bisphosphoglycerate

2 ADP

2 ATP

2 3-Phosphoglycerate

Figure 23.3 Hexose phosphate formation. 3-Phosphoglycerate is converted into fructose 6-phosphate in a pathway parallel to that of gluconeogenesis.

Hexose Phosphates Are Made from Phosphoglycerate, and Ribulose 1,5-bisphosphate Is Regenerated

In the second stage of the Calvin cycle, the 3-phosphoglycerate products of rubisco are converted into a hexose phosphate. The steps in this conversion (**Figure 23.3**) are like those of the gluconeogenic pathway (p. 314), except that glyceraldehyde 3-phosphate dehydrogenase in chloroplasts, which generates glyceraldehyde 3-phosphate (GAP) in the process of hexose formation, is specific for NADPH rather than NADH. The hexose phosphate product of the Calvin cycle exists in three isomeric forms: glucose 1-phosphate, glucose 6-phosphate, and fructose 6-phosphate, together called the *hexose monophosphate pool*. Recall that these isomers are readily interconvertible (pp. 286 and 297). Alternatively, the glyceraldehyde 3-phosphate can be transported to the cytoplasm and metabolized to fructose 6-phosphate and glucose 1-phosphate by using the gluconeogenic pathway and the enzyme phosphoglucomutase. Fructose 6-phosphate and glucose 1-phosphate are used for sucrose synthesis (Figure 23.8). These reactions and the reaction catalyzed by rubisco bring CO_2 to the oxidation level of a hexose, converting CO_2 into a chemical fuel at the expense of the NADPH and ATP that were generated in the light reactions.

The third and final stage of the Calvin cycle is the regeneration of ribulose 1,5-bisphosphate, the acceptor of CO_2 in the first stage. This regeneration is not as simply done as the regeneration of oxaloacetate in the citric acid cycle. The challenge is to construct a five-carbon sugar from a six-carbon member of the hexose monophosphate pool and three-carbon molecules, such as glyceraldehyde 3-phosphate. A *transketolase*, which we will encounter again in the pentose phosphate pathway, and an *aldolase*, an enzyme in glycolysis and gluconeogenesis, have major roles in the rearrangement of the carbon atoms. With these enzymes, the construction of the five-carbon sugar proceeds as shown in **Figure 23.4**. The sum of these reactions is

Fructose 6-phosphate + 2 glyceraldehyde 3-phosphate

+ dihydroxyacetone phosphate + 3 ATP ⟶

3 ribulose 1,5-bisphosphate + 3 ADP

This series of reactions completes the Calvin cycle (**Figure 23.5**). Figure 23.5 presents the required reactions with the proper stoichiometry to convert three molecules of CO_2 into one molecule of dihydroxyacetone phosphate (DHAP). However, two molecules of DHAP are required for the synthesis of a member of the hexose monophosphate pool. Consequently, the cycle as presented must take place twice to yield a hexose monophosphate. The outcome of the Calvin cycle is the generation of a hexose and the regeneration of the starting compound, ribulose 1,5-bisphosphate. In essence, *ribulose 1,5-bisphosphate acts catalytically, similarly to oxaloacetate in the citric acid cycle.*

Three Molecules of ATP and Two Molecules of NADPH Are Used to Bring Carbon Dioxide to the Level of a Hexose

What is the energy expenditure for synthesizing a hexose? Six rounds of the Calvin cycle are required, because one carbon atom is reduced in each round. Twelve molecules of ATP are expended in phosphorylating 12 molecules of 3-phosphoglycerate to 1,3-bisphosphoglycerate, and 12 molecules of NADPH are consumed in reducing 12 molecules of 1,3-bisphosphoglycerate to glyceraldehyde 3-phosphate. An additional 6 molecules of ATP are spent in regenerating ribulose 1,5-bisphosphate. The balanced equation for the net reaction of the Calvin cycle is then

$$6\,CO_2 + 18\,ATP + 12\,NADPH + 12\,H_2O \longrightarrow$$
$$C_6H_{12}O_6 + 18\,ADP + 18\,P_i + 12\,NADP^+ + 6\,H^+$$

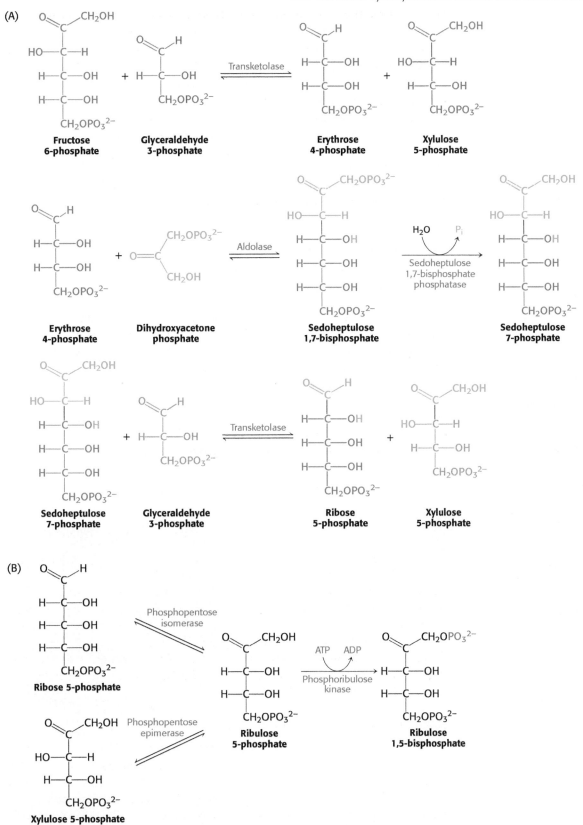

Figure 23.4 The regeneration of ribulose 1,5-bisphosphate. (A) A transketolase and aldolase generate the five-carbon sugars ribose 5-phosphate and xylulose 5-phosphate from six-carbon and three-carbon sugars. (B) Both ribose 5-phosphate and xylulose 5-phosphate are converted into ribulose 5-phosphate, which is then phosphorylated to complete the regeneration of ribulose 1,5-bisphosphate.

Thus, 3 molecules of ATP and 2 molecules of NADPH are consumed in incorporating a single CO_2 molecule into a hexose such as glucose or fructose. Biochemically, the synthesis of glucose from CO_2 is energetically expensive, but the ultimate energy source—sunlight—is abundant.

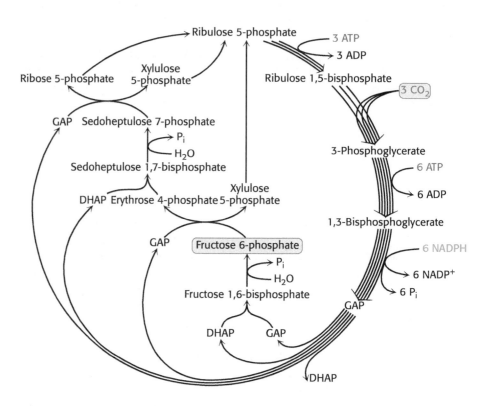

Figure 23.5 Stoichiometry of the Calvin cycle. The diagram shows the reactions necessary with the correct stoichiometry to convert three molecules of CO_2 into one molecule of dihydroxyacetone phosphate (DHAP). Two molecules of DHAP are subsequently converted into a member of the hexose monophosphate pool, shown here as fructose 6-phosphate. The cycle is not as simple as presented in Figure 23.1; rather, it entails many reactions that lead ultimately to the synthesis of glucose and the regeneration of ribulose 1,5-bisphosphate. [Information from J. R. Bowyer and R. C. Leegood, Photosynthesis. In *Plant Biochemistry*, P. M. Dey and J. B. Harborne, Eds. (Academic Press, 1997), p. 85.]

BIOLOGICAL INSIGHT

A Volcanic Eruption Can Affect Photosynthesis Worldwide

On June 12, 1991, Mount Pinatubo, a volcano on the island of Luzon in the Philippines, erupted (**Figure 23.6**). The eruption, which was the largest on Earth in a hundred years, killed more than 700 people. The eruption also spewed approximately 15 million tons of sulfur dioxide (SO_2) into the atmosphere in the form of sulfate aerosol, encircling Earth in a massive aerosol blanket in about 3 weeks. This blanket led to a decrease in total sunlight striking Earth (direct sunlight plus diffuse sunlight) but increased the proportion of diffuse sunlight.

At the same time as the sulfate aerosol was enveloping the planet, the rate of growth of atmospheric CO_2 declined sharply. Moreover, an

Figure 23.6 The eruption of Mount Pinatubo. [©InterNetwork Media/Getty Images.]

experimental station in the northeastern United States reported that noontime photosynthesis increased by almost 25%. Could these events have been related?

If the increase in photosynthesis observed in the northeastern United States were happening globally, as might have been expected, then more CO_2 would have been removed from the atmosphere by rubisco, accounting for the decrease in the rate of growth of CO_2. But why did it happen? Although still not firmly established, the diffuse light caused by the sulfate aerosol may have actually increased the amount of photosynthesis taking place. **Figure 23.7** shows that the rate of photosynthesis reaches a limiting value when light intensity is increased. Consequently, in bright direct sunlight, the top leaves of a forest may be receiving more light than they can biochemically use. However, in direct sunlight, the lower leaves in the shadows of the upper leaves will be light deprived. Diffuse light, on the other hand, will penetrate more deeply into the forest. More leaves will receive light, more rubisco will be active, and more CO_2 will be converted into sugar.

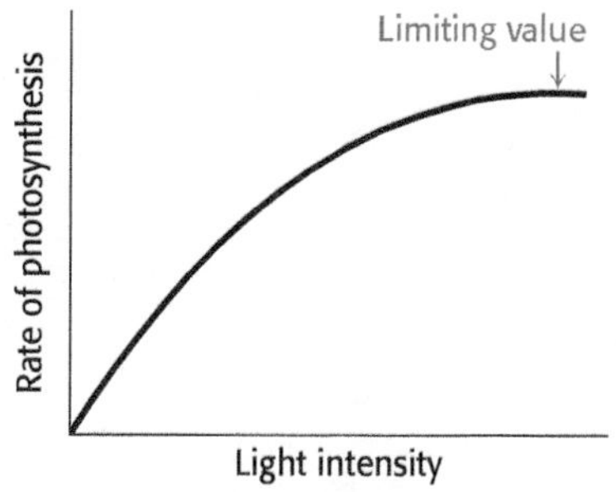

Figure 23.7 Photosynthesis has a maximum rate. The rate of photosynthesis reaches a limiting value when the light-harvesting complexes are saturated with light.

Starch and Sucrose Are the Major Carbohydrate Stores in Plants

What are the fates of the carbon atoms fixed and processed by the enzymes of the Calvin cycle? These molecules are used in a variety of ways, but there are two prominent fates: the synthesis of starch and sucrose, storage forms of carbohydrates. *Starch,* like its animal counterpart glycogen (p. 175), is a polymer of glucose residues. Starch comes in two forms: amylose, which consists of glucose molecules joined together by α-1,4 linkages only; and amylopectin, which is branched because of the presence of α-1,6 linkages. Starch is synthesized and stored in chloroplasts.

In contrast, *sucrose* (common table sugar), a disaccharide, is synthesized in the cytoplasm. Plants lack the ability to transport hexose phosphates across the chloroplast membrane, but an abundant triose phosphate-phosphate antiporter (p. 215) mediates the transport of triose phosphates from chloroplasts to the cytoplasm in exchange for phosphate. Fructose 6-phosphate is formed from these triose phosphates in the cytoplasm and joins the glucose unit of cytoplasmic UDP-glucose (activated glucose, p. 174) to form sucrose 6-phosphate (**Figure 23.8**). Sucrose 6-phosphate is then hydrolyzed by sucrose 6-phosphatase to yield sucrose.

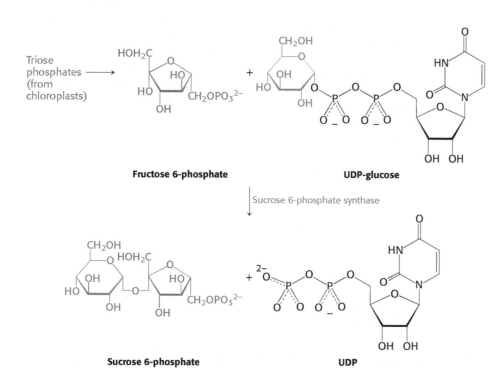

Triose phosphates (from chloroplasts) →

Fructose 6-phosphate   **UDP-glucose**

Sucrose 6-phosphate synthase

Sucrose 6-phosphate   **UDP**

Figure 23.8 The synthesis of sucrose. Sucrose 6-phosphate is formed by the reaction between fructose 6-phosphate and the activated intermediate uridine diphosphate glucose (UDP-glucose). Sucrose phosphatase subsequently generates free sucrose (not shown).

Figure 23.9 Sucrose is transported through plants as an ingredient of sap. Maple syrup, derived from the sap of the sugar maple by boiling and evaporation, is a solution of approximately 65% sucrose with some glucose and fructose. [bncc369/Getty Images.]

? QUICK QUIZ 1 Why is the Calvin cycle crucial to the functioning of all life forms?

Sucrose is transported from the leaves as a component of sap to the nonphotosynthetic parts of the plant, such as the roots, and is stored in many plant cells, as in sugar beets and sugar cane (**Figure 23.9**).

✻ BIOLOGICAL INSIGHT

Why Bread Becomes Stale: The Role of Starch

Bread staling (the process of becoming stale) is a complex process in which all of the components of bread take part—including proteins, lipids, and starch and their complex interactions. Let us consider the role of starch. A granule of starch is normally a strictly organized crystal structure rather than an amorphous form. This order is imparted to the starch granule by hydrogen bonds that form between chains of amylose and amylopectin molecules as they line up next to one another. However, when the starch is heated during baking, the hydrogen bonds of the crystal structure are disrupted, allowing water to be absorbed. The water breaks the crystal structure completely. The starch granules thus become more amorphous, contributing to the taste and texture of the bread crumb (the inside of the loaf of bread).

As the bread cools after it has been removed from the oven (**Figure 23.10**), amylose molecules separate from the amorphous starch structure and begin to associate with one another; that is, the amylose crystallizes. This crystallization takes place most prominently on the surface of the bread, where it is facilitated by the evaporation of surface water. This crystallization contributes to the crispness of the surface.

Meanwhile, inside the bread, crystallization of the amylopectin of the crumb is slower, possibly because of its branched structure, and continues for several days. Water bound to the amylopectin is released as the crystals form, which leads to a deterioration of the texture and taste of the crumb. Water is still present, but no longer in association with the amylopectin molecules. In fact, staling can be reversed at this stage if the bread is reheated to rehydrate the internal amylose and amylopectin crystals. However, if the bread is left uncovered, water will move to the surface, destroying the crispness of the bread and subsequently evaporating. Further evaporation leads to further crystallization, and, eventually, we are left with a loaf of bread more suitable for use as a doorstop.

Figure 23.10 Bread staling. Starch crystallization is responsible for the crust of bread. [Lukasphoto/FeaturePics.]

✓ **5** Describe how the light reactions and the Calvin cycle are coordinated.

23.2 The Calvin Cycle Is Regulated by the Environment

How do the light reactions communicate with the dark reactions to regulate the crucial process of fixing CO_2 into biomolecules? *The principal means of regulation is alteration of the stromal environment by the light reactions.* The light reactions lead to an increase in pH and stromal concentrations of Mg^{2+}, NADPH, and reduced ferredoxin—all of which contribute to the activation of certain Calvin-cycle enzymes (**Figure 23.11**).

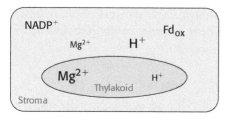

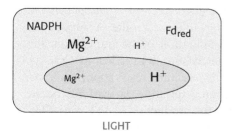

DARK

LIGHT

Figure 23.11 Light regulation of the Calvin cycle. The light reactions of photosynthesis transfer electrons out of the thylakoid lumen into the stroma, and they transfer H^+ from the stroma into the thylakoid lumen. As a consequence of these processes, the concentrations of NADPH, reduced ferredoxin (Fd), and Mg^{2+} in the stroma are higher in the light than in the dark, whereas the concentration of H^+ is lower in the light. Each of these concentration changes helps couple the Calvin-cycle reactions to the light reactions.

As stated on page 429, the rate-limiting step in the Calvin cycle is the carboxylation of ribulose 1,5-bisphosphate to form two molecules of 3-phosphoglycerate. This step is catalyzed by rubisco, a very slowly acting enzyme. *The activity of rubisco increases markedly on exposure to light.* How do the light reactions modify the activity of rubisco? Recall that CO_2 must be added to lysine 201 of rubisco to form the carbamate that is essential for catalytic activity, a process facilitated by rubisco activase (p. 429). Carbamate formation is favored by alkaline pH and high concentrations of Mg^{2+} ion in the stroma, both of which are consequences of the light-driven pumping of protons from the stroma into the thylakoid space. Magnesium ion concentration rises because Mg^{2+} ions from the thylakoid space are released into the stroma to compensate for the influx of protons. Thus, light leads to the generation of regulatory signals as well as ATP and NADPH.

Thioredoxin Plays a Key Role in Regulating the Calvin Cycle

Light-driven reactions lead to electron transfer from water to ferredoxin and, eventually, to NADPH. *Both reduced ferredoxin and NADPH regulate enzymes from the Calvin cycle.* An important protein in these regulatory processes is *thioredoxin*, a 12-kDa protein containing neighboring cysteine residues. Thioredoxin cycles between an oxidized form with a disulfide bond between the two cysteines and a reduced sulfhydryl form (**Figure 23.12**). In chloroplasts, oxidized thioredoxin is reduced by ferredoxin in a reaction catalyzed by *ferredoxin–thioredoxin reductase*. The reduced form of thioredoxin activates rubisco and other Calvin-cycle enzymes, as well as many enzymes in other metabolic pathways in the cell by reducing disulfide bridges that control their activity including the chloroplast ATP synthase (p. 419) (**Table 23.1**). Thus, *the activities of the light and dark*

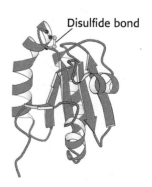

Disulfide bond

Figure 23.12 Thioredoxin. The oxidized form of thioredoxin contains a disulfide bond. When thioredoxin is reduced by reduced ferredoxin, the disulfide bond is converted into two free sulfhydryl groups. Reduced thioredoxin can cleave disulfide bonds in enzymes, activating certain Calvin-cycle enzymes and inactivating some degradative enzymes. [Drawn from 1F9M.pdb.]

Table 23.1 Enzymes regulated by thioredoxin

Enzyme	Pathway
Rubisco	Carbon fixation in the Calvin cycle
Fructose 1,6-bisphosphatase	Gluconeogenesis
Glyceraldehyde 3-phosphate dehydrogenase	Calvin cycle, gluconeogenesis, glycolysis
Sedoheptulose 1,7-bisphosphatase	Calvin cycle
Glucose 6-phosphate dehydrogenase	Pentose phosphate pathway
Phenylalanine ammonia lyase	Lignin synthesis
Ribulose 5′-phosphate kinase	Calvin cycle
$NADP^+$-malate dehydrogenase	C_4 pathway

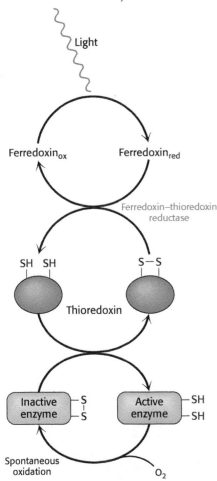

Figure 23.13 Enzyme activation by thioredoxin. Reduced thioredoxin activates certain Calvin-cycle enzymes by cleaving regulatory disulfide bonds.

reactions of photosynthesis are coordinated through electron transfer from reduced ferredoxin to thioredoxin and then to component enzymes containing regulatory disulfide bonds (**Figure 23.13**).

Rubisco Also Catalyzes a Wasteful Oxygenase Reaction

Rubisco is among the most important enzymes in life because it provides organic carbon for the biosphere. However, this enzyme has a wasteful side. Rubisco sometimes reacts with O_2 instead of CO_2, catalyzing a useless oxygenase reaction. The products of this reaction are *phosphoglycolate* and *3-phosphoglycerate* (**Figure 23.14**). Phosphoglycolate is not a versatile metabolite. A salvage pathway recovers part of its carbon skeleton, recycling three of the four carbon atoms of two molecules of glycolate. However, one carbon atom is lost as CO_2. This process is called *photorespiration* because, like cellular respiration, O_2 is consumed and CO_2 is released. However, photorespiration is wasteful because organic carbon is converted into CO_2 without the production of ATP, NADPH, or another energy-rich metabolite. This wastefulness raises the question, what is the biochemical basis of this inefficiency? The results of structural studies show that, when the reactive enediolate intermediate is formed, loops close over the active site to protect the enediolate. A channel to the environment is maintained to allow access by CO_2. However, like CO_2, O_2 is a linear molecule that also fits the channel. In essence, the problem lies not with the enzyme but in the unremarkable structure of CO_2. Carbon dioxide lacks chemical features that would allow discrimination between it and other gases such as O_2, and thus the oxygenase activity is an inevitable failing of the enzyme. Although evolutionary processes have enhanced the preference of rubisco for carboxylation—for instance, the rubisco of higher plants is eightfold as specific for carboxylation as that of photosynthetic bacteria—photorespiration still accounts for the loss of up to 25% of the carbon fixed. Another possibility exists, however. The oxygenase activity may not be an imperfection of the enzyme, but rather an imperfection in our understanding. Perhaps the oxygenase activity performs a biochemically important role that we have not yet discovered.

The oxygenase activity increases more rapidly with temperature than the carboxylase activity, presenting a problem for tropical plants. How do plants, such as sugarcane and corn that grow in hot climates, prevent very high rates of photorespiration? The solution to the problem illustrates one means by which photosynthesis responds to environmental conditions.

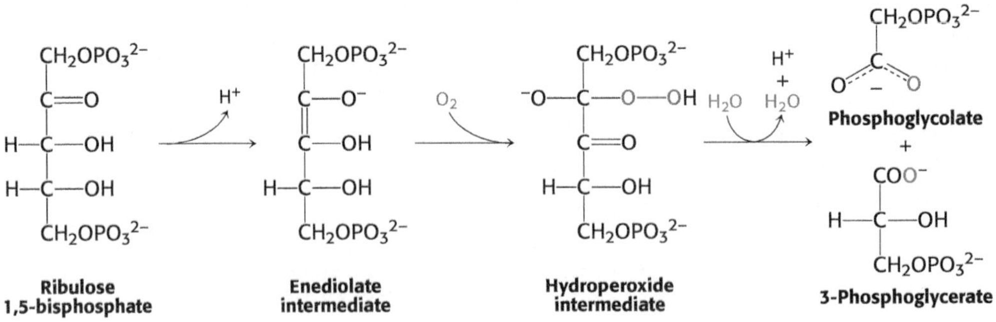

Figure 23.14 A wasteful side reaction. The reactive enediolate intermediate on rubisco also reacts with molecular oxygen to form a hydroperoxide intermediate, which then proceeds to form one molecule of 3-phosphoglycerate and one molecule of phosphoglycolate.

The C_4 Pathway of Tropical Plants Accelerates Photosynthesis by Concentrating Carbon Dioxide

One means of overcoming the inherent oxygenase activity of rubisco is generating a high local concentration of CO_2 at the site of the Calvin cycle in the

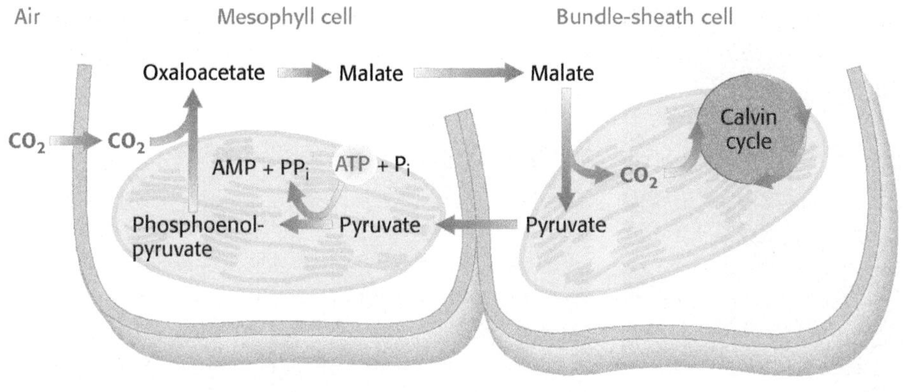

Figure 23.15 The C_4 pathway. Carbon dioxide is concentrated in bundle-sheath cells by the expenditure of ATP in mesophyll cells.

photosynthetic cells. The key to accomplishing this is *the synthesis of four-carbon (C₄) compounds such as oxaloacetate that carry CO_2 from mesophyll cells on the surfaces of leaves to interior bundle-sheath cells, which are the major sites of photosynthesis* (**Figure 23.15**). Decarboxylation of the four-carbon compound in a bundle-sheath cell maintains a high concentration of CO_2 at the site of the Calvin cycle. The resulting three-carbon compound, pyruvate, returns to the mesophyll cell for another round of carboxylation.

The C_4 *pathway* for the transport of CO_2, also called the *Hatch–Slack pathway* after its discoverers Marshall Davidson Hatch and Charles Roger Slack, starts in a mesophyll cell with the condensation of CO_2 and phosphoenolpyruvate to form *oxaloacetate,* in a reaction catalyzed by *phosphoenolpyruvate carboxylase.* Oxaloacetate is then converted into *malate* by an $NADP^+$-linked malate dehydrogenase. Malate enters the bundle-sheath cell and is decarboxylated within the chloroplasts by a different isozymic form of $NADP^+$-linked malate dehydrogenase. The released CO_2 enters the Calvin cycle in the usual way by condensing with ribulose 1,5-bisphosphate. Pyruvate formed in this decarboxylation reaction returns to the mesophyll cell. Finally, phosphoenolpyruvate is formed from pyruvate by *pyruvate-P_i dikinase.* The net reaction of this C_4 pathway is

$$CO_2 \,(\text{in mesophyll cell}) \,+\, ATP \,+\, H_2O \longrightarrow$$
$$CO_2 \,(\text{in bundle-sheath cell}) \,+\, AMP \,+\, 2\,P_i \,+\, H^+$$

Thus, *the energetic equivalent of two ATP molecules is consumed in transporting CO_2 to the chloroplasts of the bundle-sheath cells.* In essence, this process is active transport: the pumping of CO_2 into the bundle-sheath cell is driven by the hydrolysis of one molecule of ATP to one molecule of AMP and two molecules of orthophosphate. The CO_2 concentration can be 20-fold greater in the bundle-sheath cells than in the mesophyll cells as a result of this transport.

When the C_4 pathway and the Calvin cycle operate together, the net reaction is

$$6\,CO_2 \,+\, 30\,ATP \,+\, 12\,NADPH \,+\, 12\,H_2O \longrightarrow$$
$$C_6H_{12}O_6 \,+\, 30\,ADP \,+\, 30\,P_i \,+\, 12\,NADP^+ \,+\, 18\,H^+$$

Note that 30 molecules of ATP are consumed per hexose molecule formed when the C_4 pathway delivers CO_2 to the Calvin cycle, in contrast with 18 molecules of ATP per hexose molecule in the absence of the C_4 pathway. The elevated CO_2 concentration resulting from the expenditure of an extra molecule of ATP enhances the rate of photosynthesis in tropical plants where light is abundant and minimizes the energy loss caused by photorespiration. Plants that rely on the C_4 pathway are called C_4 *plants.*

Tropical plants with a C_4 pathway do little photorespiration because the high concentration of CO_2 in their bundle-sheath cells accelerates the carboxylase

(A)   (B)

Figure 23.16 C₃ and C₄ plants. (A) C₃ plants, such as trees, account for 95% of plant species. (B) Corn is a C₄ plant of tremendous agricultural importance. [(A) Audrey Ustuzhanih/FeaturePics; (B) Elena Elisseeva/FeaturePics.]

QUICK QUIZ 2 Why is the C₄ pathway valuable for tropical plants?

Figure 23.17 Desert plants. Because of crassulacean acid metabolism, cacti are well suited to life in the desert. [Yenwen Lu/Getty Images.]

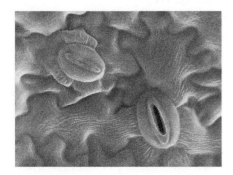

Figure 23.18 Stomata. An electron micrograph of an open stoma and a closed stoma. [Power and Syred/Science Source.]

reaction relative to the oxygenase reaction. The geographical distribution of plants having this pathway (C₄ plants) and those lacking it (C₃ plants) can now be understood in molecular terms. C₄ *plants* have the advantage in a hot environment and under high illumination, which accounts for their prevalence in the tropics. C₃ *plants*, which consume only 18 molecules of ATP per hexose molecule formed in the absence of photorespiration (compared with 30 molecules of ATP for C₄ plants), are more efficient than C₄ plants in temperate environments (**Figure 23.16**).

Crassulacean Acid Metabolism Permits Growth in Arid Ecosystems

Water is an important player in photosynthesis; if there is no water to provide electrons, oxygenic photosynthesis stops. Water availability is not a problem in temperate areas or hot, wet tropics, but it is not abundant in deserts (**Figure 23.17**). A crucial adaptation for such plants, and many non-desert plants as well, is the controlled opening and closing of the *stomata* (singular, *stoma*), microscopic pores located on the underside of plant leaves that are used for gas exchange. Such plants keep the stomata of their leaves closed in the heat of the day to prevent water loss (**Figure 23.18**). As a consequence, CO_2 cannot be absorbed during the daylight hours when it is needed for hexose synthesis. Rather, CO_2 enters the leaf when the stomata open at the cooler temperatures of night. To store the CO_2 until it can be used during the day, such plants make use of an adaptation called *crassulacean acid metabolism* (CAM), named after the family Crassulaceae (including many succulents). Carbon dioxide is fixed by the C₄ pathway into malate, which is stored in vacuoles. During the day, malate is decarboxylated and the CO_2 becomes available to the Calvin cycle. In contrast with C₄ plants, CAM plants separate CO_2 accumulation from CO_2 utilization temporally rather than spatially.

Although CAM plants do prevent water loss, the use of malate as the sole source of carbon dioxide comes at a metabolic cost. Because the malate storage is limited, CAM plants cannot generate CO_2 as rapidly as it can be imported by C₃ and C₄ plants. Consequently, the growth rate of CAM plants is slower that that of C₃ and C₄ plants. The saguaros cactus, which can live up to 200 years and reach a height of 60 feet, can take 15 years to grow to only one foot in height.

SUMMARY

23.1 The Calvin Cycle Synthesizes Hexoses from Carbon Dioxide and Water

ATP and NADPH formed in the light reactions of photosynthesis are used to convert CO_2 into hexoses and other organic compounds. The dark phase of photosynthesis, called the Calvin cycle, starts with the reaction of CO_2 and ribulose 1,5-bisphosphate to form two molecules of 3-phosphoglycerate. The steps in the conversion of 3-phosphoglycerate into fructose 6-phosphate and glucose 6-phosphate are like those of gluconeogenesis, except that glyceraldehyde 3-phosphate dehydrogenase in chloroplasts is specific for NADPH rather than NADH. Ribulose 1,5-bisphosphate is regenerated from fructose 6-phosphate, glyceraldehyde 3-phosphate, and dihydroxyacetone phosphate by a complex series of reactions. Three molecules of ATP and two molecules of NADPH are consumed for each molecule of CO_2 converted into a hexose. Starch in chloroplasts and sucrose in the cytoplasm are the major carbohydrate stores in plants.

23.2 The Calvin Cycle Is Regulated by the Environment

The activity of the Calvin cycle is coordinated with the light reaction of photosynthesis. Reduced thioredoxin formed by the light-driven transfer of electrons from ferredoxin activates enzymes of the Calvin cycle by reducing disulfide bridges. The light-induced increase in pH and Mg^{2+} levels of the stroma is important in stimulating the carboxylation of ribulose 1,5-bisphosphate by rubisco.

Rubisco also catalyzes a competing oxygenase reaction, which produces phosphoglycolate and 3-phosphoglycerate. The recycling of phosphoglycolate leads to the release of CO_2 and further consumption of O_2 in a process called photorespiration. This wasteful side reaction is minimized in tropical plants, which have an accessory pathway—the C_4 pathway—for concentrating CO_2 at the site of the Calvin cycle. This pathway enables tropical plants to take advantage of high levels of light and minimize the oxygenation of ribulose 1,5-bisphosphate. Other plants, including some that grow in arid ecosystems, employ crassulacean acid metabolism to prevent dehydration. In CAM plants, the C_4 pathway is active during the night, when the plant exchanges gases with the air. During the day, gas exchange is eliminated and CO_2 is generated from malate stored in vacuoles.

KEY TERMS

Calvin cycle (dark reactions) (p. 428)
rubisco (ribulose 1,5-bisphosphate carboxylase/oxygenase) (p. 429)
starch (p. 433)

sucrose (p. 433)
thioredoxin (p. 435)
C_4 pathway (p. 436)
C_4 plant (p. 437)

C_3 plant (p. 438)
stomata (p. 438)
crassulacean acid metabolism (CAM) (p. 438)

? Answers to QUICK QUIZZES

1. The Calvin cycle is the primary means of converting gaseous CO_2 into organic matter—biomolecules. Essentially, every carbon atom in your body passed through rubisco and the Calvin cycle at some point in the past.

2. The C_4 pathway allows the CO_2 concentration to increase at the site of carbon fixation. High concentrations of CO_2 inhibit the oxygenase reaction of rubisco. This inhibition is important for tropical plants because the oxygenase activity increases more rapidly with temperature than does the carboxylase activity.

PROBLEMS

1. *A vital cycle.* Why is the Calvin cycle crucial to the functioning of all life forms? ✓ 4

2. *Be nice to plants.* Differentiate between autotrophs and heterotrophs. ✓ 4

3. *Cabalistic reactions?* Why are the reactions of the Calvin cycle sometimes referred to as the dark reactions? Do they take place only at night, or are they grim, secret reactions? ✓ 4

4. *Three-part harmony.* The Calvin cycle can be thought of as taking place in three stages. Describe the stages. ✓ 4

5. *Like green eggs and ham.* Match each term with its description. ✓ 4

(a) Calvin cycle _____   1. CO_2 fixation

(b) Rubisco _____   2. Storage form of carbohydrates

(c) Carbamate _____   3. α-1,4 linkages only

(d) Starch _____   4. 3-Phosphoglycerate is formed after carbon fixation

(e) Sucrose _____   5. The dark reactions

(f) Amylose _____   6. Includes α-1,6 linkages

(g) Amylopectin _____   7. Required for rubisco activity

(h) C_3 plants _____   8. Carbon fixation results in oxaloacetate formation

(i) C_4 plants _____   9. Allow exchange of gases

(j) Stomata _____   10. Transport form of carbohydrates

6. *Not always to the swiftest.* Suggest a reason why rubisco might be the most abundant enzyme in the world. ✓ 4

7. *A requirement.* In an atmosphere devoid of CO_2 but rich in O_2, the oxygenase activity of rubisco disappears. Why? ✓ 5

8. *No free lunch.* Explain why the maintenance of a high concentration of CO_2 in the bundle-sheath cells of C_4 plants is an example of active transport. How much ATP is required per molecule of CO_2 to maintain a high CO_2 concentration?

9. *Reduce locally.* Glyceraldehyde 3-phosphate dehydrogenase in chloroplasts uses NADPH to participate in the synthesis of glucose. In gluconeogenesis in the cytoplasm, the isozyme of the dehydrogenase uses NADH. Why is the use of NADPH by the chloroplast enzyme advantageous? ✓ 4

10. *Light and dark talk.* Rubisco requires a molecule of CO_2 covalently bound to lysine 201 for catalytic activity. The carboxylation of rubisco is favored by high pH and high Mg^{2+} concentration in the stroma. Why does it make good

physiological sense for these conditions to favor rubisco carboxylation? ✓ 5

11. *Communication.* What are the light-dependent changes in the stroma that regulate the Calvin cycle? ✓ 5

12. *When one equals two.* In the C_4 pathway, one ATP molecule is used in combining the CO_2 with phosphoenolpyruvate to form oxaloacetate, but, in the computation of energetics bookkeeping, two ATP molecules are said to be consumed. Explain.

13. *Dog days of August.* Before the days of pampered lawns, most homeowners practiced horticultural Darwinism. A result was that the lush lawns of early summer would often convert into robust cultures of crabgrass in the dog days of August. Provide a possible biochemical explanation for this transition. ✓ 4

14. *Breathing pictures?* What is photorespiration, what is its cause, and why is it believed to be wasteful?

15. *Competition is good.* Why do high concentrations of CO_2 inhibit photorespiration?

16. *Global warming.* C_3 plants are most common in higher latitudes and become less common at latitudes near the equator. The reverse is true of C_4 plants. How might global warming affect this distribution?

17. *Cost effective?* C_3 plants require 18 molecules of ATP to synthesize 1 molecule of glucose. C_4 plants, on the other hand, require 30 molecules of ATP to synthesize 1 molecule of glucose. Why would any plant use C_4 metabolism instead of C_3 metabolism given that C_3 metabolism is so much more efficient?

Chapter Integration Problems

18. *Better to focus on one.* Rubisco catalyzes both a carboxylation reaction and a wasteful oxygenase reaction. Below are the kinetic parameters for the two reactions.

$K_M^{CO_2} (\mu M)$	$K_M^{O_2} (\mu M)$	$K_{cat}^{CO_2} (s^{-1})$	$K_{cat}^{O_2} (s^{-1})$
10	500	3	2

(a) Determine the values of $k_{cat}^{CO_2}/K_M^{CO_2}$ and $k_{cat}^{O_2}/K_M^{O_2}$ as $s^{-1}M^{-1}$.

(b In light of the k_{cat}/K_M values for the two reactions, why does the oxygenation reaction occur?

19. *Compare and contrast.* Identify the similarities and differences between the Krebs cycle and the Calvin cycle. ✓ 4

20. *Photosynthetic efficiency.* Use the following information to estimate the efficiency of photosynthesis.

The $\Delta G^{\circ\prime}$ for the reduction of CO_2 to the level of hexose is $+477$ kJ mol^{-1} ($+114$ kcal mol^{-1}).

A mole of 600-nm photons has an energy content of 199 kJ (47.6 kcal).

Assume that the proton gradient generated in producing the required NADPH is sufficient to drive the synthesis of the required ATP.

Data Interpretation Problem

21. *Deciding between three and four.* Graph A shows the photosynthetic activity of two species of plant, one a C_4 plant and the other a C_3 plant, as a function of leaf temperature.

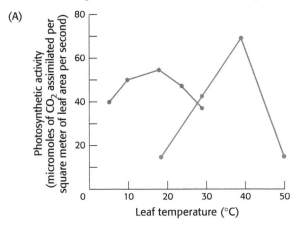

(a) Which data were most likely generated by the C_4 plant and which by the C_3 plant? Explain.

(b) Suggest some possible explanations for why the photosynthetic activity falls at higher temperatures.

Graph B illustrates how the photosynthetic activity of C_3 and C_4 plants varies with CO_2 concentration when temperature (30°C) and light intensity (high) are constant.

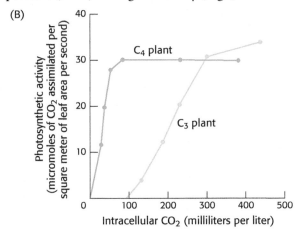

(c) Why can C_4 plants thrive at CO_2 concentrations that do not support the growth of C_3 plants?

(d) Suggest a plausible explanation for why C_3 plants continue to increase photosynthetic activity at higher CO_2 concentrations, whereas C_4 plants reach a plateau.

Challenge Problems

22. *Labeling experiments.* When Melvin Calvin performed his initial experiments on carbon fixation, he exposed algae to radioactive carbon dioxide. After 5 seconds, only a single organic compound contained radioactivity but, after 60 seconds, many compounds had incorporated radioactivity.

(a) What compound initially contained the radioactivity?

(b) What compounds contained radioactivity after 60 seconds?

23. *Total eclipse.* An illuminated suspension of *Chlorella*, a genus of single-celled green algae, is actively carrying out photosynthesis. Suppose that the light is suddenly switched off. How will the levels of 3-phosphoglycerate and ribulose 1,5-bisphosphate change in the next minute? ✓ 5

24. *CO_2 deprivation.* An illuminated suspension of *Chlorella* is actively carrying out photosynthesis in the presence of 1% CO_2. The concentration of CO_2 is abruptly reduced to 0.003%. What effect will this reduction have on the levels of 3-phosphoglycerate and ribulose 1,5-bisphosphate in the next minute? ✓ 5

25. *A potent analog.* 2-Carboxyarabinitol 1,5-bisphosphate (CABP) has been useful in studies of rubisco.

(a) Which catalytic intermediate does CABP resemble?

(b) Predict the effect of CABP on rubisco.

Selected Readings for this chapter can be found online at www.whfreeman.com/tymoczko3e.

CHAPTER 24
Glycogen Degradation

CHAPTER 25
Glycogen Synthesis

CHAPTER 26
The Pentose Phosphate Pathway

Glycogen Metabolism and the Pentose Phosphate Pathway

Cyclists sometimes call it "bonking"; distance runners, "hitting the wall." Both expressions describe a state of exhaustion in which further exercise is all but impossible. What is the biochemical basis for this condition?

After final exams are finished, tired students will sometimes sleep for 12 or more hours. Although other tissues can use fat stores as a fuel, the brain requires glucose as a fuel. How is the brain supplied with glucose during this long fast?

The answers to both of these questions entail the biomolecule *glycogen*. Glycogen is *a storage form of glucose that can be readily broken down to yield glucose molecules when energy is needed*. The depletion of muscle glycogen accounts in part for the feeling of exhaustion after intensive exercise, whereas the parceling out of the liver glycogen stores during a night's fast allows the brain to continue functioning.

As with any precious resource, glucose should be stored when plentiful. Much of the glucose consumed after an exercise bout or after a night's sleep is stored as glycogen. The interplay between glycogen breakdown and glycogen synthesis must be highly coordinated to ensure that an organism has glucose when needed.

Not all glycogen metabolism is related to energy needs. The ultimate product of glycogen breakdown—glucose 6-phosphate—can also be processed by a pathway common to all organisms, known variously as the pentose phosphate pathway, the hexose monophosphate pathway, the phosphogluconate pathway, or the pentose shunt. This pathway allows all organisms to oxidize glucose to

443

generate biosynthetic reducing power, NADPH, which is used for the biosynthesis of many biomolecules, including fats. As stated in Section 10, NADPH is also a product of photosystem I in photosynthesis, though this process provides NADPH for plants only. The pentose phosphate pathway can also be used for the metabolism of pentose sugars from the diet, the synthesis of pentose sugars for nucleotide synthesis, and the catabolism and synthesis of less common four- and seven-carbon sugars.

We begin this section with an examination of how glycogen stores are degraded, or mobilized, and how this mobilization is regulated. Glycogen mobilization is activated during exercise or fasting. We next consider the reverse process: in times of low energy demand and glucose excess, glycogen is synthesized. We will see how glycogen synthesis and degradation are coordinated. Finally, we look at how glucose 6-phosphate, the ultimate product of glycogen breakdown, can be metabolized to provide reducing power and five-carbon sugars.

✓By the end of this section, you should be able to:

✓ 1 List and describe the steps of glycogen breakdown, and identify the enzymes required.

✓ 2 Explain the regulation of glycogen breakdown.

✓ 3 Describe the steps of glycogen synthesis, and identify the enzymes required.

✓ 4 Explain the regulation of glycogen synthesis.

✓ 5 Describe how glycogen degradation and synthesis are coordinated.

✓ 6 Identify the two stages of the pentose phosphate pathway, and explain how the pathway is coordinated with glycolysis and gluconeogenesis.

✓ 7 Identify the enzyme that controls the pentose phosphate pathway.

Glycogen Degradation

Glycogen is a key source of energy for runners. Glycogen mobilization—the conversion of glycogen into glucose—is highly regulated. [Jim Rogash/Getty Images.]

G lycogen is a very large, branched polymer of glucose residues (Figure 24.1). Most of the glucose residues in glycogen are linked by α-1,4-glycosidic bonds, and branches at about every 10th residue are created by α-1,6-glycosidic bonds. Glycogen is not as reduced as fatty acids are and, consequently, not as energy rich. So why isn't all excess fuel stored as fatty acids rather than as glycogen? The readily mobilized glucose from glycogen is a good source of energy for sudden, strenuous activity. Unlike fatty acids, the released glucose can provide energy in the absence of oxygen and can thus supply energy for anaerobic activity. Moreover, the controlled release of glucose from glycogen

Figure 24.1 Glycogen. (A) Glucose units joined by α-1,4 linkages are shown as straight lines. The nonreducing ends of the glycogen molecule form the surface of the glycogen granule. At the core of the glycogen molecule is the protein glycogenin (yellow, Chapter 25). Degradation takes place at this surface. (B) A cross section of a glycogen molecule shows the branching caused by the α-1,6 linkages. The glycogenin is identified as G. [(A) Information from R. Melendez, E. Melendez-Hevia, and E. T. Canela, Fractal structure of glycogen: A clever solution to optimize cell metabolism. *Biophys. J.* 77:1327–1332, 1999.]

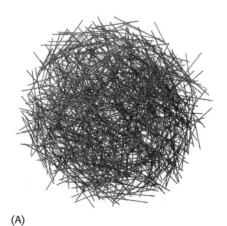

(A)

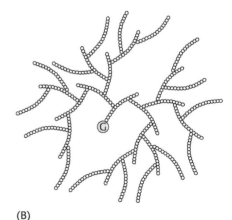

(B)

maintains the blood-glucose concentration between meals. The circulating blood keeps the brain supplied with glucose, which is virtually the only fuel used by the brain, except during prolonged starvation.

Glycogen is present in bacteria, archae, and eukaryotes. Recall that plants store glucose as starch, a similar biomolecule. Thus, storing energy as glucose polymers is common to all forms of life. In humans, most tissues have some glycogen, but the two major sites of glycogen storage are the liver and skeletal muscle. The concentration of glycogen is higher in the liver than in muscle (10% versus 2% by weight), but more glycogen is stored in skeletal muscle overall because there is more skeletal muscle in the body than there is liver tissue. Glycogen is present in the cytoplasm in the form of granules ranging in diameter from 10 to 40 nm, containing about 55,000 glucose molecules. *In the liver, glycogen synthesis and degradation are regulated to maintain the concentration of glucose in the blood required to meet the needs of the organism as a whole.* The glucose is parceled out from the liver during a nocturnal fast, maintaining brain function throughout the night. In contrast, in muscle, these processes are regulated to meet the energy needs of the muscle itself. Glycogen breakdown takes place to fuel the ATP needs of muscle contraction. The depletion of muscle glycogen is thought to be a major component of exhaustion—"bonking" or "hitting the wall."

✓ 1 List and describe the steps of glycogen breakdown, and identify the enzymes required.

24.1 Glycogen Breakdown Requires Several Enzymes

The efficient breakdown of glycogen to provide glucose 6-phosphate for further metabolism requires four enzyme activities: one to degrade glycogen, two to remodel glycogen so that it remains a substrate for degradation, and one to convert the product of glycogen breakdown into a form suitable for further metabolism. We will examine each of these activities in turn.

Phosphorylase Cleaves Glycogen to Release Glucose 1-phosphate

Glycogen phosphorylase, the key regulatory enzyme in glycogen breakdown, cleaves its substrate by the addition of orthophosphate (P_i) to yield *glucose 1-phosphate*. The cleavage of a bond by the addition of orthophosphate is referred to as *phosphorolysis*:

$$\text{Glycogen} + P_i \rightleftharpoons \text{glucose 1-phosphate} + \text{glycogen}$$
$$(n \text{ residues}) \qquad\qquad\qquad\qquad (n - 1 \text{ residues})$$

Phosphorylase catalyzes the sequential removal of glucosyl residues from the nonreducing ends of the glycogen molecule, as illustrated in **Figure 24.2** (the ends with a free OH group on carbon 4; p. 171). Orthophosphate splits the glycosidic linkage between C-1 of the terminal residue and C-4 of the adjacent one. Specifically, it cleaves the bond between the C-1 carbon atom and the glycosidic oxygen atom, and the α configuration at C-1 of the newly released glucose 1-phosphate is retained:

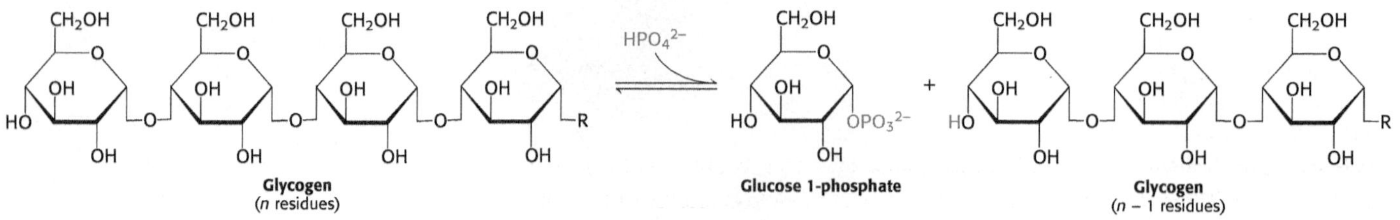

Glucose 1-phosphate released from glycogen can be readily converted into glucose 6-phosphate, an important metabolic intermediate, by the enzyme phosphoglucomutase.

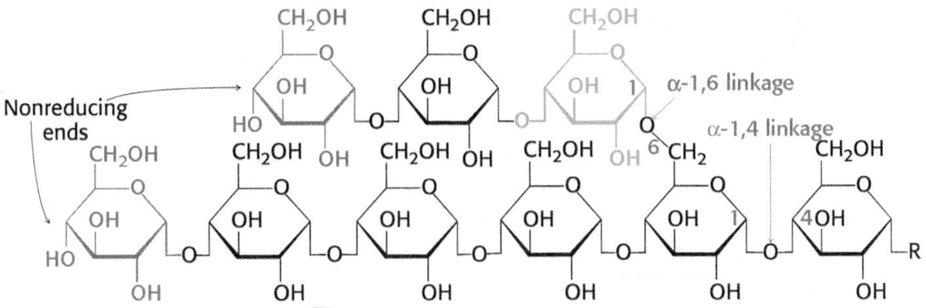

Figure 24.2 Glycogen structure. In this structure of two outer branches of a glycogen molecule, the residues at the nonreducing ends are shown in red and the residue that starts a branch is shown in green. The rest of the glycogen molecule is represented by R.

The phosphorolytic cleavage of glycogen is energetically advantageous because the released sugar is already phosphorylated. In contrast, a hydrolytic cleavage would yield glucose, which would then have to be phosphorylated at the expense of a molecule of ATP to enter the glycolytic pathway. An additional advantage of phosphorolytic cleavage for muscle cells is that no transporters exist for glucose 1-phosphate, which would be negatively charged under physiological conditions, and so it cannot be transported out of the cell.

A Debranching Enzyme Also Is Needed for the Breakdown of Glycogen

Glycogen phosphorylase can carry out the process of degrading glycogen by itself only to a limited extent before encountering an obstacle. The α-1,6-glycosidic bonds at the branch points are not susceptible to cleavage by phosphorylase. Indeed, phosphorylase stops cleaving α-1,4 linkages when it reaches a residue four residues away from a branch point. Because about 1 in 10 residues is branched, glycogen degradation by the phosphorylase alone would halt after the release of six glucose molecules per branch.

How can the remainder of the glycogen molecule be degraded for use as a fuel? Two additional enzymes, a *transferase* and α-1,6-glucosidase, remodel the glycogen for continued degradation by the phosphorylase. *The transferase shifts a block of three glucosyl residues from one outer branch to another* (**Figure 24.3**). This transfer exposes a single glucose residue joined by an α-1,6-glycosidic linkage. α-1,6-Glucosidase, also known as the *debranching enzyme*, then hydrolyzes the α-1,6-glycosidic bond, resulting in the release of a free glucose molecule:

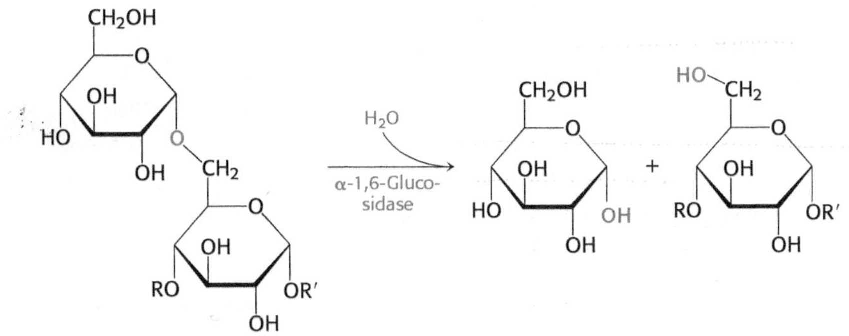

Glycogen
(*n* residues)

Glucose

Glycogen
(*n* − 1 residues)

This free glucose molecule is phosphorylated by the glycolytic enzyme hexokinase (p. 284). Thus, the transferase and α-1,6-glucosidase convert the branched structure into a linear one, which paves the way for further cleavage by phosphorylase. In eukaryotes, the transferase and the α-1,6-glucosidase activities are present in a single 160-kDa polypeptide chain, providing yet another example of a bifunctional enzyme (p. 321).

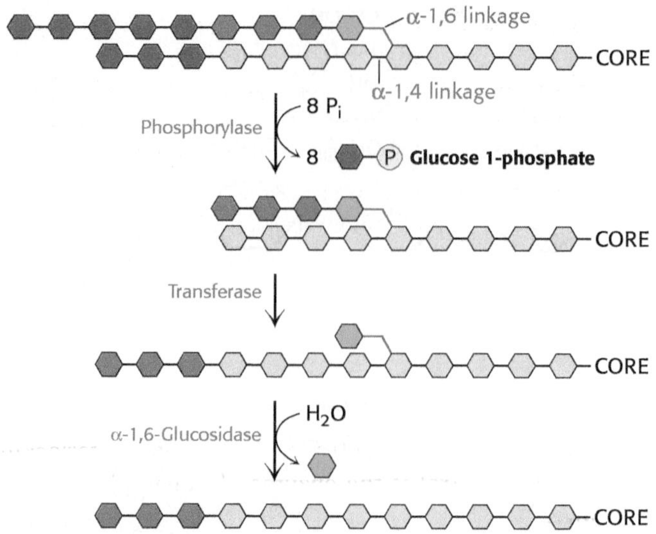

Figure 24.3 Glycogen remodeling. First, α-1,4-glycosidic bonds on each branch are cleaved by phosphorylase, leaving four residues along each branch. The transferase shifts a block of three glucosyl residues from one outer branch to the other. In this reaction, the α-1,4-glycosidic link between the blue and the green residues is broken and a new α-1,4 link between the blue and the yellow residues is formed. The green residue is then removed by α-1,6-glucosidase, leaving a linear chain with all α-1,4 linkages, suitable for further cleavage by phosphorylase.

α-1,6 linkage

α-1,4 linkage

CORE

Phosphorylase

8 P_i

8 ⓟ **Glucose 1-phosphate**

CORE

Transferase

CORE

α-1,6-Glucosidase

H_2O

CORE

Phosphoglucomutase Converts Glucose 1-phosphate into Glucose 6-phosphate

Glucose 1-phosphate formed in the phosphorolytic cleavage of glycogen must be converted into glucose 6-phosphate to enter the metabolic mainstream. This shift of a phosphoryl group is catalyzed by *phosphoglucomutase*. Recall that this enzyme is also used in galactose metabolism (p. 297). To catalyze this shift, the enzyme exchanges a phosphoryl group with the substrate (**Figure 24.4**). The active site of the mutase contains a phosphorylated serine residue. The phosphoryl group is transferred from the serine residue to the C-6 hydroxyl group of glucose 1-phosphate to form glucose 1,6-bisphosphate. The C-1 phosphoryl group of this intermediate is then shuttled to the same serine residue, resulting in the formation of glucose 6-phosphate and the regeneration of the phosphoenzyme.

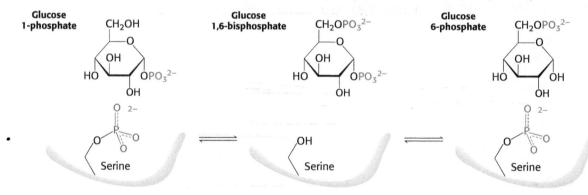

Figure 24.4 The reaction catalyzed by phosphoglucomutase. (A phosphoryl group is transferred from the enzyme to the substrate, and a different phosphoryl group is transferred back to restore the enzyme to its initial state.)

Liver Contains Glucose 6-phosphatase, a Hydrolytic Enzyme Absent from Muscle

A major function of the liver is to maintain a nearly constant concentration of glucose in the blood. The liver releases glucose into the blood during muscular activity and between meals. The released glucose is taken up by the brain, skeletal muscle, and red blood cells. In contrast with unmodified glucose, however, the phosphorylated glucose produced by glycogen breakdown is not transported out of cells. The liver contains a hydrolytic enzyme, *glucose 6-phosphatase*, that enables glucose to leave that organ. This enzyme cleaves the phosphoryl group to form free glucose and orthophosphate:

$$\text{Glucose 6-phosphate} + H_2O \longrightarrow \text{glucose} + P_i$$

This glucose 6-phosphatase is the same enzyme that releases free glucose at the conclusion of gluconeogenesis. It is located on the lumenal side of the smooth endoplasmic reticulum membrane. Recall that glucose 6-phosphate is transported into the endoplasmic reticulum; glucose and orthophosphate formed by hydrolysis are then shuttled back into the cytoplasm (p. 319). Glucose 6-phosphatase is absent from most other tissues. These tissues retain glucose 6-phosphate for the generation of ATP. In contrast, glucose is not a major fuel for the liver.

? QUICK QUIZ 1 What enzymes are required for the liver to release glucose into the blood when an organism is asleep and fasting?

24.2 Phosphorylase Is Regulated by Allosteric Interactions and Reversible Phosphorylation

✓ 2 Explain the regulation of glycogen breakdown.

Glycogen degradation is precisely controlled by multiple interlocking mechanisms. The focus of this control is the enzyme glycogen phosphorylase. *Phosphorylase is regulated by several allosteric effectors that signal the energy state of the cell, as well as by reversible phosphorylation, which is responsive to hormones such as epinephrine, glucagon, and insulin.* We will examine the differences in the control of two isozymic forms of glycogen phosphorylase: one specific to liver and one specific to skeletal muscle. These differences are due to the fact that *the liver maintains glucose homeostasis of the organism as a whole, whereas the muscle uses glucose to produce energy for itself.*

DID YOU KNOW?

Isozymes, or isoenzymes, are enzymes that are encoded by different genes yet catalyze the same reaction. Usually, isozymes display different kinetic parameters or regulatory properties.

Liver Phosphorylase Produces Glucose for Use by Other Tissues

The dimeric phosphorylase exists in two interconvertible forms: a *usually active* phosphorylase *a* and a *usually inactive* phosphorylase *b* (**Figure 24.5**). Each of these two forms exists in equilibrium between an active relaxed (R) state and a much less active tense (T) state, but the equilibrium for phosphorylase *a* favors the active R state, whereas the equilibrium for phosphorylase *b* favors the less active T state (**Figure 24.6**). The role of glycogen degradation in the liver is to

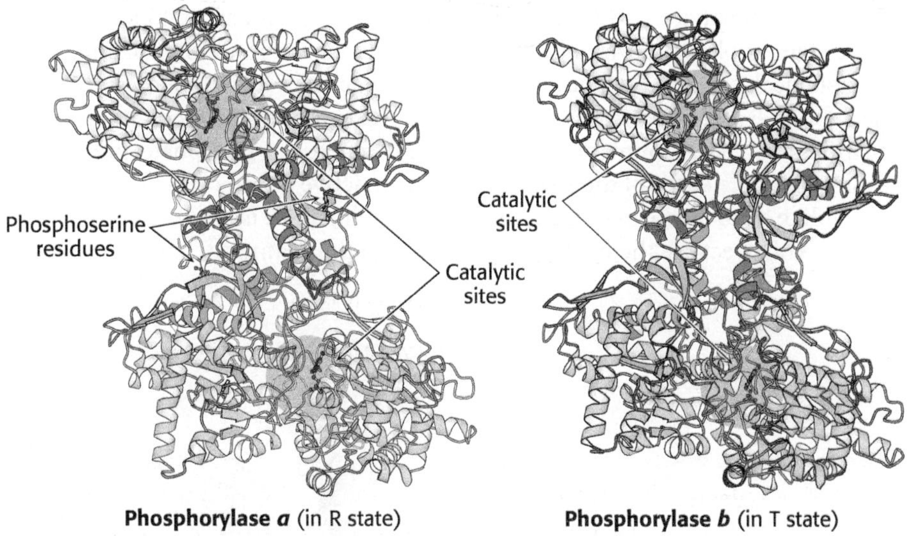

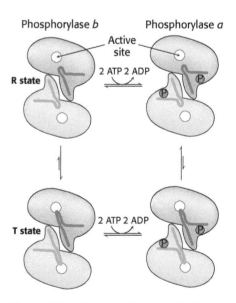

Figure 24.6 Phosphorylase regulation. Both phosphorylase *b* and phosphorylase *a* exist in equilibrium between an active R state and a less active T state. Phosphorylase *b* is usually inactive because the equilibrium favors the T state. Phosphorylase *a* is usually active because the equilibrium favors the R state. In the T state, the active site is partly blocked by a regulatory structure. The active site is unobstructed in the R state. Regulatory structures are shown in blue and green.

Phosphoserine residues

Catalytic sites

Catalytic sites

Phosphorylase *a* (in R state)

Phosphorylase *b* (in T state)

Figure 24.5 Structures of phosphorylase *a* and phosphorylase *b*. Phosphorylase *a* is phosphorylated on serine 14 of each subunit. This modification favors the structure of the more active R state. One subunit is shown in white, with helices and loops important for regulation shown in blue and red. The other subunit is shown in yellow, with the regulatory structures shown in orange and green. Phosphorylase *b* is not phosphorylated and exists predominantly in the T state. Notice that the catalytic sites are partly occluded in the T state. [Drawn from 1GPA.pdb and 1NOJ.pdb.]

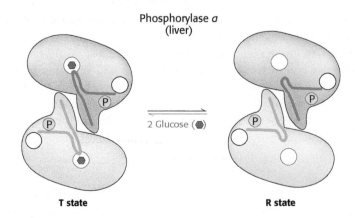

Figure 24.7 The allosteric regulation of liver phosphorylase. The binding of glucose to phosphorylase *a* shifts the equilibrium to the T state and inactivates the enzyme. Thus, glycogen is not mobilized when glucose is already abundant. Regulatory structures are shown in blue and green.

form glucose for export to other tissues when the blood-glucose concentration is low. Consequently, we can think of the default state of liver phosphorylase as being the *a* form: glucose is to be generated unless the enzyme is signaled otherwise. The liver phosphorylase *a* form thus exhibits the most responsive R ↔ T transition (**Figure 24.7**). The binding of glucose to the active site shifts the *a* form from the active R state to the less active T state. In essence, the enzyme reverts to the low-activity T state only when it detects the presence of sufficient glucose. If glucose is present in the diet, there is no need to degrade glycogen.

Muscle Phosphorylase Is Regulated by the Intracellular Energy Charge

In contrast to the liver isozyme, the default state of muscle phosphorylase is the *b* form, owing to the fact that, for muscle, phosphorylase needs to be active primarily during muscle contraction. Muscle phosphorylase *b* is activated by the presence of high concentrations of AMP, which binds to a nucleotide-binding site and stabilizes the conformation of phosphorylase *b* in the active R state (**Figure 24.8**). Recall that phosphofructokinase, the allosteric enzyme that controls glycolysis, is also activated by AMP (p. 299). ATP acts as a negative allosteric effector by competing with AMP. Thus, *the transition of phosphorylase b between the active R state and the less active T state is controlled by the energy charge of the muscle cell.* Glucose 6-phosphate also binds at the same site as ATP and stabilizes the less active state of phosphorylase *b*, an example of feedback inhibition. Under most physiological conditions, *phosphorylase b is inactive because of the inhibitory effects of ATP and glucose 6-phosphate. In contrast, phosphorylase a is fully active,* regardless of the amount of AMP, ATP, and glucose 6-phosphate present. In resting muscle, nearly all the enzyme is in the inactive *b* form.

Unlike the enzyme in muscle, the liver phosphorylase is insensitive to regulation by AMP because the liver does not undergo the dramatic changes in energy

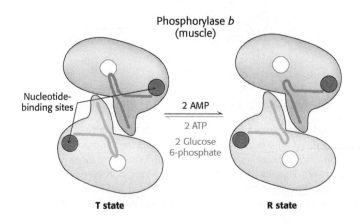

Figure 24.8 The allosteric regulation of muscle phosphorylase. A low energy charge, represented by high concentrations of AMP, favors the transition to the R state. Glucose 6-phosphate and ATP stabilize the T state.

charge seen in a contracting muscle. We see here a clear example of the use of isozymes to establish the tissue-specific biochemical properties of the liver and muscle. In human beings, liver phosphorylase and muscle phosphorylase are approximately 90% identical in amino acid sequence, yet the 10% difference results in subtle but important shifts in the stability of various forms of the enzyme.

Biochemical Characteristics of Muscle Fiber Types Differ

Not only do the biochemical needs of liver and muscle differ with respect to glycogen metabolism, but the biochemical needs of different muscle fiber types also vary. Skeletal muscle consists of three different fiber types. Type I or slow-twitch muscle, type IIb (also called type IIx) or fast-twitch fibers, and type IIa fibers, which have properties intermediate between type I and type IIb. Type I fibers, which power endurance activities, rely predominantly on cellular respiration to derive energy. These fibers are powered by fatty acid degradation and are rich in mitochondria, the site of fatty acid degradation and the citric acid cycle. As we will see in Chapter 27, fatty acids are an excellent energy source, but generating ATP from fatty acids is slower than from glycogen. Glycogen is not an important fuel for type I fibers, and consequently the amount of glycogen phosphorylase is low. Type IIb fibers use glycogen as their main fuel. Consequently, the glycogen and glycogen phosphorylase are abundant. These fibers are also rich in glycolytic enzymes needed to process glucose quickly in the absence of oxygen and poor in mitochondria. Type IIb fibers power burst activities such as sprinting and weight lifting. No amount of training can interconvert type I fibers and type IIb fibers. However, there is some evidence that type IIa fibers are "trainable"; that is, endurance training enhances the oxidative capacity of type IIa fibers while burst activity training enhances the glycolytic capacity. Table 24.1 shows the biochemical profile of the fiber types.

Phosphorylation Promotes the Conversion of Phosphorylase *b* to Phosphorylase *a*

In both liver and muscle, phosphorylase *b* is converted into phosphorylase *a* by the phosphorylation of a single serine residue (serine 14) in each subunit. This conversion is initiated by hormones. Low blood-glucose concentration leads to the secretion of the hormone glucagon. Fear or the excitement of exercise will cause release of the hormone epinephrine. The rise in glucagon and epinephrine concentration results in phosphorylation of the enzyme to the phosphorylase *a*

Table 24.1 Biochemical characteristics of muscle fiber types

Characteristic	Type I	Type IIa	Type IIb
Fatigue resistance	High	Intermediate	Low
Mitochondrial density	High	Intermediate	Low
Metabolic type	Oxidative	Oxidative/glycolytic	Glycolytic
Myoglobin content	High	Intermediate	Low
Glycogen content	Low	Intermediate	High
Triacylglycerol content	High	Intermediate	Low
Glycogen phosphorylase activity	Low	Intermediate	High
Phosphofructokinase activity	Low	Intermediate	High
Citrate synthase activity	High	Intermediate	Low

form in liver and muscle, respectively. The regulatory enzyme phosphorylase kinase catalyzes this covalent modification.

Comparison of the structures of phosphorylase *a* in the R state and phosphorylase *b* in the T state reveals that subtle structural changes at the subunit interfaces are transmitted to the active sites (Figure 24.5). The transition from the T state (the prevalent state of phosphorylase *b*) to the R state (the prevalent state of phosphorylase *a*) is associated with structural changes in α helices that move a loop out of the active site of each subunit. Thus, the T state is less active because the catalytic site is partly blocked. In the R state, the catalytic site is more accessible and a binding site for orthophosphate is well organized.

Phosphorylase Kinase Is Activated by Phosphorylation and Calcium Ions

Phosphorylase kinase activates phosphorylase *b* by attaching a phosphoryl group. The subunit composition of phosphorylase kinase in skeletal muscle is $(\alpha\beta\gamma\delta)_4$, and the mass of this very large protein is 1300 kDa. The enzyme consists of two $(\alpha\beta\gamma\delta)_2$ lobes that are joined by a β_4 bridge that is the core of the enzyme and serves as a scaffold for the remaining subunits. The γ subunit contains the active site, while all of the remaining subunits (≈90% by mass) play regulatory roles. The δ subunit is the calcium-binding protein *calmodulin*, a calcium sensor that stimulates many enzymes in eukaryotes (p. 238). The α and β subunits are targets of *protein kinase A*. The β subunit is phosphorylated first, followed by the phosphorylation of the α subunit.

Activation of phosphorylase kinase is initiated when Ca^{2+} binds to the δ subunit. This mode of activation of the kinase is especially noteworthy in muscle, where contraction is triggered by the release of Ca^{2+} from the sarcoplasmic reticulum (**Figure 24.9**). Maximal activation is achieved with the phosphorylation of the β and α subunits of the Ca^{2+}-bound kinase. The stimulation of phosphorylase kinase is one step in a signal-transduction cascade initiated by signal molecules such as glucagon and epinephrine.

QUICK QUIZ 2 Compare the allosteric regulation of phosphorylase in the liver and in muscle, and explain the significance of the difference.

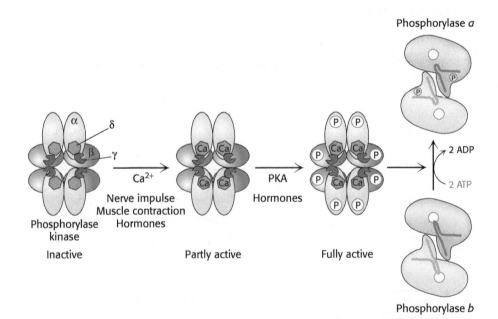

Figure 24.9 The activation of phosphorylase kinase. Phosphorylase kinase, an $(\alpha\beta\gamma\delta)_4$ assembly, is partly activated by Ca^{2+} binding to the δ subunit. Activation is maximal when the β and α subunits are phosphorylated in response to hormonal signals. When active, the enzyme converts phosphorylase *b* into phosphorylase *a*.

Hers Disease Is Due to a Phosphorylase Deficiency

Hers disease, a hereditary disorder, is a glycogen-storage disease that emphasizes the importance of the isozymes of glycogen phosphorylase. Hers disease is caused by a deficiency in the liver isozyme of glycogen phosphorylase. Because glycogen cannot be degraded, it accumulates, leading to an enlargement of the liver (hepatomegaly) that, in some cases, causes the abdomen to protrude. In extreme cases, liver damage may result. Patients with Hers disease display a low concentration of blood glucose (hypoglycemia) because their livers are not able to degrade glycogen. Clinical manifestations of Hers disease vary widely. In some patients, the disease is undetectable and, in others, it causes liver damage and growth retardation. However, in most cases the prognosis is good, and the clinical manifestations improve with age. See problem 6 on page 457 to learn the effects of a deficiency of muscle glycogen phosphorylase.

24.3 Epinephrine and Glucagon Signal the Need for Glycogen Breakdown

Protein kinase A activates phosphorylase kinase, which in turn activates glycogen phosphorylase. How is protein kinase A activated? What is the signal that ultimately triggers an increase in glycogen breakdown?

G Proteins Transmit the Signal for the Initiation of Glycogen Breakdown

As already mentioned, glucagon and epinephrine trigger the breakdown of glycogen. Muscular activity or its anticipation leads to the release of epinephrine, which markedly stimulates glycogen breakdown in muscle and, to a lesser extent, in the liver. The liver is more responsive to glucagon, which signifies the starved state (Figure 24.10).

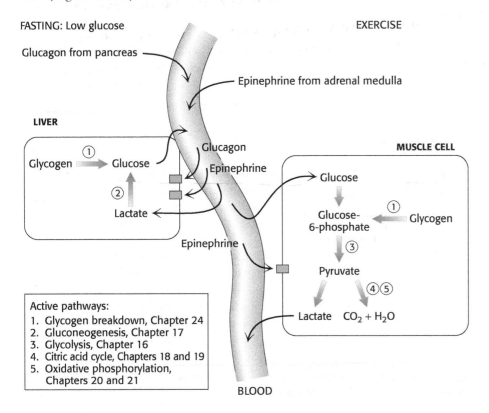

Figure 24.10 The hormonal control of glycogen breakdown. The left side of the illustration shows the hormonal response to fasting. Glucagon stimulates glycogen breakdown in the liver when blood glucose is low. The right side of the illustration shows the hormonal response to exercise. Epinephrine enhances glycogen breakdown in muscle and the liver to provide fuel for muscle contraction.

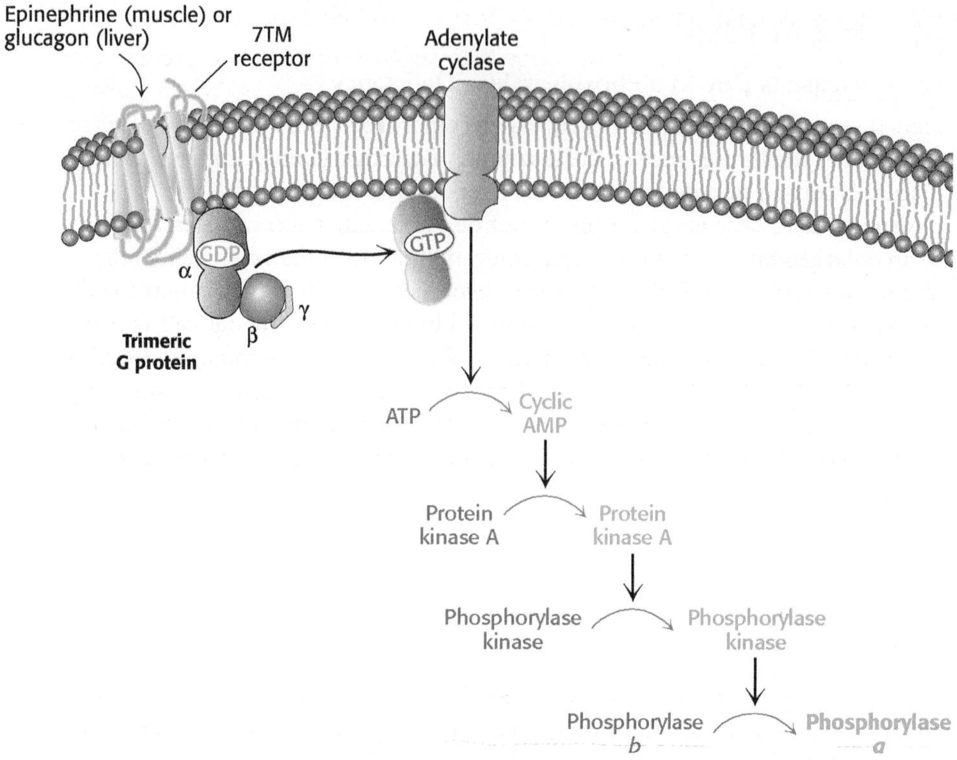

Figure 24.11 The regulatory cascade for glycogen breakdown. Glycogen degradation is stimulated by hormone binding to 7TM receptors. Hormone binding initiates a G-protein-dependent signal-transduction pathway that results in the phosphorylation and activation of glycogen phosphorylase.

How do hormones trigger the breakdown of glycogen? They initiate a cyclic AMP (cAMP) signal-transduction cascade, already discussed in Chapter 13 (Figure 24.11). This cascade leads to the activation of phosphorylase in four steps:

1. The signal molecules epinephrine and glucagon bind to specific seven-transmembrane (7TM) receptors in the plasma membranes of target cells. Epinephrine binds to the β-adrenergic receptor in muscle, whereas glucagon binds to the glucagon receptor in the liver. These binding events activate the $G_{\alpha s}$ protein.

2. The GTP-bound subunit of $G_{\alpha s}$ activates the transmembrane protein adenylate cyclase. This enzyme catalyzes the formation of the second messenger cAMP from ATP.

3. The elevated cytoplasmic concentration of cAMP activates protein kinase A.

4. Protein kinase A phosphorylates phosphorylase kinase, which subsequently activates glycogen phosphorylase.

The cAMP cascade highly amplifies the effects of hormones. The binding of a small number of hormone molecules to cell-surface receptors leads to the release of a very large number of sugar units. Indeed, much of the stored glycogen in the body would be mobilized within seconds were it not for the regulatory elements that will be discussed shortly.

The signal-transduction processes in the liver are more complex than those in muscle. Epinephrine can elicit glycogen degradation in the liver. However, in addition to binding to the β-adrenergic receptor as it does in skeletal muscle, it binds to the 7TM α-adrenergic receptor, which then initiates the *phosphoinositide cascade* (p. 232) that induces the release of Ca^{2+} from endoplasmic reticulum stores. Recall that the δ subunit of phosphorylase kinase is the Ca^{2+} sensor calmodulin. The binding of Ca^{2+} to calmodulin leads to a partial activation of phosphorylase kinase. Stimulation by both glucagon and epinephrine leads to maximal mobilization of liver glycogen.

Glycogen Breakdown Must Be Rapidly Turned Off When Necessary

There must be a rapid way to shut down glycogen breakdown to prevent the wasteful depletion of glycogen after energy needs have been met. When glucose needs have been satisfied, phosphorylase kinase and glycogen phosphorylase are dephosphorylated and inactivated. Simultaneously, glycogen synthesis is activated (Chapter 25).

The signal-transduction pathway leading to the activation of glycogen phosphorylase is shut down automatically when the initiating hormone is no longer present. The inherent GTPase activity of the G protein converts the bound GTP into inactive GDP, and phosphodiesterases always present in the cell convert cAMP into AMP, which does not activate PKA. *Protein phosphatase 1* (PP1) removes the phosphoryl groups from phosphorylase kinase, thereby inactivating the enzyme. Finally, protein phosphatase 1 also removes the phosphoryl group from glycogen phosphorylase, converting the enzyme into the usually inactive *b* form.

BIOLOGICAL INSIGHT

Glycogen Depletion Coincides with the Onset of Fatigue

Although most of us have experienced fatigue, it is a multifaceted and difficult condition to define. There are metabolic, neurological, and psychological components to fatigue, but a common definition is simply the inability to maintain the required energy output. **Figure 24.12** A shows the decrease in glycogen content of the vastus lateralis muscle (a component of the quadriceps) of cyclists who were exercising at 80% of their maximum workload as a function of time. Muscle glycogen phosphorylase was activated through allosteric mechanisms and by hormonally induced covalent modification to such an extent that, after 75 minutes, all of the glycogen in the vastus lateralis muscle was consumed. Moreover, glycogen depletion coincided with the onset of fatigue, or the inability to maintain the required effort. In other words, the cyclists "hit the wall" or "bonked" (Figure 24.12B). However, these experiments do not show that glycogen depletion *causes* fatigue;

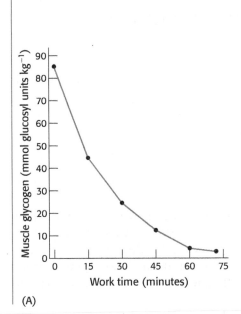

Figure 24.12 Glycogen depletion as a result of exercise. (A) Glycogen content of the vastus lateralis decreases as a function of time at 80% effort. (B) The French cyclist Tony Gallopin slumps in exhaustion after winning a stage in the 2014 edition of the Tour de France. [(A) Data from J. Bergström and E. Hultman, A study of the glycogen metabolism during exercise in man, *Scand. J. Clin. Lab. Invest.* 19:218–226, 1967. (B) ERIC LALMAND/Photoshot/Newscom.]

(A)

(B)

correlation does not mean causation. Some experiments suggest that the increase in ADP concentration resulting from glycogen depletion may be a more direct cause of the fatigue. It is fascinating that an explanation for such a common feeling still eludes scientists and demonstrates once again the scope of how little we know and the opportunities for more exciting research.

SUMMARY

Glycogen, a readily mobilized fuel store, is a branched polymer of glucose residues. Most of the glucose units in glycogen are linked by α-1,4-glycosidic bonds. At about every 10th residue, a branch is created by an α-1,6-glycosidic bond. Glycogen is present in large amounts in muscle cells and in liver cells, where it is stored in the cytoplasm in the form of hydrated granules.

24.1 Glycogen Breakdown Requires Several Enzymes

Most of the glycogen molecule is degraded to glucose 1-phosphate by the action of glycogen phosphorylase, the key enzyme in glycogen breakdown. The glycosidic linkage between C-1 of a terminal residue and C-4 of the adjacent one is split by orthophosphate to give glucose 1-phosphate, which can be reversibly converted into glucose 6-phosphate by phosphoglucomutase. Branch points are degraded by the concerted action of a transferase and an α-1,6-glucosidase.

24.2 Phosphorylase Is Regulated by Allosteric Interactions and Reversible Phosphorylation

Phosphorylase b, which is usually inactive, is converted into active phosphorylase a by the phosphorylation of a single serine residue in each subunit. This reaction is catalyzed by phosphorylase kinase. The b form in muscle can also be activated by the binding of AMP, an effect counteracted by ATP and glucose 6-phosphate. The a form in the liver is inhibited by glucose. Liver phosphorylase is activated to liberate glucose for export to other organs, such as skeletal muscle and the brain. In contrast, muscle phosphorylase is activated to generate glucose for use inside the cell as a fuel for contractile activity.

24.3 Epinephrine and Glucagon Signal the Need for Glycogen Breakdown

Epinephrine and glucagon stimulate glycogen breakdown through specific 7TM receptors. Muscle is the primary target of epinephrine, whereas the liver is responsive to glucagon. Both signal molecules initiate a kinase cascade that leads to the activation of glycogen phosphorylase.

KEY TERMS

glycogen phosphorylase (p. 446)
phosphorolysis (p. 446)
epinephrine (adrenaline) (p. 449)

glucagon (p. 449)
phosphorylase kinase (p. 452)

protein kinase A (p. 452)
calmodulin (p. 452)

? Answers to QUICK QUIZZES

1. Phosphorylase, transferase, glucosidase, phosphoglucomutase, and glucose 6-phosphatase.

2. In muscle, the b form of phosphorylase is activated by AMP. In the liver, the a form is inhibited by glucose. The difference corresponds to the difference in the metabolic role of glycogen in each tissue. Muscle uses glycogen as a fuel for contraction, whereas the liver uses glycogen to maintain proper blood-glucose concentration.

PROBLEMS

1. *Step-by-step degradation.* What are the three steps in glycogen degradation, and what enzymes catalyze each step? ✓ 1

2. *Tweedledum and Tweedledee.* Match each term with its description. ✓ 1

(a) Glycogen phosphorylase _____

(b) Phosphorolysis _____

(c) Transferase _____

(d) α-1,6-Glucosidase _____

(e) Phosphoglucomutase _____

(f) Phosphorylase kinase _____

(g) Protein kinase A _____

(h) Calmodulin _____

(i) Epinephrine _____

(j) Glucagon _____

1. Calcium-binding subunit of phosphorylase kinase
2. Activates glycogen phosphorylase
3. Removal of a glucose residue by the addition of phosphate
4. Stimulates glycogen breakdown in muscle
5. Liberates a free glucose residue
6. Shifts the location of several glucose residues
7. Stimulates glycogen breakdown in the liver
8. Catalyzes phosphorolytic cleavage
9. Prepares glucose 1-phosphate for glycolysis
10. Phosphorylates phosphorylase kinase

3. *For the greater good.* Why is the control of glycogen different in muscle and the liver? ✓ 2

4. *Get out of the way!* What structural difference accounts for the fact that the T state of phosphorylase kinase is less active than the R state? ✓ 2

5. *The regulator's regulator.* What factors result in maximal activation of phosphorylase kinase? ✓ 2

6. *Not all absences are equal.* Hers disease results from an absence of liver glycogen phosphorylase and may result in serious illness. In McArdle disease, muscle glycogen phosphorylase is absent. Although exercise is difficult for patients suffering from McArdle disease, the disease is rarely life threatening. ✓ 2

(a) Account for the different manifestations of the absence of glycogen phosphorylase in the two tissues.

(b) What does the existence of these two different diseases indicate about the genetic nature of the phosphorylase?

7. *Dare to be different.* Compare the allosteric regulation of phosphorylase in the liver and in muscle, and explain the significance of the difference. ✓ 2

8. *An appropriate inhibitor.* What is the biochemical rationale for the inhibition of muscle glycogen phosphorylase by glucose 6-phosphate when glucose 1-phosphate is the product of the phosphorylase reaction? ✓ 2

9. *Metamorphoses.* What is the predominant form of glycogen phosphorylase in resting muscle? Immediately after exercise begins, this form is activated. How does this activation take place? ✓ 2

10. *Passing along the information.* Outline the signal-transduction cascade for glycogen degradation in muscle. ✓ 2

11. *Double activation.* What path in addition to the cAMP-induced signal transduction is used in the liver to maximize glycogen breakdown? ✓ 2

12. *Slammin' on the brakes.* There must be a way to shut down glycogen breakdown quickly to prevent the wasteful depletion of glycogen after energy needs have been met. What mechanisms are employed to turn off glycogen breakdown? ✓ 2

13. *Choice is good.* Glycogen is not as reduced as fatty acids are and consequently not as energy rich. Why do animals store any energy as glycogen? Why not convert all excess fuel into fatty acids?

14. *Feeling depleted.* Glycogen depletion resulting from intense, extensive exercise can lead to exhaustion and the inability to continue exercising. Some people also experience dizziness, an inability to concentrate, and a loss of muscle control. Account for these symptoms.

15. *Family resemblance.* In problem 23 of Chapter 16, you were asked to consider the effects of exposing glycolytically active cells to arsenate. Recall that arsenate can substitute for phosphate, but that arsenate esters are unstable and spontaneously decompose to arsenate and a carboxylic acid. What will the energetic consequences be if glycogen phosphorylase uses arsenate instead of phosphate? ✓ 1

16. *Working together.* One of the liver's key roles is the maintenance of blood-glucose concentration when an organism is fasting, such as during a night's sleep. Mobilizing liver glycogen requires enzymatic teamwork. Identify the enzymes that are required for the liver to release glucose into the blood. ✓ 1

17. *Everyone has a job to do.* What accounts for the fact that liver phosphorylase is a glucose sensor, whereas muscle phosphorylase is not? ✓ 2

18. *If a little is good, a lot is better.* Amylose is an unbranched glucose polymer. Why would this polymer not be as effective a storage form of glucose as glycogen?

19. *R and T, a and b.* Glycogen phosphorylase can exist in the following states. ✓ 2

A. Phosphorylase *a*   T state
B. Phosphorylase *a*   R state
C. Phosphorylase *b*   T state
D. Phosphorylase *b*   R state

(a) Which forms of the enzyme are most active?
(b) What enzyme catalyzes the C-to-A conversion?
(c) In muscle, high concentrations of AMP cause a transition between what two forms?
(d) In liver, the transition between what two forms is stimulated by glucose?
(e) In muscle, which transition is stimulated by glucose 6-phosphate?
(f) What enzyme converts A into C?

20. *Two in one.* A single polypeptide chain houses the transferase and debranching enzyme. What is a potential advantage of this arrangement? ✓ 2

21. *How did they do that?* A strain of mice has been developed that lack the enzyme phosphorylase kinase. Yet, after strenuous exercise, the glycogen stores of a mouse of this strain are depleted. Explain how this depletion is possible. ✓ 2

Chapter Integration and Challenge Problems

22. *A shattering experience.* Crystals of phosphorylase *a* grown in the presence of glucose shatter when a substrate such as glucose 1-phosphate is added. Why? ✓ 2

23. *Two for the binding of one.* Glycogen breakdown in the liver is stimulated by glucagon. What other carbohydrate-metabolism pathway in the liver is stimulated by glucagon?

24. *An ATP saved is an ATP earned.* The complete oxidation of glucose 6-phosphate derived from free glucose yields 30 molecules of ATP, whereas the complete oxidation of glucose 6-phosphate derived from glycogen yields 31 molecules of ATP. Account for this difference. ✓ 1

25. *A thumb on the balance.* The reaction catalyzed by phosphorylase is readily reversible in vitro. At pH 6.8, the equilibrium ratio of orthophosphate to glucose 1-phosphate is 3.6. The value of $\Delta G^{\circ\prime}$ for this reaction is small because a glycosidic bond is replaced by a phosphoryl ester bond that has a nearly equal transfer potential. However, phosphorolysis proceeds far in the direction of glycogen breakdown in vivo. Suggest one means by which the reaction can be made irreversible in vivo. ✓ 1

26. *Hydrophobia.* Why is water excluded from the active site of phosphorylase? Predict the effect of a mutation that allows water molecules to enter.

27. *Quenching release.* Type 2 diabetes is a condition characterized by insulin resistance and high blood-glucose concentration. Research is underway to develop inhibitors of glycogen phosphorylase as a possible treatment for type 2 diabetes. What is the rationale for this strategy, and what is one potential problem with the approach?

Data Interpretation Problem

28. *An authentic replica.* Experiments were performed in which serine (S) 14 of glycogen phosphorylase was replaced by glutamate (E). The V_{max} of the mutant enzyme was then compared with the wild-type phosphorylase in both the *a* and the *b* forms.

	V_{max} μmol of glucose 1-PO$_4$ released min^{-1} mg^{-1}
Wild-type phosphorylase *b*	25 ± 0.4
Wild-type phosphorylase *a*	100 ± 5
S to E mutant	60 ± 3

(a) Explain the results obtained with the mutant.
(b) Predict the effect of substituting aspartic acid for the serine.

Selected Readings for this chapter can be found online at www.whfreeman.com/tymoczko3e.

CHAPTER 25

Glycogen Synthesis

Pasta and pizza are good energy sources for a host of athletic contests. Such carbohydrate-rich meals, consumed in the days preceding an endurance event such as a marathon, ensure the presence of the muscle glycogen required for a strong race. [AP Photo/Steven Senne.]

25.1 Glycogen Is Synthesized and Degraded by Different Pathways

25.2 Metabolism in Context: Glycogen Breakdown and Synthesis Are Reciprocally Regulated

As we have seen in glycolysis and gluconeogenesis, biosynthetic and degradative pathways rarely operate by precisely the same reactions in the forward and reverse directions. Glycogen metabolism provided the first known example of this important principle. *Separate pathways afford much greater flexibility, both in energetics and in control.*

We begin by learning about the substrate for glycogen biosynthesis and the enzymes required to synthesize this branched polymer. We then investigate how glycogen synthesis is controlled, paying particular attention to the coordinated regulation of glycogen synthesis and degradation. Finally, we see how the liver uses glycogen metabolism to maintain the required blood-glucose concentration.

✓ 3 Describe the steps of glycogen synthesis, and identify the enzymes required.
✓ 4 Explain the regulation of glycogen synthesis.

25.1 Glycogen Is Synthesized and Degraded by Different Pathways

A common theme for the biosynthetic pathways that we will encounter in our study of biochemistry is the requirement for an activated precursor. This axiom holds true for glycogen synthesis. Glycogen is synthesized by a pathway that utilizes *uridine diphosphate glucose* (UDP-glucose) rather than glucose 1-phosphate as the

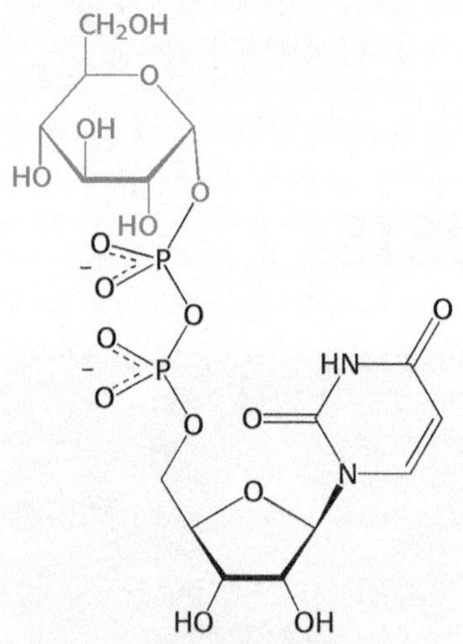

Uridine diphosphate glucose (UDP-glucose)

activated glucose donor. Recall that glucose 1-phosphate is also the product of glycogen phosphorylase. We have already encountered UDP-glucose in our consideration of galactose metabolism (p. 296). The C-1 carbon atom of the glucosyl unit of UDP-glucose is activated because its hydroxyl group is esterified to the diphosphate of UDP:

$$\text{Synthesis: Glycogen}_n + \text{UDP-glucose} \longrightarrow \text{glycogen}_{n+1} + \text{UDP}$$

$$\text{Degradation: Glycogen}_{n+1} + \text{P}_i \longrightarrow \text{glycogen}_n + \text{glucose 1-phosphate}$$

UDP-Glucose Is an Activated Form of Glucose

UDP-glucose, the glucose donor in the biosynthesis of glycogen, is an activated form of glucose, just as ATP and acetyl CoA are activated forms of orthophosphate and acetate, respectively. UDP-glucose is synthesized from glucose 1-phosphate and the nucleotide uridine triphosphate (UTP) in a reaction catalyzed by *UDP-glucose pyrophosphorylase*. This reaction liberates the outer two phosphoryl residues of UTP as pyrophosphate:

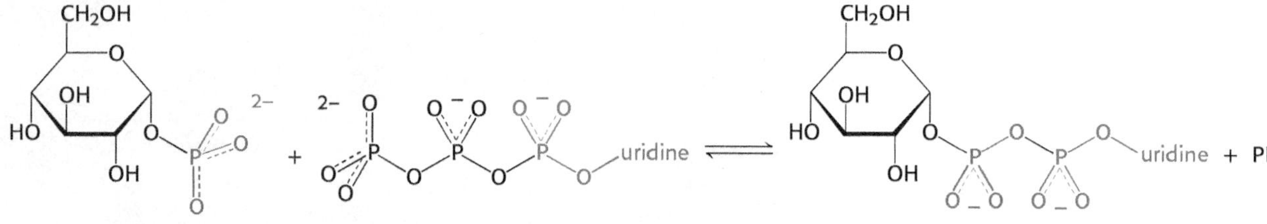

Glucose 1-phosphate   **UTP**   **UDP-glucose**

This reaction is readily reversible. However, pyrophosphate is rapidly hydrolyzed in vivo to orthophosphate by an inorganic pyrophosphatase. The essentially irreversible hydrolysis of pyrophosphate drives the synthesis of UDP-glucose:

$$\text{Glucose 1-phosphate} + \text{UTP} \rightleftharpoons \text{UDP-glucose} + \text{PP}_i$$

$$\underline{\text{PP}_i + \text{H}_2\text{O} \longrightarrow 2\,\text{P}_i}$$

$$\text{Glucose 1-phosphate} + \text{UTP} \longrightarrow \text{UDP-glucose} + 2\,\text{P}_i$$

The synthesis of UDP-glucose exemplifies another recurring theme in biochemistry: *many biosynthetic reactions are driven by the hydrolysis of pyrophosphate.*

Glycogen Synthase Catalyzes the Transfer of Glucose from UDP-Glucose to a Growing Chain

New glucosyl units are added to the nonreducing terminal residues of glycogen. The activated glucosyl unit of UDP-glucose is transferred to the hydroxyl group at C-4 of a terminal residue within a chain of glycogen to form an α-1,4-glycosidic linkage. UDP is displaced by the terminal hydroxyl group of the growing glycogen molecule. This reaction is catalyzed by *glycogen synthase, the key regulatory enzyme in glycogen synthesis*. Humans have two isozymic forms of glycogen synthase: one is specific to the liver while the other is expressed in muscle and other tissues.

Glycogen synthase can add glucosyl residues only to a polysaccharide chain already containing more than four residues. Thus, glycogen synthesis requires a *primer*. This priming function is carried out by *glycogenin*, an enzyme composed of two identical 37-kDa subunits. Each subunit of glycogenin catalyzes the formation of α-1,4-glucose polymers, 10 to 20 glucosyl units in length, on its partner subunit. The glucosyl units are covalently attached to the hydroxyl group of a specific tyrosine residue. UDP-glucose is the donor in this reaction. At this point, glycogen synthase takes over to extend the glycogen molecule. Thus, buried deeply inside each glycogen molecule lies a kernel of glycogenin (**Figure 25.1**).

A Branching Enzyme Forms Alpha-1,6 Linkages

Glycogen synthase catalyzes only the synthesis of α-1,4 linkages. Another enzyme is required to form the α-1,6 linkages that make glycogen a branched polymer. Branching takes place after a number of glucosyl residues are joined in α-1,4 linkages by glycogen synthase. A branch is created by the breaking of an α-1,4 link and the formation of an α-1,6 link. A block of residues, typically 7 in number, is transferred to a more interior site. The *branching enzyme* that catalyzes this reaction is quite exacting (**Figure 25.2**). The block of 7 or so residues must include the nonreducing terminus and come from a chain at least 11 residues long. In addition, the new branch point must be at least 4 residues away from a preexisting one.

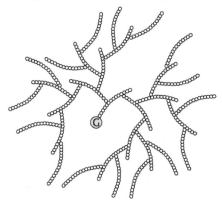

Figure 25.1 A cross section of a glycogen molecule. The component identified as G is glycogenin.

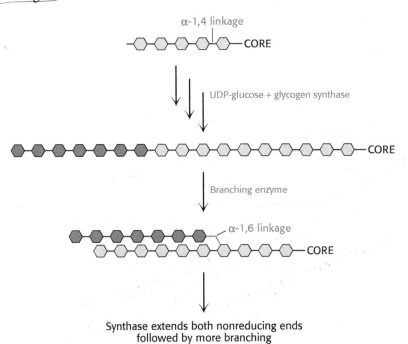

Synthase extends both nonreducing ends
followed by more branching

Figure 25.2 Branching reaction. The branching enzyme removes an oligosaccharide of approximately 7 residues from the nonreducing end and creates an internal α-1,6 linkage.

Branching is important because it increases the solubility of glycogen. Furthermore, branching creates a large number of terminal residues, the sites of action of glycogen phosphorylase and synthase (Figure 25.1). Thus, *branching increases the rate of glycogen synthesis and degradation.*

Glycogen Synthase Is the Key Regulatory Enzyme in Glycogen Synthesis

Glycogen synthase, like phosphorylase, exists in two forms: an active nonphosphorylated *a* form and a usually inactive phosphorylated *b* form. Again, like the phosphorylase, the interconversion of the two forms is regulated by covalent modification. However, the key means of regulating glycogen synthase is by allosteric regulation of the phosphorylated form of the enzyme, glycogen synthase *b*. Glucose 6-phosphate is a powerful activator of the enzyme, stabilizing the R state of the enzyme relative to the T state.

The covalent modification of glycogen synthase appears to play more of a fine-tuning role. The synthase is phosphorylated at multiple sites by several protein kinases—notably, *glycogen synthase kinase* (GSK), which is under the control of insulin (p. 465), and *protein kinase A.* The function of the multiple phosphorylation sites is still under investigation. Note that *phosphorylation has opposite effects on the enzymatic activities of glycogen synthase and glycogen phosphorylase.*

QUICK QUIZ 1 Why is the fact that phosphorylation has opposite effects on glycogen synthesis and breakdown advantageous?

Glycogen Is an Efficient Storage Form of Glucose

What is the cost of converting dietary glucose into glycogen and then into glucose 6-phosphate? Before we make this calculation, we need to introduce another enzyme, *nucleoside diphosphokinase*. This enzyme catalyzes the regeneration of UTP from UDP, a product released when glycogen grows by the addition of glucose from UDP-glucose. ATP is used by the diphosphokinase to phosphorylate UDP. The summation of the reactions in glycogen synthesis and degradation is

$$\text{Glucose} + \text{ATP} \longrightarrow \text{glucose 6-phosphate} + \text{ADP} \tag{1}$$

$$\text{Glucose 6-phosphate} \longrightarrow \text{glucose 1-phosphate} \tag{2}$$

$$\text{Glucose 1-phosphate} + \text{UTP} \longrightarrow \text{UDP-glucose} + \text{PP}_i \tag{3}$$

$$\text{PP}_i + \text{H}_2\text{O} \longrightarrow 2\,\text{P}_i \tag{4}$$

$$\text{UDP-glucose} + \text{glycogen}_n \longrightarrow \text{glycogen}_{n+1} + \text{UDP} \tag{5}$$

$$\text{UDP} + \text{ATP} \longrightarrow \text{UTP} + \text{ADP} \tag{6}$$

$$\text{Sum: Glucose} + 2\,\text{ATP} + \text{glycogen}_n + \text{H}_2\text{O} \longrightarrow$$
$$\text{glycogen}_{n+1} + 2\,\text{ADP} + 2\,\text{P}_i$$

Thus, 2 molecules of ATP are hydrolyzed to incorporate dietary glucose into glycogen. The energy yield from the breakdown of glycogen formed from dietary glucose is highly efficient. About 90% of the residues are cleaved by phosphorolysis to glucose 1-phosphate, which is converted into glucose 6-phosphate without expending an ATP molecule. The other 10% are branch residues that are hydrolytically cleaved. One molecule of ATP is then used to phosphorylate each of these glucose molecules to glucose 6-phosphate. As we saw in Chapters 16 through 21, the complete oxidation of glucose 6-phosphate yields about 31 molecules of ATP, and storage consumes slightly more than 2 molecules of ATP per molecule of glucose 6-phosphate; thus, only 2 molecules of ATP are required to store glucose as glycogen, but glycogen-derived glucose generates 31 molecules of ATP; so *the overall efficiency of storage is nearly 94%.*

✓ 5 Describe how glycogen mobilization and synthesis are coordinated.

25.2 Metabolism in Context: Glycogen Breakdown and Synthesis Are Reciprocally Regulated

An important control mechanism prevents glycogen from being synthesized at the same time as it is being broken down. *The same glucagon- and epinephrine-triggered cAMP cascades that initiate glycogen breakdown in the liver and muscle, respectively, also shut off glycogen synthesis.* Glucagon and epinephrine control both glycogen breakdown and glycogen synthesis through protein kinase A (Figure 25.3). Recall that protein kinase A adds a phosphoryl group to phosphorylase kinase, activating that enzyme and initiating glycogen breakdown. Glycogen synthase kinase and protein kinase A add a phosphoryl group to glycogen synthase, but this phosphorylation leads to a *decrease* in enzymatic activity. In this way, glycogen breakdown and synthesis are reciprocally regulated. How is the enzymatic activity of glycogen phosphorylase reversed so that glycogen breakdown halts and glycogen synthesis begins?

Protein Phosphatase 1 Reverses the Regulatory Effects of Kinases on Glycogen Metabolism

After a bout of exercise, muscle must shift from a glycogen-degrading mode to one of glycogen replenishment. A first step in this metabolic task is to shut down the phosphorylated proteins that stimulate glycogen breakdown. This step is

DURING EXERCISE OR FASTING

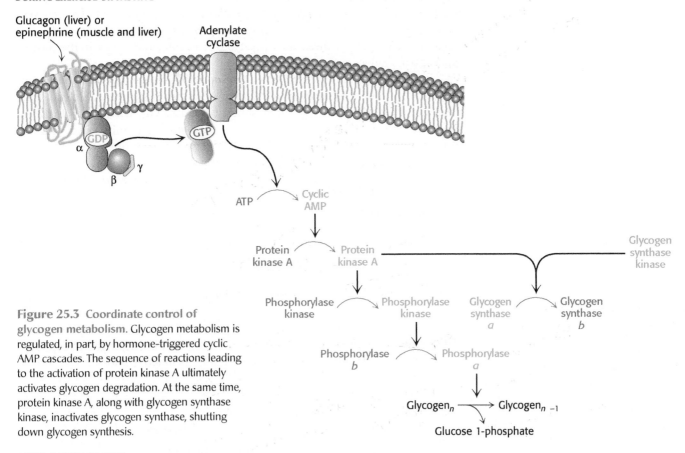

Figure 25.3 Coordinate control of glycogen metabolism. Glycogen metabolism is regulated, in part, by hormone-triggered cyclic AMP cascades. The sequence of reactions leading to the activation of protein kinase A ultimately activates glycogen degradation. At the same time, protein kinase A, along with glycogen synthase kinase, inactivates glycogen synthase, shutting down glycogen synthesis.

AFTER A MEAL OR REST

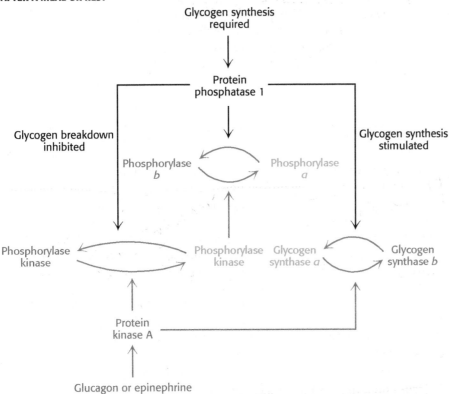

Figure 25.4 The regulation of glycogen synthesis by protein phosphatase 1. PP1 stimulates glycogen synthesis while inhibiting glycogen breakdown.

accomplished by *protein phosphatases* that catalyze the hydrolysis of phosphorylated serine and threonine residues in proteins. *Protein phosphatase 1 (PP1) plays key roles in regulating glycogen metabolism* (Figure 25.4). PP1 inactivates phosphorylase *a* and phosphorylase kinase by dephosphorylating them. PP1 thereby decreases the rate of glycogen breakdown; it reverses the effects of the phosphorylation cascade. Moreover, *PP1 removes phosphoryl groups from glycogen synthase b to convert it into the more active glycogen synthase a form.* Here, PP1 also accelerates glycogen synthesis. PP1 is yet another molecular device in the coordinate control of glycogen metabolism.

The catalytic subunit of PP1 is a single-domain protein. This subunit is usually bound to one of a family of regulatory subunits; in skeletal muscle and the heart, the most prevalent regulatory subunit is called G_M, whereas, in the liver, the most prevalent subunit is G_L. These regulatory subunits have multiple domains that participate in interactions with glycogen, with the catalytic subunit of the protein phosphatase, and with target enzymes. Thus, *these regulatory subunits act as scaffolds, bringing together the protein phosphatase and its substrates in the context of a glycogen particle.*

What prevents the phosphatase activity of PP1 from always inhibiting glycogen degradation? When glycogen degradation is called for, epinephrine or glucagon initiates the cAMP cascade that activates protein kinase A (**Figure 25.5**). Protein kinase A reduces the activity of PP1 by two mechanisms. First, in muscle, G_M is phosphorylated, resulting in the release of the catalytic subunit. Dephosphorylation is therefore greatly reduced. Second, almost all tissues contain small proteins that, when phosphorylated, inhibit the catalytic subunit of PP1. Thus, when glycogen degradation is switched on by cAMP, the phosphorylation of these inhibitors shuts off protein phosphatases, keeping glycogen phosphorylase in its active *a* form and glycogen synthase in its inactive *b* form.

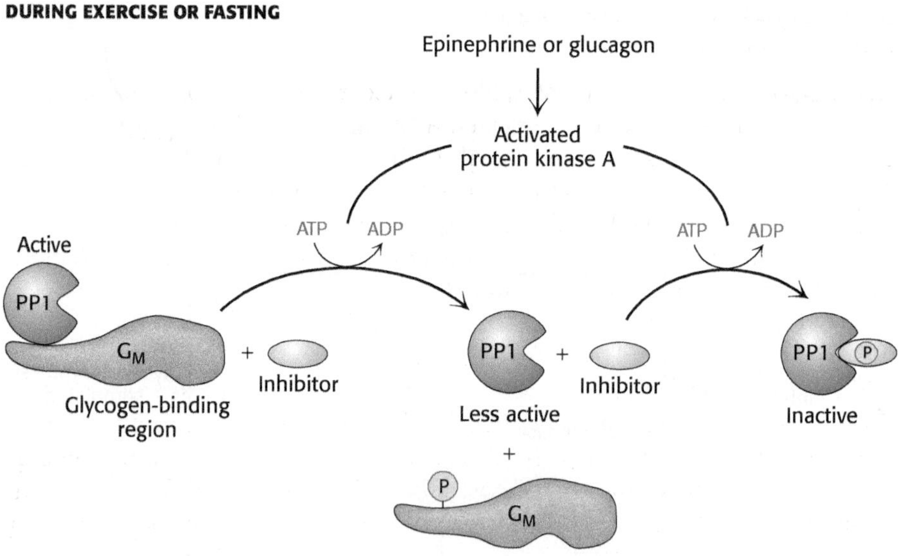

Figure 25.5 The regulation of protein phosphatase 1 in muscle takes place in two steps. Phosphorylation of G_M by protein kinase A dissociates the catalytic subunit from its substrates in the glycogen particle. Phosphorylation of the inhibitor subunit by protein kinase A inactivates the catalytic unit of PP1.

Insulin Stimulates Glycogen Synthesis by Inactivating Glycogen Synthase Kinase

After exercise, people often consume carbohydrate-rich foods to restock their glycogen stores. Indeed, the primary means of clearing glucose from the blood in human beings is its conversion into muscle glycogen. How is glycogen synthesis stimulated? When blood-glucose concentration is high, *insulin stimulates the synthesis of glycogen.* This stimulation has two components. First, recall that insulin leads to an increase in the amount of glucose in the cell by increasing the number of glucose transporters (GLUT4) in the cell membrane (p. 303). The entry of glucose and its subsequent conversion into glucose 6-phosphate allosterically activates glycogen synthase *b*. Second, *insulin leads to the inactivation of glycogen synthase kinase,* the enzyme that maintains glycogen synthase in its phosphorylated, less active state (**Figure 25.6**).

How does insulin exert its effects? The first step in the action of insulin is binding to its receptor tyrosine kinase in the plasma membrane (p. 235). The binding of insulin activates the tyrosine kinase activity of the receptor so that it phosphorylates insulin-receptor substrates. These phosphorylated proteins trigger signal-transduction pathways that eventually lead to the movement of glucose transporters to the cell membrane and to the activation of protein kinases that phosphorylate and inactivate glycogen synthase kinase. The inactive kinase can no longer maintain glycogen synthase in its phosphorylated, less active state. In the meantime, protein phosphatase 1 dephosphorylates glycogen synthase, further stimulating the enzyme and restoring glycogen reserves. The net effect of insulin is the replenishment of glycogen stores.

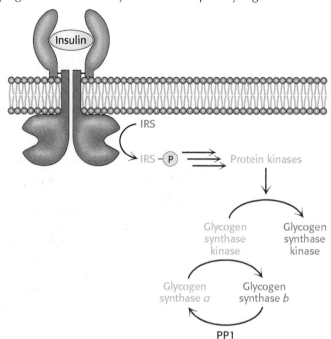

Figure 25.6 Insulin inactivates glycogen synthase kinase. Insulin triggers a cascade that leads to the phosphorylation and inactivation of glycogen synthase kinase and prevents the phosphorylation of glycogen synthase. Protein phosphatase 1 (PP1) removes the phosphoryl groups from glycogen synthase, thereby activating the enzyme and allowing glycogen synthesis. Abbreviation: IRS, insulin–receptor substrate.

Glycogen Metabolism in the Liver Regulates the Blood–Glucose Concentration

After a meal rich in carbohydrates, blood-glucose concentration rises, and glycogen synthesis is increased in the liver. Although insulin is the primary signal for glycogen synthesis, another signal is the concentration of glucose in the blood, which normally ranges from 4.4 to 6.7 mM. The liver senses the concentration of glucose in the blood and takes up or releases glucose to maintain a normal concentration. The amount of liver phosphorylase *a* decreases rapidly when glucose is infused into the blood (**Figure 25.7**). In fact, *phosphorylase a is the glucose sensor in liver cells.* The binding of glucose to phosphorylase *a* shifts its allosteric equilibrium from the active R form to the inactive T form. This conformational change *renders the phosphoryl group on serine 14 a substrate for protein phosphatase 1.* PP1 binds tightly to phosphorylase *a* only when the phosphorylase is in the R state, but PP1 is inactive when bound. When glucose induces the transition to the T form, PP1 dissociates from the phosphorylase and becomes active, converting liver phosphorylase *a* into the *b* form. Recall that the R $\rightleftharpoons$ T transition of muscle phosphorylase *a* is unaffected by glucose and is thus unaffected by the rise in blood-glucose concentration (p. 450).

After a period in which the amount of glycogen phosphorylase *a* is decreasing, the amount of glycogen synthase *a* increases, resulting in glycogen being synthesized from the glucose that has now entered the liver. How does glucose activate glycogen synthase? Phosphorylase *b*, in contrast with phosphorylase *a*, does not bind the phosphatase. Consequently, *the conversion of a into b is accompanied by the release of PP1, which is then free to activate glycogen synthase and inactivate glycogen phosphorylase* (**Figure 25.8**). The removal of the phosphoryl group of inactive glycogen synthase *b* converts it into the active *a* form. Initially, there are about 10 phosphorylase *a* molecules per molecule of phosphatase. Consequently, most of the phosphorylase *a* is converted into *b* before any

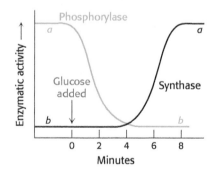

Figure 25.7 Blood glucose regulates liver-glycogen metabolism. The infusion of glucose into the bloodstream leads to the inactivation of phosphorylase followed by the activation of glycogen synthase in the liver. [Data from W. Stalmans, H. De Wulf, L. Hue, and H.-G. Hers, *Eur. J. Biochem.* 41:117–134, 1974.]

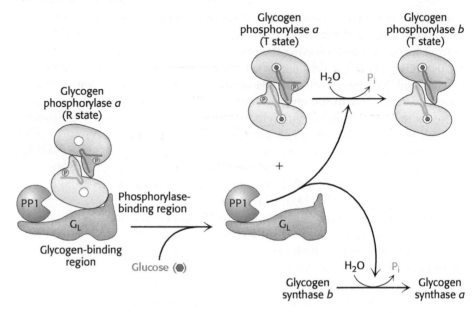

Figure 25.8 Glucose regulates liver-glycogen metabolism. In the liver, glucose binds to glycogen phosphorylase *a* and inhibits it, facilitating the formation of the T state of phosphorylase *a*. The T state of phosphorylase *a* does not bind protein phosphatase 1 (PP1), leading to the dissociation of PP1 from glycogen phosphorylase *a* and its subsequent activation. The free PP1 dephosphorylates glycogen phosphorylase *a* and glycogen synthase *b*, leading to the inactivation of glycogen breakdown and the activation of glycogen synthesis.

phosphatase is released. Hence, *the activity of glycogen synthase begins to increase only after most of the phosphorylase is inactivated* (Figure 25.7). The lag between degradation and synthesis prevents the two pathways from operating simultaneously. This remarkable glucose-sensing system depends on three key elements: (1) communication between the allosteric site for glucose and the serine phosphate, (2) the use of PP1 to inactivate phosphorylase and activate glycogen synthase, and (3) the binding of the phosphatase to phosphorylase *a* to prevent the premature activation of glycogen synthase.

❓ QUICK QUIZ 2 What accounts for the fact that liver phosphorylase is a glucose sensor, whereas muscle phosphorylase is not?

⚕ CLINICAL INSIGHT

Diabetes Mellitus Results from Insulin Insufficiency and Glucagon Excess

Diabetes mellitus (usually referred to simply as *diabetes*) is a complex disease characterized by grossly abnormal fuel usage: *glucose is overproduced by the liver and underutilized by other organs*. The incidence of diabetes mellitus is about 5% of the world's population. Indeed, diabetes is the most common serious metabolic disease in the world; it affects hundreds of millions of people. *Type 1 diabetes*, or *insulin-dependent diabetes mellitus* (IDDM), is caused by the autoimmune destruction of the insulin-secreting β cells in the pancreas and usually begins before age 20. Insulin dependency means that the affected person requires the administration of insulin to live. In contrast, most diabetics have a normal or even higher than normal concentration of insulin in their blood, but these diabetics are quite unresponsive to the hormone, a condition called *insulin resistance*. As discussed in Chapter 17, this form of the disease, known as *type 2*, or *non-insulin-dependent diabetes mellitus* (NIDDM), typically arises later in life than does the insulin-dependent form.

In type 1 diabetes, insulin production is insufficient and consequently glucagon is present at higher-than-normal concentrations. Because insulin is deficient, *the entry of glucose into adipose and muscle cells is impaired*. The liver becomes stuck in a gluconeogenic and ketogenic, or fat-utilizing, state (Chapter 27). In essence, the diabetic person is in biochemical starvation mode despite a high concentration of blood glucose. The excessive amount of glucagon relative to that of insulin leads to a decrease in the concentration of fructose 2,6-bisphosphate (F-2,6-BP) in the liver. Hence, glycolysis is inhibited and gluconeogenesis is stimulated because of the reciprocal effects of F-2,6-BP on phosphofructokinase and fructose 1,6-bisphosphatase (p. 321). The high glucagon-to-insulin ratio in diabetes also promotes glycogen breakdown. Hence, *an excessive amount of glucose is produced by the liver and released into the blood*. Glucose is excreted in the urine when its concentration in the blood exceeds the reabsorptive capacity of the renal tubules. Water accompanies the excreted glucose, and so an untreated diabetic in the acute phase of the disease is abnormally hungry and thirsty.

⚕ CLINICAL INSIGHT

A Biochemical Understanding of Glycogen-Storage Diseases Is Possible

Edgar von Gierke described the first glycogen-storage disease in 1929. A patient with this disease has a huge abdomen caused by a massive enlargement of the liver. There is a pronounced *hypoglycemia* (low blood-glucose concentration) between meals. Furthermore, the blood-glucose concentration does not rise on the administration of epinephrine and glucagon. An infant with this glycogen-storage disease may have convulsions because of hypoglycemia.

The enzymatic defect in von Gierke disease was elucidated in 1952 by Carl and Gerty Cori, the husband and wife team who were awarded the Nobel Prize in 1947 for their work on glycogen metabolism. They found that glucose 6-phosphatase is missing from the liver of a patient with this disease. This finding was the first demonstration of an inherited deficiency of a liver enzyme. The liver glycogen is normal in structure but present in abnormally large amounts. The absence of glucose 6-phosphatase in the liver causes hypoglycemia because glucose cannot be formed from glucose 6-phosphate. This phosphorylated sugar does not leave the liver, because it cannot cross the plasma membrane. The presence of excess glucose 6-phosphate triggers an increase in glycolysis in the liver, leading to high concentrations of lactate and pyruvate in the blood. Patients who have von Gierke disease also have an increased dependence on fat metabolism. This disease can also be produced by a mutation in the gene that encodes the *glucose 6-phosphate transporter*. Recall that glucose 6-phosphate must be transported into the lumen of the endoplasmic reticulum to be hydrolyzed by phosphatase (p. 319). Mutations in the other three essential proteins of this system can likewise lead to von Gierke disease.

Seven other glycogen-storage diseases have been characterized since von Gierke's first characterization, and the biochemical reasons for the symptoms of these diseases have been elucidated (Table 25.1). In Pompe disease (type II), lysosomes become engorged with glycogen because they lack α-1,4-glucosidase, a hydrolytic enzyme confined to these organelles. The lysosomes sometimes burst, releasing glycogen into the cytoplasm (Figure 25.9). The Coris elucidated the biochemical defect in another glycogen-storage disease (type III) that now bears their name. In type III disease,

Table 25. 1   Glycogen-storage diseases

	Type	Defective enzyme	Organ affected	Glycogen in the affected organ	Clinical features
I	von Gierke disease	Glucose 6-phosphatase or transport system	Liver and kidney	Increased amount; normal structure	Massive enlargement of the liver. Failure to thrive. Severe hypoglycemia, ketosis, hyperuricemia, hyperlipemia.
II	Pompe disease	α-1,4-Glucosidase (lysosomal)	All organs	Massive increase in amount; normal structure	Cardiorespiratory failure causes death, usually before age 2.
III	Cori disease	α-1,6-glucosidase (debranching enzyme)	Muscle and liver	Increased amount; short outer branches	Like type I, but milder course.
IV	Andersen disease	Branching enzyme (α-1,4 $\longrightarrow$ α-1,6)	Liver and spleen	Normal amount; very long outer branches	Progressive cirrhosis of the liver. Liver failure causes death, usually before age 2.
V	McArdle disease	Phosphorylase	Muscle	Moderately increased amount; normal structure	Limited ability to perform strenuous exercise because of painful muscle cramps. Otherwise patient is normal and well developed.
VI	Hers disease	Phosphorylase	Liver	Increased amount	Like type l, but milder course.
VII		Phosphofructokinase	Muscle	Increased amount; normal structure	Like type V.
VIII		Phosphorylase kinase	Liver	Increased amount; normal structure	Mild liver enlargement. Mild hypoglycemia.

Note: Types I through VII are inherited as autosomal recessives. Type VIII is sex linked.

the structure of liver and muscle glycogen is abnormal and the amount is markedly increased. Most striking, the outer branches of the glycogen are very short. *Patients having this type of glycogen-storage disorder lack the debranching enzyme (α-1,6-glucosidase),* and so only the outermost branches of glycogen can be effectively utilized. Thus, only a small fraction of this abnormal glycogen is functionally active as an accessible store of glucose.

A defect in glycogen metabolism confined to muscle is found in McArdle disease (type V). Muscle phosphorylase activity is absent, and a patient's capacity to perform strenuous exercise is limited because of painful muscle cramps. The patient is otherwise normal and well developed. Thus, the effective utilization of muscle glycogen is not essential for life.

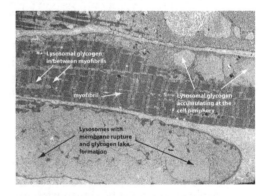

Figure 25.9 Electron micrograph of muscle cell from a patient with Pompe disease. Glycogen-engorged lysosomes are seen throughout the cell, including the myofibrils. As the disease progresses, lysosomes may rupture, releasing large amounts of glycogen into the cytoplasm. Such accumulations of cytoplasmic glycogen are called glycogen lakes. [Reproduced with permission of the author, from B. L. Thurberg et. al., Characterization of pre- and post-treatment pathology after enzyme replacement for Pompe disease. *Lab. Invest.* 86(12):1208–1220, 2006.]

SUMMARY

25.1 Glycogen Is Synthesized and Degraded by Different Pathways

Glycogen is synthesized by a different pathway from that of glycogen break-down. UDP-glucose, the activated intermediate in glycogen synthesis, is formed from glucose 1-phosphate and UTP. Glycogen synthase catalyzes the transfer of glucose from UDP-glucose to the C-4 hydroxyl group of a terminal residue in the growing glycogen molecule. Synthesis is primed by glycogenin, an autoglycosylating protein that contains a covalently attached oligosaccharide unit on a specific tyrosine residue. A branching enzyme converts some of the α-1,4 linkages into α-1,6 linkages to increase the number of ends so that glycogen can be synthesized and degraded more rapidly.

25.2 Metabolism in Context: Glycogen Breakdown and Synthesis Are Reciprocally Regulated

Glycogen synthesis and degradation are coordinated by several amplifying reaction cascades. Epinephrine and glucagon stimulate glycogen break-down and inhibit its synthesis by increasing the cytoplasmic concentration of cAMP, which activates protein kinase A. Protein kinase A activates gly-cogen breakdown by attaching a phosphoryl group to phosphorylase kinase and inhibits glycogen synthesis by phosphorylating glycogen synthase.

The glycogen-mobilizing actions of protein kinase A are reversed by protein phosphatase 1, which is regulated by several hormones. Epineph-rine inhibits this phosphatase by blocking its attachment to glycogen mol-ecules and by activating an inhibitor. Insulin, in contrast, triggers a cascade that phosphorylates and inactivates glycogen synthase kinase, one of the enzymes that inhibits glycogen synthase. Hence, glycogen syn-thesis is decreased by epinephrine and increased by insulin. Glycogen synthase and glycogen phosphorylase are also regulated by noncovalent allosteric interactions. Glycogen synthase *b* is activated by glucose 6-phos-phate, whereas glycogen phosphorylase is a key part of the glucose-sens-ing system of liver cells. Glycogen metabolism exemplifies the power and precision of reversible phosphorylation in regulating biological processes.

KEY TERMS

uridine diphosphate glucose
 (UDP-glucose) (p. 459)
glycogen synthase (p. 460)

glycogenin (p. 460)
glycogen synthase kinase
 (GSK) (p. 461)

protein phosphatase 1
 (PP1) (p. 462)

❓ Answers to QUICK QUIZZES

1. It prevents synthesis and breakdown from taking place simultaneously, which would lead to a useless expenditure of energy. See problem 9 in the Problems section.

2. Liver phosphorylase *a* is inhibited by glucose, which facilitates the R $\longrightarrow$ T transition. This transition releases protein phosphatase 1, which inactivates glycogen break-down and stimulates glycogen synthesis. Muscle phosphor-ylase is insensitive to glucose.

PROBLEMS

1. *Yin and Yang.* Match the terms on the left with the descriptions on the right. ✓ 3

(a) UDP-glucose
_____

(b) UDP-glucose pyrophos-
phorylase _____

(c) Glycogen synthase
_____

(d) Glycogenin _____

(e) Branching enzyme
_____

(f) Glucose 6-phosphate
_____

(g) Glycogen synthase
kinase _____

(h) Protein phosphatase 1
_____

(i) Insulin _____

(j) Glycogen phosphorylase
a _____

1. Glucose 1-phosphate is one of its substrates.
2. Potent activator of glycogen synthase *b*.
3. Glucose sensor in the liver.
4. Activated substrate for glycogen synthesis.
5. Synthesizes α-1,4 linkages between glucose molecules.
6. Leads to the inactivation of glycogen synthase kinase.
7. Synthesizes α-1,6 linkages between glucose molecules.
8. Catalyzes the formation of glycogen synthase *b*.
9. Catalyzes the formation of glycogen synthase *a*.
10. Synthesizes the primer for glycogen synthesis.

2. *Team effort.* What enzymes are required for the synthesis of a glycogen particle starting from glucose 6-phosphate? ✓ 3 ✓ 4

3. *ATP is behind everything!* UDP-glucose is the activated precursor for glycogen synthesis, but ultimately ATP is the power behind glycogen synthesis. Prove it by showing the reactions required to convert glucose 6-phosphate into a unit of glycogen with the concomitant regeneration of UTP. ✓ 3

4. *Force it forward.* The following reaction accounts for the synthesis of UDP-glucose. This reaction is readily reversible. How is it made irreversible in vivo? ✓ 3

Glucose 1-phosphate + UTP $\rightleftharpoons$ UDP-glucose + PP$_i$

5. *If you insist.* Why does activation of the phosphorylated *b* form of glycogen synthase by high concentrations of glucose 6-phosphate make good biochemical sense? ✓ 4

6. *Initiate and extend.* Describe the separate roles of glycogenin and glycogen synthase in glycogen synthesis. ✓ 4

7. *An ATP saved is an ATP earned.* The complete oxidation of glucose 6-phosphate derived from free glucose yields 30 molecules ATP, whereas the complete oxidation of glucose 6-phosphate derived from glycogen yields 31 molecules of ATP. Account for this difference. ✓ 5

8. *Dual roles.* Phosphoglucomutase is crucial for glycogen breakdown as well as for glycogen synthesis. Explain the role of this enzyme in each of the two processes.

9. *Working at cross-purposes.* Write a balanced equation showing the effect of the simultaneous activation of glycogen phosphorylase and glycogen synthase. Include the reactions catalyzed by phosphoglucomutase and UDP-glucose pyrophosphorylase. ✓ 5

10. *Achieving immortality.* Glycogen synthase requires a primer. The primer was once thought to be provided when the existing glycogen granules are divided between the daughter cells produced by cell division. In other words, parts of the original glycogen molecule were simply passed from generation to generation. Would this strategy have been successful in passing glycogen stores from generation to generation? How are new glycogen molecules now known to be synthesized? ✓ 4

11. *Synthesis signal.* How does insulin stimulate glycogen synthesis? ✓ 4

12. *Excessive storage.* Suggest an explanation for the fact that the amount of glycogen in type I glycogen-storage disease (von Gierke disease) is increased. ✓ 4

Chapter Integration Problems

13. *Metabolic mutants.* Predict the major consequence of each of the following mutations. ✓ 5

(a) Loss of the AMP-binding site in muscle phosphorylase.
(b) Mutation of Ser 14 to Ala 14 in liver phosphorylase.
(c) Overexpression of phosphorylase kinase in the liver.
(d) Loss of the gene that encodes the inhibitor of protein phosphatase 1.
(e) Loss of the gene that encodes the glycogen-targeting subunit of protein phosphatase 1.
(f) Loss of the gene that encodes glycogenin.

14. *More metabolic mutants.* Briefly predict the major consequences of each of the following mutations affecting glycogen utilization. ✓ 5

(a) Loss of GTPase activity of the G-protein α subunit.
(b) Loss of phosphodiesterase activity.

15. *Same symptoms, different cause.* Von Gierke disease is frequently the result of a defect in glucose 6-phosphatase. Suggest another mutation in glucose metabolism that causes symptoms similar to those of von Gierke disease.

16. *Again, von Gierke.* People suffering from von Gierke disease release a small amount of glucose into the blood after the injection of glucagon. How is this result possible?

17. *I know I've seen that face before.* UDP-glucose is the activated form of glucose used in glycogen synthesis. However, we have already met other similar activated forms of carbohydrate in our consideration of metabolism. Where else have we seen UDP-carbohydrate?

18. *Carbohydrate conversion.* Write a balanced equation for the formation of glycogen from galactose. ✓ 3

Chapter Integration, Data Interpretation, and Challenge Problems

19. *Removing all traces.* In human liver extracts, the catalytic activity of glycogenin was detectable only after treatment with α-amylase, an enzyme that hydrolyzes α-1,4-glucosidic bonds. Why was α-amylase necessary to reveal the glycogenin activity?

20. *Telltale products.* A sample of glycogen from a patient with liver disease is incubated with orthophosphate, phosphorylase, the transferase, and the debranching enzyme (α-1,6-glucosidase). The ratio of glucose 1-phosphate to glucose formed in this mixture is 100. What is the most likely enzymatic deficiency in this patient? ✓ 3

21. *Glycogen isolation 1.* The liver is a major storage site for glycogen. Purified from two samples of human liver, glycogen was either treated or not treated with α-amylase and subsequently analyzed by SDS-PAGE and western blotting with the use of antibodies to glycogenin (Chapter 5). The results are presented in the following illustration:

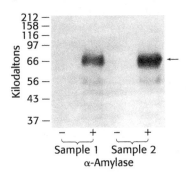

[Courtesy of Dr. Peter J. Roach, Indiana University School of Medicine.]

(a) Why are no proteins visible in the lanes without amylase treatment?

(b) What is the effect of treating the samples with α-amylase? Explain the results.

(c) List other proteins that you might expect to be associated with glycogen. Why are other proteins not visible?

22. *Glycogen isolation 2.* The gene for glycogenin was transfected into a cell line that normally stores only small amounts of glycogen. The cells were then manipulated according to the following protocol, and glycogen was isolated and analyzed by SDS-PAGE and western blotting by using an antibody to glycogenin with and without α-amylase treatment (Chapter 5). The results are presented in the following illustration.

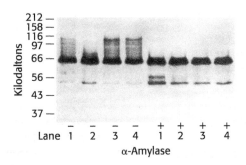

[Courtesy of Dr. Peter J. Roach, Indiana University School of Medicine.]

The protocol: Cells cultured in growth medium and 25 mM glucose (lane 1) were switched to medium containing no glucose for 24 hours (lane 2). Glucose-starved cells were re-fed with medium containing 25 mM glucose for 1 hour (lane 3) or 3 hours (lane 4). Samples (12 mg of protein) were either treated or not treated with α-amylase, as indicated, before being loaded on the gel.

(a) Why did the western analysis produce a "smear"—that is, the high-molecular-weight staining in lane 1(-)?

(b) What is the significance of the decrease in high-molecular-weight staining in lane 2(-)?

(c) What is the significance of the difference between lanes 2(-) and 3(-)?

(d) Suggest a plausible reason why there is essentially no difference between lanes 3(-) and 4(-)?

(e) Why are the bands at 66 kDa the same in the lanes treated with α-amylase, despite the fact that the cells were treated differently?

Selected Readings for this chapter can be found online at www.whfreeman.com/tymoczko3e.

CHAPTER 26

The Pentose Phosphate Pathway

Growth is an awesome biochemical feat. Two key biochemical components required for growth—ribose sugars and biochemical reducing power—are provided by the pentose phosphate pathway. [Comstock Images/AgeFotostock.]

Thus far, we have considered glycogen metabolism with a focus on energy production and storage. Another important fate for the ultimate breakdown product of glycogen degradation—glucose 6-phosphate—is as a substrate for the pentose phosphate pathway, a remarkably versatile set of reactions. The pentose phosphate pathway is an important source of NADPH, biosynthetic reducing power. Moreover, the pathway catalyzes the interconversion of the three- and six-carbon intermediates of glycolysis with five-carbon carbohydrates. These interconversions enable the synthesis of pentose sugars required for DNA and RNA synthesis as well as the metabolism of five-carbon sugars consumed in the diet.

We begin by examining the generation of NADPH, followed by an investigation of the carbohydrate interconversions. We end with a consideration of the versatility of the reactions of the pentose phosphate pathway.

✓ 6 Identify the two stages of the pentose phosphate pathway, and explain how the pathway is coordinated with glycolysis and gluconeogenesis.

26.1 The Pentose Phosphate Pathway Yields NADPH and Five-Carbon Sugars

Recall that, in photosynthesis (Chapter 22), NADPH is a key product of the light reactions. *NADPH is the source of biosynthetic reducing power in all organisms* and is used in a host of biochemical processes (Table 26.1). How do nonphotosynthetic cells and organisms generate NADPH? The metabolism of glucose 6-phosphate by the *pentose phosphate pathway* generates NADPH for nonphotosynthetic organisms. This pathway consists of two phases: (1) the oxidative generation of NADPH and (2) the nonoxidative interconversion of sugars (Figure 26.1). In the oxidative phase, NADPH is generated when glucose 6-phosphate is oxidized to ribulose 5-phosphate:

Glucose 6-phosphate $+ 2\,NADP^+ + H_2O \longrightarrow$
$$\text{ribulose 5-phosphate} + 2\,NADPH + 2\,H^+ + CO_2$$

Ribulose 5-phospate is subsequently converted into ribose 5-phosphate, which is a precursor to RNA and DNA as well as to ATP, NADH, FAD, and coenzyme A. In the nonoxidative phase, the pathway catalyzes the interconversion of three-, four-, five-, six-, and seven-carbon sugars in a series of nonoxidative reactions. Excess five-carbon sugars may be converted into intermediates of the glycolytic pathway. All these reactions take place in the cytoplasm.

Table 26.1   Pathways requiring NADPH

Synthesis
Fatty acid biosynthesis
Cholesterol biosynthesis
Neurotransmitter biosynthesis
Nucleotide biosynthesis
Detoxification
Reduction of oxidized glutathione
Cytochrome P450 monooxygenases

Two Molecules of NADPH Are Generated in the Conversion of Glucose 6-phosphate into Ribulose 5-phosphate

The oxidative phase of the pentose phosphate pathway starts with the oxidation of glucose 6-phosphate at carbon 1 with the concomitant formation of NADPH, a reaction catalyzed by *glucose 6-phosphate dehydrogenase* (Figure 26.2, on p. 476). This dehydrogenase is highly specific for $NADP^+$; the K_M for NAD^+ is about a thousand times as great as that for $NADP^+$. The product of the dehydrogenation is *6-phosphoglucono-δ-lactone*, which has an ester between the C-1 carboxyl group and the C-5 hydroxyl group. The next step is the hydrolysis of 6-phosphoglucono-δ-lactone by a specific *lactonase* to give *6-phosphogluconate*. This six-carbon acid is then oxidatively decarboxylated by *6-phosphogluconate dehydrogenase* to yield *ribulose 5-phosphate*. $NADP^+$ is again the electron acceptor. This reaction completes the oxidative phase.

The Pentose Phosphate Pathway and Glycolysis Are Linked by Transketolase and Transaldolase

The preceding reactions yield two molecules of NADPH and one molecule of ribulose 5-phosphate for each molecule of glucose 6-phosphate oxidized. The ribulose 5-phosphate is subsequently isomerized to ribose 5-phosphate by phosphopentose isomerase.

Ribulose 5-phosphate   **Ribose 5-phosphate**

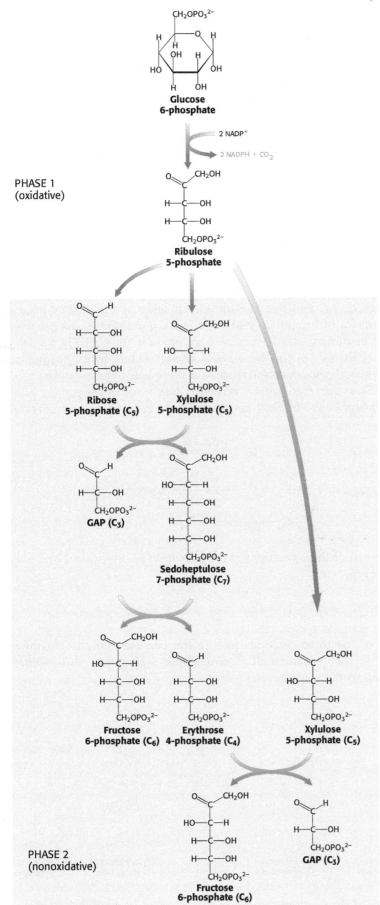

**PHASE 1
(oxidative)**

**PHASE 2
(nonoxidative)**

Figure 26.1 The pentose phosphate pathway. The pathway consists of (1) an oxidative phase that generates NADPH and (2) a nonoxidative phase that interconverts phosphorylated sugars.

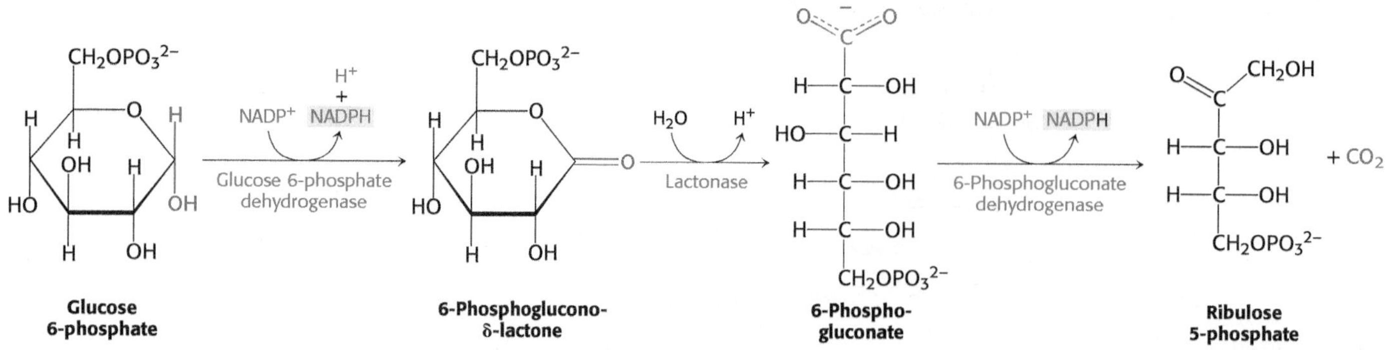

Figure 26.2 The oxidative phase of the pentose phosphate pathway. Glucose 6-phosphate is oxidized to 6-phosphoglucono-δ-lactone to generate one molecule of NADPH. The lactone product is hydrolyzed to 6-phosphogluconate, which is oxidatively decarboxylated to ribulose 5-phosphate with the generation of a second molecule of NADPH.

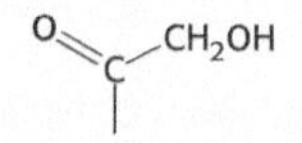

Transferred by transketolase

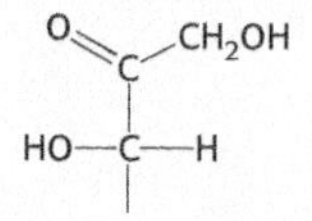

Transferred by transaldolase

Although ribose 5-phosphate is a precursor to many biomolecules, many cells need NADPH for reductive biosyntheses much more than they need ribose 5-phosphate for incorporation into nucleotides. For instance, adipose tissue, the liver, and mammary glands require large amounts of NADPH for fatty acid synthesis (Chapter 28). In these cases, ribose 5-phosphate is converted into glyceraldehyde 3-phosphate and fructose 6-phosphate by *transketolase* and *transaldolase*. *These enzymes create a reversible link between the pentose phosphate pathway and glycolysis by catalyzing these three successive reactions* (Figure 26.1):

$$C_5 + C_5 \xrightleftharpoons{\text{Transketolase}} C_3 + C_7$$

$$C_3 + C_7 \xrightleftharpoons{\text{Transaldolase}} C_6 + C_4$$

$$C_4 + C_5 \xrightleftharpoons{\text{Transketolase}} C_6 + C_3$$

The net result of these reactions is the *formation of two hexoses and one triose from three pentoses*:

$$3\ C_5 \rightleftharpoons 2\ C_6 + C_3$$

The first of the three reactions linking the pentose phosphate pathway and glycolysis is the formation of *glyceraldehyde 3-phosphate* and *sedoheptulose 7-phosphate* from two pentoses:

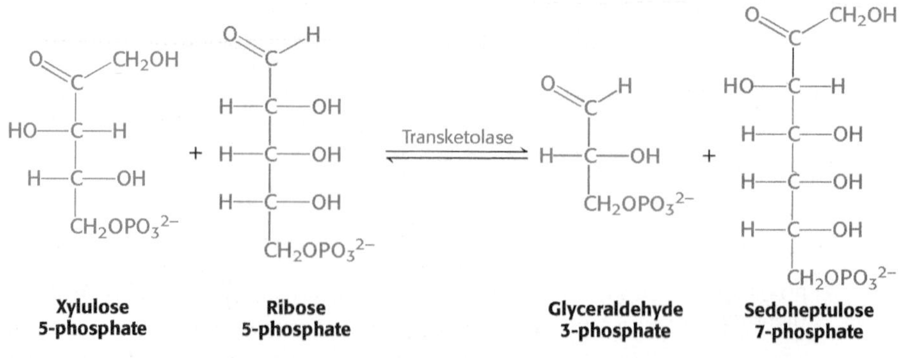

| Xylulose 5-phosphate | Ribose 5-phosphate | | Glyceraldehyde 3-phosphate | Sedoheptulose 7-phosphate |

The donor of the two-carbon unit in this reaction is xylulose 5-phosphate, an epimer of ribulose 5-phosphate. Ribulose 5-phosphate is converted into the appropriate epimer for the transketolase reaction by *phosphopentose epimerase*:

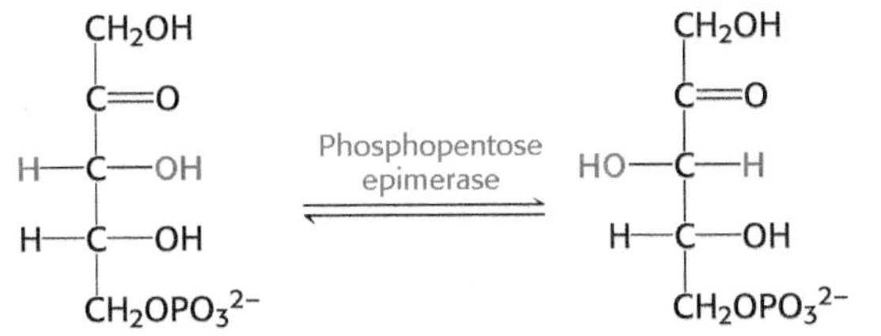

Ribulose 5-phosphate   **Xylulose 5-phosphate**

In the second reaction linking the pentose phosphate pathway and glycolysis, glyceraldehyde 3-phosphate and sedoheptulose 7-phosphate react to form *fructose 6-phosphate* and *erythrose 4-phosphate*:

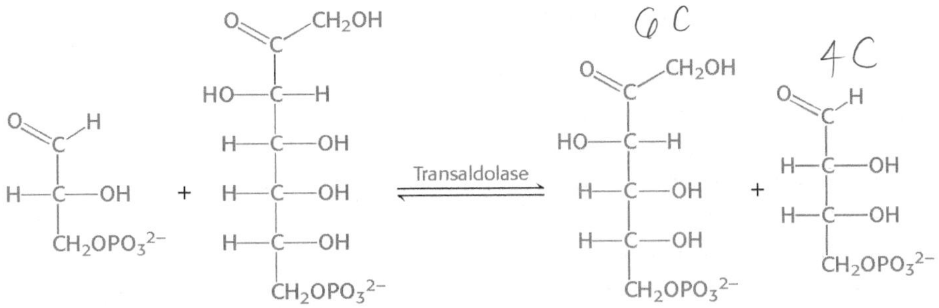

Glyceraldehyde 3-phosphate   **Sedoheptulose 7-phosphate**   **Fructose 6-phosphate**   **Erythrose 4-phosphate**

This synthesis of a four-carbon sugar and a six-carbon sugar is catalyzed by transaldolase.

In the third reaction, transketolase catalyzes the synthesis of *fructose 6-phosphate* and *glyceraldehyde 3-phosphate* from erythrose 4-phosphate and xylulose 5-phosphate:

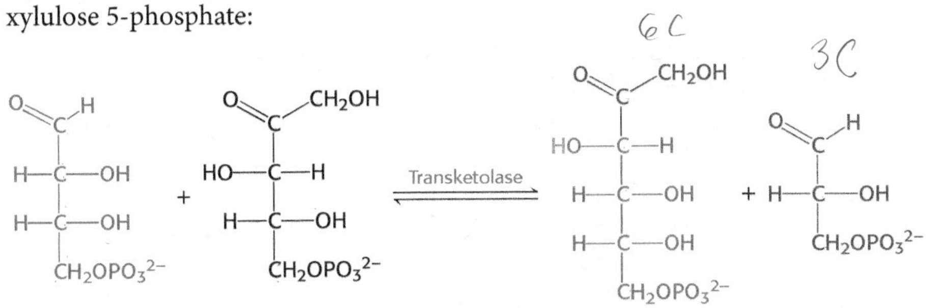

Erythrose 4-phosphate   **Xylulose 5-phosphate**   **Fructose 6-phosphate**   **Glyceraldehyde 3-phosphate**

The sum of these reactions is

2 Xylulose 5-phosphate + ribose 5-phosphate ⇌
    2 fructose 6-phosphate + glyceraldehyde 3-phosphate

Xylulose 5-phosphate can be formed from ribose 5-phosphate by the sequential action of phosphopentose isomerase and phosphopentose epimerase, and so the net reaction starting from ribose 5-phosphate is

3 Ribose 5-phosphate ⇌
    2 fructose 6-phosphate + glyceraldehyde 3-phosphate

Thus, excess ribose 5-phosphate formed by the pentose phosphate pathway can be completely converted into glycolytic intermediates. Moreover, any ribose ingested in the diet can be processed into glycolytic intermediates by this pathway. Evidently, the carbon skeletons of sugars can be extensively rearranged to meet physiologic needs (Table 26.2).

QUICK QUIZ Describe how the pentose phosphate pathway and glycolysis are linked by transaldolase and transketolase.

Table 26.2   The pentose phosphate pathway

Reaction	Enzyme
Oxidative phase	
Glucose 6-phosphate + NADP$^+$ $\longrightarrow$ 6-phosphoglucono-δ-lactone + NADPH + H$^+$	Glucose 6-phosphate dehydrogenase
6-Phosphoglucono-δ-lactone + H$_2$O $\longrightarrow$ 6-phosphogluconate + H$^+$	Lactonase
6-Phosphogluconate + NADP$^+$ $\longrightarrow$ ribulose 5-phosphate + CO$_2$ + NADPH	6-Phosphogluconate dehydrogenase
Nonoxidative phase	
Ribulose 5-phosphate $\rightleftharpoons$ ribose 5-phosphate	Phosphopentose isomerase
Ribulose 5-phosphate $\rightleftharpoons$ xylulose 5-phosphate	Phosphopentose epimerase
Xylulose 5-phosphate + ribose 5-phosphate $\rightleftharpoons$ sedoheptulose 7-phosphate + glyceraldehyde 3-phosphate	Transketolase
Sedoheptulose 7-phosphate + glyceraldehyde 3-phosphate $\rightleftharpoons$ fructose 6-phosphate + erythrose 4-phosphate	Transaldolase
Xylulose 5-phosphate + erythrose 4-phosphate $\rightleftharpoons$ fructose 6-phosphate + glyceraldehyde 3-phosphate	Transketolase

✓ 7 Identify the enzyme that controls the pentose phosphate pathway.

26.2 Metabolism in Context: Glycolysis and the Pentose Phosphate Pathway Are Coordinately Controlled

Glucose 6-phosphate is metabolized by both the glycolytic pathway and the pentose phosphate pathway. How is the processing of this important metabolite partitioned between these two metabolic routes? The cytoplasmic concentration of NADP$^+$ plays a key role in determining the fate of glucose 6-phosphate.

The Rate of the Pentose Phosphate Pathway Is Controlled by the Concentration of NADP$^+$

The first reaction in the oxidative branch of the pentose phosphate pathway, the dehydrogenation of glucose 6-phosphate by glucose 6-phosphate dehydrogenase, is irreversible. In fact, this reaction is rate limiting in vivo and serves as the control site for the oxidative branch of the pathway. The most important regulatory factor is the concentration of NADP$^+$, the electron acceptor in the oxidation steps of the pentose phosphate pathway. When there is an excess of NADPH, there is a small amount of NADP$^+$ in cells. At low concentrations of NADP$^+$, the activity of the three enzymes participating in the oxidation step is low because there are fewer oxidizing agents to move the reaction forward. The inhibitory effect of low concentrations of NADP$^+$ is enhanced by the fact that NADPH, which is present in 70-fold greater concentration than is NADP$^+$ in liver cytoplasm, competes with NADP$^+$ in binding to the enzyme, thereby preventing the oxidation of glucose 6-phosphate. The effect of the NADP$^+$ concentration on the rate of the oxidative phase ensures that the generation of NADPH is tightly coupled to its use in reductive biosyntheses or protection against oxidative stress. The nonoxidative phase of the pentose phosphate pathway is controlled primarily by the availability of substrates.

The Fate of Glucose 6-phosphate Depends on the Need for NADPH, Ribose 5-phosphate, and ATP

We can grasp the "Swiss Army knife-like" versatility of the pentose phosphate pathway, as well as its interplay with glycolysis and gluconeogenesis, by examining the metabolism of glucose 6-phosphate in four different metabolic situations (**Figure 26.3**).

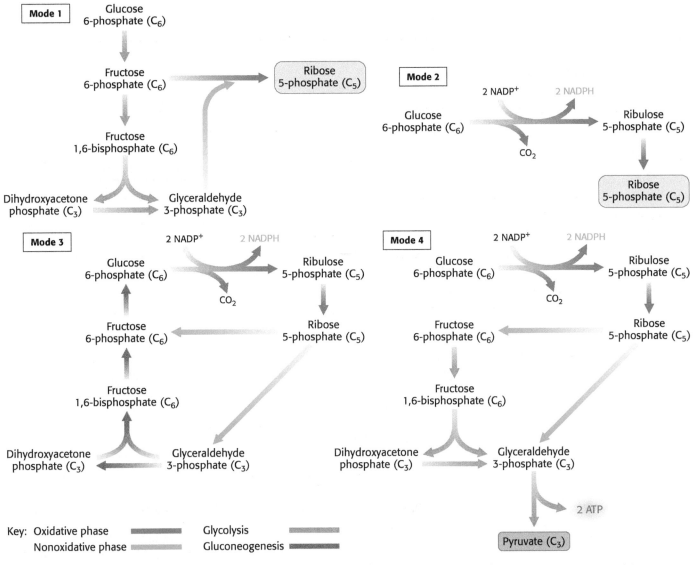

Figure 26.3 Four modes of the pentose phosphate pathway. Major products are shown in color.

Mode 1. *The need for ribose 5-phosphate is greater than the need for NADPH.* For example, rapidly dividing cells need ribose 5-phosphate for the synthesis of nucleotide precursors of DNA. Most of the glucose 6-phosphate is converted into fructose 6-phosphate and glyceraldehyde 3-phosphate by the glycolytic pathway. The nonoxidative phase of the pentose phosphate pathway then converts two molecules of fructose 6-phosphate and one molecule of glyceraldehyde 3-phosphate into three molecules of ribose 5-phosphate by a reversal of the reactions described earlier. The stoichiometry of mode 1 is

$$5 \text{ Glucose 6-phosphate} + \text{ATP} \longrightarrow 6 \text{ ribose 5-phosphate} + \text{ADP} + \text{H}^+$$

Mode 2. *The needs for NADPH and ribose 5-phosphate are balanced.* Under these conditions, glucose 6-phosphate is processed to one molecule of ribulose 5-phosphate while generating two molecules of NADPH. Ribulose 5-phosphate is then converted into ribose 5-phosphate. The stoichiometry of mode 2 is

$$\text{Glucose 6-phosphate} + 2 \text{ NADP}^+ + \text{H}_2\text{O} \longrightarrow$$
$$\text{ribose 5-phosphate} + 2 \text{ NADPH} + 2 \text{ H}^+ + \text{CO}_2$$

Table 26.3   Tissues with active pentose phosphate pathways

Tissue	Function
Adrenal gland	Steroid synthesis
Liver	Fatty acid and cholesterol synthesis
Testes	Steroid synthesis
Adipose tissue	Fatty acid synthesis
Ovary	Steroid synthesis
Mammary gland	Fatty acid synthesis
Red blood cells	Maintenance of reduced glutathione

Mode 3. *The need for NADPH is greater than the need for ribose 5-phosphate.* For example, the liver requires a large amount of NADPH for the synthesis of fatty acids (Table 26.3). In this case, glucose 6-phosphate is completely oxidized to CO_2. Three groups of reactions are active in this situation. First, the oxidative phase of the pentose phosphate pathway forms two molecules of NADPH and one molecule of ribose 5-phosphate. Then, ribose 5-phosphate is converted into fructose 6-phosphate and glyceraldehyde 3-phosphate by the nonoxidative phase. Finally, glucose 6-phosphate is resynthesized from fructose 6-phosphate and glyceraldehyde 3-phosphate by the gluconeogenic pathway. In essence, ribose 5-phosphate produced by the pentose phosphate pathway is recycled into glucose 6-phosphate by the nonoxidative phase of the pathway and by some of the enzymes of the gluconeogenic pathway. The stoichiometries of these three sets of reactions are

$$6 \text{ Glucose 6-phosphate} + 12\,NADP^+ + 6\,H_2O \longrightarrow$$
$$6 \text{ ribose 5-phosphate} + 12\,NADPH + 12\,H^+ + 6\,CO_2$$

$$6 \text{ Ribose 5-phosphate} \longrightarrow$$
$$4 \text{ fructose 6-phosphate} + 2 \text{ glyceraldehyde 3-phosphate}$$

$$4 \text{ Fructose 6-phosphate} + 2 \text{ glyceraldehyde 3-phosphate} + H_2O \longrightarrow$$
$$5 \text{ glucose 6-phosphate} + P_i$$

The sum of the mode 3 reactions is

$$\text{Glucose 6-phosphate} + 12\,NADP^+ + 7\,H_2O \longrightarrow$$
$$6\,CO_2 + 12\,NADPH + 12\,H^+ + P_i$$

Thus, *the equivalent of glucose 6-phosphate can be completely oxidized to CO_2 with the concomitant generation of NADPH.*

Mode 4. *Both NADPH and ATP are required.* Alternatively, ribulose 5-phosphate formed from glucose 6-phosphate can be converted into pyruvate when reducing power and ATP are needed. Fructose 6-phosphate and glyceraldehyde 3-phosphate derived from the nonoxidative reactions of the pentose phosphate pathway enter the glycolytic pathway rather than reverting to glucose 6-phosphate. In this mode, *ATP and NADPH are concomitantly generated, and five of the six carbon atoms of glucose 6-phosphate emerge in pyruvate.* The stoichiometry of mode 4 is

$$3 \text{ Glucose 6-phosphate} + 6\,NADP^+ + 5\,NAD^+ + 5\,P_i + 8\,ADP \longrightarrow$$
$$5 \text{ pyruvate} + 3\,CO_2 + 6\,NADPH + 5\,NADH +$$
$$8\,ATP + 2\,H_2O + 8\,H^+$$

Pyruvate formed by these reactions can be oxidized to generate more ATP or it can be used as a building block in a variety of biosyntheses.

CLINICAL INSIGHT

The Pentose Phosphate Pathway Is Required For Rapid Cell Growth

Rapidly dividing cells, such as cancer cells, require ribose 5-phosphate for nucleic acid synthesis and NADPH for reductive biosynthesis. Recall that rapidly dividing cells switch to aerobic glycolysis to meet their ATP needs (p. 304). Glucose 6-phosphate and glycolytic intermediates are then used to generate NADPH and ribose 5-phosphate using the pentose phosphate pathway as described in modes 1 and 3 above. The diversion of glycolytic intermediates into the nonoxidative phase is facilitated by the expression of the gene for a pyruvate kinase isozyme PKM. Recall that pyruvate kinase catalyzes the conversion of phosphoenolpyruvate into pyruvate with the synthesis of a molecule of ATP. PKM has a low catalytic activity creating a bottleneck in the glycolytic pathway. Glycolytic intermediates accumulate and are then used by the pentose phosphate pathway to synthesize ribose 5-phosphate. The shunting of phosphorylated intermediates into the nonoxidative phase of the pentose phosphate pathway is further enabled by the inhibition of triose phosphate isomerase by phosphoenolpyruvate, the substrate of PKM.

26.3 Glucose 6-phosphate Dehydrogenase Lessens Oxidative Stress

The NADPH generated by the pentose phosphate pathway plays a vital role in protecting the cells from reactive oxygen species (ROS). Reactive oxygen species generated in oxidative metabolism inflict damage on all classes of macromolecules and can ultimately lead to cell death. Indeed, ROS are implicated in a number of human diseases (p. 377). Reduced *glutathione* (GSH), a tripeptide with a free sulfhydryl group, combats oxidative stress by reducing ROS to harmless forms and in the process is oxidized to GSSG, two molecules of glutathione joined by a disulfide bond. GSSG must be reduced to regenerate GSH. The reducing power is supplied by the NADPH generated by glucose 6-phosphate dehydrogenase in the pentose phosphate pathway. Indeed, cells with reduced levels of glucose 6-phosphate dehydrogenase activity are especially sensitive to oxidative stress.

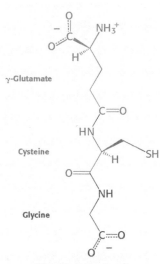

Glutathione (reduced)
(γ-Glutamylcysteinylglycine)

CLINICAL INSIGHT

Glucose 6-phosphate Dehydrogenase Deficiency Causes a Drug-Induced Hemolytic Anemia

The importance of the pentose phosphate pathway is highlighted by the anomalous response of people to certain drugs. For instance, pamaquine, the first synthetic antimalarial drug introduced in 1926, was associated with the appearance of severe and mysterious ailments. Most patients tolerated the drug well, but a few developed severe symptoms within a few days after therapy was started. Their urine turned black, jaundice developed, and the hemoglobin content of the blood dropped sharply. In some cases, massive destruction of red blood cells caused death.

This drug-induced *hemolytic anemia* was shown 30 years later to be caused by a *deficiency of glucose 6-phosphate dehydrogenase*, the enzyme catalyzing the first step in the oxidative branch of the pentose phosphate pathway. The result is a dearth of NADPH in all cells, but this deficiency is most acute in red blood cells because they lack mitochondria and have no alternative means of generating reducing power. This defect, which is inherited on the X chromosome, is the most common disease that results from an enzyme malfunction, affecting hundreds of millions of people.

Pamaquine sensitivity is not simply a historical oddity about malaria treatment many decades ago. Primaquine, an antimalarial closely related to pamaquine, is widely used in malaria-infested regions of the world today. Vicine, a purine glycoside of fava beans, a bean that is consumed in countries surrounding the Mediterranean, also induces hemolysis. People deficient in glucose 6-phosphate dehydrogenase suffer hemolysis from eating fava beans or inhaling the pollen of the fava flowers, a response called favism.

How can we explain hemolysis caused by pamaquine, primaquine, and vicine biochemically? These chemicals are oxidative agents that generate peroxides, reactive oxygen species that can damage membranes as well as other biomolecules. Peroxides are normally eliminated by the enzyme *glutathione peroxidase*, which uses reduced glutathione as a reducing agent:

$$2 \text{ GSH} + \text{ROOH} \xrightleftharpoons{\text{Glutathione peroxidase}} \text{GSSG} + \text{H}_2\text{O} + \text{ROH}$$

The major role of NADPH in red cells is to reduce the disulfide form of glutathione to the sulfhydryl form. The enzyme that catalyzes the regeneration of reduced glutathione is *glutathione reductase*:

γ-Glu—Cys—Gly
          |
          S
          | Glutathione
                               reductase

+ NADPH + H⁺ ⇌ 2 γ-Glu—Cys—Gly + NADP⁺
                                |
                               SH

γ-Glu—Cys—Gly

Oxidized glutathione (GSSG)   **Reduced glutathione (GSH)**

Red blood cells with a low concentration of reduced glutathione are more susceptible to hemolysis. In the absence of glucose 6-phosphate dehydrogenase, peroxides damage membranes because no NADPH is being produced to restore reduced glutathione. Thus, the answer to our question about the cause of hemolysis is that *glucose 6-phosphate dehydrogenase is required to maintain reduced glutathione concentration to protect against oxidative stress*. In the absence of oxidative stress, however, the deficiency is quite benign. The sensitivity to oxidative agents of people having this dehydrogenase deficiency also clearly demonstrates that *atypical reactions to drugs may have a genetic basis*.

Reduced glutathione is also essential for maintaining the normal structure of red blood cells by preserving the structure of hemoglobin. The reduced form of glutathione serves as a sulfhydryl buffer that keeps the residues of hemoglobin in the reduced sulfhydryl form. Without adequate amounts of reduced glutathione, the hemoglobin sulfhydryl groups can no longer be maintained in the reduced form. Hemoglobin molecules then cross-link with one another to form aggregates called *Heinz bodies* on cell membranes (**Figure 26.4**). Membranes damaged by Heinz bodies and reactive oxygen species become deformed, and the cell is likely to undergo lysis.

Figure 26.4 Red blood cells with Heinz bodies. The light micrograph shows red blood cells obtained from a person deficient in glucose 6-phosphate dehydrogenase. The dark particles, called Heinz bodies, inside the cells are clumps of denatured hemoglobin that adhere to the plasma membrane and stain with basic dyes. Red blood cells in such people are highly susceptible to oxidative damage. [CNRI/Science Source.]

BIOLOGICAL INSIGHT

A Deficiency of Glucose 6-phosphate Dehydrogenase Confers an Evolutionary Advantage in Some Circumstances

The incidence of the most common form of glucose 6-phosphate dehydrogenase deficiency, characterized by a 10-fold reduction in enzymatic activity in red blood cells, is 11% among Americans of African heritage. This high frequency suggests that the deficiency may be advantageous under certain environmental conditions. Indeed, *glucose 6-phosphate dehydrogenase deficiency protects against falciparum malaria* (**Figure 26.5**). The parasites causing this disease require NADPH for optimal growth. Moreover, infection by the parasites induces oxidative stress in the infected cell. Because the pentose pathway is compromised, the cell and parasite die from oxidative damage. Thus, glucose 6-phosphate dehydrogenase deficiency is a mechanism of protection against malaria, which accounts for its high frequency in malaria-infested regions of the world.

The ability of glucose 6-phosphate dehydrogenase deficiency to protect against malaria does, however, create a public health conundrum. Primaquine is a commonly used and highly effective antimalarial drug. However, indiscriminate use of primaquine will cause hemolysis in individuals deficient in glucose 6-phosphate dehydrogenase. A solution to this conundrum may be in the offing as recent work shows that an antimalaria vaccine may be within reach.

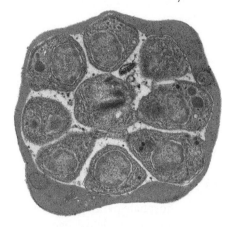

Figure 26.5 *Plasmodium falciparum–*infected red blood cell. *Plasmodium falciparum* is the protozoan parasite that causes malaria. Red blood cells are an important site of infection by *P. falciparum*. Here, the parasites are colored green. The growing parasite consumes the red-blood-cell protein, notably hemoglobin, leading to cell death and causing anemia in the host. [Omikron/Science Source.]

SUMMARY

26.1 The Pentose Phosphate Pathway Yields NADPH and Five-Carbon Sugars

The pentose phosphate pathway, present in all organisms, generates NADPH and ribose 5-phosphate in the cytoplasm. NADPH is used in reductive biosyntheses, whereas ribose 5-phosphate is used in the synthesis of RNA, DNA, and nucleotide coenzymes. The pentose phosphate pathway starts with the dehydrogenation of glucose 6-phosphate to form a lactone, which is hydrolyzed to give 6-phosphogluconate and then oxidatively decarboxylated to yield ribulose 5-phosphate. $NADP^+$ is the electron acceptor in both of these oxidations. Ribulose 5-phosphate (a ketose) is subsequently isomerized to ribose 5-phosphate (an aldose). A different mode of the pathway is active when cells need much more NADPH than ribose 5-phosphate. Under these conditions, ribose 5-phosphate is converted into glyceraldehyde 3-phosphate and fructose 6-phosphate by transketolase and transaldolase. These two enzymes create a reversible link between the pentose phosphate pathway and gluconeogenesis. Xylulose 5-phosphate, sedoheptulose 7-phosphate, and erythrose 4-phosphate are intermediates in these interconversions. In this way, 12 molecules of NADPH can be generated for each molecule of glucose 6-phosphate that is completely oxidized to CO_2.

26.2 Metabolism in Context: Glycolysis and the Pentose Phosphate Pathway Are Coordinately Controlled

Only the nonoxidative branch of the pathway is significantly active when much more ribose 5-phosphate than NADPH needs to be synthesized. Under these conditions, fructose 6-phosphate and glyceraldehyde 3-phosphate (formed by the glycolytic pathway) are converted into ribose 5-phosphate without the formation of NADPH. Alternatively, ribose 5-phosphate formed by the oxidative branch can be converted into pyruvate through fructose 6-phosphate and glyceraldehyde 3-phosphate. In this mode, ATP and NADPH are generated, and five of the six carbon atoms of glucose 6-phosphate emerge in pyruvate. The interplay of the glycolytic and pentose phosphate pathways enables the concentrations of NADPH, ATP, and building blocks such as ribose 5-phosphate and pyruvate to be continually adjusted to meet cellular needs.

26.3 Glucose 6-phosphate Dehydrogenase Lessens Oxidative Stress

NADPH generated by glucose 6-phosphate dehydrogenase maintains the appropriate concentration of reduced glutathione required to combat oxidative stress and maintain the proper reducing environment in the cell. Cells with diminished glucose 6-phosphate dehydrogenase activity are especially sensitive to oxidative stress.

KEY TERMS

pentose phosphate pathway (p. 474)
glucose 6-phosphate dehydrogenase (p. 474)

glutathione (p. 481)
glutathione reductase (p. 482)

glutathione peroxidase (p. 482)

? Answer to QUICK QUIZ

The enzymes catalyze the transformation of the five-carbon sugar formed by the oxidative phase of the pentose phosphate pathway into fructose 6-phosphate and glyceraldehyde 3-phosphate, intermediates in glycolysis (and gluconeogenesis).

PROBLEMS

1. *Biochemical taxonomy.*

(a) Identify 6-phosphglucono-δ-lactone. _____
(b) Which reactions produce NADPH? _____
(c) Identify ribulose 5-phosphate. _____
(d) What reaction generates CO_2? _____
(e) Identify 6-phosphogluconate. _____
(f) Which reaction is catalyzed by phosphopentose isomerase? _____
(g) Identify ribose 5-phosphate _____
(h) Which reaction is catalyzed by lactonase? _____
(i) Identify glucose 6-phosphate. _____
(j) Which reaction is catalyzed by 6-phosphogluconate dehydrogenase? _____

(k) Which reaction is catalyzed by glucose 6-phosphate dehydrogenase? _____

2. *Phase shift.* The pentose phosphate pathway is composed of two distinct phases. What are the two phases, and what are their roles? ✓ 6

3. *Designed to control or govern.* What is the key regulatory enzyme in the pentose phosphate pathway, and what is its most prominent regulatory signal? ✓ 7

4. *No respiration.* Glucose is normally completely oxidized to CO_2 in the mitochondria. Under what circumstance can glucose be completely oxidized to CO_2 in the cytoplasm? ✓ 6

5. *Watch your diet, Doctor.* Noted psychiatrist Hannibal Lecter once remarked to FBI Agent Starling that he enjoyed liver with some fava beans and a nice Chianti. Why might this diet be dangerous for some people? ✓ 7

6. *Offal or awful?* Liver and other organ meats contain large quantities of nucleic acids. In the course of digestion, RNA is hydrolyzed to ribose, among other chemicals. Explain how ribose can be used as a fuel. ✓ 6

7. *A required ATP.* The metabolism of glucose 6-phosphate into ribose 5-phosphate by the joint efforts of the pentose phosphate pathway and glycolysis can be summarized by the following equation:

5 glucose 6-phosphate + ATP $\longrightarrow$

6 ribose 5-phosphate + ADP

Which reaction requires the ATP?

8   *Tracing glucose.* Glucose labeled with ^{14}C at C-6 is added to a solution containing the enzymes and cofactors of the oxidative phase of the pentose phosphate pathway. What is the fate of the radioactive label?

9. *No redundancy.* Why do deficiencies in glucose 6-phosphate dehydrogenase frequently present as anemia? ✓ 7

Chapter Integration Problems

10. *Through the looking glass.* Explain why the Calvin cycle and the pentose phosphate are almost mirror images of each other.

11. *Recurring decarboxylations.* Which reaction in the citric acid cycle is most analogous to the oxidative decarboxylation of 6-phosphogluconate to ribulose 5-phosphate?

Challenge Problems

12. *You do what you can do.* Red blood cells lack mitochondria. These cells process glucose to lactate, but they also generate CO_2. What is the purpose of producing lactate? How can red blood cells generate CO_2 if they lack mitochondria? ✓ 6

13. *Carbon shuffling.* Ribose 5-phosphate labeled with ^{14}C at C-1 is added to a solution containing transketolase, transaldolase, phosphopentose epimerase, phosphopentose isomerase, and glyceraldehyde 3-phosphate. What is the distribution of the radioactive label in the erythrose 4-phosphate and fructose 6-phosphate that are formed in this reaction mixture? ✓ 6

14. *Synthetic stoichiometries.* What is the stoichiometry of the synthesis of (a) ribose 5-phosphate from glucose 6-phosphate without the concomitant generation of NADPH? (b) NADPH from glucose 6-phosphate without the concomitant formation of pentose sugars? ✓ 6

15. *Reductive power.* What ratio of NADPH to $NADP^+$ is required to sustain [GSH] = 10 mM and [GSSG] = 1 mM? Use the redox potentials given in Table 20.1.

16. *Catching carbons.* Radioactive-labeling experiments can yield estimates of how much glucose 6-phosphate is metabolized by the pentose phosphate pathway and how much is metabolized by the combined action of glycolysis and the citric acid cycle. Suppose that you have samples of two different tissues as well as two radioactively labeled glucose samples, one with glucose labeled with ^{14}C at C-1 and the other with glucose labeled with ^{14}C at C-6. Design an experiment that would enable you to determine the relative activity of the aerobic metabolism of glucose compared with metabolism by the pentose phosphate pathway. ✓ 6

Selected Readings for this chapter can be found online at www.whfreeman.com/tymoczko3e.

Fatty Acid and Lipid Metabolism

CHAPTER 27
Fatty Acid Degradation

CHAPTER 28
Fatty Acid Synthesis

CHAPTER 29
Lipid Synthesis: Storage Lipids, Phospholipids, and Cholesterol

Some birds can fly thousands of miles over water without stopping to eat. Bears can hibernate for months without the need to wake and forage. Boundaries allow for the existence of cells and, by extension, organisms. All of these circumstances are possible because of a crucial class of biomolecules—lipids. We have already encountered lipids as a storage form of energy (Chapter 11) and as components of membranes (Chapter 12). Energy is stored as triacylglycerols, a glycerol molecule esterified to three fatty acids. When energy is required, the fatty acids are liberated and oxidized to provide ATP. When fuel is abundant, fatty acids are synthesized and incorporated into triacylglycerols and stored in adipose tissue.

Triacylglycerols are the most efficient fuels because they are more reduced than carbohydrates. However, this increase in energy efficiency comes at the cost of biochemical versatility. Unlike carbohydrates, lipid metabolism requires the presence of molecular oxygen to form ATP.

Not only are lipids important fuel molecules, but they also serve a structural purpose. The common forms of membrane lipids are phospholipids, glycolipids, and cholesterol. A phospholipid is built on a backbone of either glycerol or sphingosine, an amino alcohol. These lipids contain a phosphoryl group and an alcohol, in addition to fatty acids. Sphingosine-based lipids are further decorated with carbohydrates to form glycolipids. Cholesterol is another key membrane lipid. Cholesterol does not contain fatty acids; rather, it is built on a steroid nucleus. Cholesterol is crucial for membrane structure and function as well as being a precursor to the steroid hormones. Triacylglycerols and cholesterol are transported in the blood throughout the body as lipoprotein particles.

In Chapter 27, we will examine how triacylglycerols are processed to yield fatty acids and how the fatty acids are degraded in a process called β oxidation that ultimately results in the synthesis of much ATP. In Chapter 28, we will see how fatty acids are synthesized and how fatty acid degradation and synthesis are coordinated. In Chapter 29, we will study lipid synthesis and transport, with a particular emphasis on cholesterol synthesis and its regulation.

✓ By the end of this section, you should be able to:

✓ 1 Identify the repeated steps of fatty acid degradation.

✓ 2 Describe ketone bodies and their role in metabolism.

✓ 3 Explain how fatty acids are synthesized.

✓ 4 Explain how fatty acid metabolism is regulated.

✓ 5 Describe the relation between triacylglycerol synthesis and phospholipid synthesis.

✓ 6 List the regulatory steps in the control of cholesterol synthesis.

CHAPTER 27

Fatty Acid Degradation

Many mammals, such as this house mouse, hibernate over the long winter months. Although their metabolism slows during hibernation, the energy needs of the animal must still be met. Fatty acid degradation is a key energy source for this need. [Juniors Bildarchiv/AgeFotostock.]

27.1 Fatty Acids Are Processed in Three Stages

27.2 The Degradation of Unsaturated and Odd–Chain Fatty Acids Requires Additional Steps

27.3 Ketone Bodies Are Another Fuel Source Derived from Fats

27.4 Metabolism in Context: Fatty Acid Metabolism Is a Source of Insight into Various Physiological States

Fatty acids are stored as *triacylglycerols* in adipose tissue. This fuel-rich tissue is located throughout the body, notably under the skin (subcutaneous fat) and surrounding the internal organs (visceral fat). In this chapter, we will first examine how triacylglycerols are mobilized for use by *lipolysis*, the degradation of the triacylglycerol into free fatty acids and glycerol. Next, we will investigate how the fatty acids are oxidized to acetyl CoA in the process of *β oxidation*. We will also study the formation of ketone bodies, a fat-derived fuel source especially important during fasting. Finally, we will investigate the role of fatty acid metabolism in diabetes and starvation.

✓ 1 Identify the repeated steps of fatty acid degradation.

27.1 Fatty Acids Are Processed in Three Stages

In Chapter 14, we examined how dietary triacylglycerols are digested, absorbed, transported, and stored. Now, we will examine how the stored triacylglycerols are made biochemically accessible. Peripheral tissues, such as muscle, gain access to the lipid energy reserves stored in adipose tissue through three stages of

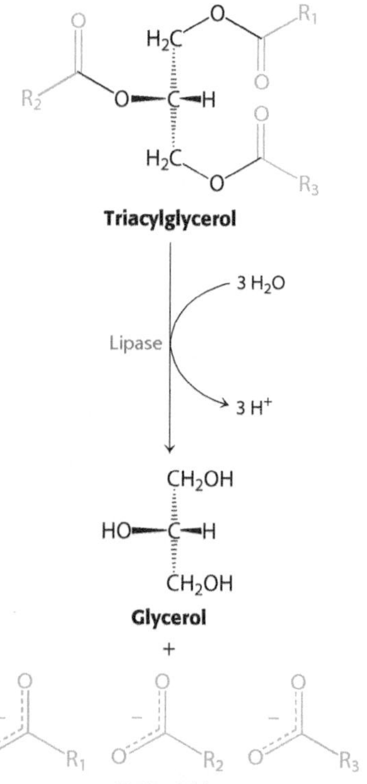

Triacylglycerol

Lipase

3 H₂O

3 H⁺

Glycerol

+

Fatty acids

Figure 27.1 Lipid degradation. Lipids are hydrolyzed by lipases in three steps to yield fatty acids and glycerol. The fatty acids are taken up by cells and used as a fuel. Glycerol also enters the liver, where it can be metabolized by the glycolytic or gluconeogenic pathways.

processing. First, the lipids must be mobilized. In this process of lipolysis, triacylglycerols are degraded to fatty acids and glycerol, which are released from the adipose tissue and transported to the energy-requiring tissues (**Figure 27.1**). Second, at these tissues, the fatty acids must be activated and transported into mitochondria for degradation. Third, the fatty acids are broken down in a step-by-step fashion into acetyl CoA, which is then processed in the citric acid cycle.

⚕ CLINICAL INSIGHT

Triacylglycerols Are Hydrolyzed by Hormone-Stimulated Lipases

Consider someone who has just awakened from a night's sleep and begins a bout of exercise. After the night's fast, glycogen stores are low, but lipids are readily available. How are these lipid stores mobilized to provide fuel for muscles and other tissues?

Triacylglycerols are stored inside a fat cell (adipocyte) as a lipid droplet, an intracellular compartment surrounded by a single layer of phospholipids and the numerous proteins required for fatty acid metabolism (Figure 11.3). Originally believed to be inert-lipid deposits, lipid droplets are now understood to be dynamic organelles essential for the regulation of lipid metabolism. Triacylglycerol mobilization and deposition take place on the surface of the droplet. Before fats can be used as fuels, the triacylglycerol storage form must be hydrolyzed to yield isolated fatty acids. Under the physiological conditions facing an early-morning runner, glucagon and epinephrine will be present. In adipose tissue, these hormones trigger 7TM receptors that activate adenylate cyclase (p. 228). The increased level of cyclic AMP then stimulates protein kinase A, which phosphorylates two key proteins: *perilipin*, a fat-droplet-associated protein, and *hormone-sensitive lipase* (**Figure 27.2**). The phosphorylation of perilipin has two crucial effects. First, it restructures the fat droplet so that the triacylglycerols are more accessible to degradation. Second, the phosphorylation

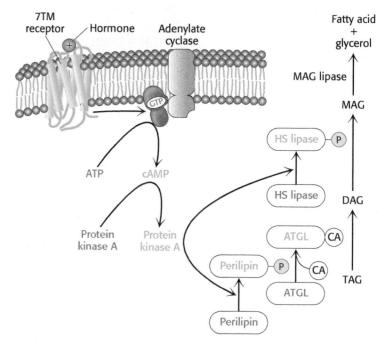

Figure 27.2 Triacylglycerols in adipose tissue are converted into free fatty acids in response to hormonal signals. The phosphorylation of perilipin restructures the lipid droplet and releases the coactivator of ATGL. The activation of ATGL by binding with its coactivator initiates the mobilization. Hormone-sensitive lipase releases a fatty acid from diacylglycerol. Monoacylglycerol lipase completes the mobilization process. Abbreviations: 7TM, seven transmembrane; ATGL, adipose triglyceride lipase; CA, coactivator; HS lipase, hormone-sensitive lipase; MAG lipase, monoacylglycerol lipase; DAG, diacylglycerol; TAG, triacylglycerol.

of perilipin triggers the release of a coactivator for *adipose triglyceride lipase* (ATGL). Once bound to the cofactor, ATGL initiates the mobilization of triacylglycerols by releasing a fatty acid from triacylglycerol, forming diacylglycerol. Diacylglycerol is converted into a free fatty acid and monoacylglycerol by the hormone-sensitive lipase, which has been activated by phosphorylation. Finally, a *monoacylglycerol lipase* completes the mobilization of fatty acids with the production of a free fatty acid and glycerol. Thus, *epinephrine and glucagon induce lipolysis.* Although their role in muscle is not as firmly established, these hormones probably also regulate the use of triacylglycerol stores in that tissue.

If the coactivator required by ATGL is missing or defective, a rare condition called Chanarin-Dorfman syndrome results. Fats accumulate throughout the body because they cannot be released by ATGL. Other symptoms include dry skin (ichthyosis), enlarged liver and muscle, and mild cognitive disability.

Free Fatty Acids and Glycerol Are Released Into the Blood

Fatty acids are not soluble in aqueous solutions. In order to reach tissues that require fatty acids, the released fatty acids bind to the blood protein albumin, which delivers them to tissues in need of fuel.

Glycerol formed by lipolysis is absorbed by the liver and phosphorylated. It is then oxidized to dihydroxyacetone phosphate, which is isomerized to glyceraldehyde 3-phosphate. This molecule is an intermediate in both the glycolytic and the gluconeogenic pathways.

Hence, glycerol can be converted into pyruvate or glucose in the liver, which contains the appropriate enzymes (**Figure 27.3**). The reverse reaction can take place just as readily. Thus, glycerol and glycolytic intermediates are interconvertible.

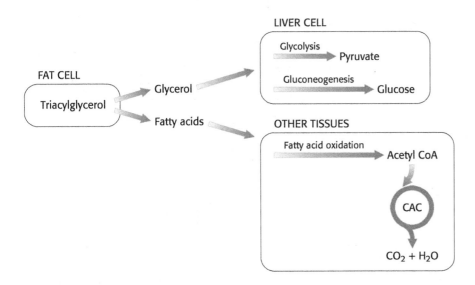

Figure 27.3 Lipolysis generates fatty acids and glycerol. The fatty acids are used as fuel by many tissues. The liver processes glycerol by either the glycolytic or the gluconeogenic pathway, depending on its metabolic circumstances. Abbreviation: CAC, citric acid cycle.

Fatty Acids Are Linked to Coenzyme A Before They Are Oxidized

Fatty acids separate from the albumin in the bloodstream and diffuse across the cell membrane with the assistance of transport proteins. In the cell, fatty acids are shuttled about in association with fatty-acid-binding proteins.

Fatty acid oxidation takes place in mitochondria, so how do these fuels gain access to the site of degradation? First, fatty acids must be activated by reacting with coenzyme A to form acyl CoA. This activation reaction takes place on the outer mitochondrial membrane, where it is catalyzed by *acyl CoA synthetase*.

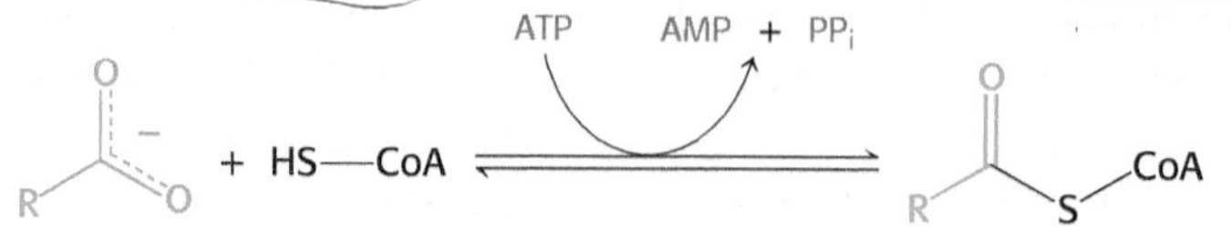

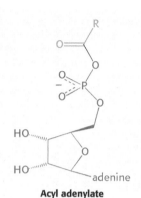

Acyl adenylate

Fatty acid   **Acyl CoA**

The activation takes place in two steps:

1. The fatty acid reacts with ATP to form an *acyl adenylate*, and the other two phosphoryl groups of the ATP substrate are released as pyrophosphate:

$$\text{Fatty acid} + \text{ATP} \rightleftharpoons \text{Acyl adenylate} + \text{PP}_i \qquad (1)$$

2. The sulfhydryl group of CoA then attacks the acyl adenylate to form acyl CoA and AMP:

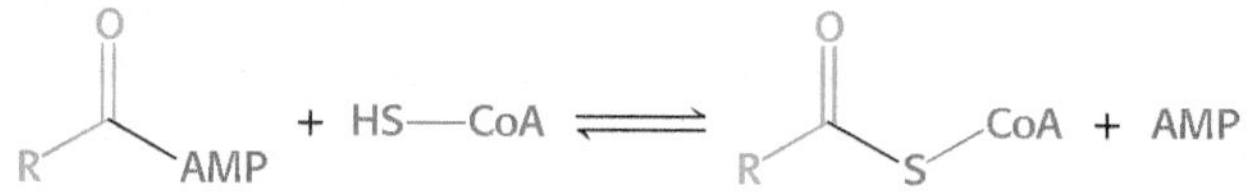

$$\text{Acyl adenylate} + \text{HS—CoA} \rightleftharpoons \text{Acyl CoA} + \text{AMP} \qquad (2)$$

Acyl CoA

These partial reactions are freely reversible. In fact, the equilibrium constant for the sum of these reactions is close to 1, meaning that the energy levels of the reactants and products are about equal. The reaction is driven forward by the hydrolysis of pyrophosphate by pyrophosphatase:

$$\underset{\text{Pyrophosphate}}{\text{PP}_i} \;+\; \text{H}_2\text{O} \;\xrightarrow{\text{Pyrophosphatase}}\; \underset{\text{Phosphate}}{2\,\text{P}_i}$$

We see here another example of a recurring theme in biochemistry: *many biosynthetic reactions are made irreversible by the hydrolysis of inorganic pyrophosphate.* Thus, the complete reaction for fatty acid activation is

$$\text{RCOO}^- + \text{CoA} + \text{ATP} + \text{H}_2\text{O} \longrightarrow \text{RCO-CoA} + \text{AMP} + 2\,\text{P}_i$$

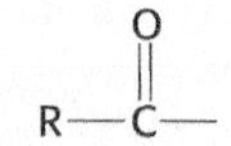

Activation is not the only step necessary to move fatty acids into the mitochondrial matrix. Activated fatty acids can cross the outer mitochondrial membrane through the *voltage-dependent ion channels* (p. 365), also called porin channels. However, transport across the inner mitochondrial membrane requires that the fatty acids be linked to the alcohol *carnitine*. The acyl group is transferred from the sulfur atom of CoA to the hydroxyl group of carnitine to form *acyl carnitine*. This reaction is catalyzed by *carnitine acyltransferase I* (also called *carnitine palmitoyl transferase I*), which is bound to the outer mitochondrial membrane:

An acyl group

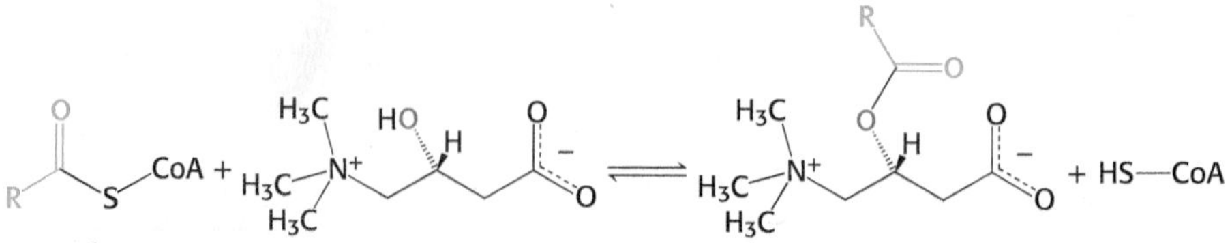

Acyl CoA   **Carnitine**   **Acyl carnitine**

Acyl carnitine is then shuttled across the inner mitochondrial membrane by a translocase (**Figure 27.4**). The acyl group is transferred back to CoA by *carnitine acyltransferase II* (carnitine palmitoyl transferase II) on the matrix side of the membrane. Finally, the translocase returns carnitine to the cytoplasmic side in exchange for an incoming acyl carnitine, allowing the process to continue.

🩺 CLINICAL INSIGHT

Pathological Conditions Result if Fatty Acids Cannot Enter the Mitochondria

A number of diseases have been traced to a deficiency of carnitine, carnitine transferase, or translocase. The symptoms of carnitine deficiency range from mild muscle cramping to severe weakness and even death. Inability to synthesize carnitine may be a contributing factor to the development of autism in males. In general, muscle, kidney, and heart are the tissues primarily impaired. Muscle weakness during prolonged exercise is a symptom of a deficiency of carnitine acyltransferases because muscle relies on fatty acids as a long-term source of energy. These diseases illustrate that *the impaired flow of a metabolite from one compartment of a cell to another can lead to a pathological condition*. Carnitine is now popular as a dietary supplement, and its proponents claim that it increases endurance, enhances brain function, and promotes weight loss. The actual effectiveness of carnitine as a dietary supplement remains to be established.

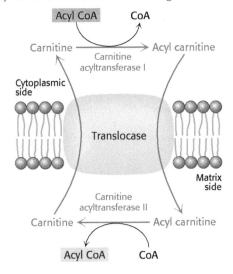

Figure 27.4 Acyl carnitine translocase. The entry of acyl carnitine into the mitochondrial matrix is mediated by a translocase. Carnitine returns to the cytoplasmic side of the inner mitochondrial membrane in exchange for acyl carnitine.

Acetyl CoA, NADH, and FADH₂ Are Generated by Fatty Acid Oxidation

After the activated fatty acid is in mitochondria, it is ready for metabolism. The goal of fatty acid degradation is to oxidize the fatty acid—two carbon atoms at a time—to acetyl CoA and to gather the released high-energy electrons to power oxidative phosphorylation. A saturated acyl CoA is degraded by a recurring sequence of four reactions: oxidation by flavin adenine dinucleotide (FAD), hydration, oxidation by nicotinamide adenine dinucleotide (NAD$^+$), and thiolysis by coenzyme A (Figure 27.5). The fatty acid chain is shortened by two carbon atoms as a result of these reactions, and FADH₂, NADH, and acetyl CoA are generated. Because oxidation takes place at the β-carbon atom, this series of reactions is called the *β-oxidation pathway*.

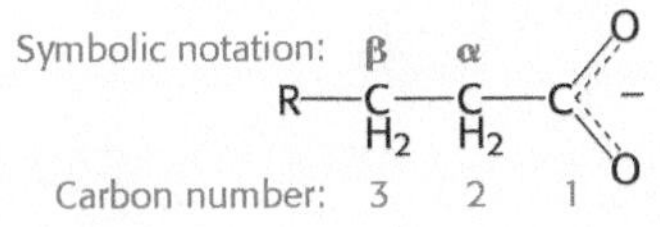

The first reaction in each round of degradation is the *oxidation* of acyl CoA by an *acyl CoA dehydrogenase* to give an enoyl CoA with a trans double bond between C-2 and C-3:

$$\text{Acyl CoA} + \text{E-FAD} \longrightarrow \textit{trans-}\Delta^2\text{-enoyl CoA} + \text{E-FADH}_2$$

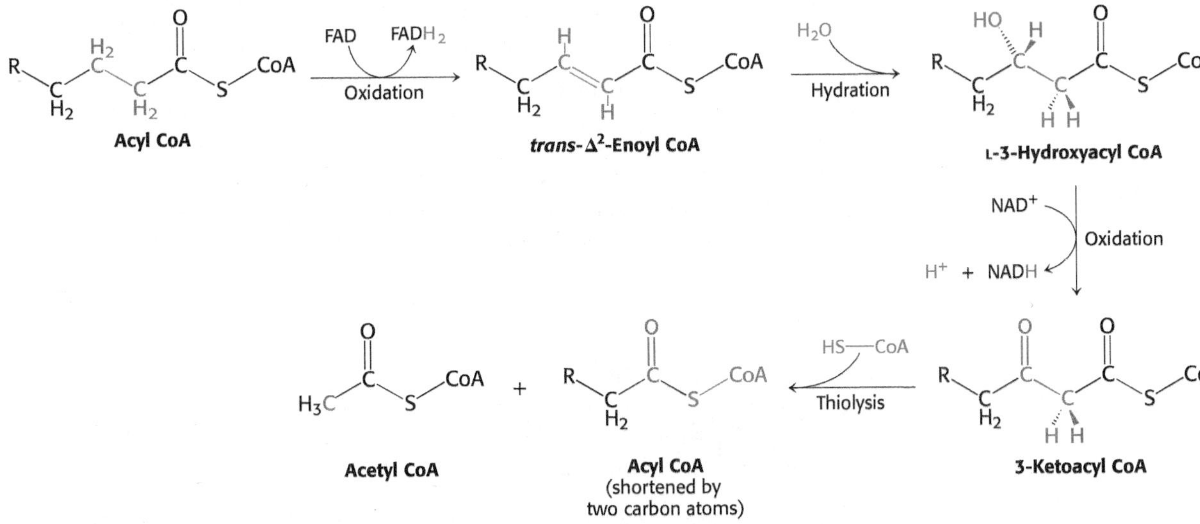

Symbolic notation:

$$\begin{array}{cc} \beta & \alpha \\ R\!-\!\underset{H_2}{C}\!-\!\underset{H_2}{C}\!-\!C\!\!\!\begin{array}{c} O \\ \diagup \\ \diagdown \\ O \end{array}^{-} \end{array}$$

Carbon number: 3   2   1

Figure 27.5 The reaction sequence for the degradation of fatty acids. Fatty acids are degraded by the repetition of a four-reaction sequence consisting of oxidation, hydration, oxidation, and thiolysis.

DID YOU KNOW?

The symbol Δ^n (capital delta, superscript number) is used to denote the position of the first carbon atom participating in a double bond. Thus, Δ^2 designates a double bond between carbon 2 and carbon 3.

As in the dehydrogenation of succinate in the citric acid cycle, FAD rather than NAD^+ is the electron acceptor because the ΔG for this reaction is insufficient to drive the reduction of NAD^+. Electrons picked up by FAD are transferred to the electron-transport chain and, ultimately, to ubiquinone, which is thereby reduced to ubiquinol. Then, ubiquinol delivers its high-potential electrons to the second proton-pumping site of the respiratory chain (p. 373).

The next step is the *hydration* of the double bond between C-2 and C-3 by *enoyl CoA hydratase:*

$$trans\text{-}\Delta^2\text{-Enoyl CoA} + H_2O \rightleftharpoons \text{L-3-hydroxyacyl CoA}$$

The hydration of enoyl CoA is stereospecific. Only the L isomer of 3-hydroxyacyl CoA is formed when the *trans*-Δ^2 double bond is hydrated.

The hydration of enoyl CoA is a prelude to the second *oxidation* reaction, which converts the hydroxyl group at C-3 into a keto group and generates NADH. This oxidation is catalyzed by *L-3-hydroxyacyl CoA dehydrogenase:*

$$\text{L-3-Hydroxyacyl CoA} + NAD^+ \rightleftharpoons \text{3-ketoacyl CoA} + NADH + H^+$$

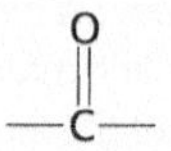

A keto group

The preceding reactions have oxidized the methylene group ($—CH_2—$) at C-3 to a keto group. The final step is the *cleavage* of 3-ketoacyl CoA by the thiol group of a second molecule of coenzyme A, which yields acetyl CoA and an acyl CoA shortened by two carbon atoms. This thiolytic cleavage is catalyzed by *β-ketothiolase*:

$$\underset{(n \text{ carbon atoms})}{\text{3-Ketoacyl CoA}} + \text{HS-CoA} \longrightarrow \text{acetyl CoA} + \underset{(n-2 \text{ carbon atoms})}{\text{acyl CoA}}$$

Table 27.1 summarizes the reactions in fatty acid degradation. The shortened acyl CoA then undergoes another cycle of oxidation, starting with the reaction catalyzed by acyl CoA dehydrogenase (Figure 27.6).

Recent evidence suggests that the enzymes of fatty acid oxidation are associated with one another to form a super complex. Moreover, this super complex is in turn associated with the inner mitochondrial membrane, the site of the electron-transport chain and ATP synthesis. This organization allows the rapid movement of substrates from enzyme to enzyme and gives the high-energy electrons generated by fatty acid oxidation immediate access to the electron-transport chain.

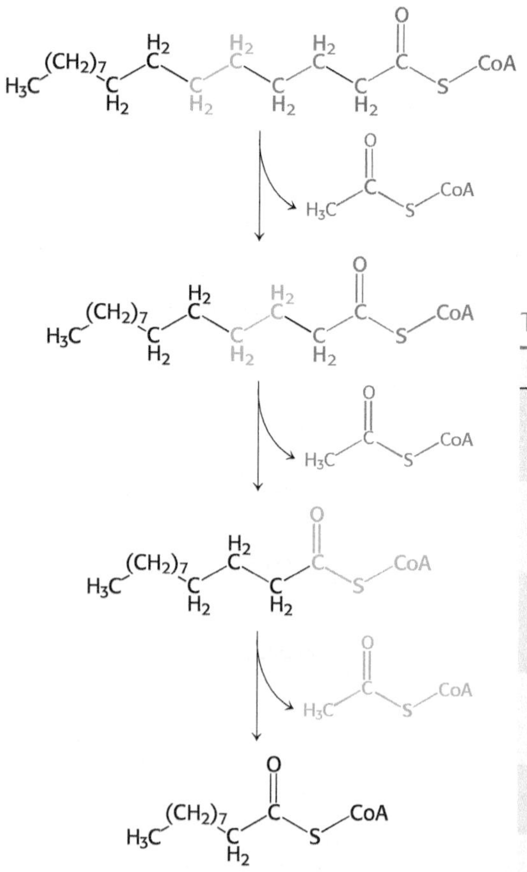

Figure 27.6 The first three rounds in the degradation of palmitate. Two carbon units are sequentially removed from the carboxyl end of the fatty acid.

Table 27.1 Principal reactions required for fatty acid degradation

Step	Reaction	Enzyme
1	Fatty Acid + CoA + ATP ⇌ acyl CoA + AMP + PP_i	Acyl CoA synthetase (also called fatty acid thiokinase and fatty acid: CoA ligase)*
2	Carnitine + acyl CoA ⇌ acyl carnitine + CoA	Carnitine acyltransferase I and II (also called carnitine palmitoyl transferase I and II)
3	Acyl CoA + E-FAD ⟶ *trans*-Δ^2-enoyl CoA + E-$FADH_2$	Acyl CoA dehydrogenases (several isozymes having different chain-length specificity)
4	*trans*-Δ^2-Enoyl CoA + H_2O ⇌ L-3-hydroxyacyl CoA	Enoyl CoA hydratase (also called crotonase or 3-hydroxyacyl CoA hydrolyase)
5	L-3-Hydroxyacyl CoA + NAD^+ ⇌ 3-Ketoacyl CoA + NADH + H^+	L-3-Hydroxyacyl CoA dehydrogenase
6	3-Ketoacyl CoA + CoA ⟶ acetyl CoA + acyl CoA (shortened by two carbon atoms)	β-Ketothiolase (also called thiolase)

*An AMP-forming ligase.

The Complete Oxidation of Palmitate Yields 106 Molecules of ATP

We can now calculate the energy yield derived from the oxidation of a fatty acid. In each reaction cycle, an acyl CoA is shortened by two carbon atoms, and one molecule each of $FADH_2$, NADH, and acetyl CoA is formed:

$$C_n\text{-acyl CoA} + FAD + NAD^+ + H_2O + CoA \longrightarrow$$
$$C_{n-2}\text{-acyl CoA} + FADH_2 + NADH + \text{acetyl CoA} + H^+$$

The degradation of palmitoyl CoA (C_{16}-acyl CoA) requires seven reaction cycles. In the seventh cycle, the C_4-ketoacyl CoA is cleaved by the thiol group of coenzyme A to two molecules of acetyl CoA. Hence, the stoichiometry of the oxidation of palmitoyl CoA is

$$\text{Palmitoyl CoA} + 7\,FAD + 7\,NAD^+ + 7\,CoA + 7\,H_2O \longrightarrow$$
$$8\,\text{acetyl CoA} + 7\,FADH_2 + 7\,NADH + 7\,H^+$$

Approximately 2.5 molecules of ATP are generated when the respiratory chain oxidizes each of these NADH molecules, whereas 1.5 molecules of ATP are formed for each $FADH_2$ because their electrons enter the chain at the level of ubiquinol. Recall that the oxidation of acetyl CoA by the citric acid cycle yields 10 molecules of ATP. Hence, the number of ATP molecules formed in the oxidation of palmitoyl CoA is 10.5 from the 7 molecules of $FADH_2$, 17.5 from the 7 molecules of NADH, and 80 from the 8 molecules of acetyl CoA, which gives a total of 108. The equivalent of 2 molecules of ATP is consumed in the activation of palmitate, in which ATP is split into AMP and two molecules of orthophosphate. Thus, *the complete oxidation of a molecule of palmitate yields 106 molecules of ATP.*

? QUICK QUIZ 1 Describe the repetitive steps of β oxidation. Why is the process called β oxidation?

27.2 The Degradation of Unsaturated and Odd-Chain Fatty Acids Requires Additional Steps

The β-oxidation pathway accomplishes the complete degradation of saturated fatty acids having an even number of carbon atoms. Most fatty acids have such structures because of their mode of synthesis (Chapter 28). However, not all fatty acids are so simple. The oxidation of fatty acids containing double bonds requires additional steps. Likewise, fatty acids containing an odd number of carbon atoms require additional enzyme reactions to yield a metabolically useful molecule.

An Isomerase and a Reductase Are Required for the Oxidation of Unsaturated Fatty Acids

Many unsaturated fatty acids are available in our diet. Indeed, we are encouraged to eat foods that are rich in certain types of polyunsaturated fatty acids, such as the ω-3 fatty acid linolenic acid, which is prominent in safflower and corn oils. Polyunsaturated fatty acids are important for a number of reasons, not the least of which is that they offer some protection from heart attacks. How are excess amounts of these fatty acids oxidized?

Consider the oxidation of palmitoleate. This C_{16} unsaturated fatty acid, which has one double bond between C-9 and C-10, is activated to palmitoleoyl CoA and transported across the inner mitochondrial membrane in the same way as saturated fatty acids. Palmitoleoyl CoA then undergoes three cycles of degradation, which are carried out by the same enzymes as those in the oxidation of saturated fatty acids. However, the *cis*-Δ^3-enoyl CoA formed in the third round is not a substrate for acyl CoA dehydrogenase. The presence of a double bond between C-3 and C-4 prevents the formation of another double bond between C-2 and C-3. This impasse is resolved by a new reaction that shifts the position and configuration of the *cis*-Δ^3 double bond. *cis*-Δ^3-*Enoyl CoA isomerase* converts this double bond into a *trans*-Δ^2 double bond (**Figure 27.7**). The subsequent

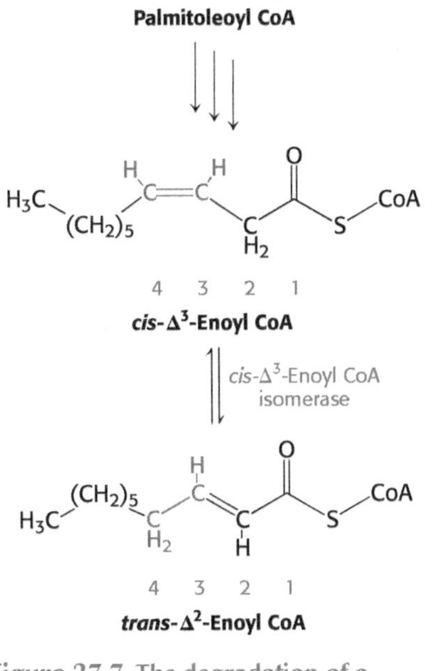

Figure 27.7 The degradation of a monounsaturated fatty acid. *cis*-Δ^3-Enoyl CoA isomerase allows the β oxidation of fatty acids with a single double bond to continue.

reactions are those of the saturated fatty acid oxidation pathway, in which the *trans*-Δ^2-enoyl CoA is a regular substrate.

Human beings require polyunsaturated fatty acids, which have multiple double bonds, as important precursors of signal molecules, but excess polyunsaturated fatty acids are degraded by β oxidation. However, when these fats are subjected to β oxidation, molecules result that cannot themselves be degraded by β oxidation. To prevent a wasteful buildup of these molecules, the initial degradation products of the polyunsaturated fatty acids must first be modified. Consider linoleate, a C_{18} polyunsaturated fatty acid with *cis*-Δ^9 and *cis*-Δ^{12} double bonds (**Figure 27.8**). The *cis*-Δ^3 double bond formed after three rounds of β oxidation is converted into a *trans*-Δ^2 double bond by the aforementioned isomerase. The acyl CoA produced by another round of β oxidation contains a *cis*-Δ^4 double bond. The dehydrogenation of this species by acyl CoA dehydrogenase yields a *2,4-dienoyl intermediate*, which is not a substrate for the next enzyme in the β-oxidation pathway. This impasse is circumvented by *2,4-dienoyl CoA reductase*, an enzyme that uses NADPH to reduce the 2,4-dienoyl intermediate to *trans*-Δ^3-enoyl CoA. *cis*-Δ^3-Enoyl CoA isomerase then converts *trans*-Δ^3-enoyl CoA into the *trans*-Δ^2 form, a normal intermediate in the β-oxidation pathway. These catalytic strategies are elegant and economical. Only two extra enzymes are needed for the oxidation of *any* polyunsaturated fatty acid. *In polyunsaturated fatty acids, odd-numbered double bonds are handled by the isomerase alone, and even-numbered ones are handled by the reductase and the isomerase.*

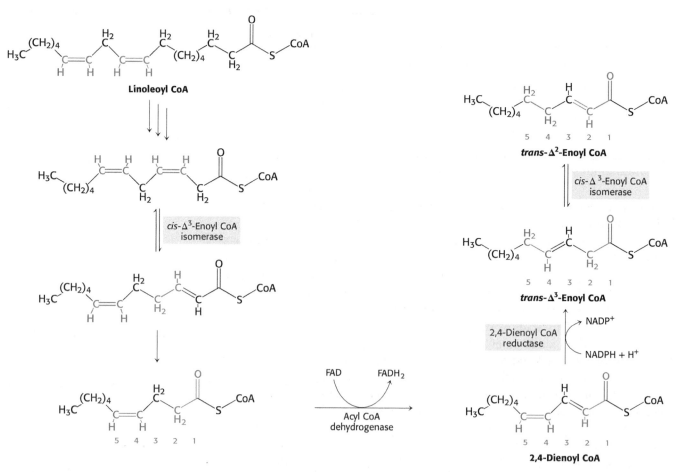

Figure 27.8 The oxidation of linoleoyl CoA. The complete oxidation of the diunsaturated fatty acid linoleate is facilitated by the activity of enoyl CoA isomerase and 2,4-dienoyl CoA reductase.

Odd-Chain Fatty Acids Yield Propionyl CoA in the Final Thiolysis Step

Fatty acids having an odd number of carbon atoms are a minor class found in small amounts in vegetables. They are oxidized in the same way as fatty acids having an even number of carbon atoms, except that propionyl CoA and acetyl CoA, rather than two molecules of acetyl CoA, are produced in the final round of degradation. The activated three-carbon unit in propionyl CoA is converted into succinyl CoA, at which point it enters the citric acid cycle.

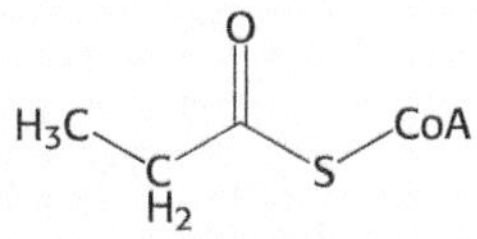

The pathway from propionyl CoA to succinyl CoA is especially interesting because it entails a rearrangement that requires *vitamin B_{12}* (also known as *cobalamin*). Propionyl CoA is carboxylated by propionyl CoA carboxylase (a biotin enzyme) at the expense of the hydrolysis of a molecule of ATP to yield the D isomer of methylmalonyl CoA (**Figure 27.9**). The D isomer of methylmalonyl CoA is converted into the L isomer, the substrate for a mutase that converts it into *succinyl CoA* by an *intramolecular rearrangement*. The —CO—S—CoA group migrates from C-2 to the methyl group in exchange for a hydrogen atom. This very unusual isomerization is catalyzed by *methylmalonyl CoA mutase*, which contains vitamin B_{12} as its coenzyme.

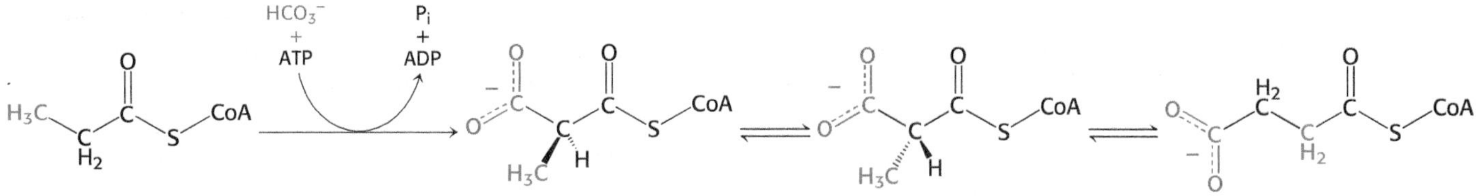

Propionyl CoA

Figure 27.9 The conversion of propionyl CoA into succinyl CoA. Propionyl CoA, generated from fatty acids having an odd number of carbon atoms as well as from some amino acids, is converted into the citric acid cycle intermediate succinyl CoA.

27.3 Ketone Bodies Are Another Fuel Source Derived from Fats

✓ 2 Describe ketone bodies and their role in metabolism.

Most acetyl CoA produced by fatty acid degradation enters the citric acid cycle. However, some acetyl CoA units are used to form an alternative fuel source called *ketone bodies*—namely, acetoacetate, D-3-hydroxybutyrate (β-hydroxybutyrate), and acetone. Although ketone bodies do not generate as much ATP as do the fatty acids from which they are derived, they have the advantage of being water soluble, and so they are an easily transportable form of acetyl units. Originally thought to be indicators of impaired metabolism, acetoacetate and D-3-hydroxybutyrate are now known to be normal fuels of respiration. Indeed, heart muscle and the renal cortex of the kidney may use acetoacetate in preference to glucose. Acetone is produced by the slow, spontaneous decarboxylation of acetoacetate. Under starvation conditions, acetone may be captured to synthesize glucose.

Ketone-Body Synthesis Takes Place in the Liver

The major site of *ketogenesis*—the production of acetoacetate and D-3-hydroxybutyrate— is the mitochondria of the liver. Acetoacetate is formed from acetyl CoA in three steps (**Figure 27.10**). The sum of these reactions is

$$2 \text{ Acetyl CoA} + \text{H}_2\text{O} \longrightarrow \text{acetoacetate} + 2 \text{ CoA} + \text{H}^+$$

A fourth step is required to form D-3-hydroxybutyrate: the reduction of acetoacetate in the mitochondrial matrix. The ketone bodies then exit the liver mitochondria into the blood, with the assistance of specific carrier proteins, and are transported to peripheral tissues.

QUICK QUIZ 2 Why might D-3-hydroxybutyrate be considered a superior ketone body compared with acetoacetate?

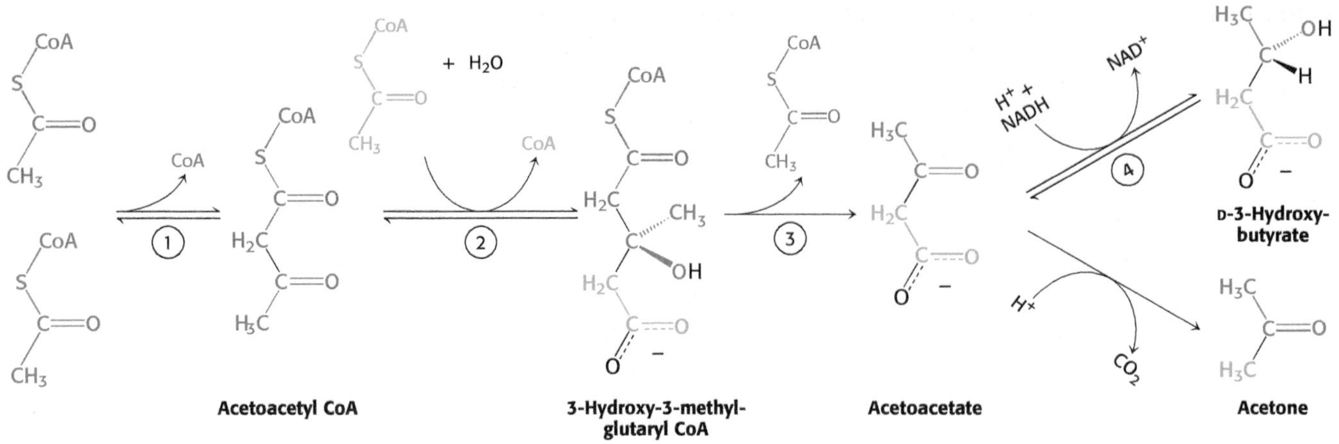

Acetoacetyl CoA   **3-Hydroxy-3-methyl-**   **Acetoacetate**   **Acetone**
                        glutaryl CoA

Figure 27.10 The formation of ketone bodies. The ketone bodies—acetoacetate, D-3-hydroxybutyrate, and acetone—are formed from acetyl CoA primarily in the liver. Enzymes catalyzing these reactions are (1) 3-ketothiolase, (2) hydroxymethylglutaryl CoA synthase, (3) hydroxymethylglutaryl CoA cleavage enzyme, and (4) D-3-hydroxybutyrate dehydrogenase. Acetoacetate spontaneously decarboxylates to form acetone.

How are ketone bodies used as a fuel? **Figure 27.11** shows how ketone bodies are metabolized to generate NADH as well as acetyl CoA, the fuel for the citric acid cycle. D-3-Hydroxybutyrate is oxidized to acetoacetate, which is then activated by a specific CoA transferase. Finally, acetoacetyl CoA is cleaved by a thiolase to yield two molecules of acetyl CoA, which enter the citric acid cycle. Liver cells, however, lack the CoA transferase, thus allowing acetoacetate to escape the liver and be transported out of the cell and to other tissues, rather than being metabolized in the liver. Acetoacetate also has a regulatory role. *High levels of acetoacetate in the blood signify an abundance of acetyl units and lead to a decrease in the rate of lipolysis in adipose tissue.*

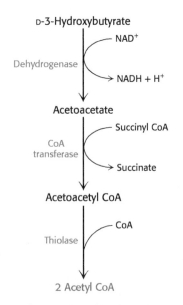

Figure 27.11 The utilization of D-3-hydroxybutyrate and acetoacetate as a fuel. D-3-Hydroxybutyrate is oxidized to acetoacetate with the formation of NADH. Acetoacetate is then converted into two molecules of acetyl CoA, which then enter the citric acid cycle.

CLINICAL INSIGHT

Ketogenic Diets May Have Therapeutic Properties

Interestingly, diets that promote ketone-body formation, called ketogenic diets, are frequently used as a therapeutic option for children with drug-resistant epilepsy. Ketogenic diets are rich in fats and low in carbohydrates, with adequate amounts of protein. In essence, the body is forced into starvation mode, where fats and ketone bodies become the main fuel source. How such diets reduce the seizures suffered by the children is currently unknown.

Animals Cannot Convert Fatty Acids into Glucose

A typical human being has far greater fat stores than glycogen stores. However, glycogen is necessary to fuel very active muscle as well as the brain, which normally uses only glucose as a fuel. When glycogen levels are low, why can't the body make use of fat stores instead and convert fatty acids into glucose? Because *animals are unable to effect the net synthesis of glucose from fatty acids.* Specifically, the acetyl CoA generated by fatty acid degradation cannot be converted into pyruvate or oxaloacetate in animals. Recall that the reaction that generates acetyl CoA from pyruvate is irreversible (p. 333). The two carbon atoms of the acetyl group of acetyl CoA enter the citric acid cycle, but two carbon atoms leave the cycle in decarboxylations catalyzed by isocitrate dehydrogenase and α-ketoglutarate dehydrogenase. Consequently, oxaloacetate is regenerated, *but it is not formed de novo when the acetyl unit of acetyl CoA is oxidized by the citric acid cycle.* In essence, two carbon atoms enter the cycle as an acetyl group, but two carbons leave the cycle as CO_2 before oxaloacetate is generated. Consequently, no net synthesis of oxaloacetate is possible. In contrast, plants have two additional enzymes enabling them to convert the carbon atoms of acetyl CoA into oxaloacetate (p. 355).

27.4 Metabolism in Context: Fatty Acid Metabolism Is a Source of Insight into Various Physiological States

Fatty acids are our most prominent fuel. Not surprisingly then, the metabolism of these key molecules is altered under different physiological conditions. We will now examine the role of fatty acid metabolism in two such conditions: diabetes and starvation.

🩺 CLINICAL INSIGHT

Diabetes Can Lead to a Life-Threatening Excess of Ketone-Body Production

Although ketone bodies are a normal fuel, excess amounts of these acids can be dangerous. Excess production can result from an imbalance in the metabolism of carbohydrates and fatty acids, as seen in the disease *diabetes mellitus,* also known as diabetes, a condition characterized by the absence of insulin or resistance to insulin (p. 466).

Why does the disruption of insulin function result in disease? Insulin has two major roles in regulating metabolism. First, insulin normally stimulates the absorption of glucose by the liver. If this absorption does not take place, oxaloacetate cannot be produced to react with acetyl CoA, the product of fatty acid degradation. Recall that animals synthesize oxaloacetate from pyruvate, a product of the glycolytic processing of glucose (p. 316). Indeed, in diabetics, oxaloacetate from the citric acid cycle is actually *consumed* to form glucose through gluconeogenesis. Second, insulin normally curtails fatty acid mobilization by adipose tissue. In the absence of functional insulin, fatty acids are released and large amounts of acetyl CoA are consequently produced by β oxidation. However, much of the acetyl CoA cannot enter the citric acid cycle, because of insufficient oxaloacetate for condensation. A striking feature of diabetes is a shift of fuel usage from carbohydrates to fats; glucose, more abundant than ever, is left unused in the bloodstream.

The liver also releases large amounts of ketone bodies into the blood because it is degrading fatty acids but lacks the glucose required to replenish the citric acid cycle (**Figure 27.12**). Ketone bodies are moderately strong

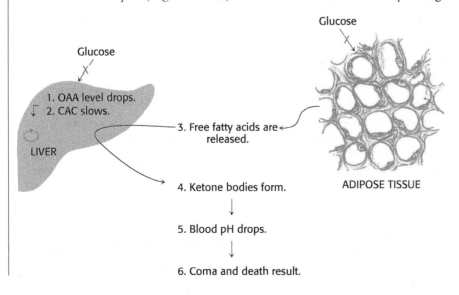

Figure 27.12 Diabetic ketosis results when insulin is absent. In the absence of insulin, fats are released from adipose tissue, and glucose cannot be absorbed by the liver or adipose tissue. The liver degrades the fatty acids by β oxidation but cannot process the acetyl CoA, because of a lack of glucose-derived oxaloacetate (OAA). Excess ketone bodies are formed and released into the blood. Abbreviation: CAC, citric acid cycle.

acids, and the result of their overabundance is severe acidosis because of the inability of the kidneys to maintain acid–base balance. The increased blood acidity impairs tissue function, most importantly in the central nervous system. An untreated diabetic who has not eaten can go into a coma because of a lowered blood-pH level, dehydration, and a lack of glucose. Before diabetes was well understood, diabetics suffering from diabetic ketosis were sometimes diagnosed as being drunk. The lack of glucose for the brain led to odd behavior, and the spontaneous decarboxylation of acetoacetate to acetone was confused with the smell of alcohol.

⚕ CLINICAL INSIGHT

Ketone Bodies Are a Crucial Fuel Source During Starvation

For some people, the use of ketone bodies as fuel is a matter of life or death. The bodies of starving people automatically resort to ketone bodies as a primary fuel source. Why does this adaptation take place? Let's consider the biochemical changes that take place in the course of prolonged fasting. A typical well-nourished 70-kg (154-lb) man has fuel reserves totaling about 670,000 kJ (161,000 kcal; Table 27.2). His energy needs for a 24-hour period range from about 6700 kJ (1600 kcal) to 25,000 kJ (6000 kcal), depending on the extent of his activity; so stored fuels meet his energy needs in starvation for 1 to 3 months. However, the carbohydrate reserves are exhausted in only a day. The first priority of metabolism is to provide sufficient glucose to the brain and other tissues (such as red blood cells) that are absolutely dependent on this fuel. However, because glycogen-derived glucose is depleted and fats cannot be converted into glucose, the only potential sources of glucose are the carbon skeletons of amino acids derived from the breakdown of proteins. Because proteins are not stored, any breakdown will necessitate a loss of function. Survival for most animals depends on being able to move rapidly, which requires a large muscle mass, and so muscle loss must be minimized.

Table 27.2   Fuel reserves in a typical 70-kg (154-lb) man

	Available energy in kilojoules (kcal)					
Organ	Glucose or glycogen		Triacylglycerols		Mobilizable proteins	
Blood	250	(60)	20	(45)	0	(0)
Liver	1,700	(400)	2,000	(450)	1,700	(400)
Brain	30	(8)	0	(0)	0	(0)
Muscle	5,000	(1,200)	2,000	(450)	100,000	(24,000)
Adipose tissue	330	(80)	560,000	(135,000)	170	(40)

Source: Data from G. F. Cahill, Jr., *Clin. Endocrinol. Metab.* 5:398, 1976.

Thus, the second priority of metabolism in starvation is to preserve muscle protein by shifting the fuel being used from glucose to fatty acids and ketone bodies, particularly in organs that normally rely on glucose (**Figure 27.13**). How is the loss of muscle protein curtailed? First of all, muscle—the largest fuel consumer in the body—shifts from glucose to fatty acids for fuel. This switch lessens the need to degrade protein for glucose formation. The degradation of fatty acids by muscle halts the conversion of pyruvate into acetyl CoA, because acetyl CoA derived from fatty acids inhibits pyruvate dehydrogenase, the enzyme that converts pyruvate into acetyl CoA (p. 338). Any available pyruvate, lactate, and alanine are then exported to the liver for conversion into glucose for use by the brain. Second, after about three days of starvation, the liver forms large amounts of the ketone bodies acetoacetate

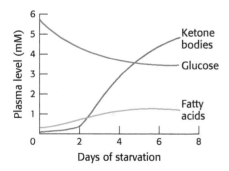

Figure 27.13 Fuel choice during starvation. The plasma levels of fatty acids and ketone bodies increase in starvation, whereas that of glucose decreases.

and D-3-hydroxybutyrate. Their synthesis from acetyl CoA increases markedly because the citric acid cycle is unable to oxidize all the acetyl units generated by the degradation of fatty acids. Gluconeogenesis depletes the supply of oxalo-acetate, which is essential for the entry of acetyl CoA into the citric acid cycle. Consequently, the liver produces large quantities of ketone bodies, which are released into the blood. At this time, *the brain begins to consume appreciable amounts of ketone bodies in place of glucose.* After three days of starvation, about a third of the energy needs of the brain are met by ketone bodies (Table 27.3).

Table 27.3 Fuel metabolism in starvation

Fuel exchanges and consumption	Amount formed or consumed in 24 hours (g)	
	3rd day	40th day
Fuel use by the brain		
Glucose	100	40
Ketone bodies	50	100
All other use of glucose	50	40
Fuel mobilization		
Adipose-tissue lipolysis	180	180
Muscle-protein degradation	75	20
Fuel output of the liver		
Glucose	150	80
Ketone bodies	150	150

After several weeks of starvation, ketone bodies become the major fuel of the brain. Only 40 g of glucose is then needed per day for the brain, compared with about 120 g in the first day of starvation. The effective conversion of fatty acids into ketone bodies by the liver and their use by the brain markedly diminishes the need for glucose. Hence, less muscle is degraded than in the first days of starvation. The breakdown of 20 g of muscle daily compared with 75 g early in starvation is crucial for survival. A person's survival time is mainly determined by the size of the triacylglycerol depot.

What happens when the lipid stores are gone? The only source of fuel that remains is protein. Protein degradation accelerates, and death inevitably results from a loss of heart, liver, or kidney function.

 CLINICAL INSIGHT

Some Fatty Acids May Contribute to the Development of Pathological Conditions

Certain polyunsaturated fatty acids are essential for life, serving as precursors to various signal molecules (p. 514). Vegetable oils, used commonly in food preparation, are rich in polyunsaturated fatty acids. However, polyunsaturated fatty acids are unstable and are readily oxidized. This tendency to become rancid reduces their shelf life and renders them undesirable for cooking. To circumvent this problem, polyunsaturated fatty acids are hydrogenated, converting them to saturated and trans unsaturated fatty acids (popularly known as "trans fat"), a variety of fat that is rare in nature. Epidemiological evidence suggests that consumption of large amounts of saturated fatty acids and trans fat promotes obesity, type 2 diabetes, and atherosclerosis. The mechanism by which these fats exert these effects is under active investigation. Some evidence suggests that they promote an inflammatory response and may mute the action of insulin and other hormones.

SUMMARY

27.1 Fatty Acids Are Processed in Three Stages

Triacylglycerols can be mobilized by the hydrolytic action of lipases that are under hormonal control. Glucagon and epinephrine stimulate triacylglycerol breakdown by activating the lipases. Insulin, in contrast, inhibits lipolysis. Fatty acids are activated to acyl CoAs, transported across the inner mitochondrial membrane by carnitine, and degraded in the mitochondrial matrix by a recurring sequence of four reactions: oxidation by FAD, hydration, oxidation by NAD^+, and thiolysis by coenzyme A. The $FADH_2$ and NADH formed in the oxidation steps transfer their electrons to O_2 by means of the respiratory chain, whereas the acetyl CoA formed in the thiolysis step normally enters the citric acid cycle by condensing with oxaloacetate. Mammals are unable to convert fatty acids into glucose because they lack a pathway for the net production of oxaloacetate, pyruvate, or other gluconeogenic intermediates from acetyl CoA.

27.2 The Degradation of Unsaturated and Odd-Chain Fatty Acids Requires Additional Steps

Fatty acids that contain double bonds or odd numbers of carbon atoms require ancillary steps to be degraded. An isomerase and a reductase are required for the oxidation of unsaturated fatty acids, whereas propionyl CoA derived from chains with odd numbers of carbon atoms requires a vitamin B_{12}-dependent enzyme to be converted into succinyl CoA.

27.3 Ketone Bodies Are Another Fuel Source Derived from Fats

The primary ketone bodies—acetoacetate and D-3-hydroxybutyrate—are formed in the liver by the condensation of acetyl CoA units. Ketone bodies are released into the blood and are an important fuel source for a number of tissues. After their uptake, ketone bodies are converted into acetyl CoA and processed by the citric acid cycle.

27.4 Metabolism in Context: Fatty Acid Metabolism Is a Source of Insight into Various Physiological States

Diabetes is characterized by the inability of cells to take up glucose. The lack of glucose as a fuel results in a greater demand for fats as a fuel. Ketone bodies may be produced in such excess as to acidify the blood, a potentially lethal condition called diabetic ketosis. Ketone bodies are also an especially important source of fuel for the brain when glucose is limited, as in prolonged fasting.

KEY TERMS

triacylglycerol (p. 489)
acyl adenylate (p. 492)

carnitine (p. 492)
β-oxidation pathway (p. 493)

ketone body (p. 497)

? Answers to QUICK QUIZZES

1. The steps are (1) oxidation by FAD; (2) hydration; (3) oxidation by NAD^+; (4) thiolysis to yield acetyl CoA. In symbolic notation, the β-carbon atom is oxidized.

2. D-3-Hydroxybutyrate is more energy rich because its oxidation potential is greater than that of acetoacetate. After having been absorbed by a cell, D-3-hydroxybutyrate is oxidized to acetoacetate, generating high-energy electrons in the form of NADH. The acetoacetate is then cleaved to yield to acetyl CoA.

PROBLEMS

1. *Stages of processing.* What are the three stages of triacylglycerol utilization? ✓ 1

2. *Control matters.* Outline the control of triacylglycerol mobilization. ✓ 1

3. *Forms of energy.* The partial reactions leading to the synthesis of acyl CoA (equations 1 and 2, p. 492) are freely reversible. The equilibrium constant for the sum of these reactions is close to 1, meaning that the energy levels of the reactants and products are about equal, even though a molecule of ATP has been hydrolyzed. Explain why these reactions are readily reversible. ✓ 1

4. *In its entirety.* Write the complete reaction for fatty acid activation. ✓ 1

5. *Activation fee.* The reaction for the activation of fatty acids before degradation is

$$R-\overset{\overset{\text{O}}{\|}}{C}-O^- + CoA + ATP + H_2O \longrightarrow$$

$$R-\overset{\overset{\text{O}}{\|}}{C}-SCoA + AMP + 2\,P_i + 2\,H^+$$

This reaction is quite favorable because the equivalent of two molecules of ATP is hydrolyzed. Explain why, from a biochemical bookkeeping point of view, the equivalent of two molecules of ATP is used despite the fact that the left side of the equation has only one molecule of ATP. ✓ 1

6. *Repeating reactions.* What are the recurring reactions of the oxidation of saturated fatty acids? ✓ 1

7. *Like Simon and Garfunkel.* Match each term with its description. ✓ 1

(a) Triacylglycerol _____
(b) Perilipin _____
(c) Adipose triglyceride lipase _____
(d) Glucagon _____
(e) Acyl CoA synthetase _____
(f) Carnitine _____
(g) β-Oxidation pathway _____
(h) Enoyl CoA isomerase _____
(i) 2,4-Dienoyl CoA reductase _____
(j) Methylmalonyl CoA mutase _____
(k) Ketone body _____

1. The enzyme that initiates lipid degradation
2. Activates fatty acids for degradation
3. Converts a *cis*-Δ^3 double bond into a *trans*-Δ^2 double bond
4. Reduces 2,4-dienoyl intermediate to *trans*-Δ^3-enoyl CoA
5. Storage form of fats
6. Required for entry into mitochondria
7. Requires vitamin B_{12}
8. Acetoacetate
9. Means by which fatty acids are degraded
10. Stimulates lipolysis
11. Lipid-droplet-associated protein

8. *Proper sequence.* Place the following list of reactions or relevant locations in the β oxidation of fatty acids in the proper order. ✓ 1

(a) Reaction with carnitine
(b) Fatty acid in the cytoplasm
(c) Activation of fatty acid by joining to CoA
(d) Hydration
(e) NAD^+-linked oxidation
(f) Thiolysis
(g) Acyl CoA in mitochondrion
(h) FAD-linked oxidation

9. *Too tired to exercise.* Explain why people with a hereditary deficiency of carnitine acyltransferase II have muscle weakness. Why are the symptoms more severe during fasting? ✓ 1

10. *A phantom acetyl CoA?* In the equation for fatty acid degradation shown here, only seven molecules of CoA are required to yield eight molecules of acetyl CoA. How is this difference possible? ✓ 1

$$\text{Palmitoyl CoA} + 7\,\text{FAD} + 7\,\text{NAD}^+ \\ + 7\,\text{CoA} + 7\,\text{H}_2\text{O} \longrightarrow \\ 8\,\text{Acetyl CoA} + 7\,\text{FADH}_2 + 7\,\text{NADH} + 7\,\text{H}^+$$

11. *Comparing yields.* Compare the ATP yields from palmitic acid and palmitoleic acid. ✓ 1

12. *Counting ATPs 1.* What is the ATP yield for the complete oxidation of C_{17} (heptadecanoic) fatty acid? Assume that the propionyl CoA ultimately yields oxaloacetate in the citric acid cycle. ✓ 1

13. *Sweet temptation.* Stearic acid is a C_{18} fatty acid component of chocolate. Suppose you had a depressing day and decided to settle matters by gorging on chocolate. How much ATP would you derive from the complete oxidation of stearic acid to CO_2? ✓ 1

14. *Buff and lean.* It has been suggested that if you are interested in losing body fat, the best time to do strenuous aerobic exercise is in the morning immediately after waking up; that is, after fasting. Don't eat breakfast before exercising, but have a cup of caffeinated coffee. Caffeine is an inhibitor of cAMP phosphodiesterase. Explain why this suggestion might work biochemically. ✓ 1

15. *The best storage form.* Compare the ATP yield from the complete oxidation of glucose, a six-carbon carbohydrate, and hexanoic acid, a six-carbon fatty acid. Hexanoic acid is also called caproic acid and is responsible for the "aroma" of goats. Why are fats better fuels than carbohydrates? ✓ 1

16. *From fatty acid to ketone body.* Write a balanced equation for the conversion of stearate into acetoacetate. ✓ 2

17. *Generous, but not to a fault.* Liver is the primary site of ketone-body synthesis. However, ketone bodies are not used by the liver but are released for other tissues to use. The liver does gain energy in the process of synthesizing and releasing ketone bodies. Calculate the number of molecules of ATP generated by the liver in the conversion of palmitate, a C-16 fatty acid, into acetoacetate. ✓ 2

18. *Counting ATPs 2.* How much energy is attained with the complete oxidation of the ketone body D-3-hydroxybutyrate? ✓ 2

19. *Another view.* Why might someone argue that the answer to problem 18 is wrong? (Consider the uses of succinyl CoA.) ✓ 2

20. *An accurate adage.* An old biochemistry adage is that *fats burn in the flame of carbohydrates.* What is the molecular basis of this adage? ✓ 2

21. *Missing acyl CoA dehydrogenases.* A number of genetic deficiencies in acyl CoA dehydrogenases have been described. A deficiency in acyl CoA dehydrogenase presents itself early in life or after a period of fasting. Symptoms include vomiting, lethargy, and, sometimes, coma. Not only are blood levels of glucose low (hypoglycemia), but also starvation-induced ketosis is absent. Provide a biochemical explanation for the last two observations.

22. *Missing ingredient.* Why are liver cells not capable of using ketone bodies as a fuel? ✓ 2

23. *Finding triacylglycerols in all the wrong places.* Insulin-dependent diabetes is often accompanied by high levels of triacylglycerols in the blood. Suggest a biochemical explanation for the high blood levels of triacylglycerols. ✓ 2

Chapter Integration Problems

24. *Leaner times might follow.* Why can't animals convert fats into glucose? Why are plants capable of such a conversion?

25. *Losing protein.* What is the purpose of protein degradation during the initial stages of starvation?

26. *Stop losing protein.* How is the loss of muscle protein delayed during starvation?

27. *After lipolysis.* During fatty acid mobilization, glycerol is produced. This glycerol is not wasted. Write a balanced equation for the conversion of glycerol into pyruvate. What enzymes are required in addition to those of the glycolytic pathway?

28. *Remembrance of reactions past.* We encountered reactions similar to the oxidation, hydration, and oxidation reactions of fatty acid degradation earlier in our study of biochemistry. What other pathway employs this set of reactions?

29. *Ill-advised diet.* Suppose that, for some bizarre reason, you decided to exist on a diet of whale and seal blubber, exclusively.

(a) How would a lack of carbohydrates affect your ability to utilize fats?

(b) What would your breath smell like?

(c) One of your best friends, after trying unsuccessfully to convince you to abandon this diet, makes you promise to consume a healthy dose of odd-chain fatty acids. Does your friend have your best interests at heart? Explain.

Data Interpretation Problem

30. *Mutant enzyme.* Carnitine palmitoyl transferase I (CPTI) catalyzes the conversion of long-chain acyl CoA into acyl carnitine, a prerequisite for transport into mitochondria and subsequent degradation. A mutant enzyme was constructed with a single amino acid change at position 3 of glutamic acid for alanine. Graphs A through C are based on data from studies performed to identify the effect of the mutation (data from J. Shi, H. Zhu, D. N. Arvidson, and G. J. Wodegiorgis, *J. Biol. Chem.* 274:9421–9426, 1999). ✓ 1

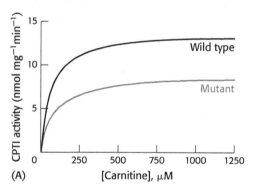

(A)

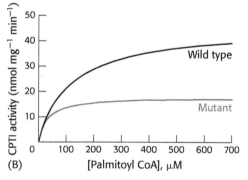

(B)

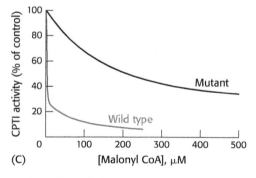

(C)

(a) What is the effect of the mutation on enzyme activity when the concentration of carnitine is varied? What are the K_M and V_{max} values for the wild-type (normal) and mutant enzymes?

(b) What is the effect when the experiment is repeated with varying concentrations of palmitoyl CoA? What are the K_M and V_{max} values for the wild-type and mutant enzymes?

(c) Graph C shows the inhibitory effect of malonyl CoA on the wild-type and mutant enzymes. Which enzyme is more sensitive to malonyl CoA inhibition? Note: Malonyl CoA is a substrate for fatty acid synthesis (p. 509).

(d) Suppose that the concentration of palmitoyl CoA is 100 μM, that of carnitine is 100 μM, and that of malonyl CoA is 10 μM. Under these conditions, what is the most prominent effect of the mutation on the properties of the enzyme?

(e) What can you conclude about the role of glutamate 3 in carnitine acyltransferase I function?

Challenge Problems

31. *Refsum disease.* Phytanic acid is a branched-chain fatty acid component of chlorophyll and is a significant component of milk. In susceptible people, phytanic acid can accumulate, leading to neurological problems. This syndrome is called Refsum disease or phytanic acid storage disease. ✓ 1

$$H_3C-CH(CH_3)-(CH_2)_3-CH(CH_3)-(CH_2)_3-CH(CH_3)-(CH_2)_3-CH(CH_3)-CH_2-COO^-$$

Phytanic acid

(a) Why does phytanic acid accumulate?

(b) What enzyme activity would you invent to prevent its accumulation?

32. *Sleight of hand.* Animals cannot affect the net synthesis of glycogen from fatty acids. Yet, if animals are fed radioactive lipids (^{14}C), over time, some radioactive glycogen appears. How is the appearance of radioactive glycogen possible in these animals?

33. *Necessary diversion.* When acetyl CoA produced by β-oxidation exceeds the capacity of the citric acid cycle, ketone bodies are produced. Although acetyl CoA is not toxic, mitochondria must divert acetyl CoA to ketone bodies to keep functioning. Explain why. What would happen if ketone bodies were not generated?

34. *A hot diet.* Tritium is a radioactive isotope of hydrogen and can be readily detected. Fully tritiated, six-carbon saturated fatty acid is administered to a rat, and a muscle biopsy of the rat is taken by concerned, sensitive, and discreet technical assistants. These assistants carefully isolate all of the acetyl CoA generated from the β oxidation of the radioactive fatty acid and remove the CoA to form acetate. What will be the overall tritium-to-carbon ratio of the isolated acetate? ✓ 1

Selected Readings for this chapter can be found online at www.whfreeman.com/tymoczko3e.

CHAPTER 28

Fatty Acid Synthesis

To prepare for their winter hibernation, bears feed into the fall, storing excess energy as fat in the form of triacylglycerols. These energy stores sustain a bear during hibernation. [Paul Souders/Getty Images.]

As we have seen, fatty acids play a variety of crucial roles in biological systems. For instance, they serve as fuel reserves, signal molecules, and components of membrane lipids. Because our diet meets our physiological needs for fats and lipids, adult human beings have little need for fatty acid synthesis. However, many tissues, such as liver and adipose tissue, are capable of synthesizing fatty acids, and this synthesis is required under certain physiological conditions. For instance, fatty acid synthesis is necessary during embryonic development and during lactation in mammary glands. Too much fatty acid synthesis in an alcoholic's liver contributes to liver failure (p. 517).

Acetyl CoA, the end product of fatty acid degradation, is the precursor for virtually all fatty acids. The biochemical challenge is to link the two-carbon units together and reduce the carbon atoms to produce palmitate, a C_{16} fatty acid. Palmitate then serves as a precursor for the variety of other fatty acids.

✓ 3 Explain how fatty acids are synthesized.

28.1 Fatty Acid Synthesis Takes Place in Three Stages

As is the case for fatty acid degradation, *fatty acid synthesis* occurs as three stages of processing:

1. In a preparatory step, acetyl CoA is transferred from mitochondria, where it is produced, to the cytoplasm, the site of fatty acid synthesis. Acetyl CoA is transported in the form of citrate, which is cleaved to yield acetyl CoA and oxaloacetate.

507

2. Fatty acid synthesis begins in the cytoplasm with the activation of acetyl CoA, in a two-step reaction, to malonyl CoA.

3. The reaction intermediates are attached to an acyl carrier protein, which serves as the molecular foundation for the fatty acid being constructed. Fatty acid is synthesized, two carbon atoms at a time, in a five-step elongation cycle.

Citrate Carries Acetyl Groups from Mitochondria to the Cytoplasm

The first biochemical hurdle in fatty acid synthesis is that synthesis takes place in the cytoplasm, but acetyl CoA, the raw material for fatty acid synthesis, is formed in mitochondria. The mitochondria are not readily permeable to acetyl CoA. How are acetyl CoA molecules transferred to the cytoplasm? The problem is solved by the transport of acetyl CoA out of the mitochondria in the form of citrate. Citrate is formed in the mitochondrial matrix by the condensation of acetyl CoA with oxaloacetate (**Figure 28.1**). This reaction begins the citric acid cycle when energy is required. When the energy needs of a cell have been met, citrate is relocated to the cytoplasm by a transport protein, where it is cleaved by *ATP-citrate lyase* at the cost of a molecule of ATP to yield cytoplasmic acetyl CoA and oxaloacetate:

$$\text{Citrate} + \text{ATP} + \text{CoA} + \text{H}_2\text{O} \xrightarrow{\text{ATP-citrate lyase}} \text{acetyl CoA} + \text{ADP} + \text{P}_i + \text{oxaloacetate}$$

This reaction occurs in three steps: (1) the formation of a phospho-enzyme with the donation of a phosphoryl group from ATP, (2) binding of citrate and CoA followed by the formation of citroyl CoA and release of the phosphate, and (3) cleavage of citroyl CoA to yield acetyl CoA and oxaloacetate. The transport and cleavage reactions must take place eight times to provide all of the carbon atoms required for palmitate synthesis. ATP-citrate lyase is stimulated by insulin, which initiates a signal transduction pathway that ultimately results in the phosphorylation and activation of the lyase by Akt (also called protein kinase B).

In addition to being a precursor for fatty acid synthesis, citrate serves as a signal molecule. It inhibits phosphofructokinase, which controls the rate of glycolysis (p. 300). Citrate in the cytoplasm indicates an energy-rich state, signaling that there is no need to oxidize glucose.

Several Sources Supply NADPH for Fatty Acid Synthesis

The synthesis of palmitate requires 14 molecules of NADPH as well as the expenditure of ATP. Some of the reducing power is generated when oxaloacetate, formed in the transfer of acetyl groups to the cytoplasm, is returned to the

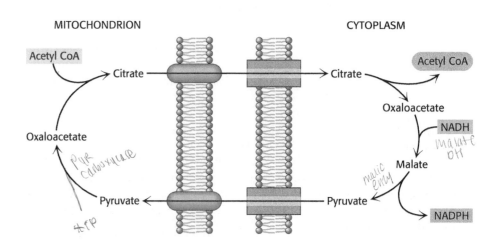

Figure 28.1 The transfer of acetyl CoA to the cytoplasm. Acetyl CoA is transferred from mitochondria to the cytoplasm, and the reducing potential of NADH is concomitantly converted into that of NADPH by this series of reactions.

mitochondria. The inner mitochondrial membrane is impermeable to oxaloace-tate. Hence, a series of bypass reactions are needed. First, oxaloacetate is reduced to malate by NADH. This reaction is catalyzed by a cytoplasmic *malate dehydrogenase*:

$$\text{Oxaloacetate} + \text{NADH} + \text{H}^+ \rightleftharpoons \text{malate} + \text{NAD}^+$$

Second, malate is oxidatively decarboxylated by an *NADP$^+$-linked malate enzyme* (also called *malic enzyme*).

$$\text{Malate} + \text{NADP}^+ \longrightarrow \text{pyruvate} + \text{CO}_2 + \text{NADPH}$$

The pyruvate formed in this reaction readily enters mitochondria, where it is carboxylated to oxaloacetate by *pyruvate carboxylase*:

$$\text{Pyruvate} + \text{CO}_2 + \text{ATP} + \text{H}_2\text{O} \longrightarrow \text{oxaloacetate} + \text{ADP} + \text{P}_i + 2\,\text{H}^+$$

The sum of these three reactions is

$$\text{NADP}^+ + \text{NADH} + \text{ATP} + \text{H}_2\text{O} \longrightarrow$$
$$\text{NADPH} + \text{NAD}^+ + \text{ADP} + \text{P}_i + \text{H}^+$$

Thus, *one molecule of NADPH is generated for each molecule of acetyl CoA that is transferred from mitochondria to the cytoplasm.* Hence, eight molecules of NADPH are formed when eight molecules of acetyl CoA are transferred to the cytoplasm for the synthesis of palmitate. *The additional six molecules of NADPH required for the synthesis of palmitate come from the pentose phosphate pathway* (Chapter 26).

The accumulation of the precursors for fatty acid synthesis is a wonderful example of the coordinated use of multiple pathways. The citric acid cycle, the transport of citrate from mitochondria, and the pentose phosphate pathway provide the carbon atoms and reducing power, whereas glycolysis and oxidative phosphorylation provide the ATP to meet the needs for fatty acid synthesis (**Figure 28.2**).

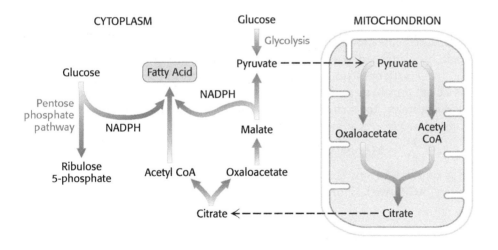

Figure 28.2 PATHWAY INTEGRATION: Fatty acid synthesis. Fatty acid synthesis requires the cooperation of various metabolic pathways located in different cellular compartments.

The Formation of Malonyl CoA Is the Committed Step in Fatty Acid Synthesis

Like the synthesis of all biopolymers, fatty acid synthesis requires an activation step. Fatty acid synthesis starts with the carboxylation of acetyl CoA to *malonyl CoA*, the activated form of acetyl CoA. This reaction is catalyzed by *acetyl CoA carboxylase 1*, a regulatory enzyme in fatty acid metabolism. As we will see shortly, malonyl CoA is the actual carbon donor for all but two of the carbon atoms of palmitic acid.

Biotin Bacteria residing in the large intestine produce biotin for human use. Biotin is also found in a wide variety of foods, such as liver, eggs, cereals, and nuts. Biotin deficiency is rare, but symptoms include lethargy, muscle pain, nausea, and dermatitis. [Photograph from Kheng Gkuan Toh/ FeaturePics.]

Acetyl CoA is combined with HCO_3^-, the form of CO_2 in aqueous solutions. This irreversible reaction is the committed step in fatty acid synthesis:

$$\text{Acetyl CoA} + \text{ATP} + HCO_3^- \longrightarrow \text{Malonyl CoA} + \text{ADP} + P_i + H^+$$

The synthesis of malonyl CoA is a two-step process:

1. Acetyl CoA carboxylase contains a biotin prosthetic group. In the first step, a carboxybiotin intermediate is formed at the expense of the hydrolysis of a molecule of ATP:

$$\text{Biotin-enzyme} + \text{ATP} + HCO_3^- \rightleftharpoons CO_2\text{-biotin} + \text{ADP} + P_i + H^+$$

2. The activated CO_2 group is then transferred to acetyl CoA to form malonyl CoA:

$$CO_2\text{-biotin enzyme} + \text{acetyl CoA} \longrightarrow \text{malonyl CoA} + \text{biotin-enzyme}$$

Acetyl CoA carboxylase 2, an isozyme of carboxylase 1 located in the mitochondria, is the essential regulatory enzyme for fatty acid degradation (p. 516). Mechanistically, it functions like carboxylase 1.

Fatty Acid Synthesis Consists of a Series of Condensation, Reduction, Dehydration, and Reduction Reactions

The enzyme system that catalyzes the synthesis of saturated long-chain fatty acids from acetyl CoA, malonyl CoA, and NADPH is called *fatty acid synthase*. The synthase is actually a complex of distinct enzymes, each of which has a different function in fatty acid synthesis. In bacteria, the enzyme complex catalyzes all but the activation step of fatty acid synthesis (Table 28.1). The elongation phase of fatty acid synthesis in bacteria starts when acetyl CoA and malonyl CoA react with a scaffold protein called *acyl carrier protein* (ACP), forming acetyl ACP and malonyl ACP. Just as most construction projects require foundations on which the structures are built, fatty acid synthesis needs a molecular foundation. The intermediates in fatty acid synthesis are linked to the sulfhydryl end of a phosphopantetheine group of ACP—the same "business end" as that of CoA—which is, in turn, attached to a serine residue of the acyl carrier protein (Figure 28.3).

Acyl carrier protein, a single polypeptide chain of 77 residues, can be regarded as a giant prosthetic group, a "macro CoA." *Acetyl transacylase and malonyl transacylase* catalyze the formation of acetyl ACP and malonyl ACP, respectively:

$$\text{Acetyl CoA} + \text{ACP} \rightleftharpoons \text{acetyl ACP} + \text{CoA}$$
$$\text{Malonyl CoA} + \text{ACP} \rightleftharpoons \text{malonyl ACP} + \text{CoA}$$

The synthesis of fatty acids with an odd number of carbon atoms starts with propionyl ACP, which is formed from propionyl CoA by acetyl transacylase.

Table 28.1 Principal reactions in fatty acid synthesis in bacteria

Step	Reaction	Enzyme
1	$\text{Acetyl CoA} + HCO_3^- + \text{ATP} \longrightarrow \text{malonyl CoA} + \text{ADP} + P_i + H^+$	Acetyl CoA carboxylase
2	$\text{Acetyl CoA} + \text{ACP} \rightleftharpoons \text{acetyl ACP} + \text{CoA}$	Acetyl transacylase
3	$\text{Malonyl CoA} + \text{ACP} \rightleftharpoons \text{malonyl ACP} + \text{CoA}$	Malonyl transacylase
4	$\text{Acetyl ACP} + \text{malonyl ACP} \longrightarrow \text{acetoacetyl ACP} + \text{ACP} + CO_2$	β-ketoacyl synthase
5	$\text{Acetoacetyl ACP} + \text{NADPH} + H^+ \rightleftharpoons \text{D-3-hydroxybutyryl ACP} + NADP^+$	β-ketoacyl reductase
6	$\text{D-3-Hydroxybutyryl ACP} \rightleftharpoons \text{crotonyl ACP} + H_2O$	3-Hydroxyacyl dehydratase
7	$\text{Crotonyl ACP} + \text{NADPH} + H^+ \longrightarrow \text{butyryl ACP} + NADP^+$	Enoyl reductase

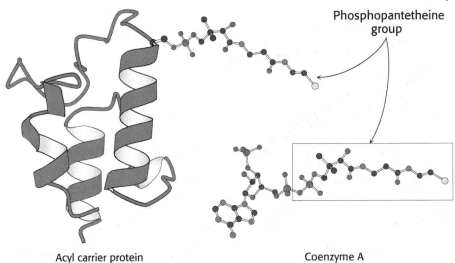

Phosphopantetheine group

Acyl carrier protein

Coenzyme A

Figure 28.3 Phosphopantetheine. Both acyl carrier protein and coenzyme A include phosphopantetheine as their reactive units.

Acetyl ACP and malonyl ACP react to form acetoacetyl ACP (**Figure 28.4**). *β-Ketoacyl synthase*, also called the condensing enzyme, catalyzes this condensation reaction.

$$\text{Acetyl ACP + malonyl ACP} \longrightarrow \text{acetoacetyl ACP + ACP + } CO_2$$

In the condensation reaction, a four-carbon unit is formed from a two-carbon unit and a three-carbon unit, and CO_2 is released. Why is the four-carbon unit not formed from two two-carbon units—say, two molecules of acetyl ACP? The equilibrium for the synthesis of acetoacetyl ACP from two molecules of acetyl ACP is highly unfavorable. In contrast, *the equilibrium is favorable if malonyl ACP, the activated form of acetyl CoA, is a reactant, because its decarboxylation contributes to a substantial decrease in free energy.* In effect, ATP drives the condensation reaction, though ATP does not directly participate in that reaction. Instead, ATP is used to carboxylate acetyl CoA to malonyl CoA, the activated form of acetyl CoA. The free energy thus stored in malonyl CoA is released in the decarboxylation accompanying the formation of acetoacetyl ACP. Although HCO_3^- is required for fatty acid synthesis, its carbon atom does not appear in the product. Rather, *all the carbon atoms of fatty acids containing an even number of carbon atoms are derived from acetyl CoA.*

The next three steps in fatty acid synthesis reduce the keto group (shown in the margin) at C-3 to a methylene group ($-CH_2-$; Figure 28.4). First, acetoacetyl ACP is reduced to D-3-hydroxybutyryl ACP. This reaction differs from the corresponding oxidation in fatty acid degradation in two respects: (1) the D rather than the L isomer is formed and (2) NADPH is the reducing agent, whereas NAD^+ is the oxidizing agent in β oxidation. This difference exemplifies the general principle that *NADPH is consumed in biosynthetic reactions, whereas NADH is generated in energy-yielding reactions.* D-3-Hydroxybutyryl ACP is then *dehydrated* to form crotonyl ACP, which is a *trans-Δ^2-enoyl ACP*. The final step in the cycle *reduces* crotonyl ACP to butyryl ACP. NADPH is again the reductant, whereas FAD is the oxidant in the corresponding reaction in β oxidation. The enzyme that catalyzes this step, *enoyl ACP reductase*, is inhibited by *triclosan*, a broad-spectrum antibacterial agent that is added to a variety of products such as toothpaste, soaps, and skin creams. These last three reactions—a reduction, a dehydration, and a second reduction—convert acetoacetyl ACP into butyryl ACP, which completes the first elongation cycle.

In the second round of fatty acid synthesis, butyryl ACP condenses with another malonyl ACP to form a C_6-β-ketoacyl ACP. This reaction is like the one in the first round, in which acetyl ACP condenses with malonyl ACP to form a

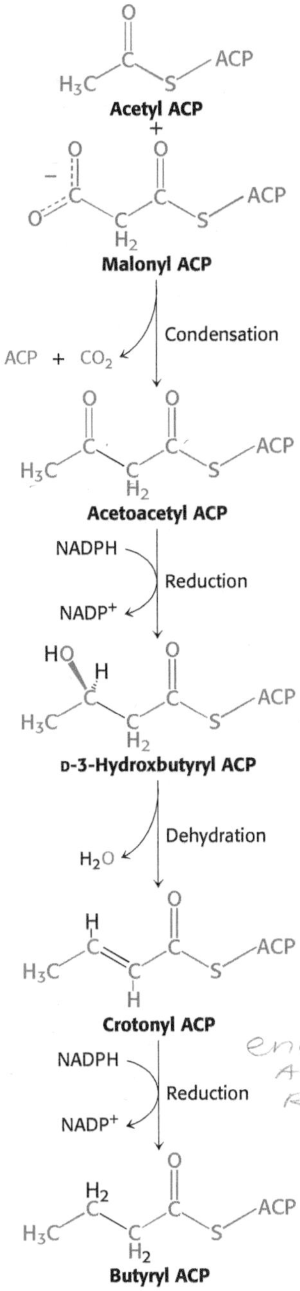

Acetyl ACP
+
Malonyl ACP

Condensation

ACP + CO_2

Acetoacetyl ACP

NADPH
Reduction
$NADP^+$

D-3-Hydroxbutyryl ACP

H_2O
Dehydration

Crotonyl ACP

enoyl ACP R.ase

NADPH
Reduction
$NADP^+$

Butyryl ACP

Figure 28.4 Fatty acid synthesis. Fatty acids are synthesized by the repetition of the following reaction sequence: condensation, reduction, dehydration, and reduction. The intermediates shown here are produced in the first round of synthesis.

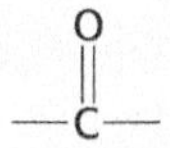

A keto group

C_4-β-ketoacyl ACP. Reduction, dehydration, and a second reduction convert the C_6-β-ketoacyl ACP into a C_6-acyl ACP, which is ready for a third round of elongation. The elongation cycles continue until C_{16}-acyl ACP is formed. This intermediate is a good substrate for a thioesterase that hydrolyzes C_{16}-acyl ACP to release the fatty acid chain from the carrier protein. Because it acts selectively on C_{16}-acyl ACP, *the thioesterase acts as a ruler to determine fatty acid chain length*. The synthesis of longer-chain fatty acids is discussed in Section 28.2.

The Synthesis of Palmitate Requires 8 Molecules of Acetyl CoA, 14 Molecules of NADPH, and 7 Molecules of ATP

Now that we have seen all of the individual reactions of fatty acid synthesis, let us determine the overall reaction for the synthesis of the C_{16} fatty acid palmitate. The stoichiometry of the synthesis of palmitate is

$$\text{Acetyl CoA} + 7\,\text{malonyl CoA} + 14\,\text{NADPH} + 7\,\text{H}^+ \longrightarrow$$
$$\text{palmitate} + 7\,\text{CO}_2 + 14\,\text{NADP}^+ + 8\,\text{CoA} + 6\,\text{H}_2\text{O}$$

The equation for the synthesis of the malonyl CoA used in the preceding reaction is

$$7\,\text{acetyl CoA} + 7\,\text{CO}_2 + 7\,\text{ATP} \longrightarrow$$
$$7\,\text{malonyl CoA} + 7\,\text{ADP} + 7\,\text{P}_i + 7\,\text{H}^+$$

Hence, the overall stoichiometry for the synthesis of palmitate is

$$8\,\text{acetyl CoA} + 7\,\text{ATP} + 14\,\text{NADPH} \longrightarrow$$
$$\text{palmitate} + 14\,\text{NADP}^+ + 8\,\text{CoA} + 6\,\text{H}_2\text{O} + 7\,\text{ADP} + 7\,\text{P}_i$$

Fatty Acids Are Synthesized by a Multifunctional Enzyme Complex in Animals

Although the basic biochemical reactions in fatty acid synthesis are similar in *E. coli* and eukaryotes, the structure of the synthase varies considerably. The component enzymes of animal fatty acid synthases, in contrast with those of *E. coli* and plants, are linked in a large polypeptide chain.

The structure of a large part of the mammalian fatty acid synthase has recently been determined, with the acyl carrier protein and thioesterase remaining to be resolved. The enzyme is a dimer of identical 270-kDa subunits. Each chain contains all of the active sites required for activity, as well as an acyl carrier protein tethered to the complex (**Figure 28.5**A). Despite the fact that each chain

(A)

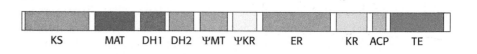

(B)

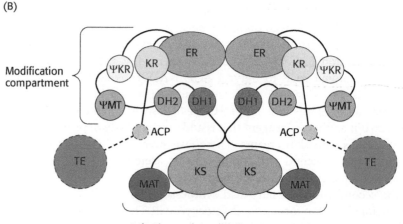

Selecting and condensing compartment

Figure 28.5 A schematic representation of a single chain of animal fatty acid synthase. (A) The arrangement of the catalytic activities present in a single polypeptide chain. (B) A cartoon of the dimer based on an x-ray crystallographic result. The ψMT and ψKR are inactive domains similar to methyl transferase and ketoreductase sequences. Although there are two domains for DH, only one is active. The inactive domains are presented in faded colors. Dotted lines outline domains for which the structure has not yet been determined. Abbreviations: KS, β-ketoacyl synthase; MAT, malonyl–acetyl transferase; DH, dehydratase; ψMT, methyl transferase (inactive); ψKR, ketoreductase (inactive); ER, enoyl reductase; KR, ketoreductase; ACP, acyl carrier protein; TE, thioesterase.

possesses all of the enzymes required for fatty acid synthesis, the monomers are not active. A dimer is required.

The two component chains interact such that the enzyme activities are partitioned into two distinct compartments (Figure 28.5B). The selecting and condensing compartment binds the acetyl and malonyl substrates and condenses them to form the growing chain. Interestingly, the mammalian fatty acid synthase has one active site, malonyl-acetyl transacylase, that adds both acetyl CoA and malonyl CoA. In contrast, most other fatty acid synthases have two separate enzyme activities, one for acetyl CoA and one for malonyl CoA. The modification compartment is responsible for the reduction and dehydration activities that result in the saturated fatty acid product.

Many eukaryotic multienzyme complexes are multifunctional proteins in which different enzymes are linked covalently. An advantage of this arrangement is that the synthetic activity of different enzymes is coordinated. In addition, intermediates can be efficiently handed from one active site to another without leaving the assembly.

? **QUICK QUIZ 1** What are the raw materials for fatty acid synthesis, and how are they obtained?

CLINICAL INSIGHT

Fatty Acid Metabolism Is Altered in Tumor Cells

We have previously seen that cancer cells alter glucose metabolism to meet the needs of rapid cell growth (p. 304 and 481). Cancer cells must also increase fatty acid synthesis for use as signal molecules as well as for incorporation into membrane phospholipids. Many of the enzymes of fatty acid synthesis are overexpressed in most human cancers, and this expression is correlated with tumor malignancy. Recall that normal cells do little de novo fatty acid synthesis, relying instead on dietary intake to meet their fatty needs.

The dependence of de novo fatty acid synthesis provides possible therapeutic targets to inhibit cancer cell growth. Inhibition of β-ketoacyl ACP synthase, the enzyme that catalyzes the condensation step of fatty acid synthesis, does indeed inhibit phospholipid synthesis and subsequent cell growth in some cancers, apparently by inducing programmed cell death (apoptosis). However, another startling observation was made: *mice treated with inhibitors of the condensing enzyme showed remarkable weight loss* because they ate less. Thus, fatty acid synthase inhibitors are exciting candidates both as antitumor and as antiobesity drugs.

Acetyl CoA carboxylase is also being investigated as a possible target for inhibiting cancer cell growth. Inhibition of the carboxylase in prostate and breast cancer cell lines induces apoptosis in the cancer cells, and yet is without effect in normal cells (problem 26). Understanding the alteration of fatty acid metabolism in cancer cells is a developing area of research that holds promise of generating new cancer therapies.

CLINICAL INSIGHT

A Small Fatty Acid That Causes Big Problems

γ-Hydroxybutyric acid (GHB) is a short-chain fatty acid that is an isomer of β-hydroxybutyric acid. The acylated version of β-hydroxybutyric acid is a metabolite in fatty acid degradation and synthesis. The ionized form of this molecule is a ketone body.

β-Hydroxybutyric acid   **γ-Hydroxybutyric acid**

This small chemical difference of where the alcohol group (—OH) is located has a great effect on the actions of these two chemicals. Small amounts of GHB are found in the brain, where it is thought to be a neurotransmitter. GHB has been used clinically as an anesthetic and for the treatment of narcolepsy and alcoholism, but it entered recreational drug use when body builders found that it stimulated the release of growth hormone. It became a popular drug because of claims that it reduces inhibition and heightens sexual awareness, and it is notorious as a date-rape drug. GHB, which is also called G, liquid ecstasy, gib, and liquid X, among other names, was banned for nonprescription uses in 1990.

28.2 Additional Enzymes Elongate and Desaturate Fatty Acids

The major product of fatty acid synthase is palmitate, a 16-carbon fatty acid. However, all cells require longer-chain fatty acids for a variety of purposes, including the synthesis of signaling molecules. In eukaryotes, enzymes on the cytoplasmic face of the endoplasmic reticulum membrane catalyze the elongation reactions in which longer fatty acids are formed. These reactions use malonyl CoA to add two-carbon units sequentially to the carboxyl ends of both saturated and unsaturated acyl CoA substrates.

Membrane-Bound Enzymes Generate Unsaturated Fatty Acids

Endoplasmic reticulum systems also introduce double bonds into long-chain acyl CoAs, an important step in the synthesis of vital signal molecules such as *prostaglandins*. For example, in the conversion of stearoyl CoA into oleoyl CoA, a *cis*-Δ^9 double bond is inserted by an oxidase that employs *molecular oxygen* and *NADH* (or *NADPH*):

$$\text{Stearoyl CoA} + \text{NADH} + \text{H}^+ + \text{O}_2 \longrightarrow \text{oleoyl CoA} + \text{NAD}^+ + 2\,\text{H}_2\text{O}$$

Unsaturated fatty acids in mammals are derived from either palmitoleate (16:1, 16 carbon atoms, 1 double bond), oleate (18:1), linoleate (18:2), or linolenate (18:3). *Mammals lack the enzymes to introduce double bonds at carbon atoms beyond C-9 in the fatty acid chain.* Hence, mammals cannot synthesize linoleate (18:2 *cis*-Δ^9, Δ^{12}) and linolenate (18:3 *cis*-Δ^9, Δ^{12}, Δ^{15}). *Linoleate and linolenate are the two essential fatty acids*, meaning that they must be supplied in the diet because they are required by an organism and cannot be endogenously synthesized. Linoleate (ω-6) and linolenate (ω-3) are the omega (ω) fatty acids that we read so much about. Linoleate and linolenate furnished by the diet are the starting points for the synthesis of a variety of other unsaturated fatty acids, including certain hormones. Safflower and corn oil are particularly rich sources of linoleate, whereas canola and soybean oil provide linolenate.

Eicosanoid Hormones Are Derived from Polyunsaturated Fatty Acids

Arachidonate, a 20:4 fatty acid derived from linoleate, is the major precursor of several classes of signal molecules: prostaglandins, prostacyclins, thromboxanes, and leukotrienes (**Figure 28.6**). Prostaglandins and related signal molecules are called *eicosanoids* (from the Greek *eikosi*, meaning "twenty") because they contain 20 carbon atoms.

Prostaglandins and other eicosanoids are *local hormones*. They are short-lived and alter the activities only of the cells in which they are synthesized and of cells in the immediate vicinity by binding to membrane receptors. These effects

$$\text{CH}_3(\text{CH}_2)_{16}\text{COO}^-$$

Stearate

$$\text{CH}_3(\text{CH}_2)_7\text{CH} = \text{CH}(\text{CH}_2)_7\text{COO}^-$$

Oleate

Precursor	Formula
Linolenate (ω-3)	$\text{CH}_3 - (\text{CH}_2)_2 = \text{CH} - \text{R}$
Linoleate (ω-6)	$\text{CH}_3 - (\text{CH}_2)_5 = \text{CH} - \text{R}$
Palmitoleate (ω-7)	$\text{CH}_3 - (\text{CH}_2)_6 = \text{CH} - \text{R}$
Oleate (ω-9)	$\text{CH}_3 - (\text{CH}_2)_8 = \text{CH} - \text{R}$

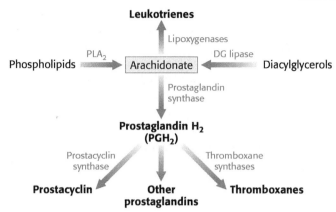

Figure 28.6 Arachidonate is the major precursor of eicosanoid hormones. Prostaglandin synthase catalyzes the first step in a pathway leading to prostaglandins, prostacyclins, and thromboxanes. Lipoxygenases catalyze the initial step in a pathway leading to leukotrienes. Abbreviations: PLA₂, phospholipase A₂; DG, diacylglycerol.

may vary from one type of cell to another. Among other effects, prostaglandins stimulate inflammation, regulate blood flow to particular organs, control ion transport across membranes, modulate synaptic transmission, and induce sleep (Figure 28.7).

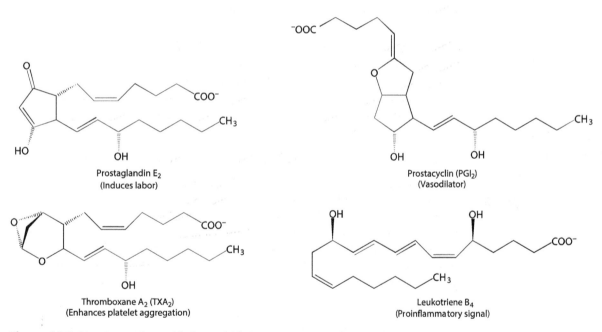

Figure 28.7 Structures of several eicosanoids.

CLINICAL INSIGHT

Aspirin Exerts Its Effects by Covalently Modifying a Key Enzyme

Aspirin (acetylsalicylate) blocks access to the active site of the enzyme that converts arachidonate into prostaglandin H₂ (p. 211). Because arachidonate is the precursor of other prostaglandins, prostacyclins, and thromboxanes, blocking this step affects many signaling pathways. It accounts for the wide-ranging effects that aspirin and related compounds have on inflammation, fever, pain, and blood clotting.

✓ 4 Explain how fatty acid metabolism is regulated.

28.3 Acetyl CoA Carboxylase Is a Key Regulator of Fatty Acid Metabolism

Fatty acid metabolism is stringently controlled so that synthesis and degradation are highly responsive to physiological needs. Fatty acid synthesis is maximal when carbohydrates and energy are plentiful and when fatty acids are scarce. *Acetyl CoA carboxylase 1 and 2 play essential roles in regulating fatty acid synthesis and degradation.* Recall that acetyl CoA carboxylase 1 catalyzes the committed step in fatty acid synthesis: the production of malonyl CoA (the activated two-carbon donor). This enzyme is subject to both local and hormonal regulation. We will examine each of these levels of regulation in turn.

Acetyl CoA Carboxylase Is Regulated by Conditions in the Cell

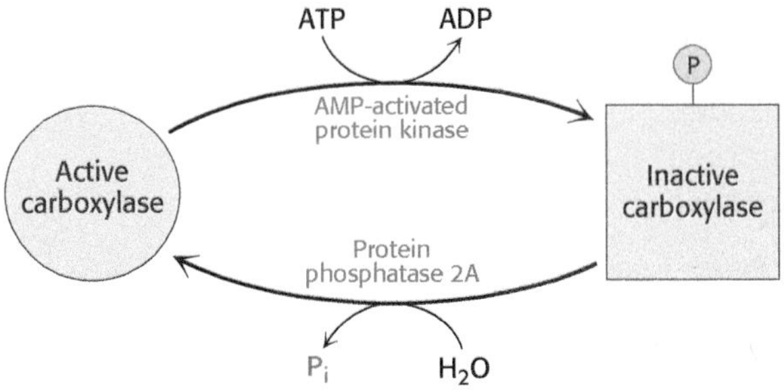

Acetyl CoA carboxylase 1 responds to changes in its immediate environment, and *is switched off by phosphorylation and activated by dephosphorylation* (**Figure 28.8**). *AMP-activated protein kinase* (AMPK) converts the carboxylase into an inactive form by modifying a single serine residue. AMPK is essentially a fuel gauge; it is activated by AMP and inhibited by ATP. Thus, the carboxylase is inactivated when the energy charge is low. Fats are not synthesized when energy is required.

The carboxylase is also allosterically stimulated by citrate. The level of citrate is high when both acetyl CoA and ATP are abundant, signifying that raw materials and energy are available for fatty acid synthesis. Citrate acts in an unusual manner on inactive acetyl CoA carboxylase, which exists as inactive isolated dimers. Citrate facilitates the polymerization of the dimers into active filaments (**Figure 28.9**). However, polymerization by citrate alone requires much higher concentrations than are normally present in the cell. Under physiological conditions, citrate-induced polymerization is facilitated by the protein MIG12, which greatly reduces the amount of citrate required. Citrate-induced polymerization can partly reverse the inhibition produced by phosphorylation (**Figure 28.10**). The level of citrate is high when both acetyl CoA and ATP are abundant, signifying that raw materials and energy are available for fatty acid synthesis. The stimulatory effect of citrate on the carboxylase is counteracted by *palmitoyl CoA*, which is abundant when there is an excess of fatty acids. Palmitoyl CoA causes the filaments to disassemble into the inactive subunits. Palmitoyl CoA also inhibits the translocase that transports citrate from mitochondria to the cytoplasm, as well as glucose 6-phosphate dehydrogenase, the regulatory enzyme in the oxidative phase of the pentose phosphate pathway (p. 478).

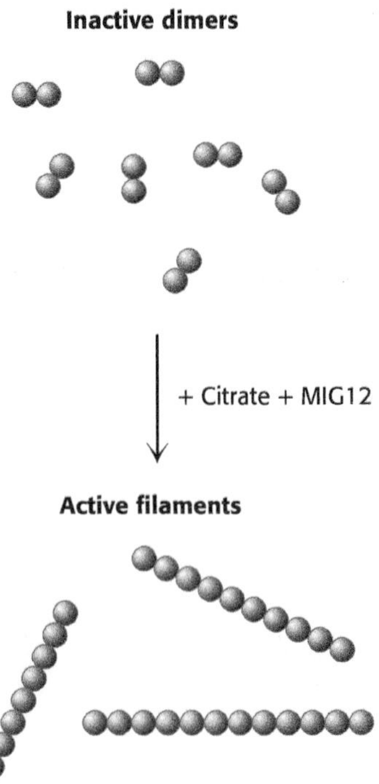

The isozyme, acetyl CoA carboxylase 2, located in the mitochondria, plays a role in the regulation of fatty acid degradation. Malonyl CoA, the product of the carboxylase reaction, is present at a high level when fuel molecules are abundant. *Mitochondrial malonyl CoA inhibits carnitine acyltransferase I, preventing the entry of fatty acyl CoAs into the mitochondrial matrix in times of plenty.* Malonyl CoA is an especially effective inhibitor of carnitine acyltransferase I in heart and muscle, tissues that have little fatty-acid-synthesis capacity of their own. In these tissues, acetyl CoA carboxylase may be a purely regulatory enzyme.

Figure 28.8 The control of acetyl CoA carboxylase. Acetyl CoA carboxylase is inhibited by phosphorylation.

Figure 28.9 Filaments of acetyl CoA carboxylase. The inactive form of acetyl CoA carboxylase is a dimer of 265-kDa subunits. Citrate and the protein MIG12 facilitate the formation of the enzymatically active filamentous form of the enzyme.

Acetyl CoA Carboxylase Is Regulated by a Variety of Hormones

Carboxylase is also controlled by the hormones glucagon, epinephrine, and insulin, which indicate the overall energy status of the organism. *Insulin stimulates fatty acid synthesis by activating the carboxylase 1, whereas glucagon and epinephrine have the reverse effect.*

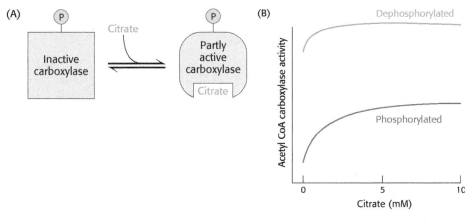

Figure 28.10 Dependence of the catalytic activity of acetyl CoA carboxylase on the concentration of citrate. (A) Citrate can partly activate the phosphorylated carboxylase. (B) The dephosphorylated form of the carboxylase is highly active even when citrate is absent. Citrate partly overcomes the inhibition produced by phosphorylation. [Data from G. M. Mabrouk, I. M. Helmy, K. G. Thampy, and S. J. Wakil. *J. Biol. Chem.* 265:6330–6338, 1990.]

Regulation by glucagon and epinephrine Consider, as we did in Chapter 27, a person who has just awakened from a night's sleep and begins to exercise. Glycogen stores will be low, but lipids are readily available for mobilization. The hormones glucagon and epinephrine, present under conditions of fasting and exercise, will stimulate the mobilization of fatty acids from triacylglycerols in fat cells, which will be released into the blood, and probably in muscle cells, where the fatty acids will be used immediately as fuel. These same hormones will inhibit fatty acid synthesis by inhibiting acetyl CoA carboxylase 1. Although the exact mechanism by which these hormones exert their effects is complex, the net result is to augment the inhibition by the AMP-activated kinase. This result makes sound physiological sense: when the energy level of a cell is low, as signified by a high concentration of AMP, and the energy level of the organism is low, as signaled by glucagon, fats should not be synthesized. Epinephrine, which signals the need for immediate energy, enhances this effect. Hence, *these catabolic hormones switch off fatty acid synthesis by keeping the carboxylase in the inactive phosphorylated state.*

Regulation by insulin Now consider the situation after the exercise has ended and the exerciser has had a meal. In this case, the hormone insulin inhibits the mobilization of fatty acids and stimulates their accumulation as triacylglycerols by muscle and adipose tissue. Insulin also stimulates fatty acid synthesis by activating acetyl CoA carboxylase 1. Insulin activates the carboxylase by enhancing the phosphorylation of AMPK by Akt, which inhibits AMPK, as well as by stimulating the activity of a protein phosphatase that dephosphorylates and activates acetyl CoA carboxylase. Thus, the signal molecules glucagon, epinephrine, and insulin act in concert on triacylglycerol metabolism and acetyl CoA carboxylase 1 to carefully regulate the utilization and storage of fatty acids.

Response to diet *Long-term control is mediated by changes in the rates of synthesis and degradation of the enzymes participating in fatty acid synthesis.* Animals that have fasted and are then fed high-carbohydrate, low-fat diets show marked increases in their amounts of acetyl CoA carboxylase 1 and fatty acid synthase within a few days. This type of regulation is known as *adaptive control.* This regulation, which is mediated both by insulin and by glucose, is at the level of gene transcription.

QUICK QUIZ 2 How are fatty acid synthesis and degradation coordinated?

28.4 Metabolism in Context: Ethanol Alters Energy Metabolism in the Liver

Ethanol has been a part of the human diet for centuries, partly because of its intoxicating effects and partly because alcoholic beverages provided a safe means of hydration when pure water was scarce (**Figure 28.11**). Indeed, throughout the

Figure 28.11 Alcoholic beverages. The cultural importance of wine is illustrated by this detail of a fourth-century vault mosaic from the Santa Costanza Mausoleum in Italy. The mausoleum was built by the Roman emperor Constantine as a burial site for his daughter Costanza. Putti, cupid-like creatures, are gathering grapes for wine-making while one of Constantine's daughters, possibly Costanza, looks on from above. [The Art Archive/Corbis.]

world, only water and tea are consumed more frequently than beer. However, ethanol consumption in excess can result in a number of health problems, most notably liver damage. What is the biochemical basis of these health problems?

Ethanol cannot be excreted and must be metabolized, primarily by the liver. There are several pathways for the metabolism of ethanol. One pathway consists of two steps (p. 114). The first step takes place in the cytoplasm:

$$CH_3CH_2OH + NAD^+ \xrightarrow{\text{Alcohol dehydrogenase}} CH_3CHO + NADH + H^+$$
$$\text{Ethanol} \qquad\qquad\qquad\qquad\qquad \text{Acetaldehyde}$$

The second step takes place in mitochondria:

$$CH_3CHO + NAD^+ + H_2O \xrightarrow{\text{Aldehyde dehydrogenase}} CH_3COO^- + NADH + H^+$$
$$\text{Acetaldehyde} \qquad\qquad\qquad\qquad\qquad \text{Acetate}$$

Note that ethanol consumption leads to an accumulation of NADH. This high concentration of NADH inhibits gluconeogenesis by preventing the oxidation of lactate to pyruvate. In fact, the high concentration of NADH will cause the reverse reaction to predominate: lactate will accumulate. The consequences may be hypoglycemia (low concentration of blood glucose) and lactic acidosis.

The NADH glut also inhibits fatty acid oxidation. The metabolic purpose of fatty acid oxidation is to generate NADH for ATP generation by oxidative phosphorylation (Chapter 27). However, an alcohol consumer's NADH needs are met by ethanol metabolism. In fact, the excess NADH signals that conditions are right for fatty acid synthesis. Hence, triacylglycerols accumulate in the liver, leading to a condition known as "fatty liver."

What are the effects of the other metabolites of ethanol? Liver mitochondria can convert acetate into acetyl CoA in a reaction requiring ATP. The enzyme is the one that normally activates fatty acids—acyl CoA synthetase.

$$\text{Acetate} + \text{CoA} + \text{ATP} \longrightarrow \text{acetyl CoA} + \text{AMP} + \text{PP}_i$$
$$\text{PP}_i \longrightarrow 2\,\text{P}_i$$

However, further processing of the acetyl CoA by the citric acid cycle is blocked because NADH inhibits two important citric acid cycle regulatory enzymes—isocitrate dehydrogenase and α-ketoglutarate dehydrogenase. The accumulation of acetyl CoA has several consequences. First, ketone bodies will form and be released into the blood, exacerbating the acidic condition already resulting from the high lactate concentration. The processing of the acetate in the liver becomes inefficient, leading to a buildup of acetaldehyde. This very reactive compound forms covalent bonds with many important functional groups in proteins, impairing protein function. If ethanol is consistently consumed at high levels, the acetaldehyde can significantly damage the liver, eventually leading to cell death.

SUMMARY

28.1 Fatty Acid Synthesis Takes Place in Three Stages

Fatty acids are synthesized in the cytoplasm by a different pathway from that of β oxidation. A reaction cycle based on the formation and cleavage of citrate carries acetyl groups from mitochondria to the cytoplasm. NADPH needed for synthesis is generated in the transfer of reducing equivalents from mitochondria by the concerted action of malate dehydrogenase and $NADP^+$-linked malate enzyme, as well as by the pentose phosphate pathway.

Synthesis starts with the carboxylation of acetyl CoA to malonyl CoA, the committed step. This ATP-driven reaction is catalyzed by acetyl CoA carboxylase 1, a biotin enzyme. The intermediates in fatty acid synthesis are linked to an acyl carrier protein. Acetyl ACP is formed from acetyl CoA, and malonyl ACP is formed from malonyl CoA. Acetyl ACP and malonyl ACP condense to form acetoacetyl ACP, a reaction driven by the release of CO_2 from the activated malonyl unit. A reduction, a dehydration, and a second reduction follow. NADPH is the reductant in these steps. The butyryl ACP formed in this way is ready for a second round of elongation, starting with the addition of a two-carbon unit from malonyl ACP. Seven rounds of elongation yield palmitoyl ACP, which is hydrolyzed to palmitate. In higher organisms, the enzymes catalyzing fatty acid synthesis are covalently linked in a multifunctional enzyme complex.

28.2 Additional Enzymes Elongate and Desaturate Fatty Acids

Fatty acids are elongated and desaturated by enzyme systems in the endoplasmic reticulum membrane. Desaturation requires NADH (or NADPH) and O_2. Mammals lack the enzymes to introduce double bonds distal to C-9, and so they require linoleate and linolenate in their diets.

Arachidonate, an essential precursor of prostaglandins and other signal molecules, is derived from linoleate. This 20:4 polyunsaturated fatty acid is the precursor of several classes of signal molecules—prostaglandins, prostacyclins, thromboxanes, and leukotrienes—that act as messengers and local hormones because of their transience. They are called eicosanoids because they contain 20 carbon atoms. Aspirin, an antiinflammatory and antithrombotic drug, irreversibly blocks the synthesis of these eicosanoids.

28.3 Acetyl CoA Carboxylase Is a Key Regulator of Fatty Acid Metabolism

Fatty acid synthesis and degradation are reciprocally regulated so that both are not simultaneously active. Acetyl CoA carboxylase 1, the essential control site, is phosphorylated and inactivated by AMP-activated kinase. The phosphorylation is reversed by a protein phosphatase. Citrate, which signals an abundance of building blocks and energy, partly reverses the inhibition by phosphorylation. Carboxylase activity is stimulated by insulin and inhibited by glucagon and epinephrine. In times of plenty, fatty acyl CoAs do not enter the mitochondrial matrix, because malonyl CoA synthesized by mitochondrial acetyl CoA carboxylase 2 inhibits carnitine acyltransferase I.

28.4 Metabolism in Context: Ethanol Alters Energy Metabolism in the Liver

Ethanol cannot be excreted and thus must be metabolized. Ethanol metabolism generates large quantities of NADH. The NADH glut inhibits fatty acid degradation and stimulates fatty acid synthesis, leading to an accumulation of fat in the liver. Excess ethanol is metabolized to acetyl CoA, which results in ketosis, and acetaldehyde, a reactive compound that modifies proteins and impairs their function.

KEY TERMS

fatty acid synthesis (p. 507)
malonyl CoA (p. 509)
acetyl CoA carboxylase (p. 509)

acyl carrier protein (ACP) (p. 510)
prostaglandin (p. 514)
arachidonate (p. 514)

eicosanoid (p. 514)
AMP-activated protein kinase
   (AMPK) (p. 516)

Answers to QUICK QUIZZES

1. Acetyl CoA is the basic substrate for fatty acid synthesis. It is transported out of mitochondria in the form of citrate. After the formation of acetyl CoA, the resulting pyruvate is transported back into the mitochondria with a concomitant formation of NADPH, the reducing power for fatty acid synthesis. Additional NADPH can be generated by the pentose phosphate pathway. Malonyl CoA, the ultimate substrate for fatty acid synthesis is formed by the carboxylation of acetyl CoA.

2. Malonyl CoA, the substrate for fatty acid synthesis, inhibits carnitine acyl transferase I, thus preventing the transport of fatty acids into mitochondria for degradation. Palmitoyl CoA inhibits acetyl CoA carboxylase, the transport of citrate into the cytoplasm, and glucose 6-phosphate dehydrogenase, the controlling enzyme of the pentose phosphate pathway.

PROBLEMS

1. *Making a fat.* Palmitate is a common fatty acid. What is the overall stoichiometry for the synthesis of palmitate from acetyl CoA? ✓ 3

2. *NADH to NADPH.* What are the three reactions that allow the conversion of cytoplasmic NADH into NADPH? What enzymes are required? Show the sum of the three reactions. ✓ 3

3. *A pledge to move forward.* What is the committed step in fatty acid synthesis, and which enzyme catalyzes the step? ✓ 4

4. *Like sugar 'n' spice.* Match each term (a–j) with its description (1–10).

(a) ATP-citrate lyase
   _____

(b) Malic enzyme _____

(c) Malonyl CoA _____

(d) Acetyl CoA
   carboxylase 1 _____

(e) Acyl carrier
   protein _____

(f) β-Ketoacyl synthase
   _____

(g) Palmitate _____

(h) Eicosanoids _____

(i) Arachidonate _____

(j) AMP-activated protein
   kinase _____

1. Helps to generate
   NADPH from NADH

2. Inactivates acetyl CoA
   carboxylase 1

3. Molecule on which
   fatty acids are
   synthesized

4. A precursor of
   prostaglandins

5. Activated acetyl CoA

6. The end product of
   fatty acid synthase

7. Fatty acids containing
   20 carbon atoms

8. Catalyzes the
   committed step in
   fatty acid synthesis

9. Catalyzes the reaction
   of acetyl CoA and
   malonyl CoA

10. Generates cytoplasmic
   acetyl CoA

5. *A supple synthesis.* Myristate, a saturated C_{14} fatty acid, is used as an emollient for cosmetics and topical medicinal preparations. Write a balanced equation for the synthesis of myristate. ✓ 3

6. *The cost of cleanliness.* Lauric acid is a 12-carbon saturated fatty acid. The sodium salt of lauric acid (sodium laurate) is a common detergent used in a variety of products, including laundry detergent, shampoo, and toothpaste. How many molecules of ATP and NADPH are required to synthesize lauric acid? ✓ 3

7. *Proper organization.* Arrange the following steps in fatty acid synthesis in their proper order. ✓ 3

(a) Dehydration

(b) Condensation

(c) Release of a C_{16} fatty acid

(d) Reduction of a carbonyl

(e) Formation of malonyl ACP

8. *No access to assets.* What would be the effect on fatty acid synthesis of a mutation in ATP-citrate lyase that reduces the enzyme's activity? Explain. ✓ 3

9. *The truth and nothing but.* Indicate whether each of the following statements is true or false. If false, explain.

(a) Biotin is required for fatty acid synthase activity.

(b) The condensation reaction in fatty acid synthesis is powered by the decarboxylation of malonyl CoA.

(c) Fatty acid synthesis does not depend on ATP.

(d) Palmitate is the end product of fatty acid synthesis.

(e) All of the enzyme activities required for fatty acid synthesis in mammals are contained in a single polypeptide chain.

(f) Fatty acid synthase in mammals is active as a monomer.

(g) The fatty acid arachidonate is a precursor for signal molecules.

(h) Acetyl CoA carboxylase 1 is inhibited by citrate.

10. *Odd fat out.* Suggest how fatty acids with odd numbers of carbon atoms are synthesized.

11. *Alpha or omega?* Only one acetyl CoA molecule is used directly in fatty acid synthesis. Identify the carbon atoms in palmitic acid that were donated by acetyl CoA. ✓ 3

12. *Now you see it, now you don't.* Although HCO_3^- is required for fatty acid synthesis, its carbon atom does not appear in the product. Explain. ✓ 4

13. *The fizz in fatty acid synthesis.* During the initial in vitro studies on fatty acid synthesis, experiments were performed to determine which buffer would result in optimal activity. Bicarbonate buffer proved far superior to phosphate buffer. The researchers could not initially account for this result. From your perspective of many decades later, explain why the result is not surprising.

14. *Tracing carbon atoms.* Consider a cell extract that actively synthesizes palmitate. Suppose that a fatty acid synthase in this preparation forms one molecule of palmitate in about 5 minutes. A large amount of malonyl CoA labeled with ^{14}C in each carbon atom of its malonyl unit is suddenly added to this system, and fatty acid synthesis is stopped a minute later by altering the pH. The fatty acids are analyzed for radioactivity. Which carbon atom of the palmitate formed by this system is more radioactive—C-1 or C-14?

15. *Driven by decarboxylation.* What is the role of decarboxylation in fatty acid synthesis? Name another key reaction in a metabolic pathway that employs this mechanistic motif.

16. *An unaccepting mutant.* The serine residue in acetyl CoA carboxylase 1 that is the target of the AMP-activated protein kinase is mutated to alanine. What is a likely consequence of this mutation? ✓ 4

17. *All for one, one for all.* What is a potential disadvantage of having many catalytic sites together on one very long polypeptide chain?

18. *Six of one, half a dozen of the other.* People who consume little fat but excess carbohydrates can still become obese. How is this result possible?

19. *No traffic allowed.* Both fatty acid synthesis and fatty acid degradation are regulated, at least in part, by controlling the movement of molecules into and out of mitochondria. Describe two such examples. ✓ 4

Chapter Integration Problems

20. *A tight embrace.* Avidin, a glycoprotein protein found in eggs, has a high affinity for biotin. Avidin can bind biotin and prevent its use by the body. How might a diet rich in raw eggs affect fatty acid synthesis? What will be the effect on fatty acid synthesis of a diet rich in cooked eggs? Explain.

21. *All about communication.* Why is citrate an appropriate inhibitor of phosphofructokinase?

22. *Counterpoint.* Compare and contrast fatty acid oxidation and synthesis with respect to (a) site of the process; (b) acyl carrier; (c) reductants and oxidants; (d) stereochemistry of the intermediates; (e) direction of synthesis or degradation; (f) organization of the enzyme system.

23. *Familiars.* One of the reactions critical for the generation of cytoplasmic NADPH from cytoplasmic NADH is also important in gluconeogenesis. What is the reaction and what is the immediate fate of the product of the reaction in gluconeogenesis?

24. *A foot in both camps.* What role does acetyl CoA carboxylase play in the regulation of fatty acid degradation? ✓ 4

Data Interpretation Problem

25. *Coming together.* The graph below shows the response of phosphorylated acetyl CoA carboxylase 1 to varying amounts of citrate. Explain this effect in light of the allosteric effects that citrate has on the enzyme. Predict the effects of increasing concentrations of palmitoyl CoA. ✓ 4

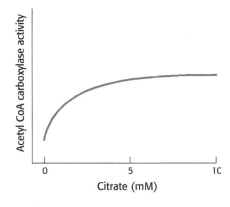

26. *Nip it in the bud.* Soraphen A is a natural antifungal agent isolated from a myxobacterium. Soraphen A is also a potent inhibitor of acetyl CoA carboxylase 1, the regulatory

enzyme that initiates fatty acid synthesis. As discussed earlier (p. 513), cancer cells must produce large amounts of fatty acids to generate phospholipids for membrane synthesis. Thus, acetyl CoA carboxylase 1 might be a target for anticancer drugs. Below are the results of experiments testing this hypothesis on a cancer line.

Figure A shows the effect of various amounts of soraphen A on fatty acid synthesis as measured by the incorporation of radioactive acetate (^{14}C) into fatty acids.

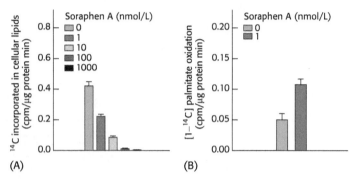

(A)   (B)

(a) How did soraphen A alter fatty acids synthesis? How was synthesis affected by increasing concentrations of soraphen A?

Figure B shows the results obtained when fatty acid oxidation was measured, as the release of radioactive CO_2 from added radioactive (^{14}C) palmitate, in the presence of soraphen A.

(b) How did soraphen A alter fatty acid oxidation?
(c) Explain the results obtained in B in light of the fact that soraphen A inhibits acetyl CoA carboxylase 1.

More experiments were undertaken to assess whether carboxylase inhibition did indeed prevent phospholipid synthesis. Again, cells were grown in the presence of radioactive acetate, and phospholipids were subsequently isolated and the amount of ^{14}C incorporated into the phospholipid was determined. The effect of soraphen A on phospholipid synthesis is shown in Figure C.

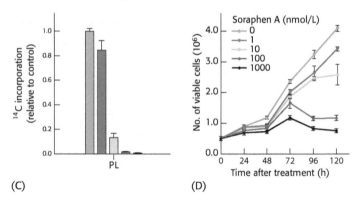

(C)   (D)

(d) Did the inhibition of carboxylase 1 by the drug alter phospholipid synthesis?
(e) How might phospholipid synthesis effect cell viability?

Finally, as shown in Figure D, experiments were performed to determine if the drug inhibits cancer growth in vitro by determining the number of cancer cells surviving upon exposure to soraphen A.

(f) How did soraphen A effect cancer cell viability?

[Data for graphs A–D from A. Beckers et al., *Cancer Res.* 67: 8180–8187, 2007.]

Challenge Problems

27. *Labels.* Suppose that you had an in vitro fatty-acid-synthesizing system that had all of the enzymes and cofactors required for fatty acid synthesis except for acetyl CoA. To this system, you added acetyl CoA that contained radioactive hydrogen (^3H, tritium) and carbon 14 (^{14}C) as shown here:

$$
\begin{array}{c}
^3H \quad\quad O \\
| \quad\quad\; \| \\
^3H\!-\!\!^{14}C\!-\!C\!-\!SCoA \\
| \\
^3H
\end{array}
$$

The ratio of $^3H/^{14}C$ is 3. What would the $^3H/^{14}C$ ratio be after the synthesis of palmitic acid (C_{16}) with the use of the radioactive acetyl CoA?

28. *If a little is good, a lot is not better.* Ethanol can be converted into acetate in the liver by alcohol dehydrogenase and aldehyde dehydrogenases:

$$CH_3CH_2OH + NAD^+ \xrightarrow{\text{Alcohol dehydrogenase}}$$
Ethanol
$$CH_3CHO + NADH + H^+$$
Acetaldehyde

$$CH_3CHO + NAD^+ + H_2O \xrightarrow{\text{Aldehyde dehydrogenase}}$$
Acetaldehyde
$$CH_3COO^- + NADH + H^+$$
Acetate

These reactions alter the $NAD^+/NADH$ ratio in the liver. What effect will this alteration have on glycolysis, gluconeogenesis, fatty acid metabolism, and the citric acid cycle? ✓ 4

Selected Readings for this chapter can be found online at www.whfreeman.com/tymoczko3e.

CHAPTER 29

Lipid Synthesis: Storage Lipids, Phospholipids, and Cholesterol

Fats are converted into triacylglycerol molecules, which are widely used to store excess energy for later use and to fulfill other purposes, illustrated by the insulating blubber of whales. The natural tendency of fats to exist in nearly water-free forms makes these molecules well suited to these roles. [Michael S. Nolan/Age Fotostock.]

W e now turn from the metabolism of fatty acids to the metabolism of lipids, which are built from fatty acids or their breakdown products. We will consider three classes of lipids: *triacylglycerols*, which are the storage form of fatty acids, *membrane lipids*, which are made up of phospholipids and sphingolipids, and *cholesterol*, a membrane component and a precursor to the steroid hormones.

✓ 5 Describe the relation between triacylglycerol synthesis and phospholipid synthesis.

29.1 Phosphatidate Is a Precursor of Storage Lipids and Many Membrane Lipids

Figure 29.1 provides a broad view of lipid synthesis. Both triacylglycerol synthesis and phospholipid synthesis begin with the precursor *phosphatidate* (diacylglycerol 3-phosphate). Phosphatidate is formed by the addition of two fatty acids to *glycerol 3-phosphate*. In most cases, glycerol 3-phosphate is first acylated by a saturated acyl CoA to form *lysophosphatidate*, which is, in turn, commonly acylated by unsaturated acyl CoA to yield phosphatidate:

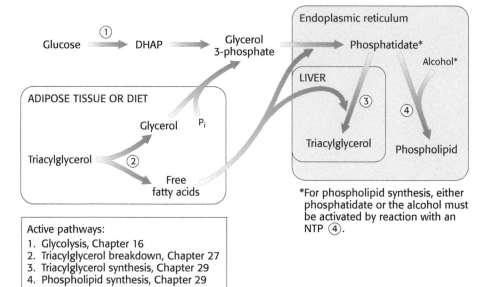

Figure 29.1 PATHWAY INTEGRATION: Sources of intermediates in the synthesis of triacylglycerols and phospholipids. Phosphatidate, synthesized from dihydroxyacetone phosphate (DHAP) produced in glycolysis and fatty acids, can be further processed to produce triacylglycerol or phospholipids. Phospholipids and other membrane lipids are continually produced in all cells.

Active pathways:
1. Glycolysis, Chapter 16
2. Triacylglycerol breakdown, Chapter 27
3. Triacylglycerol synthesis, Chapter 29
4. Phospholipid synthesis, Chapter 29

*For phospholipid synthesis, either phosphatidate or the alcohol must be activated by reaction with an NTP ④.

Glycerol 3-phosphate → **Lysophosphatidate** → **Phosphatidate**

Usually saturated
Usually unsaturated

Triacylglycerol Is Synthesized from Phosphatidate in Two Steps

The pathways diverge at phosphatidate. The synthesis of triacylglycerol is completed by a *triacylglycerol synthetase complex* that is bound to the endoplasmic reticulum membrane. Phosphatidate is hydrolyzed to give diacylglycerol (DAG), which is then acylated to a triacylglycerol:

Phosphatidate → **Diacylglycerol (DAG)** → **Triacylglycerol**

The liver is the primary site of triacylglycerol synthesis. From the liver, triacylglycerols are transported to muscles for use as a fuel or to adipose tissue for storage. Approximately 85% of a nonobese person's energy is stored as triacylglycerols, mainly in adipose tissue.

Phospholipid Synthesis Requires Activated Precursors

Phosphatidate is also a precursor for *phospholipids*. Phospholipid synthesis, which takes place in the endoplasmic reticulum, requires the combination of a diacylglycerol with an alcohol. As in most anabolic reactions, one of the components must be activated. In this case, either of the two components may be activated, depending on the source of the reactants.

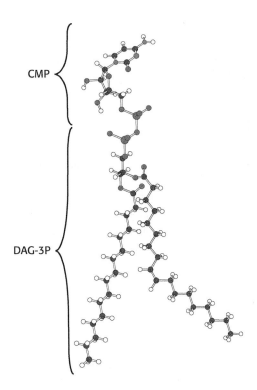

CMP

DAG-3P

Figure 29.2 The structure of CDP-diacylglycerol. A key intermediate in the synthesis of phospholipids consists of phosphatidate (diacylglycerol 3-phosphate, or DAG-3P) and cytidine monophosphate (CMP).

Synthesis from an activated diacylglycerol This pathway starts with the reaction of phosphatidate with cytidine triphosphate (CTP) to form *cytidine diphosphodiacylglycerol* (CDP-diacylglycerol; **Figure 29.2**). This reaction, like those of many biosyntheses, is driven forward by the hydrolysis of pyrophosphate.

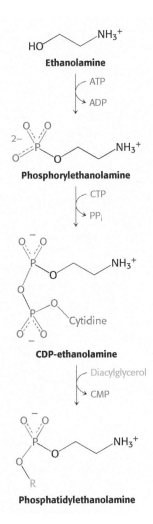

Ethanolamine

ATP
ADP

Phosphorylethanolamine

CTP
PP$_i$

CDP-ethanolamine

Diacylglycerol
CMP

Phosphatidylethanolamine

Phosphatidate   **CDP-diacylglycerol**

The activated phosphatidyl unit then reacts with the hydroxyl group of an alcohol. If the alcohol is inositol, the products are *phosphatidylinositol* and cytidine monophosphate (CMP). Subsequent phosphorylations of phosphatidylinositol catalyzed by specific kinases lead to the synthesis of *phosphatidylinositol 4,5-bisphosphate*, a membrane lipid that is also an important molecule in signal transduction (p. 232).

Synthesis from an activated alcohol Phosphatidylethanolamine in mammals can be synthesized from the alcohol ethanolamine through the formation of CDP-ethanolamine. In this case, ethanolamine is phosphorylated by ATP to form the precursor, *phosphorylethanolamine*. This precursor then reacts with CTP to form the activated alcohol, *CDP-ethanolamine*. The phosphorylethanolamine unit of CDP-ethanolamine is subsequently transferred to a diacylglycerol to form *phosphatidylethanolamine*.

CLINICAL INSIGHT

Phosphatidylcholine Is an Abundant Phospholipid

The most common phospholipid in mammals is *phosphatidylcholine,* comprising approximately 50% of the membrane mass. Dietary choline is activated in a series of reactions analogous to those in the activation of ethanolamine. CTP-phosphocholine cytidylyltransferase (CCT) catalyzes the formation of CDP-choline, the rate-limiting step in phosphatidylcholine synthesis. CCT is an amphitropic enzyme, a class of enzymes whose regulator ligand is the membrane itself. A portion of the enzyme, normally associated with the membrane, detects a fall in phosphatidylcholine as an alteration in the physical properties of the membrane. When this occurs, another portion of the enzyme is inserted into the membrane leading to enzyme activation. Indeed, the k_{cat}/K_M value (p. 116) increases three orders of magnitude upon activation, resulting in the restoration of phosphatidylcholine levels. Earlier (p. 513) we examined how cancer cells increase fatty acid synthesis to meet the fatty acid needs for membrane synthesis. Evidence is accumulating that CCT is specifically activated in some cancers to generate the required phosphocholine.

The importance of phosphatidylcholine is attested to by the fact that the liver possesses an enzyme, *phosphatidylethanolamine methyltransferase,* that synthesizes phosphatidylcholine from phosphatidylethanolamine when dietary choline is insufficient. The amino group of this phosphatidylethanolamine is methylated three times to form phosphatidylcholine. *S*-Adenosylmethionine is the methyl donor (Chapter 31):

Phosphatidyl-ethanolamine   → (3 *S*-Adenosyl-methionine, 3 *S*-Adenosyl-homocysteine) → **Phosphatidyl-choline**

Thus, phosphatidylcholine can be produced by two distinct pathways in mammals, ensuring that this phospholipid can be synthesized even if the components for one pathway are in limited supply.

Choline

Sphingosine

R_2 = H, *N*-acetylneuraminate
R_2 = OH, *N*-glycolylneuraminate

$R_1 = $

Sialic acids

Sphingolipids Are Synthesized from Ceramide

Phospholipids, with their glycerol backbones, are not the only type of membrane lipid. *Sphingolipids,* with a backbone of *sphingosine* rather than glycerol, are found in the plasma membranes of all eukaryotic cells, although the concentration is highest in the cells of the central nervous system. To synthesize a sphingolipid, palmitoyl CoA and serine condense to form 3-ketosphinganine, which is then reduced to dihydrosphingosine (**Figure 29.3**). Dihydrosphingosine is converted into *ceramide* with the addition of a long-chain acyl CoA to the amino group of dihydrosphingosine followed by an oxidation reaction to form the trans double bond.

The terminal hydroxyl group of ceramide is also substituted to form a variety of sphingolipids (**Figure 29.4**):

- In *sphingomyelin,* a component of the myelin sheath covering many nerve fibers, the substituent is phosphorylcholine.

- In a *cerebroside,* also a component of myelin, the substituent is glucose or galactose.

- In a *ganglioside,* an oligosaccharide containing at least one sialic acid is linked to the terminal hydroxyl group of ceramide by a glucose residue.

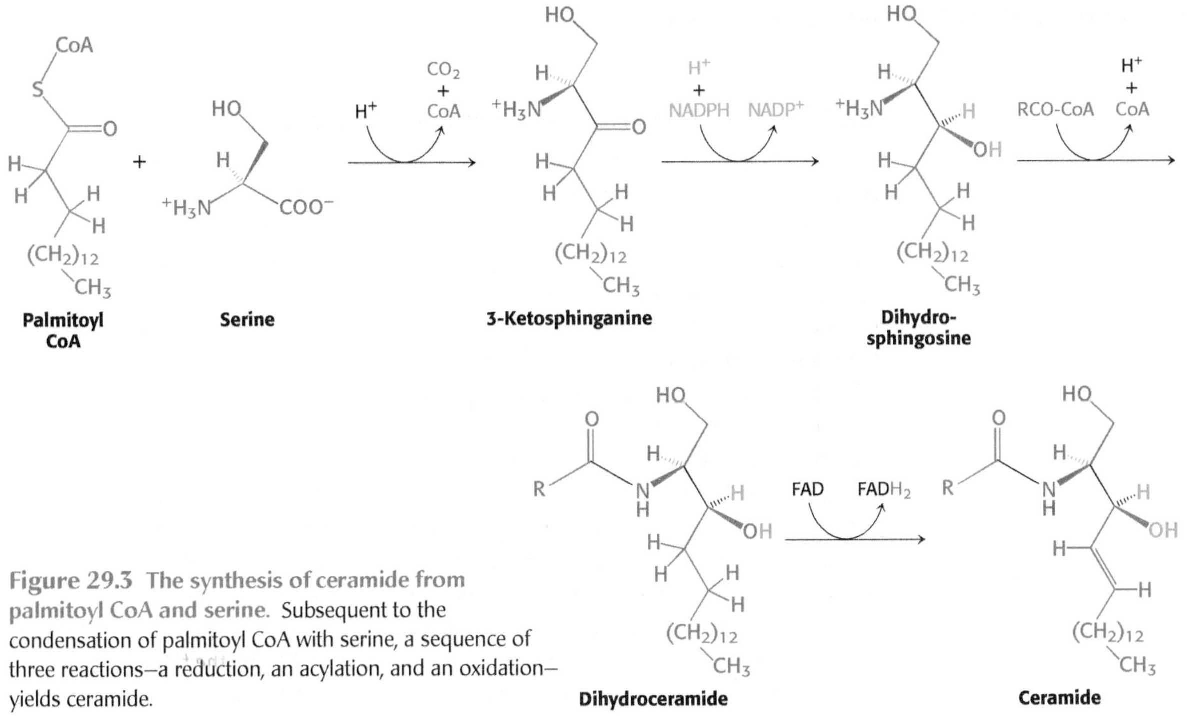

Figure 29.3 The synthesis of ceramide from palmitoyl CoA and serine. Subsequent to the condensation of palmitoyl CoA with serine, a sequence of three reactions—a reduction, an acylation, and an oxidation—yields ceramide.

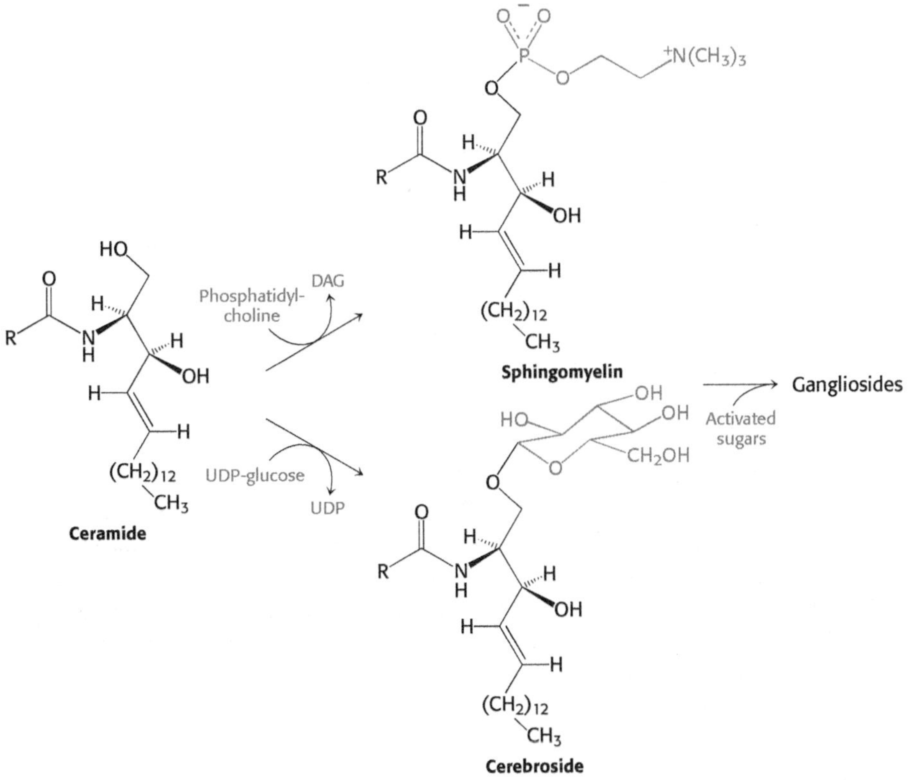

Figure 29.4 Sphingolipid synthesis. Ceramide is the starting point for the formation of sphingomyelin and gangliosides.

CLINICAL INSIGHT

Gangliosides Serve as Binding Sites for Pathogens

Ganglioside-binding by cholera toxin is the first step in the development of cholera, a pathological condition characterized by severe diarrhea (p. 231). Enterotoxigenic *E. coli*, the most common cause of diarrhea, including traveler's diarrhea, produces a toxin that also gains access to the cell by first binding to gangliosides. Gangliosides are also crucial for binding immune-system cells to sites of injury in the inflammatory response.

QUICK QUIZ 1 Describe the roles of glycerol 3-phosphate, phosphatidate, and diacylglycerol in triacylglycerol synthesis and phospholipid synthesis.

⚕ CLINICAL INSIGHT

Disrupted Lipid Metabolism Results in Respiratory Distress Syndrome and Tay–Sachs Disease

Disruptions in lipid metabolism are responsible for a host of diseases. We will briefly examine two such conditions. *Respiratory distress syndrome* is a pathological condition resulting from a failure in the biosynthesis of dipalmitoylphosphatidylcholine. This phospholipid, in conjunction with specific proteins and other phospholipids, is found in the extracellular fluid that surrounds the alveoli of the lung, where it decreases the surface tension of the fluid to prevent lung collapse at the end of the expiration phase of breathing. Premature infants may suffer from respiratory distress syndrome because their immature lungs do not synthesize enough dipalmitoylphosphatidylcholine.

Whereas respiratory distress syndrome results from failure in biosynthesis, *Tay–Sachs disease,* a congenital disease that afflicts infants soon after birth, is caused by a failure of lipid degradation: specifically, an inability to degrade gangliosides. Gangliosides are found in highest concentration in the nervous system, particularly in gray matter, where they constitute 6% of the lipids. Gangliosides are normally degraded inside lysosomes by the sequential removal of their terminal sugars, but, in Tay–Sachs disease, this degradation does not take place. The terminal residue of the ganglioside is removed very slowly or not at all. The missing or deficient enzyme is a specific *β-N-acetylhexosaminidase* (**Figure 29.5**).

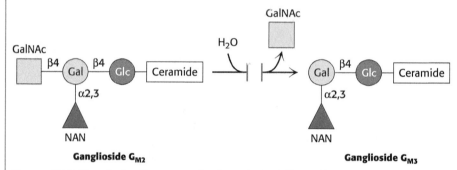

Figure 29.5 Tay–Sachs disease results from the inability to degrade a ganglioside. A particular ganglioside, G_{M2}, accumulates in Tay–Sachs patients because a key step in its degradation, conversion into ganglioside G_{M3}, cannot take place, because of insufficient β-N-acetylhexosaminidase. Abbreviations: GalNAc, N-acetylgalactosamine; NAN, *N-acetylneuraminate*; Gal, galactose; Glc, glucose.

As a consequence, neurons become significantly swollen with lipid-filled lysosomes (**Figure 29.6**). An affected infant displays weakness and retarded psychomotor skills before 1 year of age. The child is demented and blind by age 2 and usually dies before age 3.

Figure 29.6 Lysosome with lipids. An electron micrograph of a lysosome engorged with lipids. Such lysosomes are sometimes described as being "onion-skin"-like because the layers of undigested lipids resemble a sliced onion. [Graphics & Photography Service/Custom Medical Stock Photo—All rights reserved.]

Tay–Sachs disease was especially prominent among Ashkenazi Jews (descendants of Jews from central and eastern Europe). A genetic testing program, initiated in the early 1970s upon the development of a simple blood test to identify carriers, has virtually eliminated the disease in the population. Tay–Sachs disease can also be diagnosed in the course of fetal development.

Phosphatidic Acid Phosphatase Is a Key Regulatory Enzyme in Lipid Metabolism

Although the details of the regulation of lipid synthesis remain to be elucidated, evidence suggests that *phosphatidic acid phosphatase* (PAP), working in concert with diacylglycerol kinase, plays a key role in lipid synthesis regulation. Phosphatidic acid phosphatase, also called lipin 1 in mammals, controls the extent to which triacylglycerols are synthesized relative to phospholipids and regulates the type of phospholipid synthesized (**Figure 29.7**). For instance, when PAP activity

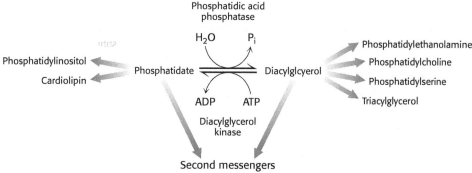

Figure 29.7 Regulation of lipid synthesis. Phosphatidic acid phosphatase (PAP) is the key regulatory enzyme in lipid synthesis. When active, PAP generates diacylglycerol, which can react with activated alcohols to form phospholipids or with fatty acyl CoA to form triacylglycerols. When PAP is inactive, phosphatidate is converted into CMP-DAG for the synthesis of different phospholipids. PAP also controls the amount of DAG and phosphatidate, both of which function as second messengers.

is high, phosphatidate is dephosphorylated and diacylglycerol is produced, which can react with the appropriate activated alcohols to yield phosphatidylethanolamine, phosphatidylserine, or phosphatidylcholine. Diacylglycerols can also be converted into triacylglycerols, and evidence suggests that the formation of triacylglycerols may act as a fatty acid buffer, which helps to regulate the levels of diacylglycerol and sphingolipids, both of which serve signaling functions.

When PAP activity is lower, phosphatidate is used as a precursor for different phospholipids, such as phosphatidylinositol and cardiolipin. Moreover, phosphatidate is a signal molecule itself. Phosphatidate regulates the growth of endoplasmic reticulum and nuclear membranes and acts as a cofactor that stimulates the expression of genes in phospholipid synthesis.

What are the signal molecules that regulate the activity of PAP? CDP-diacylglycerol, phosphatidylinositol, and cardiolipin enhance PAP activity, and sphingosine and dihydrosphingosine inhibit it.

Studies in mice clearly show the importance of PAP for the regulation of fatty acid synthesis. The loss of PAP function prevents normal adipose-tissue development, leading to lipodystrophy (severe loss of body fat) and insulin resistance. Excess PAP activity results in obesity. Understanding the regulation of phospholipid synthesis is an exciting area of research that will be active for some time to come.

29.2 Cholesterol Is Synthesized from Acetyl Coenzyme A in Three Stages

We now turn our attention to the synthesis of a different kind of lipid, which lacks the long hydrocarbon chains characteristic of triacylglycerols and membrane lipids—*cholesterol*. Cholesterol has a much higher profile than the other lipids because of its association with heart disease. Despite its lethal reputation with the public, cholesterol is vital to the body: it maintains proper fluidity of animal cell membranes (p. 197) and is the precursor of steroid hormones such as progesterone, testosterone, estradiol, and cortisol.

Cholesterol is synthesized in the liver and, to a lesser extent, in other tissues. The rate of its synthesis is highly responsive to the cellular level of cholesterol. All 27 carbon atoms of cholesterol are derived from acetyl CoA in a three-stage synthetic process:

1. Stage one is the synthesis of isopentenyl pyrophosphate, an activated isoprene unit that is the key building block of cholesterol.

2. Stage two is the condensation of six molecules of isopentenyl pyrophosphate to form squalene.

3. In stage three, squalene cyclizes and the tetracyclic product is subsequently converted into cholesterol.

The first stage takes place in the cytoplasm, and the second two in the lumen of the endoplasmic reticulum.

The Synthesis of Mevalonate Initiates the Synthesis of Cholesterol

The first stage in the synthesis of cholesterol is the formation of isopentenyl pyrophosphate from acetyl CoA. This set of reactions starts with the formation of 3-hydroxy-3-methylglutaryl CoA (HMG CoA) from acetyl CoA and acetoacetyl CoA. This intermediate is reduced to *mevalonate* for the synthesis of cholesterol.

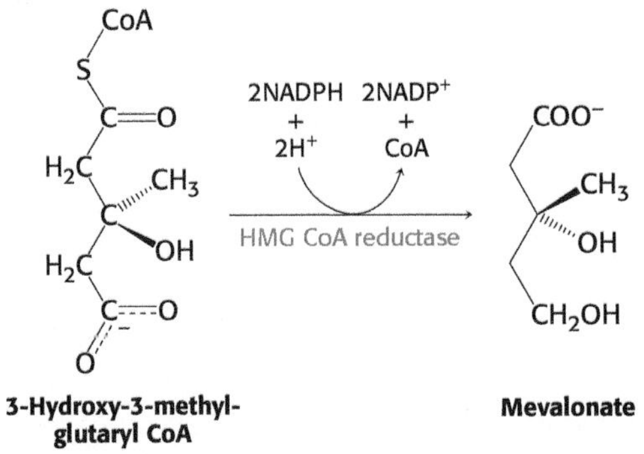

3-Hydroxy-3-methyl-glutaryl CoA → **Mevalonate** (via HMG CoA reductase, 2NADPH + 2H⁺ → 2NADP⁺ + CoA)

The synthesis of mevalonate is the committed step in cholesterol formation. The enzyme catalyzing this irreversible step, *3-hydroxy-3-methylglutaryl CoA reductase* (HMG-CoA reductase), is the key control site in cholesterol biosynthesis, as will be discussed shortly. The importance of cholesterol is vividly illustrated by the observation that mice lacking HMG-CoA reductase die very early in development.

Mevalonate is converted into *3-isopentenyl pyrophosphate* in three consecutive reactions requiring ATP (**Figure 29.8**). Stage one ends with the production of isopentenyl pyrophosphate, an activated 5-carbon *isoprene* unit.

Squalene (C₃₀) Is Synthesized from Six Molecules of Isopentenyl Pyrophosphate (C₅)

The next major precursor on the path to cholesterol is *squalene*, which is synthesized from isopentenyl pyrophosphate by the following reaction sequence:

$$C_5 \longrightarrow C_{10} \longrightarrow C_{15} \longrightarrow C_{30}$$

Before the condensation reactions take place, *isopentenyl pyrophosphate* isomerizes to *dimethylallyl pyrophosphate*:

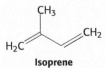

Isoprene

Isopentenyl pyrophosphate ⇌ **Dimethylallyl pyrophosphate**

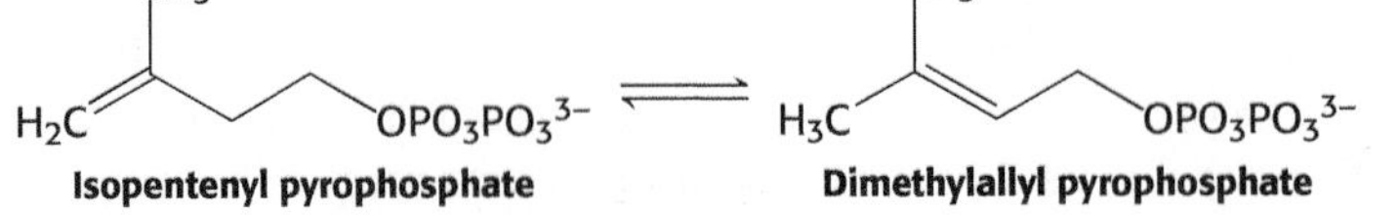

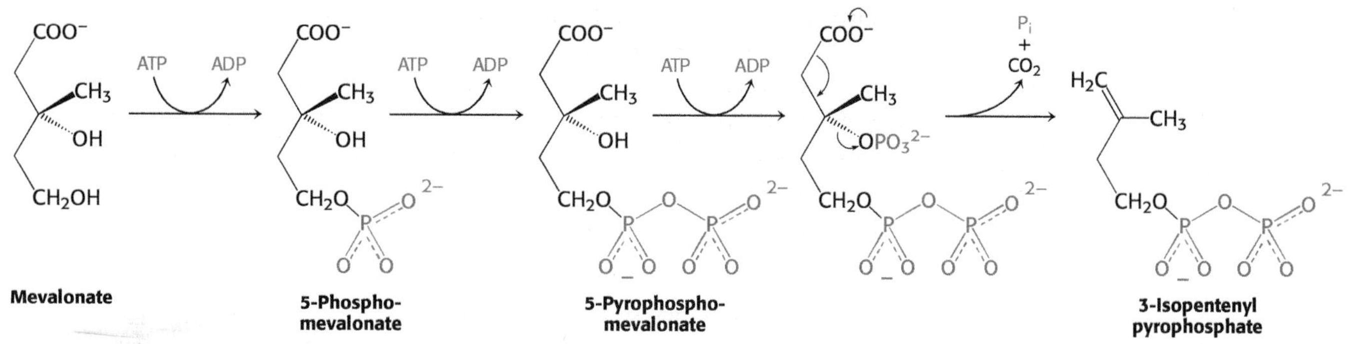

Mevalonate　　　**5-Phospho-mevalonate**　　　**5-Pyrophospho-mevalonate**　　　**3-Isopentenyl pyrophosphate**

The two isomer C_5 units (one of each) condense to begin the formation of squalene. The reactions leading from the six C_5 units to squalene, a C_{30} isoprenoid, are summarized in (**Figure 29.9**).

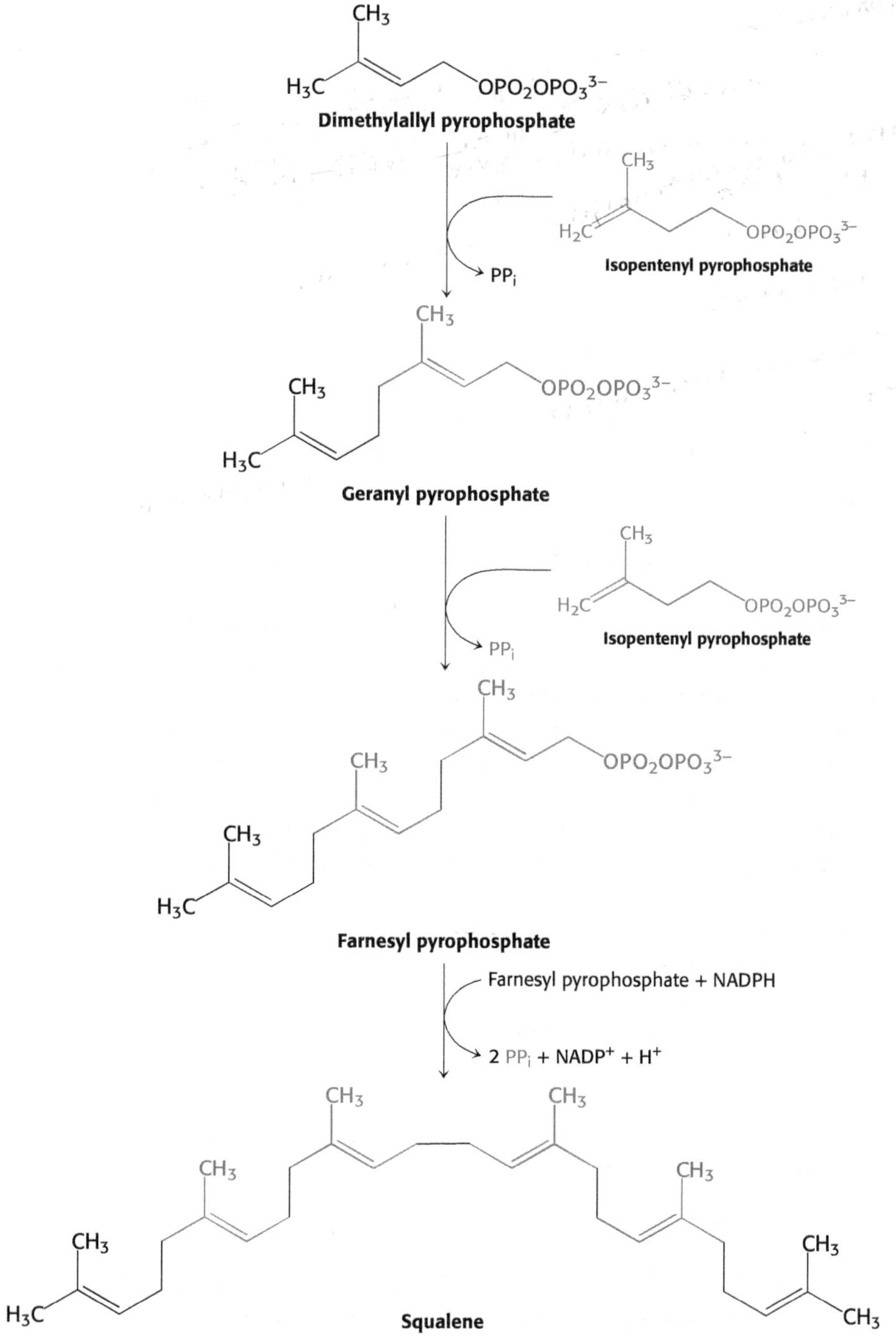

Figure 29.8 Isopentenyl pyrophosphate synthesis. This activated intermediate is formed from mevalonate in three steps, the last of which includes a decarboxylation.

Dimethylallyl pyrophosphate

Isopentenyl pyrophosphate

Geranyl pyrophosphate

Isopentenyl pyrophosphate

Farnesyl pyrophosphate

Farnesyl pyrophosphate + NADPH

2 PP_i + $NADP^+$ + H^+

Squalene

Figure 29.9 Squalene synthesis. One molecule of dimethylallyl pyrophosphate and two molecules of isopentenyl pyrophosphate condense to form farnesyl pyrophosphate. The tail-to-tail coupling of two molecules of farnesyl pyrophosphate yields squalene.

Squalene Cyclizes to Form Cholesterol

In the final stage of cholesterol biosynthesis, squalene cyclizes to form a ringlike structure (**Figure 29.10**). Squalene is first activated by conversion into squalene epoxide (2,3-oxidosqualene) in a reaction that uses O_2 and NADPH. Squalene epoxide is then cyclized to *lanosterol*. Lanosterol (C_{30}) is subsequently converted into cholesterol (C_{27}) in a multistep process, during which three carbon units are removed (Figure 29.10).

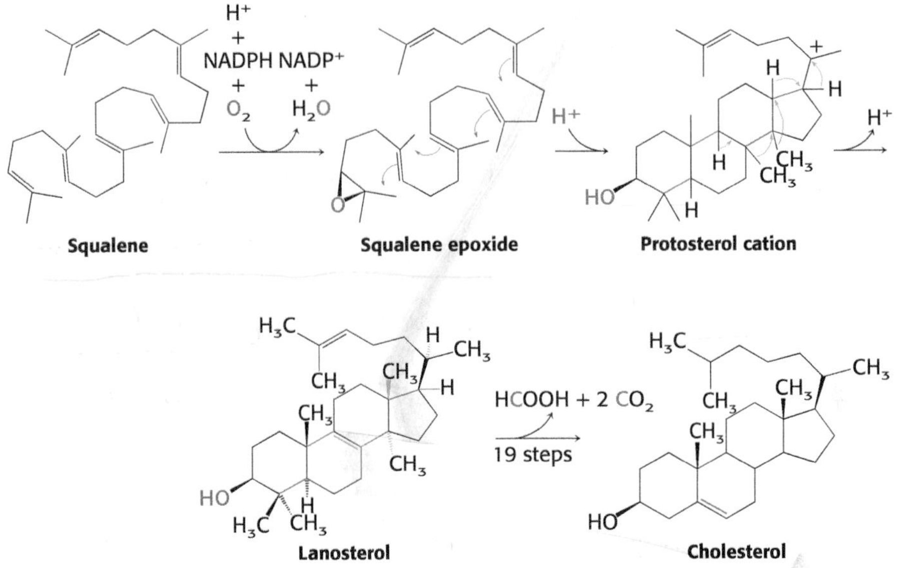

Squalene   **Squalene epoxide**   **Protosterol cation**

H^+
+
NADPH   NADP$^+$
+   +
O_2 H_2O

H^+

H^+

HCOOH + 2 CO_2

19 steps

Lanosterol   **Cholesterol**

Figure 29.10 Squalene cyclization. The formation of the steroid nucleus from squalene begins with the formation of squalene epoxide. This intermediate is protonated to form a carbocation that cyclizes to form a tetracyclic structure, which rearranges to form lanosterol. Then, lanosterol is converted into cholesterol in a complex process.

✓ 6 List the regulatory steps in the control of cholesterol synthesis.

29.3 The Regulation of Cholesterol Synthesis Takes Place at Several Levels

Cholesterol can be obtained from the diet or it can be synthesized de novo. An adult on a low-cholesterol diet typically synthesizes about 800 mg of cholesterol per day. The liver is a major site of cholesterol synthesis in mammals (**Figure 29.11**), although the intestine also forms significant amounts. The rate of cholesterol formation by these organs is highly responsive to the cellular level of cholesterol. *This feedback regulation is mediated primarily by changes in the amount and activity of 3-hydroxy-3-methylglutaryl CoA reductase (HMG-CoA reductase).* As described earlier (p. 530), this enzyme catalyzes the formation of mevalonate, the committed step in cholesterol biosynthesis. HMG-CoA reductase is controlled in multiple ways:

1. *The rate of synthesis of HMG-CoA reductase mRNA is controlled* by the sterol regulatory element binding protein (SREBP). This transcription factor binds

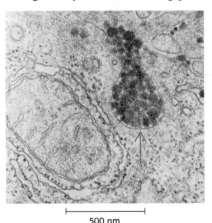

500 nm

Figure 29.11 The site of cholesterol synthesis. An electron micrograph of a part of a liver cell actively engaged in the synthesis and secretion of very low density lipoprotein (VLDL), a lipoprotein that transports lipids synthesized in the liver to elsewhere in the body (Table 29.1 and Figure 29.13). The arrow points to a vesicle that is releasing its content of VLDL particles. [Courtesy of Dr. George Palade.]

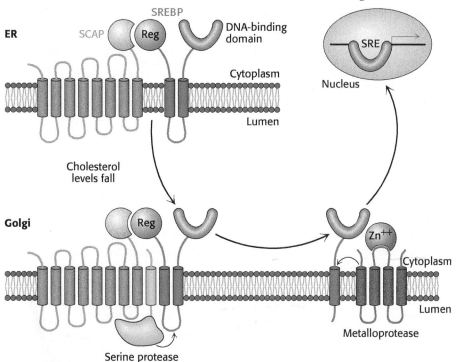

Figure 29.12 The SREBP pathway. SREBP resides in the endoplasmic reticulum, where it is bound to SCAP by its regulatory (Reg) domain. When cholesterol levels fall, SCAP and SREBP move to the Golgi complex, where SREBP undergoes successive proteolytic cleavages by a serine protease and a metalloprotease. The released DNA-binding domain moves to the nucleus to alter gene expression. [Information from an illustration provided by Dr. Michael Brown and Dr. Joseph Goldstein.]

to a short DNA sequence called the *sterol regulatory element* (SRE) on the 5′ side of the reductase gene. It binds to the SRE when cholesterol levels are low and enhances transcription. In its inactive state, the SREBP resides in the endoplasmic reticulum membrane, where it is associated with the SREBP cleavage activating protein (SCAP), an integral membrane protein. SCAP is the cholesterol sensor. When cholesterol levels fall, SCAP escorts SREBP in small membrane vesicles to the Golgi complex, where it is released from the membrane by two specific proteolytic cleavages (**Figure 29.12**). The released protein migrates to the nucleus and binds the SRE of the HMG-CoA reductase gene, as well as several other genes in the cholesterol biosynthetic pathway, to enhance transcription. When cholesterol levels rise, the proteolytic release of the SREBP is blocked, and the SREBP in the nucleus is rapidly degraded. These two events halt the transcription of genes of the cholesterol biosynthetic pathways.

What is the molecular mechanism that retains SCAP-SREBP in the endoplasmic reticulum when cholesterol is present but allows movement to the Golgi complex when the cholesterol concentration is low? When cholesterol is low, SCAP binds to vesicular proteins that facilitate the transport of SCAP-SREBP to the Golgi apparatus. When cholesterol is present, SCAP binds cholesterol, which causes a structural change in SCAP so that it binds to Insig (*insulin-induced gene*), another endoplasmic reticulum protein. Insig is the anchor that retains SCAP and thus SREBP in the endoplasmic reticulum in the presence of cholesterol.

2. *The rate of translation of HMG-CoA reductase mRNA is inhibited* by nonsterol metabolites derived from mevalonate as well as by dietary cholesterol.

3. *The degradation of HMG-CoA reductase is stringently controlled.* The enzyme has two domains: its cytoplasmic domain carries out catalysis, and its membrane domain senses signals that lead to its degradation. The membrane domain may undergo structural changes in response to increasing concentrations of sterols such as cholesterol that make the enzyme more susceptible to proteolysis. A combination of these three regulatory devices can alter the amount of enzyme in the cell more than 200-fold.

4. *Phosphorylation decreases the activity of HMG-CoA reductase.* This enzyme, like acetyl CoA carboxylase (which catalyzes the committed step in fatty acid synthesis, p. 516), is switched off by an AMP-activated protein kinase. Thus, cholesterol synthesis ceases when the ATP level is low.

We will return to the control of cholesterol in a clinical context after we consider a related topic—triacylglycerol transport.

? **QUICK QUIZ 2** Outline the mechanisms of regulation of cholesterol biosynthesis.

29.4 Lipoproteins Transport Cholesterol and Triacylglycerols Throughout the Organism

Cholesterol and triacylglycerols are made primarily in the liver, but they are used by tissues throughout the body. How are these important but hydrophobic biochemicals shuttled through the bloodstream? Cholesterol and triacylglycerols are packaged into *lipoprotein particles* for transport through bodily fluids. Each particle consists of a core of hydrophobic lipids surrounded by a shell of more-polar lipids and proteins. The protein components (called *apoproteins*) have two roles: they *solubilize hydrophobic lipids and contain cell-targeting signals.* Some of the important lipoproteins and their properties are shown in Table 29.1.

Table 29.1   Properties of plasma lipoproteins

Plasma lipoproteins	Density (g ml⁻¹)	Diameter (nm)	Apolipoprotein	Physiological role	Composition (%)				
					TAG	CE	C	PL	P
Chylomicron	<0.95	75–1200	B48, C, E	Dietary-fat transport	86	3	1	8	2
Very low density lipoprotein (VLDL)	0.95–1.006	30–80	B100, C, E	Endogenous-fat transport	52	14	7	18	8
Intermediate-density lipoprotein (IDL)	1.006–1.019	15–35	B100, E	LDL precursor	38	30	8	23	11
Low-density lipoprotein (LDL)	1.019–1.063	18–25	B100	Cholesterol transport	10	38	8	22	21
High-density lipoprotein (HDL)	1.063–1.21	7.5–20	A	Reverse cholesterol transport	5–10	14–21	3–7	19–29	33–57

Abbreviations: TAG, triacylglycerol; CE, cholesterol ester; C, free cholesterol; PL, phospholipid; P, protein.

Lipoprotein particles are classified according to increasing density: chylomicrons, chylomicron remnants, very low density lipoproteins (VLDLs), intermediate-density lipoproteins (IDLs), low-density lipoproteins (LDLs), and high-density lipoproteins (HDLs). We considered chylomicrons, which transport dietary lipids, in the context of digestion (p. 253).

Triacylglycerols and cholesterol in excess of the liver's own needs are exported into the blood in the form of *very low density lipoprotein.* Triacylglycerols in VLDL are hydrolyzed by lipases on capillary surfaces, and the freed fatty acids are taken into the cells. The cholesterol-rich remnants, called *intermediate-density* lipoproteins, can be taken up by the liver for processing or converted into *low-density lipoprotein* by the removal of more triacylglycerols (Figure 29.13).

Low-density lipoprotein is the major carrier of cholesterol in blood. This lipoprotein particle contains a core of cholesterol molecules linked by ester bonds to fatty acids. This core is surrounded by a shell of phospholipids and unesterified cholesterol. The shell also has a single copy of apoprotein B-100, which directs

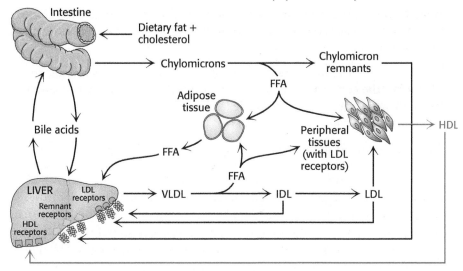

Figure 29.13 An overview of lipoprotein-particle metabolism. Fatty acids are abbreviated FFA. [Information from J. G. Hardman (Ed.), L. L. Limbrid (Ed.), and A. G. Gilman (Consult. Ed.), *Goodman and Gilman's The Pharmacological Basis of Therapeutics*, 10th ed. (McGraw–Hill, 2001), p. 975, Fig. 36.1.]

LDL to the proper cells (**Figure 29.14**). The role of LDL is to transport cholesterol to peripheral tissues and regulate de novo cholesterol synthesis at these sites, as described in the next section.

A different purpose is served by *high-density lipoprotein*, which picks up cholesterol released into the plasma from dying cells and from membranes undergoing turnover, a process termed *reverse cholesterol transport*. An acyltransferase in HDL esterifies these cholesterols, which are then returned by HDL to the liver.

Low-Density Lipoproteins Play a Central Role in Cholesterol Metabolism

Cholesterol metabolism must be precisely regulated to prevent atherosclerosis, the thickening of arterial walls with a subsequent loss of elasticity. The mode of control in the liver, the primary site of cholesterol synthesis, has already been discussed (p. 532). The primary source of cholesterol for peripheral tissues is the low-density lipoprotein. As will be discussed in the Clinical Insight on page 536, high concentrations of LDL in the blood play a role in setting the conditions for a heart attack. LDL is normally removed from the blood in a process called *receptor-mediated endocytosis*, which serves as a paradigm for the uptake of many molecules.

Receptor-mediated endocytosis of LDL is accomplished in three steps (**Figure 29.15**):

1. *Low-density lipoprotein binds to a receptor protein on the cell surface.* Apoprotein B-100 on the surface of an LDL particle binds to a specific receptor

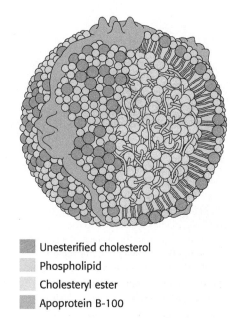

- ▮ Unesterified cholesterol
- ▮ Phospholipid
- ▮ Cholesteryl ester
- ▮ Apoprotein B-100

Figure 29.14 A schematic model of low-density lipoprotein. The LDL particle is approximately 22 nm (220 Å) in diameter.

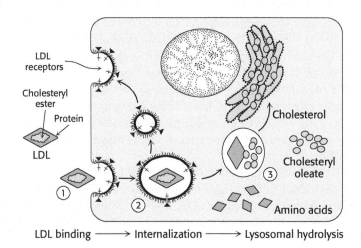

LDL binding ——→ Internalization ——→ Lysosomal hydrolysis

Figure 29.15 Receptor-mediated endocytosis. The process of receptor-mediated endocytosis is illustrated for low-density lipoprotein (LDL): (1) LDL binds to a specific receptor, the LDL receptor; (2) this complex invaginates to form an internal vesicle; (3) after separation from its receptor, the LDL-containing vesicle fuses with a lysosome, leading to the degradation of the LDL and the release of the cholesterol.

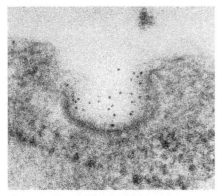

(A)

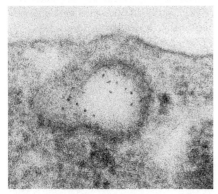

(B)

Figure 29.16 Endocytosis of low-density lipoprotein bound to its receptor. Micrographs of (A) LDL (conjugated to iron-laden ferritin for visualization, dark spots) bound to a coated-pit region on the surface of a cultured human fibroblast cell and (B) this region invaginating and fusing to form an endocytic vesicle. [Reprinted from *Cell* 10, R. G. W. Anderson, M. S. Brown, and J. L. Goldstein, Role of the coated endocytic vesicle in the uptake of receptor-bound low density lipoprotein in human fibroblasts, pp. 351–364 ©1977, with permission from Elsevier.]

protein on the plasma membrane of nonliver cells. The receptors for LDL are localized in specialized regions called *coated pits,* surrounded by a specialized protein called *clathrin.*

2. *The cell internalizes the receptor–LDL complex.* The plasma membrane in the vicinity of the complex folds in on itself (invaginates). The membrane then fuses to form an endocytic vesicle (called an endosome), enclosing the receptor–LDL complex (*endocytosis*) (Figure 29.16).

3. *Low-density lipoprotein is hydrolyzed in lysosomes.* The vesicles containing LDL subsequently fuse with *lysosomes,* acidic vesicles that carry a wide array of degradative enzymes. The protein component of the LDL is hydrolyzed to free amino acids. The cholesteryl esters in the LDL are hydrolyzed by a lysosomal acid lipase. The LDL receptor itself usually returns unscathed to the plasma membrane. The round-trip time for a receptor is about 10 minutes; in its lifetime of about a day, it brings many LDL particles into the cell.

The released unesterified cholesterol can then be used for membrane biosynthesis or reesterified for storage inside the cell. The stored cholesterol must be reesterified because high concentrations of unesterified cholesterol disrupt the integrity of cell membranes.

The synthesis of the LDL receptor is itself subject to feedback regulation. Studies show that, when cholesterol is abundant inside the cell, new LDL receptors are not synthesized, blocking the uptake of additional cholesterol from plasma LDL.

CLINICAL INSIGHT

The Absence of the LDL Receptor Leads to Familial Hypercholesterolemia and Atherosclerosis

High cholesterol levels promote atherosclerosis, which is the leading cause of death in industrialized societies. Cholesterol's role in the development of atherosclerosis was elucidated by the study of *familial hypercholesterolemia,* a genetic disorder. Familial hypercholesterolemia is characterized by high concentrations of cholesterol and LDL in the plasma, about three to four times the desired amount. In familial hypercholesterolemia, cholesterol is deposited in various tissues because of the high concentration of LDL cholesterol in the plasma. Nodules of cholesterol called *xanthomas* are prominent in skin and tendons in those having high levels of LDL. LDL also accumulates under the endothelial cells lining the blood vessels. Of particular concern is the oxidation of the excess LDL to form oxidized LDL (oxLDL), which can instigate the inflammatory response by the immune system, a response that has been implicated in the development of cardiovascular disease. The oxLDL is taken up by immune-system cells called macrophages, which become engorged to form foam cells. These foam cells become trapped in the walls of the blood vessels and contribute to the formation of atherosclerotic plaques that cause arterial narrowing and lead to heart attacks (Figure 29.17).

The molecular defect in most cases of familial hypercholesterolemia is an absence or deficiency of functional receptors for LDL. Homozygotes have almost no functional receptors for LDL, whereas heterozygotes have about half the normal number. Consequently, the entry of LDL into liver and other cells

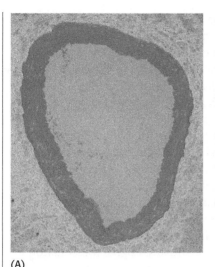

(A)

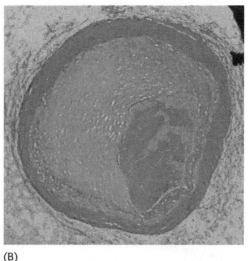

(B)

Figure 29.17 The effects of excess cholesterol. Cross section of (A) a normal artery and (B) an artery blocked by a cholesterol-rich plaque. [SPL/Science Source.]

is impaired, leading to an increased plasma level of LDL. Most homozygotes die of coronary artery disease in childhood. The disease in heterozygotes (1 in 500 people) has a milder and more variable clinical course.

🩺 CLINICAL INSIGHT

Cycling of the LDL Receptor Is Regulated

PCSK9 (*p*roprotein *c*onvertase *s*ubtilisin/*k*exin type 9) is a protease, secreted by the liver, that plays a crucial role in the regulation of cycling of the LDL receptor. Despite the fact that PCSK9 is a protease, enzymatic activity of the protein is not required for cycling regulation. PCSK9 in the blood binds to a domain on the receptor that prevents the receptor from returning to the plasma membrane, and it is degraded in the lysosome along with its cargo.

Individuals having a mutation that reduces the amount of PCSK9 in the blood have greatly reduced levels of LDL in the blood and display an almost 90% reduction in the rate of cardiovascular disease. Presumably, reduced levels of PCSK9 allow more receptor cycling and more efficient removal of LDL from the blood. Much research is now being directed at inhibiting PCSK9 activity in individuals with high cholesterol levels.

🩺 CLINICAL INSIGHT

HDL Seems to Protect Against Atherosclerosis

Although the events that result in atherosclerosis take place rapidly in patients with familial hypercholesterolemia, a similar sequence of events take place in people who develop atherosclerosis over decades. In particular, the formation of foam cells and plaques are especially hazardous occurrences. HDL and its role in returning cholesterol to the liver are important in mitigating these life-threatening circumstances.

HDL has a number of antiatherogenic properties, including the inhibition of LDL oxidation, but the best-characterized property is the removal of cholesterol from cells, especially macrophages. Earlier, we learned that HDL retrieves cholesterol from other tissues in the body to return the cholesterol to the liver for excretion as bile or in the feces. This transport, called *reverse cholesterol transport,* is especially important in regard to macrophages. Indeed, when the transport fails, macrophages become foam cells and facilitate the formation of plaques. Macrophages that collect cholesterol from LDL normally transport the cholesterol to HDL particles. The more HDL, the more readily this transport takes place and the less likely that the macrophages will develop into foam cells. Presumably, this robust reverse cholesterol transport accounts for the observation that higher HDL levels confer protection against arthrosclerosis.

The importance of reverse cholesterol transport is illustrated by the occurrence of mutations that inactivate a cholesterol-transport proteins in endothelial cells and macrophages, ABCA1 (ATP-binding cassette transporter, subfamily A1) (Figure 12.17) Loss of activity of cholesterol-transport protein ABCA1 results in a very rare condition called *Tangier disease*, which is characterized by HDL deficiency, accumulation of cholesterol in macrophages, and premature atherosclerosis. Under normal conditions, the apoprotein component of HDL, apoA-I, binds to ABCA1 to facilitate LDL transport. Moreover, the interaction between apoA-I and ABCA1 initiates a signal transduction pathway in the endothelial cells that inhibits the inflammatory response. Another antiatherogenic property of HDL is due to the association of a serum esterase, *paraoxanase*, with HDL. Paraoxanase may destroy oxLDL, accounting for some of HDL's ability to protect against coronary disease.

Until recently, high levels of HDL-bound cholesterol ("good cholesterol") relative to LDL-bound cholesterol ("bad cholesterol") were believed to protect against cardiovascular disease. This belief was based on epidemiological studies. However, a number of recent clinical trials revealed that increased levels of HDL-bound cholesterol had no protective effects at all. These studies do not discount the protective effects of HDL alone and illustrate the danger of equating free HDL and cholesterol-bound HDL.

CLINICAL INSIGHT

The Clinical Management of Cholesterol Levels Can Be Understood at a Biochemical Level

Homozygous familial hypercholesterolemia can be treated only by a liver transplant. A more generally applicable therapy is available for heterozygotes and others with high levels of cholesterol. *The goal is to reduce the amount of cholesterol in the blood by stimulating the single normal gene to produce more than the customary number of LDL receptors.* We have already observed that the production of LDL receptors is controlled by the cell's need for cholesterol. Therefore, the strategy is to deprive the cell of ready sources of cholesterol. When cholesterol is required, the amount of mRNA for the LDL receptor rises and more receptor is found on the cell surface. This state can be induced by a two-pronged approach. First, the reabsorption of bile salts from the intestine is inhibited. Bile salts are cholesterol derivatives that promote the absorption of dietary cholesterol and dietary fats (p. 539). Second, de novo synthesis of cholesterol is blocked.

The reabsorption of bile is impeded by the oral administration of positively charged polymers, such as cholestyramine, that bind negatively charged bile salts and are not themselves absorbed. Cholesterol synthesis can be effectively blocked by a class of compounds called *statins*. A well-known example of such a compound is lovastatin, which is also called Mevacor (**Figure 29.18**). These compounds are potent competitive inhibitors of HMG-CoA reductase, the essential control point in the biosynthetic pathway. Plasma cholesterol levels decrease by 50% in many patients given both lovastatin and inhibitors of bile-salt reabsorption. Lovastatin and other inhibitors of HMG-CoA reductase are

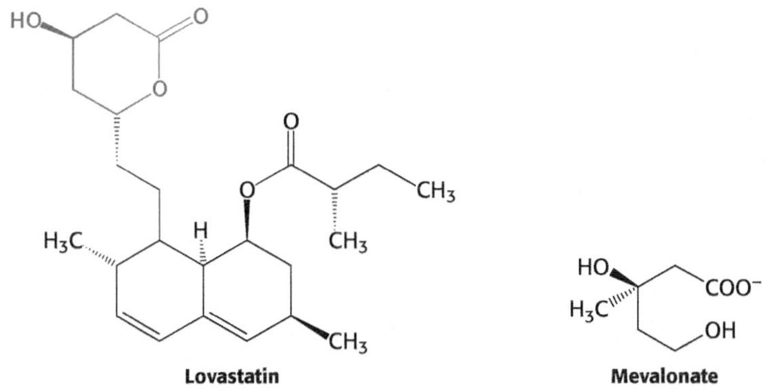

Figure 29.18 Lovastatin, a competitive inhibitor of HMG–CoA reductase. The part of the structure that resembles mevalonate is shown in red.

widely used to lower the plasma-cholesterol level in people who have atherosclerosis. Preliminary studies suggest that reducing levels of PCSK9 (p. 537) and HMG-CoA reductase activity may be an especially effective means of reducing cholesterol levels.

29.5 Cholesterol Is the Precursor of Steroid Hormones

Although cholesterol is well known in its own right as a contributor to the development of heart disease, metabolites of cholesterol—the steroid hormones—also frequently appear in the news. Indeed, steroid-hormone abuse seems to be as prominent in the sports pages as any athlete is. Cholesterol is also a precursor for two other important molecules: bile salts and vitamin D. We begin with a look at the bile salts, molecules crucial for the uptake of lipids in the diet.

🩺 CLINICAL INSIGHT

Bile Salts Facilitate Lipid Absorption

Bile salts, the major constituent of bile, are highly effective *detergents* because they contain both polar and nonpolar regions. They solubilize dietary lipids (p. 253) so that they are readily digested by lipases and absorbed by the intestine. Bile salts are synthesized in the liver from cholesterol, stored and concentrated in the gall bladder, and then released into the small intestine. Glycocholate is the major bile salt (**Figure 29.19**).

In addition to bile salts, bile is composed of cholesterol, phospholipids, and the breakdown products of heme. If too much cholesterol is present in the bile, it will precipitate to form gallbladder stones. These stones may block bile secretion and inflame the gall bladder, a condition called cholelithiasis. Symptoms include pain in the upper right abdomen and nausea. If need be, the gall bladder is removed, and bile flows from the liver through the bile duct directly into the intestine.

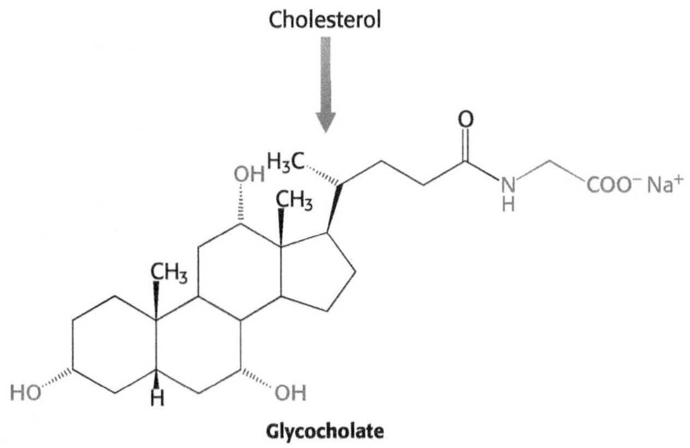

Figure 29.19 Bile–salt synthesis. The OH groups in red are added to cholesterol, as are the groups shown blue.

Steroid Hormones Are Crucial Signal Molecules

Cholesterol is the precursor of *steroid hormones* (**Figure 29.20**). These hormones are powerful signal molecules that regulate a host of organismal functions. There

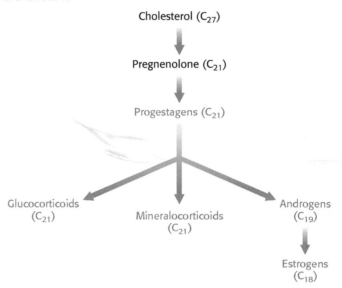

Figure 29.20 Biosynthetic relations of classes of steroid hormones and cholesterol.

are five major classes of steroid hormones: progestagens, glucocorticoids, mineralocorticoids, androgens, and estrogens.

Progesterone, a *progestagen,* prepares the lining of the uterus for the implantation of an ovum and is essential for the maintenance of pregnancy. *Androgens* are responsible for the development of male secondary sex characteristics. *Testosterone,* an important androgen, leads to the development of male sexual behavior. It is also important for the maintenance of the testes and the development of muscle mass. Owing to the latter activity, testosterone is referred to as an *anabolic steroid.* Testosterone is reduced by *5 α-reductase* to yield *dihydrotestosterone* (DHT), a powerful embryonic androgen that instigates the development and differentiation of the male phenotype:

Testosterone $\xrightarrow[\text{5α-Reductase}]{\text{NADPH + H}^+ \quad \text{NADP}^+}$

5α-Dihydrotestosterone

Estrogens such as *estradiol* are required for the development of female secondary sex characteristics. Estrogens, along with progesterone, also participate in the ovarian cycle. *Glucocorticoids* (such as *cortisol*) promote gluconeogenesis and the formation of glycogen, enhance the degradation of fat and protein, and inhibit the inflammatory response. They enable animals to respond to stress—indeed, the absence of glucocorticoids can be fatal. *Mineralocorticoids* (*primarily aldosterone*) act on the kidney to regulate salt balance, volume, and pressure of the blood.

All steroid hormones function in a similar manner. These powerful signal molecules bind to and activate receptor proteins that serve as transcription factors, proteins that regulate the expression of target genes (Chapter 37).

Vitamin D Is Derived from Cholesterol by the Energy of Sunlight

Cholesterol is also the precursor of vitamin D, which plays an essential role in the control of calcium and phosphorus metabolism. The ultraviolet light of sunlight breaks a bond in 7-dehydrocholesterol to form previtamin D_3, initiating the synthesis of *calcitriol* (1,25-dihydroxycholecalciferol), the active hormone (**Figure 29.21**). Although not a steroid, vitamin D is synthesized from cholesterol and acts in an analogous fashion to true steroids; thus, it is considered an "honorary steroid."

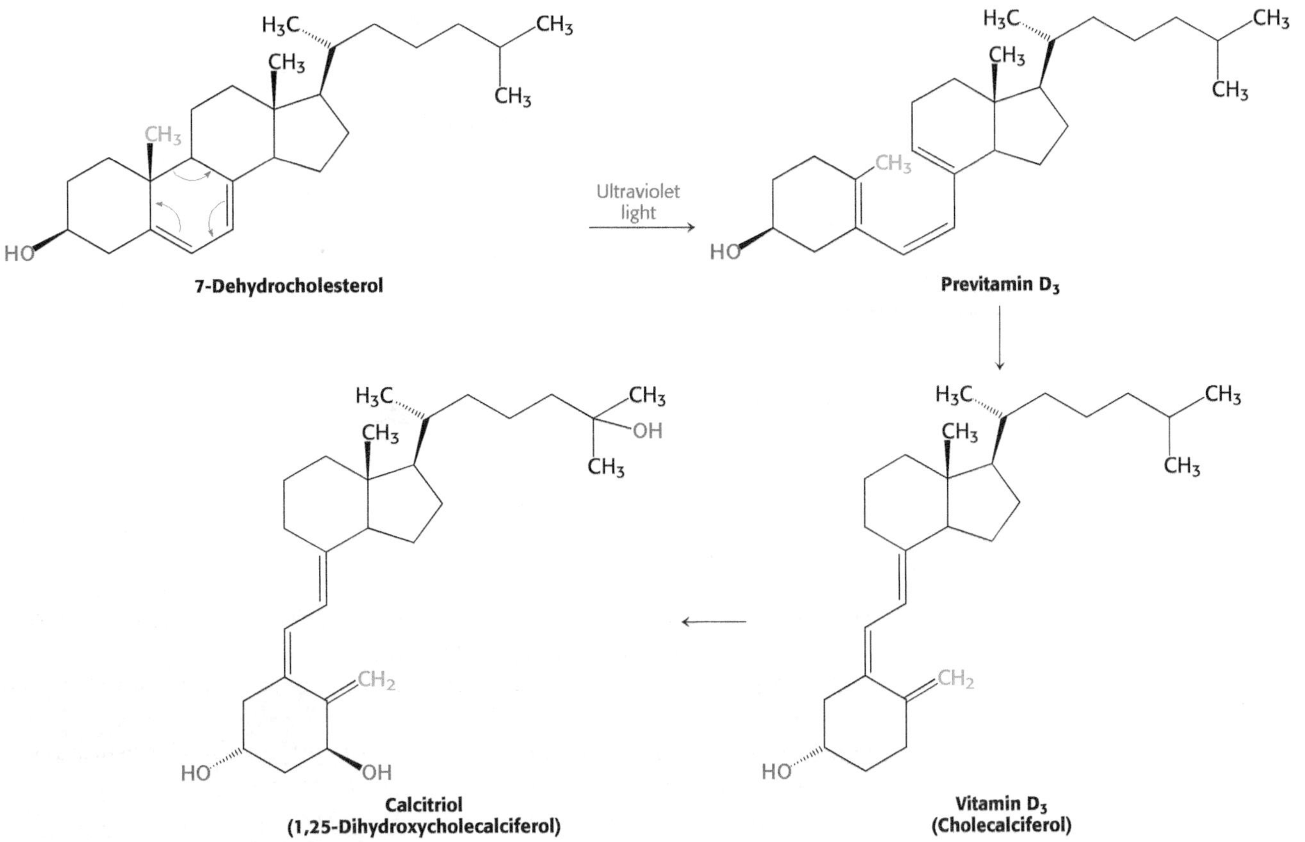

Figure 29.21 Vitamin D synthesis. The pathway for the conversion of 7-dehydrocholesterol into vitamin D$_3$ and then into calcitriol, the active hormone.

CLINICAL INSIGHT

Vitamin D Is Necessary for Bone Development

Vitamin D deficiency in childhood produces rickets, a disease characterized by inadequate calcification of cartilage and bone. Rickets was so common in seventeenth-century England that it was called the "children's disease of the English." The 7-dehydrocholesterol in the skin of these children was not cleaved to previtamin D$_3$, because there was little sunlight for many months of the year. Furthermore, their diets provided little vitamin D because most naturally occurring foods have a low content of this vitamin. Today, reliable dietary sources of vitamin D are fortified foods: milk, for example, is fortified to a level of 10 mg per quart. The recommended daily intake of vitamin D is 10 mg, irrespective of age. In adults, vitamin D deficiency leads to the softening and weakening of bones, a condition called *osteomalacia*. The occurrence of osteomalacia in Muslim women who are clothed so that only their eyes are exposed to sunlight is a striking reminder that vitamin D is needed by adults as well as by children.

Research done in the past few years indicates that vitamin D may play a much larger biochemical role than simply the regulation of bone metabolism. Muscle seems to be a target of vitamin D action. In muscle, vitamin D appears to affect a number of biochemical processes, with the net effect being enhanced muscle performance. Studies also suggest that vitamin D prevents cardiovascular disease, reduces the incidence of a variety of cancers, and protects against autoimmune diseases including diabetes. Moreover, vitamin D deficiency appears to be more common than thought. People living in northern climes may not be exposed to enough sunlight during certain times of the year to synthesize adequate amounts of vitamin D. Since skin pigmentation blocks UV radiation, people with dark complexions living in such areas may be especially susceptible to vitamin D deficiency. For example, in the United States, some studies suggest that 75% of Blacks and many Hispanics and Asians have insufficient blood levels of vitamin D. This recent research on vitamin D shows again the dynamic nature of biochemical investigations. Vitamin D, a chemical whose biochemical role was believed to be well established, now offers new frontiers of biomedical research.

🩺 CLINICAL INSIGHT

Androgens Can Be Used to Artificially Enhance Athletic Performance

Some athletes take androgens because the anabolic effects of androgens increase lean muscle mass (Figure 29.22). Synthetic steroids, such as dianabol, have been developed in an attempt to separate the anabolic properties (the growth of lean muscle mass) from the androgenic properties (the development of male sexual characteristics) of endogenous androgens.

However, to clearly separate these two activities has proved to be impossible. The use of anabolic steroids in males leads to a decrease in the secretion of testosterone, testicular atrophy, and, sometimes, breast enlargement (gynecomastia) owing to the conversion of some of the excess androgen into estrogen. In women, excess testosterone causes a decrease in ovulation and estrogen secretion, breast regression, and the growth of facial hair.

Figure 29.22 **Testosterone increases muscle mass.** In many sports, including body building, some athletes illegally use steroids to build muscle. [Brad Perks Lightscapes/Alamy.]

Androstendione
(a natural androgen)

Dianabol
(methandrostenolone)
(a synthetic androgen)

Oxygen Atoms Are Added to Steroids by Cytochrome P450 Monooxygenases

One of the chemical features that distinguish the various steroids from one another is the number and location of oxygen atoms. Carbon 3 is the only carbon atom bound to oxygen in cholesterol. All other oxygen atoms in steroid hormones are introduced by hydroxylation reactions that require *NADPH and O$_2$.* The enzymes catalyzing these reactions are called *monooxygenases* or *mixed-function oxygenases.*

$$RH + O_2 + NADPH + H^+ \longrightarrow ROH + H_2O + NADP^+$$

Hydroxylation requires the activation of oxygen. In the synthesis of steroid hormones, activation is accomplished by a *cytochrome P450 monooxygenase,* a family of membrane-anchored cytochromes, sometimes called the cytochrome P450 system, that contain a heme prosthetic group. The cytochrome P450 system also plays a role in the metabolism of drugs and other foreign substances, such as ibuprofen and caffeine.

Metabolism in Context: Ethanol Also Is Processed by the Cytochrome P450 System

In Chapter 28, we considered the consequences of excess ethanol consumption. Many of the deleterious biochemical effects stem from the overproduction of NADH by alcohol dehydrogenase and acetaldehyde dehydrogenase. Ethanol can also be processed by an ethanol-inducible cytochrome P450 system called the *microsomal ethanol-oxidizing system* (MEOS). This pathway generates acetaldehyde and, subsequently, acetate while oxidizing biosynthetic reducing power, NADPH, to NADP$^+$. Because it uses oxygen, this pathway generates free radicals that damage tissues. Moreover, because the system consumes NADPH, the antioxidant glutathione cannot be regenerated (p. 481), exacerbating the oxidative stress.

The adverse effects of ethanol are not limited to the metabolism of ethanol itself. Vitamin A (retinol) is converted into retinoic acid, an important signal molecule for growth and development in vertebrates, by the same dehydrogenases that metabolize ethanol. Consequently, this activation does not take place in the presence of ethanol, which acts as a competitive inhibitor. Moreover, the P450 enzymes induced by ethanol inactivate retinoic acid. These disruptions in the retinoic acid signaling pathway are believed to be responsible, at least in part, for fetal alcohol syndrome and for the development of a variety of cancers.

SUMMARY

29.1 Phosphatidate Is a Precursor of Storage Lipids and Many Membrane Lipids

Phosphatidate is formed by successive acylations of glycerol 3-phosphate by acyl CoA. The hydrolysis of its phosphoryl group followed by acylation yields a triacylglycerol. CDP-diacylglycerol, the activated intermediate in the de novo synthesis of several phospholipids, is formed from phosphatidate and CTP. The activated phosphatidyl unit is then transferred to the hydroxyl group of a polar alcohol, such as inositol, to form a phospholipid such as phosphatidylinositol. In mammals, phosphatidylethanolamine is formed by CDP-ethanolamine and diacylglycerol. Phosphatidylethanolamine is methylated by *S*-adenosylmethionine to form phosphatidylcholine. In mammals, this phospholipid can also be synthesized by a pathway that utilizes dietary choline. CDP-choline is the activated intermediate in this route.

Sphingolipids are synthesized from ceramide, which is formed by the acylation and reduction of dihydrosphingosine. Gangliosides are sphingolipids that contain an oligosaccharide unit having at least one residue of *N*-acetylneuraminate or a related sialic acid. They are synthesized by the step-by-step addition of activated sugars, such as UDP-glucose, to ceramide.

29.2 Cholesterol Is Synthesized from Acetyl Coenzyme A in Three Stages

Cholesterol is a steroid component of animal membranes and a precursor of steroid hormones. The committed step in its synthesis is the formation of mevalonate from 3-hydroxy-3-methylglutaryl CoA (derived from acetyl CoA and acetoacetyl CoA). Mevalonate is converted into isopentenyl pyrophosphate (C_5), which condenses with its isomer, dimethylallyl pyrophosphate (C_5), to form geranyl pyrophosphate (C_{10}). The addition of a second molecule of isopentenyl pyrophosphate yields farnesyl pyrophosphate (C_{15}), which condenses with itself to form squalene (C_{30}). This intermediate cyclizes to lanosterol (C_{30}), which is modified to yield cholesterol (C_{27}).

29.3 The Regulation of Cholesterol Synthesis Takes Place at Several Levels

In the liver, cholesterol synthesis is regulated by changes in the amount and activity of 3-hydroxy-3-methylglutaryl CoA reductase. Transcription of the gene, translation of the mRNA, and degradation of the enzyme are stringently controlled. In addition, the activity of the reductase is regulated by phosphorylation.

29.4 Lipoproteins Transport Cholesterol and Triacylglycerols Throughout the Organism

Cholesterol and other lipids in excess of those needed by the liver are exported in the form of very low density lipoprotein. After delivering its content of triacylglycerols to adipose tissue and other peripheral tissue, VLDL is converted into intermediate-density lipoprotein and then into low-density lipoprotein, both of which carry cholesteryl esters. Liver and peripheral-tissue cells take up LDL by receptor-mediated endocytosis. The LDL receptor, a protein spanning the plasma membrane of the target cell, binds LDL and mediates its entry into the cell. Absence of the LDL receptor in the homozygous form of familial hypercholesterolemia leads to a markedly elevated plasma level of LDL cholesterol and the deposition of cholesterol on blood-vessel walls, which, in turn, may result in childhood heart attacks. High-density lipoproteins transport cholesterol from the peripheral tissues to the liver.

29.5 Cholesterol Is the Precursor of Steroid Hormones

Five major classes of steroid hormones are derived from cholesterol: progestagens, glucocorticoids, mineralocorticoids, androgens, and estrogens. Hydroxylations by P450 monooxygenases that use NADPH and O_2 play an important role in the synthesis of steroid hormones and bile salts from cholesterol. Vitamin D, which is important in the control of calcium and phosphorus metabolism, is formed from a derivative of cholesterol by the action of light.

KEY TERMS

phosphatidate (p. 523)
triacylglycerol (p. 524)
phospholipid (p. 524)
cytidine diphosphodiacylglycerol
  (CDP-diacylglycerol) (p. 525)
sphingolipid (p. 526)
ceramide (p. 526)
sphingomyelin (p. 526)
cerebroside (p. 526)
ganglioside (p. 526)

phosphatidic acid phosphatase (PAP)
  (p. 529)
cholesterol (p. 529)
mevalonate (p. 530)
3-hydroxy-3-methylglutaryl CoA
  reductase (HMG-CoA reductase)
  (p. 530)
3-isopentenyl pyrophosphate (p. 530)
sterol regulatory element binding
  protein (SREBP) (p. 530)

lipoprotein particle (p. 534)
low-density lipoprotein (LDL) (p. 534)
high-density lipoprotein (HDL)
  (p. 535)
reverse cholesterol transport (p. 535)
receptor-mediated endocytosis (p. 535)
bile salt (p. 539)
steroid hormone (p. 539)
cytochrome P450 monooxygenase
  (p. 542)

? Answers to QUICK QUIZZES

1. Glycerol 3-phosphate is the foundation for both triacylglycerol synthesis and phospholipid synthesis. Glycerol 3-phosphate is acylated twice to form phosphatidate. In triacylglycerol synthesis, the phosphoryl group is removed from glycerol 3-phosphate to form diacylglycerol, which is then acylated to form triacylglycerol. In phospholipid synthesis, phosphatidate commonly reacts with CTP to form CDP-diacylglycerol, which then reacts with an alcohol to form a phospholipid. Alternatively, diacylglycerol may react with a CDP-alcohol to form a phospholipid.

2. The amount of reductase and its activity control the regulation of cholesterol biosynthesis. Transcriptional control is mediated by SREBP. Translation of the reductase mRNA also is controlled. The reductase itself may undergo regulated proteolysis. Finally, the activity of the reductase is inhibited by phosphorylation by AMP-activated kinase when ATP levels are low.

PROBLEMS

1. *Making fat.* Write a balanced equation for the synthesis of a triacylglycerol, starting from glycerol and fatty acids. ✓ 5

2. *Making a phospholipid.* Write a balanced equation for the synthesis of phosphatidylserine by the de novo pathway, starting from serine, glycerol, and fatty acids. ✓ 5

3. *Needed supplies.* How is the glycerol 3-phosphate required for phosphatidate synthesis generated? ✓ 5

4. *ATP needs.* How many high-phosphoryl-transfer-potential molecules are required to synthesize phosphatidylethanolamine from ethanolamine and diacylglycerol? Assume that the ethanolamine is the activated component. ✓ 5

5. *Identifying differences.* Differentiate among sphingomyelin, a cerebroside, and a ganglioside. ✓ 5

6. *Like Wilbur and Orville.* Match each term with its description. ✓ 5

(a) Phosphatidate _____
(b) Triacylglycerol _____
(c) Phospholipid _____
(d) Sphingolipid _____
(e) Cerebroside _____
(f) Ganglioside _____
(g) Cholesterol _____
(h) Mevalonate _____
(i) Lipoprotein particle _____
(j) Steroid hormone _____

1. Glycerol-based membrane lipid
2. Product of the committed step in cholesterol synthesis
3. Ceramide with either glucose or galactose attached
4. Storage form of fatty acids
5. Squalene is a precursor to this molecule
6. Transports cholesterol and lipids
7. Derived from cholesterol
8. Precursor to both phospholipids and triacylglycerols
9. Formed from ceramide by the attachment of phosphocholine
10. Ceramide with multiple carbohydrates attached

7. *Activated donors.* What is the activated reactant in each of the following biosyntheses? ✓ 5

(a) Phosphatidylinositol from inositol
(b) Phosphatidylethanolamine from ethanolamine
(c) Ceramide from sphingosine
(d) Sphingomyelin from ceramide
(e) Cerebroside from ceramide
(f) Farnesyl pyrophosphate from geranyl pyrophosphate

8. *Let's count the ways.* There may be 50 ways to leave your lover, but, in principle, there are only two ways to make a glycerol-based phospholipid. Describe the two pathways. ✓ 5

9. *The Decider.* What enzyme plays the key role in the regulation of lipid synthesis, and how is the regulation manifested? ✓ 5

10. *No DAG, no TAG.* What would be the effect of a mutation that decreased the activity of phosphatidic acid phosphatase? ✓ 5

11. *Turn 10 around.* What would be the effect of a mutation that increased the activity of phosphatidic acid phosphatase? ✓ 5

12. *The Law of Three Stages.* What are the three stages required for the synthesis of cholesterol? ✓ 6

13. *Telltale labels.* What is the distribution of isotopic labeling in cholesterol synthesized from each of the following precursors? ✓ 6

(a) Mevalonate labeled with ^{14}C in its carboxyl carbon atom
(b) Malonyl CoA labeled with ^{14}C in its carboxyl carbon atom

14. *Too much, too soon.* What is familial hypercholesterolemia, and what are its causes? ✓ 6

15. *Many regulations to follow.* Outline the mechanisms of the regulation of cholesterol biosynthesis. ✓ 6

16. *A good thing.* What are statins? What is their pharmacological function? ✓ 6

17. *Too much of a good thing.* Would the development of a "super statin" that inhibited all HMG-CoA reductase activity be a useful drug? Explain. ✓ 6

18. *Controlling agents.* What are the five major classes of steroid hormones?

19. *Developmental catastrophe.* Propecia (finasteride) is a synthetic steroid that functions as a competitive and specific inhibitor of 5α-reductase, the enzyme responsible for the synthesis of dihydrotestosterone from testosterone (p. 540).

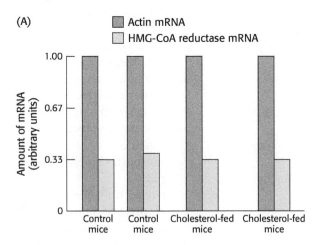

Finasteride

Propecia is now widely used to retard the development of male pattern baldness. Pregnant women are advised to avoid even handling this drug. Why is it vitally important that pregnant women avoid contact with Propecia?

20. *Breakfast conversation.* You and a friend are eating breakfast together. While eating, your friend is reading the back of her cereal box and comes across the following statement: "Cholesterol plays beneficial roles in your body, making cells, hormones, and tissues." Knowing that you are taking biochemistry, she asks if the statement makes sense. What do you reply? ✓ 6

21. *Let the sun shine in.* At a biochemical level, vitamin D functions like a steroid hormone. Therefore, it is sometimes referred to as an honorary steroid. Why is vitamin D not an actual steroid?

22. *A means of entry.* Describe the process of receptor-mediated endocytosis by using LDL as an example. ✓ 6

23. *Personalized medicine.* The cytochrome P450 system metabolizes many medicinally useful drugs. Although all human beings have the same number of P450 genes, individual polymorphisms exist that alter the specificity and efficiency of the proteins encoded by the genes. How could knowledge of individual polymorphisms be useful clinically?

24. *Honeybee crisis.* In 2006, honeybee colonies suddenly and unexplainably died off throughout the United States. The die-off was economically significant because one-third of the human diet comes from insect-pollinated plants and honeybees are responsible for 80% of the pollination. In October of 2006, the sequence of the honeybee genome was reported. Interestingly, the genome was found to contain far fewer cytochrome P450 genes than do the genomes of other insects. Suggest how the die-off and the paucity of P450 genes may be related.

Chapter Integration Problems

25. *Hold on tight or you might be thrown to the cytoplasm.* Many proteins are modified by the covalent attachment of a farnesyl (C_{15}) or a geranylgeranyl (C_{20}) unit to the carboxyl-terminal cysteine residue of the protein. Suggest why this modification might be the case.

26. *Similarities.* Compare the role of CTP in phosphoglyceride synthesis with the role of UTP in glycogen synthesis.

27. *ATP requirements.* Explain how cholesterol synthesis depends on the activity of ATP-citrate lyase.

28. *Fork in the road.* 3-Hydroxy-3-methylglutaryl CoA is on the pathway for cholesterol biosynthesis. It is also a component of another pathway. Name the pathway. What determines which pathway 3-hydroxy-3-methylglutaryl CoA follows? ✓ 6

29. *Drug resistance.* Dichlorodiphenyltrichloroethane (DDT) is a potent insecticide rarely used today because of its effects on other forms of life. In insects, DDT disrupts sodium channel function and leads to eventual death. Mosquitos have developed resistance to DDT and other insecticides that function in a similar fashion. Suggest two means by which DDT resistance might develop.

Data Interpretation Problem

30. *Cholesterol feeding.* Mice were divided into four groups, two of which were fed a normal diet and two of which were fed a cholesterol-rich diet. HMG-CoA reductase mRNA and protein from their livers were then isolated and quantified. Graph A shows the results of the mRNA isolation.

(a) What is the effect of cholesterol feeding on the amount of HMG-CoA reductase mRNA?

(b) What is the purpose of also isolating the mRNA for the protein actin, which is not under the control of the sterol response element?

HMG-CoA reductase protein was isolated by precipitation with a monoclonal antibody to HMG-CoA reductase. The amount of HMG-CoA protein in each group is shown in graph B.

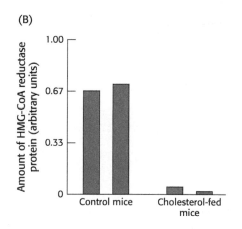

(B)

(c) What is the effect of the cholesterol diet on the amount of HMG-CoA reductase protein?

(d) Why is this result surprising in light of the results in graph A?

(e) Suggest possible explanations for the results shown in graph B. ✓ 6

Challenge Problems

31. *Familial hypercholesterolemia.* Several classes of LDL-receptor mutations have been identified as causes of familial hypercholesterolemia. Suppose that you have been given cells from patients with different mutations, an antibody specific for the LDL receptor that can be seen with an electron microscope, and access to an electron microscope.

What differences in antibody distribution might you expect to find in the cells from different patients? ✓ 6

32. *Inspiration for drug design.* Some actions of androgens are mediated by dihydrotestosterone, which is formed by the reduction of testosterone. This finishing touch is catalyzed by an NADPH-dependent 5α-reductase (p. 540). Chromosomal XY males with a genetic deficiency of this reductase are born with a male internal urogenital tract but predominantly female external genitalia. These people are usually reared as girls. At puberty, they masculinize because the testosterone level rises. The testes of these reductase-deficient men are normal, whereas their prostate glands remain small. How might this information be used to design a drug to treat *benign prostatic hypertrophy*, a common consequence of the normal aging process in men? A majority of men older than age 55 have some degree of prostatic enlargement, which often leads to urinary obstruction.

33. *Life-style consequences.* The cytochrome P450 system is a ubiquitous superfamily of monooxygenases that is present in plants, animals, and prokaryotes. The human genome encodes more than 50 members of the family, whereas the genome of the plant *Arabidopsis* encodes more than 250 members. These enzymes are responsible for the synthesis of toxins in plants. Why does the fact that plants have more of these enzymes make sense?

34. *Removal of odorants.* Many odorant molecules are highly hydrophobic and concentrate within the olfactory epithelium. They would give a persistent signal independent of their concentration in the environment if they were not rapidly modified. Propose a mechanism for converting hydrophobic odorants into water-soluble derivatives that can be rapidly eliminated.

Selected Readings for this chapter can be found online at www.whfreeman.com/tymoczko3e.

The Metabolism of Nitrogen–Containing Molecules

In this section, we examine the metabolism of nitrogen-containing compounds—specifically, amino acids and nucleoside bases. This examination will demonstrate, once again, how we are inextricably linked to our fellow inhabitants on planet Earth. As human beings, we are used to thinking of ourselves as the center of the universe, at least biologically speaking. Yet, we may have been, or should have been, humbled by the realization described in Section 10 that, without photosynthesis to introduce carbon atoms into biochemistry, we would quite literally not be here. It is humbling to realize that our lives are tied to the lives of such photosynthetic creatures as the majestic sequoias, luscious orchids, or even the ficus tree in the corner of the room, as well as to the ancestors of these living plants. However, carbon is not the only element that we depend on other organisms to provide. We, as well as all other organisms, need nitrogen. Almost every molecule that we have examined, with the exception of fuels, requires nitrogen. The primary source of this nitrogen is a special class of microorganisms—the *nitrogen-fixing bacteria*.

We begin our examination of the metabolism of nitrogen-containing molecules with amino acid degradation. Amino acids in excess of those needed for biosynthesis can be neither stored nor excreted. Rather, surplus amino acids are used as fuels. The α-amino group is first removed and converted into urea by the urea cycle and subsequently excreted. The resulting carbon skeletons are converted into metabolic intermediates, thus contributing to the generation of energy in the cell.

We then move on to amino acid synthesis. We will first see how the essentially invisible inhabitants of our world—the nitrogen-fixing prokaryotes—make life possible by removing N_2 from the atmosphere and reducing it to ammonia. We will see how the nitrogen is added to fundamental metabolites to yield amino acids. Next, we will examine how amino acids are used as the building blocks for a host of other important nitrogen-containing molecules. Finally, we will focus on the amino-acid-dependent synthesis of nucleotides, the precursors of the crucial information molecules DNA and RNA.

✓ **By the end of this section, you should be able to:**

✓ 1 Describe the fate of nitrogen that is removed when amino acids are used as fuels.

✓ 2 Explain how the carbon skeletons of the amino acids are metabolized after nitrogen removal.

✓ 3 Explain the centrality of nitrogen fixation to life, and describe how atmospheric nitrogen is converted into a biologically useful form of nitrogen.

✓ 4 Identify the sources of carbon atoms for amino acid synthesis.

✓ 5 Describe how nucleotides are synthesized.

✓ 6 Explain how nucleotide synthesis is regulated.

Amino Acid Degradation and the Urea Cycle

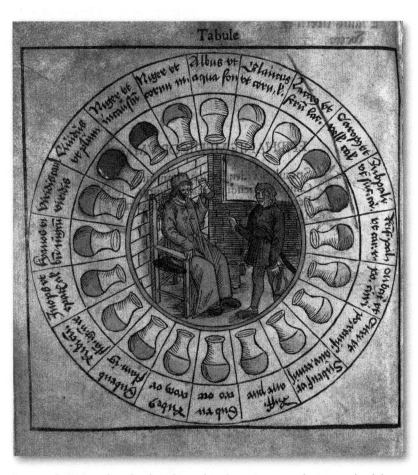

This fourteenth-century hand-colored woodcut from Germany depicts a wheel that classifies urine samples according to their color and consistency. In the middle of the wheel, a doctor inspects a patient's urine by sight, smell, and taste. The vials on the wheel aided physicians in diagnosing diseases. [Wellcome Library, London.]

Dietary protein is a vital source of amino acids. Proteins ingested in the diet are digested into amino acids or small peptides that can be absorbed by the intestine and transported in the blood (p. 250). Another source of amino acids is the degradation of defective or unneeded cellular proteins. Protein turnover—the degradation and resynthesis of proteins—takes place constantly in cells. Although some proteins, such as the proteins that make up the lens of the eye, are very stable, many proteins are short-lived, particularly those that are important in metabolic regulation. Altering the amounts of these proteins can rapidly change metabolic patterns. In addition, cells have mechanisms for detecting and removing damaged proteins. A significant proportion of newly synthesized protein molecules are defective because of errors in translation. Even proteins that are normal when first synthesized may undergo oxidative damage or be altered in other ways with the passage of time. What is the fate of the amino acids released when proteins are degraded? Amino acids are not stored like carbohydrates or fats, and so they must be processed biochemically. Excess amino acids are first used as building blocks for anabolic reactions, such as protein synthesis and nucleotide synthesis. If these needs are met, the amino acids are degraded and the carbon skeletons are used in catabolism or anabolism. What happens to the nitrogen of amino acids? Its safe removal is important because excess nitrogen in the form of ammonia is toxic.

In this chapter, we examine how nitrogen is removed from amino acids and subsequently eliminated as urea, the excreted form of NH_4^+ in most vertebrates. We then consider the fates of the carbon skeletons of the amino acids.

✓ 1 Describe the fate of nitrogen that is removed when amino acids are used as fuels.

30.1 Nitrogen Removal Is the First Step in the Degradation of Amino Acids

As already mentioned, amino acids are not stockpiled for use later. Any amino acids not needed as building blocks are degraded to various compounds, depending on the type of amino acid and the tissue from which it originates. The major site of amino acid degradation in mammals is the liver. The first step is the removal of nitrogen. The resulting α-ketoacids are then metabolized so that the carbon skeletons can enter the metabolic mainstream as precursors of glucose or of citric acid cycle intermediates. We will consider the fate of the α-amino group first, followed by that of the carbon skeleton.

Alpha-Amino Groups Are Converted into Ammonium Ions by the Oxidative Deamination of Glutamate

The α-amino group of many amino acids is transferred to α-ketoglutarate to form *glutamate*, which is then oxidatively deaminated to yield ammonium ion (NH_4^+):

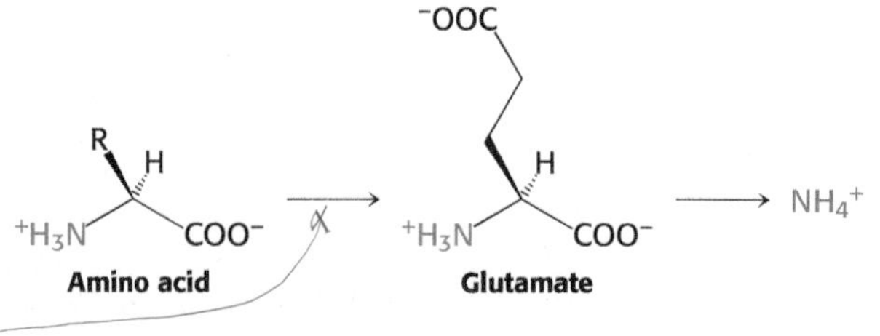

Aminotransferases catalyze the transfer of an α-amino group from an α-amino acid to an α-ketoacid. These enzymes, also called *transaminases*, generally funnel α-amino groups from a variety of amino acids to α-ketoglutarate for conversion into NH_4^+. Note that these very same transaminase enzymes can be used in the synthesis of amino acids.

Aspartate aminotransferase, one of the most important of these enzymes, catalyzes the transfer of the α-amino group of aspartate to α-ketoglutarate:

$$\text{Aspartate} + \alpha\text{-ketoglutarate} \rightleftharpoons \text{oxaloacetate} + \text{glutamate}$$

Alanine aminotransferase catalyzes the transfer of the amino group of alanine to α-ketoglutarate:

$$\text{Alanine} + \alpha\text{-ketoglutarate} \rightleftharpoons \text{pyruvate} + \text{glutamate}$$

The nitrogen atom that is transferred to α-ketoglutarate in the transamination reaction is converted into free ammonium ion by the oxidative deamination of glutamate, regenerating α-ketoglutarate:

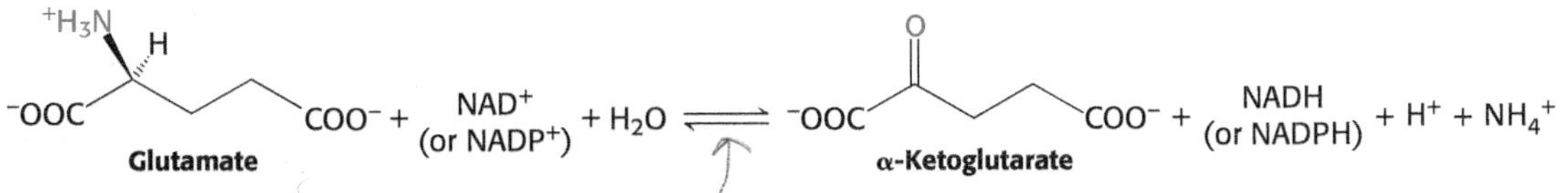

This reaction is catalyzed by *glutamate dehydrogenase*, an enzyme that we will encounter again in the context of the incorporation of ammonia into biomolecules (Chapter 31). This reaction equilibrium constant is close to 1 in the liver, so the direction of the reaction is determined by the concentrations of reactants and products. Normally, the reaction is driven forward by the rapid removal of ammonium ion. Glutamate dehydrogenase is unusual in that it can use either NADH or NADPH as reducing power. The dehydrogenase, a liver-specific enzyme, is located in mitochondria, as are some of the other enzymes required for the synthesis of urea. This compartmentalization sequesters free ammonia, which is highly toxic.

In mammals, but not in other organisms, glutamate dehydrogenase is allosterically inhibited by GTP, and stimulated by ADP. These nucleotides exert their regulatory effects in a unique manner. An *abortive complex* is formed on the enzyme when a product is replaced by substrate before the reaction is complete. For instance, the enzyme bound to glutamate and NAD(P)H is an abortive complex. GTP facilitates the formation of such complexes while ADP destabilizes the complexes.

The sum of the reactions catalyzed by aminotransferases and glutamate dehydrogenase is

$$\alpha\text{-Amino acid} + \begin{matrix} \text{NAD}^+ \\ (\text{or NADP}^+) \end{matrix} + H_2O \rightleftharpoons$$

$$\alpha\text{-ketoacid} + NH_4^+ + \begin{matrix} \text{NADH} \\ (\text{or NADPH}) \end{matrix} + H^+$$

CLINICAL INSIGHT

Blood Levels of Aminotransferases Serve a Diagnostic Function

The presence of alanine and aspartate aminotransferase in the blood is an indication of liver damage. Liver damage can occur for a number of reasons, including viral hepatitis, long-term excessive alcohol consumption, and reaction to drugs such as acetaminophen. Under these conditions, liver cell membranes are damaged and some cellular proteins, including the aminotransferases, leak into the blood. Normal blood values for alanine and aspartate aminotransferase activity are 5–30 units/l and 40–125 units/l, respectively. Depending on the extent of liver damage, the values will reach 200–300 units/l.

Serine and Threonine Can Be Directly Deaminated

Although the nitrogen atoms of most amino acids are transferred to α-ketoglutarate before their removal, the α-amino groups of serine and threonine can be directly converted into NH_4^+. These direct deaminations are catalyzed by *serine dehydratase* and *threonine dehydratase*, in which pyridoxal phosphate (PLP) is the prosthetic group.

$$\text{Serine} \longrightarrow \text{pyruvate} + NH_4^+$$

$$\text{Threonine} \longrightarrow \alpha\text{-ketobutyrate} + NH_4^+$$

In most terrestrial vertebrates, NH_4^+ is then converted into urea for excretion, a process that we will consider shortly:

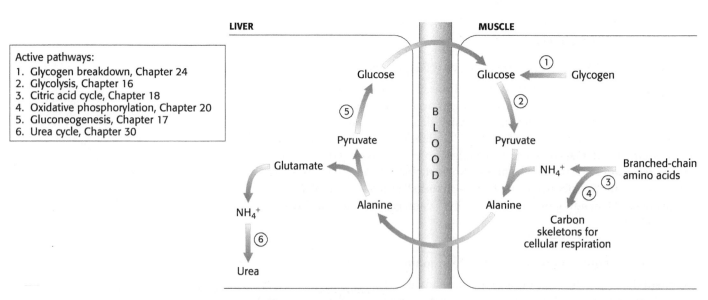

Peripheral Tissues Transport Nitrogen to the Liver

Although much amino acid degradation takes place in the liver, the liver cannot deaminate the branched-chain amino acids: leucine, valine, and isoleucine. Other tissues, most notably muscle, use the branched-chain amino acids as a source of fuel. How is the nitrogen processed in these other tissues? As in the liver, the first step is the removal of nitrogen from the amino acid. However, muscle lacks the enzymes of the urea cycle, the set of reactions that prepares nitrogen for excretion; so, in muscle, the nitrogen must be released in a form that can be absorbed by the liver and converted into urea.

Nitrogen is transported from muscle to the liver in two principal transport forms: alanine and glutamine. Glutamate is formed by transamination reactions, but the nitrogen is then transferred to pyruvate to form alanine, which is released into the blood (**Figure 30.1**). The liver takes up the alanine and converts it back into pyruvate by transamination. The pyruvate can be used for gluconeogenesis, and the amino group eventually appears as urea. This transport is referred to as the *glucose–alanine cycle*. It is reminiscent of the Cori cycle discussed earlier (Chapter 17). However, in contrast with the Cori cycle, pyruvate is not reduced to lactate, and thus more high-energy electrons are available for oxidative phosphorylation in muscle.

Figure 30.1 PATHWAY INTEGRATION: The glucose–alanine cycle. During prolonged exercise and fasting, muscle uses branched-chain amino acids as fuel. The nitrogen removed is transferred (through glutamate) to alanine, which is released into the bloodstream. In the liver, alanine is taken up and converted into pyruvate for the subsequent synthesis of glucose.

Glutamine is also a key transport form of nitrogen. *Glutamine synthetase* catalyzes the synthesis of glutamine from glutamate and NH_4^+ in an ATP-dependent reaction:

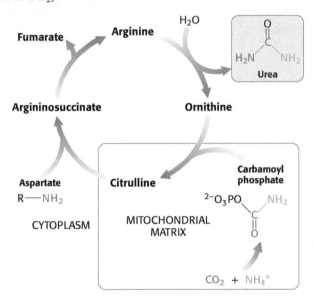

Glutamate + NH_4^+ → (Glutamine synthetase; ATP → ADP + P_i) → **Glutamine**

The nitrogen atoms of glutamine can be converted into urea in the liver.

QUICK QUIZ 1 How do aminotransferases and glutamate dehydrogenase cooperate in the metabolism of the α-amino group of amino acids?

30.2 Ammonium Ion Is Converted into Urea in Most Terrestrial Vertebrates

Some of the NH_4^+ formed in the breakdown of amino acids is consumed in the biosynthesis of nitrogen compounds. In most terrestrial vertebrates, the excess NH_4^+ is converted into urea by the *urea cycle* and then excreted (**Figure 30.2**). Such organisms are referred to as *ureotelic.* One of the nitrogen atoms of the urea is transferred from an amino acid, aspartate. The other nitrogen atom is derived directly from free NH_4^+, and the carbon atom comes from HCO_3^- (derived from the hydration of CO_2).

Figure 30.2 The urea cycle.

Fumarate **Arginine** H_2O

Urea: $H_2N-C(=O)-NH_2$ **Urea**

Argininosuccinate **Ornithine**

Aspartate **Citrulline** **Carbamoyl phosphate**

$R—NH_2$ $^{2-}O_3PO-C(NH_2)=O$

CYTOPLASM MITOCHONDRIAL MATRIX

CO_2 + NH_4^+

The urea cycle begins in mitochondria with the coupling of free NH_4^+ and HCO_3^- to form carbamoyl phosphate. This reaction, catalyzed by *carbamoyl phosphate synthetase I* (CPS I), is the committed reaction of the urea cycle. Although carbamoyl phosphate is a simple molecule, its energy-requiring biosynthesis is complex, requiring three steps:

$H_2N-C(=O)-NH_2$ **Urea**

Bicarbonate → (ATP → ADP) ① → **Carboxyphosphate** → (NH_3 → P_i) ② → **Carbamic acid** → (ATP → ADP) ③ → **Carbamoyl phosphate**

The reaction begins with the phosphorylation of HCO_3^- to form carboxyphosphate (1), which then reacts with ammonia to form carbamic acid (2). Finally, a second molecule of ATP phosphorylates carbamic acid to carbamoyl phosphate (3).

The consumption of two molecules of ATP makes this synthesis of carbamoyl phosphate irreversible. Note that NH_3, because it is a strong base, normally exists as NH_4^+ in aqueous solution. However, the carbamoyl phosphate synthetase deprotonates the ion and uses NH_3 as a substrate.

Carbamoyl Phosphate Synthetase Is the Key Regulatory Enzyme for Urea Synthesis

Carbamoyl phosphate synthetase is regulated allosterically so that it is maximally active when amino acids are being metabolized for fuel use. The allosteric regulator *N-acetylglutamate* (NAG) is required for synthetase activity. This molecule is synthesized by *N-acetylglutamate synthase*:

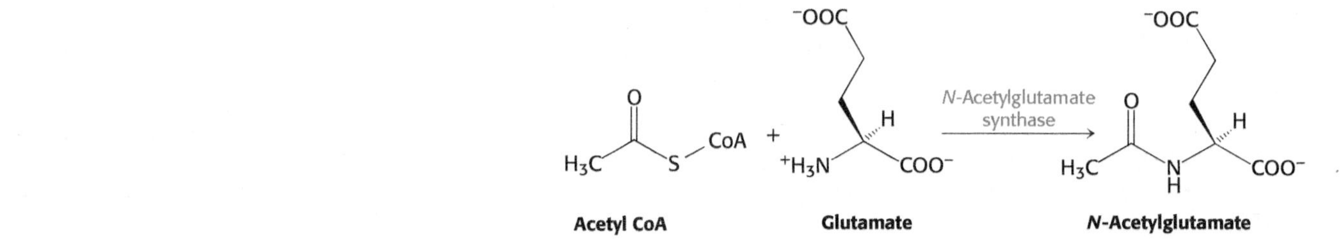

Acetyl CoA　　　**Glutamate**　　　***N*-Acetylglutamate**

N-acetylglutamate synthase is itself activated by arginine. Thus, NAG is synthesized when amino acids, as represented by arginine and glutamate, are readily available, and carbamoyl phosphate synthetase is then activated to process the generated ammonia.

Carbamoyl Phosphate Reacts with Ornithine to Begin the Urea Cycle

The carbamoyl group of carbamoyl phosphate is transferred to *ornithine* to form *citrulline*, in a reaction catalyzed by *ornithine transcarbamoylase*:

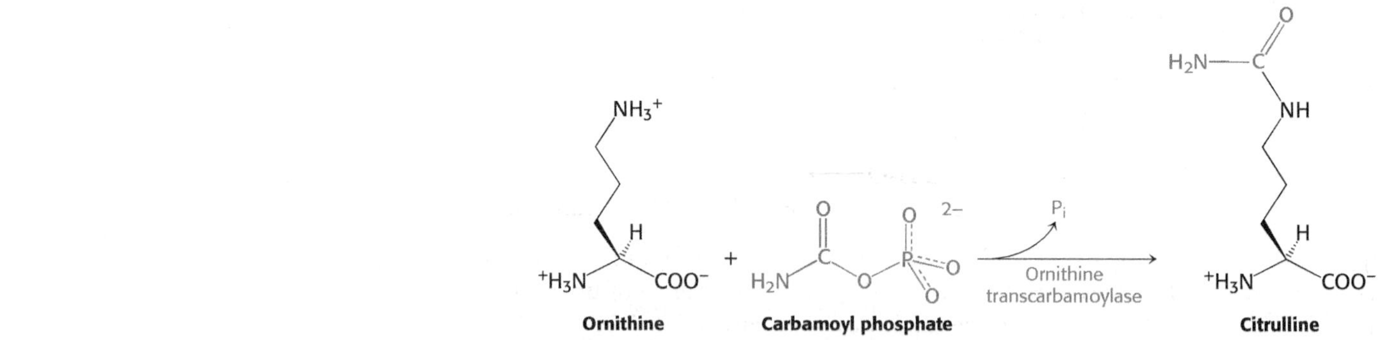

Ornithine　　　**Carbamoyl phosphate**　　　**Citrulline**

Ornithine and citrulline are amino acids, but they are not among the "alphabet" of 20 amino acids used as the building blocks of proteins. The formation of NH_4^+ by glutamate dehydrogenase, its incorporation into carbamoyl phosphate, and the subsequent synthesis of citrulline take place in the mitochondrial matrix. In contrast, the next three reactions of the urea cycle, which lead to the formation of urea, take place in the cytoplasm.

Citrulline, transported to the cytoplasm in exchange for ornithine, condenses with aspartate, the donor of the second amino group of urea. This synthesis of *argininosuccinate*, catalyzed by *argininosuccinate synthetase*, is driven by the cleavage of ATP into AMP and pyrophosphate and by the subsequent hydrolysis of pyrophosphate:

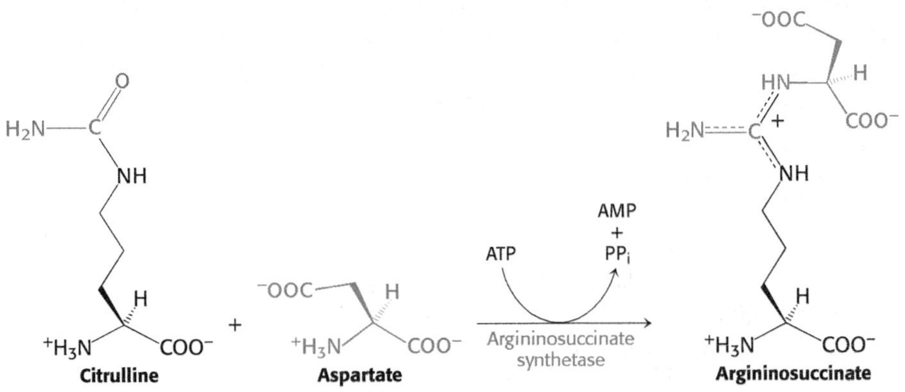

Citrulline + **Aspartate** →(ATP → AMP + PP$_i$, Argininosuccinate synthetase) **Argininosuccinate**

Argininosuccinase (also called argininosuccinate lyase) then cleaves argininosuccinate into *arginine* and *fumarate*. Thus, the carbon skeleton of aspartate is preserved in the form of fumarate:

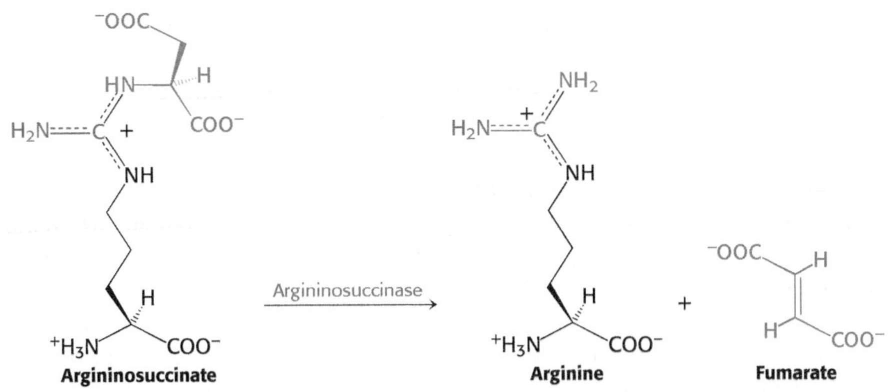

Argininosuccinate →(Argininosuccinase) **Arginine** + **Fumarate**

Finally, arginine is hydrolyzed to generate urea and ornithine in a reaction catalyzed by *arginase*. Ornithine is then transported back into a mitochondrion to begin another cycle. The urea is excreted. Indeed, human beings excrete about 10 kg (22 pounds) of urea per year.

The Urea Cycle Is Linked to Gluconeogenesis

The stoichiometry of urea synthesis is

$$CO_2 + NH_4^+ + 3\,ATP + \text{aspartate} + 2\,H_2O \longrightarrow$$
$$\text{urea} + 2\,ADP + 2\,P_i + AMP + PP_i + \text{fumarate}$$

Pyrophosphate is rapidly hydrolyzed, and so the equivalent of four molecules of ATP are consumed in these reactions to synthesize one molecule of urea. The synthesis of fumarate by the urea cycle is important because it is a precursor for glucose synthesis (**Figure 30.3**). Fumarate is hydrated to malate, an intermediate

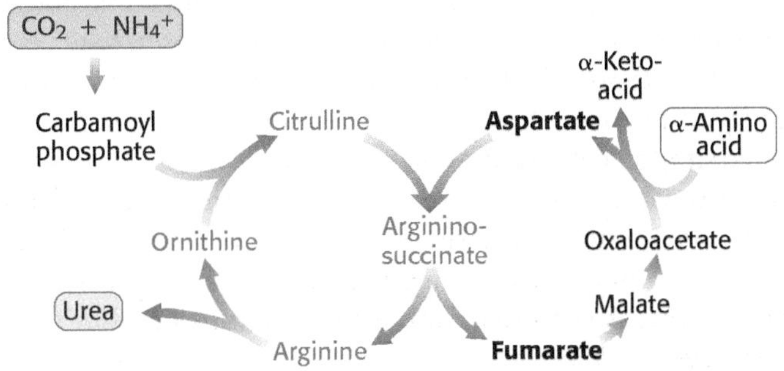

QUICK QUIZ 2 What are the immediate biochemical sources for the two nitrogen atoms in urea?

Figure 30.3 The metabolic context of nitrogen metabolism. The urea cycle, the citric acid cycle, and the transamination of oxaloacetate are linked by fumarate and aspartate.

of the citric acid cycle, which is in turn oxidized to oxaloacetate. Oxaloacetate can be converted into glucose by gluconeogenesis or transaminated to aspartate.

 CLINICAL INSIGHT

Metabolism in Context: Inherited Defects of the Urea Cycle Cause Hyperammonemia

The synthesis of urea in the liver is the major route of removal of NH_4^+. Urea cycle disorders occur with a prevalence of about 1 in 15,000 births. A blockage of carbamoyl phosphate synthesis or of any of the four steps of the urea cycle has devastating consequences because there is no alternative pathway for the synthesis of urea. *All defects in the urea cycle lead to an elevated level of NH_4^+ in the blood (hyperammonemia).* Some of these genetic defects become evident a day or two after birth, when the afflicted infant becomes lethargic and vomits periodically. Coma and irreversible brain damage may soon follow, a condition called *hepatic encephalopathy.*

Excessive alcohol consumption also can result in hyperammonemia. Earlier, we examined the effects of ethanol consumption on the liver (p. 517). Much of the damage is due to the excessive production of NADH. Liver damage from excessive ethanol consumption takes place in three stages. In stage one, a fatty liver develops. In stage two—alcoholic hepatitis—groups of cells die and inflammation results, an outcome that may be due to the overproduction of acetaldehyde and mitochondrial damage. This stage can itself be fatal. In stage three—cirrhosis—fibrous structure and scar tissue are produced around the dead cells (Figure 30.4). The cirrhotic liver is unable to convert ammonia into urea, and blood levels of ammonia rise. Ammonia is toxic to the nervous system and can cause coma and death. Cirrhosis of the liver arises in about 25% of alcoholics, and about 75% of all cases of liver cirrhosis are the result of alcoholism. Viral hepatitis is a nonalcoholic cause of liver cirrhosis.

Why are high levels of NH_4^+ toxic? The answer to this question is not yet known. Recent work, however, suggests that NH_4^+ may inappropriately activate a sodium–potassium–chloride cotransporter. This activation disrupts the osmotic balance of the nerve cell, causing swelling that damages the cell and results in neurological disorders.

Figure 30.4 Liver destruction. A healthy liver is shown at the left. The specimen in the middle shows excess fat accumulation due to the NADH glut caused by the metabolism of ethanol (p. 517). The liver on the right is a cirrhotic liver showing extensive damage, in part due to severe malfunction of the urea cycle. [Arthur Glauberman/Science Source. Enhancement by Mary Martin.]

🌸 BIOLOGICAL INSIGHT

Hibernation Presents Nitrogen Disposal Problems

A variety of animals hibernate and, during hibernation, biochemical pathways continue to function to keep the animal alive. Let's consider a hibernating bear, which may sleep for 3 to 5 months. During this time, the bear becomes a closed, self-sustained metabolic system: there is no carbon or water intake and no release of urine or digestive matter. Earlier, we considered the use of fat stores to keep the basal metabolism functioning, with ketone bodies serving as a fuel for the brain (p. 497). Protein turnover continues to take place in a hibernating bear's tissues; so how does the bear prevent the accumulation of dangerous levels of NH_4^+ if urea cannot be excreted as urine?

The nitrogen is still salvaged to produce urea, which is passed to the bladder. In the hibernating bear, the urea is then absorbed into the blood and released into the intestine. Bacteria in the intestine hydrolyze the urea, generating NH_4^+, which is used to synthesize amino acids and proteins. When the microorganisms die, the released amino acids are absorbed by the bear and used for biosynthesis. Obviously, both the bear and the bacteria benefit. The bear's urea is disposed of and returned as needed amino acids, and the bacteria have a nitrogen source for their biosynthesis. Interestingly, some of the bacteria in the bear can use the hydrolysis of urea to generate an electrochemical gradient that is used to synthesize ATP. The bear may be hibernating, but the bear and its inhabitants are a cauldron of biochemical activity.

BIOLOGICAL INSIGHT

Urea Is Not the Only Means of Disposing of Excess Nitrogen

As stated earlier, most terrestrial vertebrates are ureotelic; they excrete excess nitrogen as urea. However, urea is not the only excretable form of nitrogen. *Ammoniotelic organisms, such as aquatic vertebrates and invertebrates, release nitrogen as* NH_4^+ and rely on the aqueous environment to dilute this toxic substance. So, such organisms do not need to process nitrogen, as in the urea cycle, but can excrete it as it is removed from amino acids. Interestingly, normally ammoniotelic lungfish become ureotelic in times of drought, when they live out of the water.

Both ureotelic and ammoniotelic organisms depend on water, to varying degrees, for nitrogen excretion. *In contrast, uricotelic organisms, which secrete nitrogen as the purine uric acid, require little water.* The disposal of excess nitrogen as uric acid is especially valuable in animals, such as birds, that produce eggs having impermeable membranes that accumulate waste products. The pathway for nitrogen excretion clearly depends on the habitat of the organism.

30.3 Carbon Atoms of Degraded Amino Acids Emerge As Major Metabolic Intermediates

✓ 2 Explain how the carbon skeletons of the amino acids are metabolized after nitrogen removal.

We now turn to the fates of the carbon skeletons of amino acids after the removal of the α-amino group. *The strategy of amino acid degradation is to transform the carbon skeletons into major metabolic intermediates that can be converted into glucose or oxidized by the citric acid cycle.* The conversion pathways range from extremely simple to quite complex. The carbon skeletons of the diverse set of 20 fundamental amino acids are funneled into only seven molecules: *pyruvate, acetyl CoA, acetoacetyl CoA, α-ketoglutarate, succinyl CoA, fumarate,* and *oxaloacetate.* The conversion of 20 amino acid carbon skeletons into only seven molecules illustrates the remarkable economy of metabolic conversions, as well as the importance of certain metabolites.

Amino acids that are degraded to acetyl CoA or acetoacetyl CoA are termed *ketogenic amino acids* because they can give rise to ketone bodies or fatty acids but cannot be used to synthesize glucose. Recall that mammals lack a pathway for the net synthesis of glucose from acetyl CoA or acetoacetyl CoA (p. 498). Amino acids that are degraded to pyruvate, α-ketoglutarate, succinyl CoA, fumarate, or oxaloacetate are termed *glucogenic amino acids*. Oxaloacetate, generated from pyruvate and other citric acid cycle intermediates, and pyruvate can be converted into phosphoenolpyruvate and then into glucose through gluconeogenesis.

Of the basic set of 20 amino acids, only leucine and lysine are solely ketogenic (**Figure 30.5**). Threonine, isoleucine, phenylalanine, tryptophan, and tyrosine are both ketogenic and glucogenic. Some of their carbon atoms emerge in acetyl CoA or acetoacetyl CoA, whereas others appear in potential precursors of glucose. The other 14 amino acids are classified as solely glucogenic. We will identify the degradation pathways by the entry point into metabolism. Additionally,

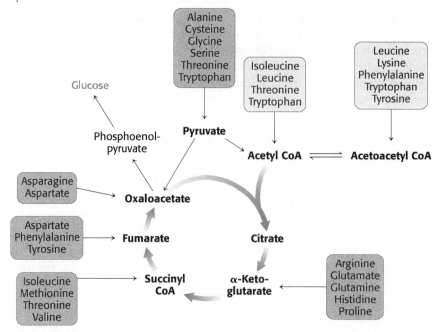

Figure 30.5 Fates of the carbon skeletons of amino acids. Glucogenic amino acids are shaded red, and ketogenic amino acids are shaded yellow. Several amino acids are both glucogenic and ketogenic.

because some degradation pathways are quite complex, they will be presented in outline form, with only some of the key intermediates highlighted.

Pyruvate Is a Point of Entry into Metabolism

Pyruvate is the entry point of the three-carbon amino acids—alanine, serine, and cysteine—into the metabolic mainstream (**Figure 30.6**). As stated earlier in the chapter, the transamination of alanine directly yields pyruvate:

$$\text{Alanine} + \alpha\text{-ketoglutarate} \rightleftharpoons \text{pyruvate} + \text{glutamate}$$

Another simple reaction in the degradation of amino acids is the *deamination of serine to pyruvate* by *serine dehydratase*:

$$\text{Serine} \rightarrow \text{pyruvate} + \text{NH}_4^+$$

The conversion of cysteine into pyruvate is more complex. The conversion can take place through several pathways, with cysteine's sulfur atom emerging in H_2S, SCN^-, or SO_3^{2-}.

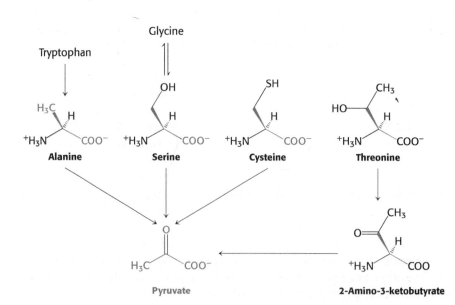

Figure 30.6 Pyruvate formation from amino acids. Pyruvate is the point of entry for alanine, serine, cysteine, glycine, threonine, and tryptophan.

The carbon atoms of three other amino acids also can be converted into pyruvate. *Glycine* can be converted into serine by the enzymatic addition of a hydroxymethyl group, or it can be cleaved to yield CO_2, NH_4^+, and the activated one-carbon unit tetrahydrofolate (Chapter 31). *Threonine* gives rise to 2-amino-3-ketobutyrate, which yields pyruvate or acetyl CoA. Three carbon atoms of *tryptophan* can emerge in alanine, which can be converted into pyruvate.

Oxaloacetate Is Another Point of Entry into Metabolism

Aspartate and asparagine are converted into oxaloacetate, a citric acid cycle intermediate. *Aspartate,* a four-carbon amino acid, is directly *transaminated to oxaloacetate*:

$$\text{Aspartate} + \alpha\text{-ketoglutarate} \rightleftharpoons \text{oxaloacetate} + \text{glutamate}$$

Asparagine is hydrolyzed by *asparaginase* to NH_4^+ and aspartate, which is then transaminated, as shown in the preceding reaction.

Recall that aspartate can also be converted into *fumarate* by the urea cycle (p. 557). Fumarate is also a point of entry for half the carbon atoms of tyrosine and phenylalanine, as will be discussed shortly.

Alpha-Ketoglutarate Is Yet Another Point of Entry into Metabolism

The carbon skeletons of several five-carbon amino acids enter the citric acid cycle at α-ketoglutarate. These amino acids are first converted into *glutamate,* which, as described earlier in the chapter (p. 552), is then oxidatively deaminated by glutamate dehydrogenase to yield α-ketoglutarate (Figure 30.7).

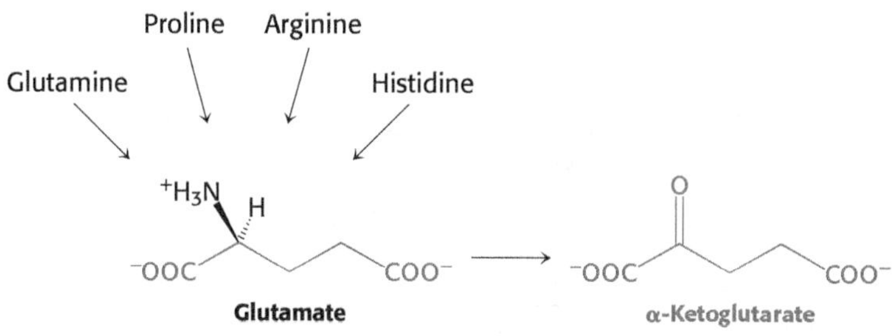

Figure 30.7 α–Ketoglutarate formation from amino acids. α–Ketoglutarate is the point of entry of several five-carbon amino acids that are first converted into glutamate.

Histidine is converted into glutamate through the intermediate 4-imidazolone 5-propionate (Figure 30.8). The amide bond in the ring of this intermediate is hydrolyzed to the *N*-formimino derivative of glutamate, which is then converted into glutamate by transfer of its formimino group to tetrahydrofolate.

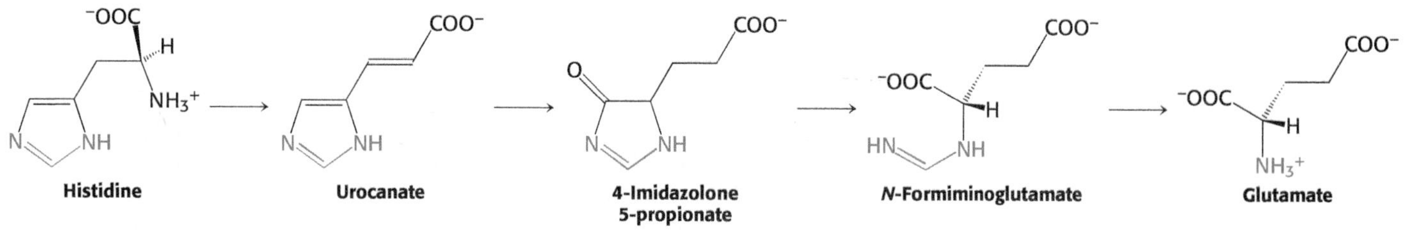

Figure 30.8 Histidine degradation. The conversion of histidine into glutamate.

Glutamine is hydrolyzed to glutamate and NH_4^+ by *glutaminase*. *Proline* and *arginine* are each converted into glutamate γ-semialdehyde, which is then oxidized to glutamate (Figure 30.9).

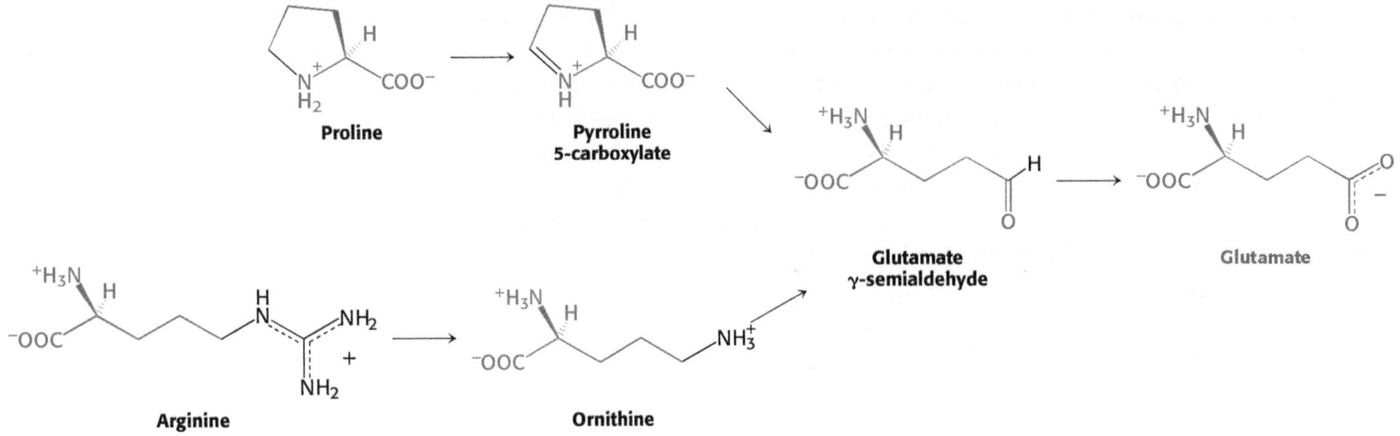

Figure 30.9 Glutamate formation. The conversion of proline and arginine into glutamate.

Succinyl Coenzyme A Is a Point of Entry for Several Nonpolar Amino Acids

Succinyl CoA is a point of entry for some of the carbon atoms of methionine, isoleucine, and valine. Propionyl CoA and then methylmalonyl CoA are intermediates in the breakdown of these three nonpolar amino acids (**Figure 30.10**). This pathway from propionyl CoA to succinyl CoA is also used in the oxidation of odd-chain fatty acids (p. 497). Because the conversion of methylmalonyl CoA into succinyl CoA is a vitamin B_{12}–dependent reaction, accumulation of methylmalonic acid is a clinical indicator of possible vitamin B_{12} deficiency.

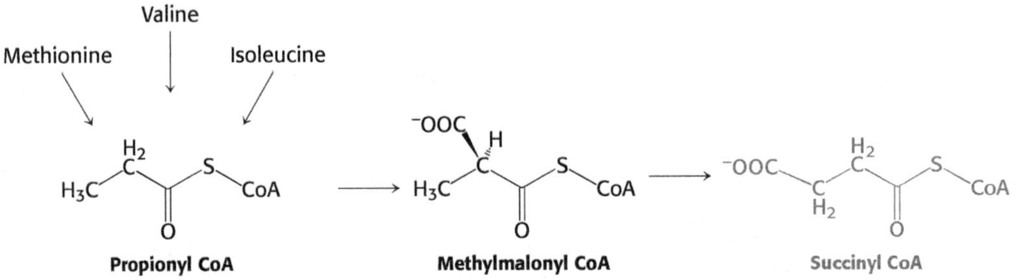

Figure 30.10 Succinyl coenzyme A formation. The conversion of methionine, isoleucine, and valine into succinyl CoA.

The Branched-Chain Amino Acids Yield Acetyl Coenzyme A, Acetoacetate, or Succinyl Coenzyme A

The degradation of the branched-chain amino acids (leucine, isoleucine, and valine) employs reactions that we have already encountered in the citric acid cycle and fatty acid oxidation. Leucine is transaminated to the corresponding α-ketoacid, α-ketoisocaproate. This α-ketoacid is oxidatively decarboxylated to *isovaleryl CoA* by the *branched-chain α-ketoacid dehydrogenase complex*. The α-ketoacids of valine and isoleucine, the other two branched-chain aliphatic amino acids, as well as α-ketobutyrate derived from methionine, also are substrates. The oxidative decarboxylation of these α-ketoacids is analogous to that of pyruvate to acetyl CoA (p. 332) and of α-ketoglutarate to succinyl CoA (p. 347).

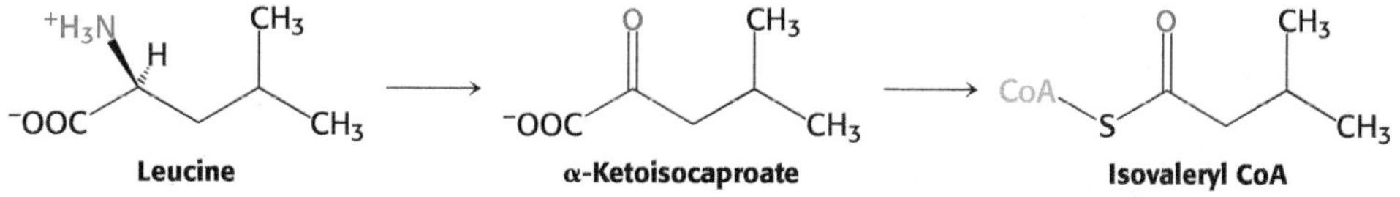

The isovaleryl CoA derived from leucine is dehydrogenated to yield *β-methylcrotonyl CoA*. The hydrogen acceptor is FAD, as in the analogous reaction in fatty acid oxidation (p. 493). *β-Methylglutaconyl CoA* is then formed by the *carboxylation* of *β*-methylcrotonyl CoA at the expense of the hydrolysis of a molecule of ATP in a reaction similar to that of pyruvate carboxylase and acetyl CoA carboxylase (pp. 316, 509):

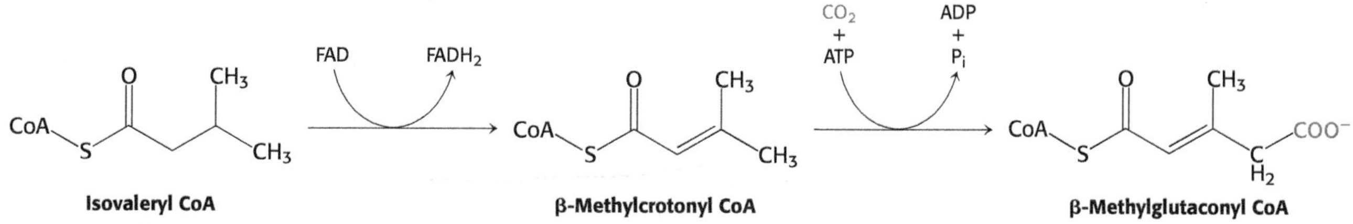

β-Methylglutaconyl CoA is hydrated to form *3-hydroxy-3-methylglutaryl CoA*, which is cleaved into *acetyl CoA* and *acetoacetate*. This reaction has already been discussed in regard to the formation of ketone bodies from fatty acids (p. 497).

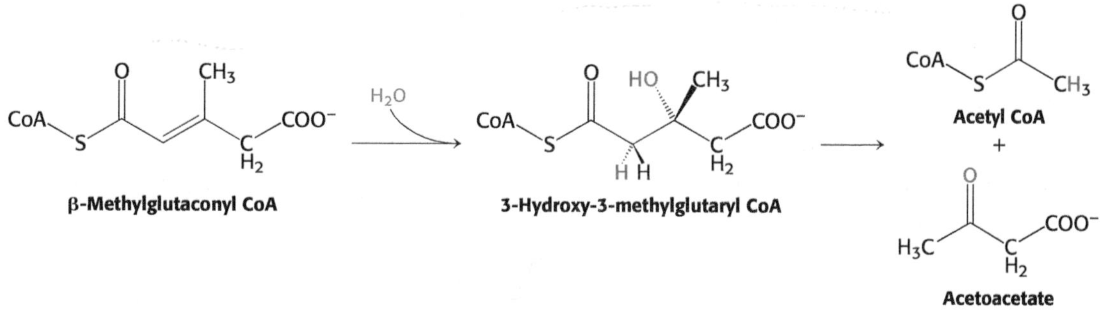

The degradative pathways of valine and isoleucine resemble that of leucine. After transamination and oxidative decarboxylation to yield a coenzyme A derivative, the subsequent reactions are like those of fatty acid oxidation. Isoleucine yields acetyl CoA and propionyl CoA, whereas valine yields CO_2 and propionyl CoA. Propionyl CoA can be converted into the citric acid cycle intermediate succinyl CoA. The degradation of leucine, valine, and isoleucine confirm a point made earlier (Chapter 15): the number of reactions in metabolism is large, but the number of *kinds* of reactions is relatively small. The degradation of leucine, valine, and isoleucine provides a striking illustration of the underlying simplicity and elegance of metabolism.

Oxygenases Are Required for the Degradation of Aromatic Amino Acids

The degradation of the aromatic amino acids is not as straightforward as that of the amino acids considered so far, although the final products—acetoacetate, fumarate, and pyruvate—are common intermediates. For the aromatic amino acids, *molecular oxygen is used to break the aromatic ring*.

The degradation of phenylalanine begins with its hydroxylation to tyrosine, a reversible reaction catalyzed by *phenylalanine hydroxylase*. This enzyme is called a *monooxygenase* or *mixed-function oxygenase* because one atom of the O_2 appears in the product and the other in H_2O (p. 542):

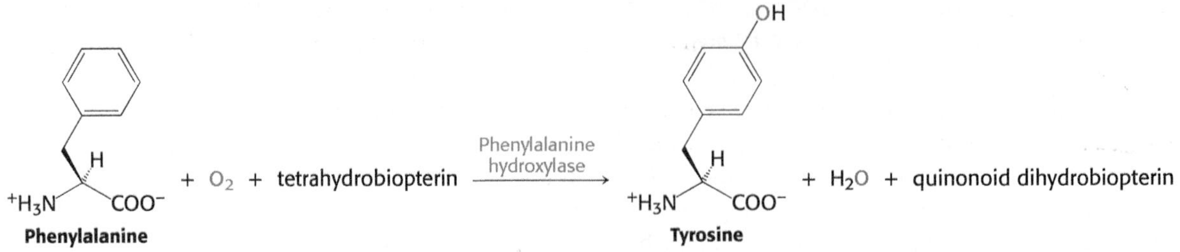

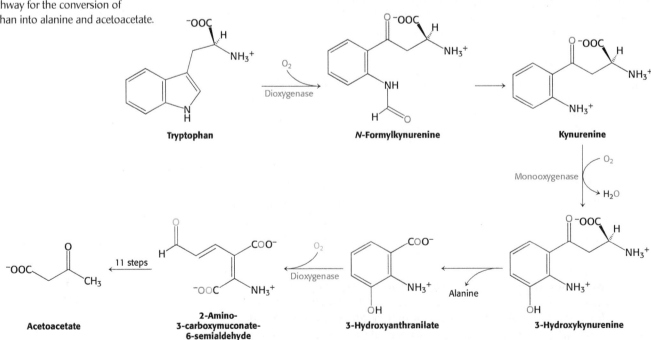

The reductant here is *tetrahydrobiopterin,* a cofactor of phenylalanine hydroxylase derived from the cofactor *biopterin.* NADH is used to regenerate tetrahydrobiopterin from the quinonoid form that was produced in the hydroxylation reaction. The sum of the reactions is

$$\text{Phenylalanine} + H_2O + NADH + H^+ \rightarrow \text{tyrosine} + NAD^+ + H_2O$$

Note that these reactions can also be used to synthesize tyrosine from phenylalanine.

The next step in the degradation of phenylalanine and tyrosine is the transamination of tyrosine to *p-hydroxyphenylpyruvate* (**Figure 30.11**). This α-ketoacid then reacts with O_2 to form *homogentisate*, to eventually yield *fumarate* and *acetoacetate*.

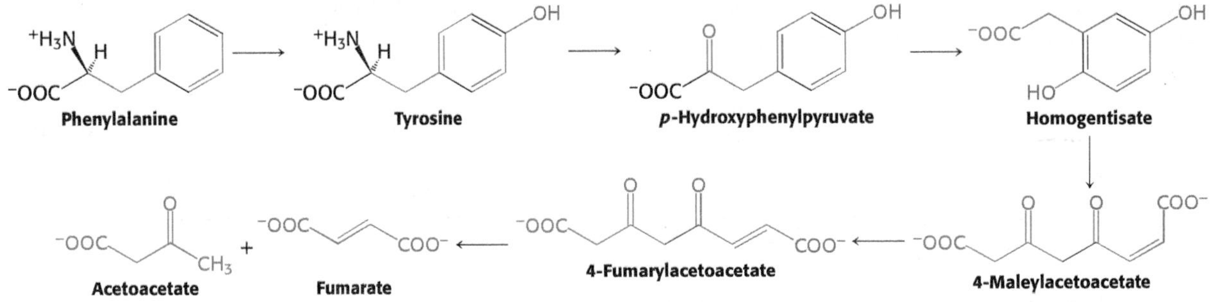

Figure 30.11 Phenylalanine and tyrosine degradation. The pathway for the conversion of phenylalanine into acetoacetate and fumarate.

Tryptophan degradation requires several oxygenases to cleave its two rings to ultimately yield acetoacetate (**Figure 30.12**). *Nearly all cleavages of aromatic rings in biological systems are catalyzed by dioxygenases*. In contrast with monooxygenases, dioxygenases incorporate both oxygen atoms into the reaction product.

Figure 30.12 Tryptophan degradation. The pathway for the conversion of tryptophan into alanine and acetoacetate.

Methionine Is Degraded into Succinyl Coenzyme A

Methionine is converted into succinyl CoA in nine steps (**Figure 30.13**). The first step is the adenylation of methionine to form *S*-adenosylmethionine (SAM). Methyl donation and deadenylation yield homocysteine, which is metabolized to propionyl CoA, which, in turn, is processed to succinyl CoA, as was described on page 562. Recall that *S*-adenosylmethionine plays a key role in cellular biochemistry as a common methyl donor in the cell (Chapter 29).

QUICK QUIZ 3 What are the common features of the breakdown products of the carbon skeletons of amino acids?

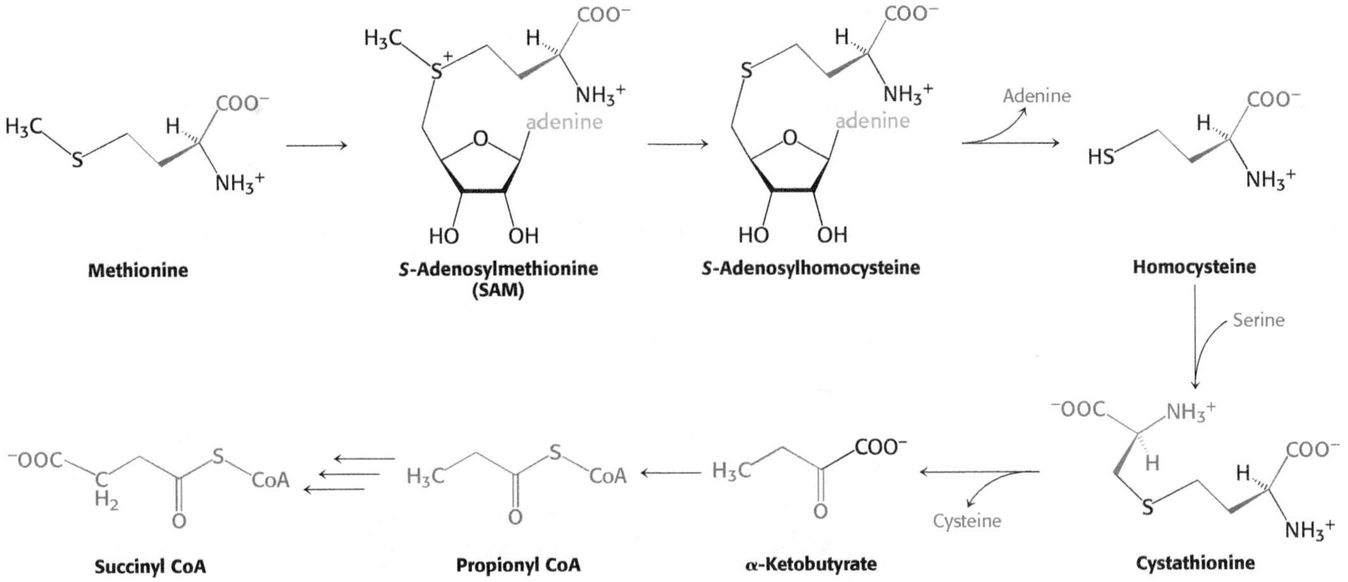

Figure 30.13 Methionine metabolism. The pathway for the conversion of methionine into succinyl CoA. *S*-Adenosylmethionine, formed along this pathway, is an important molecule for transferring methyl groups.

❊ CLINICAL INSIGHT

Inborn Errors of Metabolism Can Disrupt Amino Acid Degradation

Errors in amino acid metabolism provided some of the first correlations between biochemical defects and pathological conditions (**Table 30.1**). *Phenylketonuria*, which occurs with a prevalence of 1 in 10,000 births, is perhaps the best known of the diseases of amino acid metabolism.

Table 30.1 Inborn errors of amino acid metabolism

Disease	Enzyme deficiency	Symptoms
Citrullinema	Argininosuccinase	Lethargy, seizures, reduced muscle tension
Tyrosinemia	Various enzymes of tyrosine degradation	Weakness, liver damage, mental retardation
Albinism	Tyrosinase	Absence of pigmentation
Homocystinuria	Cystathionine β-synthase	Scoliosis, muscle weakness, mental retardation, thin blond hair
Hyperlysinemia	α-Aminoadipic semialdehyde dehydrogenase	Usually benign; in rare cases: seizures, mental retardation, lack of muscle tone, ataxia

Phenylketonuria is caused by an absence or deficiency of phenylalanine hydroxylase or, more rarely, of its tetrahydrobiopterin cofactor. Phenylalanine accumulates in all body fluids because it cannot be converted into tyrosine for complete degradation. Normally, three-quarters of phenylalanine molecules are

Phenylalanine

α-Ketoacid

α-Amino acid

Phenylpyruvate

converted into tyrosine, and the other quarter become incorporated into proteins. Because the major outflow pathway is blocked in phenylketonuria, the blood level of phenylalanine is typically at least 20-fold higher than in unaffected people. Minor fates of phenylalanine in unaffected people, such as the formation of phenylpyruvate, become major fates in phenylketonurics. Indeed, the initial description of phenylketonuria in 1934 was made by observing phenylpyruvate in the urine of phenylketonurics undergoing reaction with $FeCl_3$, the product of which turns the urine olive green.

Almost all untreated phenylketonurics are severely mentally retarded. The brain weight of these people is below normal, myelination of their nerves is defective, and their reflexes are hyperactive. The life expectancy of untreated phenylketonurics is drastically shortened. Half die by age 20 and, by age 30, three-quarters of the untreated phenylketonurics die.

Phenylketonurics appear normal at birth but are severely defective by age 1 if untreated. The therapy for phenylketonuria is a low-phenylalanine diet. The aim is to provide just enough phenylalanine to meet the needs for growth and replacement. Proteins that have a low content of phenylalanine, such as casein from milk, are hydrolyzed and phenylalanine is removed by adsorption. A low-phenylalanine diet must be started very soon after birth to prevent irreversible brain damage. In one study, the average IQ of phenylketonurics treated within a few weeks after birth was 93; a control group treated starting at age 1 had an average IQ of 53. Even as adults, phenylketonurics must restrict phenylalanine consumption (**Figure 30.14**).

Figure 30.14 Warning labels. Diet drinks in which aspartame, a phenylalanine-containing artificial sweetener, replaces sugar must have a warning on their containers that alerts phenylketonurics to the presence of phenylalanine in the drinks. [AP Photo/Rob Carr.]

Early diagnosis of phenylketonuria is essential and has been accomplished by mass screening programs. The phenylalanine level in the blood is the preferred diagnostic criterion because it is more sensitive and reliable than the $FeCl_3$ test. Prenatal diagnosis of phenylketonuria with DNA probes has become feasible because the gene has been cloned and the exact locations of many mutations have been discovered in the protein. Interestingly, whereas some mutations lower the activity of the enzyme, others decrease the enzyme concentration instead.

CLINICAL INSIGHT

Determining the Basis of the Neurological Symptoms of Phenylketonuria Is an Active Area of Research

The biochemical basis of retardation is not firmly established, but one hypothesis suggests that the lack of hydroxylase reduces the amount of tyrosine, an important precursor to neurotransmitters such as dopamine. Moreover, high concentrations of phenylalanine prevent amino acid transport of any tyrosine present as well as tryptophan, a precursor to the neurotransmitter serotonin, into the brain. Because all three of the amino acids are transported by the same carrier, phenylalanine will saturate the carrier, preventing access to tyrosine and tryptophan. Finally, high blood levels of phenylalanine result in higher levels of phenylalanine in the brain, and evidence suggests this elevated concentration inhibits glycolysis at pyruvate kinase, disrupts myelination of nerve fibers, and reduces the synthesis of several neurotransmitters.

SUMMARY

30.1   Nitrogen Removal Is the First Step in the Degradation of Amino Acids

Surplus amino acids are used as metabolic fuel. The first step in their degradation is the removal of their α-amino groups by transamination to α-ketoacids. The α-amino group funnels into α-ketoglutarate to form glutamate, which is then oxidatively deaminated by glutamate dehydrogenase to give NH_4^+ and α-ketoglutarate. NAD^+ or $NADP^+$ is the electron acceptor in this reaction.

30.2   Ammonium Ion Is Converted into Urea in Most Terrestrial Vertebrates

The first step in the synthesis of urea is the formation of carbamoyl phosphate, which is synthesized from HCO_3^-, NH_3, and two molecules of ATP by carbamoyl phosphate synthetase. Ornithine is then carbamoylated to citrulline by ornithine transcarbamoylase. These two reactions take place in mitochondria. Citrulline leaves the mitochondrion and condenses with aspartate to form argininosuccinate, which is cleaved into arginine and fumarate. The other nitrogen atom of urea comes from aspartate. Urea is formed by the hydrolysis of arginine, which also regenerates ornithine.

30.3   Carbon Atoms of Degraded Amino Acids Emerge As Major Metabolic Intermediates

The carbon atoms of degraded amino acids are converted into pyruvate, acetyl CoA, acetoacetate, or an intermediate of the citric acid cycle. Most amino acids are solely glucogenic, two are solely ketogenic, and a few are both ketogenic and glucogenic. Alanine, serine, cysteine, glycine, threonine, and tryptophan are degraded to pyruvate. Asparagine and aspartate are converted into oxaloacetate. α-Ketoglutarate is the point of entry for glutamate and four amino acids (glutamine, histidine, proline, and arginine) that can be converted into glutamate. Succinyl CoA is the point of entry for some of the carbon atoms of three amino acids (methionine, isoleucine, and valine) that are degraded through the intermediate methylmalonyl CoA. Leucine is degraded to acetoacetate and acetyl CoA. The breakdown of valine and isoleucine is like that of leucine. Their α-ketoacid derivatives are oxidatively decarboxylated by the branched-chain α-ketoacid dehydrogenase.

The rings of aromatic amino acids are degraded by oxygenases. Phenylalanine hydroxylase, a monooxygenase, uses tetrahydrobiopterin as the reductant. Four of the carbon atoms of phenylalanine and tyrosine are converted into fumarate, and four emerge in acetoacetate.

Errors in amino acid metabolism were sources of some of the first insights into the correlation between pathology and biochemistry. Phenylketonuria results from the accumulation of high levels of phenylalanine in the body fluids. By unknown mechanisms, this accumulation leads to mental retardation unless the afflicted are placed on low-phenylalanine diets immediately after birth.

KEY TERMS

aminotransferase (transaminase) (p. 552)
glutamate dehydrogenase (p. 553)
glucose–alanine cycle (p. 554)

urea cycle (p. 555)
carbamoyl phosphate synthetase I (CPS I) (p. 555)
N-acetylglutamate (p. 556)

ketogenic amino acid (p. 559)
glucogenic amino acid (p. 559)
phenylketonuria (p. 565)

Answers to QUICK QUIZZES

1. Aminotransferases transfer the α-amino group to α-ketoglutarate to form glutamate. Glutamate is oxidatively deaminated to form an ammonium ion.

2. Carbamoyl phosphate and aspartate.

3. They are either fuels for the citric acid cycle, components of the citric acid cycle, or molecules that can be converted into a fuel for the citric acid cycle in one step.

PROBLEMS

1. *Keto counterparts.* Name the α-ketoacid formed by the transamination of each of the following amino acids. ✓ 1

(a) Alanine   (d) Leucine
(b) Aspartate   (e) Phenylalanine
(c) Glutamate   (f) Tyrosine

2. *A versatile building block.* (a) Write a balanced equation for the conversion of aspartate into glucose through the intermediate oxaloacetate. Which coenzymes participate in this transformation? (b) Write a balanced equation for the conversion of aspartate into oxaloacetate through the intermediate fumarate. ✓ 2

3. *Not very discriminating.* Glutamate dehydrogenase is considered unusual in that it does not discriminate between NADH and NADPH, at least in some species. Explain why this failure to discriminate is unusual. ✓ 1

4. *Cooperation.* How do aminotransferases and glutamate dehydrogenase cooperate in the metabolism of the amino group of amino acids? ✓ 1

5. *Taking away the nitrogen.* What amino acids yield citric acid cycle components and glycolysis intermediates when deaminated? ✓ 1

6. *One reaction only.* What amino acids can be deaminated directly? ✓ 1

7. *Nitrogen sources.* What are the immediate biochemical sources for the two nitrogen atoms in urea? ✓ 1

8. *Counterparts.* Match the biochemical on the right with the property on the left. ✓ 1

(a) Formed from NH_4^+ _____   1. Aspartate

(b) Hydrolyzed to yield urea   2. Urea
_____

(c) A second source of nitrogen   3. Ornithine
_____

(d) Reacts with aspartate   4. Carbamoyl
_____   phosphate

(e) Cleavage yields fumarate   5. Arginine
_____

(f) Accepts the first nitrogen   6. Citrulline
_____

(g) Final product _____   7. Argininosuccinate

9. *Completing the cycle.* Four high-transfer-potential phosphoryl groups are consumed in the synthesis of urea according to the stoichiometry given on page 557. In this reaction, aspartate is converted into fumarate. Suppose that fumarate is converted back into aspartate. What is the resulting stoichiometry of urea synthesis? How many high-transfer-potential phosphoryl groups are spent? ✓ 1

10. *A good bet.* A friend bets you a bazillion dollars that you can't prove that the urea cycle is linked to the citric acid cycle and other metabolic pathways. Can you collect? ✓ 1

11. *A precise diagnosis.* The result of a reaction between an infant's urine and 2,4-dinitrophenylhydrazine is positive. 2,4-Dinitrophenylhydrazine reacts with α-ketoacids and suggests the presence of a high concentration of α-ketoacids. Further analysis shows abnormally high blood levels of pyruvate, α-ketoglutarate, and the α-ketoacids of valine, isoleucine, and leucine. Identify a likely molecular defect, and propose a definitive test of your diagnosis. ✓ 2

12. *Line up.* Identify structures A–D, and place them in the order that they appear in the urea cycle. ✓ 1

13. *Sweet hazard.* Why should phenylketonurics avoid the ingestion of aspartame, an artificial sweetener? (Hint: Aspartame is L-aspartyl-L-phenylalanine methyl ester.) ✓ 1

14. *Déjà vu.* N-Acetylglutamate is required as a cofactor in the synthesis of carbamoyl phosphate. How is *N*-acetylglutamate synthesized from glutamate? ✓ 1

15. *Precursors.* Differentiate between ketogenic amino acids and glucogenic amino acids. ✓ 2

16. *Supply lines.* The carbon skeletons of the 20 common amino acids can be degraded into a limited number of end products. What are the end products, and in what metabolic pathway are they commonly found? ✓ 2

17. *A sleight of hand.* The end products of tryptophan are acetyl CoA and acetoacetyl CoA, yet tryptophan is a gluconeogenic amino acid in animals. Explain.

18. *Negative nitrogen balance.* A deficiency of even one amino acid results in a negative nitrogen balance. In this state, more protein is degraded than is synthesized, and so more nitrogen is excreted than is ingested. Why would protein be degraded if one amino acid were missing?

19. *Argininosuccinic aciduria.* Argininosuccinic aciduria is a condition that results when the urea-cycle enzyme argininosuccinase is deficient. Argininosuccinate is present in the blood and urine. Suggest how this condition might be treated while still removing nitrogen from the body. ✓ 1

Chapter Integration Problems

20. *Multiple substrates.* In Chapter 8, we learned that there are two types of bisubstrate reactions, sequential and double-displacement. Which type characterizes the action of aminotransferases?

21. *Closely related.* Pyruvate dehydrogenase complex and α-ketoglutarate dehydrogenase complex are huge enzymes consisting of three discrete enzymatic activities. Which amino acids require a related enzyme complex for degradation, and what is the name of the enzyme? ✓ 2

22. *Ammonia toxicity.* Glutamate is an important neurotransmitter whose levels must be carefully regulated in the brain. Explain how a high concentration of ammonia might disrupt this regulation. How might a high concentration of ammonia alter the citric acid cycle? ✓ 1

23. *Damaged liver.* As stated in Chapter 28, excess alcoholic consumption can cause liver damage (cirrhosis). A consequence of liver damage is often ammonia poisoning. Explain. ✓ 1

Challenge Problems

24. *Inhibitor design.* Compound A has been synthesized as a potential inhibitor of a urea-cycle enzyme. ✓ 1

Compound A

Which enzyme might compound A inhibit?

25. *Fuel choice.* Within a few days after a fast begins, nitrogen excretion accelerates to a higher-than-normal level. After a few weeks, the rate of nitrogen excretion falls to a lower level and continues at this low rate. However, after the fat stores have been depleted, nitrogen excretion rises to a high level. ✓ 2

(a) What events trigger the initial surge of nitrogen excretion?

(b) Why does nitrogen excretion fall after several weeks of fasting?

(c) Explain the increase in nitrogen excretion when the lipid stores have been depleted.

26. *Isoleucine degradation.* Isoleucine is degraded to acetyl CoA and succinyl CoA. Suggest a plausible reaction sequence, on the basis of reactions discussed in the text, for this degradation pathway. ✓ 2

27. *Enough cycles to have a race.* The glucose–alanine cycle is reminiscent of the Cori cycle, but the glucose–alanine cycle can be said to be more energy efficient. Explain.

28. *A serious situation.* Pyruvate carboxylase deficiency is a fatal disorder. Patients with pyruvate carboxylase deficiency sometimes display some or all of the following symptoms: lactic acidosis, hyperammonemia (excess NH_4^+ in the blood), hypoglycemia, and demyelination of the regions of the brain due to insufficient lipid synthesis. Provide a possible biochemical rationale for each of these observations.

Selected Readings for this chapter can be found online at www.whfreeman.com/tymoczko3e.

CHAPTER 31

Amino Acid Synthesis

Nitrogen is a key component of amino acids. The atmosphere is rich in nitrogen gas (N_2), a very unreactive molecule. Certain organisms, such as bacteria that live in the root nodules of yellow clover, can convert nitrogen gas into ammonia, which can be used to synthesize, first, glutamate and, then, the other amino acids. [Hugh Spencer/Science Source.]

31.1 The Nitrogenase Complex Fixes Nitrogen

31.2 Amino Acids Are Made from Intermediates of Major Pathways

31.3 Feedback Inhibition Regulates Amino Acid Biosynthesis

The major source of nitrogen for the biosphere is the gaseous form of nitrogen, N_2, which makes up about 80% of Earth's atmosphere. However, this form of nitrogen is virtually unusable by most forms of life. Only a few prokaryotes, most notably the *nitrogen-fixing bacteria*, are able to convert N_2 gas into NH_3 (ammonia), a form of nitrogen that the rest of the biosphere can use. This process, called *nitrogen fixation*, is one of the most remarkable reactions in biochemistry.

What is so difficult about fixing nitrogen? The extremely strong $N \equiv N$ bond, which has a bond energy of 940 kJ mol^{-1} (225 kcal mol^{-1}), is highly resistant to chemical attack. Indeed, the eighteenth-century French chemist Antoine Lavoisier named nitrogen gas "azote," meaning "without life," because it is so unreactive. Nitrogen can, however, be reduced industrially. Industrial processing is in fact the source of the majority of nitrogen used for fertilizer. The industrial process for nitrogen fixation devised by Fritz Haber in 1910 is still being used today.

$$N_2 + 3\,H_2 \rightleftharpoons 2\,NH_3$$

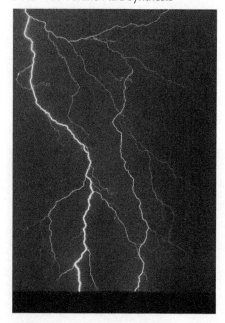

Figure 31.1 Lightning nitrogen fixation. [Royce Bair/Getty Images.]

The fixation of N_2 is typically carried out by mixing it with H_2 gas over an iron catalyst at about 500°C and a pressure of 300 atm, conditions with which even our hardiest microoganismic colleagues cannot cope. In fact, 25% of Earth's yearly fixed nitrogen is obtained by industrial processes. Lightning and ultraviolet radiation fix another 15% in the form of nitrogen oxides (**Figure 31.1**).

Interestingly, despite the severe conditions of the Haber process and of lightning strikes, the conversion of nitrogen and hydrogen to ammonia is actually thermodynamically favorable; the reaction is difficult kinetically because intermediates along the reaction pathway are unstable and rapidly decay to biochemically useless molecules. Some bacteria and archaea, *diazotrophic (nitrogen-fixing) microorganisms*, have ways of reducing the inert molecule $N \equiv N$ (nitrogen gas) to two molecules of ammonia under conditions compatible with life. An important diazotrophic microorganism is the symbiotic *Rhizobium* bacterium, which invades the roots of leguminous plants and forms root nodules in which they fix nitrogen, supplying both the bacteria and the plants. The amount of N_2 fixed by nitrogen-fixing microorganisms has been estimated to be 10^{11} kg per year, about 60% of Earth's newly fixed nitrogen.

In this chapter, we will first examine the process of nitrogen fixation and how the nitrogen is incorporated into glutamate. We will then examine how glutamate becomes a nitrogen source for the other amino acids as well as how the carbon skeletons for amino acid synthesis are derived.

✓ 3 Explain the centrality of nitrogen fixation to life, and describe how atmospheric nitrogen is converted into a biologically useful form of nitrogen.

31.1 The Nitrogenase Complex Fixes Nitrogen

To meet the kinetic challenge of nitrogen fixation, nitrogen-fixing organisms employ a complex enzyme with multiple oxidation–reduction centers that enable it to reduce the inactive N_2 gas. *The nitrogenase complex*, which carries out this fundamental transformation, consists of two proteins: a *reductase*, which provides electrons with high reducing power, and *nitrogenase*, which uses these electrons to reduce N_2 to NH_3. The transfer of electrons from the reductase to the nitrogenase component requires the hydrolysis of ATP by the reductase (**Figure 31.2**).

In principle, the reduction of N_2 to NH_3 is a six-electron process:

$$N_2 + 6\,e^- + 6\,H^+ \rightleftharpoons 2\,NH_3$$

However, the biological reaction always generates at least 1 mol of H_2 in addition to 2 mol of NH_3 for each mole of $N \equiv N$. Hence, an input of two additional electrons is required:

$$N_2 + 8\,e^- + 8\,H^+ \rightleftharpoons 2\,NH_3 + H_2$$

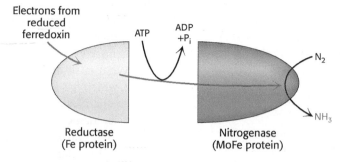

Figure 31.2 Nitrogen fixation. Electrons flow from ferredoxin to the reductase (iron protein, or Fe protein) to nitrogenase (molybdenum–iron protein, or MoFe protein) to reduce nitrogen to ammonia. ATP hydrolysis within the reductase drives conformational changes necessary for the efficient transfer of electrons.

In most nitrogen-fixing microorganisms, *the eight high-potential electrons come from reduced ferredoxin.* Two molecules of ATP are hydrolyzed for each electron transferred. Thus, *at least 16 molecules of ATP are hydrolyzed for each molecule of N_2 reduced*:

$$N_2 + 8\,e^- + 8\,H^+ + 16\,ATP + 16\,H_2O \rightleftharpoons$$
$$2\,NH_3 + H_2 + 16\,ADP + 16\,P_i$$

The product of the reduction, NH_3, is a base in aqueous solutions, attracting a proton to form NH_4^+.

Note that O_2 is required for oxidative phosphorylation to generate the ATP necessary for nitrogen fixation. However, the nitrogenase complex is exquisitely sensitive to inactivation by O_2. To allow ATP synthesis and nitrogenase to function simultaneously, leguminous plants maintain a very low concentration of free O_2 in their root nodules, the location of the nitrogenase. This is accomplished by binding O_2 to *leghemoglobin,* a homolog of hemoglobin.

The Molybdenum–Iron Cofactor of Nitrogenase Binds and Reduces Atmospheric Nitrogen

Both the reductase and the nitrogenase components of the complex are *iron–sulfur proteins,* a type of electron carrier that we have seen many times in our study of biochemistry—for instance, in the electron-transport chains of oxidative phosphorylation and photosynthesis. The *reductase* (also called the *iron protein* or the *Fe protein*) is a dimer of identical 30-kDa subunits bridged by a 4Fe-4S cluster (**Figure 31.3**). The role of the reductase is to transfer electrons from a suitable donor, such as reduced ferredoxin, to the nitrogenase component.

The nitrogenase component, an $\alpha_2\beta_2$ tetramer (240 kDa), requires the *FeMo cofactor,* which consists of $[Fe_4\text{-}S_3]$ and $[Mo\text{-}Fe_3\text{-}S_3]$ subclusters joined by three disulfide bonds. A carbon atom (the interstitial carbon) sits at the interstices of the iron atoms of the FeMo cofactor. The FeMo cofactor is also coordinated to a homocitrate moiety and to the α subunit through one histidine residue and one cysteinate residue (**Figure 31.4**).

Electrons flow from the reductase to the nitrogenase at a part of the nitrogenase called the P cluster, an iron- and sulfur-rich site. From there, the electrons progress to the FeMo cofactor, the redox center in the nitrogenase. Because molybdenum is present in this cluster, the nitrogenase component is also called the *molybdenum–iron protein* or *MoFe protein. The MoFe cofactor is the site of nitrogen fixation—the conversion of N_2 into ammonia.*

Ammonium Ion Is Incorporated into an Amino Acid Through Glutamate and Glutamine

The next task in the assimilation of nitrogen into biomolecules is to incorporate ammonium ion (NH_4^+) into the biochemically versatile amino acids. α-Ketoglutarate and glutamate play pivotal roles as the acceptor of ammonium ion, forming glutamate and glutamine, respectively. Glutamate is

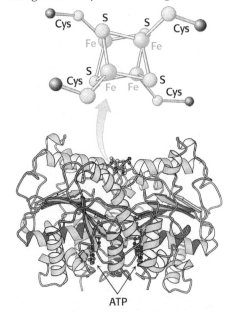

Figure 31.3 The Fe protein. This protein is a dimer composed of two polypeptide chains linked by a 4Fe–4S cluster. [Drawn from 1N2C.pdb.]

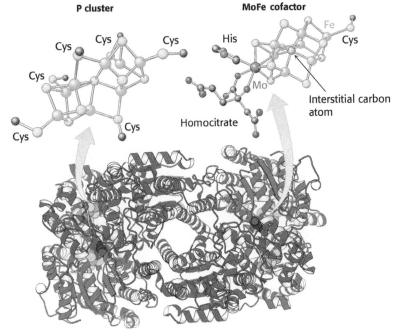

Figure 31.4 The MoFe protein. This protein is a heterotetramer composed of two α subunits (red) and two β subunits (blue). Notice that the protein contains two copies each of two types of clusters: P clusters and MoFe cofactors. Each P cluster contains eight iron atoms (green) and seven sulfides linked to the protein by six cysteinate residues. Each MoFe cofactor contains one molybdenum atom, seven iron atoms, nine sulfides, a central atom, and a homocitrate and is linked to the protein by one cysteinate residue and one histidine residue. [Drawn from 1M1N.pdb.]

synthesized from NH_4^+ and α-ketoglutarate, a citric acid cycle intermediate, by the action of *glutamate dehydrogenase*:

$$NH_4^+ + {}^-OOC\text{—}\cdots\text{—}C(=O)\text{—}COO^- + NAD(P)H + H^+ \rightleftharpoons {}^-OOC\text{—}\cdots\text{—}C(^+H_3N)(H)\text{—}COO^- + NAD(P)^+ + H_2O$$

α-Ketoglutarate   **Glutamate**

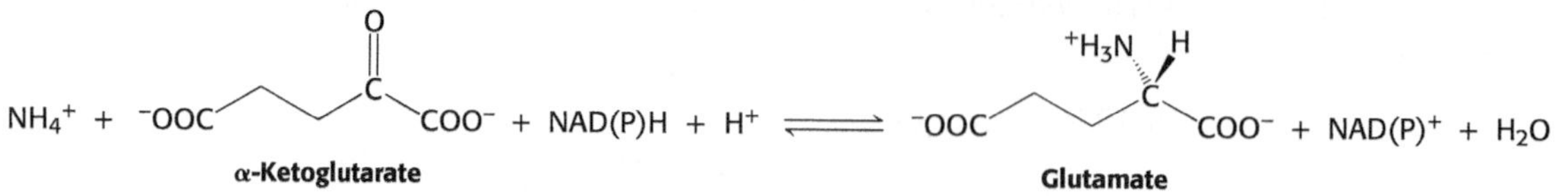

Recall that glutamate dehydrogenase also plays a key role in the removal of nitrogen from biological systems (p. 552).

A second ammonium ion is incorporated into glutamate to form glutamine by the action of *glutamine synthetase*. This amidation is driven by the hydrolysis of ATP:

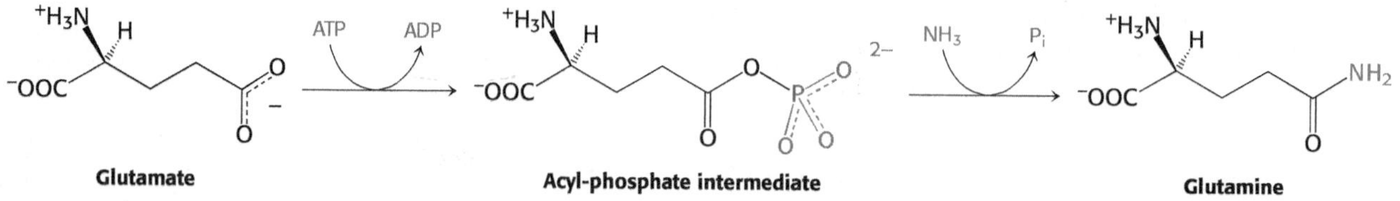

Glutamate   **Acyl-phosphate intermediate**   **Glutamine**

(ATP → ADP)   (NH_3 → P_i)

QUICK QUIZ 1 Trace the flow of nitrogen from atmospheric N_2 to glutamine.

ATP participates directly in the reaction by phosphorylating the side chain of glutamate to form an acyl-phosphate intermediate, which then reacts with ammonia to form glutamine.

Glutamate subsequently donates its α-amino group to various ketoacids by transamination reactions to form most of the amino acids. Glutamine, the other major nitrogen donor, contributes its side-chain nitrogen atom in the biosynthesis of a wide range of important compounds.

✓ **4** Identify the sources of carbon atoms for amino acid synthesis.

31.2 Amino Acids Are Made from Intermediates of Major Pathways

Thus far, we have considered the conversion of N_2 into NH_4^+ and the assimilation of NH_4^+ into glutamate and glutamine. We now turn to the biosynthesis of the other amino acids. The pathways for the biosynthesis of amino acids are diverse. However, they have an important common feature: *their carbon skeletons come from only a few sources: intermediates of glycolysis, the citric acid cycle, or the pentose phosphate pathway*. On the basis of these starting materials, amino acids can be grouped into six biosynthetic families (**Figure 31.5**).

Human Beings Can Synthesize Some Amino Acids but Must Obtain Others from the Diet

Most microorganisms such as *E. coli* can synthesize the entire basic set of 20 amino acids, but human beings can make only 11 of them. The amino acids that *must* be supplied in the diet are called *essential amino acids*, whereas the others, which can be synthesized if dietary content is insufficient, are termed *nonessential amino acids* (Table 3.2). A deficiency of even one amino acid compromises the ability of an organism to synthesize all of the proteins required for life.

The nonessential amino acids are synthesized by quite simple reactions, whereas the pathways for the formation of the essential amino acids are quite complex. For example, the nonessential amino acids *alanine* and *aspartate* are synthesized in a single step from pyruvate and oxaloacetate, respectively. In contrast, the pathways for the essential amino acids require from 5 to 16 steps (**Figure 31.6**). The sole exception to this pattern is arginine, inasmuch as the synthesis of this nonessential amino acid de novo requires 10 steps. Typically, though, arginine is made in only 3 steps from ornithine as part of the urea cycle

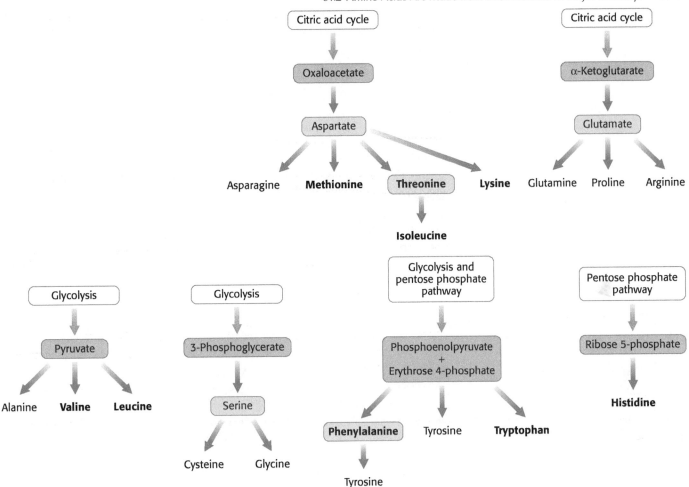

Figure 31.5 Biosynthetic families of amino acids in bacteria and plants. Major metabolic precursors are shaded blue. Amino acids that give rise to other amino acids are shaded yellow. Essential amino acids are in boldface type.

(p. 555). Tyrosine, classified as a nonessential amino acid because it can be synthesized in 1 step from phenylalanine, requires 10 steps to be synthesized from scratch and is essential if phenylalanine is not abundant. We will examine some important features of amino acid synthesis.

Some Amino Acids Can Be Made by Simple Transamination Reactions

Three α-ketoacids—α-ketoglutarate, oxaloacetate, and pyruvate—can be converted into amino acids in one step through the addition of an amino group. We have seen that α-ketoglutarate can be converted into glutamate by reductive amination (p. 574). The amino group from glutamate can be transferred to other α-ketoacids by *transamination reactions*. Thus, aspartate and alanine can be made from the addition of an amino group to oxaloacetate and pyruvate, respectively:

$$\text{Oxaloacetate} + \text{glutamate} \rightleftharpoons \text{aspartate} + \alpha\text{-ketoglutarate}$$

$$\text{Pyruvate} + \text{glutamate} \rightleftharpoons \text{alanine} + \alpha\text{-ketoglutarate}$$

Transaminations are carried out by *pyridoxal phosphate-dependent aminotransferases*. Transamination reactions participate in the synthesis of most amino acids. All aminotransferases contain the prosthetic group *pyridoxal phosphate* (PLP), which is derived from *pyridoxine* (vitamin B$_6$). In transamination, pyridoxal phosphate accepts an amino group to form a cofactor prominent in many enzymes, pyridoxamine phosphate (PMP).

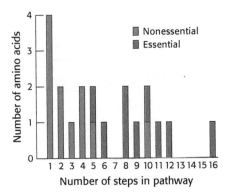

Figure 31.6 Essential and nonessential amino acids. Some amino acids are nonessential to human beings because they can be biosynthesized in a small number of steps. Amino acids requiring a large number of steps for their synthesis are essential in the diet because some of the enzymes for these steps have been lost in the course of evolution.

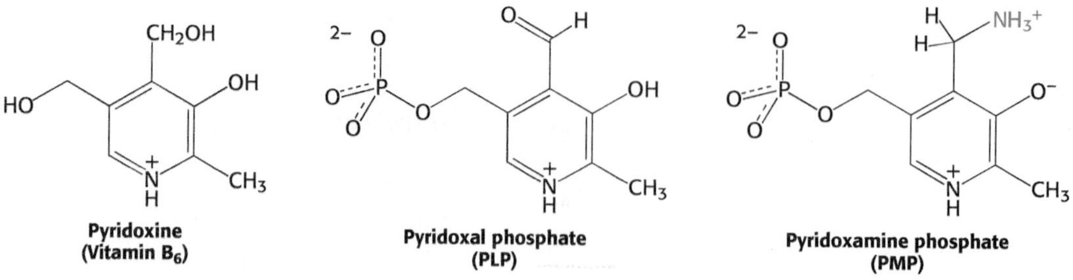

Pyridoxine (Vitamin B₆)   **Pyridoxal phosphate (PLP)**   **Pyridoxamine phosphate (PMP)**

Serine, Cysteine, and Glycine Are Formed from 3-Phosphoglycerate

Serine is synthesized from 3-phosphoglycerate, an intermediate in glycolysis. The first step is an oxidation to 3-phosphohydroxypyruvate. This α-ketoacid is transaminated to 3-phosphoserine, which is then hydrolyzed to serine. Serine is the precursor of *glycine* and *cysteine*. In the formation of glycine, the side-chain methylene group of serine is transferred to tetrahydrofolate, a carrier of one-carbon units.

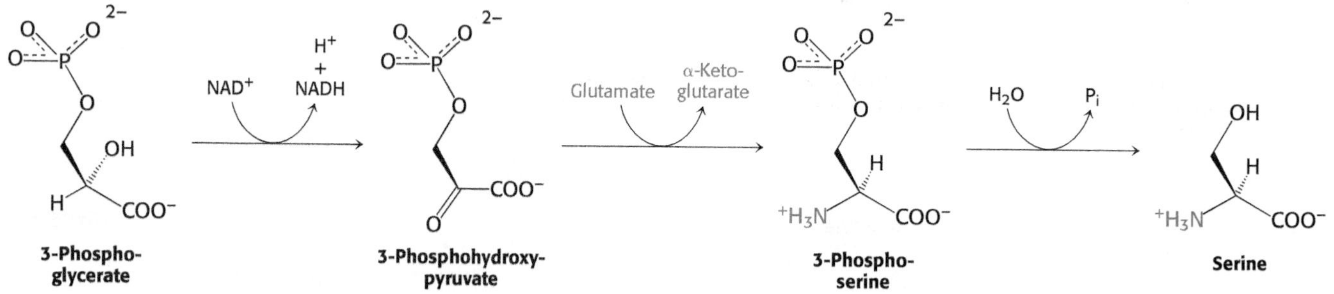

3-Phospho-glycerate   3-Phosphohydroxy-pyruvate   3-Phospho-serine   Serine

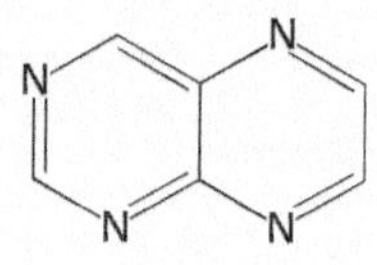

🩺 CLINICAL INSIGHT

Tetrahydrofolate Carries Activated One-Carbon Units

Tetrahydrofolate is a coenzyme essential for the synthesis of many amino acids and nucleotides. This coenzyme, a highly versatile carrier of activated one-carbon units, consists of three groups: a substituted pteridine, *p*-aminobenzoate, and a chain of one or more glutamate residues (**Figure 31.7**). Mammals can synthesize the pteridine ring, but they are unable to conjugate it to the other two units and so must obtain tetrahydrofolate from their diets or from microorganisms in their intestinal tracts.

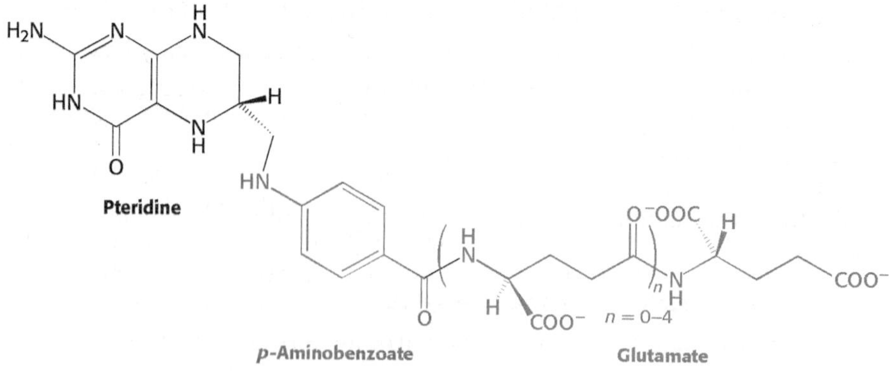

Pteridine

Pteridine   *p*-Aminobenzoate   Glutamate

Figure 31.7 Tetrahydrofolate. This cofactor includes three components: a pteridine ring, *p*-aminobenzoate, and one or more glutamate residues.

The one-carbon group carried by tetrahydrofolate is bonded to its N-5 or N-10 nitrogen atom (denoted as N^5 or N^{10}) or to both (**Figure 31.8**). This unit can exist in three oxidation states (**Table 31.1**). The most reduced form carries a *methyl* group, whereas the intermediate form carries a *methylene* group.

Tetrahydrofolate

N^5,N^{10}-**Methylene-tetrahydrofolate**

N^5-**Methyl-tetrahydrofolate**

N^{10}-**Formyl-tetrahydrofolate**

N^5,N^{10}-**Methenyl-tetrahydrofolate**

N^5-**Formimino-tetrahydrofolate**

N^5-**Formyl-tetrahydrofolate**

Figure 31.8 Conversions of one-carbon units attached to tetrahydrofolate.

Table 31.1   One-carbon groups carried by tetrahydrofolate

Oxidation state	Group	
	Formula	Name
Most reduced (= methyl)	$-CH_3$	Methyl
Intermediate (= formaldehyde)	$-CH_2-$	Methylene
Most oxidized (= formic acid)	$-CHO$	Formyl
	$-CHNH$	Formimino
	$-CH=$	Methenyl

More oxidized forms carry a *formyl, formimino,* or *methenyl* group. The fully oxidized one-carbon unit, CO_2, is carried by biotin rather than by tetrahydrofolate. The importance of tetrahydrofolate to DNA replication and cell growth is attested to by the fact that drugs that inhibit the regeneration of tetrahydrofolate are effective in the inhibition of cancer-cell growth (Chapter 32).

Tetrahydrofolate, derived from folic acid (vitamin B_9), plays an especially important role in the development of the fetal nervous system during early pregnancy. Folic acid deficiency can result in failure of the neural tube to close, which results in conditions such as spina bifida (defective closure of the vertebral column) and anencephaly (lack of a brain). The neural tube closes by about the 28th day of pregnancy, usually before a woman knows that she is pregnant. Consequently, some physicians recommend that all women of childbearing age take folic acid supplements.

NUTRITION FACTS

Folic acid (vitamin B_9) Another of the B vitamins, folic acid is especially important for growth, and a lack of folic acid during prenatal development results in spinal-cord defects. Folic acid deficiency can result in megaloblastic anemia, characterized by the release of large immature red blood cells into the blood. Green vegetables, such as spinach and broccoli, are good sources of folic acid. [Photograph from subjug/Getty Images.]

S-Adenosylmethionine Is the Major Donor of Methyl Groups

Tetrahydrofolate can carry a methyl group on its N-5 atom, but its transfer potential is not sufficiently high for most biosynthetic methylations. Rather, the activated methyl donor in such reactions is usually *S-adenosylmethionine* (SAM), which is synthesized by the transfer of an adenosyl group from ATP to the sulfur atom of methionine.

Methionine   **S-Adenosylmethionine (SAM)**

The methyl group of the methionine unit is activated by the positive charge on the adjacent sulfur atom, which makes the molecule much more reactive than N^5-methyltetrahydrofolate. Recall that *S*-adenosylmethionine is an activated methyl donor in the synthesis of phosphatidylcholine from phosphatidylethanolamine (p. 526).

The synthesis of *S*-adenosylmethionine is unusual in that the triphosphate group of ATP is split into pyrophosphate and orthophosphate; the pyrophosphate is subsequently hydrolyzed to two molecules of P_i. *S*-Adenosylhomocysteine is formed when the methyl group of *S*-adenosylmethionine is transferred to an acceptor. *S*-Adenosylhomocysteine is then hydrolyzed to *homocysteine* and adenosine:

S-Adenosylmethionine (SAM)   **S-Adenosylhomocysteine**   **Homocysteine**

Methionine can be regenerated by the transfer of a methyl group to homocysteine from N^5-methyltetrahydrofolate, a reaction catalyzed by *methionine synthase* (also known as *homocysteine methyltransferase*). The coenzyme that mediates this transfer of a methyl group is *methylcobalamin,* derived from vitamin B_{12} (p. 496).

These reactions constitute the *activated methyl cycle* (**Figure 31.9**). Methyl groups enter the cycle in the conversion of homocysteine into methionine and are then made highly reactive by the addition of an adenosyl group. The high transfer potential of the methyl group of *S*-adenosylmethionine enables it to be transferred to a wide variety of acceptors.

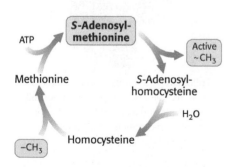

Figure 31.9 The activated methyl cycle. The methyl group of methionine is activated by the formation of *S*-adenosylmethionine.

🩺 CLINICAL INSIGHT

High Homocysteine Levels Correlate with Vascular Disease

People with elevated serum levels of homocysteine (homocysteinemia) or the disulfide-linked dimer homocystine (homocystinuria) have an unusually high risk for coronary heart disease and arteriosclerosis. The most common genetic cause of high homocysteine levels is a mutation within the gene encoding *cystathionine β-synthase*, the enzyme that combines homocysteine and serine to form cystathionine, an intermediate in the synthesis of cysteine:

Homocysteine   Serine   Cystathionine   α-Ketobutyrate   Cysteine

High levels of homocysteine appear to damage cells lining blood vessels and to increase the growth of vascular smooth muscle. The amino acid raises oxidative stress as well and has also been implicated in the development of type 2 diabetes (p. 323). The molecular basis of homocysteine's action has not been clearly identified, but may result from stimulation of the inflammatory response. Vitamin treatments are sometimes effective in reducing homocysteine levels in some people. These treatments maximize the activity of the two major metabolic pathways processing homocysteine. Pyridoxal phosphate, a vitamin B_6 derivative, is necessary for the activity of cystathionine β-synthase, and tetrahydrofolate, as well as vitamin B_{12}, are required for the methylation of homocysteine to methionine.

? QUICK QUIZ 2 Identify the six biosynthetic families of amino acids.

31.3 Feedback Inhibition Regulates Amino Acid Biosynthesis

The rate of synthesis of amino acids depends mainly on the *amounts* of the biosynthetic enzymes and on their *activities*. We now consider the control of enzymatic activity. The regulation of enzyme synthesis will be discussed in Section 15.

The Committed Step Is the Common Site of Regulation

As we have seen in every metabolic pathway that we have studied so far, the first irreversible reaction, or the *committed step*, is usually an important regulatory site. *The final product of the pathway (Z) often inhibits the enzyme that catalyzes the committed step* (A → B).

Inhibited
by Z

A ⊣⇒ B ⟹ C ⟹ D ⟹ E ⟹ Z

This kind of control is essential for the conservation of building blocks and metabolic energy. Consider the biosynthesis of serine. The committed step in this pathway is the oxidation of 3-phosphoglycerate (p. 576), catalyzed by the enzyme *3-phosphoglycerate dehydrogenase*. The *E. coli* enzyme is a tetramer of four identical subunits, each comprising a catalytic domain and a serine-binding regulatory domain (**Figure 31.10**). The binding of serine to a regulatory site reduces the value of V_{max} for the enzyme; an enzyme bound to four molecules of serine is essentially inactive. Thus, if serine is abundant in the cell, the enzyme activity is inhibited, and so 3-phosphoglycerate, a key building block that can be used for other processes, is not wasted.

Branched Pathways Require Sophisticated Regulation

The regulation of branched pathways is more complicated because the concentration of two products must be accounted for. In fact, several intricate feedback mechanisms have been found in branched biosynthetic pathways.

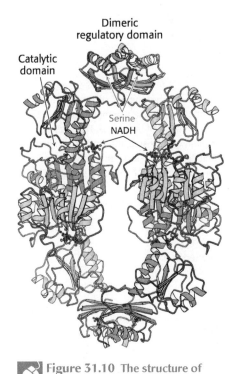

Dimeric
regulatory domain

Catalytic
domain

Serine
NADH

Figure 31.10 The structure of 3-phosphoglycerate dehydrogenase. This enzyme, which catalyzes the committed step in the serine biosynthetic pathway, is inhibited by serine. Notice the two serine-binding dimeric regulatory domains—one at the top and the other at the bottom of the structure. NADH is a required cofactor. [Drawn from 1PSD.pdb.]

Feedback inhibition and activation   Two pathways with a common initial step may each be inhibited by its own product and activated by the product of the other pathway. Consider, for example, the biosynthesis of the amino acids valine, leucine, and isoleucine (**Figure 31.11**). A common intermediate, hydroxyethyl thiamine pyrophosphate (hydroxyethyl-TPP), initiates the pathways leading to all three of these amino acids. Hydroxyethyl-TPP reacts with α-ketobutyrate in the initial step of the pathway leading to the synthesis of isoleucine. Hydroxyethyl-TPP can also react with pyruvate in the committed step for the pathways leading to valine and leucine. Thus, the relative concentrations of α-ketobutyrate and pyruvate determine how much isoleucine is produced compared with valine and leucine. How are these competing reactions regulated so that equal amounts of isoleucine, valine, and leucine are synthesized? *Threonine deaminase*, the PLP enzyme that catalyzes the formation of α-ketobutyrate, is allosterically inhibited by isoleucine (Figure 31.11). This enzyme is also allosterically activated by valine. Thus, this enzyme is inhibited by the end product of the pathway that it initiates and is activated by the end product of a competitive pathway. This mechanism balances the amounts of different amino acids that are synthesized.

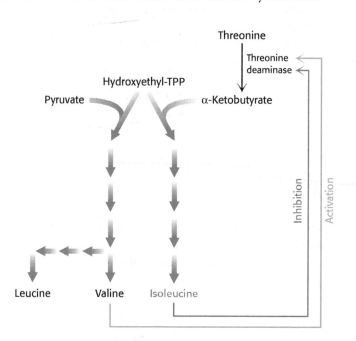

Figure 31.11 Regulation of threonine deaminase. Threonine is converted into α-ketobutyrate in the committed step, leading to the synthesis of isoleucine. The enzyme that catalyzes this step, threonine deaminase, is inhibited by isoleucine and activated by valine, the product of a parallel pathway.

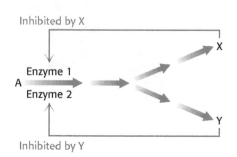

Figure 31.12 Enzyme multiplicity. Multiple enzymes that are catalytically identical or similar but have different allosteric properties may catalyze the committed step of a metabolic pathway.

Enzyme multiplicity   The committed step can be catalyzed by two or more enzymes with different regulatory properties, a strategy referred to as *enzyme multiplicity* (**Figure 31.12**). For example, the phosphorylation of aspartate is the committed step in the biosynthesis of threonine, methionine, and lysine. Three distinct aspartokinases catalyze this reaction in *E. coli*. Although the mechanisms of catalysis are essentially identical, their activities are regulated differently: one enzyme is not subject to feedback inhibition, another is inhibited by threonine, and the third is inhibited by lysine.

Cumulative feedback inhibition   A common step is partly inhibited by each of the final products, acting independently, in *cumulative feedback inhibition*. The regulation of glutamine synthetase in *E. coli* is a striking example of this inhibition. Recall that glutamine is synthesized from glutamate, NH_4^+, and ATP (p. 574). *Glutamine synthetase* regulates the flow of nitrogen and hence plays a key role in controlling bacterial metabolism. The amide group of glutamine is a source of nitrogen in the biosyntheses of a variety of compounds, such as tryptophan, histidine, carbamoyl phosphate, glucosamine 6-phosphate, cytidine triphosphate, and adenosine monophosphate. Glutamine synthetase is cumulatively inhibited by each of these final products of glutamine metabolism, as well as by alanine and glycine. *In cumulative inhibition, each inhibitor can reduce the activity of the enzyme, even when other inhibitors are exerting their own maximal inhibition*. The enzymatic activity of glutamine synthetase is switched off almost completely only when all final products are bound to the enzyme.

SUMMARY

31.1 The Nitrogenase Complex Fixes Nitrogen

Microorganisms use ATP and reduced ferredoxin, a powerful reductant, to reduce N_2 to NH_3. A molybdenum–iron cluster in nitrogenase catalyzes the fixation of N_2, a very inert molecule. Higher organisms consume the fixed nitrogen to synthesize amino acids, nucleotides, and other nitrogen-containing biomolecules. Ammonia exists as ammonium ion (NH_4^+) in aqueous solutions. The major points of entry of NH_4^+ into metabolism are glutamine or glutamate.

31.2 Amino Acids Are Made from Intermediates of Major Pathways

Human beings can synthesize 11 of the basic set of 20 amino acids. These 11 amino acids are called nonessential, in contrast with the essential amino acids, which must be supplied in the diet. The pathways for the synthesis of nonessential amino acids are quite simple. Glutamate dehydrogenase catalyzes the reductive amination of α-ketoglutarate to glutamate. A transamination reaction takes place in the synthesis of most amino acids. Alanine and aspartate are synthesized by the transamination of pyruvate and oxaloacetate, respectively.

Tetrahydrofolate, a carrier of activated one-carbon units, plays an important role in the metabolism of amino acids and nucleotides. This coenzyme carries one-carbon units at three oxidation states, which are interconvertible. The major donor of activated methyl groups is *S*-adenosylmethionine, which is synthesized by the transfer of an adenosyl group from ATP to the sulfur atom of methionine. *S*-Adenosylhomocysteine is formed when the activated methyl group is transferred to an acceptor. It is hydrolyzed to adenosine and homocysteine, and the latter is then methylated to methionine to complete the activated methyl cycle.

31.3 Feedback Inhibition Regulates Amino Acid Biosynthesis

Most of the pathways of amino acid biosynthesis are regulated by feedback inhibition, in which the committed step is allosterically inhibited by the final product. The regulation of branched pathways requires extensive interaction among the branches and includes both negative and positive regulation.

KEY TERMS

nitrogen fixation (p. 571)
nitrogenase complex (p. 572)
essential amino acids (p. 574)
nonessential amino acids (p. 574)
aminotransferase (p. 575)

pyridoxal phosphate (PLP) (p. 575)
tetrahydrofolate (p. 576)
S-adenosylmethionine (SAM) (p. 578)

activated methyl cycle (p. 578)
committed step (p. 579)
enzyme multiplicity (p. 580)
cumulative feedback inhibition (p. 580)

❓ Answers to QUICK QUIZZES

1. $N_2 \xrightarrow[\text{complex}]{\text{Nitrogenase}} NH_3/NH_4^+ \xrightarrow{\text{Glutamate dehydrogenase}}$

    $\text{glutamate} \xrightarrow{\text{Glutamate synthetase}} \text{glutamine}$

2. Asparagine, aspartate, methionine, threonine, isoleucine, and lysine constitute one family derived from oxaloacetate. Phenylalanine, tyrosine, and tryptophan constitute another family derived from phosphoenolpyruvate and erythrose 4-phosphate. The pyruvate family consists of alanine, valine, and leucine. Histidine, the only member of its family, comes from ribose 5-phosphate. The α-ketoglutarate family consists of glutamate, glutamine, proline, and arginine. The final family consists of serine, cysteine, and glycine and is based on 3-phosphoglycerate.

PROBLEMS

1. *Out of thin air.* Define nitrogen fixation. What organisms are responsible for nitrogen fixation? ✓ 3

2. *Fixing a problem.* Write the equation for biological nitrogen fixation, and explain the role of ATP. ✓ 3 ✓ 4

3. *Like Starsky and Hutch.* Match each term with its description. ✓ 3

    (a) Nitrogen fixation
    _____
    (b) Nitrogenase complex
    _____
    (c) Glutamate _____
    (d) Essential amino acids
    _____
    (e) Nonessential amino acids _____
    (f) Aminotransferase
    _____
    (g) Pyridoxal phosphate
    _____
    (h) Tetrahydrofolate
    _____
    (i) S-Adenosylmethionine
    _____
    (j) Homocysteine _____

    1. Methylated to form methionine
    2. An important methyl donor
    3. Coenzyme required by aminotransferases
    4. Conversion of N_2 into NH_3
    5. A carrier of various one-carbon units
    6. Amino acids that are dietary requirements
    7. Amino acids that are readily synthesized
    8. Responsible for nitrogen fixation
    9. Transfers amino groups between keto acids
    10. A common amino acid donor

4. *Teamwork.* Identify the two components of the nitrogenase complex, and describe their specific tasks. ✓ 3

5. *The fix is in.* "The mechanistic complexity of nitrogenase is necessary because nitrogen fixation is a thermodynamically unfavorable process." True or false? Explain. ✓ 3

6. *Siphoning resources.* Nitrogen-fixing bacteria on the roots of some plants can consume as much as 20% of the ATP produced by the plant—consumption that does not seem very beneficial to the plant. Explain why plants tolerate this loss of valuable resources and what the bacteria are doing with the ATP. ✓ 3

7. *Vital, in the truest sense.* Why are certain amino acids defined as essential for human beings? ✓ 4

8. *From few, many.* What are the seven precursors of the 20 amino acids? ✓ 4

9. *Common component.* What cofactor is required by all aminotransferases? ✓ 4

10. *Common resource.* If an animal is fed ^{15}N-labeled aspartate, many amino acids bearing the ^{15}N label subsequently appear. What reactions take part in the transfer of the label? ✓ 4

11. *One carbon at a time.* What is the role of tetrahydrofolate in biochemical systems?

12. *The same but different.* Differentiate between S-adenosylmethionine and tetrahydrofolate.

13. *Telltale tag.* In the reaction catalyzed by glutamine synthetase, an oxygen atom is transferred from the side chain of glutamate to orthophosphate, as shown by the results of ^{18}O-labeling studies. Account for this finding.

14. *Direct synthesis.* Which of the 20 amino acids can be synthesized directly from a common metabolic intermediate by a transamination reaction? ✓ 4

15. *Lines of communication.* For the following example of a branched pathway, propose a feedback-inhibition scheme that would result in the production of equal amounts of Y and Z. ✓ 4

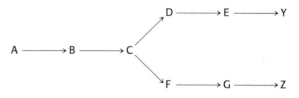

16. *Cumulative feedback inhibition.* Consider the branched pathway in problem 15. The first common step (A → B) is

partly inhibited by both of the final products, each acting independently of the other. Suppose that a high level of Y alone decreased the rate of the A → B step from 100 to 60 s^{-1} and that a high level of Z alone decreased the rate from 100 to 40 s^{-1}. What would the rate be in the presence of high levels of both Y and Z?

Chapter Integration and Data Interpretation Problems

17. *One carbon, and only one carbon.* We have identified three biomolecules that carry activated one-carbon units of some sort. Name these three carriers.

18. *I've seen that face before.* Vitamin B_{12} is required by methionine synthase to regenerate methionine from homocysteine. What other enzyme that we have encountered in our studies requires vitamin B_{12}?

19. *Further ramifications.* A person on a diet lacking in methionine would not be able to synthesize adequate amounts of proteins. However, insufficient protein synthesis would not be the only biochemical problem such a person would face. What other biosynthesis would be affected by a lack of dietary methionine?

20. *Light effects.* The adjoining graph shows the concentration of several free amino acids in light- and dark-adapted plants.

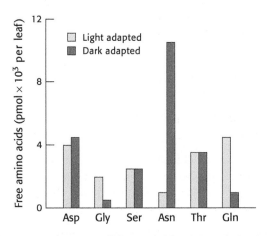

[Data from B. B. Buchanan, W. Gruissem, and R. L. Jones, *Biochemistry and Molecular Biology of Plants* (American Society of Plant Physiology, 2000), p. 363, Fig. 8.3.]

(a) Of the amino acids shown, which are most affected by light–dark adaptation?

(b) Suggest a plausible biochemical explanation for the difference observed.

(c) White asparagus, a culinary delicacy, is the result of growing asparagus plants in the dark. What chemical might you think enhances the taste of white asparagus?

Challenge Problems

21. *From sugar to amino acid.* Write a balanced equation for the synthesis of alanine from glucose. ✓ 4

22 *Connections.* How might increased synthesis of aspartate and glutamate affect energy production in a cell? How would the cell respond to such an effect? ✓ 4

23. *Comparing K_M.* Glutamate dehydrogenase (p. 574) and glutamine synthetase (p. 574) are present in all organisms. Most bacteria also contain another enzyme, *glutamate synthase*, which catalyzes the reductive amination of α-ketoglutarate with the use of glutamine as the nitrogen donor:

$$\alpha\text{-Ketoglutarate} + \text{glutamine} + \text{NADPH} + \text{H}^+$$

$$\xrightleftharpoons[\text{synthase}]{\text{Glutamate}} 2\text{ glutamate} + \text{NADP}^+$$

The side-chain amide of glutamine is hydrolyzed to generate ammonia within the enzyme. When NH_4^+ is limiting, most of the glutamate is made by the sequential action of glutamine synthetase and glutamate synthase. The sum of these reactions is

$$NH_4^+ + \alpha\text{-ketoglutarate} + \text{NADPH} + \text{ATP} \longrightarrow$$

$$\text{glutamate} + \text{NADP}^+ + \text{ADP} + \text{P}_i$$

Note that this stoichiometry differs from that of the glutamate dehydrogenase reaction in that ATP is hydrolyzed. Why do bacteria sometimes use this more expensive pathway? (Hint: The K_M value for NH_4^+ of glutamate dehydrogenase is higher than that of glutamine synthetase.) ✓ 4

Selected Readings for this chapter can be found online at www.whfreeman.com/tymoczko3e.

Nucleotide Metabolism

Charles V, Holy Roman Emperor (1519–1558), who also ruled as Charles I, king of Spain (1516–1556), was one of Europe's most powerful rulers. Under his reign, the Aztec and Inca empires of the New World fell to the Spanish. Charles V also suffered from severe gout, a pathological condition caused by the disruption of nucleotide metabolism. This painting by Titian shows the emperor's swollen left hand, which he rests gingerly in his lap. Recent analysis of a part of his finger, saved as a religious relic, confirms the diagnosis of gout. [Erich Lessing/Art Resource, NY.]

I n addition to being the building blocks of proteins and peptides, amino acids serve as precursors of many kinds of small molecules that have important and diverse biological roles. Let us briefly survey some of the biomolecules that are derived from amino acids (**Figure 32.1**).

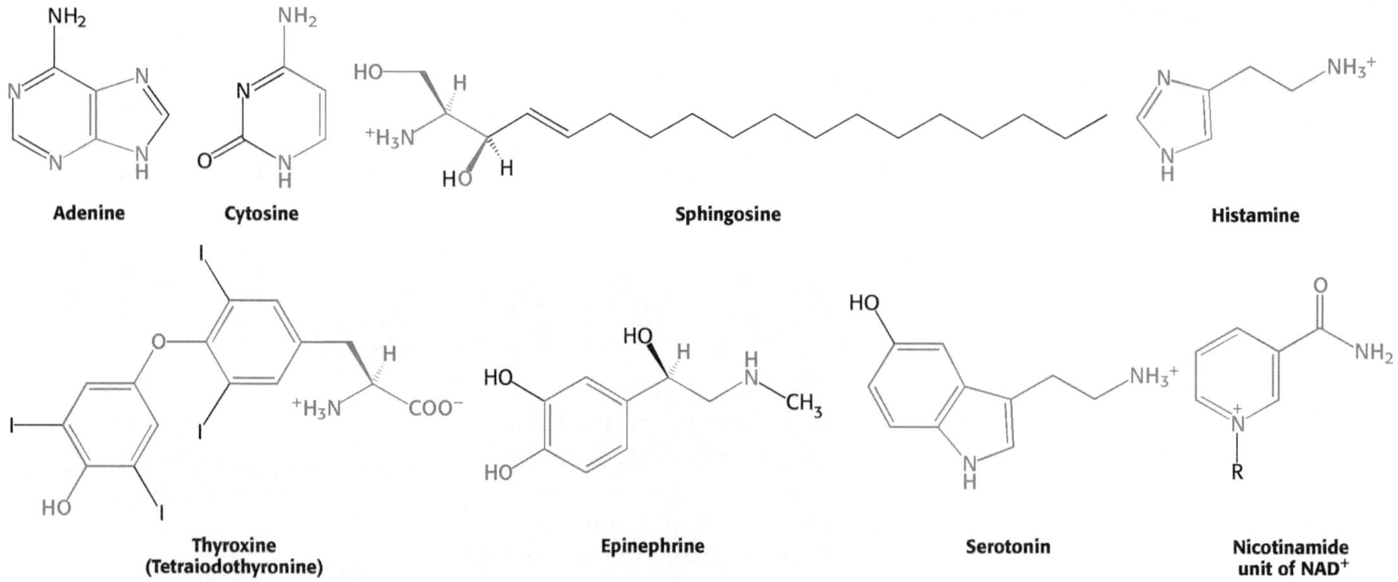

Adenine   **Cytosine**   **Sphingosine**   **Histamine**

Thyroxine (Tetraiodothyronine)   **Epinephrine**   **Serotonin**   **Nicotinamide unit of NAD$^+$**

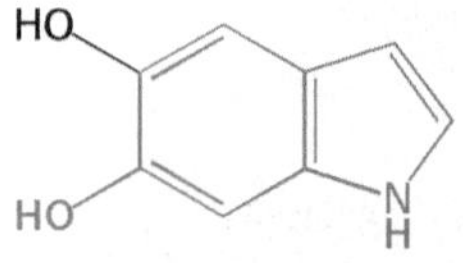

5,6-Dihydroxyindole

Figure 32.1 Selected biomolecules derived from amino acids. The atoms contributed by amino acids are shown in blue.

Histamine, a component of the immune system whose overproduction results in an allergic response, is derived from histidine by decarboxylation. Tyrosine is a precursor of the hormones *thyroxine* (tetraiodothyronine), which controls the rate of metabolic processes throughout the body, and *epinephrine* (adrenaline), the "fight or flight" hormone. Epinephrine is metabolized to 5,6-dihydroxyindole, which is a precursor of *melanin,* a complex polymeric pigment that is formed on exposure to sunlight and is the chemical basis of tanning. The neurotransmitter *serotonin* (5-hydroxytryptamine) and the *nicotinamide ring* of NAD$^+$ are synthesized from tryptophan. The reactive terminus of *sphingosine,* an intermediate in the synthesis of sphingolipids, comes from serine. We now consider in more detail especially prominent biochemicals derived from amino acids—the nucleotide bases. *Purines* and *pyrimidines* are derived largely from amino acids and are the precursors of DNA, RNA, and various coenzymes.

We begin by examining nucleotide synthesis. The initial products of nucleotide synthesis are ribonucleotides, and we will see how ribonucleotides are converted into deoxyribonucleotides, the substrates for DNA synthesis. We then turn to the regulation of nucleotide synthesis. Finally, we consider pathological conditions that result from perturbations in nucleotide metabolism.

✓ 5 Describe how nucleotides are synthesized.

32.1   An Overview of Nucleotide Biosynthesis and Nomenclature

The pathways for the biosynthesis of nucleotides fall into two classes: *de novo* pathways and *salvage pathways* (**Figure 32.2**). In de novo pathways, the nucleotide bases are assembled from simpler compounds. The framework for a *pyrimidine* base is assembled first and then attached to ribose. In contrast, the framework for a *purine* base is synthesized piece by piece directly onto a ribose-based molecule. In salvage pathways, preformed bases are recovered and reconnected to a ribose unit.

Table 32.1   Nomenclature of bases, nucleosides, and nucleotides

RNA			DNA		
Base	Ribonucleoside	Ribonucleotide (5′-monophosphate)	Base	Deoxyribonucleoside	Deoxyribonucleotide (5′-monophosphate)
Adenine (A)	Adenosine	Adenylate (AMP)	Adenine (A)	Deoxyadenosine	Deoxyadenylate (dAMP)
Guanine (G)	Guanosine	Guanylate (GMP)	Guanine (G)	Deoxyguanosine	Deoxyguanylate (dGMP)
Uracil (U)	Uridine	Uridylate (UMP)	Thymine (T)	Thymidine	Thymidylate (TMP)*
Cytosine (C)	Cytidine	Cytidylate (CMP)	Cytosine (C)	Deoxycytidine	Deoxycytidylate (dCMP)

*Thymidylate rarely exists in the ribose form, and so, by convention, the lowercase "d" is omitted in the abbreviation for thymidylate.

De novo synthesis and most salvage pathways lead to the synthesis of *ribo*nucleotides. However, DNA is built from *deoxyribo*nucleotides. All deoxyribonucleotides are synthesized from the corresponding ribonucleotides. The deoxyribose sugar is generated by the reduction of ribose within a fully formed nucleotide. Furthermore, the methyl group that distinguishes the thymine of DNA from the uracil of RNA is added at the last step in the pathway.

A *nucleoside* consists of a purine or pyrimidine base linked to a sugar, and a *nucleotide* is a phosphate ester of a nucleoside. For instance, adenosine is a nucleoside composed of the base adenine and the sugar ribose. ATP is a nucleotide, the phosphate ester (in this case, a triphosphate) of adenosine. The names of the major bases of RNA and DNA, as well as those of their nucleoside and nucleotide derivatives, are given in **Table 32.1**.

32.2 The Pyrimidine Ring Is Assembled and Then Attached to a Ribose Sugar

The precursors of pyrimidine rings are bicarbonate, aspartic acid, and ammonia usually produced from the hydrolysis of the side chain of glutamine. In the de novo synthesis of pyrimidines, the ring is synthesized first and then it is attached to ribose to form a *pyrimidine nucleotide* (**Figure 32.3**).

The first step in de novo pyrimidine biosynthesis is the synthesis of *carbamoyl phosphate* from bicarbonate and ammonia in a three-step process requiring the cleavage of two molecules of ATP. This reaction is catalyzed by *carbamoyl phosphate synthetase II* (CPS II). Recall that the carbamoyl phosphate synthetase I facilitates ammonia incorporation into urea (p. 555).

In the first step of the carbamoyl phosphate synthesis pathway, bicarbonate is phosphorylated by ATP to form carboxyphosphate, an activated form of CO_2, and ADP. In the second step, glutamine is hydrolyzed by CPS to yield glutamate and ammonia, which then reacts with carboxyphosphate to form carbamic acid and inorganic phosphate:

DE NOVO PATHWAY

Activated ribose (PRPP) + amino acids + ATP + CO_2 + . . .

↓

Nucleotide

SALVAGE PATHWAY

Activated ribose (PRPP) + base

↓

Nucleotide

Figure 32.2 Salvage and de novo pathways. In de novo synthesis, the base itself is synthesized from simpler starting materials, including amino acids. ATP hydrolysis is required for de novo synthesis. In a salvage pathway, a base is reattached to a ribose, activated in the form of 5-phosphoribosyl-1-pyrophosphate (PRPP).

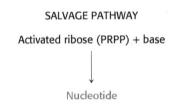

Bicarbonate   **Carboxyphosphate**   **Carbamic acid**

Carbamic acid is then phosphorylated by another molecule of ATP to form carbamoyl phosphate:

Carbamic acid   **Carbamoyl phosphate**

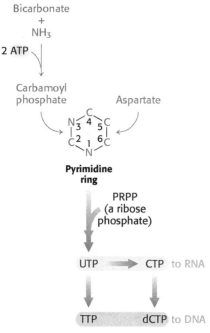

Figure 32.3 De novo pathway for pyrimidine nucleotide synthesis. The C-2 and N-3 atoms in the pyrimidine ring come from carbamoyl phosphate, whereas the other atoms of the ring come from aspartate.

The reaction catalyzed by carbamoyl phosphate synthetase II (CPS II) is identical with that of carbamoyl phosphate synthetase I (CPS I) required for urea synthesis (p. 555). CPS II, however, uses glutamine as a nitrogen source rather than free ammonia and does not require *N*-acetylglutamate.

Carbamoyl phosphate reacts with the amino acid aspartate to form carbamoylaspartate in a reaction catalyzed by *aspartate transcarbamoylase* (ATCase). This allosteric enzyme regulates the synthesis of pyrimidine nucleotides, as we will see later (p. 596).

Carbamoyl phosphate   **Carbamoylaspartate**

Carbamoylaspartate then cyclizes to form dihydroorotate, a reaction catalyzed by *dihydroorotase*. Dihydroorotate is subsequently oxidized by NAD^+ to form orotate:

Carbamoylaspartate   **Dihydroorotate**   **Orotate**

In mammals, three of the enzymes of orotate synthesis are part of a single large polypeptide chain called CAD, for *c*arbamoyl phosphate synthetase, *a*spartate transcarbamoylase, and *d*ihydroorotase.

At this stage, orotate couples to *5-phosphoribosyl-1-pyrophosphate* (PRPP), the activated form of ribose. PRPP is synthesized from ribose 5-phosphate, formed by the pentose phosphate pathway, by the addition of pyrophosphate from ATP in a reaction catalyzed by *PRPP synthetase*:

Ribose 5-phosphate   **PRPP**

Orotate reacts with PRPP to form *orotidylate*, a pyrimidine nucleotide, in a reaction catalyzed by *oroate phosphoribosyltransferase*. The hydrolysis of pyrophosphate, as shown in the margin, renders the reaction irreversible. Orotidylate is then decarboxylated to form uridine monophosphate (UMP, or uridylate), one of two major pyrimidine nucleotides that are precursors of RNA. This reaction is catalyzed by *orotidylate decarboxylase* also called orotidine-5′ phosphate decarboxylase:

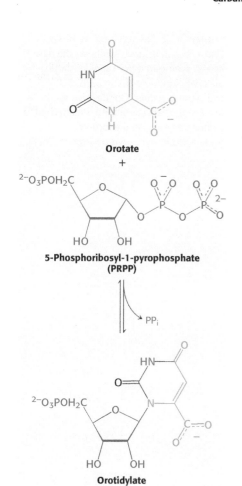

Orotidylate   **Uridylate**

The phosphoribosyltransferase and decarboxylase activities are located on the same polypeptide chain, providing another example of a bifunctional enzyme. The bifunctional enzyme is called *uridine monophosphate synthetase*.

CTP Is Formed by the Amination of UTP

How is the other major pyrimidine ribonucleotide, cytidine monophosphate (CMP, or cytidylate), formed? It is synthesized from the uracil base of UMP, but UMP must be converted into UTP before the synthesis can take place.

Kinases Convert Nucleoside Monophosphates into Nucleoside Triphosphates

Recall that the diphosphates and triphosphates are the active forms of nucleotides in biosynthesis and energy conversions. Nucleoside monophosphates are converted into nucleoside triphosphates in a step-by-step fashion. First, nucleoside monophosphates are converted into diphosphates by specific *nucleoside monophosphate kinases* that utilize ATP as the phosphoryl-group donor. For example, UMP is phosphorylated to UDP by *UMP kinase*:

$$UMP + ATP \rightleftharpoons UDP + ADP$$

Second, nucleoside diphosphates and triphosphates are interconverted by *nucleoside diphosphate kinase,* an enzyme that has broad specificity, in contrast with the monophosphate kinases. The letters X and Y can represent any of several ribonucleosides or even deoxyribonucleosides.

$$XDP + YTP \rightleftharpoons XTP + YDP$$

After uridine triphosphate has been formed, it can be transformed into *cytidine triphosphate* (CTP) by the replacement of a carbonyl group by an amino group. Like the synthesis of carbamoyl phosphate, this reaction requires ATP and uses glutamine as the source of the amino group. CTP can then be used in many biochemical processes, including RNA synthesis and the activation of molecules for phospholipid synthesis (p. 525).

> **? QUICK QUIZ 1** What are the precursors for the de novo synthesis of pyrimidines?

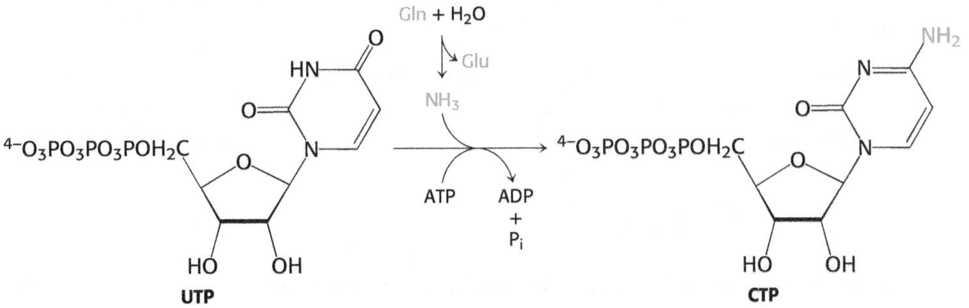

UTP   → (Gln + H$_2$O → Glu; NH$_3$; ATP → ADP + P$_i$)   CTP

🩺 CLINICAL INSIGHT

Salvage Pathways Recycle Pyrimidine Bases

Pyrimidine bases can be recovered from the breakdown products of DNA and RNA by the use of *salvage pathways*. In these pathways, a preformed base is reincorporated into a nucleotide. We will consider the salvage pathway for the pyrimidine base thymine. Thymine is found in DNA and base-pairs with adenine in the DNA double helix. Thymine released from degraded DNA is salvaged in two steps. First, thymine is converted into the nucleoside thymidine by *thymidine phosphorylase*:

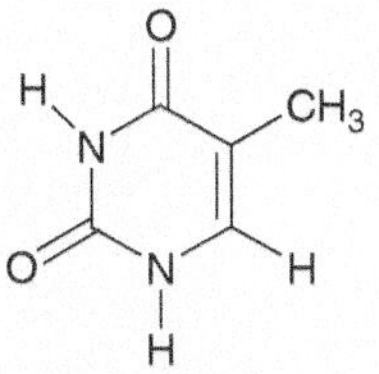

$$\text{Thymine} + \text{deoxyribose-1-phosphate} \rightleftharpoons \text{thymidine} + P_i$$

Thymine

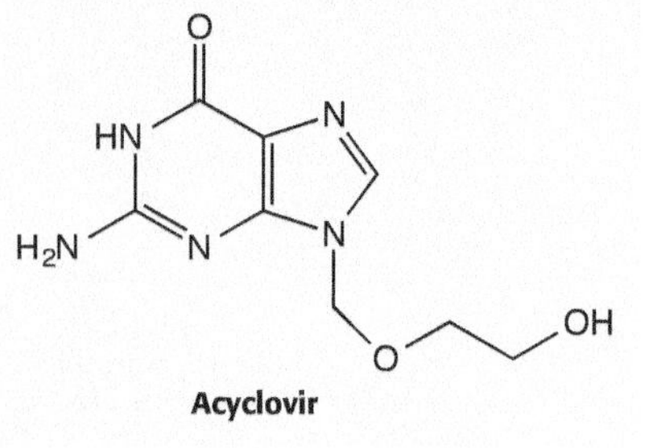

Acyclovir

Thymidine is then converted into a nucleotide by *thymidine kinase*:

$$\text{Thymidine} + \text{ATP} \rightleftharpoons \text{TMP} + \text{ADP}$$

The activity of thymidine kinase fluctuates with the cell cycle, displaying peak activity during the S phase, when DNA synthesis takes place.

Viral thymidine kinase differs from the mammalian enzyme and thus provides a therapeutic target. For instance, herpes simplex infections are treated with acyclovir, which viral thymidine kinase converts into a suicide inhibitor that terminates DNA synthesis. As we will see shortly, thymidine kinase also plays a role in the de novo synthesis of thymidylate.

32.3 The Purine Ring Is Assembled on Ribose Phosphate

Purines, like pyrimidines, are synthesized de novo, beginning with simple starting materials (**Figure 32.4**). In contrast with pyrimidine synthesis, the first step in purine assembly begins with the attachment to ribose. De novo purine biosynthesis, like pyrimidine biosynthesis, requires PRPP, but, for purines, PRPP provides the foundation on which the bases are constructed step by step. The initial step is the displacement of the pyrophosphate of PRPP by ammonia, rather than by a preassembled base, to produce *5-phosphoribosyl-1-amine*, with the amine in the β configuration. Glutamine again provides the ammonia. *Glutamine phosphoribosyl amidotransferase* catalyzes this reaction, which is the committed step in purine synthesis.

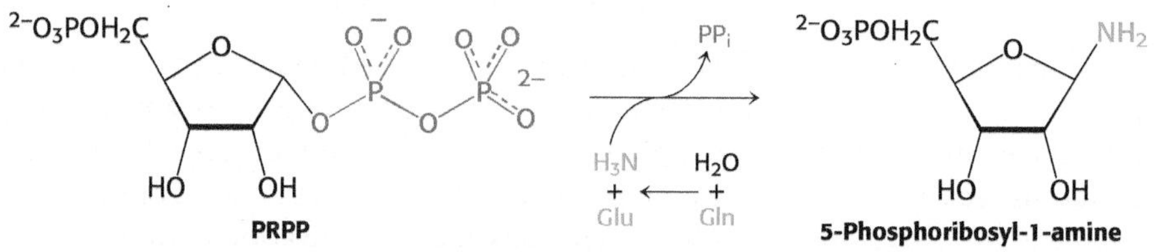

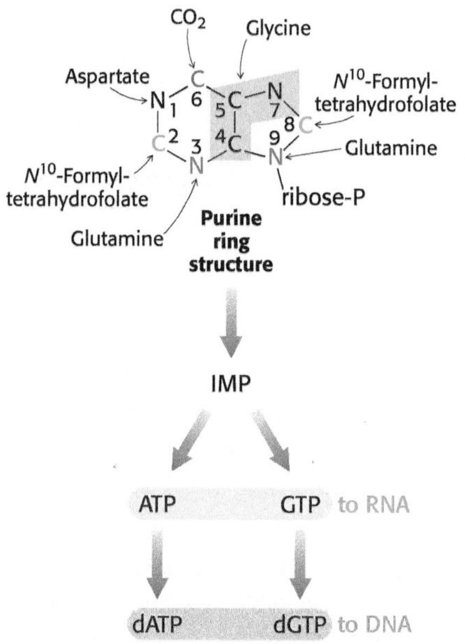

Figure 32.4 De novo pathway for purine nucleotide synthesis. The origins of the atoms in the purine ring structure are indicated.

Nine additional steps are required to assemble the purine ring. De novo purine biosynthesis proceeds by successive steps of activation by phosphorylation followed by displacement, as shown in (**Figure 32.5**). The final product is the nucleotide inosine monophosphate (IMP, or inosinate). The amino acids glycine, glutamine, and aspartate are required precursors. The synthesis of purines depends on tetrahydrofolate, a prominent carrier of activated one-carbon units (p. 576).

AMP and GMP Are Formed from IMP

Inosinate, although a component of some RNA molecules, serves primarily as a precursor to the other purines. Inosinate is at the base of a branched pathway that leads to both AMP and GMP (**Figure 32.6**). *Adenylate* is synthesized from inosinate by the substitution of an amino group for the carbonyl oxygen atom at C-6. The addition of aspartate, followed by the departure of fumarate, contributes the amino group. GTP, rather than ATP, is required for the synthesis of the adenylosuccinate intermediate from inosinate and aspartate.

Guanosine monophosphate (GMP, or guanylate) is synthesized by the oxidation of inosinate to xanthine monophosphate (XMP, or xanthylate), followed by the incorporation of an amino group at C-2. NAD^+ is the hydrogen acceptor in the oxidation of inosinate. The carbonyl group of xanthylate is activated by the transfer of an AMP group from ATP. Ammonia, generated by the hydrolysis of glutamine, then displaces the AMP group to form guanylate, in a reaction catalyzed by *GMP synthetase*. Note that the synthesis of adenylate requires GTP, whereas the synthesis of guanylate requires ATP, a contrast that, as we will see (p. 596), enables the flux down each branch to be balanced.

DID YOU KNOW?

In the ring form of ribose sugars, the β configuration means that the group attached at C-1 is on the same side of the ring as the —CH_2OH group. In the α configuration, the group attached to C-1 and —CH_2OH are on opposite sides of the ring.

? QUICK QUIZ 2 Identify the sources of all of the atoms in a purine ring.

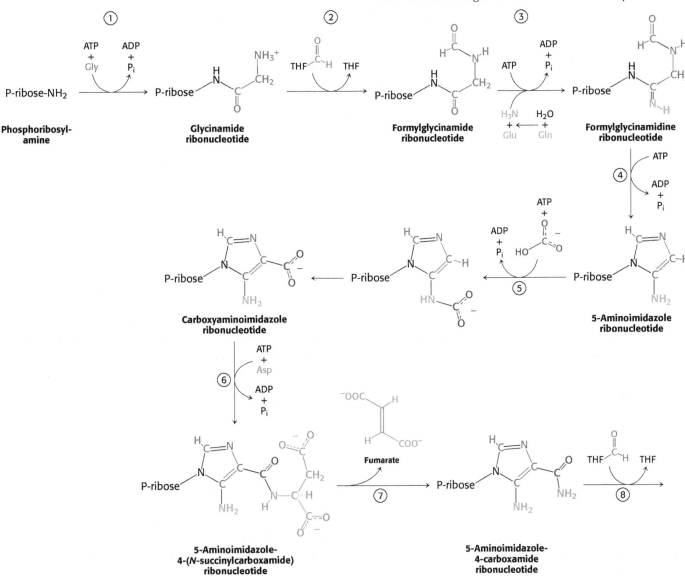

Figure 32.5 De novo purine biosynthesis. (1) Glycine is coupled to the amino group of phosphoribosylamine. (2) N^{10}-Formyltetrahydrofolate (THF) transfers a formyl group to the amino group of the glycine residue. (3) The inner amide group is phosphorylated and then converted into an amidine by the addition of ammonia derived from glutamine. (4) An intramolecular coupling reaction forms a five-membered imidazole ring. (5) Bicarbonate adds first to the exocyclic amino group and then to a carbon atom of the imidazole ring. (6) The imidazole carboxylate is phosphorylated, and the phosphoryl group is displaced by the amino group of aspartate. (7) Fumarate leaves, followed by (8) the addition of a second formyl group from N^{10}-formyltetrahydrofolate. (9) Cyclization completes the synthesis of inosinate, a purine nucleotide. Abbreviation: P stands for phosphoryl group.

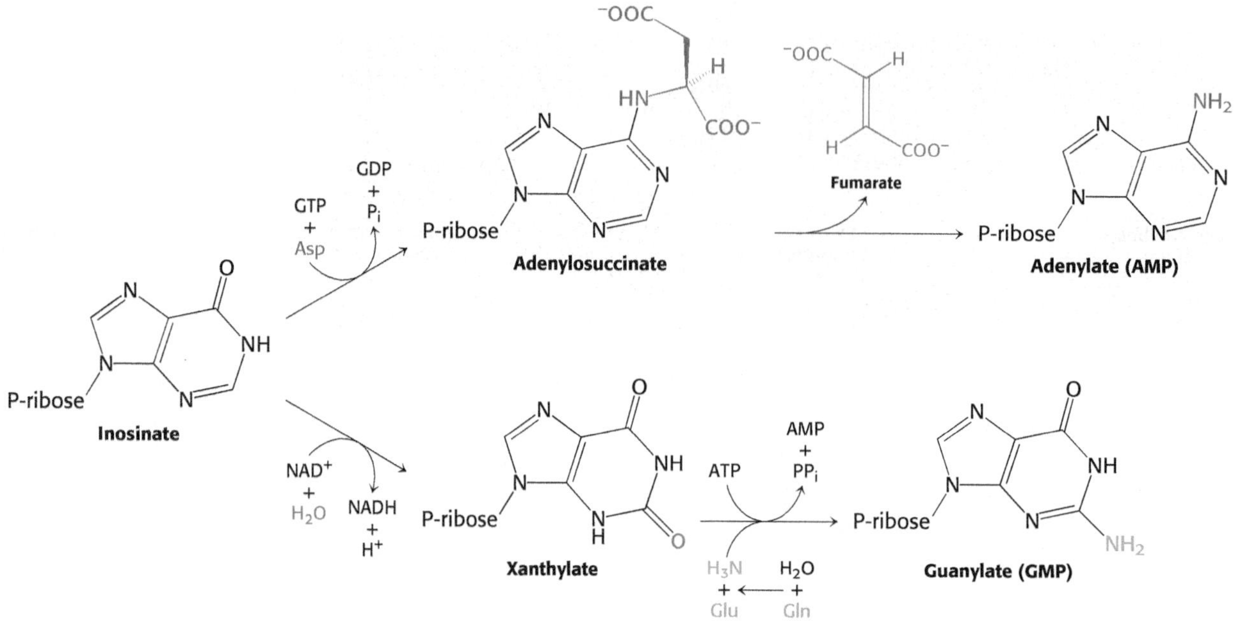

Figure 32.6 Generating AMP and GMP. Inosinate is the precursor of AMP and GMP. AMP is formed by the addition of aspartate followed by the release of fumarate. GMP is generated by the addition of water, dehydrogenation by NAD$^+$, and the replacement of the carbonyl oxygen atom by —NH$_2$ derived by the hydrolysis of glutamine.

BIOLOGICAL INSIGHT

Enzymes of the Purine-Synthesis Pathway Are Associated with One Another in Vivo

Biochemists believe that the enzymes of many metabolic pathways, such as glycolysis and the citric acid cycle, are physically associated with one another. Such associations would increase the efficiency of pathways by facilitating the movement of the product of one enzyme to the active site of the next enzyme in the pathway. The evidence for such associations comes primarily from experiments in which one component of a pathway, carefully isolated from the cell, is found to be bound to other components of the pathway. However, these observations raise the question, do enzymes associate with one another in vivo or do they spuriously associate during the isolation procedure? Recent in vivo evidence shows that the enzymes of the purine-synthesis pathway associate with one another when purine synthesis is required. Various enzymes of the pathway were fused with green fluorescent protein (GFP), which renders the enzymes visible with a fluorescent microscope, and transfected into cells. When cells were grown in the presence of purine, GFP was spread diffusely throughout the cytoplasm (Figure 32.7A). When the cells were switched to growth media lacking purines, purine synthesis began and the enzymes became associated with one another, forming complexes dubbed *purinosomes* (Figure 32.7B). The experiments were repeated with other enzymes of the purine-synthesis pathway, and the results were the same: purine synthesis takes place when the enzymes form the purinosomes. What causes complex formation? Although the results are not yet established, it appears that several G-protein coupled receptors, including those responding to epinephrine as well as ATP and ADP (purinergic receptors), instigate complex formation, while human creatine kinase II (hCK2), responding to the presence of purines, causes disassembly of the purinosome.

(A)

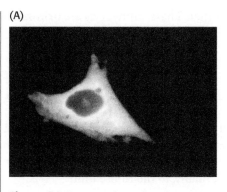

(B)

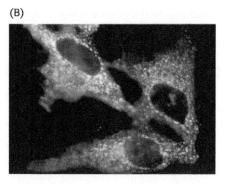

Figure 32.7 Formation of purinosomes. A gene construct encoding a fusion protein consisting of formylglycinamidine synthase and green fluorescent protein (GFP) was transfected into Hela cells (a human cell line) and expressed in them. (A) In the presence of purines (the absence of purine synthesis), the GFP was seen as a diffuse stain throughout the cytoplasm. (B) When the cells were placed in a purine-free medium, purinosomes formed, seen as cytoplasmic granules, and purine synthesis took place. [From S. An, R. Kumar, E. D. Sheets, and S. J. Benkovic, *Science* 320:103–106, 2008, Fig. 2, parts C and D.]

Bases Can Be Recycled by Salvage Pathways

Purine nucleotides, like pyrimidine nucleotides (p. 589), can also be synthesized by salvaging and recycling intact purines released by the hydrolytic degradation of nucleic acids and nucleotides. Such salvage pathways skip most of the energy-requiring steps of de novo nucleotide synthesis, thereby saving substantial amounts of ATP.

Two salvage enzymes with different specificities recover purine bases. *Adenine phosphoribosyltransferase* catalyzes the formation of adenylate,

$$\text{Adenine} + \text{PRPP} \longrightarrow \text{adenylate} + \text{PP}_i$$

whereas *hypoxanthine-guanine phosphoribosyltransferase* (HGPRT) catalyzes the formation of guanylate as well as inosinate, which, you will recall, is a precursor of guanylate and adenylate:

$$\text{Guanine} + \text{PRPP} \longrightarrow \text{guanylate} + \text{PP}_i$$
$$\text{Hypoxanthine} + \text{PRPP} \longrightarrow \text{inosinate} + \text{PP}_i$$

Purine salvage pathways are especially noteworthy in light of the amazing consequences of their absence (p. 600).

32.4 Ribonucleotides Are Reduced to Deoxyribonucleotides

We now turn to the synthesis of deoxyribonucleotides. These precursors of DNA are formed by the reduction of ribonucleotides. The 2′-hydroxyl group on the ribose moiety is replaced by a hydrogen atom. The substrates are ribonucleoside diphosphates, and the ultimate reductant is NADPH (**Figure 32.8**). The same enzyme, *ribonucleotide reductase*, acts on all four ribonucleotides. The overall stoichiometry is

$$\text{Ribonucleoside diphosphate} + \text{NADPH} + \text{H}^+ \xrightarrow{\text{Ribonucleotide reductase}}$$
$$\text{deoxyribonucleotide diphosphate} + \text{NADP}^+ + \text{H}_2\text{O}$$

The reaction mechanism is more complex than implied by this equation. Ribonucleotide reductase catalyzes only the last step in this reaction. The reductase itself must be reduced to perform another catalytic cycle, and the electrons for this reduction come from NADPH, but not directly. One carrier of reducing power linking NADPH with the reductase is *thioredoxin*, a 12-kDa protein with

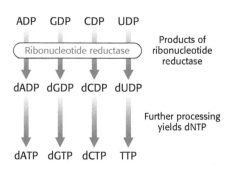

Figure 32.8 The formation of deoxyribonucleotides. Ribonucleotide reductase catalyzes the formation of deoxyribonucleotides from ribonucleotides.

two exposed cysteine residues near each other. These sulfhydryls are oxidized to a disulfide in the reaction catalyzed by ribonucleotide reductase:

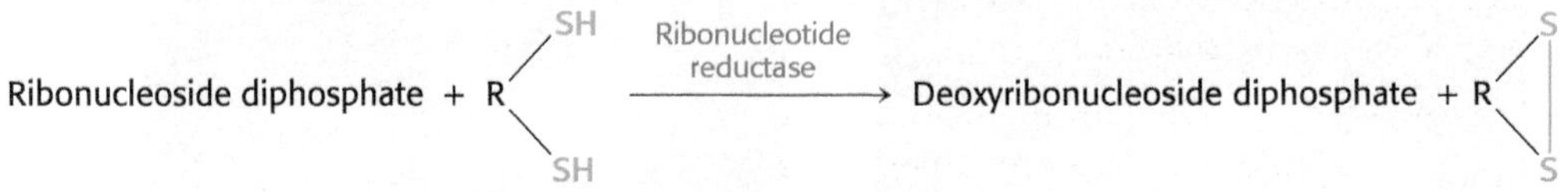

In turn, reduced thioredoxin is regenerated by electron flow from NADPH. This reaction is catalyzed by *thioredoxin reductase,* a flavoprotein. Electrons flow from NADPH to bound FAD of the reductase, to the disulfide of oxidized thioredoxin, and then to ribonucleotide reductase and, finally, to the ribose unit:

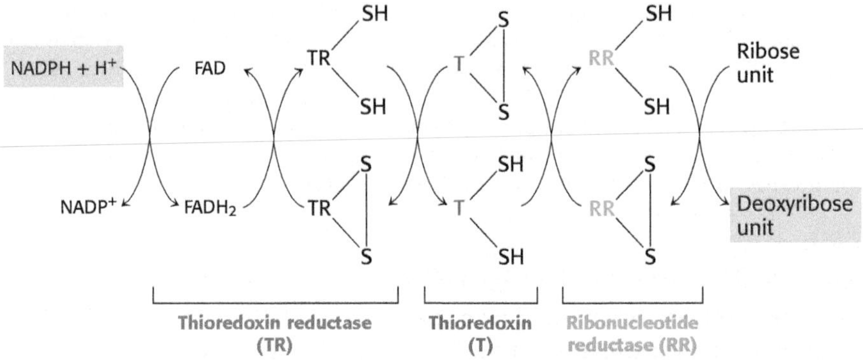

Thymidylate Is Formed by the Methylation of Deoxyuridylate

The preceding reactions have generated deoxyribonucleotides, including deoxyuridine diphosphate (dUDP). However, uracil-containing nucleotides are not components of DNA. Rather, DNA contains *thymine,* a methylated analog of uracil. Another step is required to generate thymidylate from uracil. *Thymidylate synthase* catalyzes this finishing touch: deoxyuridine monophosphate (dUMP) is methylated to thymine monophosphate (TMP, or thymidylate). The methyl donor in this reaction is N^5,N^{10}-methylenetetrahydrofolate (p. 577):

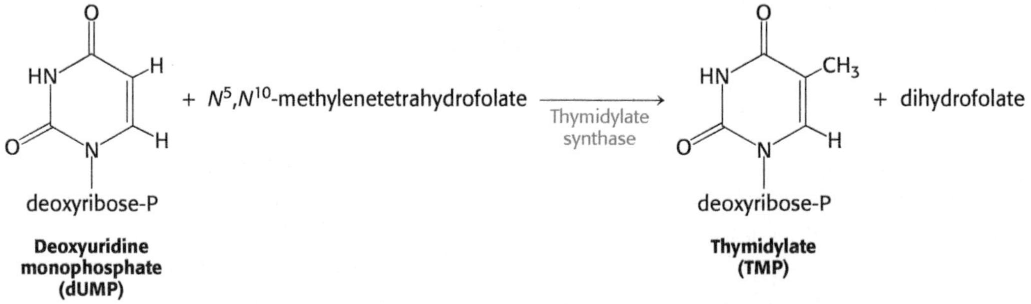

Tetrahydrofolate is regenerated from the dihydrofolate produced in the synthesis of thymidylate. This regeneration is accomplished by *dihydrofolate reductase* with the use of NADPH as the reductant:

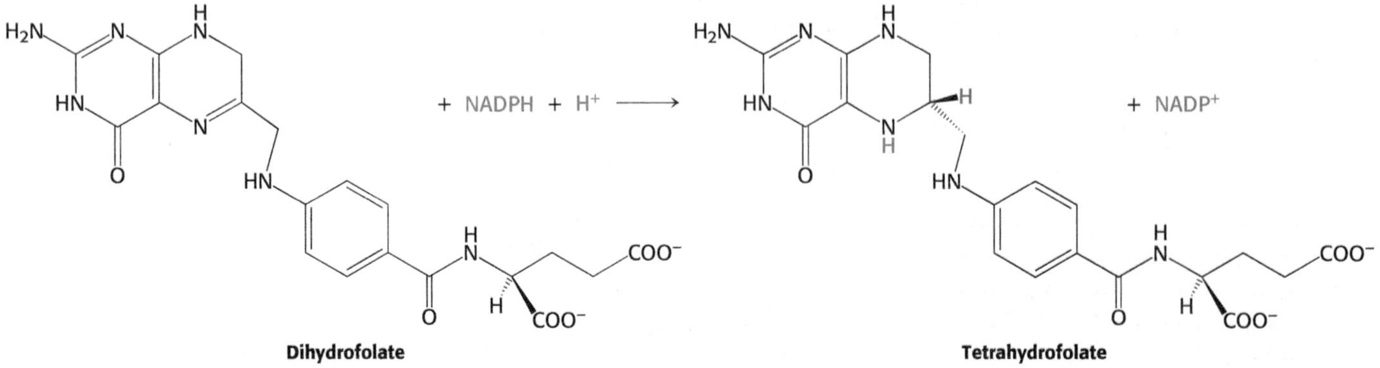

CLINICAL INSIGHT

Several Valuable Anticancer Drugs Block the Synthesis of Thymidylate

Rapidly dividing cells require an abundant supply of thymidylate, the only nucleotide specific to DNA, for DNA synthesis. The vulnerability of these cells to the inhibition of TMP synthesis has been exploited in cancer chemotherapy. Thymidylate synthase and dihydrofolate reductase are choice targets of chemotherapy (Figure 32.9).

Fluorouracil, an anticancer drug, is converted in vivo into *fluorodeoxyuridylate* (F-dUMP). This analog of dUMP acts as a normal substrate of TMP synthesis before irreversibly inhibiting thymidylate synthase. F-dUMP provides an example of *suicide inhibition,* in which an enzyme converts a substrate into a reactive inhibitor that halts the enzyme's catalytic activity (p. 138).

The synthesis of TMP can also be blocked by inhibiting the regeneration of tetrahydrofolate. Analogs of dihydrofolate, such as *aminopterin* and *methotrexate* (amethopterin), are potent competitive inhibitors of dihydrofolate reductase:

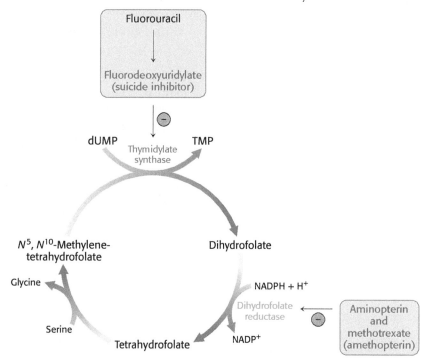

Figure 32.9 Anticancer drug targets. Thymidylate synthase and dihydrofolate reductase are choice targets in cancer chemotherapy because the generation of large quantities of precursors for DNA synthesis is required for rapidly dividing cancer cells.

Aminopterin (R = H) or methotrexate (R = CH₃)

Methotrexate is a valuable drug in the treatment of many rapidly growing tumors, such as those in acute leukemia and choriocarcinoma, a cancer derived from placental cells. However, methotrexate kills rapidly replicating cells whether they are malignant or not. Stem cells in bone marrow, epithelial cells of the intestinal tract, and hair follicles are vulnerable to the action of this folate antagonist, accounting for its toxic side effects, which include weakening of the immune system, nausea, and hair loss.

Folate analogs such as *trimethoprim* have potent antibacterial and antiprotozoal activity. Trimethoprim binds 10^5-fold less tightly to mammalian dihydrofolate reductase than it does to reductases of susceptible microorganisms. Small differences in the active-site clefts of these enzymes account for trimethoprim's highly selective antimicrobial action. The combination of trimethoprim and sulfamethoxazole (an inhibitor of folate synthesis) is widely used to treat infections such as bronchitis, traveler's diarrhea, and urinary tract infections.

Trimethoprim

✓ 6 Explain how nucleotide synthesis is regulated.

32.5 Nucleotide Biosynthesis Is Regulated by Feedback Inhibition

Nucleotide biosynthesis is regulated by feedback inhibition in a manner similar to the regulation of amino acid biosynthesis (p. 579). These regulatory pathways ensure that the various nucleotides are produced in the required quantities.

Pyrimidine Biosynthesis Is Regulated by Aspartate Transcarbamoylase

Aspartate transcarbamoylase (ATCase) is a key enzyme for the regulation of pyrimidine biosynthesis in bacteria. *ATCase is inhibited by CTP, the final product of pyrimidine biosynthesis*, and stimulated by ATP. This coupling of inhibition by a pyrimidine nucleotide with stimulation by a purine nucleotide serves to balance the two nucleotide pools.

$$\text{Aspartate} + \text{carbamoyl phosphate} \xrightarrow[\text{ATCase}]{} \text{carbamoylaspartate} \rightarrow \rightarrow \rightarrow \text{UMP} \longrightarrow \text{UDP} \longrightarrow \text{UTP} \longrightarrow \text{CTP}$$

⊖ (inhibition from CTP)   ⊕↑ ATP (stimulation)

Carbamoyl phosphate synthetase also is a site of feedback inhibition in both bacteria and eukaryotes.

The Synthesis of Purine Nucleotides Is Controlled by Feedback Inhibition at Several Sites

The regulatory scheme for purine nucleotides is more complex than that for pyrimidine nucleotides (**Figure 32.10**).

1. The committed step in purine nucleotide biosynthesis is the conversion of PRPP into phosphoribosylamine by *glutamine phosphoribosyl amidotransferase*. This important enzyme is feedback-inhibited by many purine ribonucleotides. It is noteworthy that AMP and GMP, the final products of the pathway, synergistically inhibit the amidotransferase.

2. Inosinate is the branch point in the synthesis of AMP and GMP. *The reactions leading away from inosinate are sites of feedback inhibition*. AMP inhibits the conversion of inosinate into adenylosuccinate, its immediate precursor. Similarly, GMP inhibits the conversion of inosinate into xanthylate, its immediate precursor.

3. As already noted, GTP is a substrate in the synthesis of AMP, whereas ATP is a substrate in the synthesis of GMP (p. 590). This reciprocal substrate relation tends to balance the synthesis of adenine and guanine ribonucleotides.

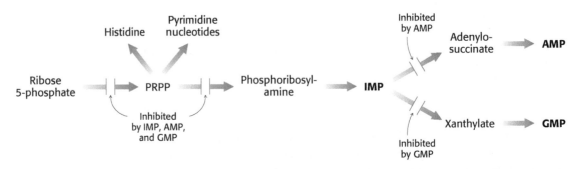

Figure 32.10 The control of purine biosynthesis. Feedback inhibition controls both the overall rate of purine biosynthesis and the balance between AMP and GMP production.

⚕ CLINICAL INSIGHT

The Synthesis of Deoxyribonucleotides Is Controlled by the Regulation of Ribonucleotide Reductase

The reduction of ribonucleotides to deoxyribonucleotides is precisely controlled by allosteric interactions. *E. coli* ribonucleotide reductase is a homodimeric enzyme. Each subunit contains two allosteric sites: one of them controls the *overall activity* of the enzyme, whereas the other regulates *substrate specificity* (**Figure 32.11**). The overall catalytic activity of ribonucleotide reductase is diminished by the binding of dATP, which signals an abundance of deoxyribonucleotides. The binding of ATP reverses this feedback inhibition. The binding of dATP or ATP to the substrate-specificity control site enhances the reduction of the pyrimidine nucleotides, UDP and CDP, to their deoxynucleotide forms. The binding of thymidine triphosphate (TTP) promotes the reduction of GDP and inhibits the further reduction of pyrimidine ribonucleotides. The subsequent increase in the level of dGTP stimulates the reduction of ADP to dADP. This complex pattern of regulation supplies the appropriate balance of the four deoxyribonucleotides needed for the synthesis of DNA.

Ribonucleotide reductase is an attractive target for cancer therapy, and a number of clinically approved anticancer drugs mimic substrates and regulators of the enzyme. The pyrmidine analog gemcitabine, once converted to the diphosphate form in vivo, becomes a suicide inhibitor of the reductase used to treat advanced pancreatic cancer. The purine analogs clofarabine and cladribine, upon conversion to their triphosphate forms in vivo, are dATP analogs and act as allosteric inhibitors of the enzyme. Clofarabine is used to treat pediatric acute myeloid leukemia while cladribine is effective against some forms of chronic lymphoid leukemia.

Gemcitabine

Clofarabine

Cladribine

(A)

Active site

Allosteric site (activity)

Allosteric site (specificity)

(B) Regulation of overall activity

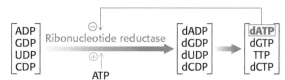

Regulation of substrate specificity

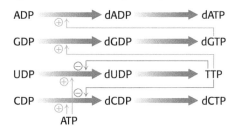

Figure 32.11 The regulation of ribonucleotide reductase. (A) Each subunit in the reductase dimer contains two allosteric sites in addition to the active site. One of the allosteric sites regulates the overall activity, and the other regulates substrate specificity. (B) The patterns of regulation of ribonucleotide reductase by the different nucleoside diphosphates.

32.6 Disruptions in Nucleotide Metabolism Can Cause Pathological Conditions

The nucleotides of a cell undergo continual turnover. Nucleotides are hydrolytically degraded to nucleosides by *nucleotidases*. The phosphorolytic cleavage of nucleosides to free bases and ribose 1-phosphate (or deoxyribose 1-phosphate) is catalyzed by *nucleoside phosphorylases*. As already discussed (p. 589 and p. 593), some of the bases are reused to form nucleotides by salvage pathways. Others are degraded to products that are excreted (**Figure 32.12**). A deficiency of an enzyme can disrupt these pathways, leading to a pathological condition.

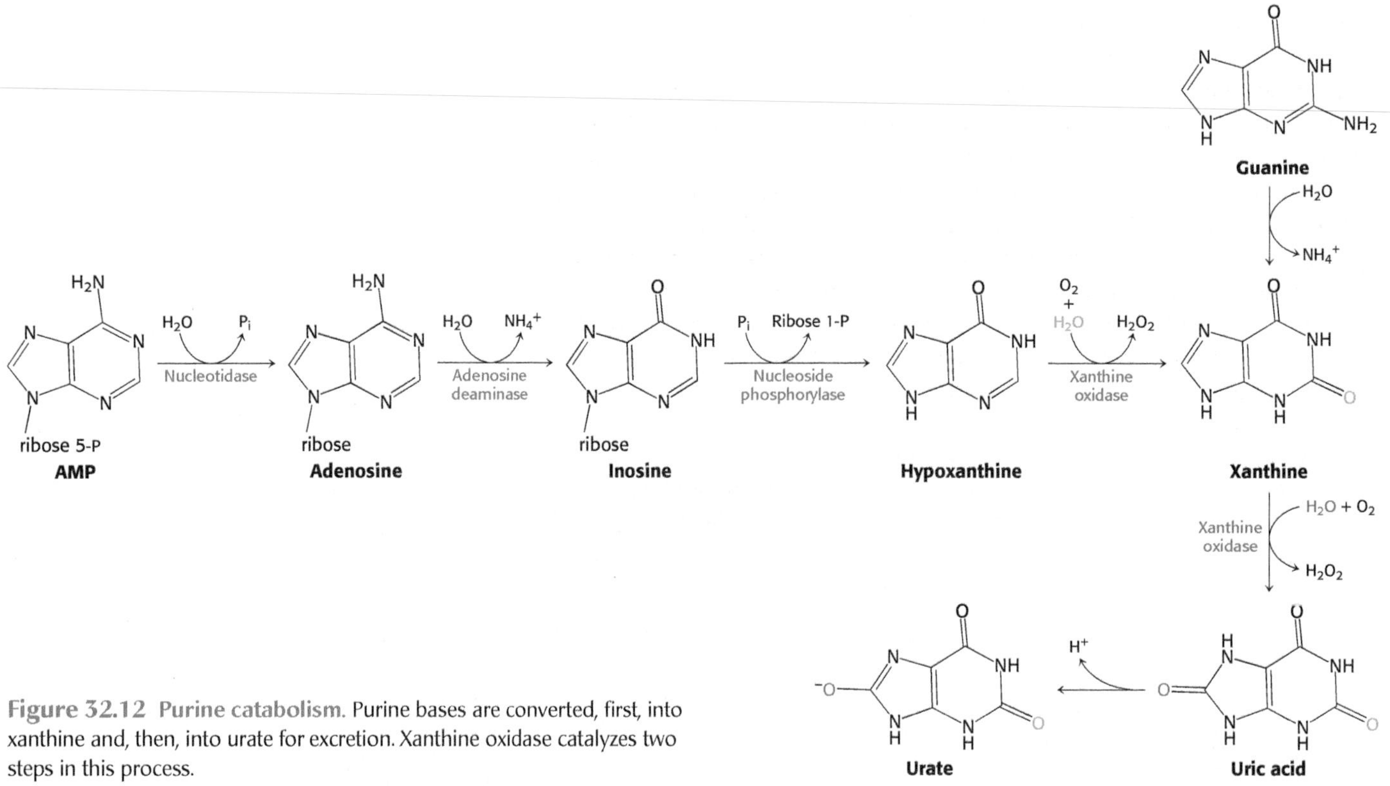

Figure 32.12 Purine catabolism. Purine bases are converted, first, into xanthine and, then, into urate for excretion. Xanthine oxidase catalyzes two steps in this process.

🜊 CLINICAL INSIGHT

The Loss of Adenosine Deaminase Activity Results in Severe Combined Immunodeficiency

The pathway for the degradation of AMP includes an extra step relative to that of the other nucleotides. First, the phosphoryl group is removed by a nucleotidase to yield the nucleoside adenosine (Figure 32.12). In the extra step, adenosine is deaminated by adenosine deaminase to form inosine. Finally, the ribose is removed by nucleoside phosphorylase, generating hypoxanthine and ribose 1-phosphate.

A deficiency in adenosine deaminase activity is associated with some forms of *severe combined immunodeficiency* (SCID), an immunological disorder. Persons with the disorder have acute recurring infections, often leading to death at an early age. SCID is characterized by a loss of T cells, which are crucial to the immune response. Although the biochemical basis of the disorder has not been clearly established, a lack of adenosine deaminase results in an increase of 50 to 100 times the normal level of dATP, which inhibits ribonucleotide reductase and, consequently, DNA synthesis. SCID is often called the "bubble boy disease" because some early treatments included complete isolation of the patient from the environment. The current treatment is bone-marrow transplantation. Adenosine deaminase deficiency was the first genetic disease to be treated by gene therapy.

CLINICAL INSIGHT

Gout Is Induced by High Serum Levels of Urate

After the production of hypoxanthine in the degradation of AMP, *xanthine oxidase* oxidizes hypoxanthine to *xanthine* and then to *uric acid*. Molecular oxygen, the oxidant in both reactions, is reduced to H_2O_2, which is decomposed to H_2O and O_2 by catalase. Uric acid loses a proton at physiological pH to form *urate* (Figure 32.12). *In human beings, urate is the final product of purine degradation and is excreted in the urine.*

High serum levels of urate (hyperuricemia) induce the painful joint disease *gout*. In this disease, the sodium salt of urate crystallizes in the fluid and lining of the joints (**Figure 32.13**). The small joint at the base of the big toe is a common site for sodium urate buildup, although the salt accumulates at other joints also. Painful inflammation results when cells of the immune system engulf the sodium urate crystals. The kidneys, too, may be damaged by the deposition of urate crystals. Gout is a common medical problem, affecting 1% of the population of Western countries. It is nine times as common in men as in women.

(A)

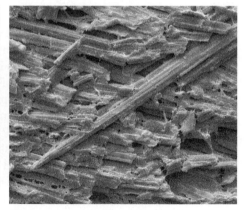

(B)

Figure 32.13 Gout. (A) Gout, "the king of diseases and the disease of kings," has been recognized as a malady since antiquity. Gout was believed to result from excessive food and drink consumption, excesses available only to the rich. Because of the pain in his inflamed leg, the corpulent musician feels as if he is being tortured by the devil. (B) Micrograph of sodium urate crystals. The accumulation of these crystals damages joints and kidneys. [(A) The color etching is from The Granger Collection. The quotation is from R. L. Wortmann, Gout and hyperuricemia. In *Kelley's Textbook of Rheumatology*, 8th ed., G. S. Firestein, Ed. (Saunders Elsevier, 2008), pp. 1481–1524. (B) Steve Gschmeissner/Science Source.]

The administration of *allopurinol*, an analog of hypoxanthine, is one treatment for gout. The mechanism of action of allopurinol is interesting: it acts first as a substrate and then as an inhibitor of xanthine oxidase. The oxidase hydroxylates allopurinol to *alloxanthine* (oxipurinol), which then remains tightly bound to the active site. We see here another example of *suicide inhibition*.

Allopurinol

The synthesis of urate from hypoxanthine and xanthine decreases soon after the administration of allopurinol. The serum concentrations of hypoxanthine and xanthine rise, and that of urate drops.

The average serum level of urate in human beings is close to the solubility limit and is higher than levels found in other primates. What is the selective advantage of a urate level so high that it teeters on the brink of gout in many people? Urate turns out to have a markedly beneficial action: it is a highly effective scavenger of reactive oxygen species. Indeed, urate is about as effective as ascorbate (vitamin C) as an antioxidant. The increased level of urate in human beings may protect against reactive oxygen species that are implicated in a host of pathological conditions.

 CLINICAL INSIGHT

Lesch–Nyhan Syndrome Is a Dramatic Consequence of Mutations in a Salvage-Pathway Enzyme

Mutations in genes that encode enzymes for nucleotide synthesis can reduce levels of needed nucleotides and can lead to an accumulation of intermediates. A nearly total absence of hypoxanthine-guanine phosphoribosyltransferase, an enzyme in the salvage pathway for guanylate and inosinate (p. 593), has unexpected and devastating consequences. The most striking expression of this inborn error of metabolism, called the *Lesch–Nyhan syndrome*, is compulsive self-destructive behavior. At age 2 or 3, children with this disease begin to bite their fingers and lips and will chew them off if unrestrained. These children also behave aggressively toward others. Mental deficiency and spasticity are other characteristics of Lesch–Nyhan syndrome. Elevated levels of urate in the serum lead to the formation of kidney stones early in life, followed by the symptoms of gout years later. The disease is inherited as an X-linked recessive disorder and predominantly affects males.

What is the connection between the absence of HGPRT activity and the behavioral characteristic of Lesch–Nyhan syndrome? The answer is not clear, but it is possible to speculate. The brain has limited capacity for de novo purine synthesis. Consequently, lack of HGPRT results in a deficiency of purine nucleotides. ATP and ADP, formed from inosinate, are especially important in the brain as signal molecules. These nucleotides bind to and activate G-protein coupled receptors that regulate the dopamine secreting neurons. Thus, the lack of hypoxanthine-guanine phosphoribosyl-transferase results in an imbalance of key neurotransmitters. Moreover, the guanosine nucleotides required for G-protein function may also be in short supply. The Lesch–Nyhan syndrome demonstrates that the salvage pathway for the synthesis of IMP and GMP is not of minor importance. Moreover, the Lesch–Nyhan syndrome reveals that *abnormal behavior such as self-mutilation and extreme hostility can be caused by the absence of a single enzyme.* Psychiatry will no doubt benefit from the unraveling of the molecular basis of such mental disorders.

 CLINICAL INSIGHT

Folic Acid Deficiency Promotes Birth Defects Such As Spina Bifida

Spina bifida is one of a class of birth defects characterized by the incomplete or incorrect formation of the neural tube early in development. In the United States, the prevalence of *neural-tube defects* is approximately 1 case per 1000 births. A variety of studies have demonstrated that the prevalence of neural-tube defects is reduced by as much as 70% when women take folic acid as a dietary supplement before and during the first trimester of pregnancy. One hypothesis is that more folate derivatives are needed for the synthesis of DNA precursors when cell division is frequent and substantial amounts of DNA must be synthesized.

SUMMARY

32.1 An Overview of Nucleotide Biosynthesis and Nomenclature

Nucleotides can be synthesized entirely from small precursor molecules in de novo pathways. Alternatively, intact bases can be recycled and attached to activated ribose in salvage pathways. A nucleoside consists of a base attached to a ribose or deoxyribose sugar, whereas a nucleotide is a phosphate ester of a nucleoside.

32.2 The Pyrimidine Ring Is Assembled and Then Attached to a Ribose Sugar

The pyrimidine ring is assembled first and then linked to ribose phosphate to form a pyrimidine nucleotide. 5-Phosphoribosyl-1-pyrophosphate (PRPP) is the donor of the ribose phosphate moiety. The synthesis of the pyrimidine ring starts with the formation of carbamoylaspartate from carbamoyl phosphate and aspartate, a reaction catalyzed by aspartate transcarbamoylase. Dehydration, cyclization, and oxidation yield orotate, which reacts with PRPP to give orotidylate. Decarboxylation of this pyrimidine nucleotide yields UMP. CTP is then formed by the amination of UTP.

32.3 The Purine Ring Is Assembled on Ribose Phosphate

The purine ring is assembled from a variety of precursors: glutamine, glycine, aspartate, N^{10}-formyltetrahydrofolate, and CO_2. The committed step in the de novo synthesis of purine nucleotides is the formation of 5-phosphoribosylamine from PRPP and glutamine. The purine ring is assembled on ribose phosphate, in contrast with the addition of the ribose to the preformed pyrimidine ring. The addition of glycine, followed by formylation, amination, and ring closure, yields 5-aminoimidazole ribonucleotide. This intermediate contains the completed five-membered ring of the purine skeleton. The addition of CO_2, the nitrogen atom of aspartate, and a formyl group, followed by ring closure, yields inosinate, a purine ribonucleotide. AMP and GMP are formed from IMP. Purine ribonucleotides can also be synthesized by a salvage pathway in which a preformed base reacts directly with PRPP.

32.4 Ribonucleotides Are Reduced to Deoxyribonucleotides

Deoxyribonucleotides, the precursors of DNA, are formed by the reduction of ribonucleoside diphosphates. These conversions are catalyzed by ribonucleotide reductase. Thioredoxin transfers electrons from NADPH to sulfhydryl groups at the active sites of this enzyme. TMP is formed by the methylation of dUMP. The donor of a methylene group and a hydride in this reaction is N^5,N^{10}-methylenetetrahydrofolate, which is converted into dihydrofolate. Tetrahydrofolate is regenerated by the reduction of dihydrofolate by NADPH. Dihydrofolate reductase, which catalyzes this reaction, is inhibited by folate analogs such as aminopterin and methotrexate. These compounds and fluorouracil, an inhibitor of thymidylate synthase, are used as anticancer drugs.

32.5 Nucleotide Biosynthesis Is Regulated by Feedback Inhibition

Pyrimidine biosynthesis in *E. coli* is regulated by the feedback inhibition of aspartate transcarbamoylase, the enzyme that catalyzes the committed step. CTP inhibits and ATP stimulates this enzyme. The feedback inhibition of glutamine-PRPP amidotransferase by purine nucleotides is important in regulating their biosynthesis. Disruption of the regulation and activity of ribonucleotide reductase by drugs is an effective chemotherapy for some types of cancer.

32.6 Disruptions in Nucleotide Metabolism Can Cause Pathological Conditions

Purines are degraded to urate in human beings. Severe combined immunodeficiency is caused by an inability to deaminate adenosine. Gout, a disease that affects joints and leads to arthritis, is associated with an excessive accumulation of urate. Lesch–Nyhan syndrome, a genetic disease characterized by self-mutilation, mental deficiency, and gout, is caused by a deficiency of hypoxanthine-guanine phosphoribosyltransferase. This enzyme is essential for the synthesis of purine nucleotides by the salvage pathway. Neural-tube defects are more frequent when a pregnant woman is deficient in folate derivatives early in pregnancy, possibly because of the important role of these derivatives in the synthesis of DNA precursors.

KEY TERMS

nucleoside (p. 587)
nucleotide (p. 587)
pyrimidine nucleotide (p. 587)
carbamoyl phosphate synthetase II
 (CPS II) (p. 587)
aspartate transcarbamoylase (p. 588)
5-phosphoribosyl-1-pyrophosphate
 (PRPP) (p. 588)

orotidylate (p. 588)
purine (p. 590)
inosinate (p. 590)
purinosomes (p. 592)
hypoxanthine-guanine
 phosphoribosyltransferase (HGPRT)
 (p. 593)
ribonucleotide reductase (p. 593)

thymidylate synthase (p. 594)
dihydrofolate reductase (p. 594)
severe combined immunodeficiency
 (SCID) (p. 598)
gout (p. 599)
Lesch–Nyhan syndrome (p. 600)
spina bifida (p. 600)

? Answers to QUICK QUIZZES

1. Bicarbonate, ammonia, and aspartate (Figure 32.3).

2. Carbon dioxide, aspartate, glutamine, N^{10}-formyltetrahydrofolate, glycine (Figure 32.4).

PROBLEMS

1. *From the beginning or extract and save and reuse.* Differentiate between the de novo synthesis of nucleotides and salvage-pathway synthesis. ✓ 5

2. *Activated ribose phosphate.* Write a balanced equation for the synthesis of PRPP from glucose through the oxidative branch of the pentose phosphate pathway.

3. *Making a pyrimidine.* Write a balanced equation for the synthesis of orotate from glutamine, CO_2, and aspartate. ✓ 5

4. *Identifying the donor.* What is the activated reactant in the biosynthesis of each of the following compounds? ✓ 5

(a) Phosphoribosylamine
(b) Carbamoylaspartate
(c) Orotidylate (from orotate)

5. *An "s" instead of a "t"?* Differentiate between a nucleoside and a nucleotide.

6. *Associate 'em.* Match the term on the right with the description on the left. ✓ 5

(a) Excessive urate _____
(b) Lack of adenosine
  deaminase _____
(c) Lack of HGPRT _____
(d) Carbamoyl phosphate
  synthetase _____
(e) Inosinate _____
(f) Ribonucleotide reductase
  _____
(g) Lack of folic acid _____
(h) Glutamine phosphoribosyl
  transferase _____
(i) Single ring _____
(j) Bicyclic ring _____
(k) Precursor to CTP _____

1. Spina bifida
2. Precursor to both
  ATP and GTP
3. Purine
4. Deoxynucleotide
  synthesis
5. UTP
6. Lesch–Nyhan
  syndrome
7. Immunodeficiency
8. Pyrimidine
9. Gout
10. First step in pyrimidine synthesis
11. Committed step in
  purine synthesis

7. *The price of methylation.* Write a balanced equation for the synthesis of TMP from dUMP that is coupled to the conversion of serine into glycine.

8. *Sulfa action.* Bacterial growth is inhibited by sulfanilamide and related sulfa drugs, and there is a concomitant accumulation of 5-aminoimidazole-4-carboxamide ribonucleotide. This inhibition is reversed by the addition of *p*-aminobenzoate.

$$H_2N-\bigcirc\!\!\!-SO_2NH_2$$

Sulfanilamide

Propose a mechanism for the inhibitory effect of sulfanilamide.

9. *A generous donor.* What major biosynthetic reactions utilize PRPP?

10. *Bringing equilibrium.* What is the reciprocal substrate relation in the synthesis of ATP and GTP? ✓ 6

11. *Calculate the ATP footprint.* How many molecules of ATP are required to synthesize one molecule of CTP from scratch?

12. *Find the label.* Suppose that cells are grown on amino acids that have all been labeled at the α carbons with ^{13}C. Identify the atoms in cytosine and guanine that will be labeled with ^{13}C. ✓ 5

13. *On the trail of carbons.* Tissue-culture cells were incubated with glutamine labeled with ^{15}N in the amide group. Subsequently, IMP was isolated and found to contain some ^{15}N. Which atoms in IMP were labeled? ✓ 5

14. *Blockages.* What intermediate in purine synthesis will accumulate if a strain of bacteria is lacking each of the following biochemicals? ✓ 5

(a) Aspartate
(b) Tetrahydrofolate
(c) Glycine
(d) Glutamine

15. *Mechanism of action.* What is the biochemical basis of allopurinol treatment for gout? ✓ 6

16. *Adjunct therapy.* Allopurinol is sometimes given to patients with acute leukemia who are being treated with anticancer drugs. Why is allopurinol used? ✓ 6

17. *Folate deficiency.* Suppose that someone was suffering from a folate deficiency. What cells would you think might be most affected? Symptoms may include diarrhea and anemia.

18. *Correcting deficiencies.* Suppose that a person is found who is deficient in an enzyme required for IMP synthesis. How might this person be treated?

19. *Labeled nitrogen.* Purine biosynthesis is allowed to take place in the presence of [^{15}N]aspartate, and the newly synthesized GTP and ATP are isolated. What positions are labeled in the two nucleotides? ✓ 5

20. *Needed supplies.* Why are cancer cells especially sensitive to inhibitors of TMP synthesis? ✓ 6

Chapter Integration Problems

21. *They're everywhere!* Nucleotides play a variety of roles in the cell. Give an example of a nucleotide that acts in each of the following roles or processes.

(a) Second messenger
(b) Phosphoryl-group transfer
(c) Activation of carbohydrates
(d) Activation of acetyl groups
(e) Transfer of electrons
(f) Chemotherapy
(g) Allosteric effector

22. *Exercising muscle.* Some interesting reactions take place in muscle tissue to facilitate the generation of ATP for contraction. ✓ 6

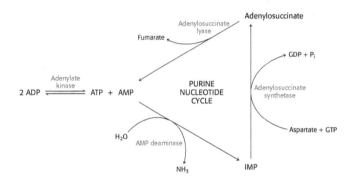

In muscle contraction, ATP is converted into ADP. Adenylate kinase converts two molecules of ADP into a molecule of ATP and AMP.

(a) Why is this reaction beneficial to contracting muscle?
(b) Why is the equilibrium for the adenylate kinase approximately equal to 1?

Muscle can metabolize AMP by using the purine nucleotide cycle. The initial step in this cycle, catalyzed by AMP deaminase, is the conversion of AMP into IMP.

(c) Why might the deamination of AMP facilitate ATP formation in muscle?
(d) How does the purine nucleotide cycle assist the aerobic generation of ATP?

23. *Different strokes.* Human beings contain two different carbamoyl phosphate synthetase enzymes. One uses glutamine as a substrate, whereas the other uses ammonia. What are the functions of these two enzymes?

24. *Your pet duck.* You suspect that your pet duck has gout. Why should you think twice before administering a dose of allopurinol-laced bread?

Challenge Problems

25. *Inhibiting purine biosynthesis.* Amidotransferases are inhibited by the antibiotic and antitumor agent azaserine (*O*-diazoacetyl-L-serine), which is an analog of glutamine. ✓ 5

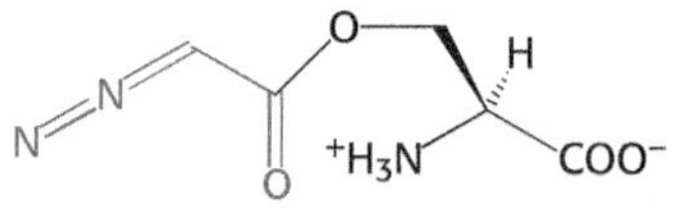

Azaserine

Which intermediates in purine biosynthesis would accumulate in cells treated with azaserine?

26. *HAT medium.* Mutant cells unable to synthesize nucleotides by salvage pathways are very useful tools in molecular and cell biology. Suppose that cell A lacks thymidine kinase, the enzyme catalyzing the phosphorylation of thymidine to thymidylate, and that cell B lacks hypoxanthine-guanine phosphoribosyl transferase.

(a) Cell A and cell B do not proliferate in a HAT medium containing *h*ypoxanthine, *a*minopterin or *a*methopterin (methotrexate), and *t*hymine. However, cell C formed by the fusion of cells A and B grows in this medium. Why?

(b) Suppose that you wanted to introduce foreign genes into cell A. Devise a simple means of distinguishing between cells that have taken up foreign DNA and those that have not. ✓ 5

27. *Hyperuricemia.* Many patients with glucose 6-phosphatase deficiency have high serum levels of urate. Hyperuricemia can be induced in normal people by the ingestion of alcohol or by strenuous exercise. Propose a common mechanism that accounts for these findings.

28. *Labeled carbon.* Succinate uniformly labeled with ^{14}C is added to cells actively engaged in pyrimidine biosynthesis. Propose a pathway by which carbon atoms from succinate can be incorporated into a pyrimidine. At what positions is the pyrimidine labeled? ✓ 5

29. *Something funny going on here.* Cells were incubated with glucose labeled with ^{14}C in carbon 2, shown in red in the following structure. Later, uracil was isolated and found to contain ^{14}C in carbons 4 and 6. Account for this labeling pattern. ✓ 5

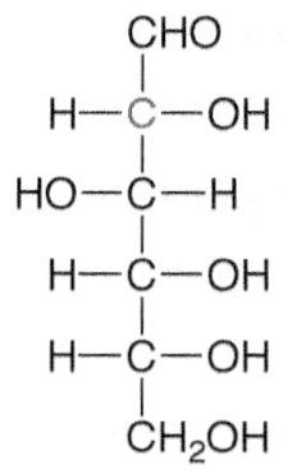

Selected Readings for this chapter can be found online at www.whfreeman.com/tymoczko3e.

CHAPTER 33
The Structure of Informational Macromolecules: DNA and RNA

CHAPTER 34
DNA Replication

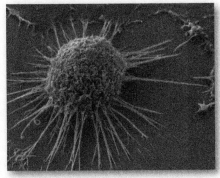

CHAPTER 35
DNA Repair and Recombination

Nucleic Acid Structure and DNA Replication

Thus far in our study of biochemistry, we have focused on the extraction of energy from fuels and the synthesis of fuel stores for later use. We have examined the metabolism of many small molecules, including carbohydrates, amino acids, and nucleotides. In this section and the following ones, we will examine how energy and small molecules are used to construct the macromolecules of biochemistry. Nucleic acids—DNA and RNA—function as the informational molecules in the cell. DNA is the stable genetic information that is passed from one generation to the next, whereas RNA, a biologically accessible copy of DNA, is a more transient copy of this information. Information carried by nucleic acids is represented by a sequence of only four letters: A, G, C, and T for DNA; and A, G, C, and U for RNA. The arrangement of these four letters into sequences of various lengths provides the blueprint for every living organism on Earth.

To preserve the information encoded in DNA through many cell divisions and prevent the introduction of mutations, the copying of the genetic information must be extremely faithful: an error rate of less than 1 base pair (bp) per 3×10^9 bp is required. Such remarkable accuracy is achieved through a multilayered system of accurate DNA synthesis, proofreading during DNA synthesis, and repair after replication.

Even after DNA has been replicated, the genome is still not safe. A variety of environmental elements can damage DNA, introducing changes in the DNA sequence (mutations), or cause lesions that can block DNA replication. All organisms contain DNA-repair systems that detect DNA damage and act to preserve the original sequence. Mutations in genes that encode components of DNA-repair systems are key factors in the development of cancer.

We begin our study of nucleic acids by focusing on structure. Especially noteworthy is the double-helical structure of DNA and how this structure facilitates the transfer of genetic information from one generation to the next. After this structural examination, we will undertake a mechanistic examination of the process of DNA replication and repair.

✓ **By the end of this section, you should be able to:**

✓ 1 List the components of nucleic acids.

✓ 2 Describe how DNA is packaged to fit inside the cell.

✓ 3 Identify the enzymes that take part in the process of DNA replication.

✓ 4 Describe how replication is organized, and distinguish between eukaryotic replication and bacterial replication.

✓ 5 Explain the mechanisms that ensure the fidelity of DNA replication.

CHAPTER 33

The Structure of Informational Macromolecules: DNA and RNA

Family resemblance, very evident in this photograph of four sisters, results from having genes in common. Genes must be expressed to exert an effect, and proteins regulate such expression. [© Nicholas Nixon, courtesy Fraenkel Gallery, San Francisco.]

By the 1950s, the composition of DNA was well known. Scientists knew that DNA is composed of four bases—A,T,G and C—and that the ratio of adenine (A) to thymine (T) and that of guanine (G) to cytosine (C) are always one. The actual structure of the DNA molecule, however, was a mystery. The elucidation of the structure of DNA is one of the key findings in the history of biological sciences. Perhaps the most exciting aspect of the structure of DNA deduced by James Watson and Francis Crick in 1953 was, as expressed in their words, that the "specific pairing we have postulated immediately suggests a possible copying mechanism for the genetic material." A *double helix* separated into two single strands can be replicated because each strand serves as a template on which its complementary strand can be assembled (**Figure 33.1**). In this chapter, we will examine the structure of DNA and how it is packaged inside the cell.

✓ 1 List the components of nucleic acids.

33.1 A Nucleic Acid Consists of Bases Linked to a Sugar–Phosphate Backbone

The nucleic acids DNA and RNA are well suited to functioning as the carriers of genetic information by virtue of their structures. These macromolecules are *linear polymers* built up from similar units connected end to end (**Figure 33.2**). Each monomer unit within the polymer consists of three components: a sugar, a phosphate, and a base. *The sequence of bases uniquely characterizes a nucleic acid and is a form of linear information.*

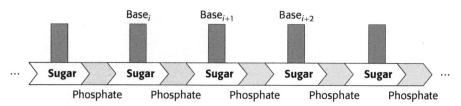

Figure 33.2 The polymeric structure of nucleic acids.

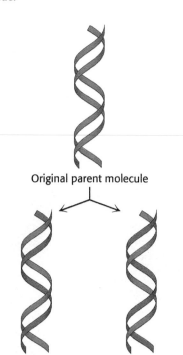

Original parent molecule

First-generation daughter molecules

Figure 33.1 DNA replication. Each strand of one double helix (shown in blue) acts as a template for the synthesis of a new complementary strand (shown in red).

DNA and RNA Differ in the Sugar Component and One of the Bases

The sugar in *deoxyribonucleic acid* (DNA) is *deoxyribose*. The "deoxy" prefix refers to the fact that the 2′-carbon atom of the sugar lacks the oxygen atom that is linked to the 2′-carbon atom of *ribose* (the sugar in ribonucleic acid, or RNA), as shown in **Figure 33.3**. The sugars in nucleic acids are linked to one another by phosphodiester bridges. Specifically, the 3′-hydroxyl (3′-OH) group of the sugar component of one nucleotide is bonded to a phosphoryl group, and the phosphoryl group is, in turn, joined to the 5′-hydroxyl group of the adjacent sugar, forming a phosphodiester linkage. The strand of sugars linked by phosphodiester bridges is referred to as the *backbone* of the nucleic acid (**Figure 33.4**). The backbone is constant in DNA and RNA, but the bases vary from one monomer to the next. Two of the bases are derivatives of *purine*—adenine and guanine—and two are derivatives of *pyrimidine*—cytosine and thymine (DNA only) or uracil (RNA only), as shown in **Figure 33.5**.

Ribonucleic acid (RNA), like DNA, is a long unbranched polymer consisting of nucleotides joined by 3′ → 5′ phosphodiester linkages (Figure 33.4). The structure of RNA differs from that of DNA in two respects: (1) the sugar units in RNA are riboses rather than deoxyriboses and (2) one of the four major bases in RNA is uracil instead of thymine (Figure 33.5).

Note that each phosphodiester linkage in the backbone of both DNA and RNA has a negative charge. This negative charge repels nucleophilic species such as hydroxide ion, which are capable of hydrolytically cleaving the phosphodiester linkages of the nucleic acid backbone. This resistance is crucial for maintaining the integrity of information stored in nucleic acids. The absence of the 2′-hydroxyl group in DNA further increases its resistance to hydrolysis. In the presence of nucleophilic species, a 2′-hydroxyl group would hydrolyze the phosphodiester linkage and cause a break in the nucleic acid backbone. The greater stability of DNA is one of the reasons for its use rather than RNA as the hereditary material in all modern cells and in many viruses.

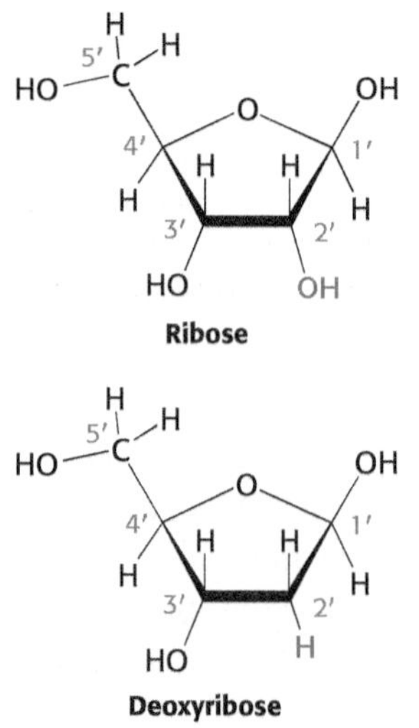

Ribose

Deoxyribose

Figure 33.3 Ribose and deoxyribose. Atoms in sugar units are numbered with primes to distinguish them from atoms in bases.

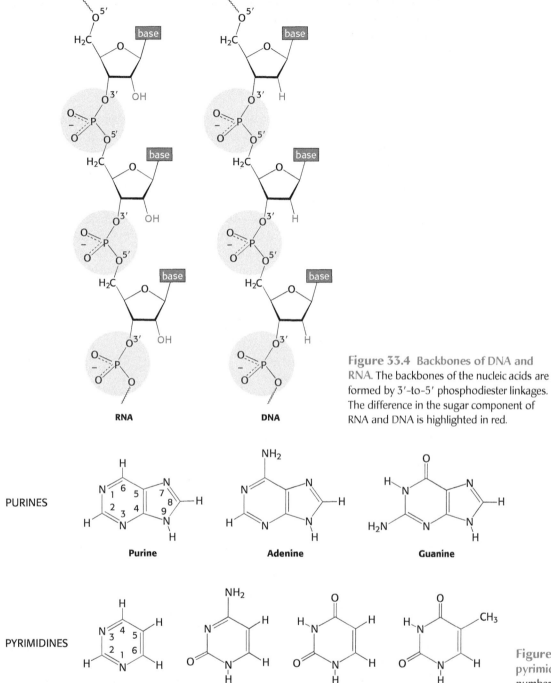

Figure 33.4 Backbones of DNA and RNA. The backbones of the nucleic acids are formed by 3′-to-5′ phosphodiester linkages. The difference in the sugar component of RNA and DNA is highlighted in red.

PURINES

Purine   Adenine   Guanine

PYRIMIDINES

Pyrimidine   Cytosine   Uracil   Thymine

Figure 33.5 Purines and pyrimidines. Atoms within bases are numbered without primes. Uracil is present in RNA instead of thymine.

Nucleotides Are the Monomeric Units of Nucleic Acids

We considered the synthesis of nucleotides, the building blocks of nucleic acids, in Chapter 32. Let's review the nomenclature of these crucial biomolecules. A unit consisting of a base bonded to a sugar is referred to as a *nucleoside*. The four nucleoside units in DNA are called *deoxyadenosine, deoxyguanosine, deoxycytidine,* and *thymidine.* Note that thymidine contains deoxyribose; however, by convention, the prefix "deoxy" is not added, because thymine-containing nucleotides are found only rarely in RNA. The nucleoside units in RNA are called *adenosine, guanosine, cytidine,* and *uridine.* In each case, N-9 of a purine or N-1 of a pyrimidine is attached to the 1′-carbon atom (C-1′) of the sugar (**Figure 33.6**). The base lies above the plane of the sugar when the structure is written in the standard orientation; that is, the configuration of the *N*-glycosidic linkage is *β*.

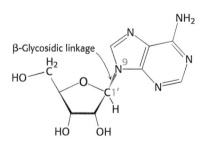

Figure 33.6 *β*-Glycosidic linkage in a purine nucleoside.

A *nucleotide* is a nucleoside joined to one or more phosphoryl groups by an ester linkage and is most commonly referred to as a nucleoside with the number of attached phosphoryl groups noted. For instance, a nucleoside monophosphate is a nucleotide. *Nucleoside triphosphates are the monomers—the building blocks—that are linked to form RNA and DNA.* The four nucleotide units that link to form DNA are nucleoside monophosphates called *deoxyadenylate, deoxyguanylate, deoxycytidylate,* and *thymidylate.*

The base name with the suffix "ate" alone is a nucleotide, but with an unknown number of phosphoryl groups attached at an undesignated carbon atom of the sugar. A more precise nomenclature also is commonly used. A compound formed by the attachment of a phosphoryl group to C-5′ of a nucleoside sugar (the most common site of phosphate esterification) is called a *nucleoside 5′-phosphate* or a *5′-nucleotide.* In this naming system for nucleotides, the number of phosphoryl groups and the attachment site are designated. For example, ATP is *adenosine 5′-triphosphate.* Another example of a nucleotide is deoxyguanosine 3′-monophosphate (3′-dGMP; Figure 33.7). This nucleotide differs from ATP in several ways: it contains guanine rather than adenine and deoxyribose rather than ribose (indicated by the prefix "d"), it has one phosphoryl group rather than three, and it has the phosphoryl group esterified to the hydroxyl group in the 3′ rather than the 5′ position.

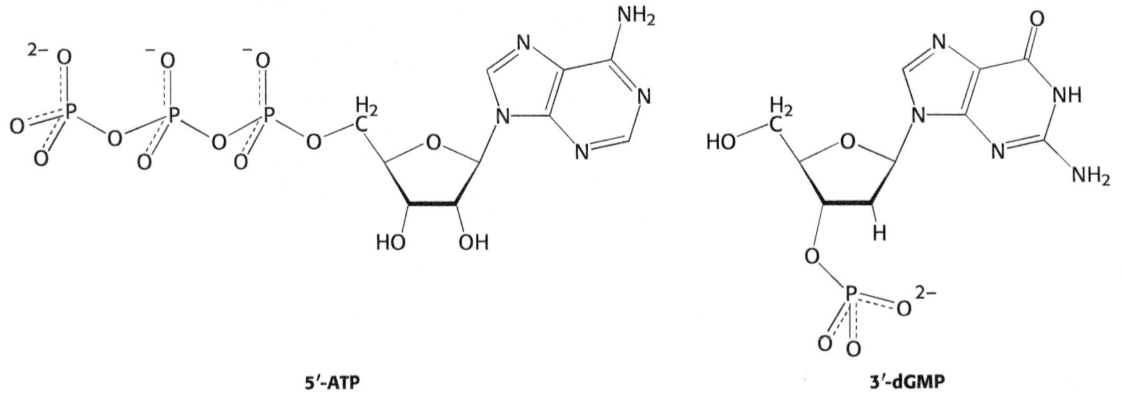

5′-ATP

3′-dGMP

Figure 33.7 Nucleotides adenosine 5′-triphosphate (5′-ATP) and deoxyguanosine 3′-monophosphate (3′-dGMP).

DNA Molecules Are Very Long and Have Directionality

Scientific communication frequently requires the sequence of a nucleic acid—in some cases, a sequence thousands of nucleotides in length—to be written. Rather than writing the cumbersome chemical structures, scientists have adopted the use of abbreviations. The abbreviated notations pApCpG, pACG, or, most commonly, ACG denote a trinucleotide. In regard to DNA, the trinucleotide consists of deoxyadenylate monophosphate, deoxycytidylate monophosphate, and deoxyguanylate monophosphate joined together by phosphodiester linkages, in which "p" denotes a phosphoryl group. The phosphoryl groups and sugar components of the trinucleotide are abbreviated as well, as shown in Figure 33.8. The 5′ end will often have a phosphoryl group attached to the 5′-OH group, which means that *a DNA or RNA strand has directionality.* One end of the strand has a free 5′-OH group (or a 5′-OH group attached to a phosphoryl group), whereas the other end has a 3′-OH group, and neither end is linked to another nucleotide. By convention, *the base sequence is written in the 5′-to-3′ direction.* Thus, the symbol ACG indicates that the phosphorylated or unlinked 5′-OH group is on deoxyadenylate (or adenylate), whereas the unlinked 3′-OH group is on deoxyguanylate (or guanylate). Because of this directionality, ACG and GCA correspond to different compounds.

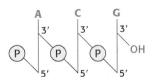

Figure 33.8 The structure of a DNA strand. The strand has a 5′ end, which is usually attached to a phosphoryl group, and a 3′ end, which is usually a free hydroxyl group.

A striking characteristic of naturally occurring DNA molecules is their length. A DNA molecule must comprise many nucleotides to carry the genetic information necessary for even the simplest organisms. For example, the DNA of a virus such as polyoma, which can cause cancer in certain organisms, is 5100 nucleotides in length. The *E. coli* genome is a single DNA molecule consisting of two strands of 4.6 million nucleotides (**Figure 33.9**).

DNA molecules from higher organisms can be much larger. The human genome comprises approximately 3 billion base pairs distributed among 24 distinct DNA molecules—22 autosomes plus two sex chromosomes (X and Y)—of different sizes. One of the largest known DNA molecules is found in the Indian muntjac, an Asiatic deer; its genome is nearly as large as the human genome but is distributed among only 3 chromosomes (**Figure 33.10**). The largest of these chromosomes has strands of more than 1 billion base pairs. If such a DNA molecule could be fully extended, it would stretch more than 1 foot in length. Some plants contain even larger DNA molecules.

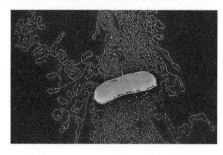

Figure 33.9 An electron micrograph of part of the *E. coli* genome. The *E. coli* was lysed, extruding the DNA. [Dr. Gopal Murti/ Science Source.]

 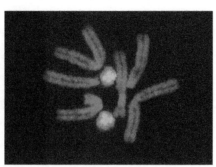

Figure 33.10 The Indian muntjac and its chromosomes. Cells from a female Indian muntjac contain three pairs of very large chromosomes (stained orange in the micrograph). The cell shown is a hybrid containing a pair of human chromosomes (stained green) for comparison. [(Left) Hugh Lansdown/Shutterstock. (Right) Reprinted by permission from Macmillan Publishers Ltd: *Nature Genetics*, J.–Y. Lee, M. Koi, E. J. Stanbridge, M. Oshimura, A. T. Kumamoto, & A. P. Feinberg, Simple purification of human chromosomes to homogeneity using muntjac hybrid cells, vol. 7, p.30, ©1994.]

33.2 Nucleic Acid Strands Can Form a Double-Helical Structure

The covalent structure of nucleic acids accounts for their ability to carry information in the form of a sequence of bases. Other features of nucleic acid structure facilitate the process of *replication*—that is, the generation of two copies of a nucleic acid from one. DNA replication is the basis of cell duplication, growth, and, ultimately, reproduction. Replication depends on the ability of the bases found in nucleic acids to form *specific base pairs* in such a way that a helical structure consisting of two strands is formed.

The Double Helix Is Stabilized by Hydrogen Bonds and the Hydrophobic Effect

The discovery of the DNA double helix has been chronicled in many articles, books, and even a movie. James Watson and Francis Crick, using data obtained by Maurice Wilkins and Rosalind Franklin and simple molecular models, inferred a structural model for DNA that was the source of some remarkable insights into the functional properties of nucleic acids (**Figure 33.11**).

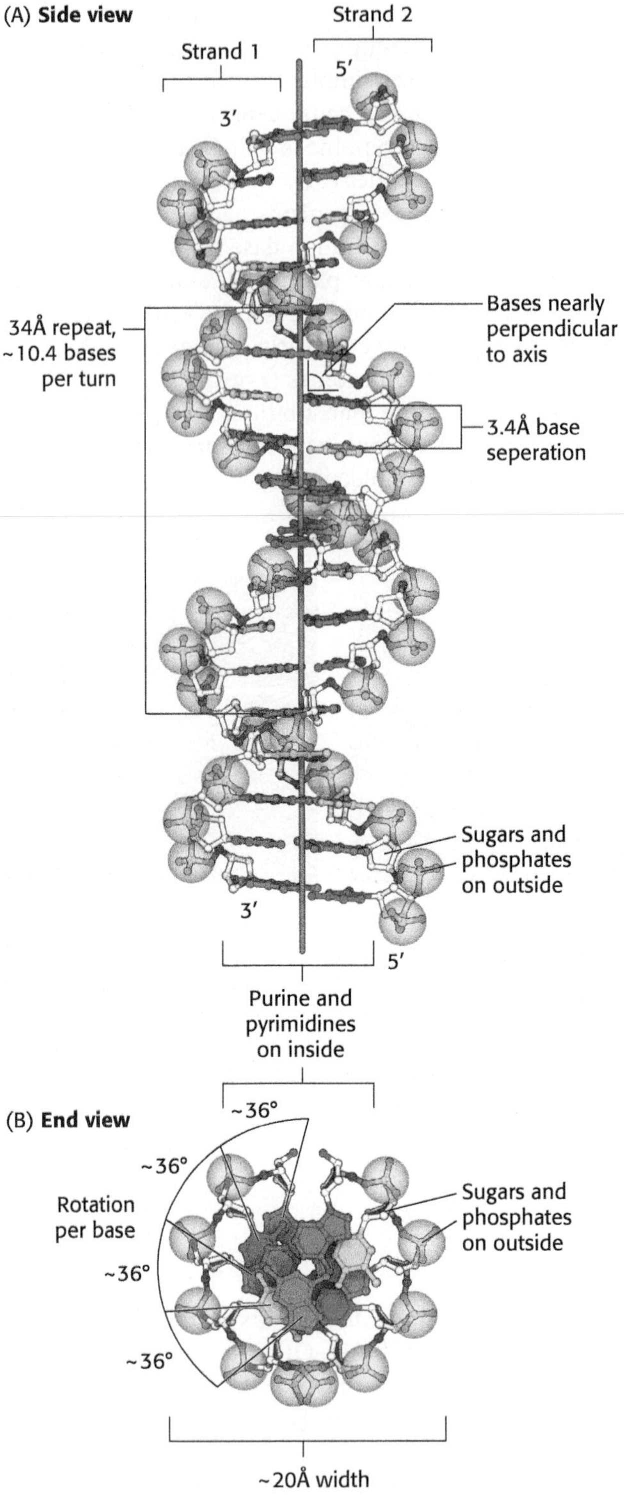

(A) Side view

Strand 1

Strand 2

5′

3′

34Å repeat, ~10.4 bases per turn

Bases nearly perpendicular to axis

3.4Å base seperation

Sugars and phosphates on outside

3′

3′

5′

Purine and pyrimidines on inside

(B) End view

~36°

~36°

Rotation per base

~36°

~36°

Sugars and phosphates on outside

~20Å width

Figure 33.11 A skeletal representation of the double helix illustrates the helix properties. (A) Side view. Adjacent bases are separated by 3.4 Å. The structure repeats along the helical axis (vertical) at intervals of 34 Å, which corresponds to 10 nucleotides on each strand. (B) End (axial) view, looking down the helix axis reveals a rotation of about 36° per base and shows that the bases are stacked one on top of another. [Drawn by Adam Steinberg.]

The features of the Watson–Crick model of DNA are

1. Two helical DNA strands are coiled around a common axis, forming a right-handed double helix. The strands run in opposite directions; that is, they have opposite directionality. One strand has a 5′-to-3′ direction and pairs with the other strand, which has a 3′-to-5′ orientation.

2. The sugar–phosphate backbones are on the outside of the helix; therefore, the purine and pyrimidine bases lie on the inside.

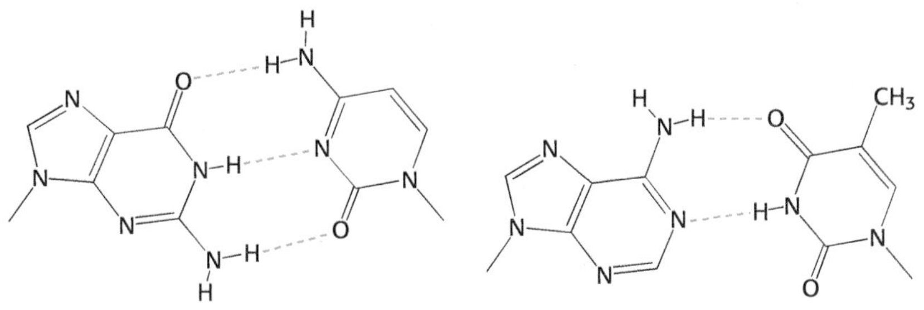

Figure 33.12 Structures of the base pairs proposed by Watson and Crick.

3. The bases are nearly perpendicular to the helix axis, and adjacent bases are separated by approximately 3.4 Å. The helical structure repeats on the order of every 34 Å, with about 10.4 bases per turn of helix. There is a rotation of nearly 36 degrees per base (360 degrees per full turn/10.4 bases per turn).

4. The diameter of the helix is about 20 Å.

How is such a regular structure able to accommodate an arbitrary sequence of bases, given the different sizes and shapes of the purines and pyrimidines? Restricting the combinations in which bases on opposite strands can be paired—A with T and C with G—alleviates this problem. When guanine is paired with cytosine and adenine with thymine to form base pairs, these pairs have essentially the same shape (**Figure 33.12**). These base pairs, often called *Watson–Crick base pairs*, are held together by hydrogen bonds. These base-pairing rules mean that, *if the sequence of one strand is known, the sequence of the other strand is defined.* We will see the implications of this below and in Chapter 34. The base-pairing also helps to stabilize the double helix.

The stacking of bases one on top of another also contributes to the stability of the double helix in two ways (**Figure 33.13**). First, the double helix is stabilized by the hydrophobic effect (p. 23): hydrophobic interactions between the bases drive the bases to the interior of the helix, resulting in the exposure of the more polar surfaces to the surrounding water. This arrangement is reminiscent of protein folding in which hydrophobic amino acids are interior in the protein and hydrophilic amino acids are exterior (p. 24). Second, stacked bases attract one another through van der Waals forces (p. 21), a phenomenon called *base stacking*. Energies associated with van der Waals interactions are quite small, such that typical interactions contribute from 2 to 4 kJ mol^{-1} (0.5 to 1.0 kcal mol^{-1}) per atom pair. In the double helix, however, a large number of atoms are in van der Waals contact, and the net effect, summed over these atom pairs, is substantial. As stated earlier, the base sequence is the information content of nucleic acids. Note that, in DNA, this information is tucked into the relative safety of the interior of the double helix.

Base stacking
(van der Waal interactions)

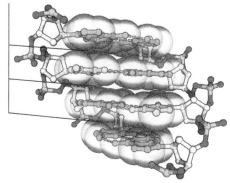

Figure 33.13 A side view of DNA. Base pairs are stacked nearly one on top of another in the double helix. The stacked bases interact with van der Waals forces. Such stacking forces help stabilize the double helix. [Drawn by Adam Steinberg.]

The Double Helix Facilitates the Accurate Transmission of Hereditary Information

The double-helical model of DNA and the presence of specific base pairs immediately suggested how the genetic material might replicate. *The sequence of bases of one strand of the double helix precisely determines the sequence of the other strand.* Thus, the separation of a double helix into its two component strands would yield two single-stranded templates onto which new double helices could be constructed, each of which would have the same sequence of bases as the parent double helix. Consequently, as DNA is replicated, one of the strands of each daughter DNA molecule would be newly synthesized, whereas the other would be passed unchanged from the parent DNA molecule. This distribution of

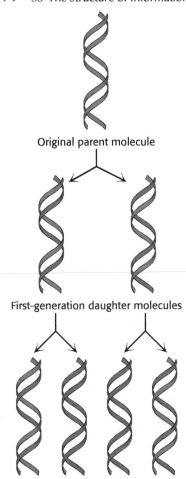

Original parent molecule

First-generation daughter molecules

Second-generation daughter molecules

Figure 33.14 A diagram of semiconservative replication. Parental DNA is shown in blue and newly synthesized DNA in red. [Information from M. Meselson and F. W. Stahl. *Proc. Natl. Acad. Sci. U. S. A.* 44:671–682, 1958.]

parental atoms is achieved by *semiconservative replication* (**Figure 33.14**). We will investigate DNA replication in detail in Chapter 34.

Meselson and Stahl Demonstrated That Replication Is Semiconservative

How do we know that DNA replication is semiconservative? Matthew Meselson and Franklin Stahl carried out an ingenious experiment to establish this fundamental point. They labeled the parent DNA with ^{15}N, a heavy isotope of nitrogen, to make it denser than ordinary DNA. The labeled DNA was generated by growing *E. coli* for many generations in a medium that contained $^{15}NH_4Cl$ as the sole nitrogen source. After the incorporation of heavy nitrogen was complete, the bacteria were abruptly transferred to a medium that contained ^{14}N, the ordinary, lighter isotope of nitrogen. The question asked was: What is the distribution of ^{14}N and ^{15}N in the DNA molecules after successive rounds of replication?

The distribution of ^{14}N and ^{15}N was revealed by the technique of *density-gradient equilibrium sedimentation*. A small amount of DNA was dissolved in a concentrated solution of cesium chloride having a density close to that of the DNA (1.7 g cm^{-3}). This solution was centrifuged until it was nearly at equilibrium, where the opposing forces of sedimentation and diffusion created a gradient in the concentration of cesium chloride across the centrifuge tube. The result was a stable density gradient, ranging from 1.66 to 1.76 g cm^{-3}. The DNA molecules in this density gradient were driven by centrifugal force into the region where the solution's density was equal to their own, yielding a narrow band of DNA that was detected by its absorption of ultraviolet light. ^{14}N DNA and ^{15}N DNA molecules in a control mixture differ in density by about 1%, enough to clearly distinguish the two types of DNA (**Figure 33.15**).

DNA was then extracted from the bacteria at various times after they had been transferred from a ^{15}N to a ^{14}N medium and centrifuged. Analysis of these samples showed that there was a single band of DNA after one generation. The density of this band was precisely halfway between the densities of the ^{14}N DNA and ^{15}N DNA

Figure 33.15 The resolution of ^{14}N DNA and ^{15}N DNA by density-gradient centrifugation. (A) A schematic representation of the centrifuge tube after density-gradient centrifugation. CsCl is cesium chloride. (B) Ultraviolet-absorption photograph of a centrifuge tube showing the two distinct bands of DNA. (C) Densitometric tracing of the absorption photograph. [(B and C) From M. Meselson and F. W. Stahl. *Proc. Natl. Acad. Sci.* 44(1958):671.]

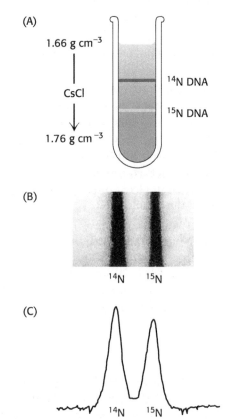

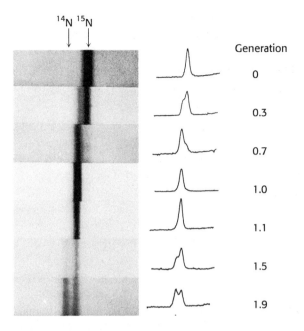

Figure 33.16 The detection of semiconservative replication of *E. coli* DNA by density-gradient centrifugation. The position of a band of DNA depends on its content of ^{14}N and ^{15}N. After 1.0 generation, all of the DNA molecules were hybrids containing equal amounts of ^{14}N and ^{15}N. [After M. Meselson and F. W. Stahl. *Proc. Natl. Acad. Sci. U. S. A.* 44:671–682, 1958.]

bands (**Figure 33.16**). The absence of ^{15}N DNA indicated that parental DNA was not preserved as an intact unit after replication. The absence of ^{14}N DNA indicated that all of the daughter DNA derived some of their atoms from the parent DNA. This proportion had to be half, because the density of the hybrid DNA band was halfway between the densities of the ^{14}N DNA and ^{15}N DNA bands.

After two generations, there were equal amounts of two bands of DNA. One was hybrid DNA, and the other was ^{14}N DNA. Meselson and Stahl concluded from these incisive experiments "that the nitrogen in a DNA molecule is divided equally between two physically continuous subunits; that following duplication, each daughter molecule receives one of these; and that the subunits are conserved through many duplications." Their results agreed perfectly with the Watson–Crick model for DNA replication (Figure 33.14).

The Strands of the Double Helix Can Be Reversibly Separated

During DNA replication and transcription, the two strands of the double helix must be separated from one another, at least in a local region. In other words, sometimes the archival information must be accessible to be useful. In the laboratory, the two strands of the double helix can be separated by heating a solution of DNA. The thermal energy causes the DNA molecules to move to such a degree that the weak forces holding the helix together, such as the hydrogen bonds between the bases, break apart. The dissociation of the double helix is called *melting* or *denaturation*, and it occurs relatively abruptly at a certain temperature. The *melting temperature* T_m is defined as the temperature at which half the helical structure is lost. Strands can also be separated by adding acid or alkali to ionize the nucleotide bases and disrupt base-pairing.

Separated complementary strands of nucleic acids spontaneously reassociate to form a double helix when the temperature is lowered below T_m. This renaturation process is sometimes called *annealing*. The facility with which double helices can be melted and then reassociated is crucial for the biological functions of nucleic acids.

QUICK QUIZ 1 What are the chemical forces that stabilize the double helix?

||||||||||||||||||||||||||||||||||||||||||||||||||||||||||||||||||||||||||||||||||||||||||||||||||||||||||||||||||||||||||||||||||||||||

33.3 DNA Double Helices Can Adopt Multiple Forms

The Watson-and-Crick model of DNA is called *B-DNA*. Under physiological conditions, most DNA is in the B form. However, this structure is not the only one that DNA can form. Studies of DNA revealed a different form called *A-DNA*,

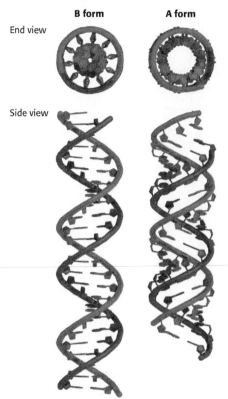

Figure 33.17 B-form and A-form DNA. Models of 10 base pairs of B-form and A-form DNA depict their right-handed helical structures. Notice that the A-form helix is shorter than the B-form helix and that the base pairs are tilted rather than perpendicular with respect to the helix axis. [Drawn by Adam Steinberg, after 1BNA.pdb and 1DNZ.pdb.]

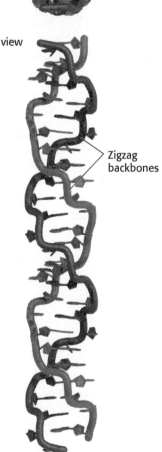

Figure 33.18 Z-DNA. DNA can adopt an alternative conformation under some conditions. This conformation is called Z-DNA because the phosphoryl groups zigzag along the backbone. [Drawn by Adam Steinberg, after 131D.pdb.]

which appears when DNA is less hydrated. A-DNA, like B-DNA, is a right-handed double helix made up of antiparallel strands held together by Watson–Crick base-pairing. The A helix is wider and shorter than the B helix, and its base pairs are tilted rather than perpendicular to the helix axis (Figure 33.17).

If the A helix were simply a property of dehydrated DNA, it would be of little significance. However, double-stranded regions of RNA and at least some RNA–DNA hybrids adopt a double-helical form very similar to that of A-DNA. As we will see (pp. 622 and 664), double-stranded regions of RNA and RNA–DNA hybrids are important players in RNA function and information processing.

Z-DNA Is a Left-Handed Double Helix in Which Backbone Phosphoryl Groups Zigzag

A third type of double helix is *left-handed*, in contrast with the *right-handed* screw sense of the A and B helices. Furthermore, the phosphoryl groups in the backbone are *zigzagged*; hence, this form of DNA is called *Z-DNA* (Figure 33.18).

Although the biological role of Z-DNA is still under investigation, Z-DNA–binding proteins, which have a highly conserved Z-DNA binding domain, are found in many organisms including humans. The existence of Z-DNA shows that *DNA is a flexible, dynamic molecule whose parameters are not as fixed as depictions suggest.* The properties of A-, B-, and Z-DNA are compared in Table 33.1.

The Major and Minor Grooves Are Lined by Sequence-Specific Hydrogen-Bonding Groups

An examination of the DNA molecule in the B form, as shown in Figure 33.17, reveals the presence of two distinct grooves, called the *major groove* and the *minor groove*. These grooves arise because the glycosidic bonds of a base pair are not diametrically opposite each other (Figure 33.19). In B-DNA, the major groove is wider (12 Å versus 6 Å) and deeper (8.5 Å versus 7.5 Å) than the minor groove (Figure 33.20).

Table 33.1 Comparison of A-, B-, and Z-DNA

	Helix type		
	A	B	Z
Shape	Broadest	Intermediate	Narrowest
Rise per base pair	2.3 Å	3.4 Å	3.8 Å
Helix diameter	~26 Å	~20 Å	~18 Å
Screw sense	Right-handed	Right-handed	Left-handed
Glycosidic bond*	*anti*	*anti*	Alternating *anti* and *syn*
Base pairs per turn of helix	11	10.4	12
Pitch per turn of helix	25.3 Å	35.4 Å	45.6 Å
Tilt of base pairs from perpendicular to helix axis	19 degrees	1 degree	9 degrees

Syn and *anti* refer to the orientation of the *N*-glycosidic bond between the base and deoxyribose. In the *syn* orientation, the base is above the deoxyribose. Pyrimidines can be in *anti* orientations only, whereas purines can be *anti* or *syn*.

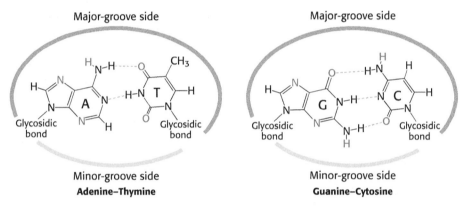

Figure 33.19 Major- and minor-groove sides. Because the two glycosidic bonds are not diametrically opposite each other, each base pair has a larger side that defines the major groove and a smaller side that defines the minor groove. The grooves are lined by potential hydrogen-bond donors (blue) and acceptors (red).

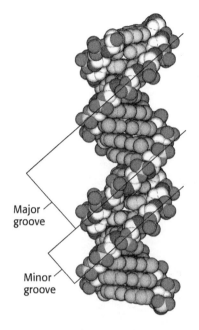

Figure 33.20 Major and minor grooves in B-DNA. Notice the presence of the major groove (depicted in orange) and the narrower minor groove (depicted in yellow). The carbon atoms of the backbone are shown in white. [Drawn by Adam Steinberg.]

Each groove is lined by potential hydrogen-bond donor and acceptor atoms that enable interactions with proteins. These interactions are essential for replication and transcription because particular proteins bind to DNA, recognizing specific hydrogen-bond donors and acceptors on the surfaces of the grooves, to catalyze these processes (Chapters 34, 36, and 37). In the minor groove, N-3 of adenine or guanine and O-2 of thymine or cytosine can serve as hydrogen acceptors, and the amino group attached to C-2 of guanine can be a hydrogen donor. In the major groove, N-7 of guanine or adenine is a potential acceptor, as are O-4 of thymine and O-6 of guanine. The amino groups attached to C-6 of adenine and C-4 of cytosine can serve as hydrogen donors. The methyl group of thymine also lies in the major groove. Note that the major groove displays more features that distinguish one base pair from another than does the minor groove. The larger size of the major groove in B-DNA makes it more accessible for interactions with proteins that recognize specific DNA sequences.

Double-Stranded DNA Can Wrap Around Itself to Form Supercoiled Structures

Our earlier consideration of the amount of DNA in a cell raises an important question: How does all of the DNA fit inside a cell or a cell nucleus? The DNA molecules in human chromosomes are linear, while DNA molecules from some other organisms are *circular* (**Figure 33.21**A). DNA molecules inside cells

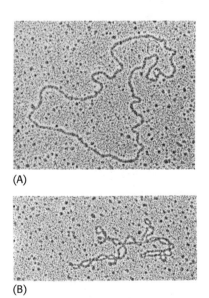

Figure 33.21 Electron micrographs of circular DNA from mitochondria. (A) Relaxed form. (B) Supercoiled form. [Courtesy of Dr. David Clayton.]

necessarily have a very compact shape. Note that the *E. coli* chromosome, fully extended, would be about 1000 times as long as the greatest diameter of the bacterium.

Bacterial chromosomes must therefore employ a new structural property to fit in the cell. The axis of the double helix can itself be twisted into a *superhelix*, a process called *supercoiling* (Figure 33.21B). Consider a linear 260-bp DNA duplex in the B-DNA form (**Figure 33.22**A). Because the number of residues per turn in a DNA molecule is 10.4, this linear DNA molecule has 25 (260/10.4) turns. The ends of this helix can be joined to produce a *relaxed* circular DNA (Figure 33.22B). A different circular DNA can be formed by unwinding the linear duplex by two turns before joining its ends (Figure 33.22C). What is the structural consequence of unwinding before ligation? Two limiting conformations are possible. First, the DNA can fold into a structure containing 23 turns of B helix and an unwound loop (Figure 33.22D). However, bases in the loop do

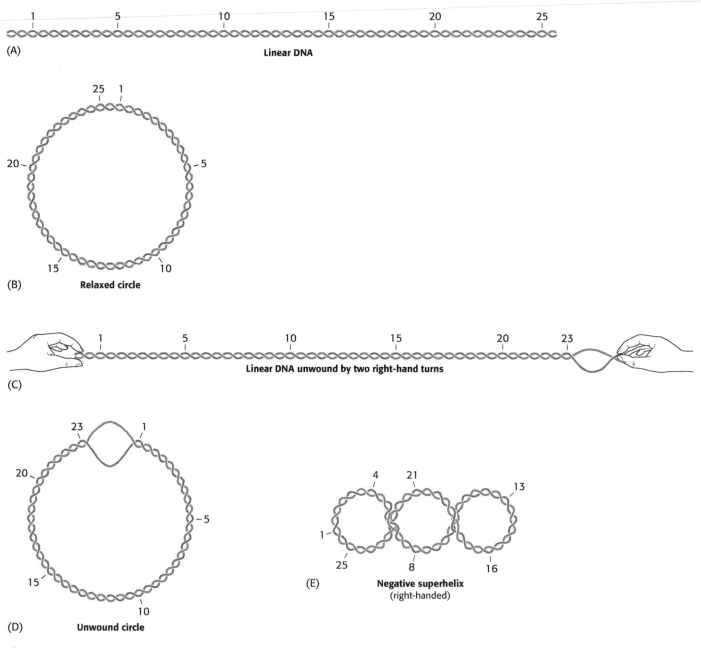

Figure 33.22 Supercoiling in DNA. Partial unwinding of a circular molecule of DNA allows supercoiling. [Information from W. Saenger, *Principles of Nucleic Acid Structure* (Springer, 1984), p. 452.]

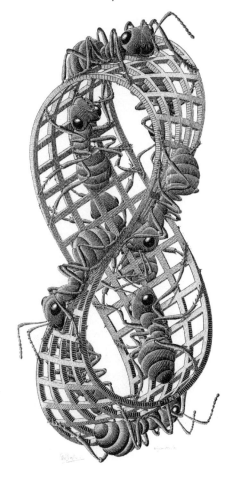

Figure 33.23 The Mobius strip. The study of DNA structures has been enriched by a branch of mathematics called topology, which deals with structural properties unchanged by deformations such as stretching and bending. The Mobius strip, which has only one side, is an example of a topological structure. The Dutch graphic artist M. C. Escher illustrated the unique property of the Mobius strip by depicting ants crawling along the strip, always returning to the starting point after having traversed every part of the strip on both sides without ever crossing an edge. [M. C. Escher's "Mobius Strip II." 2008. The M. C. Escher Company–Holland. All rights reserved. www.mcescher.com.]

not form base pairs, and base pairs stabilize the double helix; so this structure is somewhat unstable. The second structural possibility is for the DNA to adopt a *supercoiled* structure with 23 turns of B helix and 2 turns of *right-handed* (termed *negative*) superhelix (Figure 33.22E). In the superhelix, no bases are left unpaired. The unwound DNA and the supercoiled DNA are *topological isomers*, or *topoisomers*; that is, they have the same sequence but have different structures because of differences in the coiling of the DNA molecules (**Figure 33.23**). A supercoiled DNA molecule has a more compact shape than that of its relaxed counterpart.

Most naturally occurring DNA molecules are negatively supercoiled. What is the basis for this prevalence? Negative supercoiling arises from the unwinding or underwinding of the DNA. In essence, negative supercoiling prepares DNA for processes requiring separation of the DNA strands, such as replication or transcription. Positive supercoiling condenses DNA as effectively, but it makes strand separation more difficult.

Supercoiling is a property not only of circular DNA. It also applies to linear DNA molecules that have a constrained configuration, such as when DNA is packaged into chromosomes (Section 33.4). The separation of the two strands of DNA in replication requires local unwinding of the double helix. This local unwinding must lead either to the overwinding of surrounding regions of DNA or to supercoiling. To prevent the strain induced by overwinding, a specialized set of enzymes is present to introduce supercoils that favor strand separation (Chapter 34).

33.4 Eukaryotic DNA Is Associated with Specific Proteins

✓ 2 Describe how DNA is packaged to fit inside the cell.

There are about 3.6 m of DNA in each human cell, which are packaged into 46 chromosomes, all of which are in a nucleus that has a diameter of about 5 μm. Clearly, the length of DNA and the size of the nucleus present a packaging problem. Supercoiling contributes to the compaction of DNA. However, further compaction is necessary while, at the same time, maintaining the accessibility of base-sequence information. What ways other than supercoiling exist to compact the large amount of DNA in a eukaryotic nucleus? Further compaction is attained with the assistance of certain proteins. The final DNA–protein complex is called a chromosome. We will examine one important class of proteins that takes part in the initial stages of compaction.

Nucleosomes Are Complexes of DNA and Histones

Eukaryotic DNA is tightly bound to a group of small basic proteins called *histones*. In fact, histones constitute half the mass of a eukaryotic chromosome. The entire complex of a cell's DNA and associated protein is called *chromatin*. Chromatin is essentially isolated chromosomes, although the exact relation between the two remains to be determined. Five major histones are present in chromatin: four histones, called H2A, H2B, H3, and H4, associate with one another; the other histone is called H1. Histones have strikingly basic properties because a quarter of the residues in each histone are either arginine or lysine.

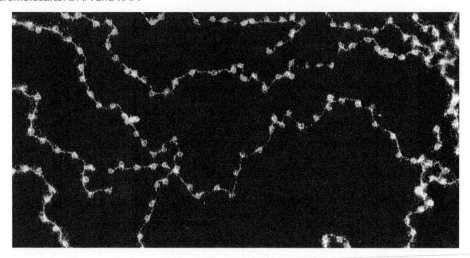

Figure 33.24 Chromatin structure. An electron micrograph of chromatin showing its "beads on a string" character. [©Don. W. Fawcett/Science Source.]

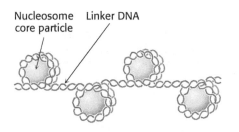

Figure 33.25 Linked core particles. Core particles are joined to one another by linker DNA. [Information from D. L. Nelson and M. M. Cox, *Lehninger Principles of Biochemistry*, 5th ed. (W. H. Freeman and Company, 2008), p. 963.]

Chromatin is made up of repeating units, each containing 200 bp of DNA and two copies each of H2A, H2B, H3, and H4, called the *histone octamer*. These repeating units are known as *nucleosomes*. When chromatin is stretched out so that the DNA and associated nucleosomes can be seen by an electron microscope, it has the appearance of beads on a string; each bead has a diameter of approximately 100 Å (**Figure 33.24**). Partial digestion of chromatin with DNAse, an enzyme that hydrolytically cleaves the phosphodiester backbone of the DNA, yields the isolated beads. These particles consist of fragments of DNA ≈200 bp in length bound to the eight histones. More extensive digestion yields a DNA fragment of 145 bp bound to the histone octamer. This smaller complex of the histone octamer and the 145-bp DNA fragment is the *nucleosome core particle*. The DNA connecting core particles in undigested chromatin is called *linker DNA* (**Figure 33.25**). Histone H1 binds, in part, to the linker DNA.

Eukaryotic DNA Is Wrapped Around Histones to Form Nucleosomes

The eight histones in the core particle are arranged into an $(H3)_2(H4)_2$ tetramer and a pair of H2A–H2B dimers (**Figure 33.26**). The tetramer and dimers come together to form a complex around which the DNA wraps. In addition, each histone has an amino-terminal tail that extends out from the core structure. These tails are flexible and contain a number of lysine and arginine residues. As we shall

Figure 33.26 A nucleosome core particle. Schematic representations of a core of eight histone proteins surrounded by DNA. (A) A view showing the DNA wrapping around the histone core. (B) A view related to that in part A by a 90-degree rotation.

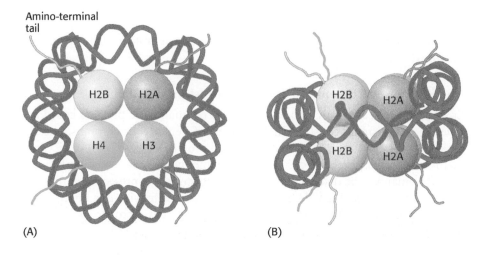

see, *covalent modifications of these tails play an essential role in modulating the accessibility of DNA for transcription* (Chapter 37).

The DNA forms a left-handed superhelix as it wraps around the outside of the histone octamer. The protein core forms contacts with the inner surface of the superhelix at many points, particularly along the phosphodiester backbone and the minor groove of the DNA. Histone H1, which has a different structure from that of the other histones, seals off the nucleosome at the location at which the linker DNA enters and leaves the nucleosome. The amino acid sequences of histones, including their amino-terminal tails, are remarkably conserved from yeast through human beings.

The winding of DNA around the nucleosome core contributes to DNA's packing by decreasing its linear extent. An extended 200-bp stretch of DNA would have a length of about 680 Å. Wrapping this DNA around the histone octamer reduces the length to approximately 100 Å along the long dimension of the nucleosome. Thus, the DNA is compacted by a factor of 7. However, human chromosomes in metaphase, which are highly condensed, are compacted by a factor of 10^4. Clearly, the nucleosome is just the first step in DNA compaction. What is the next step? The nucleosomes themselves are arranged in 30-nm fibers. The nature of the arrangement is under investigations, and more than one arrangement is likely. In one model, the nucleosomes are packed into two interwound left-handed helical stacks. The linker DNA connects successive nucleosomes, crossing the interior of the fiber (**Figure 33.27**). The folding of these fibers of nucleosomes into loops further compacts DNA (**Figure 33.28**).

? QUICK QUIZ 2 Why are supercoiling and nucleosome formation important for the structure of DNA in a cell?

(A)

(B)

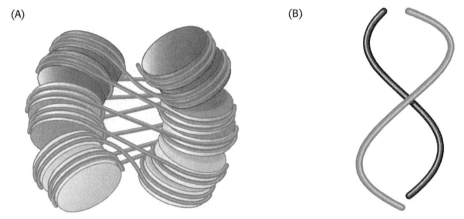

Figure 33.27 Higher-order chromatin structure. (A) A proposed model for chromatin in which the nucleosomes are packed into two interwound left-handed helical stacks. (B) The paths of two helical stacks are illustrated. [Information from F. Song et al. *Science* 344:376–380, 2014 (Figure 1D, p. 377).]

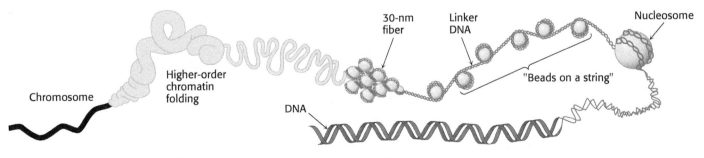

Figure 33.28 The compaction of DNA into a eukaryotic chromosome. [Information from H. Lodish, A. Berk, P. Matsudaria, C. A. Kaiser, M. Krieger, M. P. Scott, S. L. Zipursky, and J. Darnell, *Molecular Cell Biology*, 5th ed. (W. H. Freeman and Company, 2004), p. 406.]

Figure 33.29 Cisplatin alters the structure of DNA. The anticancer drug cisplatin binds to DNA and disrupts its structure, leading to cell death. [Drawn by Adam Steinberg; after Michael J. Hannon, *Chem. Soc. Rev.* 36:280–295, 2007.]

CLINICAL INSIGHT

Damaging DNA Can Inhibit Cancer-Cell Growth

Some cancer chemotherapy drugs work by modifying the structure of DNA, thereby inhibiting replication and transcription. Because cancer cells grow and divide faster than most nonmalignant cells, chemotherapy drugs have a more pronounced effect on cancer cells. An example of a chemotherapy drug is cisplatin, which is used to treat a variety of cancers, including testicular cancer and ovarian cancer. When cisplatin reacts with DNA, the chloride ligands are displaced by purine nitrogen atoms, most commonly in guanine, on two adjacent bases on the same strand. The formation of these bonds causes a severe kink in the structure of the DNA (**Figure 33.29**). This structural disruption prevents replication and transcription. Moreover, research results suggest that certain nuclear proteins bind to the cisplatin-damaged DNA and prevent access to DNA-repair enzymes (Chapter 35). The net effect of the

Cisplatin

cisplatin treatment is that the cell undergoes programmed cell death, or apoptosis, killing the cancer cell. Unfortunately, cisplatin is not specific for cancer cells and consequently many normal cells die, causing serious side effects such as nausea, vomiting, and kidney and nerve damage.

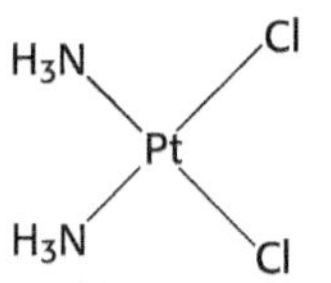

33.5 RNA Can Adopt Elaborate Structures

As we have seen, DNA is usually not a simple helix but is compacted into a more complex structure. Even single-stranded nucleic acids, most commonly RNA, often fold back on themselves to form well-defined and often complex structures. These complex structures allow RNA to perform a host of functions that the double-stranded DNA molecule cannot. Indeed, the complexity of some RNA molecules rivals that of proteins, and these RNA molecules perform a number of functions that had formerly been thought to be the exclusive role of proteins. For instance, one of the RNA components of ribosomes—a large complex of RNAs and proteins on which proteins are synthesized—is the actual catalyst for protein synthesis (Chapter 40).

The simplest and most common structural motif formed is a *stem-loop*, created when two complementary sequences within a single strand come together to form a double-helical structure (**Figure 33.30**). In many cases, these double helices are made up entirely of Watson–Crick base pairs. In other cases, however, the structures include mismatched base pairs or unmatched bases that bulge out from the helix. Such mismatches destabilize the local structure but introduce deviations from the standard double-helical structure that can be important for higher-order folding and for function.

Single-stranded nucleic acids can adopt structures more complex than simple stem-loops through the interaction of more widely separated bases. Often, three or more bases interact to stabilize these structures. In such cases, hydrogen-bond donors and acceptors that do not participate in Watson–Crick base pairs form hydrogen bonds in nonstandard pairings (**Figure 33.31**). Metal ions such as magnesium ion (Mg^{2+}) often assist in the stabilization of these more elaborate structures.

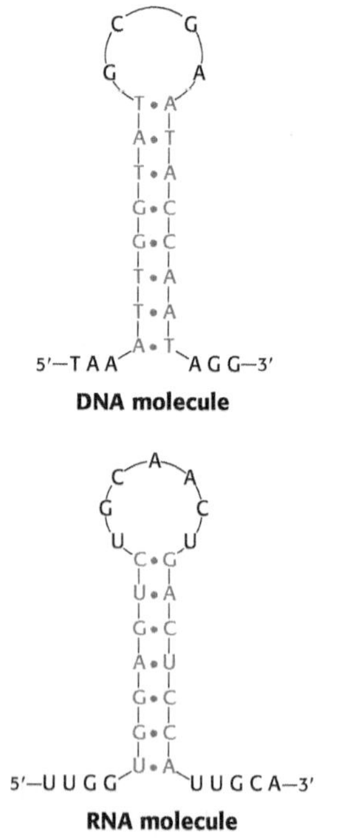

DNA molecule

RNA molecule

Figure 33.30 Stem-loop structures. A stem-loop structure can be formed from single-stranded DNA or from RNA.

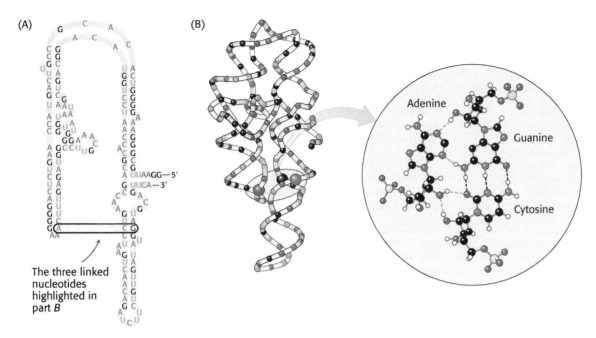

Figure 33.31 The complex structure of an RNA molecule. A single-stranded RNA molecule can fold back on itself to form a complex structure. (A) The nucleotide sequence showing Watson–Crick base pairs and other nonstandard base pairings in stem-loop structures. (B) The three-dimensional structure and one important long-range interaction between three bases. In the three-dimensional structure to the left, cytidine nucleotides are shown in blue, adenosine in red, guanosine in black, and uridine in green. In the blown-up view, hydrogen bonds within the Watson–Crick base pair are shown as dashed black lines; additional hydrogen bonds are shown as dashed green lines.

SUMMARY

33.1 A Nucleic Acid Consists of Bases Linked to a Sugar–Phosphate Backbone

DNA and RNA are linear polymers of a limited number of monomers. In DNA, the repeating units are nucleotides with the sugar being a deoxyribose and the bases being adenine, thymine, guanine, and cytosine. In RNA, the sugar is a ribose and the base uracil is used in place of thymine. DNA is the molecule of heredity in all prokaryotic and eukaryotic organisms.

33.2 Nucleic Acid Strands Can Form a Double-Helical Structure

All cellular DNA consists of two very long, helical polynucleotide strands coiled around a common axis. The sugar–phosphate backbone of each strand is on the outside of the double helix, whereas the purine and pyrimidine bases are on the inside. The two strands of the double helix run in opposite directions. The two strands are held together by hydrogen bonds between pairs of bases: adenine is always paired with thymine, and guanine is always paired with cytosine. The sequence of one strand thus determines the sequence of the other. This sequence determination makes DNA an especially appropriate molecule for storing genetic information, which is the precise sequence of bases along a strand. The strands can be separated, and each separated strand can be used to make a double helix identical with the original molecule.

33.3 DNA Double Helices Can Adopt Multiple Forms

DNA is a structurally dynamic molecule that can exist in a variety of helical forms: A-DNA, B-DNA (the classic Watson–Crick helix), and Z-DNA. DNA can be bent, kinked, and unwound. In A-, B-, and Z-DNA, two antiparallel strands are held together by Watson–Crick base pairs and by stacking forces between bases in the same strand. The sugar–phosphate backbone is on the outside of the helix, and the bases are inside. A- and B-DNA are right-handed helices. In B-DNA, the base pairs are nearly perpendicular to the helix axis. An important structural feature of the B helix is the presence of major and minor grooves, which display different potential hydrogen-bond acceptors and donors according to the base sequence. Z-DNA is a left-handed helix. Most of the DNA in a cell is in the B form.

Double-stranded DNA can also wrap around itself to form supercoiled structures. The supercoiling of DNA has two important consequences. Supercoiling compacts the DNA and, because supercoiling is left-handed, the DNA is partly unwound and more accessible for processes such as replication and transcription.

33.4 Eukaryotic DNA Is Associated with Specific Proteins

Eukaryotic DNA is tightly bound to proteins, most notably to a group of small basic proteins called histones. The entire complex of a cell's DNA and associated protein is called chromatin. Five major histones are present in chromatin: four histones—H2A, H2B, H3, and H4—associate with one another; the other histone is H1. Chromatin is made up of nucleosomes—repeating units, each containing 200 bp of DNA and two copies each of H2A, H2B, H3, and H4. The DNA forms a left-handed superhelix as it wraps around the outside of a nucleosome. The winding of DNA around the nucleosome core contributes to DNA's packing. The nucleosomes themselves are arranged into two intertwined left-handed helical stacks.

33.5 RNA Can Adopt Elaborate Structures

Although RNA is usually single stranded, the single strand folds back on itself to form elaborate structures. The simplest structure is the stem-loop stabilized by Watson–Crick base pairs. In other cases, hydrogen-bond donors and acceptors that are not normal participants in Watson–Crick base pairs form hydrogen bonds in nonstandard pairings.

KEY TERMS

double helix (p. 607)
deoxyribonucleic acid (DNA) (p. 608)
deoxyribose (p. 608)
ribose (p. 608)
purine (p. 608)
pyrimidine (p. 608)
ribonucleic acid (RNA) (p. 608)

nucleoside (p. 609)
nucleotide (p. 610)
Watson-Crick base pair (p. 613)
semiconservative replication (p. 614)
B-DNA (p. 615)
A-DNA (p. 615)
Z-DNA (p. 616)

major groove (p. 616)
minor groove (p. 616)
supercoiling (p. 618)
histone (p. 619)
chromatin (p. 619)
nucleosome (p. 620)
nucleosome core particle (p. 620)
stem-loop (p. 622)

? Answers to QUICK QUIZZES

1. Hydrogen bonds between base pairs and van der Waals interactions among the bases. The van der Waals interactions come into play because of the hydrophobic effect, which forces the bases to the interior of the helix.

2. Both supercoiling and nucleosome formation help to compact the DNA.

PROBLEMS

1. *A "t" instead of an "s"?* Differentiate between a nucleoside and a nucleotide. ✓ 1

2. *A lovely pair.* What is a Watson–Crick base pair? ✓ 1

3. *Almost like base pairs.* Match each term with its description.

(a) Base stacking _____

(b) B-DNA _____
(c) A-DNA _____
(d) Z-DNA _____
(e) Topoisomers _____

(f) Supercoiling _____

(g) Histone _____
(h) Chromatin _____

(i) Nucleosome _____

(j) Nucleosome core particle _____

1. DNA molecules with the same sequence but differ in coiling
2. DNA in a left-handed helix
3. Twisting of the axis of a DNA helix into a superhelix
4. Constitute half the mass of a chromosome
5. Composed of 200 base pairs of DNA and 8 histones
6. Stabilization of the double helix due to van der Waals forces
7. Results from extensive digestion of chromatin by DNAse
8. Most common form of DNA
9. DNA and associated proteins
10. Form of DNA found under dehydrating conditions

4. *Chargaff rules!* Biochemist Erwin Chargaff was the first to note that, in DNA, [A] = [T] and [G] = [C], equalities now called Chargaff's rule. With the use of this rule, determine the percentages of all the bases in DNA that is 20% thymine. ✓ 1

5. *But not always.* A single strand of RNA is 20% U. What can you predict about the percentages of the remaining bases?

6. *Size matters.* Why are GC and AT the only base pairs permissible in the double helix? ✓ 1

7. *Complements.* Write the complementary sequence (in the standard 5′ → 3′ notation) for (a) GATCAA, (b) TCGAAC, (c) ACGCGT, and (d) TACCAT.

8. *Compositional constraint.* The composition (in mole-fraction units) of one of the strands of a double-helical DNA molecule is [A] = 0.30 and [G] = 0.24. (a) What can you say about [T] and [C] for the same strand? (b) What can you say about [A], [G], [T], and [C] of the complementary strand? ✓ 1

9. *Inside out.* Single-stranded DNA absorbs more ultraviolet light than does double-stranded DNA. Suggest why this might be the case.

10. *Axial ratio.* What is the axial ratio (length:diameter) of a DNA molecule 20 μm long?

11. *Strong, but not strong enough.* Why does heat denature, or melt, DNA in solution?

12. *Coming and going.* What does it mean to say that the DNA strands in a double helix have opposite directionality?

13. *Lost DNA.* The DNA of a deletion mutant of λ bacteriophage has a length of 15 μm instead of 17 μm. How many base pairs are missing from this mutant?

14. *An unseen pattern.* What result would Meselson and Stahl have obtained if the replication of DNA were conservative (i.e., the parental double helix stayed together)? Give the expected distribution of DNA molecules after 1.0 and 2.0 generations for conservative replication.

15. *Overcharged.* DNA in the form of a double helix must be associated with cations, usually Mg^{2+}. Why is this requirement the case? ✓ 2

16. *Packing it in.* Does packing DNA into nucleosomes account for the compaction found in a metaphase chromosome (the most condensed form of a chromosome)? ✓ 2

17. *Resistance is futile.* Chromatin viewed with the electron microscope has the appearance of beads on a string. Partial digestion of chromatin with DNAse yields the isolated beads, containing fragments of DNA approximately 200 bp in length bound to the eight histones. More extensive digestion yields a reduced DNA fragment of 145 bp bound to the histone octamer. Why is more extensive digestion required to yield the 145-bp fragment? ✓ 2

18. *Charge neutralization.* Given the histone amino acid sequences illustrated here, estimate the charge of a histone octamer at pH 7. Assume that histidine residues are uncharged at this pH value. How does this charge compare with the charge on 150 base pairs of DNA? ✓ 2

Histone H2A

MSGRGKQGGKARAKAKTRSSRAGLQFPVGRVHRLLRKGNYSERVGAGAPVYLAAVLEYLTAEILELAGNA
ARDNKKTRIIPRHLQLAIRNDEELNKLLGRVTIAQGGVLPNIQAVLLPKKTESHHKAKGK

Histone H2B

MPEPAKSAPAPKKGSKKAVTKAQKKDGKKRKRSRKESYSVYVYKVLKQVHPDTGISSKAMGIMNSFVNDI
FERIAGEASRLAHYNKRSTITSREIQTAVRLLLPGELAKHAVSEGTKAVTKYTSSK

Histone H3

MARTKQTARKSTGGKAPRKQLATKAARKSAPSTGGVKKPHRYRPGTVALREIRRYQKSTELLIRKLPFQR
LVREIAQDFKTDLRFQSAAIGALQEASEAYLVGLFEDTNLCAIHAKRVTIMPKDIQLARRIRGERA

Histone H4

MSGRGKGGKGLGKGGAKRHRKVLRDNIQGITKPAIRRLARRGGVKRISGLIYEETRGVLKVFLENVIRDA
VTYEHAKRKTVTAMDVVYALKRQGRTLYGFGG

19. *Around we go.* Assuming that 145 base pairs of DNA wrap around the histone octamer 1¾ times, estimate the radius of the histone octamer. Assume 3.4 Å per base pair, and simplify the calculation by assuming that the wrapping is in two rather than three dimensions and neglecting the thickness of the DNA. ✓ 2

Data Interpretation Problems

20. *3 is greater than 2.* The illustration below graphs the relation between the percentage of GC base pairs in DNA and the melting temperature. Suggest a possible explanation for these results.

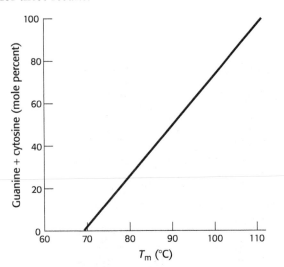

[Data from R. J. Britten and D. Kohne, *Science* 161–540, 1968.]

21. *Blast from the past.* The illustration below is a graph called a C_0t curve (pronounced "cot"). The *y* axis shows the percentage of DNA that is double stranded. The *x* axis is the product of the concentration of DNA and the time required for the double-stranded molecules to form. Explain why the mixture of poly A and poly U and the three DNAs shown vary in the C_0t value (the point at which half the molecules are double stranded). MS-2 and T4 are bacterial viruses (bacteriophages) with genome sizes of 3569 and 168,903 bp, respectively. The *E. coli* genome is 4.6×10^6 bp.

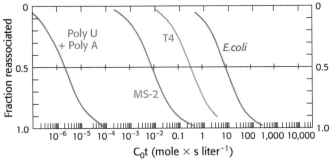

[Data from J. Marmur and P. Doty, *J. Mol. Biol.* 5:120, 1962.]

22. *Salt to taste.* The graph below shows the effect of salt concentration on the melting temperature of bacterial DNA. How does salt concentration affect the melting temperature of DNA? Account for this effect.

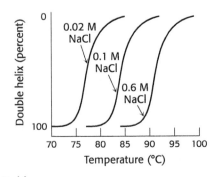

Challenge Problems

23. *Uniqueness.* The human genome contains 3 billion base pairs arranged in a vast array of sequences. What is the minimum length of a DNA sequence that will, in all probability, appear only once in the human genome? You need consider only one strand and may assume that all four nucleotides have the same probability of appearance.

24. *Information content.* (a) How many different 8-mer sequences of DNA are there? (Hint: There are 16 possible dinucleotides and 64 possible trinucleotides.) We can quantify the information-carrying capacity of nucleic acids in the following way. Each position can be one of four bases, corresponding to two bits of information ($2^2 = 4$). Thus, a strand of 5100 nucleotides corresponds to $2 \times 5100 = 10,200$ bits, or 1275 bytes (1 byte = 8 bits). (b) How many bits of information are stored in an 8-mer DNA sequence? In the *E. coli* genome? In the human genome? (c) Compare each of these values with the amount of information that can be stored on a 700 megabytes CD.

25. *A tougher strand.* RNA is readily hydrolyzed by alkali, whereas DNA is not. Why?

26. *A picture is worth a thousand words.* Write a reaction sequence showing why RNA is more susceptible to nucleophilic attack than DNA.

Selected Readings for this chapter can be found online at www.whfreeman.com/tymoczko3e.

CHAPTER 34

DNA Replication

Faithful copying is essential to the storage of genetic information. With the precision of a diligent 15th-century monk copying an illuminated manuscript, a DNA polymerase copies DNA strands, preserving the precise sequence of bases with very few errors. [Ramon Manent/The Art Archive at Art Resource, NY.]

In Chapter 33, we learned that DNA is the archival form of genetic information. An archive is a historical record or document, which makes referring to DNA as an archive or archival information particularly appropriate; DNA contains the information to generate a new individual organism—be it a bacterium, a ficus, or a human being, depending on its source. In other words, DNA contains the blueprint for the development of the organism as well as the evolutionary history of an organism.

If DNA is the archival information, who are the archivists of the cell? Such vital information must be handled carefully. Just as the monk depicted in the illustration above carefully copied manuscripts for preservation, molecular

627

equivalents copy and preserve DNA. The molecular equivalents of these archivists are enzymes called DNA polymerases. As discussed in Chapter 33, the DNA double helix is a stable structure, as should be the case for archival information. Because of this stability, the rate of spontaneous strand separation is negligible. However, strand separation is necessary to synthesize copies of genetic information for the next cell generation. Thus, energy is required for this process, and other molecules are required to assist the DNA polymerases to gain access to the sequence information. Other enzymes and proteins separate the strands and prepare the DNA for copying by the master archivists, the DNA polymerases.

We begin this chapter with an examination of the enzymes taking part in DNA replication. We then see how these enzymes are coordinated to complete the complex process of DNA replication. Finally, we examine some of the special difficulties in eukaryotic replication.

✓ 3 Identify the enzymes that take part in the process of DNA replication.

34.1 DNA Is Replicated by Polymerases

The full replication machinery in a cell comprises more than 20 proteins engaged in intricate and coordinated interplay. The key enzymes are called *DNA polymerases,* which promote the formation of the phosphodiester linkages joining units of the DNA backbone. *E. coli* has five DNA polymerases, designated by roman numerals, that participate in DNA replication and repair (Table 34.1). We will focus our attention on two of the better understood polymerases: DNA polymerase I and DNA polymerase III.

Table 34.1 *E. coli* DNA polymerases

Enzyme	Function	Additional enzyme activities	Type of DNA damage
Polymerase I	Primer removal and DNA repair	$5' \rightarrow 3'$ exonuclease	
		$3' \rightarrow 5'$ exonuclease	
Polymerase II	Repair	$3' \rightarrow 5'$ exonuclease	Attachment of bulky hydrocarbons to bases
Polymerase III	Replication	$3' \rightarrow 5'$ exonuclease	
Polymerase IV	Repair		Attachment of bulky hydrocarbons to bases
Polymerase V	Repair		Attachment of bulky hydrocarbons to bases
			Sites missing bases
			Covalently joined bases

Note: Polymerases II, IV, and V can replicate through regions of damaged DNA (Chapter 35). They are called translesion polymerases or error-prone polymerases.

DNA Polymerase Catalyzes Phosphodiester-Linkage Formation

DNA polymerases catalyze the step-by-step addition of deoxyribonucleotides to a DNA strand (Figure 34.1). The reaction, in its simplest form, is

$$(DNA)_n + dNTP \rightleftharpoons (DNA)_{n+1} + PP_i$$

where dNTP stands for any deoxyribonucleotide and PP_i is a pyrophosphate ion. DNA synthesis has the following characteristics:

1. The reaction requires all four activated precursors—that is, *the deoxynucleoside 5'-triphosphates dATP, dGTP, dCTP, and TTP*—as well as the Mg^{2+} ion.

2. *The new DNA strand is assembled directly on a preexisting DNA template.* DNA polymerases catalyze the formation of a phosphodiester linkage efficiently only if the base on the incoming nucleoside triphosphate is complementary to the base on the *template* strand. Thus, *DNA polymerase is a*

DID YOU KNOW?

A template is a sequence of nucleic acids that determines the sequence of a complementary nucleic acid.

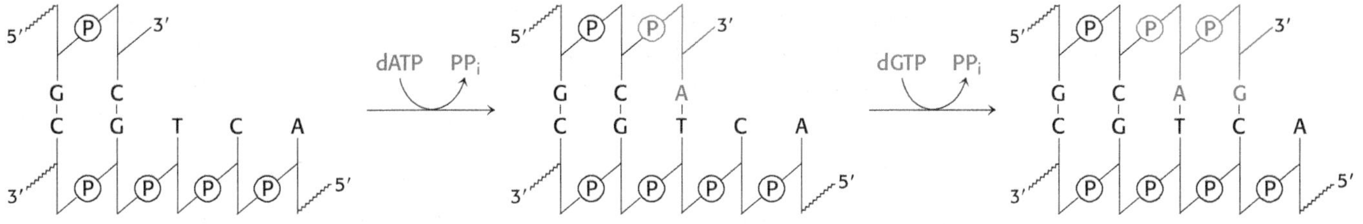

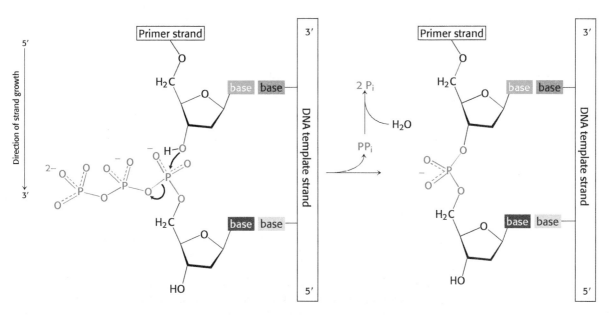

Figure 34.1 A polymerization reaction catalyzed by DNA polymerases.

template-directed enzyme that synthesizes a product with a base sequence complementary to that of the template.

3. *DNA polymerases require a primer to begin synthesis.* A *primer* strand having a free 3'-OH group must be already bound to the template strand. The strand-elongation reaction catalyzed by DNA polymerases is a nucleophilic attack by the 3'-OH end of the growing strand on the innermost phosphorus atom of deoxynucleoside triphosphate (**Figure 34.2**). A phosphodiester bridge is formed and pyrophosphate is released. The subsequent hydrolysis of pyrophosphate to yield two ions of orthophosphate (P_i) by pyrophosphatase drives the polymerization forward. *Elongation of the DNA strand proceeds in the 5'-to-3' direction.*

4. *Many DNA polymerases are able to correct mistakes in DNA by removing mismatched nucleotides.* These polymerases have a distinct nuclease activity that allows them to excise incorrect bases by a separate reaction. For instance, DNA polymerase I has three distinct active sites: the polymerase site, a $3' \rightarrow 5'$ exonuclease site, and a $5' \rightarrow 3'$ exonuclease site. The $3' \rightarrow 5'$ nuclease activity contributes to the remarkably high fidelity of DNA replication, which has an error rate of less than 10^{-8} per base pair. We will consider the function of the $5' \rightarrow 3'$ nuclease activity shortly.

The three-dimensional structures of a number of DNA polymerase enzymes are known. The first such structure to be elucidated was a fragment of *E. coli* DNA polymerase I, called the Klenow fragment, that consisted of the polymerase and the $3' \rightarrow 5'$ exonuclease. The shape approximates the shape of a right hand with

> **DID YOU KNOW?**
>
> A primer is the initial segment of a polymer that is to be extended on which elongation depends.

Figure 34.2 Strand–elongation reaction. DNA polymerases catalyze the formation of a phosphodiester bridge. Elongation of the DNA strand proceeds in the 5'-to-3' direction.

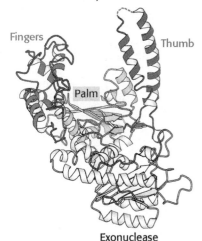

Figure 34.3 DNA polymerase structure. The first DNA polymerase structure determined was that of a fragment of E. coli DNA polymerase I called the Klenow fragment. Notice that, like other DNA polymerases, the polymerase unit resembles a right hand with fingers (blue), palm (yellow), and thumb (red). The Klenow fragment also includes an exonuclease domain that removes incorrect nucleotide bases. [Drawn from 1DPI.pdb.]

domains that are referred to as the fingers, the thumb, and the palm (Figure 34.3). The finger and thumb domains wrap around DNA—much as your own fingers wrap around a baseball bat—and hold it across the enzyme's active site, which is located in the palm domain.

The Specificity of Replication Is Dictated by the Complementarity of Bases

Because DNA is the repository of genetic information, it must be replicated with high accuracy. Each base added to the growing strand should be the Watson–Crick complement of the base in the corresponding position in the template strand. The binding of the dNTP containing the proper base is favored by the formation of a base pair, which is stabilized by specific hydrogen bonds. The binding of a noncomplementary base is less likely because the interactions are energetically weaker. Although the hydrogen bonds linking two complementary bases make a significant contribution to the fidelity of DNA replication, overall shape complementarity also is crucial. Studies show that a nucleotide with a base that is very similar in shape to adenine but lacks the ability to form base-pairing hydrogen bonds can still direct the incorporation of thymidine, both in vitro and in vivo (Figure 34.4). However, DNA polymerases replicate DNA even more faithfully than can be accounted for by these interactions alone.

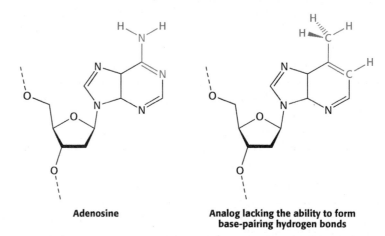

Adenosine   Analog lacking the ability to form
                              base-pairing hydrogen bonds

Figure 34.4 Shape complementarity. The base analog on the right has the same shape as adenosine, but groups that form hydrogen bonds between base pairs have been replaced by groups (shown in red) not capable of hydrogen bonding. Nonetheless, studies reveal that, when incorporated into the template strand, this analog directs the insertion of thymidine in DNA replication.

What is the basis of DNA polymerase's low error rate? The answer to this question is complex, but one important factor is induced fit—the change in the structure of the enzyme when it binds the correct nucleotide. DNA polymerases close down around the incoming nucleoside triphosphate (dNTP), as shown in Figure 34.5. The binding of a dNTP into the active site of a DNA polymerase triggers a conformational change: the finger domain rotates to form a tight pocket into which only a properly shaped base pair will readily fit.

CLINICAL INSIGHT

The Separation of DNA Strands Requires Specific Helicases and ATP Hydrolysis

For a double-stranded DNA molecule to replicate, the two strands of the double helix must be separated from each other, at least at the site of replication. This separation allows each strand to act as a template on which a new DNA strand can be assembled. Specific enzymes, termed *helicases*, utilize

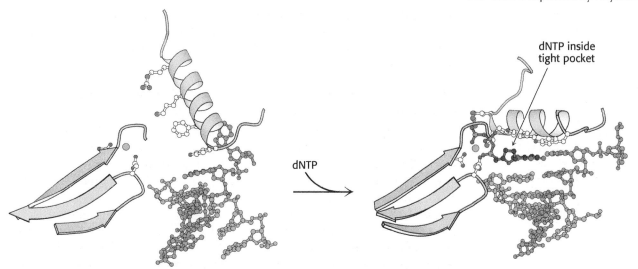

dNTP inside tight pocket

dNTP

Figure 34.5 Shape selectivity. The binding of a deoxyribonucleoside triphosphate (dNTP) to DNA polymerase induces a conformational change, generating a tight pocket for the base pair consisting of the dNTP and its partner on the template strand. Such a conformational change is possible only when the dNTP corresponds to the Watson–Crick partner of the template base. Only the part of the polymerase directly participating in nucleotide binding is displayed (yellow ribbons). The green ball represents the Mg^{2+} ion. [Drawn from 2BDP.pdb and 1T7P.pdb.]

the energy of ATP hydrolysis to power strand separation. Pathological conditions that result from defects in helicase activity attest to the importance of helicases. For instance, Werner syndrome, which is characterized by premature aging, is due to a defect in helicase activity (Figure 34.6).

Helicases are a large and diverse family of enzymes taking part in many biological processes. The helicases in DNA replication are typically oligomers containing six subunits that form a ring structure. Each subunit has a loop that extends toward the center of the ring structure and interacts with DNA. A possible mechanism for the action of a helicase is shown in Figure 34.7. Two subunits are bound to ATP, two to ADP and P_i, and two are initially free of nucleotides. One of the strands of the double helix passes through the hole in the center of the helicase, bound to the loops on two adjacent subunits, one of which has bound ATP and the other of which has bound ADP + P_i. The binding of ATP to the subunit that initially had no bound nucleotides results

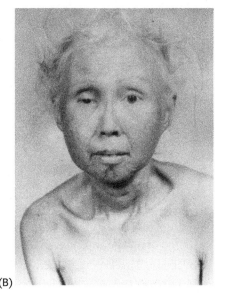

(A) (B)

Figure 34.6 Werner syndrome. A very rare disorder, Werner syndrome is characterized by premature aging. (A) People afflicted by Werner syndrome develop normally until adolescence, at which time they begin to age rapidly. (B) By age 40, they look several decades older. [Courtesy of the International Registry of Werner Syndrome.]

Figure 34.7 Helicase mechanism. One of the strands of the double helix passes through the hole in the center of the helicase, bound to the loops of two adjacent subunits. Two of the subunits do not contain bound nucleotides. On the binding of ATP to these two subunits and the release of ADP + P$_i$ from two other subunits, the helicase hexamer undergoes a conformational change, pulling the DNA through the helicase. The helicase acts as a wedge to force separation of the two strands of DNA.

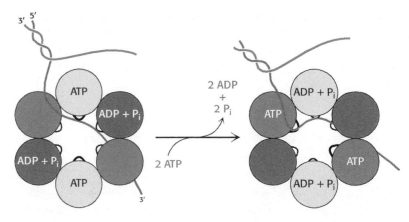

in a conformational change within the entire hexamer, leading to the release of ADP + P$_i$ from two subunits and the binding of the single-stranded DNA by one of the subunits that has just bound ATP. This conformational change pulls the DNA through the center of the hexamer. The protein acts as a wedge, forcing the two strands of the double helix apart. This cycle then repeats itself, moving two bases along the DNA strand with each cycle.

Topoisomerases Prepare the Double Helix for Unwinding

Most naturally occurring DNA molecules are negatively supercoiled. What is the basis for this prevalence? As discussed in Chapter 33, negative supercoiling arises from the unwinding or underwinding of DNA. In essence, negative supercoiling prepares DNA for processes requiring separation of the DNA strands, such as replication. The presence of supercoils in the immediate area of unwinding would, however, make unwinding difficult (Figure 34.8). Therefore, negative supercoils must be continuously removed, and the DNA relaxed, as the double helix unwinds.

Unwound   Overwound

Figure 34.8 Consequences of strand separation. DNA must be locally unwound to expose single-stranded templates for replication. This unwinding puts a strain on the molecule by causing the overwinding of nearby regions.

Enzymes called *topoisomerases* introduce or eliminate supercoils by temporarily cleaving DNA. *Type I topoisomerases* catalyze the relaxation of supercoiled DNA, a thermodynamically favorable process. *Type II topoisomerases* utilize free energy from ATP hydrolysis to add negative supercoils to DNA. In bacteria, type II topoisomerase is called *DNA gyrase.*

⚕ CLINICAL INSIGHT

Bacterial Topoisomerase Is a Therapeutic Target

DNA gyrase is the target of several antibiotics that inhibit this bacterial topoisomerase much more than the eukaryotic one. *Novobiocin* blocks the binding of ATP to gyrase. *Nalidixic acid* and *ciprofloxacin,* in contrast, interfere with the breakage and rejoining of DNA strands. These two gyrase inhibitors are widely used to treat urinary-tract and other infections.

Ciprofloxacin, more commonly known as "cipro," became a "celebrity" in the United States, owing to the anthrax poisonings in the fall of 2001 (**Figure 34.9**). It is a potent broad-spectrum antibiotic that prevents anthrax poisoning by preventing the growth of *Bacillus anthracis* if taken early enough after infection.

Many Polymerases Proofread the Newly Added Bases and Excise Errors

Many polymerases further enhance the fidelity of replication by the use of proof-reading mechanisms. One polymerase from *E. coli*, *DNA polymerase I* used in DNA replication and repair, displays an exonuclease activity in addition to the polymerase activity. The *exonuclease* removes mismatched nucleotides from the 3' end of DNA by hydrolysis. If the wrong nucleotide is inserted, the malformed product is not held as tightly in the polymerase active site. It is likely to flop about because of the weaker hydrogen bonding and to find itself in the exonuclease active site, where the trespassing nucleotide is removed (**Figure 34.10**). This flopping is the result of Brownian motion (p. 18).

How does the enzyme sense whether a newly added base is correct? First, an incorrect base will not pair correctly with the template strand and will be unlikely to be linked to the new strand. Second, even if an incorrect base is inserted into the new strand, it is likely to be deleted. After the addition of a new nucleotide, the DNA is pulled by one base pair into the enzyme. If an incorrect base is incorporated, the enzyme stalls owing to the structural disruption caused by the presence of a non-Watson–Crick base pair in the enzyme, and the pause provides additional time for the strand to wander into the exonuclease site. There is a cost to this editing function, however: DNA polymerase I removes

Figure 34.9 Anthrax poisoning in the United States in 2001. A mailbox in Hamilton, New Jersey, is covered in plastic because of possible contamination with *B. anthracis* spores. In the background, the flag flies at half-staff to honor two Hamilton postal workers who died from anthrax poisoning. [William Thomas Cain/Getty Images.]

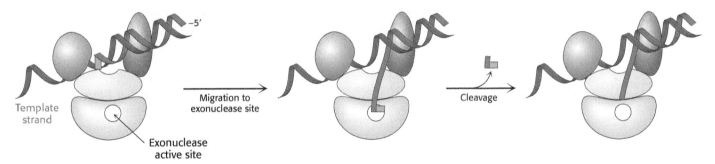

Figure 34.10 Proofreading. The growing DNA strand occasionally leaves the polymerase site, especially if there is improper base-pairing, and migrates to the active site of exonuclease. There, one or more nucleotides are excised from the newly synthesized strand, removing potentially incorrect bases.

approximately 1 correct nucleotide in 20. Although the removal of correct nucleotides is slightly wasteful energetically, proofreading increases the accuracy of replication by a factor of approximately 1000.

? QUICK QUIZ 1 What are the three key enzymes required for DNA synthesis, and what biochemical challenges to replication do they address?

34.2 DNA Replication Is Highly Coordinated

DNA replication must be very rapid, given the sizes of the genomes and the rates of cell division. The *E. coli* genome contains 4.6 million base pairs and is copied in less than 40 minutes. Thus, 2000 bases are incorporated per second. Enzyme activities need to be highly coordinated to replicate entire genomes precisely and rapidly.

✓ 4 Describe how replication is organized, and distinguish between eukaryotic replication and bacterial replication.

We begin our consideration of the coordination of DNA replication by looking at *E. coli,* which has been extensively studied. For this organism, with a small genome, replication begins at a single site and continues around the circular chromosome. The coordination of eukaryotic DNA replication is more complex than bacterial replication because there are many initiation sites throughout the genome and because an additional enzyme is needed to replicate the ends of linear chromosomes.

DNA Replication in *E coli* Begins at a Unique Site

In *E. coli,* DNA replication starts at a unique site within the entire 4×10^6 bp genome. This *origin of replication,* called the *oriC locus,* is a 245-bp region that has several unusual features.

The *oriC* locus contains four copies of a sequence that are preferred binding sites for the origin-recognition protein *DnaA.* In addition, the locus contains a tandem array of 13-bp sequences that are rich in AT base pairs (**Figure 34.11**A). The binding of DnaA molecules to the oriC locus begins the building of the replication complex. The DnaA proteins then oligomerize, wrapping the origin around themselves (Figure 34.11B). Additional proteins, such as *DnaB,* then join DnaA. The DnaB protein is a helicase that utilizes ATP hydrolysis to unwind the duplex, including the AT-rich regions. *Single-strand-binding proteins* (SSB) then bind to the newly generated single-stranded regions, preventing the two complementary strands from re-forming the double helix (Figure 34.11C). The result of this process is the generation of a structure called the *prepriming complex,* which makes single-stranded DNA accessible for other enzymes to begin the synthesis of the complementary strands.

An RNA Primer Synthesized by Primase Enables DNA Synthesis to Begin

Even with the DNA template exposed and the prepriming complex assembled, there are still obstacles to DNA synthesis. DNA polymerases can add nucleotides only to a free hydroxyl group; they cannot start a strand de novo. Therefore, a primer is required. How is this primer formed? An important clue came from the observation that RNA synthesis is required for the initiation of DNA synthesis. In fact, *RNA primes the synthesis of DNA.* A specialized RNA polymerase called *primase* joins the prepriming complex in a multisubunit assembly called the *primosome.* Primase synthesizes a short stretch of RNA (about 10 nucleotides) that is complementary to one of the template DNA strands (**Figure 34.12**). The RNA primer is removed by a $5' \rightarrow 3'$ exonuclease; in *E. coli,* the exonuclease is present as the third active site in DNA polymerase I.

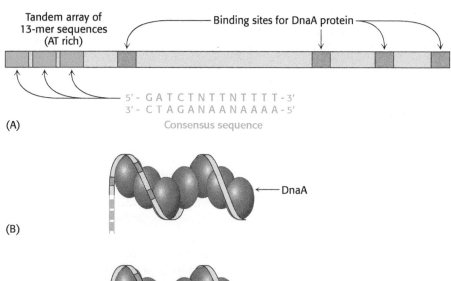

(A)

5'- G A T C T N T T N T T T -3'
3'- C T A G A N A A N A A A A -5'
Consensus sequence

(B) ←— DnaA

(C) ←— SSB
←— DnaB

Figure 34.11 The origin of replication in *E. coli* and formation of the prepriming complex. (A) The *oriC* locus has a length of 245 bp. It contains a tandem array of three nearly identical AT-rich regions (green) and four binding sites (orange) for the DnaA protein. (B) Monomers of DnaA bind to their binding sites in *oriC* and oligomerize, wrapping the DNA around the oligomer. This structure marks the origin of replication and favors DNA strand separation in the AT-rich sites (green). (C) The AT-rich regions are unwound by DnaB and trapped by the single-stranded-binding protein (SSB). At this stage, the complex is ready for the synthesis of the RNA primers and assembly of the DNA polymerase III holoenzyme.

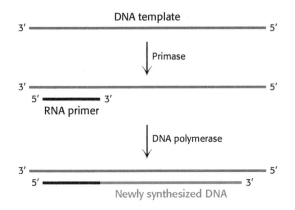

Figure 34.12 Priming. DNA replication is primed by a short stretch of RNA that is synthesized by primase, an RNA polymerase. The RNA primer is removed at a later stage of replication.

One Strand of DNA Is Made Continuously and the Other Strand Is Synthesized in Fragments

Both strands of parental DNA serve as templates for the synthesis of new DNA. The site of DNA synthesis is called the *replication fork* because the complex formed by the newly synthesized daughter strands arising from the parental duplex resembles a two-pronged fork (**Figure 34.13**). Recall that the two strands are antiparallel; that is, they run in opposite directions. On cursory examination, both daughter strands appear to grow in the same direction, as shown in Figure 34.13. However, this appearance of same-direction growth presents a conundrum, because all known DNA polymerases synthesize DNA in the $5' \rightarrow 3'$ direction but not in the $3' \rightarrow 5'$ direction. How then does one of the daughter DNA strands appear to grow in the $3' \rightarrow 5'$ direction?

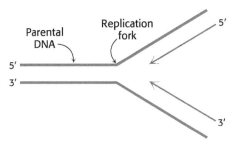

Figure 34.13 DNA replication at low resolution. On cursory examination, both strands of a DNA template appear to replicate continuously in the same direction.

This dilemma was resolved when careful experimentation found that *a significant proportion of newly synthesized DNA exists as small fragments.* These units of about a thousand nucleotides, called *Okazaki fragments* (after Reiji Okazaki, who first identified them), are present briefly in the vicinity of the replication fork (**Figure 34.14**). As replication proceeds, these fragments become covalently joined through the action of *DNA ligase,* an enzyme that uses ATP hydrolysis to power the joining of DNA fragments to form one of the daughter strands. The other new strand is synthesized continuously. The strand formed from Okazaki fragments is termed the *lagging strand,* whereas the one synthesized without interruption is the leading strand. Both the Okazaki fragments and the *leading strand* are synthesized in the $5' \rightarrow 3'$ direction. *The discontinuous assembly of the lagging strand enables $5' \rightarrow 3'$ polymerization at the nucleotide level to give rise to overall growth in the $3' \rightarrow 5'$ direction.*

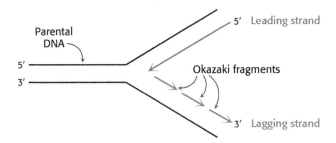

Figure 34.14 Okazaki fragments. At a replication fork, both strands are synthesized in the $5' \rightarrow 3'$ direction. The leading strand is synthesized continuously, whereas the lagging strand is synthesized in short pieces termed Okazaki fragments.

DNA Replication Requires Highly Processive Polymerases

We have been introduced to the enzymes that participate in DNA replication. They include helicases and topoisomerases, and we have examined the general properties of polymerases by focusing on DNA polymerase I. We now examine the enzyme responsible for the rapid and accurate synthesis of DNA

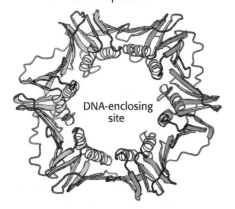

Figure 34.15 The structure of a sliding DNA clamp. The dimeric β_2 subunit of DNA polymerase III forms a ring that surrounds the DNA duplex. Notice the central cavity through which the DNA template slides. Clasping the DNA molecule in the ring, the polymerase enzyme is able to move without falling off the DNA substrate. [Drawn from 2POL.pdb.]

in *E. coli*—the holoenzyme *DNA polymerase III*. The hallmarks of this multi-subunit assembly are not only its fidelity, but also its very high catalytic potency and processivity. The term *processivity* refers to the ability of an enzyme to catalyze many consecutive reactions without releasing its substrate. The holoenzyme catalyzes the formation of many thousands of phosphodiester linkages before releasing its template, compared with only 20 for DNA polymerase I. The DNA polymerase III holoenzyme grasps its template and does not let go until the template has been completely replicated. Another distinctive feature of the holoenzyme is its catalytic prowess: 1000 nucleotides are added per second compared with only 10 per second for DNA polymerase I. This acceleration is accomplished with no loss of accuracy. The greater catalytic prowess of polymerase III is largely due to its processivity; no time is lost in repeatedly stepping on and off the template.

The source of the processivity—the ability to stay "on task"—is the β_2 subunit, which has the form of a star-shaped ring (**Figure 34.15**). A 35-Å-diameter hole in its center can readily accommodate a duplex DNA molecule yet leaves enough space between the DNA and the protein to allow rapid sliding and turning in the course of replication. A catalytic rate of 1000 nucleotides polymerized per second requires the sliding of 100 turns of duplex DNA (a length of 3400 Å, or 0.34 mm) through the central hole of β_2 per second. Thus, β_2 *plays a key role in replication by serving as a sliding DNA clamp.*

How does DNA become trapped inside the *sliding clamp*? Polymerases also include assemblies of subunits that function as clamp loaders. These enzymes grasp the sliding clamp and, utilizing the energy of ATP, pull apart the two subunits of the sliding clamp on one side. DNA can move through the gap, inserting itself through the central hole. ATP hydrolysis then releases the clamp, which closes around the DNA.

The Leading and Lagging Strands Are Synthesized in a Coordinated Fashion

Replicative polymerases such as DNA polymerase III synthesize the leading and lagging strands simultaneously at the replication fork (**Figure 34.16**). DNA polymerase III begins the synthesis of the leading strand starting from the RNA primer formed by primase. The duplex DNA ahead of the polymerase

Figure 34.16 The replication fork. A schematic representation of the arrangement of DNA polymerase III and associated enzymes and proteins present in replicating DNA. The helicase separates the two strands of the parent double helix, allowing DNA polymerases to use each strand as a template for DNA synthesis. Abbreviation: SSB, single-strand-binding protein.

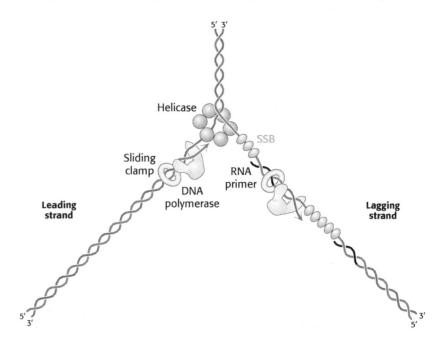

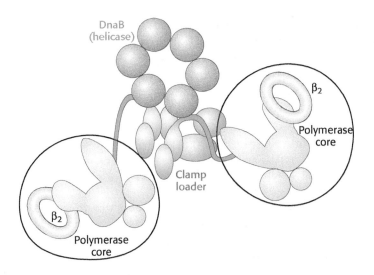

Figure 34.17 The DNA polymerase holoenzyme. Each holoenzyme consists of two copies of the polymerase core enzyme linked to a central structure. The central structure includes the clamp-loader complex, which binds to the hexameric helicase DnaB.

is unwound by a helicase. Single-strand-binding proteins bind to the unwound strands, keeping the strands separated so that both strands can serve as templates. The leading strand is synthesized continuously by polymerase III. Topoisomerase II concurrently introduces negative supercoils to avert a topological crisis.

The mode of synthesis of the lagging strand is necessarily more complex. As mentioned earlier, the lagging strand is synthesized in fragments so that $5' \rightarrow 3'$ polymerization leads to overall growth in the $3' \rightarrow 5'$ direction. Yet the synthesis of the lagging strand is coordinated with the synthesis of the leading strand. How is this coordination accomplished? Examination of the subunit composition of the DNA polymerase III holoenzyme reveals an elegant solution (Figure 34.17). The holoenzyme includes two copies of the polymerase core enzyme, which consists of the DNA polymerase itself, a $3'$-to-$5'$ proofreading exonuclease, two copies of the dimeric β_2-subunit sliding clamp, and several other proteins. The core enzymes are linked to a central structure, which consists of the clamp loader, and subunits that interact with the single-strand-binding protein. The entire central structure interacts with helicase DnaB.

The lagging-strand template is looped out so that it passes through the polymerase site in one subunit of polymerase III in the $3' \rightarrow 5'$ direction. After adding about 1000 nucleotides, DNA polymerase III lets go of the lagging-strand template by releasing the sliding clamp. A new loop is then formed, a sliding clamp is added, and primase again synthesizes a short stretch of RNA primer to initiate the formation of another Okazaki fragment. This mode of replication has been termed the *trombone model* because the size of the loop lengthens and shortens like the slide on a trombone (Figure 34.18). The gaps between fragments of the nascent lagging strand are filled by DNA polymerase I. This essential enzyme also uses its $5' \rightarrow 3'$ exonuclease activity to remove the RNA primer lying ahead of the polymerase site. The primer cannot be erased by DNA polymerase III, because the enzyme lacks $5' \rightarrow 3'$ editing capability. Finally, DNA ligase connects the fragments. *DNA ligase catalyzes the formation of a phosphodiester linkage between the $3'$-hydroxyl group at*

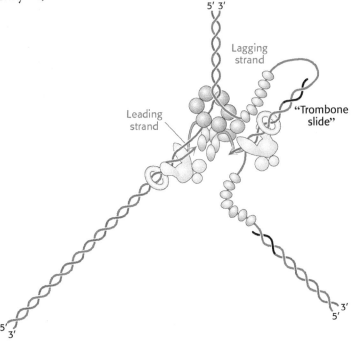

Figure 34.18 The trombone model. The replication of the leading and lagging strands is coordinated by the looping out of the lagging strand to form a structure that acts somewhat as a trombone slide does, growing as the replication fork moves forward. When the polymerase on the lagging strand reaches a region that has been replicated, the sliding clamp is released and a new loop is formed.

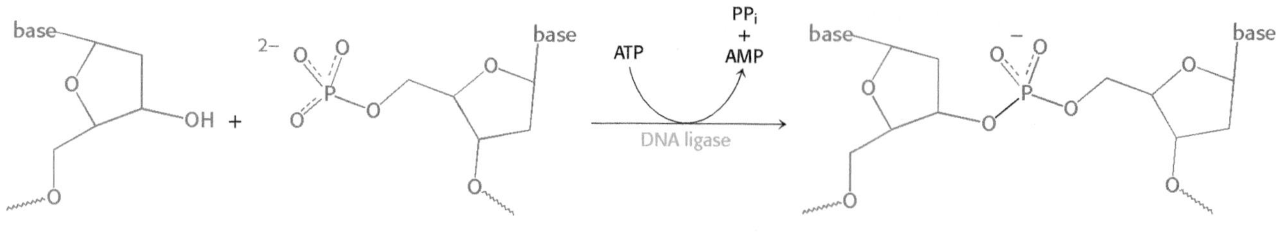

Figure 34.19 The DNA ligase reaction. DNA ligase catalyzes the joining of one DNA strand with a free 3'-hydroxyl group to another with a free 5'-phosphate group. In eukaryotes and archaea, ATP is cleaved to AMP and PP$_i$ to drive this reaction.

QUICK QUIZ 2 Why is processivity a crucial property for DNA polymerase III?

the end of one DNA chain and the 5'-phosphate group at the end of the other (Figure 34.19). ATP is commonly used as the energy source to drive this thermodynamically uphill reaction.

DNA Synthesis Is More Complex in Eukaryotes Than in Bacteria

Replication in eukaryotes is mechanistically similar to replication in bacteria but is more challenging in three ways. One challenge is sheer size: *E. coli* must replicate 4.6 million base pairs, whereas a human diploid cell must replicate 6 billion base pairs. The second challenge is the fact that, while the genetic information for *E. coli* is contained on 1 chromosome, human beings have 23 pairs of chromosomes that must be replicated. Finally, whereas the *E. coli* chromosome is circular, human chromosomes are linear. The third challenge arises because of the nature of DNA synthesis on the lagging strand. Linear chromosomes are subject to shortening with each round of replication unless countermeasures are taken.

The first two challenges are met by the use of multiple origins of replication, which are located between 30 and 300 kilobase pairs (kbp) apart. In human beings, replication of the entire genome requires about 30,000 origins of replication, with each chromosome containing several hundred. Each origin of replication represents a replication unit, or *replicon*. The use of multiple origins of replication requires mechanisms for ensuring that each sequence is replicated once and only once. How are replicons controlled so that each replicon is replicated only once in each cell division? Proteins, called *licensing factors* because they permit (license) the formation of the DNA synthesis initiation complex, bind to the origin of replication. These proteins ensure that each replicon is replicated only once in each round of DNA synthesis. After the licensing factors have established the initiation complex, these factors are subsequently destroyed. The license expires after one use.

Two distinct polymerases are needed to copy a eukaryotic replicon. An initiator polymerase called *polymerase α* begins replication. This enzyme includes a primase subunit, used to synthesize the RNA primer, as well as an active DNA polymerase. After this polymerase has added a stretch of about 20 deoxynucleotides to the primer, it is replaced by *DNA polymerase δ,* a more processive enzyme and the principal replicative polymerase in eukaryotes. This process is called *polymerase switching* because one polymerase has replaced another.

Telomeres Are Unique Structures at the Ends of Linear Chromosomes

Whereas the genomes of essentially all bacteria are circular, the chromosomes of human beings and other eukaryotes are linear. The free ends of linear DNA molecules introduce several complications. First, the unprotected termini of the DNA at the end of a chromosome are likely to be more susceptible to digestion by exonucleases if they are left to freely dangle at the end of the chromosome during replication. Second, the complete replication of DNA ends is difficult because polymerases act only in the $5' \rightarrow 3'$ direction. The lagging strand would have an incomplete 5' end after the removal of the RNA primer. Each round of replication would further shorten the chromosome, which would, like the Cheshire cat, slowly disappear (Figure 34.20).

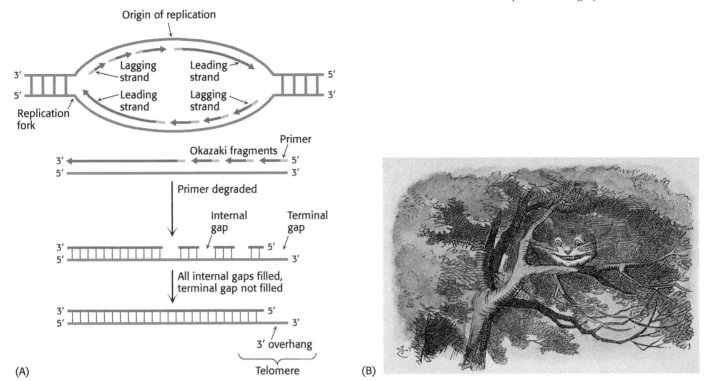

(A)

(B)

Figure 34.20 Telomere shortening. (A) Over many cycles of replication, the DNA at the telomeres would continuously shorten. (B) Without countermeasures, telomeres would vanish like the Cheshire cat. [(A) Information from A. J. F. Griffiths, S. R. Wessler, R. C. Lewontin, and S. B. Carroll, *Introduction to Genetic Analysis*, 9th ed. (W. H. Freeman and Company, 2008), p. 288. (B) The Granger Collection.]

What characteristics of the chromosome ends, called *telomeres* (from the Greek *telos*, meaning "an end"), mitigate these two problems? The most notable feature of telomeric DNA is that it contains hundreds of tandem repeats of a hexanucleotide sequence. One of the strands is G-rich at the 3′ end, and it is slightly longer than the other strand. In human beings, the repeating G-rich sequence is AGGGTT. This simple repeat precludes the likelihood that the sequence encodes any information, but the simple repeat does facilitate the formation of large duplex loops (**Figure 34.21**). The single-stranded region at the very end of the structure has been proposed to loop back to form a DNA duplex with another part of the repeated sequence, displacing a part of the original telomeric duplex. This loop-like structure is formed and stabilized by specific telomere-binding proteins. Such structures would nicely protect the end of the chromosome from degradation.

G-rich strand

Figure 34.21 A proposed model for telomeres. A single-stranded segment of the G-rich strand extends from the end of the telomere. In one model for telomeres, this single-stranded region invades the duplex to form a large duplex loop.

✚ CLINICAL INSIGHT

Telomeres Are Replicated by Telomerase, a Specialized Polymerase That Carries Its Own RNA Template

We now see how telomeres help to protect the ends of the DNA, but the problem of how the ends are replicated remains. What steps are taken to prevent the lagging strand from disappearing after repeated cycles of replication? The solution is to add nucleotides to the leading strand so that the lagging strand will always maintain its approximate length. This task falls to the enzyme *telomerase*, which contains an RNA molecule that acts as a

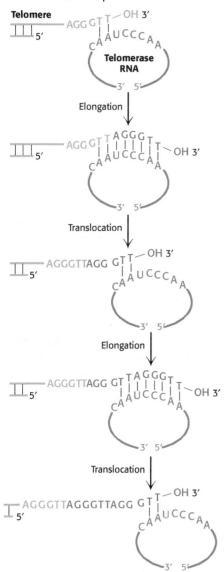

Figure 34.22 Telomere formation. The mechanism of synthesis of the G-rich strand of telomeric DNA. The RNA template of telomerase is shown in blue and the nucleotides added to the G-rich strand of the primer are shown in red. [Information from E. H. Blackburn, *Nature* 350:569–573, 1991.]

template for extending the leading strand. The extended leading strand then acts as a template to elongate the lagging strand, thus ensuring that the chromosome will not shorten after many rounds of replication (**Figure 34.22**).

Telomerase activity is low or absent in most human cells. Consequently, as the cells divide, telomeres shorten. At some point, cell division ceases because the telomeres are too short and programmed cell death is initiated. Thus, telomere shortening serves as a counting device to limit cell proliferation. Indeed, telomere shortening may play a role in aging and the development of pathological conditions. So if telomerase activity is low in human cells, why can cancer, which is characterized by unlimited cell proliferation, develop in humans? *Telomerase reactivation is one of the hallmarks of cancer cells.* Because cancer cells express high levels of telomerase, whereas most normal cells do not, telomerase is a potential target for anticancer therapy. A variety of approaches for blocking telomerase expression or blocking its activity are under investigation for cancer treatment and prevention.

SUMMARY

34.1 DNA Is Replicated by Polymerases

DNA polymerases are template-directed enzymes that catalyze the formation of phosphodiester linkages by the 3′-hydroxyl group's nucleophilic attack on the innermost phosphorus atom of a deoxyribonucleoside 5′-triphosphate. They cannot start strands de novo; a primer with a free 3′-hydroxyl group is required. Many DNA polymerases proofread the nascent product: their 3′ → 5′ exonuclease activity potentially edits the outcome of each polymerization step. A mispaired nucleotide is excised before the next step proceeds. In *E. coli*, DNA polymerase I repairs DNA and participates in replication. Fidelity is further enhanced by an induced fit that results in a catalytically active conformation only when the complex of enzyme, DNA, and correct dNTP is formed.

Helicases prepare the way for DNA replication by using ATP hydrolysis to separate the strands of the double helix. Single-strand-binding proteins bind to the separated strands of DNA to prevent them from reannealing. Separation of the DNA into single strands introduces torsional stress in the DNA ahead of the separation. Topoisomerase relieves the stress by introducing supercoils.

34.2 DNA Replication Is Highly Coordinated

DNA replication in *E. coli* starts at a unique origin (*oriC*) and proceeds sequentially in opposite directions. More than 20 proteins are required for replication. An ATP-driven helicase unwinds the *oriC* region to create a replication fork. At this fork, both strands of parental DNA serve as templates for the synthesis of new DNA. A short stretch of RNA formed by primase, an RNA polymerase, primes DNA synthesis. One strand of DNA (the leading strand) is synthesized continuously, whereas the other strand (the lagging strand) is synthesized discontinuously, in the form of 1-kb Okazaki fragments. Both new strands are formed simultaneously by the concerted actions of the highly processive DNA polymerase III holoenzyme, an asymmetric dimer. The discontinuous assembly of the lagging strand enables 5′ → 3′ polymerization at the molecular level to give rise to overall growth of this strand in the 3′ → 5′ direction. The RNA primer stretch is hydrolyzed by the 5′ → 3′ nuclease activity of DNA polymerase I, which also fills gaps. Finally, nascent DNA fragments are joined by DNA ligase in a reaction driven by ATP.

DNA synthesis in eukaryotes is more complex than in bacteria. Eukaryotes require thousands of origins of replication to complete replication in a timely fashion. A special RNA-dependent DNA polymerase called telomerase is responsible for the replication of the ends of linear chromosomes.

KEY TERMS

DNA polymerase (p. 628)
template (p. 628)
primer (p. 629)
helicase (p. 630)
topoisomerase (p. 632)
exonuclease (p. 633)

origin of replication (p. 634)
primase (p. 634)
replication fork (p. 635)
Okazaki fragment (p. 635)
DNA ligase (p. 635)
lagging strand (p. 635)

leading strand (p. 635)
processivity (p. 636)
sliding clamp (p. 636)
trombone model (p. 637)
telomere (p. 638)
telomerase (p. 639)

? Answers to QUICK QUIZZES

1. DNA polymerases, helicases, and topoisomerases. Polymerases faithfully replicate the DNA. Helicases unwind the double helix, allowing the polymerases access to the base sequence. Topoisomerases relax the DNA by removing negative supercoils.

2. DNA polymerase III is charged with copying the entire genome in a limited time. Processivity—binding to the DNA for thousands of elongation steps—makes rapid replication feasible.

PROBLEMS

1. *Guide and starting point.* Define template and primer as they relate to DNA synthesis. ✓ 3

2. *Cooperation among polymerases.* Explain why DNA synthesis depends on RNA synthesis. ✓ 4

3. *One from many.* What is an Okazaki fragment? ✓ 4

4. *Foremost and reluctant.* Distinguish between the leading and the lagging strands in DNA synthesis. ✓ 4

5. *Like Homer and Marge.* Match each term with its description.

(a) DNA polymerase _____

(b) Template _____
(c) Primer _____
(d) Helicase _____
(e) Topoisomerase _____
(f) Exonuclease _____
(g) Primase _____
(h) Okazaki fragment

_____

(i) DNA ligase _____
(j) Telomerase _____

1. Unwinds the double helix
2. Crucial for proofreading
3. Uses ATP to join DNA fragments
4. Determines the sequence of a complementary nucleic acid
5. Faithfully replicates the DNA
6. Synthesizes a segment of RNA
7. Found on the lagging strand
8. The initial segment of a polymer that will be extended
9. Prevents the disappearance of the lagging strand
10. Relaxes or introduces supercoils

6. *Which way?* Explain, on the basis of nucleotide structure, why DNA synthesis proceeds in the 5′-to-3 direction. ✓ 4

7. *Strength in numbers.* For long double-stranded DNA molecules, the rate of spontaneous strand separation is negligibly low under physiological conditions despite the fact that only weak reversible bonds hold the strands together. Explain.

8. *Only one oar in the water.* DNA replication in *E. coli* begins at a unique site on the circular chromosome called the origin of replication and proceeds in both directions as illustrated in part A.

What would be the effect of a mutation that yielded only one functional fork per origin of replication site (part B)? ✓ 4

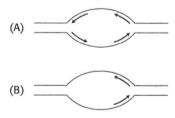

9. *Wound tighter than a drum.* Why would replication come to a halt in the absence of topoisomerase II? ✓ 3

10. *And just like that, poof. He's gone.* Telomerase is not active in most human cells. Some cancer biologists have suggested that the activation of the telomerase gene would be a requirement for a cell to become cancerous. Explain. ✓ 4

11. *I need to unwind.* With the assumption that the energy required to break an average base pair in DNA is 10 kJ mol^{-1} (2.4 kcal mol^{-1}), estimate the maximum number of base pairs that could be broken per ATP hydrolyzed by a helicase operating under standard conditions. ✓ 3

12. *The terminator.* (a) What would be the effect of adding dideoxyadenosine triphosphate (ddATP) to a DNA-replication reaction in large excess of deoxyadenosine

triphosphate (dATP)? (b) What would happen if ddATP were added at 10% of the concentration of dATP?

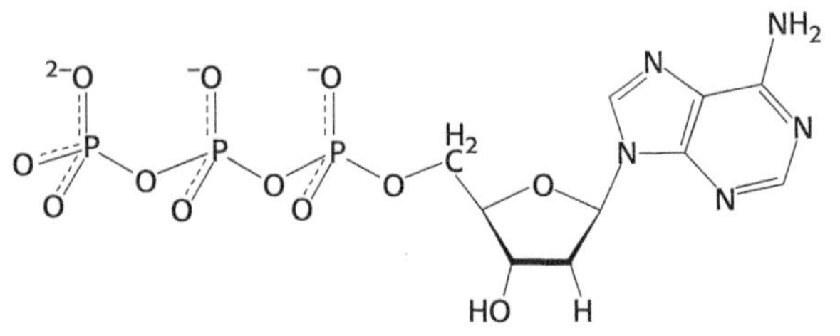

5′-dATP

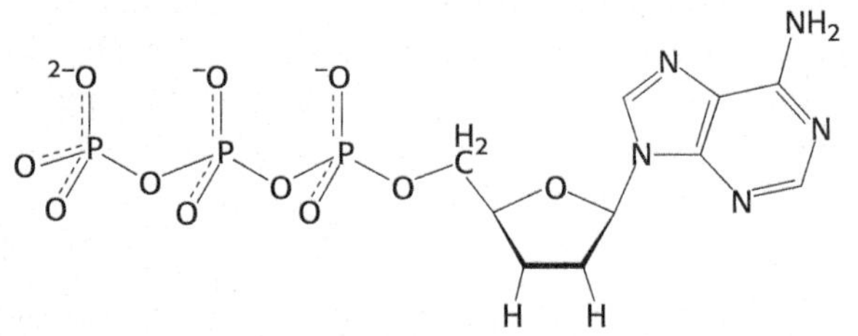

5′-ddATP

Chapter Integration Problem

13. *Life in a hot tub.* An archaeon (*Sulfolobus acidocaldarius*) found in acidic hot springs contains a topoisomerase that catalyzes the ATP-driven introduction of positive supercoils into DNA. How might this enzyme be advantageous to this unusual organism? ✓ 3

Data Interpretation and Challenge Problem

14. *Like a ladder.* Circular DNA from SV40 virus was isolated and subjected to gel electrophoresis. The results are shown below in lane A (the control) of the electrophoretic patterns. ✓ 3

[From W. Keller, *Proc. Nat. Acad. Sci. U. S. A.* 72:2553, 1975.]

(a) Why does the DNA separate in agarose gel electrophoresis? How does the DNA in each band differ?
The DNA was then incubated with topoisomerase I for 5 minutes and again analyzed by gel electrophoresis with the results shown in lane B.

(b) What types of DNA do the various bands represent?
Another sample of DNA was incubated with topoisomerase I for 30 minutes and again analyzed as shown in lane C.

(c) What is the significance of the fact that more of the DNA is in slower-moving forms?

Challenge Problems

15 *Don't rush it.* The accuracy rate for DNA polymerase falls if the next nucleotide to be polymerized is present in high concentration. Conversely, the accuracy rate rises if the next nucleotide is present in low concentration, a phenomenon termed the "next nucleotide effect." Suggest an explanation for this effect. ✓ 3

16. *Fast flying fork.* (a) How fast does template DNA spin (expressed in revolutions per second) at an *E. coli* replication fork? (b) What is the velocity of movement (in micrometers per second) of DNA polymerase III holoenzyme relative to the template?

17. *Nick translation.* Suppose that you wish to make a sample of DNA duplex highly radioactive to use as a DNA probe. You have a DNA endonuclease that cleaves the DNA internally to generate 3′-OH and 5′-phosphoryl groups, intact DNA polymerase I, and radioactive dNTPs. Suggest a means for making the DNA radioactive. ✓ 3

Selected Readings for this chapter can be found online at www.whfreeman.com/tymoczko3e.

CHAPTER 35

DNA Repair and Recombination

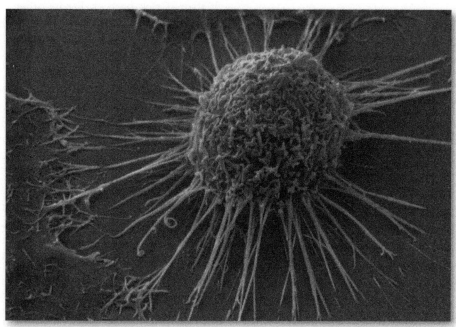

Cancer may result if DNA damage is not repaired. The Greek physician Hippocrates used the term carcinoma (from the Greek *karcinos*, meaning "crab") to describe tumors. Like crabs, some tumors have a shell-like surface and leg-like filaments. The image shows a colored scanning electron micrograph of a cervical cancer cell. This large rounded cell has an uneven surface with many cytoplasmic projections. Cervical cancer is one of the most common cancers affecting women. [STEVE GSCHMEISSNER/Getty Images.]

We have examined how even very large and complex genomes can be replicated with considerable fidelity. However, DNA does become damaged, both in the course of replication and through other processes. Damage to DNA can be as simple as the misincorporation of a single base, or it can take more complex forms such as the chemical modification of bases, chemical cross-links between the two strands of the double helix, or breaks in one or both of the phosphodiester backbones. The results may be cell transformation, changes in the DNA sequence that can be inherited by future generations, blockage of the DNA replication process itself, or even cell death. A variety of DNA-repair systems have evolved that can recognize these defects and, in many cases, restore the DNA molecule to its undamaged form. We begin with some of the sources of DNA damage and then proceed to examine the mechanisms that repair damaged DNA.

35.1 Errors Can Arise in DNA Replication

Errors introduced in the replication process are the simplest source of damage in the double helix. With the addition of each base, there is the possibility that an incorrect base might be incorporated, forming a non-Watson–Crick base pair. These non-Watson–Crick base pairs can locally distort the DNA double helix. Furthermore, such mismatches can be *mutagenic*; that is, they can result in permanent changes in the DNA sequence. When a double helix containing a non-Watson–Crick base pair is replicated, the two daughter double helices will have different sequences because the mismatched base is very likely to pair with its Watson–Crick partner. Errors other than mismatches include insertions, deletions, and breaks in one or both strands. Such errors are especially damaging because replicative polymerases can stall or even fall off a damaged template entirely. As a consequence, replication of the genome may halt before it is complete.

A variety of mechanisms have evolved to deal with such interruptions, including the recruitment of specialized DNA polymerases, called *translesion* or *error-prone polymerases,* to the region of the DNA that contains the lesions (Table 34.1). Because these molecular machines are not as precise as the replicative polymerases, they can replicate DNA across many lesions. However, a drawback to the use of these enzymes is that such polymerases are substantially more error prone than are normal replicative polymerases when replicating normal DNA. Nonetheless, these error-prone polymerases allow the completion of a draft sequence of the damaged area of the genome that can be at least partly repaired by DNA-repair processes. DNA recombination (Section 35.3) provides an additional mechanism for salvaging interruptions in DNA replication.

 CLINICAL INSIGHT

Some Genetic Diseases Are Caused by the Expansion of Repeats of Three Nucleotides

Some genetic diseases are caused by the presence of DNA sequences that are inherently prone to errors in the course of replication. One particular class of such diseases is characterized by the presence of long tandem arrays of repeats of three nucleotides. An example is *Huntington disease,* an autosomal dominant neurological disorder with a variable age of onset. The symptoms of Huntington disease include uncontrolled movements (chorea), the loss of cognitive facilities, and personality alterations. The mutated gene in this disease expresses a protein in the brain called huntingtin, which contains a stretch of consecutive glutamine residues. These glutamine residues are encoded by a tandem array of CAG sequences within the gene. In unaffected persons, this array is between 6 and 31 repeats, whereas, in those having the disease, the array is between 36 and 82 repeats or longer. Moreover, the array tends to become longer from one generation to the next. The consequence is a phenomenon called *anticipation*: the children of an affected parent tend to show symptoms of the disease at an earlier age than did the parent.

The tendency of *trinucleotide repeats* to expand is explained by the formation of alternative structures in DNA replication (**Figure 35.1**). Part of the array within a template strand can loop out without disrupting base-pairing outside this region. In replication, DNA polymerase extends this strand through the remainder

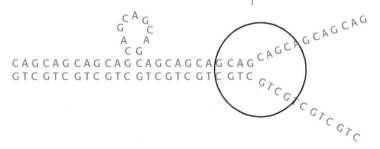

Figure 35.1 Triplet-repeat expansion. Sequences containing tandem arrays of repeated triplet sequences can be expanded to include more repeats by the looping out of some of the repeats before replication. The double helix formed from the red template strand will contain additional sequences encompassing the looped-out region. The circle represents the replication machinery.

of the array by a poorly understood mechanism, leading to an increase in the number of copies of the trinucleotide sequence. Friedreich's ataxia (p. 371) is caused by trinucleotide repeat expansion in the gene encoding frataxin, which disrupts iron–sulfur cluster synthesis and thus mitochondrial function.

A number of other neurological diseases are characterized by expanding arrays of trinucleotide repeats. How do these long stretches of repeated amino acids cause disease? For huntingtin, it appears that the polyglutamine stretches become increasingly prone to aggregate as their length increases; the additional consequences of such aggregation are still under investigation. Thus, Huntington disease and other trinucleotide expansion diseases may be yet other examples of a protein-aggregation disease (p. 63).

Bases Can Be Damaged by Oxidizing Agents, Alkylating Agents, and Light

The integrity of DNA is constantly under assault, not only from errors in replication, but also by an array of chemical agents. Indeed, an estimated 10^4 to 10^6 DNA-damaging events take place in a single human cell in a day. The chemical agents that alter specific bases within DNA after replication is complete are called *mutagens* and include reactive oxygen species such as hydroxyl radical. For example, hydroxyl radical reacts with guanine to form 8-oxoguanine. 8-Oxoguanine is mutagenic because it often pairs with adenine rather than cytosine in DNA replication (Figure 35.2).

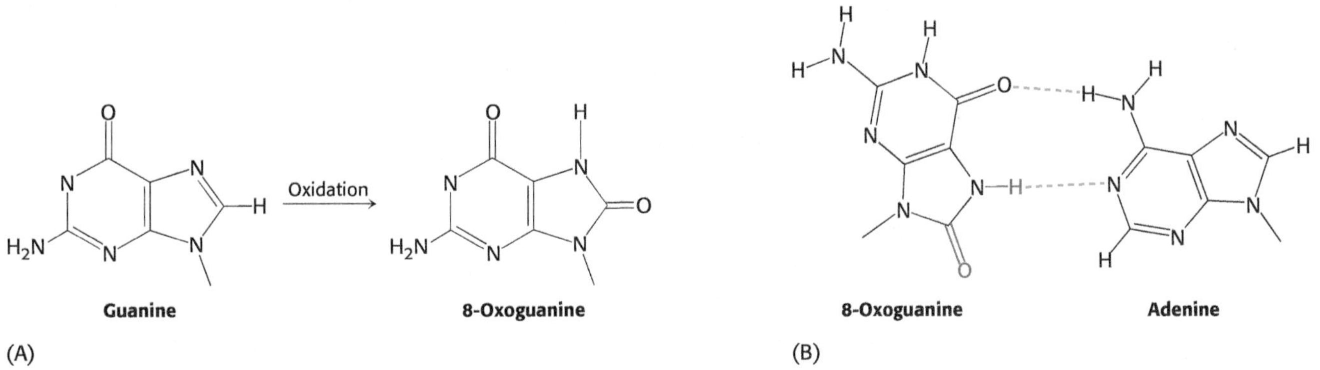

Figure 35.2 Guanine oxidation. (A) Guanine, the base component of dGMP in DNA, can be oxidized to 8-oxoguanine. (B) 8-Oxoguanine can base-pair with adenine.

After replication, one of the replicated helices will have an A–T base pair in place of a G–C base pair. Deamination is another potentially deleterious process. For example, adenine can be deaminated to form hypoxanthine (Figure 35.3). This process is mutagenic because hypoxanthine pairs with cytosine rather than thymine. Guanine and cytosine also can be deaminated to yield bases that pair differently from the parent base.

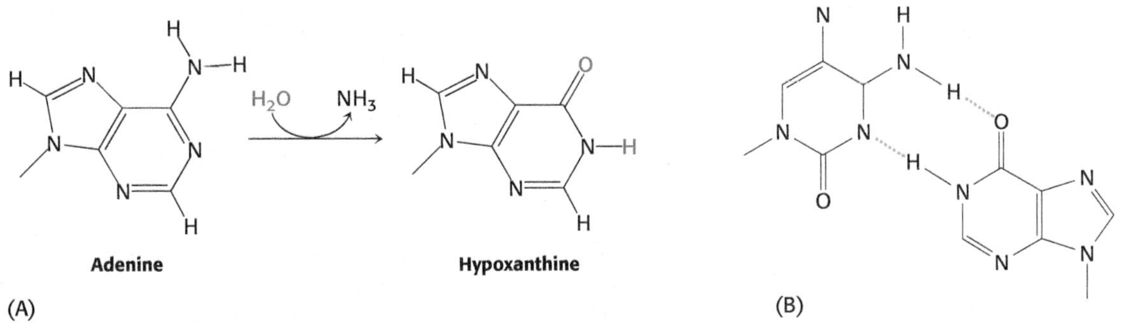

Figure 35.3 Adenine deamination. (A) The base adenine can be deaminated to form hypoxanthine. (B) Hypoxanthine forms base pairs with cytosine in a manner similar to that of guanine, and so the deamination reaction can result in mutation.

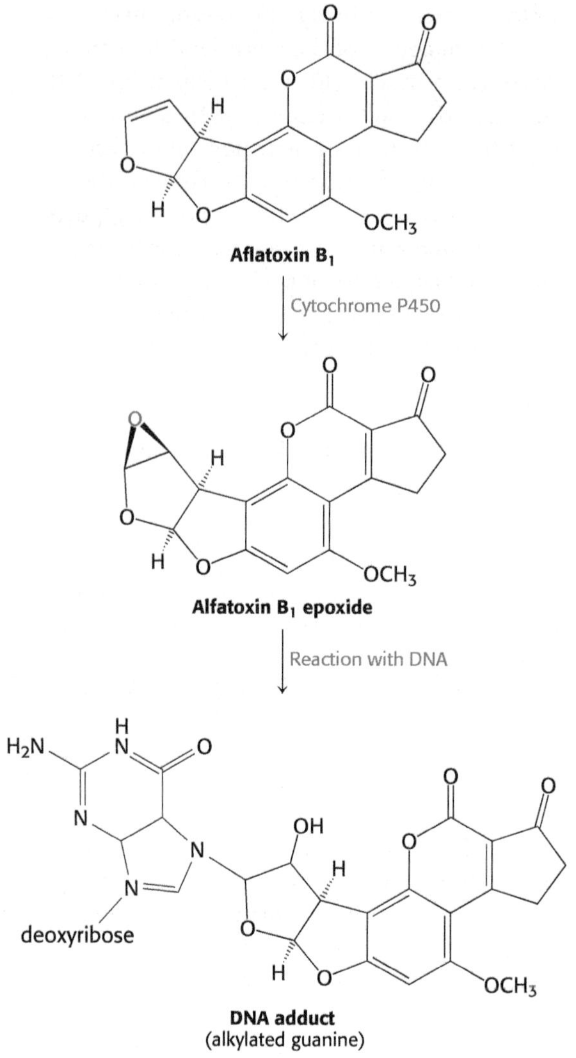

Figure 35.4 Aflatoxin B_1 activation. The compound, produced by molds that grow on peanuts and other grains, is activated by a cytochrome P450 enzyme to form a highly reactive species that modifies bases such as guanine in DNA, leading to mutations.

DID YOU KNOW?

Epoxide is an oxygen atom bonded to two carbon atoms and is also called a cyclic ether. Because of the strained ring, epoxides are especially reactive.

In addition to oxidation and deamination, nucleotide bases are subject to the addition of a hydrocarbon molecule, a reaction called alkylation. A striking example is aflatoxin B_1, a compound produced by molds that grow on peanuts and other foods. A cytochrome P450 enzyme (p. 542) converts this compound into a highly reactive epoxide (Figure 35.4). This agent reacts with the N-7 atom of guanine to form a mutagenic adduct that frequently leads to the conversion of a G–C base pair into a T–A base pair. Polycyclic aromatic hydrocarbons, present in automobile exhaust and cigarette smoke, also are converted into reactive epoxides by a cytochrome P450 enzyme. These epoxides alkylate DNA, accounting in part, for the carcinogenicity of those pollutants.

The most pervasive DNA-damaging agent is the ultraviolet component of sunlight (Figure 35.5). Its major effect is to covalently link adjacent pyrimidine residues along a DNA strand (Figure 35.6). Such a pyrimidine dimer cannot fit into a double helix, and so replication and gene expression are blocked until the lesion is removed. The tanning that occurs from exposure to sunlight is due to the formation of melanin pigments. These pigments absorb ultraviolet light and may protect against further DNA damage.

High-energy electromagnetic radiation (ionizing radiation) such as x-rays can damage DNA by producing high concentrations of reactive chemicals. X-ray exposure can also induce several types of DNA damage, including single- and double-stranded breaks in DNA.

Figure 35.5 Tanning. Ultraviolet light increases skin pigmentation and causes DNA damage. [Robert Daly/Getty Images.]

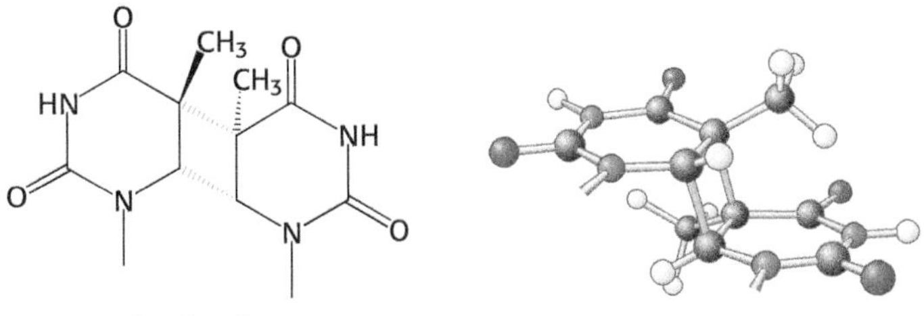

Thymine dimer

Figure 35.6 A cross-linked dimer of two thymine bases. Ultraviolet light induces cross-links between adjacent pyrimidines along one strand of DNA.

35.2 DNA Damage Can Be Detected and Repaired

✓ 5 Explain the mechanisms that ensure the fidelity of DNA replication.

To protect the integrity of the genetic message, a wide range of DNA-repair systems are present in most organisms. The bacterium *Deinococcus radiodurans* illustrates the extraordinary power of such repair systems. This bacterium was discovered in 1956 when scientists were studying the use of high doses of gamma radiation to sterilize canned meat. In some cases, the meat still spoiled due to the growth of a bacterial species that withstood doses of gamma radiation more than 1000 times larger than those that would kill a human being. Even when the bacterial chromosomes are broken into many fragments by the ionizing radiation, they can reassemble and recombine to regenerate the intact genome with essentially no loss of information. While most organisms do not have the robust repair systems of *D. radiodurans*, all organisms require the ability to repair damaged DNA.

Many systems repair DNA by using sequence information from the uncompromised strand. Such single-strand replication systems follow a similar mechanistic outline:

1. Recognize the offending base(s).

2. Remove the offending base(s).

3. Repair the resulting gap with a DNA polymerase and DNA ligase.

We will briefly consider examples of several repair pathways that follow these basic steps. Although many of these examples are taken from *E. coli*, corresponding repair systems are present in most other organisms, including human beings.

The replicative DNA polymerases are able to correct many DNA mismatches produced in the course of replication. For example, a subunit of *E. coli* DNA polymerase III functions as a 3′-to-5′ exonuclease, as described in Chapter 34. This domain removes mismatched nucleotides from the 3′ end of DNA by hydrolysis. How does the enzyme sense whether a newly added base is correct? As a new strand of DNA is synthesized, it is proofread. If an incorrect base is inserted, then DNA synthesis slows down and the mismatched pair, which is weakly bound, is able to fluctuate in position. The delay from the slowdown allows time for these fluctuations to take the newly synthesized strand out of the polymerase active site and into the exonuclease active site (Figure 34.9). There, the DNA is degraded until it moves back into the polymerase active site and synthesis continues.

A second mechanism is present in essentially all cells to correct errors made in replication that are not corrected by proofreading (**Figure 35.7**).

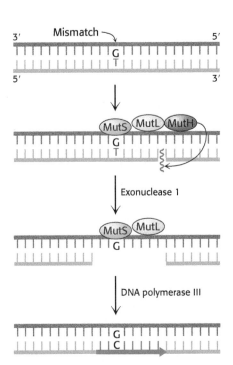

Figure 35.7 Mismatch repair. DNA mismatch repair in *E. coli* is initiated by the interplay of MutS, MutL, and MutH proteins. A G–T mismatch is recognized by MutS. MutL binds to MutS and activates the MutH nuclease. MutH cleaves the backbone in the vicinity of the mismatch. A segment of the DNA strand containing the erroneous T is removed by exonuclease I and synthesized anew by DNA polymerase III. [Information from R. F. Service, *Science* 263:1559–1560, 1994.]

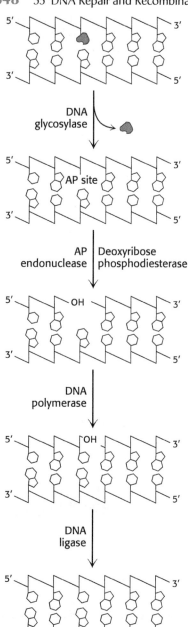

DNA
glycosylase

AP site

AP | Deoxyribose
endonuclease | phosphodiesterase

DNA
polymerase

DNA
ligase

Figure 35.8 Base-excision repair. The base-excision-repair process identifies and removes a modified base. Subsequently, the entire nucleotide is replaced. [Information from B. A. Pierce, *Genetics: A Conceptual Approach*, 3d ed. (W. H. Freeman and Company, 2008), p. 493.]

Mismatch-repair systems consist of at least two proteins, one for detecting the mismatch and the other for recruiting an endonuclease that cleaves the newly synthesized DNA strand close to the lesion. An exonuclease can then excise the incorrect base, and a polymerase fills the gap. In *E. coli*, the mismatch-repair proteins are called MutS and MutL and the endonuclease is called MutH.

How does the mismatch-repair machinery determine which of the bases is incorrect? The answer is not well established for higher organisms; however, in *E. coli*, some adenine bases of the parent strand of DNA are methylated, whereas the newly synthesized daughter strand is not yet methylated. Thus, the repair machinery recognizes that the base attached to the methylated DNA is correct and removes the offending base from the daughter strand.

Sometimes, the damage to DNA can be repaired directly, without having to remove any fragments of the DNA, a process called *direct repair*. An example of direct repair is the photochemical cleavage of pyrimidine dimers by a *photoreactivating enzyme* called *DNA photolyase*. The *E. coli* enzyme, a 35-kDa protein that contains bound N^5, N^{10}-methenyltetrahydrofolate and flavin adenine dinucleotide (FAD) cofactors, binds to the distorted region of DNA. The enzyme uses light energy—specifically, the absorption of a photon by the N^5, N^{10}-methenyltetrahydrofolate coenzyme—to form an excited state that cleaves the dimer into its component bases.

Modified bases, such as 8-oxyguanine or 3-methyladenine, are excised by the *E. coli* enzyme *AlkA*, an example of *base-excision repair*. The binding of this enzyme to damaged DNA flips the affected base out of the DNA double helix and into the active site of the enzyme. The enzyme then acts as a *glycosylase*, cleaving the glycosidic bond to release the damaged base. At this stage, the DNA backbone is intact, but a base is missing. This hole is called an *AP site* because it is *apurinic* (devoid of A or G) or *apyrimidinic* (devoid of C or T). An *AP endonuclease* recognizes this defect and nicks the backbone adjacent to the missing base. *Deoxyribose phosphodiesterase* excises the residual deoxyribose phosphate unit, and DNA polymerase I inserts an undamaged nucleotide, as dictated by the base on the undamaged complementary strand. Finally, the repaired strand is sealed by DNA ligase (**Figure 35.8**).

Base-excision repair corrects the most common point mutation in humans, C → T. Cytosine in eukaryotic DNA is frequently methylated in the 5 position as a means of transcription regulation. When 5-methylcytosine spontaneously deaminates, thymine is formed, generating a T–G base pair (**Figure 35.9**). Inasmuch as both T and G are normal bases in DNA, how does the repair machinery know which base to retain and which to remove? Because the C → T mutation is so common, the T in a T–G pair is always treated as the incorrect base and removed by the base-excision-repair proteins.

Base-excision repair requires recognition of an altered base. However, not all damaged bases are recognized by DNA glycosylase. What system recognizes improper nucleotide pairs that escape the base-excision-repair system? The

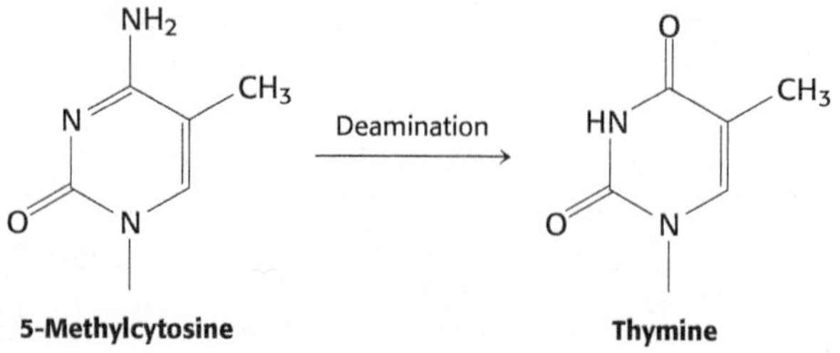

Figure 35.9 The deamination of 5-methylcytosine forms thymine.

Deamination

5-Methylcytosine   **Thymine**

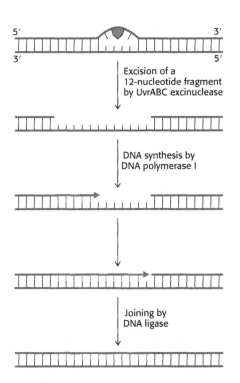

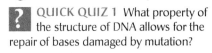

QUICK QUIZ 1 What property of the structure of DNA allows for the repair of bases damaged by mutation?

Figure 35.10 Nucleotide-excision repair. The repair of a region of DNA containing a thymine dimer by the sequential action of a specific excinuclease, a DNA polymerase, and a DNA ligase. The thymine dimer is shown in blue, and the new region of DNA is in red. [Information from P. C. Hanawalt. *Endeavour* 31:83, 1982.]

nucleotide-excision-repair system recognizes distortions in the DNA double helix caused by the presence of a damaged base. One of the best-understood examples of nucleotide-excision repair is utilized for the excision of a pyrimidine dimer. Three enzymatic activities are essential for this repair process in *E. coli* (**Figure 35.10**). First, an enzyme complex consisting of the proteins encoded by the *uvrABC* genes detects the distortion produced by the DNA damage. The UvrABC enzyme, an excinuclease (from the Latin *exci*, meaning "to cut out"), cuts the damaged DNA strand at two sites, eight nucleotides away from the damaged site on the 5′ side and four nucleotides away on the 3′ side. DNA polymerase I then enters the gap to carry out repair synthesis. The 3′ end of the nicked strand is the primer, and the intact complementary strand is the template. Finally, the 3′ end of the newly synthesized stretch of DNA and the original part of the DNA chain are joined by DNA ligase.

The Presence of Thymine Instead of Uracil in DNA Permits the Repair of Deaminated Cytosine

The presence in DNA of thymine rather than uracil was an enigma for many years. Both bases pair with adenine. The only difference between them is a methyl group in thymine in place of the C-5 hydrogen atom in uracil. Why is a methylated base employed in DNA and not in RNA? The existence of an active repair system to correct the deamination of cytosine provides a convincing solution to this puzzle.

Cytosine in DNA spontaneously deaminates at a perceptible rate to form uracil. The deamination of cytosine is potentially mutagenic because uracil pairs with adenine, and so one of the daughter strands will contain a U–A base pair rather than the original C–G base pair. This mutation is prevented by a base-excision-repair system that recognizes uracil to be foreign to DNA (**Figure 35.11**). The repair enzyme, *uracil DNA glycosylase*, hydrolyzes the glycosidic bond between the uracil and deoxyribose moieties but does not attack thymine-containing nucleotides. The AP site generated is repaired to reinsert cytosine. Thus, *the methyl group on thymine is a tag that distinguishes thymine from deaminated cytosine*. If thymine were not used in DNA, uracil correctly in

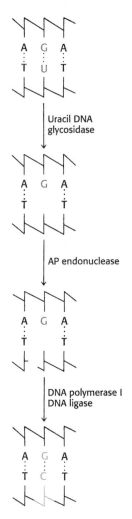

Figure 35.11 Uracil repair. Uridine bases in DNA, formed by the deamination of cytidine, are excised and replaced by cytidine.

place would be indistinguishable from uracil formed by deamination. The defect would persist unnoticed, and so a C–G base pair would necessarily be mutated to U–A in one of the daughter DNA molecules. *Thymine is used instead of uracil in DNA to preserve the fidelity of the genetic message.*

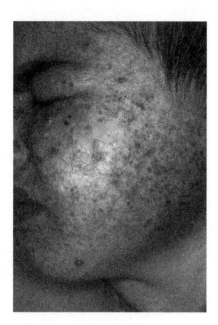

Figure 35.12 Xeroderma pigmentosa. In people suffering from xeroderma pigmentosa, numerous hyperpigmented blemishes appear most commonly on sun-exposed skin. These blemishes may progress to rough-surfaced growths (solar keratoses) and, finally, to skin malignancies. [ISM/Phototake.]

Cyclophosphamide

Cisplatin

☤ CLINICAL INSIGHT

Many Cancers Are Caused by the Defective Repair of DNA

As described in Chapter 13, cancers are caused by mutations in genes associated with growth control. Defects in DNA-repair systems increase the overall frequency of mutations and, hence, the likelihood of cancer-causing mutations. Indeed, the synergy between studies of mutations that predispose people to cancer and studies of DNA repair in model organisms have had a tremendous impact in revealing the biochemistry of DNA-repair pathways. Genes for DNA-repair proteins are often *tumor-suppressor genes;* that is, they suppress tumor development when at least one copy of the gene is free of a deleterious mutation. When both copies of a gene are mutated, however, tumors develop at rates greater than those for the population at large. People who inherit defects in a single tumor-suppressor allele do not necessarily develop cancer but are susceptible to developing the disease because only the one remaining normal copy of the gene must mutate to further the development of cancer.

Consider, for example, *xeroderma pigmentosum*, a rare human skin disease (**Figure 35.12**). The skin in an affected person is extremely sensitive to sunlight or ultraviolet light. In infancy, severe changes in the skin become evident and worsen with time. The skin becomes dry, and there is a marked atrophy of the dermis. Keratoses (skin growth derived from epidermal keratinocytes) appear, the eyelids become scarred, and the cornea ulcerates. Skin cancer usually develops at several sites. Many patients die before age 30 from metastases of these malignant skin tumors. Studies of xeroderma pigmentosum patients have revealed that mutations occur in genes for a number of different proteins. These proteins are components of the human nucleotide-excision-repair pathway, including components similar to the UvrABC subunits.

Defects in other repair systems can increase the frequency of other tumors. For example, *hereditary nonpolyposis colorectal cancer* (HNPCC, or *Lynch syndrome*) results from defective DNA mismatch repair. HNPCC is not rare—as many as 1 in 200 people will develop this form of cancer. Mutations in two genes, called *hMSH2* and *hMLH1*, account for most cases of this hereditary predisposition to cancer. These mutated genes seem likely to allow mutations to accumulate throughout the genome. In time, genes important in controlling cell proliferation become altered, resulting in the onset of cancer.

The absence of repair mechanisms and the high rates of cell duplication in cancer cells present a therapeutic opportunity. Several agents widely used in cancer chemotherapy, including cyclophosphamide and cisplatin (p. 622), act by damaging DNA. Ideally, the cancer cells are irreparably damaged and die.

☤ CLINICAL INSIGHT

Many Potential Carcinogens Can Be Detected by Their Mutagenic Action on Bacteria

Many human cancers are caused by exposure to chemicals that cause mutations. It is important to identify compounds that can cause mutations and ascertain their potency so that human exposure to them can be minimized.

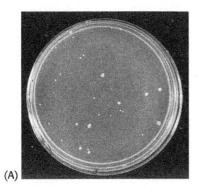

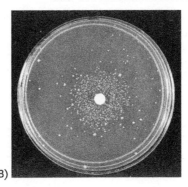

(A)   (B)

Figure 35.13 Ames test. (A) A Petri plate containing about 10^9 *Salmonella* bacteria that cannot synthesize histidine. (B) The same Petri plate with the addition of a filter-paper disc with a mutagen, which produces a large number of revertants that can synthesize histidine. After 2 days, the revertants appear as rings of colonies around the disc. The small number of visible colonies in plate A are spontaneous revertants. [Reprinted from *Mutation Research/Environmental Mutagenesis and Related Subjects*, 31, B. N. Ames, J. McCann, and E. Yamasaki, Methods for detecting carcinogens and mutagens with the salmonella/mammalian-microsome mutagenicity test, p. 347, ©1975, with permission from Elsevier.]

Bruce Ames devised a simple and sensitive test for detecting chemical mutagens. In the *Ames test,* a thin layer of agar containing about 10^9 bacteria of a specially constructed strain of *Salmonella* is placed on a Petri plate. These bacteria are unable to grow in the absence of histidine because a mutation is present in one of the genes for the biosynthesis of this amino acid. The addition of a chemical mutagen to the plate results in many new mutations. A small proportion of them reverse the original mutation, and histidine can be synthesized. These *revertants* multiply in the absence of an external source of histidine and appear as discrete colonies after the plate has been incubated at 37°C for 2 days (**Figure 35.13**). For example, a sample of 2-aminoanthracene, a known cancer-causing agent (carcinogen), gives 11,000 revertant colonies, compared with only 30 spontaneous revertants in its absence. A series of concentrations of a chemical can be readily tested to generate a dose–response curve. These curves are usually linear, which suggests that there is no threshold concentration for mutagenesis.

A key feature of this detection system is the inclusion of a *mammalian liver homogenate.* Recall that some potential carcinogens such as aflatoxin are converted into their active forms by P450 enzyme systems in the liver or other mammalian tissues (p. 646). Bacteria lack these enzymes, and so the test plate requires a few milligrams of a liver homogenate to activate this group of mutagens.

The *Salmonella* test is extensively used to help evaluate the mutagenic and carcinogenic risks of a large number of chemicals. This rapid and inexpensive bacterial assay for mutagenicity complements epidemiological surveys and animal tests that are necessarily slower, more laborious, and far more expensive.

35.3 DNA Recombination Plays Important Roles in Replication and Repair

Most processes associated with DNA replication function to copy the genetic message as faithfully as possible. However, several biochemical processes require the *recombination* of genetic material between two DNA molecules. In genetic recombination, two daughter molecules of DNA are formed by the exchange of genetic material between two parent molecules. Under what circumstance will recombination be required for DNA repair?

Common DNA lesions that can be repaired by recombination are breaks in both strands of a DNA helix. Double-strand breaks arise when replication stalls, such as when the polymerase encounters an unrepaired nick in one of the template strands at the replication fork (**Figure 35.14**). The replication fork collapses, leaving a double-strand break on one of the daughter helices. Double-strand breaks can also result from exposure to ionizing radiation. Radiation on the

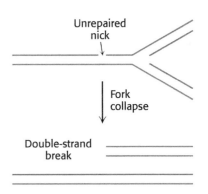

Figure 35.14 Generation of a double-strand break. DNA replication takes place at the replication fork. When the replication machinery on one strand encounters a nick in the DNA, replication stalls. The replication fork may collapse, yielding an uncompleted strand with a double-strand break.

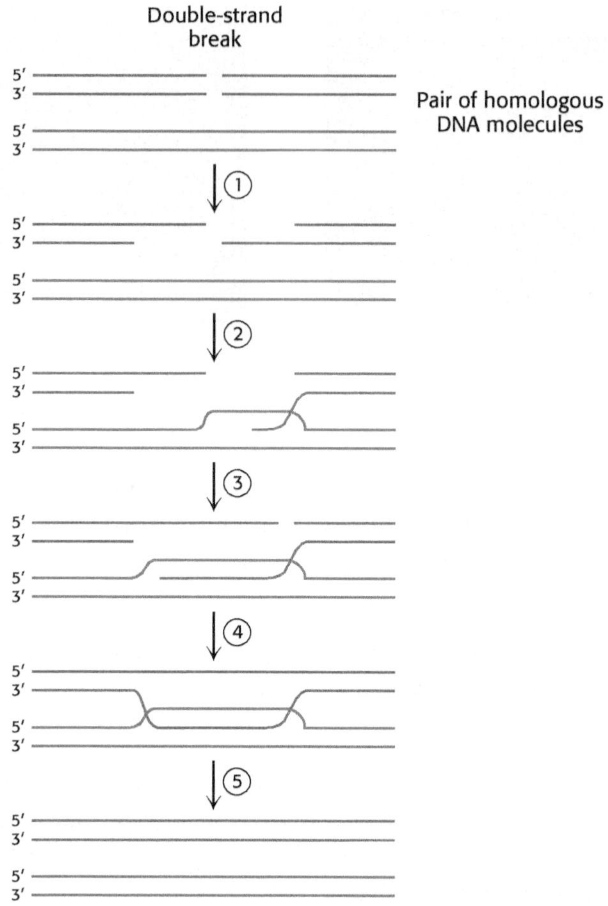

Figure 35.15 **Repair of double-strand break by using recombination.** 1. A 5′ exonuclease generates single-strand DNA at the site of the break. 2. Strand invasion takes place when the strand on the damaged DNA base-pairs with the complementary strand on the undamaged DNA, forming a displacement loop, or D-loop. 3. DNA synthesis takes place. 4. A second strand invasion takes place, resulting in the formation of two Holliday junctions. 5. The Holliday junctions are cleaved and ligated, forming two intact molecules of DNA.

short-wavelength end of the electromagnetic spectrum—notably, x-ray and gamma rays—is powerful enough to break the DNA backbone. Let's examine how a double-strand break caused by ionizing radiation is repaired by recombination.

Recombination is most efficient between stretches of DNA that are similar in sequence. In *homologous recombination*, parent DNA duplexes align at regions of sequence similarity, and new DNA molecules are formed by the breakage and joining of homologous segments.

Double-Strand Breaks Can Be Repaired by Recombination

Recombination is a complex process, requiring dozens of proteins. One key protein in recombination in humans is *RAD51*, an ATPase that binds single-stranded DNA. How are single-stranded regions generated at the site of a double-strand break?

The repair process begins with the digestion of the 5′ ends at the break site, generating single-stranded regions of DNA that are then bound by multiple copies of RAD51. The single-stranded DNA then displaces one of the strands of the undamaged double helix in an ATP-dependent process called strand invasion. The resulting three-stranded structure is called a *displacement loop* or *D-loop*. DNA synthesis takes place, with the use of the undamaged helix as a template. A second strand invasion then takes place to complete the repair of the damaged strand, resulting in the formation of two crosslike structures called *Holliday junctions* (**Figure 35.15**). Cleavage of the Holliday junctions, followed by ligation, yields two intact DNA double helices.

DNA Recombination Is Important in a Variety of Biological Processes

In the example illustrating the use of recombination for DNA repair, we assumed for simplicity's sake that the DNA sequences on the homologous strands were

(A) (B)

Figure 35.16 Consequences of gene disruption by recombination. Sections of muscle from normal (A) and gene-disrupted-by-recombination (B) mice, as viewed under the light microscope. Muscles do not develop properly in mice having both genes for myogenin disrupted. Arrows indicates bones of the pelvis, establishing that the two views show the same anatomical region. M in part B indicates a muscle fiber in the knockout mouse. [Reprinted by permission from Macmillan Publishers Ltd: Nature, v. 364, Hasty, P., Bradley, A., Morris, J. H., Edmondson, D. G., Venuti, J. M., Olson, E. N., Klein, W. H., *Muscle deficiency and neonatal death in mice with a targeted mutation in the myogenin gene*, pp. 501–506, ©1993.]

identical. However, such identical sequences are rarely the case. There are sequence differences—ranging from slight to substantial—in the alleles of the same gene on homologous DNA molecules. Consequently, when recombination takes place, new DNA sequences are generated. In meiosis, the limited exchange of genetic material between paired chromosomes provides a simple mechanism for generating genetic diversity in a population. Recombination also plays a crucial role in generating molecular diversity for antibodies and some other molecules in the immune system. Some viruses employ recombination pathways to integrate their genetic material into the DNA of a host cell.

Recombination is the foundation for an enormously powerful biochemical tool. Recombination is used to move, excise, or insert genes in, for example, the generation of "gene-knockout" mice, in which a specific gene is deleted, or "knock-in" mice, in which a specific gene is inserted. Such modified mice have proved to be valuable experimental tools.

For example, the gene-knockout approach has been applied to the genes encoding transcription factors that control the differentiation of muscle cells. When both copies of the gene for the regulatory protein *myogenin* are disrupted, an animal dies at birth because it lacks functional skeletal muscle. Microscopic inspection reveals that the tissues from which muscle normally forms contain precursor cells that have failed to differentiate fully (**Figure 35.16**). Heterozygous mice containing one normal gene for myogenin and one disrupted gene appear normal, suggesting that the level of gene expression is not essential for its function. Analogous studies have probed the function of many other genes to generate animal models for known human genetic diseases.

DID YOU KNOW?

Meiosis is a special type of cell division that is required for sexual reproduction. In animals, meiosis results in the production of sperm and eggs.

? QUICK QUIZ 2 How many strands of DNA are present in a Holliday structure?

SUMMARY

35.1 Errors Can Arise in DNA Replication

DNA can be damaged in a variety of ways. For example, mismatched bases may be incorporated in the course of DNA replication or individual bases may be damaged by oxidation or by hydrocarbon attachment after DNA replication. Other forms of damage are the formation of cross-links and the introduction of single- or double-stranded breaks in the DNA backbone.

35.2 DNA Damage Can Be Detected and Repaired

Several different repair systems detect and repair DNA damage. Repair begins with the process of proofreading in DNA replication: mismatched bases that were incorporated in the course of synthesis are excised by exonuclease activity present in replicative polymerases. Some DNA lesions such as thymine dimers can be directly reversed through the action of specific enzymes. Other DNA-repair pathways act through the excision of single damaged bases (base-excision repair) or short segments of nucleotides (nucleotide-excision repair). Defects in DNA-repair components are associated with susceptibility to many different sorts of cancer. Such defects are a common target of cancer treatments. Many potential carcinogens can be detected by their mutagenic action on bacteria (the Ames test).

35.3 DNA Recombination Plays Important Roles in Replication and Repair

Recombination is the exchange of segments between two DNA molecules. It is important in some types of DNA repair as well as other processes such as meiosis, the generation of antibody diversity, and the life cycles of some viruses. Some recombination pathways are initiated by strand invasion, in which a single strand at the end of a DNA double helix forms base pairs with one strand of DNA in another double helix and displaces the other strand. A common intermediate formed in other recombination pathways is the Holliday junction, which consists of four strands of DNA that come together to form a crosslike structure.

KEY TERMS

trinucleotide repeat (p. 644)
mutagen (p. 645)
mismatch repair (p. 648)
direct repair (p. 648)

base-excision repair (p. 648)
nucleotide-excision repair (p. 649)
tumor-suppressor gene (p. 650)
Ames test (p. 651)

RAD51 (p. 652)
Holliday junction (p. 652)

? Answers to QUICK QUIZZES

1. DNA's property of being a double helix. Damage to one strand of DNA can be repaired by the use of the other strand as a template.

2. Four.

PROBLEMS

1. *Mistakes were made.* List some of the means by which mutations can be introduced into DNA.

2. *Molecular insults.* List the factors discussed in this chapter that can cause DNA damage.

3. *As simple as 1,2,3.* What are the steps required of all DNA-repair mechanisms? ✓ 5

4. *DNA repair services.* List the repair systems responsible for maintaining the integrity of DNA. ✓ 5

5. *Sad, but true.* Generations of children have been told by their mothers: "Get out of the house and play outside. Sunshine is good for you!" Why is that now sometimes considered questionable advice?

6. *Flipped out.* Describe the process of base-excision repair, which removes modified bases.

7. *T time.* Explain the benefit of using thymine in DNA instead of uracil. ✓ 5

8. *A coincidence? I think not.* Cytosine in eukaryotic DNA is frequently methylated in the 5 position as a means of transcription regulation. Eukaryotic cells also contain a specialized repair system that recognizes G–T mismatches and repairs them to G–C base pairs. Why is this repair system useful to the cell? ✓ 5

9. *Corrections.* Most of the time, the misincorporation of a nucleotide in the course of DNA synthesis in *E. coli* does not lead to mutated progeny. What is the mechanistic basis for this good fortune?

10. *No wonder you're so tired.* The spontaneous cleavage of adenine or guanine from DNA, a process called spontaneous depurination, is a common form of DNA damage in mammalian cells. The rate of spontaneous depurination is estimated to be 3×10^{-9} per purine per minute. This damage must be repaired. A diploid human cell has 6×10^9 bp. How many spontaneous depurinations must be repaired per cell per day? ✓ 5

11. *Induced spectrum.* DNA photolyases convert the energy of light in the near-ultraviolet, or visible, region (300–500 nm) into chemical energy to break the cyclobutane ring of pyrimidine dimers. In the absence of substrate, these photoreactivating enzymes do not absorb light of wavelengths longer than 300 nm. Why is the substrate-induced absorption band advantageous? ✓ 5

12. *Like Lewis and Clark.* Match each term with its description.

(a) Trinucleotide repeat _____
(b) Mutagen _____
(c) Mismatch repair _____
(d) Direct repair _____
(e) Base-excision repair _____
(f) Nucleotide-excision repair _____
(g) Tumor-suppressor gene _____
(h) RAD51 _____
(i) Holliday junction _____
(j) Recombination _____

1. DNA repair without the removal of any fragments of DNA
2. Often a gene for DNA repair
3. An ATPase that binds single-stranded DNA
4. Exchange of genetic information
5. Chemical agents that alter DNA
6. Requires glycosylase activity
7. Repair mechanism encoded by the *uvrABC* gene in *E. coli*
8. Recombination intermediate
9. Expansion of it causes Huntington disease
10. Requires MutS, MutL, and MutH

Data Interpretation and Challenge Problems

13. *Ames test.* The adjoining illustration shows four Petri plates used for the Ames test. A piece of filter paper (white circle in the center of each plate) was soaked in one of four preparations and then placed on a Petri plate.

The four preparations contained (A) purified water (control), (B) a known mutagen, (C) a chemical whose mutagenicity is under investigation, and (D) the same chemical after treatment with liver homogenate. The number of revertants, visible as colonies on the Petri plates, was determined in each case.

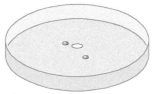

(A) Control: No mutagen

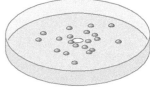

(B) + Known mutagen

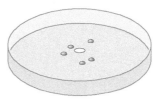

(C) + Experimental sample

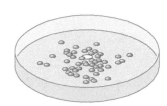

(D) + Experimental sample after treatment with liver homogenate

(a) What was the purpose of the control plate, which was exposed only to water?

(b) Why was it wise to use a known mutagen in the experimental system?

(c) How would you interpret the results obtained with the experimental compound?

(d) What liver components are likely responsible for the effects observed in preparation D?

14. *Here comes the sun, and I say, it's not alright.* Skin cancer is the most common cancer in Caucasians, accounting for nearly half of all cancers in the United States per year. More than 2 million people are expected to develop nonmelanoma skin cancer this year in the United States. Fortunately, this skin cancer is readily treated. Sun exposure is the major environmental agent implicated in the induction of nonmelanoma skin cancer. The adjoining graph shows the cumulative percentage of patients with nonmelanoma skin cancer plotted against the age of onset of skin cancers in normal and xeroderma pigmentosum (XP) skin-cancer patients. ✓ 5

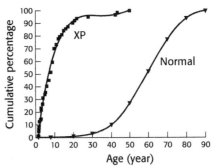

[Data from K. H. Kramer, *Proc. Natl. Acad. Sci. U. S. A.* 94:11–14, 1997.]

(a) Of the two groups of patients, which group is more susceptible to skin cancer?

(b) Explain your answer to part *a*.

(c) Hypothesize why it takes approximately six decades for skin cancer to develop in normal people.

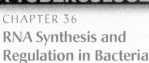

CHAPTER 36
RNA Synthesis and
Regulation in Bacteria

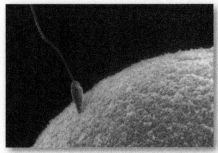

CHAPTER 37
Gene Expression in
Eukaryotes

CHAPTER 38
RNA Processing in
Eukaryotes

RNA Synthesis, Processing, and Regulation

D NA can be thought of as the archival information that is converted into a useful form by being copied as RNA. This process is called transcription or RNA synthesis. Transcription is the initial step in the expression of genomic information. It is analogous to making photocopies of particular parts of a rare out-of-print book for use in a specific class session. After having served their purpose, the copies can be destroyed, but the book itself—the genomic equivalent—is intact and undamaged.

The synthesis of accurate RNA is only part of the story. Even in an organism as simple as *E. coli*, not all of the genes are expressed at a given time. The pattern of genes that are transcribed into RNA changes with environmental conditions, such as when a bacterium finds itself in the presence of the disaccharide lactose as a carbon source instead of glucose. Transcription must be regulated to meet the biochemical needs of the cell.

This regulation becomes more complex in eukaryotes. There, transcription can change not only in response to diet, as in *E. coli*, but also in response to hormones and even behavior. For instance, exercise exerts its beneficial effects in part by altering the transcription patterns of muscle, liver, and fat cells. Perhaps the most dramatic example of the regulation of gene expression is the development of a single-cell fertilized egg into a fully functioning human being with trillions of cells, hundreds of cell types, and dozens of tissues. Each cell contains the entire repertoire of genetic information, but regulated transcription allows the process of growth and development.

The process of transcription in eukaryotes is even more complex than in bacteria in that the initial transcription product of most protein-encoding genes is not the final, functional RNA product. Most of the protein-encoding genes are actually mosaics of information-rich sequences that will be expressed (exons), interrupted by sequences that will not be expressed (introns). Subsequent to transcription, the information-rich regions must be removed from the initial RNA product of RNA synthesis and joined to form the final, usable product in a process called RNA splicing.

Transcription is thus the first step in unlocking the information in the genome. We will examine this process and its control first in bacteria and then in eukaryotes. The section ends with an examination of RNA splicing.

✓ **By the end of this section, you should be able to:**

✓ 1 Identify the key enzyme required for transcription.

✓ 2 Describe how transcription is controlled in bacteria.

✓ 3 Describe how transcription is regulated in eukaryotes.

✓ 4 Explain how RNA is processed after its transcription in eukaryotes.

RNA Synthesis and Regulation in Bacteria

A picture of the shores of the French Riviera and a poster identifying "The Great Scourge of Tuberculosis" might seem an odd pairing. However, in 1957 a strain of bacteria was isolated from a soil sample near St. Raphael on the Riviera that produced antibiotics used to treat tuberculosis. Rifampicin, derived from these antibiotics, acts by inhibiting bacterial RNA polymerase. [(Left) Niday Picture Library/Alamy. (Right) Mary Evans/ONSLOW AUCTIONS/The Image Works.]

W e begin by examining RNA synthesis and regulation in our constant traveling companion—the gut bacterium *E. coli*. As we have seen on many occasions before, studies on this model organism are an excellent source of insight into the basic biochemical processes common to all living systems.

✓ 1 Identify the key enzyme required for transcription.

36.1 Cellular RNA Is Synthesized by RNA Polymerases

The synthesis of RNA from a DNA template is called *transcription* and is catalyzed by a large enzyme common to all life forms called *RNA polymerase* (**Figure 36.1**). RNA polymerase requires the following components:

1. *A Template.* Double-stranded DNA is one of the substrates for RNA synthesis. However, usually only one of the strands is transcribed. We differentiate the two strands of DNA on the basis of their relation to the RNA product. *The sequence of the template strand of DNA is the complement of that of the RNA*

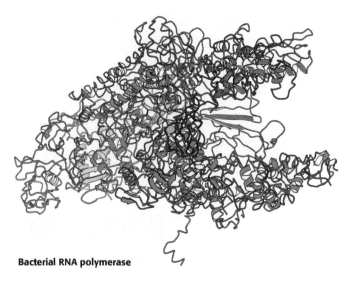

Figure 36.1 RNA polymerase. This large enzyme comprises many subunits, including β (blue) and β' (red), which form a "claw" that holds the DNA to be transcribed. Notice that the active site includes a magnesium ion (green) at the center of the structure. [Drawn from 1L9Z. pdb.]

Bacterial RNA polymerase

transcript (Figure 36.2). In contrast, *the coding strand of DNA has the same sequence as that of the RNA transcript* except for thymine (T) in place of uracil (U). The coding strand is also known as the *sense* (+) *strand* and the template strand as the *antisense* (−) *strand*.

2. *Activated Precursors.* In addition to DNA, the other required substrates are the building blocks—the *ribonucleoside triphosphates* ATP, GTP, UTP, and CTP—of the RNA product.

3. *A Divalent Metal Ion.* RNA polymerase requires a divalent cation cofactor. Either Mg^{2+} or Mn^{2+} is effective.

```
5'—GCGGCGACGCGCAGUUAAUCCCACAGCCGCCAGUUCCGCUGGCGGCAU—3'   mRNA
3'—CGCCGCTGCGCGTCAATTAGGGTGTCGGCGGTCAAGGCGACCGCCGTA—5'   Template (antisense) strand of DNA
5'—GCGGCGACGCGCAGTTAATCCCACAGCCGCCAGTTCCGCTGGCGGCAT—3'   Coding (sense) strand of DNA
```

Figure 36.2 Complementarity between mRNA and DNA. The base sequence of mRNA (green) is the complement of that of the DNA template strand. The other strand of DNA is called the coding strand because it has the same sequence as that of the RNA transcript except for thymine (T) in place of uracil (U).

RNA polymerase catalyzes the initiation and elongation of RNA strands. The reaction catalyzed by this enzyme is

$$(\text{RNA})_{n\text{ residues}} + \text{ribonucleoside triphosphate} \rightleftharpoons (\text{RNA})_{n+1\text{ residues}} + \text{PP}_i$$

The synthesis of RNA is similar to that of DNA in several respects (Figure 36.3). First, the direction of synthesis is $5' \rightarrow 3'$. Second, the mechanism of elongation is similar: the 3'-OH group at the terminus of the growing chain attacks the innermost phosphoryl group of the incoming ribonucleoside triphosphate. Third, the synthesis is driven forward by the hydrolysis of pyrophosphate. In contrast with DNA polymerase, however, RNA polymerase does not require a primer.

Genes Are the Transcriptional Units

There are many types of RNA, all of which are products of RNA polymerase. The segments of DNA that encode the various species of RNA are called *genes*. RNA polymerase recognizes the beginning and end of a gene by mechanisms that we will consider shortly.

Three major types of RNA are produced in all cells. *Messenger RNA* (mRNA) encodes the information for the synthesis of a protein, whereas *transfer RNA* (tRNA) and *ribosomal RNA* (rRNA) are crucial components in the machinery that

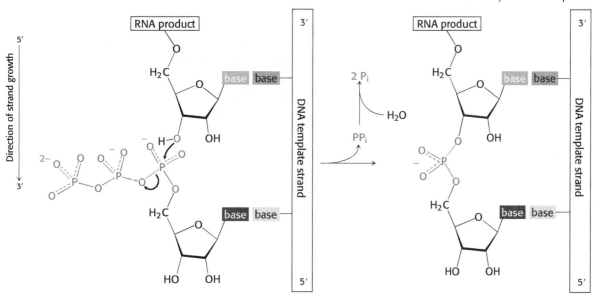

Figure 36.3 RNA strand-elongation reaction.

translates the mRNA into protein. All types of cellular RNA are synthesized in *E. coli* by the same RNA polymerase according to instructions given by a DNA template. In mammalian cells, there is a division of labor among several different kinds of RNA polymerases (Chapter 37), although the chemistry is the same for all types.

RNA Polymerase Is Composed of Multiple Subunits

As stated earlier, the enzyme responsible for transcription in all organisms is RNA polymerase. In *E. coli*, RNA polymerase is a very large (~500-kDa) and complex enzyme consisting of five kinds of subunits (Table 36.1). The subunit composition of the entire enzyme, called the *holoenzyme*, is $\alpha_2\beta\beta'\sigma\omega$. The *sigma* ($\sigma$) *subunit* helps to find a site where transcription begins, participates in the initiation of RNA synthesis, and then dissociates from the rest of the enzyme. RNA polymerase without this subunit ($\alpha_2\beta\beta'\omega$) is called the *core enzyme*. The core enzyme contains the active site.

Table 36.1 Subunits of *E. coli* RNA polymerase

Subunit	Gene	Number	Mass (kDa)	Function
α	*rpoA*	2	37	Required for assembly of core enzyme; interacts with regulatory factors
β	*rpoB*	1	151	Takes part in all stages of catalysis
β'	*rpoC*	1	155	Binds to DNA; takes part in catalysis
ω	*rpoZ*	1	10	Required to restore denatured polymerase to its native form
σ^{70}	*rpoD*	1	70	Takes part in promoter recognition

36.2 RNA Synthesis Comprises Three Stages

RNA synthesis, like nearly all biological polymerization reactions, takes place in three stages: *initiation*, *elongation*, and *termination*.

Transcription Is Initiated at Promoter Sites on the DNA Template

Let us consider the problem of initiation. There may be billions of bases in a genome, arranged into thousands of genes. How does RNA polymerase

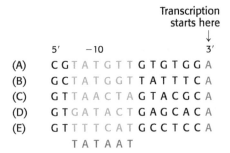

Transcription
starts here
↓

5′   −10   3′

(A)   C G T A T G T T G T G T G G A
(B)   G C T A T G G T T A T T T C A
(C)   G T T A A C T A G T A C G C A
(D)   G T G A T A C T G A G C A C A
(E)   G T T T T C A T G C C T C C A
       T A T A A T

Figure 36.4 Bacterial promoter sequences. A comparison of five sequences from bacterial promoters reveals a recurring sequence of TATAAT centered on position −10. The −10 consensus sequence (in red) was deduced from a large number of promoter sequences.

determine where to begin transcription? Special DNA sequences, called *promoters*, direct the RNA polymerase to the proper site for the initiation of transcription. Two common DNA sequences that act as the promoter for many bacterial genes are present on the 5′ (upstream) side of the start site. They are known as the −10 *site* (also called the *Pribnow box*) and the −35 *sequence* because they are centered at approximately 10 and 35 nucleotides upstream of the start site. They are each 6 bp long. Their *consensus* (average) *sequences*, deduced from the analysis of many promoters (**Figure 36.4**), are

$$5'\text{\textasciitilde\textasciitilde\textasciitilde}\overset{-35}{TTGACA}\text{\textasciitilde\textasciitilde\textasciitilde\textasciitilde\textasciitilde}\overset{-10}{TATAAT}\text{\textasciitilde\textasciitilde\textasciitilde}\overset{+1}{Start\ site}$$

The first nucleotide (the start site) of a transcribed DNA sequence (gene) is denoted as +1 and the second one as +2; the nucleotide preceding the start site is denoted as −1. These designations refer to the coding strand of DNA.

Not all promoters are equally efficient. Genes with strong promoters are transcribed frequently—as often as every 2 seconds in *E. coli*. In contrast, genes with very weak promoters are transcribed about once in 10 minutes. Thus, *the efficiency, or strength, of a promoter sequence serves to regulate transcription*. The −10 and −35 regions of the strongest promoters have sequences that correspond closely to the consensus sequences, whereas weak promoters tend to have multiple substitutions at these sites. Indeed, the mutation of a single base in either the −10 sequence or the −35 sequence can diminish promoter activity. The distance between these conserved sequences also is important; a separation of 17 nucleotides is optimal. Regulatory proteins, called *transcription factors*, that bind to specific sequences near promoter sites and interact with RNA polymerase (Chapter 37) also markedly influence the frequency of transcription of many genes.

Outside the promoter in a subset of highly expressed genes is the *upstream element*, also called the UP element (for *up*stream element). This sequence is present 40 to 60 nucleotides upstream of the transcription start site. The UP element is bound by the α subunit of RNA polymerase and increases the efficiency of transcription by creating an additional binding site for the polymerase.

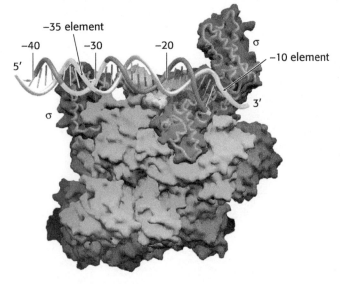

−35 element

−40   −30   −20   σ
5′   −10 element

                          3′

σ

Figure 36.5 The RNA polymerase holoenzyme complex. Notice that the σ subunit (orange) of the bacterial RNA polymerase holoenzyme makes sequence-specific contacts with the −10 and −35 promoter sequences (yellow). [From K. S. Murakami, S. Masuda, E. A. Campbell, O. Muzzin, and S. A. Darst. *Science* 296:1285–1290, 2002.]

Sigma Subunits of RNA Polymerase Recognize Promoter Sites

The $\alpha_2\beta\beta'\omega$ core of *E. coli* RNA polymerase is unable to start transcription at promoter sites. The first step in initiation is finding the start site for transcription. Proper initiation depends on the σ *subunit*, which helps the polymerase locate the correct start site. The σ subunit does so in two ways. First, paradoxically, *it decreases the affinity of RNA polymerase for general regions of DNA by a factor of 10^4*. This decrease permits the enzyme to bind to the DNA double helix and rapidly slide along it, searching for the promoter. In the absence of the σ subunit, the core enzyme binds DNA indiscriminately and tightly. *Second, the σ subunit enables RNA polymerase to recognize promoter sites* (**Figure 36.5**). When the new RNA chain (the *nascent* chain) reaches 9 or 10 nucleotides in length, it contacts the σ subunit, facilitating its ejection from the transcription complex. Release of the subunit marks the initiation to elongation transition. After its release, the σ subunit can assist initiation by another core enzyme. Thus, *the σ subunit acts catalytically* (**Figure 36.6**).

E. coli has many kinds of σ factors for recognizing several types of promoter sequences in *E. coli* DNA. The type that

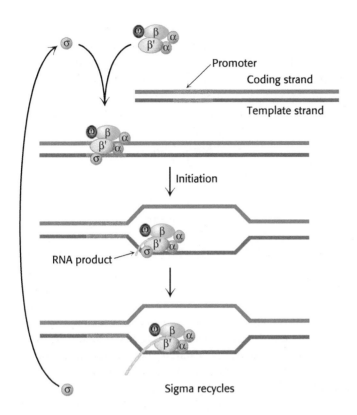

Figure 36.6 Sigma factors act catalytically. The σ factor assists the polymerase in finding the correct promoter site and then dissociates from the active enzyme to help another polymerase find the proper promoter. Thus, one σ factor can aid many polymerases in locating promoters.

recognizes the consensus sequences described earlier is called σ^{70} because it has a mass of 70 kDa. Other σ factors help the bacteria to withstand periods of elevated temperature, nitrogen starvation, and other environmental conditions.

RNA Strands Grow in the 5'-to-3' Direction

After RNA polymerase has bound to a promoter and before synthesis can begin, it must unwind a segment of the DNA double helix so that nucleotides on the template strand can direct the synthesis of the RNA product. Each bound polymerase molecule unwinds a 17-bp segment of DNA (**Figure 36.7**).

The transition from the *closed promoter complex* (in which DNA is double helical) to the *open promoter complex* (in which a DNA segment is unwound) is an essential event in transcription. After this event has taken place, the stage is set for the formation of the first phosphodiester linkage of the new RNA chain.

In contrast with DNA synthesis, _RNA synthesis can start de novo, without needing a primer_. In bacteria, most newly synthesized RNA strands carry a highly distinctive tag on the 5' end: the first base at that end is either *pppG* or *pppA*, confirming that RNA strands, like DNA strands, grow in the 5' → 3' direction (**Figure 36.8**).

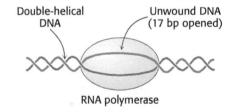

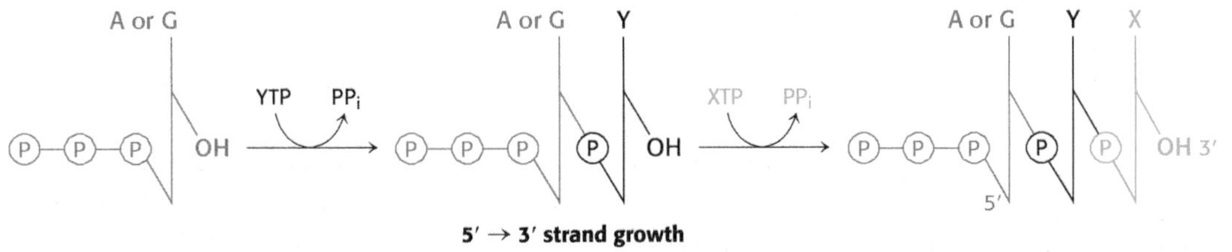

Figure 36.7 DNA unwinding. RNA polymerase unwinds about 17 base pairs of double-helical DNA to form the open promoter complex.

Figure 36.8 RNA strand growth. Newly synthesized RNA grows in the 5'-to-3' direction, with the first nucleotide being either pppA or pppG. YTP and XTP represent any other nucleotides.

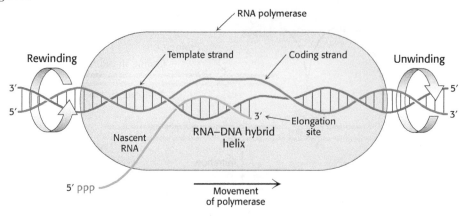

Figure 36.9 A transcription bubble. A schematic representation of a transcription bubble in the elongation of an RNA transcript. Duplex DNA is unwound at the forward end of RNA polymerase and rewound at its rear end. The RNA–DNA hybrid rotates during elongation.

Elongation Takes Place at Transcription Bubbles That Move Along the DNA Template

The elongation phase of RNA synthesis begins after the formation of the first phosphodiester linkage. An important change is the loss of the σ subunit shortly after initiation. The loss of σ enables the core enzyme to strongly bind to the DNA template (p. 662). Indeed, RNA polymerase continues transcription until a termination signal is reached. The region containing RNA polymerase, DNA, and nascent RNA is called a *transcription bubble* because it contains a locally denatured "bubble" of DNA (**Figure 36.9**). About 17 bp of DNA are unwound at a time throughout the elongation phase, as in the initiation phase. The newly synthesized RNA forms a hybrid helix with about 8 bp of the template DNA strand. The 3′-hydroxyl group of the RNA in this hybrid helix is positioned so that it can attack the α-phosphorus atom of an incoming ribonucleoside triphosphate. The transcription bubble moves a distance of 170 Å (17 nm) in a second, which corresponds to a rate of elongation of about 50 nucleotides per second, extruding the newly synthesized RNA from the polymerase (**Figure 36.10**). Although rapid, this rate is much slower than that of DNA synthesis, which is nearly 1000 nucleotides per second.

As in all real-life phenomena, mistakes are sometimes made by RNA polymerase. The error rate of the order of one mistake per 10^4 or 10^5 nucleotides is higher than that for DNA replication, including all error-correcting mechanisms. The lower fidelity of RNA synthesis can be tolerated because mistakes are not transmitted to progeny. For most genes, many RNA transcripts are synthesized; a few defective transcripts are unlikely to be harmful. However, RNA polymerases do show proofreading nuclease activity, particularly in the presence of accessory proteins. The RNA polymerase pauses when a wrong nucleotide is inserted in the nascent RNA. The enzyme then backtracks and, using its inherent nuclease activity, removes the mis-incorporated nucleotide.

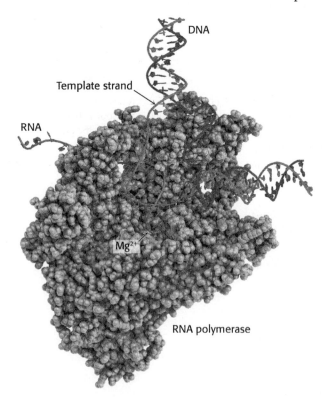

Figure 36.10 RNA–DNA hybrid separation. A model based on the crystal structure of the RNA polymerase holoenzyme shows the unwound DNA forming the transcription bubble. Notice that the RNA is peeled from the template strand and extruded from the enzyme. [Drawn by Adam Steinberg, after 1CDW.pdb.]

An RNA Hairpin Followed by Several Uracil Residues Terminates the Transcription of Some Genes

As already stated, the σ subunit enables bacterial RNA polymerase to initiate transcription at specific sites. Elongation then takes place, but how does the enzyme determine when to stop transcription? The termination of transcription is as precisely controlled as its initiation. In the termination phase of

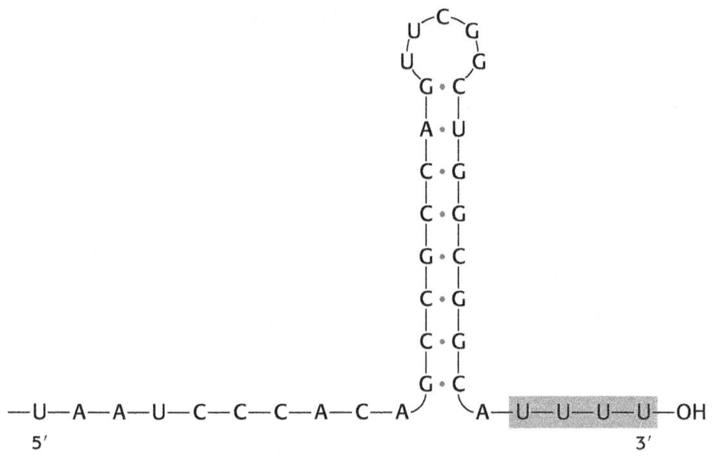

Figure 36.11 Termination signal. A termination signal found at the 3′ end of an mRNA transcript consists of a series of bases that form a stable stem-and-loop structure and a series of U residues.

transcription, the formation of phosphodiester linkages stops, the RNA–DNA hybrid dissociates, the melted region of DNA reanneals, and RNA polymerase releases the DNA. What is the signal for transcription termination? Interestingly, although the transcribed regions of DNA contain stop signals, *the transcribed product of the stop signals terminates transcription.* The simplest stop signal on DNA is a palindromic (inverted repeat) GC-rich region followed by a sequence of T residues. The RNA transcript of this DNA palindrome is self-complementary (Figure 36.11). Hence, its bases can pair to form a hairpin structure with a stem and loop. This stable hairpin is followed by a sequence of four or more uracil residues, which also are crucial for termination. The RNA transcript ends within or just after them. This type of termination is called *intrinsic termination.*

How does this combination hairpin–oligo(U) structure terminate transcription? First, RNA polymerase appears to pause immediately after it has synthesized a stretch of RNA that folds into a hairpin. Second, the RNA–DNA hybrid helix produced in the oligo(U) tail is unstable because rU–dA base pairs are the weakest of the three kinds of Watson–Crick base pairs. Hence, the pause in transcription caused by the hairpin permits the weakly bound nascent RNA to dissociate from the DNA template and then from the enzyme. The solitary DNA template strand rejoins its partner to re-form the DNA duplex, and the transcription bubble closes.

The Rho Protein Helps Terminate the Transcription of Some Genes

RNA polymerase needs no help in terminating transcription when it encounters a hairpin followed by several U residues. At other sites, however, termination requires the participation of an additional protein with ATPase activity called *rho* (ρ). This type of termination is called *protein-dependent termination.*

Hexameric ρ specifically binds a stretch of 72 nucleotides on single-stranded RNA. Rho binds the RNA in such a way that the RNA passes through the center of the structure (Figure 36.12). Rho is brought into action by sequences located in the nascent RNA that are rich in cytosine and poor in guanine. The ATPase activity of ρ enables the protein to pull on the nascent RNA, essentially racing down the RNA strand in pursuit of the RNA polymerase. When ρ collides with the RNA polymerase at the transcription bubble, it breaks the RNA–DNA hybrid helix, unwinding the hybrid helix and stopping transcription. *A common feature of protein-independent and protein-dependent termination is that the functioning signals lie in newly synthesized RNA rather than in the DNA template.*

DID YOU KNOW?

Derived from the Greek *palindromos,* meaning "running back again," a palindrome is a word, sentence, or verse that reads the same from right to left as it does from left to right: "radar" or "senile felines" are examples.

QUICK QUIZ 1 How is transcription initiation controlled in *E. coli?* What is the mechanism of termination?

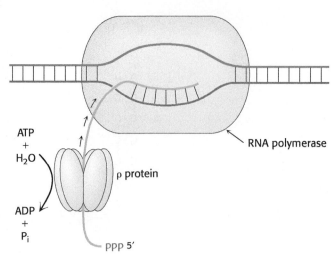

Figure 36.12 The mechanism for the termination of transcription by ρ protein. This protein is an ATP-dependent helicase that binds the nascent RNA strand and pulls it away from RNA polymerase and the DNA template.

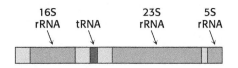

Figure 36.13 Primary transcript. Cleavage of this transcript produces 5S, 16S, and 23S rRNA molecules and a tRNA molecule. Spacer regions are shown in yellow.

Precursors of Transfer and Ribosomal RNA Are Cleaved and Chemically Modified After Transcription

The product of RNA synthesis is not always the mature RNA molecule. Although messenger RNA molecules in bacteria undergo little or no modification after synthesis by RNA polymerase and, indeed, may be translated while they are still being transcribed, it is not the case for transfer RNA molecules and ribosomal RNA molecules. *Transfer RNA and ribosomal RNA molecules are generated by cleavage and other modifications of the transcription product.* For example, in *E. coli*, three kinds of rRNA molecules and a tRNA molecule are excised from a single primary RNA transcript that also contains noncoding regions called *spacer regions* (**Figure 36.13**). Other transcripts contain arrays of several kinds of tRNA or of several copies of the same tRNA. The nucleases that cleave and trim these precursors of rRNA and tRNA are highly precise. *Ribonuclease P*, for example, generates the correct 5′ terminus of all tRNA molecules in *E. coli*. This interesting enzyme contains a catalytically active RNA molecule embedded in its polypeptide chains (Chapter 38). *Ribonuclease III* excises 5S, 16S, and 23S rRNA precursors from the primary transcript by cleaving double-helical hairpin regions at specific sites. The RNAs are named for how fast they move in a centrifugal field; the larger the S value, the faster the molecule moves (Chapter 5).

Excision from a precursor is not the only way in which rRNAs and tRNAs are processed. *A second type of processing, common for tRNA molecules, is the addition of nucleotides to the termini of some RNA strands.* For example, CCA, a terminal sequence required for the function of all tRNAs, is added to the 3′ ends of tRNA molecules that do not already possess this terminal sequence. *A third way to alter rRNA and tRNA is the modification of bases and ribose units.* In bacteria, some bases of rRNA are methylated. Similarly, unusual bases are formed in all tRNA molecules by the enzymatic modification of a standard ribonucleotide in a tRNA precursor. For example, uridylate residues are modified after transcription to form *ribothymidylate* and *pseudouridylate* (**Figure 36.14**). These modifications generate diversity, allowing greater structural and functional versatility.

(A)

6-Dimethyladenylate

(B)

Ribothymidylate

Uridylate

Pseudouridylate

Figure 36.14 Base modifications in RNA. (A) A modified adenylate. (B) Two examples of modified uridylate.

CLINICAL INSIGHT

Some Antibiotics Inhibit Transcription

We are all aware of the benefits of antibiotics, highly specific inhibitors of biological processes that are synthesized by many bacteria. Rifampicin and actinomycin are two antibiotics that inhibit transcription, although in quite different ways. *Rifampicin* is a semisynthetic derivative of *rifamycins*, which are isolated from a strain of *Amycolatopsis* that is related to the bacterium that causes strep throat.

Rifampicin

This antibiotic specifically inhibits the initiation of RNA synthesis. Rifampicin interferes with the formation of the first few phosphodiester linkages in the RNA strand by blocking the channel into which the RNA–DNA hybrid generated by the enzyme must pass (**Figure 36.15**).

Actinomycin D, a polypeptide-containing antibiotic from a strain of *Streptomyces,* inhibits transcription by an entirely different mechanism. *Actinomycin D binds tightly and specifically to double-helical DNA and thereby prevents it from being an effective template for RNA synthesis.* A part of the actinomycin molecule slips in

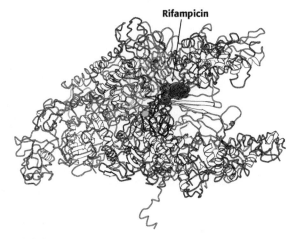

Figure 36.15 Antibiotic action. Rifampicin, shown in red, binds to a pocket in the channel that is normally occupied by the newly formed RNA–DNA hybrid. Thus, the antibiotic blocks elongation after only two or three nucleotides have been added.

between neighboring base pairs in DNA, a mode of binding called *intercalation*. At low concentrations, actinomycin D inhibits transcription without significantly affecting DNA replication. Hence, *actinomycin D is extensively used as a highly specific inhibitor of the formation of new RNA in both bacterial and eukaryotic cells*. Its ability to inhibit the growth of rapidly dividing cells makes it an effective therapeutic agent in the treatment of some cancers, including breast and testicular cancer.

✓ 2 Describe how transcription is controlled in bacteria.

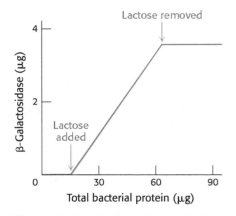

Figure 36.16 β-Galactosidase induction. The addition of lactose to an *E. coli* culture causes the production of β-galactosidase to increase from very low amounts to much larger amounts. The increase in the amount of enzyme parallels the increase in the number of cells in the growing culture. β-Galactosidase constitutes 6.6% of the total protein synthesized in the presence of lactose.

36.3 The *lac* Operon Illustrates the Control of Bacterial Gene Expression

Let us now examine how transcription in bacteria responds to environmental information. An especially clear example is the regulation of genes for the enzymes that catabolize the disaccharide lactose. Bacteria such as *E. coli* usually rely on glucose as their source of carbon and energy. However, when glucose is scarce, *E. coli* can use lactose as its carbon source even though this disaccharide does not lie on any major metabolic pathways. An essential enzyme in the metabolism of lactose is *β-galactosidase*, which hydrolyzes lactose into galactose and glucose. These products are then metabolized by pathways discussed in Chapter 16.

An *E. coli* cell growing on a carbon source such as glucose or glycerol contains fewer than 10 molecules of β-galactosidase. In contrast, the same cell will contain several thousand molecules of the enzyme when grown on lactose (**Figure 36.16**).

Interestingly, two other proteins are synthesized whenever β-galactosidase is synthesized—namely, *galactoside permease* and *thiogalactoside transacetylase*. The permease is required for the transport of lactose across the bacterial cell membrane. The transacetylase is not essential for lactose metabolism but may play a role in the detoxification of compounds that also may be transported by the permease. *The expression levels of a set of enzymes that all contribute to adaptation to a specific change in the environment change together*. Such a coordinated unit of gene expression is called an *operon*.

An Operon Consists of Regulatory Elements and Protein-Encoding Genes

The parallel regulation of β-galactosidase, the permease, and the transacetylase suggests that the expression of genes encoding these enzymes is controlled by a single mechanism (**Figure 36.17**). The DNA components of the regulatory system are a *regulator gene*, *an operator site*, and a *set of structural genes* that, in the galactosidase example, encode the three enzymes. The regulator gene encodes a *repressor* protein that binds to the operator site. As the name suggests, binding of the repressor to the operator represses transcription of the structural genes. The operator and its associated structural genes constitute the operon.

For the *lactose (lac) operon*, the gene encoding the repressor is designated *i*, the operator site is *o*, and the structural genes for β-galactosidase, the permease, and the transacetylase are called *z*, *y*, and *a*, respectively. The operon also contains a promoter site (denoted

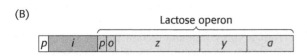

Figure 36.17 Operons. (A) The general structure of an operon. (B) The structure of the lactose operon. In addition to the promoter, *p*, in the operon, a second promoter is present in front of the regulator gene, *i*, to drive the synthesis of the regulator.

by *p*), which directs the RNA polymerase to the correct transcription-initiation site. The *z, y,* and *a* genes are transcribed to yield a single mRNA molecule that encodes all three proteins. An mRNA molecule encoding more than one protein is known as a *polygenic* or *polycistronic* transcript.

How does the *lac repressor* inhibit the expression of the *lac* operon? In the absence of lactose, the repressor binds very tightly and rapidly to the operator, blocking the bound RNA polymerase from using the DNA as a template. The biochemical rationale is that, in the absence of lactose, there is no need to transcribe the genes that encode enzymes required for lactose degradation.

Ligand Binding Can Induce Structural Changes in Regulatory Proteins

How is the repressor removed from the operator? This question is tantamount to asking what signal triggers the expression of the *lac* operon. A signal molecule called an *inducer* binds to the *lac* repressor, *causing a structural change in the repressor that greatly reduces the affinity of the repressor for the operator DNA.* The repressor–inducer complex leaves the DNA and allows transcription of the operon (Figure 36.18).

What is the inducer? Intuitively, we would expect it to be lactose itself. Interestingly, however, it is not lactose that announces its own presence; rather, it is *allolactose*, a combination of galactose and glucose with an α-1,6 rather than an α-1,4 linkage, as in lactose. Allolactose is a side product of the β-galactosidase reaction produced at low levels by the few molecules of β-galactosidase that are present before induction. Many other gene-regulatory networks function in ways analogous to those of the *lac* operon.

Transcription Can Be Stimulated by Proteins That Contact RNA Polymerase

The *lac* repressor functions by inhibiting transcription until lactose is present. There are also sequence-specific DNA-binding proteins that stimulate the transcription of the *lac* operon when glucose is in short supply. One particularly well-studied example is the *catabolite activator protein* (CAP), which is also known as the cyclic AMP (cAMP) response protein (CRP). When bound to cAMP, CAP stimulates the transcription of lactose genes. When the concentration of glucose decreases, the concentration of cAMP increases, thus activating the *lac* operon to help *E. coli* metabolize lactose for fuel.

Within the *lac* operon, the CAP–cAMP complex binds to a site that is centered near position –61 relative to the start site for transcription. Binding stimulates the initiation of transcription by approximately 50-fold by recruiting RNA polymerase to promoters to which CAP is bound (Figure 36.19). The CAP–cAMP complex forms protein–protein contacts with the RNA polymerase that increases transcription initiation. Thus, in regard to the *lac* operon, gene expression is maximal when the binding of allolactose relieves the inhibition by the *lac* repressor, and the CAP–cAMP complex stimulates the binding of RNA polymerase. Glucose, on the other hand, inhibits the expression of the *lac* operon in a process called *catabolite repression*. An increase in cellular glucose concentration leads to a decrease in the intracellular concentration of cAMP.

How is the level of cAMP controlled in bacteria? Enzyme IIA (EIIA) is phosphorylated at the expense of the glycolytic intermediate phosphoenolpyruvate. Phosphorylated EIIA then transfers a phosphate to a glucose molecule, generating glucose 6-phosphate. However, if glucose is absent, phosphorylated EIIA activates adenylate cyclase, leading to an increase in cAMP and enhanced transcription of the *lac* operon.

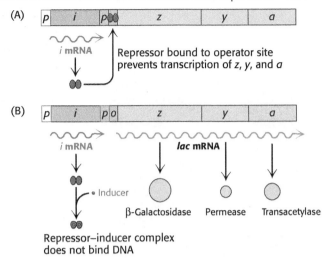

(A)

(B)

i mRNA

Repressor bound to operator site prevents transcription of *z, y,* and *a*

i mRNA

lac mRNA

• Inducer

β-Galactosidase   Permease   Transacetylase

Repressor–inducer complex does not bind DNA

Figure 36.18 The induction of the *lac* operon. (A) In the absence of lactose, the *lac* repressor binds DNA and represses transcription from the *lac* operon. (B) Allolactose or another inducer binds to the *lac* repressor, leading to its dissociation from DNA and to the production of *lac* mRNA.

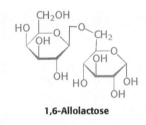

1,6-Allolactose

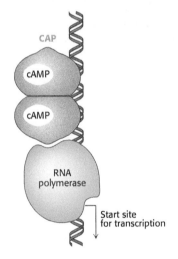

CAP

cAMP

cAMP

RNA polymerase

Start site for transcription

Figure 36.19 The binding site for catabolite activator protein (CAP). This protein binds as a dimer to an inverted repeat that is at the position –61 relative to the start site of transcription. The CAP-binding site on DNA is adjacent to the position at which RNA polymerase binds.

? QUICK QUIZ 2 Describe how allolactose and cAMP combine to regulate the *lac* operon.

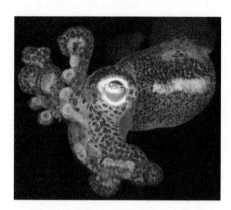

Figure 36.20 Bobtail squid. The bobtail squid, only 1 to 8 cm in length, inhabits the coastal waters of the Pacific and provides safe haven for the bacterium *Vibrio fischeri*. [David Fleetham/Alamy.]

 CLINICAL AND BIOLOGICAL INSIGHT

Many Bacterial Cells Release Chemical Signals that Regulate Gene Expression in Other Cells

Bacterial cells have been traditionally viewed as solitary single cells. However, it is becoming increasingly clear that, in many circumstances, bacterial cells live in complex communities, interacting with other cells of their own species and of different species. These social interactions change the patterns of gene expression within the cells.

A prominent type of interaction is called *quorum sensing*. This phenomenon was discovered in *Vibrio fischeri*, a species of bacterium that can live inside a specialized light organ in the bobtail squid (**Figure 36.20**). The bobtail squid inhabits shallow waters and hunts at night. On moonlit nights, the squid would be visible to predators prowling below it. The squid adjusts the light emitted by the light organ to match the background, thus becoming less visible to predators. In this symbiotic relation, the bacteria produce the enzyme luciferase, which generates bioluminescence, thereby providing protection for the squid. In return, the bacteria have a protected place to live and reproduce. Interestingly, when these bacteria are grown in culture at low density, they are not bioluminescent. However, when the cell density reaches a critical level, the gene for luciferase is expressed and the cells bioluminesce. How do these cells sense the density of their population?

Cells of *V. fischeri* release an *autoinducer* into their environment and other *V. fischeri* cells take up the chemical. After the inducer concentration inside the cell has increased to an appropriate level, the inducer binds to a regulatory protein. The complex of inducer–regulatory protein then binds to promoter sites, increasing the rate of transcription of the operon that encodes the luciferase gene as well as the gene for the autoinducer. Because each cell produces only a small amount of the autoinducer, this regulatory system allows each *V. fischeri* cell to determine the density of the *V. fischeri* population in its environment—hence, the term *quorum sensing* for this process.

Quorum sensing appears to play a major role in the formation of bacterial communities with particular species compositions. Many species of bacteria can be found in specialized structures termed *biofilms* that can form on surfaces. Some genes controlled by quorum-sensing mechanisms promote the formation of specific molecules, which include proteins, polysaccharides, and nucleic acids, that serve as a matrix for the biofilm. Biofilms are of considerable medical importance. Organisms within them are more than 1000-fold more resistant to antibiotics and more able to evade the host immune response than free-living bacteria. Patients implanted with prosthetic devices, including simple bladder catheters, are especially at risk for biofilm formation. Dental plaque and shower scum are common examples of biofilms.

Some Messenger RNAs Directly Sense Metabolite Concentrations

A newly discovered control mechanism for the regulation of gene expression depends on the remarkable ability of some mRNA molecules to form special secondary structures, some of which are capable of directly binding small molecules. These structures are termed *riboswitches*. Consider a riboswitch-containing RNA that encodes the genes that participate in the biosynthesis of riboflavin (p. 269) in bacteria. As synthesis is taking place, the RNA can adopt two alternative structures—one that is capable of binding flavin mononucleotide (FMN), a key intermediate in riboflavin synthesis, or one that is not. The binding of FMN traps the RNA transcript in a conformation that favors the termination of further RNA synthesis, preventing the production of functional mRNA. This strategy makes good biochemical sense: if the precursor is present, it is not necessary to make the enzymes that synthesize the precursor. However, when FMN is absent, an alternative conformation forms that allows the production of the full-length mRNA. Riboswitches serve as a vivid illustration of how RNAs are capable of forming elaborate, functional structures, though we tend to depict them as simple lines in the absence of specific information.

SUMMARY

36.1 Cellular RNA Is Synthesized by RNA Polymerases

All cellular RNA molecules are synthesized by RNA polymerases according to instructions given by DNA templates. The activated monomer substrates are ribonucleoside triphosphates. The direction of RNA synthesis is $5' \rightarrow 3'$, as in DNA synthesis. RNA polymerases, unlike DNA polymerases, do not need a primer. RNA polymerase in *E. coli* is a multisubunit enzyme. The subunit composition of the $\sim$500-kDa holoenzyme is $\alpha_2\beta\beta'\sigma\omega$ and that of the core enzyme is $\alpha_2\beta\beta'\omega$.

36.2 RNA Synthesis Comprises Three Stages

Transcription is initiated at promoter sites consisting of two sequences, one centered near -10 and the other near -35; that is, 10 and 35 nucleotides away from the start site in the 5′ (upstream) direction, respectively. The consensus sequence of the -10 region is TATAAT. The σ subunit enables the holoenzyme to recognize promoter sites. RNA polymerase must unwind the template double helix for transcription to take place. Unwinding exposes some 17 bases on the template strand and sets the stage for the formation of the first phosphodiester linkage. RNA chains usually start with pppG or pppA. The σ subunit dissociates from the holoenzyme after the initiation of the new chain. Elongation takes place at transcription bubbles that move along the DNA template at a rate of about 50 nucleotides per second. The nascent RNA chain contains stop signals that end transcription. One stop signal is an RNA hairpin, which is followed by several U residues. A different stop signal is read by the rho protein, an ATPase.

Although mRNA undergoes little posttranscriptional modification in bacteria, rRNA and tRNA are heavily processed. In *E. coli*, precursors of transfer RNA and ribosomal RNA are cleaved and chemically modified after transcription.

36.3 The *lac* Operon Illustrates the Control of Bacterial Gene Expression

The *lac* operon is a well-studied example of gene regulation in *E. coli*. An operon is a coordinated unit of gene expression, consisting of regulatory elements and protein-encoding genes. In regard to the *lac* operon, three enzymes—β-galactosidase, a permease, and a transacetylase—facilitate the use of lactose as a fuel. The regulatory components include a regulatory gene that encodes the *lac* repressor protein and a DNA sequence called the operator site. In the absence of lactose, the repressor binds to the operator site, preventing the transcription of the protein-encoding genes. When lactose is present, a metabolite of lactose, allolactose, binds the repressor, causing it to leave the operator and permitting transcription.

The *lac* operon is stimulated when the catabolite activator protein, also known as the cAMP response protein, binds near the *lac* operon promoter and stimulates the initiation of transcription. The binding takes place only when glucose is scarce.

Another example of the regulation of gene expression observed in some bacterial species is quorum sensing. This process includes the release of chemicals called autoinducers into the medium surrounding the cells. Autoinducers are taken up by surrounding cells and activate the expression of genes, including those that promote the synthesis of more autoinducer. These chemically mediated social interactions allow these bacteria to change their gene-expression patterns in response to the number of other cells in their environments. Biofilms are complex communities of bacteria that are promoted by quorum-sensing mechanisms.

KEY TERMS

transcription (p. 659)
RNA polymerase (p. 659)
sigma (σ) subunit (p. 661)
promoter (p. 662)
consensus sequence (p. 662)
transcription factor (p. 662)
transcription bubble (p. 664)

rho (ρ) (p. 665)
β-galactosidase (p. 668)
operon (p. 668)
operator (p. 668)
repressor (p. 668)
lac repressor (p. 669)
inducer (p. 669)

catabolite activator protein (CAP) (p. 669)
catabolite repression (p. 669)
quorum sensing (p. 670)
riboswitch (p. 670)

? Answers to QUICK QUIZZES

1. The σ factor recognizes the promoter site. When transcription starts, σ leaves the enzyme to assist in the initiation by another polymerase. Termination takes place in one of two common ways. If the RNA product forms a hairpin that is followed by several uracil residues, the polymerase pauses and falls from the template. Alternatively, the ATPase rho binds the RNA products and pushes the polymerase off the template.

2. Allolactose binds to the repressor protein, preventing the repressor from binding to the operator and allowing the polymerase access to the gene so that transcription can take place. Cyclic AMP binds to CAP. This complex binds to the *lac* operon upstream of the promoter and stimulates the binding of polymerase to the promoter.

PROBLEMS

1. *The process of recording.* Define transcription. ✓ 1

2. *Different strands.* Explain the difference between the coding strand and the template strand in DNA. ✓ 1

3. *True of many endeavors.* What are the three stages of RNA synthesis? ✓ 1

4. *Necessary ingredients.* What components does RNA polymerase require for the synthesis of RNA? ✓ 1

5. *Location, location, location.* What is the function of the sigma subunit in transcription? ✓ 1

6. *Like stalagmites and stalactites.* Match each term with its description. ✓ 1

(a) Transcription _____
(b) RNA polymerase
_____
(c) Sigma (σ) factor
_____
(d) Promoter _____
(e) Consensus sequence
_____
(f) Rho (ρ) _____
(g) Operon _____
(h) Operator _____
(i) Repressor _____
(j) Inducer _____

1. Recognizes promoter sites
2. The average order of nucleotides
3. Repressor binding site
4. RNA synthesis
5. DNA sequences that determine the site of transcription initiation
6. Consists of regulatory elements and protein-encoding genes
7. In the lac operon, allolactose
8. Prevents the transcription of structural genes
9. RNA-synthesis-termination protein
10. Transcriptional machinery

7. *Comparisons.* Compare transcription with replication.

8. *A gene advocate.* What is a promoter?

9. *Cerrada/Abierta.* Distinguish between a closed promoter complex and an open promoter complex.

10. *Complements.* The sequence of part of an mRNA is

5'-AUGGGGAACAGCAAGAGUGGGGCCCUGUC-CAAGGAG-3'

What is the sequence of the DNA coding strand? Of the DNA template strand?

11. *Checking for errors.* Why is RNA synthesis not as carefully monitored for errors as is DNA synthesis?

12. *Speed is not of the essence.* Why is it advantageous for DNA synthesis to be more rapid than RNA synthesis?

13. *A loose cannon.* The σ protein by itself does not bind to promoter sites. Predict the effect of a mutation enabling σ to bind to the −10 region in the absence of other subunits of RNA polymerase. ✓ 1

14. *Stuck sigma.* What would be the likely effect of a mutation that would prevent σ from dissociating from the RNA polymerase core? ✓ 1

15. *Transcription time.* What is the minimum length of time required for the synthesis by *E. coli* polymerase of an mRNA encoding a 100-kDa protein? ✓ 1

16. *Between bubbles.* How far apart are transcription bubbles on *E. coli* genes that are being transcribed at a maximal rate?

17. *Anything like Robocop?* Define riboswitch, and explain its role in the regulation of RNA synthesis. ✓ 2

18. *Missing genes.* Predict the effects of deleting the following regions of DNA: (a) the gene encoding *lac* repressor; (b) the *lac* operator; (c) the gene encoding CAP. ✓ 2

19. *Unavailable resources. Lac* operon mutants that lack permease activity exist. Even though the gene for β-galactosidase is intact, the enzyme is not induced when lactose is added to the bacterial growth medium. Why? ✓ 2

20. *Not such a random walk.* RNA polymerase is a large enzyme, and one of its substrates, DNA, is a large biomolecule with thousands to millions of bases. Yet, RNA polymerase can locate a promoter sequence in DNA faster than two small molecules can collide in solution. Explain this apparent paradox. ✓ 1

21. *Alterations.* What are the three types of tRNA modifications seen in *E. coli*?

22. *Christmas trees.* The adjoining image depicts several bacterial genes undergoing transcription. Identify the DNA. What are the strands of increasing length? Where is the beginning of transcription? The end of transcription? On the page, what is the direction of RNA synthesis? What can you conclude about the number of enzymes participating in RNA synthesis on a given gene? ✓ 1

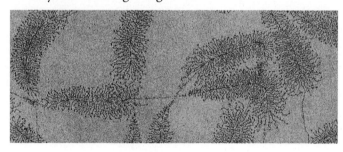

[Thomas Broker/Phototake]

Challenge Problems

23. *Potent inhibitor.* Heparin, a glycosaminoglycan (p. 177), inhibits transcription by binding to RNA polymerase. What properties of heparin allow it to bind so effectively to RNA polymerase?

24. *Minimal concentration.* Calculate the concentration of *lac* repressor, assuming that one molecule is present per cell. Assume that each *E. coli* cell has a volume of 10^{-12} cm^3. Would you expect the single molecule to be free or bound to DNA? The dissociation constant for the repressor–DNA complex is 10^{-13} M. ✓ 2

25. *The opposite direction.* Some compounds, called anti-inducers, bind to repressors such as the *lac* repressor and inhibit the action of inducers; that is, transcription is repressed and higher concentrations of inducer are required to induce transcription. Propose a mechanism of action for anti-inducers. ✓ 2

26. *Abortive cycling.* Di- and trinucleotides are occasionally released from RNA polymerase at the very start of transcription, a process called abortive cycling. This process requires the restart of transcription. Suggest a plausible explanation for abortive cycling. ✓ 1

27. *Scientist at work.* You have just performed an experiment in which you investigated RNA synthesis in vitro. Interestingly, the RNA molecules synthesized in vitro by RNA polymerase acting alone are *longer* than those made in vivo. Being the bright scientist that you are, you hypothesized that something was missing from the in vitro system that was present in vivo. To check for this possibility, you added some cell extract and found that the correct termination took place. You purified this protein, which you named "boat" after your interest in sailing. You performed an experiment in which boat was added to the RNA synthesis mixture at various times after the initiation of RNA synthesis. The results are illustrated below. Explain these results. ✓ 2

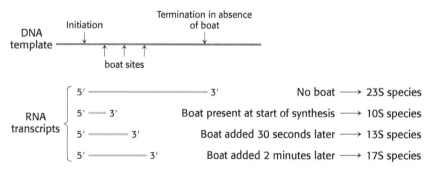

Selected Readings for this chapter can be found online at www.whfreeman.com/tymoczko3e.

CHAPTER 37

Gene Expression in Eukaryotes

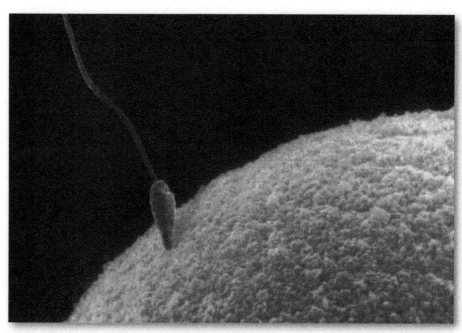

Two haploid cells join to form a single diploid cell at fertilization. This joining initiates an astounding biochemical process—the parceling out of genetic information, in the process of development, to construct, for example, the person reading these words. The regulation of gene transcription is a key component of development. [Don W. Fawcett/Science Source.]

The basics of RNA synthesis presented in Chapter 36 apply across all kingdoms of life. We now turn to *transcription* in eukaryotes, which is much more complicated than in bacteria because of a much larger genome and the resulting increased complexity of transcriptional regulation. Although eukaryotic cells from any organism have the same DNA, they have a remarkable ability to precisely regulate which genes are expressed, the time at which each gene is transcribed, and how much RNA is produced. This ability has allowed some eukaryotes to evolve into multicellular organisms, with distinct tissues. That is, *multicellular eukaryotes differentially use transcriptional regulation of DNA common to all cells to create different cell types*.

Gene expression is influenced by three important characteristics unique to eukaryotes: more complex transcriptional regulation, RNA processing, and the nuclear membrane.

1. *More Complex Transcriptional Regulation.* RNA synthesis in eukaryotes is carried out by three distinct *RNA polymerases*. As in bacteria, these RNA polymerases rely on promoter sequences in DNA to regulate the initiation of transcription. But not all *promoters* are recognized by each of the polymerases. Adding complexity to regulation is the presence of many different types of promoter

675

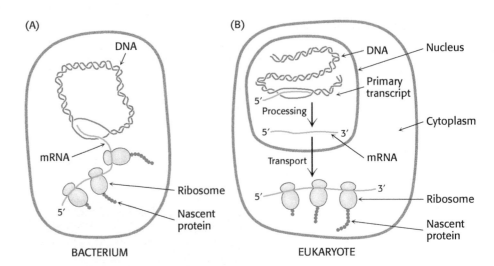

Figure 37.1 Transcription and translation. (A) In bacteria, the primary transcript serves as mRNA and is used immediately as the template for protein synthesis. (B) In eukaryotes, mRNA precursors are processed and spliced in the nucleus before being transported to the cytoplasm for translation. [Information from J. Darnell, H. Lodish, and D. Baltimore, *Molecular Cell Biology*, 2d ed. (Scientific American Books, 1990), p. 230.]

elements. Bacteria, in contrast, have just three promoter elements (the -10, -35, and UP elements). The eukaryotic promoter elements can combine in a multitude of ways, greatly increasing the number of types of promoters.

2. *RNA Processing.* Although both bacteria and eukaryotes modify RNA, *eukaryotes very extensively process nascent RNA destined to become mRNA*. This processing includes modifications to both ends and, most significantly, splicing out segments of the primary transcript. RNA processing is described in Chapter 38.

3. *The Nuclear Membrane. In eukaryotes, transcription and translation take place in different cellular compartments:* transcription takes place in the membrane-bounded nucleus, whereas translation takes place outside the nucleus in the cytoplasm. In bacteria, the two processes are closely coupled (**Figure 37.1**). Indeed, the translation of bacterial mRNA begins while the transcript is still being synthesized. The spatial and temporal separation of transcription and translation enables eukaryotes to regulate gene expression in much more intricate ways, contributing to the richness of eukaryotic form and function.

37.1 Eukaryotic Cells Have Three Types of RNA Polymerases

In bacteria, RNA is synthesized by a single kind of polymerase. In contrast, the nucleus of a eukaryote contains three types of RNA polymerase differing in a number of properties, including template specificity and susceptibility to α-amanitin (**Table 37.1**). α-Amanitin is a toxin produced by the mushroom *Amanita phalloides* (**Figure 37.2**) that binds very tightly to RNA polymerase II and thereby blocks the elongation phase of RNA synthesis. Higher concentrations of α-amanitin inhibit polymerase III, whereas polymerase I is insensitive to this toxin. This pattern of sensitivity is highly conserved throughout the animal and plant kingdoms.

    RNA polymerase I is located in nucleoli, where it transcribes the three ribosomal RNAs 18S, 5.8S, and 28S ribosomal RNA as a single transcript. *RNA polymerase II* is located in the nucleoplasm, the fluid and structural components in

Figure 37.2 RNA polymerase poison. α-Amanitin is produced by the poisonous mushroom *Amanita phalloides,* also called the *death cap* or the *destroying angel*. More than a hundred deaths result worldwide each year from the ingestion of poisonous mushrooms. [Jacana/Science Source.]

Table 37.1 Eukaryotic RNA polymerases

Type	Location	Transcripts	Subunits	Mass (kDa)	α-Amatin sensitivity
I	Nucleolus	18S, 5.8S, and 28S rRNA	14	514	Insensitive
II	Nucleoplasm	mRNA precursors and snRNA	12	588	Very sensitive
III	Nucleoplasm	tRNA and 5S rRNA	17	693	Moderately sensitive

Table 37.2 Additional classes of RNA

RNA	Size (nucleotides)	Function
Small nuclear RNA (snRNA)	Less than 300	Components of RNA splicing machinery
Small nucleolar RNA (snoRNA)	Less than 300	rRNA biogenesis and modification
MicroRNA (miRNA)	20–25	Regulates use of mRNA
Small interfering RNA (siRNA)	20–25	Antiviral defense mRNA degradation
Piwi-interacting RNA (piRNA)	29–30	Gene regulation
Long noncoding RNA (lncRNA)	Greater than 200	Gene regulation

the nucleus that surround the chromosomes and nucleoli. RNA polymerase II synthesizes the precursors of messenger RNA as well as several small RNA molecules, such as those of the splicing apparatus. The other ribosomal RNA molecule (5S rRNA, Chapter 39) and all the transfer RNA molecules are synthesized by *RNA polymerase III*, which also is located in the nucleoplasm. In addition to mRNA, rRNA, and tRNA, there are several other classes of RNA (Table 37.2). The function of some of these RNAs will be considered in Section 16.

Although all eukaryotic RNA polymerases are similar to one another and to bacterial RNA polymerase in structure and function, RNA polymerase II contains a unique *carboxyl-terminal domain* (CTD) on the 220-kDa subunit; this domain is unusual because it contains multiple repeats of a YSPTSPS consensus amino acid sequence. The activity of RNA polymerase II is regulated by phosphorylation mainly on the serine residues of the CTD. Phosphorylation of the CTD enhances transcription and recruits other factors required to process the RNA polymerase II product.

Eukaryotic genes, like bacterial genes, require promoters for transcription initiation. Like bacterial promoters, eukaryotic promoters consist of sequences that serve to attract the polymerases to the start sites. As is the case for bacteria, a promoter and the gene that it regulates are always on the same molecule of DNA. Consequently, promoters are referred to as cis-acting elements. Eukaryotic promoters differ distinctly in sequence and position relative to the regulated genes, depending on the type of RNA polymerase to which they bind (Figure 37.3).

RNA polymerase I promoter

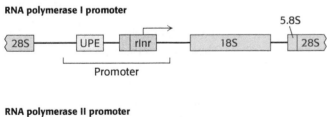

RNA polymerase II promoter

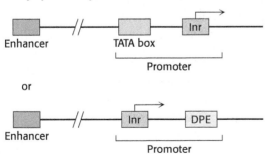

RNA polymerase III promoter

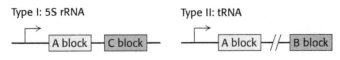

Figure 37.3 Common eukaryotic promoter elements. Each eukaryotic RNA polymerase recognizes a set of promoter elements—sequences in DNA that promote transcription. The RNA polymerase I promoter consists of a ribosomal initiator (rInr) and an upstream promoter element (UPE). The RNA polymerase II promoter includes an initiator element (Inr) and may also include either a TATA box or a downstream promoter element (DPE). Separate from the promoter region, enhancer elements bind specific transcription factors. RNA polymerase III promoters consist of conserved sequences that lie within the transcribed genes.

DID YOU KNOW?

Cis-acting elements are DNA sequences that regulate the expression of a gene located on the same molecule of DNA.

Trans-acting elements are proteins that recognize cis-acting elements and regulate RNA synthesis. They are commonly called transcription factors.

The ability to recognize specific promoters accounts for the template specificity of the various eukaryotic RNA polymerases. Some characteristics of polymerase-promoter interaction for each of the RNA polymerases are

1. *RNA Polymerase I.* The ribosomal DNA (rDNA) transcribed by polymerase I is arranged in several hundred tandem repeats, each containing a copy of the three rRNA genes. The promoter sequences are located in stretches of DNA separating the gene repeats. At the transcriptional start site lies a sequence called the *ribosomal initiator element* (rInr). Farther upstream, 150 to 200 bp from the start site, is the *upstream promoter element* (UPE). Both elements aid transcription by binding proteins that recruit RNA polymerase I.

2. *RNA Polymerase II.* Promoters for RNA polymerase II, like bacterial promoters, include a set of conserved-sequence elements that define the start site and recruit the polymerase. However, the promoter can contain any combination of possible elements. The diverse array of promoter elements also includes enhancer elements, unique to eukaryotes, that can be very distant (more than 1 kb) from the start site (p. 679).

3. *RNA Polymerase III.* Promoters for RNA polymerase III are *within* the transcribed sequence, downstream of the start site. There are two types of promoters for RNA polymerase III. Type I promoters, found in the 5S rRNA gene, contain two short sequences known as the A block and the C block. Type II promoters, found in tRNA genes, consist of two 11-bp sequences, the A block and the B block, situated about 15 bp from either end of the gene.

✓ 3 Describe how transcription is regulated in eukaryotes.

37.2 RNA Polymerase II Requires Complex Regulation

The elaborate regulation of RNA polymerase II accounts for cell differentiation and development in higher organisms. Consequently, we will focus our attention on this vital enzyme. Promoters for RNA polymerase II, like those for bacterial polymerases, are generally located on the 5′ side of the start site for transcription. The most commonly recognized cis-acting element for genes transcribed by RNA polymerase II is called the *TATA box* on the basis of its consensus sequence (**Figure 37.4**). The TATA box is usually centered at ~ −25. Note that the eukaryotic TATA box closely resembles the bacterial −10 sequence (TATAAT) but is farther from the start site. The mutation of a single base in the TATA box markedly impairs promoter activity. Thus, the precise sequence, not just a high content of AT pairs, is essential.

5′ T_{82} A_{97} T_{93} A_{85} A_{63} A_{88} A_{50} 3′

TATA box

Figure 37.4 The TATA box. Comparisons of the sequences of more than 100 eukaryotic promoters led to the consensus sequence shown. The subscripts denote the frequency (%) of the base at that position.

The TATA box is often paired with an *initiator element* (Inr), a sequence found at the transcriptional start site located at ~ +1. This sequence defines the start site because the other promoter elements are at variable distances from that site. Its presence increases transcriptional activity.

A third element, the *downstream core promoter element* (DPE), is commonly found in conjunction with the Inr in transcripts that lack the TATA box. In contrast with the TATA box, the DPE is found downstream of the start site, at ~ +30.

Additional regulatory sequences are located between −40 and −150. Many promoters contain a *CAAT box*, and some contain a *GC box* (**Figure 37.5**). Constitutive genes (genes that are continuously expressed rather than regulated) tend to have GC boxes in their promoters. The positions of these upstream sequences vary from one promoter to another, in contrast with the location of the −35 region in bacteria. Another difference is that the CAAT box and the GC box can be effective when present on the template (antisense) strand, unlike the −35 region, which must be present on the coding (sense) strand. These differences between bacteria and eukaryotes correspond to fundamentally different mechanisms for the recognition of cis-acting elements. The −10 and −35 sequences in

5′ GGNCAATCT 3′

CAAT box

5′ GGGCGG 3′

GC box

Figure 37.5 The CAAT and GC boxes. Consensus sequences for the CAAT and GC boxes of eukaryotic promoters for mRNA precursors. N signifies that any nucleotide can occupy the position.

bacterial promoters are binding sites for RNA polymerase and its associated σ factor. In contrast, the TATA, CAAT, and GC boxes and other cis-acting elements in eukaryotic promoters are recognized by proteins other than RNA polymerase itself.

The Transcription Factor IID Protein Complex Initiates the Assembly of the Active Transcription Complex

Cis-acting elements constitute only part of the puzzle of eukaryotic gene expression. *Transcription factors* that bind to these elements also are required. For example, RNA polymerase II is guided to the start site by a set of transcription factors known collectively as *TFII* (*TF* stands for transcription factor, and *II* refers to RNA polymerase II). Individual TFII factors are called TFIIA, TFIIB, and so on. Initiation begins with the binding of TFIID to the TATA box (Figure 37.6). These general transcription factors are found in all eukaryotes, suggesting that the fundamentals of transcription are conserved in higher organisms.

In TATA-box promoters, the key initial event is the recognition of the TATA box by the TATA-box-binding protein (TBP), a 30-kDa component of the 700-kDa TFIID complex. TBP is a saddle-shaped protein consisting of two similar domains (Figure 37.7). The TATA box of DNA binds to the concave surface of TBP, inducing large conformational changes in the bound DNA. The double helix is substantially unwound to widen its *minor groove*, enabling it to make extensive contact with the concave side of TBP with hydrophobic interactions.

TBP bound to the TATA box nucleates the formation of the preinitiation complex (PIC) (Figure 37.6). The surface of the TBP saddle provides docking sites for the binding of other components. TFIIA and TFIIB bind next, stabilizing the binding of TBP to the DNA. With the arrival of TFIIB, the complex recruits RNA polymerase, which is escorted to the promoter site by TFIIE. Binding by TFIIH completes the formation of the PIC.

TFIIH has two essential catalytic activities. First, it is an ATP-dependent helicase that unwinds the DNA as a prelude to transcription. Second, the protein is also a kinase that phosphorylates the CTD of the polymerase. In the formation of the PIC, the CTD is unphosphorylated and binds to other proteins that facilitate transcription. *Phosphorylation of the CTD by TFIIH marks the transition from initiation to elongation.* The phosphorylated CTD stabilizes transcription elongation by RNA polymerase II and recruits RNA-processing enzymes that act in the course of elongation (Chapter 38). Most of the factors are released before the polymerase leaves the promoter and can then participate in another round of initiation.

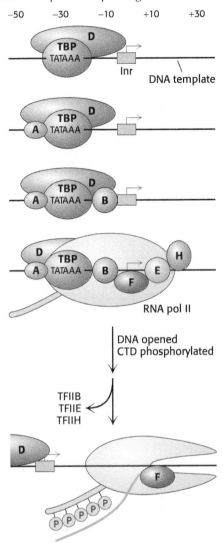

Figure 37.6 Transcription initiation. Transcription factors TFIIA, B, D, E, F, and H are essential in initiating transcription by RNA polymerase II.

Figure 37.7 The complex formed by the TATA-box-binding protein and DNA. The saddlelike structure of the protein sits atop a DNA fragment. Notice that the DNA is significantly unwound and bent. [Drawn from 1CDW.pdb by Adam Steinberg.]

Enhancer Sequences Can Stimulate Transcription at Start Sites Thousands of Bases Away

The activities of many promoters in higher eukaryotes are greatly increased by another type of cis-acting element called an *enhancer*. *Enhancer sequences have no promoter activity of their own yet can exert their stimulatory actions over distances of several thousand base pairs.* They can be upstream, downstream, or even in the midst of a transcribed gene. Moreover, *enhancers are effective when present on either DNA strand*. Like promoter sequences, enhancers are bound by proteins called transcription activators that participate in the regulation of transcription.

 CLINICAL INSIGHT

Inappropriate Enhancer Use May Cause Cancer

Enhancer sequences are important in establishing the tissue specificity of gene expression because *a particular enhancer is effective only in certain cells*. For example, the immunoglobulin enhancer functions in B lymphocytes but not elsewhere. Cancer can result if the relation between genes and enhancers is disrupted. In Burkitt lymphoma and B-cell leukemia, a chromosomal translocation brings the proto-oncogene *myc* (which encodes a transcription factor) under the control of a powerful immunoglobulin enhancer. The consequent dysregulation of the *myc* gene is believed to play a role in the progression of the cancer.

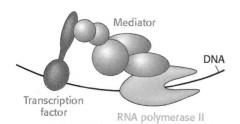

? QUICK QUIZ 1 Differentiate between a promoter and an enhancer.

Figure 37.8 Mediator. A large complex of protein subunits, mediator acts as a bridge between transcription factors bearing activation domains and RNA polymerase II. These interactions help recruit and stabilize RNA polymerase II near specific genes that are then transcribed.

Multiple Transcription Factors Interact with Eukaryotic Promoters and Enhancers

The basal transcription complex described on page 679 initiates transcription at a low frequency. Additional transcription factors that bind to other sites are required to achieve a high rate of mRNA synthesis and to selectively stimulate specific genes. Indeed, many transcription factors have been isolated. These transcription factors are often expressed in a tissue-specific manner.

In contrast with those of bacterial transcription, few eukaryotic transcription factors have any effect on transcription on their own. Instead, *each factor recruits other proteins to build up large complexes that interact with the transcriptional machinery to activate or repress transcription*. These intermediary proteins act as a bridge between the transcription factors and the polymerase. An important target intermediary for many transcription factors is *mediator*, a huge complex of 25 to 30 subunits with a mass of more than 1-MDa, that joins the transcription machinery before initiation takes place. Mediator acts as a bridge between enhancer-bound activators, other transcription factors, and promoter-bound RNA polymerase II (Figure 37.8).

Transcription factors and other proteins that bind to regulatory sites on DNA can be regarded as passwords that cooperatively open multiple locks, giving RNA polymerase access to specific genes. A major advantage of this mode of regulation is that a given regulatory protein can have different effects, depending on what other proteins are present in the same cell. This phenomenon, called *combinatorial control*, is crucial to multicellular organisms that have many different cell types. A comparison between human beings and the simple roundworm *Caenorhabditis elegans* provides an example of the power of combinatorial control. Human beings have only one-third more genes than the worm, but we have many more cell types, and the dramatic increase in complexity without a corresponding increase in gene number may be due to a more sophisticated regulation allowed by combinatorial control.

DID YOU KNOW?

Pluripotent cells are stem cells that can develop into any type of fetal or adult cell. Totipotent stems cells not only can develop into any fetal or adult cell type, but also can develop extraembryonic tissues and thus grow into an entire organism.

 CLINICAL INSIGHT

Induced Pluripotent Stem Cells Can Be Generated by Introducing Four Transcription Factors into Differentiated Cells

An important application illustrating the power of transcription factors is the development of *induced pluripotent stem* (iPS) *cells*. Pluripotent stem cells, which can be derived from embryos, have the ability to differentiate into many different cell types on appropriate treatment. Recent experiments have identified just four genes, encoding transcription factors, that can induce pluripotency in already-differentiated skin cells. When these four genes are introduced into skin cells called fibroblasts, the fibroblasts de-differentiate into cells that appear to have characteristics very nearly identical with those of embryonic stem cells.

These iPS cells are powerful new research tools and, potentially, a new class of therapeutic agents. Ideally, a sample of a patient's fibroblasts could be readily isolated and converted into iPS cells. These iPS cells could then be treated to differentiate into a desired cell type that could be transplanted into the patient, repairing tissue damage. Although the field of iPS-cell research is still in its very early stages, it holds great promise as a possible approach to treatment for many common and difficult-to-treat diseases.

37.3 Gene Expression Is Regulated by Hormones

Just as bacteria can adjust their patterns of gene expression in response to chemicals in their environment, eukaryotes have many systems for responding to environmental stimuli. As the first step in the control of gene expression, the initiation of transcription by RNA polymerase II is the focal point for many signal-transduction pathways. As an example, we will examine a system that detects and responds to estrogens. Synthesized and released by the ovaries, *estrogens*, such as estradiol, are cholesterol-derived steroid hormones (p. 539). They are required for the development of female secondary sex characteristics and, along with progesterone, participate in the ovarian cycle.

Because they are hydrophobic signal molecules, estrogens easily diffuse across cell membranes. When inside a cell, estrogens bind to highly specific receptor proteins. Estrogen receptors are members of a large family of proteins that act as receptors for a wide range of hydrophobic molecules. The steroid receptors are different from the receptors discussed thus far in that they are soluble and located in the cytoplasm or nucleoplasm, rather than being bound to membranes. This receptor family includes receptors for other steroid hormones, such as the androgen testosterone, as well as thyroid hormones, and vitamin A-derivative retinoids. All these receptors, common in multicellular organisms, have a similar mode of action. On binding of the signal molecule (called, generically, a *ligand*), the ligand–receptor complex modifies the expression of specific genes by binding to control elements in the DNA. The human genome encodes approximately 50 members of this family, often referred to as *nuclear hormone receptors*.

Estradiol
(an estrogen)

Nuclear Hormone Receptors Have Similar Domain Structures

Nuclear hormone receptors bind to specific DNA sites referred to as *response elements*. In regard to the estrogen receptor, the *estrogen response elements* (EREs) contain the consensus sequence 5′-**AGGTCANNNTGACCT**-3′. As expected from the symmetry of this sequence, an estrogen receptor binds to such sites as a dimer.

A comparison of the amino acid sequences of members of the nuclear hormone-receptor family reveals two highly conserved domains: a DNA-binding domain and a ligand-binding domain (**Figure 37.9**). The DNA-binding domain lies toward the center of the primary structure and includes nine conserved cysteine residues. This domain provides these receptors with sequence-specific DNA-binding activity. Eight of the cysteine residues bind zinc ions to form DNA-binding domains that are called *zinc-finger domains*.

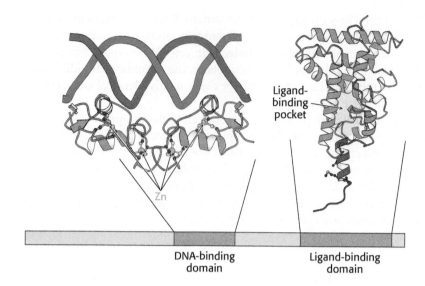

Figure 37.9 The structure of two nuclear hormone–receptor domains. Nuclear hormone receptors contain two crucial conserved domains: (1) a DNA-binding domain toward the center of the sequence and (2) a ligand-binding domain toward the carboxyl terminus. The structure of a dimer of the DNA-binding domain bound to DNA is shown, as is one monomer of the normally dimeric ligand-binding domain. [Drawn from 1HCQ and 1LBD.pdb.]

The second highly conserved region of a nuclear hormone receptor lies near the carboxyl terminus. This area of the receptor forms a hydrophobic pocket that is the ligand-binding site. Ligand binding leads to substantial structural rearrangement (**Figure 37.10**). How does ligand binding by the receptor lead to changes in gene expression? Intuitively, we might think that ligand binding allows the receptor to bind to DNA. Interestingly, ligand binding has little effect on the ability of the receptor to bind its response element. Rather, ligand binding allows the receptor to recruit other proteins that facilitate transcription.

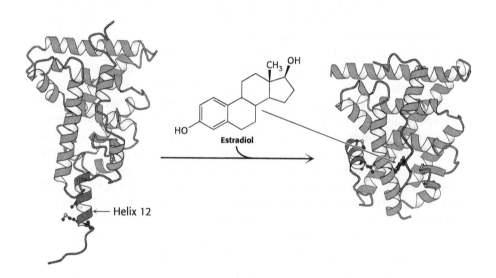

Figure 37.10 Ligand binding to nuclear hormone receptor. The ligand binding causes structural alteration in the receptor so that the hormone lies completely surrounded within a pocket in the ligand-binding domain. Notice that helix 12 (shown in purple) folds into a groove on the side of the structure on ligand binding. [Drawn from 1LDB and 1ERE.pdb.]

Nuclear Hormone Receptors Recruit Coactivators and Corepressors

Although the receptor can bind DNA without a ligand, the receptor cannot serve its recruitment function without first binding the ligand. Proteins that bind to the receptor only after it has bound to the steroid are called *coactivators*. The site for the interaction between the nuclear hormone–receptor complex and the coactivators is fully formed only when the ligand is bound (**Figure 37.11**). These coactivators are referred to as the p160 family because of their size (~160 kDa). Coactivators work in concert with other proteins to form large multicomponent complexes that modify chromatin and the transcription machinery to regulate gene expression.

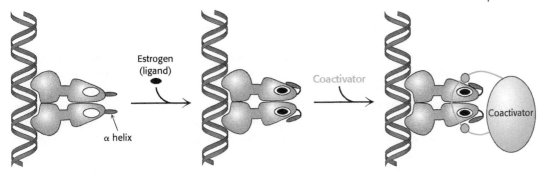

Figure 37.11 Coactivator recruitment. The binding of ligand to a nuclear hormone receptor induces a conformational change in the ligand-binding domain. This change in conformation generates favorable sites for the binding of a coactivator.

Although the estrogen receptor is inactive in the absence of estrogen, other members of the nuclear hormone-receptor family, such as the receptors for thyroid hormone and retinoic acid, repress transcription in the absence of ligand. This repression also is mediated by the ligand-binding domain. In their unbound forms, the ligand-binding domains of these receptors bind to *corepressor proteins*. Corepressors bind to a site in the ligand-binding domain that overlaps the coactivator binding site. Ligand binding triggers the release of the corepressor and frees the ligand-binding domain, enabling it to bind to a coactivator.

CLINICAL INSIGHT

Steroid-Hormone Receptors Are Targets for Drugs

Molecules such as estradiol that bind to a receptor and trigger signaling pathways are called *agonists*. As discussed in Chapter 29, some athletes take natural and synthetic agonists of the androgen receptor, a member of the nuclear hormone-receptor family, because their binding to the androgen receptor stimulates the expression of genes that enhance the development of lean muscle mass. Referred to as *anabolic steroids*, such compounds used in excess have substantial negative side effects (p. 542).

Other molecules bind to nuclear hormone receptors but do not effectively trigger signaling pathways. Such compounds are called *antagonists* and are, in many ways, like competitive inhibitors of enzymes. Some important drugs are antagonists that target the estrogen receptor. For example, *tamoxifen* and *raloxifene* are used in the treatment and prevention of breast cancer, because some breast tumors rely on estrogen-mediated pathways for growth. These compounds are sometimes called *selective estrogen receptor modulators* (SERMs).

Tamoxifen

Raloxifene

QUICK QUIZ 2 Outline the mechanism of action of estradiol.

37.4 Histone Acetylation Results in Chromatin Remodeling

We have seen that nuclear receptors respond to signal molecules by recruiting coactivators. Now we can ask, how do coactivators modulate transcriptional activity? Recall that the template for RNA synthesis in eukaryotes is not simply naked DNA; rather, it is a complex of DNA and *histones* called *chromatin*. *Some proteins that stimulate transcription act to loosen the histone complex from the DNA, exposing additional DNA regions to the transcription machinery.* One means by which loosening takes place is with the enzymatic attachment of acetyl groups to histones.

Metabolism in Context: Acetyl CoA Plays a Key Role in the Regulation of Transcription

So far in our study of biochemistry, we have seen acetyl CoA only in the context of intermediary metabolism, for instance, as a fuel for the citric acid cycle or as a precursor for fatty acid synthesis and for steroid synthesis. However, acetyl CoA also plays a role in the regulation of gene expression. Recall that citrate is transported out of mitochondria and is cleaved into oxaloacetate and acetyl CoA in the cytoplasm by ATP-citrate lyase. Recent research shows that citrate can also enter the nucleus, where a nuclear ATP-citrate lyase generates acetyl CoA for use as a substrate for histone-modifying enzymes. For instance, some of the p160 coactivators and the proteins that they recruit covalently modify the amino-terminal tails of histones by catalyzing the transfer of acetyl groups from acetyl CoA to specific lysine residues in these amino-terminal tails.

Lysine in histone tail   **Acetyl CoA**

Enzymes that catalyze such reactions are called *histone acetyltransferases* (HATs). The histone tails are readily extended, so they can fit into the HAT active site and become acetylated (**Figure 37.12**).

What are the consequences of histone acetylation? Lysine bears a positively charged ammonium group at neutral pH. The addition of an acetyl group neutralizes the ammonium group to an amide group while adding a negative charge.

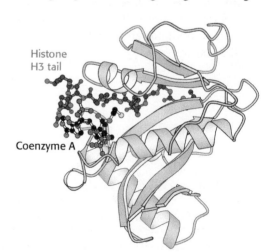

Histone
H3 tail

Coenzyme A

Figure 37.12 The structure of histone acetyltransferase. The amino-terminal tail of histone H3 extends into a pocket in which a lysine side chain can accept an acetyl group from acetyl CoA bound in an adjacent site. [Drawn from 1QSN.pdb.]

This change dramatically reduces the affinity of the tail for DNA and decreases the affinity of the entire histone complex for DNA, loosening the histone complex from the DNA.

In addition, the acetylated lysine residues interact with a specific *acetyllysine-binding domain*, termed a *bromodomain*, that is present in many proteins that regulate eukaryotic transcription. These domains serve as docking sites to recruit proteins that play a variety of roles in transcription and chromatin remodeling. Bromodomain-containing proteins are components of two large complexes essential for transcription. One is a complex of more than 10 polypeptides that binds to the TATA-box-binding protein. Recall that TBP is an essential transcription factor for many genes (p. 679). Proteins that bind to TBP are called *TAFs* (for *TATA-box-binding protein associated factors*). In particular, TAF1 contains a pair of bromodomains near its carboxyl terminus. The two domains are oriented such that each can bind one of two acetyllysine residues at positions 5 and 12 in the histone H4 tail. Thus, *acetylation of the histone tails provides a mechanism for recruiting other components of the transcriptional machinery*.

Bromodomains are also present in some components of large complexes known as *chromatin-remodeling engines*. These complexes, which also contain domains similar to those of helicases, utilize the free energy of ATP hydrolysis to shift the positions of *nucleosomes* along the DNA and to induce other conformational changes in chromatin, potentially exposing binding sites for other factors (**Figure 37.13**). Thus, *histone acetylation can activate transcription through a combination of three mechanisms*: by reducing the affinity of the histones for DNA, by recruiting other components of the transcriptional machinery, and by initiating the remodeling of the chromatin structure. Acetylation is not the only means of modifying histones. Other means include phosphorylation and methylation (**Table 37.3**).

Recall that nuclear hormone receptors also include regions that interact with components of coactivators. Thus, two mechanisms of gene regulation can work in concert. The modification of histones and chromatin remodeling can open up

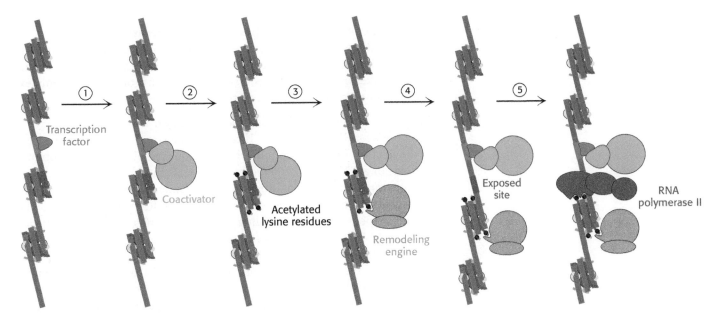

Figure 37.13 Chromatin remodeling. Eukaryotic gene regulation begins with an activated transcription factor bound to a specific site on DNA. One scheme for the initiation of transcription by RNA polymerase II requires five steps: (1) recruitment of a coactivator, (2) acetylation of lysine residues in the histone tails, (3) binding of a remodeling-engine complex to the acetylated lysine residues, (4) ATP-dependent remodeling of the chromatin structure to expose a binding site for RNA polymerase or for other factors, and (5) recruitment of RNA polymerase II. Only two subunits are shown for each complex, although the actual complexes are much larger.

Table 37.3 Selected histone modifications

Histone modified*		Amino acid modification	Effect
H4	K8	Acetylation	Transcription activation
H3	K14	Acetylation	Transcription activation
H3	K27	Methylation	Transcription activation
H3	R17	Methylation	Transcription activation
H2B	S14	Phosphorylation	Apoptosis initiation

*The letters K, R, and S are the single-letter abbreviations for lysine (K), arginine (R), and serine (S). The numbers with these letters refer to the location in the primary structure.

regions of chromatin into which the transcription complex can be recruited through protein–protein interactions.

Histone Deacetylases Contribute to Transcriptional Repression

Just as in bacteria, some changes in a cell's environment lead to the repression of genes that had been active. The modification of histone tails again plays an important role. However, in repression, a key reaction appears to be the deacetylation of acetylated lysine, catalyzed by specific *histone deacetylase* enzymes. Indeed, all covalent modifications of histone tails are reversible.

In many ways, the acetylation and deacetylation of lysine residues in histone tails is analogous to the phosphorylation and dephosphorylation of serine, threonine, and tyrosine residues in other stages of signaling processes. Like the addition of phosphoryl groups, the addition of acetyl groups can induce conformational changes and generate novel binding sites. Without a means of removing these groups, however, these signaling switches will become stuck in one position and lose their effectiveness. Like phosphatases, deacetylases help reset the switches.

SUMMARY

37.1 Eukaryotic Cells Have Three Types of RNA Polymerases

There are three types of RNA polymerase in the nucleus: RNA polymerase I makes ribosomal RNA precursors, II makes messenger RNA precursors, and III makes transfer RNA precursors. Eukaryotic promoters are complex, being composed of several different elements.

37.2 RNA Polymerase II Requires Complex Regulation

Promoters for RNA polymerase II can be located on the 5′ side or the 3′ side of the start site for transcription. One common type of eukaryotic promoter consists of a TATA box centered between −30 and −100 and paired with an initiator element. Eukaryotic promoter elements are recognized by proteins called transcription factors rather than by RNA polymerase II. The saddle-shaped TATA-box-binding protein unwinds and sharply bends DNA at TATA-box sequences and serves as a focal point for the assembly of transcription complexes. The TATA-box-binding protein initiates the assembly of the active transcription complex. The activity of many promoters is greatly increased by enhancer sequences that have no promoter activity of their own. Enhancers are DNA elements that can modulate gene expression from more than 1000 bp away from the start site of transcription. Enhancers are often specific for certain cell types, depending on which DNA-binding proteins are present.

Most eukaryotic genes are not expressed unless they are activated by the binding of specific transcription factors to sites on the DNA. These specific DNA-binding proteins interact directly or indirectly with RNA polymerases or their associated proteins.

37.3 Gene Expression Is Regulated by Hormones

The estradiol receptor is a member of the nuclear hormone-receptor family, a group of receptors that regulate gene expression in response to hormone binding. These receptors are modular: they consist of separate DNA-binding and hormone-binding domains. The receptor recognizes particular DNA sequences, called response elements, upstream of the genes that they regulate. On hormone binding, the hormone receptor recruits coactivators that alter the activity of RNA polymerase II.

37.4 Histone Acetylation Results in Chromatin Remodeling

Eukaryotic DNA is tightly bound to basic proteins called histones; the combination is called chromatin. Changes in chromatin structure play a major role in regulating gene expression. Among the most important functions of coactivators is to catalyze the addition of acetyl groups to lysine residues in the tails of histones. Histone acetylation decreases the affinity of the histones for DNA, making additional genes accessible for transcription. In addition, acetylated histones are targets for proteins containing specific binding units called bromodomains. Bromodomains are components of two classes of large complexes: (1) chromatin-remodeling engines and (2) factors associated with RNA polymerase II. These complexes open up sites on chromatin and initiate transcription.

KEY TERMS

transcription (p. 675)
RNA polymerase (p. 675)
carboxyl-terminal domain (CTD)
  (p. 677)
TATA box (p. 678)
transcription factor (p. 679)
enhancer (p. 679)
mediator (p. 680)
combinatorial control (p. 680)
nuclear hormone receptor (p. 681)

estrogen response element (ERE)
  (p. 681)
zinc-finger domain (p. 681)
coactivator (p. 682)
agonist (p. 683)
antagonist (p. 683)
selective estrogen receptor modulator
  (SERM) (p. 683)
histone (p. 684)
chromatin (p. 684)

histone acetyltransferase (HAT)
  (p. 684)
acetyllysine-binding domain (p. 685)
bromodomain (p. 685)
TATA-box-binding protein associated
  factor (TAF) (p. 685)
chromatin-remodeling engine (p. 685)
nucleosome (p. 685)
histone deacetylase (p. 686)

? Answers to QUICK QUIZZES

1. A promoter is a DNA sequence that attracts the polymerase to the start site for transcription. Promoters are usually located just upstream of the gene. An enhancer is a DNA sequence that has no promoter activity itself but greatly enhances the activity of an associated promoter. Enhancers can exert their effects over a distance of several thousand base pairs.

2. Estradiol diffuses into the cell. It binds to a specific receptor protein. When bound to the hormone, the receptor recruits other proteins called coactivators that stimulate RNA polymerase.

PROBLEMS

1. *Defining features.* Identify three characteristics that differentiate eukaryotic transcription from bacterial transcription.

2. *Three of a kind.* Differentiate among the three types of eukaryotic RNA polymerases.

3. *Promoting for two.* Describe the major features of a promoter for RNA polymerase II.

4. *Distinguishing characteristic.* What structural feature of RNA polymerase II distinguishes it from RNA polymerases I and III?

5. *Different sides of a bond?* Explain the difference between cis-acting and trans-acting elements.

6. *Does it require spandex?* What is an enhancer, and how does it function?

7. *Listening to its world.* Why are promoters for RNA polymerase II necessarily more complex than promoters for RNA polymerases I and III? ✓ 3

8. *Making noncombustible matches.* Pair the properties with the appropriate RNA polymerase.

(a) RNA polymerase I
_____

(b) RNA polymerase II
_____

(c) RNA polymerase III
_____

1. Synthesizes mRNA precursors
2. Insensitive to α-amanitin
3. Synthesizes tRNA
4. Located in the nucleoplasm
5. Contains a carboxyl-terminal domain
6. Located in the nucleolus
7. TATA is a promoter sequence
8. Very sensitive to α-amanitin
9. Synthesized rRNA
10. Moderately sensitive to α-amanitin

9. *A fitting end.* Cordycepin inhibits RNA synthesis.

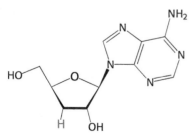

Cordycepin (3'-deoxyadenosine)

Cordycepin is an analog of which ribonucleotide? Why does it inhibit RNA synthesis?

10. *Leaving the station.* What is the functional consequence of phosphorylation of the carboxyl-terminal domain of RNA polymerase II?

11. *Bubble translation.* The lengths of the RNA–DNA hybrid and of the unwound region of DNA stay rather constant as RNA polymerase moves along the DNA template. What does this finding show about the rate of unwinding and rewinding of the DNA in the course of transcription?

12. *Like Harry Potter and Lord Voldemort.* Match each term with its description.

(a) Transcription factor
_____
(b) Enhancer _____
(c) Mediator _____
(d) Combinatorial control
_____
(e) Zinc-finger domain
_____
(f) Coactivator _____
(g) Selective estrogen-receptor modulator
_____
(h) Histone acetyltransferase _____
(i) Acetyllysine-binding domain _____
(j) Chromatin-remodeling engine _____

1. Depends on ATP-citrate lyase for its substrate
2. Acts as a bridge between transcription factors and RNA polymerase II
3. Based on the fact that the regulatory properties of a protein depend on what other proteins are present
4. Facilitates estrogen-receptor action
5. Binds to cis-acting elements
6. Bromodomain
7. Shifts the location of nucleosomes
8. Tamoxifen
9. DNA-binding domain in some proteins
10. Cis-acting element that may be far from the gene regulated

13. *Teamwork.* What is the relation between the nuclear hormone receptors and the coactivators? ✓ 3

14. *Creating complexity.* What is the advantage of combinatorial control? ✓ 3

15. *Charge changes.* In light of the fact that histones are highly basic proteins, how does acetylation affect these molecules? ✓ 3

16. *Destroying-angel toxin.* α-Amanitin is a deadly poison and potent inhibitor of RNA polymerase II. The toxin is produced by a species of *Amanita*. Initial symptoms of

α-amanitin poisoning are gastrointestinal distress, but these symptoms usually subside. Liver-cell cytolysis is evident by the fourth day after ingestion. Death caused by liver failure is within 10 days. Suggest some reasons why it takes days for this powerful poison to cause liver failure and death.

17. *Hybrid receptor.* Through recombinant DNA methods, a modified steroid-hormone receptor was prepared that consists of an estrogen receptor with its ligand-binding domain replaced by the ligand-binding domain from the progesterone receptor. Predict the expected responsiveness of gene expression for cells treated with estrogen or with progesterone. ✓ 3

Chapter Integration Problems

18. *Hormone action.* How does estradiol action differ from that of hormones that bind G-protein-coupled receptors? ✓ 3

19. *Specificity.* If estradiol can diffuse through cell membranes, explain why all cell types are not sensitive to the presence of estradiol. ✓ 3

20. *RDA.* The recommended daily allowance (RDA) for zinc is between 9 and 11 mg per day. Suggest one crucial use for zinc in human beings.

Data Interpretation and Challenge Problems

21. *Accessibility matters.* Chromatin accessibility varies with cell type. For instance, if cells synthesizing globin mRNA are treated briefly with DNase (an enzyme that digests DNA), the genes that encode globin are destroyed but the gene for ovalbumin remains intact. Conversely, if the oviduct is similarly treated, the gene for ovalbumin is destroyed but the genes that encode globin are undamaged. Explain these results. ✓ 3

22. *Run-off experiment.* Nuclei were isolated from the liver, muscle, and the brain. The nuclei were then incubated with α-[^{32}P]UTP under conditions that allow RNA synthesis, except that an inhibitor of RNA initiation was present. The radioactive RNA was isolated and annealed to various DNA sequences that had been attached to a gene chip (Chapter 41). In the adjoining graphs, the intensity of the shading indicates roughly how much RNA was attached to each DNA sequence. ✓ 3

Liver

Muscle

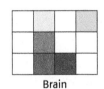
Brain

(a) Why does the intensity of hybridization differ between genes?

(b) What is the significance of the fact that some of the RNA molecules display different hybridization patterns in different tissues?

(c) Some genes are expressed in all three tissues. What would you guess is the nature of these genes?

(d) Suggest a reason why an initiation inhibitor was included in the reaction mixture.

Selected Readings for this chapter can be found online at www.whfreeman.com/tymoczko3e.

CHAPTER 38

RNA Processing in Eukaryotes

Vincent van Gogh, a post-Impressionist painter (Dutch, 1853–1890) shown here in a self-portrait, suffered bouts of madness, and scholars have long debated the extent to which his madness affected his painting. His metal health problems may have been caused by acute intermittent porphyria, a disease that can result from splicing defects in an enzyme in heme metabolism. [BuyenlargeUIG/age fotostock.]

Virtually all the initial products of transcription are further processed in eukaryotes. For example, primary transcripts (pre-mRNA molecules), the products of RNA polymerase II action, acquire a cap at their 5′ ends and a poly(A) tail at their 3′ ends. Most importantly, *nearly all mRNA precursors in higher eukaryotes are spliced*. Most genes in eukaryotes are discontinuous mosaics of regions that encode amino acids, called *exons,* and noncoding regions, called

691

introns, that are removed by splicing. Introns can make up most of the primary transcript; some mRNAs end up being only a tenth of the size of their precursors.

We begin by examining the processing of rRNA and tRNA. We then move on to the processing of mRNA, including how alternative splicing generates multiple mRNAs from a single gene. The chapter ends with a consideration of the catalytic properties of RNA.

38.1 Mature Ribosomal RNA Is Generated by the Cleavage of a Precursor Molecule

RNA polymerase I transcription results in a single precursor (45S in mammals) that encodes three RNA components of the ribosome: the 18S rRNA, the 28S rRNA, and the 5.8S rRNA (**Figure 38.1**). The 18S rRNA is the RNA component of the small ribosomal subunit (40S), and the 28S and 5.8S rRNAs are two RNA components of the large ribosomal subunit (60S). The other RNA component of the large ribosomal subunit, the 5S rRNA, is transcribed by RNA polymerase III as a separate transcript.

Figure 38.1 The processing of eukaryotic pre-rRNA. The mammalian pre-rRNA transcript contains the RNA sequences destined to become the 18S, 5.8S, and 28S rRNAs of the small and large ribosomal subunits. First, nucleotides are modified (indicated by red lines). Next, the pre-rRNA is cleaved and packaged to form mature ribosomes in a highly regulated process in which more than 200 proteins take part.

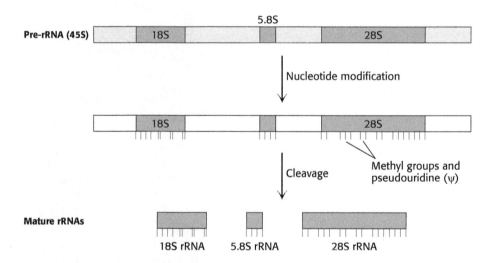

The cleavage of the precursor into three separate rRNAs is actually the final step in its processing. Prior to cleavage, two other modifications occur. First, the nucleotides of the pre-rRNA sequences destined for the ribosome undergo extensive modification, on both ribose and base components, directed by many *small nucleolar ribonucleoproteins* (snoRNPs), each of which consists of one small nucleolar RNA (snoRNA) and several proteins. Next, the pre-rRNA is assembled with ribosomal proteins in a large ribonucleoprotein. Cleavage of the pre-rRNA (sometimes coupled with additional processing steps) releases the mature rRNAs assembled with ribosomal proteins as ribosomes. Like those of RNA polymerase I transcription itself, most of these processing steps take place in the cell nucleolus, a nuclear subcompartment.

38.2 Transfer RNA Is Extensively Processed

Transfer RNA precursors, which are synthesized by RNA polymerase III, are processed by several means to yield the mature tRNA molecule. As with bacterial tRNAs, nucleotides from the 5′ end on the precursor (the 5′ leader) are cleaved by RNase P, the nucleotides on the 3′ end (the 3′ trailer) are removed by RNase Z, and nucleotides CCA are added by tRNA nucleotidyltransferase (**Figure 38.2**). Eukaryotic tRNAs are also heavily modified on base and ribose moieties; these modifications are important for function. Many eukaryotic pre-tRNAs also

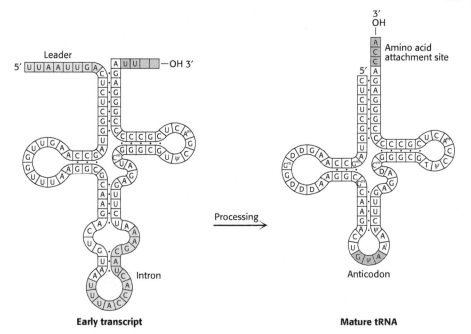

Leader

Intron

Early transcript

Processing

Anticodon

Mature tRNA

3'
OH

Amino acid
attachment site

Figure 38.2 Transfer RNA precursor processing. The conversion of a yeast tRNA precursor into a mature tRNA requires the removal of a 14-nucleotide intron (yellow), the cleavage of a 5' leader (green), the removal of the 3' trailer (blue), and the attachment of CCA at the 3' end (red). In addition, several bases are modified.

contain an intron, which is removed by an endonuclease, with the resulting products joined by a ligase.

38.3 Messenger RNA Is Modified and Spliced

The most extensively modified transcription product is that of RNA polymerase II: most of this RNA will be processed to mRNA. The immediate product of RNA polymerase II is sometimes referred to as *pre-mRNA*. Both the 5' and the 3' ends of the pre-mRNA are modified, and most pre-mRNA molecules are spliced to remove the introns.

The 5' triphosphate end of the nascent RNA strand is modified shortly after initiation of RNA synthesis, when the strand is about 25 nucleotides in length. The modification occurs in three steps. First, a phosphoryl group is removed by hydrolysis by RNA triphosphatase. Second, the diphosphate 5' end of the RNA attacks the α-phosphorus atom of a molecule of GTP to form an unusual 5'–5' triphosphate linkage, a reaction catalyzed by guanylyltransferase. This distinctive terminus is called a *5' cap* (**Figure 38.3**). Third, the N-7 nitrogen atom of the terminal guanine is then methylated by RNA N-7 guanine methyltransferase, which uses *S*-adenosylmethionine as the methyl donor (p. 578) to form *cap 0*. The triphosphatase and guanylyltranseferase are present on the same polypeptide chain, providing another example of a bifunctional enzyme. The adjacent riboses may be subsequently methylated to form *cap 1* or *cap 2*. Caps contribute to the stability of mRNAs by protecting their 5' ends from phosphatases and nucleases. In addition, caps enhance the translation of mRNA by eukaryotic protein-synthesizing systems (Chapter 40).

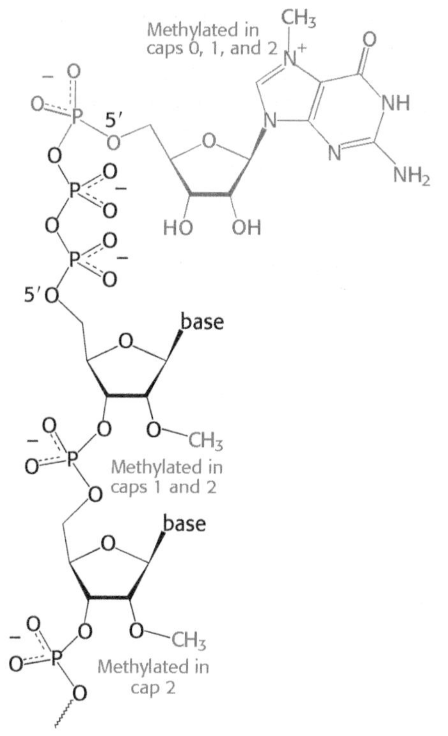

As mentioned earlier, pre-mRNA is also modified at the 3' end. *Most eukaryotic mRNAs contain a polyadenylate, or poly(A), tail at that end*, added after transcription has ended. Thus, DNA does not encode this *poly(A) tail*. Indeed, the nucleotide preceding the poly(A) attachment site is not the last nucleotide to be transcribed. Some primary transcripts contain hundreds of nucleotides beyond the 3' end of the mature mRNA, all of which are removed before the poly(A) tail is added.

How is the final form of the 3' end of the pre-mRNA created? *The 3' end of the pre-mRNA is generated by a complex that contains a specific endonuclease (the cleavage and polyadenylation specificity factor, CPSF) that recognizes the*

✓ 4 Explain how RNA is processed after its transcription in eukaryotes.

Methylated in caps 0, 1, and 2

base

Methylated in caps 1 and 2

base

Methylated in cap 2

Figure 38.3 Capping the 5' end. Caps at the 5' end of eukaryotic mRNA include 7-methylguanylate (red) attached by a triphosphate linkage to the ribose at the 5' end. None of the riboses are methylated in cap 0, one is methylated in cap 1, and both are methylated in cap 2.

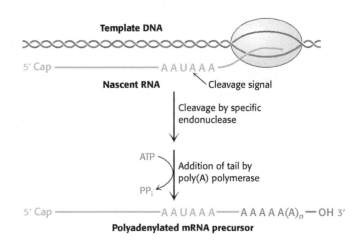

Figure 38.4 Polyadenylation of a primary transcript. A specific endonuclease cleaves the RNA downstream of AAUAAA. Poly(A) polymerase then adds about 250 adenylate residues.

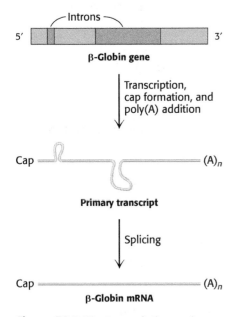

Figure 38.5 The transcription and processing of the β-globin gene. The gene is transcribed to yield the primary transcript, which is modified by cap and poly(A) addition. The introns in the primary RNA transcript are removed to form the mRNA.

sequence *AAUAAA* (Figure 38.4). The presence of internal AAUAAA sequences in some mature mRNAs indicates that AAUAAA is only part of the cleavage signal; its context also is important. After cleavage of the pre-RNA by the endonuclease, a poly(A) polymerase adds about 250 adenylate residues to the 3′ end of the transcript; ATP is the donor in this reaction.

The role of the poly(A) tail is still not firmly established despite much effort. However, evidence that it enhances translation efficiency and the stability of mRNA is accumulating. Messenger RNA devoid of a poly(A) tail can be transported out of the nucleus, but such RNA is usually a much less effective template for protein synthesis than is one with a poly(A) tail. The half-life of an mRNA molecule may be determined in part by the rate of degradation of its poly(A) tail.

Sequences at the Ends of Introns Specify Splice Sites in mRNA Precursors

Processing at the ends of the pre-mRNA molecule is not the only modification that these molecules must undergo. More than 90% of human protein-encoding genes are composed of exons and introns—coding regions and noncoding regions, respectively. The introns must be removed, and the exons must be linked to form the final mRNA in a process called *splicing* (Figure 38.5). To achieve the accurate removal of introns, the correct splice site must be clearly marked. Does a particular sequence denote the splice site? The sequences of thousands of intron–exon junctions within RNA transcripts are known. In eukaryotes ranging from yeast to mammals, these sequences have a common structural motif: *the intron begins with GU and ends with AG*. The consensus sequence at the 5′ splice site in vertebrates is AGGUAAGU, where the GU is invariant (Figure 38.6). At the 3′ end of an intron, the consensus sequence is a stretch of *10 pyrimidines* (U or C, termed the *polypyrimidine tract*), followed by any base and then by C, and ending with the invariant AG. Introns also have an important internal site located between 20 and 50 nucleotides upstream of the 3′ splice site; it is called the *branch site* for reasons that will be evident shortly.

The 5′ and 3′ splice sites and the branch site are essential for determining where splicing takes place. Mutations in each of these three critical regions lead to aberrant splicing. Introns vary in length from 50 to 10,000 nucleotides, and so the splicing machinery may have to find the 3′ site several thousand nucleotides away. This splicing must be exquisitely sensitive because the nucleotide information is

Figure 38.6 Splice sites. Consensus sequences for the 5′ splice site and the 3′ splice site are shown. Py stands for pyrimidine.

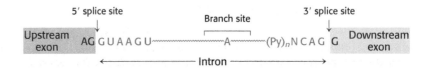

converted into protein information in three nonoverlapping nucleotide sequences called codons (Chapter 39). Thus, a one-nucleotide shift in the course of splicing would alter the nucleotide information on the 3′ side of the splice to give an entirely different amino acid sequence, likely including a Stop codon that would prematurely stop protein synthesis. Specific sequences near the splice sites (in both the introns and the exons) play an important role in splicing regulation, particularly in designating splice sites when there are many alternatives (p. 697). Despite our knowledge of splice-site sequences, predicting pre-mRNAs and their protein products from genomic DNA sequence information remains a challenge.

Small Nuclear RNAs in Spliceosomes Catalyze the Splicing of mRNA Precursors

What are the molecular machines that so precisely excise introns and join exons? A group of special RNAs and more than 300 proteins combine with pre-mRNA to form a large splicing complex called the *spliceosome*. The splicesome has a mass of ~ 4.8 MDa, and may be the most complicated molecular machine in the cell. Let's consider some of the components of the spliceosome.

The nucleus contains many types of small RNA molecules (Table 37.2). One particular class of such RNAs, referred to as *snRNAs* (small *n*uclear RNAs) and designated U1, U2, U4, U5, and U6, is essential for splicing mRNA precursors. These RNA molecules and their associated proteins are key components of the splicesome. The RNA-protein complexes are termed *snRNPs* (small *n*uclear *r*ibo-*n*ucleoprotein *p*articles); investigators often speak of them as "snurps." In mammalian cells, the first step in the splicing is the recognition of the 5′ splice site by the U1 snRNP (Figure 38.7). In fact, the RNA component of U1 snRNP contains a six-nucleotide sequence that base-pairs to the 5′ splice site of the pre-mRNA. This binding initiates spliceosome assembly on the pre-mRNA molecule. U2 snRNP then binds the branch site in the intron. A preassembled U4-U5-U6 tri-snRNP joins this complex of U1, U2, and the mRNA precursor to form the spliceosome.

The splicing process continues when U5 interacts with exon sequences in the 5′ splice site and subsequently with the 3′ exon. Next, U6 disengages from U4 and interacts with U2 and with the 5′ end of the intron, displacing U1 from the spliceosome. The *U2 and U6 snRNAs form the catalytic center of the spliceosome* (Figure 38.8). U4 serves as an inhibitor that masks U6 until the specific splice sites are aligned. These rearrangements bring the ends of the intron together and

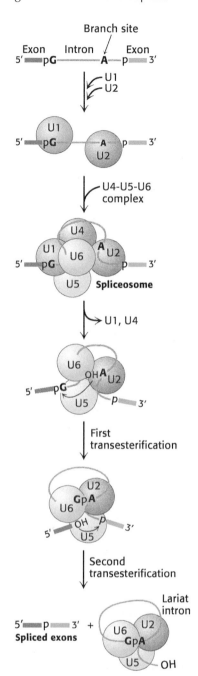

Figure 38.7 Spliceosome assembly and action. U1 binds the 5′ splice site, and U2 binds to the branch point. A preformed U4-U5-U6 complex then joins the assembly to form the complete spliceosome. Extensive interactions between U6 and U2 displace U1 and U4. Then, in the first transesterification step, the branch-site adenosine attacks the 5′ splice site, making a lariat intermediate. U5 holds the two exons in close proximity, and the second transesterification takes place, with the 5′ splice-site hydroxyl group attacking the 3′ splice site. These reactions result in the mature spliced mRNA and a lariat form of the intron bound by U2, U5, and U6. [Information from T. Villa, J. A. Pleiss, and C. Guthrie, *Cell* 109:149–152, 2002.]

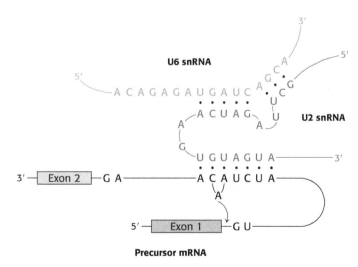

Figure 38.8 The splicing catalytic center. The catalytic center of the spliceosome is formed by U2 snRNA (red) and U6 snRNA (green), which are base-paired. U2 is also base-paired to the branch site of the mRNA precursor. [Information from H. D. Madhani and C. Guthrie, *Cell* 71:803–817, 1992.]

DID YOU KNOW?

Transesterification is the reaction of an alcohol with an ester to form a different alcohol and a different ester.

Transesterification

QUICK QUIZ Describe the roles of the various snRNPs in the splicing process.

result in a *transesterification reaction*. The 5′ exon is cleaved, and the remaining pre-mRNA molecule forms a lariat intermediate in which the first nucleotide of the intron (G) is joined to an A in the branch region in an unusual 5′–2′ phosphodiester linkage.

Further rearrangements of RNA in the spliceosome facilitate the second transesterification. In these rearrangements, U5 holds the free 5′ exon and the 3′ exon close together to facilitate the second transesterification, generating the spliced product. U2, U5, and U6 bound to the excised lariat intron are released to complete the splicing reaction.

Many of the steps in the splicing process require ATP hydrolysis. How is the free energy associated with ATP hydrolysis used to power splicing? To achieve the well-ordered rearrangements necessary for splicing, ATP-powered RNA helicases must unwind RNA helices and allow alternative base-pairing arrangements to form. Thus, two features of the splicing process are noteworthy. First, *RNA molecules play key roles in directing the alignment of splice sites and in carrying out catalysis*. Second, *ATP-powered helicases unwind RNA duplex intermediates that facilitate catalysis and induce the release of snRNPs from the mRNA*.

CLINICAL INSIGHT

Mutations that Affect Pre-mRNA Splicing Cause Disease

Mutations in either the pre-mRNA (cis-acting) or the splicing factors (trans-acting) can cause defective pre-mRNA splicing. Mutations in the pre-mRNA cause some forms of thalassemia, a group of hereditary anemias characterized by the defective synthesis of hemoglobin. The gene for the hemoglobin β chain consists of three exons and two introns. Cis-acting mutations that cause aberrant splicing can arise at the 5′ or 3′ splice site in either of the two introns or in its exons. The mutations usually result in an incorrectly spliced pre-mRNA that, because of a premature termination signal, cannot encode a full-length protein. The defective mRNA is normally degraded rather than translated (**Figure 38.9**). *Mutations affecting splicing have been estimated to cause at least 15% of all genetic diseases*.

Disease-causing mutations may also appear in protein splicing factors. Retinitis pigmentosa is a disease of acquired blindness, first described in 1857,

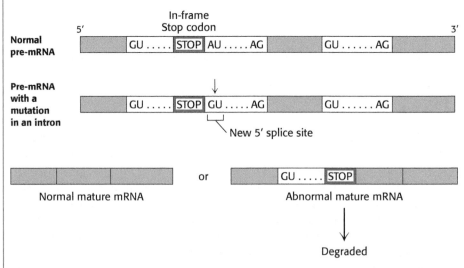

Figure 38.9 A splicing mutation that causes thalassemia. An A-to-G mutation within the first intron of the gene for the human hemoglobin β chain creates a new 5′ splice site (GU). The abnormal mature mRNA now has a premature Stop codon and is degraded.

with an incidence of 1/3500 individuals. About 5% of the autosomal dominant form of retinitis pigmentosa is likely due to mutations in a pre-mRNA splicing factor that is a component of the U4-U5-U6 tri-snRNP. How a mutation in a splicing factor that is present in all cells causes disease only in the retina is unclear; nevertheless, retinitis pigmentosa is a good example of how mutations that disrupt spliceosome function can cause disease.

⚕ CLINICAL INSIGHT

Most Human Pre-mRNAs Can Be Spliced in Alternative Ways to Yield Different Proteins

Alternative splicing is a widespread mechanism for generating protein diversity. Current estimates are that more than 70% of human genes that encode proteins are alternatively spliced. Different combinations of exons in the same gene may be spliced into a mature RNA, producing distinct forms of a protein for specific tissues, developmental stages, or signaling pathways. For example, a precursor of an antibody-producing cell forms an antibody that is anchored in the cell's plasma membrane (**Figure 38.10**). The attached antibody recognizes a specific foreign antigen, which leads to cell differentiation and proliferation. The activated antibody-producing cells then splice their nascent RNA transcript in an alternative manner to form soluble antibody molecules that are secreted rather than retained on the cell surface. We see here a clear-cut example of a benefit conferred by the complex arrangement of introns and exons in higher organisms.

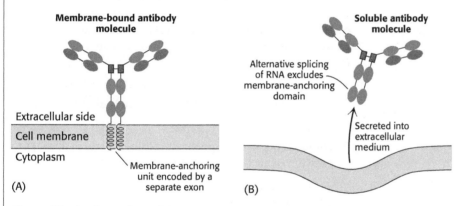

Figure 38.10 Alternative splicing. Alternative splicing generates mRNAs that are templates for different forms of a protein: (A) a membrane-bound antibody on the surface of a lymphocyte and (B) its soluble counterpart, exported from the cell. The membrane-bound antibody is anchored to the plasma membrane by a helical segment (highlighted in yellow) that is encoded by its own exon.

What controls which splicing sites are selected? The selection is determined by the binding of trans-acting splicing factors to cis-acting sequences in the pre-mRNA. Most alternative splicing leads to changes in the coding sequence, resulting in proteins with different functions. To better understand the power of alternative splicing, let's consider a gene with five positions at which alternative splicing can take place. With the assumption that these alternative splicing pathways can be regulated independently, a total of $2^5 = 32$ different mRNAs can be generated. *Alternative splicing provides a powerful mechanism for expanding the versatility of genomic sequences through combinatorial control.* Several human diseases that can be attributed to defects in alternative splicing are listed in **Table 38.1**.

Table 38.1 Selected human diseases attributed to defects in alternative splicing

Disorder	Gene or its product
Acute intermittent porphyria	Porphobilinogen deaminase
Breast and ovarian cancer	*BRCA*1
Cystic fibrosis	CFTR
Frontotemporal dementia	τ protein
Hemophilia A	Factor VIII
HGPRT deficiency (Lesch-Nyhan syndrome)	Hypoxanthine-guanine phosphoribosyltransferase
Leigh encephalomyelopathy	Pyruvate dehydrogenase E1 α
Severe combined immunodeficiency	Adenosine deaminase
Spinal muscle atrophy	*SMN*1 or *SMN*2

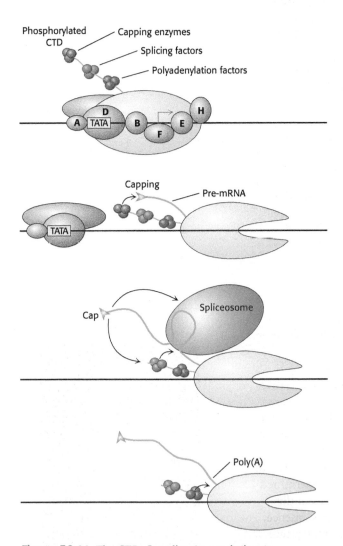

Figure 38.11 The CTD: Coupling transcription to pre-mRNA processing. The transcription factor TFIIH phosphorylates the carboxyl-terminal domain (CTD) of RNA polymerase II, signaling the transition from transcription initiation to elongation. The phosphorylated CTD binds factors required for pre-mRNA capping, splicing, and polyadenylation. These proteins are brought in close proximity to their sites of action on the nascent pre-mRNA as it is transcribed in the course of elongation. [Information from P. A. Sharp. *Trends Biochem. Sci.* 30:279–281, 2005.]

The Transcription and Processing of mRNA Are Coupled

Although the transcription and processing of mRNAs have been described herein as separate events in gene expression, experimental evidence suggests that the two steps are coordinated by the carboxyl-terminal domain of RNA polymerase II. We have seen that the CTD consists of a unique repeated seven-amino-acid sequence, YSPTSPS. Either the second serine, the fifth serine, or both may be phosphorylated in the various repeats. The phosphorylation state of the CTD is controlled by a number of kinases and phosphatases and leads the CTD to bind many of the proteins having roles in RNA transcription and processing. The CTD contributes to efficient transcription by recruiting these proteins to the pre-mRNA (**Figure 38.11**), including

1. capping enzymes, which methylate the 5′ guanine nucleotide base on the pre-mRNA immediately after transcription begins;

2. components of the splicing machinery, which initiate the excision of each intron as it is synthesized; and

3. an endonuclease that cleaves the transcript at the poly(A) addition site, creating a free 3′-OH group that is the target for 3′ adenylation.

These events take place sequentially, directed by the phosphorylation state of the CTD.

BIOLOGICAL INSIGHT

RNA Editing Changes the Proteins Encoded by mRNA

Alternative splicing is not the only means of generating diverse proteins from one gene. Remarkably, the amino-acid-sequence information encoded by some mRNAs is altered after transcription. *RNA editing* is the term for a change in the nucleotide sequence of RNA after transcription by processes other than RNA splicing. RNA editing is prominent in the synthesis of apolipoproteins (Table 29.1). *Apolipoprotein B* (apo B) plays a role in the transport of triacylglycerols and cholesterol as a component of lipoprotein particles. Apo B exists in two forms, a

512-kDa *apo B-100* and a 240-kDa *apo B-48*. The larger form, synthesized by the liver, participates in the transport of lipids synthesized in the cell. The smaller form, synthesized by the small intestine, carries dietary fat in the form of chylomicrons. Apo B-48 contains the 2152 N-terminal residues of the 4536-residue apo B-100. What is the relation between these two forms of apo B? Experiments revealed that a totally unexpected mechanism for generating diversity is at work: *the changing of the nucleotide sequence of mRNA after its synthesis* (**Figure 38.12**). A specific cytidine residue of mRNA is deaminated to uridine, which changes the codon at residue 2153 from CAA (Gln) to UAA (stop). The deaminase that catalyzes this reaction is present in the small intestine, but not in the liver, and is expressed only at certain developmental stages.

RNA editing is likely much more common than was formerly thought. Antarctic and Arctic octopuses provide a fascinating example of RNA editing. In very cold temperatures, nerve impulses in these creatures would slow because the K^+ channels (p. 216) required for nerve impulse transmission cannot open and close rapidly enough. To prevent this situation, RNA editing is used to generate channels that function rapidly enough to maintain nerve functions in the extreme cold. Thus, temperature adaptation in octopus neurons occurs by RNA editing.

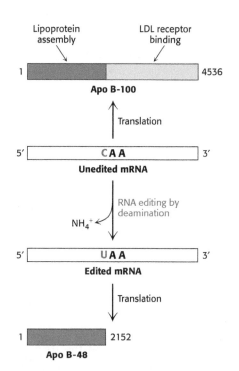

Figure 38.12 RNA editing. Enzyme-catalyzed deamination of a specific cytidine residue in the mRNA for apolipoprotein B-100 changes a codon for glutamine (CAA) to a Stop codon (UAA). Apolipoprotein B-48, a truncated version of the protein lacking the LDL receptor-binding domain, is generated by this posttranscriptional change in the mRNA sequence. [Information from P. Hodges and J. Scott. *Trends Biochem. Sci.* 17:77–81, 1992.]

38.4 RNA Can Function as a Catalyst

RNAs are a surprisingly versatile class of molecules. As we have seen, splicing is catalyzed largely by RNA molecules, with proteins playing a secondary role. As we shall see in Chapter 40, the RNA component of ribosomes is the catalyst that carries out protein synthesis. The discovery of *catalytic RNA* revolutionized the concept of biological catalysis.

The versatility of RNA first became clear from observations of the processing of ribosomal RNA in a single-cell eukaryote, *Tetrahymena*—a ciliated protozoan (**Figure 38.13**). In this reaction, a 414-nucleotide intron is removed from a precursor to yield the mature rRNA molecule (**Figure 38.14**). In an elegant series of studies of this splicing reaction, Thomas Cech and his coworkers established that the RNA spliced itself to precisely excise the intron in the absence of protein. Indeed, the RNA alone is catalytic under certain conditions and thus can function as an enzyme. Catalytic RNAs are called *ribozymes*. Catalytic biomolecules, formerly believed to be limited to proteins, now include certain RNA molecules. More than 1500 similar introns have since been found in species as widely dispersed as bacteria and eukaryotes, though not in vertebrates. Collectively, they are referred to as *group I introns*.

The *self-splicing* reaction in the group I intron requires an added guanosine or guanine nucleotide. Guanosine serves not as an energy source but as an attacking group that becomes transiently incorporated into the RNA (Figure 38.14). Guanosine associates with the RNA and then attacks the 5′ splice site to form a

Figure 38.13 Tetrahymena. [From R. Robinson (2006), Ciliate genome sequence reveals unique features of a model eukaryote. *PLoS Biol.* 4(9): e304. doi:10.1371/journal. pbio.0040304.]

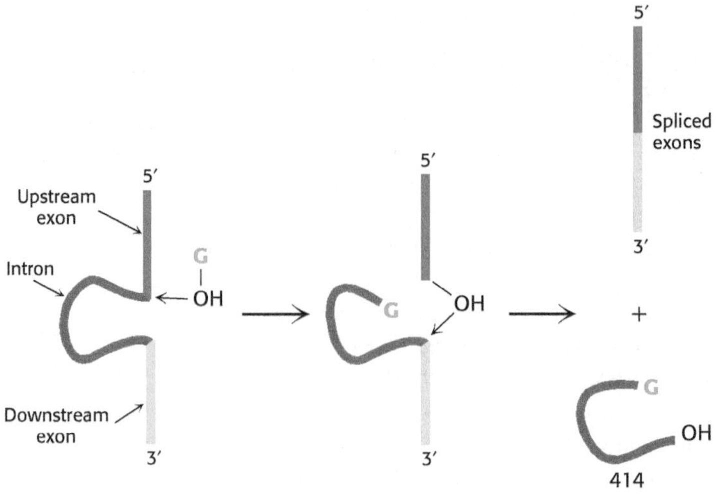

Figure 38.14 Self-splicing. A ribosomal RNA precursor from *Tetrahymena* splices itself in the presence of a guanosine cofactor (G, shown in green). An intron (red) is released in the first splicing reaction. This intron then splices itself twice again to produce a linear RNA. [Information from T. Cech, RNA as an enzyme. Copyright © 1986 by Scientific American, Inc. All rights reserved.]

phosphodiester linkage with the 5′ end of the intron, generating a 3′-OH group at the end of the upstream exon. This newly generated 3′-OH group then attacks the 3′ splice site, joining the two exons and releasing the intron. Self-splicing depends on the structural integrity of the RNA precursor. This molecule, like many RNAs, has a folded structure formed by many double-helical stems and loops (**Figure 38.15**), with a well-defined pocket for binding the guanosine.

Messenger RNA precursors in the mitochondria of yeast and fungi also undergo self-splicing, as do some RNA precursors in the chloroplasts of unicellular organisms such as *Chlamydomonas*. Self-splicing reactions can be classified according to the nature of the unit that attacks the upstream splice site. Group I self-splicing is mediated by a guanosine cofactor, as in *Tetrahymena*. The attacking moiety in group II splicing is the 2′-OH group of a specific adenylate of the intron.

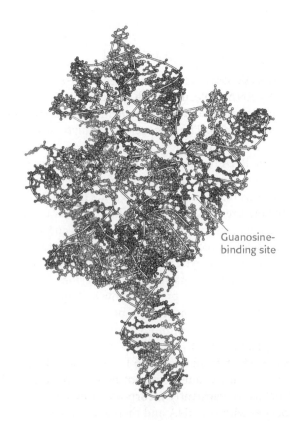

Figure 38.15 The structure of a self-splicing intron. The structure of a large fragment of the self-splicing intron from *Tetrahymena* reveals a complex folding pattern of helices and loops. Bases are shown in green, A; yellow, C; purple, G; and orange, U

SUMMARY

38.1 Mature Ribosomal RNA Is Generated by the Cleavage of a Precursor Molecule

RNA polymerase I transcribes the genes for three of the four rRNA molecules. The 18S RNA, 28S RNA, and 5.8S RNA are formed by the processing of an initial 45S transcript. In addition, the rRNAs are modified on the ribose and base components. Synthesis and processing take place in the nucleolus. The remaining rRNA, 5S RNA, is synthesized by RNA polymerase III.

38.2 Transfer RNA Is Extensively Processed

Eukaryotic tRNAs are the products of RNA polymerase III. Both the 5′ and the 3′ ends are cleaved, and CCA is added to the 3′ end. Introns, present in some pre-tRNA molecules, are removed by an endonuclease and a ligase. Additionally, bases and riboses are modified.

38.3 Messenger RNA Is Modified and Spliced

The 5′ ends of mRNA precursors become capped and methylated in the course of transcription. A 3′ poly(A) tail is added to most mRNA precursors after the nascent chain has been cleaved by an endonuclease.

The splicing of mRNA precursors is carried out by spliceosomes, which consist of small nuclear ribonucleoprotein particles. Splice sites in mRNA precursors are specified by sequences at the ends of introns and by branch sites near their 3′ ends. The 2′-OH group of an adenosine residue in the branch site attacks the 5′ splice site to form a lariat intermediate. The newly generated 3′-OH terminus of the upstream exon then attacks the 3′ splice site to become joined to the downstream exon. Splicing thus consists of two transesterification reactions, with the number of phosphodiester linkages remaining constant during reactions. Small nuclear RNAs in spliceosomes catalyze the splicing of mRNA precursors. In particular, U2 and U6 snRNAs form the active centers of spliceosomes.

The events in the posttranscriptional processing of mRNA are controlled by the phosphorylation state of the carboxyl-terminal domain, part of RNA polymerase II. RNA editing, which changes the mRNA nucleotide sequence, is another means of modifying mRNA.

38.4 RNA Can Function as a Catalyst

Some RNA molecules, such as those containing the group I intron, undergo self-splicing in the absence of protein. A self-modified version of this rRNA intron displays true catalytic activity and is thus a ribozyme.

KEY TERMS

exon (p. 691)
intron (p. 692)
small nucleolar ribonucleoprotein (snoRNP) (p. 692)
pre-mRNA (p. 693)
5′ cap (p. 693)

poly(A) tail (p. 693)
splicing (p. 694)
spliceosome (p. 695)
small nuclear RNA (snRNA) (p. 695)
small nuclear ribonucleoprotein (snRNP) particles (p. 695)

transesterification (p. 696)
alternative splicing (p. 697)
RNA editing (p. 698)
catalytic RNA (p. 699)
self-splicing (p. 699)

? Answer to QUICK QUIZ

U1 snRNP binds to the 5′ splice site, initiating spliceosome formation. U2 snRNP then binds at the branch site. The tri-snRNP U4-U5-U6 then joins, completing spliceosome formation. U5 interacts with the 5′ splice site and the 3′ exon. U6 disengages from U4 and pairs with U2. These rearrangements result in the cleavage of the 5′ exon and the formation of the lariat. U5 facilitates the alignment of the 5′ exon to attack the 3′ splice site, which forms the final product.

PROBLEMS

1. *Like Jordan and Pippen.* Match each term with its description. ✓ 4

(a) Exon _____
(b) Intron _____
(c) Pre-mRNA _____
(d) 5′ Cap _____
(e) Poly A tail _____
(f) Splicing _____
(g) Spliceosome _____
(h) snRNA _____
(i) snRNP _____
(j) Alternative splicing

   _____

1. Removed from initial transcript
2. Has a distinctive 5′–5′ linkage
3. A complex that coordinates two transesterification reactions
4. Allows one gene to encode several mRNAs
5. Removes introns and joins exons
6. Protein–RNA particles important for splicing
7. The initial product of RNA polymerase II
8. A 3′ posttranscriptional addition
9. Encodes part of the final mRNA
10. Catalytic component of the splicing machinery

2. *Nice hat.* Describe the 5′ cap.

3. *Tales of the end.* What are the steps required for the formation of a poly(A) tail on an mRNA molecule?

4. *Maturing.* What are the three most common modifications by which primary transcripts are converted into mature mRNA?

5. *A coupling device.* What structural component of RNA polymerase II that is lacking in RNA polymerases I and III allows modification of the newly synthesized RNA?

6. *Polymerase inhibition.* Cordycepin inhibits poly(A) synthesis at low concentrations and RNA synthesis at higher concentrations.

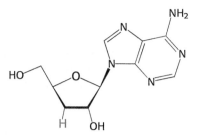

Cordycepin (3′-deoxyadenosine)

(a) What is the basis of inhibition by cordycepin?
(b) Why is poly(A) synthesis more sensitive to the presence of cordycepin?
(c) Does cordycepin need to be modified to exert its effect?

7. *Teamwork.* What is a spliceosome, and of what is it composed?

8. *Recruiting the right team.* What is the role of the carboxyl-terminal domain of RNA polymerase II in coupling RNA synthesis and splicing?

9. *From a few, many.* Explain why alternative splicing effectively enlarges the size of the genome.

10. *Alternative splicing.* A gene contains eight sites where alternative splicing is possible. With the assumption that the splicing pattern at each site is independent of those at all other sites, how many splicing products are possible? ✓ 4

11. *A good idea at the time.* George Beadle and Edward Tatum were awarded the Nobel Prize for their research on the relation of genes to enzymes. In a paper published in 1941, they proposed the "one gene, one enzyme hypothesis," which postulated that each gene encodes a single enzyme or, more generally, a single protein. Although this hypothesis was very influential, we now know that it is an oversimplification. Explain why.

12. *Searching for similarities.* Compare self-splicing with spliceosome-catalyzed splicing.

Chapter Integration Problems

13. *Proteome complexity.* What process considered in this chapter makes the proteome more complex than the genome? What processes might further enhance this complexity? ✓ 4

14. *Separation technique.* Suggest a means by which you could separate mRNA from the other types of RNA in a eukaryotic cell.

15. *A cell-cycle matter.* Histone genes differ from most genes in eukaryotes in several ways. The genes are arranged in repeated arrays, with each array containing a gene for each of the five histones. Moreover, the histone genes do not contain introns. Finally, messenger RNAs encoded by histone genes do not have poly(A) tails attached after transcription. Propose an explanation for these remarkable characteristics in regard to histone function.

Challenge Problems

16. *An extra piece.* In one type of mutation leading to a form of thalassemia, the mutation of a single base (G → A) generates a new 3′ splice site (blue in the adjoining illustration) like the normal one (yellow) but farther upstream.

Normal 3′ end of intron

5′ CCTATT**GG**TCTATTTTCCACCC**TTAG**GCTGCTG 3′

5′ CCT**ATTAG**TCTATTTTCCACCCTTAGGCTGCTG 3′

What is the amino acid sequence of the extra segment of protein synthesized in a thalassemic patient having a mutation leading to aberrant splicing? The reading frame after the splice site begins with TCT.

17. *A long-tailed messenger.* Another thalassemic patient had a mutation leading to the production of an mRNA for the β chain of hemoglobin that was 900 nucleotides longer than the normal one. The poly(A) tail of this mutant mRNA was located a few nucleotides after the only AAUAAA sequence in the additional sequence. Propose a mutation that would lead to the production of this altered mRNA.

Selected Readings for this chapter can be found online at www.whfreeman.com/tymoczko3e.

Protein Synthesis and Recombinant DNA Techniques

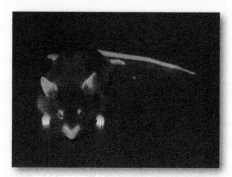

In Sections 14 and 15, we saw how genetic information is replicated and transcribed. We will complete our examination of the central dogma by investigating the last leg: *protein synthesis*, or *translation*. Protein synthesis is called translation because it is the biochemical process that translates nucleic acid information into amino-acid-sequence information. The genetic code is the Rosetta stone of the translation process, a code that correlates a sequence of three nucleotide bases, called a codon, with a particular amino acid. Translation is a crucial step in the central dogma because proteins perform virtually all biochemical functions, including the synthesis of DNA and RNA.

Befitting its position linking the nucleic acid and protein languages, the process of protein synthesis depends critically on both nucleic acid and protein factors. Protein synthesis takes place on *ribosomes*—enormous complexes containing three large RNA molecules and more than 50 proteins. Interestingly, *the ribosome is a ribozyme*; that is, the RNA components play the most fundamental roles.

Transfer RNA molecules, messenger RNA, and many proteins participate in protein synthesis along with ribosomes. The link between amino acids and nucleic acids is first made by enzymes called aminoacyl-tRNA synthetases. By specifically linking a particular amino acid to each tRNA, these enzymes perform the translation. Indeed, the enzymes that catalyze this crucial reaction are the actual translators of the genetic code.

We begin our examination of translation by first investigating the code that links the nucleic acid alphabet to the amino acid alphabet. Next, we will study some of the key players in the process of translation. We complete our study of translation by learning how all of the components cooperate to orchestrate protein synthesis in an accurate and controlled fashion.

In the final chapter of this section, and of the textbook itself, we complete our examination of the techniques of biochemistry, a task that we began in Chapter 5, with an investigation of recombinant DNA technology. Although only two chapters explore the tools of biochemistry, all that we have learned, and all that future students of biochemistry will learn, is knowledge revealed by the creative use of techniques like those that we examined in Chapter 5 and will examine in Chapter 41.

✓ By the end of this section, you should be able to:

✓ 1 Describe the genetic code.

✓ 2 Identify the step in protein synthesis in which translation takes place.

✓ 3 List the key components of the protein-synthesis machinery.

✓ 4 Describe the roles of RNA and proteins in protein synthesis.

✓ 5 Explain the role of the ribosome in protein synthesis.

✓ 6 Compare and contrast bacterial and eukaryotic protein synthesis.

✓ 7 Explain how protein synthesis can be regulated.

✓ 8 List the key tools of recombinant DNA technology, and explain how they are used to clone DNA.

CHAPTER 39

The Genetic Code

The Rosetta stone is inscribed with a decree issued by the Ptolemaic dynasty, which ruled Egypt from 305 B.C. to 30 B.C. The text is made up of three translations—Greek and two Egyptian scripts (hieroglyphics and demotic script)—of the same passage and enabled scholars to decipher hieroglyphics. The genetic code, deciphered by biochemists, allows the translation of nucleic acid information into protein sequence. [Mary Evans Picture Library Ltd/age fotostock.]

Many people spend hours weight training, with the goal of sculpting their bodies. Vigorous weight training can double or triple the size of a muscle. These people are altering their behavior so as to alter the biochemistry of their muscles; they are stimulating muscle-protein synthesis.

Protein synthesis is called *translation* because the four-letter alphabet of nucleic acids is translated into the entirely different twenty-letter alphabet of proteins. We begin our study of protein synthesis by examining the code that links these two alphabets and the machinery that is required to interpret the code.

✓ 1 Describe the genetic code.

39.1 The Genetic Code Links Nucleic Acid and Protein Information

For any sort of translation to take place, there must be a lexicon—a Rosetta stone—that links the two languages. The *genetic code* is the relation between the sequence of bases in DNA (or its RNA transcripts) and the sequence of amino acids in proteins. What are the characteristics of this code?

1. *Three nucleotides encode an amino acid.* Proteins are built from 20 amino acids, but there are only four bases in nucleic acids. Simple calculations show that a minimum of three bases is required to encode at least 20 amino acids. Genetic experiments showed that *an amino acid is in fact encoded by a group of three bases, called a codon.*

2. *The code is nonoverlapping.* Consider a base sequence ABCDEF. In an overlapping code, ABC specifies the first amino acid, BCD the next, CDE the next, and so on. In a nonoverlapping code, ABC designates the first amino acid, DEF the second, and so forth. Genetics experiments established the code to be nonoverlapping.

3. *The code has no punctuation.* In principle, one base (denoted as Q) might serve as a "comma" between codons:

   … QABCQDEFQGHIQJKLQ …

   However, it is not the case. Rather, *the sequence of bases is read sequentially from a fixed starting point without punctuation.*

4. *The genetic code has directionality.* The code is read from the 5′ end of the messenger RNA to its 3′ end.

5. *The genetic code is degenerate.* In the context of the genetic code, degeneracy means that some amino acids are encoded by more than one codon, inasmuch as there are 64 possible base triplets and only 20 amino acids. In fact, 61 of the 64 possible triplets specify particular amino acids and 3 triplets (called Stop codons) designate the termination of translation. Thus, *for most amino acids, there is more than one codon.* Codons that specify the same amino acid are called *synonyms.* For example, CAU and CAC are synonyms for histidine.

All 64 codons have been deciphered (Table 39.1). Only tryptophan and methionine are encoded by just one triplet each. The other 18 amino acids are each encoded by two or more. Indeed, leucine, arginine, and serine are specified by six codons each.

What is the biological significance of the extensive degeneracy of the genetic code? If the code were not degenerate, 20 codons would designate amino acids and 44 would lead to chain termination. The probability of mutating to chain termination would therefore be much higher with a nondegenerate code. Chain-termination mutations usually lead to inactive proteins. Degeneracy also allows for mutations that will not change the encoded amino acid, such as when a codon mutates to a synonym or mutates to the codon for another amino acid. The latter mutation is called a substitution, many of which are harmless. Thus, *degeneracy minimizes the deleterious effects of mutations.*

The Genetic Code Is Nearly Universal

Most organisms use the same genetic code. This universality accounts for the fact that human proteins, such as insulin, can be synthesized in the bacterium *E. coli* and harvested from it for the treatment of diabetes. However, genome-sequencing studies have shown that not all genomes are translated by the

Table 39.1   The genetic code

First position (5' end)	Second position				Third position (3' end)
	U	C	A	G	
U	Phe	Ser	Tyr	Cys	U
	Phe	Ser	Tyr	Cys	C
	Leu	Ser	Stop	Stop	A
	Leu	Ser	Stop	Trp	G
C	Leu	Pro	His	Arg	U
	Leu	Pro	His	Arg	C
	Leu	Pro	Gln	Arg	A
	Leu	Pro	Gln	Arg	G
A	Ile	Thr	Asn	Ser	U
	Ile	Thr	Asn	Ser	C
	Ile	Thr	Lys	Arg	A
	Met	Thr	Lys	Arg	G
G	Val	Ala	Asp	Gly	U
	Val	Ala	Asp	Gly	C
	Val	Ala	Glu	Gly	A
	Val	Ala	Glu	Gly	G

Note: This table identifies the amino acid encoded by each triplet. For example, the codon 5'-AUG-3' on mRNA specifies methionine, whereas CAU specifies histidine. UAA, UAG, and UGA are termination signals. AUG is part of the initiation signal, in addition to coding for internal methionine residues.

same code. Ciliated protozoa, for example, differ from most organisms in that UAA and UAG are read as codons for amino acids rather than as stop signals; UGA is their sole termination signal. The first variations in the genetic code were found in mitochondria from a number of species, including human beings (Table 39.2). The genetic code of mitochondria can differ from that of the rest of the cell because mitochondrial DNA encodes a distinct set of transfer RNAs, adaptor molecules that recognize the alternative codons. Thus, the genetic code is nearly but not absolutely universal.

Table 39.2   Distinctive codons of human mitochondria

Codon	Standard code	Mitochondrial code
UGA	Stop	Trp
UGG	Trp	Trp
AUA	Ile	Met
AUG	Met	Met
AGA	Arg	Stop
AGG	Arg	Stop

Transfer RNA Molecules Have a Common Design

The fidelity of protein synthesis requires the accurate recognition of three-base codons on messenger RNA. An amino acid itself is not structurally complex enough to recognize a codon. Consequently, some sort of adaptor is required. *Transfer RNA (tRNA) serves as the adapter molecule between the codon and its specified amino acid.* The tRNA acts as an adaptor by binding to a specific codon and brings with it an amino acid for incorporation into the polypeptide chain.

There is at least one tRNA molecule for each of the amino acids. These molecules have many common structural features, as might be expected because all tRNA molecules must be able to interact in nearly the same way with the ribosomes, mRNAs, and protein factors that participate in translation.

All known transfer RNA molecules have the following features:

1. Each is a single strand containing between 73 and 93 ribonucleotides (~25 kDa).

2. The three-dimensional molecule is L-shaped (**Figure 39.1**).

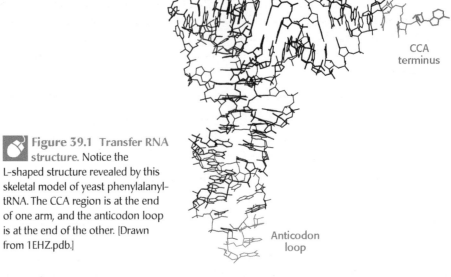

Inosine

ribose

5-Methylcytidine (mC)

ribose

Dihydrouridine (UH₂)

ribose

3. They contain many unusual bases, typically between 7 and 15 per tRNA. Some are methylated or dimethylated derivatives of A, U, C, and G. Methylation prevents the formation of certain base pairs, thereby rendering some of the bases accessible for interactions with other components of the translation machinery. In addition, methylation imparts a hydrophobic character to some regions of tRNAs, which may be important for their interaction with proteins required for protein synthesis. Modified bases, such as inosine, also are components of tRNA. The inosines in tRNA are formed by deamination of adenosine after the synthesis of the primary transcript.

4. When depicted on a two-dimensional surface, all tRNA molecules can be arranged in a cloverleaf pattern, with about half the nucleotides in tRNAs base-paired to form double helices (Figure 39.2). Five groups of bases are not base-paired in this way: the 3′ *CCA terminal region*, which is part of a region called the *acceptor stem*; the *TψC loop*, which acquired its name from

Figure 39.2 The general structure of transfer RNA molecules. The structure of the tRNA molecule is shown in the cloverleaf pattern. Comparison of the base sequences of many tRNAs reveals a number of conserved features.

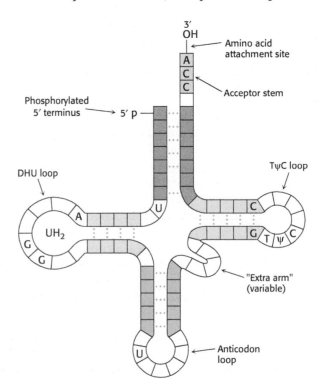

the sequence ribothymine-pseudouracil-cytosine; the *"extra arm,"* which contains a variable number of residues; the *DHU loop*, which contains several dihydrouracil residues; and the *anticodon loop*. The structural diversity generated by this combination of helices and loops containing modified bases ensures that the tRNAs can be uniquely distinguished, though structurally similar overall.

5. The 5′ end of a tRNA is phosphorylated. The 5′ terminal residue is usually pG.

6. The activated amino acid is attached to a hydroxyl group of the adenosine residue located at the end of the 3′ CCA component of the acceptor stem. This region is a flexible single strand at the 3′ end of mature tRNAs.

7. The *anticodon* is present in a loop near the center of the sequence.

Some Transfer RNA Molecules Recognize More Than One Codon Because of Wobble in Base-Pairing

What are the rules that govern the recognition of a codon by the anticodon of a tRNA? A simple hypothesis is that each of the bases of the codon forms a Watson–Crick type of base pair with a complementary base on the anticodon. The codon and anticodon would then be lined up in an antiparallel fashion. Recall that, by convention, nucleotide sequences are written in the 5′ → 3′ direction unless otherwise noted. Hence, the anticodon to AUG is written as CAU, but the actual base-pairing with the codon would be

$$
\begin{array}{c}
\textbf{Anticodon} \\
\overset{3'}{-\text{U}}-\text{A}-\overset{5'}{\text{C}}- \\
\overset{}{-\underset{5'}{\text{A}}}-\text{U}-\underset{3'}{\text{G}}- \\
\textbf{Codon}
\end{array}
$$

According to this model, a particular anticodon can recognize only one codon.

However, things are not so simple. *Some tRNA molecules can recognize more than one codon.* For example, consider the yeast alanyl-tRNA, with the anticodon IGC, where I is the nucleoside inosine. This anticodon binds to *three* codons: GCU, GCC, and GCA. The first two bases of these codons are the same, whereas the third is different. Could it be that recognition of the third base of a codon is sometimes less discriminating than recognition of the other two? The pattern of degeneracy of the genetic code indicates that it might be so. Look again at Table 39.1. Generally speaking, XYU and XYC always encode the same amino acid; XYA and XYG usually do. These data suggest that the steric criteria might be less stringent for pairing of the third base than for the other two. In other words, there is some steric freedom ("*wobble*") in the pairing of the third base of the codon. With this steric freedom, tRNA anticodons can bond to mRNA codons as shown in Table 39.3.

Table 39.3 Allowed pairings at the third base of the codon according to the wobble hypothesis

First base of anticodon	Third base of codon
C	G
A	U
U	A or G
G	U or C
I	U, C, or A

Two generalizations concerning the codon–anticodon interaction can be made:

1. The first two bases of a codon pair in the standard way. Recognition is precise. Hence, *codons that differ in either of their first two bases must be recognized by different tRNAs.* For example, both UUA and CUA encode leucine but are read by different tRNAs.

2. The first base of an anticodon determines whether a particular tRNA molecule reads one, two, or three kinds of codons: C or A (one codon), U or G (two codons), or I (three codons). Thus, *part of the degeneracy of the genetic code*

arises from imprecision in the pairing of the third base of the codon with the first base of the anticodon. We see here a strong reason for the frequent appearance of inosine, one of the unusual nucleosides, in anticodons. *Inosine maximizes the number of codons that can be read by a particular tRNA molecule.*

The Synthesis of Long Proteins Requires a Low Error Frequency

The process of transcription is analogous to copying, word for word, a page from a book. There is no change of alphabet or vocabulary; so the likelihood of a change in meaning is small. Translating the base sequence of an mRNA molecule into a sequence of amino acids is similar to translating the page of a book into another language. Translation is a complex process, entailing many steps and dozens of molecules. The potential for error exists at each step. The complexity of translation creates a conflict between two requirements: the process must be not only accurate, but also fast enough to meet a cell's needs. How fast is "fast enough"? In *E. coli*, translation can take place at a rate of 40 amino acids per second, a truly impressive speed considering the complexity of the process.

How accurate must protein synthesis be? The average *E. coli* protein is about 300 amino acids in length, with several dozen greater than 1000 amino acids. Let us consider possible error rates when synthesizing proteins of this size. As Table 39.4 shows, an error frequency of 10^{-2} (one incorrect amino acid for every 100 correct ones incorporated into a protein) would be intolerable, even for small proteins. An error value of 10^{-3} would usually lead to the error-free synthesis of a 300-residue protein ($\sim$33 kDa) but not of a 1000-residue protein ($\sim$110 kDa). Thus, the error frequency must not exceed approximately 10^{-4} to produce the larger proteins effectively. Lower error frequencies are conceivable; however, except for the largest proteins, they will not dramatically increase the percentage of proteins with accurate sequences. In addition, such lower error rates are likely to be possible only by a reduction in the rate of protein synthesis because additional time for proofreading will be required. *In fact, the observed error values are close to 10^{-4}.* An error frequency of about 10^{-4} per amino acid residue was selected in the course of evolution to accurately produce proteins consisting of as many as 1000 amino acids while maintaining a remarkably rapid rate for protein synthesis.

Table 39.4   Accuracy of protein synthesis

Frequency of inserting an incorrect amino acid	Probability of synthesizing an error-free protein		
	Number of amino acid residues		
	100	300	1000
10^{-2}	0.366	0.049	0.000
10^{-3}	0.905	0.741	0.368
10^{-4}	0.990	0.970	0.905
10^{-5}	0.999	0.997	0.990

Note: The probability p of forming a protein with no errors depends on n, the number of amino acids, and ε, the frequency of insertion of a wrong amino acid; $p = (1 - \varepsilon)^n$.

✓ 2 Identify the step in protein synthesis in which translation takes place.

39.2   Amino Acids Are Activated by Attachment to Transfer RNA

Before codon and anticodon meet, the amino acids required for protein synthesis must first be attached to specific tRNA molecules, linkages that are crucial for two reasons. *First, the attachment of a given amino acid to a particular tRNA establishes the genetic code.* When an amino acid has been linked to a tRNA, it will be incorporated into a growing polypeptide chain at a position dictated by the anticodon of the tRNA. *Second, because the formation of a peptide bond between free amino acids is not thermodynamically favorable, the amino acid must first be activated in order for protein synthesis to proceed.* The activated intermediates in protein synthesis are amino acid esters, in which the carboxyl group of an amino acid is linked to either the 2′- or the 3′-hydroxyl group of the ribose unit at the 3′ end of tRNA. An amino acid ester of tRNA is called an *aminoacyl-tRNA* or sometimes a *charged tRNA* (**Figure 39.3**). For a specific amino acid attached to its cognate tRNA—for instance, threonine—the charged tRNA is designated Thr-tRNA^{Thr}.

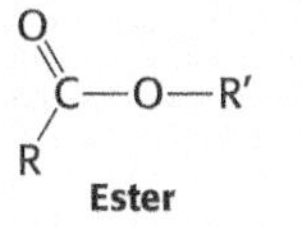

Amino Acids Are First Activated by Adenylation

The activation of amino acids is catalyzed by specific *aminoacyl-tRNA synthetases.* The first step is the formation of an *aminoacyl adenylate* from an amino acid and ATP. In this molecule, the carboxyl group of the amino acid is linked to the phosphoryl group of AMP; hence, it is also known as *aminoacyl-AMP.*

$$\text{Amino acid} + \text{ATP} \rightleftharpoons \text{aminoacyl-AMP} + \text{PP}_i$$

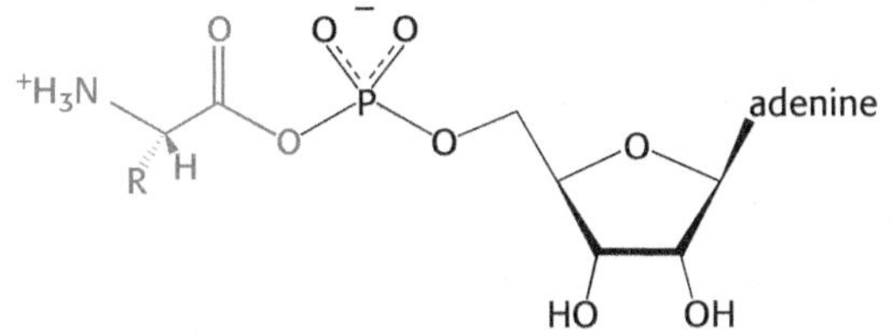

Aminoacyl adenylate

The next step is the transfer of the aminoacyl group of aminoacyl-AMP to a particular tRNA molecule to form *aminoacyl-tRNA*:

$$\text{Aminoacyl-AMP} + \text{tRNA} \rightleftharpoons \text{aminoacyl-tRNA} + \text{AMP}$$

The sum of the activation and transfer steps is

$$\text{Amino acid} + \text{ATP} + \text{tRNA} \rightleftharpoons \text{aminoacyl-tRNA} + \text{AMP} + \text{PP}_i$$

The $\Delta G°'$ of this reaction is close to zero because the free energy of hydrolysis of the ester bond of aminoacyl-tRNA is similar to that for the hydrolysis of ATP to AMP and PP$_i$. As we have seen many times, the reaction is driven by the hydrolysis of pyrophosphate. The sum of these three reactions is highly exergonic:

$$\text{Amino acid} + \text{ATP} + \text{tRNA} + \text{H}_2\text{O} \longrightarrow \text{aminoacyl-tRNA} + \text{AMP} + 2\,\text{P}_i$$

Thus, *the equivalent of two molecules of ATP are consumed in the synthesis of each aminoacyl-tRNA.*

The activation and transfer steps for a particular amino acid are catalyzed by the same aminoacyl-tRNA synthetase. Indeed, *the aminoacyl-AMP intermediate does not dissociate from the synthetase.* Rather, it is tightly bound to the active site of the enzyme by noncovalent interactions. Translation takes place with the formation of the ester linkage between an amino acid and a specific tRNA. Thus aminoacyl-tRNA synthetases are the actual translators of the genetic code.

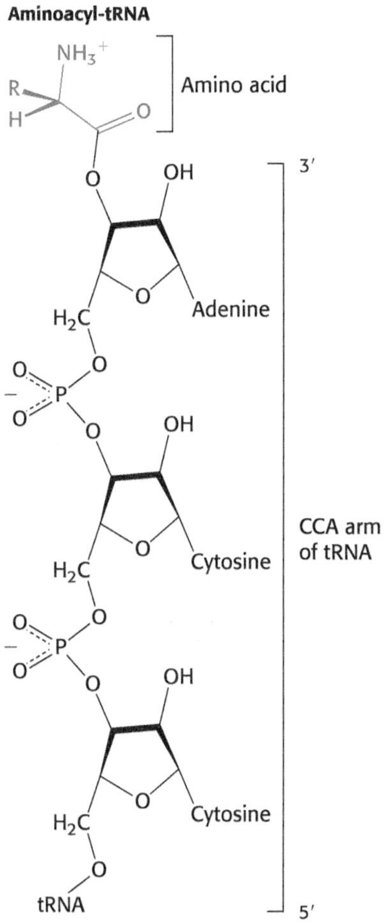

Ester

Figure 39.3 Aminoacyl-tRNA. Amino acids are coupled to tRNAs through ester linkages to either the 2′- or the 3′-hydroxyl group of the 3′-adenosine residue. A linkage to the 3′-hydroxyl group is shown.

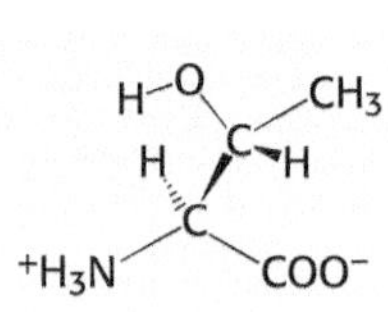

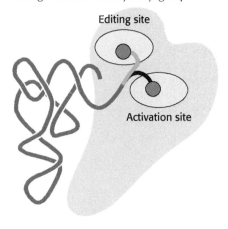

Threonine

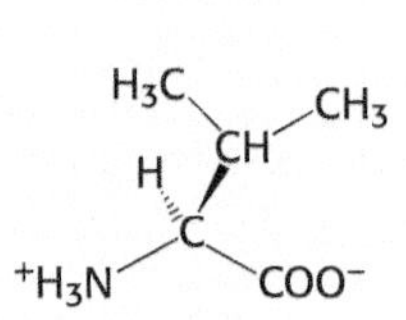

Valine

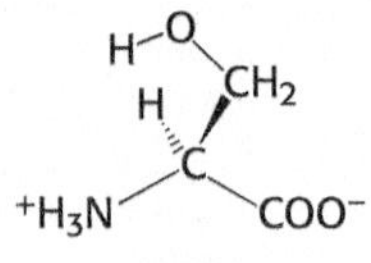

Serine

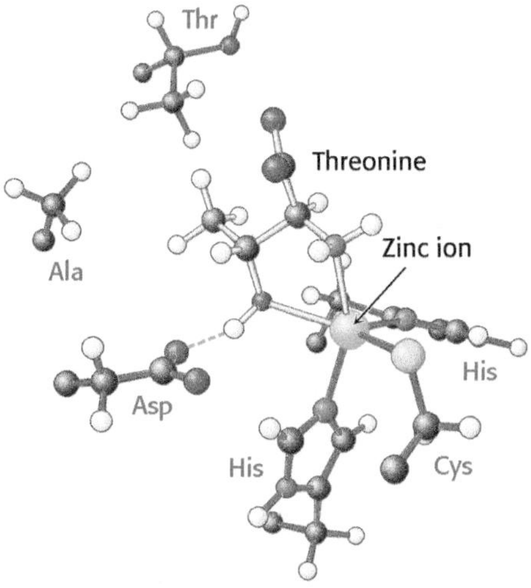

Figure 39.4 The active site of threonyl-tRNA synthetase. Notice that the site for binding the amino acid includes a zinc ion (green ball) that coordinates threonine through its amino and hydroxyl groups.

Figure 39.5 The editing of aminoacyl-tRNA. The flexible CCA arm of an aminoacyl-tRNA can move the amino acid between the activation site and the editing site. If the amino acid fits well into the editing site, the amino acid is removed by hydrolysis.

Aminoacyl-tRNA Synthetases Have Highly Discriminating Amino Acid Activation Sites

For protein synthesis to be accurate, the correct amino acid must bond to the correct tRNA. Each aminoacyl-tRNA synthetase is highly specific for a given amino acid. Indeed, a synthetase will incorporate the incorrect amino acid only once in 10^4 or 10^5 reactions. How is this level of specificity achieved? Each aminoacyl-tRNA synthetase takes advantage of the properties of its amino acid substrate. Let us consider the challenge faced by threonyl-tRNA synthetase, as an example. Threonine is similar to two other amino acids—namely, valine and serine. Valine has almost exactly the same shape as that of threonine, except that it has a methyl group in place of a hydroxyl group. Like threonine, serine has a hydroxyl group but lacks the methyl group. How can the threonyl-tRNA synthetase prevent the coupling of these amino acids to threonyl-tRNA (abbreviated tRNA^{Thr})?

The structure of threonyl-tRNA synthetase's binding site for the amino acid reveals how coupling with valine is prevented (**Figure 39.4**). The enzyme's active site contains a zinc ion. Threonine binds to the zinc ion through its amino group and its side-chain hydroxyl group. The side-chain hydroxyl group is further recognized by an aspartate residue to which it hydrogen bonds. The methyl group present in valine in place of this hydroxyl group cannot participate in these interactions; it will not bind at the active site and, hence, does not become adenylated and transferred to threonyl-tRNA.

The zinc site is less well suited to discrimination against serine because this amino acid does have a hydroxyl group that can bind to the zinc. Indeed, with only this mechanism available, threonyl-tRNA synthetase does mistakenly couple serine to threonyl-tRNA at a rate 10^{-2} to 10^{-3} times that for threonine. As noted (p. 712), this error rate is likely to lead to many translation errors. How is the formation of Ser-tRNA^{Thr} prevented?

Proofreading by Aminoacyl-tRNA Synthetases Increases the Fidelity of Protein Synthesis

In addition to its active site, threonyl-tRNA synthetase has an editing site located 20 Å from the active site. This site accepts serine attached to tRNA^{Thr} and hydrolyzes the bond between serine and tRNA^{Thr}, providing the synthetase an opportunity to correct its mistakes and improve its accuracy to less than one mistake in 10^4. The CCA arm with serine attached can swing out of the activation site and into the editing site (**Figure 39.5**). Thus, the aminoacyl-tRNA can be edited without dissociating from the synthetase.

What prevents the hydrolysis of Thr-tRNA^{Thr}? The discrimination of serine from threonine is relatively easy because threonine contains an *extra* methyl group, and this methyl group prevents Thr-tRNA^{Thr} from fitting into the editing site. Most aminoacyl-tRNA synthetases contain editing sites in addition to acylation sites. These complementary pairs of sites function as a *double sieve* to ensure very high fidelity. In general, the active site rejects amino acids that are larger than the correct one because there is insufficient room for them, whereas the editing site cleaves activated species that are *smaller* than the correct one.

Synthetases Recognize the Anticodon Loops and Acceptor Stems of Transfer RNA Molecules

How do synthetases choose their tRNA partners? This enormously important step is the point at which "translation" takes place—at which the correlation between the amino acid and the nucleic acid vocabularies is made. In a sense, aminoacyl-tRNA synthetases are the only molecules in biology that "know" the genetic code. Their precise recognition of tRNAs is as important for high-fidelity protein synthesis as is the accurate selection of amino acids.

A priori, the anticodon of tRNA would seem to be a good identifier of the appropriate tRNA partner for a synthetase because each type of tRNA has a different one. Indeed, some synthetases recognize their tRNA partners primarily on the basis of their anticodons, although they may also recognize other aspects of tRNA structure. The most direct evidence for how recognition takes place comes from crystallographic studies of complexes formed between synthetases and their cognate tRNAs. **Figure 39.6** shows the aspects of the tRNA molecule that are recognized by tRNA synthetases. Note that many of the recognition sites are loops rich in unusual bases that can provide structural identifiers.

QUICK QUIZ 1 Why is the synthesis of aminoacyl–tRNA the crucial step in protein synthesis?

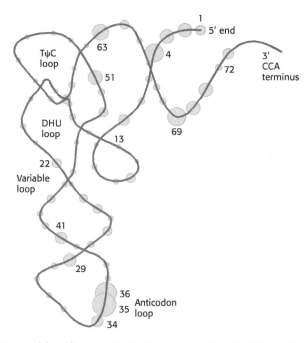

Figure 39.6 Recognition sites on tRNA. Circles represent nucleotides, and the sizes of the circles are proportional to the frequency with which they are used as recognition sites by aminoacyl-tRNA synthetases. The numbers indicate the positions of the nucleotides in the base sequence, beginning from the 5' end of the tRNA molecule. [Information from M. Ibba and D. Söll, *Annu. Rev. Biochem.* 69:617–650, 1981, p. 636.]

39.3 A Ribosome Is a Ribonucleoprotein Particle Made of Two Subunits

We now turn to ribosomes, the molecular machines that coordinate the interplay of aminoacyl-tRNAs, mRNA, and proteins that results in protein synthesis. An *E. coli* ribosome is a ribonucleoprotein assembly with a mass of about 2.7 MDa, a diameter of approximately 250 Å, and a sedimentation coefficient (p. 78) of 70S. The 20,000 ribosomes in a bacterial cell constitute nearly a fourth of its mass.

A *ribosome* can be dissociated into a *large subunit* (50S) and a *small subunit* (30S) (**Figure 39.7**). These subunits can be further dissociated into their constituent proteins and RNAs. The small subunit contains 21 different proteins (referred to as S1 through S21) and a 16S RNA molecule. The large subunit contains 34 different proteins (L1 through L34) and two RNA molecules, a 23S and a 5S species.

Ribosomal RNAs Play a Central Role in Protein Synthesis

The prefix *ribo* in the name *ribosome* is apt, because RNA constitutes nearly two-thirds of the mass of these large molecular assemblies. The three RNAs present—5S, 16S, and 23S—are critical for ribosomal architecture and function. These

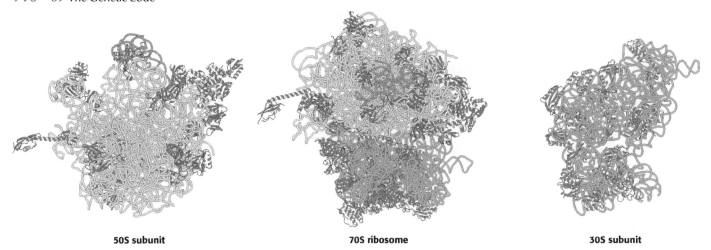

50S subunit   **70S ribosome**   **30S subunit**

Figure 39.7 The ribosome at high resolution. Detailed models of the ribosome based on the results of x-ray crystallographic studies of the 70S ribosome and the 30S and 50S subunits. (Left) View of the part of the 50S subunit that interacts with the 30S subunit; (center) side view of the 70S ribosome; (right) view of the part of the 30S subunit that interacts with the 50S subunit. 23S RNA is shown in yellow, 5S RNA in orange, 16S RNA in green, proteins of the 50S subunit in red, and proteins of the 30S subunit in blue. Notice that the interface between the 50S and the 30S subunits consists entirely of RNA. [Drawn from 1GIX.pdb and 1GIY.pdb.]

ribosomal RNAs (rRNAs) are folded into complex structures with many short duplex regions (**Figure 39.8**).

For many years, the protein components of ribosomes were thought to orchestrate protein synthesis, with the RNA components serving primarily as structural scaffolding. But this idea raised a perplexing evolutionary "chicken-and-egg" question: Namely, how can complex proteins be synthesized if complex proteins are required for protein synthesis? The discovery of catalytic RNA (p. 699) made biochemists receptive to the possibility that RNA plays a much more active role in ribosomal function. Detailed structural analyses make it clear that the key catalytic sites in the ribosome are composed almost entirely of RNA. Contributions from the proteins are minor. The almost inescapable conclusion is that, early in the evolution of life, the ribosome initially consisted only of RNA and that the proteins were added later to fine-tune its functional properties. This conclusion has the pleasing consequence of dodging the chicken-and-egg question.

Messenger RNA Is Translated in the 5′-to-3′ Direction

The sequence of amino acids in a protein is translated from the nucleotide sequence in mRNA, and *the direction of translation is 5′ → 3′*. The direction of translation has significant consequences. Recall that transcription also is in the 5′ → 3′ direction (p. 660). If the direction of translation were opposite that of transcription, only fully synthesized mRNA could be translated. In contrast, because the directions are the same, mRNA can be translated while it is being synthesized. In bacteria, almost no time is lost between transcription and translation. The 5′ end of mRNA interacts with ribosomes very soon after it is made, much before the 3′ end of the mRNA molecule is finished. *A key feature of bacterial gene expression is that translation and transcription are closely coupled in space and time.* Many ribosomes can be translating an mRNA molecule simultaneously. This parallel synthesis markedly increases the efficiency of mRNA translation. The group of ribosomes bound to an mRNA molecule is called a *polyribosome* or a *polysome* (**Figure 39.9**).

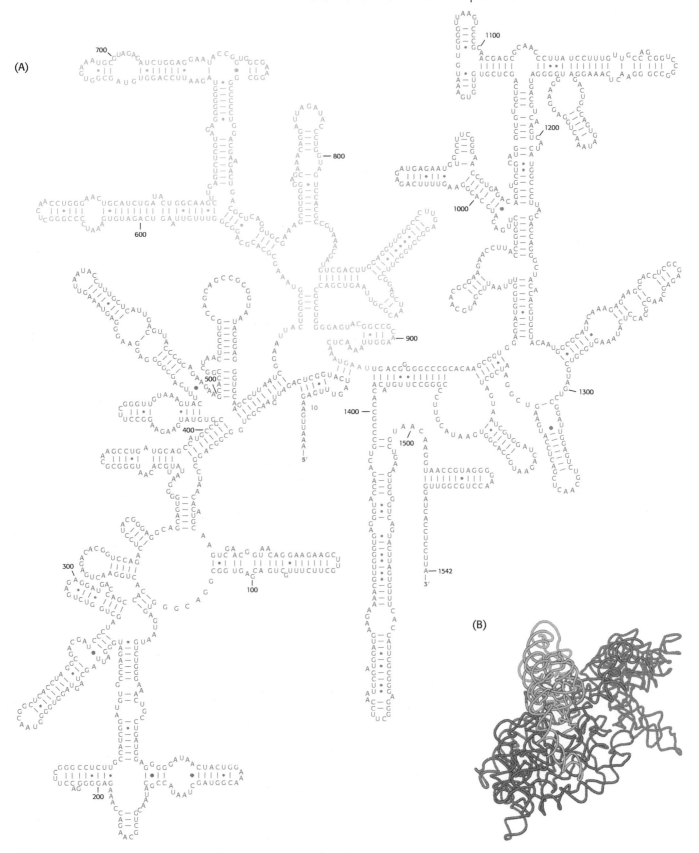

Figure 39.8 Ribosomal RNA folding pattern. (A) The secondary structure of 16S ribosomal RNA deduced from sequence comparison and the results of chemical studies. (B) The tertiary structure of 16S RNA determined by x-ray crystallography. [(A) Courtesy of Dr. Bryn Weiser and Dr. Harry Noller; (B) drawn from 1FJG.pdb.]

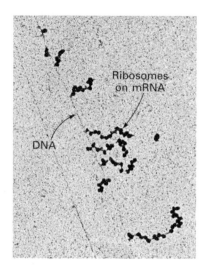

Figure 39.9 Polysomes. Transcription of a segment of DNA from *E. coli* generates mRNA molecules that are immediately translated by multiple ribosomes. [From O. L. Miller, Jr., et al., Visualization of bacterial genes in action, *Science* 169:392–395, 1970. *Reprinted with permission from AAAS.*]

? QUICK QUIZ 2 What are the chief components required for protein synthesis that are discussed in this chapter?

SUMMARY

39.1 The Genetic Code Links Nucleic Acid and Protein Information

Protein synthesis is called translation because information present as a nucleic acid sequence is translated into a different language, the sequence of amino acids in a protein. The genetic code is the relation between the sequence of bases in DNA (or its RNA transcript) and the sequence of amino acids in proteins. Amino acids are encoded by groups of three bases (called codons) starting from a fixed point. Sixty-one of the 64 codons specify particular amino acids, whereas the other 3 codons are signals for chain termination. Thus, for most amino acids, there is more than one code word. In other words, the code is degenerate. The genetic code is nearly the same in all organisms. Messenger RNAs contain start and stop signals for translation, just as genes do for directing where transcription begins and ends.

Translation is mediated by the coordinated interplay of more than a hundred macromolecules, including mRNA, rRNAs, tRNAs, aminoacyl-tRNA synthetases, and protein factors. Given that proteins typically consist of 100 to 1000 amino acids, the frequency at which an incorrect amino acid is incorporated in the course of protein synthesis must be less than 10^{-4}. Transfer RNAs are the adaptors that make the link between a nucleic acid and an amino acid. These molecules, single strands of about 80 nucleotides, have an L-shaped structure.

39.2 Amino Acids Are Activated by Attachment to Transfer RNA

Each amino acid is activated and linked to a specific transfer RNA by an enzyme called an aminoacyl-tRNA synthetase. Such an enzyme links the carboxyl group of an amino acid to the 2′- or 3′-hydroxyl group of the adenosine unit of a CCA sequence at the 3′ end of the tRNA by an ester linkage. There is at least one specific aminoacyl-tRNA synthetase and at least one specific tRNA for each amino acid. A synthetase utilizes both the functional groups and the shape of its cognate amino acid to prevent the attachment of the incorrect amino acid to a tRNA. Some synthetases have a separate editing site at which incorrectly linked amino acids are removed by hydrolysis. A synthetase recognizes the anticodon, the acceptor stem, and sometimes other parts of its tRNA substrate. By specifically recognizing both amino acids and tRNAs, aminoacyl-tRNA synthetases implement the instructions of the genetic code.

39.3 A Ribosome Is a Ribonucleoprotein Particle Made of Two Subunits

Protein synthesis takes place on ribosomes—ribonucleoprotein particles (about two-thirds RNA and one-third protein) consisting of large and small subunits. In *E. coli*, the 70S ribosome (2.7 MDa) is made up of 30S and 50S subunits. The 30S subunit consists of 16S ribosomal RNA and 21 different proteins; the 50S subunit consists of 23S and 5S rRNA and 34 different proteins.

Proteins are synthesized in the amino-to-carboxyl direction, and the mRNA is read by the ribosome in the 5′ → 3′ direction. Because RNA is transcribed in the 5′ → 3′ direction also, the translation of mRNA in bacteria can begin before transcription is complete.

KEY TERMS

translation (p. 707)
genetic code (p. 708)
codon (p. 708)
transfer RNA (tRNA) (p. 709)

anticodon loop (p. 711)
wobble (p. 711)
aminoacyl-tRNA synthetase (p. 713)
ribosome (p. 715)

50S subunit (p. 715)
30S subunit (p. 715)
ribosomal RNA (rRNA) (p. 716)
polysome (p. 716)

? Answers to QUICK QUIZZES

1. Aminoacyl-tRNA molecules are the actual substrates for protein synthesis. Furthermore, the synthesis of the aminoacyl-tRNA is the step at which the translation of nucleic acid information into amino acid information takes place.

2. Ribosomes, tRNA, mRNA, and aminoacyl-tRNA synthetases.

PROBLEMS

1. *Minimal code.* What is the minimum number of contiguous bases required to encode 20 amino acids? Explain your answer. ✓ 1

2. *Nearly universal.* Why has the code remained nearly invariant through billions of years of evolution, from bacteria to human beings? ✓ 1

3. *Like Pooh and Christopher Robin.* Match each term with its description.

(a) Translation _____
(b) Genetic code _____
(c) Codon _____
(d) tRNA _____
(e) Anticodon loop _____
(f) Wobble _____
(g) Aminoacyl-tRNA synthetase _____
(h) Ribosome _____
(i) 50S subunit _____
(j) 30S subunit _____

1. Code reader
2. Pairing freedom
3. Protein synthesis
4. Three nucleotides
5. Contains 16S RNA
6. Interacts with the codon
7. Carrier of activated amino acids
8. Contains 28S and 5S RNA
9. Site of protein synthesis
10. Nucleic acid information to protein information

4. *A code you can live by.* What are the key characteristics of the genetic code? ✓ 1

5. *Degeneracy.* Usually degeneracy should be avoided, except perhaps in well-controlled circumstances. However, it is advantageous for the genetic code to be degenerate. What does it mean to say that the code is degenerate, and explain why a degenerate code is valuable? ✓ 1

6. *Odds of getting it right.* Calculate the probability of synthesizing an error-free protein of 50 amino acids and one of 300 amino acids when the frequency of inserting an incorrect amino acid is 10^{-2}. Repeat the calculations with error frequencies of 10^{-4} and 10^{-6}.

7. *Careful, but not too careful.* Why is it crucial that protein synthesis has an error frequency of 10^{-4}?

8. *Going wobbly.* Explain how it is possible that some tRNA molecules recognize more than one codon. ✓ 1

9. *Encoded sequences.* (a) Write the sequence of the mRNA molecule synthesized from a DNA template strand having the following sequence. ✓ 1

5'-ATCGTACCGTTA-3'

(b) What amino acid sequence is encoded by the following base sequence of an mRNA molecule? Assume that the reading frame starts at the 5' end.

5'-UUGCCUAGUGAUUGGAUG-3'

(c) What is the sequence of the polypeptide formed on addition of poly(UUAC) to a cell-free protein-synthesizing system that does not require a start codon?

10. *Deciphering the code, 1.* Synthetic RNA molecules of defined sequence were instrumental in deciphering the genetic code. Their synthesis first required the synthesis of DNA molecules to serve as a template. H. Gobind Khorana synthesized, by organic-chemical methods, two complementary deoxyribonucleotides, each with nine residues: d(TAC)₃ and d(GTA)₃. Partly overlapping duplexes that formed on mixing these oligonucleotides then served as templates for the synthesis by DNA polymerase of long, repeating double-helical DNA strands. The next step was to obtain long polyribonucleotide strands with a sequence complementary to only one of the two DNA strands. How did Khorana obtain only poly(UAC)? Only poly(GUA)? ✓ 1

11. *Deciphering the code, 2.* The code word GGG cannot be deciphered in the same way as can UUU, CCC, and AAA, because poly(G) does not act as a template for protein synthesis. Poly(G) forms a triple-stranded helical structure. Why is it an ineffective template? ✓ 1

12. *Overlapping or not.* In a nonoverlapping triplet code, each group of three bases in a sequence ABCDEF . . . specifies only one amino acid—ABC specifies the first, DEF the second, and so forth—whereas, in a completely overlapping triplet code, ABC specifies the first amino acid, BCD the second, CDE the third, and so forth. Assume that you can mutate an individual nucleotide of a codon and detect the mutation in the amino acid sequence. Design an experiment that would establish whether the genetic code is overlapping or nonoverlapping. ✓ 1

13. *Triple entendre.* The RNA transcript of a region of T4 phage DNA contains the sequence 5′-AAAUGAGGA-3′. In theory, this sequence is capable of encoding three different polypeptides. What are they? ✓ 1

14. *Valuable synonyms.* Proteins generally have low contents of Met and Trp, intermediate ones of His and Cys, and high ones of Leu and Ser. What does this observation suggest about the relation between the number of codons of an amino acid and its frequency of occurrence in proteins? What might be the selective advantage of this relation?

15. *A new translation.* A transfer RNA with a UGU anticodon is enzymatically conjugated to ^{14}C-labeled cysteine. The cysteine unit is then chemically modified to alanine. The altered aminoacyl-tRNA is added to a protein-synthesizing system containing normal components except for this tRNA. The mRNA added to this mixture contains the following sequence:

5′-UUUUGCCAUGUUUGUGCU-3′

What is the sequence of the corresponding radiolabeled peptide?

16. *Commonalities.* What features are common to all tRNA molecules? ✓ 2

17. *The ol' two step.* What two reaction steps are required for the formation of an aminoacyl-tRNA? ✓ 2

18. *The same but different.* Why must tRNA molecules have both unique structural features and common structural features? ✓ 2

19. *Charge it.* In the context of protein synthesis, what is meant by an activated amino acid? ✓ 2

20. *1 = 2, for sufficiently large values of 1.* The energetic equivalent of two molecules of ATP is used to activate an amino acid, yet only one molecule of ATP is used. Explain.

21. *Sieves.* Using threonyl-tRNA synthetase as an example, account for the specificity of threonyl-tRNA formation. ✓ 2

22. *Knowledge rich.* Aminoacyl-tRNA synthetases are the only components of gene expression that know the genetic code. Explain. ✓ 2

Challenge Problems

23. *A tougher strand.* RNA is readily hydrolyzed by alkali, whereas DNA is not. Why?

24. *Synthetase mechanism.* The formation of isoleucyl-tRNA proceeds through the reversible formation of an enzyme-bound Ile-AMP intermediate. Predict whether ^{32}P-labeled ATP is formed from ^{32}PP$_i$ when each of the following sets of components is incubated with the specific activating enzyme. ✓ 2

(a) ATP and ^{32}PP$_i$

(b) tRNA, ATP, and ^{32}PP$_i$

(c) Isoleucine, ATP, and ^{32}PP$_i$

25. *Finding direction.* A series of experiments were performed to establish the direction of chain growth in protein synthesis. Reticulocytes (young red blood cells) that were actively synthesizing hemoglobin were treated with [^{3}H]leucine. In a period of time shorter than that required to synthesize a complete chain, samples of hemoglobin were taken, separated into α and β chains, and analyzed for the distribution of ^{3}H within their sequences. In the earliest samples, only regions near the carboxyl ends contained radioactivity. In later samples, radioactivity was present closer to the amino terminus as well. Explain how these results determine the direction of chain growth in protein synthesis.

26. *Mission possible.* Your mission, should you accept it, is to determine the direction in which the mRNA is read during protein synthesis. You are provided with synthetic polynucleotide

$$\overset{5'}{A}—A—A\!\!-\!\!\left(A—A—A\right)_{\!n}\!\!-\!\!A\quad A\overset{3'}{—C}$$

as the template in a cell-free protein-synthesizing system, which is also kindly provided to you. The polypeptide product was

$$^{+}H_3N—Lys—(Lys)_{\!n}—Asn—\overset{\displaystyle O}{\underset{\displaystyle O}{C}}—$$

What does the nature of this product say about the direction in which the mRNA is read? ✓ 2

Selected Readings for this chapter can be found online at www.whfreeman.com/tymoczko3e.

The Mechanism of Protein Synthesis

Henry Ford developed the assembly-line approach to manufacturing in the early 1900s. The assembly line allows the rapid and precise production of many items, including automobiles. Living systems have used this approach for billions of years for protein synthesis, where even very long polypeptide chains can be assembled rapidly and with impressive accuracy. [Mary Evans Picture Library/The Image Works.]

The synthesis of proteins is an energetically expensive, complicated process consisting of three parts: initiation, elongation, and termination. In initiation, the translation machinery must locate the correct Start codon. In elongation, the codons are read sequentially in the 5'-to-3' direction as the protein is synthesized from the amino end to the carboxyl end. When a termination codon is reached, special proteins hydrolyze the polypeptide from the last transfer RNA, releasing the completed protein. The translation machinery comes apart and is ready to begin another round of protein synthesis.

We begin our consideration of protein synthesis by examining this process in bacteria. We then compare protein synthesis in bacteria and in eukaryotes. Finally, we consider how protein synthesis is affected by environmental factors, such as inhibitors and cell structure, and how the process is regulated.

✓ 3 List the key components of the protein-synthesis machinery.
✓ 4 Describe the roles of RNA and proteins in protein synthesis.

40.1 Protein Synthesis Decodes the Information in Messenger RNA

The first step in protein synthesis is initiation. Protein-synthesis initiation requires the cooperation of the ribosome, tRNA, mRNA, and various protein factors. We begin our consideration of this process with an examination of the tRNA-binding sites on ribosomes.

Ribosomes Have Three tRNA-Binding Sites That Bridge the 30S and 50S Subunits

The three tRNA-binding sites on ribosomes are arranged to allow the sequential formation of peptide bonds between amino acids encoded by the codons on mRNA (**Figure 40.1**). The mRNA fragment being translated at a given moment is bound within the 30S subunit. Each of the tRNA molecules is in contact with both the *30S subunit* and the *50S subunit*. At the 30S end, two of the three tRNA molecules are bound to the mRNA through anticodon–codon base pairs. These binding sites are called the A site (for *a*minoacyl) and the P site (for *p*eptidyl). The third tRNA molecule is bound to an adjacent site called the E site (for *e*xit).

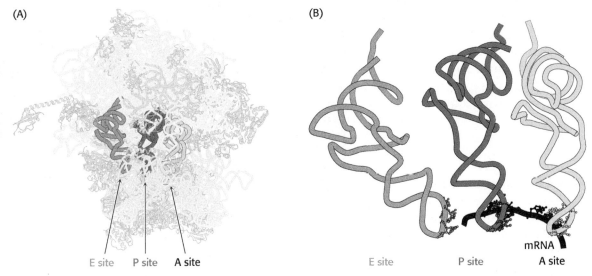

(A)

(B)

E site   P site   A site

E site   P site   A site   mRNA

Figure 40.1 Binding sites for transfer RNA. (A) Three tRNA molecules are shown in the tRNA-binding sites on the 70S ribosome. They are called the A (for aminoacyl), P (for peptidyl), and E (for exit) sites. Each tRNA molecule contacts both the 30S and the 50S subunit. (B) The tRNA molecules in sites A and P are base-paired with mRNA. [(B) Drawn from 1JGP.pdb.]

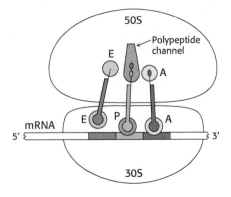

Figure 40.2 An active ribosome. This schematic representation shows the relations among the key components of the translation machinery.

The other end of each tRNA molecule interacts with the 50S subunit. The acceptor stems of the tRNA molecules occupying the A site and the P site converge at a site where a peptide bond is formed. A tunnel connects this site to the back of the ribosome, through which the polypeptide chain passes during synthesis (**Figure 40.2**).

The Start Signal Is AUG Preceded by Several Bases That Pair with 16S Ribosomal RNA

How does protein synthesis start? In other words, how are the codons appropriately positioned in the A and P sites? The simplest possibility would be for the first three nucleotides of each mRNA to serve as the first codon; no special start signal would then be needed. However, the experimental fact is that translation in bacteria does not begin immediately at the 5′ terminus of mRNA. Indeed, the first translated codon is nearly always more than 25 nucleotides away from the 5′ end. Furthermore, in bacteria, many mRNA molecules are *polycistronic*—that is, they

```
5'                                          3'
AGCACGAGGGGAAAUCUGAUGGAACGCUAC    E. coli trpA

UUUGGAUGGAGUGAAACGAUGGCGAUUGCA    E. coli araB

GGUAACCAGGUAACAACCAUGCGAGUGUUG    E. coli thrA

CAAUUCAGGGUGGUGAAUGUGAAACCAGUA    E. coli lacI

AAUCUUGGAGGCUUUUUUUAUGGUUCGUUCU   φX174 phage A protein

UAACUAAGGAUGAAAUGCAUGUCUAAGACA    Qβ phage replicase

UCCUAGGAGGUUUGACCUAUGCGAGCUUUU    R17 phage A protein

AUGUACUAAGGAGGUUGUAUGGAACAACGC    λ phage cro
```

Pairs with   Pairs with
16S rRNA   initiator tRNA

Figure 40.3 Initiation sites. Sequences of mRNA initiation sites for protein synthesis in some bacterial and viral mRNA molecules. Comparison of these sequences reveals some recurring features.

encode two or more polypeptide chains. Recall that, in the *lac* operon, the mRNA product encodes three enzymes (p. 668). Each of these three proteins has its own start and stop signals on mRNA. In fact, *all known mRNA molecules contain signals that define the beginning and the end of each encoded polypeptide chain.*

A clue to the mechanism of initiation was the finding that nearly half the amino-terminal residues of proteins in *E. coli* are methionine. In fact, the initiating codon in mRNA is usually AUG (methionine). However, the initiating codon alone is not up to the task of protein-synthesis initiation. In addition, each initiator region contains a purine-rich sequence, called the *Shine–Dalgarno sequence* (named after John Shine and Lynn Dalgarno, who first described the sequence), centered about 10 nucleotides on the 5′ side of the initiator codon (**Figure 40.3**). Nucleotide sequences in mRNA, such as the Shine–Dalgarno sequence, that are not translated are called *untranslated regions* (UTRs) and may be at the 5′ end of the mRNA (5′-UTR) or at the 3′ end (3′-UTR). Untranslated regions usually function to regulate the usage of mRNA molecules.

The Shine–Dalgarno sequence interacts with a complementary sequence on the 3′ end of the 16S rRNA component of the ribosome's small subunit. Thus, *two kinds of interactions determine where protein synthesis starts in bacteria: (1) the pairing of mRNA bases with the 3′ end of 16S rRNA and (2) the pairing of the initiator codon on mRNA with the anticodon of an initiator tRNA molecule.*

Bacterial Protein Synthesis Is Initiated by Formylmethionyl Transfer RNA

As stated earlier, methionine is the first amino acid in many *E. coli* proteins. However, the methionine residue found at the amino-terminal end of *E. coli* proteins is usually modified. In fact, *protein synthesis in bacteria starts with the modified amino acid N-formylmethionine* (fMet). A special tRNA brings formylmethionine to the ribosome to initiate protein synthesis. This *initiator tRNA* (abbreviated as tRNA$_f$) differs from the tRNA that inserts methionine in internal positions (abbreviated as tRNA$_m$). The subscript "f" indicates that methionine attached to the initiator tRNA can be formylated, whereas it cannot be formylated when attached to tRNA$_m$. Although virtually all proteins synthesized in *E. coli* begin with formylmethionine, in approximately one-half of the proteins, *N*-formylmethionine is removed when the nascent chain is 10 amino acids long.

Methionine is linked to these two kinds of tRNAs by the same aminoacyl-tRNA synthetase. A specific enzyme then formylates the amino group of the methionine molecule that is attached to tRNA$_f$ (**Figure 40.4**). The activated formyl donor in this reaction is *N*^{10}-formyltetrahydrofolate, a folate derivative that carries activated one-carbon units (p. 576). Free methionine and methionyl-tRNA$_m$ are not substrates for this transformylase.

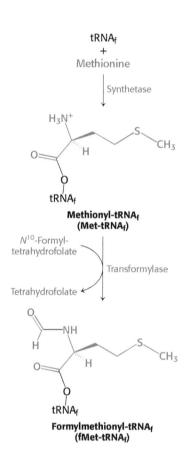

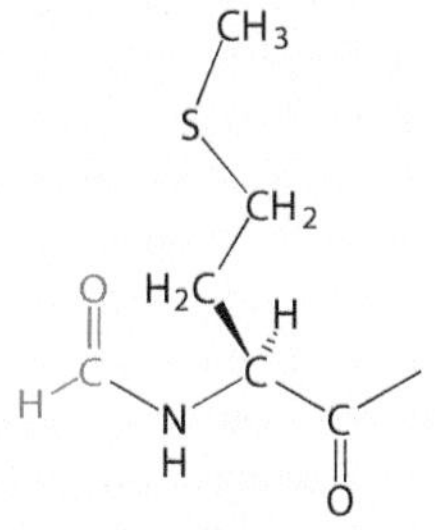

Figure 40.4 Formylation of methionyl-tRNA. Initiator tRNA (tRNA$_f$) is first charged with methionine, and then a formyl group is transferred to the methionyl-tRNA$_f$ from *N*^{10}-formyltetrahydrofolate.

Formylmethionyl-tRNA$_f$ Is Placed in the P Site of the Ribosome in the Formation of the 70S Initiation Complex

Messenger RNA and formylmethionyl-tRNA$_f$ must be brought to the ribosome for protein synthesis to begin. How is this task accomplished? Three protein *initiation factors* (IF1, IF2, and IF3) are essential. The 30S ribosomal subunit first forms a complex with IF1 and IF3 (**Figure 40.5**). The binding of these factors to the 30S subunit prevents it from prematurely joining the 50S subunit to form a dead-end 70S complex, devoid of mRNA and fMet-tRNA$_f$. Initiation factor 2, a GTPase, binds GTP, and the concomitant conformational change enables IF2 to associate with formylmethionyl-tRNA$_f$. The IF2–GTP–initiator-tRNA complex binds with mRNA (correctly positioned by the interaction of the Shine–Dalgarno sequence with the 16S rRNA) and the 30S subunit to form the *30S initiation complex*. Structural changes then lead to the ejection of IF1 and IF3. IF2 stimulates the association of the 50S subunit to the complex. The GTP bound to IF2 is hydrolyzed upon arrival of the 50S subunit, releasing IF2. The result is a *70S initiation complex*. The formation of the 70S initiation complex is the rate-limiting step in protein synthesis.

When the 70S initiation complex has been formed, the ribosome is ready for the elongation phase of protein synthesis. The fMet-tRNA$_f$ molecule occupies the P site on the ribosome, positioned so that its anticodon pairs with the initiating codon on mRNA. The other two sites for tRNA molecules, the A site and the E site, are empty. This interaction establishes the *reading frame* for the translation of the entire mRNA. After the initiator codon has been located, groups of three nonoverlapping nucleotides are defined and subsequently translated as protein synthesis takes place.

Elongation Factors Deliver Aminoacyl-tRNA to the Ribosome

At this point, fMet-tRNA$_f$ occupies the P site, and the A site is vacant, as stated. The particular species inserted into the empty A site depends on the mRNA codon in the A site. However, the appropriate aminoacyl-tRNA does not simply leave the synthetase and diffuse to the A site. Rather, it is delivered to the A site in association with a 43-kDa protein called *elongation factor Tu* (EF-Tu), which requires GTP for activity. Elongation factor Tu, the most abundant bacterial protein, binds aminoacyl-tRNA only in the GTP form (**Figure 40.6**). The binding of EF-Tu to aminoacyl-tRNA serves two functions. First, EF-Tu protects the delicate ester linkage in aminoacyl-tRNA from hydrolysis. Second, EF-Tu contributes to the accuracy of protein synthesis because GTP hydrolysis and expulsion of the EF-Tu-GDP complex from the ribosome occur only if the pairing between the

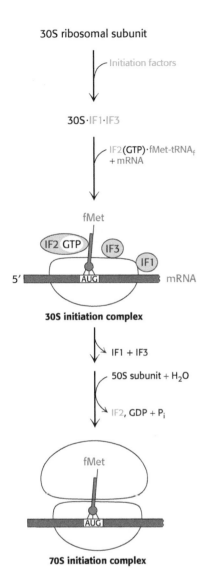

30S ribosomal subunit

Initiation factors

30S·IF1·IF3

IF2 (**GTP**)·fMet-tRNA$_f$ + mRNA

fMet

IF2 GTP   IF3   IF1

5′   AUG   mRNA

30S initiation complex

IF1 + IF3

50S subunit + H$_2$O

IF2, GDP + P$_i$

fMet

AUG

70S initiation complex

Figure 40.5 Translation initiation in bacteria. Initiation factors aid the assembly, first, of the 30S initiation complex and, then, of the 70S initiation complex.

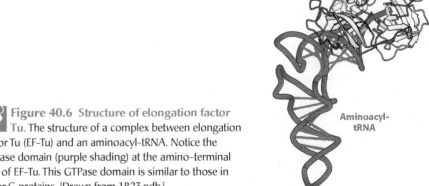

Figure 40.6 Structure of elongation factor Tu. The structure of a complex between elongation factor Tu (EF-Tu) and an aminoacyl-tRNA. Notice the GTPase domain (purple shading) at the amino-terminal end of EF-Tu. This GTPase domain is similar to those in other G proteins. [Drawn from 1B23.pdb.]

anticodon and the codon is correct. EF-Tu interacts with the 16S RNA. Correct codon recognition induces structural changes in the 30S subunit that activates the GTPase activity of EF-Tu, releasing EF-Tu-GDP from the ribosome. These same structural changes also rotate the aminoacyl-tRNA in the A site so that the amino acid is brought into proximity with the aminoacyl-tRNA in the P site, a process called *accommodation*. Accommodation aligns the amino acids for peptide bond formation.

Released EF-Tu is then reset to its GTP form by a second elongation factor, *elongation factor Ts*. EF-Ts induces the dissociation of GDP. GTP binds to EF-Tu, and EF-Ts is concomitantly released. It is noteworthy that *EF-Tu does not interact with fMet-tRNA$_f$*. Hence, this initiator tRNA is not delivered to the A site. In contrast, Met-tRNA$_m$, like all other aminoacyl-tRNAs, does bind to EF-Tu. These findings account for the fact that *internal AUG codons are not read by the initiator tRNA*. Conversely, IF2 recognizes fMet-tRNA$_f$ but no other tRNA. The cycle of elongation continues until a termination codon is met.

40.2 Peptidyl Transferase Catalyzes Peptide-Bond Synthesis

✓ 5 Explain the role of the ribosome in protein synthesis.

With both the P site and the A site occupied by aminoacyl-tRNA, the stage is set for the formation of a peptide bond: the formylmethionine molecule linked to the initiator tRNA will be transferred to the amino group of the amino acid in the A site. The formation of the peptide bond, one of the most important reactions in life, is a thermodynamically spontaneous reaction catalyzed by a site on the 23S rRNA of the 50S subunit called the *peptidyl transferase center*. This catalytic center is located deep in the 50S subunit near the tunnel that allows the nascent peptide to leave the ribosome.

The ribosome, which enhances the rate of peptide bond synthesis by a factor of 10^7 over the uncatalyzed reaction ($\sim 10^{-4}$ M^{-1} s^{-1}), derives much of its catalytic power from *catalysis by proximity and orientation*. The ribosome positions and orients the two substrates so that they are situated to take advantage of the inherent reactivity of an amine group (on the aminoacyl-tRNA in the A site) with an ester (on the initiator tRNA in the P site).

The reaction begins when the amino group of the aminoacyl-tRNA in the A site, in its unprotonated state, makes a nucleophilic attack on the ester linkage between the initiator tRNA and the formylmethionine molecule in the P site (**Figure 40.7**A). The nature of the transition state that follows the attack is not established, and several models are plausible. One model proposes roles for the 2′OH of the adenosine of the tRNA in the P site and a molecule of water at the peptidyl transferase center (Figure 40.7B). The nucleophilic attack of the α-amino group generates an eight-membered transition state in which three protons are shuttled about in a concerted manner. The proton of the attacking amino group hydrogen bonds to the 2′ oxygen of ribose of the tRNA. The hydrogen of 2′OH in turn interacts with the oxygen of the water molecule at the center, which then donates a proton to the carbonyl oxygen. Collapse of the transition state with the formation of the peptide bond allows protonation of the 3′OH of the now-empty tRNA in the P site (Figure 40.7C). The stage is now set for translocation and formation of the next peptide bond.

The Formation of a Peptide Bond Is Followed by the GTP-Driven Translocation of tRNAs and mRNA

With the formation of the peptide bond, the peptide chain is now attached to the tRNA whose anticodon is in the A site on the 30S subunit. The two ribosomal subunits rotate with respect to one another, and this structural change places the CCA end of the same tRNA and its peptide in the P site of the large

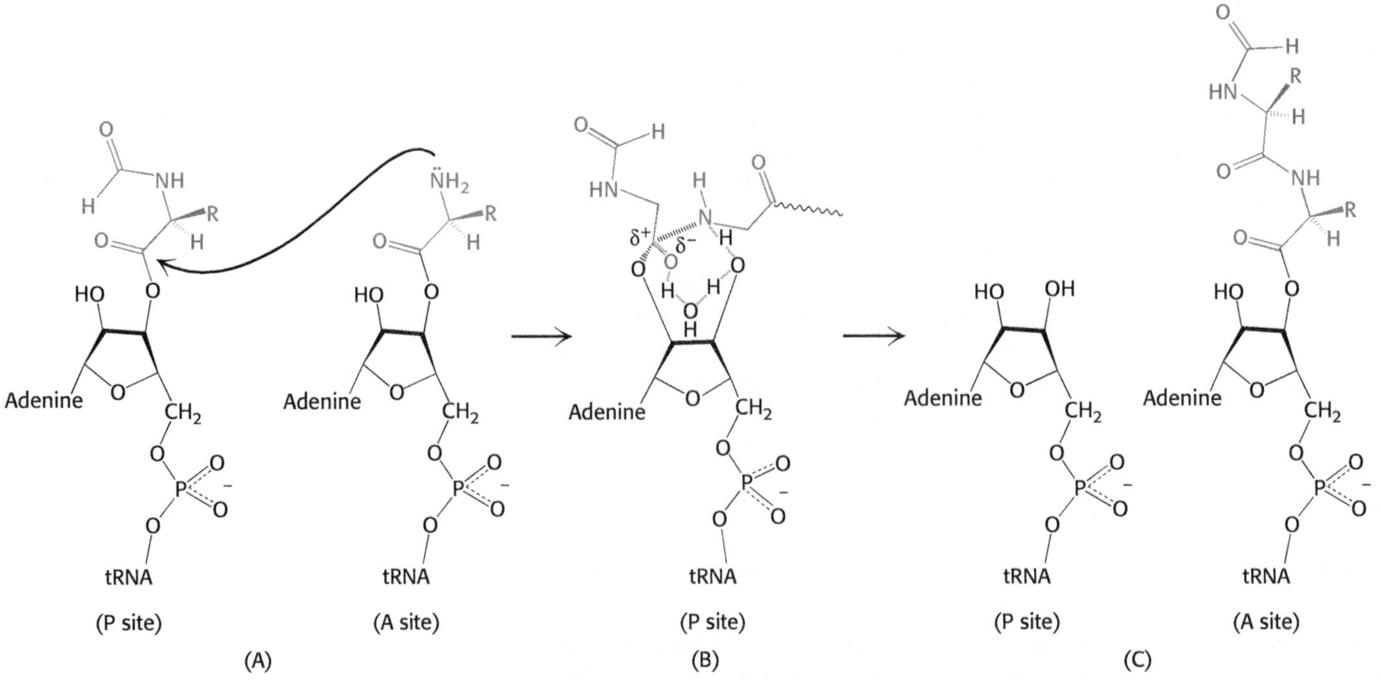

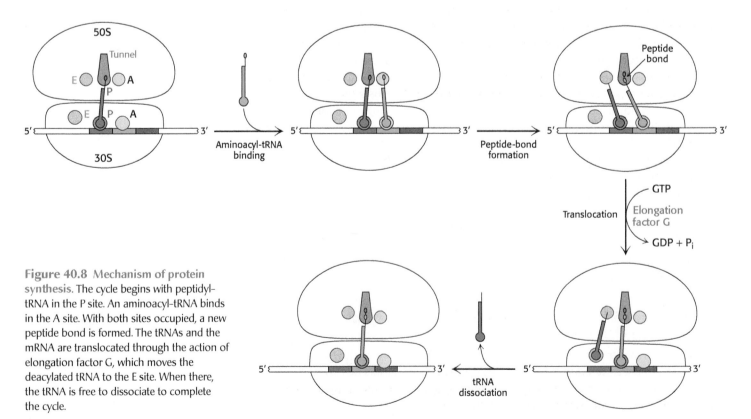

Figure 40.7 Peptide-bond formation. (A) The amino group of the aminoacyl–tRNA attacks the carbonyl group of the ester linkage of the peptidyl–tRNA. (B) An eight-membered transition state is formed. Note: Not all atoms are shown, and some bond lengths are exaggerated for clarity. (C) This transition state collapses to form the peptide bond and release the deacylated tRNA.

subunit (**Figure 40.8**). However, protein synthesis cannot continue without the translocation of the mRNA and the tRNAs within the ribosome. The mRNA must move by a distance of three nucleotides so that the next codon is positioned in the A site for interaction with the incoming aminoacyl-tRNA. The peptidyl-tRNA moves out of the A site into the P site on the 30S subunit and at the same time, the

Figure 40.8 Mechanism of protein synthesis. The cycle begins with peptidyl–tRNA in the P site. An aminoacyl-tRNA binds in the A site. With both sites occupied, a new peptide bond is formed. The tRNAs and the mRNA are translocated through the action of elongation factor G, which moves the deacylated tRNA to the E site. When there, the tRNA is free to dissociate to complete the cycle.

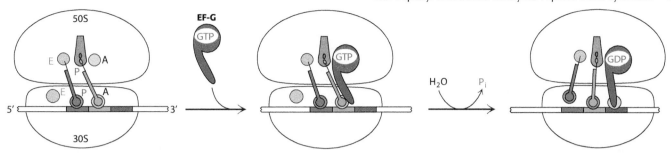

Figure 40.9 Translocation mechanism. In the GTP form, EF-G binds to the EF-Tu-binding site on the 50S subunit. This binding stimulates GTP hydrolysis, inducing a conformational change in EF-G that forces the tRNAs and the mRNA to move through the ribosome by a distance corresponding to one codon.

deacylated tRNA moves out of the P site into the E site and is subsequently released from the ribosome. The movement of the peptidyl-tRNA into the P site shifts the mRNA by one codon, exposing the next codon to be translated in the A site.

The three-dimensional structure of the ribosome undergoes significant change during translocation, and evidence suggests that translocation may result from properties of the ribosome itself. However, protein factors accelerate the process. Translocation is enhanced by *elongation factor G* (EF-G, also called *translocase*). A possible mechanism for accelerating the translocation process is shown in **Figure 40.9**. First, EF-G in the GTP form binds to the ribosome near the A site, interacting with the 23S rRNA of the 50S subunit. The binding of EF-G to the ribosome stimulates the GTPase activity of EF-G. On GTP hydrolysis, EF-G undergoes a conformational change that displaces the peptidyl-tRNA in the A site to the P site, which carries the mRNA and the deacylated tRNA with it. The dissociation of EF-G leaves the ribosome ready to accept the next aminoacyl-tRNA into the A site.

Note that *the peptide chain remains in the P site on the 50S subunit throughout this cycle*, growing into the exit tunnel. This cycle is repeated as new aminoacyl-tRNAs move into the A site, allowing the polypeptide to be elongated indefinitely. Keep in mind that all of this activity is taking place 40 times per second. Note that *the polypeptide is synthesized in the amino-terminal-to-carboxyl-terminal direction* (Figure 40.10). Many ribosomes can be translating an mRNA molecule simultaneously, forming a *polyribosome*, or a *polysome* (Figure 39.9).

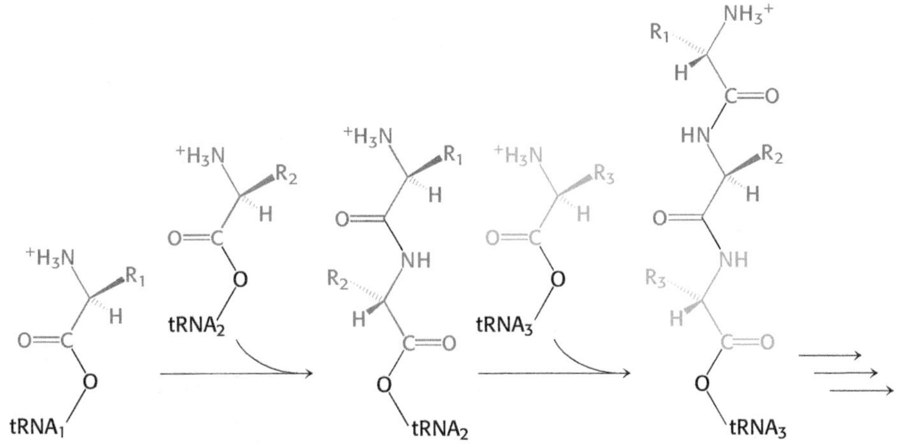

Figure 40.10 Polypeptide-chain growth. Proteins are synthesized by the successive addition of amino acids to the carboxyl terminus.

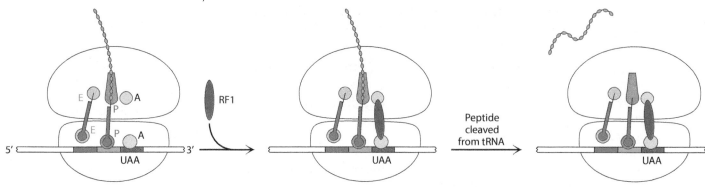

Figure 40.11 Termination of protein synthesis. A release factor recognizes a Stop codon in the A site and stimulates the release of the completed protein from the tRNA in the P site.

Protein Synthesis Is Terminated by Release Factors That Read Stop Codons

The final phase of translation is termination. How does the synthesis of a polypeptide chain come to an end when a Stop codon is encountered? No tRNAs with anticodons complementary to the Stop codons—UAA, UGA, or UAG—exist in normal cells. Instead, these *Stop codons are recognized by proteins called release factors* (RFs). One of these release factors, RF1, recognizes UAA or UAG. A second factor, RF2, recognizes UAA or UGA. A third factor, RF3, another GTPase, catalyzes the removal of RF1 or RF2 from the ribosome upon release of the newly synthesized protein.

RF1 and RF2 are compact proteins that, when bound to the ribosome, unfold to bridge the gap between the Stop codon on the mRNA and the peptidyl transferase center on the 50S subunit (**Figure 40.11**). The RF interacts with the peptidyl transferase center with a loop containing a highly conserved glycine-glycine-glutamine (GGQ) sequence, with the glutamine methylated on the amide nitrogen atom of the R group. This modified glutamine is crucial in promoting, assisted by the peptidyl transferase, a water molecule's attack on the ester linkage between the tRNA and the polypeptide chain, freeing the polypeptide chain. The detached polypeptide leaves the ribosome. Transfer RNA and messenger RNA remain briefly attached to the 70S ribosome until the entire complex is dissociated through the hydrolysis of GTP in response to the binding of EF-G and another factor, called the *ribosome release factor* (RRF).

? QUICK QUIZ 1 What are the roles of the protein factors required for protein synthesis?

✓ 6 Compare and contrast bacterial and eukaryotic protein synthesis.

40.3 Bacteria and Eukaryotes Differ in the Initiation of Protein Synthesis

The basic plan of protein synthesis in eukaryotes and archaea is similar to that in bacteria. The major structural and mechanistic themes recur in all domains of life. However, eukaryotic protein synthesis requires more protein components than does bacterial protein synthesis, and some steps are more complicated. Some noteworthy similarities and differences are as follows:

1. *Ribosomes.* Eukaryotic ribosomes are larger. They consist of a 60S large subunit and a 40S small subunit, which come together to form an 80S complex having a mass of 4.2 MDa, compared with 2.7 MDa for the bacterial 70S ribosome. The 40S subunit contains an 18S RNA that is homologous to the bacterial 16S RNA. The 60S subunit contains three RNAs: the 5S RNA, which is homologous to the bacterial 5S rRNA; the 28S RNA, which is homologous to the bacterial 23S molecule; and the 5.8S RNA, which is homologous to the 5′ end of the bacterial 23S RNA.

2. *Initiator tRNA.* In eukaryotes, the initiating amino acid is methionine rather than *N*-formylmethionine. However, as in bacteria, a special tRNA participates

in initiation. This aminoacyl-tRNA is called Met-tRNA$_i$ or Met-tRNA$_f$ (the subscript "i" stands for initiation, and "f" indicates that it can be formylated in vitro).

3. *Initiation.* The initiating codon in eukaryotes is always AUG. Unlike bacteria, eukaryotes do not have a specific purine-rich sequence on the 5′ side to distinguish initiator AUGs from internal ones. Instead, the AUG nearest the 5′ end of mRNA is usually selected as the start site. Eukaryotes utilize many more initiation factors than do bacteria, and their interplay is much more complex. The prefix *eIF* denotes a eukaryotic initiation factor.

   Initiation begins with the formation of a ternary complex consisting of the 40S ribosome and Met-tRNA$_i$ in association with eIF-2. The complex is called the 43S preinitiation complex (PIC). The PIC binds to the 5′ end of mRNA and begins searching for an AUG codon by moving step-by-step in the 3′ direction. Initiation factor eIF-4E binds to the 5′ cap of the mRNA (p. 693) and facilitates binding of PIC to the mRNA (**Figure 40.12**). This scanning process is catalyzed by helicases that move along the mRNA powered by ATP hydrolysis. Pairing of the anticodon of Met-tRNA$_i$ with the AUG codon of mRNA signals that the target has been found. In almost all cases, eukaryotic mRNA has only one start site and, hence, is the template for a single protein. In contrast, a bacterial mRNA can have multiple Shine–Dalgarno sequences and, hence, multiple start sites, and it can serve as a template for the synthesis of several proteins.

   The difference in initiation mechanisms between bacteria and eukaryotes is, in part, a consequence of the difference in RNA processing. The 5′ end of mRNA is readily available to ribosomes immediately after transcription in bacteria. In contrast, in eukaryotes pre-mRNA must be processed and transported to the cytoplasm before translation is initiated. The 5′ cap provides an easily recognizable starting point.

4. *The Structure of mRNA.* Eukaryotic mRNA is circular (**Figure 40.13**). Soon after the PIC binds the mRNA, eIF-4G links eIF-4E to a protein associated with the poly (A) tail, the poly (A)-binding protein I (PABPI). Cap and tail are thus brought together to form a circle of mRNA. The benefits of circularization of mRNA remain to be determined, but a requirement for circularization may prevent translation of mRNA molecules that have lost their poly A tails.

5. *Elongation and Termination.* Eukaryotic elongation factors EF1α and EF1$\beta\gamma$ are the counterparts of bacterial EF-Tu and EF-Ts, whereas eukaryotic EF2 corresponds to the EF-G (translocase) in bacteria. Termination in eukaryotes is carried out by a single release factor, eRF1, compared with two in bacteria. Initiation factor eIF-3 accelerates the activity of eRF-1.

6. *Organization.* The components of the translation machinery in higher eukaryotes are organized into large complexes associated with the cytoskeleton. This association is believed to facilitate the efficiency of protein synthesis. Recall that the organization of elaborate biochemical processes into physical complexes is a recurring theme in biochemistry.

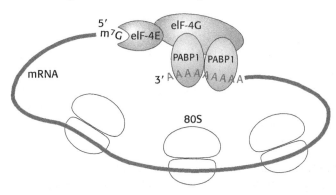

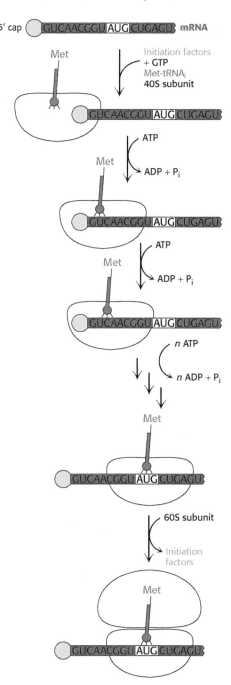

Figure 40.12 Eukaryotic translation initiation. In eukaryotes, translation initiation starts with the assembly of a complex on the 5′ cap that includes the 40S subunit and Met-tRNA$_i$. Driven by ATP hydrolysis, this complex scans the mRNA until the first AUG is reached. The 60S subunit is then added to form the 80S initiation complex.

Figure 40.13 Protein interactions circularize eukaryotic mRNA. [Information from H. Lodish et al., *Molecular Cell Biology*, 6th ed. (W. H. Freeman and Company, 2008), Fig. 4.28.]

Mutations in Initiation Factor 2 Cause a Curious Pathological Condition

Mutations in eukaryotic initiation factor 2 result in a mysterious disease, called vanishing white matter (VWM) disease, in which nerve cells in the brain disappear and are replaced by cerebrospinal fluid (**Figure 40.14**). The white matter of the brain consists predominately of nerve axons that connect the gray matter of the brain to the rest of the body. Death, resulting from fever or extended coma, is anywhere from a few years to decades after the onset of the disease. An especially puzzling aspect of the disease is its tissue specificity. A mutation in a biochemical process as fundamental to life as protein-synthesis initiation would be predicted to be lethal or to at least affect all tissues of the body. Diseases such as VWM graphically show that, although much progress has been made in biochemistry, much more research will be required to understand the complexities of health and disease.

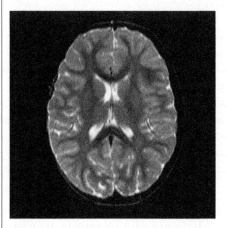
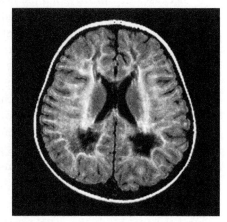

Figure 40.14 The effects of vanishing white matter disease. (A) In the normal brain, magnetic resonance imaging (MRI) visualizes the white matter as dark gray. (B) In the diseased brain, MRI reveals that white matter is replaced by cerebrospinal fluid, seen as white. [(A) Du Cane Medical Imaging Ltd./Science Source. (B) Reprinted from M.C. van der Knaap, et al., Vanishing white matter disease, *Lancet Neurology* 5:413–423, ©2006, with permission from Elsevier.]

40.4 A Variety of Biomolecules Can Inhibit Protein Synthesis

Many chemicals that inhibit various aspects of protein synthesis have been identified. These chemicals are powerful experimental tools and clinically useful drugs.

Some Antibiotics Inhibit Protein Synthesis

The differences between eukaryotic and bacterial ribosomes can be exploited for the development of antibiotics (**Table 40.1**). For example, the antibiotic *streptomycin*, a highly basic trisaccharide, interferes with the binding of formylmethionyl-tRNA to ribosomes and thereby prevents the correct initiation of protein synthesis. Other *aminoglycoside antibiotics* such as neomycin, kanamycin, and gentamycin interfere with the interaction between tRNA and the 16S rRNA of the 30S subunit (p. 715). *Chloramphenicol* acts by inhibiting peptidyl transferase activity. *Erythromycin* binds to the 50S subunit and blocks translocation.

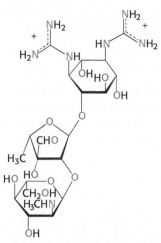

Streptomycin

Table 40.1 Antibiotic inhibitors of protein synthesis

Antibiotic	Action
Streptomycin and other aminoglycosides	Inhibit initiation and cause the misreading of mRNA (bacteria)
Tetracycline	Binds to the 30S subunit and inhibits the binding of aminoacyl-tRNAs (bacteria)
Chloramphenicol	Inhibits the peptidyl transferase activity of the 50S ribosomal subunit (bacteria)
Cycloheximide	Inhibits translocation (eukaryotes)
Erythromycin	Binds to the 50S subunit and inhibits translocation (bacteria)
Puromycin	Causes premature chain termination by acting as an analog of aminoacyl-tRNA (bacteria and eukaryotes)

The antibiotic *puromycin* inhibits protein synthesis in both bacteria and eukaryotes by causing nascent polypeptide chains to be released before their synthesis is completed. Puromycin is an analog of the terminal part of aminoacyl-tRNA (**Figure 40.15**). It binds to the A site on the ribosome and blocks the entry of aminoacyl-tRNA. Furthermore, puromycin contains an α-amino group. This amino group, like the one on aminoacyl-tRNA, forms a peptide bond with the carboxyl group of the growing peptide chain. The product, a peptide having a covalently attached puromycin residue at its carboxyl end, dissociates from the ribosome. No longer used medicinally, puromycin remains an experimental tool for the investigation of protein synthesis. *Cycloheximide,* another antibiotic, blocks translocation in eukaryotic ribosomes, making it a useful laboratory tool for blocking protein synthesis in eukaryotic cells.

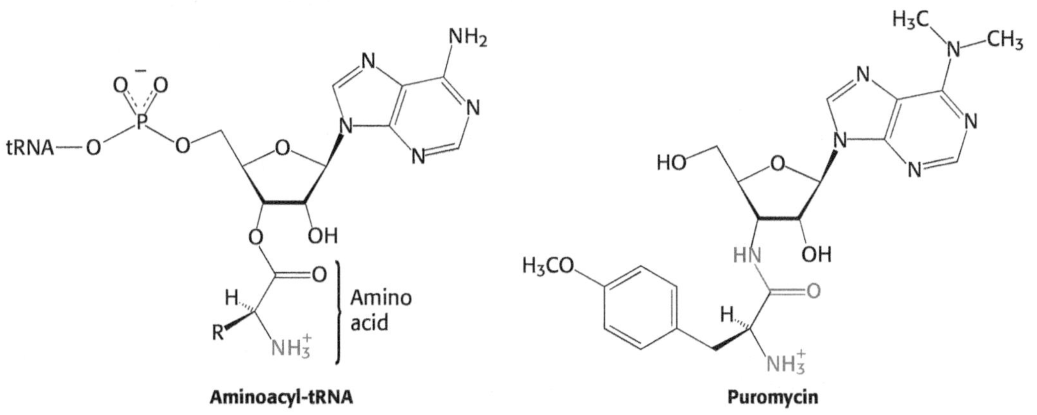

Aminoacyl-tRNA

Puromycin

Figure 40.15 Antibiotic action of puromycin. Puromycin resembles the aminoacyl terminus of an aminoacyl-tRNA. Its amino group joins the carboxyl group of the growing polypeptide chain to form an adduct that dissociates from the ribosome. This adduct is stable because puromycin has an amide (shown in red) rather than an ester linkage.

CLINICAL INSIGHT

Diphtheria Toxin Blocks Protein Synthesis in Eukaryotes by Inhibiting Translocation

Many antibiotics, harvested from bacteria for medicinal purposes, are inhibitors of bacterial protein synthesis. However, some bacteria produce protein-synthesis inhibitors that inhibit eukaryotic protein synthesis, leading to diseases such as diphtheria, which was a major cause of death in childhood before the advent of effective immunization. Symptoms include painful sore throat, hoarseness, fever, and difficulty breathing. The lethal effects of this disease are due mainly to a protein toxin produced by a bacteriophage infecting *Corynebacterium diphtheriae,* a bacterium that grows in the upper

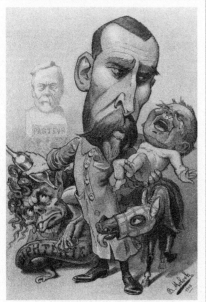

ADP-ribose

Figure 40.16 The blocking of translocation by diphtheria toxin. Diphtheria toxin blocks protein synthesis in eukaryotes by catalyzing the transfer of an ADP-ribose unit from NAD$^+$ to diphthamide, a modified amino acid residue in elongation factor 2 (translocase). Diphthamide is formed by the posttranslational modification (blue) of a histidine residue.

respiratory tract of an infected person. A few micrograms of diphtheria toxin is usually lethal in an unimmunized person because the toxin inhibits protein synthesis. The toxin consists of a single polypeptide chain that binds to the cell, enabling the toxin to enter the cytoplasm of its target cell. Shortly after entering a target cell, the toxin is cleaved into a 21-kDa A fragment and a 40-kDa B fragment. The A fragment of the toxin catalyzes the covalent modification of elongation factor 2, the elongation factor catalyzing translocation in eukaryotic protein synthesis, resulting in protein synthesis inhibition and cell death.

A single A fragment of the toxin in the cytoplasm can kill a cell. Why is it so lethal? EF2 contains *diphthamide*, an unusual amino acid residue that enhances the fidelity of codon shifting during translocation. Diphthamide is formed by a highly conserved complicated pathway that posttranslationally modifies histidine. The A fragment of the diphtheria toxin catalyzes the transfer of the ADP ribose unit of NAD$^+$ to the diphthamide ring (**Figure 40.16**).

This ADP ribosylation of a single side chain of EF2 blocks EF2's capacity to carry out the translocation of the growing polypeptide chain. Protein synthesis ceases, accounting for the remarkable toxicity of diphtheria toxin.

Pierre Paul Émile Roux (1853–1933), French physician, collaborator of Louis Pasteur, and cofounder of the Pasteur Institute, developed one of the first successful vaccines against diphtheria. [Photograph from Bibliotheque de la Faculte de Medecine, Paris, France/ Archives Charmet/Bridgeman Images.]

Figure 40.17 Castor beans. The seeds of castor beans from *Ricinus communis* are a rich source of oils with a wide variety of uses, including the production of biodiesel fuels. The seeds are also rich in the toxin ricin. [Ted Kinsman/Science Source.]

CLINICAL INSIGHT

Ricin Fatally Modifies 28S Ribosomal RNA

Ricin is a biomolecule frequently in the news because of its potential use as a bioterrorism agent. Ricin is a small protein (65 kDa) found in the seeds (the beans) of the castor plant, *Ricinus communis* (**Figure 40.17**). It is indeed a deadly molecule because as little as 500 mg is lethal for an adult human being and a single molecule can inhibit all protein synthesis in a cell, resulting in cell death.

Ricin is a heterodimeric protein composed of a catalytic A chain joined by a single disulfide bond to a B chain. The B chain allows the toxin to bind to the target cell, and this binding leads to an endocytotic uptake of the dimer and the eventual release of the A chain into the cytoplasm. The A chain is an

N-glycosidase that cleaves adenine from a particular adenosine nucleotide on the 28S rRNA that is found in all eukaryotic ribosomes. Removal of the adenine base completely inactivates the ribosome by preventing the binding of elongation factors. Thus, ricin and diphtheria toxin both act by inhibiting protein-synthesis elongation; ricin does so by covalently modifying rRNA, and diphtheria toxin does so by covalently modifying the elongation factor.

40.5 Ribosomes Bound to the Endoplasmic Reticulum Manufacture Secretory and Membrane Proteins

Not all newly synthesized proteins are destined to function in the cytoplasm. A newly synthesized protein in *E. coli* can stay in the cytoplasm or it can be sent to the plasma membrane, the outer membrane, the space between them, or the extracellular medium. Eukaryotic cells can direct proteins to internal sites such as mitochondria, the nucleus, and the endoplasmic reticulum, a process called *protein targeting* or *protein sorting*. How is sorting accomplished? There are two general mechanisms by which sorting takes place. In one mechanism, the protein is synthesized in the cytoplasm, and then the completed protein is delivered to its intracellular location posttranslationally. Proteins destined for the nucleus, chloroplast, mitochondria, and peroxisomes are delivered by this general process. The other mechanism, termed the *secretory pathway*, directs proteins into the *endoplasmic reticulum* (ER), the extensive membrane system that comprises about half the total membrane of a cell, cotranslationally—that is, while the protein is being synthesized. Proteins sorted by the secretory pathway include secreted proteins, residents of the ER, the Golgi complex, lysosomes, and integral membrane proteins of these organelles as well as integral plasma-membrane proteins. We will focus our attention on the secretory pathway only.

In eukaryotic cells, a ribosome remains free in the cytoplasm unless it is directed to the endoplasmic reticulum. The region that binds ribosomes is called the *rough ER* because of its studded appearance, as stated in Chapter 1, in contrast with the *smooth ER,* which is devoid of ribosomes (**Figure 40.18**).

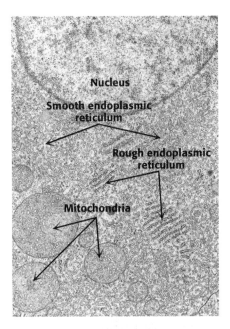

Figure 40.18 Ribosomes bound to the endoplasmic reticulum. In this electron micrograph, ribosomes appear as small black dots binding to the cytoplasmic side of the endoplasmic reticulum to give a rough appearance. In contrast, the smooth endoplasmic reticulum is devoid of ribosomes. [Don W. Fawcett/Science Source.]

Protein Synthesis Begins on Ribosomes That Are Free in the Cytoplasm

The synthesis of proteins sorted by the secretory pathway begins on a free ribosome, but shortly after synthesis begins, it is halted until the ribosome is directed to the cytoplasmic side of the endoplasmic reticulum. When the ribosome docks with the membrane, protein synthesis begins again. As the newly forming peptide chain exits the ribosome, it is transported through the membrane into the lumen of the endoplasmic reticulum. Free ribosomes that are synthesizing proteins for use in the cell are identical with those attached to the ER. What is the process that directs the ribosome synthesizing a protein destined to enter the ER to bind to the ER?

Signal Sequences Mark Proteins for Translocation Across the Endoplasmic Reticulum Membrane

The machinery required to direct a ribosome to the ER and to translocate the nascent protein across the ER consists of four components:

1. *The Signal Sequence. The signal sequence is a sequence of 9 to 12 hydrophobic amino acid residues, sometimes containing positively charged amino acids.* This sequence, which adopts an α-helical structure, is usually near the amino terminus of the nascent polypeptide chain. The presence of the signal sequence identifies the nascent peptide as one that must cross the ER membrane. Some signal

sequences are maintained in the mature protein, whereas others are cleaved by a *signal peptidase* on the lumenal side of the ER membrane (**Figure 40.19**).

2. *The Signal-Recognition Particle.* The signal-recognition particle (SRP) is a ribonucleoprotein consisting of a 7S RNA and six different proteins. The SRP recognizes the signal sequence and binds the sequence and the ribosome as soon as the signal sequence exits the ribosome. After the SRP is bound to the signal sequence, interactions between the ribosome and the SRP occlude the EF-binding site, thereby halting protein synthesis. The SRP then shepherds the ribosome and its nascent polypeptide chain to the ER membrane. Like the G proteins considered earlier, the SRP is a GTP-binding protein with GTPase activity. Thus, the SRP samples ribosomes until it locates one exhibiting a signal sequence.

3. *The SRP Receptor.* The SRP–ribosome complex travels to the endoplasmic reticulum, where the SRP binds the SRP receptor (SR), an integral membrane protein consisting of two subunits, SRα and SRβ. SRα is, like the SRP, a GTPase.

4. *The Translocon.* The SRP–SR complex delivers the ribosome to the translocation machinery, called the *translocon,* a multisubunit assembly of integral and peripheral membrane proteins. *The translocon is a protein-conducting channel.* This channel opens when the translocon and ribosome bind to each other. Protein synthesis resumes, with the growing polypeptide chain passing through the translocon channel into the lumen of the ER.

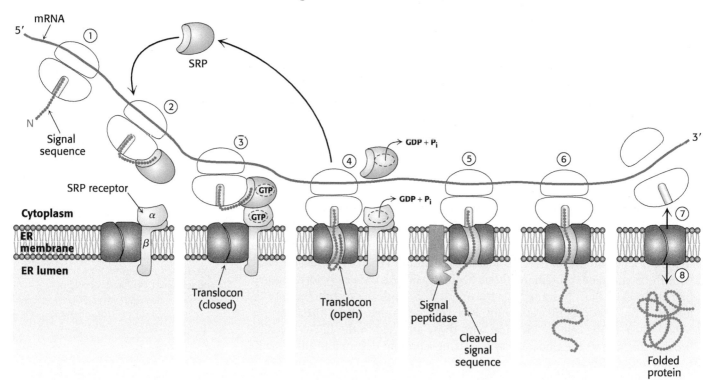

Figure 40.19 The SRP targeting cycle. (1) Protein synthesis begins on free ribosomes. (2) After the signal sequence has exited the ribosome, it is bound by the signal-recognition particle (SRP), and protein synthesis halts. (3) The SRP–ribosome complex docks with the SRP receptor in the ER membrane. (4) The ribosome–nascent polypeptide is transferred to the translocon. The SRP and the SRP receptor simultaneously hydrolyze bound GTPs. Protein synthesis resumes, and the SRP is free to bind another signal sequence. (5) The signal peptidase may remove the signal sequence as it enters the lumen of the ER. (6) Protein synthesis continues as the protein is synthesized directly into the ER. (7) On completion of protein synthesis, the ribosome is released and the protein tunnel in the translocon closes. [Information from H. Lodish et al., *Molecular Cell Biology,* 6th ed. (W. H. Freeman and Company, 2008), Fig. 13.6.]

The interactions of the components of the translocation machinery are shown in Figure 40.19. Both the SRP and the SRα subunit of the SR must bind GTP to facilitate the formation of the SRP-SR complex. For the SRP–SR complex to then deliver the ribosome to the translocon, the two GTP molecules—one in the SRP and the other in the SR—are aligned in what is essentially an active site shared by the two proteins. The formation of the alignment is catalyzed by the 7S RNA of the SRP. After the ribosome has been passed along to the translocon, the GTPs are hydrolyzed, the SRP and the SR dissociate, and the SRP is free to search for another signal sequence to begin the cycle anew. Thus, the SRP acts catalytically. The signal peptidase, which is associated with the translocon in the lumen of the ER, removes the signal sequence from most proteins. The proteins that are now in the lumen of the endoplasmic reticulum are subsequently packaged into transport vesicles. The transport vesicles bud-off the endoplasmic reticulum and are shuttled to various locations in the cell.

? QUICK QUIZ 2 What four components are required for the translocation of proteins across the endoplasmic reticulum membrane?

40.6 Protein Synthesis Is Regulated by a Number of Mechanisms

✓ 7 Explain how protein synthesis can be regulated.

Protein synthesis can be controlled at various stages in the translation process. We will first examine the control of protein synthesis by iron. An essential nutrient, iron is required for the synthesis of hemoglobin, cytochromes, and many other proteins. Recall that iron is a key component of the respiratory chain, the primary source of ATP in aerobic organisms (Chapter 20). However, excess iron can be quite harmful because, untamed by a suitable protein environment, iron can initiate a range of free-radical reactions that damage proteins, lipids, and nucleic acids. Animals have evolved sophisticated systems for the accumulation of iron in times of scarcity and for the safe storage of excess iron for later use. Key proteins include *transferrin*, a transport protein that carries iron in the blood serum, *transferrin receptor*, a membrane protein that binds iron-loaded transferrin and facilitates its entry into cells by the process of receptor mediated endocytosis (p. 535), and *ferritin*, an iron-storage protein found primarily in the liver and kidneys.

Messenger RNA Use Is Subject to Regulation

Ferritin and transferrin-receptor expression levels are reciprocally related in their responses to changes in iron levels. When iron is scarce, the amount of transferrin receptor increases and little or no new ferritin is synthesized. Interestingly, the extent of mRNA synthesis for these proteins does not change correspondingly. Instead, regulation takes place at the level of translation.

Let us consider ferritin first. Ferritin mRNA includes a stem-loop structure termed an *iron-response element* (IRE) in its 5′ untranslated region (**Figure 40.20**). This stem-loop structure binds a protein, called an *IRE-binding protein* (IRE-BP),

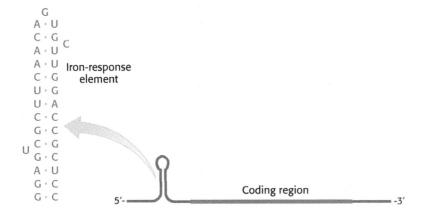

Figure 40.20 The iron-response element. Ferritin mRNA includes a stem-loop structure, termed an iron-response element (IRE), in its 5′ untranslated region. The IRE binds a specific protein that blocks the translation of this mRNA under low-iron conditions.

which blocks the initiation of translation. When the iron level increases, the IRE-BP binds iron. The IRE-BP bound to iron cannot bind RNA, because the binding sites for iron and RNA substantially overlap. Thus, in the presence of iron, ferritin mRNA is released from the IRE-BP and translated to produce ferritin, which sequesters the excess iron.

The Stability of Messenger RNA Also Can Be Regulated

Transferrin-receptor mRNA also has several IRE-like regions. Unlike those in ferritin mRNA, these regions are located in the 3′ untranslated region rather than in the 5′ untranslated region (Figure 40.21). In the absence of iron, IRE-BP binds to these IREs. However, given the location of these binding sites, the transferrin-receptor mRNA can still be translated. What happens when the iron level increases? IRE-BP bound to iron no longer binds transferrin-receptor mRNA. Freed from the IRE-BP, transferrin-receptor mRNA is rapidly degraded. Thus, an increase in the cellular iron level leads to the destruction of transferrin-receptor mRNA and, hence, a reduction in the production of transferrin-receptor protein. The IRE-BP serves as an iron sensor. If there is enough iron present to bind the IRE-BP, then ferritin will be synthesized to store the iron. Likewise, if enough iron is present to be stored, there is no need to produce the receptor for its uptake. An environmental signal—the concentration of iron—controls the translation of proteins required for the metabolism of this metal.

Figure 40.21 Transferrin-receptor mRNA. This mRNA has a set of iron-response elements (IREs) in its 3′ untranslated region. The binding of the IRE-binding protein to these elements stabilizes the mRNA but does not interfere with translation.

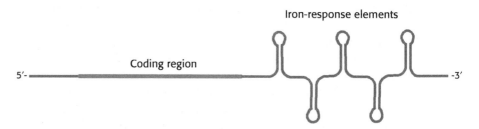

Small RNAs Can Regulate mRNA Stability and Use

In recent years, an entirely new means of regulating protein synthesis was discovered. *RNA interference* (RNAi) was originally identified as a process that leads to mRNA degradation, induced by the presence of double-stranded RNA (dsRNA). RNAi, observed in all eukaryotes, may have evolved as a protective mechanism against viruses that employ dsRNA at some point in the viral life cycle. In RNAi, *Dicer,* an RNase, cleaves double-stranded RNAs into 21-nucleotide fragments. Single-stranded components of the cleavage products, called *small interfering RNA* (siRNA), are bound by members of a class of proteins called the *Argonaute* family to form an *RNA induced silencing complex* (RISC). This complex uses the single-stranded siRNA to locate and degrade complementary mRNAs (Figure 40.22).

Figure 40.22 MicroRNA action. MicroRNAs bind to members of the Argonaute family where they serve to target specific mRNA molecules for cleavage.

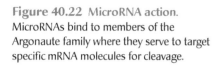 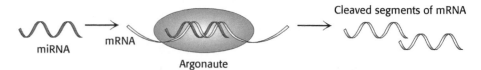

The investigation of RNAi led to the discovery of a new class of RNAs that regulate gene expression posttranscriptionally. Whereas siRNAs are derived from exogenous RNA (either from a virus or introduced experimentally), *microRNAs* (miRNAs) are a class of noncoding RNAs generated from larger transcripts produced by RNA polymerase II and, in some cases, RNA polymerase III. These transcripts are cleaved by specific nucleases to yield fragments of double-stranded RNA (dsRNA) ~21 base pairs in length. Further processing leads to the degradation of one of the strands of this ~21-base-pair dsRNA; the other strand associates with

Argonaute to form RISC, also called a microRNP (miRNP). The complex is believed to function in one of two ways. If the small RNA component of the complex binds to an mRNA by precise Watson–Crick base-pairing, the mRNA is destroyed. If the complementarity is not precise, the mRNA is simply not translated. In either mode of action, protein synthesis is inhibited.

Gene regulation by miRNAs was originally thought to be limited to a small number of species. However, subsequent studies have revealed that this mode of gene regulation is nearly ubiquitous in eukaryotes. Indeed, more than 700 miRNAs encoded by the human genome have been identified. Each miRNA can regulate many different genes because many different target sequences are present in each mRNA. An estimated 60% of all human genes are regulated by one or more miRNAs. MicroRNA has been shown to regulate a wide variety of biochemical processes, including development, cell differentiation, and oncogenesis—the development of the cancer.

The discovery of small RNAs that regulate gene expression is among the most exciting biochemical discoveries in recent years. Elucidating the precise mechanism of action of these RNAs is an active area of biochemical research.

SUMMARY

40.1 Protein Synthesis Decodes the Information in Messenger RNA

Protein synthesis takes place in three phases: initiation, elongation, and termination. In bacteria, mRNA, formylmethionyl-tRNA$_f$ (the special initiator tRNA that recognizes AUG), and a 30S ribosomal subunit come together with the assistance of initiation factors to form a 30S initiation complex. A 50S ribosomal subunit then joins this complex to form a 70S initiation complex, in which fMet-tRNA$_f$ occupies the P site of the ribosome.

Elongation factor Tu delivers the appropriate aminoacyl-tRNA to the ribosome's A (aminoacyl) site as an EF-Tu–aminoacyl-tRNA–GTP ternary complex. EF-Tu serves both to protect the aminoacyl-tRNA from premature cleavage and to increase the fidelity of protein synthesis by ensuring that the correct anticodon–codon pairing has taken place before hydrolyzing GTP and releasing aminoacyl-tRNA into the A site.

40.2 Peptidyl Transferase Catalyzes Peptide-Bond Synthesis

A peptide bond is formed when the amino group of the aminoacyl-tRNA attacks the ester linkage of the peptidyl-tRNA. The ribosome catalyzes the formation of the peptide bond by using proximity and orientation catalysis as well as by stabilizing the transition state.

On peptide-bond formation, the tRNAs and mRNA must be translocated for the next cycle to begin. The deacylated tRNA moves to the E site and then leaves the ribosome, and the peptidyl-tRNA moves from the A site into the P site. Elongation factor G uses the free energy of GTP hydrolysis to drive translocation. Protein synthesis is terminated by release factors, which recognize the termination codons UAA, UGA, and UAG and cause the hydrolysis of the ester bond between the polypeptide and tRNA.

40.3 Bacteria and Eukaryotes Differ in the Initiation of Protein Synthesis

The basic plan of protein synthesis in eukaryotes is similar to that of bacteria, but there are some significant differences. Eukaryotic ribosomes (80S) consist of a 40S small subunit and a 60S large subunit. The initiating amino acid is again methionine, but it is not formylated. The initiation of protein synthesis is more complex in eukaryotes than in bacteria. The AUG codon closest to the 5′ end of mRNA is nearly always the start site. The 40S ribosome finds this site by binding to the 5′ cap and then scanning the RNA until AUG is reached. Eukaryotic mRNA is circularized by interaction between eukaryotic initiation factor 4 and the poly(A)-binding protein.

40.4   A Variety of Biomolecules Can Inhibit Protein Synthesis

Protein synthesis is the site of action of an array of antibiotics. Virtually all aspects of protein synthesis are sensitive to one antibiotic or another. The fact that many antibiotics target protein synthesis in bacteria but not eukaryotes makes them clinically useful. Toxins such as diphtheria toxin and ricin are lethal because they inhibit protein synthesis. Diphtheria toxin modifies and inactivates a protein factor, and ricin removes a base from 28S rRNA, eliminating its catalytic activity.

40.5   Ribosomes Bound to the Endoplasmic Reticulum Manufacture Secretory and Membrane Proteins

Proteins contain signals that determine their ultimate destination. In eukaryotes, some proteins are transported into organelles, such as mitochondria and nuclei, posttranslationally. Other proteins, following the secretory pathway, are inserted into the endoplasmic reticulum cotranslationally. Protein synthesis continues in the cytoplasm unless the nascent chain contains a signal sequence that directs the ribosome to the endoplasmic reticulum. Amino-terminal signal sequences consist of 9 to 12 hydrophobic amino acids. A signal-recognition particle (SRP) recognizes signal sequences and brings ribosomes bearing them to the ER. A GTP–GDP cycle releases the signal sequence from the SRP and then detaches the SRP from its receptor. The nascent chain is then translocated across the ER membrane.

40.6   Protein Synthesis Is Regulated by a Number of Mechanisms

In eukaryotes, genes encoding proteins that transport and store iron are regulated at the translational level. Iron-response elements (IREs), structures that are present in certain mRNAs, are bound by an IRE-binding protein when this protein is not binding iron. Whether the expression of a gene is stimulated or inhibited in response to changes in the iron status of a cell depends on the location of the IRE within the mRNA. Recently discovered microRNAs also can regulate protein synthesis either by causing RNA degradation or by inhibiting the translation of mRNA.

KEY TERMS

30S subunit (p. 722)
50S subunit (p. 722)
Shine–Dalgarno sequence (p. 723)
initiation factor (p. 724)
elongation factor Tu (EF-Tu) (p. 724)
accommodation (p. 725)
elongation factor Ts (EF-Ts) (p. 725)
peptidyl transferase center (p. 725)
elongation factor G (EF-G) (p. 727)

polysome (p. 727)
release factor (p. 728)
protein targeting (p. 733)
signal sequence (p. 733)
signal peptidase (p. 734)
signal-recognition particle (SRP) (p. 734)
SRP receptor (SR) (p. 734)
translocon (p. 734)

transferrin (p. 735)
transferrin receptor (p. 735)
ferritin (p. 735)
iron-response element (IRE) (p. 735)
IRE-binding protein (IRE-BP) (p. 735)
RNA interference (RNAi) (p. 736)
microRNA (miRNA) (p. 736)

❓ Answers to QUICK QUIZZES

1. Protein factors modulate the initiation of protein synthesis. The role of IF1 and IF3 is to prevent premature binding of the 30S and 50S ribosomal subunits, whereas IF2 delivers Met-tRNA$_f$ to the ribosome. Protein factors are also required for elongation (EF-G and EF-Tu), for termination (release factors, RF), and for ribosome dissociation (ribosome release factors, RRFs).

2. The signal sequence, the signal-recognition particle (SRP), the SRP receptor, and the translocon.

PROBLEMS

1. *Babel fish.* Why is protein synthesis also called translation?

2. *Correct phrasing.* What is meant by the phrase *reading frame*?

3. *Match 'em.* Match each term on the right with the terms in parts *a, b,* and *c.* ✓ 3 ✓ 4

    (a) Initiation _____
    (b) Elongation _____
    (c) Termination _____

    1. GTP
    2. AUG
    3. fMet
    4. RRF
    5. IF2
    6. Shine–Dalgarno
    7. EF-Tu
    8. Peptidyl transferase
    9. UGA
    10. Transformylase

4. *Wasted effort?* Transfer RNA molecules are quite large, given that the anticodon consists of only three nucleotides. What is the purpose of the rest of the tRNA molecule? ✓ 3

5. *Light and heavy ribosomes.* Density-gradient centrifugation is a technique that allows the separation of biological molecules and molecular complexes by differences in density. Ribosomes were isolated from bacteria grown in a "heavy" medium (^{13}C and ^{15}N) and from bacteria grown in a "light" medium (^{12}C and ^{14}N). These 70S ribosomes were added to an in vitro system undergoing protein synthesis. An aliquot removed several hours later was analyzed by density-gradient centrifugation. How many types of 70S ribosomes differing in density would you expect to see in the density gradient? ✓ 4 ✓ 5

6. *The price of protein synthesis.* What is the smallest number of molecules of ATP and GTP consumed in the synthesis of a 200-residue protein, starting from amino acids? Assume that the hydrolysis of PP_i is equivalent to the hydrolysis of ATP for this calculation.

7. *Viral mutation.* An mRNA transcript of a T7 phage gene contains the base sequence

$$\downarrow$$

5′–AACUGCACGAGGUAACACAAGAUGGCU–3′

Predict the effect of a mutation that changes the G identified by the arrow to A. ✓ 3

8. *Enhancing fidelity.* Compare the accuracy of (a) DNA replication, (b) RNA synthesis, and (c) protein synthesis. What mechanisms are used to ensure the fidelity of each of these processes? ✓ 4

9. *You have to know where to look.* Bacterial messenger RNAs usually contain many AUG codons. How does the ribosome identify the AUG specifying initiation? ✓ 4 ✓ 5

10. *Triggered GTP hydrolysis.* Ribosomes markedly accelerate the hydrolysis of GTP bound to the complex of EF-Tu and aminoacyl-tRNA. What is the biological significance of this enhancement of GTPase activity by ribosomes? ✓ 4

11. *Blocking translation.* Devise an experimental strategy for switching off the expression of a specific mRNA without changing the gene encoding the protein or the gene's control elements. ✓ 3

12. *Directional problem.* Suppose that you have a protein-synthesis system that is synthesizing a protein designated A. Furthermore, you know that protein A has four trypsin-sensitive sites, equally spaced in the protein, that, on digestion with trypsin, yield the peptides A_1, A_2, A_3, A_4, and A_5. Peptide A_1 is the amino-terminal peptide, and A_5 is the carboxyl-terminal peptide. Finally, you know that your system requires 4 minutes to synthesize a complete protein A. At $t = 0$, you add all 20 amino acids, each carrying a ^{14}C label. ✓ 4

(a) At $t = 1$ minute, you isolate intact protein A from the system, cleave it with trypsin, and isolate the five peptides. Which peptide is most heavily labeled?
(b) At $t = 3$ minutes, what will be the order of the labeling of peptides from heaviest to lightest?
(c) What does this experiment tell you about the direction of protein synthesis?

13. *A timing device.* EF-Tu, a member of the G-protein family, plays a crucial role in the elongation process of translation. Suppose that a slowly hydrolyzable analog of GTP were added to an elongating system. What would be the effect on the rate of protein synthesis? ✓ 4

14. *Fundamentally the same, yet . . .* List the differences between bacterial and eukaryotic protein synthesis. ✓ 6

15. *Membrane transport.* What four components are required for the translocation of proteins across the endoplasmic reticulum membrane? ✓ 4

16. *Like a border collie.* What is the role of the signal-recognition particle in protein translocation? ✓ 4 ✓ 5

17. *Push, don't pull.* What is the energy source that powers the cotranslational movement of proteins across the endoplasmic reticulum? ✓ 5

18. *An assembly line.* Why is the fact that protein synthesis takes place on polysomes advantageous? ✓ 5

19. *Iron regulation.* What effect would you expect from the addition of an iron-response element (IRE) to the 5′ end of a gene that is not normally regulated by iron levels? To the 3′ end? ✓ 7

20. *Predicting microRNA regulation.* Suppose that you have identified an miRNA that has the sequence 5'-GCCUAGCCUUAGCAUUGAUUGG-3'. Propose a strategy for identifying mRNA that might be regulated by this miRNA, given the sequences of all mRNAs encoded by the human genome. ✓ 7

Chapter Integration Problems

21. *Déjà vu.* Which protein in G-protein cascades plays a role similar to that of elongation factor Ts? ✓ 4

22. *Family resemblance.* Eukaryotic elongation factor 2 is inhibited by ADP ribosylation catalyzed by diphtheria toxin. What other G proteins are sensitive to this mode of inhibition?

23. *Contrasting modes of elongation.* The two basic mechanisms for the elongation of biomolecules are represented in the adjoining illustration. In type 1, the activating group (X) is released from the growing chain. In type 2, the activating group is released from the incoming unit as it is added to the growing chain. Indicate whether each of the following biosyntheses is by means of a type 1 or a type 2 mechanism:

(a) Glycogen synthesis
(b) Fatty acid synthesis
(c) $C_5 \rightarrow C_{10} \rightarrow C_{15}$ in cholesterol synthesis
(d) DNA synthesis
(e) RNA synthesis
(f) Protein synthesis

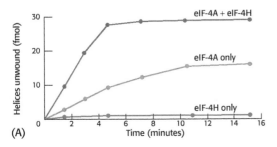

24. *The final step.* What aspect of primary structure allows the transfer of linear nucleic acid information into the functional three-dimensional structure of proteins?

Data Interpretation and Challenge Problems

25. *Helicase helper.* The initiation factor eIF-4 displays ATP-dependent RNA helicase activity. Another initiation factor, eIF-4H, has been proposed to assist the action of eIF-4. Graph A shows some of the experimental results from an assay that can measure the activity of eIF-4 helicase in the presence of eIF-4H. ✓ 4

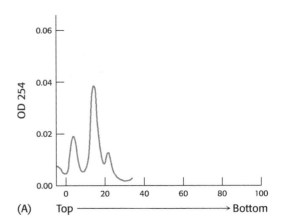

(a) What are the effects on eIF-4 helicase activity in the presence of eIF-4H?
(b) Why did measuring the helicase activity of eIF-4H alone serve as an important control?
(c) The initial rate of helicase activity of $0.2\ \mu M$ of eIF-4 was then measured with varying amounts of eIF-4H (graph B). What ratio of eIF-4H to eIF-4 yielded optimal activity?

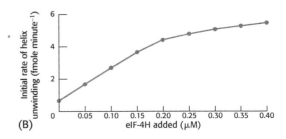

(d) Next, the effect of RNA–RNA helix stability on the initial rate of unwinding in the presence and absence of eIF-4H was tested (graph C). How does the effect of eIF-4H vary with helix stability?

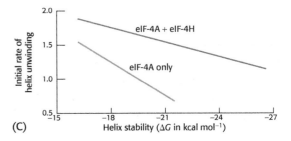

(e) How might eIF-4H affect the helicase activity of eIF-4A?

[Data from N. J. Richter, G. W. Rodgers, Jr., J. O. Hensold, and W. C. Merrick. Further biochemical and kinetic characterization of human eukaryotic initiation factor 4H. *J. Biol. Chem.* 274:35415–35424, 1999.]

26. *Size separation.* The protein-synthesizing machinery was isolated from eukaryotic cells and briefly treated with a low concentration of RNase. The sample was then subjected to sucrose gradient centrifugation. The gradient was fractionated and the absorbance, or optical density (OD), at 254 nm was recorded for each fraction. The plot in graph A was obtained. ✓ 4

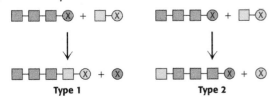

(a) What do the three peaks of absorbance in graph A represent?

The experiment was repeated except that, this time, the RNase treatment was omitted.

(b) Why is the centrifugation pattern in graph B more complex? What do the series of peaks near the bottom of the centrifuge tube represent?

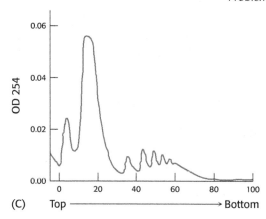

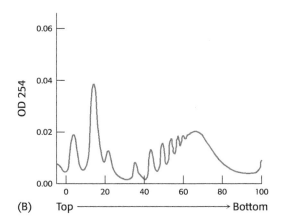

(c) What is the effect of growing cells under hypoxic conditions?

[Data from M. Koritzinsky et al. *EMBO J.* 25:1114–1125, 2006.]

27. *Suppressing frameshifts.* The insertion of a base in a coding sequence leads to a shift in the reading frame, which in most cases produces a nonfunctional protein. Propose a mutation in a tRNA that might suppress frameshifting.

28. *The exceptional E. coli.* In contrast with *E. coli,* most bacteria do not have a full complement of aminoacyl-tRNA synthetases. For instance, *Helicobacter pylori,* the cause of stomach ulcers, has tRNA^{Gln}, but no Gln-tRNA synthetase. However, glutamine is a common amino acid in *H. pylori* proteins. Suggest a means by which glutamine can be incorporated into proteins in *H. pylori.* (Hint: Glu-tRNA synthetase can misacylate tRNA^{Gln}.) ✓ 4

Before the isolation of the protein-synthesizing machinery, the cells were grown in low concentrations of oxygen (hypoxic conditions). Again the experiment was repeated without RNase treatment (graph C).

Selected Readings for this chapter can be found online at www.whfreeman.com/tymoczko3e.

Recombinant DNA Techniques

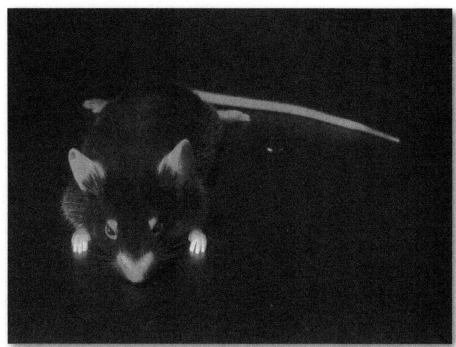

Inherited diseases often result from a defective or missing gene. A goal of biochemistry is to eventually treat such diseases by replacing the defective gene with a functioning version. In the example of gene manipulation illustrated here, a recombinant gene consisting of the gene for the cytoskeleton protein actin linked to the gene for green fluorescent protein was inserted into a mouse embryo. The fluorescence photograph of the mouse shows the location of green fluorescent protein (GFP) attached to actin molecules. Hair has no actin and thus does not fluoresce. [Dr. Charles Mazel.]

We focused on protein purification, immunological techniques, and amino-acid-sequence determination when we last explored experimental techniques in Chapter 5. Here, we will expand our repertoire of procedures to include those used with nucleic acids—in particular, the techniques of recombinant DNA technology. We will continue our study of the estrogen receptor, a transcription factor that regulates certain genes in response to the presence of the steroid hormone estradiol (p. 78). Here, we will use the purified receptor protein to isolate DNA encoding the receptor protein, to express the receptor in bacteria, to investigate the nature of its gene, and then to find out whether there are relatives of this protein in other organisms or similar proteins in the same organism.

41.1 Nucleic Acids Can Be Synthesized from Protein-Sequence Data

In Chapter 5, we purified the receptor by using monoclonal antibodies. What is the next step in characterizing the receptor? There are many possible answers to the question, but an especially common one is to isolate the DNA that encodes the receptor. There are "libraries" of DNA sequences that we can search for sequences corresponding to the receptor. For instance, a *complementary DNA* (cDNA) *library* is a collection of DNA sequences representing all of the mRNA expressed by a cell. The enzyme *reverse transcriptase* uses mRNA as a template to make a DNA copy, or cDNA. If we can isolate the cDNA for the estrogen receptor, we can insert the DNA into bacteria, and the bacteria will produce the estrogen receptor for our experiments. How can we find the estrogen receptor in the cDNA library? We can use our purified estrogen receptor to generate a probe that will allow us to identify the DNA sequences in the library that correspond to the estrogen receptor.

Protein Sequence Is a Guide to Nucleic Acid Information

Chapter 5 examined the techniques used to sequence an entire protein. To search DNA libraries, we need a probe—a chemical of some sort that will recognize DNA sequences that correspond to the estrogen receptor. To generate probes to search for the estrogen-receptor gene in a cDNA library, we need to know only a part of the primary structure of the receptor. Knowledge of a protein's primary structure permits the synthesis of DNA sequences that correspond to a part of the amino acid sequence on the basis of the genetic code. These DNA sequences can be used as probes to isolate the gene encoding the protein or the cDNA corresponding to the mRNA. For instance, the single-stranded probe can be made radioactive and then mixed with DNA libraries. The probe will pair, or hybridize, with a complementary DNA molecule, identifying the DNA molecule as the one encoding the sequence corresponding to the protein that was used to generate the probe.

Using purified receptor as starting material, we can use mass spectroscopy (p. 88) to determine the amino acid sequence of a part of the receptor. With this information and a copy of the genetic code (p. 709) in hand, we can convert the protein-sequence information into nucleic-acid-sequence information and then into a DNA probe.

DNA Probes Can Be Synthesized by Automated Methods

A DNA strand can be synthesized by the sequential addition of activated monomers to a growing strand that is linked to an insoluble support. The activated monomers are protonated *deoxyribonucleoside 3'-phosphoramidites* in which all but one of the reactive groups of the nucleotide are protected by the temporary attachment of nonreactive groups.

In step 1, the 3'-phosphorus atom of this incoming unit joins the growing strand to form a *phosphite triester* (Figure 41.1). In step 2, the phosphite triester is oxidized by iodine to stabilize the new linkage. In step 3, the protecting group on the 5'-OH group of the growing strand is removed by the addition of dichloroacetic acid. The DNA strand is now elongated by one unit and ready for another cycle of addition. Each cycle only takes a few minutes and usually elongates more than 99% of the strands.

A deoxyribonucleoside 3'-phosphoramidite with DMT and βCE attached

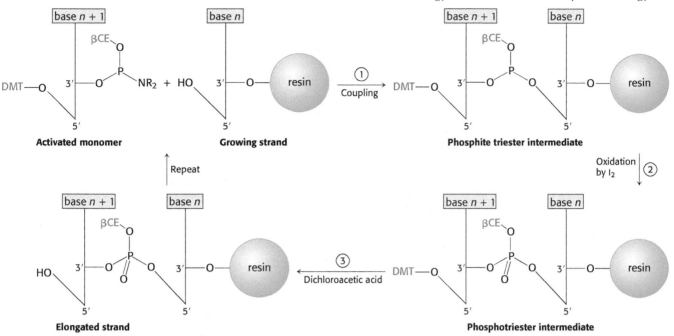

Figure 41.1 The synthesis of a DNA strand by the phosphite triester method. The activated monomer added to the growing strand is a deoxyribonucleoside 3′-phosphoramidite containing a dimethoxytrityl (DMT) protecting group on its 5′-oxygen atom, a β-cyanoethyl (βCE) protecting group on its 3′-phosphoryl oxygen atom, and a protecting group on the base.

The ability to rapidly synthesize DNA strands of any selected sequence opens many experimental avenues. For example, a synthesized oligonucleotide labeled at one end with ^{32}P or a fluorescent tag can be used to search for a complementary sequence in a collection of DNA sequences. In our experiment, we will make a radioactive DNA probe corresponding to a part of the estrogen receptor's primary structure. We will use this probe to screen collections of DNA sequences that encode the estrogen receptor. However, before we do so, we must examine the practical characteristics of the DNA that we are looking for and, in the process, we will examine some of the tools for recombinant DNA technology.

41.2 Recombinant DNA Technology Has Revolutionized All Aspects of Biology

✓ 8 List the key tools of recombinant DNA technology, and explain how they are used to clone DNA.

Recombinant DNA techniques, developed in the early 1970s, have taken biology from an exclusively analytical science to a synthetic one. New combinations of unrelated genes can be constructed in the laboratory by applying recombinant DNA techniques. These novel combinations can be cloned—amplified many times—by introducing them into suitable cells, where they are replicated by the DNA-synthesizing machinery of the host. The inserted genes are often transcribed and translated in their new setting, producing proteins that would not ordinarily be found in the host cell. Recall also that recombinant DNA techniques allow the researcher to remove (knock-out) or add (knock-in) specific genes from or to a genome (p. 652).

Restriction Enzymes Split DNA into Specific Fragments

Restriction enzymes are perhaps the tools that made the development of recombinant DNA technology possible. *Restriction enzymes,* also called *restriction*

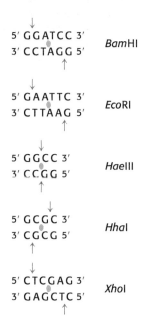

Figure 41.2 Specificities of some restriction endonucleases. The sequences recognized by these enzymes contain a twofold axis of symmetry about the green dot. The cleavage sites are denoted by red arrows. The abbreviated name of each restriction enzyme is given at the right of the sequence that it recognizes. The cuts can be staggered or even.

endonucleases, recognize specific base sequences in double-helical DNA and cleave, at specific places, both strands of that duplex. To biochemists, these exquisitely precise scalpels are marvelous gifts of nature. They are indispensable for analyzing chromosome structure, sequencing very long DNA molecules, isolating genes, and creating new DNA molecules that can be cloned.

Restriction enzymes are found in a wide variety of bacteria. Their biological role is to cleave and thereby destroy foreign DNA molecules, such as the DNA of viruses that attack bacteria (bacteriophages). The cell's own DNA is not degraded, because the sites recognized by its own restriction enzymes are methylated, which prevents DNA cleavage. Many restriction enzymes recognize specific sequences of four to eight base pairs, called cleavage sites, and hydrolyze a phosphodiester linkage at a specific site in each strand in this region. A striking characteristic of these cleavage sites is that they almost always possess *twofold rotational symmetry*. In other words, the recognized sequence is *palindromic*, or an inverted repeat, and the cleavage sites are symmetrically positioned. For example, the sequence recognized by a restriction enzyme from *Streptomyces achromogenes* is

In each strand, the enzyme cleaves the C–G phosphodiester linkage on the 3′ side of the symmetry axis.

Hundreds of restriction enzymes have been purified and characterized. Their names consist of a three-letter abbreviation for the host organism (*e.g., Eco* for *Escherichia coli, Hin* for *Haemophilus influenzae, Hae* for *Haemophilus aegyptius*), followed by a strain designation (if needed) and a roman numeral (if more than one restriction enzyme from the same strain has been identified). The specificities of several of these enzymes are shown in **Figure 41.2**. Restriction enzymes are used to cleave DNA molecules into specific fragments that are more readily analyzed and manipulated than the entire parent molecule.

Restriction Fragments Can Be Separated by Gel Electrophoresis and Visualized

Small differences between related DNA molecules can be readily detected because their restriction fragments can be separated and displayed by gel electrophoresis. In Chapter 5, we considered the use of gel electrophoresis to separate protein molecules. Gel electrophoresis of nucleic acids is similar in principle to gel electrophoresis of proteins. When working with nucleic acids, however, the sample is not denatured and the gel is often made of agarose instead of polyacrylamide. Because the phosphodiester backbone of DNA is highly negatively charged, this technique is also suitable for the separation of nucleic acid fragments. For most gels, the shorter the DNA fragment, the farther the migration. Polyacrylamide gels are used to separate, by size, fragments containing as many as 1000 base pairs, whereas more porous agarose gels are used to resolve mixtures of larger fragments (as large as 20 kb). An important feature of these gels is their high resolving power. In certain kinds of gels, fragments differing in length by just one nucleotide of several hundred can be distinguished. Bands or spots of DNA in gels can be visualized by staining with a dye such as ethidium bromide, which fluoresces an intense orange under irradiation by ultraviolet light when bound to a double-helical DNA molecule (**Figure 41.3**). A band containing only 50 ng of DNA can be readily seen.

A restriction fragment containing a specific base sequence can be identified by hybridizing it with a labeled complementary DNA strand, such as our probe for the estrogen receptor (**Figure 41.4**). A mixture of restriction fragments is separated by electrophoresis through an agarose gel, denatured to form single-stranded DNA, and transferred to a nitrocellulose sheet. The positions of the DNA fragments in the gel are preserved on the nitrocellulose sheet, where they are exposed to a ^{32}P-labeled single-stranded DNA probe. A sheet of x-ray film is then placed over the blot. The radioactivity of the probe will expose (darken) the x-ray film. This process, called *autoradiography*, reveals the position of the restriction-fragment–probe duplex. A particular fragment amid a million others can be readily identified in this way. This powerful technique is named *Southern blotting*, for its inventor Edwin Southern.

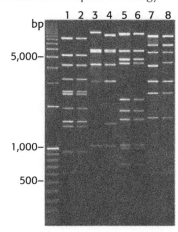

Figure 41.3 Gel-electrophoresis pattern of a restriction digest. This gel shows the fragments produced by cleaving DNA from two viral strains (odd- vs. even-numbered lanes) with each of four restriction enzymes. These fragments were made fluorescent by staining the gel with ethidium bromide. [Carr et al., *Emerging Infectious Diseases*, 17(8): 1402–1408, August 2011, www.cdc.gov/eid]

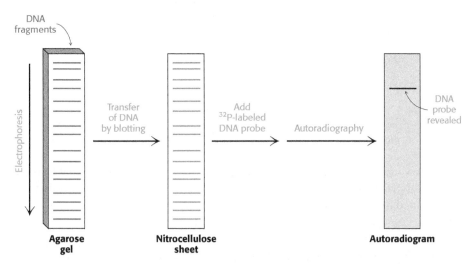

Figure 41.4 Southern blotting. A DNA fragment containing a specific sequence can be identified by separating a mixture of fragments by electrophoresis, transferring them to nitrocellulose, and hybridizing with a ^{32}P-labeled probe complementary to the sequence. The fragment containing the sequence is then visualized by autoradiography.

Similarly, RNA molecules can be separated by gel electrophoresis, and specific sequences can be identified by hybridization subsequent to their transfer to nitrocellulose. This analogous technique for the analysis of RNA has been whimsically termed *northern blotting*. A further play on words accounts for the term *western blotting*, which refers to a technique for detecting a particular protein by staining with specific antibody (Chapter 5). Southern, northern, and western blots are also known respectively as *DNA, RNA,* and *protein blots.*

Restriction Enzymes and DNA Ligase Are Key Tools for Forming Recombinant DNA Molecules

Let us examine how novel DNA molecules can be constructed in the laboratory as preparation for isolating DNA encoding the estrogen receptor. Our immediate goal is to insert DNA encoding the receptor into a piece of DNA, called a *vector*, that is readily taken up and replicated by bacteria. The bacteria containing the foreign DNA can be isolated, or cloned. The cloned bacteria can then produce large amounts of receptor proteins with which we can perform experiments.

How do we construct a recombinant DNA molecule? A DNA fragment of interest is covalently joined to a DNA vector. The essential feature of a vector is that it can replicate autonomously in an appropriate host. *Plasmids* (naturally occurring circles of DNA that act as accessory chromosomes in bacteria) and

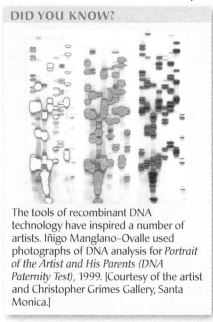
bacteriophage lambda (λ phage), a virus, are commonly used vectors for cloning in *E. coli*. The vector can be prepared for accepting a new DNA fragment by cleavage at a single specific site with a restriction enzyme. The staggered cuts made by this enzyme produce *complementary single-stranded ends,* which have specific affinity for each other and hence are known as *cohesive* or *sticky ends.* Any DNA fragment can be inserted into this plasmid if the fragment has the same cohesive ends. Such a fragment can be prepared from a larger piece of DNA by using the same restriction enzyme as was used to open the plasmid DNA (**Figure 41.5**).

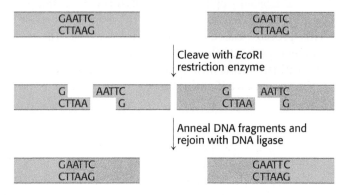

Figure 41.5 The joining of DNA molecules by the cohesive-end method. Two DNA molecules, cleaved by a common restriction enzyme such as *Eco*RI, can be ligated to form recombinant molecules.

The single-stranded ends of the fragment are then complementary to those of the cut plasmid. The DNA fragment and the cut plasmid can be annealed and then joined by *DNA ligase.*

41.3 Eukaryotic Genes Can Be Manipulated with Considerable Precision

Eukaryotic genes can be introduced into bacteria by using the techniques of recombinant DNA technology as heretofore discussed. The bacteria can then be used as factories to produce the desired gene product, usually a protein. Producing such a molecular factory will be our plan of attack for the estrogen receptor. First, however, we must generate DNA encoding the estrogen receptor.

Complementary DNA Prepared from mRNA Can Be Expressed in Host Cells

How can mammalian DNA be cloned and expressed by *E. coli?* Recall that most mammalian genes are mosaics of introns and exons (p. 694). These interrupted genes cannot be expressed by bacteria, which lack the machinery to splice introns out of the primary transcript. However, this difficulty can be circumvented by introducing recombinant DNA that is complementary to mature mRNA, or cDNA, into the bacteria. For example, proinsulin, a precursor of insulin, is synthesized by bacteria harboring plasmids that contain DNA complementary to mRNA for proinsulin (**Figure 41.6**). Indeed, bacteria produce much of the insulin used today by millions of diabetics.

The key to forming complementary DNA is the enzyme *reverse transcriptase,* an RNA-directed DNA polymerase isolated from retroviruses. Reverse transcriptase synthesizes a DNA strand complementary to an RNA template if the transcriptase is provided with a DNA primer that is base-paired to the RNA and contains a free 3′-OH group. We can use a simple sequence of linked thymidine, or oligo(T), residues as the primer. This oligo(T) sequence pairs with the poly(A)

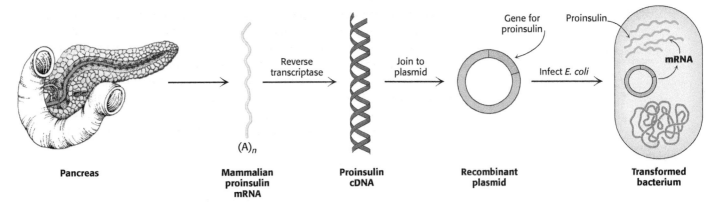

Figure 41.6 Synthesis of proinsulin by bacteria. Proinsulin, a precursor of insulin, can be synthesized by transformed (genetically altered) clones of *E. coli*. The clones contain the mammalian DNA encoding proinsulin.

sequence at the 3′ end of most eukaryotic mRNA molecules (p. 693), as shown in **Figure 41.7**. Reverse transcriptase then synthesizes the rest of the cDNA strand in the presence of the four deoxyribonucleoside triphosphates. The RNA strand of this RNA–DNA hybrid is subsequently hydrolyzed by raising the pH. Unlike RNA, DNA is resistant to alkaline hydrolysis. The single-stranded DNA is converted into double-stranded DNA by creating another primer site. The enzyme *terminal transferase* adds nucleotides—for instance, several residues of dG—to the 3′ end of the single-stranded DNA to act as a platform for the primer. Oligo(dC) can bind to dG residues and prime the synthesis of the second DNA strand. The second DNA strand is synthesized as it would be in nature, with DNA polymerase and dNTPs. Synthetic linkers, segments of DNA synthesized to contain several restriction sites, can be added to this double-helical DNA for ligation to a suitable vector. Complementary DNA for all mRNA that a cell contains can be made, inserted into such vectors, and then inserted into bacteria. Thus, each bacterium will contain a vector with an inserted piece of DNA as well as its own circular chromosomes. Such a collection of vectors constitutes a *cDNA library*.

> **DID YOU KNOW?**
>
> Retroviruses contain an RNA genome but replicate through a DNA intermediate. The conversion of RNA information into DNA information is catalyzed by reverse transcriptase. Human immunodeficiency virus (HIV), the cause of AIDS, is a retrovirus.

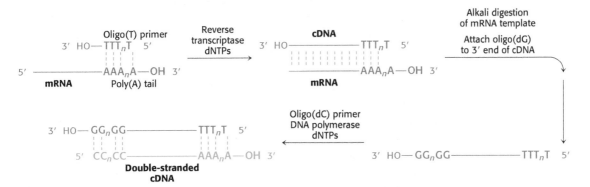

Figure 41.7 Formation of a cDNA duplex. A complementary DNA (cDNA) duplex is created from mRNA by using reverse transcriptase to synthesize a cDNA strand—first, along the mRNA template and, then, after digestion of the mRNA, along that same newly synthesized cDNA strand.

? QUICK QUIZ Why are restriction enzymes such vital tools for recombinant DNA technology?

Estrogen-Receptor cDNA Can Be Identified by Screening a cDNA Library

With these techniques at our disposal, let us return to our experiments with the estrogen receptor. We will use the probe that we synthesized earlier to screen a cDNA library generated from an estrogen-responsive tissue, such as a rat uterus. For our experiments, we will use a λ phage cDNA library.

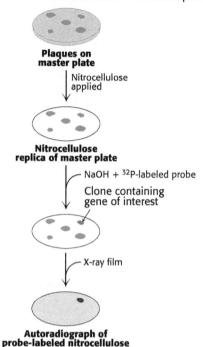

Figure 41.8 Screening a cDNA library for a specific gene. Here, a plate is tested for plaques containing cDNA for the estrogen receptor.

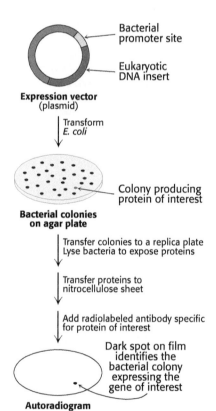

Figure 41.9 Screening expression vectors for the presence of the estrogen receptor. A method of screening for cDNA clones is to identify expressed protein products by staining with specific antibody.

A dilute suspension of the recombinant phages is first plated on a lawn of bacteria (**Figure 41.8**). Where each phage particle has landed and infected a bacterium, a *plaque* containing identical phages develops on the plate. A replica of this master plate is then made by applying a sheet of nitrocellulose. Infected bacteria and phage DNA released from lysed cells adhere to the sheet in a pattern of spots corresponding to the plaques. Intact bacteria on this sheet are lysed with NaOH, which also serves to denature the DNA so that it becomes accessible for hybridization with a ^{32}P-labeled single-stranded probe. The probe will bind only to the DNA sequence encoding the estrogen-receptor cDNA. Autoradiography of the replica then reveals the positions of spots harboring recombinant DNA. The corresponding plaques are picked out of the intact master plate and grown. A single investigator can readily screen a million clones in a day.

The vector containing the cDNA for the estrogen receptor can be isolated and transcribed. The resulting mRNA can be translated in vitro to produce receptor for experiments.

Complementary DNA Libraries Can Be Screened for Synthesized Protein

The vectors discussed so far simply carry the incorporated DNA and allow for the transcription of the inserted DNA. However, with the use of a specially prepared vector, bacteria that are actually expressing the estrogen-receptor protein can be isolated. Complementary DNA molecules can be inserted into specially engineered vectors that favor the efficient expression of these molecules in hosts such as *E. coli*. Such plasmids or phages are called *expression vectors*. These vectors maximize the transcription of the inserted DNA by using a powerful promoter. Expression vectors also contain a segment of DNA that encodes a ribosome-binding site on the mRNA that is transcribed from the inserted cDNA. Thus, cDNA molecules that are inserted into these vectors are not only transcribed but also translated. Clones containing the cDNA can be screened on the basis of their capacity to direct the synthesis of a foreign protein in bacteria. Spots of bacteria on a replica plate are lysed to release proteins, which bind to an applied nitrocellulose filter. With the use of immunological techniques similar to those mentioned in Chapter 5, the monoclonal antibody for the estrogen receptor can be used to identify colonies of bacteria that harbor the corresponding cDNA vector (**Figure 41.9**).

Having a cDNA for the estrogen receptor enables us to perform a number of experiments to determine the biochemical properties of the protein. For instance, we learned earlier that the estrogen receptor is a transcription factor that functions by binding to the DNA of select genes (p. 681). By using the cloned receptor, we can perform experiments to determine the DNA sequence to which the receptor binds most tightly. We could investigate whether the receptor reacts with other proteins when binding to DNA. Indeed, the knowledge to be gained is limited only by our imagination and experimental skill. However, having cDNA for the receptor tells us little about the gene that encodes the receptor itself. Does the gene contain introns? What regulatory sequences control its expression? To answer these questions and similar ones, we must isolate the gene that encodes the receptor. To do so, we return to a library, but this time to a genomic library.

Specific Genes Can Be Cloned from Digests of Genomic DNA

Let us see how we can clone a gene that is present just once in a haploid genome, such as the gene encoding the estrogen receptor. The approach is to prepare a large collection (library) of fragments of genomic DNA and then to identify those members of the collection that have the gene of interest.

A sample containing many copies of total genomic DNA—in our case, rat DNA—is first mechanically sheared or partly digested by restriction enzymes into large fragments. This process yields a nearly random population of overlapping DNA fragments. These fragments are then separated by gel electrophoresis to isolate the set of all fragments that are about 15 kb long because this size is convenient for insertion into vectors. Synthetic linkers are attached to the ends of these fragments, cohesive ends are formed, and the fragments are then inserted into a vector, such as λ phage DNA, prepared with the same cohesive ends (**Figure 41.10**). *E. coli* bacteria are then infected with these recombinant phages. These phages replicate themselves and then lyse their bacterial hosts. The resulting lysate contains fragments of rat DNA housed in a sufficiently large number of virus particles to ensure that the entire genome is represented. These phages constitute a *genomic library*. This genomic library is then screened in a similar fashion to the screening of a cDNA library.

The gene of interest is unlikely to be found in one piece of DNA, because genes are usually larger than 15 kb, the size of the fragments used to make the genomic library. Consequently, several clones from the genomic library will harbor different parts of the gene for the estrogen receptor. These clones must be isolated and sequenced to determine the sequence of the entire gene.

DNA Can Be Sequenced by the Controlled Termination of Replication

The analysis of DNA structure and its role in gene expression also have been markedly facilitated by the development of powerful techniques for the *sequencing* of DNA molecules. The key to DNA sequencing is the generation of DNA fragments whose length is determined by the last base in the sequence. Collections of such fragments can be generated through the *controlled termination of replication* (Sanger dideoxy method), a method developed by Frederick Sanger and his coworkers. The same procedure is performed on four reaction mixtures at the same time. In all these mixtures, a DNA polymerase is used to make the complement of a short sequence within a single-stranded DNA molecule that is being sequenced. The synthesis is primed by a chemically synthesized fragment that is complementary to a part of the sequence. In addition to the four deoxyribonucleoside triphosphates (radioactively labeled), each reaction mixture contains a small amount of the *2′, 3′-dideoxy analog* of one of the nucleotides, a different nucleotide for each reaction mixture.

2′,3′-Dideoxy analog

The incorporation of this analog blocks further growth of the new strand because the dideoxy analog lacks the 3′-hydroxyl terminus needed to form the next phosphodiester linkage. The concentration of the analog is low enough that strand termination will take place only occasionally. The polymerase will insert the correct nucleotide sometimes and the dideoxy analog other times, stopping the reaction. For instance, if the dideoxy analog of dATP (ddATP) is present, fragments of various lengths are produced, but all will be terminated by ddATP (**Figure 41.11**). Importantly, ddATP will be inserted only where a T was located in the DNA being sequenced. Thus, the fragments of different length will correspond to the positions of T. Four such sets of *strand-terminated fragments* (one for each dideoxy analog) then undergo electrophoresis, and the base sequence of the new DNA is read from the autoradiogram of the four lanes.

Fluorescence detection is a highly effective alternative to autoradiography. A fluorescent tag is incorporated into each dideoxy analog—a differently colored

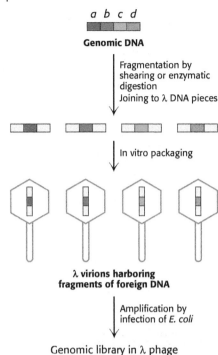

Figure 41.10 Creating a genomic library. A genomic library can be created from a digest of a whole complex genome.

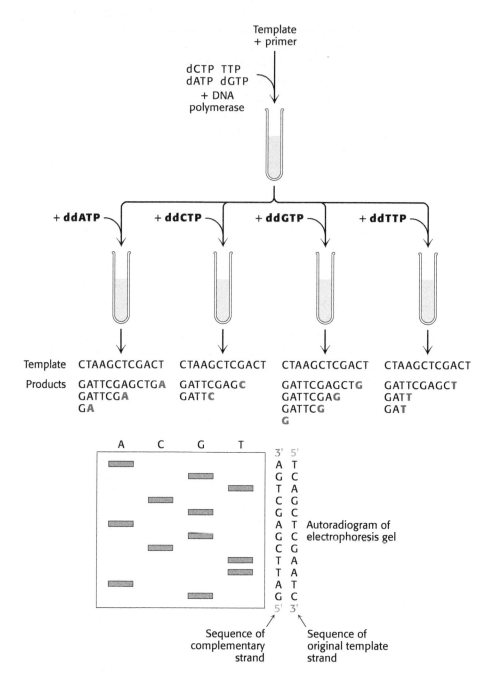

Figure 41.11 Strategy of the strand-termination method for sequencing DNA. Fragments are produced by adding the 2',3'-dideoxy analog of a dNTP to each of four polymerization mixtures. For example, the addition of the dideoxy analog of dATP results in fragments ending in A. The strand cannot be extended past the dideoxy analog. [Information from B. A. Pierce, *Genetics*, 3d ed. (W. H. Freeman and Company, 2008), p. 526.]

one for each of the four strand terminators (e.g., a blue emitter for termination at A and a red one for termination at C). With the use of a mixture of terminators, a single reaction can be performed and the resulting fragments are then subjected to electrophoresis. The separated bands of DNA are detected by their fluorescence as they emerge subsequent to electrophoresis; the sequence of their colors yields the base sequence directly (**Figure 41.12**). Fluorescence detection is attractive because it eliminates the use of radioactive reagents and can be readily automated. Indeed, modern DNA-sequencing instruments can sequence millions of bases per day with the use of this method.

Applying such sequencing tools to the investigation of the gene for the estrogen receptor reveals that the gene is more than 140 kb in length and contains eight exons. In addition to TATA and CAAT boxes (p. 678), the upstream region of the gene contains a P1 promoter that is activated by the transcription factor AP2γ. Interestingly, certain breast cancers depend on the presence of the estrogen receptor for malignant growth, and AP2γ may play a critical role in the regulation of the gene for the estrogen receptor in cancer cells.

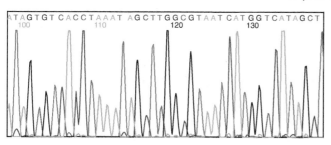

ATAGTGT CACCTAAAT AGCTTGGCGTAAT CAT GGTCATAGCT
100   110   120   130

Figure 41.12 Fluorescence detection of oligonucleotide fragments produced by the dideoxy method. A sequencing reaction is performed with four strand-terminating dideoxy nucleotides, each labeled with a tag that fluoresces at a different wavelength (e.g., red for T). Each of the four colors represents a different base in a chromatographic trace produced by fluorescence measurements at four wavelengths. [Data from A. J. F. Griffiths et al., *An Introduction to Genetic Analysis*, 8th ed. (W. H. Freeman and Company, 2005).]

CLINICAL AND BIOLOGICAL INSIGHT

Next-Generation Sequencing Methods Enable the Rapid Determination of a Complete Genome Sequence

Since the introduction of the Sanger dideoxy method in the mid-1970s, significant advances have been made in DNA-sequencing technologies, enabling the readout of progressively longer sequences with higher fidelity and shorter run times. The development of *next-generation sequencing* (NGS) platforms has extended this capability to formerly unforeseen levels. By combining technological breakthroughs in the handling of very small amounts of liquid, high-resolution optics, and computing power, these methods have already made a significant impact on the ability to obtain whole genome sequences rapidly and cheaply.

Next-generation sequencing refers to a family of technologies, each of which utilizes a unique approach for the determination of a DNA sequence. All of these methods are *highly parallel*: from 1 million to 1 billion DNA fragment sequences are acquired in a single experiment. How are NGS methods capable of attaining such a high number of parallel runs? Individual DNA fragments are amplified by polymerase chain reaction (PCR) (p. 754) on a solid support, such as a single bead or a small region of a glass slide, such that clusters of identical DNA fragments are distinguishable by high-resolution imaging. These fragments then serve as templates for DNA polymerase, where the addition of nucleotide triphosphates is converted to a signal that can be detected in a highly sensitive manner. The technique used to detect individual base incorporation varies among NGS methods. However, most of these can be understood simply by considering the overall reaction of chain elongation catalyzed by DNA polymerase (**Figure 41.13**). In the *reversible terminator method*, the four nucleotides are added to the template DNA, with each base tagged with a unique fluorescent label and a reversibly blocked 3′ end. The blocked end assures that only one phosphodiester linkage will form. Once the nucleotide is incorporated into the growing strand, it is identified by its fluorescent tag, the blocking agent is removed and the process is repeated.

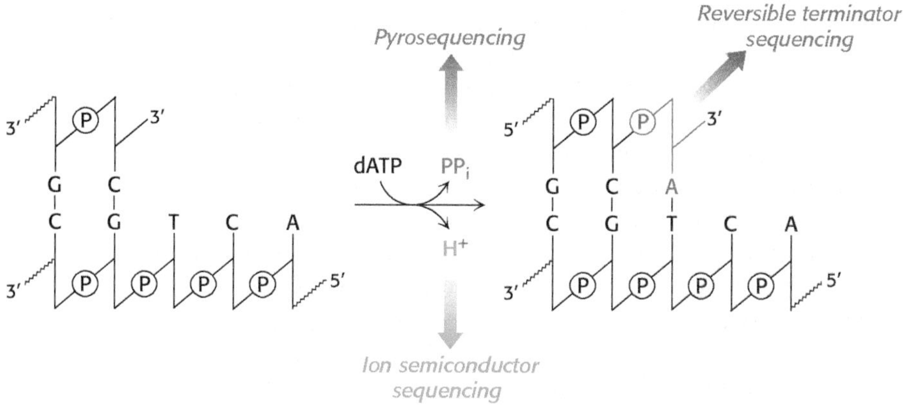

Pyrosequencing

Reversible terminator sequencing

dATP   PP$_i$

H$^+$

Ion semiconductor sequencing

Figure 41.13 Detection methods in next-generation sequencing. Measurement of base incorporation in next-generation sequencing methods relies on the detection of the various products of the DNA polymerase reaction. Reversible terminator sequencing measures the nucleotide incorporation in a manner similar to Sanger sequencing, while pyrosequencing and ion semiconductor sequencing detect the release of pyrophosphate and protons, respectively.

In *pyrosequencing*, nucleotides are added to the template DNA, one at a time in a defined order. One of the nucleotides will be incorporated into the growing strand, releasing a pyrophosphate that is detected by coupling the formation of pyrophosphate with the production of light by the sequential action of the enzymes ATP sulfurylase and luciferase:

$$PP_i + \text{adenylyl sulfate} \xrightleftharpoons{\text{ATP sulfurylase}} ATP + \text{sulfate}$$

$$ATP + \text{luciferin} \xrightleftharpoons{\text{Luciferase}} \text{oxyluciferin} + \text{light}$$

The protocol for *ion semiconductor sequencing* is similar to pyrosequencing except that nucleotide incorporation is detected by sensitively measuring the very small changes in pH of the reaction mixture due to the release of proton upon nucleotide incorporation.

Regardless of the sequencing method, the technology exists to quantify the signal produced by millions of DNA fragment templates simultaneously. However, for many approaches, as few as 50 bases are read per fragment. Hence, significant computing power is required to both store the massive amounts of sequence data and perform the necessary alignments required to assemble a complete sequence. Next-generation sequencing methods are being used to answer an ever-growing number of questions in genomics, transcriptomics, and evolutionary biology, to name a few. Additionally, individual genome sequences will provide information about genetic variation within populations and may usher in an era of personalized medicine, when these data can be used to guide treatment decisions.

Selected DNA Sequences Can Be Greatly Amplified by the Polymerase Chain Reaction

Let us summarize our research accomplishments thus far. We have purified the estrogen receptor by using the monoclonal antibody that we generated (Chapter 5). We have synthesized a DNA probe that allowed us to isolate the cDNA of the receptor as well as the gene for the receptor. Finally, we have deduced the DNA sequence of the gene. Although there are many possible experiments to perform on the basis of what we have accomplished so far, let us start a new research project that will introduce us to one of the most powerful techniques in experimental biochemistry. Our experimental system thus far has been with the rat uterus. We can ask whether the receptor gene is transcribed in other tissues, such as the brain and the liver.

We could, in fact, screen cDNA libraries from these tissues, searching for a clone that contains the receptor cDNA as heretofore described. However, we will use a much more rapid means of detection. We will study cDNA prepared from brain and liver tissues as well as other tissues and will determine, with the use of the *polymerase chain reaction* (PCR), whether cDNA (and, by implication, the mRNA) for the receptor is present.

Consider a DNA duplex consisting of a target sequence surrounded by nontarget DNA. In our example, the target would be the putative receptor cDNA in the brain, the liver, or muscle. If the target DNA is present, we can detect it if we first amplify the amount of DNA present. Millions of copies of the target sequences can be readily obtained by PCR if the flanking sequences of the target are known, and we know what the flanking sequences are because we have the DNA sequence of the receptor. PCR is carried out by adding the following components to a solution containing the target sequence: (1) a pair of primers that hybridize with the flanking sequences of the target, (2) all four deoxyribonucleoside triphosphates (dNTPs), and (3) a heat-stable DNA polymerase. A PCR cycle consists of three steps (Figure 41.14):

1. *Strand Separation.* The two strands of the parent DNA molecule are separated by heating the solution to 95°C for 15 s.

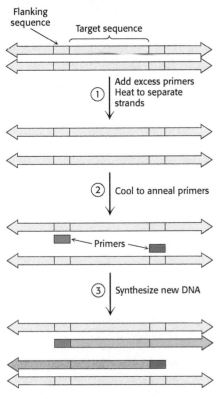

Figure 41.14 The first cycle in the polymerase chain reaction (PCR). A cycle consists of three steps: strand separation, the hybridization of primers, and the extension of primers by DNA synthesis.

2. *Hybridization of Primers.* The solution is then abruptly cooled to 54°C to allow each primer to hybridize to a DNA strand. One primer hybridizes to the 3′ end of the target on one strand, and the other primer hybridizes to the 3′ end on the complementary target strand. Parent DNA duplexes do not form, because the primers are present in large excess. Primers are typically from 20 to 30 nucleotides long.

3. *DNA Synthesis.* The solution is then heated to 72°C, the optimal temperature for *Taq* DNA polymerase. This heat-stable polymerase comes from *Thermus aquaticus*, a thermophilic bacterium that lives in hot springs. The polymerase elongates both primers in the direction of the target sequence because DNA synthesis is in the 5′-to-3′ direction. DNA synthesis takes place on both strands but extends beyond the target sequence.

These three steps—strand separation, the hybridization of primers, and DNA synthesis—constitute one cycle of PCR amplification and can be carried out repetitively just by changing the temperature of the reaction mixture. The thermostability of the polymerase makes it feasible to carry out PCR in a closed container; no reagents are added after the first cycle. The duplexes are heated to begin the second cycle, which produces four duplexes, and then the third cycle is initiated (**Figure 41.15**). At the end of the third cycle, two short strands appear that constitute only the target sequence—the sequence including and bounded by the primers. Subsequent cycles will amplify the target sequence exponentially. The larger strands increase in number arithmetically and serve as a source for the synthesis of more short strands. Ideally, after n cycles, the desired sequence is amplified 2^n-fold. The amplification is a millionfold after 20 cycles and a billionfold after 30 cycles, which can be carried out in less than an hour.

Several features of this remarkable method for amplifying DNA are noteworthy. First, the sequence of the target need not be known. All that is required is knowledge of the flanking sequences. Second, the target can be much larger than the primers. Targets larger than 10 kb have been amplified by PCR. Third, primers do not have to be perfectly matched to flanking sequences to amplify targets. With the use of primers derived from a gene of known sequence, it is possible to search for variations on the theme. In this way, families of genes are being discovered with the use of PCR. Fourth, PCR is highly specific because of the stringency of hybridization at relatively high temperature. *Stringency* is the required closeness of the match between primer and target, which can be controlled by temperature and salt. At high temperatures, the only DNA that is amplified is that situated between primers. A gene constituting less than a millionth of the total DNA of a higher organism is accessible by PCR. Fifth, PCR is exquisitely sensitive. A single DNA molecule can be amplified and subsequently visualized in gel electrophoresis. Indeed, the amplified DNA can be isolated from the gel and inserted into a vector and cloned if so desired.

PCR examination for the presence of the estrogen receptor in cDNA libraries from various tissues reveals that significant amounts of receptor mRNA are present in pituitary, bone, liver, and muscle cells, as well as in the reproductive tissues, including ovary, mammary gland, and uterus. Further studies using the same techniques show that the estrogen receptor is found in all vertebrates.

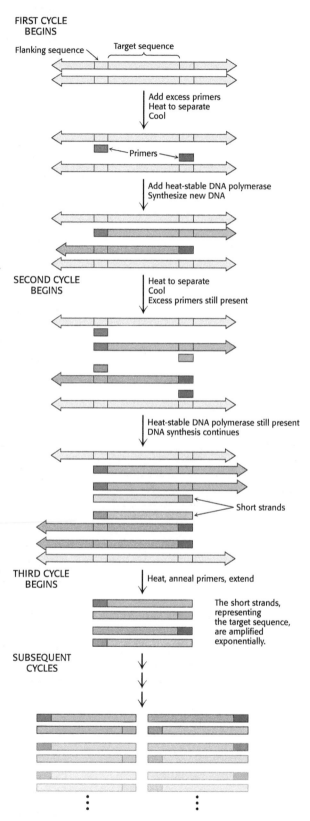

Figure 41.15 Multiple cycles of the polymerase chain reaction. The two short strands produced at the end of the third cycle (along with longer strands not shown) represent the target sequence. Subsequent cycles will amplify the target sequence exponentially and the parent sequence arithmetically.

PCR Is a Powerful Technique in Medical Diagnostics, Forensics, and Studies of Molecular Evolution

PCR can provide valuable diagnostic information in medicine. Bacteria and viruses can be readily detected with the use of specific primers. For example, PCR can reveal the presence of human immunodeficiency virus in people who have not mounted an immune response to this pathogen and would therefore be missed with an antibody assay. Finding *Mycobacterium tuberculosis* bacilli, the cause of tuberculosis, in tissue specimens is slow and laborious. With PCR, as few as 10 tubercle bacilli per million human cells can be readily detected. PCR is a promising method for the early detection of certain cancers. This technique can identify mutations of certain growth-control genes, such as the *ras* genes (p. 236). The capacity to greatly amplify selected regions of DNA can also be highly informative in monitoring cancer chemotherapy. Tests using PCR can detect when cancerous cells have been eliminated and treatment can be stopped; they can also detect a relapse and the need to immediately resume treatment. PCR is ideal for detecting leukemias caused by chromosomal rearrangements.

PCR is also having an effect in forensics and legal medicine. An individual DNA profile is highly distinctive because many genetic loci are highly variable within a population. For example, variations at specific loci determine a person's HLA type (human-leukocyte-antigen type); organ transplants are rejected when the HLA types of the donor and recipient are not sufficiently matched. PCR amplification of multiple genes is being used to establish biological parentage in disputed paternity and immigration cases. Analyses of blood stains and semen samples by PCR have implicated guilt or innocence in numerous assault and rape cases. The root of a single shed hair found at a crime scene contains enough DNA for typing by PCR (**Figure 41.16**).

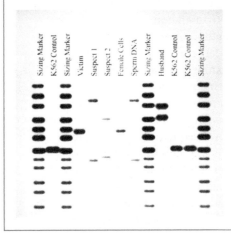

Figure 41.16 DNA and forensics. DNA isolated from sperm obtained during the examination of a rape victim was amplified by PCR, then compared with DNA from the victim and three potential suspects—the victim's husband and two additional individuals—using gel electrophoresis and autoradiography. Sperm DNA matched the pattern of Suspect 1, but not that of Suspect 2 or the victim's husband. Sizing marker and K562 lanes refer to control DNA samples. [Martin Shields/Science Source.]

Gene-Expression Levels Can Be Comprehensively Examined

Let us look now at a final technique, one that enables us to see how environmental signals, such as the presence of hormones, or pathological conditions, such as cancer, alter the expression of an array of genes in a tissue. Most genes are present in the same quantity in every cell—namely, one copy per haploid cell or two copies per diploid cell. However, the level at which a gene is expressed, as indicated by mRNA quantities, can vary widely, ranging from no expression to hundreds of mRNA copies per cell. Gene-expression patterns vary from cell type to cell type, distinguishing, for example, a muscle cell from a nerve cell. Even within the same cell, gene-expression levels can vary as the cell responds to changes in physiological circumstances. Note that mRNA levels sometimes correlate with the levels of proteins expressed, but this correlation does not always hold. Thus, care must be exercised when interpreting the results of mRNA levels alone.

The quantity of individual mRNA transcripts can be determined by *quantitative PCR* (qPCR), or real-time PCR. RNA is first isolated from the cell or tissue of interest. With the use of reverse transcriptase, cDNA is prepared from this RNA sample. In one qPCR approach, the transcript of interest is PCR amplified with the appropriate primers in the presence of the dye SYBR Green I, which fluoresces brightly when bound to double-stranded DNA. In the initial PCR cycles, not enough duplex is present to allow a detectable fluorescence signal. However, after repeated PCR cycles, the fluorescence intensity exceeds the detection threshold and continues to rise as the number of duplexes corresponding to the transcript of interest increases (**Figure 41.17**). Importantly, the cycle number at which the fluorescence becomes detectable over a defined threshold (or C_T) is indirectly proportional to the number of copies of the original template. After the relation between the original copy number and the C_T has been established with the use of a known standard, subsequent qPCR experiments can be used to determine the number of copies of any desired transcript in the original sample, provided the appropriate primers are available.

Although qPCR is a powerful technique for quantitation of a small number of transcripts in any given experiment, we can now use our knowledge of complete genome sequences to investigate an entire *transcriptome*, the pattern and level of expression of all genes in a particular cell or tissue. One of the most powerful methods developed to date for this purpose is based on hybridization. Oligonucleotides or cDNAs are affixed to a solid support such as a microscope slide, creating a *DNA microarray,* or *gene chip*. Fluorescently labeled cDNA is then hybridized to the chip to reveal the expression level for each gene, identifiable by its known location on the chip. The intensity of the fluorescent spot on the chip reveals the extent of the transcription of a particular gene. **Figure 41.18** shows the pattern of genes that are induced or repressed in various breast-cancer tumors, and **Figure 41.19** shows how yeast transcription varies under different conditions. An analysis of mRNA pools with the use of these chips revealed, for example, that approximately 50% of all yeast genes are expressed at steady-state levels of 0.1 to 1.0 mRNA copy per cell. This method readily detected variations in expression levels displayed by specific genes under different growth conditions.

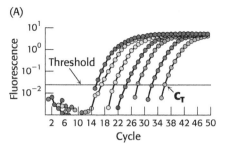

(A)

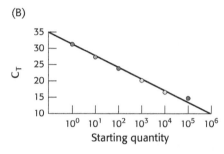

(B)

Figure 41.17 Quantitative PCR. (A) In qPCR, fluorescence is monitored in the course of PCR amplification to determine C_T, the cycle at which this signal exceeds a defined threshold. Each color represents a different starting quantity of DNA. **(B)** C_T values are inversely proportional to the number of copies of the original cDNA template. [Data from N. J. Walker, *Science* 296:557–559, 2002.]

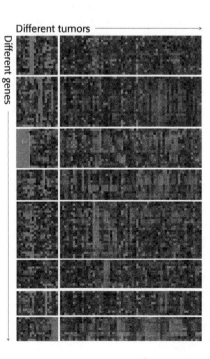

Figure 41.18 Gene-expression analysis with microarrays. The expression levels of thousands of genes can be simultaneously analyzed by using DNA microarrays (gene chips). Here, the analysis of 1733 genes in 84 breast-tumor samples reveals that the tumors can be assorted into distinct classes on the basis of their gene-expression patterns. Red corresponds to gene induction, and green corresponds to gene repression. [Reprinted by permission from Macmillan Publishers Ltd: *Nature*, 406:747. C. M. Perou et al., Molecular portraits of human breast tumours. ©2000.]

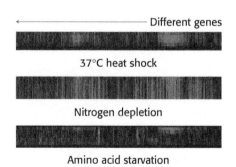

Figure 41.19 Monitoring changes in gene expression in yeast. This microarray analysis shows levels of gene expression for yeast genes under different conditions. [Adapted by permission from Macmillan Publishers Ltd: *Nature* 409:533, Iyer, V.R. et al., Genomic binding sites of the yeast cell-cycle transcription factors SBF and MBF, ©2001.]

SUMMARY

41.1 Nucleic Acids Can Be Synthesized from Protein-Sequence Data

Amino-acid-sequence information obtained from the purified receptor can be used to synthesize DNA probes that correspond to the amino acid sequence. The technique is to add modified deoxynucleotides to one another to form a growing strand that is linked to an insoluble support. DNA strands as long as 100 nucleotides can be readily synthesized.

41.2 Recombinant DNA Technology Has Revolutionized All Aspects of Biology

The recombinant DNA revolution in biology is rooted in the repertoire of enzymes that act on nucleic acids. Restriction enzymes are a key group among them. These endonucleases recognize specific base sequences in double-helical DNA and cleave both strands of the duplex, forming specific fragments of DNA. These restriction fragments can be separated, displayed by gel electrophoresis, and inserted in vectors to generate recombinant DNA molecules.

41.3 Eukaryotic Genes Can Be Manipulated with Considerable Precision

New genes can be constructed in the laboratory, introduced into host cells, and expressed. Novel DNA molecules are made by joining fragments that have complementary cohesive ends produced by the action of a restriction enzyme. DNA ligase seals breaks in DNA chains. Vectors for propagating the DNA include plasmids and λ phage. With the use of these techniques, cDNA libraries for any tissue and genomic libraries from any organism can be generated. These libraries can then be screened by using DNA probes. Foreign DNA can be expressed after insertion into bacteria by the appropriate vector. Specific genes can be cloned from a genomic library by using a DNA probe.

Rapid sequencing techniques have been developed to further the analysis of DNA molecules. DNA can be sequenced by the controlled interruption of replication. The fragments produced are separated by gel electrophoresis and visualized by autoradiography of a ^{32}P label at the 5′ end or by fluorescent tags.

The polymerase chain reaction makes it possible to greatly amplify specific segments of DNA in vitro. The region amplified is determined by the placement of a pair of primers that are added to the target DNA along with a thermostable DNA polymerase and deoxyribonucleoside triphosphates.

DNA microarrays, or gene chips, allow the detection of changes in the transcription of hundreds of genes simultaneously. Microarray analysis shows how transcription responds to various physiological and pathological conditions, such as the presence of a hormone or the transformation into a cancerous state.

KEY TERMS

complementary DNA (cDNA) library (p. 744)
reverse transcriptase (p. 744)
complementary DNA (cDNA) (p. 744)
restriction enzyme (p. 745)
Southern blotting (p. 747)
northern blotting (p. 747)

vector (p. 747)
plasmid (p. 747)
bacteriophage lambda (λ phage) (p. 748)
sticky ends (p. 748)
DNA ligase (p. 748)
expression vector (p. 750)
genomic library (p. 751)

controlled termination of replication (Sanger dideoxy method) (p. 751)
polymerase chain reaction (PCR) (p. 754)
quantitative PCR (qPCR) (p. 757)
DNA microarray (gene chip) (p. 757)

? Answer to QUICK QUIZ

Because of their high degree of specificity, restriction enzymes allow precise cleavage of double-stranded DNA. They are essentially molecular scalpels.

PROBLEMS

1. *Building a library.* What is a cDNA library? How it is constructed? ✓ 8

2. *Different libraries.* Differentiate between a cDNA library and a genomic library. ✓ 8

3. *It's not the heat, . . .* Why is *Taq* polymerase especially useful for PCR? ✓ 8

4. *Like Marie and Pierre.* Match each term with its description. ✓ 8

(a) cDNA library _____
(b) Reverse transcriptase _____
(c) Restriction enzyme _____
(d) Vector _____
(e) Sticky ends _____
(f) DNA ligase _____
(g) Genomic library _____
(h) Sanger dideoxy method _____
(i) Polymerase chain reaction _____
(j) DNA microarray _____

1. Cleave double-helical DNA at specific sequences
2. Measures the expression of many genes
3. Sequencing by strand termination
4. DNA sequences representing mRNA
5. Piece of DNA taken up and replicated by bacteria
6. Fragments of DNA, housed in phages, that represent an entire genome
7. Joins two DNA molecules
8. Copies RNA into DNA
9. Amplifies selected DNA sequences
10. Complementary single strands of DNA

5. *Probe design.* Which of the following amino acid sequences would yield the most optimal oligonucleotide probe? ✓ 8

Ala-Met-Ser-Leu-Pro-Trp

Gly-Trp-Asp-Met-His-Lys

Cys-Val-Trp-Asn-Lys-Ile

Arg-Ser-Met-Leu-Gln-Asn

6. *The right template.* Ovalbumin is the major protein of egg white. The chicken ovalbumin gene contains eight exons separated by seven introns. Which should be used to form the protein in *E. coli*: ovalbumin cDNA or ovalbumin genomic DNA? Why?

7. *Cleavage frequency.* The restriction enzyme *Alu*I cleaves at the sequence 5'-AGCT-3', and *Not*I cleaves at 5'-GCGGCCGC-3'. What would be the average distance between cleavage sites for each enzyme on digestion of double-stranded DNA? Assume that the DNA contains equal proportions of A, G, C, and T.

8. *The right cuts.* Suppose that a human genomic library is prepared by exhaustive digestion of human DNA with the *Eco*RI restriction enzyme. Fragments averaging about 4 kb in length will be generated. Is this procedure suitable for cloning large genes? Why or why not? ✓ 8

9. *Reading sequences.* An autoradiogram of a sequencing gel containing four lanes of DNA fragments is shown in the adjoining illustration. ✓ 8

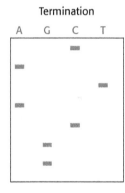

Termination

(a) What is the sequence of the DNA fragment?
(b) Suppose that the Sanger dideoxy method shows that the template-strand sequence is 5'-TGCAATGGC-3'. Sketch the gel pattern that would lead to this conclusion.

10. *A revealing cleavage.* Sickle-cell anemia arises from a mutation in the gene for the β chain of human hemoglobin. The change from GAG to GTG in the mutant eliminates a cleavage site for the restriction enzyme *Mst*II, which recognizes the target sequence CCTGAGG. These findings form the basis of a diagnostic test for the sickle-cell gene. Propose a rapid procedure for distinguishing between the normal and the mutant gene. Would a positive result prove that the mutant contains GTG in place of GAG?

11. *A blessing and a curse.* The power of PCR can create problems. Suppose someone claims to have isolated dinosaur DNA by using PCR. What questions might you ask to determine if it is indeed dinosaur DNA? ✓ 8

12. *Questions of accuracy.* The stringency (p. 755) of PCR amplification can be controlled by altering the temperature at which the primers and the target DNA undergo hybridization. How would altering the temperature of hybridization affect the amplification? Suppose that you have a particular yeast gene *A* and that you wish to see if it has a counterpart in human beings. How would controlling the stringency of the hybridization help you? ✓ 8

13. *A puzzling ladder.* A gel pattern displaying PCR products shows four strong bands. The four pieces of DNA have lengths that are approximately in the ratio of 1:2:3:4. The largest band is cut out of the gel, and PCR is repeated with the same primers. Again, a ladder of four bands is evident in the gel. What does this result reveal about the structure of the encoded protein?

14. *Man's best friend.* Why might the genomic analyses of dogs be particularly useful for investigating the genes responsible for body size and other physical characteristics?

Chapter Integration Problem

15. *Designing primers.* A successful PCR experiment often depends on designing the correct primers. In particular, the T_m, the melting temperature of a double helix, for each primer should be approximately the same. What is the basis of this requirement? ✓ 8

Data Interpretation Problems

16. *DNA diagnostics.* Representations of sequencing gels for variants of the α chain of human hemoglobin are shown here. What is the nature of the amino acid change in each of the variants? The first triplet encodes valine.

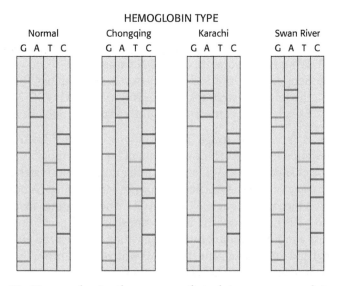

17. *Two peaks.* In the course of studying a gene and its possible mutation in humans, you obtain genomic DNA samples from a collection of persons and PCR amplify a region of interest within this gene. For one of the samples, you obtain the sequencing chromatogram shown here. Explain the appearance of these data at position 49 (indicated by the arrow).

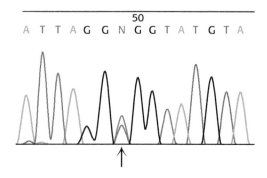

18. *Any direction but east.* A series of people are found to have difficulty eliminating certain types of drugs from their bloodstreams. The problem has been linked to a gene *X*, which encodes an enzyme Y. Six people were tested with the use of various techniques of molecular biology. Person A is a normal control, person B is asymptomatic but some of his children have the metabolic problem, and persons C through F display the trait. Tissue samples from each person were obtained. Southern analysis was performed on the DNA after digestion with the restriction enzyme *Hind*III. Northern analysis of mRNA also was done. In both types of analysis, the gels were probed with labeled *X* cDNA. Finally, a western blot with an enzyme-linked monoclonal antibody was used to test for the presence of protein Y. The results are shown here. Why is person B without symptoms? Suggest possible defects in the other people. ✓ 8

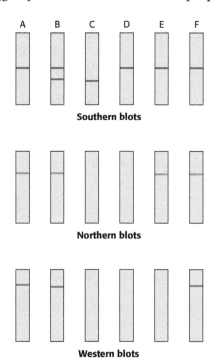

Challenge Problems

19. *Many melodies from one cassette.* Suppose that you have isolated an enzyme that digests paper pulp and have obtained its cDNA. The goal is to produce a mutant that is effective at high temperature. You have engineered a pair of unique restriction sites in the cDNA that flank a 30-bp coding region. Propose a rapid technique for generating many different mutations in this region.

20. *Terra incognita.* PCR is typically used to amplify DNA that lies between two known sequences. Suppose that you want to explore DNA on both sides of a single known sequence. Devise a variation of the usual PCR protocol that would enable you to amplify entirely new genomic terrain. ✓ 8

21. *Chromosome walking.* Propose a method for isolating a DNA fragment that is adjacent in the genome to a previously isolated DNA fragment. Assume that you have access to a complete library of DNA fragments in a vector but that the sequence of the genome under study has not yet been determined. ✓ 8

Selected Readings for this chapter can be found online at www.whfreeman.com/tymoczko3e.

APPENDIX A: Physical Constants and Conversion of Units

Values of physical constants

Physical constant	Symbol or abbreviation	Value
Atomic mass unit (dalton)	amu	1.660×10^{-24} g
Avogadro's number	N	6.022×10^{23} mol^{-1}
Boltzmann's constant	k	1.381×10^{-23} J K^{-1} 3.298×10^{-24} cal K^{-1}
Electron volt	eV	1.602×10^{-19} J 3.828×10^{-20} cal
Faraday constant	F	9.649×10^{4} J V^{-1} mol^{-1} 9.649×10^{4} C mol^{-1}
Curie	Ci	3.70×10^{10} disintegrations s^{-1}
Gas constant	R	8.315 J mol^{-1} K^{-1} 1.987 cal mol^{-1} K^{-1}
Planck's constant	h	6.626×10^{-34} J s 1.584×10^{-34} cal s
Speed of light in a vacuum	c	2.998×10^{10} cm s^{-1}

Abbreviations: C, coulomb; cal, calorie; cm, centimeter; K, kelvin; eq, equivalent; g, gram; J, joule; mol, mole; s, second; V, volt.

Mathematical constants

$\pi = 3.14159$

$e = 2.71828$

$\log_e x = 2.303 \log_{10} x$

Conversion factors

Physical quantity	Equivalent
Length	$1\ cm = 10^{-2}\ m = 10\ mm = 10^4\ \mu m = 10^7\ nm$
	$1\ cm = 10^8\ \text{Å} = 0.3937\ inch$
Mass	$1\ g = 10^{-3}\ kg = 10^3\ mg = 10^6\ \mu g$
	$1\ g = 3.527 \times 10^{-2}\ ounce\ (avoirdupois)$
Volume	$1\ cm^3 = 10^{-6}\ m^3 = 10^3\ mm^3$
	$1\ ml = 1\ cm^3 = 10^{-3}\ liter = 10^3\ \mu l$
	$1\ cm^3 = 6.1 \times 10^{-2}\ inch^3 = 3.53 \times 10^{-5}\ ft^3$
Temperature	$K = {}^\circ C + 273.15$
	${}^\circ C = (5/9)({}^\circ F - 32)$
Energy	$1\ J = 107\ erg = 0.239\ cal = 1\ watt\ s$
Pressure	$1\ torr = 1\ mm\ Hg\ (0{}^\circ C)$
	$= 1.333 \times 10^2\ newtons\ m^{-2}$
	$= 1.333 \times 10^2\ pascals$
	$= 1.316 \times 10^{-3}\ atmospheres$

Standard prefixes

Prefix	Abbreviation	Factor
kilo	k	10^3
hecto	h	10^2
deca	da	10^1
deci	d	10^{-1}
centi	c	10^{-2}
milli	m	10^{-3}
micro	μ	10^{-6}
nano	n	10^{-9}
pico	p	10^{-12}

APPENDIX B: Acidity Constants

pK_a values of some acids

Acid	pK_a (at 25°C)	Acid	pK_a (at 25°C)
Acetic acid	4.76	Malic acid, pK_1	3.40
Acetoacetic acid	3.58	pK_2	5.11
Ammonium ion	9.25	Phenol	9.89
Ascorbic acid, pK_1	4.10	Phosphoric acid, pK_1	2.12
pK_2	11.79	pK_2	7.21
Benzoic acid	4.20	pK_3	12.67
n-Butyric acid	4.81	Pyridinium ion	5.25
Cacodylic acid	6.19	Pyrophosphoric acid, pK_1	0.90
Citric acid, pK_1	3.14	pK_2	2.00
pK_2	4.77	pK_3	6.60
pK_3	6.39	pK_4	9.40
Ethylammonium ion	10.81	Succinic acid, pK_1	4.21
Formic acid	3.75	pK_2	5.64
Glycine, pK_1	2.35	Trimethylammonium ion	9.79
pK_2	9.78	Tris (hydroxymethyl) aminomethane	8.08
Imidazolium ion	6.95	Water*	15.74
Lactic acid	3.86		
Fumaric acid, pK_1	3.03		
pK_2	4.44		

*$[H^+][OH^-] = 10^{-14}$; $[H_2O] = 55.5$ M.

pK$_a$ values of some ionizable groups in proteins

Group	Acid $\rightleftharpoons$ Base		Typical pK$_a$
Terminal α-carboxyl group			3.1
Aspartic acid Glutamic acid			4.1
Histidine			6.0
Terminal α-amino group			8.0
Cysteine			8.3
Tyrosine			10.4
Lysine			10.0
Arginine			12.5

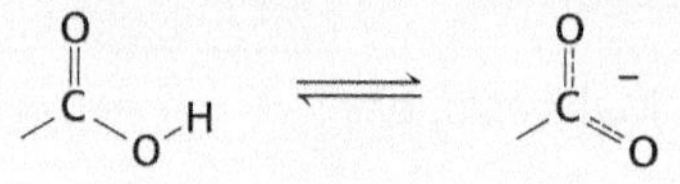

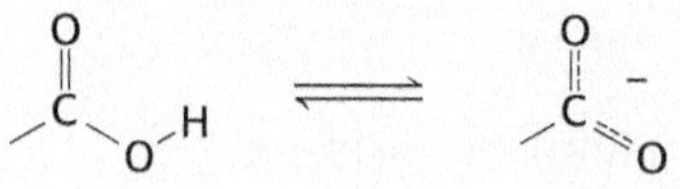

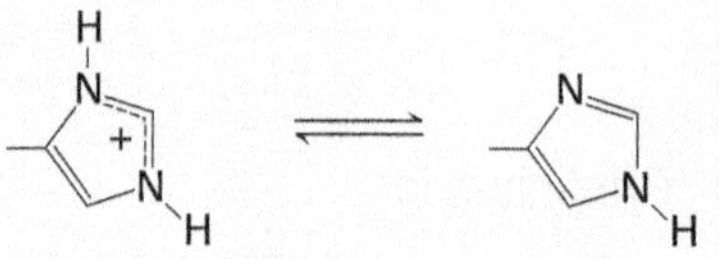

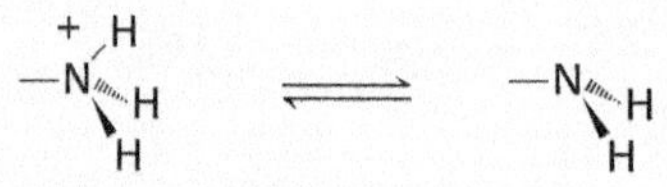

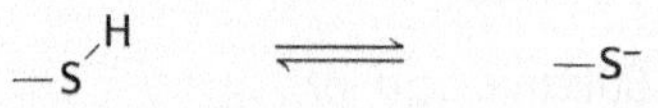

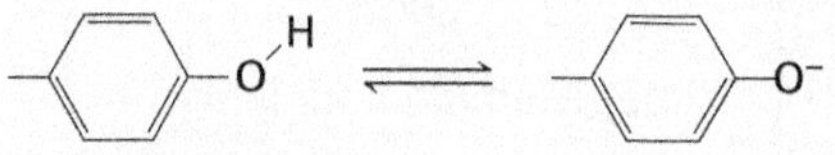

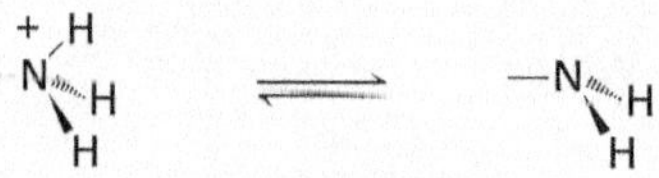

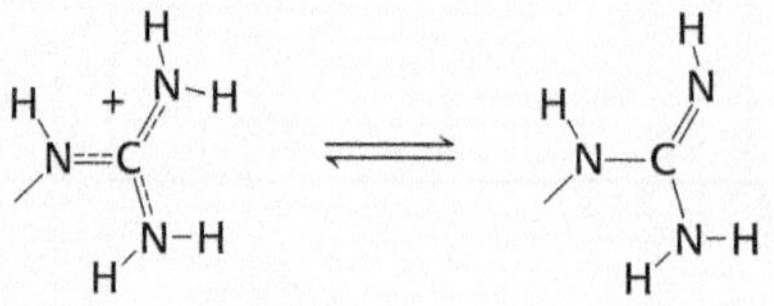

Note: pK$_a$ values depend on temperature, ionic strength, and the microenvironment of the ionizable group.

APPENDIX C: Standard Bond Lengths

Bond	Structure	Length (Å)	Bond	Structure	Length (Å)
C — H	R_2CH_2	1.07	C = O	Aldehyde	1.22
	Aromatic	1.08		Amide	1.24
	RCH_3	1.10	C — S	R_2S	1.82
C — C	Hydrocarbon	1.54	N — H	Amide	0.99
	Aromatic	1.40	O — H	Alcohol	0.97
C = H	Ethylene	1.33	O — O	O_2	1.21
C ≡ C	Acetylene	1.20	P — O	Ester	1.56
C — N	RNH_2	1.47	S — H	Thiol	1.33
	O = C — N	1.34	S — S	Disulfide	2.05
C — O	Alcohol	1.43			
	Ester	1.36			

APPENDIX D: Water-Soluble Vitamins

With the exception of vitamin C, the vitamins listed in this table are the B vitamins, which are also components of coenzymes. Some vitamins are substantially modified to become coenzymes, which are vitamin-derived cofactors that certain enzymes require for their activity.

Vitamin B₁ (Thiamine)

Vitamin	Coenzyme
Vitamin B₁ (Thiamine)	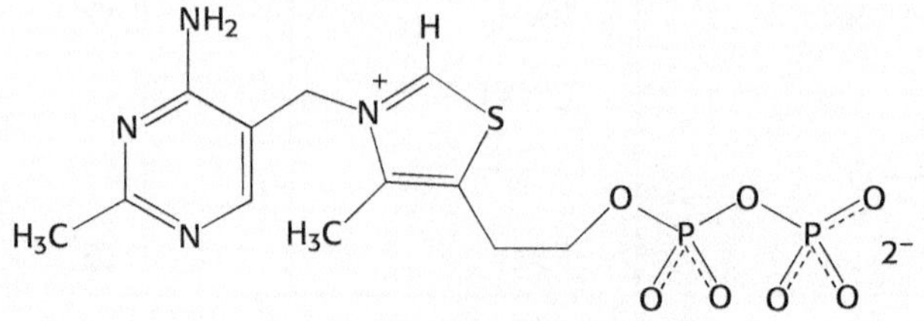 Thiamine pyrophosphate (TPP)

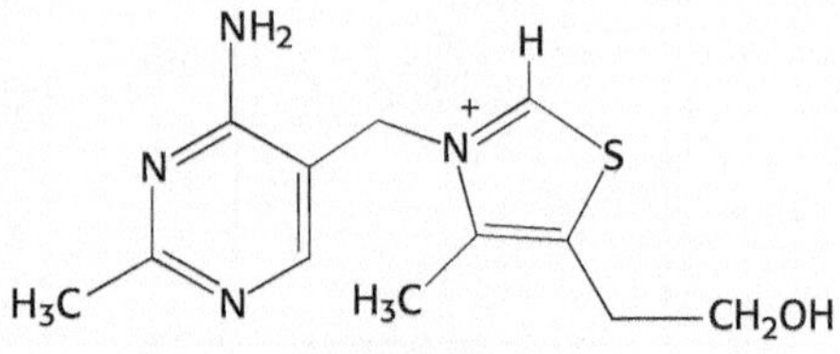

Vitamin B₂ (Riboflavin)

Vitamin	Coenzyme
Vitamin B₂ (Riboflavin)	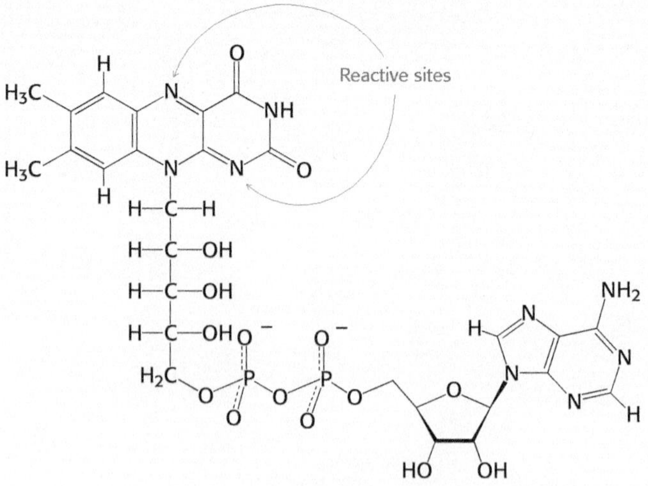 Flavin adenine dinucleotide (FAD)

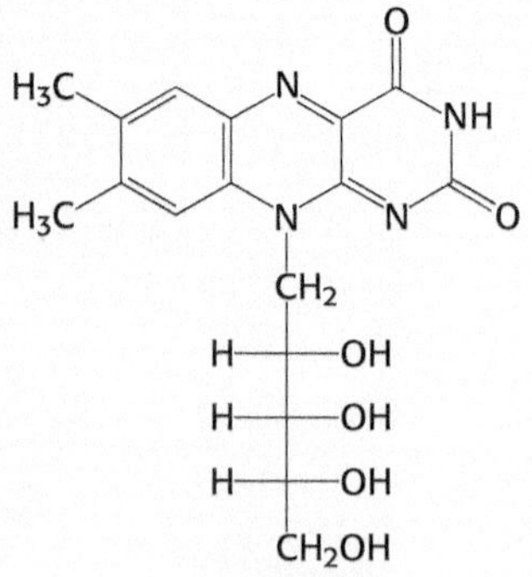

Abbreviation: **RDA** = recommended daily allowance.

Source	Deficiency	Typical reaction	RDA	Page
Pork, legumes, bread and bread products 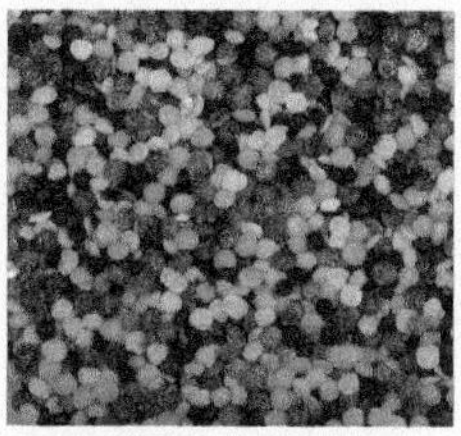 Ruslan/AgeFotostock	Beriberi (muscle weakness, anorexia, weak heart)	Aldehyde transfer	Males 19 years and older: 1.2 mg Females 19 years and older: 1.1 mg	293

Source	Deficiency	Typical reaction	RDA	Page
Almonds, organ meats, whole grains, broccoli 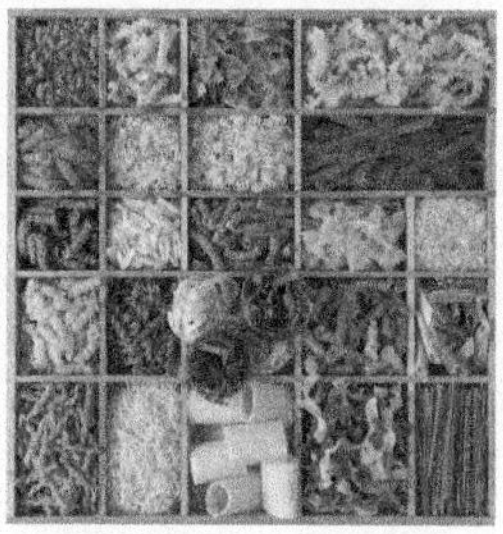 Elena Schweitzer/Shutterstock	Fissures at the corners of the mouth (cheilosis), lesions of the mouth (angular stomatitis), dermatitis	Oxidation–reduction	Males 19 years and older: 1.3 mg Females 19 years and older: 1.1 mg	175

continued on next page

Vitamin B₃ (Niacin)

Vitamin	Coenzyme
Vitamin B₃ (Niacin)	Reactive site **Nicotinamide adenine dinucleotide reaction (phosphate)** R = H (NAD⁺) R = PO₃²⁻ (NADP⁺)

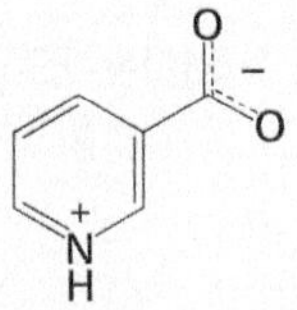

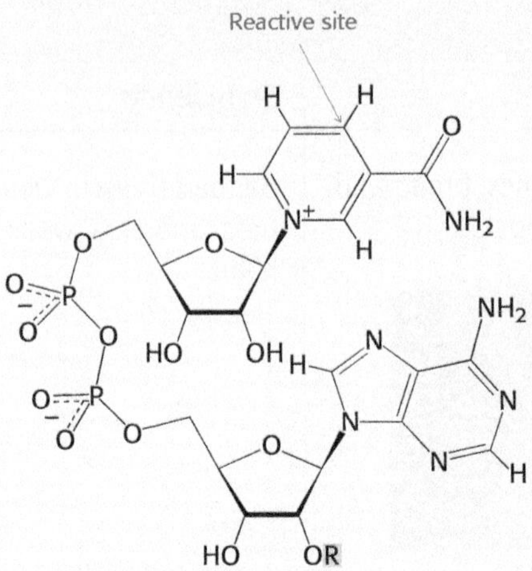

Vitamin B₅ (Pantothenate)

Vitamin	Coenzyme
Vitamin B₅ (Pantothenate)	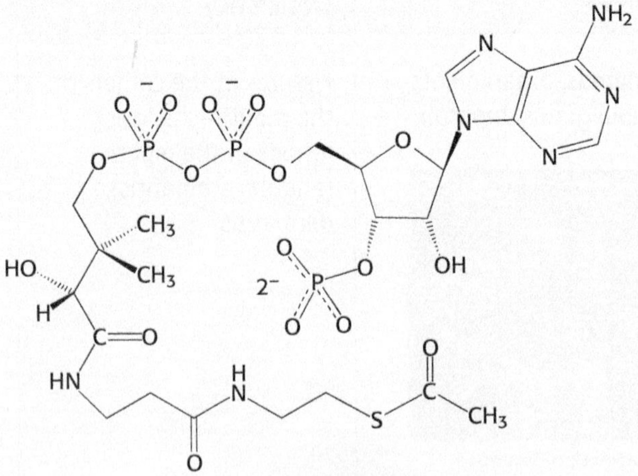 **Acetyl coenzyme A (Acetyl CoA)**

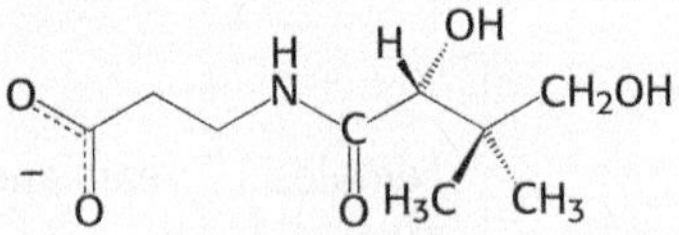

Abbreviations: **RDA** = recommended daily allowance. **AI** = adequate intake; no RDA has been established.

Source	Deficiency	Typical reaction	RDA	Page
Chicken breast, dairy products, legumes, nuts  Brand X Pictures	Pellagra (dermatitis, dementia, diarrhea)	Oxidation-reduction	Males 19 years and older: 16 mg Females 19 years and older: 14 mg	292

Source	Deficiency	Typical reaction	AI	Page
Egg yolk, chicken, beef, broccoli  FoodCollection/AgeFotostock	A deficiency has never been clearly documented. However, some reports suggest numbness, tingling of the skin (paresthesia), muscle cramps	Acyl-group carrier	Males 19 years and older: 5 mg Females 19 years and older: 5 mg	271

continued on next page

Vitamin B$_6$ (Pyridoxine)

Vitamin	Coenzyme

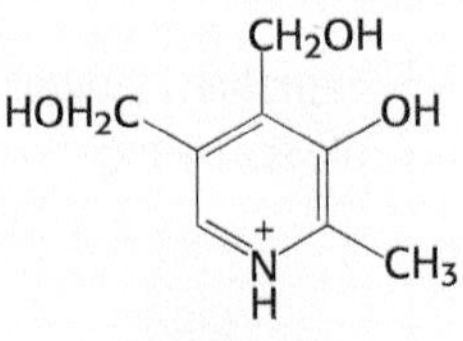

Vitamin B$_6$
(Pyridoxine)

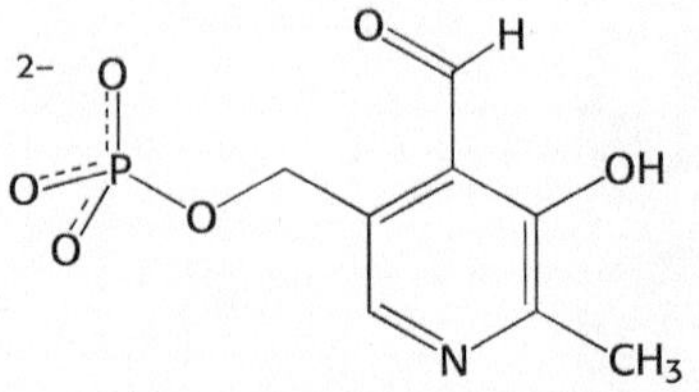

Pyridoxal phosphate (PLP)

Vitamin B$_7$ (Biotin)

Vitamin	Coenzyme

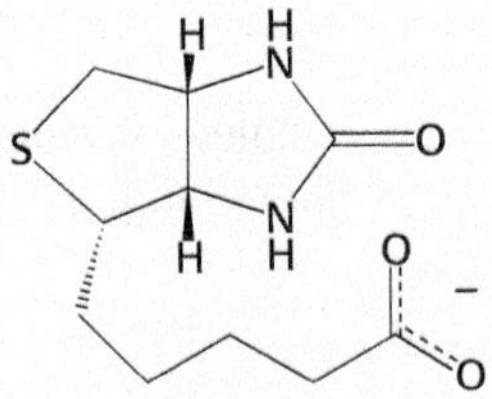

Vitamin B$_7$
(Biotin)

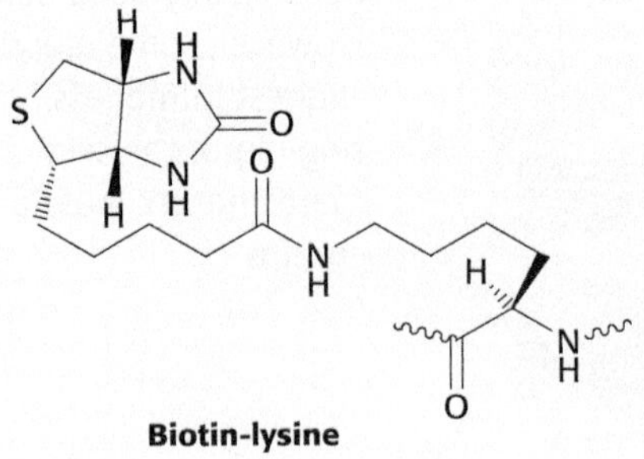

Biotin-lysine

Abbreviations: **RDA** = recommended daily allowance. **AI** = adequate intake; no RDA has been established.

Source	Deficiency	Typical reaction	RDA	Page
Beef, salmon, banana, whole grains  Og-vision/Dreamstime.com	Fatigue, fissures at the corners of the mouth (cheilosis), inflammation of the tongue (glossitis), inflammation of the lining of the mouth (stomatitis)	Group transfer	Males 19 years and older: 1.3 mg Females 19 years and older: 1.3 mg	574

Source	Deficiency	Typical reaction	AI	Page
Liver, soybean, nuts  Charles Brutlag/FeaturesPics	Muscle pain, lethargy, anorexia, depression	ATP-dependent carboxylation and carboxyl-group transfer	Males 19 years and older: 30 μg Females 19 years and older: 30 μg	317

continued on next page

Vitamin B$_9$ (Folic acid)

Vitamin	Coenzyme

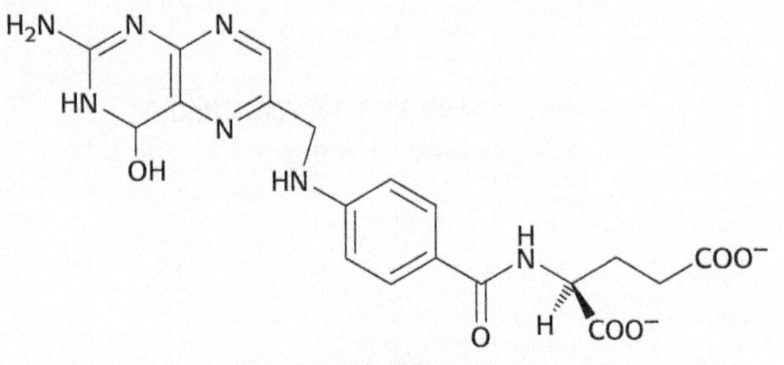

Vitamin B$_9$
(Folic acid)

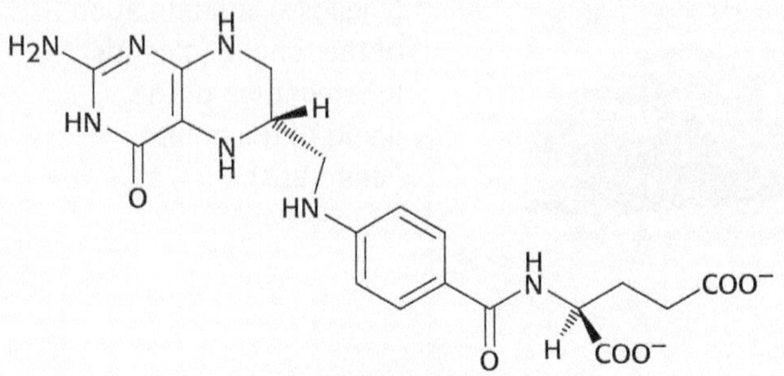

Tetrahydrofolate

Vitamin B$_{12}$ (Cobalamin)

Vitamin	Coenzyme

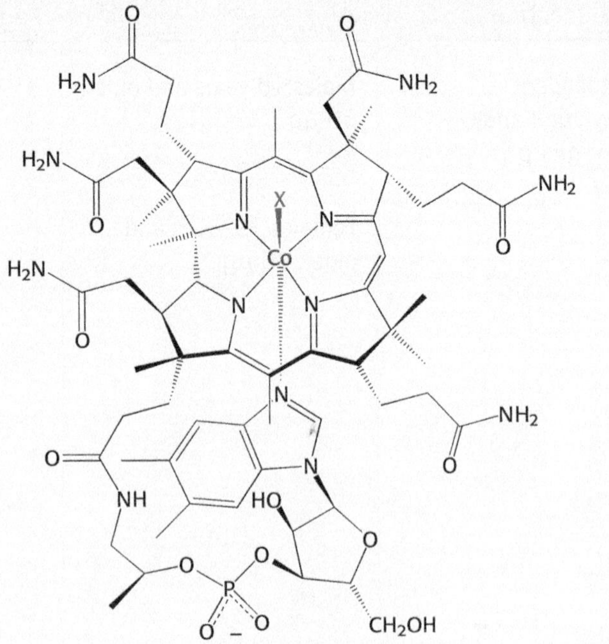

Vitamin B$_{12}$
(Hydroxycobalamin)
X = OH

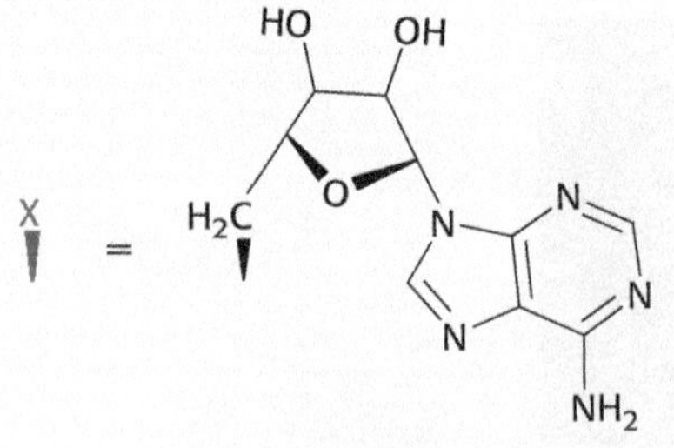

Coenzyme B$_{12}$
(5-Deoxyadenosylcobalamin)

Abbreviation: **RDA** = recommended daily allowance.

Source	Deficiency	Typical reaction	RDA	Page
Green vegetables such as spinach, asparagus, and broccoli  subjug/Getty Images	Diarrhea, fatigue, depression, the presence of fewer but larger red blood cells (megaloblastic anemia)	Transfer of one-carbon compounds	Males 19 years and older: 400 μg Females 19 years and older: 400 μg	577

Source	Deficiency	Typical reaction	RDA	Page
Meat, fish, poultry, and eggs  Chepe Nicoli/FeaturePics	Enlarged but fewer red blood cells (megaloblastic anemia), fatigue, shortness of breath, numbness in the extremities	Intramolecular rearrangements, methylations, reduction of ribonucleotides to deoxyribonucleotides	Males 19 years and older: 2.4 μg Females 19 years and older: 2.4 μg	496

continued on next page

Vitamin C (Ascorbic acid)

Vitamin

Vitamin C
(Ascorbic acid)

Abbreviation: **RDA** = recommended daily allowance.

Source	Deficiency	Typical reaction	RDA	Page
Citrus fruits, tomatoes, broccoli 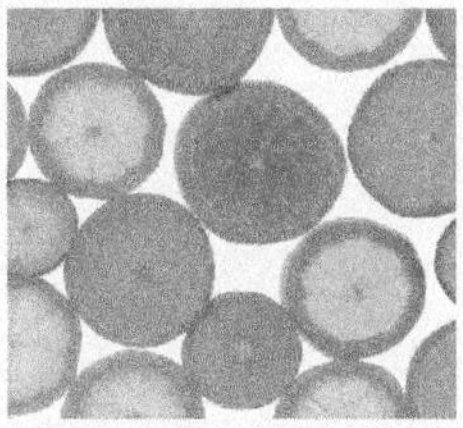 Don Farrell/Digital Vision/Getty Images	Scurvy (bleeding gums, rupture of capillaries, slow wound healing)	Antioxidant	Males 19 years and older: 90 mg Females 19 years and older: 75 mg	57

Glossary

accommodation A process that rotates the aminoacyl-tRNA in the A site so that the amino acid is brought into proximity with the aminoacyl-tRNA in the P site on the ribosome.

acetyl coenzyme A (acetyl CoA) An activated two-carbon acetyl unit formed by the thioester linkage to coenzyme A.

acetyl CoA carboxylase An enzyme that catalyzes the ATP-dependent synthesis of malonyl CoA from acetyl CoA and carbon dioxide, the committed step in fatty acid synthesis.

acetyllipoamide A compound that is an intermediate in the synthesis of acetyl CoA from pyruvate. The acetyl group is transferred to coenzyme A.

acetyllysine-binding domain A domain consisting of a four-helix bundle that binds peptides containing acetyllysine. Also called bromodomain.

activated carrier A small molecule carrying activated functional groups that can be donated to another molecule; for instance, ATP carries activated phosphoryl groups and coenzyme A carries activated acyl groups.

activated methyl cycle A series of reactions in which methyl groups from methionine are converted into a biochemically reactive form through the formation of S-adenosylmethionine an active methyl group can be transferred from S-adenosylmethionine to acceptor molecules such as phosphatidylethanolamine. The remaining part of the cycle includes the regeneration of methionine from homocysteine and N^5- methyltetrahydrofolate.

active site A specific region of an enzyme that binds the enzyme's substrate and carries out catalysis.

active transport The transport of an ion or a molecule against a concentration gradient, where ΔG for the transported species is positive; the process must be coupled to an input of free energy from a source such as ATP, an electrochemical gradient, or light.

acyl adenylate A mixed anhydride in which the carboxyl group of a molecule is linked to the phosphoryl group of AMP; the formation of acyl adenylates is a means of activating carboxyl groups in biochemical reactions, such as the formation of fatty acyl CoA molecules from a free fatty acid and coenzyme A.

acyl carrier protein (ACP) A polypeptide that is linked to phosphopantetheine and acts as a carrier of the growing fatty acyl chain in fatty acid biosynthesis.

adenine (A) A base in DNA and RNA.

adenosine triphosphate (ATP) A nucleotide that consists of adenine, a ribose, and a triphosphate unit and is the free-energy donor in most energy-requiring processes such as motion, active transport, or biosynthesis. Most of catabolism consists of reactions that extract energy from fuels such as carbohydrates and fats and convert it into ATP.

adenylate cyclase An enzyme that generates cyclic adenosine monophosphate, a second messenger, from ATP.

A-DNA A form of DNA that is wider and shorter than B-DNA. A-DNA is a right-handed double helix made up of antiparallel strands held together by A–T and G–C base-pairing and is seen in dehydrated DNA as well as in double-stranded regions of RNA and in RNA–DNA helices.

affinity chromatography A biochemical purification technique based on a specific interaction between a protein and a ligand. The ligand is immobilized on inert material and a protein mixture is passed over the ligand. Only the protein with the specific interaction will bind.

affinity label A substrate analog used to map the active site of an enzyme by binding to the active site and forming a covalent bond with a nearby amino acid.

agonist A molecule that binds to a receptor protein and triggers a signaling pathway.

alcoholic fermentation The anaerobic conversion of glucose into ethanol with the concomitant production of ATP.

aldose A monosaccharide whose C-1 carbon atom contains an aldehyde group.

allosteric enzyme An enzyme having multiple active sites as well as distinct regulatory sites that control the flux of biochemicals through a metabolic pathway. The regulation of catalytic activity is by environmental signals, including the final product of the metabolic pathway regulated by the enzyme; its kinetics are more complex than those of Michaelis–Menten enzymes.

α-amylase A pancreatic enzyme that digests starch and glycogen by cleaving the α-1,4 bonds of starch and glycogen but not the α-1,6 bonds. The products are the di- and trisaccharides maltose and maltotriose as well as the material not digestible by the enzyme.

$\alpha\beta$ dimer In hemoglobin, consists of an α subunit and a β subunit. The three-dimensional structure of hemoglobin consists of a pair of identical $\alpha\beta$ dimers ($\alpha_1\beta_1$ and $\alpha_2\beta_2$) that associate to form the hemoglobin tetramer.

α helix A common structural motif in proteins, in which a polypeptide main chain forms the inner part of a right-handed helix, with the side chains extending outward; the helix is stabilized by intrachain hydrogen bonds between amino and carboxyl groups of the main chain.

α-ketoglutarate dehydrogenase complex A citric acid cycle enzyme that catalyzes the oxidative decarboxylation of α-ketoglutarate to yield succinyl CoA. This enzyme, which helps to regulate the rate of the citric acid cycle, is structurally and mechanistically similar to the pyruvate dehydrogenase complex.

α subunit A subunit of human adult hemoglobin A, which consists of four subunits: two α subunits and two β subunits.

alternative splicing The generation of unique but related mRNA molecules by the differential splicing of the pre-mRNA transcript. By allowing the synthesis of more than one mRNA molecule from a pre-mRNA transcript, alternative splicing increases the encoding potential of the genome.

Ames test A simple, rapid means of detecting carcinogens by measuring a chemical's ability to induce mutations in *Salmonella* bacteria.

L amino acid One of two mirror-image forms of α-amino acids.

aminoacyl-tRNA synthetase An enzyme that activates an amino acid and then links it to transfer RNA. Also known as an activating enzyme, each aminoacyl-tRNA synthetase is specific for a particular amino acid.

aminotransferases A class of enzymes that transfer an α-amino group from an α-amino acid to an α-ketoacid. Also called transaminases.

AMP-activated protein kinase (AMPK) A protein kinase, conserved among eukaryotes, that is activated on binding AMP and inhibited by ATP; consequently, it functions as a cellular fuel gauge, inhibiting certain processes by phosphorylating key enzymes when the energy supply is low.

amphibolic pathway Consists of metabolic reactions that can be anabolic or catabolic, depending on the energy conditions in the cell.

amphipathic molecule A molecule, such as a membrane lipid, that contains both a hydrophobic and a hydrophilic moiety.

anabolism The set of metabolic reactions that require energy to synthesize molecules from simpler precursors.

anapleurotic reaction A reaction that leads to the net synthesis, or replenishment, of pathway components.

anomers Isomers of cyclic hemiacetals or hemiketals, with different configurations only at the carbonyl carbon atom; that carbon atom is known as the anomeric carbon atom.

antagonist A molecule that binds to a receptor protein but does not trigger the signaling pathway. Such molecules are like competitive inhibitors for enzymes.

antibody A protein synthesized by an animal in response to a foreign substance, called an antigen. Antibodies have specific and high affinity for their antigens. The binding of antibody to antigen is a key step in the immune response.

anticodon Three-nucleotide sequence of tRNA that base-pairs with a codon in mRNA.

anticodon loop One of five groups of bases that are not base-paired to form double helices.

antigen A foreign substance that generates an immune response in an organism.

antigenic determinant A specific group or cluster of amino acids on an antigen that is recognized by an antibody. Also called an epitope.

antiporter A transport system in which a molecule is carried across a membrane in the direction opposite that of an ion, which in turn is pumped back across the membrane through active transport linked to ATP hydrolysis.

apoenzyme An enzyme without its cofactor.

arachidonate Derived from linoleate, a 20:4 fatty acid that is a major precursor to several classes of signal molecules, including prostaglandins.

aspartate transcarbamoylase An allosteric enzyme that regulates the synthesis of pyrimidine nucleotides.

assay Any procedure to measure the activity of a biomolecule, such as an enzyme.

ATP-ADP translocase A transport protein in the inner mitochondrial membrane that carries ADP into the mitochondria and ATP out in a coupled fashion. Also called adenine nucleotide translocase (ANT).

ATP-binding cassette (ABC) domain The ATP-binding domain characteristic of specific ATP-driven pumps, called ABC transporters, that provide the energy needed for the active transport of ions into and out of cells; these transporters also contain a membrane-spanning region.

ATP synthase Molecular assembly of the inner mitochondrial membrane responsible for the respiratory-chain-driven synthesis of ATP. Also called Complex V, mitochondrial ATPase, H^+-ATPase, or F_0F_1-ATPase. The corresponding enzyme in chloroplasts is called the CF_1-CF_0 complex.

bacteriophage lambda (λ phage) A bacterial virus that is frequently used as a cloning vector. λ phage can multiply within a host and lyse it (lytic pathway) or its DNA can become integrated into the host genome (lysogenic pathway) where it is dormant until activated.

base-excision repair A means of repairing DNA in which the damaged base is removed and replaced by a base complementary to the undamaged DNA strand.

B-DNA The form in which most DNA exists under physiological conditions: the two strands are antiparallel; the bases are inside the helix, and the phosphates and deoxyribose sugars are on the outside; adenine forms hydrogen bonds with thymine, and guanine forms them with cytosine; the bases in each pair are coplanar; there are 10.4 residues per turn, with a pitch of 35 Å. It is the form depicted in the Watson-and-Crick model of DNA.

beriberi A neurological and cardiovascular disorder caused by a dietary deficiency of thiamine (vitamin B_1).

β-galactosidase An essential enzyme in lactose metabolism that hydrolyzes lactose into galactose and glucose.

β-oxidation pathway In the degradation of a fatty acyl CoA molecule, the sequence of oxidation, hydration, and oxidation reactions that converts a methylene group at the C-3 carbon atom (also called the β-carbon atom) into a β-keto group, which is subsequently cleaved to yield acetyl CoA.

β pleated sheet A common structural motif in proteins, in which a number of β strands are associated as stacks of chains, stabilized by interchain hydrogen bonds and running in the same direction. Stacks of β strands running in opposite directions form an antiparallel pleated sheet.

β strand A polypeptide chain that is almost fully extended rather than being tightly coiled as in the α helix.

β subunit A subunit of human adult hemoglobin A, which consists of four subunits: two α subunits and two β subunits.

bifunctional enzyme An enzyme with two different, often opposing, catalytic activities on one polypeptide chain. For instance, phosphofructokinase 2 synthesizes fructose 2,6-bisphosphate and fructose 2,6-bisphosphatase hydrolyzes it, yet both active sites are on the same polypeptide chain.

bile salt A polar derivative of cholesterol that is made in the liver, stored in the gall bladder, and released into the small intestine, where it acts as a detergent to solubilize dietary lipids, facilitating their digestion and absorption.

binding energy The free energy released in the formation of the weak interactions between enzyme and substrate that facilitate the formation of the transition state.

biotin (vitamin B_7) A vitamin that plays a role in carboxylation and decarboxylation reactions.

2,3-bisphosphoglycerate (2,3-BPG) An allosteric regulator of oxygen binding by hemoglobin. Also called 2,3-diphosphoglycerate (2,3-DPG).

Bohr effect The observation made by Christian Bohr that H^+ and CO_2 promote the release of oxygen from oxyhemoglobin.

bromodomain A domain consisting of a four-helix bundle that binds peptides containing acetyllysine. Also called acetyllysine-binding domain.

Brownian motion The random movement of gases and liquids powered by the background thermal energy. Brownian motion inside the cell supplies the energy for many of the interactions required for a functioning biochemical system.

buffer An acid–base conjugate pair that resists changes in the pH of a solution.

calmodulin (CaM) In vertebrates, a ubiquitous protein that, when bound to calcium, stimulates many enzymes and transporters.

calmodulin-dependent protein (CaM) kinase A protein kinase that is activated by the binding of a Ca^{2+}–calmodulin complex.

Calvin cycle In plants, a cyclic metabolic pathway in which carbon dioxide is incorporated into ribulose 1,5-bisphosphate to yield compounds that can be used for the synthesis of glucose. Also called dark reactions.

5′ cap A structure at the 5′ end of eukaryotic mRNA that stabilizes the mRNA and enhances its translation. The cap contains a 7-methyl guanylate residue attached by a triphosphate linkage to the sugar at the 5′ end of the mRNA in a rare 5′–5′ linkage.

carbamate A negatively charged group formed between an amino-terminal group and carbon dioxide. In hemoglobin, carbamate formation stimulates oxygen release.

carbamoyl phosphate synthetase (CPS) An enzyme that begins the urea cycle by catalyzing the synthesis of carbamoyl phosphate from bicarbonate, ammonium ion, and ATP. An isozymic form of the enzyme also catalyzes the initial reaction in pyrimidine biosynthesis.

carbohydrate A saccharide, which can be an aldehyde or a ketone compound having multiple hydroxyl groups. Also defined as an organic compound with the empirical formula $(CH_2O)_n$.

carbonic anhydrase An enzyme abundant in red blood cells that hydrates carbon dioxide in the blood to form bicarbonate (HCO_3^-), the transport form of carbon dioxide.

carboxyl-terminal domain (CTD) A regulatory domain in RNA polymerase II. Phosphorylation of this domain regulates aspects of transcription and splicing.

carnitine A zwitterionic alcohol formed from lysine that acts as a carrier of long-chain fatty acids from the cytoplasm to the mitochondrial matrix.

carotenoid An extended polyene that absorbs light between 400 and 500 nm and serves as an accessory pigment in photosynthesis by funneling the energy to the photosynthetic reaction center.

catabolism The set of metabolic reactions that transform fuels into cellular energy.

catabolite activator protein (CAP) The cAMP-response protein; when bound to cyclic AMP, CAP binds to an inverted repeated of the *lac* operon, near position -61 relative to the start site of transcription, to stimulate transcription.

catabolite repression The repression by glucose of catabolic enzymes required for the catabolism of carbohydrates other than glucose.

catalase A ubiquitous heme protein that catalyzes the dismutation of hydrogen peroxide into molecular oxygen and water.

catalysis by approximation Enhances the rate of a reaction by bringing multiple substrates together along a single binding surface of an enzyme.

catalyst Agent that enhances the rate of a chemical reaction without being permanently affected itself.

catalytic RNA One of a class of RNA molecules that display enzymatic activity.

catalytic triad A constellation of three residues, found in many proteolytic enzymes, in which two of the residues convert the remaining residue, usually a serine or cysteine residue, into a potent nucleophile.

cellular respiration The generation of high-transfer-potential electrons by the citric acid cycle, their flow through the respiratory chain to reduce oxygen to water, and the accompanying synthesis of ATP. Also called respiration.

cellulose An unbranched homopolysaccharide in plants, composed of glucose residues in β-1,4 linkage; the major structural polysaccharide in plants.

central dogma States that information flows from DNA to RNA and then to protein, a scheme that underlies information processing at the level of gene expression that was first proposed by Francis Crick in 1958.

ceramide A sphingosine with a long-chain acyl group attached to the amino group.

cerebroside A sphingolipid in which glucose or galactose is linked to the terminal hydroxyl group of a ceramide.

channel A protein passage that is continuous and that allows ions to flow rapidly through a membrane from a compartment of higher to a compartment of lower concentration. Channels are generally composed of four to six subunits, or domains, and are gated by membrane potential, allosteric effectors, or covalent modification.

chemotroph An organism that obtains energy by the oxidation of foodstuffs. See also **phototroph**.

chlorophyll A substituted tetrapyrrole that is the principal photoreceptor in plants.

chloroplast The plant organelle in which photosynthesis takes place.

cholecystokinin (CCK) A peptide hormone secreted by the endocrine cells of the upper intestine. CCK stimulates the release of digestive enzymes by the pancreas and of bile salts by the gall bladder.

cholesterol A sterol that is an important constituent of eukaryotic membranes and lipoproteins; also a precursor of steroid hormones.

chromatin Nucleoprotein chromosomal material consisting mainly of DNA and histones.

chromatin-remodeling engine A complex of proteins that contain domains homologous to helicases and use the energy of ATP hydrolysis to shift the positions of nucleosomes and induce other conformational changes in chromatin.

chylomicron A lipoprotein particle that transports dietary triacylglycerol from the intestine to other tissues; apolipoprotein B-48 is a protein component of chylomicrons.

citrate synthase An enzyme that catalyzes the condensation of acetyl CoA with oxaloacetate to form citrate, initiating the citric acid cycle.

citric acid cycle A cyclic series of metabolic reactions that completely oxidize acetyl units to carbon dioxide. Also known as the tricarboxylic acid cycle (after citrate) or the Krebs cycle, after Hans Krebs, who elucidated the cyclic nature of the pathway.

coactivator A protein that enhances transcription by binding to a transcription factor. Coactivators themselves do not bind to DNA.

codon A nucleotide triplet in mRNA that encodes a particular amino acid.

coenzyme A small organic molecule required for the activity of many enzymes; vitamins are often components of coenzymes.

coenzyme A (CoA) A vitamin-derived carrier of acyl groups.

coenzyme Q A mobile electron carrier that is a component of the respiratory chain; it shuttles between the oxidized ubiquinone form to the reduced ubiquinol form through a semiquinone intermediate; accepts electrons from NADH-Q oxidoreductase as well as succinate-Q reductase.

cofactor A small molecule on which the catalytic activity of an enzyme may depend. The cofactor's precise role varies with the cofactor and the enzyme. Cofactors can be subdivided into two groups: (1) small organic molecules called coenzymes and (2) metals.

coiled coil Two right-handed α helices intertwined to form a type of left-handed superhelix.

combinatorial control A means of controlling gene expression in eukaryotes in which each transcription factor, rather than acting on its own to effect transcription, recruits other proteins to build up large complexes that regulate the transcription machinery.

committed step The first irreversible step in a metabolic pathway under physiological conditions; this step is catalyzed by an allosteric enzyme and commits the product to a particular chemical fate.

competitive inhibition The reduction in the rate of enzyme activity observed when the enzyme can bind the substrate or the inhibitor but not both. Many competitive inhibitors resemble the substrate and compete with it for binding to the enzyme's active site. Relief from inhibition by increasing substrate concentration is a kinetic hallmark of competitive inhibition.

complementary DNA (cDNA) DNA that is complementary to mRNA. Complementary DNA is synthesized by reverse transcriptase with the use of mRNA as a template.

complementary DNA (cDNA) library A collection of cDNA molecules, inserted into cloning vectors, made from all the mRNA molecules that a cell is expressing.

consensus sequence Idealized base sequence that represents common features of a promoter site.

controlled termination of replication A DNA-sequencing technique based on the generation of DNA fragments whose length depends on the last base in the sequence. The same procedure is performed on four reaction mixtures at the same time. DNA polymerase is used to make the complement of a short sequence within a single-stranded DNA molecule that is being sequenced. The synthesis is primed by a fragment that is complementary to a part of the sequence. In addition to the four deoxyribonucleoside triphosphates (radioactively labeled), each reaction mixture contains a small amount of the $2',3'$-dideoxy analog of one of the nucleotides, a different nucleotide for each reaction mixture. The incorporation of this analog blocks further growth of the new strand. The concentration of the analog is low enough that strand termination will take place only occasionally. In all of the reaction mixtures, fragments of various lengths are produced, but all will be terminated by the dideoxy analog. Four such sets of strand-terminated fragments then undergo electrophoresis, and the base sequence of the new DNA is read from the autoradiogram of the four lanes. Also called the Sanger dideoxy method.

cooperative effect Enhanced activity resulting from cooperation between subunits of an allosteric molecule.

Cori cycle A cyclic interorgan metabolic pathway in which lactate from active muscle is converted into glucose by the liver, which in turn supplies newly synthesized glucose to muscle and other tissues.

covalent catalysis Catalysis in which the active site contains a reactive group that becomes temporarily covalently modified in the course of catalysis.

covalent modification The attachment of chemical groups to an enzyme or their removal from an enzyme and the consequent change in the catalytic properties of that enzyme. The catalytic properties of many enzymes are altered by the covalent attachment and removal of phosphoryl groups.

C_4 pathway A means by which four-carbon compounds, such as oxaloacetate and malate, carry carbon dioxide from mesophyll cells in contact with the air to bundle-sheath cells, which are the major sites of photosynthesis. The pathway accelerates photosynthesis by concentrating carbon dioxide in photosynthetic cells.

C_3 plant A plant that lacks the C_4 pathway.

C_4 plant A plant that utilizes the C_4 pathway.

crassulacean acid metabolism (CAM) An adaptation by plants living in an arid environment; the C_4 pathway concentrates carbon dioxide at night, and gas exchange with the environment is curtailed during the heat of the day by closure of the stomata.

cumulative feedback inhibition A regulatory strategy in which the enzyme catalyzing the committed step common to several pathways is incrementally inhibited by the products of each of the pathways. Thus, each inhibitor can reduce the activity of the enzyme even if other inhibitors are bound at saturating levels.

cyclic photophosphorylation In photosynthesis, the generation of ATP without the concomitant formation of NADPH; electron cycling from the reaction center of photosystem I to ferredoxin and then back to the reaction center through cytochrome b_6f and plastoquinone, generating a proton gradient that is used to drive ATP formation.

cytidine diphosphodiacylglycerol (CDP-diacylglycerol) An activated precursor for the synthesis of many phospholipids formed by the reaction of phosphatidate with cytidine triphosphate.

cytochrome b_6f A cytochrome complex that links photosystem II and photosystem I in green plants; cytochrome b_6f contributes to the proton gradient by oxidizing plastoquinol to plastoquinone.

cytochrome c A water-soluble, highly conserved cytochrome component of the respiratory chain that accepts electrons from cytochrome c reductase and is in turn oxidized by cytochrome c oxidoreductase.

cytochrome c oxidase The final complex of the respiratory chain, cytochrome c oxidase transfers electrons from cytochrome c to molecular oxygen and concomitantly pumps protons across the inner mitochondrial membrane to generate the proton-motive force. Also called Complex IV.

cytochrome P450 monooxygenase An enzyme that uses O_2 to add oxygen atoms to biochemicals such as steroids. One oxygen atom is bound to the substrate and the other one is reduced to water.

cytoplasm The biochemical material between the plasma membrane and the nuclear membrane that is not enclosed by any other membrane. The cytoplasm is the site of a host of biochemical processes, including the initial stage of glucose metabolism, fatty acid synthesis, and protein synthesis.

cytosine (C) A base in DNA and RNA.

cytoskeleton The internal scaffolding of cells that is made up of actin filaments, intermediate filaments, and microtubules and that enables cells to transport vesicles, change shape, and migrate.

de novo pathway A biosynthetic pathway in which a molecule is synthesized from simple precursors.

deoxyribonucleic acid (DNA) A nucleic acid constructed of four different deoxyribonucleotides; contains the genome and exists as a double-stranded helix in all higher organisms.

deoxyribose A five-carbon monosaccharide ($C_5H_9O_5$) that constitutes the carbohydrate moiety of a deoxynucleotide; the deoxyribose commonly found in deoxynucleotides is 2-deoxyribose.

dialysis A process of removing small molecules from a mixture of solutes. The mixture of solutes is placed in a semipermeable bag, which in turn in placed in a large volume of solvent. Small molecules diffuse from the bag.

diastereoisomers A pair of molecules, each with more than one asymmetric center, that have opposite configurations at one such center but are not mirror images of each other; in the aldotetrose series, D-erythrose and D-threose are diastereoisomers.

differential centrifugation A procedure that yields several fractions of decreasing density, each still containing hundreds of different proteins, which are assayed for the activity being purified.

digestion The process of breaking large molecules in food down into smaller units by hydrolytic enzymes. Proteins are hydrolyzed to the 20 different amino acids, polysaccharides are hydrolyzed to simple sugars such as glucose, and fats are hydrolyzed to fatty acids.

dihydrofolate reductase An enzyme that catalyzes the regeneration of tetrahydrofolate from dihydrofolate formed in the synthesis of thymidylate.

dihydrolipoyl dehydrogenase (E_3) One of three distinct enzymes in the pyruvate dehydrogenase complex, dihydrolipoyl dehydrogenase regenerates the oxidized form of lipoamide, which enables the pyruvate dehydrogenase complex to complete another catalytic cycle.

dihydrolipoyl transacetylase (E_2) One of three distinct enzymes in the pyruvate dehydrogenase complex, dihydrolipoyl transacetylase catalyzes the transfer of the acetyl group from acetyllipoamide to CoA in the conversion of pyruvate into acetyl CoA.

dipolar ion An ion carrying both a positive and a negative charge. Also called a zwitterion.

direct repair A means of repairing damaged DNA in which the damaged region is corrected in place. For example, pyrimidine dimers are simply cleaved to restore the original nucleotides.

disaccharide A carbohydrate consisting of two sugars joined by an O-glycosidic bond.

distal histidine A histidine located near the heme group in myoglobin and hemoglobin that helps maintain the heme iron in the Fe^{2+} oxidation state and inhibits carbon monoxide binding. It is on the opposite side of the heme from the proximal histidine.

disulfide bond A covalent bond formed by the oxidation of two sulfhydryl groups; the oxidation of cysteine residues in a polypeptide yields a disulfide bond linking the two residues.

DNA ligase An enzyme that catalyzes the formation of a phosphodiester linkage between the 3'-hydroxyl group at the end of one DNA strand and the 5'-phosphoryl group at the end of another strand; it takes part in the synthesis, repair, and splicing of DNA.

DNA microarray A collection of known DNA sequences attached to a specific location on a solid surface. Fluorescently labeled cDNA is hybridized to the surface-bound DNA to reveal the expression level of each gene, identifiable by its known position in the microarray. Also called a gene chip.

DNA polymerases Enzymes that catalyze the template-directed, primer-dependent addition of deoxynucleotide units, using deoxynucleoside triphosphates as substrates, to the 3' end of a DNA chain; chain growth is in the 5'-to 3'-direction; such enzymes replicate and repair DNA.

domain An independently folded unit in the tertiary structure of a polypeptide chain; may contain a number of supersecondary structures. In multienzyme complexes, each domain may carry out one or more catalytic reactions. In proteins, a compact globular unit of 100 to 400 residues, possibly joined to other domains by a flexible polypeptide segment; often encoded by a specific exon in the gene encoding the protein.

double-displacement reaction A reaction having multiple substrates in which one or more products are released before all substrates bind the enzyme. The defining feature of these reactions is the formation of a substituted-enzyme intermediate. Also called ping-pong reaction.

double helix In DNA, a right-handed double helix is formed by two helical polynucleotide chains coiled around a common axis. The chains run in opposite directions; that is, they have opposite polarity.

Edman degradation A protein-sequencing method that sequentially removes one residue at a time from the amino end of a peptide. Phenyl isothiocyanate reacts with the terminal amino group of the peptide, which then cyclizes and breaks off the peptide, yielding an intact peptide shortened by one amino acid. The cyclic compound is a phenylthiohydantoin (PTH) amino acid, which can be identified by chromatographic procedures. The Edman procedure can then be repeated sequentially to yield the amino acid sequence of the peptide.

EF hand A helix-loop-helix motif that forms a binding site for calcium; found in many calcium-sensitive proteins.

eicosanoid A carbon compound containing 20 carbon atoms; prostaglandins are examples. Eicosanoids often act as local hormones.

electron-transport chain Four large protein complexes embedded in the inner mitochondrial membrane. The oxidation–reduction reactions that allow the flow of electrons from NADH and $FADH_2$ to oxygen take place in these complexes. Also called respiratory chain.

electrostatic interaction A weak interaction between ions having opposite charges. Also called ionic bond or salt bridge.

elongation factor G (EF-G) A member of the G-protein family required for protein synthesis in eukaryotes. The hydrolysis of GTP by EF-G causes the tRNAs and mRNA to move through the ribosome a distance corresponding to one codon.

elongation factor Ts (EF-Ts) A protein that binds to the GDP-bound form of EF-Tu and induces the release of GDP, thereby enabling EF-Tu to participate in another elongation step.

elongation factor Tu (EF-Tu) A member of the G-protein family that delivers aminoacyl-tRNAs to the A site of the ribosome with the concomitant hydrolysis of GTP to GDP.

emulsion A mixture of two immiscible substances.

enantiomers A pair of molecules, each with one or more chiral centers, that are mirror images of each other.

endocytosis The process of the internalization of extracellular material by the invagination and budding of the cell membrane.

endoplasmic reticulum (ER) An extensive system of cytoplasmic membranes that comprises about half the total cell membrane. The region of the ER that binds ribosomes is called the rough ER, and the region that is devoid of ribosomes is called the smooth ER.

endosome A component of the receptor-mediated endocytotic pathway in which sorting decisions about the endocytosized material are made; endosomes are derived from coated vesicles that lose the clathrin coat.

energy charge A means of determining the energy status of the cell, equal to the concentration of ATP plus one-half the concentration of ADP, all divided by the total adenine nucleotide concentration.

enhancer A DNA sequence that has no promoter activity itself but can greatly enhance the activity of other promoters; enhancers can exert their stimulatory effect over a distance of several thousand nucleotides.

enol phosphate A compound with a high phosphoryl-transfer potential because the phosphoryl group traps the molecule in an unstable enol form. On transfer of the phosphate, the molecule converts into the more-stable ketone form.

enteropeptidase A proteolytic enzyme, secreted by the epithelial cells of the small intestine, that activates the pancreatic zymogen trypsinogen to form trypsin. Also called enterokinase.

entropy A measure of the degree of randomness or disorder in a system; denoted by the symbol S in thermodynamics, the change in entropy (ΔS) increases when a system becomes more disordered and decreases when the system becomes more ordered, or less random.

enzyme A biological macromolecule that acts as a catalyst for a biochemical reaction; although almost all enzymes are composed of protein, some RNA molecules are catalytically active.

enzyme-linked immunosorbent assay (ELISA) An immunological-based means of detecting and quantifying proteins. An enzyme that reacts with a colorless substrate to produce a colored product is covalently linked to a specific antibody that recognizes a target antigen. If the antigen is present, the antibody–enzyme complex will bind to it; on the addition of the substrate, the enzyme will catalyze the reaction, generating the colored product revealing the presence of the antigen.

enzyme multiplicity A regulatory strategy in which the committed step common to several pathways is catalyzed by different enzymes with the same catalytic properties but different regulatory properties. Each enzyme thus responds to the final product of one of the pathways having the committed step in common.

enzyme–substrate (ES) complex The product of specific binding between the active site of an enzyme and its substrate.

epimers A pair of molecules, each with more than one asymmetric center, that differ in configuration at only one such center; for example, glucose and galactose are epimers, differing only in the configuration at carbon 4 (C-4).

epinephrine A catecholamine released by the adrenal medulla in response to muscle activity or its anticipation that stimulates the breakdown of muscle glycogen. Also called adrenaline.

essential amino acid An amino acid that cannot be synthesized de novo and therefore must be acquired from the diet; in adult mammals, at least nine amino acids are considered essential.

estrogen response element (ERE) A DNA sequence to which an estrogen receptor binds as a dimer.

eukaryote An organism whose cells contain a nucleus and other membrane-bounded compartments.

exocytosis The process by which secretory granules fuse with the plasma membrane and dump their cargo into the extracellular environment.

exon Region of pre-mRNA that is retained in mature mRNA.

exonuclease An enzyme that digests nucleic acids from the ends of the molecules, rather than at an internal site; exonucleases can be specific for digestion from the $3'$ or $5'$ ends of the nucleic acid.

expression vector A cloning vector that allows the translation of the inserted DNA. These vectors not only maximize the transcription with the use of a powerful promoter, but also contain a segment of DNA that encodes a ribosome-binding site on the mRNA that is transcribed from the inserted cDNA. Thus, cDNA molecules that are inserted into these vectors are not only transcribed but also translated.

facilitated diffusion The transport of an ion or a molecule down a concentration gradient, where ΔG for the transported species is negative. Also called passive transport.

fatty acid A carboxylic acid containing a long hydrocarbon chain. Fatty acids are an important fuel source as well as a key component of membrane lipids.

fatty acid synthase An enzyme system that catalyzes the synthesis of saturated long-chain fatty acids from acetyl CoA, malonyl CoA, and NADPH; in bacteria, the constituent enzymes of the synthase complex can be dissociated when cell extracts are prepared; in mammals, all constituent enzyme activities in fatty acid synthase are part of the same polypeptide.

fatty acid synthesis The synthesis of saturated long-chain fatty acids from acetyl CoA, malonyl CoA, and NADPH by a sequence of reduction, dehydration, and reduction reactions that convert a β-keto group into a methylene group.

feedforward stimulation The activation of an allosteric enzyme in a later stage of a pathway by the product of a reaction that takes place earlier in the reaction pathway.

ferritin An iron-storage protein found primarily in the liver and kidneys.

fetal hemoglobin A tetrameric molecule comprising two α subunits and two γ subunits. The γ subunit is 72% identical in amino acid sequence with the adult hemoglobin β subunit.

50S subunit The large subunit of the bacterial 70S ribosome; the site of peptide-bond synthesis, it contains 34 different proteins, a 5S RNA species, and a 23S RNA species.

Fischer projection A two-dimensional depiction of a biomolecule in which every atom is identified and the bonds to the central carbon atom are represented by horizontal and vertical lines. By convention, the horizontal bonds are assumed to project out of the page toward the viewer, whereas the vertical bonds are assumed to project behind the page away from the viewer.

flavin adenine dinucleotide (FAD) A coenzyme for oxidation–reduction reactions composed of the vitamin riboflavin bonded to adenosine diphosphate. The electron acceptor of FAD is the isoalloxazine ring of riboflavin.

$5'$-phosphoribosyl-1-pyrophosphate (PRPP) The activated form of ribose.

flavin mononucleotide (FMN) A coenzyme for oxidation–reduction reactions derived from the vitamin riboflavin. The electron acceptor of FMN, the isoalloxazine ring, is identical with that of the oxidized form of flavin adenine dinucleotide (FAD), but FMN lacks the adenyl nucleotide component of FAD.

flavoprotein A protein tightly associated with FAD or FMN; flavoproteins play important roles in many oxidation–reduction reactions.

folding funnel Visualizes the folding of proteins. The breadth of the funnel represents all possible conformations of an unfolded protein, and the depth represents the energy difference between the unfolded and the native protein. Each point on the surface represents a possible three-dimensional structure and its energy value. The funnel suggests that there are alternative pathways to the native structure. Also called energy landscape. See also **molten globule**.

free energy A form of energy capable of doing work under conditions of constant temperature and pressure. Also, a measure of the usable energy generated in a chemical reaction; denoted by the symbol G in thermodynamics. The change in free energy (ΔG) of a system undergoing transformation at constant pressure is equal to the change in enthalpy (ΔH) minus the product of the absolute temperature (T) and the change in entropy (ΔS).

free energy of activation The energy required to form the transition state from the substrate of a reaction.

fructose 1,6-bisphosphatase An enzyme responsible for the conversion of fructose 1,6-bisphosphate into fructose 6-phosphate and orthophosphate.

functional group A group of atoms having distinct chemical properties.

G_α The guanyl nucleotide-binding subunit of heterotrimeric G proteins.

$G_{\beta\gamma}$ The dimeric component of G proteins that helps to maintain the G_α component in the inactive state. In some signal-transduction pathways, the $G_{\beta\gamma}$ component can also function as a second messenger.

ganglioside A ceramide, common in membranes of the nervous system, in which an oligosaccharide is linked to the ceramide by a glucose residue.

gel electrophoresis A technique for separating biomolecules, such as nucleic acids and proteins, on the basis of charge and size. When placed under an electric field, the biomolecules move through the gel (a cross-linked polymer such as polyacrylamide), with larger molecules moving more slowly than smaller molecules.

gel-filtration chromatography See **molecular exclusion chromatography**.

general acid–base catalysis Catalysis in which a molecule other than water plays the role of a proton donor or acceptor.

genetic code The relation between nucleic acid sequence information and protein sequence information.

genome The heritable information in DNA.

genomic library A collection of DNA fragments, inserted into a cloning vector, that represents the entire genome of an organism.

glucagon A polypeptide hormone that is secreted by the α cells of the pancreas when the blood-glucose level is low and leads to glycogen breakdown in the liver and the release of glucose to the blood.

glucogenic amino acid An amino acid whose carbon skeleton, entirely or in part, can be converted into substrates for gluconeogenesis.

gluconeogenesis The synthesis of glucose from noncarbohydrate precursors, including lactate, glycerol, and amino acids.

glucose–alanine cycle The cycle in which nitrogen is transported from muscle to the liver as alanine.

glucose 6-phosphatase A membrane protein of the lumenal side of the liver endoplasmic reticulum that catalyzes the formation of free glucose from glucose 6-phosphate. The enzyme plays a key role in maintaining blood-glucose levels.

glucose 6-phosphate dehydrogenase An enzyme that initiates the oxidative phase of the pentose phosphate pathway by oxidizing glucose 6-phosphate to 6-phosphoglucono-δ-lactone to generate one molecule of NADPH.

glutamate dehydrogenase An enzyme that catalyzes the oxidative deamination of glutamate, yielding ammonium ion and α-ketoglutarate.

glutathione A tripeptide playing a role in combating oxidative stress by maintaining the reduced state of the cell. Glutathione cycles between the reduced (GSH) and oxidized (GSSG) state.

glutathione peroxidase An enzyme that eliminates peroxides, reactive oxygen species that can damage membranes as well as other biomolecules.

glutathione reductase An enzyme that catalyzes the regeneration of reduced glutathione with the use of NADPH.

glycerol 3-phosphate shuttle A pathway that transfers electrons from cytoplasmic NADH into the mitochondria; dihydroxyacetone phosphate (DHAP) is reduced by NADH to glycerol 3-phosphate, which enters the mitochondria and is oxidized to yield $FADH_2$ and DHAP, which leaves the mitochondria.

glycogen A readily mobilized storage form of glucose in which the glucose monomers are linked by α-1,4-glycosidic bonds and in which there are branches (α-1,6-glycosidic bonds) at about every tenth residue.

glycogenin A protein that bears an oligosaccharide of α-1,4 glucose units and is the primer for glycogen synthase. Glycogenin uses UDP-glucose to catalyze its own autoglycosylation.

glycogen phosphorylase An enzyme that catalyzes the phosphorolysis of glycogen to yield glucose 1-phosphate; an allosteric enzyme whose activity is further regulated by reversible covalent modification.

glycogen synthase An allosteric enzyme that can be reversibly covalently regulated and is responsible for the synthesis of glycogen; it transfers glucose from UDP-glucose to the hydroxyl group at a C-4 terminus of glycogen.

glycogen synthase kinase (GSK) An enzyme that phosphorylates glycogen synthase, keeping glycogen synthase in its inactive state.

glycolipid A sugar-containing lipid derived from sphingosine; the sugar moiety is attached at the alcohol on sphingosine.

glycolysis A sequence of reactions that convert glucose into pyruvate with the concomitant generation of energy.

glycoprotein A proteins that has a specific carbohydrate moiety attached.

glycosaminoglycan A heteropolysaccharide made of repeating disaccharide units and containing the amino sugar glucosamine or galactosamine.

glycosidic bond A covalent bond between an aldehyde or ketone function of a monosaccharide and an oxygen, nitrogen, or sulfur atom of another molecule. The most common glycosidic linkages are O-links between the anomeric carbon atom of a sugar and a hydroxyl group of another saccharide.

glycosyltransferase Any of a number of specific enzymes that catalyze the formation of glycosidic bonds.

glyoxylate cycle A metabolic pathway that converts two-carbon acetyl units into succinate for energy production and biosyntheses; found primarily in bacteria and plants, the cycle bypasses two decarboxylation steps in the citric acid cycle and allows the net formation of glucose and other molecules through oxaloacetate from acetate or acetyl CoA.

glyoxysome A plant organelle in which enzymes of the glyoxylate pathway are present.

Golgi complex In the cytoplasm, a stack of membranous sacs that constitute the major sorting center for proteins that reside in cell membranes and the lumen of organelles.

gout A disease characterized by inflammation of the joints and kidneys due to the precipitation of abnormally high levels of sodium urate, a breakdown product of purines.

G protein A guanyl nucleotide-binding protein that is a component of intracellular signaling pathways. In the inactive state, the G protein (sometimes called a heterotrimeric G protein) is a trimeric protein consisting of $\alpha\beta\gamma$ subunits, with the GDP bound to the α subunit. In the active state, the α protein exchanges GDP for GTP and dissociates from the $\beta\gamma$ subunits. The GTP-bound α subunit propagates the signal. Signal propagation is terminated when the α subunit hydrolyzes GTP to GDP and reassociates with the $\beta\gamma$ subunits.

G-protein-coupled receptor (GPCR) See **seven-transmembrane-helix (7TM) receptors.**

gradient centrifugation An ultracentrifugation technique used to separate large biomolecules or molecular complexes. A linear gradient of viscous solution is created in a centrifuge tube—for instance, a gradient of 5% to 20% sucrose. The sample is layered on the top of the gradient, and the tube is centrifuged at high speeds. Large, dense molecules or complexes move through the gradient faster than smaller complexes. The separated complexes are harvested by making a hole in the bottom of the tube and collecting drops. Also called zonal centrifugation.

granum A pile or stack of thylakoid membranes in the chloroplast.

group-specific reagent A reagent that reacts with specific R groups (side chains) of amino acids.

guanine (G) A base in DNA and RNA.

helicase An enzyme that catalyzes the ATP-driven unwinding of nucleic acids; DNA helicases are important in DNA replication.

heme The prosthetic group of myoglobin and hemoglobin as well as other proteins; consists of an organic constituent, protoporphyrin, and an iron atom.

hemiacetal A compound formed by the reaction of an aldehyde functional group and a hydroxyl group; for example, the C-1 group of the open-chain form of glucose reacts with the C-5 hydroxyl group to form an intermolecular hemiacetal.

hemiketal A compound formed by the reaction of a ketone group and a hydroxyl group; for example, the C-2 keto group of the open-chain form of fructose reacts with the C-5 hydroxyl group to form an intermolecular hemiketal.

hexokinase A kinase that phosphorylates six-carbon sugars, usually glucose, at the expense of ATP.

high-density lipoprotein (HDL) A lipoprotein that picks up cholesterol from dying cells and from membranes undergoing turnover, esterifies it, and then transfers the cholesterol esters to the liver and other steroid-synthesizing tissues.

high-pressure liquid chromatography (HPLC) An enhanced column chromatography technique. The separation materials are more finely divided and thus possess more interaction sites, resulting in greater resolution. Because the column material is so fine, pressure must be applied to obtain adequate flow rates.

histone Member of a highly conserved group of small basic proteins found in eukaryotes in association with DNA to form nucleosomes.

histone acetyltransferase (HAT) An enzyme that catalyzes the attachment of acetyl groups from acetyl CoA to specific lysine residues in the amino-terminal domains of histones. These enzymes play crucial roles in the transcription-enhancing modification of chromatin structure.

histone deacetylase An enzyme that contributes to transcriptional repression by the deacetylation of acetylated lysine residues in histones.

Holliday junction A crosslike structure, formed by four polynucleotide chains, that is a key intermediate in the recombination process.

holoenzyme An enzyme that consists of the protein component forming the main body of the enzyme (the apoenzyme) and any necessary, usually small, cofactors.

homogenate The mixture formed when a tissue is treated in such a way that the plasma membranes are disrupted.

hydrogen bond A bond formed when two relatively electronegative atoms, such as oxygen or nitrogen, unequally share a hydrogen atom that is covalently bonded to one of the electronegative atoms.

hydrophobic effect The process in which nonpolar molecules in aqueous solutions are driven together because of the resulting increase in entropy of water molecules.

3-hydroxy-3-methylglutaryl CoA reductase (HMG-CoA reductase) A highly regulated enzyme that catalyzes the committed step in cholesterol synthesis—the formation of mevalonate from 3-hydroxy-3-methylglutaryl CoA.

hypoxanthine-guanine phosphoribosyltransferase (HGPRT) An enzyme that catalyzes the formation of guanylate or inosinate by attaching guanine or hypoxanthine, respectively, to 5-phosphoribosyl-1-pyrophosphate (PRPP); the reaction is part of the salvage pathway for purine nucleotides.

immunoprecipitation A precipitation technique in which a protein reacts with an antibody to the protein. The antibody is in turn linked to an insoluble matrix.

induced fit The modification of the shape of an active site in an enzyme after the substrate has been bound.

inducer A small molecule that binds to a repressor and alters its interaction with an operator.

initiation factor (IF) One of a set of proteins that assist in the association of the ribosome, mRNA, and initiator tRNA to initiate the process of protein synthesis.

inosinate A purine nucleotide formed by the reaction of hypoxanthine with 5-phosphoribosyl-1-pyrophosphate (PRPP); a precursor to both AMP and GMP.

insulin A polypeptide hormone, secreted by the β cells of the pancreas, that stimulates fuel storage and protein synthesis.

integral membrane proteins Proteins found in membranes that interact extensively with the hydrocarbon chains of the membrane lipids and usually span the membrane.

intermediary metabolism Consists of a collection of defined pathways that process a biochemical from a starting point to an end point without the generation of wasteful or harmful side products. These pathways are interdependent—a biochemical ecosystem—and their activities are coordinated by exquisitely sensitive means of communication in which allosteric enzymes are predominant.

intrinsically unstructured protein (IUP) A protein that, completely or in part, does not have a discrete three-dimensional structure under physiological conditions until it interacts with another molecule.

intron Region of the primary transcript that is removed in the mature mRNA. Also called intervening sequence.

ion channel A passive-transport system for ions capable of very high transport rates; ion channels often display a high degree of specificity for the transported ion.

ion-exchange chromatography A means of protein purification that takes advantage of the differences in net charge among proteins. A mixture of proteins is percolated through a column of beads that contain a charged group, such as a carboxylate group. Proteins bearing charges opposite those on the bead will bind. The bound proteins can be eluted (released) by washing with a buffer containing salt.

ionic bond See **electrostatic interaction.**

IRE-binding protein (IRE-BP) An iron-sensitive protein that regulates the translational capability of ferritin mRNA and the stability of transferrin-receptor mRNA by binding to a stem-loop structure called an iron-response element (IRE) in the mRNA molecule.

iron-response element (IRE) A stem-loop structure found in the mRNAs for ferritin and transferrin receptor that interacts with the IRE-binding protein and regulates the translation of the mRNAs.

iron–sulfur proteins Proteins that contain clusters of iron and sulfur that play a role in electron-transfer reactions; iron cycles between the Fe^{2+} and Fe^{3+} state. Also called nonheme-iron proteins.

isocitrate dehydrogenase An enzyme that catalyzes the oxidative decarboxylation of isocitrate to form α-ketoglutarate in the citric acid cycle; plays a role in controlling the rate of the citric acid cycle.

isocitrate lyase An enzyme of the glyoxylate cycle, isocitrate lyase cleaves isocitrate into succinate and glyoxylate.

isoelectric focusing A means of separating proteins electrophoretically based on the relative content of acidic and basic residues of the proteins. A mixture of proteins undergoes electrophoresis in a pH gradient. Each protein will migrate in the gel until it reaches the pH equal to its pI, the pH at which the protein has no net charge.

3-isopentenyl pyrophosphate Activated isoprene; the basic building block of cholesterol.

isozymes Enzymes encoded by different genes and catalyze the same reaction, yet display different kinetic parameters and respond to different regulatory molecules.

k_{cat}/K_M The specificity constant; directly relates the rate of a reaction to the concentration of the reactant and allows comparison of the efficiency of an enzyme capable of reacting with different substrates.

ketogenic amino acids Amino acids whose carbon skeletons, entirely or in part, are degraded into acetyl CoA or acetoacetyl CoA; only leucine and lysine are solely ketogenic.

ketone body Refers to acetoacetate, β-hydroxybutyrate, and acetone, produced when acetyl CoA is diverted from the citric acid cycle to the formation of acetoacetyl CoA in the liver; subsequent reactions generate the three compounds, known as ketone bodies.

ketose A monosaccharide that has a ketone group as its most oxidized carbon atom.

kinase An enzyme that catalyzes the attachment of a phosphoryl group to a substrate by using ATP as a phosphoryl donor.

kinetics The study of the rates of chemical reactions.

lac **repressor** The regulator protein that binds to the operator site of the *lac* operon and thereby inhibits the expression of the structural genes of the operon; inhibition is relieved when the repressor protein binds allolactose, an inducer of the *lac* operon.

lactic acid fermentation The anaerobic metabolism of glucose to yield lactic acid with the concomitant production of ATP.

lagging strand A newly synthesized strand of DNA at the replication fork that is initially synthesized as Okazaki fragments. See also **leading strand**.

L amino acid One of two mirror-image forms, called the L isomer and the D isomer, of α-amino acids. Only L amino acids are constituents of proteins.

lateral diffusion The ability of lipid and protein molecules to move laterally in the membrane rapidly and spontaneously.

leading strand A newly synthesized strand of DNA at the replication fork that is synthesized continuously. See also **lagging strand**.

lectin A protein with a high affinity for a specific sugar residue; as such, lectins are important probes of carbohydrate-containing molecules.

Lesch–Nyhan syndrome A disease resulting from the loss of a single enzyme in the salvage pathway for purines; marked by mental retardation, extreme hostility, and self-mutilation.

ligand A small molecule that binds to a protein, inducing a specific structural change. For instance, a steroid is a ligand for a steroid-hormone receptor.

light-harvesting complex A complex of light-absorbing pigments and protein that completely surrounds the reaction center of photosynthesis; funnels the energy of absorbed light to the reaction center.

light reactions In chloroplasts, the reactions in which light is used to create reducing potential, to form ATP, and to generate oxygen.

Lineweaver–Burk equation A plot of $1/V_0$ versus $1/[S]$ yields a straight line with a y-intercept of $1/V_{max}$ and a slope of K_M/V_{max}. Also called a double-reciprocal plot.

lipase An enzyme that degrades a triacylglycerol into free fatty acids and monoacylglycerol.

lipid A key biomolecule that is both hydrophilic and hydrophobic, enabling it to form barriers that delineate the cell and cellular compartments. Also serves as a storage form of energy and as a signal molecule.

lipid bilayer A bimolecular sheet formed by amphipathic molecules in which the hydrophobic moieties are on the inside of the sheet and the hydrophilic ones are on the aqueous outside.

lipid raft A complex formed by cholesterol and some phospholipids, notably sphingolipids. Such complexes may concentrate in specific regions within membranes. The resulting structures are thicker, more stable, and less fluid than the rest of the membrane. Lipid rafts may also play a role in concentrating proteins that participate in signal-transduction pathways. Also called membrane raft.

lipoic acid An acyl-group carrier that functions as a cofactor in some dehydrogenase enzymes; linked covalently to specific lysine residues in enzyme proteins, it can exist as the reduced open-chain form or the closed-ring disulfide form, undergoing interconversion in a catalytic cycle.

lipoprotein particle Consists of a core of hydrophobic lipids surrounded by a shell of polar lipids and specific proteins and plays a role in the transport of cholesterol and triacylglycerols.

liposome An artificial lipid vesicle that consists of an aqueous compartment enclosed by a lipid bilayer.

low-density lipoprotein (LDL) The major carrier of cholesterol in the blood; consists of a core of esterified cholesterol molecules surrounded by a shell of phospholipids, unesterified cholesterol, and apoprotein B-100; primary source of cholesterol for cells other than those in the liver or intestine.

lysosome An organelle containing a wide array of digestive enzymes that degrades and recycles damaged cellular components or material brought into the cell by endocytosis.

major groove A 12-Å-wide, 8.5-Å-deep groove in B-DNA resulting from the fact that the glycosidic bonds of a base pair are not diametrically opposite each other.

malate–aspartate shuttle A reversible shuttle, found in the liver and heart, used to transport electrons from cytoplasmic NADH to mitochondrial NAD^+.

malate synthase An enzyme of the glyoxylate cycle that catalyzes the formation of oxaloacetate from glyoxylate and acetyl CoA.

MALDI-TOF "Matrix-assisted laser desorption–ionization–time of flight": a technique for determining a protein's mass.

malonyl CoA Carboxylated acetyl CoA; the activated form of acetyl CoA used in fatty acid synthesis.

manganese center See **water-oxidizing complex**.

mass spectrometry A technique for determining a protein's mass and identity and for sequencing a protein.

mechanism-based inhibition An inhibitor binds to an enzyme as a substrate and is initially processed by the normal catalytic mechanism. The mechanism of catalysis then generates a chemically reactive intermediate that inactivates the enzyme through covalent modification. The fact that the enzyme participates in its own irreversible inhibition strongly suggests that the covalently modified group on the enzyme is catalytically vital. Also called suicide inhibition.

membrane A lipid bilayer.

mediator A complex of 25 to 30 subunits that is part of the preinitiation complex of the eukaryotic transcription machinery. Mediator acts as a bridge between enhancer-bound activators and promoter-bound RNA polymerase II.

messenger RNA (mRNA) A special class of RNA molecules that serve as templates for the synthesis of proteins.

metabolism A highly integrated network of chemical pathways that enables a cell to extract energy from the environment and use this energy for biosynthetic purposes.

metal ion catalysis Catalysis in which a metal acts as an electrophilic catalyst by stabilizing a negative charge on a reaction intermediate, generates a nucleophile by increasing the acidity of nearby molecules, or increases the binding energy of the enzyme–substrate interaction by binding to substrates.

metamorphic protein Belongs to a class of proteins that exist in an ensemble of structures of approximately equal energy that are in equilibrium.

mevalonate A precursor for the synthesis of cholesterol; its formation by 3-hydroxy-3-methylglutaryl CoA reductase constitutes the committed step in cholesterol biosynthesis.

micelle A globular structure formed by amphipathic molecules in which the hydrophilic part is exposed to water and the hydrophobic part is sequestered inside, away from the water.

Michaelis constant The concentration of substrate at which half the active sites of an enzyme are filled; a ratio of rate constants for the reaction model.

$$E + S \underset{k_{-1}}{\overset{k_1}{\rightleftharpoons}} ES \overset{k_2}{\longrightarrow} E + P$$

The constant is defined as $K_M = (k_{-1} + k_2)/k_1$.

Michaelis–Menten equation An equation that expresses the velocity (V) of an enzyme-catalyzed reaction in terms of maximum velocity (V_{max}), substrate concentration (S), and the Michaelis–Menten constant (K_M). The equation accounts for the hyperbolic kinetics observed when V is plotted as a function of S; the equation is $V = V_{max} [S]/([S] + K_M)$.

microRNA (miRNA) Belongs to a class of noncoding RNAs found in plants, invertebrates, and vertebrates. These ~21-nucleotide-long RNA molecules are excised from longer double-stranded RNAs and play a role in posttranscriptional regulation.

minor groove A 6-Å-wide, 7.5-Å-deep groove in B-DNA that arises because the glycosidic bonds of a base pair are not diametrically opposite each other.

mismatch repair A system of DNA repair that consists of at least two proteins, one for detecting the mismatch and the other for recruiting an endonuclease that cleaves the newly synthesized DNA strand close to the lesion. An exonuclease can then excise the incorrect base, and a polymerase fills the gap.

mitochondrion An oval-shaped organelle, about 2 μm in length and 0.5 μm in diameter, that is the site of oxidative phosphorylation, the enzymes of the citric acid cycle, and the enzymes of fatty acid oxidation.

molecular exclusion chromatography (also called gel filtration chromatography) A technique for separating proteins on the basis of size. A mixture of proteins is percolated through a column of porous beads. Large proteins are excluded from the beads and exit the column first. Smaller proteins, because they enter the beads and take a longer path, exit subsequently.

molten globule A stage in protein folding. The hydrophobic effect brings together hydrophobic amino acids that are far apart in the amino acid sequence. The drawing together of hydrophobic amino acids in the interior leads to the formation of a globular structure. The hydrophobic interactions are presumed to be dynamic, allowing the protein to form progressively more stable interactions.

monoclonal antibody An antibody, produced by a cloned hybridoma cell, that binds to only one epitope on a specific antigen.

monosaccharide A single aldehyde or ketone that has two or more hydroxyl groups; monosaccharides are the simplest carbohydrates.

motif Refers to certain combinations of secondary structure present in many proteins, frequently exhibiting similar functions. Also called supersecondary structure.

mucin Belongs to a class of glycoproteins, consisting mainly of carbohydrate, in which the protein component is extensively glycosylated to serine or threonine by *N*-acetylgalactosamine. Also called a mucoprotein.

multidrug-resistance (MDR) protein A protein that acts as an ATP-dependent pump that extrudes a wide range of small molecules from cells that express it. When cells are exposed to a drug, the MDR protein pumps the drug out of the cell before the drug can exert its effects. The MDR protein comprises four domains: two membrane-spanning domains and two ATP-binding cassette domains. Also called P glycoprotein.

mutagen Perturbs the base sequence of DNA and causes a mutation; often chemical but can also be energy sources such as ultraviolet light.

mutase An enzyme that catalyzes the intramolecular shift of a chemical group.

N-acetylglutamate An allosteric activator of mammalian carbamoyl phosphate synthetase, which initiates the synthesis of urea.

NADH-Q oxidoreductase A large component of the respiratory chain that transfers electrons from NADH to ubiquinone and, in the process, pumps protons across the inner mitochondrial membrane to generate the proton-motive force. Also called NADH dehydrogenase or Complex I.

Na$^+$–K$^+$ pump A membrane-bound enzyme that uses the energy of ATP hydrolysis to pump Na$^+$ out of the cell and K$^+$ into the cell, against their concentration gradients. Also called Na$^+$–K$^+$ ATPase.

nitrogenase complex An enzyme complex that catalyzes the reduction of N$_2$ to ammonia; found in bacteria and the blue-green algae.

nitrogen fixation The conversion of nitrogen gas (N$_2$) into ammonia; the first step in the flow of nitrogen into amino acids, nucleotides, and other nitrogen-containing compounds.

noncompetitive inhibition The reduction in the rate of enzyme activity observed when an enzyme can bind its substrate and its inhibitor simultaneously. Noncompetitive inhibitors decrease the turnover number for an enzyme but do not alter the K_M of the enzyme; their effects are not overcome by increased substrate concentration.

nonessential amino acids Amino acids that can be synthesized by an organism and are thus not a dietary requirement.

northern blotting A technique that allows the identification of a specific RNA molecule. A mixture of RNA molecules is separated by gel electrophoresis and then transferred (blotted) to nitrocellulose. The blot is then hybridized with a ^{32}P-labeled probe complementary to the RNA molecule of interest. The RNA molecule is visualized by autoradiography.

nuclear hormone receptor A member of a large family of transcription factors that, on the binding of a signal molecule such as a steroid hormone, modify the expression of specific genes by binding to control elements in the DNA.

nucleic acid A macromolecule composed of nucleotide monomers that stores and transfers information in cells. DNA and RNA are nucleic acids.

nucleoside A purine or pyrimidine base linked to a sugar.

nucleosome The repeating unit of chromatin that consists of 200 base pairs of DNA and two each of the histones H2A, H2B, H3, and H4.

nucleosome core particle A complex of the histone octamer and the 145-bp DNA fragment in eukaryotic DNA.

nucleotide A nitrogenous purine or pyrimidine base linked to a sugar, which is in turn linked to one or more phosphoryl groups.

nucleotide-excision repair A means of repairing DNA in which a stretch of DNA around the site of damage is removed and replaced.

nucleus A double-membrane-enclosed compartment that contains most of a eukaryotic cell's DNA.

obligate anaerobe An organism that cannot survive in the presence of oxygen and is thus usually dependent on fermentation as a source of cellular energy.

Okazaki fragments Small fragments of DNA (approximately 1000 nucleotides) that are formed on the lagging strand at the replication fork of DNA synthesis and later joined; enable $5' \rightarrow 3'$ polymerization at the nucleotide level while overall growth is in the $3' \rightarrow 5'$ direction.

oligosaccharide A carbohydrate composed of 2 to 12 monosaccharide units.

oncogene A gene whose expression contributes to the development of cancer.

operator A DNA segment that is adjacent to a group of structural genes and is the target sequence for a repressor protein.

operon In bacteria, a coordinated unit of gene expression that includes structural genes and regulatory elements recognized by one or more regulatory-gene products.

organelle Any intracellular membrane-bounded component of a cell.

origin of replication A particular sequence in a genome at which replication begins.

orotidylate A nucleotide precursor to uridylate and cytidylate formed by the reaction of orotate with 5-phosphoribosyl-1-pyrophosphate (PRPP).

oxidative phosphorylation The process in which ATP is formed as a result of the transfer of electrons from NADH or $FADH_2$ to O_2 by a series of electron carriers.

oxyanion hole A region on certain proteolytic enzymes that stabilizes the oxyanion constituent of the tetrahedral intermediate of the reaction.

pentose phosphate pathway A metabolic pathway that generates NADPH and five-carbon sugars such as ribose 5-phosphate from glucose 6-phosphate; it includes oxidative reactions that produce NADPH and ribose 5-phosphate as well as nonoxidative reactions that together convert five-carbon sugar phosphates into gluconeogenic precursors of glucose 6-phosphate. Also referred to as the hexose monophosphate shunt or the phosphogluconate pathway.

pepsin A proteolytic enzyme that begins the process of protein degradation in the stomach. Pepsin's action yields protein fragments that will be further degraded by the proteases of the intestine.

peptide bond A covalent linkage formed between the α-carboxyl group of one amino acid and the α-amino group of another. Also called an amide bond.

peptidyl transferase center A region of the large ribosomal subunit that catalyzes peptide-bond formation between the aminoacyl (or peptidyl) tRNA in the P site and the aminoacyl tRNA in the A site.

peripheral membrane protein A protein associated with the surface of a membrane by electrostatic and hydrogen-bond interactions.

pH A measure of the hydrogen ion concentration of a solution.

phagocytosis The process of taking large amounts of material into a cell.

phenylketonuria A disease caused by the inability to convert phenylalanine into tyrosine, which results in excess phenylalanine and its secondary metabolites; the disease is characterized by severe retardation.

phi angle The angle of rotation about the bond between the nitrogen atom and the α-carbon atom of an amino acid in a polypeptide chain.

phosphatidate (diacylglycerol 3-phosphate) A precursor to triacylglycerols and to many phospholipids.

phosphatidic acid phosphatase (PAP) The key regulatory enzyme in lipid synthesis. The enzyme controls the extent to which triacylglycerols are synthesized relative to phospholipids and regulates the type of phospholipid synthesized.

phosphofructokinase (PFK) A kinase that phosphorylates fructose 6-phosphate to fructose 1,6-bisphosphate; PFK, an allosteric enzyme, is the major control point for flux through the glycolytic pathway.

phosphoglycerol A phospholipid derived from glycerol and composed of a glycerol backbone to which are attached two fatty acid chains and a phosphorylated alcohol.

phosphoinositide cascade A set of reactions that convert an extracellular signal into an intracellular one; the conversion entails the cleavage of the phospholipid phosphatidylinositol 4,5-bisphosphate into two second messengers: inositol 1,4,5-trisphosphate and diacylglycerol.

phospholipase C An enzyme that catalyzes the degradation of phospholipids in signal-transduction pathways. Phospholipase C catalyzes the formation of two second messengers: inositol 1,4,5-trisphosphate, a soluble molecule that can diffuse from the membrane, and diacylglycerol, which stays in the membrane.

phospholipid An important constituent of membranes that is composed of three components: a backbone (usually glycerol or sphingosine), two fatty acid chains, and a phosphorylated alcohol.

5-phosphoribosyl-1-pyrophosphate (PRPP) A form of ribose activated to accept nucleotide bases.

phosphorolysis The cleavage of a bond by orthophosphate, as in the degradation of glycogen to glucose 1-phosphate.

phosphorylase kinase An enzyme that activates phosphorylase b by attaching a phosphoryl group in the regulation of glycogen degradation. This kinase is itself under dual control: it is activated both by phosphorylation and by increases in Ca^{2+} levels.

phosphorylation potential A means of measuring the energy status of a cell that is derived by dividing the concentration of ATP by the product of the concentrations of ADP and P_i.

phosphoryl-transfer potential A measure of the tendency of a phosphorylated compound to transfer a phosphoryl group to another compound; presented as the $\Delta G^{\circ\prime}$ of hydrolysis of the phosphorylated compound; the more negative the $\Delta G^{\circ\prime}$ of hydrolysis, the greater the phosphoryl-group-transfer potential.

photoinduced charge separation The excitation of an electron from its ground state to a higher energy level by light absorption and the subsequent movement of the excited electron from the initial molecule to an acceptor, resulting in a positive charge on the initial molecule and a negative charge on the acceptor molecule.

photorespiration The conversion of organic carbon into carbon dioxide without the production of energy-rich metabolites; the result of the oxygenase reaction catalyzed by rubisco and the subsequent synthesis of glycine from two molecules of glycolate, with the release of carbon dioxide and ammonia and the consumption of ATP.

photosynthesis The conversion of sunlight into chemical energy.

photosystem I (PS I) In chloroplasts, a photosynthetic unit that includes a light-harvesting complex, a reaction center, and an electron-transport chain. The system catalyzes the light-driven transfer of electrons from reduced plastoquinone to ferredoxin, which in turn drives the formation of NADPH; it requires light of wavelength shorter than 700 nm.

photosystem II (PS II) In chloroplasts, a photosynthetic unit that includes a light-harvesting complex, a reaction center, and an electron-transport chain. The system catalyzes the light-driven transfer of electrons from water to plastoquinone, with the concomitant generation of oxygen; it requires light of wavelength shorter than 680 nm.

phototroph An organism that can meet its energy needs by converting light energy into chemical energy.

plasma membrane Separates the inside of a cell, whether prokaryotic or eukaryotic, from the outside. Although impermeable to most substances, thus serving as a barrier, the plasma membrane has selective permeability owing to the presence of proteins that permit the entry and exit of certain molecules and information.

plasmid In bacteria, a small circular DNA separate from the bacterial chromsomal DNA. Plasmids can replicate independently of the chromosomal DNA and often carry genes that confer antibiotic resistance. Plasmids are also used as cloning vectors.

poly(A) tail A long (as many as 250 nucleotides) polyadenylate segment added posttranscriptionally to the 3′ end of most eukaryotic mRNA.

polymerase chain reaction (PCR) A technique used to amplify a target DNA sequence over several orders of magnitude. A PCR cycle consists of three steps—strand separation, hybridization of primers, and DNA synthesis—that are repeated 20 or 30 times.

polysaccharide A carbohydrate composed of a large number of linear or branched monosaccharide units; homopolysaccharides are composed of large numbers of one type of sugar, whereas heteropolysaccharides contain more than one type.

polysome A group of ribosomes bound to an mRNA molecule and simultaneously carrying out translation. Also called polyribosome.

pre-mRNA The unspliced immediate product of RNA polymerase II in eukaryotes.

primary messenger The information embodied in the interaction between a ligand and its receptor molecule.

primary structure Usually refers to the linear sequence of amino acids in a protein; more generally, the linear sequence of units that form a polymer.

primase A specialized RNA polymerase that synthesizes the RNA primers for DNA synthesis.

primer In the elongation of polymers, the initial segment of the polymer that is to be extended. In DNA synthesis, elongation depends on the primer.

prion A brain protein that can fold into an abnormal structure. Prions aggregate and cause spongioform encephalopathies, such as mad cow disease; these diseases are transmissible, and thus prions are infectious proteins.

processivity A property of an enzyme that enables it to catalyze multiple rounds of elongation or the digestion of a polymer while the polymer stays bound to the enzyme.

prokaryote A single-celled organism lacking a nucleus and membrane-bounded compartments. Prokaryotes consist of two taxonomic domains: the bacteria and the archaea.

promoter A specific sequence of DNA, usually just upstream of a gene, that specifies the start site and extent of transcription of the associated gene.

prostaglandin Belongs to a class of short-lived signal molecules that are 20-carbon fatty acids containing a five-membered ring.

protein A biological macromolecule composed of a linear array of amino acids joined by peptide bonds; roles of proteins in biological processes include catalysis, transport and storage, motion, mechanical support, immune protection, the generation and transmission of nerve impulses, and the control of growth and differentiation.

protein kinase A (PKA) A protein kinase that consists of two catalytic subunits and two regulatory subunits that inhibit the catalytic subunits; on binding of cyclic AMP, the regulatory subunits dissociate from the catalytic subunits, which then become active.

protein kinase C (PKC) A protein kinase that is activated by the binding of diacylglycerol.

protein phosphatase 1 (PP1) A protein phosphatase stimulated by insulin that inhibits glycogen degradation and stimulates glycogen synthesis.

protein targeting A means by which proteins are directed to specific cellular compartments or processed for secretion. Also called protein sorting.

proteoglycan A protein containing one or more covalently linked glycosaminoglycan chains; cartilage proteoglycan contains keratan sulfate and chondroitin chains linked to a polypeptide backbone.

proteolytic enzyme Belongs to a class of enzymes that hydrolyze the peptide bonds between amino acids, thus digesting proteins. Also called protease.

proteome Refers to the proteins expressed by the genome. It encompasses the types, functions, and interactions of proteins that yield a functional unit. Unlike the genome, the proteome varies with cell type, developmental state, and environmental conditions.

proton-motive force The energy inherent in the proton gradient established during the functioning of the respiratory chain; consists of a membrane potential as well as a chemical gradient.

proto-oncogene A signal-transduction protein that usually regulates cell growth in some fashion; when proto-oncogenes mutate, they become oncogenes and contribute to the development of cancer.

protoporphyrin An organic constituent of the heme prosthetic group; consists of four pyrrole rings joined by methine bridges and contains various side chains.

proximal histidine The residue occupying the fifth coordination site to which iron can bind in hemoglobin and myoglobin.

P700 A special pair of molecules in photosystem I in green plants; absorption of light by P700 results in the transfer of electrons that generates ferredoxin and, ultimately, NADPH.

psi angle The angle of rotation about the bond between the α-carbon atom and the carbonyl carbon atom of an amino acid in a polypeptide chain.

P680 A special pair of molecules in photosystem II in green plants; absorption of light by P680 results in the transfer of electrons from water to plastoquinone, which generates a proton gradient.

P-type ATPases A family of enzymes that use the energy of hydrolysis to move ions across membranes; called "P-type ATPases" because the reaction mechanism includes a phosphorylaspartate intermediate.

purine A nitrogenous base that includes a pyrimidine ring fused with a five-membered imidazole ring; the purine derivatives adenine and guanine are found in nucleotides and nucleic acids.

purinosome The complex formed upon the association of the enzymes of purine synthesis.

pyridoxal phosphate (PLP) A prosthetic group derived from vitamin B_6 (pyridoxine) that plays a key role in transamination reactions.

pyrimidine A nitrogenous base that is a six-membered heterocyclic ring containing two nitrogen atoms and four carbon atoms; the pyrimidine derivatives cytosine, uracil, and thymine are found in nucleotides and nucleic acids.

pyrimidine nucleotide A nucleotide in which the base is a pyrimidine.

pyruvate carboxylase A biotin-dependent enzyme that catalyzes the formation of oxaloacetate from pyruvate and CO_2 at the expense of ATP; important in gluconeogenesis as well as in the replenishment of the citric acid cycle.

pyruvate dehydrogenase (E_1) One of three distinct enzymes in the pyruvate dehydrogenase complex; E_1 catalyzes the decarboxylation of pyruvate and the formation of acetyllipoamide.

pyruvate dehydrogenase complex A large, complex mitochondrial enzyme that catalyzes the oxidative decarboxylation of pyruvate to form acetyl CoA; this irreversible reaction is the link between glycolysis and the citric acid cycle.

pyruvate kinase An enzyme in the glycolytic pathway that catalyzes the virtually irreversible transfer of a phosphoryl group from phosphoenolpyruvate to ADP.

Q cycle A set of reactions in which coenzyme Q cycles between the fully reduced state and the fully oxidized state through one-electron transfer reactions in which one of the electrons is temporarily stored in Q$^{\bar{\bullet}}$; provides a means of passing the two electrons of coenzyme Q to the single-electron carrier cytochrome *c*, one electron at a time.

Q-cytochrome *c* oxidoreductase A component of the respiratory chain, this oxidoreductase carries electrons from coenzyme Q to cytochrome *c* and, in the process, pumps protons out of the mitochondrial matrix to generate the proton-motive force. Also called cytochrome *c* reductase or Complex III.

Q pool A pool of coenzyme Q (Q) and ubiquinol (QH$_2$) thought to exist in the inner mitochondrial membrane. Because ubiquinone is soluble in the hydrophobic interior of the membrane, the components of the Q pool can readily shuttle between the larger electron carriers in the membrane.

quantitative PCR (qPCR) A polymerase chain reaction technique capable of determining the quantity of individual mRNA transcripts. Also called real-time PCR.

quaternary structure In proteins containing more than one polypeptide chain, the spatial arrangements of those chains (subunits) and the nature of contacts among them.

quorum sensing A signal-transduction system in bacteria that allows the coordination of gene expression in a population of bacteria as a function of population density.

RAD51 A key ATPase in human recombination that binds single-stranded DNA.

Ramachandran plot A steric contour diagram that depicts allowed ranges of the angles ψ and ϕ for amino acid residues in polypeptide chains; for each residue, its conformation in the main chain of a polypeptide can be completely defined by ψ (the degree of rotation at the bond between the nitrogen atom and the α-carbon atom) and ϕ (the degree of rotation at the bond between the carbonyl carbon atom and the α-carbon atom).

Ras A small G protein that stimulates cell growth and differentiation and possesses intrinsic GTPase activity.

reaction center In a photosynthetic unit, a specialized chlorophyll molecule that collects excitation energy from other chlorophyll molecules and mediates the transformation of light into chemical energy.

receptor-mediated endocytosis A means of importing specific proteins into a cell by their binding to plasma-membrane receptors and their subsequent endocytosis and inclusion into vesicles.

receptor tyrosine kinase (RTK) A transmembrane receptor protein that, when bound to the appropriate signal molecule, displays intracellular protein kinase activity, phosphorylating proteins at tyrosine residues.

reducing sugar A sugar that converts into a form with a free aldehyde group that is readily oxidized and can thus reduce another compound.

reduction potential The electron-transfer potential, expressed symbolically as E_0'. Also called redox potential or oxidation–reduction potential.

release factor (RF) One of a set of proteins that recognize Stop codons on mRNA at the A site on the ribosome, which leads to the release of the completed protein from the tRNA in the P site of the ribosome.

replication The process of copying the genome.

replication fork The site of DNA synthesis where the parental strands are separated and daughter strands complementary to each parent are synthesized.

repressor A protein that binds to an operator sequence and inhibits the transcription of the structural genes in the operon.

respiratory control Tight coupling or coordination of the oxidation of reduced cofactors (NADH and FADH$_2$) in the electron-transport chain and the phosphorylation of ADP to yield ATP in the mitochondrion; such control ensures that the rate of the citric acid cycle, where reduced cofactors are generated, corresponds to the demand for ATP. Also called acceptor control.

restriction enzyme Bacterial enzymes that recognize specific base sequences in double-helical DNA and cleave, at specific places, both strands of DNA. Restriction enzymes are among the key tools of recombinant DNA technology. Also called restriction endonuclease.

reverse cholesterol transport A process in which high-density lipoproteins pick up cholesterol released into the plasma from dying cells and from membranes undergoing turnover and transport them to the liver.

reverse transcriptase A viral polymerase that uses RNA as template to synthesize DNA.

R group Side chain of an amino acid.

rho An ATP-dependent bacterial helicase that breaks the RNA–DNA hybrid at the transcription bubble and thereby terminates transcription.

ribonucleic acid (RNA) A long unbranched polymer consisting of ribonucleotides joined by $3' \rightarrow 5'$ phosphodiester linkages. This single-stranded molecule is similar to DNA in composition except that it contains the base uridine instead of thymine and its sugar component has an extra hydroxyl group. RNA plays many roles in the cell, including that of a catalyst.

ribonucleotide reductase An enzyme that catalyzes the reduction of all four ribonucleotides to deoxyribonucleotides.

ribose A five-carbon monosaccharide ($C_5H_{10}O_5$) that constitutes the carbohydrate moiety of ATP, other ribonucleosides and ribonucleotides, and cofactors such as NAD$^+$ and coenzyme A.

ribosomal RNA (rRNA) The RNA components of ribosomes. In eukaryotes, 18S RNA is a component of the 40S subunit, and 5S RNA, 28S RNA and 5.8S RNA are components of the 60S subunit. A site on the 28S RNA catalyzes peptide-bond formation.

ribosome A large ribonucleoprotein assembly that catalyzes the formation of peptide bonds; a molecular machine that coordinates protein synthesis.

riboswitch A molecular structure that is a component of some bacterial mRNA molecules that can bind to specific small molecules and thereby prevent further transcription of the mRNA.

RNA editing A process, distinct from RNA splicing, that changes the nucleotide sequence of mRNA after transcription.

RNA interference (RNAi) Refers to the regulation of gene expression posttranscriptionally by small molecules of RNA.

RNA polymerase Belongs to a class of enzymes that synthesize RNA molecules complementary to DNA templates.

rubisco (ribulose 1,5-bisphosphate carboxylase/oxygenase) An enzyme that catalyzes the reaction of carbon dioxide with ribulose 1,5-bisphosphate to form two molecules of 3-phosphoglycerate.

S-adenosylmethionine (SAM) An activated methyl donor that consists of an adenosyl group linked to the sulfur atom of methionine.

salting out Protein precipitation caused by an increase in the salt concentration.

salvage pathway In general, a pathway that synthesizes the final product from preformed components; purine nucleotides can be synthesized in a salvage pathway by attaching purine bases to phosphoribosylpyrophosphate (PRPP).

secondary structure In a protein, the spatial arrangement of amino acid residues that are near one another in the linear sequence; the α helix and the β strand are both elements of secondary structure.

secondary transporter A transporter that uses the energy of the downhill (exergonic) flow of one ion or molecule to power the uphill (endergonic) flow of another. Also called a cotransporter.

second messenger A small intracellular signal molecule whose concentration changes in response to a primary messenger.

secretin A hormone, released by the cells of the small intestine, that stimulates the release of sodium bicarbonate, which neutralizes the low pH of the partly digested proteins, carbohydrates, and lipids coming from the stomach.

secretory granule A vesicle filled with the proteins destined for secretion. Also called a zymogen granule.

sedimentation coefficient (*s*) The rate at which a particle will move in a liquid medium when subjected to centrifugal force. It is a function of the mass, shape, and density of the particle as well as the density of the medium. Sedimentation coefficients are usually expressed as Svedberg units (S), equal to 10^{-13} s.

selectin A carbohydrate-binding protein that constrains immune-system cells to the site of injury in an inflammatory response by mediating cell–cell interactions.

selective estrogen receptor modulator (SERM) An antagonist, such as a drug used in the treatment and prevention of breast cancer, that targets the estrogen receptor; some breast tumors rely on estrogen-mediated pathways for growth.

selectivity filter A region of ion-channel proteins that determines the specificity of a particular channel.

self-splicing Refers to introns that have the ability to remove themselves from precursor RNA and assist in the splicing of exons to form mature RNA.

semiconservative replication In the duplication of DNA, one of the strands of each daughter molecule is newly synthesized, whereas the other is unchanged from the parental DNA double helix.

sequential reaction A reaction having multiple reactants, in which all substrates bind to the enzyme before any product is released. Thus, in a reaction with two substrates, a ternary complex of the enzyme and both substrates forms.

seven-transmembrane-helix (7TM) receptor Belongs to a class of integral membrane proteins in which the intramembrane part consists of seven helical regions; these receptors are always coupled to G proteins.

severe combined immunodeficiency (SCID) An immunological disorder characterized by a loss of T cells and severe recurring infections, often leading to death at an early age. A deficiency in adenosine deaminase activity is associated with some forms of the disorder.

Shine–Dalgarno sequence In bacterial messenger RNA, a purine-rich region about 10 nucleotides on the 5′ side of an initiator codon that pairs with the 3′ end of 16S RNA in the 30S ribosomal subunit; helps to determine where translation is initiated on an mRNA molecule.

sickle-cell anemia A blood disease caused by a single amino acid substitution in one hemoglobin chain.

side chain The distinctive variable group bonded to the α-carbon atom of an amino acid. Also called R group.

sigma subunit A component of bacterial RNA polymerase that enables the core RNA polymerase to recognize promoter sites.

sigmoidal curve An S-shaped curve. The sigmoidal curve for the binding of oxygen in hemoglobin indicates that the oxygen binding is cooperative. A sigmoidal curve is a key characteristic of allosteric proteins.

signal peptidase An enzyme that cleaves signal sequences in mature proteins on the lumenal side of the endoplasmic reticulum membrane.

signal-recognition particle (SRP) Recognizes the signal sequence and binds the sequence and the ribosome as soon as the signal sequence exits the ribosome. SRP then shepherds the ribosome and its nascent polypeptide chain to the endoplasmic reticulum membrane.

signal sequence A sequence of 9 to 12 hydrophobic amino acid residues, sometimes containing positively charged amino acids and usually near the amino terminus of a nascent polypeptide chain. The presence of the signal sequence identifies the nascent peptide as one that must cross the endoplasmic reticulum membrane.

simple diffusion The movement of molecules down a concentration gradient.

sliding clamp The dimeric β_2 subunit of DNA polymerase III through which the DNA template slides. The enzyme is able to move without falling off its substrate by clasping the DNA molecule.

small nuclear ribonucleoprotein particles (snRNPs) Complexes of small nuclear RNA and specific proteins that catalyze the splicing of mRNA precursors.

small nuclear RNA (snRNA) Belongs to a class of small RNAs confined to the nucleus; some play a role in splicing.

small nucleolar ribonucleoprotein (snoRNP) Directs the extensive modification of the ribose and base components of nucleotides of pre-rRNA sequences destined for the ribosome, processing that takes place in the nucleolus. Each snoRNP consists of one small nucleolar RNA and several proteins.

Southern blotting A technique that allows the identification of a specific DNA sequence subsequent to the digestion of the DNA by restriction enzymes. Digestion fragments are separated by gel electrophoresis and then transferred (blotted) to nitrocellulose. The blot is then hybridized with a ^{32}P-labeled probe complementary to the DNA sequence of interest. The DNA fragment containing the sequence of interest is visualized by autoradiography.

special pair In a photosynthetic reaction center, the pair of chlorophyll molecules that collect excitation energy from antenna chlorophyll molecules and then transfer high-energy electrons to other electron acceptors.

sphingolipid A lipid having sphingosine rather than glycerol as its backbone. Sphingolipids are found in the plasma membranes of all eukaryotic cells, although the concentration is highest in the cells of the central nervous system.

sphingomyelin Common in brain tissue, a sphingolipid in which the terminal hydroxyl group of ceramide has a phosphorylcholine substituent.

sphingosine An amino alcohol containing a long, unsaturated hydrocarbon chain that is a component of the phospholipid sphingomyelin and of glycolipids; serves a role analogous to that of glycerol in phosphoglycerides.

spina bifida A class of birth defects characterized by incomplete or incorrect formation of the neural tube early in development.

spliceosome An assembly of proteins and small nuclear RNAs that splices primary transcripts to form mature mRNA.

splicing The removal of introns and the ligation of exons from precursors of RNA to form mature RNA.

SRP receptor (SR) An integral membrane protein consisting of two subunits to which a complex consisting of a signal-recognition particle and a ribosome binds.

starch A homopolysaccharide that is a storage form of glucose in plant chloroplasts; amylopectin, the branched form of starch, has approximately one α-1,6 linkage per thirty α-1,4 linkages, whereas amylose is unbranched, composed of glucose residues in α-1,4 linkage.

stem-loop A nucleic acid structure, commonly seen in RNA, created when two complementary sequences within a single strand come together to form a double-helical structure, which constitutes the stem. The two strands of the stem are linked by the loop.

steroid hormone A hormone, such as androgen or estrogen, derived from cholesterol.

sterol regulatory element binding protein (SREBP) A transcription factor that binds to the sterol regulatory element of the HMG-CoA reductase gene and other genes in cholesterol and fatty acid metabolism to stimulate their transcription.

sticky ends Complementry strands produced when a restriction enzyme cleaves the DNA target sequence in a staggered fashion.

stomata Microscopic pores located on the underside of plants leaves that are used for gas exchange. Singular, stoma.

stroma The matrix of the chloroplast; contains thylakoids as well as soluble enzymes and is enclosed by the inner membrane of the chloroplast.

substrate A reactant in a chemical reaction. An enzyme catalyzes a single chemical reaction or a set of closely related reactions, and the components of those reactions are called substrates.

substrate cycle A pair of thermodynamically irreversible biochemical reactions that simultaneously produce and consume a pair of metabolic intermediates; these paired pathways may amplify metabolic signals and, in some cases, can generate heat for the maintenance of temperature in tissues. Also called a futile cycle.

substrate-level phosphorylation The formation of ATP from ADP in which the phosphate donor is a substrate with a higher phosphoryl-transfer potential than that of ATP.

subunit Any of the polypeptide chains in a protein that contains more than one such chain.

succinate-Q reductase An integral-membrane-protein complex of the inner mitochondrial membrane that transfers electrons from $FADH_2$ formed in the citric acid cycle to coenzyme Q. Also called Complex II.

sucrose A disaccharide of glucose and fructose (commonly known as table sugar) that is readily transportable and stored in many plant cells.

supercoiling Refers to the ability of closed, circular DNA to coil upon itself.

superoxide dismutase An enzyme that scavenges superoxide radicals by catalyzing the conversion of two of these radicals into hydrogen peroxide and molecular oxygen; protects against damage by reactive oxygen species.

symporter A transport system in which a molecule is carried across a membrane against its concentration gradient in the same direction as an ion moving down its concentration gradient.

TATA box Found in nearly all eukaryotic genes, a promoter element giving rise to mRNA. The TATA box is centered between 30 and 90 residues upstream of the transcription start site and has the consensus sequence 5′-TATAAAA-3′.

TATA-box-binding protein associated factor (TAF) One of a group of proteins, many of which contain bromodomains, that bind to the TATA-box-binding protein to form a complex required for RNA transcription.

telomerase A reverse transcriptase that contains its own template; a highly processive enzyme that elongates the 3′-ending strand of a telomere.

telomere Each end of a chromosome; the DNA at the telomere consists of hundreds of repeats of a hexanucleotide sequence characteristic of the organism.

template In DNA or RNA, a sequence that directs the production of a complementary sequence.

tertiary structure In proteins, the spatial arrangement of amino acid residues that are far from each other in the linear sequence, as well as the pattern of disulfide bonds.

tetrahydrofolate A highly versatile carrier of activated one-carbon units.

thalassemia An inherited disorder of hemoglobin caused by the loss or substantial reduction of a single hemoglobin chain.

thiamine pyrophosphate (TPP) The coenzyme form of thiamine (vitamin B_1), composed of a modified thiazole ring linked by a methylene bridge to a substituted pyrimidine; a cofactor in enzymatic reactions in which bonds to carbonyl carbon atoms are cleaved or synthesized.

thioester intermediate An ester in which the noncarbonyl oxygen atom is replaced by a sulfur atom; thioesters are energy-rich intermediates in a number of biochemical reactions.

thioredoxin A protein with exposed cysteine residues that can be reversibly oxidized and reduced; an important electron carrier in the reduction of ribonucleotides and in photosynthesis.

30S subunit The small subunit of the bacterial 70S ribosome; composed of 21 different proteins and a 16S RNA molecule.

thylakoid In chloroplasts, a membranous sac, or vesicle, that contains the energy-transducing machinery of photosynthesis, including light-harvesting proteins, reaction centers, electron-transport chains, and ATP synthase.

thymidylate synthase An enzyme that catalyzes the methylation of deoxyuridylate (dUMP) to form thymidylate (TMP).

thymine (T) A base in DNA.

titration curve A plot of how the pH of a solution changes with the amount of OH^- added.

topoisomerase An enzyme that catalyzes the interconversion of topoisomers of DNA; can relax supercoiled DNA.

transcription DNA-directed synthesis of RNA catalyzed by RNA polymerase.

transcription bubble The site of RNA synthesis or transcription; it contains RNA polymerase, a locally melted "bubble" of DNA, and hybrid helices consisting of the template strand and the newly synthesized RNA.

transcription factor A protein that assists RNA polymerase in the initiation of RNA synthesis; binds to a specific promoter element.

transesterification The reaction of an alcohol with an ester to form a different alcohol and a different ester.

transferrin A transport protein that carries iron in the blood serum.

transferrin receptor A membrane protein that binds iron-loaded transferrin and initiates its entry into cells.

transfer RNA (tRNA) The adaptor molecule in protein synthesis; contains an amino-acid-recognition site as well as a template-recognition site, or anticodon.

transition state A chemical species that has the highest free energy and the lowest concentration of those on the pathway from a substrate to a product.

transition-state analog A compound resembling the transition state of a catalyzed reaction; such compounds are often potent inhibitors of enzyme-catalyzed reactions.

translation Cellular protein synthesis, so named because the four-letter alphabet of nucleic acids is translated into the different amino acids that make up proteins.

translocon A multisubunit assembly of integral and peripheral membrane proteins constituting the machinery that translocates a nascent protein across the endoplasmic reticulum; forms a protein-conducting channel that opens when the translocon and ribosome bind to each other.

triacylglycerol A glycerol that has fatty acyl chains esterified to each of its hydroxyl groups; storage form of fats. Also called neutral fat or triglyceride.

trinucleotide repeat Found in stretches of DNA in which a trinucleotide sequence is repeated many times; these segments of DNA can expand in the course of DNA replication, causing such genetic diseases as Huntington disease.

trombone model In DNA replication, the lagging strand loops out to form a structure that acts somewhat as a trombone slide does, growing as the replication fork moves forward. When the polymerase on the lagging strand reaches a region that has been replicated, the sliding clamp is released and a new loop is formed.

trypsin A pancreatic enzyme that digests dietary proteins and activates most of the pancreatic zymogens.

trypsinogen A pancreatic zymogen that is the precursor of trypsin.

tumor-suppressor gene A gene that encodes a protein, such as a DNA-repair protein, that suppresses tumor development when at least one copy of the gene is free of a deleterious mutation.

turnover number The number of substrate molecules converted into product by an enzyme molecule in a unit time when the enzyme is fully saturated with substrate; equal to the kinetic constant k_2 (see **Michaelis constant**).

two-dimensional electrophoresis A means of obtaining high-resolution separations of proteins in a mixture. The protein mixture is first separated in one dimension by isoelectric focusing in a single-lane gel. The gel is then is then placed horizontally on top of an SDS-acrylamide gel and electrophoresis is performed in the second dimension, perpendicular to the first direction.

uncompetitive inhibition Inhibition distinguished by the fact that the inhibitor binds only to the enzyme–substrate complex.

uncoupling protein 1 (UCP-1) A mitochondrial membrane protein that plays a role in thermogenesis by forming a pathway for the flow of protons into the mitochondria, thereby generating heat without synthesizing ATP. Also called thermogenin.

unity of biochemistry The uniformity of organisms at the molecular level.

uracil (U) A base in RNA.

urea cycle A cyclic pathway that converts excess ammonia into urea for secretion; the first metabolic pathway to be discovered.

uridine diphosphate glucose (UDP-glucose) The activated form of glucose. For instance, UDP-glucose is used in the synthesis of glycogen; formed from glucose 1-phosphate and UTP.

van der Waals interaction The attraction between two molecules based on transient electron asymmetry around an atom that induces a complementary asymmetry in a nearby molecule.

vector A small DNA molecule that can replicate autonomously in an appropriate organism. Vectors, such as plasmids and λ phage, are used to replicate inserted foreign DNA in recombinant DNA technology.

vitamin An organic substance required in trace amounts for a number of essential biochemical reactions.

water-oxidizing complex (WOC) The site of oxygen generation in photosynthesis in green plants. The complex, which includes four manganese ions, donates electrons to positively charged P680. After donating four electrons, the complex oxidizes two molecules of water to replenish its electrons and thus forms a single molecule of molecular oxygen and four protons. Also called the manganese center.

Watson-Crick base pair Guanine paired with cytosine and adenosine paired with thymine.

western blotting A means of detecting small quantities of a protein of interest in a cell or bodily fluid. The sample is subjected to SDS-gel electrophoresis, and then the proteins in the gel are transferred (blotted) to a polymer sheet pressed against the gel. The blot is then treated with a labeled antibody specific for the protein of interest, allowing for the identification and quantitation of the protein. Also called immunoblotting.

wobble Refers to steric freedom in the pairing of the third base of an mRNA codon with the anticodon of a transfer RNA molecule, which allows more than one codon to be recognized by a particular tRNA molecule.

Z-DNA A left-handed double helix in which the backbone phosphates zigzag; can be formed by oligonucleotides with alternating sequences of purines and pyrimidines.

zinc-finger domain A conserved sequence-specific DNA-binding domain, found in members of the nuclear hormone-receptor family, that consists of eight cysteine residues: the first four bind one zinc ion and the second four bind another.

Z scheme of photosynthesis The pathway of electron flow between photosystem I and photosystem II; so called because the redox diagram from P680 to P700 looks like the letter Z.

zwitterion A dipolar ion.

zymogen A catalytically inactive precursor of an enzyme. Also called a proenzyme.

Answers to Problems

Chapter 1

1. The phrase refers to the fact that all organisms are remarkably similar at a biochemical level, which strongly suggests that all organisms on Earth are derived from a common ancestor.
2. DNA is double stranded, and its sugar is deoxyribose. DNA contains the base thymine. RNA is usually single stranded, and its sugar is ribose. RNA uses uracil in place of thymine.
3. Proteins are linear polymers composed from 20 different amino acids. Glycogen is a branched polymer composed only of glucose.
4. The central dogma describes the fundamental information flow in biological systems. DNA is replicated to form new DNA, which is transcribed into RNA. The RNA is translated into protein.
5. Replication is the generation of two daughter double helices from a single parent double helix. The process is catalyzed by DNA polymerase. Transcription is the process of copying DNA information into RNA and is catalyzed by RNA polymerase. Translation converts the sequence information of RNA into proteins and takes place on ribosomes.
6. An enzyme is usually a protein catalyst, although some types of RNA also function as catalysts. Catalysts enhance the rate of a chemical reaction without themselves being permanently altered.
7. Eukaryotic cells contain a nucleus and a complex of membrane-bounded internal structures called organelles. Prokaryotic cells do not have a nucleus and lack the complex internal organization of eukaryotic cells.
8. An organelle can be any of a number of membrane-bounded structures inside eukaryotic cells.
9. Mitochondria, chloroplasts, and nuclei.
10. The nuclear membrane is not continuous. It is a set of closed membranes that come together at pores.
11. The gene is transcribed into RNA, which is translated into the protein on ribosomes bound to the endoplasmic reticulum. The protein enters the lumen of the endoplasmic reticulum, is sequestered into transport vesicles, and moves to the Golgi complex, where the protein is modified. The protein is packaged into secretory vesicles that fuse with the plasma membrane, resulting in the exocytosis of the protein.
12. (a) 9; (b) 12; (c) 5; (d) 10; (e) 11; (f) 8; (g) 4; (h) 7; (i) 2; (j) 6; (k) 1; (l) 3

Chapter 2

1. Brownian motion is the random movement of molecules in a fluid or gas powered by the background thermal energy.
2. Water is polar in that the hydrogen atoms bear a partial positive charge, whereas the oxygen atom has a partial negative charge owing to the electronegative nature of the oxygen atom. However, the total charge on the molecule is zero; that is, the positive charges are equal to the negative charges.
3. Many weak bonds allow for highly specific yet transient interactions.
4. Ionic bonds, hydrogen bonds, and van der Waals interactions. Water disrupts ionic bonds and hydrogen bonds. Because van der Waals interactions are most common between hydrophobic groups, water can be said to strengthen these bonds by facilitating their formation through the hydrophobic effect.
5. Lowering the temperature would reduce the motion of the water molecules and allow the formation of more hydrogen bonds, which

is indeed the case: each molecule of water in ice is hydrogen bonded to approximately 3.7 molecules of water. The opposite takes place as the water is heated, and fewer hydrogen bonds would be expected to form. At 100°C, a molecule of water is hydrogen bonded to 3.2 water molecules.
6. Electrostatic interactions would be stronger in an organic solvent relative to a polar solvent because there would be no competition from the solvent for the components of the electrostatic interaction.
7. An electronegative atom is one that has a high affinity for electrons. Consequently, when bonded to a hydrogen atom, the electronegative atom assumes a partial negative charge and the hydrogen atom assumes a partial positive charge. Such polarity allows the formation of hydrogen bonds.
8. The hydrophobic effect is the tendency of nonpolar molecules to interact with one another in the presence of water. The interaction is powered by the increase in entropy of water molecules when the nonpolar molecules are removed from the watery environment.
9. The Second Law of Thermodynamics states that the entropy of a system and its surroundings always increases in a spontaneous process. When hydrophobic molecules are sequestered away from water, the entropy of the water increases. Such sequestration, called the hydrophobic effect, also leads to the formation of biochemical structures.
10. 10^{-9} M
11. 10^{-12} M
12. The Henderson–Hasselbalch equation is pH = pK_a + log[A^-]/[HA]. If [A^-] = [HA], then the equation becomes pH = pK_a + log 1. But log 1 = 0. Thus, pH = pK_a under the conditions stated.
13. The lower the pK_a, the greater the K_a. The greater the K_a, the stronger the acid.
14. Recall that pK_a = log 1/K_a. The antilog of 4.76, the pK_a of acetic acid, is 57,543, and the K_a is the inverse. Thus, the K_a for acetic acid = 1.7×10^{-5}.
The pK_a for trichloroacetic acid is 0.7. Calculating as above, the K_a is 2.0×10^{-1}.
Trichloroacetic acid is a stronger acid; that is, it is more completely dissociated.
15. As with many problems in life, this one can be solved with the Henderson–Hasselbalch equation:

$$pH = pK_a + \log\left(\frac{[A^-]}{[HA]}\right)$$

Pyruvic acid:

$$7.4 = 2.50 + \log [pyruvate]/[pyruvic\ acid]$$
$$4.5 = \log [pyruvate]/[pyruvic\ acid]$$

The antilog of 4.5 = 3.2×10^5 = [pyruvate]/[pyruvic acid].

Lactic acid:

$$7.4 = 3.86 + \log [lactate]/[lactic\ acid]$$
$$3.54 = \log [lactate]/[lactic\ acid]$$

The antilog of 3.54 = 3.5×10^3 = [lactate]/lactic acid].

Thus, these organic acids, like most organic acids that we will encounter in our study of biochemistry, are extensively ionized at physiological pH.

16. 6.48
17. 7.8
18. 100
19.

Initial acetate (M)	Initial HCl (M)	pH
0.1	0.0025	6.3
0.1	0.005	6.0
0.1	0.01	5.7
0.1	0.05	4.8
0.01	0.0025	5.2
0.01	0.005	4.8
0.01	0.01	3.4
0.01	0.05	1.4

20. Acetate ion = 0.128 M; acetic acid = 0.072 M.

21. First, take a deep breath. You need to determine the concentration of base and acid that will be present in the 0.2 M acetate buffer. Fortunately, you have just completed the previous problem (problem 20), so you know that in the final buffer,

$$[acetate] = 0.128 \text{ M and } [acetic\ acid] = 0.072 \text{ M}$$

Because you are making half a liter of buffer, you will need

$$0.5\ l \times 0.128 \text{ mol } l^{-1} = 0.064 \text{ mol of sodium acetate}$$
$$0.064 \text{ mol} \times 82 \text{ g mol}^{-1} = 5.25 \text{ g of sodium acetate}$$

You weigh the required amount of sodium acetate. For the 500-ml buffer, you will require 0.036 mol of acetic acid. Your acid solution is 1 mol l^{-1} [1 M], so

$$1 \text{ mol } l^{-1} \times 0.036 \text{ mol} = 0.036\ l \text{ or } 36 \text{ ml}$$

To produce the final buffer, dissolve 5.25 g of acetate in some water, add the 36 ml of 1M acetic acid, dilute to 500 ml, and bask in the adoring gaze of your lab mates.

22. To solve this problem, we resort to the Henderson–Hasselbalch equation:

$$pH = pK_a + \log\left(\frac{[A^-]}{[HA]}\right)$$

Substituting the given values,

$$7.1 = 6.1 + \log 8 \text{ mM}/[CO_2]$$
$$1 = \log 8 \text{ mM}/[CO_2]$$

The antilog of 1 is 10; thus,

$$10 = 8 \text{ mM}/[CO_2]$$

and

$$[CO_2] = 8 \text{ mM}/10 = 0.8 \text{ mM}$$

23. (a) Again, we fall back on the Henderson–Hasselbalch equation:

$$pH = pK_a + \log\left(\frac{[A^-]}{[HA]}\right)$$

Substituting the values, we arrive at

$$0.93 = \log [HCO_3^-]/1.10 \text{ mM}$$

The antilog of 0.93 = 8.5, so

$$8.5 = [HCO_3^-]/1.10 \text{ mM}$$

and

$$[HCO_3^-] = 9.4, \text{ a dangerously low value.}$$

(b) Intravenous administration of sodium bicarbonate is a common treatment.

(c) Middle-distance running involves aspects of endurance running as well as sprinting. Some studies have shown that drinking a sodium bicarbonate solution prior to the run mitigates a drop in pH and improves performance.

Chapter 3

1. ELVISISLIVINGINLASVEGAS
2. TWITCHYLITTLEFERRETARENTYOUMALFOY (Twitchy little ferret, aren't you Malfoy?)
3. A: Proline, Pro, P; B: tyrosine, Tyr, Y; C: leucine, Leu, L; D: lysine, Lys, K
4. (a) C, A; (b) D; (c) B, D; (d) B, D; (e) B
5. (a) 6; (b) 2; (c) 3; (d) 1; (e) 4; (f) 5
6. (a) Ala; (b) Tyr; (c) Ser; (d) His
7. pH 7: no net charge. pH 12: −1.
8. Calculate the pI by taking the average of the two pK_a values. Thus, $pI = (pK_a^{acid} + pK_a^{amino})/2 = (2.72 + 9.60)/2 = 6.16$.
9. Arginine, lysine, and histidine
10. Nonessential amino acids can be synthesized from other molecules in the body. Essential amino acids cannot and thus must be consumed in the diet.
11. Phenylalanine, tyrosine, and tryptophan
12. The side chains of aspartate, glutamate, histidine, cysteine, tyrosine, lysine, and arginine
13.

14. Tyrosine
15. Ser, Glu, Tyr, and Thr
16. Using the Henderson–Hasselbalch equation (p. 28), we find the ratio of alanine-COOH to alanine-COO$^-$ at pH 7 to be 10^{-4}. The ratio of alanine-NH$_2$ to alanine-NH$_3^+$, determined in the same fashion, is 10^{-1}. Thus, the ratio of neutral alanine to the zwitterionic species is $10^{-4} \times 10^{-1} = 10^{-5}$.
17. Recall that the pK_a is the pH at which the concentration of the unionized form of a molecule is equal to the ionized form. The carboxylic acid group in Figure 3.2 has a pK_a of slightly more than 2, whereas the amino groups have a pK_a of 9.
18. To answer this, we need only apply the Henderson–Hasselbalch equation to the ionization of NH$_3^+$. NH$_3^+$ is the acid and NH$_2$ the conjugate base.

For pH 9.8,

$$9.8 = 10.8 + \log [NH_2]/[NH_3^+]$$
$$-1 = \log [NH_2]/[NH_3^+]$$
$$10^{-1} = [NH_2]/[NH_3^+]$$

Thus, the fraction of protonated molecules = 10/(10 + 1) = 91%.
For pH 11.8,

$$1 = \log [NH_2]/[NH_3^+]$$

$$10 = [NH_2]/[NH_3^+]$$

Thus, the fraction of protonated molecules 1/(10 + 1) = 9%.
19. We need to apply the Henderson–Hasselbalch equation to the ionization of the carboxylic acid.
For pH 2.1,

$$2.1 = 4.1 + \log [COO^-]/[COOH]$$

$$-2 = \log [COO^-]/[COOH]$$

$$10^{-2} = [COO^-]/[COOH]$$

Thus, the fraction of protonated molecules = 100/(100 + 1) = 99%.
For pH 7.1,

$$3 = \log [COO^-]/[COOH]$$

$$10^3 = [COO^-]/[COOH]$$

Thus, the fraction of protonated molecules = 1/(1000 + 1) = 10^{-3}.

Chapter 4

1. The energy barrier that must be crossed to go from the polymerized state to the hydrolyzed state is large even though the reaction is thermodynamically favorable.
2. (a) Alanine-glycine-serine; (b) Alanine; (c and d):

3.

At pH 4.0, the net charge is +1.

At pH 7.5, the net charge is 0.

4. (a) TEPIVAPMEYGK; (b) −1 at pH 7; (c) −4 at pH 12
5. This observation demonstrates that pK_a values are affected by the environment. A given amino acid can have a variety of pK_a values, depending on the chemical environment inside the protein.
6. The peptide bond has partial double-bond character, which prevents rotation. This lack of rotation constrains the conformation of the peptide backbone and limits possible structures.

7. (a) 8; (b) 7; (c) 1; (d) 3; (e) 9; (f) 10; (g) 2; (h) 5; (i) 4; (j) 6
8. The translation of an α helix is 1.5 Å/amino acid. Therefore, the axial length of the helix would be 1.5 Å/amino acid × 120 amino acids = 180 Å. In a fully extended polypeptide chain, the distance between amino acids is 3.5 Å, so the length of the fully extended chain is 420 Å.
9. There are 20 choices for each of the 50 amino acids: 20^{50}, or 1.13×10^{65}, a very large number.
10.

11. The (nitrogen–α carbon–carbonyl carbon) repeating unit
12. Side chain is the functional group attached to the α-carbon atom of an amino acid.
13. Amino acid composition refers simply to the amino acids that make up the protein. The order is not specified. Amino acid sequence is the same as the primary structure—the sequence of amino acids from the amino terminal to the carboxyl terminal of the protein. Different proteins may have the same amino acid composition, but amino acid sequence identifies a unique protein.
14. The primary structure determines the tertiary structure. Knowing the primary structure helps to elucidate the function of the protein. Knowledge of the primary structure of mutated proteins enables an understanding of the biochemical basis of some diseases. Primary structure can reveal the evolutionary history of the protein.
15. The helix is a condensed, coiled structure, with the R groups bristling outward from the axis of the helix. The distance between two adjacent amino acids is 1.5 Å. The strand is a fully extended polypeptide chain, and the side chains of adjacent amino acids point in opposite directions. The distance between adjacent amino acids is 4.5 Å. Both structures are stabilized by hydrogen bonding between components of the polypeptide backbone.
16. Primary structure—peptide bond; secondary structure—local hydrogen bonds between components of the polypeptide backbone; tertiary structure—various types of noncovalent bonds between R groups that are far apart in the primary structure; quaternary structure—various noncovalent bonds between R groups on the surface of subunits.
17. (a) 5; (b) 10; (c) 7; (d) 1; (e) 8; (f) 2; (g) 3; (h) 6; (i) 4; (j) 9
18. No, the Pro-X bond would have the characteristics of any other peptide bond. The steric hindrance in X–Pro arises because the R group of Pro is bonded to the amino group. Hence, in X–Pro, the proline R group is near the R group of X, which would not be the case in Pro–X.
19. The methyl group attached to the β-carbon atom of isoleucine sterically interferes with α-helix formation. In leucine, this methyl group is attached to the γ-carbon atom, which is farther from the main chain and hence does not interfere.
20. The first mutation destroys activity because valine occupies more space than alanine does, and so the protein must take a different shape, assuming that this residue lies in the closely packed interior. The second mutation restores activity because of a compensatory reduction of volume; glycine is smaller than isoleucine.

21. Loops invariably are on the surface of proteins, exposed to the environment. Because many proteins exist in aqueous environments, the exposed loops are hydrophilic so as to interact with water.

22. (a) Heat would increase the thermal energy of the chain. The weak bonds holding the chain in its correct three-dimensional structure would not be able to withstand the wiggling of the backbone, and the tertiary structure would be lost. Often, the denatured chains would interact with each other, forming large complexes that precipitate out of solution.

(b) Detergents would denature the protein by essentially turning it inside out. The hydrophobic residues in the interior of the protein would interact with the detergent, whereas the hydrophilic residues would interact with one another and not with the environment.

(c) All ionic interaction, including hydrogen bonds, would be disrupted, resulting in protein denaturation.

23. Glycine has the smallest side chain of any amino acid. Its size is often critical in allowing polypeptide chains to make tight turns or to approach one another closely.

24. Glutamate, aspartate, and the terminal carboxylate can form salt bridges with the guanidinium group of arginine. In addition, this group can be a hydrogen-bond donor to the side chains of glutamine, asparagine, serine, threonine, aspartate, glutamate, and tyrosine and to the main-chain carbonyl group. At pH 7, histidine also can hydrogen bond with arginine.

25. Disulfide bonds in hair are broken by adding a thiol-containing reagent and applying gentle heat. The hair is curled, and an oxidizing agent is added to re-form disulfide bonds to stabilize the desired shape.

26. Some proteins that span biological membranes are "the exceptions that prove the rule" because they have the reverse distribution of hydrophobic and hydrophilic amino acids. For example, consider porins, proteins found in the outer membranes of many bacteria. Membranes are built largely of hydrophobic chains. Thus, porins are covered on the outside largely with hydrophobic residues that interact with the neighboring hydrophobic chains. In contrast, the center of the protein contains many charged and polar amino acids that surround a water-filled channel running through the middle of the protein. Thus, because porins function in hydrophobic environments, they are "inside out" relative to proteins that function in aqueous solution.

27. The amino acids will be hydrophobic in nature. An α helix is especially suitable for crossing a membrane because all of the amide hydrogen atoms and carbonyl oxygen atoms of the peptide backbone take part in intrachain hydrogen bonds, thus stabilizing these polar atoms in a hydrophobic environment.

28. Recall that hemoglobin exists as a tetramer, whereas myoglobin is a monomer. Consequently, the hydrophobic residues on the surface of hemoglobin subunits probably take part in van der Waals interactions with similar regions of the other subunits and are shielded from the aqueous environment by these interactions.

29. A possible explanation is that the severity of the symptoms corresponds to the degree of structural disruption. Hence, the substitution of alanine for glycine might result in mild symptoms, but the substitution of the much larger tryptophan might prevent little or no collagen triple-helix formation.

30. The reason is that the wrong disulfides formed pairs in urea. There are 105 different ways of pairing eight cysteine molecules to form four disulfides; only one of these combinations is enzymatically active. The 104 wrong pairings have been picturesquely termed "scrambled" ribonuclease.

31. The added β-mercaptoethanol catalyzed the rearrangement of disulfide pairings until the native structure was regained. This process was driven by the decrease in free energy as the scrambled conformations were converted into the stable, native conformation of the enzyme. The native disulfide pairings of ribonuclease thus contribute to the stabilization of the thermodynamically preferred structure.

32. The native conformation of insulin is not the thermodynamically most stable form, because it contains two separate chains linked by disulfide bonds. Insulin is formed from proinsulin, a single-chain precursor, which is cleaved to form insulin, a 51-residue molecule, after the disulfide bonds have formed.

33. A segment of the main chain of the protease could hydrogen bond to the main chain of the target protein to form an extended parallel or antiparallel pair of β strands.

34. As the size of the protein increases, the surface-to-volume ratio decreases. Consequently, the ratio of hydrophilic amino acids to hydrophobic amino acids also decreases.

35. Each strand is 35 kDa and hence has about 318 residues (the mean residue mass is 110 Da). Because the rise per residue in an α helix is 1.5 Å, the length is 477 Å. More precisely, for an α-helical coiled coil, the rise per residue is 1.46 Å; so the length is 464 Å.

36. Most likely, the RNase molecules would have become tangled with one another to form a large, insoluble aggregate. For example, suppose a positively charged R group interacted with a negatively charged R group in the native structure. If the protein concentration were high upon denaturation, groups from different molecules might interact to form a large, insoluble aggregate.

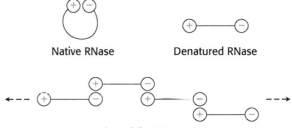

Native RNase   Denatured RNase

A portion of the RNase aggregate

Chapter 5

1. An assay identifies the desired protein. The ability to identify the protein is important in determining if particular purification steps are effective in isolating the protein from the other cellular material.

2. (a) 10; (b) 1; (c) 6; (d) 9; (e) 2; (f) 8; (g) 5; (h) 7; (i) 3; (j) 4

3. If the salt concentration becomes too high, the salt ions interact with the water molecules. Eventually, there are not enough water molecules to interact with the protein, and the protein precipitates.

4. If there is a lack of salt in a protein solution, the proteins may interact with one another—the positive charges on one protein with the negative charges on another or several others. Such an aggregate becomes too large to be solubilized by water alone. If salt is added, it neutralizes the charges on the proteins, preventing protein–protein interactions.

5. Charged and polar R groups on the surface of an enzyme

6. (a) Trypsin cleaves after arginine (R) and lysine (K), generating AVGWR, VK, and S. Because they differ in size, these products could be separated by molecular exclusion chromatography.

(b) Chymotrypsin, which cleaves after large aliphatic or aromatic R groups, generates two peptides of equal size (AVGW) and (RVKS). Separation based on size would not be effective. The peptide RVKS has two positive charges (R and K), whereas the other peptide is neutral. Therefore, the two products could be separated by ion-exchange chromatography.

7. The long hydrophobic tail on the SDS molecule (p. 73) disrupts the hydrophobic interactions in the interior of the protein. The protein unfolds, with the hydrophobic R groups now interacting with SDS rather than with one another.

8. An inhibitor of the enzyme being purified might have been present and subsequently removed by a purification step. This removal would lead to an apparent increase in the total amount of enzyme present.

9.

Purification procedure	Total protein (mg)	Total activity (units)	Specific activity (units mg^{-1})	Purification level	Yield (%)
Crude extract	20,000	4,000,000	200	1	100
$(NH_4)_2SO_4$ precipitation	5,000	3,000,000	600	3	75
DEAE–cellulose chromatography	1,500	1,000,000	667	3.3	25
Molecular exclusion chromatography	500	750,000	1,500	7.5	19
Affinity chromatography	45	675,000	15,000	75	17

10. (a) Because one SDS molecule binds to a protein for every two amino acids in the proteins, in principle, all proteins will have the same charge-to-mass ratio. For instance, a protein consisting of 200 amino acids will bind 100 SDS molecules, whereas a protein consisting of 400 amino acids will bind 200 SDS molecules. The average mass of an amino acid is 110, and there is one negative charge per SDS molecule. Thus, the charge-to-mass ratio of both proteins is the same—0.0045.

(b) The statement might be incorrect if the protein contains many charged amino acids.

(c) The protein may be modified. For instance, serine, threonine, and tyrosine may have phosphoryl groups attached.

11. The term $\bar{v}$ is the partial specific volume, the reciprocal of the particle density. Thus, the denser a particle, the smaller $\bar{v}$. The smaller $\bar{v}$ means that the opposing force is less, so the denser particle moves faster.

12. The estrogen receptor has a unique, high affinity for the estrogen estradiol.

13. Polyclonal antibodies are a collection of antibodies that bind to multiple epitopes on an antigen. Monoclonal antibodies constitute a collection of antibodies that bind to a single epitope on an antigen.

14. If an antibody to a protein of interest exists, the antibody can be attached to an insoluble bead of some sort. A mixture of proteins that includes the protein of interest is mixed with the antibody. Only the protein of interest will bind to the antibody. The mixture is centrifuged, and the supernatant is discarded. The protein of interest is then released from the antibody, often by adding a protein denaturant.

15. An enzyme-linked immunoabsorbant assay. ELISA is used for quantitating the presence of an antigen by using an enzyme linked to an antibody to the antigen.

16. Western blotting is an immunological technique used to detect a specific protein in a cell or in a body fluid. A sample is subjected to SDS–polyacrylamide electrophoresis. The resolved proteins are transferred, or blotted, to a polymer sheet, and then an antibody specific for the protein of interest is incubated with the blotted sample. Other, enzyme-linked antibodies can then be used to visualize the desired antigen–antibody complex.

17. Keep in mind that we are sequencing a large population of identical molecules, not a single molecule, and that the cleavage reaction is not 100% effective. Consequently, after many repetitions (approximately 50), many different peptides are releasing different amino acids at the same time. To illustrate this point, assume that each sequencing step is 98% efficient. The proportion of correct amino acids released after 50 rounds is 0.98^{50}, or 0.4—a hopelessly impure mix.

18. Many proteins have similar masses but different sequences and different patterns when digested with trypsin. The set of masses of tryptic peptides forms a detailed "fingerprint" of a protein that is very unlikely to appear at random in other proteins regardless of size.

19. Treatment with urea disrupts noncovalent bonds. Thus, the original 60-kDa protein must be made of two 30-kDa subunits. When these subunits are treated with urea and β-mercaptoethanol, a single 15-kDa species results, suggesting that disulfide bonds link the 30-kDa subunits.

20. (a) At the pI, the protein has no net charge, so the repulsive forces between protein molecules are minimal. This lack of repulsion allows individual proteins to interact, forming large complexes that cannot be solvated; that is, the complexes precipitate.

(b) Salt binds to the charges on the protein, preventing protein-protein interactions that lead to precipitation. This is the process of salting in (Figure 5.2 and problem 4).

21. Amino terminal: A

Trypsin digestion: Cleaves at R. Only two peptides are produced. Therefore, one R must be internal and the other must be the carboxyl-terminal amino acid. Because A is amino terminal, the sequence of one of the peptides is AVR.

Carboxypeptidase digestion: No digestion confirms that R is the carboxyl-terminal amino acid.

Chymotrypsin digestion: Cleaves only at Y. Combined with the preceding information, chymotrypsin digestion tells us that the sequences of the two peptides are AVRY and SR.

Thus, the complete peptide is AVRYSR.

22. First amino acid: S

Last amino acid: L

Cyanogen bromide cleavage: M is 10th position; carboxyl-terminal residues are (2S,L,W).

Amino-terminal residues: (G,K,S,Y), tryptic peptide, ends in K

Amino-terminal sequence: SYGK

Chymotryptic peptide order: (S,Y), (G,K,L), (F,I,S), (M,T), (S,W), (S,L)

Sequence: SYGKLSIFTMSWSL

23. The sample was diluted 1000-fold. The concentration after dialysis is thus 0.001 M, or 1 mM. You could reduce the salt concentration by dialyzing your sample, now 1 mM, in more buffer free of $(NH_4)_2SO_4$.

Chapter 6

1. Rate enhancement and substrate specificity

2. The active site is a three-dimensional crevice or cleft; it makes up only a small part of the total volume of the enzyme. Active sites have unique microenvironments. A substrate binds to the active site with multiple weak interactions. The specificity of the active site depends on the active site's precise three-dimensional structure.

3. A cofactor

4. Coenzymes and metals

5. Vitamins are converted into coenzymes that are required for most biochemical reactions.

6. Enzymes facilitate the formation of the transition state.

7.   The intricate three-dimensional structure of proteins allows the construction of active sites that will recognize only specific substrates.

8.   Binding energy is the free energy released when two molecules bind together, such as when an enzyme and a substrate interact.

9.   Binding energy is maximized when an enzyme interacts with the transition state, thereby facilitating the formation of the transition state and enhancing the rate of the reaction.

10.   There would be no catalytic activity. If the enzyme–substrate complex is more stable than the enzyme–transition-state complex, the transition state would not form and catalysis would not take place.

11.   (a) 4; (b) 7; (c) 8; (d) 3; (e) 9; (f) 10; (g) 5; (h) 1; (i) 2; (j) 6

12.   The energy required to reach the transition state (the activation energy) is returned when the transition state proceeds to product.

13.   The product is more stable than the substrate in graph A; so ΔG is negative and the reaction is exergonic. In graph B, the product has more energy than the substrate has; ΔG is positive, meaning that the reaction is endergonic.

A

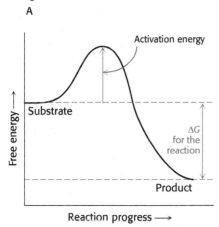

B

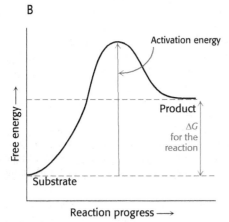

14.   Protein hydrolysis has a large activation energy. Protein synthesis requires energy to proceed.

15.   Lysozyme helps protect the fluid that surrounds eyes from bacterial infection.

16.   Transition states are very unstable. Consequently, molecules that resemble transition states are themselves likely to be unstable and, hence, difficult to synthesize.

17.   (a) 0; (b) 28.53; (c) −22.84; (d) −11.42; (e) 5.69

18. $K_{eq} = 19$, $\Delta G^{\circ\prime} = -7.41$ kJ mol^{-1} (−1.77 kcal mol^{-1})

19.   This reaction takes place in glycolysis (Chapter 16). At equilibrium, the ratio of GAP to DHAP is 0.0475 at 25°C (298 K) and pH 7. Hence, $K'_{eq} = 0.0475$. The standard free-energy change for this reaction is then calculated from equation 5:

$$\Delta G^{\circ\prime} = -RT \ln K'_{eq}$$
$$= -8.315 \times 10^{-3} \times 298 \times \ln (0.0475)$$
$$= +7.53 \text{ kJ mol}^{-1}(+1.80 \text{ kcal mol}^{-1})$$

Under these conditions, the reaction is endergonic. DHAP will not spontaneously convert into GAP.

   Substituting these values into equation 1 gives

$$\Delta G = 7.53 \text{ kJ mol}^{-1} + RT \ln \frac{3 \times 10^{-6} \text{ M}}{2 \times 10^{-4} \text{ M}}$$
$$= 7.53 \text{ kJ mol}^{-1} - 10.42 \text{ kJ mol}^{-1}$$
$$= -2.89 \text{ kJ mol}^{-1}(-0.69 \text{ kcal mol}^{-1})$$

This negative value for ΔG indicates that the isomerization of DHAP to GAP is exergonic and can take place spontaneously when these species are present at the preceding concentrations. Note that ΔG for this reaction is negative, although $\Delta G^{\circ\prime}$ is positive.

20.   The mutation slows the reaction by a factor of 100. The activation-free energy is increased by 11.4 kJ mol^{-1} (2.73 kcal mol^{-1}). Strong binding of the substrate relative to the transition state slows catalysis.

21.   (a) Incubating the enzyme at 37°C leads to the denaturation of enzyme structure and a loss of activity. For this reason, most enzymes must be kept cool if they are not actively catalyzing their reactions.

   (b) The coenzyme apparently helps to stabilize the enzyme's structure, because enzyme from PLP-deficient cells denatures faster. Cofactors often help to stabilize enzyme structure.

22.   (a)

$$\Delta G^{\circ\prime} = -RT \ln K'_{eq}$$
$$+1.8 = -(1.98 \times 10^{-3} \text{ K}^{-1} \text{ deg}^{-1} \text{ mol}^{-1})$$
$$(298 \text{ K})(\ln [\text{G1P}]/[\text{G6P}])$$
$$-3.05 = \ln [\text{G1P}]/[\text{G6P}]$$
$$+3.05 = \ln [\text{G6P}]/[\text{G1P}]$$
$$K'^{-1}_{eq} = 21 \text{ or } K'_{eq} = 4.8 \times 10^{-2}$$

Because [G6P]/[G1P] = 21, there is 1 molecule of G1P for every 21 molecules of G6P. Because we started with 0.1 M, the [G1P] is 1/22(0.1 M) = 0.0045 M and [G6P] must be 21/22(0.1 M), or 0.096 M. The reaction does not proceed to a significant extent as written.

   (b) Supply G6P at a high rate, and remove G1P at a high rate by other reactions. In other words, make sure that the [G6P]/[G1P] ratio is kept large.

23.   Potential hydrogen-bond donors at pH 7 are the side chains of the following residues: arginine, asparagine, glutamine, histidine, lysine, serine, threonine, tryptophan, and tyrosine.

24.   (a) $K = \dfrac{[\text{P}]}{[\text{S}]} = \dfrac{k_F}{k_R} = \dfrac{10^{-4}}{10^{-6}} = 100$. Using equation 5 in the text,

$\Delta G^{\circ\prime} = -11.42$ kJ mol^{-1} (−2.73 kcal mol^{-1}).

   (b) $k_F = 10^{-2}$ s^{-1} and $k_R = 10^{-4}$ s^{-1}. The equilibrium constant and $\Delta G^{\circ\prime}$ values are the same for both the uncatalyzed and catalyzed reactions.

Chapter 7

1.   A first-order rate constant is a proportionality constant for reactions having only one reactant; it relates the rate of a reaction to the concentration of the sole reactant. First-order rate constants have the unit s^{-1}. A second-order rate constant is a proportionality constant for reactions having two reactants; it relates the rate of a reaction to the concentration of both reactants. Second-order rate constants have the unit M^{-1} s^{-1}.

2. A second-order reaction that appears to be first order. If the concentration of one reactant is much greater than that of the second reactant, the velocity will appear to be first order with respect to the reactant present in the lower concentration.

3. (a)

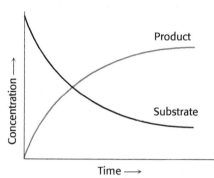

(b)

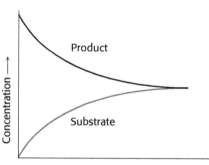

4. At substrate concentrations near the K_M, the enzyme displays significant catalysis yet is sensitive to changes in substrate concentration.

5. Sequential reactions are characterized by the formation of a ternary complex consisting of the enzyme and both substrates. Double-displacement reactions always require the formation of a temporarily substituted enzyme intermediate.

6. No, K_M is not equal to the dissociation constant, because the numerator also contains k_2, the rate constant for the conversion of the enzyme–substrate complex into enzyme and product. If, however, k_2 is much smaller than k_{-1}, $K_M \approx K_D$.

7. (a) Yes, $K_M = 5.2 \times 10^{-6}$ M; (b) $V_{max} = 6.8 \times 10^{-10}$ mol minute^{-1}; (c) 337 s^{-1}

8. (a) 31.1 μmol; (b) 0.05 μmol; (c) 622 s^{-1}, which is a midrange value for enzymes (Table 6.2)

9. (a) $V = V_{max} - (V/[S])K_M$
    (b) Slope $= -K_M$, y-intercept $= V_{max}$, x-intercept $= V_{max}/K_M$

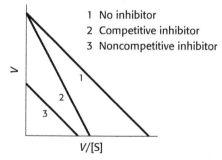

1 No inhibitor
2 Competitive inhibitor
3 Noncompetitive inhibitor

10. 1.1 μmol minute^{-1}

11.

$$f_{es} = \frac{V_0}{V_{max}} = \frac{[S]}{K_M + [S]}$$

12. (a) 7; (b) 4; (c) 5; (d) 1; (e) 8; (f) 2; (g) 9; (h) 6; (i) 10; (j) 3

13.

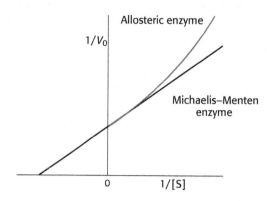

14.

$$V_0 = \frac{V_{max}[S]}{K_M + [S]} \qquad \frac{V_0}{V_{max}} = \frac{[S]}{K_M + [S]}$$

When $[S] = 10\ K_M$, $V_0/V_{max} = 0.91$, and when $[S] = 20\ K_M$, $V_0/V_{max} = 0.95$. So, any Michaelis–Menten curves showing that the enzyme actually attains V_{max} are pernicious lies.

15. The inhibition of an allosteric enzyme by the end product of the pathway controlled by the enzyme. Feedback inhibition prevents the production of too much end product and the consumption of substrates when product is not required.

16. The enzyme would show simple Michaelis–Menten kinetics because it is essentially always in the R state.

17. Homotropic effectors are the substrates of allosteric enzymes. Heterotropic effectors are the regulators of allosteric enzymes. Homotropic effectors account for the sigmoidal nature of the velocity-versus-substrate-concentration curve, whereas heterotropic effectors alter the midpoint of K_M of the curve. Ultimately, both types of effectors work by altering the T/R ratio.

18. The reconstitution shows that the complex quaternary structure and the resulting catalytic and regulatory properties are ultimately encoded in the primary structure of individual components.

19. The sequential model more readily accounts for negative cooperativity than does the concerted model.

20. (a) K_M is a measure of affinity *only* if k_2 is rate limiting, which is the case here. Therefore, the lower K_M means higher affinity. The mutant enzyme has a higher affinity.
    (b) 50 μmol minute^{-1}. 10 mM is K_M, and K_M yields $1/2V_{max}$. V_{max} is 100 μmol minute^{-1}, and so . . .
    (c) Enzymes do not alter the equilibrium of the reaction.

21. Enzyme 2. Despite the fact that enzyme 1 has a higher V_{max} than enzyme 2, enzyme 2 shows greater activity at the concentration of the substrate in the environment because enzyme 2 has a lower K_M for the substrate.

22. If the total amount of enzyme $[E]_T$ is increased, V_{max} will increase, because $V_{max} = k_2[E_T]$. But $K_M = (k_{-1} + k_2)/k_1$; that is, it is independent of substrate concentration. The middle graph describes this situation.

23. The first step will be the rate-limiting step. Enzymes E_B and E_C are operating at $1/2V_{max}$, whereas the K_M for enzyme E_A is greater than the substrate concentration. E_A would be operating at approximately $10^{-2}\ V_{max}$.

24. (a) The most effective means of measuring the efficiency of any enzyme–substrate complex is to determine the k_{cat}/K_M values. For the three substrates in question, the respective values of k_{cat}/K_M are 6, 15, and 36. Thus, the enzyme exhibits a strong preference for cleaving peptide bonds in which the second amino acid is a large hydrophobic amino acid and will cleave such peptides most rapidly.

(b) The k_{cat}/K_M for this substrate is 2. Not very effective. This value suggests that the enzyme prefers to cleave peptide bonds with the following specificity:

small R group—large hydrophobic R group

25. The rates of utilization of substrates A and B are given by

$$V_A = \left[\frac{k_2}{K_M}\right]_A [E][A]$$

and

$$V_B = \left[\frac{k_2}{K_M}\right]_B [E][B]$$

Hence, the ratio of these rates is

$$V_A/V_B = \left[\frac{k_2}{K_M}\right]_A [A] \Big/ \left[\frac{k_2}{K_M}\right]_B [B]$$

Thus, an enzyme discriminates between competing substrates on the basis of their values of k_2/K_M rather than that of K_M alone.

26. The fluorescence spectroscopy reveals the existence of an enzyme–serine complex and of an enzyme–serine–indole complex.

27. (a)

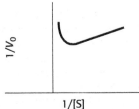

(b) This behavior is substrate inhibition: at high concentrations, the substrate forms unproductive complexes at the active site. The adjoining drawing shows what might happen. Substrate normally binds in a defined orientation, shown in the drawing as red to red and blue to blue. At high concentrations, the substrate may bind at the active site such that the proper orientation is met for each end of the molecule, but two different substrate molecules are binding.

Normal substrate binding at the active site. Substrate will be cleaved into red balls and blue balls.

Substrate inhibition

28. The binding of PALA switches ATCase from the T to the R state because PALA acts as a substrate analog. An enzyme molecule containing bound PALA has fewer free catalytic sites than does an unoccupied enzyme molecule. However, the PALA-containing enzyme will be in the R state and hence have higher affinity for the substrates. The dependence of the degree of activation on the concentration of PALA is a complex function of the allosteric constant L_0 and of the binding affinities of the R and T states for the analog and the substrates.

29. The simple sequential model predicts that the fraction of catalytic chains in the R state, f_R, is equal to the fraction containing bound substrate, Y. The concerted model, in contrast, predicts that f_R increases more rapidly than Y as the substrate concentration is increased. The change in f_R leads to the change in Y on addition of substrate, as predicted by the concerted model.

30. The binding of succinate to the functional catalytic sites of the native catalytic trimer changed the visible absorption spectrum of nitrotyrosine residues in the *other*, modified trimer of the hybrid enzyme. Thus, the binding of substrate analog to the active sites of one trimer altered the structure of the other trimer.

31. According to the concerted model, an allosteric activator shifts the conformational equilibrium of all subunits toward the R state, whereas an allosteric inhibitor shifts it toward the T state. Thus, ATP (an allosteric activator) shifted the equilibrium to the R form, resulting in an absorption change similar to that obtained when substrate is bound. CTP had a different effect. Hence, this allosteric inhibitor shifted the equilibrium to the T form. Thus, the concerted model accounts for the ATP-induced and CTP-induced (heterotropic), as well as for the substrate-induced (homotropic), allosteric interactions of ATCase.

Chapter 8

1. Covalent catalysis; general acid–base catalysis; metal ion catalysis; catalysis by approximation and orientation

2. The three-dimensional structure of an enzyme is stabilized by interactions with the substrate, reaction intermediates, and products. This stabilization minimizes thermal denaturation.

3. (a) This piece of information is necessary for determining the correct dosage of succinylcholine to administer.

(b) The duration of the paralysis depends on the ability of the serum cholinesterase to clear the drug. If there were one-eighth the amount of enzyme activity, paralysis could last eight times as long, which is undesirable for a number of reasons. First, the respirator might break from extended use, which would not be good for the patient on the respirator; second, the doctors might miss their golf game.

(c) K_M is the concentration needed by the enzyme to reach ½ V_{max}. Consequently, for a given concentration of substrate, the reaction catalyzed by the enzyme with the lower K_M will have the higher rate. The mutant patient with the higher K_M will clear the drug at a much lower rate. Part *b* describes the consequences.

4. If a particular amino acid side chain is suspected of participating in a catalytic mechanism, covalent modification of the residue by a group-specific reagent may alter it sufficiently that the enzyme activity is altered or inhibited.

5. Competitive inhibition: 2, 3, 9; uncompetitive: 4, 5, 6; noncompetitive: 1, 7, 8

6. (a) In the absence of inhibitor, V_{max} is 47.6 μmol minute^{-1} and K_M is 1.1×10^{-5} M. In the presence of inhibitor, V_{max} is the same and the apparent K_M is 3.1×10^{-5} M.

(b) Competitive

7. (a) V_{max} is 9.5 μmol minute^{-1}. K_M is 1.1×10^{-5} M, the same as without inhibitor.

(b) Noncompetitive

8. Group-specific inhibitors; affinity analogs; suicide inhibitors; transition-state analogs

9. The lactam ring of penicillin reacts with a serine residue in glycopeptide transpeptidase, an enzyme that stabilizes the bacterial cell wall. If the lactam were destroyed, penicillin would be ineffective. Indeed, the presence of β-lactamase confers penicillin resistance.

10. The catalytic triad, composed of the amino acids serine 195, histidine 57, and aspartate 102, resides at the active site of chymotrypsin. The histidine residue serves to position the serine side chain and to polarize its hydroxyl group so that it is poised for deprotonation. In the presence of the substrate, histidine 57 accepts the proton from the serine-195 hydroxyl group. The withdrawal

of the proton from the hydroxyl group generates an alkoxide ion, which is a much more powerful nucleophile than a hydroxyl group is. The aspartate residue helps orient the histidine residue and make it a better proton acceptor through hydrogen bonding and electrostatic effects.

11. The oxyanion hole is a structure at the active site of chymotrypsin that stabilizes the tetrahedral intermediate in the proteolysis reaction and facilitates the formation of the acyl-enzyme intermediate.

12. Chymotrypsin cleaves peptide bonds in a two-step reaction, in which the first step, the formation of the acyl-enzyme intermediate, is faster than the second step, hydrolysis.

13. Chymotrypsin recognizes large hydrophobic groups, which are usually buried in the enzyme's core owing to the hydrophobic effect.

14.

Experimental condition	V_{max}	K_M
a. Twice as much enzyme is used.	Doubles	No change
b. Half as much enzyme is used.	Half as large	No change
c. A competitive inhibitor is present.	No change	Increases
d. An uncompetitive inhibitor is present.	Decreases	Decreases
e. A pure noncompetitive is present.	Decreases	No change

15. Because catalysis by chymotrypsin involves a substituted enzyme intermediate, the reaction is a double-displacement (ping-pong) reaction.

16. (a) When $[S^+]$ is much greater than the value of K_M, pH will have a negligible effect on the enzyme because S^+ will interact with E^- as soon as the enzyme becomes available (left-hand graph).

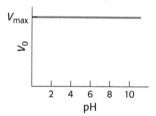
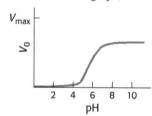

(b) When $[S^+]$ is much less than the value of K_M, the plot of V_0 versus pH becomes essentially a titration curve for the ionizable groups, with enzyme activity being the titration marker. At low pH, the high concentration of H^+ will keep the enzyme in the EH form and inactive. As the pH rises, more and more of the enzyme will be in the E^- form and active. At high pH (low H^+), all of the enzyme is E^- (right-hand graph).

(c) The midpoint on this curve will be the pK_a of the ionizable group, which is stated to be pH 6.

17. The negative charge on the aspartic acid helps orient histidine 57 so that it acts as a general base catalyst to assist in the formation of the reactive alkoxide ion on serine. Asparagine, lacking a charge, would be less effective at orienting histidine 57. Indeed, chymotrypsin with this mutation has 10,000-fold lower activity than that of the wild-type enzyme.

18. The formation of the acyl-enzyme intermediate is slower than the hydrolysis of this amide substrate, and so no burst is observed. For ester substrates, the formation of the acyl-enzyme intermediate is faster.

19. (a) Tosyl-L-lysine chloromethyl ketone (TLCK). (b) Determine whether substrates protect trypsin from inactivation. Ascertain whether the D isomer of TLCK is an effective inhibitor.

Chapter 9

1. (a) 6; (b) 10; (c) 5; (d) 7; (e) 9; (f) 3; (g) 4; (h) 2; (i) 1; (j) 8

2. (a) 2.96×10^{-11} g
   (b) 2.71×10^8 molecules
   (c) No. There would be 3.22×10^8 hemoglobin molecules in a red cell if they were packed in a cubic crystalline array. Hence, the actual packing density is about 84% of the maximum possible.

3. 2.65 g (4.75×10^{-2} mol) of Fe

4. (a) In human beings, 1.44×10^{-2} g (4.49×10^{-4} mol) of O_2 per kilogram of muscle. In sperm whales, 0.144 g (4.49×10^{-3} mol) of O_2 per kilogram.
   (b) 128

5. The cooperativity allows hemoglobin to become saturated in the lungs, where oxygen pressure is high. When the hemoglobin moves to tissues, the lower oxygen pressure induces it to release oxygen and thus deliver oxygen where it is needed. Thus, the cooperative release favors a more complete unloading of oxygen in the tissues.

6. Deoxyhemoglobin is in the T state. The presence of oxygen disrupts the R $\rightleftharpoons$ T equilibrium in favor of the R state. The structural changes are significant enough to cause the crystal to come apart.

7. Hemoglobin with oxygen bound to only one of four sites remains primarily in the T-state quaternary structure, an observation consistent with the sequential model. On the other hand, hemoglobin behavior is concerted in that hemoglobin with three sites occupied by oxygen is almost always in the quaternary structure associated with the R state.

8. Fetal hemoglobin does not bind 2,3-BPG as well as maternal hemoglobin does. Recall that the tight binding of 2,3-BPG by hemoglobin reduces the oxygen affinity of hemoglobin.

9. Hemoglobin S molecules bind together to form large fibrous aggregates that extend across the cell, deforming the red blood cells and giving them their sickle shape. This aggregation takes place predominantly in the deoxygenated form of Hb S. Small blood vessels are blocked because of the deformed cells, which creates a local region of low oxygen concentration. Hence, more hemoglobin changes into the deoxy form and so more cells undergo sickling. Sickled red cells become trapped in the small blood vessels, impairing circulation and leading to the damage of many tissues. Sickled cells, which are more fragile than normal red blood cells, rupture (hemolyze) readily to produce severe anemia.

10. Deoxyhemoglobin A contains a complementary site, and so it can add to a fiber of deoxyhemoglobin S. The fiber cannot then grow further, because the terminal deoxy Hb A molecule lacks a sticky patch.

11. The whale swims long distances between breaths. A high concentration of myoglobin in the whale muscle maintains a ready supply of oxygen for the muscle between breathing episodes.

12. The presence of 2,3-BPG shifts the equilibrium toward the T state. 2,3-BPG binds only to the center cavity of deoxyhemoglobin (T state). The size of the center cavity decreases on the change to the R form, expelling the 2,3-BPG and thus facilitating the formation of the R state.

13. A higher concentration of 2,3-BPG would shift the oxygen-binding curve to the right. The rightward shift of the oxygen-binding curve would promote the dissociation of oxygen in the tissues and would thereby increase the percentage of oxygen delivered to the tissues.

14. (a) The transfusion would increase the number of red blood cells, which increases the oxygen-carrying capacity of the blood, allowing more sustained effort.

(b) BPG stabilizes the T state of hemoglobin, which results in a more efficient release of oxygen. If BPG is depleted, the oxygen will not be released even though the red blood cells are carrying more oxygen.

15. The Bohr effect, not to be confused with the boring effect of a monotonous lecture, is the regulation of hemoglobin oxygen binding by hydrogen ions and carbon dioxide. Deoxyhemoglobin is stabilized by ionic bonds that stabilize the T state. One of these bonds forms between carboxyl-terminal His $\beta146$ and Asp $\beta94$. As the pH increases, this stabilizing salt bridge is broken because His $\beta146$ becomes deprotonated and loses its positive charge, facilitating the formation of the R state. At lower pH values, His $\beta146$ is positively charged. The formation of the ionic bonds shifts the equilibrium from the R state to the T state, thus releasing oxygen.

16. Oxygen binding appears to cause the copper ions, along with their associated histidine ligands, to move closer to each other, thereby also moving the helices to which the histidines are attached (in similar fashion to the conformational change in hemoglobin).

17. Inositol pentaphosphate in part c. Inositol pentaphosphate has negative charges, similarly to 2,3-BPG.

18. (a) 2; (b) 4; (c) 2; (d) 1

19. The electrostatic interactions between 2,3-BPG and hemoglobin would be weakened by competition with water molecules. The T state would not be stabilized.

20. (a) $k_{off} = k_{on}$, $K = 20$ s^{-1}. (b) Mean duration is 0.05 s (the reciprocal of k_{off}).

21. The pK_a is (a) lowered; (b) raised; and (c) raised.

22. The tight binding of carbon monoxide forces the tetramer into the R state even under oxygen partial pressures that should lead to the release of oxygen. In essence, the bound carbon monoxide shifts the oxygen saturation curve to the left.

23. 62.7% oxygen-carrying capacity

24. The modified hemoglobin would not show cooperativity. Although the imidazole in solution will bind to the heme iron atom (in place of histidine) and will facilitate oxygen binding, the imidazole lacks the crucial connection to the particular α helix that must move so as to transmit the change in conformation.

25. The release of acid will lower the pH. A lower pH promotes oxygen dissociation in the tissues. However, the enhanced release of oxygen in the tissues will increase the concentration of deoxyhemoglobin, thereby increasing the likelihood that the cells will sickle.

Chapter 10

1. Carbohydrates were originally regarded as *hydrates* of *carbon* because the empirical formula of many of them is $(CH_2O)_n$.

2. Three amino acids can be linked by peptide bonds in only six different ways. However, three different monosaccharides can be linked in a plethora of ways. The monosaccharides can be linked in a linear or a branched manner, with α or β linkages, with bonds between C-1 and C-3, between C-1 and C-4, between C-1 and C-6, and so forth. There are 12,288 possible trisaccharides but only 6 different tripeptides.

3. (a) 10; (b) 6; (c) 8; (d) 9; (e) 2; (f) 4; (g) 1; (h) 5; (i) 7; (j) 3

4. (a) Aldose–ketose; (b) epimers; (c) aldose–ketose; (d) anomers; (e) aldose–ketose; (f) epimers

5. Erythrose: tetrose aldose
Ribose: pentose aldose
Glyceraldehyde: triose aldose
Dihydroxyacetone: triose ketose
Erythrulose: tetrose ketose
Ribulose: pentose ketose
Fructose: hexose ketose

6.

D-Allose   **D-Altrose**   **D-Mannose**   **D-Gulose**

D-Idose   **D-Galactose**   **D-Talose**

7. Glucose is reactive because of the presence of an aldehyde group in its open-chain form. The aldehyde group slowly condenses with amino groups to form an amino ketone.

8. No, sucrose is not a reducing sugar. The anomeric carbon atoms of glucose and fructose can act as the reducing agent, but, in sucrose, the two anomeric carbon atoms are joined by a covalent bond and thus not available to react.

9. From methanol

10. (a) β-D-Mannose; (b) β-D-galactose; (c) β-D-fructose; (d) β-D-glucosamine

11. Synthesize the trisaccharide, and test its effect on cell–cell interaction. The trisaccharide itself should be a competitive inhibitor of cell adhesion if the trisaccharide unit of the glycoprotein is critical for the interaction.

12. (a) Not a reducing sugar; no open-chain forms are possible. (b) D-Galactose, D-glucose, and D-fructose. (c) D-Galactose and sucrose (glucose + fructose).

13. The hemiketal linkage of the α anomer is broken to form the open form. Rotation about the C-1 and C-2 bonds allows the formation of the β anomer, and a mixture of isomers results.

β-D-**Glucose**

14. Heating converts the very sweet pyranose form into the more stable but less sweet furanose form. Consequently, the sweetness of the preparation is difficult to accurately control, which also accounts for why honey loses sweetness with time. Figure 10.6 shows the structures.

15. (a) Each glycogen molecule has one reducing end, whereas the number of nonreducing ends is determined by the number of branches, or α-1,6 linkages. (b) Because the number of nonreducing ends greatly exceeds the number of reducing ends in a collection of glycogen molecules, all of the degradation and synthesis of glycogen takes place at the nonreducing ends, thus maximizing the rate of degradation and synthesis.

16. Glycogen is a polymer of glucose linked by α-1,4-glycosidic bonds with branches formed approximately every 10 glucose units by α-1,6-glycosidic bonds. Starch consists of two polymers of glucose. Amylose is a straight-chain polymer formed by α-1,4-glycosidic bonds. Amylopectin is similar to glycogen, but amylopectin has fewer branches, one branch per 30 or so glucose units.

17. Cellulose is a linear polymer of glucose joined by β-1,4 linkages. Glycogen is a branched polymer with the main chain being formed by α-1,4-glycosidic bonds. The β-1,4 linkages allow the formation of a linear polymer ideal for structural roles. The α-1,4 linkages of glycogen form a helical structure, which allows the storage of many glucose moieties in a small space.

18. Simple glycoproteins are often secreted proteins and thus play a variety of roles. For example, the hormone EPO is a glycoprotein. Usually, the protein component constitutes the bulk of the glycoprotein by mass. In contrast, proteoglycans and mucoproteins are predominantly carbohydrates. Proteoglycans have glycosaminoglycans attached and play structural roles, as in cartilage and the extracellular matrix. Mucoproteins often serve as lubricants and have multiple carbohydrates attached through an N-acetylgalactosamine moiety.

19. The attachment of the carbohydrate allows EPO to stay in circulation longer and thus to function for longer periods of time than would a carbohydrate-free EPO.

20. A glycoprotein is a protein that is decorated with carbohydrates. A lectin is a protein that specifically recognizes carbohydrates. A lectin can also be a glycoprotein.

21. The glycosaminoglycan, because it is heavily charged, binds many water molecules. When cartilage is stressed, such as when your heel hits the ground, the water is released, thus cushioning the impact. When you lift your heel, the water rebinds.

22. It suggests that carbohydrates are on the cell surfaces of all organisms for the purpose of recognition by other organisms or by the environment.

23. The lectin that binds the mannose 6-phosphate might be defective and not recognize a correctly addressed protein.

24. 64 proteins. Each site either is or is not glycosylated, and so there are $2^6 = 64$ possible proteins.

25. The wide array of possible linkages between carbohydrates in concert with the wide variety of monosaccharides and their many isomeric forms makes complex carbohydrates information-rich molecules.

26. As discussed in Chapter 6, many enzymes display stereochemical specificity. Clearly, the enzymes of sucrose synthesis are able to distinguish between the isomers of the substrates and link only the correct pair.

27. (a) Aggrecan is heavily decorated with glycosaminoglycans. If glycosaminoglycans are released into the media, aggrecan must be undergoing degradation.

(b) Another enzyme might be present that cleaves glycosaminoglycans from aggrecan without degrading aggrecan. Other experiments not shown established that glycosaminoglycan release is an accurate measure of aggrecan destruction.

(c) The control provides a baseline of "background" degradation inherent in the assay.

(d) Aggrecan degradation is greatly enhanced.

(e) Aggrecan degradation is reduced to the background system.

(f) It is an in vitro system in which not all the factors contributing to cartilage stabilization in vivo are present.

28. The proportion of the α anomer is 0.36, and that of the β anomer is 0.64.

29. A pyranoside reacts with two molecules of periodate; formate is one of the products. A furanoside reacts with only one molecule of periodate; formate is not formed.

30. Reducing ends would form 1,2,3,6-tetramethylglucose. The branch points would yield 2,3-dimethylglucose. The remainder of the molecule would yield 2,3,6-trimethylglucose.

Chapter 11

1. Lipids are water-insoluble molecules that are highly soluble in organic solvents.

2. (a) 5; (b) 10; (c) 8; (d) 7; (e) 1; (f) 2; (g) 4; (h) 3; (i) 6; (j) 9

3. (a)

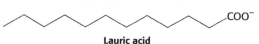

Lauric acid

(b)

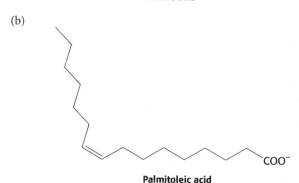

Palmitoleic acid

(c)

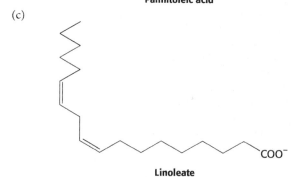

Linoleate

4. Triacylglycerols from plants may have many cis double bonds or have shorter fatty acid chains than those from animals.

5. Triacylglycerols consist of three fatty acid chains attached to a glycerol backbone. Triacylglycerols are a storage form of fuel. Phosphoglycerides consist of two fatty acid chains attached to a glycerol backbone. The remaining alcohol of the glycerol is bonded to a phosphate, which is in turn bonded to an alcohol. Phosphoglycerides are membrane components.

6. The backbone in phosphoglycerides is glycerol, whereas that in sphingolipids is sphingosine. In sphingolipids, one of the fatty acids is linked to the sphingosine by an amide bond.

7. Examples of head groups include serine, ethanolamine, choline, glycerol, and inositol.

8.

Palmitic acid $CH_3(CH_2)_{14}\overset{\overset{O}{\|}}{C}-O-CH_2$

Stearic acid $CH_3(CH_2)_{16}\overset{\overset{O}{\|}}{C}-O-CH$

Oleic acid $CH_3(CH_2)_7CH=CH(CH_2)_7\overset{\overset{O}{\|}}{C}-O-CH_2$

9.   Lipids are primarily hydrophobic molecules. For instance, triacylglycerols have three fatty acid side chains. This predominately hydrophobic nature accounts for their solubility in organic solvents and their lack of solubility in aqueous solvents.

10.   The hydrophobic chains would shun the water, interacting with similar chains in other molecules. Meanwhile, the hydrophilic head groups would readily interact with the water, resulting in the formation of a membrane or a small membrane vesicle called a liposome (p. 207).

11.   Steroids are cyclical rather than linear.

12.   The C_{16} fatty acid is attached by an ether linkage. The C-2 carbon atom of glycerol has only an acetyl group attached by an ester linkage instead of a fatty acid, as is the case with most phospholipids.

13.

Sodium stearate

The sodium stearate will form a micelle, with the hydrophilic head groups (red) exposed to water and the fatty acid chains (green) in the interior. When the sodium stearate is worked into the clothes by agitation or onto the skin by rubbing in the presence of water, the grease, which is hydrophobic, will localize in the hydrophobic interior of the micelle and be washed down the drain with rinsing.

14.   Instead of forming a soluble micelle that will be washed down the drain, the magnesium or calcium salts will precipitate, forming a scum-like bathtub ring. You should clean the bathtub immediately because, when the scum dries, it is more difficult to remove.

15.   Lipids are more reduced than glycogen, and they are stored in an anhydrous form.

16.   Hibernators selectively feed on plants that have a high proportion of polyunsaturated fatty acids with lower melting temperature.

17.   (a) 4; (b) 1; (c) 5; (d) 3; (e) 7; (f) 2; (g) 6

18.   The presence of a cis double bond introduces a kink in the fatty acid chain that prevents tight packing and reduces the number of atoms in van der Waals contact. The kink will lower the melting point compared with that of a saturated fatty acid. Trans fatty acids do not have the kink, and so their melting temperatures will be higher, more similar to those of saturated fatty acids.

19.   Palmitic acid is shorter than stearic acid. Thus, when the chains pack together, there will be less opportunity for van der Waals interactions and the melting point will thus be lower than that of the longer stearic acid.

Chapter 12

1.   (1) Membranes are sheetlike structures that are two molecules thick. (2) Membranes are composed of lipids and proteins, both of which may be decorated by carbohydrates. (3) Membrane lipids are amphipathic molecules, composed of hydrophilic and hydrophobic components, that spontaneously form closed bimolecular sheets in aqueous solution. (4) Proteins, unique to each membrane, mediate the transfer of molecules and information across the membrane. (5) Membranes are noncovalent assemblies. (6) The leaflets of the membrane bilayers are different; that is, membranes are asymmetric. (7) Membranes are fluid, rather than rigid structures. (8) Membranes are electrically polarized, with the inside of the cell negative with respect to the outside.

2.   First, the molecule must be lipophilic, and second, the concentration of the molecule must be greater on one side of the membrane than on the other.

3.   (a) 3 ; (b) 5; (c) 6 ; (d) 1; (e) 7 ; (f) 10; (g) 2; (h) 4; (i) 8 ; (j) 9

4.   c, a, e, b, d

5.   In simple diffusion, the molecule in question can diffuse down its concentration gradient through the membrane. In facilitated diffusion, the molecule is not lipophilic and cannot directly diffuse through the membrane. A channel or carrier is required to facilitate movement down the gradient.

6.   In passive transport (facilitated diffusion), a substance moves down its concentration gradient through a channel or transporter. In active transport, a concentration gradient is generated at the expense of another source of energy, such as the hydrolysis of ATP.

7.   Recall that most proteins are not static structures and require conformational changes to perform their biochemical tasks. If the membrane were too rigid, the required structural conformation could not be obtained. If the membrane were too fluid, the interactions with the environment (hydrophobic core of the membrane) that the protein needs to maintain its structure would be disrupted.

8.   The heart contraction (the heartbeat) is initiated by the release of calcium from calcium stores. The contraction is terminated by the removal of calcium from the cytoplasm. This removal is accomplished, in part, by a sodium–calcium antiporter, which moves calcium out of the cell, against its concentration gradient, by allowing sodium to flow into the cell down its concentration gradient. The sodium gradient is established by the Na^+–K^+ ATPase. Cardiotonic steroids function by inhibiting the Na^+–K^+ ATPase, which in turn inhibits the sodium–calcium antiporter. As a result, calcium, the signal for contraction, remains in the heart cell longer, allowing for a more robust heartbeat.

9.   Ouabain, like digitalis, inhibits the Na^+–K^+ ATPase. The Na^+–K^+ ATPase is crucial to maintaining the sodium gradient that renders neurons and muscle cells electrically excitable. Inhibition of the enzyme shuts down a host of biochemical process required for life, such as cardiac and respiratory function.

10.   Inhibition of the symporter would eventually lead to the inhibition of the ATPase. Because the sodium gradient would not be dissipated by the symporter, the sodium concentration outside the cell would become so great that the hydrolysis of ATP by the ATPase would not provide sufficient energy to pump against such a large gradient.

11.   Selectivity and the rapid transport of ions

12.   Ligand-gated channels open in response to the binding of a molecule by the channel, whereas voltage-gated channels open in response to changes in the membrane potential.

13.   False. Although the cotransporter does not directly depend on ATP, the formation of the Na^+ gradient that powers glucose uptake depends on ATP hydrolysis.

14.   The two forms are (1) ATP hydrolysis and (2) the movement of one molecule down its concentration gradient coupled with the movement of another molecule up its concentration gradient.

15.   Databases could be searched for proteins with stretches of 20 hydrophobic amino acids.

16.   The hydrophobic effect. If there is a hole, the hydrophobic tails of the phospholipids will come together, freeing any associated water.

17.   Peripheral proteins are attached to the phospholipid head groups of membrane lipids or the exposed portions of integral membrane proteins. Integral membrane proteins are embedded in the membrane.

18.   For both sides of a membrane to become identical, the hydrophilic parts of the lipids, proteins, and carbohydrates would have to pass through the hydrophobic interior of the membrane. Such movement is energetically unfavorable.

19. Establish a glucose gradient across vesicle membranes that contain a properly oriented Na^+–glucose linked transporter. Initially, Na^+ concentration should be the same on both sides of the membrane. As the glucose flows "in reverse" through the transporter, down its concentration gradient, a Na^+ concentration gradient becomes established as the glucose gradient is dissipated.

20. An ion channel must transport ions in either direction at the same rate. The net flow of ions is determined only by the composition of the solutions on either side of the membrane.

21. All of the amide hydrogen atoms and carbonyl oxygen atoms are stabilized in the hydrophobic environment by intrachain hydrogen bonds. If the R groups are hydrophobic, they will interact with the hydrophobic interior of the membrane, further stabilizing the helix.

22. The catalytic prowess of acetylcholinesterase ensures that the duration of the nerve stimulus will be short.

23. Ibuprofen is a competitive inhibitor of the synthase.

24. (a) The graph shows that, as temperature increases, the phospholipid bilayer becomes more fluid. T_m is the temperature of the transition from the predominantly less fluid state to the predominantly more fluid state. Cholesterol broadens the transition from the less fluid to the more fluid state. In essence, cholesterol makes membrane fluidity less sensitive to temperature changes.

(b) This effect is important because the presence of cholesterol tends to stabilize membrane fluidity by preventing sharp transitions. Because protein function depends on the proper fluidity of the membrane, cholesterol maintains the proper environment for membrane-protein function.

25. Glucose displays a transport curve that suggests the participation of a carrier, because the initial rate is high but then levels off at higher concentrations, consistent with saturation of the carrier, which is reminiscent of Michaelis–Menten enzymes (p. 113). Indole shows no such saturation phenomenon, which implies that the molecule is lipophilic and simply diffuses across the membrane. Ouabain is a specific inhibitor of the Na^+–K^+ pump. If ouabain were to inhibit glucose transport, then a Na^+– glucose linked transporter would be assisting in transport.

26. During the day, the lipids are likely to be long saturated hydrocarbon chains. With the onset of night, the chains may be shorter or contain cis double bonds or both.

27. Cells may be exposed to many environmental chemicals, called xenobiotics. Many of these chemicals are likely to be toxic. Cells that can remove such chemicals will survive longer.

28. Essentially a reverse membrane. The hydrophilic groups would come together on the interior of the structure, away from the solvent, whereas the hydrocarbon chains would interact with the solvent.

29. (a) Only ASIC1a is inhibited by the toxin. (b) Yes; when the toxin was removed, the activity of the acid-sensing channel began to be restored. (c) 0.9 nM.

30. This mutation is one of a class of mutations that result in slow-channel syndrome (SCS). The results suggest a defect in channel closing; so the channel remains open for prolonged periods. Alternatively, the channel may have a higher affinity for acetylcholine than does the control channel.

31. The recordings would show the channel opening only infrequently. The mutation reduces the affinity of acetylcholine for the receptor.

32. The blockage of ion channels inhibits action potentials, leading to a loss of nervous function. These toxin molecules are useful for isolating and specifically inhibiting particular ion channels.

33. For either ion to pass through the channel, it must shed the water of solvation, which is an endergonic process. For potassium, the energy to compensate for the loss of water is provided by interaction between the ion and the selectivity filter. Because sodium is smaller than potassium, the energetic compensation between the ion and the selectivity filter is too small to compensate the sodium ions for the loss of the water of solvation. Thus, sodium cannot pass through the channel.

Chapter 13

1. (a) 7; (b) 13; (c) 6; (d) 1; (e) 10; (f) 2; (g) 14; (h) 3; (i) 12; (j) 8; (k) 4; (l) 9; (m) 11; (n) 5

2. (1) G-protein coupled (seven-transmembrane-helix) receptors; (2) receptors that dimerize on ligand binding and recruit tyrosine kinases; (3) receptors that dimerize on ligand binding that are tyrosine kinases (receptor tyrosine kinases)

3. The initial signal—the binding of the hormone by a receptor—is amplified by enzymes and channels.

4. The receptor must have a site on the extracellular side of the membrane to which the signal molecule can bind and must have an intracellular domain. Binding of the signal to the receptor must induce structural changes on the intracellular domain so that the signal can be transmitted.

5. The GTPase activity terminates the signal. Without such activity, after a pathway has been activated, it remains activated and is unresponsive to changes in the initial signal.

6. The presence of the appropriate receptor

7. (a) 7; (b) 3; (c) 1; (d) 10; (e) 2; (f) 9; (g) 4; (h) 6; (i) 5; (j) 8

8. The insulin receptor and the EGF receptor employ a common mechanism of signal transmission across the plasma membrane.

9. Growth-factor receptors can be activated by dimerization. If an antibody causes the growth-factor receptor to dimerize, the signal-transduction pathway in a cell will be activated.

10. Heterotrimeric G proteins are composed of $\alpha\beta\gamma$ subunits. The α subunit contains the GTP-binding site. On activation by the signal-receptor event, the GDP is exchanged for GTP, and the $\beta\gamma$ subunits dissociate from the α subunit bound to GTP, which then activates other pathway components such as adenylate cyclase. Small G proteins, such as Ras, are single-subunit proteins. They are activated by proteins such as Sos in the EGF signal pathway. The activation causes the exchange of GDP for GTP to activate Ras, which in turn, actives specific kinases.

11. The mutated α subunit would always be in the GTP form and, hence, in the active form, which would stimulate its signaling pathway.

12. Calcium ions diffuse slowly because they bind to many protein surfaces within a cell, impeding their free motion. Cyclic AMP does not bind as frequently, and so it diffuses more rapidly.

13. $G_{\alpha s}$ stimulates adenylate cyclase, leading to the generation of cAMP. This signal then leads to glucose mobilization (Chapter 24). If cAMP phosphodiesterase were inhibited, then cAMP levels would remain high even after the termination of the epinephrine signal, and glucose mobilization would continue.

14. The full network of pathways initiated by insulin includes a large number of proteins and is substantially more elaborate than indicated in Figure 13.18. Furthermore, many additional proteins take part in the termination of insulin signaling. A defect in any of the proteins in the insulin-signaling pathways or in the subsequent termination of the insulin response could potentially cause problems. Therefore, it is not surprising that many different gene defects can cause type 2 diabetes.

15. The binding of growth hormone causes its monomeric receptor to dimerize. The dimeric receptor can then activate a separate tyrosine kinase to which the receptor binds. The signaling pathway can then continue in similar fashion to the pathways that are activated by the insulin receptor or other mammalian EGF receptors.

16. Proto-oncogenes are normally expressed versions of genes that encode proteins that usually regulate cell growth. Oncogenes are proto-oncogenes that are mutated or overexpressed such that the encoded protein always enhances growth. Tumor-suppressor genes encode proteins that inhibit cell growth or induce cell death.

17. Proto-oncogenes often initiate or are components of pathways leading to cell growth and division in response to some sort of signal. If only one gene is mutated, the cell will be continuously stimulated to grow, even if the other gene continues to function normally. On the other hand, tumor-suppressor genes inhibit the growth signals in some fashion. Thus, even if one gene is nonfunctional, the remaining normal gene is usually sufficient to inhibit unrestricted growth.

18. Other potential drug targets within the EGF signaling cascade include, but are not limited to, the kinase active sites of the EGF receptor, Grb-2, Sos, Ras or any other downstream components of the signaling pathway.

19. Like the receptors discussed in this chapter, ligand-gated channels bind signal molecules, which alters their activity so as to propagate a signal. For example, the IP_3-activated calcium channel is closed until it binds IP_3.

20. Recall that hydrophobic residues are rarely exposed to the aqueous environment of the cells. The exposure of such residues allows calmodulin to bind other proteins and thus propagate the signal.

21. A G protein was a component of the signal-transduction pathway. The terminal sulfate must not be an effective substrate for the GTPase activity of the G protein.

22. Two identical receptors must recognize different aspects of the same signal molecule.

23. The negatively charged glutamate residues mimic the negatively charged phosphoserine or phosphothreonine residues and stabilize the active conformation of the enzyme.

24. Calcium ion levels are kept low by transport systems that extrude Ca^{2+} from the cell. Given this low steady-state level, transient Ca^{2+} increases in intracellular concentration produced by signaling events can be readily sensed.

25. The truncated receptor will dimerize with the full-length monomers on EGF binding, but cross-phosphorylation cannot take place, because the truncated receptor possesses neither the substrate for the neighboring kinase domain nor its own kinase domain to phosphorylate the C-terminal tail of the other monomer. Hence, these mutant receptors will block normal EGF signaling.

26. 10^5

27. (a) $X \approx 10^{-7}$ M; $Y \approx 5 \times 10^{-6}$ M; $Z \approx 10^{-3}$ M

(b) Because much less X is required to fill half of the sites, X displays the highest affinity.

(c) The binding affinity almost perfectly matches the ability to stimulate adenylate cyclase, suggesting that the hormone–receptor complex leads to the stimulation of adenylate cyclase.

(d) Try performing the experiment in the presence of antibodies to $G_{\alpha s}$.

28. (a) The total binding does not distinguish binding to a specific receptor from binding to different receptors or from nonspecific binding to the membrane.

(b) The rationale is that the receptor will have a high affinity for the ligand. Thus, in the presence of excess nonradioactive ligand, the receptor will bind to nonradioactive ligand. Therefore, any binding of the radioactive ligand must be nonspecific.

(c) The plateau suggests that the number of receptor-binding sites in the cell membrane is limited.

29.

Number of receptors per cell =

$$\frac{10^4 \text{cpm}}{\text{mg of membrane protein}} \times \frac{\text{mg of membrane protein}}{10^{10} \text{ cells}}$$

$$\times \frac{\mu\text{mol}}{10^{12} \text{cpm}} \times \frac{6.023 \times 10^{20} \text{ molecules}}{\mu\text{mol}} = 600$$

Chapter 14

1. (a) 6; (b) 4; (c) 1; (d) 9; (e) 2; (f) 10; (g) 5; (h) 3; (i) 7; (j) 8

2. Our most important foods—lipids, complex carbohydrates, and proteins—are large macromolecules that cannot be taken up by cells of the intestine. They must be converted into small molecules that have transport systems allowing them entry to the cells of the intestine. Once inside the cells, the carbohydrates, fatty acids, and amino acids are processed by metabolic pathways to yield energy or used as building blocks.

3. Chewing well efficiently homogenizes the food, rendering it more accessible to the digestive enzymes.

4. Denaturation is the unraveling of a protein's three-dimensional structure. The denatured, extended protein is a much more efficient substrate for digestion by proteases.

5. α-Amylase hydrolyzes the α-1,4 bonds, generating limit dextrin, maltotriose, maltose, and glucose. Maltase digests maltose, whereas α-glucosidase digests maltotriose and other oligosaccharides that may have been generated by α-amylase. Dextrinase digests the limit dextrin. The simple sugars resulting from these enzyme activities are absorbed by the intestine.

6. If a small amount of trypsinogen were inappropriately activated in the pancreas or pancreatic ducts, trypsin could activate other zymogens and lead to the destruction of the pancreas.

7. Macaroni is starch. Hydrating the starch enables α-amylase to more effectively bind to the starch molecules and degrade them.

8. Unlike proteins and carbohydrates, neither lipids nor the products of lipid digestion—fatty acids—are water soluble. The lipids are converted into mixtures of lipid droplets and water (emulsions), a conversion enhanced by bile salts. The emulsions are accessible to lipases. The fatty acids generated by the lipases are carried in micelles to the intestinal membrane.

9. The formation of emulsions allows aqueous lipase access to the ester linkages of the lipids in the lipid droplets.

10. Lipid digestion and absorption would be hindered, and much lipid would be excreted in the feces.

11. Micelles transport the products of lipase digestion, fatty acids and monoacylglycerol, to the intestinal cells for absorption.

12. Secretion of the digestive enzymes as precursors reduces the likelihood that the secretory tissue will be damaged by its secretory products.

13. CCK stimulates the secretion of digestive enzymes by the pancreas and the secretion of bile salts by the gall bladder.

14. Sodium, in cooperation with the sodium-glucose linked transporter, will allow entry of glucose into the intestinal cells.

15. Activation is independent of zymogen concentration because the reaction is intramolecular.

Chapter 15

1. The highly integrated biochemical reactions that take place inside the cell

2. Anabolism is the set of biochemical reactions that use energy to build new molecules and, ultimately, new cells. Catabolism is the set of biochemical reactions that extract energy from fuel sources or breakdown biomolecules.

3. You reply that vandalism is disrespectful and expensive. Part of your tuition money will now have to pay to remove the vandalism. Plus, the fool should know that Gibbs free energy is at a minimum when a system is in equilibrium.

4. (1) Cellular movements and the performance of mechanical work; (2) active transport; (3) biosynthetic reactions

5. (a) 6; (b) 8; (c) 9; (d) 1; (e) 7; (f) 2; (g) 3; (h) 5; (i) 10; (j) 4

6. Charge repulsion, resonance stabilization, increase in entropy, and stabilization by hydration

7. Having only one nucleotide function as the energy currency of the cell enables the cell to monitor its energy status.

8. These divalent ions bind to the negatively charged oxygen atoms found on the phosphoryl groups and help stabilize the charges on ATP.

9. Increasing the concentration of ATP or decreasing the concentration of cellular ADP or P_i (by rapid removal by other reactions, for instance) would make the reaction more exergonic. Likewise, altering the Mg^{2+} concentration could raise or lower the ΔG of the reaction (problems 25 and 30).

10. Reactions in parts *a* and *c*, to the left; reactions in parts *b* and *d*, to the right

11. None whatsoever

12. (a) $\Delta G^{\circ\prime} = 31.4$ kJ mol^{-1} (7.5 kcal mol^{-1}) and $K'_{eq} = 3.06 \times 10^{-6}$; (b) 3.28×10^4

13. $\Delta G^{\circ\prime} = 7.1$ kJ mol^{-1} (1.7 kcal mol^{-1}). The equilibrium ratio is 17.8.

14. (a) Acetate + CoA + H^+ goes to acetyl CoA + H_2O, $\Delta G^{\circ\prime} = +31.4$ kJ mol^{-1} (+7.5 kcal mol^{-1}). ATP hydrolysis to AMP, $\Delta G^{\circ\prime} = 45.6$ kJ mol^{-1} (−10.9 kcal mol^{-1}). Overall reaction, $\Delta G^{\circ\prime} = -14.2$ kJ mol^{-1} (−3.4 kcal mol^{-1}).

(b) With pyrophosphate hydrolysis, $\Delta G^{\circ\prime} = -33.5$ kJ mol^{-1} (−8.0 kcal mol^{-1}). Pyrophosphate hydrolysis dramatically increases the exergonicity of the reaction.

15. The free-energy changes of the individual steps in a pathway are summed to determine the overall free-energy change of the entire pathway. Consequently, a reaction with a positive free-energy value can be powered to take place if coupled to a sufficiently exergonic reaction.

16. An ADP unit

17. Higher organisms cannot make vitamins, and thus are dependent on obtaining them from other organisms.

18. NADH and $FADH_2$ are electron carriers for catabolism; NADPH is the carrier for anabolism.

19. The electrons of the C=O bond cannot form resonance structures with the C—S bond that are as stable as those that they can form with the C—O bond. Thus, the thioester is not stabilized by resonance to the same degree as is an oxygen ester.

20. (1) Control of the amount of enzymes; (2) control of enzyme activity; (3) control of the availability of substrates

21. Recall that $\Delta G = \Delta G^{\circ\prime} + RT \ln$ [products]/[reactants]. Altering the ratio of products to reactants will cause ΔG to vary. In glycolysis, the concentrations of the components of the pathway result in a value of ΔG greater than that of $\Delta G^{\circ\prime}$.

22. A, Ethanol; B, lactate; C, succinate; D, isocitrate; E, malate

23. Unless the ingested food is converted into molecules capable of being absorbed by the intestine, no energy can ever be extracted by the body.

24. Although the reaction is thermodynamically favorable, the reactants are kinetically stable because of the large activation energy. Enzymes lower the activation energy so that the reaction takes place on time scales required by the cell.

25. (a) For an acid AH,

$$AH \rightleftharpoons A^- + H^+ \quad K = \frac{[A^-][H^+]}{[AH]}$$

The pK_a is defined as $pK_a = -\log_{10} K$.
$\Delta G^{\circ\prime}$ is the standard free-energy change at pH 7. Thus, $\Delta G^{\circ\prime} = -RT \ln K = -2.303\, RT \log_{10} K = +2.303\, RT\, pK_a$.

(b) $\Delta G^{\circ\prime} = 27.32$ kJ mol^{-1} (6.53 kcal mol^{-1})

26. The activated form of sulfate in most organisms is 3′-phosphoadenosine-5′-phosphosulfate.

27. (a) As the Mg^{2+} concentration falls, the ΔG of hydrolysis rises. Note that pMg is a logarithmic plot, and so each number on the *x* axis represents a 10-fold change in $[Mg^{2+}]$.

(b) Mg^{2+} binds to the phosphoryl groups of ATP and helps to mitigate charge repulsion. As the $[Mg^{2+}]$ falls, there is less charge stabilization of ATP, leading to greater charge repulsion and an increase in ΔG on hydrolysis.

28. Arginine phosphate in invertebrate muscle, like creatine phosphate in vertebrate muscle, serves as a reservoir of high-potential phosphoryl groups. Arginine phosphate maintains a high level of ATP in muscular exertion.

29. (a) The rationale behind creatine supplementation is that it would be converted into creatine phosphate and thus serve as a rapid means of replenishing ATP after muscle contraction.

(b) If creatine supplementation were beneficial, it would affect activities that depend on short bursts of activity; any sustained activity would require ATP generation by fuel metabolism, which, as Figure 15.8 shows, requires more time.

30. Under standard conditions, $\Delta G^{\circ\prime} = -RT \ln$ [products]/[reactants]. Substituting 23.8 kJ mol^{-1} (5.7 kcal mol^{-1}) for $\Delta G^{\circ\prime}$ and solving for [products]/[reactants] yields 7×10^{-5}. In other words, the forward reaction does not take place to a significant extent. Under intracellular conditions, ΔG is −1.3 kJ mol^{-1} (−0.3 kcal mol^{-1}). Using the equation $\Delta G = \Delta G^{\circ\prime} + RT \ln$ [product]/[reactants] and solving for [products]/[reactants] gives a ratio of 3.7×10^{-5}. Thus, a reaction that is endergonic under standard conditions can be converted into an exergonic reaction by maintaining the [products]/[reactants] ratio below the equilibrium value. This conversion is usually attained by using the products in another coupled reaction as soon as they are formed.

31. Under standard conditions,

$$K'_{eq} = \frac{[B]_{eq}}{[A]_{eq}} \times \frac{[ADP]_{eq}\,[P_i]_{eq}}{[ADP]_{eq}} = 2.67 \times 10^2$$

At equilibrium, the ratio of [B] to [A] is given by

$$\frac{[B]_{eq}}{[A]_{eq}} \times K'_{eq} \frac{[ADP]_{eq}}{[ADP]_{eq}\,[P_i]_{eq}}$$

The ATP-generating system of cells maintains the $[ATP]/[ADP][P_i]$ ratio at a high level, typically about 500 M^{-1}. For this ratio,

$$\frac{[B]_{eq}}{[A]_{eq}} = 2.67 \times 10^2 \times 500 = 1.34 \times 10^5$$

This equilibrium ratio is strikingly different from the value of 1.15×10^{-3} for the reaction A→B in the absence of ATP hydrolysis. In other words, coupling the hydrolysis of ATP with the conversion of A into B has changed the equilibrium ratio of B to A by a factor of about 10^8.

32. Liver: −45.2 kJ mol^{-1} (−10.8 kcal mol^{-1}); muscle: −48.1 kJ mol^{-1} (−11.5 kcal mol^{-1}); brain: −48.5 kJ mol^{-1} (−11.6 kcal mol^{-1}). The free energy of ATP hydrolysis is most negative in the brain.

Chapter 16

1. Glucose is formed under prebiotic conditions. It is the most stable hexose sugar and consequently, has a low tendency, relative to other monosaccharides, to nonenzymatically react with proteins.

2. (a) 4; (b) 3; (c) 1; (d) 6; (e) 8; (f) 2; (g) 10; (h) 9; (i) 7; (j) 5

3. In both cases, the electron donor is glyceraldehyde 3-phosphate. In lactic acid fermentation, the electron acceptor is pyruvate, converting it into lactate. In alcoholic fermentation, acetaldehyde is the electron acceptor, forming ethanol.

4. (a) 3 ATP; (b) 2 ATP; (c) 2 ATP; (d) 2 ATP; (e) 4 ATP

5. Glucokinase enables the liver to remove glucose from the blood when hexokinase is saturated, ensuring that glucose is captured for later use.

6. Glucokinase has a higher K_M value, which allows this enzyme to become more active at high glucose concentrations, conditions that saturate hexokinase.

7. Glucose cannot be cleaved into two three-carbon fragments, whereas fructose can, and three-carbon molecules are metabolized in the second stage of glycolysis. The conversion of fructose 6-phosphate into fructose 1,6-bisphosphate prevents the glucose isomer from being re-formed.

8. The GAP formed is immediately removed by subsequent reactions, resulting in the conversion of DHAP into GAP by the enzyme.

9. A thioester couples the oxidation of glyceraldehyde 3-phosphate to 3-phosphoglycerate with the formation of 1,3-bisphosphoglycerate. 1,3-Bisphosphoglycerate can subsequently power the formation of ATP.

10. Glycolysis is a component of alcoholic fermentation, the pathway that produces alcohol for beer and wine. The belief was that understanding the biochemical basis of alcohol production might lead to a more efficient means of producing beer.

11. The conversion of glyceraldehyde 3-phosphate into 1,3-bisphosphoglycerate would be impaired. Glycolysis would be less effective.

12. Glucose 6-phosphate must have other fates. Indeed, it can be converted into glycogen or be processed to yield reducing power for biosynthesis.

13. The energy needs of a muscle cell vary widely, from rest to intense exercise. Consequently, the regulation of phosphofructokinase by energy charge is vital. In other tissues, such as the liver, ATP concentration is less likely to fluctuate and will not be a key regulator of phosphofructokinase.

14. The $\Delta G^{\circ\prime}$ for the reverse of glycolysis is $+90$ kJ mol^{-1} ($+22$ kcal mol^{-1}), far too endergonic to take place.

15. Pyruvate can be metabolized to ethanol in alcoholic fermentation, to lactate in lactic acid fermentation, or be completely oxidized to CO_2 and H_2O in cellular respiration.

16. The conversion of glucose into glucose 6-phosphate by hexokinase; the conversion of fructose 6-phosphate into fructose 1,6-bisphosphate by phosphofructokinase; the formation of pyruvate from phosphoenolpyruvate by pyruvate kinase

17. Lactic acid is a strong acid (problem 2.15). If it remained in the cell, the pH of the cell would fall, which could lead to the denaturation of muscle protein and result in muscle damage.

18. In liver, fructokinase converts fructose into fructose 1-phosphate. Fructose 1-phosphate is cleaved by a specific aldolase to yield glyceraldehyde and dihydroxyacetone phosphate, which is a component of the glycolytic pathway. Glyceraldehyde is converted into the glycolytic intermediate glyceraldehyde 3-phosphate by triose kinase. In other tissues, fructose is converted into fructose 6-phosphate by hexokinase.

19. Without triose phosphate isomerase, only one of the two three-carbon molecules generated by aldolase could be used to generate ATP. Only two molecules of ATP would result from the metabolism of each molecule of glucose. But two molecules of ATP would still be required to form fructose 1,6-bisphosphate,

the substrate for aldolase. The net yield of ATP would be zero, a yield incompatible with life.

20. 3.06×10^{-5}

21. The equilibrium concentrations of fructose 1,6-bisphosphate, dihydroxyacetone phosphate, and glyceraldehyde 3-phosphate are 7.8×10^{-4} M, 2.2×10^{-4} M, and 2.2×10^{-4} M, respectively.

22. (a) The fructose 1-phosphate pathway forms glyceraldehyde 3-phosphate, bypassing the control by phosphofructokinase.

(b) Since phosphofructokinase is bypassed, glycolysis proceeds in an unregulated manner. Lactic acidosis may result, and fatty liver may develop (Chapter 28).

23. The net reaction in the presence of arsenate is

Glyceraldehyde 3-phosphate $+$ NAD$^+$ $+$ H$_2$O $\longrightarrow$

$$3\text{-phosphoglycerate} + \text{NADH} + 2\,\text{H}^+$$

Glycolysis proceeds in the presence of arsenate, but the ATP normally formed in the conversion of 1,3-bisphosphoglycerate into 3-phosphoglycerate is lost. Thus, arsenate uncouples oxidation and phosphorylation by forming a highly labile acyl arsenate.

24. This example illustrates the difference between the *stoichiometric* and the *catalytic* use of a molecule. If cells used NAD$^+$ stoichiometrically, a new molecule of NAD$^+$ would be required each time a molecule of lactate was produced. As we will see, the synthesis of NAD$^+$ requires ATP. On the other hand, if the NAD$^+$ that is converted into NADH could be recycled and reused, a small amount of the molecule could regenerate a vast amount of lactate, which is the case in the cell. NAD$^+$ is regenerated by the oxidation of NADH and reused. NAD$^+$ is thus used catalytically.

25. Recall from our discussion of enzyme kinetics in Chapter 7 that substrates are usually present in much higher concentration than their enzymes. Consequently, converting a small amount of substrate into a potent activator of PFK will lead to a rapid increase in the rate of ATP synthesis.

26. (a) 4; (b) 10; (c) 1; (d) 5; (e) 7; (f) 8; (g) 9; (h) 2; (i) 3; (j) 6

27. Fructose 2,6-bisphosphate stabilizes the R state of the enzyme.

28. Galactose is a component of glycoproteins. Possibly, the absence of galactose leads to the improper formation or function of glycoproteins required in the central nervous system. More generally, the fact that the symptoms arise in the absence of galactose suggests that galactose is required in some fashion.

29. A potassium channel, a ligand-gated channel, is inhibited by binding ATP. This inhibition alters the voltage across the plasma membrane, which activates a calcium channel, a voltage-gated channel, allowing an influx of calcium ions. The calcium ions stimulate the fusion of insulin-containing granules with the plasma membrane, resulting in the secretion of insulin.

30. Using the Michaelis–Menten equation to solve for [S] when $K_M = 50$ μM and $v_0 = 0.9 V_{max}$ shows that a substrate concentration of 0.45 mM yields 90% of V_{max}. Under normal conditions, the enzyme is essentially working at V_{max}.

31. (a) Curiously, the enzyme uses ADP as the phosphoryl donor rather than ATP.

(b) Both AMP and ATP behave as competitive inhibitors of ADP, the phosphoryl donor. Apparently, the *P. furiosus* enzyme is not allosterically inhibited by ATP.

32. ATP initially stimulates PFK activity, as would be expected for a substrate. Higher concentrations of ATP inhibit the enzyme. Although this effect seems counterintuitive for a substrate, recall that the function of glycolysis in muscle is to generate ATP. Consequently, high concentrations of ATP signal that the ATP needs are met and glycolysis should stop. In addition to being a substrate, ATP is an allosteric inhibitor of PFK.

33. (a) Carbon dioxide is a good indicator of the rate of alcoholic fermentation because it is, along with ethanol, a product of alcoholic fermentation.

(b) Phosphate is a required substrate for the reaction catalyzed by glyceraldehyde 3-phosphate dehydrogenase. The phosphate is incorporated into 1,3-bisphosphoglycerate.

(c) It indicates that the amount of free phosphate must have been limiting.

(d) The ratio would be expected to be 1. A phosphate would be consumed for every pyruvate decarboxylated.

(e) Lack of phosphate would inhibit glyceraldehyde 3-phosphate dehydrogenase. This inhibition would "back-up" glycolysis, resulting in the accumulation of fructose 1,6-bisphosphate.

34. Glucose is reactive because its open-chain form contains an aldehyde group. In other words, glucose is a reducing sugar.

35. (a) The label is in the methyl carbon atom of pyruvate.

(b) 5 mCi mM^{-1}. The specific activity is halved because the number of moles of product (pyruvate) is twice that of the labeled substrate (glucose).

36. (a) Glucose + 2 P$_i$ + 2 ADP → 2 lactate + 2 ATP

(b) $\Delta G^{\circ\prime} = -123$ kJ mol^{-1} (-29 kcal mol^{-1})

(c) $\Delta G = -114$ kJ mol^{-1} (-27.2 kcal mol^{-1})

37. GLUT 2 transports glucose only when the blood concentration of glucose is high, which is precisely the condition in which the β cells of the pancreas secrete insulin.

38. Fructose + ATP → fructose 1-phosphate + ADP: Fructokinase

Fructose 1-phosphate → dihydroxyacetone phosphate + glyceraldehyde: Fructose 1-phosphate aldolase

Glyceraldehyde + ATP → glyceraldehyde 3-phosphate + ADP: Triose kinase

The primary controlling step of glycolysis catalyzed by phosphofructokinase is bypassed by the preceding reactions. Glycolysis will proceed in an unregulated fashion.

39. The ATPase activity of hexokinase is low in the absence of a sugar because it is in a catalytically inactive conformation. The addition of xylose closes the cleft between the two lobes of the enzyme. However, xylose lacks a hydroxymethyl group, and so it cannot be phosphorylated. Instead, a water molecule at the site normally occupied by the C-6 hydroxymethyl group acts as the acceptor of the phosphoryl group from ATP.

40. Consider the equilibrium equation of adenylate kinase:

$$K_{eq} = [ATP][AMP]/[ADP]^2 \qquad (1)$$

or

$$[AMP] = K_{eq}[ADP]^2/[ATP] \qquad (2)$$

Recall that [ATP] > [ADP] > [AMP] in the cell. As ATP is utilized, a small decrease in its concentration will result in a larger percentage increase in [ADP] because ATP's concentration is greater than that of ADP. This larger percentage increase in [ADP] will result in an even greater percentage increase in [AMP] because the concentration of AMP is related to the square of [ADP]. In essence, equation 2 shows that monitoring the energy status with AMP magnifies small changes in [ATP], leading to tighter control.

Chapter 17

1. The reverse of glycolysis is highly endergonic under cellular conditions. The expenditure of six NTP molecules in gluconeogenesis renders gluconeogenesis exergonic.

2. (a) 6; (b) 1; (c) 7; (d) 3; (e) 2; (f) 5 ; (g) 4

3. In glycolysis, the formation of pyruvate and ATP by pyruvate kinase is irreversible. This step is bypassed by two reactions in gluconeogenesis: (1) the formation of oxaloacetate from pyruvate and CO$_2$ by pyruvate carboxylase and (2) the formation of phosphoenolpyruvate from oxaloacetate and GTP by phosphoenolpyruvate carboxykinase. The formation of fructose 1,6-bisphosphate by phosphofructokinase is bypassed by fructose 1,6-bisphosphatase in gluconeogenesis, which catalyzes the conversion of fructose 1,6-bisphosphate into fructose 6-phosphate. Finally, the hexokinase-catalyzed formation of glucose 6-phosphate in glycolysis is bypassed by glucose 6-phosphatase, but only in the liver.

4. Reactions in parts *b* and *e* would be blocked.

5. The synthesis of glucose during intense exercise provides a good example of interorgan cooperation in higher organisms. When muscle is actively contracting, lactate is produced from glucose by glycolysis. The lactate is released into the blood and absorbed by the liver, where it is converted by gluconeogenesis into glucose. The newly synthesized glucose is then released and taken up by the muscle for energy generation.

6. Muscle is likely to produce lactic acid during contraction. Lactic acid is a strong acid and must not accumulate in muscle or blood. Liver removes the lactic acid from the blood and converts it into glucose. The glucose can be released into the blood or stored as glycogen for later use.

7. Glucose produced by the liver could not be released into the blood. Tissues that rely on glucose as an energy source would not function as well unless glucose was provided in the diet.

8. Glucose is an important energy source for both tissues and is essentially the only energy source for the brain. Consequently, these tissues should never release glucose. Glucose release is prevented by the absence of glucose 6-phosphatase.

9. In gluconeogenesis, lactate dehydrogenase synthesizes pyruvate from lactate. In lactic acid fermentation, the enzyme synthesizes lactate from pyruvate.

10. (a) A, B; (b) C, D; (c) D; (d) A; (e) B; (f) C; (g) A; (h) D; (i) none; (j) A; (k) A

11. Some of the amino acids will be released into the blood. The liver will take up the amino acids and convert the carbon skeletons into glucose.

12. 6 NTP (4 ATP and 2 GTP); 2 NADH

13. (a) None; (b) none; (c) 4 (2 ATP and 2 GTP); (d) none

14. One cycle would require the glycolytic enzyme phosphofructokinase and the gluconeogenic enzyme fructose 1,6-bisphosphatase. The other cycle would require pyruvate kinase from glycolysis and pyruvate carboxylase and phosphoenolpyruvate carboxykinase from gluconeogenesis.

15. The substrate cycles regulate flux on one or the other pathway by amplifying metabolic signals.

16. The enzymes in two substrate cycles are control points. The glycolytic pathway is activated by F-2,6-BP, AMP, and F-1,6-BP, whereas ATP, alanine, citrate, and protons inhibit glycolysis. Gluconeogenesis is activated by citrate and acetyl CoA and inhibited by F-2,6-BP, AMP, and ADP.

17. (a) 2, 3, 6, 9; (b) 1, 4, 5, 7, 8

18. If the amino groups are removed from alanine and aspartate, the ketoacids pyruvate and oxaloacetate are formed. Both of these molecules are components of the gluconeogenic pathway.

19. (a) Increased; (b) increased; (c) increased; (d) decreased

20.

Glycerol   **Glycerol phosphate**

Dihydroxyacetone phosphate

Dihydroxyacetone phosphate can be used in gluconeogenesis or glycolysis.

21. For glycerol to be fermented to ethanol, two NADH are generated. However, only one of NAD^+ is regenerated during alcoholic fermentation. Thus, redox balance is not maintained and the yeast run out of NAD^+ and pass on.

22. (a) If both enzymes operated simultaneously, the following reactions would take place:

The net result would simply be

$$ATP + H_2O \rightarrow ADP + P_i$$

The energy of ATP hydrolysis would be released as heat.

(b) Not really. For the cycle to generate heat, both enzymes must be functional at the same time in the same cell.

(c) The species *B. terrestris* and *B. rufocinctus* might show some futile cycling because both enzymes are active to a substantial degree.

(d) No. These results simply suggest that simultaneous activity of phosphofructokinase and fructose 1,6-bisphosphatase is unlikely to be employed to generate heat in the species shown.

23. Lactic acid is capable of being further oxidized and is thus useful energy. The conversion of this acid into glucose saves the carbon atoms for future combustion.

24. Fructose 2,6-bisphosphate, present at high concentration when glucose is abundant, normally inhibits gluconeogenesis by blocking fructose 1,6-bisphosphatase. In this genetic disorder, the phosphatase is active irrespective of the glucose level. Hence, substrate cycling is increased. The level of fructose 1,6-bisphosphate is consequently lower than normal. Less pyruvate is formed and thus less ATP is generated.

25. There will be no labeled carbon atoms. The CO_2 added to pyruvate (formed from the lactate) to form oxaloacetate is lost with the conversion of oxaloacetate into phosphoenolpyruvate.

26. The input of four additional high-phosphoryl-transfer-potential molecules in gluconeogenesis changes the equilibrium constant by a factor of 10^{32}, which makes the conversion of pyruvate into glucose thermodynamically feasible. Without this energetic input, gluconeogenesis would not take place.

Chapter 18

1. The pyruvate dehydrogenase complex catalyzes the following reaction, linking glycolysis and citric acid cycle:

$$\text{Pyruvate} + \text{CoA} + \text{NAD}^+ \rightarrow \text{acetyl CoA} + \text{NADH} + \text{H}^+ + \text{CO}_2$$

2. Pyruvate dehydrogenase catalyzes the decarboxylation of pyruvate and the formation of acetyllipoamide. Dihydrolipoyl transacetylase catalyzes the formation of acetyl CoA. Dihydrolipoyl dehydrogenase catalyzes the reduction of the oxidized lipoic acid. PDH kinase associated with the complex phosphorylates and inactivates the complex, whereas PDH phosphatase dephosphorylates and activates the complex.

3. The remaining steps regenerate oxidized lipoamide, which is required to begin the next reaction cycle. Moreover, this regeneration results in the production of high-energy electrons in the form of NADH.

4. Decarboxylation, oxidation, and transfer of the resultant acetyl group to CoA

5. Oxidation to CO_2 by the citric acid cycle or incorporation into lipids

6. Thiamine pyrophosphate plays a role in the decarboxylation of pyruvate. Lipoic acid (as lipoamide) transfers the acetyl group. Coenzyme A accepts the acetyl group from lipoic acid to form acetyl CoA. FAD accepts the electrons and hydrogen ions when oxidized lipoic acid is reduced. NAD^+ accepts electrons and a proton from $FADH_2$.

7. Catalytic coenzymes (TPP, lipoic acid, and FAD) are modified but regenerated in each reaction cycle. Thus, they can play a catalytic role in the processing of many molecules of pyruvate. Stoichiometric coenzymes (coenzyme A and NAD^+) are used in only one reaction because they are the components of products of the reaction.

8. The electrons are transferred from reduced lipoamide to FAD initially and then to NAD^+. This transfer is unusual because the electrons are passed to NAD^+ from $FADH_2$. The transfer is usually in the other direction.

9. (a) 6; (b) 10; (c) 1; (d) 7; (e) 2; (f) 8; (g) 3; (h) 4; (i) 5; (j) 9

10. In muscle, the acetyl CoA generated by the complex is used for energy generation. Consequently, signals that indicate an energy-rich state (high ratios of ATP/ADP and NADH/NAD^+) inhibit the complex, whereas the reverse conditions stimulate the enzyme. Calcium as the signal for muscle contraction (and, hence, energy need) also stimulates the enzyme. In liver, acetyl CoA derived from pyruvate is used for biosynthetic purposes, such as fatty acid synthesis. Insulin, the hormone denoting the fed state, stimulates the complex.

11. (a) Enhanced kinase activity will result in a decrease in the activity of the PDH complex because phosphorylation by the kinase inhibits the complex.

(b) Phosphatase activates the complex by removing a phosphoryl group. If the phosphatase activity is diminished, the activity of the PDH complex also will decrease.

12. She might have been ingesting, in some fashion, the arsenite from the wallpaper. The arsenite would have inhibited enzymes that require lipoic acid—notably, the pyruvate dehydrogenase complex.

13. Acetyllipoamide and acetyl CoA

14. The mercury was inhibiting the PDH complex, specifically the transacetylase, by binding to lipoic acid. Many of the symptoms are neurological because the brain uses only glucose as a fuel.

15. Acetyl CoA will inhibit the complex. The metabolism of glucose to pyruvate will be slowed because acetyl CoA is being derived from an alternative source.

16. A thioester is a key intermediate in the formation of 1,3-bisphosphoglycerate from glyceraldehyde 3-phosphate in the reaction catalyzed by glyceraldehyde 3-phosphate dehydrogenase. 1,3-Bisphosphoglycerate is subsequently metabolized to pyruvate.

17. (a) A decrease in the amount of O_2 will necessitate an increase in anaerobic glycolysis for energy production, leading to the generation of a large amount of lactic acid. (b) Under conditions of shock, the kinase inhibitor is administered to ensure that pyruvate dehydrogenase is operating maximally.
18. (a) DCA inhibits pyruvate dehydrogenase kinase.

(b) The fact that inhibiting the kinase results in more dehydrogenase activity suggests that there must be some residual activity that is being inhibited by the kinase.
19. Pyruvate dehydrogenase kinase phosphorylates and inhibits the pyruvate dehydrogenase component of the pyruvate dehydrogenase complex. Inhibiting pyruvate dehydrogenase kinase may be an effective mechanism to increase glucose utilization, thereby lowering blood-glucose levels. Furthermore, increasing pyruvate oxidation may contribute to the lowering of the blood-glucose level by decreasing the supply of gluconeogenic substrates.
20. Thiamine thiazolone pyrophosphate is a transition-state analog. The sulfur-containing ring of this analog is uncharged, and so it closely resembles the transition state of the normal coenzyme in thiamine-catalyzed reactions (e.g., the uncharged resonance form of hydroxyethyl-TPP).

Chapter 19

1. Acetyl-CoA + 3 NAD^+ + FAD + ADP + P_i →
   2 CO_2 + 3 NADH + 3 H^+ + $FADH_2$ + ATP + CoA
2. The citric acid cycle depends on a steady supply of NAD^+ and FAD, which are typically generated from NADH and $FADH_2$ by reaction of these electron carriers with oxygen. If there is no oxygen to accept the electrons, the citric acid cycle will cease to operate.
3. In the first stage, two carbon atoms are introduced into the cycle by reacting with oxaloacetate to form citrate. Two carbon atoms are then oxidized to CO_2. In the second stage, the resulting four-carbon molecule is metabolized to regenerate oxaloacetate. High-energy electrons are generated in both stages.
4. (a) 5; (b) 7; (c) 1; (d) 10; (e) 2; (f) 4; (g) 9; (h) 3; (i) 8; (j) 6
5. The reaction is powered by the hydrolysis of a thioester. Acetyl CoA provides the thioester that is converted into citryl CoA. When this thioester is hydrolyzed, citrate is formed in an irreversible reaction.
6. Acetyl CoA does not bind citrate synthase until oxaloacetate is bound and ready for the synthesis of citrate. In addition, the catalytic groups required for the hydrolysis of the thioester are not positioned for catalysis until citryl CoA has formed.
7. (a) Isocitrate lyase and malate synthase are required in addition to the enzymes of the citric acid cycle.

(b) 2 Acetyl CoA + 2 NAD^+ + FAD + 3 H_2O →
   oxaloacetate + 2 CoA + 2 NADH + $FADH_2$ + 3 H^+

(c) No. Hence, mammals cannot carry out the net synthesis of oxaloacetate from acetyl CoA.
8. -41.0 kJ mol^{-1} (-9.8 kcal mol^{-1})
9. Enzymes or enzyme complexes are biological catalysts. Recall that a catalyst facilitates a chemical reaction without the catalyst itself being permanently altered. The citric acid cycle operates catalytically in that oxaloacetate binds to an acetyl group, leads to the oxidative decarboxylation of the two carbon atoms, and is regenerated at the completion of a cycle.
10. Succinate will increase in concentration, followed by α-ketoglutarate and the other intermediates "upstream" of the site of inhibition. Succinate has two methylene groups that are required for the dehydrogenation, whereas malonate has but one.
11. Succinate dehydrogenase is the only enzyme in the citric acid cycle that is embedded in the mitochondrial membrane in association with the electron-transport chain.

12. The reaction catalyzed by succinyl CoA synthetase:

Succinyl CoA + P_i + ADP → succinate + CoA + ATP

13. The reaction catalyzed succinyl CoA synthetase:

Succinyl CoA + P_i + ADP → succinate + CoA + ATP

14. Isocitrate dehydrogenase and α-ketoglutarate dehydrogenase
15. It enables organisms such as plants and bacteria to convert fats, through acetyl CoA, into glucose.
16. Both α-ketoglutarate and pyruvate are α-ketoacids that are decarboxylated to form thioesters of CoA. Moreover, the dihydrolipoyl dehydrogenase components of the complexes are identical and the other enzymes are similar. Finally, the reaction mechanisms are the same.
17. We cannot attain the net conversion of fats into glucose, because the only means to get the carbon atoms from fats into oxaloacetate, the precursor to glucose, is through the citric acid cycle. However, although two carbon atoms enter the cycle as acetyl CoA, two carbon atoms are lost as CO_2 before oxaloacetate is formed. Thus, although some carbon atoms from fats may end up as carbon atoms in glucose, we cannot obtain a *net* synthesis of glucose from fats.
18. Pyruvate carboxylase should be active only when the acetyl CoA concentration is high. Acetyl CoA might accumulate if the energy needs of the cell are not being met, because of a deficiency of oxaloacetate. Under these conditions the pyruvate carboxylase catalyzes an anapleurotic reaction. Alternatively, acetyl CoA might accumulate because the energy needs of the cell have been met. In this circumstance, pyruvate will be converted back into glucose, and the first step in this conversion is the formation of oxaloacetate.
19. Citrate is a symmetric molecule. Consequently, the investigators assumed that the two $-CH_2COO^-$ groups in it would react identically. Thus, for every citrate molecule undergoing the reactions shown in path 1, they thought that another citrate molecule would react as shown in path 2. If so, then only *half* the label should have emerged in the CO_2.

Path 1

Path 2 (does not occur)

20. Call one hydrogen atom A and the other B. Now suppose that an enzyme binds three groups of this substrate—X, Y, and H—at three complementary sites. The adjoining diagram shows X, Y, and H_A bound to three points on the enzyme. In contrast, X, Y, and H_B cannot be bound to this active site; two of these three groups can be bound, but not all three. Thus, H_A and H_B will have different fates.

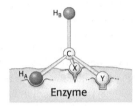

Sterically nonequivalent groups such as H_A and H_B will almost always be distinguished in enzymatic reactions. The essence of the differentiation of these groups is that the enzyme holds the substrate in a specific orientation. Attachment at three points, as depicted in the diagram, is a readily visualized way of achieving a particular orientation of the substrate, but it is not the only means of doing so.

21. (a) The complete oxidation of citrate requires 4.5 μmol of O_2 for every micromole of citrate:

$$C_6H_8O_7 + 4.5\ O_2 \rightarrow 6\ CO_2 + 4\ H_2O$$

Thus, 13.5 μmol of O_2 would be consumed by 3 μmol of citrate.

(b) Citrate led to the consumption of far more O_2 than can be accounted for simply by the oxidation of citrate itself. Citrate thus facilitated O_2 consumption.

22. (a) In the absence of arsenite, the amount of citrate remained constant. In its presence, the concentration of citrate fell, suggesting that it was being metabolized.

(b) Arsenite's action is not altered. Citrate still disappears.

(c) Arsenite is preventing the regeneration of citrate. Recall (p. 340) that arsenite inhibits the pyruvate dehydrogenase complex.

23. (a) The initial infection is unaffected by the absence of isocitrate lyase, but the absence of this enzyme inhibits the latent phase of the infection.

(b) Yes

(c) A critic could say that, in the process of deleting the isocitrate lyase gene, some other gene was damaged, and the absence of this other gene prevents latent infection. Reinsertion of the isocitrate lyase gene into the bacteria from which it had been removed renders the criticism less valid.

(d) Isocitrate lyase enables the bacteria to synthesize carbohydrates that are necessary for survival, including carbohydrate components of the cell membrane.

24. (a) After one round of the citric acid cycle, the label emerges in C-2 and C-3 of oxaloacetate. (b) The label emerges in CO_2 in the formation of acetyl CoA from pyruvate. (c) After one round of the citric acid cycle, the label emerges in C-1 and C-4 of oxaloacetate. (d and e) Same fate as that in part a.

25. (a) The steady-state concentrations of the products are low compared with those of the substrates. (b) The ratio of malate to oxaloacetate must be greater than 1.57×10^4 for oxaloacetate to be formed.

26. The energy released when succinate is reduced to fumarate is not sufficient to power the synthesis of NADH but is sufficient to reduce FAD.

27. Citrate is a tertiary alcohol that cannot be oxidized, because oxidation requires a hydrogen atom to be removed from the alcohol and a hydrogen atom to be removed from the carbon atom bonded to the alcohol. No such hydrogen exists in citrate. The isomerization converts the tertiary alcohol into isocitrate, which is a secondary alcohol that can be oxidized.

28. Pyruvate + CoA + NAD$^+$ $\xrightarrow{\text{Pyruvate dehydrogenase complex}}$ acetyl CoA + CO_2 + NADH

Pyruvate + CO_2 + ATP + H_2O $\xrightarrow{\text{Pyruvate carboxylase}}$ oxaloacetate + ADP + P_i

Oxaloacetate + acetyl CoA + H_2O $\xrightarrow{\text{Citrate synthase}}$ citrate + CoA + H$^+$

Citrate $\xrightarrow{\text{Aconitase}}$ isocitrate

Isocitrate + NAD$^+$ $\xrightarrow{\text{Isocitrate dehydrogenase}}$ α-ketoglutarate + CO_2 + NADH

Net: 2 Pyruvate + 2 NAD$^+$ + ATP + H_2O $\rightarrow$ α-ketoglutarate + CO_2 + ADP + 2 NADH + 2 H$^+$ + P_i

Chapter 20

1. An oxidizing agent, or oxidant, accepts electrons in oxidation–reduction reactions. A reducing reagent, or reductant, donates electrons in such reactions.

2. (a) 4; (b) 5; (c) 2; (d) 10; (e) 3; (f) 8; (g) 9; (h) 7; (i) 1; (j) 6

3. Biochemists use E'_0, the value at pH 7, whereas chemists use E_0, the value in 1 M H$^+$. The prime denotes that pH 7 is the standard state.

4. $\Delta G^{\circ\prime}$ is + 67 kJ mol^{-1} (+16.1 kcal mol^{-1}) for oxidation by NAD$^+$ and -3.8 kJ mol^{-1} (-0.92 kcal mol^{-1}) for oxidation by FAD. The oxidation of succinate by NAD$^+$ is not thermodynamically feasible.

5. The most bacteria-like mitochondrial genome, that of the protozoan *Reclinomonas americana*, consist of 97 genes with 62 genes encoding proteins. The protein-coding genes, which comprise only 2% of the protein-coding genes in the bacterium *E. coli*, include all of the protein-coding genes found in all of the sequenced mitochondrial genomes. Thus, 2% of bacterial genes are found in all examined mitochondria. It seems unlikely that mitochondrial genomes resulting from several endosymbiotic events could have been independently reduced to the same set of genes found in *R. americana*. The simplest explanation is that the endosymbiotic event took place just once and all existing mitochondria are descendants of that ancestor.

6. Pyruvate accepts electrons and is thus the oxidant. NADH gives up electrons and is the reductant.

7. The $\Delta E'_0$ value of iron can be altered by changing the environment of the ion.

8. $\Delta G^{\circ\prime} = -nF\Delta E'_0$

9. The 10 isoprene units render coenzyme Q soluble in the hydrophobic environment of the inner mitochondrial membrane. The two oxygen atoms can reversibly bind two electrons and two protons as the molecule transitions between the quinone form and the quinol form.

10. c, e, b, a, d

11. (a) 4; (b) 3; (c) 1; (d) 5; (e) 2

12. Hydroxyl radical (OH ·), hydrogen peroxide (H_2O_2), superoxide ion ($O_2^{\cdot -}$), and peroxide (O_2^{2-}). These small molecules react with a host of macromolecules—including proteins, nucleotides, and membranes—to disrupt cell structure and function.

13. Rotenone: NADH, NADH-Q oxidoreductase will be reduced. The remainder will be oxidized. Antimycin A: NADH, NADH-Q oxidoreductase and coenzyme Q will be reduced. The remainder will be oxidized. Cyanide: All will be reduced.

14. The respirasome is another example of the use of supramolecular complexes in biochemistry. Having the three

complexes that are proton pumps associated with one another will enhance the efficiency of electron flow from complex to complex, which in turn will cause more efficient proton pumping.

15. Triose phosphate isomerase converts dihydroxyacetone phosphate (a potential dead end) into glyceraldehyde 3-phosphate (a mainstream glycolytic intermediate).

16. (a) Vitamins C and E

(b) Exercise induces superoxide dismutase, which converts ROS into hydrogen peroxide and oxygen.

(c) The answer to this question is not fully established. Two possibilities are (1) the suppression of ROS by vitamins prevents the expression of more superoxide dismutase and (2) some ROS may be signal molecules required to stimulate the biochemical benefits of exercise.

17. Succinate dehydrogenase is a component of Complex II.

18. In fermentations, organic compounds are both the donors and the acceptors of electrons. In respiration, the electron donor is usually an organic compound, whereas the electron acceptor is an inorganic molecule, such as oxygen.

19. The $\Delta G^{\circ\prime}$ for the reduction of oxygen by $FADH_2$ is -200 kJ mol^{-1} (-48 kcal mol^{-1}).

20. This inhibitor (like antimycin A, problem 13) blocks the reduction of cytochrome c_1 by QH_2, the crossover point.

Chapter 21

1. The ATP is recycled by ATP-generating processes, most notably oxidative phosphorylation.

2. (a) 4; (b) 6; (c) 8; (d) 1; (e) 10; (f) 9; (g) 2; (h) 3; (i) 5; (j) 7

3. (a) 12.5; (b) 14; (c) 32; (d) 13.5; (e) 30; (f) 16

4. (a) Azide blocks electron transport and proton pumping at Complex IV.

(b) Atractyloside blocks electron transport and ATP synthesis by inhibiting the exchange of ATP and ADP across the inner mitochondrial membrane.

(c) Rotenone blocks electron transport and proton pumping at Complex I.

(d) DNP blocks ATP synthesis without inhibiting electron transport by dissipating the proton gradient.

(e) Carbon monoxide blocks electron transport and proton pumping at Complex IV.

(f) Antimycin A blocks electron transport and proton pumping at Complex III.

5. If the proton gradient is not dissipated by the influx of protons into a mitochondrion with the generation of ATP, the outside of the mitochondrion eventually develops such a large positive charge that the electron-transport chain can no longer pump protons against the gradient.

6. (a) No effect; mitochondria cannot metabolize glucose.

(b) No effect; no fuel is present to power the synthesis of ATP from ADP and P_i.

(c) The $[O_2]$ falls because citrate is a fuel and ATP can be formed from ADP and P_i.

(d) Oxygen consumption stops because oligomycin inhibits ATP synthesis, which is coupled to the activity of the electron-transport chain.

(e) No effect, for the reasons given in part d

(f) $[O_2]$ falls rapidly because the system is uncoupled and does not require ATP synthesis to lower the proton-motive force.

(g) $[O_2]$ falls, though at a lower rate. Rotenone inhibits Complex I, but the presence of succinate will enable electrons to enter at Complex II.

(h) Oxygen consumption ceases because Complex IV is inhibited and the entire chain backs up.

7. Such a defect (called Luft syndrome) was found in a 38-year-old woman who was incapable of performing prolonged physical work. Her basal metabolic rate was more than twice normal, but her thyroid function was normal. A muscle biopsy showed that her mitochondria were highly variable and atypical in structure. Biochemical studies then revealed that oxidation and phosphorylation were not tightly coupled in these mitochondria. In this patient, much of the energy of fuel molecules was converted into heat rather than ATP.

8. Dicyclohexylcarbodiimide reacts readily with carboxyl groups. Hence, the most likely targets are aspartate and glutamate side chains. In fact, Asp 61 of subunit **c** of *E. coli* F_0 is specifically modified by this reagent. The conversion of Asp 61 into asparagine by site-specific mutagenesis eliminates proton conduction, showing that the acid is required for proton conduction.

9. If oxidative phosphorylation were uncoupled, no ATP could be produced. In a futile attempt to generate ATP, much fuel would be consumed. The danger lies in the dose. Too much uncoupling would lead to tissue damage in highly aerobic organs such as the brain and heart, which would have severe consequences for the organism as a whole. The energy that is normally transformed into ATP would be released as heat. To maintain body temperature, sweating might increase, although the very process of sweating itself depends on ATP.

10. If the proton gradient cannot be dissipated by flow through ATP synthase, the proton gradient will eventually become so large that the energy released by the electron-transport chain will not be great enough to pump protons against the larger-than-normal gradient.

11. The proton gradient is necessary for ATP synthesis because proton flow through the enzyme causes conformational changes that convert a T subunit into an O subunit with the subsequent release of ATP. The role of the proton gradient is not to form ATP but to release it from the synthase.

12. Arg 210, with its positive charge, will facilitate proton release from aspartic acid by stabilizing the negatively charged Asp 61.

13. 2.7; 4; 5

14. Presumably, because the muscle has greater energy needs, especially during exercise, it will require more ATP. This requirement means that more sites of oxidative phosphorylation are called for, and these sites can be provided by an increase in the amount of cristae.

15. If ATP and ADP cannot exchange between the matrix and the mitochondria, ATP synthase will cease to function because its substrate ADP is absent. The proton gradient will eventually become so large that the energy released by the electron-transport chain will not be great enough to pump protons against the larger-than-normal gradient.

16. Remember that the extra negative charge on ATP relative to that on ADP accounts for ATP's more rapid translocation out of the mitochondrial matrix. If the charge differences between ATP and ADP were lessened by the binding of the Mg^{2+}, ADP might more readily compete with ATP for transport to the cytoplasm.

17. The subunits are jostled by background thermal energy (Brownian motion). The proton gradient makes clockwise rotation more likely because that direction results in protons flowing down their concentration gradient.

18. ATP export from the matrix. Phosphate import into the matrix.

19. If ADP cannot enter the mitochondria, the electron-transport chain will cease to function because there will be no acceptor for the energy. NADH will build up in the matrix. Recall that NADH inhibits some citric acid cycle enzymes, and NAD$^+$ is required by several citric acid cycle enzymes. Glycolysis will stop functioning aerobically but will switch to anaerobic glycolysis so that the NADH can be reoxidized to NAD$^+$ by lactate dehydrogenase.

20. When all of the available ADP has been converted into ATP, ATP synthase can no longer function. The proton gradient becomes

large enough that the energy of the electron-transport chain is not enough to pump against the gradient, and electron transport and, hence, oxygen consumption falls.

21. The effect on the proton gradient is the same in each case.

22. The ATP synthase would pump protons at the expense of ATP hydrolysis, thus maintaining the proton-motive force. The synthase would function as an ATPase. There is some evidence that damaged mitochondria use this tactic to maintain, at least temporarily, the proton-motive force.

23. First, a closed compartment, intrinsically impermeable to protons, is required to obtain ATP synthesis. Second, electron transport does generate a proton gradient across the inner mitochondrial membrane. Third, an artificial system representing the cellular respiration system demonstrates the basic principle of the chemiosmotic hypothesis (Figure 21.2). Synthetic vesicles containing bacteriorhodopsin and mitochondrial ATP synthase purified from beef heart were created. A proton gradient is generated by bacteriorhodopsin, a purple membrane protein from halobacteria that pumps protons when illuminated. When the vesicles were exposed to light, ATP was formed.

24. The inside of the mitoplasts would be more basic (pH 7) relative to the outside (pH 4) after the mixing. Thus, an artificial pH gradient would have been imposed. ATP synthesis would indeed be seen under these circumstances.

25. If **b** and δ subunits were absent, the γ subunit would simply rotate the $\alpha_3\beta_3$ ring rather than power the structural changes (O→L→T) that result in ATP synthesis. In other words, the proton-motive force would power rotation.

26. Recall that enzymes catalyze reactions in both directions. The hydrolysis of ATP is exergonic. Consequently, ATP synthase will catalyze the conversion of ATP into its more stable products. ATP synthase works as a synthase in vivo because the energy of the proton gradient overcomes the tendency toward ATP hydrolysis.

27. It suggests that malfunctioning mitochondria may play a role in the development of Parkinson disease. Specifically, it implicates Complex I.

28. The cytoplasmic kinases will thereby obtain preferential access to the exported ATP.

29. The organic acids in the blood are indications that the mice are deriving a large part of their energy needs through aerobic glycolysis. Lactate is the end product of aerobic glycolysis. Alanine is an aminated transport form of lactate. Alanine formation plays a role in succinate formation, which is caused by the reduced state of the mitochondria.

30. In the presence of poorly functioning mitochondria, the only means of generating ATP is by anaerobic glycolysis, which will lead to an accumulation of lactic acid in blood.

31. (a) Succinate is oxidized by Complex II, and the electrons are used to establish a proton-motive force that powers ATP synthesis.

(b) The ability to synthesize ATP is greatly reduced.

(c) Because the goal was to measure ATP hydrolysis. If succinate had been added in the presence of ATP, no reaction would have taken place, because of respiratory control.

(d) The mutation has little effect on the ability of the enzyme to catalyze the hydrolysis of ATP.

(e) They suggest two things: (1) the mutation did not affect the catalytic site on the enzyme, because ATP synthase is still capable of catalyzing the reverse reaction, and (2) the mutation did not affect the amount of enzyme present, given that the controls and patients had similar amounts of activity.

32. (a) The P : O ratio is equal to the product of $(H^+/2\ e^-)$ and (P/H^+). Note that the P : O ratio is identical with the P : 2 e^- ratio.

(b) 2.5 and 1.5, respectively

33. Cyanide can be lethal because it binds to the ferric form of cytochrome c oxidase and thereby inhibits oxidative phosphorylation. Nitrite converts ferrohemoglobin into ferrihemoglobin, which also binds cyanide. Thus, ferrihemoglobin competes with cytochrome c oxidase for cyanide. This competition is therapeutically effective because the amount of ferrihemoglobin that can be formed without impairing oxygen transport is much greater than the amount of cytochrome c oxidase.

34. The available free energy from the translocation of two, three, and four protons is -38.5, -57.7, and -77.4 kJ mol^{-1} (-9.2, -13.8, and -18.5 kcal mol^{-1}), respectively. The free energy consumed in synthesizing a mole of ATP under standard conditions is 30.5 kJ (7.3 kcal). Hence, the residual free energy of -8.1, -27.2, and -46.7 kJ mol^{-1} (-1.93, -6.5, and -11.2 kcal mol^{-1}) can drive the synthesis of ATP until the [ATP]/[ADP][P$_i$] ratio is 26.2, 6.5×10^4, and 1.6×10^8, respectively. Suspensions of isolated mitochondria synthesize ATP until this ratio is greater than 10^4, which shows that the number of protons translocated per ATP synthesized is at least three.

35. Add the inhibitor with and without an uncoupler, and monitor the rate of O$_2$ consumption. If the O$_2$ consumption increases again in the presence of inhibitor and uncoupler, the inhibitor must be inhibiting ATP synthase. If the uncoupler has no effect on the inhibition, the inhibitor is inhibiting the electron-transport chain.

Chapter 22

1. Ultimately, all of the carbon atoms of which we are made, not just carbohydrates, enter the biosphere through the process of photosynthesis. Moreover, the oxygen that we require is produced by photosynthesis.

2. $2\ NADP^+ + 3\ ADP^{3-} + 3\ P_i^{2-} + H^+ \rightarrow$
$$O_2 + 2\ NADPH + 3\ ATP^{4-} + H_2O$$

3. (a) 7; (b) 5; (c) 4; (d) 10; (e) 1; (f) 2; (g) 9; (h) 3; (i) 8; (j) 6

4. Photosystem II, in conjunction with the oxygen-generating complex, powers oxygen release. The reaction center of photosystem II absorbs light maximally at 680 nm.

5. Oxygen consumption will be maximal when photosystems I and II are operating cooperatively. Oxygen will be efficiently generated when electrons from photosystem II fill the electron holes in photosystem I, which were generated when the reaction center of photosystem I was illuminated by light of 700 nm.

6. The light reactions take place on thylakoid membranes. Increasing the membrane surface increases the number of ATP- and NADH-generating sites.

7. Photoinduced separation of charge results when a high-energy electron generated by light absorption moves to a neighboring acceptor molecule with a lower excited state. This step is in photosynthesis is fundamental because the now negatively charged acceptor molecule possesses a high-energy electron that can be used to generate a proton gradient or biosynthetic reducing power.

8. These complexes absorb more light than can a reaction center alone. The light-harvesting complexes funnel light to the reaction centers.

9. NADP$^+$ is the acceptor. H$_2$O is the donor. Light energy powers the electron flow.

10. The charge gradient, a component of the proton-motive force in mitochondria, is neutralized by the influx of Mg^{2+} into the lumen of the thylakoid membranes.

11. Chlorophyll is readily inserted into the hydrophobic interior of the thylakoid membranes.

12. Chlorophyll contains networks of alternating single and double bonds. Such networks allow electrons to resonate and thus are not

held tightly by a particular atom. This situation allows excitation of the electrons by light energy.

13. Protons released by the oxidation of water; protons pumped into the lumen by the cytochrome b_6f complex; protons removed from the stroma by the reduction of $NADP^+$ and plastoquinone

14. The electron flow from PS II to PS I is uphill, or endergonic. For this uphill flow, ATP would need to be consumed, defeating the purpose of photosynthesis.

15. $\Delta E'_0 = 10.11$ V, and $\Delta G^{\circ\prime} = -21.3$ kJ mol^{-1} (-5.1 kcal mol^{-1}).

16. (a) All ecosystems require an energy source from outside the system, because the chemical-energy sources will ultimately be limited. The photosynthetic conversion of sunlight is one example of such a process.

   (b) Not at all. Spock would point out that chemicals other than water can donate electrons and protons.

17. DCMU inhibits electron transfer in the link between photosystems II and I. O_2 can evolve in the presence of DCMU if an artificial electron acceptor such as ferricyanide can accept electrons from Q.

18. DCMU will have no effect because it blocks photosystem II, and cyclic photophosphorylation uses photosystem I and the cytochrome b_6f complex.

19. Exposure to light allowed the generation of a proton gradient, but the absence of ADP and P_i prevented the synthesis of ATP. When the chloroplasts were placed in a dark environment with ADP and P_i, ATP synthesis could take place until the gradient was depleted.

20. Oxygen generation would stop. The proton gradient generated by photosystem II would not be dissipated. Soon the gradient would become so great that the energy of electron flow in photosystem II would be incapable of pumping more protons. Oxygen generation (as well as photosystem I) would halt.

21. An uncoupler, such as dinitrophenol

22. The cristae

23. In eukaryotes, both processes take place in specialized organelles. Both depend on high-energy electrons to generate ATP. In oxidative phosphorylation, the high-energy electrons originate in fuels and are extracted as reducing power in the form of NADH. In photosynthesis, the high-energy electrons are generated by light and are captured as reducing power in the form of NADPH. Both processes use redox reactions to generate a proton gradient, and the enzymes that convert the proton gradient into ATP are similar in both processes. In both systems, electron transport takes place in membranes inside organelles.

24. Both enzymes transfer electrons from $FADH_2$ to a nicotinamide nucleotide: $NADP^+$ in the reductase and NAD^+ in the dehydrogenase complex. Such electron flow is uncommon. Usually, electrons flow from reduced nicotinamide nucleotide to FAD.

25. Both photosynthesis and cellular respiration are powered by high-energy electrons flowing toward a more stable state. In cellular respiration, the high-energy electrons are derived from the oxidation of carbon fuels as NADH and $FADH_2$. They release their energy as they reduce oxygen. In photosynthesis, high-energy electrons are generated by absorbing light energy, and they find stability in photosystem 1 and ferredoxin.

26. (a) Thioredoxin

   (b) The control enzyme is unaffected, but the mitochondrial enzyme with part of the chloroplast γ subunit increases activity as the concentration of DTT increases.

   (c) The increase was even larger when thioredoxin was present. Thioredoxin is the natural reductant for the chloroplast enzyme, and so it presumably operates more efficiently than would DTT, which probably functions to keep the thioredoxin reduced.

   (d) They seem to have done so.

   (e) The enzyme is susceptible to control by the redox state. In plant cells, reduced thioredoxin is generated by photosystem I. Thus, the enzyme is active when photosynthesis is taking place.

   (f) Cysteine

   (g) Group-specific modification or site-specific mutagenesis

27. The absorption of light by photosystem I results in a $\Delta E'_0$ of -1.0 V. Recall that $\Delta G'_0 = -nF\Delta E'_0$, where $F = 96.48$ kJ mol^{-1} V^{-1}. Under standard conditions, the energy change for the electrons is 96.5 kJ. Thus, the efficiency is 96.5/172 = 56%.

28. (a) 120 kJ einstein^{-1} (28.7 kcal einstein^{-1})

   (b) 1.24 V

   (c) One 1000-nm photon has the free-energy content of 2.4 molecules of ATP. A minimum of 0.42 photon is needed to drive the synthesis of a molecule of ATP.

29. The electrons flow through photosystem II directly to ferricyanide. No other steps are required.

30. NADPH must be factored in because it is an energy-rich molecule. Recall from Chapter 21, that NADH is worth 2.5 ATP if oxidized by the electron-transport chain. Thus, 12 NADPH = 30 ATP. Eighteen ATP are used directly, so the equivalent of 48 molecules of ATP are required for the synthesis of glucose.

Chapter 23

1. The Calvin cycle is the primary means of converting gaseous CO_2 into organic matter—that is, biomolecules. Essentially, every carbon atom in your body passed through rubisco and the Calvin cycle at some time in the past.

2. Autotrophs can use the energy of sunlight, carbon dioxide, and water to synthesize carbohydrates, which can subsequently be used for catabolic or anabolic purposes. Heterotrophs require chemical fuels and are thus ultimately dependent on autotrophs.

3. Nothing grim or secret about these reactions. They are sometimes called the dark reactions because they do not directly depend on light.

4. Stage 1 is the fixation of CO_2 with ribulose 1,5-bisphosphate and the subsequent formation of 3-phosphoglycerate. Stage 2 is the conversion of some of the 3-phosphoglycerate into hexose. Stage 3 is the regeneration of ribulose 1,5-bisphosphate.

5. (a) 5; (b) 1; (c) 7; (d) 2; (e) 10; (f) 3; (g) 6; (h) 4; (i) 8; (j) 9

6. Rubisco catalyzes a crucial reaction, but it is very slow. Consequently, it is required in large amounts to overcome its slow catalysis.

7. Because the carbamate forms only in the presence of CO_2, this property would prevent rubisco from catalyzing the oxygenase (or the carboxylase) reaction when CO_2 is absent.

8. ATP is required to form phosphoenolpyruvate (PEP) from pyruvate. PEP combines with CO_2 to form oxaloacetate and, subsequently, malate. Two ATP molecules are required because a second ATP molecule is required to phosphorylate AMP to ADP.

9. Because NADPH is generated in the chloroplasts by the light reactions

10. The stroma will accumulate Mg^{2+} and become alkaline as a consequence of the movement of protons from the stroma to the thylakoid space. Thus, rubisco is primed for activity when the light reactions are providing ATP and NADPH required for carbon fixation and glucose synthesis.

11. The light reactions lead to an increase in the stromal concentrations of NADPH, reduced ferredoxin, and Mg^{2+}, as well as an increase in pH.

12. ATP is converted into AMP. To convert this AMP back into ATP, two molecules of ATP are required: one to form ADP and another to form ATP from the ADP.

13. The oxygenase activity of rubisco increases with temperature. Crabgrass is a C_4 plant, whereas most grasses lack this capability. Consequently, the crabgrass will thrive at the hottest part of the summer because the C_4 pathway provides an ample supply of CO_2.

14. Photorespiration is the consumption of oxygen by plants with the production of CO_2, but it does not generate energy. Photorespiration is due to the oxygenase activity of rubisco. It is wasteful because, instead of fixing CO_2 for conversion into hexoses, rubisco generates CO_2.

15. High concentrations of CO_2 prevent O_2 from entering the active site of rubisco.

16. As global warming progresses, C_4 plants will invade the higher latitudes, whereas C_3 plants will retreat to cooler regions.

17. C_4 metabolism allows rubisco to function efficiently even when the temperatures are high, which favor oxygenase activity. Moreover, C_4 metabolism allows desert plants to accumulate CO_2 at night when the temperatures are cooler and water evaporation is not a problem.

18. (a) $k_{cat}^{CO_2}/K_M^{CO_2} = 3 \times 10^6 \, s^{-1} \, M^{-1}$

$k_{cat}^{O_2}/K_M^{O_2} = 4 \times 10^3 \, s^{-1} \, M^{-1}$

(b) Despite the fact that the specificity constant for CO_2 as a substrate is much greater than that of O_2, the concentration of O_2 in the atmosphere is higher than that of CO_2, allowing the oxygenation reaction to occur.

19.

Calvin cycle	Krebs cycle
Stroma	Matrix
Carbon chemistry for photosynthesis	Carbon chemistry for oxidative phosphorylation
Fixes CO_2	Releases CO_2
Requires high-energy electrons (NADPH)	Generates high-energy electrons (NADH)
Regenerates starting compound (ribulose 1,5-bisphosphate)	Regenerates starting compound (oxaloacetate)
Requires ATP	Generates ATP
Complex stoichiometry	Simple stoichiometry

20. The reduction of each mole of CO_2 to the level of a hexose requires 2 mol of NADPH. The reduction of $NADP^+$ is a two-electron process. Hence, the formation of 2 mol of NADPH requires the pumping of 4 mol of electrons by photosystem I. The electrons given up by photosystem I are replenished by photosystem II, which needs to absorb an equal number of photons. Hence, eight photons are needed to generate the required NADPH. The energy input of 8 mol of photons is 1594 kJ (381 kcal). Thus, the overall efficiency of photosynthesis under standard conditions is at least 477/1594, or 30%.

21. (a) The blue curve on the right in graph A was generated by the C_4 plant. Recall that the oxygenase activity of rubisco increases with temperature more rapidly than does the carboxylase activity. Consequently, at higher temperatures, the C_3 plants fix less CO_2. Because C_4 plants can maintain a higher CO_2 concentration, the rise in temperature is less deleterious.

(b) The oxygenase activity predominates. Additionally, when the temperature rise is very high, the evaporation of water might become a problem. The higher temperatures can begin to damage protein structures as well.

(c) The C_4 pathway is a very effective active-transport system for concentrating CO_2, even when environmental concentrations are very low.

(d) With the assumption that the plants have approximately the same capability to fix CO_2, the C_4 pathway is apparently the rate-limiting step in C_4 plants.

22. (a) 3-Phosphoglycerate; (b) the other members of the Calvin cycle

23. The concentration of 3-phosphoglycerate will increase, whereas that of ribulose 1,5-bisphosphate will decrease.

24. The concentration of 3-phosphoglycerate will decrease, whereas that of ribulose 1,5-bisphosphate will increase.

25. (a) CABP resembles the addition compound formed in the reaction of CO_2 and ribulose 1,5-bisphosphate. (b) CABP will be a potent inhibitor of rubisco.

Chapter 24

1. Step 1 is the release of glucose 1-phosphate from glycogen by glycogen phosphorylase. Step 2 is the formation of glucose 6-phosphate from glucose 1-phosphate, a reaction catalyzed by phosphoglucomutase. Step 3 is the remodeling of the glycogen by the transferase and the glucosidase.

2. (a) 8; (b) 3; (c) 6; (d) 5; (e) 9; (f) 2; (g) 10; (h) 1; (i) 4; (j) 7

3. Because muscle maintains glucose for its own use, whereas the liver maintains glucose homeostasis for the whole organism

4. The active site is partly blocked in the T state.

5. Phosphorylase kinase is maximally active when calcium is bound, and it is subsequently phosphorylated.

6. (a) The different manifestations correspond to the different roles of the liver and muscle. Liver glycogen phosphorylase plays a crucial role in the maintenance of blood-glucose concentration. Recall that glucose is the primary fuel for the brain. Muscle glycogen phosphorylase provides glucose only for the muscle and, even then, only when the energy needs of the muscle are high, as during exercise.

(b) The fact that there are two different diseases suggests that there are two different isozymic forms of the glycogen phosphorylase—a liver-specific isozyme and a muscle-specific isozyme.

7. In muscle, the b form of phosphorylase is activated by AMP. In the liver, the a form is inhibited by glucose. The difference corresponds to the difference in the metabolic role of glycogen in each tissue. Muscle uses glycogen as a fuel for contraction, whereas the liver uses glycogen to maintain blood-glucose concentration.

8. Although glucose 1-phosphate is the actual product of the phosphorylase reaction, glucose 6-phosphate, generated from glucose 1-phosphate by phosphoglucomutase, is a more versatile molecule with respect to metabolism. Among other fates, glucose 6-phosphate can be processed to yield energy or building blocks. In the liver, glucose 6-phosphate can be converted into glucose and released into the blood.

9. Glycogen phosphorylase b in the T state. AMP acts as an allosteric activator to stabilize the active R state of glycogen phosphorylase b.

10. Two signals account for the activation of muscle phosphorylase. First, the calcium released during muscle contraction activates the phosphorylase kinase, and makes it a substrate for protein kinase A. Second, epinephrine binds to its G-protein-coupled receptor. The resulting structural changes activate a $G_{\alpha s}$ protein, which in turn activates adenylate cyclase. Adenylate cyclase synthesizes cAMP, which activates protein kinase A. Protein kinase A phosphorylates phosphoryl kinase, completing the activation phosphorylase kinase. Phosphorylase kinase phosphorylates and activates glycogen phosphorylase.

11. In the liver, glucagon stimulates the cAMP-dependent pathway that activates protein kinase A. Epinephrine binds to a

7TM α-adrenergic receptor in the liver plasma membrane, which activates phospholipase C and the phosphoinositide cascade. This activation causes calcium ions to be released from the endoplasmic reticulum, which bind to calmodulin, and further stimulates phosphorylase kinase and glycogen breakdown.

12. First, the signal-transduction pathway is shut down when the initiating hormone is no longer present. Second, the inherent GTPase activity of the G protein converts the bound GTP into inactive GDP. Third, phosphodiesterases convert cAMP into AMP. Fourth, the enzyme protein phosphatase 1 (PP1) removes the phosphoryl groups from phosphorylase kinase and glycogen phosphorylase, converting the enzymes into their inactive forms.

13. Glycogen is an important fuel reserve for several reasons. The controlled breakdown of glycogen and the release of glucose increase the amount of glucose that is available between meals. Hence, glycogen serves as a buffer to maintain blood-glucose concentration. This role of glycogen is especially important because glucose is virtually the only fuel used by the brain, except during prolonged starvation. Moreover, the glucose from glycogen is readily mobilized and is therefore a good source of energy for sudden, strenuous activity. Unlike fatty acids, the released glucose can provide energy in the absence of oxygen and can thus supply energy for anaerobic activity.

14. All these symptoms suggest central nervous system problems. If exercise is exhaustive enough or the athlete has not prepared well enough or both, liver glycogen also can be depleted. The brain depends on glucose derived from liver glycogen. The symptoms suggest that the brain is not getting enough fuel.

15. Glucose 1-arsenate would be formed by the phosphorylase and would spontaneously hydrolyze to glucose and arsenate. Glucose liberated by phosphorylase in the presence of arsenate would have to be phosphorylated by hexokinase at the expense of a molecule of ATP.

16. Phosphorylase, transferase, glucosidase, phosphoglucomutase, and glucose 6-phosphatase

17. Liver phosphorylase a is inhibited by glucose, which facilitates the R → T transition. Muscle phosphorylase is insensitive to glucose.

18. As an unbranched polymer, amylose has only one nonreducing end. Therefore, only one glycogen phosphorylase molecule could degrade each amylose molecule. Because glycogen is highly branched, there are many nonreducing ends per molecule. Consequently, many phosphorylase molecules can release many glucose molecules per glycogen molecule.

19. (a) B and D; (b) phosphorylase kinase; (c) C to D; (d) B to A; (e) D to C; (f) protein phosphatase 1

20. The substrate can be handed directly from the transferase site to the debranching site.

21. During exercise, [ATP] falls and [AMP] rises. Recall that AMP is an allosteric activator of glycogen phosphorylase b. Thus, even in the absence of covalent modification by phosphorylase kinase, glycogen is degraded.

22. Glucose is an allosteric inhibitor of phosphorylase a. Hence, crystals grown in its presence are in the T state. The addition of glucose 1-phosphate, a substrate, shifts the R-to-T equilibrium toward the R state. The conformational differences between these states are sufficiently large that the crystal shatters unless it is stabilized by chemical cross-links.

23. Gluconeogenesis

24. Free glucose must be phosphorylated at the expense of a molecule of ATP. Glucose 6-phosphate derived from glycogen is formed by phosphorolytic cleavage, sparing one molecule of ATP. Thus, the net yield of ATP when glycogen-derived glucose is

processed to pyruvate is three molecules of ATP compared with two molecules of ATP from free glucose.

25. Cells maintain the $[P_i]/[\text{glucose 1-phosphate}]$ ratio at greater than 100, substantially favoring phosphorolysis. We see here an example of how the cell can alter the free-energy change to favor a reaction taking place by altering the ratio of substrate and product.

26. Water is excluded from the active site of phosphorylase to prevent hydrolysis. The entry of water could lead to the formation of glucose rather than glucose 1-phosphate. A site-specific mutagenesis experiment is revealing in this regard. In wild-type glycogen phosphorylase, Tyr 573 is hydrogen bonded to the 2'-OH group of a glucose residue. The ratio of glucose 1-phosphate to glucose product is 9000:1 for the wild-type enzyme, and 500:1 for the Phe 573 mutant. Model building suggests that a water molecule occupies the site normally filled by the OH group of tyrosine and occasionally attacks the oxocarbonium ion intermediate to form glucose.

27. As we examined previously (p. 323) and will see again in Chapter 25, one of the consequences of insulin resistance is failure to appropriately inhibit gluconeogenesis and glycogen breakdown. Consequently, the inhibition of liver glycogen phosphorylase would help to ameliorate the high blood-glucose concentration. The danger is that muscle phosphorylase, which does not contribute glucose to the blood, would also be inhibited.

28. (a) Apparently, the glutamate, with its negatively charged R group, can mimic to some extent the presence of a phosphoryl group on serine. That the stimulation is not as great is not surprising in that the carboxyl group is smaller and not as charged as the phosphate.

(b) Substitution of aspartate would give some stimulation, but being that it is smaller than the glutamate, the simulation would be smaller.

Chapter 25

1. (a) 4; (b) 1; (c) 5; (d) 10; (e) 7; (f) 2; (g) 8; (h) 9; (i) 6; (j) 3
2. Phosphoglucomutase, UDP-glucose pyrophosphorylase, pyrophosphatase, glycogenin, glycogen synthase, and branching enzyme
3.

$$\text{Glucose 6-phosphate} \longrightarrow \text{glucose 1-phosphate} \qquad (1)$$
$$\text{Glucose 1-phosphate} + \text{UTP} \longrightarrow \text{UDP-glucose} + \text{PP}_i \qquad (2)$$
$$\text{PP}_i + \text{H}_2\text{O} \longrightarrow 2\,\text{P}_i \qquad (3)$$
$$\text{UDP-glucose} + \text{glycogen}_n \longrightarrow \text{glycogen}_{n+1} + \text{UDP} \qquad (4)$$
$$\text{UDP} + \text{ATP} \longrightarrow \text{UTP} + \text{ADP} \qquad (5)$$

Sum: Glucose 6-phosphate + ATP + glycogen$_n$ + H$_2$O $\longrightarrow$
$$\text{glycogen}_{n+1} + \text{ADP} + 2\,\text{P}_i$$

4. The enzyme pyrophosphatase converts the pyrophosphate into two molecules of inorganic phosphate. This conversion renders the overall reaction irreversible.

$$\text{Glucose 1-phosphate} + \text{UTP} \rightleftharpoons \text{UDP-glucose} + \text{PP}_i$$
$$\underline{\text{PP}_i + \text{H}_2\text{O} \longrightarrow 2\,\text{P}_i}$$
$$\text{Glucose 1-phosphate} + \text{UTP} \longrightarrow \text{UDP-glucose} + 2\,\text{P}_i$$

5. The presence of high concentrations of glucose 6-phosphate indicates that glucose is abundant and that it is not being used by glycolysis. Therefore, this valuable resource is saved by incorporation into glycogen.

6. Glycogenin, a dimer, catalyzes the addition of 10–20 glucosyl units linked by α-1,4 bonds to each glycogenin subunit. The modified glycogenin then serves as a primer for glycogen synthase, which extends the glycogen chain with the formation of more α-1,4 linkages.

7. Free glucose must be phosphorylated at the expense of a molecule of ATP. Glucose 6-phosphate derived from glycogen is formed by phosphorolytic cleavage, sparing one molecule of ATP. Thus, the net yield of ATP when glycogen-derived glucose is processed to pyruvate is three molecules of ATP compared with two molecules of ATP from free glucose.

8. Breakdown: Phosphoglucomutase converts glucose 1-phosphate, liberated from glycogen breakdown, into glucose 6-phosphate, which can be released as free glucose (liver) or processed in glycolysis (muscle and liver). Synthesis: Converts glucose 6-phosphate into glucose 1-phosphate, which reacts with UTP to form UDP-glucose, the substrate for glycogen synthase.

9.
$$\text{Glycogen}_n + P_i \longrightarrow \text{glycogen}_{n-1} + \text{glucose 6-phosphate}$$
$$\text{Glucose 6-phosphate} \rightleftharpoons \text{glucose 1-phosphate}$$
$$\text{UTP} + \text{glucose 1-phosphate} \longrightarrow \text{UDP-glucose} + 2\,P_i$$
$$\text{Glycogen}_{n-1} + \text{UDP-glucose} \longrightarrow \text{glycogen}_n + \text{UDP}$$
$$\text{Sum: Glycogen}_n + \text{UTP} \longrightarrow \text{glycogen}_n + \text{UDP} + P_i$$

10. In principle, having glycogen be the only primer for the further synthesis of glycogen should be a successful strategy. However, if the glycogen granules were not evenly divided between daughter cells, glycogen stores for future generations of cells might be compromised. Glycogenin synthesizes the primer for glycogen synthase.

11. Insulin binds to its receptor and activates the tyrosine kinase activity of the receptor, which in turn triggers a pathway that activates protein kinases. The signal-transduction pathway results in an increase in the number of glucose transporters in the membrane. Moreover, the kinases phosphorylate and inactivate glycogen synthase kinase. Protein phosphatase 1 then removes the phosphate from glycogen synthase and thereby activates the synthase.

12. The high concentration of glucose 6-phosphate in von Gierke disease, resulting from the absence of glucose 6-phosphatase or the transporter, shifts the allosteric equilibrium of phosphorylated glycogen synthase toward the active form.

13. (a) Muscle phosphorylase *b* will be inactive even when the concentration of AMP is high. Hence, glycogen will not be degraded unless phosphorylase is converted into the *a* form by hormone-induced or Ca^{2+}-induced phosphorylation.

(b) Phosphorylase *b* cannot be converted into the much more active *a* form. Hence, the mobilization of liver glycogen will be markedly impaired.

(c) The elevated amount of the kinase will lead to the phosphorylation and activation of glycogen phosphorylase. Because glycogen will be persistently degraded, little glycogen will be present in the liver.

(d) Protein phosphatase 1 will be continually active. Hence, the amount of phosphorylase *b* will be higher than normal, and glycogen will be less readily degraded.

(e) Protein phosphatase 1 will be much less effective in dephosphorylating glycogen synthase and glycogen phosphorylase. Consequently, the synthase will stay in the less active *b* form, and the phosphorylase will stay in the more active *a* form. Both changes will lead to increased degradation of glycogen.

(f) The absence of glycogenin will prevent the initiation of glycogen synthesis. Very little glycogen will be synthesized in its absence.

14. (a) The α subunit will thus always be active. Cyclic AMP will always be produced. Glycogen will always be degraded, and glycogen synthesis will always be inhibited.

(b) Phosphodiesterase destroys cAMP. Therefore, glycogen degradation will always be active and glycogen synthesis will always be inhibited.

15. This disease can also be produced by a mutation in the gene that encodes the glucose 6-phosphate transporter. Recall that glucose 6-phosphate must be transported into the lumen of the endoplasmic reticulum to be hydrolyzed by phosphatase (p. 319). Mutations in the other essential proteins of this system can likewise lead to von Gierke disease.

16. Glucagon stimulates glycogen breakdown, and the product of debranching enzyme is free glucose, which is released into the blood ($\approx 10\%$ of available glucose in glycogen is contained in α-1,6 branch points).

17. Galactose is converted into UDP-galactose to eventually form glucose 6-phosphate.

18.
$$\text{Galactose} + \text{ATP} + \text{UTP} + H_2O + \text{glycogen}_n \longrightarrow$$
$$\text{glycogen}_{n+1} + \text{ADP} + \text{UDP} + 2\,P_i + H^+$$

19. The amylase activity was necessary for the removal of all of the glycogen from glycogenin. Recall that glycogenin synthesizes oligosaccharides of about 10–20 glucose units, and then activity stops. Consequently, if the glucose residues are not removed by extensive amylase treatment, glycogenin will not be detected.

20. The patient has a deficiency of the branching enzyme.

21. (a) Glycogen was too large to enter the gel, and, because analysis was by western blot with the use of an antibody specific to glycogenin, we would not expect to see background proteins.

(b) α-Amylase degrades glycogen, releasing the protein glycogenin, which can be visualized by the western blot.

(c) Glycogen phosphorylase, glycogen synthase, and protein phosphatase 1. These proteins might be visible if the gels were stained for protein, but a western analysis reveals the presence of glycogenin only.

22. (a) The smear was due to molecules of glycogenin with increasingly large amounts of glycogen attached to them.

(b) In the absence of glucose in the medium, glycogen is metabolized, resulting in a loss of the high-molecular-weight material.

(c) Glycogen could have been resynthesized and added to the glycogenin when the cells were fed glucose again.

(d) No difference between lanes 3 and 4 suggests that, by 1 hour, the glycogen molecules had attained maximum size in this cell line. Prolonged incubation does not apparently increase the amount of glycogen.

(e) α-Amylase removes essentially all of the glycogen, and so only the glycogenin remains.

Chapter 26

1. (a) C; (b) B and F; (c) G; (d) F; (e) E; (f) H; (g) I; (h) D; (i) A; (j) F; (k) B

2. The oxidative phase generates NADPH and is irreversible. The nonoxidative phase allows for the interconversion of phosphorylated sugars.

3. Glucose 6-phosphate dehydrogenase; $NADP^+$ concentration

4. When much NADPH is required. The oxidative phase of the pentose phosphate pathway is followed by the nonoxidative phase. The resulting fructose 6-phosphate and glyceraldehyde 3-phosphate are used to generate glucose 6-phosphate through gluconeogenesis, and the cycle is repeated until the equivalent of one glucose molecule is oxidized to CO_2.

5. Fava beans contain vicine, a purine glycoside that can lead to the generation of peroxides—reactive oxygen species that can damage membranes as well as other biomolecules. Glutathione is used to detoxify the ROS. The regeneration of glutathione depends on an adequate supply of NADPH, which is synthesized by the oxidative phase of the pentose phosphate pathway. People with low

levels of the dehydrogenase activity are especially susceptible to vicine toxicity.

6. The nonoxidative phase of the pentose phosphate pathway can be used to convert three molecules of ribose 5-phosphate into two molecules of fructose 6-phosphate and one molecule of glyceraldehyde 3-phosphate. These molecules are components of the glycolytic pathway.

7. The conversion of fructose 6-phosphate into fructose 1,6-bisphosphate by phosphofructokinase requires ATP.

8. The label emerges at C-5 of ribulose 5-phosphate.

9. Because red blood cells do not have mitochondria and the only means to obtain NADPH is through the pentose phosphate pathway. There are biochemical means to convert mitochondrial NADH into cytoplasmic NADPH.

10. The Calvin cycle begins with the fixation of CO_2 and proceeds to use NADPH in the synthesis of glucose. The pentose phosphate pathway begins with the oxidation of a glucose-derived carbon atom to CO_2 and concomitantly generates NADPH. The regeneration phase of the Calvin cycle converts C_6 and C_3 molecules back into the starting material—the C_5 molecule ribulose 1,5-bisphosphate. The pentose phosphate pathway converts a C_5 molecule, ribose 5-phosphate, into C_6 and C_3 intermediates of the glycolytic pathway.

11. The oxidative decarboxylation of isocitrate to α-ketoglutarate

12. Lacking mitochondria, red blood cells metabolize glucose to lactate to obtain energy in the form of ATP. The CO_2 results from extensive use of the pentose phosphate pathway coupled with gluconeogenesis. This coupling allows the generation of much NADPH with the complete oxidation of glucose by the oxidative branch of the pentose phosphate pathway.

13. C-1 and C-3 of fructose 6-phosphate are labeled, whereas erythrose 4-phosphate is not labeled.

14. (a)   5 Glucose 6-phosphate + ATP →

6 ribose 5-phosphate + ADP + H^+

(b)   Glucose 6-phosphate + 12 $NADP^+$ + 7 H_2O →

6 CO_2 + 12 NADPH + 12 H^+ + P_i

15. $\Delta E_0'$ for the reduction of glutathione by NADPH is $+0.09$ V. Hence, $\Delta G^{\circ\prime}$ is -17.4 kJ mol^{-1} (-4.2 kcal mol^{-1}), which corresponds to an equilibrium constant of 1126. The required [NADPH]/[$NADP^+$] ratio is 8.9×10^{-5}.

16. Incubate an aliquot of a tissue homogenate with glucose labeled with ^{14}C at C-1, and incubate another with glucose labeled with ^{14}C at C-6. Compare the radioactivity of the CO_2 produced by the two samples. The rationale of this experiment is that only C-1 is decarboxylated by the pentose phosphate pathway, whereas C-1 and C-6 are decarboxylated equally when glucose is metabolized by the glycolytic pathway, the pyruvate dehydrogenase complex, and the citric acid cycle. The reason for the equivalence of C-1 and C-6 in the latter set of reactions is that glyceraldehyde 3-phosphate and dihydroxyacetone phosphate are rapidly interconverted by triose phosphate isomerase.

Chapter 27

1. In stage 1, triacylglycerols are degraded to fatty acids and glycerol, which are released from the adipose tissue and transported to the energy-requiring tissues. In stage 2, the fatty acids are activated and transported into mitochondria for degradation. In stage 3, the fatty acids are broken down in a step-by-step fashion into acetyl CoA, which is then processed in the citric acid cycle.

2. Glucagon and epinephrine trigger 7TM receptors in adipose tissue that activate adenylate cyclase (p. 228). The increased level of cyclic AMP then stimulates protein kinase A, which phosphorylates two key proteins: perilipin, a fat-droplet-associated protein,

and hormone-sensitive lipase. The phosphorylation of perilipin restructures the fat droplet so that the triacylglycerols are more readily mobilized, and it triggers the release of a coactivator for the adipose triglyceride lipase (ATGL). Activated ATGL then initiates the mobilization of triacylglycerols by releasing a fatty acid from triacylglycerol, forming diacylglycerol. Diacylglycerol is converted into a free fatty acid and monoacylglycerol by the hormone-sensitive lipase. Monoacylglycerol lipase completes the mobilization of fatty acids with the production of a free fatty acid and glycerol.

3. The ready reversibility is due to the high-energy nature of the thioester in the acyl CoA.

4. $RCOO^- + CoA + ATP + H_2O \longrightarrow$

$RCO\text{-}CoA + AMP + 2 P_i$

5. To return the AMP to a form that can be phosphorylated by oxidative phosphorylation or substrate-level phosphorylation, another molecule of ATP must be expended in the reaction:

$$ATP + AMP \rightleftharpoons 2 ADP$$

6. Oxidation by flavin adenine dinucleotide (FAD), hydration, oxidation by nicotinamide adenine dinucleotide (NAD^+), and thiolysis by coenzyme A.

7. (a) 5; (b) 11; (c) 1; (d) 10; (e) 2; (f) 6; (g) 9; (h) 3; (i) 4; (j) 7; (k) 8

8. b, c, a, g, h, d, e, f

9. Fatty acids cannot be transported into mitochondria for oxidation. The muscles could not use fats as a fuel. Muscles could use glucose derived from glycogen. However, when glycogen stores are depleted, as after a fast, the effect of the deficiency is especially apparent.

10. The next-to-last degradation product, acetoacetyl CoA, yields two molecules of acetyl CoA with the thiolysis by only one molecule of CoA.

11. Palmitic acid yields 106 molecules of ATP. Palmitoleic acid has a double bond between carbons C-9 and C-10. When palmitoleic acid is processed in β oxidation, one of the oxidation steps (to introduce a double bond before the addition of water) will not take place, because a double bond already exists. Thus, $FADH_2$ will not be generated, and palmitoleic acid will yield 1.5 fewer molecules of ATP than palmitic acid, for a total of 104.5 molecules of ATP.

12.

Activation fee to form the acyl CoA	−2	ATP
Seven rounds of yield:		
7 acetyl CoA at 10 ATP/acetyl CoA	+70	ATP
7 NADH at 2.5 ATP/NADH	+17.5	ATP
7 $FADH_2$ at 1.5 ATP/$FADH_2$	+10.5	ATP
Propionyl CoA, which requires an ATP to be converted into succinyl CoA	−1	ATP
Succinyl CoA → succinate	+1	ATP
Succinate → fumarate + $FADH_2$ $FADH_2$ at 1.5 ATP/$FADH_2$	+1.5	ATP
Fumarate → malate		
Malate → oxaloacetate + NADH NADH at 2.5 ATP/NADH	+2.5	ATP
Total	100	ATP

13. You might hate yourself in the morning, but at least you won't have to worry about energy. To form stearoyl CoA requires the equivalent of 2 molecules of ATP.

$$\text{Stearoyl CoA} + 8\,\text{FAD} + 8\,\text{NAD}^+ + 8\,\text{CoA} + 8\,\text{H}_2\text{O} \longrightarrow$$
$$9\,\text{acetyl CoA} + 8\,\text{FADH}_2 + 8\,\text{NADH} + 8\text{H}^+$$

9 acetyl CoA at 10 ATP/acetyl CoA	+90	ATP
8 NADH at 2.5 ATP/NADH	+20	ATP
8 FADH$_2$ at 1.5 ATP/FADH$_2$	+12	ATP
Activation fee	−2.0	
Total	122	ATP

14. After a night's sleep, glycogen stores will be low, but fats will be plentiful. Muscles will burn fat as a fuel. Why the caffeine? Lipid mobilization is stimulated by glucagon and epinephrine, and both of these hormones can work through the cAMP cascade. cAMP stimulates protein kinase A, which stimulates the breakdown of triacylglycerols in the adipose tissue. If cAMP hydrolysis to AMP is inhibited, protein kinase A will be maximally stimulated, fat will be maximally mobilized, and you will look as buff and lean as a svelte waterfowl.

15. Keep in mind that, in the citric acid cycle, 1 molecule of FADH$_2$ yields 1.5 molecules of ATP, 1 molecule of NADH yields 2.5 molecules of ATP, and 1 molecule of acetyl CoA yields 10 molecules of ATP. Two molecules of ATP are produced when glucose is degraded to 2 molecules of pyruvate. Two molecules of NADH also are produced, but the electrons are transferred to FADH$_2$ to enter mitochondria. Each molecule of FADH$_2$ can generate 1.5 molecules of ATP. Each molecule of pyruvate will produce 1 molecule of NADH. Each molecule of acetyl CoA generates 3 molecules of NADH, 1 molecule of FADH$_2$, and 1 molecule of ATP. So, we have a total of 10 molecules of ATP per molecule of acetyl CoA, or 20 for the 2 molecules of acetyl CoA. The total for glucose is 30 ATP. Now, what about hexanoic acid? Caproic acid is activated to caproic CoA at the expense of 2 ATP, and so we are 2 ATP in the hole. The first cycle of β oxidation generates 1 FADH$_2$, 1 NADH, and 1 acetyl CoA. After the acetyl CoA has been run through the citric acid cycle, this step will have generated a total of 14 ATP. The second cycle of β oxidation generates 1 FADH$_2$ and 1 NADH but 2 acetyl CoA. After the acetyl CoA has been run through the citric acid cycle, this step will have generated a total of 24 ATP. The total is 36 ATP. Thus, the foul-smelling caproic acid has a net yield of 36 ATP. So on a per carbon basis, this fat yields 20% more ATP than does glucose, a manifestation of the fact that fats are more reduced than carbohydrates.

16. $\text{Stearate} + \text{ATP} + 13.5\,\text{H}_2\text{O} + 8\,\text{FAD} + 8\,\text{NAD}^+ \rightarrow$
$4.5\,\text{acetoacetate} + 14.5\,\text{H}^+ + 8\,\text{FADH}_2 + 8\,\text{NADH} + \text{AMP} + 2\,\text{P}_\text{i}$

17. Palmitate is activated and then processed by β oxidation according to the following reactions:

$$\text{Palmitate} + \text{CoA} + \text{ATP} \longrightarrow \text{palmitoyl CoA} + \text{AMP} + 2\,\text{P}_\text{i}$$
$$\text{Palmitoyl CoA} + 7\,\text{FAD} + 7\,\text{NAD}^+ + 7\,\text{CoASH} + \text{H}_2\text{O} \rightarrow$$
$$8\,\text{acetyl CoA} + 7\,\text{FADH}_2 + 7\,\text{NADH} + 7\,\text{H}^+$$

The 8 molecules of acetyl CoA combine to form 4 molecules of acetoacetate for release into the blood, and so they do not contribute to the energy yield in the liver. However, the FADH$_2$ and NADH generated in the preparation of acetyl CoA can be processed by oxidative phosphorylation to yield ATP.

$$1.5\,\text{ATP/FADH}_2 \times 7 = 10.5\,\text{ATP}$$
$$2.5\,\text{ATP/NADH} \times 7 = 17.5\,\text{ATP}$$

The equivalent of 2 ATP were used to form palmitoyl CoA. Thus, 26 ATP were generated for use by the liver.

18. NADH produced with the oxidation to acetoacetate = 2.5 ATP Acetoacetate is converted into acetoacetyl CoA.

Two molecules of acetyl CoA result from the hydrolysis of acetoacetyl CoA, each worth 10 ATP when processed by the citric acid cycle. Total ATP yield is 22.5.

19. Because a molecule of succinyl CoA is used to form acetoacetyl CoA. Succinyl CoA could be used to generate one molecule of ATP, and so someone could argue that the yield is 21.5.

20. For fats to be combusted, not only must they be converted into acetyl CoA, but the acetyl CoA must be processed by the citric acid cycle. In order for acetyl CoA to enter the citric acid cycle, there must be a supply of oxaloacetate. Oxaloacetate can be formed by the metabolism of glucose to pyruvate and the subsequent carboxylation of pyruvate to form oxaloacetate.

21. The absence of ketone bodies is due to the fact that the liver, the source of ketone bodies in the blood, cannot oxidize fatty acids to produce acetyl CoA. Moreover, because of the impaired fatty acid oxidation, the liver becomes more dependent on glucose as an energy source. This dependency results in a decrease in gluconeogenesis and a drop in blood-glucose levels, which is exacerbated by the lack of fatty acid oxidation in muscle and a subsequent increase in glucose uptake from the blood.

22. Liver cells lack the specific CoA transferase that converts acetoacetate into acetoacetyl CoA, which is subsequently cleaved into two molecules of acetyl CoA. The lack of the enzyme allows liver to produce ketone bodies but not to use them.

23. In the absence of insulin, lipid mobilization will take place to an extent that it overwhelms the ability of the liver to convert the lipids into ketone bodies.

24. Two carbon atoms enter the cycle as an acetyl group, but two carbons leave the cycle as CO_2 before oxaloacetate is generated. Consequently, no net synthesis of oxaloacetate is possible. In contrast, plants have two additional enzymes enabling them to convert the carbon atoms of acetyl CoA into oxaloacetate in the glyoxylate cycle (p. 355).

25. The carbon skeletons of the released amino acids are used to synthesize glucose for use by the brain and red blood cells.

26. Muscle shifts from glucose to fatty acids for fuel. This switch lessens the need to degrade protein for glucose formation. The degradation of fatty acids by muscle halts the conversion of pyruvate into acetyl CoA, because acetyl CoA derived from fatty acids inhibits pyruvate dehydrogenase, the enzyme that converts pyruvate into acetyl CoA.

27. $\text{Glycerol} + 2\,\text{NAD}^+ + \text{P}_\text{i} + \text{ADP} \rightarrow$
$\text{pyruvate} + \text{ATP} + \text{H}_2\text{O} + 2\,\text{NADH} + \text{H}^+$
Glycerol kinase and glycerol phosphate dehydrogenase

28. The citric acid cycle. The reactions that take succinate to oxaloacetate, or the reverse, are similar to those of fatty acid metabolism (pp. 493 and 510).

29. (a) Fats burn in the flame of carbohydrates. Without carbohydrates, there would be no anapleurotic reactions to replenish the components of the citric acid cycle. With a diet of fats only, the acetyl CoA from fatty acid degradation would build up.

(b) Acetone from ketone bodies

(c) Yes. Odd-chain fatty acids would lead to the production of propionyl CoA, which can be converted into succinyl CoA, a citric acid cycle component. It would serve to replenish the citric acid cycle and mitigate the halitosis.

30. (a) The V_max is decreased, and the K_M is increased. V_max (wild type) = 13 nmol minute^{-1} mg^{-1}; K_M (wild type) = 45 μM; V_max (mutant) = 8.3 nmol minute^{-1} mg^{-1}; K_M (mutant) = 74 μM.

(b) Both the V_max and the K_M are decreased. V_max (wild type) = 41 nmol minute^{-1} mg^{-1}; K_M (wild type) = 104 μM; V_max (mutant) = 23 nmol minute^{-1} mg^{-1}; K_M (mutant) = 69 μM.

(c) The wild type is significantly more sensitive to malonyl CoA.

(d) With respect to carnitine, the mutant displays approximately 65% of the activity of the wild type; with respect to palmitoyl CoA, approximately 50% activity. On the other hand, 10 μM of malonyl CoA inhibits approximately 80% of the wild type but has essentially no effect on the mutant enzyme.

(e) The glutamate appears to play a more prominent role in regulation by malonyl CoA than in catalysis.

31. (a)

$$H_3C-\overset{\overset{\displaystyle CH_3}{|}}{CH}-(CH_2)_3-\overset{\overset{\displaystyle CH_3}{|}}{CH}-(CH_2)_3-\overset{\overset{\displaystyle CH_3}{|}}{CH}-(CH_2)_3-\overset{\overset{\displaystyle CH_3}{|}}{CH}-CH_2-COO^-$$

Phytanic acid

The problem with phytanic acid is that, as it undergoes β oxidation, we encounter the dreaded pentavalent carbon atom. Because the pentavalent carbon atom doesn't exist, β oxidation cannot take place and phytanic acid accumulates.

$$R-\overset{\overset{\displaystyle CH_3}{|}}{\underset{\underset{\displaystyle H}{|}}{C}}-\overset{}{\underset{\underset{\displaystyle H_2}{|}}{C}}-COO^- \longrightarrow R-\overset{\overset{\displaystyle CH_3}{|}}{C}=C-COO^- \longrightarrow$$

$$R-\overset{\overset{\displaystyle CH_3}{|}}{C}=C-COO^- \longrightarrow R-\overset{\overset{\displaystyle CH_3}{|}}{\underset{\underset{\displaystyle OH}{|}}{C}}-\overset{}{\underset{\underset{\displaystyle H_2}{|}}{C}}-COO^- \longrightarrow$$

$$R-\overset{\overset{\displaystyle CH_3}{|}}{\underset{\underset{\displaystyle O}{\|}}{C}}-\overset{}{\underset{\underset{\displaystyle H_2}{|}}{C}}-COO^-$$

The dreaded pentavalent carbon atom

(b) How do we solve the problem? Well, the removal of methyl groups, though theoretically possible, would be time consuming and, well, lacking in elegance. What would we do with the methyl groups? We solve this problem—well, actually our livers solve the problem—by inventing α oxidation.

$$R-\overset{\overset{\displaystyle CH_3}{|}}{\underset{\underset{\displaystyle H}{|}}{C}}-\overset{}{\underset{\underset{\displaystyle H_2}{|}}{C}}-COO^- \longrightarrow R-\overset{\overset{\displaystyle CH_3}{|}}{\underset{\underset{\displaystyle H}{|}}{C}}-\overset{}{\underset{\underset{\displaystyle OH}{|}}{CH}}-COO^- \longrightarrow$$

$$R-\overset{\overset{\displaystyle CH_3}{|}}{\underset{\underset{\displaystyle H}{|}}{C}}-\overset{}{\underset{\underset{\displaystyle O}{\|}}{C}}-COO^- \longrightarrow R-\overset{\overset{\displaystyle CH_3}{|}}{\underset{\underset{\displaystyle H}{|}}{C}}-\overset{}{\underset{\underset{\displaystyle O}{\|}}{C}}-O^- + CO_2$$

One round of α oxidation rather than β oxidation converts phytanic acid into a β-oxidation substrate.

32. Radioactive lipids are combusted to acetyl CoA, which is metabolized by the citric acid cycle. However, the two carbon atoms that enter the cycle as acetyl CoA are not the two carbon atoms that leave the cycle as CO_2. Consequently, some ^{14}C will appear in oxaloacetate, which can be converted into glucose and then into glycogen.

33. Eventually all of the CoA would be in the form of acetyl CoA, and no energy could be produced. Moreover, the high concentrations of acetyl CoA would eventually inhibit fatty acid oxidation (not to mention glucose oxidation). Getting rid of the acetyl CoA (generating ketone bodies) allows for the generation of high-energy electrons by fatty acid oxidation and for the provision of fuel for other tissues.

34. The first oxidation removes two tritium atoms. The hydration adds nonradioactive H and OH. The second oxidation removes another tritium atom from the β-carbon atom. Thiolysis removes an acetyl CoA with only one tritium; so the tritium-to-carbon ratio is 1/2. This ratio will be the same for two of the acetates. The last one, however, does not undergo oxidation, and so all tritium remains. The ratio for this acetate is 3/2. The ratio for the entire molecule is then 5/6.

Chapter 28

1. 8 acetyl CoA + 7 ATP + 14 NADPH $\longrightarrow$
   palmitate + 14 NADP$^+$ + 8 CoA + 6 H$_2$O + 7 ADP + 7 P$_i$

2. The first reaction
   Oxaloacetate + NADH + H$^+$ $\rightleftharpoons$ malate + NAD$^+$
   is catalyzed by cytoplasmic malate dehydrogenase. The next reaction is catalyzed by malic enzyme:
   Malate + NADP$^+$ $\longrightarrow$ pyruvate + CO$_2$ + NADPH
   Finally, oxaloacetate is regenerated from pyruvate by pyruvate carboxylase:
   Pyruvate + CO$_2$ + ATP + H$_2$O $\longrightarrow$
   oxaloacetate + ADP + P$_i$ + 2 H$^+$
   The sum of the reactions is
   NADP$^+$ + NADH + ATP + H$_2$O $\longrightarrow$
   NADPH + NAD$^+$ + ADP + P$_i$ + H$^+$

3. The formation of malonyl CoA from acetyl CoA by acetyl CoA carboxylase 1

4. (a) 10; (b) 1; (c) 5; (d) 8; (e) 3; (f) 9; (g) 6; (h) 7; (i) 4; (j) 2

5. 7 acetyl CoA + 6 ATP + 12 NADPH + 12 H$^+$ $\rightarrow$
   myristate + 7 CoA + 6 ADP + 6 P$_i$ + 12 NADP$^+$ + 5 H$_2$O

6. Six acetyl CoA units are required. One acetyl CoA unit is used directly to become the two carbon atoms farthest from the acid end (the ω end). The other five units must be converted into malonyl CoA. The synthesis of each malonyl CoA molecule costs a molecule of ATP; so 5 molecules of ATP are required. Each round of elongation requires 2 molecules of NADPH, one to reduce the keto group to an alcohol and one to reduce the double bond. As a result, 10 molecules of NADPH are required. Therefore, 5 molecules of ATP and 10 molecules of NADPH are required to synthesize lauric acid.

7. e, b, d, a, c

8. Such a mutation would inhibit fatty acid synthesis because the enzyme cleaves cytoplasmic citrate to yield acetyl CoA for fatty acid synthesis.

9. (a) False. Biotin is required for acetyl CoA carboxylase activity.
   (b) True
   (c) False. ATP is required to synthesize malonyl CoA.
   (d) True
   (e) True
   (f) False. Fatty acid synthase is a dimer.
   (g) True
   (h) False. Acetyl CoA carboxylase 1 is stimulated by citrate, which is cleaved to yield its substrate acetyl CoA.

10. Fatty acids with odd numbers of carbon atoms are synthesized starting with propionyl ACP (instead of acetyl ACP), which is formed from propionyl CoA by acetyl transacylase.

11. The only acetyl CoA used directly, not in the form of malonyl CoA, provides the two carbon atoms at the ω end of the fatty acid chain. Because palmitic acid is a C$_{16}$ fatty acid, acetyl CoA will have provided carbons 15 and 16.

12. HCO_3^- is attached to acetyl CoA to form malonyl CoA. When malonyl CoA condenses with acetyl CoA to form the four-carbon ketoacyl CoA, the HCO_3^- is lost as CO_2.

13. Bicarbonate is required for the synthesis of malonyl CoA from acetyl CoA by acetyl CoA carboxylase.

14. C-1 is more radioactive.

15. Decarboxylation drives the condensation of malonyl ACP and acetyl ACP. In contrast, the condensation of two molecules of acetyl ACP is energetically unfavorable. In gluconeogenesis, decarboxylation drives the formation of phosphoenolpyruvate from oxaloacetate.

16. The mutant enzyme will be persistently active because it cannot be inhibited by phosphorylation. Fatty acid synthesis will be abnormally active. Such a mutation might lead to obesity.

17. The probability of synthesizing an error-free polypeptide chain decreases as the length of the chain increases (p. 712). A single mistake can make the entire polypeptide ineffective. In contrast, a defective subunit can be spurned in the formation of a noncovalent multienzyme complex; the good subunits are not wasted.

18. When glycogen stores are filled, the excess carbohydrates are metabolized to acetyl CoA, which is then converted into fats. Human beings cannot convert fats into carbohydrates, but they can certainly convert carbohydrates into fats.

19. Cytoplasmic palmitoyl CoA inhibits the translocase that shuttles citrate from mitochondria to the cytoplasm, thereby preventing fatty acid synthesis. Mitochondrial malonyl CoA inhibits carnitine acyltransferase I, preventing the entry of fatty acids into mitochondria, thereby inhibiting fatty acid degradation in times of plenty.

20. With a diet rich in raw eggs, avidin will inhibit fatty acid synthesis by reducing the amount of biotin required by acetyl CoA carboxylase. Cooking the eggs will denature avidin, and so it will no longer bind biotin.

21. Phosphofructokinase controls the flux down the glycolytic pathway. Glycolysis functions to generate ATP or building blocks for biosynthesis, depending on the tissue. The presence of citrate in the cytoplasm indicates that those needs are met, and there is no need to metabolize glucose.

22. (a) Oxidation in mitochondria; synthesis in the cytoplasm

(b) Coenzyme A in oxidation; acyl carrier protein in synthesis

(c) FAD and NAD^+ in oxidation; NADPH in synthesis

(d) L isomer of 3-hydroxyacyl CoA in oxidation; D isomer in synthesis

(e) From carboxyl to methyl in oxidation; from methyl to carboxyl in synthesis

(f) The enzymes of fatty acid synthesis, but not those of oxidation, are organized in a multienzyme complex.

23. Pyruvate $+ CO_2 + ATP + H_2O \longrightarrow$

$$oxaloacetate + ADP + P_i + 2 H^+$$

In gluconeogenesis, the oxaloacetate is converted into phosphoenolpyruvate by PEP carboxykinase at the expense of a molecule of GTP.

24. The product of the reaction catalyzed by acetyl CoA carboxylase 2, malonyl CoA, inhibits the import of fatty acids into mitochondria by inhibiting carnitine acyltransferase I.

25. Citrate is an allosteric activator of acetyl CoA carboxylase 1, causing the enzyme to form active filaments. Palmitoyl CoA will inhibit the enzyme.

26. (a) Soraphen A inhibits fatty acid synthesis in a dose-dependent manner.

(b) Fatty acid oxidation is increased in the presence of soraphen A.

(c) Recall that acetyl carboxylase 2 synthesizes malonyl CoA to inhibit the transport of fatty acids into the mitochondria, thereby preventing fatty acid oxidation. Soraphen A apparently inhibits both forms of the carboxylase.

(d) Phospholipid synthesis was inhibited in a dose-dependent manner.

(e) Phospholipids are required for membrane synthesis.

(f) Soraphen A inhibits cell proliferation, especially at higher concentrations.

27. All of the labeled carbon atoms will be retained. Because you need 8 acetyl CoA molecules and only 1 carbon atom is labeled in the acetyl group, you will have 8 labeled carbon atoms. The only acetyl CoA used directly will retain 3 tritium atoms. The 7 acetyl CoA molecules used to make malonyl CoA will lose 1 tritium atom on addition of the CO_2 and another tritium atom at the dehydration step. Each of the 7 malonyl CoA molecules will retain 1 tritium atom. Therefore, the total retained tritium is 10 atoms. The ratio of tritium to carbon is 1.25.

28. The high concentration of NADH inhibits gluconeogenesis by preventing the oxidation of lactate to pyruvate while causing the reverse reaction to predominate, leading to lactic acidosis and hypoglycemia. The NADH glut also inhibits fatty acid oxidation and the citric acid cycle because the drinker's NADH needs are met by ethanol metabolism. The excess NADH signals that conditions are right for fatty acid synthesis, which leads to the development of "fatty liver."

Chapter 29

1. Glycerol $+ 4$ ATP $+ 3$ fatty acids $+ 4 H_2O \rightarrow$

$$triacylglycerol + ADP + 3 AMP + 7 P_i + 4 H^+$$

2. Glycerol $+ 3$ ATP $+ 2$ fatty acids $+ 2 H_2O + CTP +$ serine $\rightarrow$

$$phosphatidylserine + CMP + ADP + 2 AMP + 6 P_i + 3 H^+$$

3. Glycerol 3-phosphate is formed primarily by the reduction of dihydroxyacetone phosphate, a glycolytic intermediate, and to a lesser extent by the phosphorylation of glycerol (p. 314).

4. Three. One molecule of ATP to form phosphorylethanolamine and two molecules of ATP to regenerate CTP from CMP.

5. All are synthesized from ceramide. In sphingomyelin, the terminal hydroxyl group of ceramide is modified with phosphorylcholine. In a cerebroside, the hydroxyl group has a glucose or galactose attached. In a ganglioside, oligosaccharide chains are attached to the hydroxyl group.

6. (a) 8; (b) 4; (c) 1; (d) 9; (e) 3; (f) 10; (g) 5; (h) 2; (i) 6; (j) 7

7. (a) CDP-diacylglycerol; (b) CDP-ethanolamine; (c) acyl CoA; (d) CDP-choline; (e) UDP-glucose or UDP-galactose; (f) geranyl pyrophosphate

8. No need to make a new plan, Stan. Just listen to me. (1) Activate the diacylglycerol as CDP-DAG. (2) Activate the alcohol as CDP-alcohol.

9. Phosphatidic acid phosphatase (PAP). PAP controls the extent to which triacylglycerols are synthesized relative to phospholipids and regulates the type of phospholipid synthesized.

10. Such mutations are seen in mice. The amount of adipose tissue would decrease severely because diacylglycerol could not be formed. Normally, diacylglycerol is acylated to form triacylglycerols. If there were deficient phosphatidic acid phosphatase activity, no triacylglycerols would form.

11. Excess PAP activity would increase the amount of diacylglycerol. After the phospholipid needs requiring diacylglycerol had been met, the excess diacylglycerol would be converted into triacylglycerol and obesity would result.

12. The three stages are (1) the synthesis of activated isoprene units (isopentyl pyrophosphate), (2) the condensation of six of the activated isoprene units to form squalene, and (3) cyclization of the squalene to form cholesterol.

13. (a and b) None, because the label is lost as CO_2

14. The hallmark of this genetic disease is elevated cholesterol levels in the blood of even young children. The excess cholesterol is taken up by macrophages, which eventually results in the formation of plaques and heart disease. There are many mutations that cause the disease, but all result in malfunctioning of the LDL receptor.

15. The amount of reductase and its activity control the regulation of cholesterol biosynthesis. Transcriptional control is mediated by SREBP. Translation of the reductase mRNA also is controlled. The reductase itself may undergo regulated proteolysis. Finally, the activity of the reductase is inhibited by phosphorylation by AMP-dependent kinase when ATP levels are low.

16. Statins are competitive inhibitors of HMG-CoA reductase. They are used as drugs to inhibit cholesterol synthesis in patients who have high levels of cholesterol.

17. No. Cholesterol is essential for membrane function and as a precursor for bile salts and steroid hormones. A complete lack of cholesterol would be lethal.

18. Progestagens, glucocorticoids, mineralocorticoids, androgens, and estrogens

19. Recall that dihydrotestosterone is crucial for the development of male characteristics in the embryo. If a pregnant woman were to be exposed to Propecia, the 5α-reductase of the male embryo would be inhibited, which could result in severe developmental abnormalities.

20. "None of your business" and "I don't talk biochemistry until after breakfast" are appropriate but rude and uninformative answers. A better answer might be: "Although it is true that cholesterol is a precursor to steroid hormones, the rest of the statement is oversimplified. Cholesterol is a component of membranes, and membranes literally define cells, and cells make up tissues. But to say that cholesterol 'makes' cells and tissues is wrong."

21. The core structure of a steroid is four fused rings: three cyclohexane rings and one cyclopentane ring. In vitamin D, the second cyclohexane ring from the left, called the B ring, is split by ultraviolet light.

22. The LDL contains apolipoprotein B-100, which binds to an LDL receptor on the cell surface in a region known as a coated pit. On binding, the complex is internalized by endocytosis to form an internal vesicle. The vesicle is separated into two components. One, with the receptor, is transported back to the cell surface and fuses with the membrane, allowing continued use of the receptor. The other vesicle fuses with lysosomes inside the cell. The cholesteryl esters are hydrolyzed, and free cholesterol is made available for cellular use. The LDL protein is hydrolyzed to free amino acids.

23. This knowledge would enable clinicians to characterize the likelihood of a patient having an adverse drug reaction or being susceptible to chemical-induced illnesses. It would also permit a personalized and especially effective drug-treatment regime for diseases such as cancer.

24. The honeybees may be especially sensitive to environmental toxins, including pesticides, because these chemicals are not readily detoxified, owing to the minimal P450 system.

25. The attachment of isoprenoid side chains confers hydrophobic character. Proteins having such a modification are targeted to membranes.

26. Note that a cytidine nucleotide plays the same role in the synthesis of these phosphoglycerides as a uridine nucleotide does in the formation of glycogen (p. 460). In all of these biosyntheses, an activated intermediate (UDP-glucose, CDP-diacylglycerol, or CDP-alcohol) is formed from a phosphorylated substrate (glucose 1-phosphate, phosphatidate, or a phosphorylalcohol) and a nucleoside triphosphate (UTP or CTP). The activated intermediate then reacts with a hydroxyl group (the terminus of glycogen, the side chain of serine, or a diacylglycerol).

27. Citrate is transported out of mitochondria in times of plenty. ATP-citrate lyase yields acetyl CoA and oxaloacetate. The acetyl CoA can then be used to synthesize cholesterol.

28. 3-Hydroxy-3-methylglutaryl CoA is also a precursor for ketone-body synthesis. If fuel is needed elsewhere in the body, as might be the case during a fast, 3-hydroxy-3-methylglutaryl CoA is converted into the ketone acetoacetate. If energy needs are met, the liver will synthesize cholesterol.

29. Mutations could occur in the gene encoding sodium channel that would prevent the action of DDT. Alternatively, P450 enzyme synthesis could be increased to accelerate metabolism of the insecticide to inactive metabolites. In fact, both types of responses to insecticides have been observed.

30. (a) There is no effect.

(b) Because actin is not controlled by cholesterol, the amount isolated should be the same in both experimental groups. A difference would suggest a problem in the RNA isolation.

(c) The presence of cholesterol in the diet dramatically reduces the amount of HMG-CoA reductase protein.

(d) A common means of regulating the amount of a protein present is to regulate transcription, which is clearly not the case here.

(e) Translation of mRNA could be inhibited. The protein could be rapidly degraded.

31. The categories of mutations are (1) no receptor is synthesized; (2) receptors are synthesized but do not reach the plasma membrane, because they lack the signals for intracellular transport or do not fold properly; (3) receptors reach the cell surface, but they fail to bind LDL normally because of a defect in the LDL-binding domain; (4) receptors reach the cell surface and bind LDL, but they fail to cluster in coated pits because of a defect in their carboxyl-terminal regions.

32. Benign prostatic hypertrophy can be treated by inhibiting 5α-reductase. Finasteride, a steroid analog of dihydrotestosterone, competitively inhibits the reductase but does not act on androgen receptors. Patients taking finasteride have a markedly lower plasma level of dihydrotestosterone and a nearly normal level of testosterone. The prostate gland becomes smaller, but testosterone-dependent processes such as fertility, libido, and muscle strength appear to be unaffected.

33. In cartoons, plants can run from their predators. In real life, it is not the case, and the predators might eat the plants. Consequently, as a means of self-defense, it behooves plants to produce a wide array of toxins so that they can ward off those who might eat them.

34. Many hydrophobic odorants are deactivated by hydroxylation. Molecular oxygen is activated by a cytochrome P450 monooxygenase. NADPH serves as the reductant. One oxygen atom of O_2 goes into the odorant substrate, whereas the other is reduced to water.

Chapter 30

1. (a) Pyruvate; (b) oxaloacetate; (c) α-ketoglutarate; (d) β-ketoisocaproate; (e) phenylpyruvate; (f) hydroxyphenylpyruvate

2. (a) Aspartate + α-ketoglutarate + GTP + ATP + 2 H_2O + NADH + H^+ →
   1/2 glucose + glutamate + CO_2 +
   ADP + GDP + NAD^+ + 2 P_i

The required coenzymes are pyridoxal phosphate in the transamination reaction and $NAD^+/NADH$ in the redox reactions.

(b) Aspartate + CO_2 + NH_4^+ + 3 ATP + NAD^+ + 4 H_2O → oxaloacetate + urea + 2 ADP + 4 P_i + AMP + NADH + H^+

3. Most enzymes are specific for one or the other. Enzymes in catabolic pathways use $NADH/NAD^+$, whereas enzymes in anabolic pathways use only $NADPH/NADP^+$.

4. Aminotransferases transfer the α-amino group to α-ketoglutarate to form glutamate. Glutamate is oxidatively deaminated to form an ammonium ion.

5. Aspartate (oxaloacetate), glutamate (α-ketoglutarate), alanine (pyruvate)

6. Serine and threonine

7. Carbamoyl phosphate and aspartate

8. (a) 4; (b) 5; (c) 1; (d) 6; (e) 7; (f) 3; (g) 2

9. CO_2 + NH_4^+ + 3 ATP + NAD^+ + glutamate + 3 H_2O →

urea + 2 ADP + 2 P_i + AMP + PP_i + NADH + H^+ +

α-ketoglutarate

The number of high-transfer-potential phosphoryl groups used remains four, as in the equation on page 557.

10. The synthesis of fumarate by the urea cycle is important because it links the urea cycle and the citric acid cycle. Fumarate is hydrated to malate, which, in turn, is oxidized to oxaloacetate. Oxaloacetate has several possible fates: (1) transamination to aspartate, (2) conversion into glucose by the gluconeogenic pathway, (3) condensation with acetyl CoA to form citrate, or (4) conversion into pyruvate.

11. The analytical results strongly suggest that three enzymes—pyruvate dehydrogenase, α-ketoglutarate dehydrogenase, and the branched-chain α-ketoacid dehydrogenase—are deficient. In fact, the E_3 component (Chapter 18) is common to all of these enzymes. The results suggest that E_3 is missing or defective. This proposal could be tested by purifying these three enzymes and assaying their ability to catalyze the regeneration of lipoamide.

12. A, arginine; B, citrulline; C, ornithine; D, arginosuccinate. The order of appearance: C, B, D, E.

13. Aspartame, a dipeptide ester, is hydrolyzed to L-aspartate and L-phenylalanine. High levels of phenylalanine are harmful in phenylketonurics.

14. N-Acetylglutamate is synthesized from acetyl CoA and glutamate. Once again, acetyl CoA serves as an activated acetyl donor. This reaction is catalyzed by N-acetylglutamate synthase.

15. The carbon skeletons of ketogenic amino acids can be converted into ketone bodies or fatty acids. Only leucine and lysine are purely ketogenic. Glucogenic amino acids are those whose carbon skeletons can be converted into glucose.

16. Pyruvate (glycolysis and gluconeogenesis), acetyl CoA (citric acid cycle and fatty acid synthesis), acetoacetyl CoA (ketone-body formation), α-ketoglutarate (citric acid cycle), succinyl CoA (citric acid cycle), fumarate (citric acid cycle), and oxaloacetate (citric acid cycle and gluconeogenesis)

17. As shown in Figure 30.12, alanine, a gluconeogenic amino acid, is released during the metabolism of tryptophan to acetyl CoA and acetoacetyl CoA.

18. Not all proteins are created equal; some are more important than others. Some proteins would be degraded to provide the missing amino acid. The nitrogen from the other proteins would be excreted as urea. Consequently, more nitrogen would be excreted than ingested.

19. This defect can be partly bypassed by providing a surplus of arginine in the diet and restricting the total protein intake. In the liver, arginine is split into urea and ornithine, which then reacts

with carbamoyl phosphate to form citrulline. This urea-cycle intermediate condenses with aspartate to yield argininosuccinate, which is then excreted. Note that two nitrogen atoms—one from carbamoyl phosphate and the other from aspartate—are eliminated from the body per molecule of arginine provided in the diet. In essence, argininosuccinate substitutes for urea in carrying nitrogen out of the body. The formation of argininosuccinate removes the nitrogen, and the restriction on protein intake relieves the aciduria.

20. Double-displacement. A substituted enzyme intermediate is formed.

21. The branched-chain amino acids leucine, isoleucine, and valine. The required enzyme is the branched-chain α-ketoacid dehydrogenase complex.

22. Ammonia could lead to the amination of α-ketoglutarate, producing a high concentration of glutamate in an unregulated fashion. α-Ketoglutarate for glutamate synthesis could be removed from the citric acid cycle, thereby diminishing the cell's respiration capacity.

23. The liver is the primary tissue for capturing nitrogen as urea. If the liver is damaged (for instance, by hepatitis or the excessive consumption of alcohol), free ammonia is released into the blood.

24. Ornithine transcarbamoylase

25. (a) Depletion of glycogen stores. When they are gone, proteins must be degraded to meet the glucose needs of the brain. The resulting amino acids are deaminated, and the nitrogen atoms are excreted as urea.

(b) The brain has adapted to the use of ketone bodies, which are derived from fatty acid catabolism. In other words, the brain is being powered by fatty acid breakdown.

(c) When the glycogen and lipid stores are gone, the only available energy source is protein.

26. Deamination to α-keto-β-methylvalerate; oxidative decarboxylation to α-methylbutyryl CoA; oxidation to tiglyl CoA; hydration, oxidation, and thiolysis yields acetyl CoA and propionyl CoA; propionyl CoA to succinyl CoA.

27. In the Cori cycle, the carbon atoms are transferred from muscle to the liver as lactate. For lactate to be of any use, it must be reduced to pyruvate. This reduction requires high-energy electrons in the form of NADH. When the carbon atoms are transferred as alanine, transamination yields pyruvate directly.

28. The precise cause of all of the symptoms is not firmly established, but a likely explanation depends on the centrality of oxaloacetate to metabolism. A lack of pyruvate carboxylase would reduce the amount of oxaloacetate. The lack of oxaloacetate would reduce the activity of the citric acid cycle, and so ATP would be generated by lactic acid formation. If the concentration of oxaloacetate is low, aspartate cannot be formed and the urea cycle would be compromised. Oxaloacetate is also required to form citrate, which transports acetyl CoA to the cytoplasm for fatty acid synthesis. Finally, oxaloacetate is required for gluconeogenesis.

Chapter 31

1. Nitrogen fixation is the conversion of atmospheric N_2 into NH_4^+. Diazotrophic (nitrogen-fixing) microorganisms are responsible for this reaction.

2. N_2 + 8 e^- + 8 H^+ + 16 ATP + 16 H_2O ⇌

2 NH_3 + H_2 + 16 ADP + 16 P_i

The fixation of nitrogen is an exergonic reaction. The role of ATP is to reduce the activation energy of the barriers in the reaction pathway so as to render the reaction kinetically feasible.

3. (a) 4; (b) 8; (c) 10; (d) 6; (e) 7; (f) 9; (g) 3; (h) 5; (i) 2; (j) 1

4. The reductase provides electrons with high reducing power, whereas the nitrogenase, which requires ATP hydrolysis, uses the electrons to reduce N_2 to NH_3.

5. False. Nitrogen is thermodynamically favored. ATP expenditure by the nitrogenase is required to make the reaction kinetically possible.

6. The bacteria provide the plant with ammonia by reducing atmospheric nitrogen. This reduction is energetically expensive, and the bacteria use ATP from the plant.

7. Human beings do not have the biochemical pathways to synthesize essential amino acids from simpler precursors. Consequently, they must be obtained from the diet.

8. Oxaloacetate, pyruvate, ribose-5-phosphate, phosphoenolpyruvate, erythrose-4-phosphate, α-ketoglutarate, and 3-phosphoglycerate

9. Pyridoxal phosphate (PLP)

10. A reversible transamination reaction will transfer the labeled amino group from aspartate to α-ketoglutarate to form glutamate and oxaloacetate. Glutamate is an amino-group donor for the synthesis of many amino acids from their corresponding ketoacids.

11. Tetrahydrofolate is a carrier of a variety of one-carbon units.

12. Both carry one-carbon units. S-Adenosylmethionine is a more useful methyl donor than tetrahydrofolate because it has a greater transfer potential.

13. γ-Glutamyl phosphate is a likely reaction intermediate.

14. Alanine from pyruvate; aspartate from oxaloacetate; glutamate from α-ketoglutarate

15. Y could inhibit the C → D step, Z could inhibit the C → F step, and C could inhibit A → B. This scheme is an example of sequential feedback inhibition. Alternatively, Y could inhibit the C → D step, Z could inhibit the C → F step, and the A → B step would be inhibited only in the presence of both Y and Z. This scheme is called concerted feedback inhibition.

16. The rate of the A → B step in the presence of high levels of Y and Z would be 24 s^{-1} $(0.6 \times 0.4 \times 100 \text{ s}^{-1})$.

17. Tetrahydrofolate, S-adenosylmethionine, and biotin

18. Methylmalonyl CoA mutase, an enzyme required for the degradation of odd-chain fatty acids (p. 497)

19. Methionine is a component of S-adenosylmethionine, which is a methyl donor for the synthesis of phosphatidylcholine from phosphatidylethanolamine (p. 526).

20. (a) Asparagine is much more abundant in the dark. More glutamine is present in the light. These amino acids show the most dramatic effects. Glycine also is more abundant in the light.

   (b) Glutamine is a more metabolically reactive amino acid, used in the synthesis of many other compounds. Consequently, when energy is available as light, glutamine will be preferentially synthesized. Asparagine, which carries more nitrogen per carbon atom and is thus a more efficient means of storing nitrogen when energy is short, is synthesized in the dark. Glycine is more prevalent in the light because of photorespiration.

   (c) White asparagus has an especially high concentration of asparagine, which accounts for its intense taste. All asparagus has a large amount of asparagine. In fact, as suggested by its name, asparagine was first isolated from asparagus.

21. Glucose + 2 ADP + 2 P_i + 2 NAD^+ + 2 glutamate →

   2 alanine + 2 α-ketoglutarate + 2 ATP + 2 NADH + 2 H_2O + 2 H^+

22. Synthesis from oxaloacetate and α-ketoglutarate would deplete the citric acid cycle, which would decrease ATP production. Anapleurotic reactions would be required to replenish the citric acid cycle.

23. The value of K_M of glutamate dehydrogenase for NH_4^+ is high (≫1 mM), and so this enzyme is not saturated when NH_4^+ is limiting. In contrast, glutamine synthetase has very low K_M for NH_4^+. Thus, ATP hydrolysis is required to capture ammonia when it is scarce.

Chapter 32

1. In de novo synthesis, the nucleotides are synthesized from simpler precursor compounds, in essence from scratch. In salvage pathways, preformed bases are recovered and attached to activated riboses.

2. Glucose + 2 ATP + 2 $NADP^+$ + H_2O
   PRPP + CO_2 + ADP + AMP + 2 NADPH + 3 H^+

3. Glutamine + aspartate + CO_2 + 2 ATP + NAD^+ →
   orotate + 2 ADP + 2 P_i + glutamate + NADH + H^+

4. (a and c) PRPP; (b) carbamoyl phosphate

5. A nucleoside is a base attached to ribose. A nucleotide is a nucleoside with the ribose bearing one or more phosphates.

6. (a) 9; (b) 7; (c) 6; (d) 10; (e) 2; (f) 4; (g) 1; (h) 11; (i) 8; (j) 3; (k) 5

7. dUMP + serine + NADPH + H^+ → TMP + $NADP^+$ + glycine

8. Sulfanilamide inhibits the synthesis of folate by acting as an analog of p-aminobenzoate, one of the precursors of folate. This results in a deficiency of N^{10}-formyltetrahydrofolate.

9. PRPP is the activated intermediate in the synthesis of phosphoribosylamine in the de novo pathway of purine formation; in the synthesis of purine nucleotides from free bases by the salvage pathway; and in the synthesis of orotidylate in the formation of pyrimidines.

10. The reciprocal substrate relation refers to the fact that AMP synthesis requires GTP, whereas GMP synthesis requires ATP. This relation tends to balance the synthesis of ATP and GTP.

11. The synthesis of carbamoyl phosphate requires 2 ATP	2 ATP
The formation of PRPP from ribose 5-hosphate yields 1 AMP*	2 ATP
The conversion of UMP to UTP requires 2 ATP	2 ATP
The conversion of UTP to CTP requires 1 ATP	1 ATP
Total	7 ATP

*Remember that 1 AMP is the equivalent of 2 ATP because an ATP must be expended to generate ADP, the substrate for ATP synthesis.

12. The label will be on C-6 for cytosine and on C-5 for guanine, to correspond to ^{13}C labels on the α carbons of aspartate and glycine, respectively.

13. N-3 and N-9 in the purine ring

14. (a) Carboxyaminoimidazole ribonucleotide; (b) glycinamide ribonucleotide; (c) phosphoribosylamine; (d) formylglycinamide ribonucleotide

15. Allopurinol, an analog of hypoxanthine, is a suicide inhibitor of xanthine oxidase.

16. Leukemia patients have a high level of urate because of the breakdown of nucleic acids due to cell death. Allopurinol prevents the formation of kidney stones and blocks other deleterious consequences of hyperuricemia by preventing the formation of urate (p. 599).

17. Because folate is required for nucleotide synthesis, cells that are dividing rapidly would be most readily affected. They would include cells of the intestine, which are constantly replaced, and precursors to blood cells. The lack of intestinal cells and blood cells would account for the symptoms often observed.

18. Inosine or hypoxanthine could be administered.

19. N-1 in both cases, and the amine group linked to C-6 in ATP.

20. By their nature, cancer cells divide rapidly and thus require frequent DNA synthesis. Inhibitors of TMP synthesis will impair DNA synthesis and cancer growth.

21. (a) cAMP; (b) ATP; (c) UDP-glucose; (d) acetyl CoA; (e) NAD^+, FAD; (f) fluorouracil; (g) CTP inhibits ATCase.

22. (a) Some ATP can be salvaged from the ADP that is being generated. (b) There are equal numbers of high-phosphoryl-transfer-potential groups on each side of the equation. (c) Because the adenylate kinase reaction is at equilibrium, the removal of AMP would lead to the formation of more ATP. (d) Essentially, the cycle serves as an anapleurotic reaction for the generation of the citric acid cycle intermediate fumarate.

23. Carbamoyl phosphate I uses ammonia to synthesize carbamoyl phosphate for a reaction with ornithine, the first step of the urea cycle. Carbamoyl phosphate II uses glutamine to synthesize carbamoyl phosphate for use in the first step of pyrimidine biosynthesis.

24. Allopurinol is an inhibitor of xanthine oxidase, which is on the pathway for urate synthesis. In your pet duck, this pathway is the means by which excess nitrogen is excreted. If xanthine oxidase were inhibited in your duck, nitrogen could not be excreted, with severe consequences such as a dead duck.

25. PRPP and formylglycinamide ribonucleotide

26. (a) Cell A cannot grow in a HAT medium, because it cannot synthesize TMP either from thymidine or from dUMP. Cell B cannot grow in this medium, because it cannot synthesize purines by either the de novo pathway or the salvage pathway. Cell C can grow in a HAT medium because it contains active thymidine kinase from cell B (enabling it to phosphorylate thymidine to TMP) and hypoxanthine guanine phosphoribosyltransferase from cell A (enabling it to synthesize purines from hypoxanthine by the salvage pathway).

(b) Transform cell A with a plasmid containing foreign genes of interest and a functional thymidine kinase gene. The only cells that will grow in a HAT medium are those that have acquired a thymidylate kinase gene; nearly all of these transformed cells will also contain the other genes on the plasmid.

27. The cytoplasmic level of ATP in the liver falls and that of AMP rises above normal in all three conditions. The excess AMP is degraded to urate.

28. Succinate → malate → oxaloacetate by the citric acid cycle. Oxaloacetate → aspartate by transamination, followed by pyrimidine synthesis. Carbons 4, 5, and 6 are labeled.

29. Glucose will most likely be converted into two molecules of pyruvate, one of which will be labeled in the 2 position:

$$
\begin{array}{c}
\text{O} \overset{-}{\cdots} \text{O} \\
\text{C} \\
\text{C}=\text{O} \\
\text{CH}_3
\end{array}
$$

Now consider two common fates of pyruvate—conversion into acetyl CoA and subsequent processing by the citric acid cycle or carboxylation by pyruvate carboxylase to form oxaloacetate.

The formation of citrate by condensing the labeled acetyl CoA (derived from labeled pyruvate) with oxaloacetate will yield labeled citrate:

$$
\begin{array}{c}
\text{COO}^- \\
\text{CH}_2 \\
^-\text{OOC}-\text{C}-\text{OH} \\
\text{H}_2\text{C} \\
\text{COO}^-
\end{array}
$$

The labeled carbon atom will be retained through one round of the citric acid cycle, but, on the formation of the symmetric succinate,

the label will appear in two different positions. Thus, when succinate is metabolized to oxaloacetate, which may be aminated to form aspartate, two carbons will be labeled:

$$
\begin{array}{c}
\text{O} \overset{-}{\cdots} \text{O} \\
\text{C} \\
^+\text{H}_3\text{N}-\text{CH} \\
\text{CH}_2 \\
\text{O} \overset{\text{C}}{\underset{-}{\cdots}} \text{O}
\end{array}
$$

When this aspartate is used to form uracil, the labeled COO^- attached to the α-carbon is lost and the other COO^- becomes incorporated into uracil as carbon 4.

Suppose instead that pyruvate labeled in the 2 position is carboxylated to form oxaloacetate and processed to form aspartate. In this case, the α-carbon of aspartate bears the label.

$$
\begin{array}{c}
\text{O} \overset{-}{\cdots} \text{O} \\
\text{C} \\
^+\text{H}_3\text{N}-\text{CH} \\
\text{CH}_2 \\
\text{O} \overset{\text{C}}{\underset{-}{\cdots}} \text{O}
\end{array}
$$

When this aspartate is used to synthesize uracil, carbon 6 bears the label.

Chapter 33

1. A nucleoside is a base attached to a ribose sugar. A nucleotide is a nucleoside with one or more phosphoryl groups attached to the ribose.

2. Hydrogen-bond pairing between the base A and the base T as well as hydrogen-bond pairing between the base G and the base C in DNA.

3. (a) 6; (b) 8; (c) 10; (d) 2; (e) 1; (f) 3; (g) 4; (h) 9; (i) 5; (j) 7

4. T is always equal to A, and so these two nucleotides constitute 40% of the bases. G is always equal to C, and so the remaining 60% must be 30% G and 30% C.

5. Nothing, because the base-pair rules do not apply to single-stranded nucleic acids.

6. Two purines are too large to fit inside the double helix, and two pyrimidines are too small to form base pairs with each other.

7. (a) TTGATC; (b) GTTCGA; (c) ACGCGT; (d) ATGGTA

8. (a) $[T] + [C] = 0.46$; (b) $[T] = 0.30$, $[C] = 0.24$, and $[A] + [G] = 0.46$

9. The heterocyclic bases readily absorb ultraviolet lights. The bases in the interior of a double helix are stacked and not as exposed as the bases in single-stranded DNA. Consequently, the bases of double-stranded DNA absorb light less efficiently than single-stranded DNA.

10. The diameter of DNA is 20 Å and 1 Å = 0.1 nm, so the diameter is 2 nm. Because 1 μm = 10^3 nm, the length is 2×10^4 nm. Thus, the axial ratio is 1×10^4.

11. The thermal energy causes the strands to wiggle about, which disrupts the hydrogen bonds between base pairs and the stacking forces between bases and thereby causes the strands to separate.

12. One end of a nucleic acid polymer ends with a free 5′-hydroxyl group (or a phosphoryl group esterified to the hydroxyl group), and the other end has a free 3′-hydroxyl group. Thus, the ends are different. Two strands of DNA can form a double helix only if the strands are running in different directions—that is, have opposite directionality.

13. 5.88×10^3 base pairs

14. In conservative replication, after 1.0 generation, half of the molecules would be ^{15}N-^{15}N and the other half would be ^{14}N-^{14}N. After 2.0 generations, one-quarter of the molecules would be ^{15}N-^{15}N and the other three-quarters would be ^{14}N-^{14}N. Hybrid ^{14}N-^{15}N molecules would not be observed in conservative replication.

15. There would be too much charge repulsion from the negative charges on the phosphoryl groups. These charges must be countered by the addition of cations.

16. The nucleosome structure results in a packaging ratio of only 7. Other structures—such as the 30-nm chromatin fiber—must form to attain the compaction ratio of 10^4 found in metaphase chromosomes.

17. The DNA is bound to the histones. Consequently, it is not readily accessible to DNase. However, with an extended reaction time or more enzyme, the DNA becomes available for digestion.

18. The distribution of charged amino acids is H2A (13 K, 13 R, 2 D, 7 E, charge = +15), H2B (20 K, 8 R, 3 D, 7 E, charge = +18), H3 (13 K, 18 R, 4 D, 7 E, charge = +20), H4 (11 K, 14 R, 3 D, 4 E, charge = +18). The total charge of the histone octamer is estimated to be $2 \times (15 + 18 + 20 + 18) = +142$. The total charge on 150 base pairs of DNA is -300. Thus, the histone octamer neutralizes approximately one-half of the charge.

19. The total length of the DNA is estimated to be 145 bp $\times$ 3.4 Å/bp = 493 Å, which represents 1.75 turns. Thus, the circumference is 281 Å. The formula for the circumference is $C = 2\pi r$. Solving for r yields 41 Å.

20. GC base pairs have three hydrogen bonds compared with two for AT base pairs. Thus, the higher content of GC means more hydrogen bonds and greater helix stability.

21. $C_0 t$ value essentially corresponds to the complexity of the DNA sequence—in other words, how long it will take for a sequence of DNA to find its complementary strand to form a double helix. Simpler or small DNA sequences will more rapidly reanneal and have lower $C_0 t$ values.

22. Increasing amounts of salt increase the melting temperature. Because the DNA backbone is negatively charged, there is a tendency for charge repulsion to destabilize the helix and cause it to melt. The addition of salt neutralizes the charge repulsion, thereby stabilizing the double helix. The results show that, within the parameters of the experiment, more salt results in more stabilization, which gives the DNA a higher melting temperature.

23. The probability that any sequence will appear is $1/4^n$, where 4 is the number of nucleotides and n is the length of the sequence. The probability of any 15-base sequence appearing is $1/4^{15}$, or 1/1,073,741,824. Thus, a 15-nucleotide sequence would be likely to appear approximately three times (3 billion $\times$ probability of appearance). The probability of a 16-base sequence appearing is $1/4^{16}$, which is equal to 1/4,294,967,296. Such a sequence will be unlikely to appear more than once.

24. (a) $4^8 = 65,536$. In computer terminology, there are 64K 8-mers of DNA.

   (b) A bit specifies two bases (say, A and C) and a second bit specifies the other two (G and T). Hence, two bits are needed to specify a single nucleotide (base pair) in DNA. For example, 00, 01, 10, and 11 could encode A, C, G, and T. An 8-mer stores 16 bits ($2^{16} = 65,536$), the E. coli genome (4.6×10^6 bp) stores 9.2×10^6 bits, and the human genome (3.0×10^9 bases) stores 6.0×10^9 bits of genetic information.

   (c) A 700 megabyte CD is equal to 5.6×10^9 bits. A large number of 8-mer sequences could be stored on such a CD. The DNA sequence of E. coli could be written on a single CD with room to spare for a lot of music. One CD would not be quite enough to record the entire human genome.

25. The 2′-OH group in RNA acts as an intramolecular nucleophile. In the alkaline hydrolysis of RNA, it forms a 2′-3′ cyclic intermediate.

26.

Chapter 34

1. A template is the sequence of DNA or RNA that directs the synthesis of a complementary sequence. A primer is the initial segment of a polymer that is to be extended on which elongation depends.

2. DNA polymerase cannot initiate primer synthesis. Consequently, an RNA polymerase, called primase, synthesizes a short sequence of RNA that is used as a primer by the DNA polymerase.

3. Okazaki fragments are short segments of DNA that are synthesized on the lagging stand of DNA. These fragments are subsequently joined by DNA ligase to form a continuous segment of DNA.

4. When DNA is being synthesized at the replication fork, the leading strand is synthesized continuously in the 5′-to-3′ direction as the template is read in the 3′-to-5′ direction. The lagging strand is synthesized as short Okazaki fragments.

5. (a) 5; (b) 4; (c) 8; (d) 1; (e) 10; (f) 2; (g) 6; (h) 7; (i) 3; (j) 9

6. The nucleotides used for DNA synthesis have the triphosphate attached to the 5′-hydroxyl group with free 3′-hydroxyl groups. Such nucleotides can be utilized only for 5′-to-3′ DNA synthesis.

7. The rate of strand separation is low owing to the hydrogen bonds of the helix and the stacking forces between bases. Although individually weak, the thousands or millions of such bonds that hold a helix together make spontaneous separation of the strands unlikely.

8. Replication would take twice as long.

9. Eventually, the DNA would become so tightly wound that movement of the replication complex would be energetically impossible.

10. A hallmark of most cancer cells is prolific cell division, which requires DNA replication. If the telomerase were not activated, the chromosomes would shorten until they became nonfunctional, leading to cell death. Interestingly, telomerase is often, but not always, found to be activated in cancer cells.

11. The free energy of ATP hydrolysis under standard conditions is -30.5 kJ mol^{-1} (-7.3 kcal mol^{-1}). In principle, it could be used to break three base pairs.

12. (a) The nucleotide ddATP is virtually identical in structure with dATP except that it lacks a 3'-OH. Thus, it will be incorporated into a newly synthesized DNA by DNA polymerase when the polymerase first encounters a T in the template strand. However, because the incorporated ddAMP has no 3'-OH, synthesis stops.

(b) Because the ddATP is only 10% of the concentration of dATP, there is a 10% chance that it will be incorporated into the newly synthesized DNA when a T is encountered in the template strand. Consequently, a population of fragments of DNA will be synthesized, all ending in ddAMP. The inability to extend a dideoxy nucleotide is the basis of the strand-termination, or dideoxy method, of DNA sequencing (p. 751).

13. Positive supercoiling resists the unwinding of DNA. The melting temperature of DNA increases in proceeding from negatively supercoiled to relaxed to positively supercoiled DNA. Positive supercoiling is probably an adaptation to high temperature to prevent unregulated DNA unwinding.

14. (a) Size; the top is relaxed and the bottom is supercoiled DNA.

(b) Topoisomers

(c) The DNA is becoming progressively more unwound, or relaxed, and thus slower moving.

15. The high concentration of nucleotides will have the effect of speeding up the polymerization reaction. Consequently, a misformed product may exit the polymerase active site before it has time to wander into the exonuclease activity site. The reverse is true if the next nucleotide is scarce. The polymerase pauses for a longer time, increasing the likelihood that a misformed product will visit the exonuclease site.

16. (a) 96.2 revolutions per second (1000 nucleotides per second divided by 10.4 nucleotides per turn for B-DNA gives 96.2 rps)

(b) 0.34 μm s^{-1} (1000 nucleotides per second corresponds to 3400 Å s^{-1} because the axial distance between nucleotides in B-DNA is 3.4 Å)

17. Treat the DNA briefly with endonuclease to occasionally nick each strand. Add the polymerase with the radioactive dNTPs. At the broken bond, or nick, the polymerase will degrade the existing strand with its 5'→ 3' exonuclease activity and replace it with a radioactive complementary copy by using its polymerase activity. This reaction scheme is referred to as nick translation because the nick is moved, or translated, along the DNA molecule without ever becoming sealed.

Chapter 35

1. Mismatches, insertions, deletions, and breaks

2. Oxidizing agents such as reactive oxygen species; deamination; alkylation; ultraviolet light; ionizing radiation

3. First, recognize the damaged base or bases. Second, remove the damaged base(s). Third, repair the gap with DNA polymerase I, and seal the deal with DNA ligase.

4. The proofreading ability of DNA polymerase; mismatch-repair systems; direct repair; base-excision repair; nucleotide-excision repair; DNA recombination.

5. The ultraviolet radiation of sunlight causes, in addition to tan lines, thymine dimers in DNA, which if not repaired, block replication and gene expression, possibly leading to a mutation.

6. The repair enzyme, AlkA in *E. coli*, binds the DNA and flips the base out of the helix. The enzyme cleaves the glycosidic bond to release the damaged base. The AP endonuclease nicks the backbone,

and another enzyme removes the deoxyribose phosphate unit. DNA polymerase I fills in the gap, and DNA ligase seals the strand.

7. Cytosine in DNA, which pairs with guanine, sometimes undergoes deamination to form uracil, which pairs with adenine. The result would be a mutation (a U–A base pair replacing a C–G base pair). By using thymine instead of uracil, the repair machinery immediately recognizes uracil in DNA as a mistake and replaces it with cytosine.

8. 5-Methylcytosine spontaneously deaminates to form thymine. This results in a T–G base pair. Because the C → T mutation is so common, the T in a T–G pair is always treated as the incorrect base and removed by the base-excision-repair proteins. This repair system allows the methylation of cytosine to serve a role in transcription regulation without resulting in a DNA mutation.

9. A subunit of DNA polymerase III, acting as an exonuclease, removes the offending base. The polymerase activity makes the correction, and DNA ligase seals the backbone.

10. Each base pair has one purine, which can undergo spontaneous depurination at the rate 3×10^{-9} depurinations per purine per minute. The human genome contains 6×10^9 base pairs, and so there must be 18 depurinations per minute. Multiplying 18 by 60 minutes in an hour and 24 hours in a day reveals that 26,000 repair events are required per day per cell.

11. Potentially deleterious side reactions are prevented. The enzyme itself might be damaged by light if it could be activated by light in the absence of bound DNA harboring a pyrimidine dimer.

12. (a) 9; (b) 5; (c) 10; (d) 1; (e) 6; (f) 7; (g) 2; (h) 3; (i) 8; (j) 4

13. (a) The control plate was used to determine the number of spontaneous revertants—that is, the background mutation rate.

(b) To firmly establish that the system was working. A known mutagen's failure to produce revertants would indicate that something was wrong with the experimental system.

(c) The chemical itself has little mutagenic ability but is apparently activated into a mutagen by the liver homogenate.

(d) The cytochrome P450 system or other hepatic enzymes.

14. (a) People with xeroderma pigmentosum develop skin cancer at a much earlier age than do people without it.

(b) People with xeroderma pigmentosum lack a component of the human nucleotide-excision-repair pathway. This pathway is especially important in the repair of ultraviolet-radiation-induced DNA lesions, such as thymidine dimers. Thus, skin cancer readily develops.

(c) The late appearance of skin cancer in normal people suggests that multiple mutations in the DNA must occur before skin cancer can develop.

Chapter 36

1. Transcription is DNA-directed RNA synthesis by RNA polymerase.

2. The template strand has a sequence complementary to that of the RNA transcript. The coding strand has the same sequence as that of the RNA transcript except for thymine (T) in place of uracil (U).

3. Beginning, middle, and end, although strictly true, doesn't count. The three stages are initiation, elongation, and termination.

4. DNA template, ATP, GTP, UTP, CTP, Mg^{2+}, or Mn^{2+}

5. The sigma subunit helps the RNA polymerase locate promoter sites. After the promoter is located, the sigma subunit leaves the enzyme and assists another polymerase to find a promoter.

6. (a) 4; (b) 10; (c) 1; (d) 5; (e) 2; (f) 9; (g) 6; (h) 3; (i) 8; (j) 7

7.

	Transcription	Replication
Template	DNA	DNA
Strands copied	One	Both
Enzyme	RNA polymerase	DNA polymerase
Substrates	Ribonucleotides	Deoxyribonucleotides
Primer	None	Required

8. A promoter is a DNA sequence that directs RNA polymerase to the proper initiation site for transcription.

9. In the closed promoter complex, the DNA is double helical and transcription is not possible. In the open promoter complex, the DNA is unwound, an essential requirement for transcription.

10. The sequence of the coding (+, sense) strand is

5′-ATGGGGAACAGCAAGAGTGGGGCCCTGTCCAAGGAG-3′

and the sequence of the template (−, antisense) strand is

3′-TACCCCTTGTCGTTCTCACCCCGGGACAGGTTCCTC-5′

11. In RNA synthesis, an error affects only one molecule of mRNA of many synthesized from a gene. In addition, the errors do not become a permanent part of the genomic information.

12. At any given instant, only a fraction of the genome (total DNA) is being transcribed. Consequently, speed is not necessary.

13. This mutant σ will competitively inhibit the binding of holoenzyme and prevent the specific initiation of RNA strands at promoter sites.

14. The core enzyme without σ binds more tightly to the DNA template than does the holoenzyme. The retention of σ after strand initiation would make the mutant RNA polymerase less processive. Hence, RNA synthesis would be much slower than normal.

15. A 100-kDa protein contains about 910 residues, which are encoded by 2730 nucleotides. At a maximal transcription rate of 50 nucleotides per second, the mRNA is synthesized in 54.6 s.

16. Initiation at strong promoters takes place every 2 s. In this interval, 100 nucleotides are transcribed. Hence, centers of transcription bubbles are 34 nm (340 Å) apart.

17. Riboswitches are special secondary structures formed by some mRNA molecules, capable of directly binding small molecules, that determine whether transcription will continue or cease. For instance, a riboswitch controls the synthesis of an mRNA encoding a protein required for FMN synthesis. If FMN is already present, the riboswitch binds the FMN and traps the RNA transcript in a conformation that favors the termination of further RNA synthesis, preventing the production of functional mRNA. However, if FMN is absent, an alternative conformation forms that allows the production of the full-length mRNA.

18. (a) Cells will express β-galactosidase, lac permease, and thiogalactoside transacetylase even in the absence of lactose.

(b) Cells will express β-galactosidase, lac permease, and thiogalactoside transacetylase even in the absence of lactose.

(c) The levels of catabolic enzymes such as β-galactosidase will remain low even at low levels of glucose.

19. Lactose can't get into the cell, because the permease is missing.

20. Because of the σ factor, RNA polymerase binds DNA and slides along the DNA in a one-dimensional search for the promoter. Diffusion in one dimension is faster than diffusion in three dimensions, which is how the two small molecules must find each other.

21. Cleavage of a precursor, addition of CCA to tRNA, and modification of bases

22. DNA is the single strand that forms the trunk of the tree. Strands of increasing length are RNA molecules; the beginning of transcription is where growing strands are the smallest; the end of transcription is where strand growth stops. Direction is left to right. Many enzymes are actively transcribing each gene.

23. Heparin, a glycosaminoglycan, is highly anionic. Its negative charges, like the phosphodiester bridges of DNA templates, allow it to bind to lysine and arginine residues of RNA polymerase.

24. A liter is equivalent to 1000 cm^3, so 10^{-12} cm^3 is 10^{-15} l. The concentration is $1/(6 \times 10^{23})$ mol per 10^{-15} l $= 1.7 \times 10^{-9}$ M. Because $K_d = 10^{-13}$ M, the single molecule should be bound to its specific binding site.

25. Anti-inducers bind to the conformation of repressors, such as the lac repressor, that is capable of binding DNA. They occupy a site that overlaps that for the inducer and, therefore, compete for binding to the repressor.

26. The base-pairing energy of the di- and trinucleotide DNA–RNA hybrids formed at the very beginning of transcription is not sufficient to prevent strand separation and loss of product.

27. RNAs of different sizes were obtained, designated 10S, 13S, and 17S when boat was added at initiation, a few seconds after initiation, and 2 minutes after initiation, respectively. If no boat was added, transcription yielded a 23S RNA product. Boat, like ρ, is evidently a termination factor. The template that was used for RNA synthesis contained at least three termination sites that respond to boat (yielding 10S, 13S, and 17S RNA) and one termination site that does not (yielding 23S RNA). Thus, specific termination at a site producing 23S RNA can take place in the absence of boat. However, boat detects additional termination signals that are not recognized by RNA polymerase alone. Sadly, your search of the literature reveals that someone has already characterized the factor that you named boat—the termination factor ρ.

Chapter 37

1. (1) Eukaryotes have more complex transcriptional regulation than that of bacteria. (2) RNA, especially mRNA, is more highly processed in eukaryotes than in bacteria. (3) In eukaryotes, RNA synthesis is localized to a particular organelle—the nucleus. As a result, transcription and translation take place in different cellular compartments.

2. RNA polymerase I catalyzes the synthesis of all ribosomal RNA except 5S RNA and is located in the nucleolus. RNA polymerase I is insensitive to α-amanitin inhibition. RNA polymerase II makes mRNA, is located in the nucleoplasm, and is very sensitive to α-amanitin inhibition. RNA polymerase III, also located in the nucleoplasm, synthesizes tRNA and 5S RNA. RNA polymerase III can be inhibited by high concentrations of α-amanitin. The polymerases also differ in their subunit composition and the mass of the holoenzyme.

3. Three common elements are found in the promoter for RNA polymerase II: (1) the TATA box, (2) the initiator element (Inr), and (3) a downstream promoter element (DPE). Additional elements, such as the CAAT box and GC box may be present.

4. RNA polymerase II has a domain called the carboxyl-terminal domain that plays a key role in the regulation of RNA polymerase II activity.

5. Cis-acting elements are DNA sequences that regulate the expression of a gene located on the same molecule of DNA. Trans-acting elements, also called transcription factors or transcription

activators, are proteins that bind to cis-acting elements and regulate RNA synthesis.

6. Enhancers are DNA sequences that have no promoter activity of their own; yet they can stimulate promoters located several thousand base pairs away. They can be upstream, downstream, or even in the midst of a transcribed gene and are effective when present on either DNA strand. Enhancers are bound by proteins that participate in the regulation of transcription.

7. RNA polymerases I and III produce the same products in all cells. RNA polymerase II, however, must respond to changes in environmental conditions. Thus, the promoters for the genes transcribed by RNA polymerase II must be complex enough to respond to a variety of signals.

8. (a) 2, 6, 9; (b) 1, 4, 5, 7, 8; (c) 3, 4, 10

9. Cordycepin (3′-deoxyadenosine) is an adenosine analog and, when phosphorylated, is an adenylate nucleotide analog. The absence of a 3′-hydroxyl group means that it cannot form a phosphodiester linkage with another nucleotide. If cordycepin is incorporated, RNA synthesis will cease.

10. Phosphorylation of the CTD marks the transition from initiation to elongation. The phosphorylated CTD facilitates elongation by RNA polymerase II and serves as a binding site for RNA-processing enzymes that act in the course of elongation.

11. This finding indicates that DNA is rewound at about the same rate at the rear of RNA polymerase as it is unwound at the front of the enzyme, locking the genomic information away again.

12. (a) 5; (b) 10; (c) 2; (d) 3; (e) 9; (f) 4; (g) 8; (h) 1; (i) 6; (j) 7

13. Nuclear hormone receptors bind to appropriate response elements in the DNA. When a nuclear hormone receptor binds a hormone, the receptor's structure changes so that the receptor can bind a coactivator (or corepressor), which in turn activates (or inhibits) RNA polymerase.

14. A given regulatory protein may have different effects, depending on the other regulatory proteins in the environment. Therefore, complex regulatory patterns leading to differentiation and development can be obtained with a smaller number of regulatory proteins.

15. The acetylation reduces the number of positive charges while introducing a negative charge and thus reduces the affinity for the negatively charged DNA.

16. Although RNA polymerase II might be rapidly inhibited, there are still many functional mRNAs and proteins in the liver. However, as they are damaged and replacements cannot be provided, the liver fails and death results.

17. Gene expression is not expected to respond to the presence of estrogen. However, genes for which expression normally responds to estrogen will respond to the presence of progesterone.

18. Estradiol, because of its hydrophobic nature, can diffuse into the cell, where it can bind to its receptor. The complex can bind directly to DNA. The ligands for G-protein-coupled receptors do not enter the cell. Moreover, these receptors themselves are confined to the membrane. Any effect on gene regulation must include other components of a signal-transduction pathway.

19. Estradiol exerts its effect only in the presence of the estradiol receptor. Thus, only tissues having the receptor recognize the presence of estradiol.

20. Zinc is a component of zinc fingers, the DNA-binding motifs of nuclear hormone receptors.

21. These results suggest that chromatin structure varies, depending on the transcription status of the gene. If the gene is being transcribed, chromatin packing is loosened to allow access to the DNA for transcription factors and RNA polymerase. In the presence of DNase, these regions are susceptible to digestion.

22. (a) Different amounts of RNA are present for the various genes.

(b) Although all of the tissues have the same genes, the genes are expressed to different extents in different tissues.

(c) These genes are called housekeeping genes—genes expressed by most tissues. They might include genes for glycolysis or for citric acid cycle enzymes.

(d) The point of the experiment was to determine which genes are initiated in vivo. The initiation inhibitor was added to prevent initiation at start sites that may have been activated during the isolation of the nuclei.

Chapter 38

1. (a) 9; (b) 1; (c) 7; (d) 2; (e) 8; (f) 5; (g) 3; (h) 10; (i) 6; (j) 4

2. A GTP linked to the 5′ end of the mRNA by a 5′-5′ linkage forms the basic cap. By the way, 5′ caps are never worn backward.

3. A specific endonuclease recognizes an AAUAAA sequence and cleaves the precursor. Poly(A) polymerase adds a tail of adenylate residues to the 3′ end of the transcript.

4. The addition of the 5′ cap, the addition of the poly(A) tail, and the removal of the introns by splicing

5. The carboxyl-terminal domain

6. (a) Because cordycepin lacks a 3′-OH group, it cannot participate in 3′ → 5′ bond formation. (b) Because the poly(A) tail is a long stretch of adenosine nucleotides, the likelihood that a molecule of cordycepin will become incorporated into the poly(A) tail is higher than it is for most RNA. (c) Yes, it must be converted into cordycepin 5′-triphosphate.

7. A spliceosome is the splicing machinery in the nucleus. It is composed of snRNPs (U1, U2, U4, U5, and U6) and various protein splicing factors.

8. The carboxyl-terminal domain recruits proteins required for cap formation, splicing, and polyadenylation.

9. One gene with several introns can be spliced to yield several different mRNAs, and these mRNAs will produce different proteins. In essence, one gene can encode more than one protein.

10. Because the gene contains eight sites at which alternative splicing is possible, there are $2^8 = 256$ possible products.

11. Alternative splicing allows the formation of more than one protein from a gene.

12. The splicing mechanisms are similar in two respects. First, in the initial step, a ribose hydroxyl group attacks the 5′ splice site. The newly formed 3′-OH terminus of the upstream exon then attacks the 3′ splice site to form a phosphodiester linkage with the downstream exon. Second, both reactions are transesterifications in which the phosphate moieties at each splice site are retained in the products. The number of phosphodiester linkages stays constant.

13. Alternative splicing and RNA editing. Covalent modification of the proteins subsequent to synthesis further enhances the complexity.

14. Attach an oligo(dT) or oligo(U) sequence to an inert support to create an affinity column. When RNA is passed through the column, only poly(A)-containing RNA will be retained.

15. Histones are required in large amounts and in equal quantities only when DNA is being synthesized. The multiple arrays with an equal number of histone types per array will facilitate rapid production of a large number in equal proportions. The lack of posttranscriptional processing may speed up the synthesis of the histones themselves, and the lack of a poly(A) tail may facilitate the degradation of the histone mRNA, which need be present only during DNA synthesis.

16. Ser-Ile-Phe-His-Pro-Stop

17. A mutation that disrupted the normal AAUAAA recognition sequence for the endonuclease could account for this finding. In fact, a change from U to C in this sequence caused this defect in the thalassemic patient. Cleavage was at the AAUAAA 900 nucleotides downstream from this mutant AACAAA site.

Chapter 39

1. Three contiguous bases. Because there are four bases, a code based on a two-base codon could encode only 16 amino acids. A three-base codon would allow 64 different combinations, more than enough to account for the 20 amino acids.

2. A mutation that altered the reading of mRNA would change the amino acid sequence of most, if not all, proteins synthesized by that particular organism. Many of these changes would undoubtedly be deleterious, and so there would be strong selection against a mutation with such pervasive consequences.

3. (a) 3; (b) 10; (c) 4; (d) 7; (e) 6; (f) 2; (g) 1; (h) 9; (i) 8; (j) 5

4. Three nucleotides encode an amino acid; the code is nonoverlapping; the code has no punctuation; the code is degenerate.

5. Degeneracy of the code means that, for most amino acids, there is more than one codon. This property is valuable because, if the code were not degenerate, 20 codons would encode amino acids and the rest of the codons would lead to chain termination. Most mutations would then likely lead to inactive proteins.

6. The probability is calculated with the equation $p = (1 - \varepsilon)^n$, where p is the probability of synthesizing the error-free protein, ε is the error rate, and n is the number of amino acid residues in the protein.

	Probability of synthesizing an error-free protein	
	Number of amino acid residues	
Frequency of inserting an incorrect amino acid	50	500
10^{-2}	0.605	0.0066
10^{-4}	0.995	0.951
10^{-6}	0.999	0.999

7. An error frequency of 1 incorrect amino acid every 10^4 incorporations allows for the rapid and accurate synthesis of proteins as large as 1000 amino acids. Higher error rates would result in too many defective proteins. Lower error rates would likely slow the rate of protein synthesis without a significant gain in accuracy.

8. The first two bases in a codon form Watson–Crick base pairs that are checked for fidelity by bases of the 16S rRNA. The third base is not inspected for accuracy, and so some variation is tolerated.

9. (a) 5′-UAACGGUACGAU-3′; (b) Leu-Pro-Ser-Asp-Trp-Met; (c) Poly(Leu-Leu-Thr-Tyr)

10. Incubation with RNA polymerase and only UTP, ATP, and CTP led to the synthesis of only poly(UAC). Only poly(GUA) was formed when GTP was used in place of CTP.

11. Only single-stranded RNA can serve as a template for protein synthesis.

12. These alternatives were distinguished by the results of studies of the sequence of amino acids in mutants. Suppose that the base C is mutated to C′. In a nonoverlapping code, only amino acid 1 will be changed. In a completely overlapping code, amino acids 1, 2, and 3 will all be altered by a mutation of C to C′. The results of amino-acid-sequence studies of tobacco mosaic virus mutants and abnormal hemoglobins showed that alterations usually affected only a single amino acid. Hence, the genetic code was concluded to be nonoverlapping.

13. A peptide terminating with Lys (UGA is a Stop codon), -Asn-Glu-, and -Met-Arg-

14. Highly abundant amino acid residues have the most codons (e.g., Leu and Ser each have six), whereas the least abundant amino acids have the fewest (Met and Trp each have only one). Degeneracy (1) allows variation in base composition and (2) decreases the likelihood that a substitution for a base will change the encoded amino acid. If the degeneracy were equally distributed, each of the 20 amino acids would have three codons. Both benefits (1 and 2) are maximized by the assignment of more codons to prevalent amino acids than to less frequently used ones.

15. Phe-Cys-His-Val-Ala-Ala

16. (1) Each is a single chain. (2) They contain unusual bases. (3) Approximately half of the bases are base-paired to form double helices. (4) The 5′ end is phosphorylated and is usually pG. (5) The amino acid is attached to the hydroxyl group of the A residue of the CCA sequence at the 3′ end of the tRNA. (6) The anticodon is located in a loop near the center of the tRNA sequence.

17. First is the formation of the aminoacyl adenylate, which then reacts with the tRNA to form the aminoacyl-tRNA. Both steps are catalyzed by aminoacyl-tRNA synthetase.

18. Unique features are required so that the aminoacyl-tRNA synthetases can distinguish among the tRNAs and attach the correct amino acid to the correct tRNA. Common features are required because all tRNAs must interact with the same protein-synthesizing machinery.

19. An activated amino acid is one linked to the appropriate tRNA.

20. The ATP is cleaved to AMP and PP_i. Consequently, a second ATP is required to convert AMP into ADP, the substrate for oxidative phosphorylation.

21. Amino acids larger than the correct amino acid cannot fit into the activation site of the tRNA. Smaller but incorrect amino acids that become attached to the tRNA fit into the editing site and are cleaved from the tRNA.

22. These enzymes convert nucleic acid information into protein information by interpreting the tRNA and linking it to the correct amino acid.

23. The 2′-OH group in RNA acts as an intramolecular nucleophile. In the alkaline hydrolysis of RNA, the 2′-OH group forms a 2′-3′ cyclic intermediate.

24. (a) No; (b) no; (c) yes

25. This distribution is the one expected if the amino-terminal regions of some chains had already been partly synthesized before the addition of the radioactive amino acid. Thus, protein synthesis begins at the amino terminus and extends toward the carboxyl terminus.

26. AAA encodes lysine, whereas AAC encodes asparagine. Because asparagine was the carboxyl-terminal residue, we can conclude that the codon AAC was the last to be read.

Chapter 40

1. The *Oxford English Dictionary* defines translation as the action or process of turning from one language into another. Protein synthesis converts nucleic-acid-sequence information into amino-acid-sequence information.

2. The reading frame is a set of contiguous, nonoverlapping three-nucleotide codons that encode the amino acid sequence of the protein. The reading frame begins with a Start codon and ends with a Stop codon.

3. (a) 1, 2, 3, 5, 6, 10; (b) 1, 2, 7, 8; (c) 1, 4, 8, 9

4. Transfer RNAs have roles in several recognition processes. A tRNA must be recognized by the appropriate aminoacyl-tRNA synthetase, and the tRNA must interact with the ribosome and, in particular, with the peptidyl transferase.

5. Four bands: light, heavy, a hybrid of light 30S and heavy 50S, and a hybrid of heavy 30S and light 50S

6. Two hundred molecules of ATP are converted into 200 AMP + 400 P_i to activate the 200 amino acids, which is equivalent to 400 molecules of ATP. One molecule of GTP is required for initiation, and 398 molecules of GTP are needed to form 199 peptide bonds.

7. The sequence GAGGU is complementary to a sequence of five bases at the 3′ end of 16S rRNA and is located several bases upstream of an AUG Start codon. Hence, this region is a start signal for protein synthesis. The replacement of G by A would be expected to weaken the interaction of this mRNA with the 16S rRNA and thereby diminish its effectiveness as an initiation signal. In fact, this mutation results in a 10-fold decrease in the rate of synthesis of the protein specified by this mRNA.

8. The error rates of DNA, RNA, and protein synthesis are of the order of 10^{-10}, 10^{-5}, and 10^{-4}, respectively, per nucleotide (or amino acid) incorporated. The fidelity of all three processes depends on the precision of base-pairing to the DNA or mRNA template. Few errors are corrected in RNA synthesis. In contrast, the fidelity of DNA synthesis is markedly increased by the $3′ \rightarrow 5′$ proofreading nuclease activity and by postreplicative repair. In protein synthesis, the mischarging of some tRNAs is corrected by the hydrolytic action of aminoacyl-tRNA synthetase. Proofreading also takes place when aminoacyl-tRNA occupies the A site on the ribosome; the GTPase activity of EF-Tu sets the pace of this final stage of editing.

9. The Shine–Dalgarno sequence of the mRNA base-pairs with a part of the 16S rRNA of the 30S subunit, which positions the subunit so that the initiator AUG is recognized.

10. GTP is not hydrolyzed until aminoacyl-tRNA is delivered to the A site of the ribosome. An earlier hydrolysis of GTP would be wasteful because EF-Tu–GDP has little affinity for aminoacyl-tRNA.

11. The translation of an mRNA molecule can be blocked by an RNA molecule with the complementary sequence. Such RNAs are called antisense RNAs. The antisense–sense RNA duplex cannot serve as a template for translation; single-stranded mRNA is required. Furthermore, the antisense–sense duplex is degraded by the RNA-induced silencing complex. Antisense RNA added to the external medium is spontaneously taken up by many cells. A precise quantity can be delivered by microinjection. Alternatively, a plasmid encoding the antisense RNA can be introduced into target cells.

12. (a) A_5; (b) $A_5 > A_4 > A_3 > A_2$; (c) synthesis is from the amino terminus to the carboxyl terminus.

13. The rate would fall because the elongation step requires that the GTP be hydrolyzed before any further elongation can take place.

14.

	Bacterium	Eukaryote
Ribosome size	70S	80S
mRNA	Polycistronic	Not polycistronic
Initiation	Shine–Dalgarno is required	First AUG is used
Protein factors	Required	Many more required
Relation to transcription	Translation can start before transcription is completed	Transcription and translation are spatially separated
First amino acid	fMet	Met

15. The signal sequence, the signal-recognition particle (SRP), the SRP receptor, and the translocon

16. The signal-recognition particle (SRP) binds to the signal sequence and inhibits further translation. The SRP ushers the inhibited ribosome to the ER, where it interacts with the SRP receptor (SR). The SRP–SR complex binds the translocon and simultaneously hydrolyzes GTP. On GTP hydrolysis, SRP and SR dissociate from each other and from the ribosome. Protein synthesis resumes, and the nascent protein is channeled through the translocon.

17. The formation of peptide bonds, which in turn are powered by the hydrolysis of the aminoacyl-tRNAs.

18. The alternative would be to have a single ribosome translating a single mRNA molecule. The use of polysomes allows more protein synthesis per mRNA molecule in a given period of time and thus the production of more protein.

19. The addition of an IRE to the 5′ end of the mRNA would be expected to block translation in the absence of iron. The addition of an IRE to the 3′ end of the mRNA would not be expected to block translation, but it would make the mRNA more susceptible to degradation.

20. The sequences of all of the mRNAs would be searched for sequences that are fully or nearly complementary to the sequence of the miRNA. These sequences would be candidates for regulation by this miRNA.

21. EF-Ts catalyzes the exchange of GTP for GDP bound to EF-Tu. In G-protein cascades, an activated 7TM receptor catalyzes GTP–GDP exchange in a G protein.

22. The α subunits of G proteins are inhibited by a similar mechanism in cholera and whooping cough (p. 231).

23. (a, d, and e) Type 2; (b, c, and f) type 1

24. The primary structure determines the three-dimensional structure of the protein. Thus, the final phase of information transfer from DNA to RNA to protein synthesis is the folding of the protein into its functional state.

25. (a) Factor eIF-4H has two effects: (1) the extent of unwinding is increased and (2) the rate of unwinding is increased, as indicated by the increased rise in activity at early reaction times.

(b) To firmly establish that the effect of eIF-H4 was not due to any inherent helicase activity.

(c) Half-maximal activity was achieved at 0.11 μM of eIF-4H. Therefore, maximal stimulation would be achieved at a ratio of 1:1.

(d) Factor eIF-4H enhances the rate of unwinding of all helices, but the effect is greater as the helices increase in stability.

(e) The results in graph C suggest that it increases the processivity.

26. (a) The three peaks represent, from left to right, the 40S ribosomal subunit, the 60S ribosomal subunit, and the 80S ribosome.

(b) Not only are ribosomal subunits and the 80S ribosome present, but polysomes of various lengths also are apparent. The individual peaks in the polysome region represent polysomes of discrete length.

(c) The treatment significantly inhibited the number of polysomes, whereas it increased the number of free ribosomal subunits. This outcome could be due to inhibited protein-synthesis initiation or inhibited transcription.

27. A mutation caused by the insertion of an extra base can be suppressed by a tRNA that contains a fourth base in its anticodon. For example, UUUC rather than UUU is read as the codon for phenylalanine by a tRNA that contains 3′-AAAG-5′ as its anticodon.

28. Glu-tRNA^{Gln} is formed by misacylation. The activated glutamate is subsequently amidated to form Gln-tRNA^{Gln}. Ways in which glutamine is formed from glutamate were discussed in Section 31.2. In regard to *H. pylori*, a specific enzyme, Glu-tRNA^{Gln} amidotransferase, catalyzes the following reaction:

$$\text{Gln} + \text{Glu-tRNA}^{Gln} + \text{ATP} \rightarrow \text{Gln-tRNA}^{Gln} + \text{Glu} + \text{ADP} + P_i$$

Glu-tRNA^{Glu} is not a substrate for the enzyme; so the transferase must also recognize aspects of the structure of tRNA^{Gln}.

Chapter 41

1. A cDNA library is set of DNA segments complementary to mRNA sequences from a cell type. The mRNA is isolated and converted into double-stranded DNA with the use of reverse transcriptase and DNA polymerase. The double-stranded DNA is ligated to linkers, which are then inserted into some sort of vector.

2. A cDNA library is a DNA representation of the mRNA expressed in a particular tissue under a specific set of physiological conditions. cDNA libraries vary from cell type to cell type from the same organism. A genomic library is a collection of DNA fragments, inserted into vector molecules, that represent the entire genome of an organism. Genomic libraries prepared from any diploid tissue in an organism are identical.

3. *Taq* polymerase is the DNA polymerase from the thermophilic bacterium that lives in hot springs. Consequently, it is heat stable and can withstand the high temperatures required for PCR without denaturing.

4. (a) 4; (b) 8; (c) 1; (d) 5; (e) 10; (f) 7; (g) 6; (h) 3; (i) 9; (j) 2

5. The codon(s) for each amino acid can be used to determine the number of possible nucleotide sequences that encode each peptide sequence (Table 4.5):

Ala–Met–Ser–Leu–Pro–Trp:

$4 \times 1 \times 6 \times 6 \times 4 \times 1 = 576$ total sequences

Gly–Trp–Asp–Met–His–Lys:

$4 \times 1 \times 2 \times 1 \times 2 \times 2 = 32$ total sequences

Cys–Val–Trp–Asn–Lys–Ile:

$2 \times 4 \times 1 \times 2 \times 2 \times 3 = 96$ total sequences

Arg–Ser–Met–Leu–Gln–Asn:

$6 \times 6 \times 1 \times 6 \times 2 \times 2 = 864$ total sequences

The set of DNA sequences encoding the peptide Gly-Trp-Asp-Met-His-Lys would be most ideal for probe design because it encompasses only 32 total oligonucleotides.

6. Ovalbumin cDNA should be used. *E. coli* lacks the machinery to splice the primary transcript arising from genomic DNA.

7. The presence of the *Alu*I sequence would, on average, be $(1/4)^4$, or 1/256, because the likelihood of any base being at any position is one-fourth and there are four positions. By the same reasoning, the presence of the *Not*I sequence would be $(1/4)^8$, or 1/65,536. Thus, the average product of digestion by *Alu*I would be 250 base pairs (0.25 kbp) in length, whereas that for *Not*I would be 66,000 base pairs (66 kbp) in length.

8. No, because most human genes are much longer than 4 kb. A fragment would contain only a small part of a complete gene.

9. (a) 5′-GGCATAC-3′

(b) The Sanger dideoxy method of sequencing would give the gel pattern shown here.

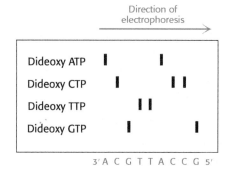

Direction of electrophoresis →

Dideoxy ATP		
Dideoxy CTP		
Dideoxy TTP		
Dideoxy GTP		

3′ A C G T T A C C G 5′

10. Southern blotting of an *Mst*II digest would distinguish between the normal and the mutant genes. The loss of a restriction site would lead to the replacement of two fragments on the Southern blot by a single longer fragment. Such a finding would not prove that GTG replaced GAG; other sequence changes at the restriction site could yield the same result.

11. Because PCR can amplify as little as one molecule of DNA, statements claiming the isolation of ancient DNA need to be greeted with some skepticism. The DNA would need to be sequenced. Is it similar to human, bacterial, or fungal DNA? If so, contamination is the likely source of the amplified DNA. Is it similar to the DNA of birds or crocodiles? This sequence similarity would strengthen the case that it is dinosaur DNA because these species are evolutionarily close to dinosaurs.

12. At high temperatures of hybridization, only very close matches between primer and target would be stable because all (or most) of the bases would need to find partners to stabilize the primer–target helix. As the temperature is lowered, more mismatches would be tolerated; so the amplification is likely to yield genes with less sequence similarity. In regard to the yeast gene, synthesize primers corresponding to the ends of the gene, and then use these primers and human DNA as the target. If nothing is amplified at 54°C, the human gene differs from the yeast gene, but a counterpart may still be present. Repeat the experiment at a lower temperature of hybridization.

13. The encoded protein contains four repeats of a specific sequence.

14. Within a single species, individual dogs show enormous variation in body size and substantial diversity in other physical characteristics. Therefore, genomic analyses of individual dogs would provide valuable clues concerning the genes responsible for the diversity within the species.

15. If the melting temperatures of the primers are too different, the extent of hybridization with the target DNA will differ during the annealing phase, which would result in differential replication of the strands.

16. Chongqing: residue 2, L → R, CTG → CGG
    Karachi: residue 5, A → P, GCC → CCC
    Swan River: residue 6, D → G, GAC → G*G*C

17. This particular person is heterozygous for this particular mutation: one allele is wild type, whereas the other carries a point mutation at this position. Both alleles are PCR amplified in this experiment, yielding the "dual peak" appearance on the sequencing chromatogram.

18. A mutation in person B has altered one of the alleles for gene *X*, leaving the other intact. The fact that the mutated allele is smaller suggests that it has undergone a deletion. The one functioning allele is transcribed and translated and apparently produces enough protein to render the person asymptomatic.

Person C has only the smaller version of the gene. This gene is neither transcribed (negative northern blot) nor translated (negative western blot).

Person D has a normal-size allele of the gene but no corresponding RNA or protein. There may be a mutation in the promoter region of the gene that prevents transcription.

Person E has a normal-size allele of the gene that is transcribed, but no protein is made, which suggests that a mutation prevents translation. There are a number of possible explanations, including a mutation that introduced a premature Stop codon in the mRNA.

Person F has a normal amount of protein but still displays the metabolic problem. This finding suggests that the mutation affects the activity of the protein—for instance, a mutation that compromises the active site of enzyme Y.

19. A simple strategy for generating many mutants is to synthesize a degenerate set of the 30-bp sequence by using a mixture of activated nucleosides in particular rounds of oligonucleotide synthesis. Suppose that the 30-bp coding region begins with GTT, which encodes valine. If a mixture of all four nucleotides is used in the first and second rounds of synthesis, the resulting oligonucleotides will begin with the sequence XYT (where X and Y denote A, C, G, or T). These 16 different versions of the cassette will encode proteins containing either Phe, Leu, Ile, Val, Ser, Pro, Thr, Ala, Tyr, His, Asn, Asp, Cys, Arg, or Gly at the first position. Likewise, degenerate cassettes can be made in which two or more codons are simultaneously varied.

20. Digest genomic DNA with a restriction enzyme, and select the fragment that contains the known sequence. Circularize this fragment. Then carry out PCR with the use of a pair of primers that serve as templates for the synthesis of DNA away from the known sequence.

21. Use chemical synthesis or the polymerase chain reaction to prepare hybridization probes that are complementary to both ends of the known (previously isolated) DNA fragment. Challenge clones representing the library of DNA fragments with both of the hybridization probes. Select clones that hybridize to one of the probes but not the other; such clones are likely to represent DNA fragments that contain one end of the known fragment along with the adjacent region of the particular chromosome.

Index

Note: Page numbers followed by f and t refer to figures and tables, respectively. Page numbers preceded by A refer to appendices. **Boldface** page numbers indicate structural formulas and ribbon diagrams.

(continued)

(continued)

Common Abbreviations in Biochemistry

A	adenine	IgG	immunoglobulin G
ACP	acyl carrier protein	Ile	isoleucine
ADP	adenosine diphosphate	IP$_3$	inositol 1,4,5-trisphosphate
Ala	alanine	ITP	inosine triphosphate
AMP	adenosine monophosphate	LDL	low-density lipoprotein
Arg	arginine	Leu	leucine
Asn	asparagine	Lys	lysine
Asp	aspartic acid	Met	methionine
ATP	adenosine triphosphate	miRNA	microRNA
ATPase	adenosine triphosphatase	mRNA	messenger RNA
C	cytosine	NAD$^+$	nicotinamide adenine dinucleotide (oxidized form)
cAMP	cyclic AMP (adenosine 3′,5′-cyclic monophosphate)	NADH	nicotinamide adenine dinucleotide (reduced form)
cDNA	complementary DNA	NADP$^+$	nicotinamide adenine dinucleotide phosphate (oxidized form)
CDP	cytidine diphosphate		
cGMP	cyclic GMP (guanosine 3′,5′-cyclic monophosphate)	NADPH	nicotinamide adenine dinucleotide phosphate (reduced form)
CMP	cytidine monophosphate	PFK	phosphofructokinase
CoA	coenzyme A	Phe	phenylalanine
CoQ	coenzyme Q (ubiquinone)	P$_i$	inorganic orthophosphate
CTP	cytidine triphosphate	PLP	pyridoxal phosphate
Cys	cysteine	PP$_i$	inorganic pyrophosphate
Cyt	cytochrome	Pro	proline
DNA	deoxyribonucleic acid	PRPP	5-phosphoribosyl-1-pyrophosphate
DNase	deoxyribonuclease	Q	ubiquinone (plastoquinone)
EcoRI	EcoRI restriction endonuclease	QH$_2$	ubiquinol (plastoquinol)
EF	elongation factor	RNA	ribonucleic acid
FAD	flavin adenine dinucleotide (oxidized form)	RNase	ribonuclease
		rRNA	ribosomal RNA
FADH$_2$	flavin adenine dinucleotide (reduced form)	scRNA	small cytoplasmic RNA
fMet	formylmethionine	Ser	serine
FMN	flavin mononucleotide (oxidized form)	siRNA	small interfering RNA
FMNH$_2$	flavin mononucleotide (reduced form)	snRNA	small nuclear RNA
G	guanine	T	thymine
GDP	guanosine diphosphate	Thr	threonine
Gln	glutamine	TPP	thiamine pyrophosphate
Glu	glutamic acid	tRNA	transfer RNA
Gly	glycine	Trp	tryptophan
GMP	guanosine monophosphate	TTP	thymidine triphosphate
GSH	reduced glutathione	Tyr	tyrosine
GSSG	oxidized glutathione	U	uracil
GTP	guanosine triphosphate	UDP	uridine diphosphate
GTPase	guanosine triphosphatase	UMP	uridine monophosphate
Hb	hemoglobin	UTP	uridine triphosphate
HDL	high-density lipoprotein	Val	valine
HGPRT	hypoxanthine-guanine phosphoribosyl-transferase	VLDL	very low density lipoprotein
His	histidine		
Hyp	hydroxyproline		